Larry Abbott, William Bloor Professor of Theoretical Neuroscience, Columbia University; co-author with Peter Dayan of the textbook *"Theoretical Neuroscience"*

"In this textbook, Xiao-Jing Wang takes the reader on a remarkable journey from cellular and synaptic biophysics to memory, decision making, psychiatry and large-scale brain structure. His insights on how to apply theoretical methods to higher cognitive functions will be valuable to anyone thinking about thinking."

Jean-Pierre Changeux, Institute Pasteur, Paris; author of *"Neuronal Man: the Biology of Mind"*

"X-J Wang book illuminates how in the past decades a new field has emerged from the theoretical modeling of a wide diversity of neural processes which has successfully contributed to our understanding of brain functions up to Cognition. The book progresses from the most elementary building blocks of the brain, neurons, synapses and their molecular components up to global executive functions and the new field of computational psychiatry. The immense knowledge recently acquired and beautifully presented in the book is a must for anybody concerned by the higher functions of the brain: a bible for the students, a grandiose panorama for the educated scientists."

Anne Churchland, Professor of Neurobiology, University of California, Los Angeles

"This compelling book explains foundational concepts in theory and modeling in a way that marries rigor with intuition. It furnishes readers with powerful theory and modeling tools that are critically needed to tackle problems in modern neuroscience. Wang's explanations of network dynamics, normalization and attractor models for cognition are certain to inspire experimentalists and theorists alike."

Robert Desimone, Director of the McGovern Institute for Brain Research, Doris and Donald Berkey Professor of Neuroscience, MIT

"In Theoretical Neuroscience: Understanding Cognition, *Xiao-Xing Wang explains that while the brain is not a computer, all of its elements, from synapses to circuits, obey computational principles that underlie our cognitive functions. As a leader in the computational field, he gives an insider's account of these principles at both a mathematical and conceptual level. This should be required reading for any student of modern neuroscience."*

Stanislas Dehaene, Chair of Experimental Cognitive Psychology, Collège de France, Paris

"In the 20th century, Molecular Biology decoded the intricate mechanisms of DNA and RNA, the "logic of life" celebrated by Nobel Prize Laureate François Jacob. Understanding the logic of neural circuits of cognition is an equivalent goal for 21st century in Neuroscience. This important book represents a fundamental step forward : it provides a synthesis of elemental circuit building blocks of core cognitive functions, how their operation can be mathematically modeled, and how their composition may account for a variety of cognitive processes and behavioral tasks."

Brent Doiron, Heinrich Kluver Professor of Neurobiology, Director of the Grossman Center for Quantitative Biology and Human Behavior, University of Chicago

"Understanding the brain needs an appreciation of its components and activity over a wide range of spatial and temporal scales: from synapses to whole brain circuits, from rapid sensory and motor responses to lifetime memory. In this book Xiao-Jing Wang beautifully shows how theoretical and mathematical thinking can provide the needed abstraction to not only describe the brain at each scale, but synthesize these across scales into a compelling and deep understanding of neural function. At whatever stage you are in your neuroscience journey, from novice to expert, this is now required reading."

John Duncan, University of Cambridge and the Medical Research Council; author of *"How Intelligence Happens"*

"Nobody understands better than Xiao-Jing Wang how the events of the mind are built from the dynamic activity of neural populations, and in this book, he explains the state of the art from top to bottom. With extraordinary depth and scope, the story moves from the fundamentals of the neuron to the highest levels of human intelligence. Anybody aiming to understand the union of mind and brain can start right here."

Ann Graybiel, Institute Professor, MIT

"Xiao-Jing Wang has written a remarkable book, melding together viewpoints of a physicist/mathematician and those of a neuroscientist to produce a unique account of how the brain and its vast number of networks could operate to produce behavior. Wang has undertaken the job of bridging these two worlds, and the result is a grand success. This volume will be invaluable not only to academics, but also to all who have an interest in how the brain does its work. Xiao-Jing Wang has created a book to read and re-read, culminating in an introduction to brain-wide modeling to address the unsolved mysteries of the normal brain and the need for computational psychiatry to alleviate mental illness."

John Krystal, Chair of the Department of Psychiatry, Robert L. McNeil, Jr. Professor of Translational Research, Yale University School of Medicine

"This book is a remarkable opus from Dr. X-J Wang, a pioneering computational neuroscientist who's work continues to surprise and inspire me. This book is logically organized. It builds from a model neuron, to networks, to cognition/behavior, and it integrates all of this in considering issues of "computational psychiatry", i.e., behavioral pathology. Wirtten with clarity and an accessible style, this wonderful book is useful for people entering the field of computational neuroscience. It is also a wonderful resource for clinical investigators hoping to learn about the formalism of computational models."

Eve Marder, University Professor, Brandeis University; Former President of the Society for Neuroscience

"Theoretical Neuroscience: Understanding Cognition *by X-J. Wang is a remarkable book. It reads easily, and puts much of the field of quantitative and computational neuroscience into frameworks that allow the reader to see the etiology and genesis of the particular methods and findings being discussed. The book is comprehensive and moves seamlessly from the details of the biophysics of single neurons and synapses to some of the most complex puzzles in the neuroscience of cognition. This book will be useful for trainees and for established investigators as its breadth and depth will be helpful for everyone as they wish to gain access to the conceptual frameworks of large areas of neuroscience and their quantitative analyses."*

William Newsome, Harman Family Provostial Professor of Neurobiology, the inaugural Director of Wu Tsai Neuroscience Institute, Stanford University, Co-chair of the US Brain Initiative

"In a rapidly advancing field, Xiao-Jing Wang provides a beautiful and timely account of where we have been, where we are now, and where we are likely heading. From biophysical models of single neurons, through network models of neural circuits, to macro-level models of communication between brain areas, Wang sure-handedly guides the reader with clear, incisive exposition of each topic. For dessert, we are treated to Wang's forward-looking views on computational psychiatry and future interactions between artificial intelligence and natural intelligence. This book is a must-read for anyone intrigued with the nexus between neuroscience, cognitive science, and computational theory."

Mu-ming Poo, Scientific Director of Institute of Neuroscience, Chinese Academy of Sciences; Paul Licht Distinguished Professor in Biology Emeritus, University of California, Berkeley

"This monumental treatise by Xiao-Jing Wang, one of the leading computational neuroscientists in the world, offers a thorough survey and critical synthesis of both experimental and theoretical works on various cognitive processes in the brain. A valuable book for consultation by researchers and students of neuroscience."

Matthew Rushworth, Chair of Psychology Department, Oxford University, Britain

"Xiao-Jing Wang has been a leader in theoretical neuroscience for many years. The approaches and models that he has devised have inspired the field. Even more importantly, however, Xiao-Jing Wang has had an impact on cognitive science and neuroscience more generally that has changed how scientists working with brain and behavioural data think about the issues that they are investigating. Theoretical Neuroscience: Understanding Cognition *provides an accessible but comprehensive guide not just to how this happened but also to the very latest ideas emerging at the field's frontier."*

Doris Tsao, University of California at Berkeley

"If a genie were to grant me a wish, I might ask for a guided tour of the brain—starting in the microscopic world of neurons and synapses, traveling through the vast landscapes of large-scale cortical interactions underlying higher cognitive functions, and culminating in our modern era of AI. A tour with a master at my side who could cut through the dense jungle of thousands of scientific papers and explain the key findings with clarity, putting them in a precise conceptual and mathematical framework. Xiao-Jing Wang's remarkable book is exactly that wish come true. This book is an invaluable guide for young researchers and anyone who wants to truly understand what is known about the brain."

Theoretical Neuroscience

This textbook is an introduction to Systems and Theoretical/Computational Neuroscience, with a particular emphasis on cognition. It consists of three parts: Part I covers fundamental concepts and mathematical models in computational neuroscience, along with cutting-edge topics. Part II explores the building blocks of cognition, including working memory (how the brain maintains and manipulates information "online" without external input), decision making (how choices are made among multiple options under conditions of uncertainty and risk) and behavioral flexibility (how we direct attention and control actions). Part III is dedicated to frontier research, covering models of large-scale multi-regional brain systems, Computational Psychiatry and the interface with Artificial Intelligence. The author highlights the perspective of neural circuits as dynamical systems, and emphasizes a cross-level mechanistic understanding of the brain and mind, from genes and cell types to collective neural populations and behavior. Overall, this textbook provides an opportunity for readers to become well versed in this highly interdisciplinary field of the twenty-first century.

Xiao-Jing Wang is Distinguished Global Professor of Neural Science at New York University, where he is also affiliated with the Courant Institute of Mathematical Sciences and Department of Physics. Before joining New York University, he was Professor at Yale University. His honors include a Guggenheim Fellowship, the Swartz Prize for Theoretical and Computational Neuroscience and the Goldman-Rakic Prize for Outstanding Achievement in Cognitive Neuroscience.

Theoretical Neuroscience

Understanding Cognition

Xiao-Jing Wang

CRC Press
Taylor & Francis Group
Boca Raton London New York

CRC Press is an imprint of the
Taylor & Francis Group, an **informa** business

First edition published 2025
by CRC Press
2385 NW Executive Center Drive, Suite 320, Boca Raton, FL 33431

and by CRC Press
4 Park Square, Milton Park, Abingdon, Oxon, OX14 4RN

CRC Press is an imprint of Taylor & Francis Group, LLC

Library of Congress Cataloging-in-Publication Data
Names: Wang, Xiao-Jing, author.
Title: Theoretical neuroscience : understanding cognition / Xiao-Jing Wang.
Description: Boca Raton : CRC Press, 2025. | Includes bibliographical references and index.
Identifiers: LCCN 2024019547 (print) | LCCN 2024019548 (ebook) | ISBN 9781032604824 (hardback) | ISBN 9781032604817 (paperback) | ISBN 9781003459361 (ebook)
Subjects: LCSH: Neural networks (Neurobiology)—Textbooks | Human information processing—Textbooks | Computational neuroscience—Textbooks | Cognition—Textbooks
Classification: LCC QP355.2 .W36 2025 (print) | LCC QP355.2 (ebook) | DDC 612.8—dc23/eng/20240430
LC record available at https://lccn.loc.gov/2024019547
LC ebook record available at https://lccn.loc.gov/2024019548

ISBN: 978-1-032-60482-4 (hbk)
ISBN: 978-1-032-60481-7 (pbk)
ISBN: 978-1-003-45936-1 (ebk)

DOI: 10.1201/9781003459361

Typeset in ITC Leawood
by Apex CoVantage, LLC

For Product Safety Concerns and Information please contact our EU representative:
GPSR@taylorandfrancis.com
Taylor & Francis Verlag GmbH, Kaufingerstraße 24, 80331 München, Germany

MIX
Paper | Supporting responsible forestry
FSC™ C013985

Printed in the United Kingdom
by Henry Ling Limited

CONTENTS

PREFACE

This book aims at a synthesis of neuroscience and cognitive science across multiple levels, grounded in recent advances in experimental and theoretical studies of basic cognitive functions. Galileo Galilei wrote more than three centuries ago:

> [The universe] cannot be read until we have learnt the language and become familiar with the characters in which it is written. It is written in mathematical language . . . without which means it is humanly impossible to comprehend a single word.

This holds true for the biological brain as well as the inanimate world. Cognition and intelligence arise from neural circuits endowed with an abundance of feedback loops, so their behavior is difficult to understand by intuition alone. Furthermore, connectome and neurophysiology, the two pillars of neuroscience, are presently undergoing a great upheaval with tremendous advances in both fields. New tools for statistical analysis and theoretical ideas are needed in order to translate big data into knowledge leading to discoveries. For these reasons, theory and modeling play an increasingly important role in neuroscience. Besides its importance in basic research, modeling translates "wet" neural tissues to "dry" computational algorithms, thereby bridging brain research and the field of artificial intelligence.

I will cover the basics of computational neuroscience required for discussions of the brain mechanisms and computational principles of cognition. Special efforts have been made, with the goal for the book to be accessible to both neuroscientists and cognitive scientists, especially young students. Neurobiological and mathematical concepts are introduced so that the book is self-contained. For more background, a reader can consult a number of excellent existing textbooks: *Principles of Neural Science* by Eric Kandel et al.; *Neuroscience: Exploring the Brain* by Mark Bear et al.; *Principles of Neurobiology* by Liqun Luo; *Cognitive Neuroscience: The Biology of the Mind* by Michael Gazzaniga et al.; and *Theoretical Neuroscience* by Peter Dayan and Larry Abbott.

A semester-long course can use Parts I and II as a textbook, with Part III for optional reading on frontier topics. The book can also be of interest as a monograph to researchers who want to become familiar with recurrent neural circuits underlying cognition. The best way to learn computational neuroscience is through hands-on practice. I have provided a number of exercises and Python codes of a dozen basic models discussed in this book, available on the book's website. Readers are encouraged to go through the exercises and model simulations to acquire in-depth understanding of the materials, explore new questions and extend the models for their own research.

The focus here is on basic cognitive functions that can be studied using nonhuman animals including monkeys and rodents and possibly much simpler species like *Drosophila* flies. This limited scope makes it possible to investigate in detail neural circuits underlying cognition. Little will be said about higher cognitive processes such as thought or language. However, a wide range of cognitive functions will be discussed, so this book should be of broad interest beyond "elemental" cognitive functions, to neurobiologists, cognitive neuroscientists and researchers in artificial intelligence.

I chose to use "we" instead of "I" for two reasons. First, it reflects my interactive style in teaching such a course where students participate in discussions as a group. Second, much of my research has been carried out in collaboration with a cohort of talented graduate students and postdoctoral fellows; models and findings belong to a community. Another pragmatic matter concerns references. The literature is so vast that some way to be selective is mandatory. I attempted to include works based on either historical significance or relevance to the materials covered. There is a "recency effect" with an increased chance of citing new publications, although it is impossible to stay fully updated as the field continues to move on many fronts. While works that I am familiar with are more likely to be included, every effort has been made to highlight key publications in the covered areas.

The root of this book can be traced back to 30 years ago, when I was a research associate with the physicist Leo Kadanoff at the University of Chicago. My fellowship had a teaching component. Having started to make a transition to neuroscience, I offered to design and teach a course on computational neuroscience in the fall of 1992. Leo not only kindly approved my plan but also provided funds to hire an undergraduate who typed up my handwritten lecture notes. After three decades, I can finally report that Leo's trust in me and his clairvoyance have not been in vain.

In the intervening years, the field has matured and my own research has evolved to focus on the brain systems underlying cognition. This book is based on my teaching such a course over the years at Brandeis University, Yale University and New York University. In addition, the same materials were used at various summer schools, especially the Methods in Computational Neuroscience course at the Marine Biological Laboratory in Woods Hole, Massachusetts, that started in 1988 and the Computational and Cognitive Neuroscience (CCN) course in China that was launched in 2010.

This project has taken multiple years to complete, supported by a Guggenheim Fellowship and other funding sources for which I am grateful. The book was partly written as a Rockefeller Fellow at the beautiful Rockefeller Foundation Bellagio Center by Lake Como in Italy and as a visiting professor at the Collège de France hosted by Stanislas Dehaene in France. I have fond memories of both experiences.

My warmest thanks go to the past and current members of my research group: Salva Ardid, Alberto Bernacchia, Aldo Battista, Gabriel Benigno, Daniel Bliss, Nicolas Brunel, Marcelo Camperi, Gene Carter, Warasinee Chaisangmongkon, Rishidev Chaudhuri, Elisa Chinigo, Albert Compte, Xingyu Ding, Tatiana Engel, Sean Froudist-Walsh, Moran Furman, Stefano Fusi, Luis Carlos Garcia, Caroline Geisler, Vishwa Goudar, Jorge Jaramillo, Junjie Jiang, Madhura Joglekar, Alan Lai, Zhenghan Liao, Feng Liu, Yinghui Liu, Yue Liu, Chung-Chuan Lo, Christian Luhmann, Loïc Magrou, Victor Matveev, Jorge F. Mejias, John Hongyu Meng, Paul Miller, Bin Min, John Murray, Jacinto Pereira, Ulises Pereira-Obilinovic, Warren Pettine, Alfonso Renart, Alireza Soltani, Francis Song, Simon Song, Jesper Tegnér, Panagiota Theodoni, Dahui Wang, Wei Wei, Kong-Fatt Wong, Guangyu Robert Yang and Manyi Yim.

I have had the good fortune to collaborate with numerous experimentalists, including Alan Anticevic, Amy Arnsten, Elisabeth Buffalo, Gyuri Buzsaki, Xinying Cai, Marvin Chun, Philip Corlett, Stanislas Dehaene, Robert Desimone, David Freedman, Dan Fulcher, Winrich Freiwald, Wen-Jun Gao, Patricia Goldman-Rakic, Henry Kennedy, Robert Knight, John Krystal, Daeyeol Lee, Songting Li, Nuo Li, Julio Martinez-Trujillo, David McCormick, Earl Miller, Sacha Nelson, Bill Newsome, Nicola Palmero-Gallagher, Ranulfo Romo, Marcello Rosa, Mavi Sanchez-Vives, Jeff Schall, Claire Sergent, Michael Shadlen, Karel Svoboda, Gina Turrigiano and Liping Wang. Close interactions between theory and experimentation through these joint efforts are most fruitful. I am also grateful to the National Institute of Mental Health, the National Science Foundation, the Office of Naval Research, the Swartz Foundation and the James Simons Foundation, for their funding support of my lab's research over the last three decades.

Sincere thanks are due to many colleagues and friends who read various chapters of this book and provided valuable feedback: Larry Abbott, Steve Baccus, David Badre, Tim Behrens, Nicolas Brunel, Rishi Chaudhuri, Anne Churchland, Albert Compte, Carl Craver, Stefano Fusi, Tamar Gendler, Josh Gold, Mark Goldman, Henry Kennedy, Liqun Luo, Victor Matveev, Camillo Padoa-Schioppa, David Redish, John Rinzel, Robb Rutledge, Jeff Schall, Ali Soltani and Robert Guangyu Yang. I am also deeply appreciative of people who helped with figures: Amy Arnsten, Aldo Battista, Nicolas Brunel, Paul Cisek, Tatiana Engel, Stefan Everling, Ev Fedorenko, Ila Fiete, Sean Froudist-Walsh, Stefano Fusi, Michael Graupner, Boris Gutkin, Laurence Hunt, Yijie Kang, Daeyeol Lee, Tony Li, Sukbin Lim, Chung-Chuan Lo, Kenway Louie, Tony Movshon, John Murray, Srdjan Ostojic, Robb Rutledge, Trevor Robbins, Terry Sejnowski, Claire Sergent, Mike Shadlen, Francis Song, Yota Theodoni, Ziqiang Wei and Guangyu Yang. Credit for the cover design goes to Qian Lin and Benjamin Marder. Inputs from them and all lab members helped to improve the manuscript considerably. My friend Liqun Luo introduced me to the editor of his textbook, *Principles of Neurobiology*, Chuck Crumly. I had the pleasure to work with Chuck and the team at CRC/Taylor & Francis through the publication process, their professionalism and high standards are much valued.

I am deeply indebted to my late parents Xiangming Wang and Shiguang Zhao, and my sister Xiaolan Wang, who have been a constant source of nurture and affection. With my profound gratitude, this book is dedicated to my wife Qian and son Lucian who give me so much love, understanding, support and joys in life.

Part I

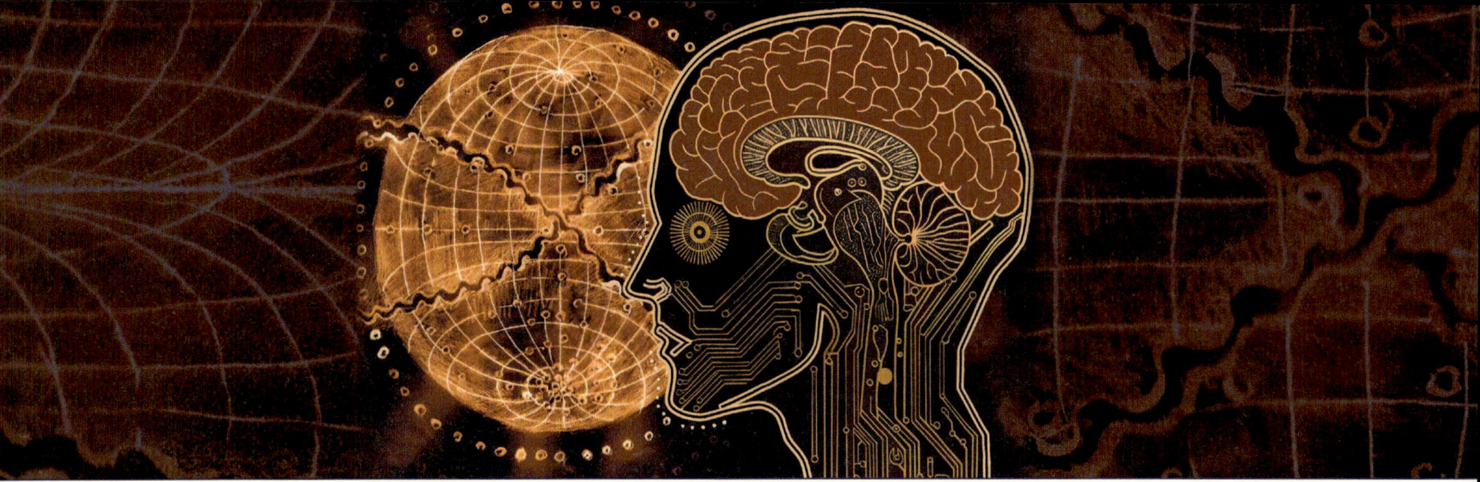

Understanding the Cognitive Brain

1

"You," your joys and your sorrows, your memories and your ambitions, your sense of personal identity and free will, are in fact no more than the behavior of a vast assembly of nerve cells and their associated molecules.

Francis Crick in *The Astonishing Hypothesis*

1.1 INTRODUCTION

When your hand touches fire and withdraws from it abruptly, a sensory stimulation triggers a reflexive motor response. This book is not about such a simple behavior. Rather, we shall primarily discuss cognition, which broadly refers to mental processes, intermediate between perception and movement, that are not enslaved by the immediate sensory environment (Shadlen and Kiani 2013). Examples include: how does the brain internally maintain and process information in the absence of any sensory stimulation (an ability called working memory)? How do we evaluate alternative options leading to a deliberate choice in decision making? How can our reflexive actions be kept in check when they are not appropriate? How does a neural system perform logical reasoning? Cognitive abilities are essential for volitional actions, thought and general intelligence (Gazzaniga et al. 2013). Neuroscience of cognition is not only a frontier of basic research, but also has potential implications for mental healthcare and future artificial intelligence. We will strive to understand cognitive functions across levels spanning the biophysics of neurons and synapses, recurrent neural circuits and computational principles. The aim is to build a scientific foundation to explain the immaterial mind in terms of its physical substrate (the biological brain), in the spirit of Crick's quote expressed in his inimitable style.

An impediment in the field has been the limited availability of experimental tools at our disposal, as is clearly illustrated by a comparison between cognitive and early sensory processes. To elucidate visual information encoding in the retina, we can use low-level animal species such as a fly or a salamander *in vivo*; we can also carry out experiments on an isolated retina *in vitro*, so that neural circuit dissection can be conducted

DOI: 10.1201/9781003459361-2

rigorously and systematically (Dowling 1987). These approaches are usually not feasible for cognitive functions such as deliberate and subjective choices by humans or nonhuman animals. The brain is organized modularly, wherein some structures are more important than others for internally governed cognitive functions. In the words of Aleksandr Romanovich Luria (1980): "the material basis of the higher nervous processes is the brain as a whole but that the brain is a highly differentiated system whose parts are responsible for different aspects of the unified whole." Only mammals have a well-developed neocortex. Today we know that the prefrontal cortex (PFC), the frontal part of the neocortex located above the eyes, is crucial for many cognitive functions (Fuster 1989; Goldman-Rakic 1987a; Miller and Cohen 2001; Stuss and Knight 2002; Passingham and Wise 2012; Passingham 2021; Chafee and Heilbronner 2022; Levy 2024). The PFC is especially well developed in primates (Preuss 1995; Wise 2008; Le Merre et al. 2021; Preuss and Wise 2022), with exceptional expansion in great apes and humans (Smaers et al. 2017).

The field of PFC research has changed drastically in the last few decades (Laubach et al. 2018; Arnsten 2023), thanks to a confluence of happenstances. First, functional brain imaging has shown that the PFC plays a central role in many higher cognitive functions. Second, neurophysiological experiments with nonhuman primates, and increasingly with rodents, have identified neural correlates of basic cognitive processes at the levels of single cells and neural populations. Third, in close interactions with experimentation, theory and models have begun to identify neural circuit mechanisms and general principles underlying cognition. Of course, mental processes work together with sensation and movement, which interact reciprocally (with action plan influencing perception and vice versa) (Gibson 1979; Prinz 1997; Warren 2006; Fuster 1989). In the brain, the PFC plays a central role in executive control of perception and action, through its interactions with other cortical (Fuster 1985; Goldman-Rakic 1987a; Petrides and Pandya 2002; Passingham and Wise 2012; Barbas 2015) and subcortical (Janacsek et al. 2022) brain structures.

The focus here is on *how* the brain works. Consider perceptual decision making where you are faced with ambiguous sensory inputs and must make judgments under uncertainty, such as steering a car while driving in a heavy snowstorm or making a medical diagnosis based on probabilistic evidence. A behavioral task commonly used in laboratory experiments for studying perceptual decision making is called the random dot direction discrimination task: In a visual motion stimulus, a fraction (motion coherence) of dots moves coherently in one of two possible directions (e.g., A, left; B, right), and subjects make a choice about the correct net motion direction (Newsome et al. 1989). When the motion coherence is small (say 5%, while 95% of dots move in random directions), the stimulus is quite weak, yet a categorical judgment must be rendered under uncertainty. This simple paradigm captures salient characteristics of daily life perceptual judgments based on noisy and incomplete information. A series of laboratory studies using monkeys performing this task, pioneered by Tony Movshon, Bill Newsome and others, have yielded a wealth of information about the neural correlates of perceptual decision making. At the behavioral level, the psychometric function (accuracy) and chronometric function (reaction time) as a function of the motion coherence are quantitatively assessed. Neurobiologically, whereas the visual middle temporal (MT) area encodes veridical motion direction, neurons in the PFC and the posterior parietal cortex (PPC) signal subjective choices (e.g., in erroneous trials where the judgment is A while the actual motion stimulus direction is B). Moreover, the temporal dynamics of decision-correlated neurons display a linear ramping over time, suggestive of a temporal integration process underlying accumulation of evidence in favor of different options (Gold and Shadlen 2007).

These experimental observations are broadly consistent with the drift-diffusion model, popular in mathematical psychology (Ratcliff 1978; Shadlen et al. 2007), which describes an ideal integration over time (in the sense of Calculus) to a decision threshold (a criterion for a sufficient cumulative amount of evidence to yield a judgment). This model is simple and elegant, the algorithm of which amounts to a few lines of programming code. However, to genuinely achieve a mechanistic explanation, theoretical neuroscience should tackle questions such as: (1) What may be the biological mechanism for a long-time integration in a neural circuit? (2) What are the sources of stochasticity in single-trial probabilistic decision making? (3) Can accumulation of evidence and decision choice be accomplished in a single neural network? (4) What is the brain substrate of a decision threshold, confidence of judgment, speed-accuracy tradeoff? (5) What is the general principle of a distributed decision process engaging multiple brain regions? etc. Going beyond abstract mathematical modeling towards cross-level mechanistic understanding not only would represent a fundamental advance at the interface of neurobiology and cognitive science, but it is also necessary for translational research in psychiatry, identifying key molecules and neural circuit pathways of cognitive deficits associated with mental disorders.

1.2 ON EPISTEMOLOGY

Our inquiry animates philosophical questions of two kinds (Churchland 1989). One aims at providing insights from a neuroscientific perspective into traditional puzzles such as consciousness (Dennett 1991; Dehaene 2014); the other addresses the field's epistemological questions like what constitutes an explanation in brain sciences (Craver 2007; Kaplan 2011; Cao and Yamins 2024). A neuroscientist may choose not to work on big questions about the mind but cannot avoid epistemological matters. What does it mean to understand the brain and its functions? How can we improve our philosophical methodologies in neuroscience?

A most important issue, at the heart of our endeavor to understand the brain mechanism of cognition and intelligence, is concerned with the relationship between behavior and neural circuits. A popular guide to tackle this knotty problem was proposed by David Marr (Marr 1982), a pioneer of computational neuroscience who died at the young age of 35 in 1980: first, formulate the problem and identify its normative solution (the way it should be, optimally); second, search for computational algorithms that accomplish the optimal solution; and third, elucidate hardware implementations of such algorithms in the brain. Over the years, Marr's legacy has been mostly remembered for his three-step recipe of brain theory-building (Marr II). Yet, it would do Marr injustice to forget his deep roots in neurobiology; his Ph.D. research (Marr I) led to his seminal trilogy of papers on the theories of cerebellum (Marr 1969), neocortex (Marr 1970) and hippocampus (Marr 1971). In the words of Francis Crick (1991):

> David [Marr]'s work clearly falls into two phases: Marr I was concerned with neural circuitry and what it might compute. Marr II (the AI phase) was more functional. The emphasis was on the theory of the process and possible algorithms, with much less attention to realistic implementations. I believe that if he had lived he would have moved to a synthesis of these two approaches.

An earlier manifesto of the three levels, in collaboration with Tommy Poggio, explicitly stressed the necessity of simultaneous investigations at all levels (Marr and Poggio 1977; Poggio 2012). Nevertheless, some, too simplistically, took Marrian "three levels" as a unidirectional function→ algorithm→implementation linear framework. As neuroscience advances, it has been argued that Marrian progression from behavior ("the outside") to the brain ("the inside") is problematic, partly due to considerable difficulties with mapping psychological constructs onto neural substrates; instead, a "from inside out" approach would be more effective (Buzsáki 2019, 2020). In support of this view, empirical evidence shows that the brain is as active "at rest" in the absence of obvious external stimulation as during performance of tasks (Raichle et al. 2001); activity patterns in a sensory area like the primary visual cortex (V1) are similar in the spontaneous state or in response to a stimulus (Tsodyks et al. 1999); the distribution of firing rates across recorded single cells of the prefrontal cortex is virtually identical before a task trial begins, during stimulus presentation and when the animal is engaged in working memory (Murray et al. 2017a; Wang 2021). Therefore, conceptualization of the brain merely as an input (stimulus)-output (behavior) information processing system seems quite limited. This is particularly relevant to cognition, which critically depends on internal states of the brain.

On the other hand, there is no lack of pushback to "the epistemological primacy of the hardware" (Poeppel and Adolfi 2020). A not uncommon concern is that neuroscience and cognitive psychology have in recent years over-emphasized neural measurements and tool developments, to the point that behavior appears to take a secondary place (Krakauer et al. 2017; Gomez-Marin and Ghazanfar 2019; Niv 2021). Back to the example of the visual motion direction discrimination, the task must be carefully designed to ascertain that an animal is indeed using temporal accumulation of information in making a decision, that recorded neural activity in a brain region is causally important for the perceptual decision, etc. Ignoring behavioral considerations would be a mistake for anyone interested in the brain basis of cognition. The notion that behavior and functions would automatically emerge from a biologically detailed computer model of the brain is misguided and could potentially lead to the field "veering off course" (Destexhe 2021; Frégnac 2021; Fairhall 2021; Jirsa 2021; Marder and Rue 2021).

Normative theories center on the optimality of behavior, which has a special appeal. This approach has been supported by some empirical observations, especially in early sensory systems (Barlow 1961; Rieke et al. 1999; Bialek 2012) and motor systems (Wolpert and Ghahramani 2000; Todorov 2004; Shadmehr and Mussa-Ivaldi 2012). In the cognitive domains, however, there is little evidence that our judgments and choices are optimal (Simon 1997; Kahneman et al. 1982; Gigerenzer et al. 2000; Rahnev and Denison 2018). Optimal behavior models can serve as a useful tool for exploring performance benchmarks. However, while adaptation leads to better performance, it does not follow logically that optimality has already been achieved for a specific behavior through biological evolution or overtraining (Doebeli et al. 2017). Importantly, optimality considerations need to take into account the facts that modifications in the brain by natural selection, development or learning are made on existing biological systems (Lamb 2013; Cisek 2019); computational resources are limited, and the neural systems are stochastic (Lieder and Griffiths 2020; Summerfield and Parpart 2021; Glimcher 2022).

The viewpoint of behavior primacy is linked to the traditional metaphor of brains as computers (von Neumann 1958). An abstract Turing machine capable of simulating any algorithm can be realized by a variety of physical devices, from the first general-purpose computer *ENIAC* made of vacuum tubes to modern supercomputers powered by integrated circuits. This is dubbed "multiple realizability" (one function realizable in many different ways), a functionalistic view advocated by Putnam (1960). However, this dogma has been thoroughly criticized on epistemological grounds (Block 1978; Bechtel and Mundale 1999; Besson 1999).

Proponents of the functionalism often cite the example of a small neural network ("central pattern generator") in the lobster or crab stomach, where multiple configurations of synaptic and cellular parameter values can achieve the same desirable functional output, namely, food grinding (Prinz et al. 2004). This system thus appears to demonstrate multiple realizability. However, in this example, the "target" behavior is quite simple, namely, a tri-phasic oscillation pattern. Moreover, redundancy becomes understandable when one considers the need to ensure the robustness of a desired behavioral output in spite of perturbations and drastic changes throughout an animal's lifetime (Marder and Goaillard 2006), or for crustacean species living at various temperatures and with different diets. Thanks to redundancy, diverse phenotypes are available, which may be essential for evolvability by natural selection (Changeux and Danchin 1976; Edelman and Gally 2001; Whitacre and Bender 2010). By contrast, for the PFC, we cannot limit ourselves to consider only one "target" function. Instead, we need to first identify a long list of various cognitive processes in which the PFC plays a central role, then investigate how many ways they can actually all be realized in a single neural system. Thus, we do not know if multiple realizability is even relevant for such a complex system as the PFC underlying a wide range of cognitive functions. A superficial take on multiple realizability implies a decreasing order of importance across Marrian levels; if a desired computation can be carried out by many hardwares, dissection of neural circuits would seem to be not so fundamental. But what if this presumption is wrong? Perhaps, those who care only about algorithms recognize that the software of deep convolutional neural networks, the workhorse of today's AI, was actually inspired by our knowledge about the hardware of the primate visual system.

It is likely that our actions and human nature cannot be scientifically understood without elucidating the brain (Badre 2020; Deisseroth 2021). As Patricia Churchland (1996) pointedly put it:

> What precisely are supposed to be the programmatic sequelae to the multiple realizability argument? Is it that neuroscience is irrelevant to understanding the nature of the human mind? Obviously not. . . . Is it perhaps that we should not get our hopes up too high? What precisely is too high here? Is it the hope that we will discover the general principles of how the brain works? Why is that too high a hope?

1.3 THE MIND-BRAIN SOCIETY

Time is ripe for a more sophisticated epistemology than a naïve unidirectional hierarchy of three levels. Minsky (1988) likened the mind to a society. Society could also be a useful metaphor for the brain, where billions of neurons are akin to individuals, and synaptic circuits are similar to social networks (assuming that relations between individuals are defined in a specific way by a mathematical graph). Like a society, the brain is characterized by multiple interacting systems. For instance, it has multiple decision centers specialized in different (potentially conflicting) considerations (Redish 2013), akin to fights among interest groups in a society.

Brain functions are underpinned by neuroanatomy (connectome), just like a human society is organized by, and thus must be understood in terms of, its structure. Furthermore, it would be absurd to claim that as far as the behavior of a society in whatever domain is concerned, observing and understanding individuals do not matter. Similarly, in a neural system, the biophysics of single synapses and neurons have functional implications for the whole animals. Thus, we cannot achieve understanding of behavior arising from the whole brain without a profound knowledge about the inner workings of its biological components.

This observation brings us to another question: what do "levels" mean exactly in the Marrian framework? How levels are defined has epistemological implications (Wimsatt 1976; Craver 2015). Behavior is observable at the macroscopic level of the whole animal, whereas in much of the twentieth century, neurophysiology has been very much preoccupied with microscopic single-neuron activity (Barlow 1972) and local neural networks (Douglas and Martin 1991). Is this mismatch of spatial scales the reason for calling the behavior-to-brain approach "top-down"? If so, it would be an accidental consequence of past experimental limitations in neuroscience, which are currently being overcome with new technologies to investigate the whole macroscopic brain. Both the behavior and the neural system exhibit multiple levels of organization. The biological brain is organized from molecules to synapses and neurons, to local circuits and large-scale systems (Churchland and Sejnowski 1992), defined in terms of multiple spatial scales (**Figure 1.1**).

Behavioral repertoires are rich (see www.cognitiveatlas.org) and also exhibit multiple levels (Von Uexküll 2010; Tinbergen 1951; Hinde 1966; Hutt and Hutt 1970; Dawkins 1976). Cognitive science and neuroscience implicitly assume a taxonomy of perception, cognition and action (**Figure 1.2a**). For instance, cognition involves attention, memory, intention, thinking and decision making; memory in turn has multiple types. This approach, originating from the psychology tradition, is useful, but functions thus defined do not easily map onto the brain system. Alternatively,

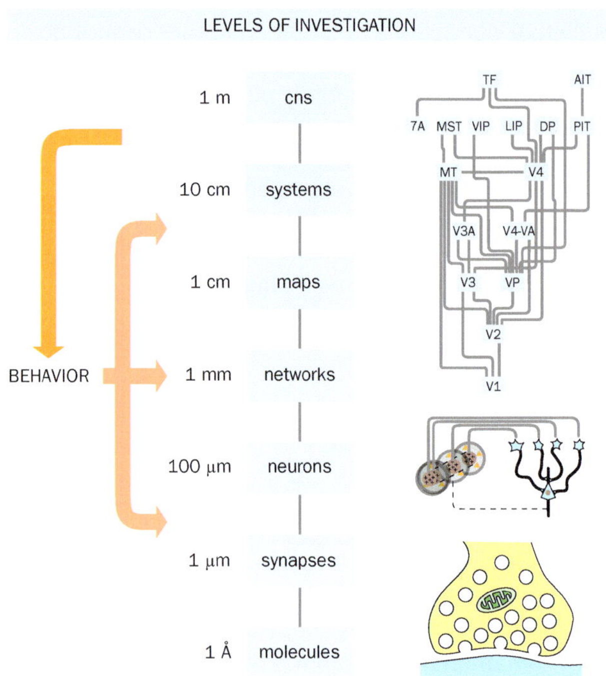

Figure 1.1 Organization of a central nervous system (CNS) on multiple spatial scales. Given an external environment including social setting, a behavior in turn leads to plastic changes in single synapses and cells as well as at the level of circuits. The system generates behavior (upper left dark orange arrow), which induces changes across several spatial scales in the brain (light orange arrows). Adapted from Churchland and Sejnowski (1992). [Courtesy of Terry Sejnowski.]

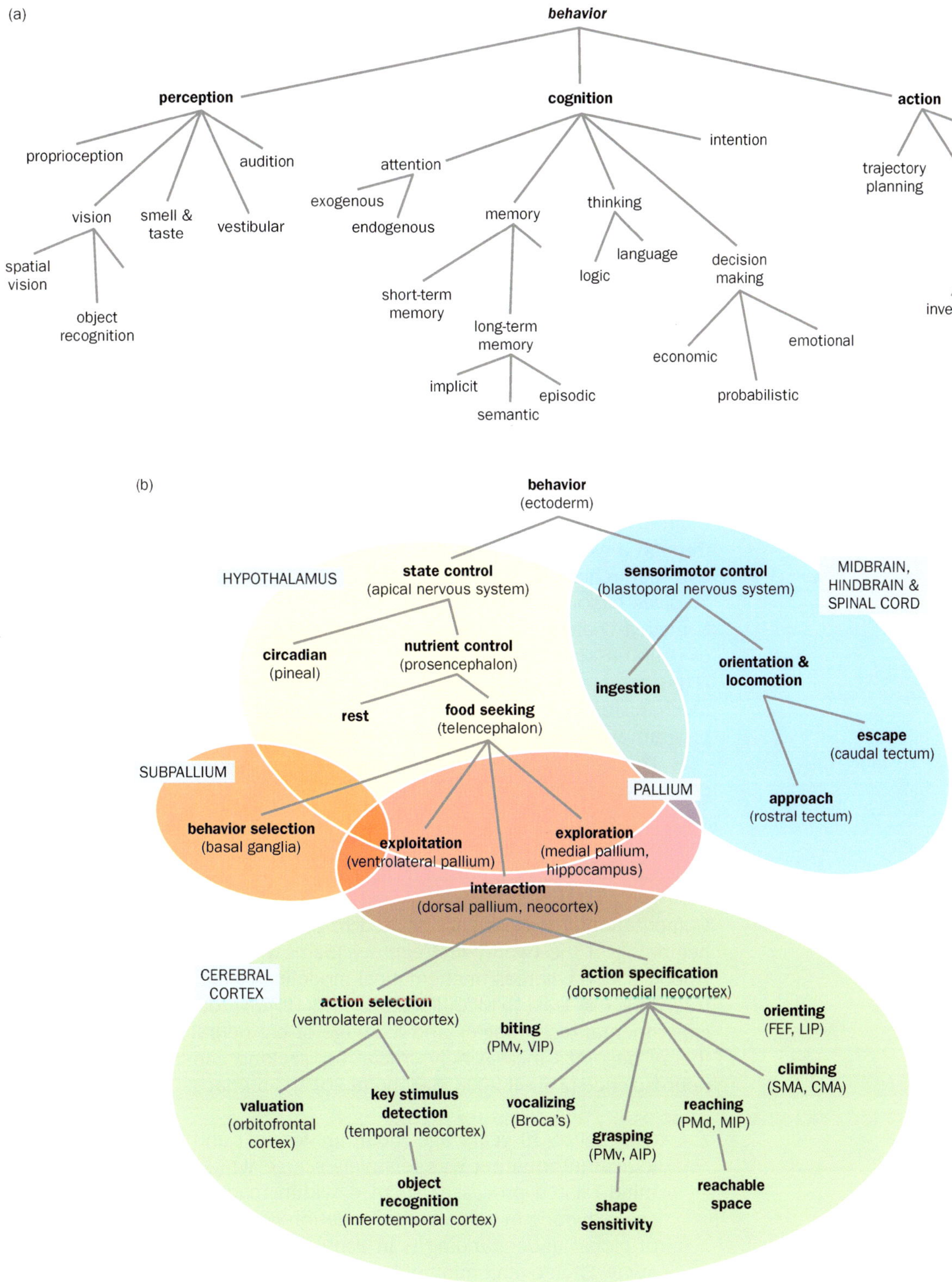

Figure 1.2 Hierarchy of behavior. (a) Partial sketch of a conceptual taxonomy implicit in mainstream cognitive science and neuroscience. (b) An alternative conceptual taxonomy resulting from following a phylogenetic approach along the vertebrate lineage. Here, each functional category is conceived as a particular specialization of the functional category above it, and each corresponds to a biological structure that emerged as a specialization within an ancestral structure. AIP, anterior intraparietal area; CMA, cingulate motor area; FEF, frontal eye fields; LIP, lateral intraparietal area; MIP, medial intraparietal area; PMd, dorsal premotor cortex; PMv, ventral premotor cortex; SMA, supplemental motor area; VIP, ventral intraparietal area. [Reproduced from Cisek (2019). Cisek, P. (2019). Resynthesizing behavior through phylogenetic refinement. Attention, Perception, & Psychophysics. 81(7), 2265–2287.]

the perspective of "phylogenetic refinement" aims at advancing a theory of behavior based on biological evolution (**Figure 1.2b**) (Cisek 2019; Pessoa et al. 2022). According to this view, behaviors began with the need for sensorimotor control in the most primitive species that move to seek food and mates, and progressively expanded to object recognition, exploration versus exploitation, climbing, reaching, vocalization, etc. They form a hierarchy, in which biological substrate(s) specialized for each of these can potentially be identified according to its emergence "at an empirically definable time in history." Presently, this idea remains qualitative and limited to simple behaviors; how the complex higher cognitive functions can be described by a quantitative ontology remains an open question, even for a particular trait such as behavioral self-regulation (Eisenberg et al. 2019). A hierarchical scheme misses the richness of relationships between behaviors (and their associated brain structures), which may share features of a heterarchical network (in which no unit dominates the rest and any unit can govern or be governed by others depending on circumstances) (Bechtel 2019).

In summary, neither the mind nor the brain can be assigned to a single level in a linear Marrian hierarchy. Instead, each is a rich world with its own multi-level structure. The connections between the two societies are defined through a web of complex interactions across many different levels of the two hierarchies of brain and behavior. For instance, learning in behavior involves biological processes ranging from plasticity that depends on the N-methyl-D-aspartate (NMDA) receptors at the synapse level to formation of neural assemblies and memory consolidation through the brain system (**Figure 1.1**). The goal of neuroscience is to understand the biological brain mechanisms that causally explain behavior and cognition; success in this program would establish a rigorous science of the mind in neuronal terms.

This book attempts to make the case for a new epistemology called cross-level mechanistic theory.

1.4 CROSS-LEVEL MECHANISTIC THEORY

Computational neuroscience started to take shape as a field around the beginning of the twenty-first century (Sejnowski et al. 1988; Dayan and Abbott 2001), whereas mathematical modeling of neural networks can be traced back at least to McCulloch and Pitts (1943). Today, with advances in experimental technologies and a deluge of big neural data (**Figure 1.3**), theory becomes increasingly crucial for seeking causal explanations in neuroscience (Abbott 2008; Carandini 2012). Frégnac (2017) observed:

> Only 20 to 30 years ago, neuroanatomical and neurophysiological information was relatively scarce while understanding mind-related processes seemed within reach. Nowadays, we are drowning in a flood of information. Paradoxically all sense of global understanding is in acute danger of getting washed away. Each overcoming of technological barriers opens a Pandora's box by revealing hidden variables, mechanisms, and nonlinearities, adding new levels of complexity.

Churchland and Abbott (2016) wrote: "Obtaining deep understanding from this onslaught will require, in addition to the skillful and creative application of experimental technologies, substantial advances in data analysis methods and intense application of theoretic concepts and models." A wide range of approaches includes behavioral theories, algorithmic models, graph analysis of connectome, biophysical models of ion channels, cell types and neural circuits.

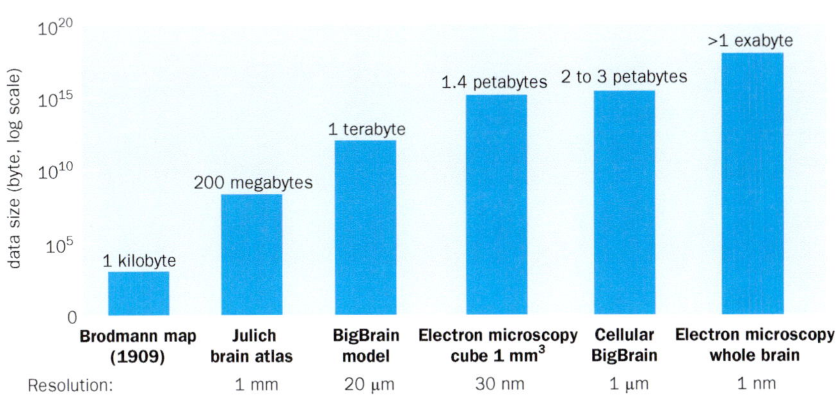

Figure 1.3 Big data from human brain research. The size of data is exponentially growing according to the resolution of studies on the human brain connectome and organization. The two largest bars are estimated sizes because such data do not exist for the human brain. Reproduced from Amunts, K., & Lippert, T. (2021). Brain research challenges supercomputing. Science, 374(6571), 1054–1055.

What are the relative merits of endeavors to solve specific brain/mind puzzles versus those to develop a grand unifying theory of the brain (Cobb 2020)? Not all methods are equally effective, and better approaches accelerate research progress (Platt 1964), but to some extent the answer depends on a scientist's taste and aesthetics. Physics is full of examples of major breakthroughs that resulted from solving specific problems. When classical Newtonian physics failed to explain the observed properties of black body radiation, Max Planck's solution ushered in the birth of quantum mechanics (Segrè 2007). Another example is the so-called Lamb shift, an observed tiny difference of two energy levels of the hydrogen atom not predicted by quantum mechanics in the 1940s, which became the harbinger of quantum electrodynamics, the relativistic quantum field theory of electromagnetism. In biology, Charles Darwin did not start his research with a grand theory; his preoccupation with explaining the diversity of beaks among finches collected at the Galápagos Islands eventually led to his theory of evolution through natural selection of random variations.

As an example in neuroscience, theoretical consideration of highly irregular spiking activity of cortical cells (Softky and Koch 1993) gave rise to the idea of synaptic excitation and inhibition balance (Shadlen and Newsome 1994; van Vreeswijk and Sompolinsky 1996), which became a general concept of neuroscience. To understand the PFC, one could attempt an all-encompassing normative theory that must explain all PFC functions. Alternatively, one can focus on one particular phenomenon at a time. Taking the latter approach, experiments by Fuster (1989) and Goldman-Rakic (1987a) focused on working memory, which can be studied using delayed response tasks where information about a brief stimulus must be held internally in the brain over a short period of time to produce a memory-based response later. Their seminal findings on mnemonic self-sustained persistent neural activity played a critical role in spurring the growth of PFC research in recent decades (Figure 1.4). This is not because working memory of sensory information can be exclusively mapped onto the PFC. Rather, by defining a specific new problem, neuroscientists were galvanized in research on the brain mechanism of active maintenance and manipulation of information in the absence of external stimulation, which enables the PFC to subserve cognition. Moreover, as we shall see in this book, theoretical considerations of information-specific persistent neural activity led to a common neural circuit mechanism for working memory and decision making, thereby suggesting a general "cognitive-type" neural circuit (Wang 2013). These examples argue in favor of the practice of focusing on concrete problems that ought to be as fruitful in neuroscience as in physics.

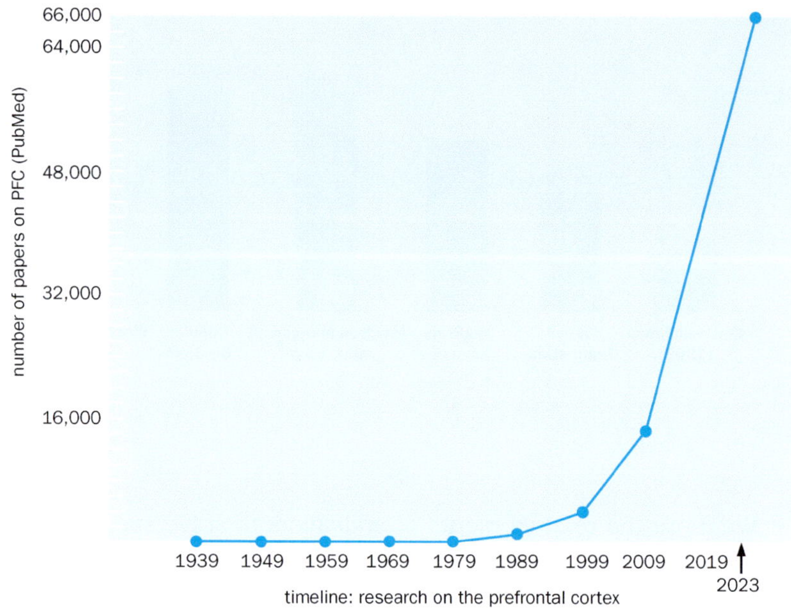

Figure 1.4 Timeline of the number of papers published on the "prefrontal cortex" (search term in PubMed). [Courtesy of Amy Arnsten.]

Theory and mathematical models in neuroscience should and eventually will play a similar role as theoretical physics does in physics. In close interplay with experiments, theory is needed to (a) identify computational algorithms carried out by a given empirically measured neural population activity and to assess whether it can explain behavior; and (b) elucidate the underlying circuit mechanism, i.e., how the observed neural signals arise from genes, cells and connectivity (**Figure 1.5**). Furthermore, internal computation underlying cognition depends on recurrent neural circuits, which are difficult to understand by experiments alone. For instance, consider three interacting brain areas A, B and C. An alternation of activity in area A causes changes in areas B and C, which in turn produce modified outputs to each other and to area A, leading to further changes in all three areas; predicting the end result is not straightforward. Interpreting experimental observations would benefit from a theoretical framework for testing scientific hypotheses.

The approach taken in this book, namely, cross-level mechanistic understanding, is by no means equivalent to the radical reductionism defined by macroscopic-to-microscopic progression of explanation. Brain functions and behavior cannot be reduced to the smallest constituents of a central nervous system. Rather, they emerge as collective phenomena of interacting neurons described as a large and complex nonlinear dynamical system. One can be both a reductionist and an emergentist (Wimsatt 1976). The physicist Leo Kadanoff was fond of saying: "The world is like an onion." What he meant was that we progress in understanding the world by peeling through layers, one at a time, going from observed phenomena at one level to the underlying mechanism at a deeper level. In neuroscience, there are many levels of investigation. Crucially, the brain is plastic; behavior generated by the brain in turn alters the nervous system on multiple spatial scales from synapses and neurons to circuits. The challenge is to achieve understanding across levels for complex systems (Anderson 1972; Changeux 1983; Churchland and Sejnowski 1992; Craver 2007).

To meet this challenge, we shall eschew the computer as a metaphor for the brain. Van Gelder (1995) wrote: "Rather than computers, cognitive

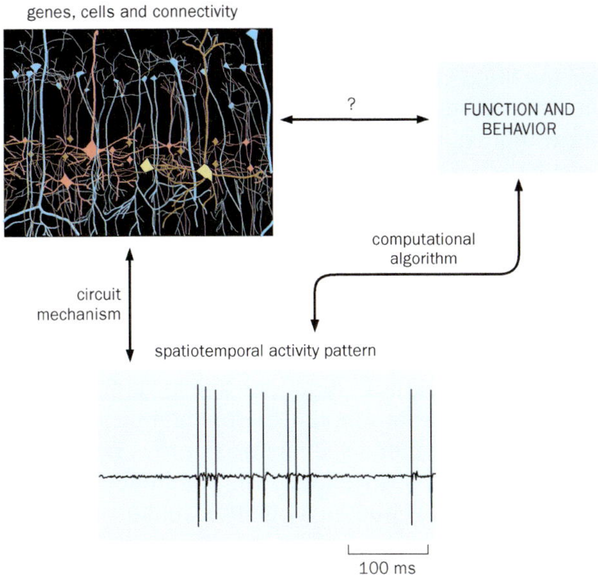

Figure 1.5 Schematic inter-relationship between brain function/behavior, neural activity pattern and biological matter. Amarasingham, A., Chen, T. L., Geman, S., Harrison, M. T., Sheinberg, D. L., Spike count reliability and the Poisson hypothesis. *J Neurosci.* 26(3): 801–9. Copyright 2006 Society for Neuroscience.

systems may be dynamical systems; rather than computation, cognitive processes may be state-space evolution within these very different kinds of systems." In a similar spirit (Kelty-Stephen et al. 2022), the theoretical emphasis of this book is on treating a neural circuit as a dynamical system. The framework of dynamical systems provides a natural mathematical language for describing complex neural population spatiotemporal activity patterns. A dynamical system describes how the current state of a system (e.g., a neural circuit) and interactions between constituent units (e.g., neurons) predict the future evolution of the system over time (Strogatz 2016). The theory of dynamical systems offers a variety of concepts and analysis tools for understanding neural population activity (Freeman 1975; Ermentrout and Terman 2010).

The dynamical system perspective highlights collective behaviors as emergent phenomena, but can also capture properties of constituent units, so therefore it is suited for cross-level understanding. In studies at the single-cell resolution, a shift of focus from individual neurons to neural populations naturally led to describing neural population activity as a dynamical system in a high-dimensional state space (Shenoy et al. 2013; Vyas et al. 2020). This perspective has gained increasing attention in recent years with the technological advances in recording simultaneously from many single cells in behaving animals, opening the door to probe the brain basis of behavior and cognition in real time. While dynamical system modeling appears to emphasize self-organized emergent behavior, it can and should take into account biological details such as biophysics and selectivity of single neurons, as we shall see throughout the book. Therefore, the supposed dichotomy between specific neuronal computation and collective network function (Barack and Krakauer 2021) will turn out to be a straw man, once we achieve cross-level understanding of cognition.

A solid foundation built along these lines is relevant for both basic research and clinical psychiatry. A recent polemic asserted: "Currently, there is no compelling evidence for the viability of reducing mental disorders to unique biological abnormalities, both in terms of enhanced etiological understanding and of improving the effectiveness of interventions" (Borsboom et al. 2019). Genomics of psychiatric disorders is seen as the poster child of "explanatory reductionism" that has not delivered, but instead of concluding that the field is set on the wrong track, we should acknowledge the huge gulf between genes and brain functions and redouble efforts in cross-level translational neuroscience. A main goal of this book is to show that cognition, and cognitive deficits associated with mental illness, can be elucidated theoretically, bridging molecular and neural circuit mechanisms. Thus, this line of research offers a path forward in integrating the biology of the brain and the psychology of the mind.

1.5 LAYOUT OF THE BOOK

This book is focused on cognitive functions that are sufficiently "elemental" to be studied using relatively simple experimental paradigms that can be performed by nonhuman animals, enabling rigorous neural circuit dissection. Lessons learnt from such efforts will potentially shed insights into higher-level, some uniquely human, cognitive abilities. Since our ultimate goal is to understand how the brain subserves cognition, our emphasis will be on "biologically based modeling," by which we mean modeling that takes into consideration knowledge from the two main pillars of neuroscience: anatomy (connectome) and neurophysiology.

We will cover basic functions such as working memory, decision making and executive control of behavior. These themes have a long history in psychology (Gazzaniga et al. 2013). Each theme as a basic "building block" of cognition has been investigated extensively in laboratory experiments with careful task designs to isolate a specific aspect of cognition, validating the "divide-and-conquer" scientific method. Consideration of each as a distinct mental function has been questioned on the epistemic grounds that they do not neatly map onto non-overlapping brain substrates; from the viewpoint of biological evolution, our ultimate goal is to acquire knowledge about behavioral flexibility and its brain basis rather than psychological constructs *per se* (Pessoa et al. 2022). However, at a minimum, the psychological description of functional taxonomy (e.g., working memory, decision making, executive control) is valuable, even though it does not constitute a mechanistic explanation. Here, we take the perspective that basic cognitive functions can be studied separately, and insights thus gained will facilitate our investigations of more elaborate behaviors through flexible combinations of these building blocks. As we shall see throughout the book, this approach in our inquiries paves the way for the nascent understanding of cognitive functions that are distributed across overlapping neural systems in the whole brain.

This book consists of three parts. Part I is devoted to models and concepts about neurons and synapses (Chapter 2), neural circuits (Chapter 3), plasticity and learning (Chapter 4). The aims are twofold. One is to provide a pedagogical introduction of theoretical neuroscience to students and researchers. Second, the neurobiology and mathematics needed to follow the rest of the book can be found therein, so that the book is essentially self-contained. In that sense, Part I is a preliminary to the coverage of cognition in Parts II and III. We shall discuss the basics, as well as cutting-edge topics.

Part II constitutes the core of this book. In order to make it accessible to a broad audience in neurobiological and cognitive sciences, mathematics is kept at its minimum here, and the reader is referred to Part I for more technical details. We shall start in Chapter 5 with working memory, which is essential for all cognitive functions while relatively simple for behavioral experiments, neural circuit dissection and cross-level theoretical investigations. Then, in Chapter 6, we will show that the model developed for working memory accounts for decision making at both the behavioral and neurophysiological levels. This theme is expanded in Chapter 7 to cover value-based choice behavior in a cross-disciplinary field called neuroeconomics. Chapter 8 is devoted to behavioral flexibility, both in the sensory domain (selective attention) and motor domain (inhibitory control of actions). Executive function is critically dependent on the PFC; elucidating this type of neural circuit holds the key to understanding flexible behavior.

In Part III, we will discuss frontier areas of neuroscience. First, most theory and models have so far been about local circuits, such as rhythmic movement from a central pattern generator, orientation selectivity of neurons in the primary visual cortex. With the advance of brain connectome and new neurophysiological tools, we are now in a position to develop theory and models of multi-regional large-scale brain systems, which are discussed in Chapter 9. Second, since the PFC and associated brain regions are implicated in virtually all psychiatric disorders, it is natural to explore how progress in basic research covered here may shed insights into clinical research on mental disorders, in a nascent field called computational psychiatry, which is covered in Chapter 10. Third, Chapter 11 focuses on the reciprocal interplay between brain science and artificial intelligence (AI). We will discuss how modern machine learning tools can be deployed to build models for increasingly more complex cognitive functions that are of central importance to intelligence, such as multi-tasking, abstract knowledge and reasoning. Conversely, recent experiments and computational theories of these brain functions have the potential to inspire the next generation of smart machine systems. Finally, in Chapter 12, we will provide a summary of the book and discuss briefly how cross-species neuroscience could close the explanatory gap between nonhuman animals and *Homo Sapiens*.

FURTHER READING

Anderson, P. W. (1972). More is different. *Science 177*, 393–396.

Barlow, H. B. (1961). Possible principles underlying the transformations of sensory messages. In W. A. Rosenblith (Ed.), *Sensory Communication*, pp. 217–234. Cambridge, MA: MIT Press.

Buzsáki, G. (2019). *The Brain from Inside Out*. New York: Oxford University Press.

Churchland, P. (1989). *Neurophilosophy: Toward a Unified Science of the Mind-Brain*. Cambridge, MA: MIT Press.

Craver, C. F. (2007). *Explaining the Brain: Mechanisms and the Mosaic Unity of Neuroscience*. Oxford: Clarendon Press.

Dayan, P. and L. F. Abbott (2001). *Theoretical Neuroscience*. Cambridge, MA: MIT Press.

Fuster, J. M. (1989). *The Prefrontal Cortex* (2nd ed.). New York: Academic Press.

Gazzaniga, M. S., R. B. Ivry, and G. R. Mangun (2013). *Cognitive Neuroscience: The Biology of the Mind* (4th ed.). New York: W. W. Norton & Company.

Goldman-Rakic, P. S. (1987). Circuitry of primate prefrontal cortex and regulation of behavior by representational memory. In F. Plum and V. Mountcastle (Eds.), *Handbook of Physiology—The nervous system V, Chapter 9*, pp. 373–417. Bethesda, MD: American Physiological Society.

Poeppel, D. and F. Adolfi (2020). Against the epistemological primacy of the hardware: The brain from inside out, turned upside down. *eNeuro 7*. https://doi.org/10.1523/ENEURO.0215-20.2020.

Vyas, S., M. D. Golub, D. Sussillo, and K. V. Shenoy (2020). Computation through neural population dynamics. *Annu. Rev. Neurosci. 43*, 249–275.

Wang, X.-J. (2013). The prefrontal cortex as a quintessential "cognitive-type" neural circuit: Working memory and decision making. In D. T. Stuss and R. T. Knight (Eds.), *Principles of Frontal Lobe Function* (2nd ed.), pp. 226–248. New York: Cambridge University Press.

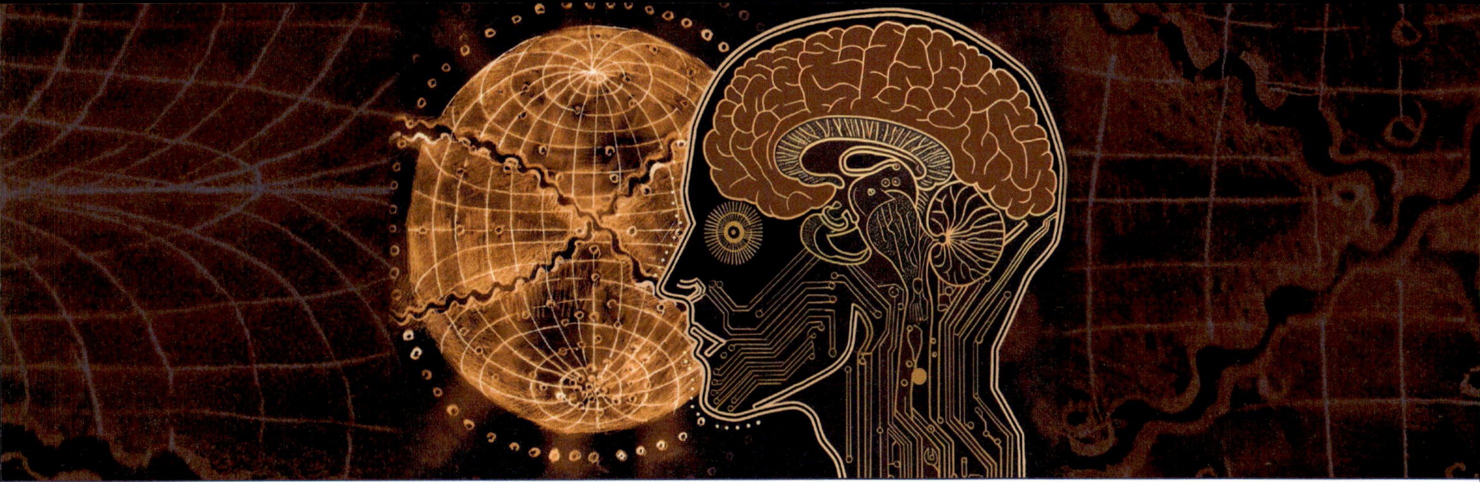

Neurons and Synapses

2

Like the entomologist hunting for brightly colored butterflies, my attention has chased in the gardens of the grey matter cells with delicate and elegant shapes, the mysterious butterflies of the soul, whose beating of wings may one day reveal to us the secrets of the mind.

Santiago Ramón y Cajal in *Recollections of My Life*

2.1 INTRODUCTION

A brain is an intricate biological system composed of nerve cells (neurons) as well as non-neuronal glia cells (Cajal 1911). Individual neurons connect each other through junctions called synapses (Sherrington 1906) (**Figure 2.1a**). A neuron receives electrical signals from other neurons at synapses distributed across its dendrite (a branched, tree-like structure). These inputs converge to the soma (the cell body of a neuron), where the inputs are summated. The overall input determines neuronal output signal in the form of stereotypical electrical pulses called action potentials or spikes. Spikes propagate along fibers called axons away from the cell body to synaptic terminals, the presynaptic side of synapses onto postsynaptic neurons. The brain size spans a wide range across species: the worm *C. elegans*, fly *Drosophila* and mouse have 302, 135,000 and 70 million neurons, respectively. The human brain is empowered with 86 billion neurons (Azevedo et al. 2009); the total length of fibers connecting neurons in our head is roughly the distance between the earth and the moon, forming a tremendously complex circuit underlying our mental life.

To face the challenge of understanding the brain dynamics and functions, a good starting point is to consider microcircuits within a brain region dedicated to a specific function such as the retina at the back of the eyeball for vision. Microcircuits come in innumerable varieties; at the same time some general principles of organization are valid across neural microcircuits ranging from sensory organs and motor pattern generators of invertebrates to the primate prefrontal cortex (PFC) often dubbed "the CEO of the brain" in primates (Shepherd 1998; Shepherd and Grillner 2018).

DOI: 10.1201/9781003459361-3

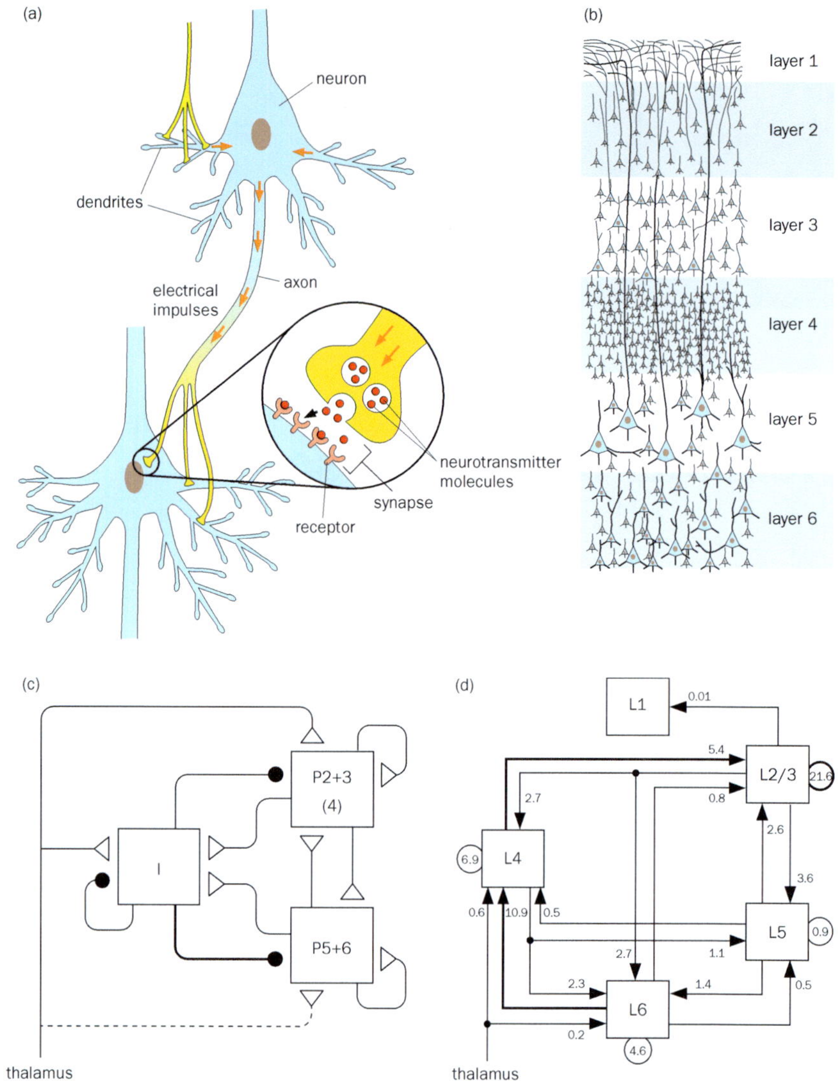

Figure 2.1 Cortical microcircuit. (a) Schematic of two pyramidal neurons, one (upper right) sends an axon to the other (lower left). Insert is the connecting synapse where an action potential in the presynaptic neuron triggers release of neurotransmitters, which travel to the other side of the (20–40 nanometer wide) synaptic cleft and change the activity of the postsynaptic neuron. (b) Neocortex displays six layers across its thickness; only excitatory neurons are shown for the sake of simplicity. (c) Interconnectivity of a layered microcircuit in the neocortex, divided into superficial layers (including also layer 4 spiny stellates) and deep layers. P denotes excitatory pyramidal neurons (also stellate cells in layer 4), I denotes inhibitory interneurons. External inputs to a microcircuit target both superficial excitatory neurons and inhibitory cells. (d) Percentage of all excitatory-to-excitatory synapses for each connection type, indicated by arrows, in the primary visual cortex of the cat. Panel (a) is reproduced from Bear et al. (2015), (b) from Barrett (2017), (c) from Nelson (2002) provided by Kevan Martin, (d) from Binzegger et al. (2004). [Courtesy of Kevan Martin.] [Binzegger, T., R. Douglas, and K. Martin (2004). A quantitative map of the circuit of cat primary visual cortex. J. Neurosci. 24(39), 8441–8453. Copyright 2004 Society for Neuroscience.]

The neocortex, a thin (1–3 mm thick) sheet of nerve tissue beneath the pia that covers the brain, is a distinct feature of mammals in the animal kingdom and plays a major role in higher brain functions. In the neocortex, excitatory neurons increase the activity of target neurons, the majority of which are pyramidal cells named for their pyramid-shaped soma (**Figure 2.1a**). The remaining nerve cells are inhibitory, acting on suppressing target neurons. They are often referred to as interneurons because virtually all of them make connections locally within a cortical region. Intriguingly, excitatory:inhibitory cell ratio varies significantly in different cortical regions; in humans, the inhibitory neurons are proportionally more numerous in the PFC (nearly 30% of all neurons) than in the primary visual area V1 (15%) (Krienen et al. 2020). In a given cortical area, primates have higher percentages of interneurons than do rodents; in the motor cortex, the excitatory to inhibitory cell ratio is 5:1 in mouse, 3:1 in marmoset monkey and 2:1 in human (Bakken et al. 2021).

A cortical microcircuit has a laminar structure subdivided into several layers (**Figure 2.1b**); their interconnections are shown in **Figure 2.1c** simplified with "superficial layers 2/3" (including layer 4) and "deep layers 5/6." It is endowed with local recurrent connection loops; for instance, layer 2/3 pyramidal cells project to neurons in layer 5/6, which in turn project back to layer 2/3. The primary visual cortex (visual area 1 or V1) is the entry point into the neocortex for visual inputs, which arrive in the retina

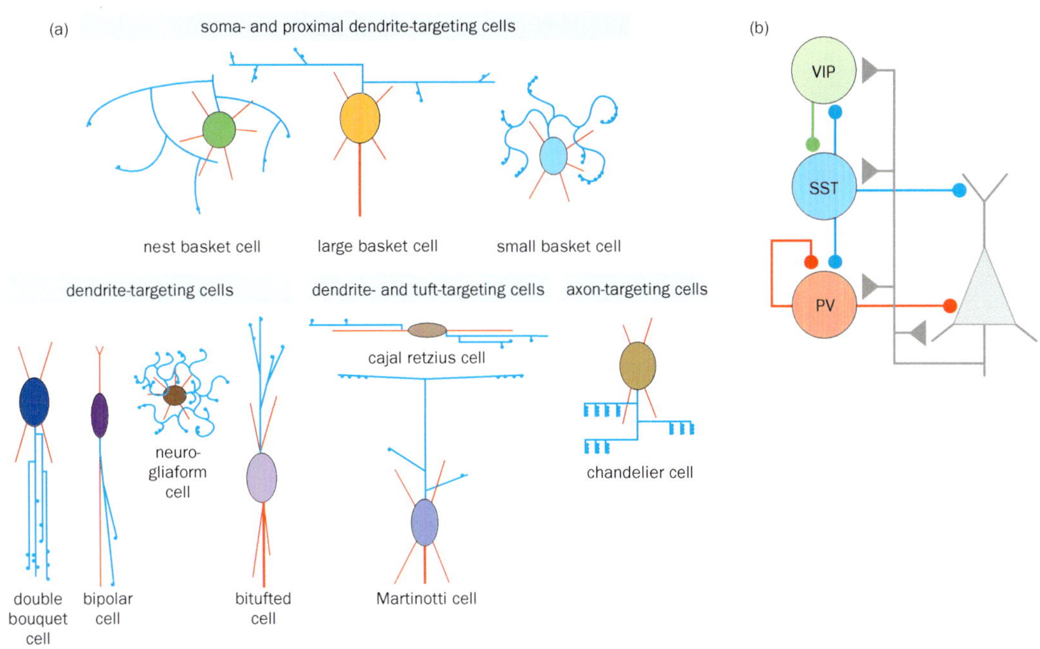

Figure 2.2 Diversity of inhibitory neurons in the neocortex. (a) Scheme summarizing the main anatomical properties of inhibitory interneurons. Each neuron type has a different colored soma; dendrites, red; axons, blue lines; axonal boutons, blue dots. Neurons are orientated with the pia facing upwards. Some interneurons have a prominent vertical dendrite. (b) A disinhibitory motif composed of three types of interneurons, where PV cells target the perisomatic region of a pyramidal cell, and SST cells target the pyramidal dendrite, whereas VIP cells target SST cells. Panel (a) is reproduced from Markram et al. (2004), (b) from Wang and Yang (2018). [Markram, H., Toledo-Rodriguez, M., Wang, Y., Gupta, A., Silberberg, G., & Wu, C. (2004). Interneurons of the neocortical inhibitory system. Nature Reviews Neuroscience, 5(10), 793–807.] [Wang, X. J., & Yang, G. R. (2018). A disinhibitory circuit motif and flexible information routing in the brain. Current Opinion in Neurobiology, 49, 75–83.]

and are relayed to the cortex through the lateral geniculate nucleus (LGN) in the thalamus. Thalamic inputs constitute about 6% of all excitatory synapses in a microcircuit of V1, although physiologically they powerfully drive the activity of target cells. Recurrent connections between excitatory neurons are especially abundant in the superficial layers 2/3, constituting more than 20% of the bulk of excitatory synapses in a given volume of the cortical tissue (**Figure 2.1d**). This feature holds true for excitatory-to-inhibitory, inhibitory-to-excitatory and inhibitory-to-inhibitory connections (Binzegger et al. 2004). In primates, the intrinsic connections in the superficial layers 2/3 display stripe-like patterns in the PFC in contrast to patch-like patterns in V1 (Levitt et al. 1993; Kritzer and Goldman-Rakic 1995). As we shall see throughout this book, feedback loops within cortical microcircuits play a key role in explaining neural population dynamics and cognitive functions.

Another characteristic of the neocortical microcircuits is a wide diversity of inhibitory cells (**Figure 2.2a**) (Kawaguchi and Kubota 1997; Markram et al. 2004; Paul et al. 2017). Three distinct types can be classified according to their projections and protein expressions (**Figure 2.2b**). The first type targets the peri-somatic region of pyramidal cells and acts to control spiking output of excitatory neurons. Soma-targeting cells (STCs) express the calcium-binding protein parvalbumin (PV) and can be labeled as such. The second type, which either express somatostatin (SST) or the calcium-binding protein calbindin (CB), target pyramidal dendrites. Thus SST/CB interneurons are well positioned to control inputs onto excitatory neurons. The third type, identifiable by expression of either vasoactive intestinal peptide (VIP) or the calcium-binding protein calretinin (CR), preferentially targets SST/CB inhibitory neurons (Freund and Buzsáki 1996; DeFelipe 2002). These three interneuron types form a disinhibitory motif, originally proposed theoretically (Wang et al. 2004), which has been shown experimentally to represent a general feature of cortical microcircuits (**Figure 2.2b**) (Kepecs and Fishell

2014; Tremblay et al. 2016). The fourth major type of inhibitory neurons does not express PV, SST or VIP (Machold et al. 2023): they constitute 90% of interneurons in layer 1 with various firing patterns (Meng et al. 2023); in other layers, they either express neuropeptide Y and mediate $GABA_B$ receptor-dependent slow inhibition (Tamás et al. 2003), or express peptide CCK and underlie the cannabinoid action in the brain (Katona et al. 1999).

We shall discuss microcircuits of interconnected neurons in Chapter 3. Here, we will cover the basics of mathematical models of single neurons and synapses, with reference to basic experimental observations of excitatory pyramidal neurons and inhibitory cells in the mammalian cortex. We will primarily describe electrical signaling, but it is important to keep in mind that neuronal operations are shaped by chemical substances. There are a plethora of chemical processes inside a nerve cell that interact with electrical signaling and may underlie interesting computations. As a simple example of this kind, we will describe single-neuron adaptation mediated by intracellular calcium ions. At a synapse, communication is mediated by the release of chemicals (neurotransmitters) from a presynaptic neuron, and binding of transmitters with a receptor in the postsynaptic neuron. Signaling processes of transmitters and neuromodulators are targets of many substances that affect the brain, including therapeutic medicine for mental illness and addictive drugs. Furthermore, the hallmark of the brain as a biological organ is its ability to adapt from experience and acquire knowledge. In Chapter 4 we shall discuss biological mechanisms of plasticity of synaptic connections believed to underlie learning.

As we shall see, biophysical properties of single neurons and synapses matter a lot to brain dynamics and functions, and they display variations across species (Elston et al. 2001; Gidon et al. 2020; Beaulieu-Laroche et al. 2021). Accurate mathematical description of single neurons affects greatly our understanding of collective network behaviors (Chapter 3); neuronal adaptation plays a key role in decision computations (Chapter 6). Biologically realistic modeling of synaptic transmission is crucial for the demonstration of "cognitive-type" neural circuits capable of both working memory and decision making (Chapters 5 and 6), as well as shedding insights into biological deficits associated with psychiatric disorders (Chapter 10). Models of the four major cell types (excitatory pyramidal cells and three inhibitory subclasses) will be used to realize a gating mechanism for a working memory system to filter out behaviorally irrelevant distracting stimuli (Chapter 5) and more generally in a multi-regional large-scale neocortex (Chapter 9). Therefore, what is discussed in this chapter forms a foundation for neural circuit models of cognitive functions covered in Parts II and III.

2.2 INTEGRATE-AND-FIRE NEURON

The electrical signaling of a neuron is determined by voltage difference across its membrane, $V = V_{in} - V_{out}$, where V_{in} and V_{out} are voltages inside and outside the neuron, respectively. The levels of V_{in} and V_{out} are determined by concentrations of ions such as Cl^-, Na^+, K^+ and Ca^{2+}, in the cytosol and in the extracellular medium (Hille 2001). A nerve cell has an elaborated spatial structure (**Figures 2.1a** and **2.2**), and the membrane potential varies from a spatial location to another. In general, $V(x,t)$ depends on both the location x of a spatially extended neuron and time t.

The choice of a single neuron model with a certain degree of complexity should be done judiciously depending on the question the model is designed to investigate. A morphologically fully reconstructed pyramidal neuron (**Figure 2.3a**) may be modeled with hundreds of compartments, each assumed to be isopotential (i.e., the transmembrane potential

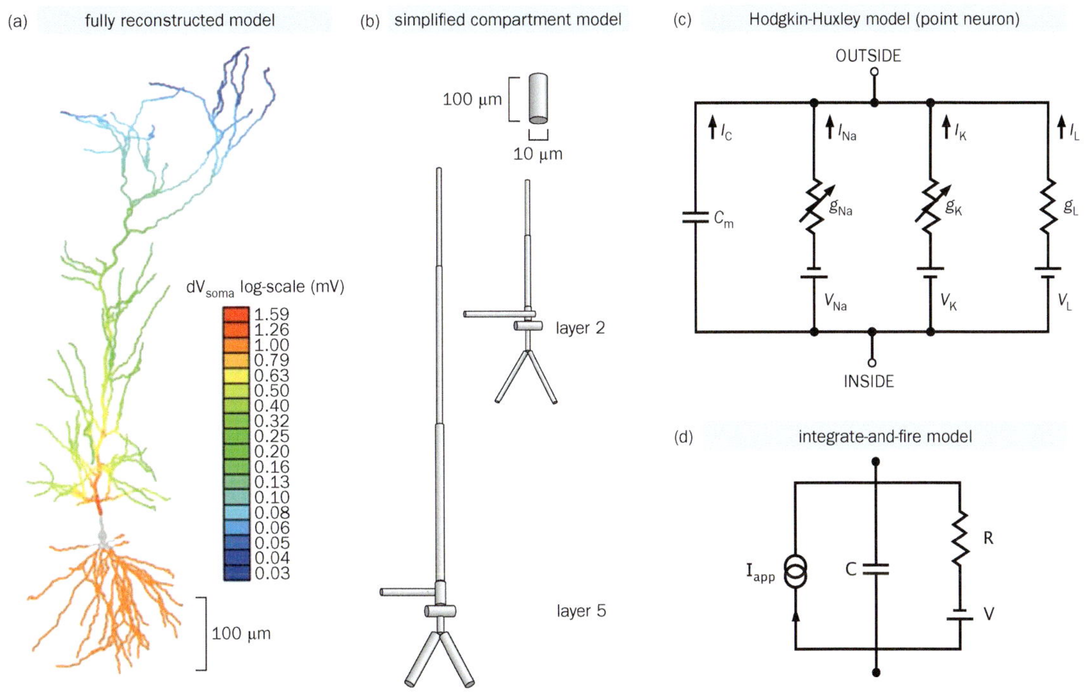

Figure 2.3 Single neuron models. (a) A pyramidal neuron model with a fully reconstructed dendritic tree. (b) A simplified multi-compartmental model of a pyramidal neuron in layer 5 (left) and another one in layer 2/3 (right). (c) A point Hogdkin-Huxley model of action potential. (d) A leaky integrate-and-fire neuron. Panel (a) is reproduced from Nicholson et al. (2006), (b) from Bush and Sejnowski (1993), (c) from Hodgkin and Huxley (1952) and (d) from Koch (1999). [Nicholson, D. A., Trana, R., Katz, Y., Kath, W. L., Spruston, N., & Geinisman, Y. (2006). Distance-dependent differences in synapse number and AMPA receptor expression in hippocampal CA1 pyramidal neurons. Neuron, 50(3), 431–442.] [Bush, P. C., & Sejnowski, T. J. (1993). Reduced compartmental models of neocortical pyramidal cells. Journal of Neuroscience Methods, 46(2), 159–166.]

everywhere within a compartment is roughly the same) (Rall 1964; Rall and Shepherd 1968; Segev et al. 1995), which is appropriate to study how synaptic inputs propagate and are spatiotemporally integrated throughout the dendrite tree (Spruston 2008; Stuart et al. 2016; Goetz et al. 2021). In neural circuit modeling, such a model is often simplified, reduced approximately to a small number of compartments (**Figure 2.3b**), which captures the basic spatial structure of a neuron with apical and basal dendrites (Bush and Sejnowski 1993), down to just three (Wang et al. 2004) or two (perisomatic and dendritic) compartments (Pinsky and Rinzel 1994; Wang 1999a; Kepecs et al. 2002). When distinct dendritic versus somatic signaling is not the main focus, it is reasonable to use a "point neuron" with a single compartment, which can be the Hodgkin-Huxley conductance-based model (**Figure 2.3c**) or the integrate-and-fire model (**Figure 2.3d**). Similarly, modeling of a synapse varies in terms of the degree of complexity. It is important to keep in mind that, on the one hand, it is not the case that the more details a model incorporates, the better; on the other hand, certain model properties of single neurons or synapses may have drastic implications for collective dynamical behavior of a network comprising thousands of neurons.

2.2.1 Neuronal Membrane as an RC Circuit

A neuronal membrane is mostly a fatty lipid bilayer, which is not permeable to charged molecules, and hence acts as a capacitor with a capacitance C by separating the charges inside and outside the neuron. Charge stored in a capacitor is $Q = CV$ and its time derivative defines a capacitive current $dQ/dt = CdV/dt$. There are various types of transmembrane ion channels, which, when opened, selectively allow particular types of ions to pass through them. For instance, a potassium channel highly prefers K^+ to Na^+ even though Na^+ has the same charge and is smaller than K^+.

Opening an ion channel leads to a current I that, according to Ohm's law, can be described by $I = g(V - V_{\text{rev}})$ (with the unit of $\mu A / cm^2$), where g is a conductance (the inverse of resistance) per unit area (in $\mu S / cm^2$) and V_{rev} is a reversal potential. Given a membrane surface area A, we can write $G = gA$ (in μS) and the total current $I = G(V - V_{\text{rev}})$ (in nA) (with a slight abuse of notation for I). The conductance G quantifies the strength of an ion current, but the size of a current is also proportional to its "driving force," $V - V_{\text{rev}}$, the difference between the membrane potential V (which varies dynamically over time) and the reversal potential V_{rev}. The value of V_{rev} depends on the selective permeability of an ion channel and the intracellular and extracellular concentrations of ions that can pass through the channel. When a channel is permeable for a single type of ions, V_{rev} is given by the Nernst equation, $V_{\text{rev}} = \left(\frac{RT}{Fz}\right)\ln([\text{C}_{\text{out}}]/[\text{C}_{\text{in}}])$, where R is the universal gas constant, T is the temperature, F is the Faraday constant, z is the charge of an ion, and C_{in} and C_{out} are the intracellular and extracellular ionic concentrations. Typically, the potassium concentration is higher inside than outside a neuron, thus V_K is negative; the opposite holds for the sodium and calcium concentrations, thus V_{Na} and V_{Ca} are positive. At room temperature, the value of reversal potential is about -80 mV for a potassium current, $+50$ mV for a sodium current, and $+120$ mV for a calcium current. The combined effects of various ion types yield a net reversal potential given by the Goldman-Hodgkin-Katz equation (Hille 2001).

When a neuron is "at rest," the voltage is lower inside compared with outside the membrane, and V is typically about -70 mV. This is conveniently modeled by a "leak" current $I_L = G_L(V - V_L)$ with a constant G_L and $V_L = -70$ mV. A minimum model for a passive neuronal membrane is described by an RC circuit, with the input resistance $R = 1/G_L$ and the capacitance C associated with the membrane's lipid bilayer. If I_{app} represents an external current injected into the membrane, then according to Kirchhoff's circuit law, the sum of the resistive and capacitive currents must equal to I_{app}, $C_m dV/dt + I_L = I_{app}$, or

$$C_m \frac{dV}{dt} = -G_L(V - V_L) + I_{app} \tag{2.1}$$

where I_{app} is an external stimulus as an injected current. Eq. (2.1) can be rewritten as

$$\frac{dV}{dt} = \frac{(V_{ss} - V)}{\tau_m} \tag{2.2}$$

where $V_{ss} = V_L + I_{app}/G_L$ is the steady-state and $\tau_m = C_m/G_L = RC_m$ is the membrane time constant. For instance, if $C_m = 0.2$ nF, $G_L = 0.02$ μS, then $\tau_m = 10$ ms.

Eq. (2.2) is a differential equation that, unlike an algebraic equation, depends on both a variable V and its time derivative. It is also called a dynamical system, which describes and predicts how the behavior unfolds in time: in a given "state" (a particular value of V), the right-hand side determines the slope (dV/dt) of its change into the future. A "steady-state" V_{ss} is determined by the value of V when it does not change in time, $dV/dt = 0$ yielding $V = V_{ss}$. In the absence of input ($I_{app} = 0$), the steady-state membrane potential is V_L. When driven by an input current I_{app}, V_{ss} increases linearly with I_{app}, and is inversely proportional to the leak conductance G_L (a given amount of excitation produces more depolarization if leak of current across the membrane is smaller). This steady-state is stable: if V is transiently depolarized or hyperpolarized by a current pulse, after the input offset it will evolve back to V_{ss} exponentially with a time constant τ_m. This is a simple example of an attractor state, which is a mathematical term to describe a state of a dynamical system that is stable under perturbations. More specifically, since the membrane is governed by the linear RC circuit equation, it can be readily solved. For an initial condition $V(t=0)=V_0$, the general solution of Eq. (2.1) is

$$V(t)=V_{ss}+(V_0-V_{ss})exp(-t/\tau_m). \tag{2.3}$$

Therefore, as shown in **Figure 2.4b** (black curve), when a current pulse is applied, V evolves over time so that its difference with V_{ss} decreases exponentially. The time it takes for V to reach the steady-state within 1% is ~5 τ_m (at $t=5\tau_m$, $\exp(-t/\tau_m)=\exp(-5)\simeq 0.7\%$). When the current pulse I_{app} is turned off, it takes a similar amount of time for V to decay back to V_L. For instance, if $\tau_m=10$ ms, the system "forgets" a previously shown input within about 50 ms.

Neurons are endowed with many active channels, and their conductances are not fixed but depend on voltage or/and intracellular ion concentrations (Llinás 1988). Therefore, generally the Kirchhoff's circuit equation has the form of $C_m dV/dt=-\sum_j I_{ion,j}+I_{app}$, where the sum is over all intrinsic ion currents $I_{ion,j}=g_j\,(V-V_{rev,j})$. The value of V_{rev} relative to V determines whether a current is positive or negative. By convention, at resting membrane $V\simeq -70$ mV, a negative current is called an "inward" current that tends to increase or depolarize the transmembrane potential V, whereas a positive current is called an outward current that tends to decrease or hyperpolarize V. This convention may be a bit confusing at first sight, but it makes sense when we notice the negative sign in front of an ion current in the membrane equation.

2.2.2 LIF as a Simple Spiking Neuron Model

Neuronal communication is conveyed by action potentials, or spikes, which are brief pulses ($\sim 1-2$ millisecond duration) of electrical signals (**Figure 2.4a**). The more spikes are emitted per unit time (second), the more active is a neuron. The leaky integrate-and-fire (LIF) model, originally proposed in 1907 by the French physiologist Louis Lapicque (see Brunel and van Rossum 2007), is built on the simplification that spikes are stereotypical and brief events. It describes a passive neuron that incorporates spikes into an RC circuit in a simplified way. Namely, the model assumes the generation of a spike as a "point event" in time whenever V exceeds a preset voltage threshold V_{th}. Specifically, if V reaches V_{th} at time t, a spike is discharged with the spike time t; V is instantaneously reset to V_{rest} and stays there for a refractory period of time τ_{ref}. The values of V_{th}, V_{rest} and τ_{ref} are parameters that can be determined by physiological measurements for a particular type of neurons that one wishes to model. For the sake of concreteness, let us set as an example $V_{th}=-50$ mV, $V_{reset}=-60$ mV and $\tau_{ref}=2$ ms.

The behavior of the LIF model neuron depends on the intensity of input current I_{app}. With a small I_{app}, $V(t)$ settles in a steady-state V_{ss} below V_{th} (**Figure 2.4b**, black curve). On the other hand, if I_{app} exceeds a threshold value I_c, the steady-state V_{ss} is larger than V_{th}. The value of I_c is given by imposing $V_{ss}=V_{th}$, or $I_c=G_L(V_{th}-V_L)=0.4$ nA. For $I_{app}>I_c$, V_{ss} is larger than V_{th}. As $V(t)$ evolves towards V_{ss}, a spike is produced when the membrane potential crosses V_{th} and is reset to V_{rest} for a refractory period τ_{ref}. Afterwards, V rises and over time reaches V_{th} again, leading to repetitive firing of spikes (**Figure 2.4b**, blue curve). With a fixed I_{app} the firing is regular in time, characterized by a constant period T or interspike interval (ISI), and firing rate $r=1/T$.

The ISI can be obtained analytically from Eq. (2.3). Immediately after a refractory period, $V=V_{reset}$. Using this value as an initial condition, $V(t)$

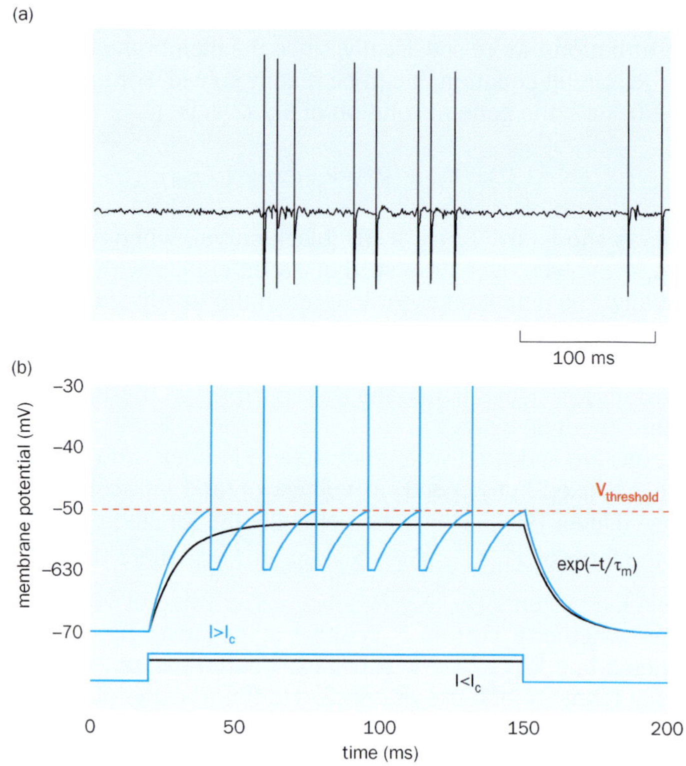

Figure 2.4 Leaky integrate-and-fire (LIF) model of a spiking neuron. (a) A sample spike train recorded from a visual cortical area of a behaving monkey. (b) Behavior of the LIF neuron model. In response to an input current I is below a threshold level I_c, the membrane potential researches a steady-state (black). With an increased I above I_c (blue), the neuron model displays repetitive firing of spikes, which are not explicitly modeled in the model and assumed to be a point process. The horizontal dashed red line indicates the voltage threshold V_{th} for spike firing. Panel (a) is reproduced from Amarasingham et al. (2006). [Amarasingham A, Chen TL, Geman S, Harrison MT, Sheinberg DL. (2006). Spike count reliability and the Poisson hypothesis. J Neurosci. 26(3):801–9. Copyright 2006 Society for Neuroscience.]

evolves in time according to Eq. (2.3) and reaches V_{th} at $t = T - \tau_{ref}$. This condition yields an algebraic equation for T. Therefore, one obtains an analytical expression for the firing rate r as a function of the current I_{app},

$$r = \begin{cases} \dfrac{1}{\tau_{ref} + \dfrac{C_m}{G_L}\ln\left[1 + \dfrac{G_L \Delta V}{(I_{app} - I_c)}\right]} & \text{if } I_{\text{app}} > I_c \\ 0 & \text{otherwise} \end{cases} \tag{2.4}$$

where $\Delta V = (V_{th} - V_{reset})$ is the voltage difference between the firing threshold and reset.

The frequency-current relationship ($f - I$ curve), also called the neuronal input-output transfer function, is a fundamental concept that describes a neuron's mapping from an input current to an output firing rate. For the LIF model, the firing rate starts at zero at $I_{app} = I_c$ and increases steeply near I_c (in fact, the slope of $r(I)$ is infinity at $I_{app} = I_c$). Well above I_c, one can show from Eq. (2.4) that r is approximately linear with I_{app}, with a slope given by $1 / C_m \Delta V$. At very large I_{app}, r cannot exceed $1 / \tau_{ref}$ because *ISI* must be longer than the refractory period; therefore, the firing rate eventually plateaus. With $\tau_{ref} = 2$ ms, the maximum firing rate is $1 / \tau_{ref} = 500$ Hz.

2.2.3 Spiking Variability

In an intact brain *in vivo*, synaptic inputs to a neuron fluctuate considerably in time (Destexhe et al. 2003). Consequently, neuronal firing of a spike train is highly irregular (see **Figure 2.4a**). For a periodic spike train, the ISI is constant. By contrast, ISI is a random variable for a stochastic point process. For instance, in a Poisson process, at any small time window dt, an event (a spike) occurs with a probability λdt, and there is no temporal correlation between events. In this case, consecutive ISIs are independent from each other, with an exponential probability density $p(ISI) = \lambda \exp(-\lambda ISI)$, and the mean firing rate is $r = \lambda = 1/\langle ISI \rangle$.

The stochasticity of a spike train can be quantified in two ways. The first is the coefficient of variation (CV) of ISIs, defined as

$$CV = \frac{\text{Standard Deviation}(\text{ISI})}{\text{Mean}(\text{ISI})} = \frac{<(\text{ISI} - \langle \text{ISI} \rangle)^2>^{1/2}}{\langle \text{ISI} \rangle} \tag{2.5}$$

where $\langle x \rangle = \int p(x) x dx$ denotes the average of x. The second measure is Fano factor of spike count $N(T)$ (number of spikes in a time window T) defined by

$$F(T) = \frac{\text{Variance}(N(T))}{\text{Mean}(N(T))} \tag{2.6}$$

For a regular periodic spike train, both CV and $F(T)$ are 0. At the other extreme, if a spike train is Poisson, we have CV and $F(T)$ equal to 1. High variability represents a salient characteristic of spike trains observed in cortical neurons *in vivo*, for which typically the spike-count variance exceeds the mean (Tolhurst et al. 1983; Softky and Koch 1993; Goris et al. 2014). An example is shown in **Figure 2.5a** for neurons in the primary visual cortex V1 of a macaque monkey in response to a visual stimulus, with a time window of $T = 1$ s, in which case the spike count is the same as firing rate per second. As it can be seen, the variance is larger than the mean, and the Fano factor is larger than 1. This excessive variability can be explained by a statistical model where spike trains is approximated by a Poisson process with the rate λ that itself varies stochastically over time (Goris et al. 2014). Similarly, the CV is typically larger than 1 for neural activity in the prefrontal cortex of a behaving macaque monkey (**Figure 2.5b**) (Compte et al. 2003).

To capture stochastic spike firing in an integrate-and-fire model, one can introduce a noise $\eta(t)$ into I_{app}, so Eq. (2.1) becomes

$$C_m \frac{dV}{dt} = -G_L(V - V_L) + \mu + \sigma\eta(t) \tag{2.7}$$

where μ is the mean of I_{app}, σ is the noise level. The simplest kind of noise has no temporal correlation and a Gaussian probability density for η of zero mean and unit variance, $p(\eta) = 1/\sqrt{2\pi}\exp(-\eta^2/2)$. This is called "white noise," because its power spectrum is constant and therefore completely devoid of dependence on frequencies. The absence of temporal correlation is mathematically described by $\langle \eta(t)\eta(t') \rangle = \delta(t - t')$, where the delta function $\delta(t)$ is infinity with $t = 0$ and zero otherwise (its integral is 1 and physical unit is one over time). Such a stochastic linear differential equation with a leak and white noise is called the Ornstein-Uhlenbeck process (Wax 1954).

The input current now is noisy with a mean of μ and a standard deviation of σ. When μ is below I_c, the deterministic steady-state V_{SS} is below

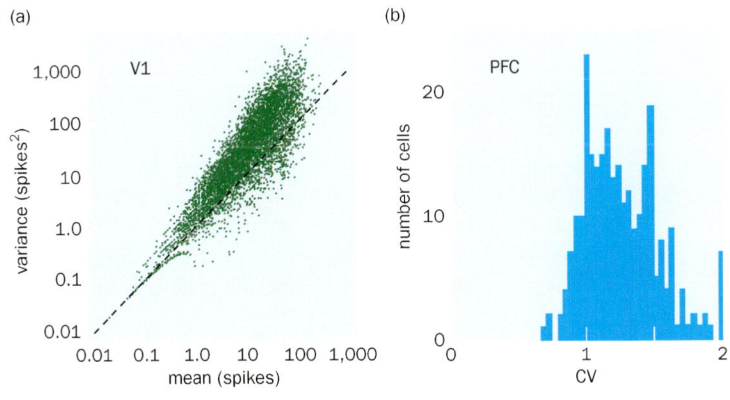

Figure 2.5 Variability of the spiking activity of cortical neurons. (a) Variance versus mean of spike count in a 1-second time window of neurons recorded from the primary visual cortex V1 of monkey (each point corresponds to one neuron under one stimulus condition). The slope is Fano factor, which is about 1 at low firing rates (around the diagonal line) and larger than 1 at high firing rates (above the diagonal line). (b) The histogram of the coefficient of variation (CV) of interspike intervals across a population of neurons recorded in monkey prefrontal cortex, showing that CV values are mostly close to or larger than 1. (a) is adapted from Goris et al. (2014), (b) from (Compte et al. 2003). [Goris, R. L., Movshon, J. A., & Simoncelli, E. P. (2014). Partitioning neuronal variability. Nature Neuroscience, 17(6), 858–865.] [Compte, A., Constantinidis, C., Tegnér, J., Raghavachari, S., Chafee, M. V., Goldman-Rakic, P. S., & Wang, X. J. (2003). Temporally irregular mnemonic persistent activity in prefrontal neurons of monkeys during a delayed response task. Journal Neurophysiology, 90(5), 3441–3454.]

the firing threshold, but spiking is now possible because V is driven by noise and may still reach V_{th} through fluctuations (**Figure 2.6a**). In the presence of noise, the interspike interval *ISI* is a random variable, following a probabilistic distribution. ISIs are statistically independent of each other, since each time a spike is fired, V is reset to V_{reset} and memory about the past history is lost.

Starting at V_{reset}, the time it takes for V to reach V_{th} for the first time is mathematically called "first passage time" (Ricciardi 1977; Tuckwell 1988); an ISI is the first passage time plus the refractory period. Therefore, the statistics of ISIs can be analyzed by the theory of first passage times of an Ornstein-Uhlenbeck process. In particular, the average $\langle ISI \rangle$ is the inverse of the mean firing rate r, and the input-output function is given by (Ricciardi 1977; Renart et al. 2003):

$$\frac{1}{r} = \tau_{ref} + \tau_m \sqrt{\pi} \int_{\frac{C_m(V_{reset}-V_{ss})}{\sqrt{\tau_m}\sigma}}^{\frac{C_m(V_{th}-V_{ss})}{\sqrt{\tau_m}\sigma}} e^{x^2}(1+\mathrm{erf}(x))dx \tag{2.8}$$

where $\mathrm{erf}\ (x) = \left(2/\sqrt{\pi}\right)\int_0^x \exp\left(-x'^2\right)dx'$ is the error function. The CV is given by (Renart et al. 2003):

$$CV^2 = 2\pi r^2 \int_{\frac{C_m(V_{reset}-V_{ss})}{\sqrt{\tau_m}\sigma}}^{\frac{C_m(V_{th}-V_{ss})}{\sqrt{\tau_m}\sigma}} e^{x^2}\left[\int_{-\infty}^{x} e^{y^2}(1+\mathrm{erf}(y))dy\right]dx \tag{2.9}$$

Figure 2.6b (left panel) shows the frequency-current relationship as a function of the mean μ, with three σ different values. As it can be seen in **Figure 2.6b** (middle panel), the LIF model driven by noise exhibits two separate regimes as far as variability is concerned. When the mean input is sub-threshold ($\mu < I_c$), spike discharge is triggered by fluctuations in the current, and spike trains are irregular. Therefore, in this regime, the CV is high, close to one. For supra-threshold mean current ($\mu > I_c$), the mean becomes dominant over the noise, and firing becomes increasingly regular. This happens when $(\mu - I_c)$ is much larger than $\sigma_I = \sigma / \sqrt{\tau_m}$ (note that σ has the unit of nA$\sqrt{ms}$, thus σ_I has the unit of nA). With large μ, CV decays to zero and spiking becomes regular. The sharpness of this transition depends on σ. When the fluctuations are small, the transition is very

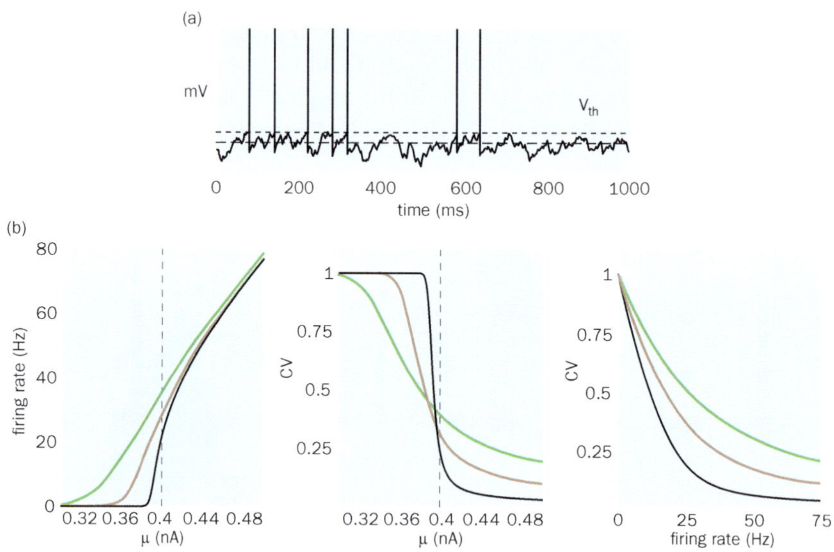

Figure 2.6 LIF model in the presence of noise. (a) Activity driven by a noisy input current with a mean μ near the threshold I_c. (b) The frequency-current curve (left) and CV (middle) as a function of the mean input current μ for three different levels of noise. CV decreases with the mean firing rate (right). Reproduced from Renart et al. (2003). Copyright 2003. From Mean-field theory of recurrent cortical networks: Working memory circuits with irregularly spiking neurons. In J. Feng (Ed.), Computational Neuroscience: A Comprehensive Approach by Renart, A., N. Brunel, and X.-J. Wang. Reproduced by permission of Taylor & Francis Group, LLC, a division of Informa plc.

sharp, so as soon as the neuron starts firing, it does so in a regular fashion. For large values of σ the transition is smooth, so the neuron fires irregularly for μ near I_c and becomes progressively more regular witn μ much larger than I_c. When the firing rate increases as a result of an increase in the mean current, CV decreases. This decrease is faster with smaller noise (**Figure 2.6b** (right panel).

Evidently, a model with a fixed noise level cannot explain the commonly observed high CV or Fano factor of cortical neurons in a broad range of firing rates (**Figure 2.5**). Neuronal variability is important to understand, because it plays a crucial role in neural computation. For instance, cognitive processes such as decision making are probabilistic. To what extent does the source of stochasticity originate from the irregular firing of neurons? A more basic physiological question is how a high variability of neural spike firing can be explained mechanistically. This phenomenon suggests that the total input to a single cortical cell must operate near the firing threshold, regardless of the firing rate level. How can this be realized in a cortical circuit? We will come back to this issue when we discuss the excitation-inhibition balance in recurrent neural networks in Chapter 3.

2.3 CONDUCTANCE-BASED MODELS OF SINGLE NEURONS

2.3.1 Hodgkin-Huxley Formalism of Action Potential

The biophysical mechanism for the generation of an action potential is well understood, starting with the seminal work of Hodgkin and Huxley (1952). The description of this model can be found in standard textbooks (Hille 2001; Nichols et al. 1992). Here we will present it succinctly. According to the Hodgkin-Huxley model, an action potential is produced by an interplay of voltage-gated Na^+ and K^+ ion currents (I_{Na} and I_K). I_{Na} is an inward current and depolarizes the neuronal membrane; by contrast, I_K is an outward current hyperpolarizes the membrane. As usual, we can write $I_{Na} = g_{Na}(V - V_{Na})$ and $I_K = g_K(V - V_K)$ with the reversal potentials V_{Na} and V_K. Unlike a passive leak conductance, the conductances g_{Na} and g_K are not constant but "gated" by membrane potential V. Specifically, the potassium conductance g_K becomes $g_K n^4$, where n is the activation gating variable. The idea is that n represents the fraction of "gates" in an open state, and $1-n$ is the fraction of gates in a closed state. The gating variable n obeys a dynamical equation according to a first-order chemical kinetics:

$$dn/dt = \alpha_n(1-n) - \beta_n n \tag{2.10}$$

with the opening and closing rates α_n and β_n dependent on V. This equation can be rewritten as $dn/dt = (n_\infty - n)/\tau_n$, with the steady-state $n_\infty = \alpha_n/(\alpha_n + \beta_n)$ and time constant $\tau_n = 1/(\alpha_n + \beta_n)$.

The sodium conductance g_{Na} has not only an activation gating variable m but also an inactivation gating variable h, such that g_{Na} becomes $g_{Na}m^3h$, with m and h described by similar first-order kinetic equations as for n but with different opening and closing rates.

Thus, the Hodgkin-Huxley model consists of four differential equations:

$$c_m\frac{dV}{dt} = -g_L(V - V_L) - g_{Na}m^3h(V - V_{Na}) - g_K n^4(V - V_K) + I_{app} \tag{2.11}$$

$$\frac{dm}{dt} = \phi_m\left(\alpha_m(V)(1-m) - \beta_m(V)m\right) = \frac{\left(m_\infty(V) - m\right)}{\tau_m(V)} \tag{2.12}$$

$$\frac{dh}{dt} = \phi_h\left(\alpha_h(V)(1-h) - \beta_h(V)h\right) = \frac{\left(h_\infty(V) - h\right)}{\tau_h(V)} \tag{2.13}$$

$$\frac{dn}{dt} = \phi_n\left(\alpha_n(V)(1-n) - \beta_n(V)n\right) = \frac{\left(n_\infty(V) - n\right)}{\tau_n(V)} \tag{2.14}$$

where I_{app} is the injected current (in $\mu A/cm^2$). The parameters ϕ_x are the temperature factors, equal to 1 in the original Hodgkin-Huxley model (with the kinetic rates measured at 6–7°C). At a higher room or body temperature, chemical reactions are faster, and ϕ_m, ϕ_h and ϕ_n are larger than 1. Note that our convention is to use lowercase g for specific conductance (per cm^2) and uppercase G for total conductance (as in the above LIF model).

The Hodgkin-Huxley model is a highly nonlinear coupled system, as the V equation depends on all four variables, and the rates α and β in the equations for the gating variables are complex functions of V. However, using an experimental technique called "voltage-clamp," one can hold V at a constant level in time, in which case α and β are mere numbers. By measuring the time course of an ion current such as I_K or I_{Na} and its conductance (equal to $I/(V - V_{Na})$ and $I/(V - V_K)$ with a fixed V, respectively), one can deduce the α and β values for each V. Using this method, Hodgkin and Huxley deduced the quantitative dependence of the rates on V. **Figure 2.7** shows the resulting steady-states and time constants for the three gating variables as a function of V. Explicitly, the rate functions are:

$$\alpha_m(V) = -0.1(V+40)/\left(\exp\left(-0.1(V+40)\right)-1\right); \tag{2.15}$$

$$\beta_m(V) = 4\exp\left(-(V+65)/18\right) \tag{2.16}$$

$$\alpha_h(V) = 0.07\exp\left(-0.05(V+65)\right); \tag{2.17}$$

$$\beta_h(V) = 1/\left(1+\exp\left(-0.1(V+35)\right)\right) \tag{2.18}$$

$$\alpha_n(V) = -0.01(V+55)/\left(\exp\left(-0.1(V+55)\right)-1\right); \tag{2.19}$$

$$\beta_n(V) = 0.125\exp\left(-(V+65)/80\right) \tag{2.20}$$

The other parameter values of the classic Hodgkin-Huxley model are: $c_m = 1\ \mu F/cm^2$; the maximum conductances $g_{Na} = 120$, $g_K = 36$, $g_L = 0.3$ (in mS/cm^2); and the reversal potentials $V_{Na} = +50$, $V_K = -77$, $V_L = -54.4$ (in mV). The passive time constant $\tau_m = c_m / g_L \simeq 3.3$ ms.

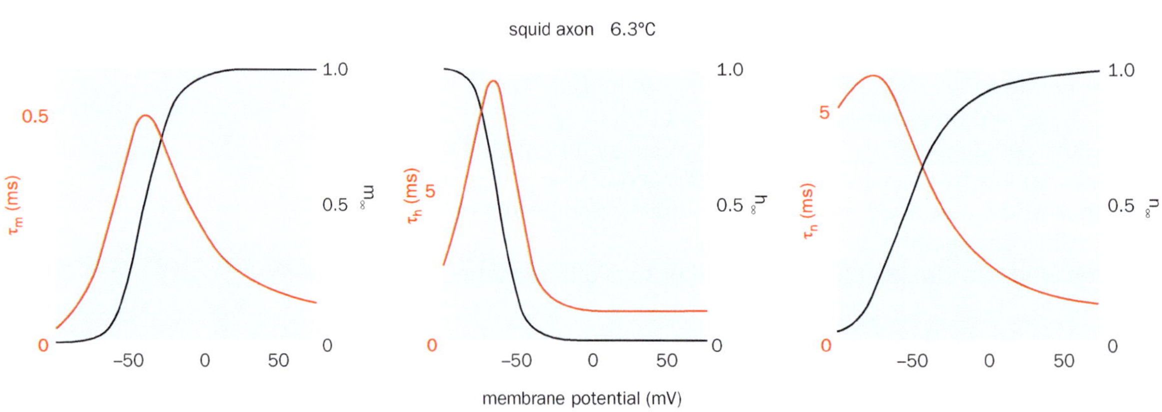

Figure 2.7 The steady-state and time constant as a function of membrane potential *V* for the gating variables *m*, *h* and *n* of the Hodgkin-Huxley model of an action potential. The steady-state is an increasing function of *V* for the activation variables *m* and *n* of the sodium and potassium conductances, respectively, whereas it is a decreasing function of *V* for the inactivation variable *h* of I_{Na}. The time constant for *m* is about tenfold shorter than those for *h* and *n*.

When the input current I_{app} is sufficiently large, above a critical value I_c, the Hodgkin-Huxley neuron model generates periodic discharges of action potentials (**Figure 2.8a**). The repetitive firing can be intuitively understood as a result of the interplay between fast positive feedback mediated by I_{Na} and slower negative feedback mediated by I_K. Because the activation variable m increases with V (**Figure 2.7**, left), depolarization leads to the opening of I_{Na}, which as an inward current further increases V. This positive feedback occurs like a chain reaction very quickly (τ_m is sub-millisecond, about 10-fold shorter than τ_h and τ_n, see **Figure 2.7**), underlying the rapid upstroke of an action potential. On the other hand, the inactivation variable h decreases with V (**Figure 2.7**, middle). Thus, the sodium conductance decreases on a slower timescale. Moreover, the activation variable n of I_K grows with depolarization (**Figure 2.7**, right; **Figure 2.8a**). This outward potassium current eventually produces the downstroke of an action potential. During a refractory period, the membrane potential is hyperpolarized, causing n (thus I_K) to decay away. At the same time, the inactivation variable h for the sodium conductance recovered to a high value. The whole cycle can start anew, leading to a periodic train of action potentials.

Oscillation underlying repetitive firing of action potentials naturally results from a fast positive feedback process and a slower negative feedback process. Indeed, if the negative feedback by activation of g_K was as fast as, or faster than, the positive feedback by activation of g_{Na}, rhythmic action potential would be impossible. Instead, the membrane potential would reach a steady-state regardless of the injected current intensity I_{app}. As a general principle, oscillation occurs in a system endowed with a fast positive feedback (here the interplay between depolarization and activation of I_{Na}) and a slower negative feedback (the back-and-forth between voltage-activated I_K and its hyperpolarizing influence on V).

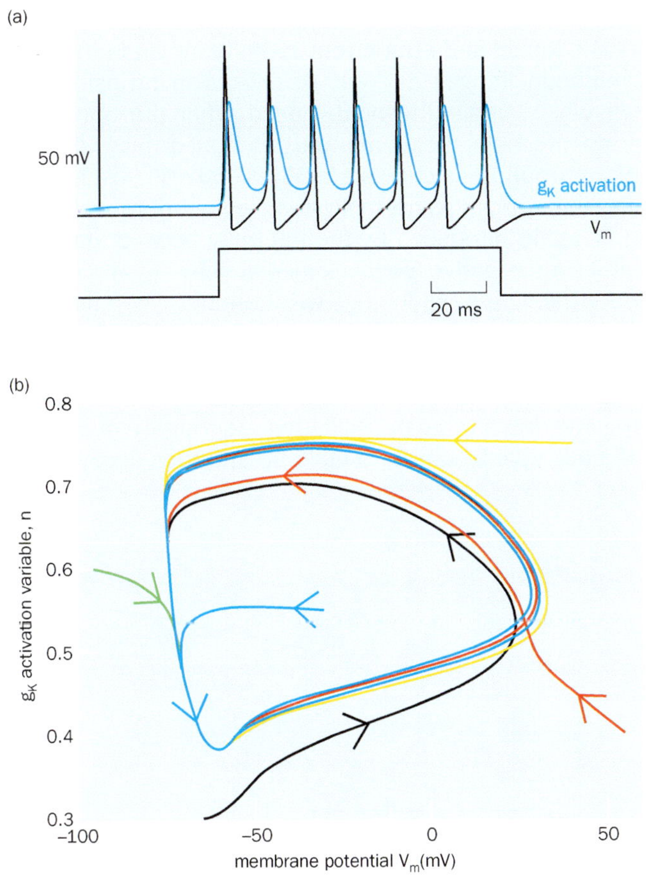

Figure 2.8 Repetitive firing of action potentials in the Hodgkin-Huxley model. (a) Membrane potential V (black) and the activation variable of the potassium conductance n (blue) as a function of time, in response to a long current pulse $I_{app} = 9\ \mu A/cm^2$. (b) The potassium activation variable n is plotted against V, showing that the oscillation forms a closed trajectory in this projected two-dimensional "state space." The trajectory roughly consists of three portions: The upstroke of an action potential (when both V and n increase), the downstroke (when V decreases while n keeps increasing then starts to decrease), and the refractory period (when n continues to decrease while V starts to increase). Different colors correspond to five different initial conditions (with different V and n values at $t = 0$); in all cases the system dynamically evolves into the limit cycle of the oscillatory attractor state.

In the Hodgkin-Huxley model, as the input current I_{app} is increased gradually across I_c, the behavior undergoes a *qualitative* change from steady-state to repetitive firing. Such sudden emergence of a novel behavior can be understood mathematically in the theory of nonlinear dynamical systems as a *bifurcation* phenomenon, and I_c is called a bifurcation point of the parameter I_{app} (Strogatz 2016). Importantly, bifurcations can never occur in linear dynamical systems such as an RC circuit.

Mathematically, a stable state is called an *attractor*, which simply means that when the system deviates slightly away from that state, for instance, by a small perturbation, it will evolve back to that state over time. We have seen the steady-state of an RC circuit as a stationary attractor. On the other hand, a regular train of action potentials in the Hodgkin-Huxley model represents an oscillatory type of attractor behavior called *limit cycle*. The attractor nature of periodic spiking is shown in **Figure 2.8b**: regardless of the initial states, the model always converges to the same periodic attractor state. If the system is perturbed by a transient stimulus, it would evolve over time and eventually resume the same firing pattern after the stimulus offset, except for a shift of spiking time, or the phase of the periodic attractor state. Thus, a periodic attractor is robust (the amplitude and periodicity are not sensitive to transient perturbations). At the same time, the phase (e.g., the time relative to the peak of an action potential in every oscillatory cycle) can be readily shifted by an external stimulation, hence the clock can be readily reset.

2.3.2 Type I and Type II Neurons

Figure 2.9 displays the frequency-current function for the classic Hodgkin-Huxley model. Contrary to the LIF model, in the Hodgkin-Huxley model the firing rate of action potentials is not zero but finite ($\sim$ 45 Hz) at the threshold input current, and the range of possible firing rates is rather limited. A finite minimum firing rate at the onset of repetitive spike firing is a characteristic of the current-frequency relationship for "type II" neurons (Hodgkin 1948; Rinzel and Ermentrout 1989; Wang and Rinzel 1995).

This minimum firing rate of $\sim$ 45 Hz reflects an intrinsic oscillatory frequency even when V is subthreshold. Indeed, when the input current I_{app} is near the firing threshold, the membrane potential displays damped temporal oscillations around its steady-state ($V_{SS} \simeq -60$ mV), just like a pendulum in a viscous fluid swinging back to its stable equilibrium state. The damped oscillation is, again, produced by the interplay between the fast positive feedback and slow negative feedback mediated by I_{Na} and I_K, as explained previously for the action potential generation; but when the input current is not sufficiently strong to drive V above the spiking threshold, oscillations are subthreshold and occur on a small scale of membrane potential variations around -60 mV. Assuming the positive feedback is very fast, the upswing and downswing of oscillation are roughly given by the time it takes for the negative feedback to build up and then decay way. The time

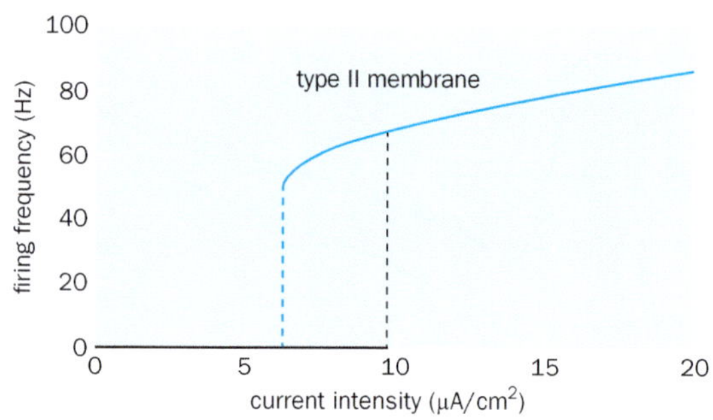

Figure 2.9 Frequency of repetitive firing as a function of input current intensity of the Hodgkin-Huxley model. Note a gap of firing rate between 0 and ~45Hz and a range of I_{app} where bistability exists between a stable steady-state (with zero firing rate) and an oscillatory attractor of repetitive firing for I_{app} between 7 and 10.

constants of the activation of I_K and inactivation of I_{Na} are $\sim 5-10$ ms (**Figure 2.7**), yielding a periodicity of ~ 20 ms in agreement with a frequency of 45 Hz. This intrinsic frequency sets a minimum for firing rate of action potentials when the input current exceeds the spiking threshold, thereby creating a gap in the frequency-current curve (**Figure 2.9**).

In contrast to the classic Hodgkin-Huxley model, many neurons in the mammalian cortex exhibit a continuous current-frequency curve, starting at zero firing rate ("Type I neuron"), and the firing rate can be as high as a few hundred hertz (McCormick et al. 1985; Koch 1999). It turns out that a few quantitative changes of the parameters from the original Hodgkin-Huxley model gives rise to "Type I" behavior (Wang and Buzsáki 1996). These changes involve shifting the activation and inactivation curves of I_{Na} and I_K to more depolarized membrane potential, by a few mV for m_∞ and h_∞ and by more than 20 mV for n_∞. As a result, the voltage-dependent conductances are no longer significant at ~ -60 mV. Furthermore, because the activation of I_K requires a lot more depolarization compared to I_{Na}, once activated the positive feedback mediated by I_{Na} inevitably produces sufficient depolarization for the generation of an action potential. Therefore, subthreshold oscillation is no longer possible, and spike firing starts at zero hertz (**Figure 2.10**). Finally, the gating kinetics of I_{Na} and I_K speed up, producing a broad range of firing rates (from 0 to 400 Hz).

The ion channel model of type I neurons (Wang and Buzsáki 1996) can be captured by a modified integrate-and-fire model. Fourcaud-Trocme et al. (2003) introduced an exponential term into the LIF equation that describes the positive feedback underlying action potential. The exponential integrate-and-fire (EIF) model obeys

$$c_m \frac{dV}{dt} = -g_L(V - V_L) + \Delta exp\left((V - V_{th})/\Delta\right) + I_{app} \tag{2.21}$$

where V_{th} is the voltage firing threshold, and Δ measures the sharpness of the action potential upstroke. Thus, when V approaches V_{th}, the inward current grows exponentially with V, providing a rapid and powerful positive feedback. Each time V reaches V_{th}, a spike is triggered and V is reset to V_{reset} for a time that includes the duration of an action potential (τ_{spike}) and a refractory period (τ_{ref}). With the appropriate choice of model parameter

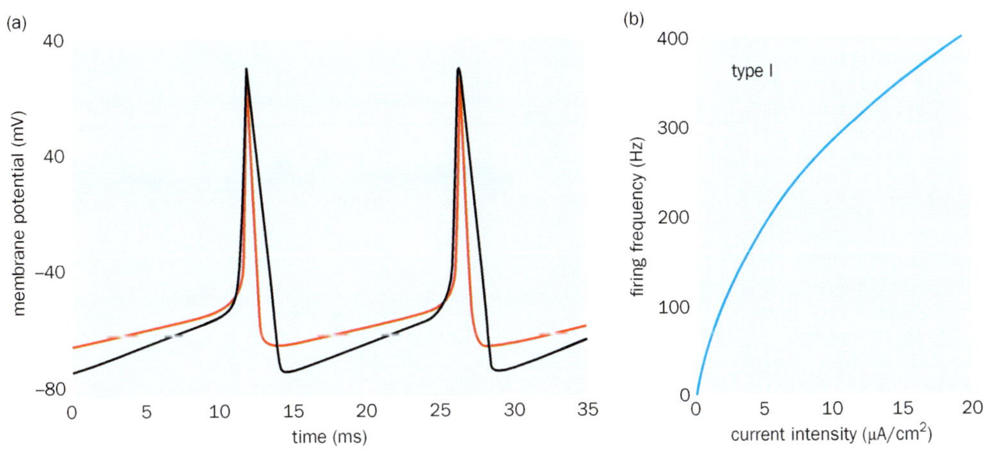

Figure 2.10 Frequency of repetitive firing as a function of input current intensity of a type I neuron. (a) Compared to the original Hodgkin-Huxley model (black), a modified Hodgkin-Huxley model displays narrower spikes and less afterhyperpolarization (red). (b) The firing rate ranges from 0 to about 400Hz. Reproduced from Wang and Buzsáki (1996). [Wang, X. J., & Buzsáki, G. (1996). Gamma oscillation by synaptic inhibition in a hippocampal interneuronal network model. Journal of Neuroscience, 16(20), 6402–6413.]

values, the EIF model reproduces quite accurately the behavior of the Wang-Buzsaki model. In particular, the current-frequency relationship is virtually indistinguishable between the two models. In the absence of noise, for input current near the firing threshold value I_c, the firing rate is predicted to behave as $r = \beta\sqrt{(I_{app} - I_c)}$ with $\beta = 38\ \text{Hz}/((\mu\text{A})^{1/2}/\text{cm})$ (Fourcaud-Trocme et al. 2003). As in the classical LIF, with the addition of a white noise to the input, neuronal firing can be triggered by fluctuating inputs for I_{app} below I_c, and the current-frequency curve of the type I neuron model is again reproduced by the EIF model.

2.4 TIME-DEPENDENT NEURONAL FIRING PATTERNS

2.4.1 Resonance in Response to Time-Dependent Noisy Inputs

The frequency-current relationship describes the neuronal response to a constant input (or a stationary input in the presence of noise). How does a neuron respond to a time-varying input? For instance, can a neuron "keep up" with increasingly rapid input fluctuations? When synaptic inputs reflect a synchronous network rhythm at frequency f, can a neuron respond effectively at a higher and higher frequency?

To address these questions, consider a noisy oscillatory input current $I_{app} = I_0 + I_1\cos(2\pi ft) + I_{noise}$, where I_0 sets the overall firing rate above 0 (say 20 Hz), I_1 is the amplitude of the sinusoidal input and I_{noise} is a white noise. **Figure 2.11** shows an integrate-and-fire neuron model generalized to include a slow outward current (Richardson et al. 2003). In response to a noisy sinusoidal input with a low (a) or high (b) noise level (upper panel in [a] and [b]), membrane potential fluctuates over time (middle panel in [a] and [b]), and spike train varies considerably from trial to trial. Across trials

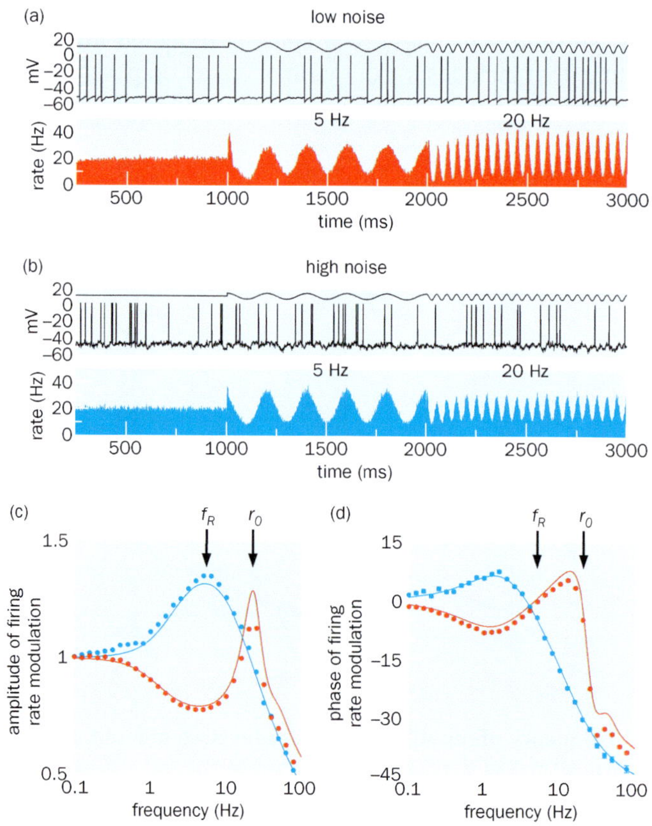

Figure 2.11 Subthreshold and spiking resonance of a model neuron in the presence of noise. The model neuron exhibits subthreshold membrane resonance at frequency $r_R \simeq 5$ Hz. (a) Response with low noise. Top: Input current with three (tonic, 5 Hz and 20 Hz sinusoidal current injection) epochs. The mean output firing rate remains constant ($r_0 = 20$ Hz) throughout. Middle: A sample membrane trace. Bottom: Post-stimulus time histogram (PSTH). (b) Response with high noise. Same format as in (a). (c) Signal gain amplitude versus input frequency f (normalized to be one at $f = 0$). Firing rate resonance occurs at the subthreshold resonance frequency ($f \simeq f_R$) with high noise (blue), but at the mean firing frequency ($f \simeq r_0$) with low noise (red). (d) Phase shift of the firing response relative to the sinusoidal input (positive: Phase advance; negative: Phase delay). Panels (a)–(c) are reproduced from Richardson et al. (2003). [Richardson, M. J., Brunel, N., & Hakim, V. (2003). From subthreshold to firing-rate resonance. Journal of Neurophysiology, 89(5), 2538–2554.

one can compute the number of spikes within each small time window, normalized by the time window size, to yield the trial-averaged firing rate $r(t)$ (post-stimulus time histogram, or PSTH) (bottom panel in [a] and [b]).

Under the condition that that rhythmic modulation is small relative to the baseline (I_1 is small compared to I_0), $r(t)$ can be described as $r(t) = r_0 + r_1(f)\cos(2\pi ft + \Phi(f))$, where $r_1(f)$ is the amplitude of rhythmic modulation and $\Phi(f)$) is a phase shift relative to the input current. At low f, the neuronal response can "keep up" with input oscillations and $r_1(f)$ is roughly independent of f. By contrast, for f above a "cutoff" value f_c, $r_1(f)$ decreases with f, reflecting the fact that neuron dynamics cannot follow effectively more rapid changes in the input. Both the type I neuron (Wang and Buzsáki 1996) and its reduced EIF model (Richardson et al. 2003) predict that $r_1(f)$ decays with f as $1/f$, whereas the phase shift $\Phi(f)$ reaches an asymptotic level of 90 degrees at high frequencies (Fourcaud-Trocme et al. 2003). In other words, neural firing activity lags input current by up to a quarter of a cycle. By contrast, for the classical LIF model driven by a sinusoidal input and white noise, the attenuation of firing response decreases less steeply, as $r_1(f) \sim 1/\sqrt{f}$, and $\Phi(f)$ plateaus at 45 degrees. Therefore, even though LIF displays a current-frequency relationship with zero onset rate, its behavior in response to temporally varying noisy inputs differs qualitatively from the EIF model and the ion channel model of type I neurons.

Some types of neurons display a resonance phenomenon, where the oscillatory component of neural responses $r_1(f)$ shows a peak at a "preferred" frequency. As shown in **Figure 2.11c**, at low noise (red) the resonance frequency is the mean firing rate r_0. On the other hand, at high noise (blue) the resonance frequency is determined by damped oscillations of subthreshold membrane potential, which is around 5 Hz in this case. Interestingly, around the resonance frequency, neural activity shows a phase advance rather than a delay with respect to the input current; i.e., it tends to peak on the upswing of the oscillatory input (**Figure 2.11c–d**) (Kepecs et al. 2002; Richardson et al. 2003).

2.4.2 Spike Rate Adaptation

In addition to the transient sodium current and non-inactivating potassium current that underlie action potential generation, neurons display a variety of other active ion currents that are gated by membrane potential or intracellular calcium ions (Llinás 1988; Koch 1999). Those currents are expressed differentially in different types of neurons, endowing them with a rich repertoire of membrane potential response dynamics and spike firing patterns. A particularly prominent example is spike rate adaptation (SRA), commonly observed in a majority of excitatory pyramidal cells and some inhibitory neurons of the mammalian cortex (McCormick et al. 1985; Lorenzon and Foehring 1992; Markram et al. 2004). SRA refers to the phenomenon of a graduate decrease of spiking discharges of a neuron in response to a long-lasting stimulus, as a form of neuronal dynamics that renders a single neuron to be more sensitive to changes in the input rather than steady-state, in order to achieve efficient coding of inputs. SRA can be quantified by the inverse of instantaneous ISI as a function of time, $r(t) = 1/ISI(t)$ (**Figure 2.12**). It was found both *in vitro* (Lorenzon and Foehring 1992) and *in vivo* (Ahmed et al. 1998) that $r(t)$ can be well fit with an exponential function of time,

$$r(t) = r_{ss} + (r_0 - r_{ss})exp(-t/\tau_{adap}) \tag{2.22}$$

The degree of adaptation is quantified by $F_{adap} = (r_0 - r_{ss})/r_0$, which is 0 without adaptation ($r_{ss} = r_0$), and 1 with maximum adaptation (in the case

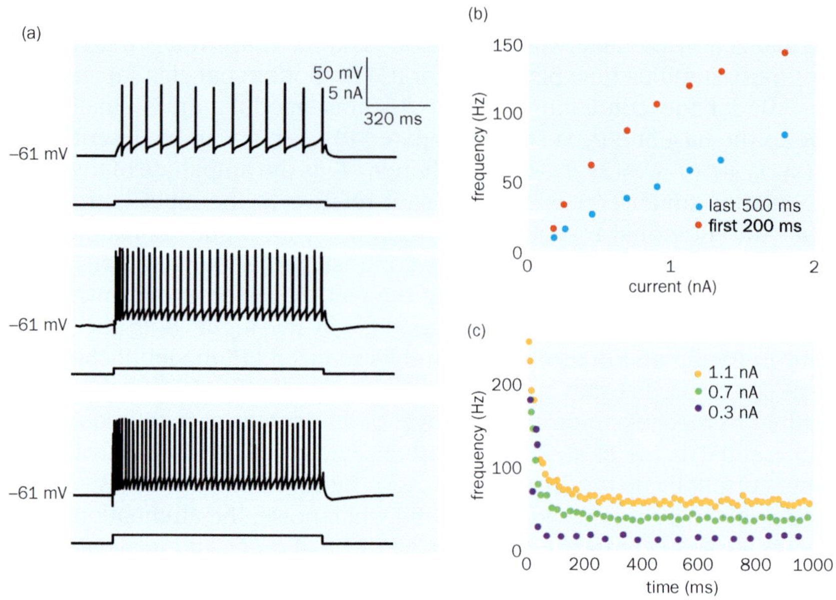

Figure 2.12 Spike frequency adaptation of an excitatory pyramidal neuron from the human cortex recorded in *in vitro* slices. (a) Membrane potential in response to a current injection at three intensity values. (b) The initial firing rate and steady-state firing rate as a function of input current intensity. (c) The instantaneous firing rate (defined as the inverse of an interspike interval) as a function of time for the three examples from (a), each can be fitted with an exponential function of time. Reproduced from Lorenzon and Foehring (1992). Lorenzon, N. M., & Foehring, R. C. (1992). Relationship between repetitive firing and afterhyperpolarizations in human neocortical neurons. Journal of Neurophysiology, 67(2), 350–363.

of $r_{ss} = 0$). Therefore, SRA can be fully described with two quantities, F_{adap} and τ_{adap}. Physiological studies have revealed that, in cortical pyramidal neurons, SRA on a timescale of a few hundreds of milliseconds is primarily mediated by a calcium-activated potassium ion current (I_{KCa}) (Sah 1996), one of many types of potassium currents (Rudy 1988). I_{KCa} is sometimes also denoted as I_{K-AHP} because it underlies slow afterhypopolarization (AHP). The way I_{KCa} works is that membrane depolarization during spikes induces calcium influx in the cell through an "L-type" calcium channel I_{Ca-L}. The intracellular calcium level $[Ca^{2+}]$ (relative to a low baseline) accumulates over time and activates I_{KCa}. This outward current reduces the neuronal excitability as a negative feedback mechanism. Thus, the firing rate decreases over time.

SRA can be incorporated in the LIF model by assuming $I_{KCa} = G_{KCa}[Ca^{2+}](V - V_K)$, where $V_K = -80$ mV is the reversal potential. A modified integrate-and-fire model with adaptation is given by (Liu and Wang 2001):

$$Cm\frac{dV}{dt} = -G_L(V - V_L) - G_{KCa}\left[Ca^{2+}\right](V - V_K) + I_{app} \tag{2.23}$$

$$\frac{d\left[Ca^{2+}\right]}{dt} = \alpha\sum_j \delta(t - t_j) - \frac{\left[Ca^{2+}\right]}{\tau_{Ca}} \tag{2.24}$$

where $\{t_j\}$ is a presynaptic spike train, I_{Ca-L} mediated calcium influx per spike is described by a delta-function, α is the amount of calcium influx per spike (~ 200 nM,) and τ_{Ca} is the decay time constant (~ 100 ms) (Helmchen et al. 1996).

An example of spike train generated by this adaptive LIF model is shown in **Figure 2.13**, where the stimulation is a stochastic Poisson input. The resulting firing rate time course $r(t)$ is well fit by Eq. (2.22) with $F_{adap} = 0.7$ and $\tau_{adap} = 30$ ms, similar to *in vivo* observation of cortical neurons (Ahmed et al. 1998).

How are F_{adap} and τ_{adap} determined by the biophysical parameters, and why is τ_{adap} much shorter than τ_{Ca} even though adaptation results from the calcium dynamics? To answer these questions, a theoretical analysis can be carried out in several steps (Wang 1998; Liu and Wang 2001).

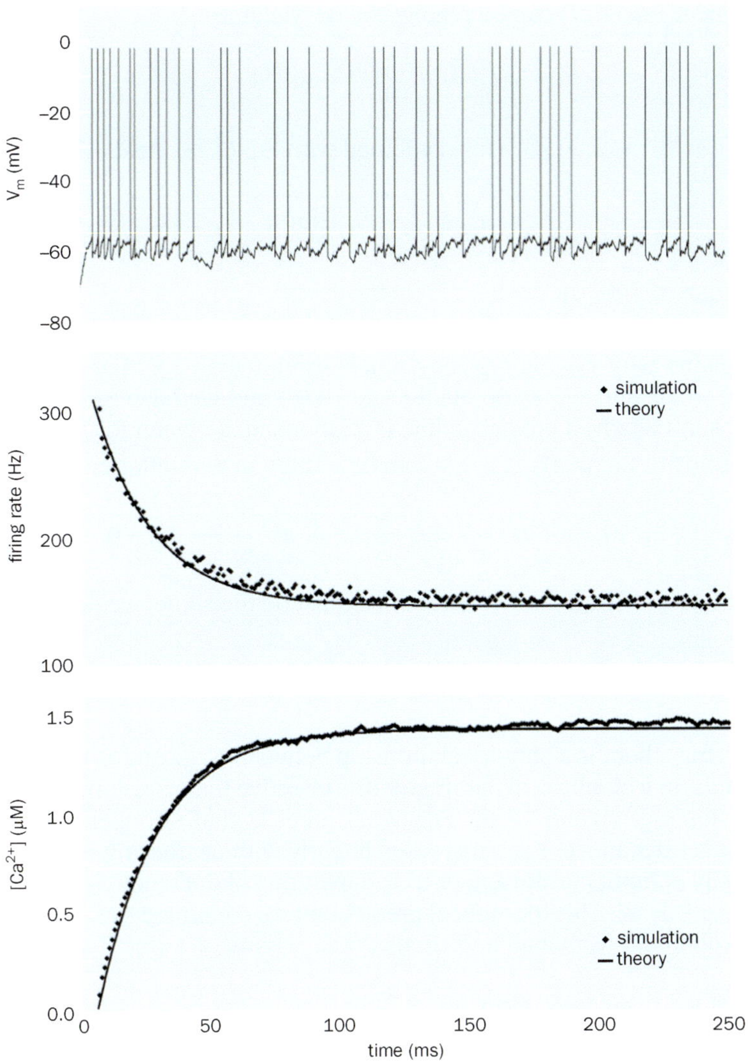

Figure 2.13 Adaptation dynamics of an adaptive LIF model neuron. Upper panel: Adaptation in response to Poisson synaptic inputs. Middle panel: The trial-averaged instantaneous firing rate decreases in time exponentially with a characteristic time constant τ_{adap}. Lower panel: The intracellular $[Ca^{2+}]$ level builds up with the same time course. Adapted with permission from Liu and Wang (2001). [Liu, Y. H., & Wang, X. J. (2001). Spike-frequency adaptation of a generalized leaky integrate-and-fire model neuron. Journal of Computational Neuroscience, 10(1), 25–45.]

First, let us assume that the firing rate $r(t) = F(I_{tot}) \simeq \beta I_{tot}$ (approximately linear with the total current $I_{tot} = I_{app} - I_{KCa}$, above the firing threshold). Second, replacing V by a constant average $\langle V \rangle$ in I_{KCa}:

$$r(t) \simeq \beta\Big(I_{app} - G_{KCa}[Ca^{2+}](\langle V \rangle - V_K)\Big) \simeq r_0 - J_{KCa}[Ca^{2+}] \tag{2.25}$$

where $r_0 = \beta I_{app}$ is the initial firing rate (without adaptation) and $J_{KCa} = \beta G_{KCa}(\langle V \rangle - V_K)$ quantifies the effective strength of the adaptation current. Third, assuming that electrical activity is faster than calcium signal, which thus averages spikes over time, so that the spike train in the $[Ca^{2+}]$ equation can be approximately replaced by the firing rate r,

$$\frac{d[Ca^{2+}]}{dt} - \alpha \sum_j \delta(t - t_j) - [Ca^{2+}] / \tau_{Ca} \tag{2.26}$$

$$\simeq \alpha r - [Ca^{2+}] / \tau_{Ca} = \alpha r_0 - [Ca^{2+}] / \tau_{adap} \tag{2.27}$$

with

$$\frac{1}{\tau_{adap}} = \alpha J_{KCa} + \frac{1}{\tau_{Ca}} \tag{2.28}$$

The linear Eq. (2.27) can be readily solved, yielding

$$\left[Ca^{2+}\right](t) = \left[Ca^{2+}\right]_{ss}\left(1 - exp\left(-t/\tau_{adap}\right)\right) \tag{2.29}$$

with $\left[Ca^{2+}\right]_{ss} = \alpha r_0 \tau_{adap}$. This combined with Eq. (2.25) yields

$$r(t) = r_{ss} + (r_0 - r_{ss})exp\left(-t/\tau_{adap}\right) \tag{2.30}$$

with $r_{ss} = r_0 - J_{KCa}\left[Ca^{2+}\right]_{ss}$.

This theory fits the numerical results quite well. In particular, it explains why τ_{adap} is typically much shorter than τ_{Ca}. For instance, if $\alpha J_{KCa} = 0.02$ and $\tau_{Ca} = 100$ ms ($1/\tau_{Ca} = 0.01$) and we have $\tau_{adap} \simeq 30$ ms. More surprisingly, it predicts a linear relationship between F_{adap} and τ_{adap}. According to Eq. (2.27), $\left[Ca^{2+}\right]_{ss}$ can be written in two different ways,

$$\alpha r_{ss} - \left[Ca^{2+}\right]_{ss}/\tau_{Ca} = \alpha r_0 - \left[Ca^{2+}\right]_{ss}/\tau_{adap} = 0 \tag{2.31}$$

Since $\left[Ca^{2+}\right]_{ss}$ from the two equations must be the same, we deduce

$$F_{adap} = 1 - \frac{\tau_{adap}}{\tau_{Ca}}. \tag{2.32}$$

Thus, there is a simple relationship between F_{adap} and τ_{adap} that only depends on a biophysical parameter (the effective time constant of calcium extrusion τ_{Ca}) (Wang 1998). Eq. (2.32) predicts that stronger adaptation has faster dynamics: F_{adap} increases linearly with decreasing τ_{adap}, as the intensity of input current I_{app} or the adaptation conductance G_{KCa} is varied (**Figure 2.14a**). This theoretical prediction was confirmed by a physiological experiment (Gabbiani and Krapp 2006) (**Figure 2.14b**).

2.4.3 Input Decorrelation

SRA takes place on multiple timescales (Marom 2010; Weber et al. 2019). A slower form of adaptation in a single neuron is realized by a potassium current activated by the intracellular sodium concentration rather than by calcium concentration (Sanchez-Vives and McCormick 2000; Sanchez-Vives et al. 2000). What happens is that prolonged firing of action potentials leads to a gradual Na^+ influx into the cell through the opening of sodium channels. Intracellular sodium accumulation eventually activates a potassium current, I_{KNa}, which gives rise to hyperpolarization of the neuronal membrane and a reduced responsiveness of the cell. After the offset of a large stimulus, sodium ions are slowly extruded from the cell through a Na^+/K^+ ionic pump. This process underlies the recovery of neuronal responsiveness with a time constant of about ten seconds.

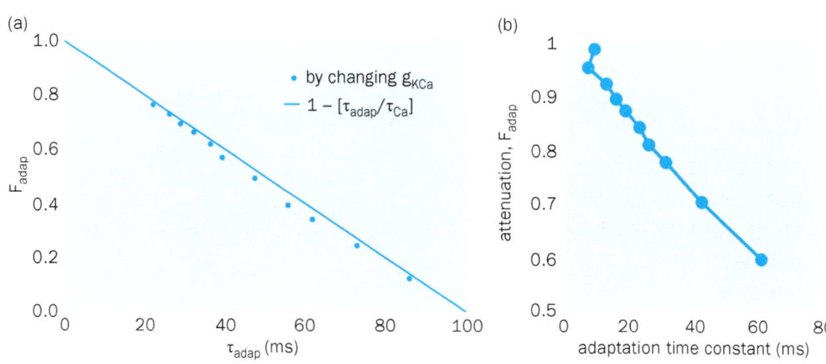

Figure 2.14 A relationship between the amount of adaptation F_{adap} and adaptation time constant τ_{adap} of spike rate adaptation by a single neuron. Left: Model prediction; right: Experimental measurement in a looming sensitive neuron. Panel (a) is reproduced from Liu and Wang (2001), (b) from Gabbiani and Krapp (2006). [Liu, Y. H., & Wang, X. J. (2001). Spike-frequency adaptation of a generalized leaky integrate-and-fire model neuron. Journal of Computational Neuroscience, 10(1), 25–45.] [Gabbiani, F., & Krapp, H. G. (2006). Spike-frequency adaptation and intrinsic properties of an identified, looming-sensitive neuron. Journal of Neurophysiology, 96(6), 2951–2962.]

This phenomenon, observed at the level of single cells, is directly relevant to perception of an animal. Our visual system has evolved to be more sensitive to changes in the visual world than stimuli that are constant in time. For instance, when we enter a dark room from a bright one, we momentarily become blind, but as our eyes adjust to the dim light, we start to see things. We also adapt to contrast, which is the difference in luminance of an object relative to its surroundings; thus, a face of the same luminance appears brighter indoor than under the sun in a park.

In neurophysiological laboratory experiments, contrast of a visual stimulus, such as sinusoidal grating, could be varied, and it was found that spike firing rate of V1 single cells display adaptation on a timescale of seconds comparable to contrast adaptation at the behavioral level (Ohzawa et al. 1985; Sanchez-Vives and McCormick 2000). *In vitro* experiments on single cells suggested that a sodium activated potassium current I_{KNa} could account for contrast adaptation (Sanchez-Vives et al. 2000).

Figure 2.15 displays a type I neuron model with the addition of an I_{KNa}, and intracellular Na^+ dynamics (influx through I_{Na} and outflux through a Na^+/K^+ pump) (Wang et al. 2003). The input is as in the experimental protocol, namely, a sinusoidal current (at 2 Hz) mimicking a sinusoidal grating (Ohzawa et al. 1985; Sanchez-Vives and McCormick 2000). Input amplitude represents stimulus contrast. Initially, the model neuron responds to a low amplitude input at a constant rate. During a high amplitude input, the firing rate transiently increases, then decays to a lower steady-state level according to a time course in parallel with that of intracellular Na^+ accumulation, hence activation of I_{KNa}. After the stimulus returns from the high to the original low amplitude, the membrane potential is hyperpolarized, and the model neuron is incapable of spiking response for seconds. Slowly, however, intracellular Na^+ decays back to baseline, and the model neuron resumes the same activity level as before. Thus, the model captures essential observations from neurons in the primary visual cortex.

What may be the behavioral impact of such neuronal adaptation? Sensory inputs are highly redundant. Removal of redundancy has long been proposed to enhance the efficiency of neural coding of sensory stimuli in the brain, and neuronal adaptation provides a mechanism for a decorrelation operation that reduces input redundancy in time (Barlow 1961; Barlow and Foldiak 1989).

If I_{KNa} subserves adaptation, is it capable of input decorrelation as well? Specifically, light intensity signals of optical flow, observed with a narrow visual field, exhibit correlations at all timescales. This is quantified by a scale-free power spectrum, with the power growing like $1/f$ at decreasing frequencies f (van Hatern and van der Schaaf 1996) (red curve in **Figure 2.15b**), characterizing correlations at increasingly longer timescales. By contrast, a completely uncorrelated signal has a constant power spectrum, i.e., the same power at all frequencies (hence the name "white noise"). How would a cell endowed with adaptation respond to an $1/f$-type input? This question can be investigated in model simulations, using stochastic inputs with strong time correlations. When an $1/f$ stochastic input is used as injected current to drive the model neuron, long-term correlations of the spiking activity are suppressed, so that the power spectrum of the output becomes essentially flat ("whitened") at frequencies below 1 Hz (blue curve, left panel in **Figure 2.15b**). Decorrelation takes place due to the negative feedback mediated by I_{KNa} : a higher firing activity leads to a larger I_{KNa}, hence a stronger negative signal to subtractively counterbalance the external input drive. Decorrelation takes place on a timescale of a second or longer, hence whitening of the power spectrum below 1 Hz. Therefore, a single neuron mechanism can subserve decorrelation function. This model prediction was directly tested in

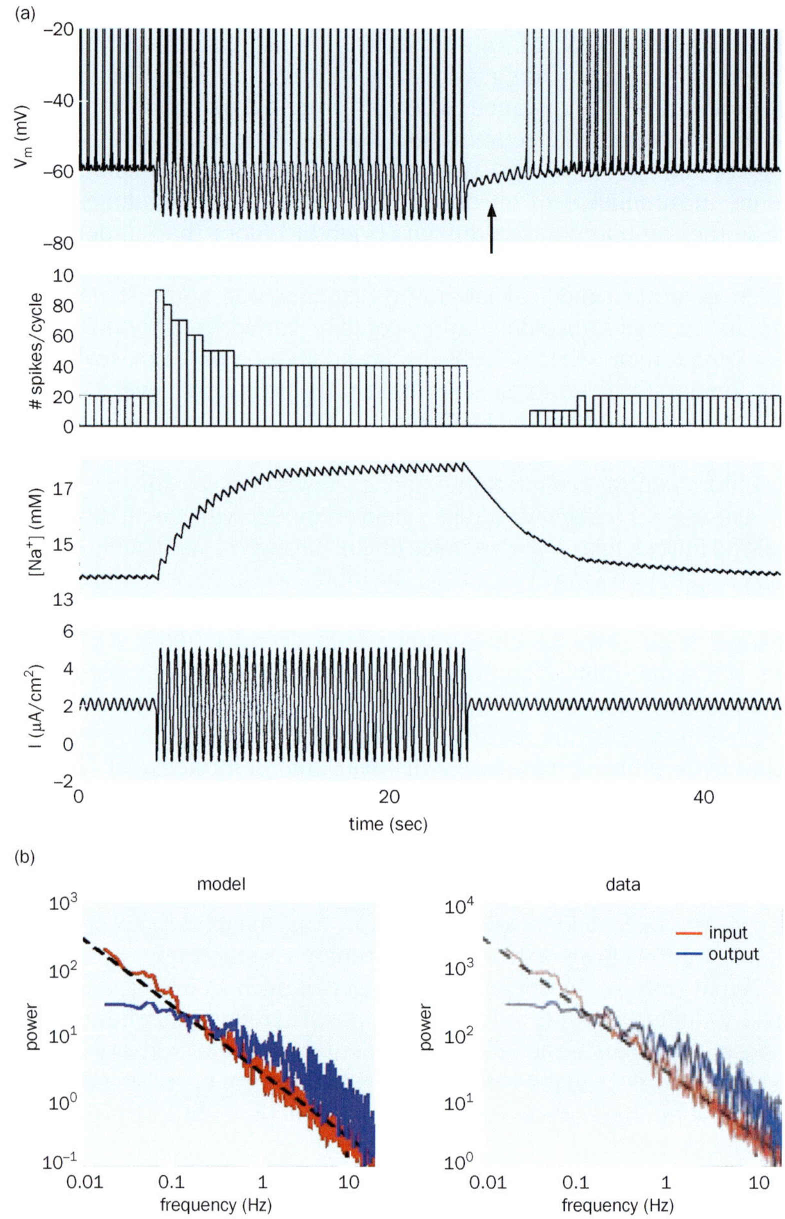

Figure 2.15 Adaptation and decorrelation by a single neuron. (a) The response of the model neuron to a low-high-low sinusoidal current input. Top to bottom: Membrane potential, number of spikes per cycle, intracellular Na^+ concentration, and 2 Hz sinusoidal injected current that varies from low to high to low in amplitude. Following the injection of high-amplitude sinusoidal current, there is a slow hyperpolarization lasting for about ten seconds (indicated by an arrow). Intracellular Na^+ accumulation mirrors the adaptation time course of the instantaneous firing rate during the high-contrast input, whereas Na^+ decay is correlated with the slow recovery from hyperpolarization during the second low-amplitude period. (b) Decorrelation of $1/f$ -type stochastic input by the model (left panel) and a visual cortical neuron recorded intracellularly in a ferret cortical slice (right panel). Reproduced from Wang et al. (2003). [Wang, X. J., Liu, Y., Sanchez-Vives, M. V., & McCormick, D. A. (2003). Adaptation and temporal decorrelation by single neurons in the primary visual cortex. Journal of Neurophysiology, 89(6), 3279–3293.]

an *in vitro* preparation of ferret cortical slices. When a real neuron of the primary visual cortex receives a stochastic $1/f$-type input, the output of the neuron exhibits the same kind of decorrelation as the model (Figure 2.15b, right panel). This experiment demonstrates that decorrelation can occur in single cortical cells.

2.5 BURST FIRING

2.5.1 Ping-Pong Interplay between Soma and Dendrite

Compartmental modeling of dendrites (Segev et al. 1995) reduces a spatially continuous dendrite into a number of discrete compartments, each is assumed to be isopotential. The interactions between two compartments j and k are mediated by a current flow through an axial resistance R_{jk}, and act to equalize the voltages V_j and V_k. The smaller the resistance R_{jk}, the more effective the current flow and the inter-compartment coupling. In principle, the value of R_{jk} (the coupling strength) may be different for any pair (j and k) of compartments, because it depends on the physical properties such as the radius and length of each cylinder-shaped compartment (Koch 1999; Dayan and Abbott 2001).

Importantly, dendrites are not passive but endowed with various types of active ion channels and show a rich repertoire of electroresponsiveness (Stuart et al. 2016). In particular, voltage-dependent ion channels at dendritic sites, once activated, last for tens of milliseconds, and this plateau potential provides a depolarizing drive during a burst firing of multiple spikes (Wong et al. 1979; Traub et al. 1991; Pinsky and Rinzel 1994; Wang 1999a; Kepecs et al. 2002). Furthermore, Hodgkin-Huxley-type sodium and potassium currents on the dendrite enable action potential (initiated at the initial segment of axon, near the soma) to propagate back into the dendrite (Stuart et al. 1997). Therefore, there is a soma-dendritic "ping-pong" interplay: an action potential backpropagates and depolarizes dendritic membrane potential, activates dendritic voltage-gated channels and prolongs dendritic depolarization, which in turn produces more spikes. In the neocortex, such burst firing is especially prominent in pyramidal cells of deep layers (Silva et al. 1991).

Since burst firing is rather common in neural systems, let us discuss it in some detail. We shall use a relatively simple model that captures the ping-pong mechanism (Pinsky and Rinzel 1994), which is deduced from a more elaborated model with hundreds of compartments (Traub et al. 1991). Consider a two-compartment model with a soma and a dendrite (**Figure 2.16a**). The somatic voltage V_s and dendritic voltage V_d obey the following current-balance equations:

$$c_m A_s \frac{dV_s}{dt} = -A_s\left(I_L + I_{Na} + I_K + I_s\right) - \frac{1}{R_c}\left(V_s - V_d\right); \tag{2.33}$$

$$c_m A_d \frac{dV_d}{dt} = -A_d\left(I_L + I_{Ca-L} + I_{K-C} + I_{KCa} + I_d\right) - \frac{1}{R_c}\left(V_d - V_s\right) \tag{2.34}$$

where $c_m = 1\,\mu\text{F/cm}^2$. The membrane surface areas are A_s and A_d (in μm^2), and the axial resistance (resistive coupling) is denoted as R_c (in MΩ). The soma contains Hodgkin-Huxley-type spike-generating currents I_{Na} and I_K (see Section 2.3). The dendrite contains three active conductances: a voltage-gated, L-type calcium current I_{Ca-L}, a calcium-activated potassium current I_{KCa} (the same current I_{KCa} underlying spike-rate adaptation [Section 2.4]), and another potassium current I_{K-C}, which depends on both dendritic voltage and intracellular calcium concentration. The injected currents to the perisomatic and dendritic compartments are I_s and I_d (all currents are in $\mu A\,/\,cm^2$),

Defining the total membrane area $A_{\text{tot}} = A_s + A_d$, and the fractional area for the perisomatic membrane $p = A_s\,/\,A_{\text{tot}}$, one can rewrite the two equations as

$$c_m \frac{dV_s}{dt} = -\left(I_L + I_{Na} + I_K + I_s\right) - \frac{g_c}{p}\left(V_s - V_d\right); \tag{2.35}$$

$$c_m \frac{dV_d}{dt} = -\left(I_L + I_{Ca-L} + I_{K-C} + I_{KCa} + I_d\right) - \frac{g_c}{(1-p)}\left(V_d - V_s\right) \tag{2.36}$$

where $g_c = 1/\left(R_c A_{\text{tot}}\right)$ (in mS/cm^2). This form of equations helps us to see better the interactions between somatic and dendritic membranes. The resistive coupling acts to equalize V_s and V_d: if V_d is higher than V_s, then the coupling term in the first equation is positive, whereas it is negative in the second equation, thereby acting to increase V_s and decrease V_d. The coupling term is zero when $V_s = V_d$. Moreover, soma-dendritic interactions are not symmetrical: if the dendritic compartment is larger than the somatic compartment, p is smaller than 0.5, $g_c\,/\,p$ is larger $g_c\,/\,(1-p)$,

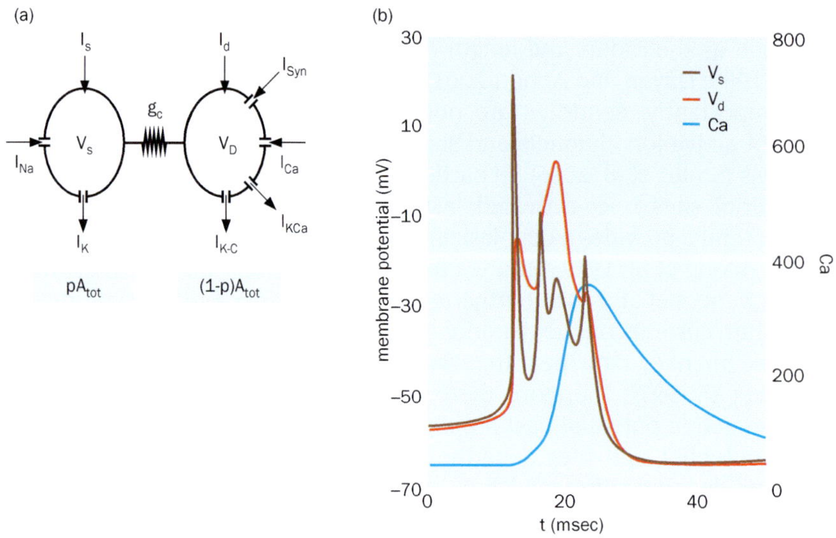

Figure 2.16 Bursting in a two-compartmental model of pyramidal neuron. (a) Schematic of the two-compartment model showing applied currents and outward and inward active currents to soma and dendrite compartments. Soma has Hodgkin-Huxley type sodium (I_{Na}) and potassium (I_K) currents; dendrite has a high-threshold L-type calcium current (I_{Ca-L}), Ca^+- and voltage-activated potassium (I_{K-C}), potassium after hyperpolarization (I_{KCa}) currents. I_s and I_d are applied currents, with the addition of a synaptic current (I_{syn}). Coupling strength between soma and dendrite is governed by parameter g_c, and A_{tot} refers to the total cell membrane area. (b) Time courses of somatic voltage, V_s, dendritic voltage, V_d, and Ca during a burst (the scale for Ca is different from voltage). Burst is initiated by a somatic action potential, which triggers a subthreshold dendritic Ca spike by I_{Ca-L}. This is followed by a full Ca spike in the dendrite, leading to somatic burst pattern. Dendritic spike, and hence burst, is terminated by I_{K-C} and I_{K-AHP}, which build up when Ca reaches appreciable levels. Reproduced from Pinsky and Rinzel (1994). [Pinsky, P. F., & Rinzel, J. (1994). Intrinsic and network rhythmogenesis in a reduced Traub model for CA3 neurons. Journal of Computational Neuroscience, 1(1), 39–60.]

reflecting the intuition that there is a stronger influence of the larger dendritic compartment to the smaller somatic compartment than in the opposite direction.

Figure 2.16b displays a burst of spikes produced by this model. During a burst, action potentials originate in the perisomatic compartment (V_S, brown line) and backpropagate into the dendritic compartment (V_d, dashed line) (Larkum et al. 1999). This interaction is described as a dynamic "ping-pong" flow of current between the soma and the dendrite. A burst occurs when a narrow somatic spike (in V_S) elicits a broad dendritic spike (in V_d, red line) that, in turn, produces a depolarizing afterpotential in the soma. The broad dendritic spike, which is due to a dendritic calcium current I_{Ca-L}, acts as a source of inward current to depolarize the soma and bring it to threshold again. During repetitive spikes, intracellular calcium accumulates ($[Ca^{2+}]$, blue), the resulting gradual activation of the potassium currents I_{K-C} and I_{KCa} eventually shut off local regeneration of dendritic spikes and a burst is terminated. During hyperpolarization $[Ca^{2+}]$ decays back to a baseline, the membrane potential recovers, possibly leading to another burst of spikes.

The two-compartment model describes how a ping-pong mechanism produces bursting firing. However, note that multiple mechanisms contribute to bursting, especially the voltage-gated synaptic conductance mediated by the N-methyl-D-aspartate (NMDA) receptors (which we will discuss in Section 2.6) (Larkum et al. 1999; Larkum 2013).

A two-compartment model may not be suitable for investigating other questions. For instance, relatively independent segments of a dendritic tree suggest a two-layered model of pyramidal cells (Poirazi et al. 2003). To describe input and output control of pyramidal cells by SST/CB-expressing and PV-expressing inhibitory neurons, respectively, a better yet minimal model for pyramidal cells may have three compartments (soma, proximate and distal dendrites), so that the distal dendrite is separated by the proximate dendrite from the soma and can act less dependently (Wang et al. 2004). As another example, the alignment of pyramidal dendrites is an important factor for summating the extracellular current flow across a population of pyramidal cells, which underlies local field potential (LFP), electroencephalography (EEG) and magnetoencephalography (MEG) (Wang 2010; Buzsaki et al. 2012). LFP is captured in a model where pyramidal cells have a more complicated dendrite (Sherman et al. 2016; Jones et al. 2009). Modeling of dendrites is an active research topic that continues to yield surprising findings (Stuart et al. 2016; Poirazi and Papoutsi 2020; Otor et al. 2022).

2.5.2 Postinhibitory Rebound

Like potassium currents, there are multiple types of calcium currents (Tsien et al. 1987). Another kind of burst firing relies on a low-threshold (T-type) calcium current I_{Ca-T} rather than a high-threshold calcium current I_{Ca-L}, and thus can be activated at membrane potentials below the spiking threshold (Llinás 1988; Wang et al. 1991; McCormick and Huguenard 1992; Wang 1994; Destexhe et al. 1996). Curiously, I_{Ca-T} activation is conditioned on a prolonged membrane hyperpolarization: its inactivation variable h has a steady-state that is virtually zero near or above the resting membrane potential; it increases when V is hyperpolarized with a time constant of $\sim$100 ms. On the other hand, its activation variable becomes significant with moderate depolarization. Therefore, when a neuron is endowed with a large I_{Ca-T}, depolarization from resting-state leads to repetitive spikes (tonic mode). On the other hand, upon release from a sufficiently long hyperpolarization, I_{Ca-T} is activated to underlie a "depolarization wave" (postinhibitory rebound) with a burst of spikes riding on top of it (burst mode) (**Figure 2.17a**). This phenomenon has been observed in several neural types, including thalamic neurons that relay information from sensory organs to the neocortex (Llinás 1988). The thalamus is a pacemaker of synchronous brain oscillations during certain (non-dreaming) phases of sleep, which depends on I_{Ca-T} underlying rhythmic bursting at a few hertz (Steriade and Llinas 1988; Steriade et al. 1993; Wang and Rinzel 1992; Wang et al. 1995; Destexhe and Sejnowski 2001). Thus, understanding the ion channel mechanism of postinhibitory bursting represents an example of how single-cell properties shape collective dynamics of the whole brain. Waking from sleep is correlated with depolarization of thalamic neurons, which in behaving conditions operate in the tonic mode, although rebound bursts may occur occasionally in wakefulness (Ramcharan et al. 2005).

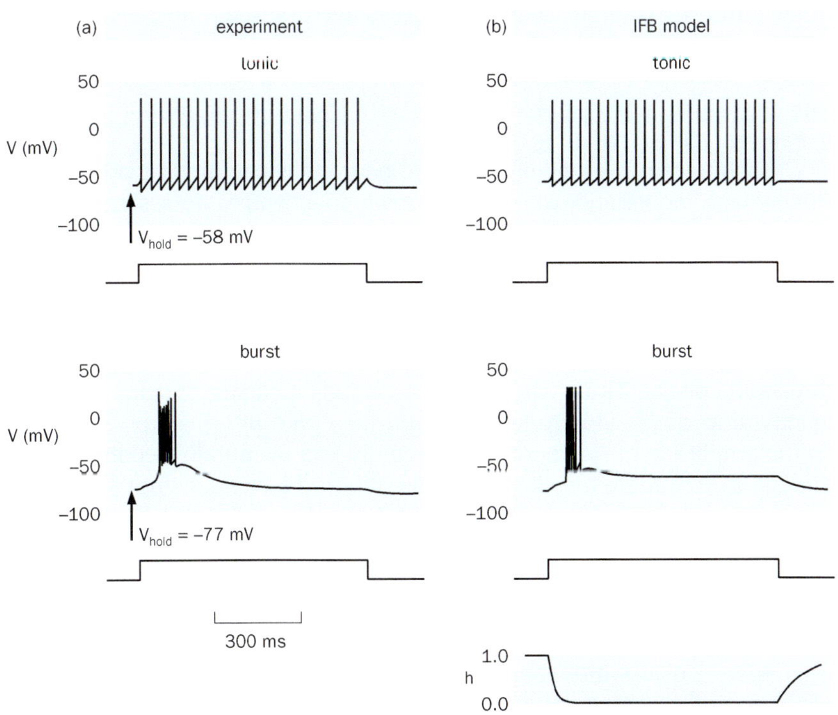

Figure 2.17 Bursting as postinhibitory rebound. Experimental records of relay neuron responses to depolarizing current steps and comparable integrate-and-fire-or-burst (IFB) model simulations. (a) A thalamic relay neuron responds in either tonic or burst mode depending on the initial membrane potential (V_{hold}). (b) The IFB model reproduces both tonic and burst responses depending on the initial membrane potential. Bottom: h, the deinactivation gating variable for I_{Ca-T}, during the IFB model burst response. Adapted with permission from Smith et al. (2000). [Smith, G. D., Cox, C. L., Sherman, S. M., & Rinzel, J. (2000). Fourier analysis of sinusoidally driven thalamocortical relay neurons and a minimal integrate-and-fire-or-burst model. Journal of Neurophysiology, 83(1), 588–610.]

Smith et al. (2000) proposed an extension of the LIF model to incorporate postinhibitory rebound. I_{Ca-T} is added to the voltage equation, a second equation describes the dynamics of its slow inactivation:

$$c_m \frac{dV}{dt} = -g_L(V - V_L) - g_{Ca-T} m_\infty(V) h(V - V_{Ca}) + I_{app} \tag{2.37}$$

$$\frac{dh}{dt} = \begin{cases} -h / \tau_h^- & \text{if} \quad V > V_h \\ (1-h) / \tau_h^+ & \text{if} \quad V < V_h \end{cases} . \tag{2.38}$$

where I_{Ca-T} has a reversal potential of $V_{Ca} = 120$ mV and activation threshold of $V_h = -60$ mV. For the sake of simplicity, this model assumes that the fast activation variable m is replaced by its steady-state (m_∞), which is zero for $V < V_h$ and 1 for $V > V_h$. The opposite holds true for the steady-state of the inactivation variable h. The time constant $\tau_h^+ = 100$ ms controls the timescale of hyperpolarization needed for postinhibitory rebound, and $\tau_h^- = 20$ ms determines the duration of depolarization underlying rebound burst. The model reproduced the salient tonic and burst modes of thalamic relay neurons (**Figure 2.17b**). This model has been used for other neuron types endowed with postinhibitory rebound, such as neurons in the subthalamic nucleus and external globus pallidus of basal ganglia (Wei et al. 2015).

2.5.3 Clustered and Irregular Spiking

Bursting denotes a brief period of spiking at a high frequency, a variant of which is clusters of spikes inter-nested in time with episodes of subthreshold membrane fluctuations (Rinzel 1987; Wang 1993). **Figure 2.18a** (upper panel) shows an example of such firing patterns of an inhibitory neuron in rat medial septum (Serafin et al. 1996). The regular repetition of spiking clusters is in the range of theta oscillations, a prominent rhythm in the hippocampus that occurs during an animal's spatial navigation and depends on inputs from the medial septum (Vanderwolf 1969; Buzsáki 2002). Interestingly, the medial septum contains two major cell types. Cholinergic cells release acetylcholine, a major neuromodulator, to neurons in the hippocampus, thereby regulating the network's excitability; on the other hand, GABAergic cells act on a faster timescale and entrain hippocampal neurons at theta oscillations (Stewart and Fox 1990). The intrinsic rhythm of spike clustering in GABAergic neurons of the medial septum thus represents a pacemaking function for the entire brain in a particular behavioral state (spatial navigation).

Such pacemaker activity of single neurons is captured by a Hodgkin-Huxley-type model endowed with a slowly inactivating potassium current (I_{KS}) (Wang 1993), as shown in **Figure 2.18a** (lower panel) (Wang 2002a). The mechanism relies on two assumptions. First, a low-threshold K^+ current I_{KS} should be transient and inactivates slowly during depolarization. Second, the neuronal spike afterhyperpolarization is strong, so that the average membrane potential is lower during spikes than during subthreshold epochs. Thus, when the neuron fires a cluster of spikes, the de-inactivation of I_{KS} slowly builds up during spike afterhyperpolarization, so the amplitude of I_{KS} increases gradually and eventually becomes large enough to terminate a spiking episode. When the cell does not fire spikes in a subsequent subthreshold epoch, the membrane potential is relatively depolarized, and I_{KS} slowly decreases due to inactivation, until the cell is sufficiently recovered and can start to fire again (**Figure 2.18b**). The subthreshold oscillations are produced by the interplay between a Na^+ current and the low-threshold activation of I_{KS}.

Spike clusters interspersed with subthreshold epochs can also take place irregularly rather than rhythmically, as observed in VIP/CR type of

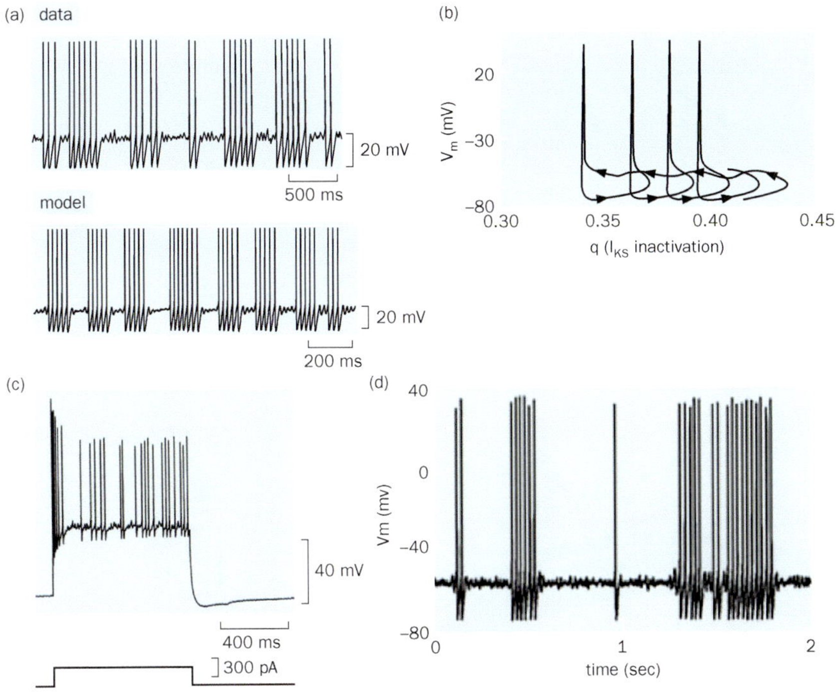

Figure 2.18 Spike clustering. (a) Upper panel: A non-cholinergic (putative inhibitory) cell in the rat medial septum displays rhythmic alternations at theta frequency between "clusters" of spikes and epochs of subthreshold membrane potential oscillations. Lower panel: A Hodgkin-Huxley-type conductance-based model of the neuron shown in the upper panel. (b) For the model, membrane potential versus the inactivation gating variable q for a low-threshold potassium channel (I_{KS}). I_{KS} gradually de-inactivates (an increase of q, hence I_{KS}) during the hyperpolarizing phases of spikes (rightward arrows), whereas it inactivates (decreasing q) during subthreshold membrane oscillations (leftward arrows). (c) A VIP/CR interneuron in the neocortex shows spike clustering interspersed with subthreshold epochs. (d) A model of VIP/CR neuron with the Hodgkin-Huxley conductances and a low-threshold potassium current, driven by noisy inputs. The upper part of panel (a) is reproduced from Serafin, M., Williams, S., Khateb, A., Fort, P., & Mühlethaler, M. (1996). Rhythmic firing of medial septum non-cholinergic neurons. Neuroscience, 75(3), 671–675; the lower part of panel (a) and panel (b) are reproduced from X. J. (2002). Pacemaker neurons for the theta rhythm and their synchronization in the septohippocampal reciprocal loop. Journal of Neurophysiology, 87(2), 889; panel (c) is reproduced from Cauli, B., Audinat, E., Lambolez, B., Angulo, M. C., Ropert, N., Tsuzuki, K., ... & Rossier, J. (1997). Molecular and physiological diversity of cortical nonpyramidal cells. Journal of Neuroscience, 17(10), 3894–3906; panel (d) is reproduced from Stiefel, K. M., Englitz, B., & Sejnowski, T. J. (2013). Origin of intrinsic irregular firing in cortical interneurons. Proceedings of the National Academy of Sciences of the United States of America, 110(19), 7886–7891.

inhibitory neurons in the neocortex (**Figure 2.18c**) (Cauli et al. 1997; Markram et al. 2004). It is interesting that single isolated cells, in the absence of synaptic inputs, can display irregular firing dynamics, which may be deterministic chaos (Strogatz 2016) or a result of intrinsic noise (White et al. 2000). A conductance-based model with a low-threshold potassium current, driven by a noisy input current, produces irregular firing patterns similar to experimental observations (**Figure 2.18d**) (Stiefel et al. 2013).

Modern genomic analysis of the neural system has made it possible to link genes to physiological actions of their encoded proteins. An early example was provided from a study on the single cells of a central pattern generator of crustaceans, showing a strong positive correlation between mRNA abundance and the corresponding ion channel conductance measured by voltage-clamp experiments (Schulz et al. 2006). **Figure 2.19a** shows the result for three types of ion channels: non-inactivating potassium current I_{Kd} described in the Hodgkin-Huxley model, an inactivating potassium current I_A, a voltage- and calcium-activated potassium current I_{BK-KCa}. From ion channels one can deduce their physiological effects on a single cell. In particular, the width of an action potential is determined by I_{Kd}, its corresponding gene is called Kv3.1 or KCNC1 (Rudy 1988). The spike width is narrower in PV-expressing interneurons than in SST-expressing ones in the cortex, thus the gene that encodes I_{Kd} should have a higher expression in PV than in SST cells. That is indeed the case, as shown in **Figure 2.19b**. However, there are several intermediate steps from a gene to its encoded protein to physiology. It remains to be seen to what extent the expression level of a gene is a good predictor for the physiological effect of its encoded protein.

A remarkable advance in this area is to use single-cell transcriptomic data for cell-type classification in the mammalian cortex, which has been recently reported for mouse (Yao et al. 2023) and macaque monkey (Chen et al. 2023a). In the macaque cortex, data were collected using single-nucleus RNA-seq measurements, and a clustering method was applied to the database, revealing 264 transcriptomic clusters for cell-type taxonomy of the macaque cortex (**Figure 2.19c**). The dendrogram depicts 264 cell types (annotated in the outer circle) based on the median expression levels

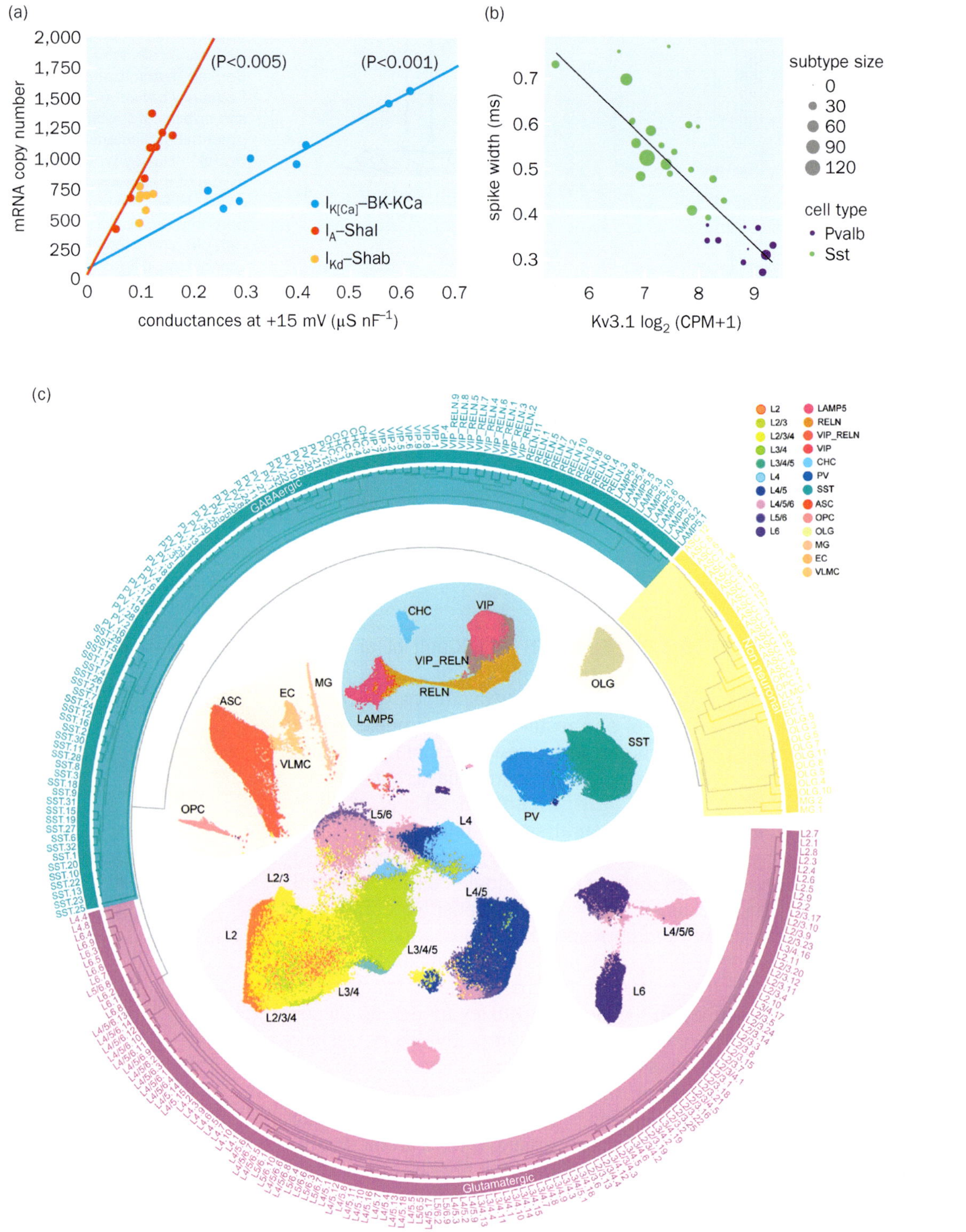

Figure 2.19 Genomic analysis for cell-type classification and single cell properties. (a) Ion channel mRNA abundance is correlated with measured membrane conductance in neurons of a crab's central pattern generator; each data point is for an individual cell. (b) The spike width (at half of the maximal voltage) is negatively correlated with the expression level of the gene Kv3.1 (KCNC1) that encodes the potassium conductance underlying spike repolarization of PV and SST positive GABAergic cells in the mouse V1. Each bubble represents a subcell type defined in Tasic et al. (2018); the size of each bubble represents the number of cells within that subcell type. (c) Transcriptomic cell-type taxonomy of 146 cortical regions in macaque brain. CHC, chandelier cells; ASC, astrocyte; MG, microglia; OLG, oligodendrocyte; OPC, oligodendrocyte precursor cells; EC, endothelial cells; VLMC, vascular leptomeningeal cell. Panel (a) is reproduced from Schulz et al. (2006), (b) is provided by Yijie Kang with data from Gouwens et al. (2020), (c) is reproduced from Chen et al. (2023a). [Schulz, David J., Jean-Marc Goaillard, & Eve Marder. Variable channel expression in identified single and electrically coupled neurons in different animals. Nature Neuroscience 9.3 (2006): 356–362.] [Chen, Ao, Yidi Sun, Ying Lei, Chao Li, Sha Liao, Juan Meng, Yiqin Bai et al. Single-cell spatial transcriptome reveals cell-type organization in the macaque cortex. Cell (2023), 186, https://doi.org/10.1016/j.cell.2023.06.009]

of top 50 marker genes, revealing the hierarchical clustering of three major classes (glutamatergic, GABAergic, and non-neuronal types; marked by background colors) and 23 cell subclasses (marked by dot colors). This database offers a wealth of information to be mined by the community. Future genomic data analysis will also help our studies of the neural circuits described in later chapters.

In summary, single neurons display a plethora of firing patterns, with potential functional implications such as spike frequency adaptation and rhythmogenesis. In the neocortex, SRA is typically weak in PV inhibitory neurons (which are hence often called "fast spiking" cells), but prominently displayed by excitatory pyramidal neurons (McCormick et al. 1985), with some in deep layers showing burst firing. SST/CB inhibitory neurons also show SRA (Kawaguchi and Kubota 1993; Gibson et al. 1999), whereas VIP/CR inhibitory neurons are characterized by irregular firing (Cauli et al. 1997; Mendonça et al. 2016). Thus, we have now become acquainted with models of major cell types of excitatory and inhibitory neurons in the neocortex. They provide neuronal units in a variety of network models. But before we discuss neural networks, it is necessary to learn how to mathematically describe synaptic interactions between nerve cells.

2.6 SINGLE SYNAPSE MODELS

A chain of physiological events takes place when a spike traveling along an axon arrives at a synapse (Katz 1966). During an action potential, the strong membrane depolarization of the axonal terminal activates voltage-gated calcium channels, resulting in a brief and large calcium influx into the presynaptic intracellular space (Katz and Miledi 1968; Llinás 1982). The latter triggers release of neurotransmitters as a stochastic process with some probability. The availability of transmitters at the presynaptic terminal and the release probability are activity dependent, which give rise to short-term synaptic plasticity (see Section 2.7). Released transmitters diffuse through the synaptic cleft and bind with receptors at the postsynaptic site, thereby imparting on the membrane of the targeted cell. In the central nervous system, at an excitatory synapse the primary transmitter is glutamate, and major types of glutamate receptors include AMPA (α-amino-3-hydroxy-5-methyl-4-isoxazolepropionic acid receptor) and NMDA (N-Methyl-D-aspartic acid) receptors as well as mGluR (G-protein-coupled metabotropic glutamate) receptors. On the other hand, at an inhibitory synapse, the primary transmitter is GABA (γ-Aminobutyric acid), and major types of GABA receptors are $GABA_A$ and $GABA_B$ receptors (Nichols et al. 1992; Koch 1999). The binding of a transmitter with its receptor opens an ion channel on the postsynaptic membrane, resulting in either an excitation (a depolarization mediated by glutamate receptors) or an inhibition (a hyperpolarization mediated by GABA receptors). Therefore, neuronal communication requires a combination of electrical signals (spikes) and chemical substances, and it is inherently unidirectional (there is no principled reason that if one neuron sends signals to another neuron, the communication is reciprocated). Moreover, in the central nervous system, a given neuron only synthesizes either glutamate or GABA. Therefore, its action on all its connected postsynaptic neurons is uniformly depolarizing or hyperpolarizing. In other words, a neuron is either excitatory or inhibitory, not both. This is the "Dale's law," named after the British physiologist Henry Hallett Dale (Nichols et al. 1992).

Just like a diversity of models for single neurons, modeling single synapses may take into account a varying degree of complexity of the synaptic transmission process. Here, we will discuss mathematical descriptions of synaptic transmission in multiple ways. In those descriptions, the focus is on the impact of a presynaptic spike onto a postsynaptic neuronal membrane. The time taken by all the intermediate steps of synaptic transmission is lumped into a single parameter called "transmission latency" Δ_L, which is small and neglected sometimes by assuming $\Delta_L = 0$.

Two major respects differentiate various models of single synapses. First, a distinction must be made between "current-based" and "conductance-based" synapse models. A conductance-based model describes a synaptic current similarly to any other ionic current, as $I_{syn} = g_{syn}s(V - V_{syn})$, where g_{syn} is the maximum conductance, and s is a synaptic gating variable that is not activated unless there is a presynaptic spike. Therefore, a synaptic current depends on both presynaptic activity and postsynaptic membrane V. The reversal potential determines whether a current is depolarizing or hyperpolarizing; typically, $V_E = 0$ mV for synaptic excitation mediated by the AMPA and NMDA receptors of the transmitter glutamate, and $V_I = -70$ mV and -80 mV for synaptic inhibition mediated by the $GABA_A$ and $GABA_B$ receptors, respectively, of the transmitter GABA.

For a neuron that receives synaptic inputs from M presynaptic neurons, each with a weight w_i and gating variable s_i (which depends on the firing of presynaptic neuron $i = 1,\dots,M$), the summated synaptic current is written as:

$$I_{syn} = \sum\nolimits_{i=1}^{M} w_i s_i (V - V_{syn}), \tag{2.39}$$

A current-based model neglects the dependence on the postsynaptic voltage, so instead of Eq. (2.39), we have:

$$I_{syn} = \sum\nolimits_{i=1}^{M} J_i s_i, \tag{2.40}$$

where synaptic weight is denoted as J_i instead of w_i with a different physical unit. A current-based model is simpler than a conductance-based model, but it is worth keeping in mind that it does not account for conductance changes due to synaptic inputs, which may vary considerably depending on the brain states and stimulation conditions (Destexhe et al. 2003).

Secondly, choosing how to model the synaptic gating variable s involves a tradeoff between the desire for mathematical simplicity and the need for certain details to capture aspects of synaptic dynamics. This will become clear when we present several types of synapse models in progressively more complicated forms.

2.6.1 Kick Synapses

We will start with the simplest description of synaptic interaction, by assuming that each presynaptic spike causes an instantaneous change (a "kick") in the postsynaptic neuron. For a current-based kick synapse model, the impact is directly on the voltage. Consider a presynaptic neuron, which a train of spikes at time $t_j, j = 1,2,3\dots$, then the synaptic current is $I_{syn} = Js$ and $s = \sum_j \delta(t - t_j)$. Note that since the physical unit of a delta function is $1/s$, the unit of J is electric charge (current has the unit of charge per unit time).

If an LIF neuron is initially at rest, and a presynaptic cell fires a single spike at time $t = 0$, then by integrating Eq. (2.1) one obtains:

$$V(t) = V_L + \frac{J}{C_m}\exp\left(-\frac{t}{\tau_m}\right)H(t), \tag{2.41}$$

where H(t) is a Heaviside function, H(t) = 0 if $t < 0$ and 1 if $t > 0$. Thus, the postsynaptic potential (PSP) produced by each presynaptic spike consists of an instantaneous kick of size $\bar{J} = J / C_m$ followed by an exponential decay with time constant τ_m. For example, if the synapse is excitatory (J is positive) and the unitary charge is $J = 0.04$ pC, and $C_m = 0.2$ nF, then the kick size is $\bar{J} = 0.04 / 0.2 = 0.2$ mV.

With a spike train fired by a presynaptic neuron, the impacts are additive. Moreover, If M neurons synapse onto a postsynaptic cell, each with a weight $J_i\,(i = 1,\ldots,M)$, then the current into the postsynaptic neuron is additive over all the presynaptic neurons, $I_{syn} = \sum_{i=1}^{M} J_i s_i$ and $s_i = \sum_j \delta\left(t - t_j^i\right)$, where t_j^i is the time of the jth spike from the ith presynaptic neuron.

As a simple example, assume that all synaptic weights are the same ($J_i = J$) and spike trains are Poisson processes, with a rate λ_i for presynaptic neuron i ($i = 1, 2,\ldots, M$). Because the sum of Poisson processes is itself a Poisson process, the spike train of overall synaptic input is still a Poisson process with a rate $\lambda = \sum_i^M \lambda_i$. When the firing rate λ is low (say 1 Hz), a typical time interval between two synaptic inputs is large (~ 1s) compared to τ_m (~ 10 ms), in which case V decays back to V_L after each synaptic input. On the other hand, with a large firing rate (e.g., $M = 100$ and $\lambda_i = 5$ Hz, therefore $\lambda = 500$ Hz), intervals between synaptic inputs are small (2ms), and V does not have time to decay to baseline between synaptic inputs. Instead, the neuron integrates synaptic inputs with a filter characterized by τ_m. With sufficiently strong excitation (determined by presynaptic activity and synaptic weight), a postsynaptic neuron can reach the firing threshold, generating an output spike train.

For a conductance-based synapse model, we have $I_{syn} = \sum_{i=1}^{M} w_i s_i \left(V - V_{syn}\right)$. Inserting this input current into an LIF model (Eq. 2.1), it is clear that, as a major difference with a current-based synapse model, synaptic bombardments can significantly affect the neuronal time constant. Specifically, if synaptic inputs have an average conductance of G_{syn}, so the average $I_{syn} \sim G_{syn}\left(V - V_{syn}\right)$, then the effective time constant is $\tau_{m,eff} = C_m / \left(G_L + G_{syn}\right)$, which can be much shorter than $\tau_m = C_m / G_L$ without synaptic inputs (Destexhe et al. 2003). One of the implications is that with increased synaptic inputs, the neuron becomes leakier (with a larger effective leak conductance), thereby less capable of temporal integration. This implies that a neuron may not be able to integrate synaptic inputs over tens of milliseconds. Instead, its firing is determined by the coincidence of multiple synaptic inputs within a very brief period of time, in which case a neuron would behave as a "coincidence detector" rather than as an "integrator" (Abeles 1991; Softky 1994).

2.6.2 Filter and Kinetic Models of Synaptic Transmission

Kick synapse modeling does not account for the temporal properties of synaptic transmission, which can be measured with voltage-clamp technique. If the membrane potential is fixed and a presynaptic neuron is triggered to fire a spike or several spikes, one can measure $I_{syn} / \left(V - V_{syn}\right)$ to deduce the time course of the gating variable $s(t)$ itself, which has an onset and a decay as well as temporal integration ability for a train of spikes (Katz 1966; Nichols et al. 1992).

A variety of models have been used to account for the temporal synaptic dynamics in addition to transmission delay. First, one can assume a temporal filter $h(t)$, for a given presynaptic train $t_j, j = 1,2,3...$, the synaptic variable $s(t)$ is given by

$$s(t) = \sum_j h(t - t_j) \tag{2.42}$$

The kick synapse model is a special case, with $h(t - t_j) = \delta(t - t_j)$. A commonly used filter is $h(t) = (t / \alpha)\exp(-\alpha t)$. This "alpha-function" rises from zero, reaches a maximum at time $t_{peak} = 1/\alpha$, then decays exponentially with a time constant $\tau_{syn} = 1/\alpha$. If we want to describe the rise time and decay time with independent parameters, we can choose a difference of two exponentials, $h(t) = A(\exp(-t/\tau_1) - \exp(-t/\tau_2))$, where τ_1 is longer than τ_2 so that $h(t)$ is positive. The filter $h(t)$ rises from zero, peaks at $t_{peak} = \ln(\tau_1/\tau_2)/(1/\tau_2 - 1/\tau_1)$ and afterwards decays as $\sim \exp(-t/\tau_1)$. **Figure 2.20** displays the time course of synaptic response to a single presynaptic spike for the AMPA receptor and NMDA receptor-mediated excitation, and $GABA_A$ receptor-mediated inhibition according to this description of a difference of two exponentials.

Synapses can also be described by a kinetic equation, assuming that channels that bind with a receptor-like glutamate or GABA obey a similar first-order chemical reaction as those for intrinsic ion channels (Eq. 2.10), except that it is activated by a presynaptic neuron that fires spikes (Wang and Rinzel 1992; Destexhe et al. 1998). For a presynaptic spike train $t_j, j = 1,2,3...$:

$$\frac{ds}{dt} = \alpha_S \sum_j \delta(t - t_j)(1 - s) - \beta_S s \tag{2.43}$$

where the opening rate $\alpha_S \sum_j \delta(t - t_j)$ is zero except when there is presynaptic spiking activity, and β is the rate of "unbinding" with the receptor leading to channel closing; in the absence of spiking, $s(t)$ decays exponentially with a time constant $\tau_S = 1/\beta_S$. Note that when s is close to 1, the first term in Eq. (2.43) approaches 0, thus s cannot exceed 1, reflecting the fact that there is only a finite amount of receptor-channel complex available at a synapse. The saturation property distinguishes the kinetic model from the filter models in which the synaptic current can grow without bound. This formulation can be extended to a sequence of two or more chemical reactions so that $s(t)$ displays different onset and decay times as well as summation to saturation properties that can be fit with experimental measurements (Destexhe et al. 1998; Wang 1999b).

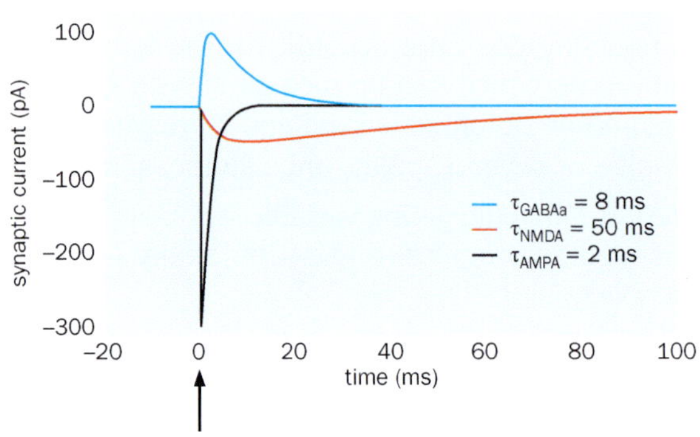

Figure 2.20 Model of an excitatory synaptic current mediated by the AMPA (black) or NMDA (red) receptor and an inhibitory synaptic current mediated by the $GABA_A$ receptor (blue). With a clamped voltage, the time course of synaptic response to a single spike at $t = 0$ is determined by the kinetics of the gating variable s of synaptic conductance. An excitatory current is negative (inward), whereas an inhibitory current is positive (outward). Note different rise time and decay time course for each of the three currents.

To understand better synaptic saturation, let us approximate the presynaptic spike train by its average rate $r = (1/T)\int_0^T \sum_j \delta(t-t_j)$. Then Eq. (2.43) becomes:

$$ds / dt = \alpha_S r(1-s) - \beta_S s. \tag{2.44}$$

Thus with repetitive stimulation, s increases according to this linear equation and eventually reaches a steady-state:

$$s_{SS} = \frac{\alpha_S \tau_S r}{\alpha_S \tau_S r + 1} \tag{2.45}$$

The level of synaptic activation increases roughly linearly with the presynaptic firing rate as long as $\alpha_S \tau_S r \ll 1$ so that the denominator is ~ 1. For the sake of simplicity, consider $\alpha_S \sim 1$, when this condition of linearity becomes $r \ll 1/\tau_S$. This holds true when the firing rate is tens of Hz, for the fast AMPA receptor-mediated excitation (with $\tau_{AMPA} \sim 2$ ms (Hestrin et al. 1990b; Spruston et al. 1995), $1/\tau_{AMPA} \sim 500$ Hz) or $GABA_A$ receptor-mediated inhibition (with $\tau_{GABA} \sim 10$ ms (Salin and Prince 1996; Xiang et al. 1998), $1/\tau_{GABA} \sim 100$ Hz). This is why we often can neglect the factor $(1-s)$ in Eq. (2.43), so it becomes $ds/dt = \alpha_S \sum_j \delta(t-t_j) - \beta_S s$ for AMPA and $GABA_A$ receptor-mediated synaptic currents, which is equivalent to a linear filter with the time constant τ_S. One advantage of the simplification is that if M presynaptic neurons converge on a postsynaptic cell, one does not need to write M equations for $s_i, i = 1,2,\ldots M$. Instead, the summated inputs can be described by a single variable $s_{tot} = \sum_{i=0}^{M} s_i$, which obeys $ds_{tot}/dt = \alpha_S \sum_i \sum_j \delta\left(t - t_j^{(i)}\right) - \beta_S s_{tot}$. The approximation (1-s)~1 is no longer valid in the case of the NMDA receptor-mediated excitation with a time constant of 50–100 ms (Hestrin et al. 1990b; Spruston et al. 1995).

2.6.3 NMDA Receptor-Mediated Synaptic Excitation

A distinguishing feature of the NMDA synaptic conductance is its nonlinear dependence on the postsynaptic voltage, so the current is written as:

$$I_{NMDA} = g_{NMDA} F(V) s(V - V_E) \tag{2.46}$$

where $F(V) = 1/\left(1 + [Mg^{2+}] exp(-0.062V)/3.57\right)$ and $[Mg^{2+}]$ is the extracellular magnesium concentration (~ 1 mM in the central nervous system). **Figure 2.21a** shows this sigmoid-shaped dependence on the postsynaptic membrane potential, derived from single-channel experiments (Jahr and Stevens 1990). Thus, the NMDA channel opening is negligible if the membrane potential is close to resting state (~ -70 mV), and becomes large only with depolarization. In other words, the NMDA conductance activation requires a coincidence of presynaptic and postsynaptic activity. This property endows the NMDARs with a crucial role in Hebbian synaptic plasticity that relies on pre- and post-synaptic co-activation (see Chapter 4).

Furthermore, NMDA receptors also play a major role in online neural computation, especially in brain structures of importance to higher cognition (Wang 1999b). The NMDAR-mediated excitation is much slower than that mediated by the AMPA receptors, with $\tau_{NMDA} \sim 50 - 100$ ms (depending on the NMDA receptor subtypes). In this case, saturation cannot be neglected even for firing rates within the commonly observed physiological range (10–50 Hz). **Figure 2.21b–d** show the behavior of

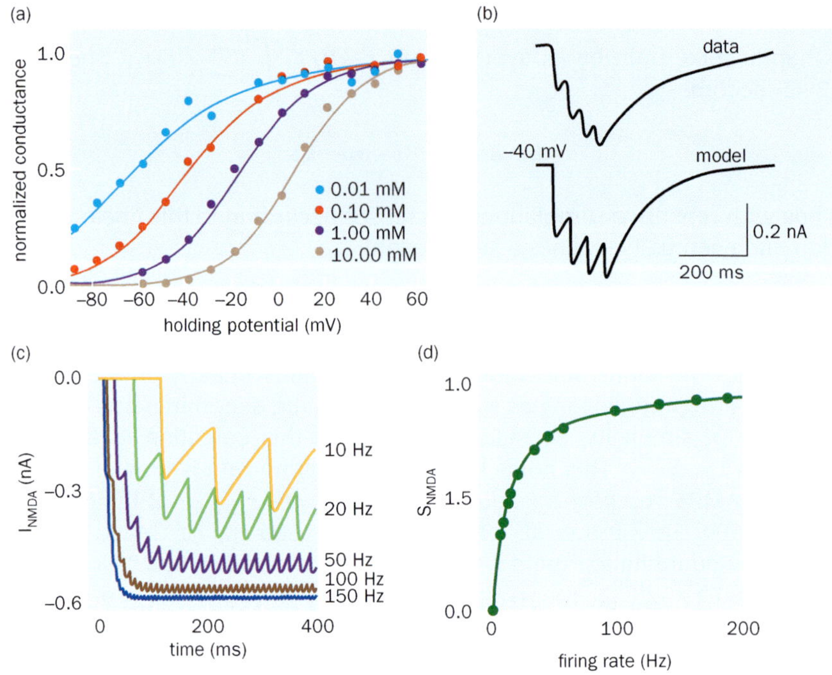

Figure 2.21 NMDA receptor-mediated current. (a) Voltage dependence. (b) Time course of the current in response to a train of four spikes. (c) Current saturates at high input firing rate. (d) Average synaptic gating variable (steady-state) as a function of input firing rate. Panel (a) is reproduced from Jahr and Stevens (1990), upper trace of (b) from Hestrin et al. (1990a), lower trace of (b) and (c–d) from Wang (1999b). [Jahr, C. E., & Stevens, C. F. (1990). Voltage dependence of NMDA-activated macroscopic conductances predicted by single-channel kinetics. Journal of Neuroscience, 10(9), 3178–3182.] [Hestrin, S., Perkel, D. J., Sah, P., Manabe, T., Renner, P., & Nicoll, R. A. (1990, January). Physiological properties of excitatory synaptic transmission in the central nervous system. In Cold Spring Harbor Symposia on Quantitative Biology (Vol. 55, pp. 87–93). Cold Spring Harbor Laboratory Press.] [Wang, X. J. (1999). Synaptic basis of cortical persistent activity: The importance of NMDA receptors to working memory. Journal of Neuroscience, 19(21), 9587–9603.]

a model of NMDAR-mediated current with saturation. In **Figure 2.21b**, the model (lower panel) reproduces a sample trace of measured NMDAR-mediated current in a pyramidal neuron (upper panel). The voltage was clamped to -40 mV, and thus the observed current is negative and proportional to s. In response to four stimuli at 25 Hz, the increment of current by each stimulus gradually decreases, because s does not have time to decay back to zero between stimuli, and the amount of increase by each stimulus is proportional to $(1-s)$. Biologically, the unbinding of channels from NMDARs is slow; subsequent transmitter release can only recruit those unbound (closed) channels. This effect is stronger with a higher stimulus rate (**Figure 2.21c**). The temporal average of s displays a firing rate dependence (**Figure 2.21d**), as predicted by Eq. (2.45). Notably, when the firing rate is higher than ~ 50 Hz, a further increase in presynaptic activity no longer recruits more excitatory drive. It is noteworthy that saturation does not occur for a single synaptic input (Mainen et al. 1999) but for a train of them at sufficiently high frequencies.

2.7 SHORT-TERM SYNAPTIC PLASTICITY

Short-term synaptic plasticity (Zucker 1989; Fisher et al. 1997; Abbott and Regehr 2004) refers to the phenomenon that synaptic transmission between two neurons is not static, but can either decrease (**Figure 2.22a**) or increase (**Figure 2.22b**) depending on the history of the presynaptic neuron's firing activity (Abbott and Regehr 2004).

Recall that synaptic transmission depends on transmitter release, but there is only a limited number of vesicles of transmitters at a presynaptic terminal. Consider a presynaptic terminal with a number $N \le N_0$ of readily releasable vesicles of neurotransmitters. Upon the arrival of an action potential, a vesicle of neurotransmitters is released with certain probability P_{rel}, hence on average $P_{rel}N$ vesicles are released, and N is reduced to $N - P_{rel}N$. Between two stimuli, the pool of vesicles is gradually refilled, with a recovery time constant τ_D. Hence, the postsynaptic potential response (proportional to $P_{rel}N$) is initially $P_{rel}N_0$ but can exhibit short-term depression (**Figure 2.22a**) as N decreases with time, typically on a

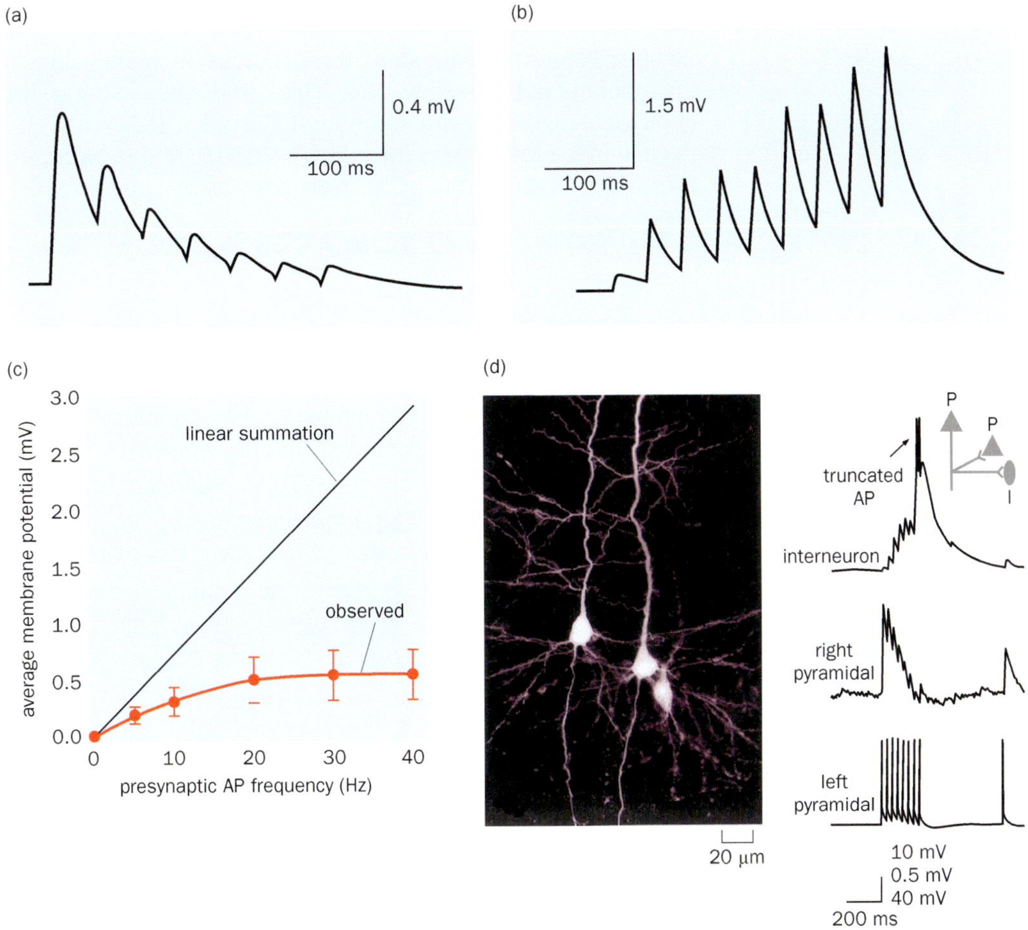

Figure 2.22 Short-term synaptic plasticity. (a) An example of short-term synaptic depression. Two connected neurons are recorded, showing the membrane potential of the postsynaptic neuron in response to a train of spikes in the presynaptic neuron. The amount of voltage increase is the largest at the first response and gradually decreases to a lower steady-state level. (b) An example of short-term synaptic facilitation. In this case, the voltage increase is the smallest at the first response and gradually grows to a higher steady-state level. (c) The average change of voltage as a function of the presynaptic firing rate. Unlike linear summation, with short-term depression measured neuronal response (red) reaches a plateau at a presynaptic firing rate above 20 Hz. (d) Triple recording from a presynaptic excitatory pyramidal neuron that projects to a postsynaptic pyramidal neuron and a postsynaptic inhibitory interneuron. Synaptic connection displays short-term depression for the former, facilitation for the latter, target cell. Panel (a) is reproduced from [Markram, H, & Tsodyks, M (1996) Redistribution of synaptic efficacy between neocortical pyramidal neurons. Nature 382(6594):807–810; panel (b) is reproduced from Markram, H, Wang, Y, & Tsodyks, MV (1998) Differential signaling via the same axon of neocortical pyramidal neurons. Proceedings of the National Academy of Sciences of the United States of America, 95(9):5323–5328. Copyright (1998) National Academy of Sciences, U.S.A.] [Tsodyks, M. V., & Markram, H. (1997). The neural code between neocortical pyramidal neurons depends on neurotransmitter release probability. Proceedings of the National Academy of Sciences of the United States of America, 94(2), 719–723. Copyright (1997) National Academy of Sciences, U.S.A.; panel (d) is reproduced from Markram, H, Wang, Y, & Tsodyks, MV (1998) Differential signaling via the same axon of neocortical pyramidal neurons. Proceedings of the National Academy of Sciences of the United States of America, 95(9):5323–5328. Copyright (1998) National Academy of Sciences, U.S.A.

timescale of hundreds of milliseconds. On the other hand, P_{rel} depends on the dynamical state of the transmitter release machinery, and can increase over time in response to a stimulus train, leading to short-term facilitation (**Figure 2.22b**).

Here we will discuss short-term synaptic depression and facilitation at the presynaptic site that occur on the timescale of hundreds of milliseconds to a few seconds. It is noteworthy that at the postsynaptic site, there exist dynamical changes such as receptor desensitization on comparable timescales (Hille 2001). Through these processes, synapses display short-term memory traces and can carry out interesting computations.

2.7.1 Short-Term Synaptic Depression

Synapses endowed with short-term depression are sensitive to changes in the input but not its steady-state. This can be seen using a simple phenomenological model (Dayan and Abbott 2001). Let $D = N / N_0$, then after each release in average D is reduced to $(1 - P_{rel})D$, and between release events D recovers towards one with a time constant τ_D. Mathematically, D obeys:

$$\frac{dD}{dt} = -P_{rel}D\sum_j \delta(t - t_j) + \frac{(1 - D)}{\tau_D} \tag{2.47}$$

where D multiplying $\delta(t - t_j)$ is understood to take the value D^- evaluated immediately before the spikes, at t_j^-. In this description, "during a spike," which is a point in time, the δ function has an infinite amplitude, and the second term can be neglected. Thus, the update by a spike is given by $D^+ - D^- = -P_{rel}D^-$, or $D^+ = (1 - P_{rel})D^-$. Between spikes, $D(t)$ recovers exponentially towards 1 with a time constant τ_D.

For a Poisson spike train at rate r, if we approximate the presynaptic spike train $\sum_j \delta(t - t_j)$ by its rate r, the dynamical equation for D can be written as

$$\frac{dD}{dt} = -P_{rel}rD + \frac{1 - D}{\tau_D} \tag{2.48}$$

The steady-state of this equation is given by $dD / dt = 0$, which yields:

$$D_{ss} = \frac{1}{1 + P_{rel}r\tau_D}.$$

According to this result, for the input rate r higher than $1 / P_{rel}\tau_D$, the term $P_{rel}r\tau_D$ is larger than 1 and dominates the denominator. Therefore, $D_{ss} \simeq 1 / P_{rel}r\tau_D$, inversely proportional to r. The transmission rate (release per unit time) is the amount of release per spike, $P_{rel}D$, multiplied by the presynaptic firing rate r. Its steady state, hence the postsynaptic response, is independent of input rate r when $r \gg 1 / P_{rel}\tau_D$. For instance, if $P_{rel} = 0.6$ and $\tau_D = 500$ ms, then $1 / P_{rel}\tau_D = 1 / 0.3 \simeq 3$ Hz. These model predictions have been confirmed by experimental observations on synapses between neocortical neurons (**Figure 2.22c**) (Abbott et al. 1997; Tsodyks and Markram 1997).

Functionally, this means that the impact on the postsynaptic cell is not sensitive to the steady-state level of presynaptic activity. On the other hand, when there is a sudden change in the input, the synapse can detect it by a transient response before the slow depression has the time to react. This is illustrated in **Figure 2.23**, where the postsynaptic response (proportional to $P_{rel}rD$) is plotted as the presynaptic firing rate undergoes a series of step changes. Note first that regardless of the presynaptic firing rate 25, 100, 10, 40 Hz, the steady-state $P_{rel}rD$ is quite similar, as predicted before. Second, when there is a sudden jump in the presynaptic rate, because the depression process takes time to react, the synapse signals input change by a rigorous but transient response, thereby performing a time derivative of the input. When r is decreased or increased to $r + \Delta r$, D briefly remains $\sim 1 / r$ (where r is the rate prior to the jump), thus the response jump is $\sim \Delta r / r$, reflecting the change in the input relative to the baseline, rather than the absolute amount of input change. This feature is apparent in **Figure 2.23**, where the transient response is similar for the transitions from 25 Hz to 100 Hz and from 10 Hz to 40 Hz. In the first step change, $r = 25$, $\Delta r = 75$; in the second step change, $r = 10$ Hz and $\Delta r = 30$ Hz. In both cases, $\Delta r / r = 3$. The transient postsynaptic response is roughly the same during

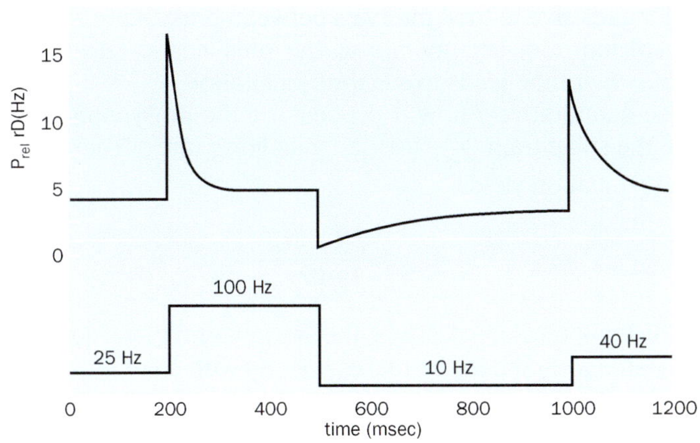

Figure 2.23 Synapse exhibiting short-term depression performs time derivative-like computation. The average postsynaptic response from a synapse model with short-term depression remains roughly the same for different long fixed presynaptic firing rate (r at 10, 25 or 100 Hz). It responds to stepwise change Δr with an amount proportional to the relative change of r, $\Delta r / r$. The parameters of the model are $P_{rel} = 0.6$ and $\tau_D = 500$ ms. Reproduced from Dayan, P. and L. F. Abbott (2001). Theoretical Neuroscience. MIT Press, Cambridge MA.

these two jumps of the presynaptic firing rate (**Figure 2.23**). Moreover, the time constant governing the decay process toward a steady-state is given by $\tau_D / (1 + P_{rel} r \tau_D)$, which is shorter at a higher input rate r. This explains why the transient response is longer lasting for the 10 to 40 Hz transition than for the 25 to 100 Hz transition. Therefore, short-term depression provides a synaptic mechanism for postsynaptic neurons to be responsive to changes in the input but not sensitive to constant input level, and contribute to adaptation in the visual system (Abbott et al. 1997).

2.7.2 Short-Term Synaptic Facilitation

Experiments showed that a presynaptic pyramidal cell connects to another pyramidal cell with short-term depression or an inhibitory neuron with short-term facilitation (**Figure 2.22d**) (Markram et al. 1998). This and other findings (Caputi et al. 2009) suggest the following rule: whether an excitatory synaptic connection displays predominantly short-term depression or facilitation is determined by the postsynaptic (excitatory or inhibitory) cell types. By contrast, short-term dynamics of an inhibitory connection depends on the presynaptic cell subclass, e.g., depression for PV+ cell-mediated inhibition and facilitation for SST+ cell-mediated inhibition (Silberberg and Markram 2007; Ma et al. 2012; Feldmeyer et al. 2018; Campagnola et al. 2022).

The release probability of transmitters P_{rel} depends on the intracellular calcium concentration at a presynaptic terminal. The latter may accumulate over repetitive stimulation due to slow residual calcium (Zucker 1989; Fisher et al. 1997), and therefore P_{rel} would increase in time, leading to gradually larger synaptic response (which on average is proportional to P_{rel}). Various models have been proposed for short-term facilitation (Bertram et al. 1996; Markram et al. 1998; Tsodyks et al. 1998; Matveev and Wang 2000). Here we will describe a phenomenological model following Tsodyks et al. (1998). Assume that the release probability $P_{rel} = u(t)$ is activity-dependent and obeys the following dynamical equation:

$$\frac{du}{dt} = U \sum_j \delta(t - t_j)(1 - u) + (U - u) / \tau_F \tag{2.49}$$

where U (less than one) is the baseline value of u or release probability. Again, $U(1-u)$ multiplying the spike train $\delta(t - t_j)$ takes the value $U(1 - u^-)$ evaluated immediately before the delta functions, at t_j^-. Thus, the update from before to after a spike is given by $u^+ - u^- = U(1 - u^-)$, or $u^+ = u^- + U(1 - u^-$. In the absence of presynaptic firing, the first term of Eq. (2.49) is zero, and u decays exponentially towards U with a time constant

τ_F. If τ_F is long relative to time intervals between presynaptic spikes, spike-triggered calcium can accumulate at the presynaptic terminal, and an increase of u over time leads to synaptic facilitation.

To see quantitatively how u depends on the presynaptic activity, let us replace the spike train $\sum_j \delta(t-t_j)$ by its firing rate r. Then, solving the steady-state equation yields:

$$u_{ss} \simeq \frac{U(1+r\tau_F)}{1+Ur\tau_F} \tag{2.50}$$

Thus, at low firing rates ($r \leq 1/U\tau_F$), the second term in the denominator is small compared to 1, and u_{ss} increases linearly with r. For firing rates much larger than $1/U\tau_F$, u_{ss} approaches 1.

A synapse can display both short-term depression and facilitation. In that case, the coupled equations for u and D are (Mongillo et al. 2008):

$$\frac{du}{dt} = U\sum_j \delta(t-t_j)(1-u) + \frac{(U-u)}{\tau_F} \tag{2.51}$$

$$\frac{dD}{dt} = -uD\sum_j \delta(t-t_j) + \frac{(1-D)}{\tau_D} \tag{2.52}$$

In the case, the change of synaptic variable s induced by a presynaptic spike (the parameter α_s in Eq. 2.43) is proportional to uD, or more precisely u^+D^- because of the subtlety in differentiating a variable before and after a spike (as a delta-function of time). Recall the sequential biological processes of synaptic transmission: after each spike-triggered calcium influx, the dynamical variable u is increased from u^- to u^+, which should be used as P_{rel} for transmitter release. The amount of transmitter released is given by D^-, and therefore the size of the postsynaptic potential is evaluated using u^+D^-. After the transmitter release, D is updated from D^- to D^+ (using $P_{rel} = u^+$) as a result of vesicle depletion. To reiterate, by a spike input $u^+ = u^- + U(1-u^-)$, $D^+ = (1-u^+)D^-$ and the synaptic variable s is updated using $\alpha_s = \alpha_0 u^+ D^-$ in Eq. (2.43) with a constant α_0.

When the spike train is replaced by a time-varying firing rate r, the approximate equations become:

$$\frac{du}{dt} = Ur(1-u) + \frac{(U-u)}{\tau_F} \tag{2.53}$$

$$\frac{dD}{dt} = -uDr + \frac{(1-D)}{\tau_D} \tag{2.54}$$

$$\frac{ds}{dt} = \alpha_0 uDr(1-s) - \frac{s}{\tau_s} \tag{2.55}$$

which are used in firing rate models endowed with short-term synaptic plasticity.

Figure 2.24 displays an example of synaptic response to a stochastic spike train that includes single spikes and brief bursts of spikes. Facilitation occurs maximally when spikes occur at a high rate within a brief burst in time. By contrast, single spikes do not facilitate synapses, and thus have a low probability of being transmitted to the postsynaptic side. In other words, a facilitating synapse can act as a detector of bursts while "ignoring" temporally isolated spikes among a mixture of single spikes and bursts (Lisman 1997; Wang 1999a; Matveev and Wang 2000).

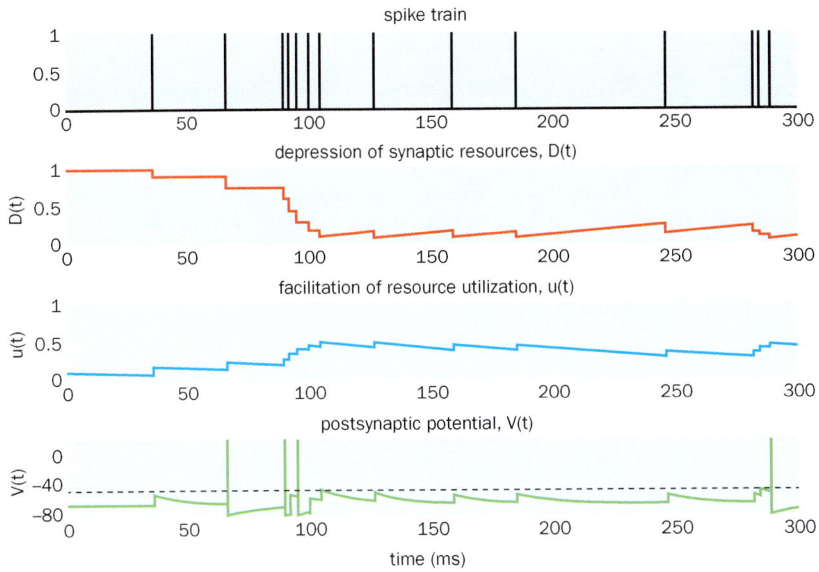

Figure 2.24 A model with a presynaptic neuron (top) and a postsynaptic neuron (bottom) connected by a synapse endowed with both short-term facilitation and depression. The depression variable $D(t)$ decreases while the facilitation variable $u(t)$ increases at each presynaptic spike. The combined effects result in a maximal synaptic drive and high probability of spiking in the postsynaptic neuron during a burst of spikes in the presynaptic neuron. $U = 0.1$, $\tau_F = 150$ ms and $\tau_D = 300$ ms. Model simulation by Victor Matveev based on Matveev and Wang (2000).

2.8 SUMMARY

In this chapter we introduced several models of neurons and synapses. We learnt some basic neurobiological concepts, such as membrane potential dynamics, ion channels, neurotransmitters and Dale's law. Single cells may exhibit various firing patterns such as adaptation and bursting. It is worth keeping in mind that single neuron properties vary from one brain region to another, as well as from one species to another, exemplified by unique features of pyramidal cells in the human cortex (Gidon et al. 2020; Galakhova et al. 2022). Synapses have their own temporal dynamics as well as short-term plasticity, which, if prominent, can accomplish interesting computation at the synaptic level, such as time-derivative of inputs or burst detection. We also covered mathematical concepts: time constant of a linear dynamical system, bifurcation in a nonlinear dynamical system, stability and attractors. These basic biological properties of neurons and synapses will serve as the building blocks of the brain. We will use them in models designed to investigate the complex dynamics of neural circuits and cognitive functions later in the book.

FURTHER READING

Books and Reviews

Abbott, L. F. and W. G. Regehr (2004). Synaptic computation. *Nature 431*, 796–803.

Bear, M. F., B. W. Connors, and M. A. Paradiso (Eds.) (2015). *Neuroscience: Exploring the Brain* (4th ed.). Alphen aan den Rijn, Netherlands: Wolters Kluwer.

Destexhe, A., Z. F. Mainen, and T. J. Sejnowski (1998). Kinetic models of synaptic transmission. In C. Koch and I. Segev (Eds.), *Methods in Neuronal Modeling* (2nd ed.), pp. 1–25. Cambridge, MA: MIT Press.

Hille, B. (2001). *Ion Channels of Excitable Membranes* (3rd ed.). Sunderland, MA: Sinauer Associates.

Katz, B. (1966). *Nerve, Muscle, and Synapse*. New York: McGraw-Hill.

Koch, C. (1999). *Biophysics of Computation: Information Processing in Single Neurons*. New York: Oxford University Press.

Llinás, R. (1988). The intrinsic electrophysiological properties of mammalian neurons: Insights into central nervous system function. *Science 242*, 1654–1664.

Markram, H., M. Toledo-Rodriguez, Y. Wang, A. Gupta, G. Silberberg, and C. Wu (2004). Interneurons of the neocortical inhibitory system. *Nat. Rev. Neurosci. 5*, 793–807.

Nichols, J. G., A. R. Martin, and B. G. Wallace (1992). *From Neuron to Brain* (3rd ed.). Sunderland, MA: Sinauer Associates.

Rinzel, J. and G. B. Ermentrout (1989). Analysis of neural excitability and oscillations. In C. Koch and I. Segev (Eds.), *Methods in Neuronal Modeling: From Synapses to Networks*, pp. 135–169. Cambridge, MA: MIT Press.

Stuart, G., N. Spruston, and M. Hausser (2016). *Dendrite* (3rd ed.). Cambridge, MA: MIT Press.

Weber, A. I., K. Krishnamurthy, and A. L. Fairhall (2019). Coding principles in adaptation. *Annu. Rev. Vis. Sci. 5*, 427–449.

Primary Research Papers

Single Neurons

Hodgkin, A. L. and A. F. Huxley (1952). A quantitative description of membrane current and its application to conductance and excitation in nerve. *J. Physiol. 117*, 500–544.

Kawaguchi, Y. and Y. Kubota (1997). GABAergic cell subtypes and their synaptic connections in rat frontal cortex. *Cereb. Cortex. 7*, 476–486.

McCormick, D., B. Connors, J. Lighthall, and D. Prince (1985). Comparative electrophysiology of pyramidal and sparsely spiny stellate neurons in the neocortex. *J. Neurophysiol. 54*, 782–806.

Traub, R. D., R. K. Wong, R. Miles, and H. Michelson (1991). A model of a ca3 hippocampal pyramidal neuron incorporating voltage-clamp data on intrinsic conductances. *J. Neurophysiol. 66*, 635–650.

Wang, X.-J. and G. Buzsáki (1996). Gamma oscillation by synaptic inhibition in a hippocampal interneuronal network model. *J. Neurosci. 16*, 6402–6413.

Wang, X.-J., Tegnér, J., Constantinidis, C. and Goldman-Rakic, P. S. (2004). Division of labor among distinct subtypes of inhibitory neurons in a cortical microcircuit of working memory. *Proc. Natl. Acad. Sci. U.S.A. 101*, 1368–1373.

Single Synapses

Abbott, L. F., J. A. Varela, K. Sen, and S. B. Nelson (1997). Synaptic depression and cortical gain control. *Science 275*, 220–224.

Jahr, C. E. and C. F. Stevens (1990). Voltage dependence of NMDA-activated macroscopic conductances predicted by single-channel kinetics. *J. Neurosci. 10*, 3178–3182.

Rall, W. (1964). Theoretical significance of dendritic trees for neuronal input-output relations. In R. F. Reiss (Ed.), *Neural Theory and Modeling*, pp. 73–97. Stanford: Stanford University Press.

Tsodyks, M. V. and H. Markram (1997). The neural code between neocortical pyramidal neurons depends on neurotransmitter release probability. *Proc. Natl. Acad. Sci. U.S.A. 94*, 719–723.

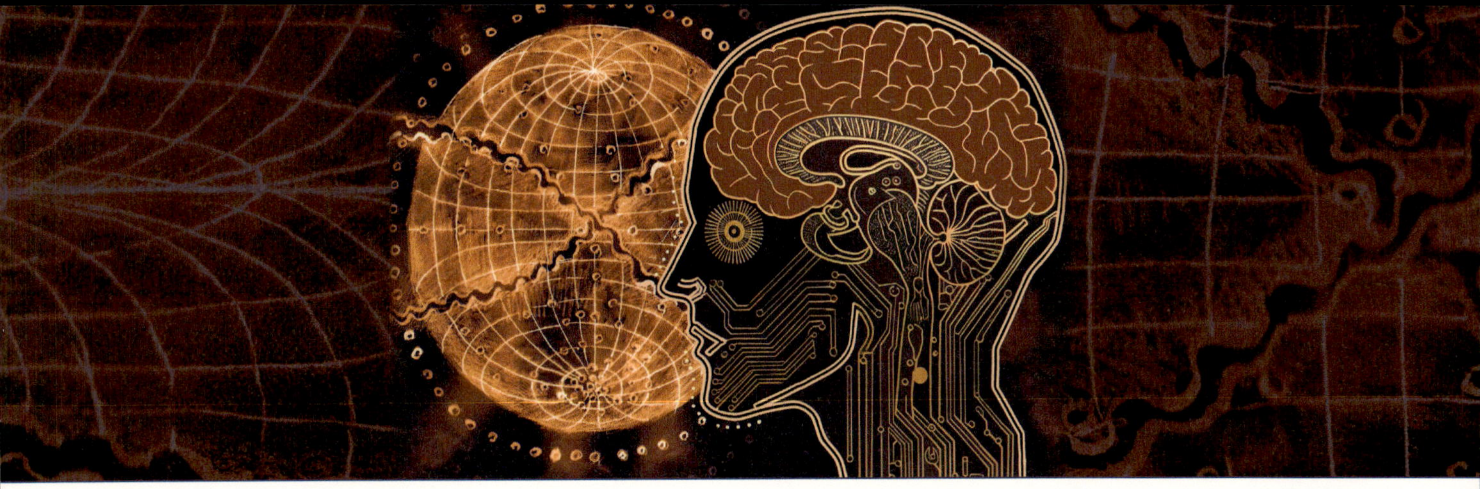

Neural Networks

3

The earth to be spann'd, connected by network,
The races, neighbors, to marry and be given in marriage,
The oceans to be cross'd, the distant brought near,
The lands to be welded together.

Walt Whitman in *Passage to India*

3.1 INTRODUCTION

A prerequisite for understanding the operation of a neural network is some knowledge about its anatomical connectivity. Neuroanatomy has a long tradition (Brodmann 1909; Cajal 1911; von Economo 1929). Today, description and analysis of connectivity of neural circuits are the subject of brain connectomics (Seung 2012; Sporns 2014; Fornito et al. 2016). Much work has been devoted to the analysis of microcircuits within a brain region (Shepherd and Grillner 2018), such as a neocortical area (White et al. 1986; Braitenberg and Schütz 1991; Lichtman et al. 2008; Helmstaedter et al. 2008), which is the focus of this chapter. Large-scale multi-regional brain systems will be the subject of Chapter 9.

A neural circuit is a network consisting of interconnected "nodes," which can be single neurons or neural groups. Broadly speaking, to describe a network structure is to map out its "wiring diagram" in which links are drawn between nodes that are connected. Obviously, the connectivity of a neural network is typically not symmetrical, thus links are "directed": A may be connected to B, but the connection in the opposite direction may or may not exist. The "topological" connectivity is quantified by a matrix with elements c_{ij}, which are equal to 1 if there is a link from node j to node i, and 0 otherwise. Modern network science (Newman 2010; Barabási 2016), built on a branch of mathematics called graph theory, provides tools to characterize statistical properties of network connectivity. For instance, given a connectivity matrix, the average probability p of connection between any two nodes informs us whether the network is sparse (with a small p) or dense (with a p close to 1). We can compute the statistical distribution of the number of nodes that are linked onto a given "target" node (called the

DOI: 10.1201/9781003459361-4

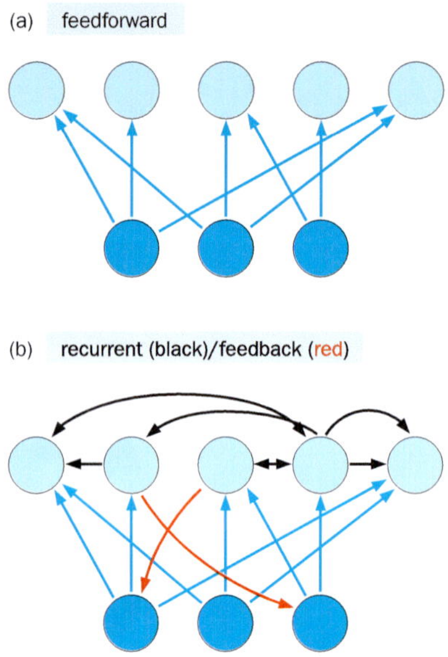

Figure 3.1 Schematics of feedforward (a) and recurrent (b) neural networks.

"in-degree") or the distribution of number of nodes that receive links from a "source" node (the "out-degree"). A node with an exceptionally large number of links may be considered a "hub" of the network.

For neural networks, we would like to know more than all-or-none binary connectivity because, as we shall see, the weights of connections are of paramount importance in determining a network's computation. In addition, as we will discuss in Chapter 4, learning and memory involve experience-dependent modifications of synaptic connection weights. Generally, we will quantitatively describe a network wiring structure by a weighted connectivity matrix W with elements w_{ij}, which are real numbers.

Figure 3.1a shows schematically a feedforward network with neurons divided into two groups or layers. Neurons in the first layer project to the second layer, but there is no projection in the opposite direction. Also, units within each layer do not interact with each other. By contrast, **Figure 3.1b** illustrates a recurrent network, with loops of connections between neurons. Because of feedback loops, when a node is perturbed by some input, it will lead to cascading changes through chains of looped connections.

In this chapter, we shall discuss two types of recurrent network models: randomly connected neural networks and networks with spatially structured connections. Random network models are a first approximation to describe a local population of neurons that encode a single variable and perform an identical computation. In particular, a minicolumn of mammalian cortex that contains about 100 neurons; it corresponds to a small volume of a cylindrical shape, with a surface diameter varying from 20 to 50 μm (depending on species), across the 2 mm cortical depth (Mountcastle 1997). In the primary visual cortex (V1), a minicolumn can be defined by neurons selective for a particular preferred orientation of visual stimulus at a location in the visual field (Hubel and Wiesel 1968; Buxhoeveden and Casanova 2002). Neurons within a minicolumn can be assumed to connect with each other randomly, while the distance dependence of connectivity is neglected. Interconnections between a number of such local populations form a larger circuit with spatially dependent connections, like a hypercolumn in V1 comprising neural populations selective for all orientations.

For a given scientific problem, a model is designed with choices of single neuron and synapse models (Chapter 2) and a network architecture. Consider a population of N neurons, each indexed by $i = 1, 2, \dots, N$. Assume the LIF model for single neurons and a conductance-based synapse model (Chapter 2), then the network obeys the following equation:

$$C_m \frac{dV_i}{dt} = -G_L(V_i - V_L) - G_{syn} \sum_{j=1}^{N} w_{ij} s_j (V_i - V_{syn}) + I_{ext,i} \tag{3.1}$$

where w_{ij} is an entry of a weighted and directed connectivity matrix, the absolute connection strength is scaled by G_{syn}. The synaptic gating variable s_j obeys a dynamical equation chosen from one of the possible descriptions (Chapter 2), and the reversal potential V_{syn} depends on the receptor type that mediates the synaptic interactions, glutamate for excitation and GABA for inhibition. Finally, $I_{ext,i}$ is the external input to neuron i. Eq. (3.1) can be readily generalized to cases where single neurons are described by the Hodgkin-Huxley conductance-based model or a model with multiple compartments (Chapter 2).

When there are more than one neural population, the network includes multiple types of connections. For instance, an excitatory-inhibitory circuit model has coupled equations with dynamical variables V_i^E and V_i^I for E and I cells, respectively. There are four types of recurrent connections: excitatory-to-excitatory (E-to-E) connections with G^{EE} and w_{ij}^{EE};

similarly for excitatory-to-inhibitory (E-to-I, G^{IE} and w_{ij}^{IE}), inhibitory-to-excitatory (I-to-E, G^{EI} and w_{ij}^{EI}) and inhibitory-to inhibitory (I-to-I, G^{II} and w_{ij}^{II}) connections.

Let us first consider the case where w_{ij} is random. To make it simple, suppose that $w_{ij} = c_{ij}$ where c_{ij} is either 0 or 1, in other words, all existing connections have the same weight, and each c_{ij} is determined by a probability p. The probability p that one neuron connects to another tells us how sparse the network connectivity is. For instance, $p \simeq 0.1$ between pyramidal cells in layer 2/3 or layer 5 of a cortical microcircuit (within a spatial extent of $\sim 200\mu$ m) (Holmgren et al. 2003; Song et al. 2005; Cossell et al. 2015). For a random network model, connections may be wired by "coin tossing" (a link between two neurons in either direction is present if the outcome is "head" with probability p, and absent otherwise with probability $1-p$). The number C of connections per neuron has a binomial distribution:

$$P(C=k) = \binom{N}{k} \cdot p^k (1-p)^{N-k} \tag{3.2}$$

with a mean $\mu = K = pN$ and variance of $\sigma^2 = Np(1-p)$ (Gardiner 1985).

When we build models for cortical circuits, typically our model has a smaller number N of neurons compared to the real biological system. For instance, in the mammalian cortex, a small patch of mammalian cortex with a linear dimension of 1 mm, or a "voxel" of functional magnetic resonance imaging (fMRI), contains about half a million neurons, whereas a model typically contains a smaller number N of neurons. We would like model results to be robust and insensitive to the precise size N used in network model simulations. For an increasingly large N, p may be fixed, in which case K grows proportionally with N, and every neuron is connected with a finite fraction (p) of all neurons in the network. With a sufficiently large N, neuron-to-neuron variability of synaptic inputs decreases relative to the mean, $\sigma / \mu \sim 1/\sqrt{N}$.

On the other hand, K may be held finite with increasing N. In that case, $p = K/N$ decreases inversely with N. The distribution of the number of connections per neuron converges to a Poisson distribution in the limit of large N:

$$P(C=k) = \frac{K^k \exp(-K)}{k!} \tag{3.3}$$

with both the mean and variance equal to K.

Next, let us consider a large enough network where distance dependence of connections needs to be taken into account. Suppose that w_{ij} is given by the product of a topological matrix c_{ij} (with 0s and 1s) and sampled weights from a distribution $p(\tilde{w})$, $w_{ij} = c_{ij}\tilde{w}_{ij}$, the connection probability p_{ij} for determining c_{ij} depends on the distance d_{ij} between a neural pair i and j. The dependence may be a Gaussian distribution:

$$p_{ij} = A\exp\left(-d_{ij}^2 / 2\sigma^2\right) \tag{3.4}$$

where σ defines the spatial footprint. In the mice cortex, excitatory-to-excitatory, excitatory-to-inhibitory and inhibitory-to-excitatory connections all have σ values of about 100 μm (Levy and Reyes 2012). Importantly, the peak probability A is in the range of 0.1 to 0.3; neuron-to-neuron connections are probabilistic and sparse in the cortex.

Another often used formula is an exponential function:

$$p_{ij} = A\exp\left(-\lambda d_{ij}\right) \tag{3.5}$$

where λ is a decay rate defining the spatial "footprint" $1/\lambda$.

Such analysis using *in vitro* cortical slices is typically restricted to a small cortical volume, and therefore not suitable to physiologically examine long-range horizontal connections of pyramidal neurons, which can extend a few millimeters in the monkey cortex (Gilbert 1983). These long-range excitatory projections target both pyramidal cells and inhibitory neurons. Further, certain types of GABAergic inhibitory interneurons also project widely, exemplified by large basket cells, which project up to $\sim 500\,\mu m$ (Kisvarday et al. 2003). An intriguing possibility is that the probability of long-distance connections falls off according to a power law such as:

$$p_{ij} = \frac{A}{1+(d_{ij} / d_0)^{\alpha}} \tag{3.6}$$

with an exponent α (Buzsáki et al. 2004). A power distribution decreases with d_{ij} much less steeply than a Gaussian distribution or an exponential distribution, yielding an excessive likelihood of connections between neurons separated by a long distance that facilitates rapid monosynaptic communication. Assume that maximal connection probability $A = 1$. According to an exponential distance rule, the probability of connection between two neurons separated by a distance $d_{ij} = 5d_0$ is negligible, $p_{ij} = \exp(-5) = 0.007$; by contrast, if the connection probability obeys a power law with $\alpha = 1$, $p_{ij} = 1/6 = 0.17$ is quite significant.

When there is a topographical organization, neurons with similar functional selectivity are close to each other physically in the neural tissue. For example, neurons in area V1 of cat or monkey cortex are selective for the direction θ of a visual stimulus orientation, and they are organized in a topographical fashion (Hubel and Wiesel 1968). A model with N neurons describes a continuous "ring network" discretized with a bin size of $\Delta\theta = 180 / N$, so that each neuron indexed by i (i=1,2, . . ., N) has the preferred direction $\theta_i = i\Delta\theta$. The distance between two neurons with preferred directions θ_i and θ_j is $d_{ij} = \theta_i - \theta_j$. Note that since θ is a circular variable (180° is the same as 0°), strictly speaking p_{ij} should be a periodic function of d, such as $\exp\left(\cos\left(2\left(\theta_i - \theta_j\right)\right) / \lambda^2\right)$ (which is maximum when $\theta_i = \theta_j$). By contrast, V1 neurons in mouse do not display a spatial map of orientation selectivity (Ohki et al. 2006), yet the connection weights still display a decaying dependence on $\theta_i - \theta_j$ (Cossell et al. 2015). This is plausible since neurons with similar functional selectivity tend to connect with each other more strongly than those with dissimilar selectivity by virtue of Hebbian plasticity that dictates that "neurons that fire together, wire together" (see Chapter 4).

Note that w_{ij} depends on both c_{ij} and $\tilde{w}_{ij}$, each with its own statistical properties. The connection probability p_{ij} decays with distance (Levy and Reyes 2012), whereas the distribution $p(\tilde{w})$ is skewed (Feldmeyer et al. 1999; Song et al. 2005; Lefort et al. 2009) and can be well fit with a log-normal distribution (**Figure 3.2**), i.e., while many connections are weak, some strong ones would not be expected by a Gaussian distribution.

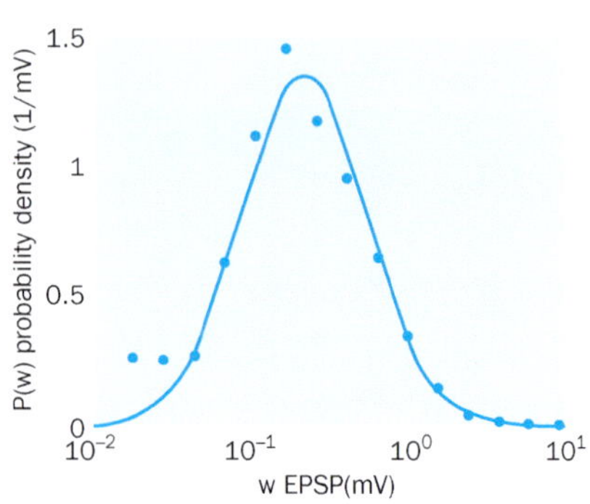

Figure 3.2 Log-normal distribution of synaptic strengths (solid curve) fitting a histogram of measured strengths of monosynaptic excitatory connections between neighboring neurons in neocortex. Reproduced from Song et al. (2005). [Song S, Sjöström PJ, Reigl M, Nelson S, Chklovskii DB (2005) Highly Nonrandom Features of Synaptic Connectivity in Local Cortical Circuits. PLoS Biol 3(3): e68. https://doi.org/10.1371/journal.pbio.0030068., https://creativecommons.org/licenses/by/4.0/]

While the knowledge of structural network connectivity is tremendously valuable, it is insufficient to predict the dynamical behavior of a neural circuit. To illustrate this point, consider two identical neurons that are connected symmetrically by mutual inhibition (**Figure 3.3**). Given this "connectome," can we deduce how the network behaves? It turns out that experiments and theory have uncovered multiple scenarios of this network's dynamical behavior. First, both neurons may simply stay silent. Second, when driven by inputs, the system may behave as a "switch," with one neuron active while inhibiting the other neuron, or vice versa, and a brief input can switch the system between the two states. Third, if neurons are endowed with slow adaptation (Chapter 2), one of the two neurons could be active and over time eventually become "fatigued," when the

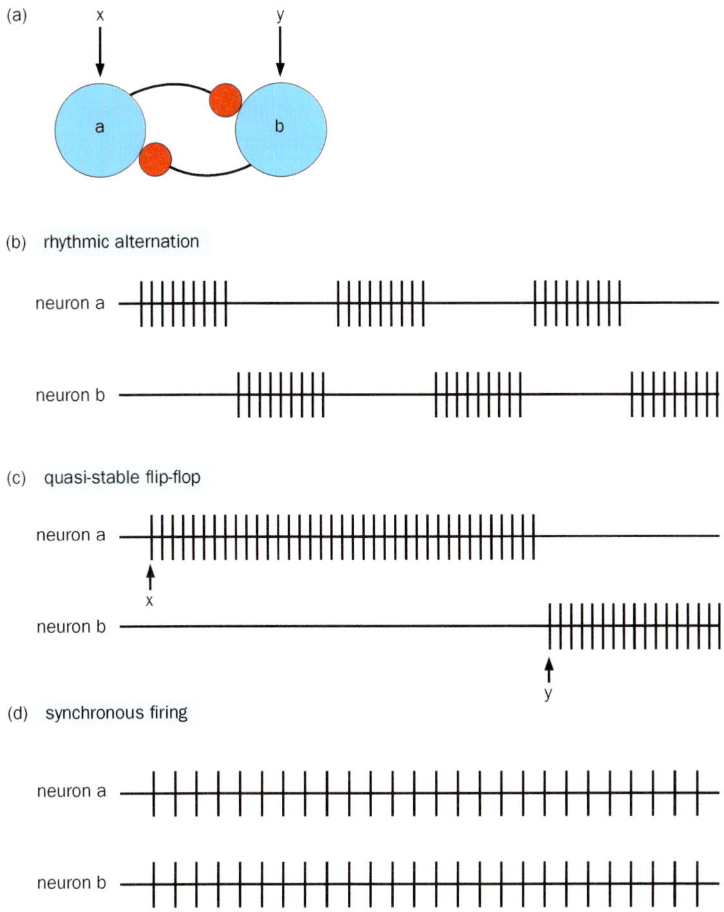

Figure 3.3 Distinct dynamical regimes in a circuit of two reciprocally connected inhibitory neurons. (a) Neurons #1 and #2 receive excitatory inputs x and y, respectively, and are connected with each other through symmetric connections. (b) Under different conditions of neuronal and synaptic properties, the circuit can exhibit qualitatively distinct dynamical regimes, including rhythmic alternation as a "half-center oscillator," quasi-stable flip-flop and spike-by-spike synchrony. Adapted from Kristan and Katz (2006).

other neuron takes over. The alternation of activity leads to a "half-center" oscillator, which is the core of rhythmic central pattern generators (Perkel and Mulloney 1974; Getting 1989; Marder and Calabrese 1996). Fourth and finally, under certain conditions, the two neurons can be perfectly synchronized, spike by spike: the two neurons fire at the same time, leading to mutual inhibition after a brief delay, and when this inhibition decays away, they can fire again together (Wang and Rinzel 1992; Wang 2010). This simple example illustrates that dynamical behavior often cannot be predicted by anatomy in a straightforward fashion; physiology and modeling are important for discovering the dynamical operations of neural circuits.

3.2 NETWORK DYNAMICS OF SPIKING NEURONS

3.2.1 Signal Propagation in a Feedforward Network

In a feedforward network, an interesting problem is concerned with how a signal propagates reliably across the system. Imagine a number of groups of neurons (neural assemblies) $i = 1, 2, \dots, M$ organized in a purely feedforward manner. Neurons in group 1 do not interact with each other; they merely project to neurons in group 2, and the same holds for neurons in group 2 projecting to group 3, etc. When neurons in group 1 fire synchronously, a brief burst of a few spikes in time, how does the signal lead to representation by neural activity in group 2, then to group 3, and so forth? Diesmann et al. (1999) built a simple model for studying signal propagation, where projections from one group to the next are random in a feedforward architecture, with a number of independently chosen neurons in group i projecting to any target neuron in group $(i+1)$ (**Figure 3.4a**). It was found that, across the network, a burst either propagates reliably (**Figure 3.4b**, upper panel) or decays away and fails to reach some neural groups (**Figure 3.4b**, lower panel).

In their model simulations, spikes in the first group were generated by a Gaussian distribution over time with a standard deviation of σ. Two quantities are defined in each neural group i: the number of spikes (a_i) within a burst and the temporal width of the burst σ_i. Diesmann et al. (1999) analyzed the network dynamics in a state space of (a,σ), where the dynamics of signal propagation corresponds to a trajectory, a discrete sequence $\{a_i,\sigma_i\}, i=1,2,\ldots,M$ (**Figure 3.4c**). The plane is divided into two regions, separated by a "separatrix" (blue line without arrow). As shown with sample trajectories, starting with a pair (a_1,σ_1) above the line, the dynamics converges to a fixed point with a high number of spikes and small temporal dispersion. This means that from group to group the "package" of spikes remains preserved and signal propagation is reliable. By contrast, starting with a pair (a_1,σ_1) below the line, the dynamics converges to another fixed point with a large σ and vanishingly small a, implying that the activity decays away across the system.

Studies of signal propagation were originally motivated by the idea of a "synfire chain" where neurons fire and communicate with each other by brief packages of spikes with a high temporal precision (the "temporal code") (Abeles 1991). On the other hand, in the cortex neurons exhibit highly irregular firing, which by some measures is close to a Poisson process (Chapter 2), hence lacks temporal precision. Signal propagation has also been studied for transmission of a "rate code" rather than temporal code (Shadlen and Newsome 1998; Litvak et al. 2003). Interestingly, even when spike firing is asynchronous at the start, neurons in successive groups tend to gradually become synchronized because of shared inputs (Litvak et al. 2003). Sparse connectivity would reduce overlap of inputs onto target neurons, thus facilitate asynchronous signal propagation. Furthermore, it is worth keeping in mind that these studies considered a purely feedforward network architecture, whereas recurrent connections commonly present in the cortex are expected to have a major impact on the collective population dynamics (see later discussion).

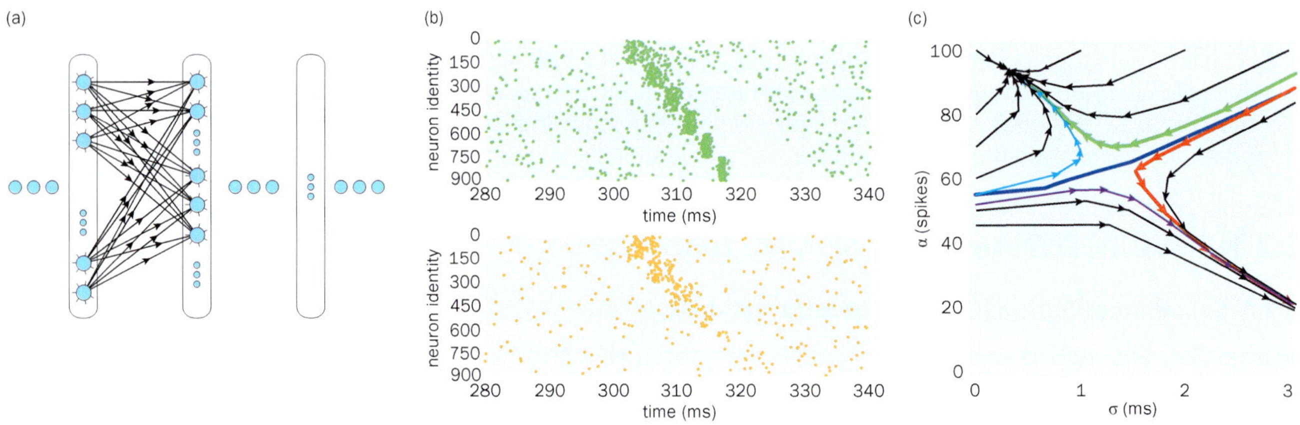

Figure 3.4 Signal propagation across a sequence of neural groups in a feedforward network model. (a) Connections are from one group to the next, without feedback in the opposite direction nor interactions among neurons within each group. (b) A package of spikes starting in the first group either propagates faithfully across the system (upper panel) or fails to reach later parts of the model system (lower panel). (c) State-space representation of pulse packet propagation. Activity in each group is quantified by the number of spikes (a) and the dispersion of spike times (σ). The upper left point, with high a and low σ, marks a fixed point of the network, corresponding to a stable propagation of the pulse packet. Another fixed point is located at the lower right corner, with low a and high σ, corresponding to failed propagation. The blue line (without arrows) separates trajectories that converge to either of the two fixed points. Reproduced from Kumar et al. (2010). [Diesmann, M., Gewaltig, MO. & Aertsen, A. (1999). Stable propagation of synchronous spiking in cortical neural networks. Nature 402(6761), 529–533; panels (b)–(c) are reproduced from Kumar, A., Rotter, S., & Aertsen, A. (2010). Spiking activity propagation in neuronal networks: Reconciling different perspectives on neural coding. Nature Reviews Neuroscience, 11(9), 615–627.]

3.2.2 Excitation and Inhibition Balance and Asynchronous State in a Recurrent Network

In the cortex, a temporal code or rate code emerges as a collective phenomenon of recurrent interactions between excitatory (E) and inhibitory (I) neurons. A local cortical circuit is typically endowed with all four (E-to-E, E-to-I, I-to-E and I-to-I) types of synaptic interactions. Even with completely random connections, such a network can display a variety of dynamical behaviors. A salient characteristic of the cortex is that each neuron receives a large number K (in the thousands) of synaptic connections, and just tens of individual excitatory postsynaptic potentials (EPSPs) would be sufficient for a neuron's membrane potential to reach the spiking threshold. Large synaptic excitation thus needs to be balanced by equally large synaptic inhibition (E/I balance), so that the net input to a neuron is in a reasonable range. Moreover, E/I balance provides a mechanism to ensure that the total input current is close to firing threshold, required to explain high variability of spike activity by cortical cells.

At first sight, the E/I balance imposes an approximate equality between average excitatory and inhibitory inputs, regardless of the firing rates, which appears to require fine-tuning of network properties. Instead, van Vreeswijk and Sompolinsky (1996) proposed a clever mechanism for achieving E/I balance dynamically, without requiring fine-tuning of parameters. The idea is that if the strength of synaptic connection strength scales like $1/\sqrt{K}$, $G_{syn} \sim g_{syn}/\sqrt{K}$ ($G^{EE} \sim g^{EE}/\sqrt{K}$, etc.), every input will remain relatively strong even with a large K value (for instance, if $K = 1{,}000$, $1/K = 0.001$, whereas $1/\sqrt{K} \simeq 0.03$). Interestingly, a recent experiment reported empirical support for $1/\sqrt{K}$ scaling of synaptic strength in cortical culture slices (Barral and Reyes 2016). The assumption is that both N and K are large, but K is smaller than N ($N \gg K \gg 1$).

Consider a neuron that receives roughly the same number K of excitatory inputs and inhibitory inputs. The excitatory current I_E is a sum of K terms, each with the strength of $\sim g_E/\sqrt{K}$ multiplied by the mean firing rate r_E. Therefore, its mean μ_E scales like $Kg_E r_E/\sqrt{K} \sim \sqrt{K} g_E r_E$. The same holds true for the inhibitory current, with a mean of $\mu_I \sim \sqrt{K} g_I r_I$ where r_I is the mean firing rate of inhibitory neurons. Consequently, the total current has an average of $\mu_E - \mu_I \sim \sqrt{K}(g_E r_E - g_I r_I + i_{ext}) + O(1)$, with the excitatory external current $I_{ext} \sim \sqrt{K} i_{ext}$. The term $O(1)$ denotes "of order 1," which means that a value is neither vanishingly small nor unboundedly large with increasing K. Both μ_E and μ_I grow with K, but a stable state can be achieved when neural firing activities r_E and r_I evolve dynamically until $g_I r_I$ largely cancels out $g_E r_E + i_{ext}$. Specifically, the argument yields two coupled equations for r_E and r_I:

$$g^{EE} r_E - g^{EI} r_I + i_{ext,E} = 0 \tag{3.7}$$

$$g^{IE} r_E - g^{II} r_I + i_{ext,I} = 0 \tag{3.8}$$

from which r_E and r_I can be solved. Consequently, the mean of the total current $\mu_E - \mu_I$ remains at a modest level (order 1).

Moreover, the variance of fluctuations in either excitatory or inhibitory current scales like $(1/\sqrt{K})^2$ (from each individual synaptic input) times the number of inputs K, or $\sim K(1/\sqrt{K})^2$, which remains finite ($O(1)$) with large K. This is in sharp contrast to the scenario where the synaptic strength scales as $1/K$, in which case the variance would scale like $(1/K)^2 K \sim 1/K$ and vanishes with large K. Importantly, the variance of the total current is additive rather than subtractive for contributions from excitation and inhibition, $\sigma_E^2 + \sigma_I^2$, and remains order one. In other words, when excitation and inhibition are balanced, the mean and standard deviation of total input to a neuron are of comparable size, and fluctuations of synaptic inputs play a important role in producing spike firing, leading to highly irregular firing of cortical neurons (**Figure 3.5a–b**).

When excitation and inhibition are balanced, a recurrent network can display a state of deterministic chaos (van Vreeswijk and Sompolinsky 1996) where neurons fire irregularly in the absence of any noise input. **Figure 3.5c** shows a random network of LIF neurons in the chaotic regime (van Vreeswijk and Sompolinsky 1996; Ostojic 2014; Kass et al. 2018), with a network of 2000 spiking neurons and the probability of connection $p = 0.2$ (so the average number of connections $K = pN = 200$). There is no obvious structure in the spatiotemporal pattern (**Figure 3.5c**); the population firing rate $r(t)$ is flat in time, which defines a state of complete asynchrony. In a balanced state, across individual neurons the firing rate distribution is broad and skewed; depending on the model details and stimulus conditions, the distribution can be exponential (Amit and Brunel 1997; Brunel and Wang 2001) (**Figure 3.5d**, left panel) or log-normal (Roxin et al. 2011), as commonly observed experimentally (**Figure 3.6**). Furthermore, the CV of ISIs is fairly large around 1 (**Figure 3.5d**, right panel). Therefore, E/I balance offers an explanation for highly irregular firing of cortical neurons (Chapter 2, **Figure 2.5**) (Shadlen and Newsome 1994; van Vreeswijk and Sompolinsky 1996).

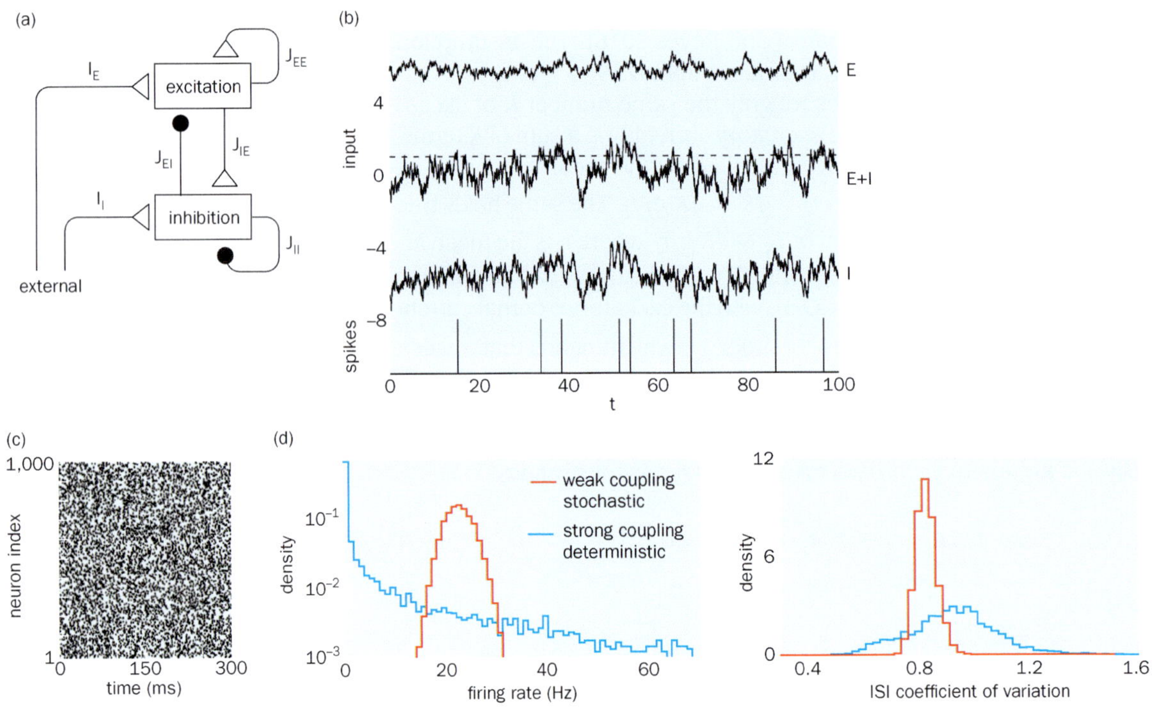

Figure 3.5 Chaotic state of a recurrent network model with E/I balance. (a) A local circuit model with all four types of connections between E and I neurons. (b) In a balanced state, both E and I currents to a neuron are large, and their average values approximately cancel each other so that the total synaptic current has a mean (solid horizontal line) near the threshold (dash-dotted horizontal line) for the onset of spiking activity. As a result, spikes are triggered by input fluctuations and display a high degree of variability (bottom). (c) Spatiotemporal pattern of a network of spiking neurons, displaying an asynchronous state where the population firing rate is constant in time. (d) Left panel: Distribution of firing rates of individual neurons is broad (blue) in a balanced state with synaptic strength $\sim 1/\sqrt{K}$, in contrast to a narrow distribution in the case the average input is not balanced with synaptic strength $\sim 1/K$ (red). Right panel: The distribution of coefficient of variation of ISIs across neurons is broad and centered around 1 in the balanced state (blue), in contrast to low variability of spike trains in an unbalanced state (red). Panels (a–b) are reproduced from van Vreeswijk and Sompolinsky (1998), (c–d) from Kass et al. (2018). [van Vreeswijk, C., & Sompolinsky, H. (1998). Chaotic balanced state in a model of cortical circuits. Neural Computation, 10(6), 1321–1371.] [Kass, R. E., Amari, S. I., Arai, K., Brown, E. N., Diekman, C. O., Diesmann, M., ... & Kramer, M. A. (2018). Computational neuroscience: Mathematical and statistical perspectives. Annual Review of Statistics and Its Application, 5, 183.]

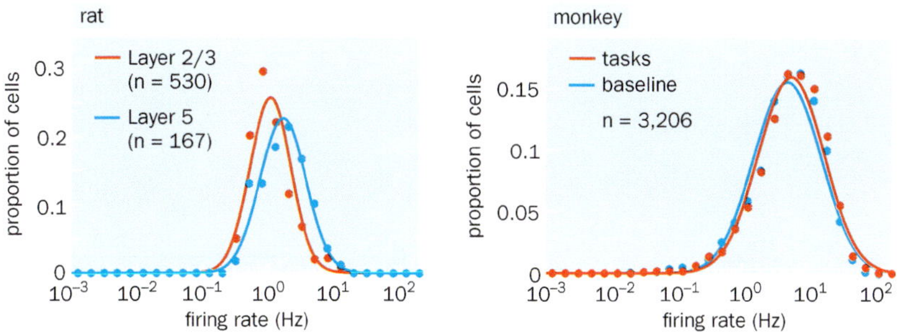

Figure 3.6 Log-normal distributions of firing rates, shown by the histograms of firing rates of cortical neurons recorded from layer 2/3 and layer 5 of rat frontal cortex (left) and from monkey posterior parietal cortex (right). Reproduced from Buzsaki and Mizuseki (2014). [Buzsáki, G., & Mizuseki, K. (2014). The log-dynamic brain: How skewed distributions affect network operations. Nature Reviews Neuroscience, 15(4), 264–278.]

This asynchronous behavior is a self-sustained state of a recurrent network endowed with E/I balance. A salient characteristic of chaotic dynamics is sensitivity to initial conditions: an infinitesimally small difference in the initial network state grows exponentially in time so that the network state is entirely different after a finite (and typically short) time (Strogatz 2016). In an *in vivo* experiment, a single cell was perturbed by an intracellular electrode in the rat cortex, resulting in cascading changes of firing in other cells monitored by extracellular recording electrodes yielded evidence for chaotic dynamics (London et al. 2010).

Presently existing models of E/I balance have certain limitations. First, the external inputs are assumed to scale also as $\sqrt{K}$; it is unclear how such large external input, say from LGN to V1, can be justified. Second, models on E/I balance tend to use a current-based "kick synapse model" (see Chapter 2) and a random network architecture (Ostojic 2014; Renart et al. 2010). In a more realistic neural circuit with conductance-based synapses, the total synaptic conductance would increase as $\sqrt{K}$, and thus single neurons would become excessively leaky (see Chapter 2, Section 6.1); whether this is biologically plausible remains unknown. Third, Eqs. (3.7–3.8) predict that average firing rates are a linear function of inputs, and the concept of E/I balance needs to be extended to be compatible with highly nonlinear input-output network computations (Renart et al. 2007). Regardless, the theoretically proposed concept of E/I balance has been recognized as a general principle of wide relevance to neuroscience (Hensch and Fagiolini 2004). Furthermore, certain abnormalities of cortical functions may be attributable to E/I imbalance, which we will discuss in Chapter 10 on psychiatric disorders.

3.2.3 Neuronal Correlations

Correlation between two neurons refers to how their activities co-vary. For instance, activities of two neurons may be correlated because of a shared synaptic input. In a network, predicting neural correlations is not straightforward: if neuron j projects to neuron k, which in turn synapses with neuron i, spike activities of neurons i and j may be correlated even though they do not directly connect with each other. Collective brain oscillations (see Section 3.4) naturally induce inter-neuronal correlations. On the other extreme, in a network with balanced excitation and inhibition, under certain conditions the average correlation between pairs of neurons can be vanishingly small even though synaptic connections are strong (Renart et al. 2010).

Extreme cases aside, firing activities of neurons are commonly correlated (Zohary et al. 1994; Buzsáki 2006; Wang 2010; Cohen and Kohn 2011; Doiron et al. 2016). Cross-correlation function $CC(\tau)$ measures correlation of spiking by one neuron at one time point t_1 and spiking by another neuron at another time t_2 separated by an interval $\tau = t_1 - t_2$ (Gerstein and Perkel 1972). $CC(\tau)$ provides a measure of functional connections of two simultaneously recorded neurons (Aertsen and Gerstein 1985). If neuron i projects to neuron j unidirectionally, firing of neuron i will have an impact

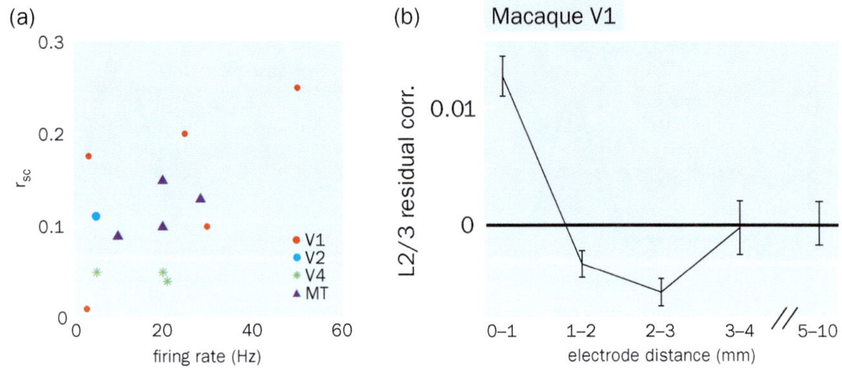

Figure 3.7 Noise correlation between pairs of neurons. (a) Noise correlation averaged over recorded neurons varies substantially depending on cortical areas and behavioral states. (b) Noise correlations display distance dependence between neuronal pairs in L2/3 of macaque primary cortex. Panel (a) is reproduced from Cohen and Kohn (2011) and (b) from Rosenbaum et al. (2017). [Cohen, M. R., & Kohn, A. (2011). Measuring and interpreting neuronal correlations. Nature Neuroscience, 14(7), 811–819.] [Rosenbaum, R., Smith, M. A., Kohn, A., Rubin, J. E., & Doiron, B. (2017). The spatial structure of correlated neuronal variability. Nature Neuroscience, 20(1), 107–114.]

on that of neuron j, and one expects that their $CC(\tau)$ has a sharp peak at a value of τ reflecting the latency of synaptic transmission. How sharp is the peak in $CC(\tau)$ measures the degree of temporal precision of neuronal correlation.

It is also important to assess how neural activities co-vary over a longer timescale under a given behavioral condition such as with repetition of a fixed sensory stimulus. This is measured by "noise correlation" or spike count correlation (r_{SC}) of two neurons (Zohary et al. 1994; Averbeck et al. 2006; Cohen and Kohn 2011). Given a time window T, r_{SC} computes the normalized co-variance of the numbers $N_1(T)$ and $N_2(T)$ of spikes within T that fluctuate from trial to trial:

$$r_{SC}(T) = \frac{\left\langle \left(N_1(T) - \langle N_1(T)\rangle\right)\left(N_2(T) - \langle N_2(T)\rangle\right)\right\rangle}{\sqrt{< \left(N_1(T) - \langle N_1(T)\rangle\right)^2 >< \left(N_2(T) - \langle N_2(T)\rangle\right)^2 >}} \tag{3.9}$$

where $\langle x \rangle$ denotes average of x; possible r_{SC} values are between -1 and 1.

As expected intuitively, noise correlation is the integrated cross-correlation over τ between $-T$ and $+T$ (Bair et al. 2001). Noise correlation depends on T, reaching a plateau at large T, which is the total integral of the cross-correlation function.

Noise correlations of neurons recorded in behaving monkeys vary widely over different recorded brain areas and in distinct behavioral conditions, with the mean ranging from 0.01 to 0.2 (**Figure 3.7a**) (Cohen and Kohn 2011; Doiron et al. 2016). Interestingly, the noise correlation of two neurons depends on the spatial separation non-monotonically (**Figure 3.7b**), presumably as a result of distance-dependent synaptic excitation and inhibition (Rosenbaum et al. 2017). Correlations represent a key for understanding dynamics and computation by a neuronal population (Cohen et al. 2009; Verhoef and Maunsell 2017; Ni et al. 2018). For example, variability in synaptic inputs onto a neuron vanishes with an increasing number of presynaptic neurons only if the latter were completely uncorrelated (according to the law of large numbers). In the presence of neuronal correlations, this is no longer the case; instead, when more neurons are "pooled" the signal-to-noise ratio plateaus (Zohary et al. 1994) while neural activity remains stochastic. Hence, neural correlations contribute to an explanation of the observed high variability of neuronal activity.

3.3 POPULATION RATE MODELS

We are often interested in a group of neurons, such as a minicolumn of the mammalian cortex, that behave more or less the same way and perform a specific computation. Since cells in such a neural pool or "assembly" are functionally homogeneous, for many purposes one can average spiking activities across neurons, yielding a population firing rate. Rate models are more accessible to mathematical analysis than spiking neuron network models, and therefore provide useful tools for our understanding of complex neural networks.

Imagine that spike trains of N neurons in a cortical minicolumn are recorded simultaneously. One can divide time into small bins of size Δt and compute "population firing rate" $r(t)$ as the number of spikes from all recorded neurons in each bin divided by N and Δt. The population rate thus estimated has a temporal resolution given by the bin size, which can be arbitrarily small. Dynamical equations for describing temporal evolution of $r(t)$ are called "population rate models."

3.3.1 Formulations of Rate Models

Theories of rate models have a long tradition in neuroscience, starting with pioneering works in the 1970s (Wilson and Cowan 1972, 1973; Freeman 1975; Amari 1977). Rate models can be constructed heuristically in different ways (Ermentrout 1998), but here we will limit ourselves to a simple formulation. Consider a recurrent network of an excitatory neural group and an inhibitory neural group, with the firing rates r_E and r_I, respectively. Moreover, let us denote the E-to-E, E-to-I, I-to-E and I-to-I connection weights by J_{EE}, J_{IE}, J_{EI} and J_{II}, respectively. Synaptic dynamical variables are denoted as s_E and s_I, which can be thought of as "synaptic drives." One can write down the following equations for the firing rates:

$$\tau_E \frac{dr_E}{dt} = -r_E + F_E\left(J_{EE}s_E - J_{EI}s_I + I_{E,ext}\right) \tag{3.10}$$

$$\tau_I \frac{dr_I}{dt} = -r_I + F_I\left(J_{IE}s_E - J_{II}s_I + I_{I,ext}\right) \tag{3.11}$$

where τ_E and τ_I are the time constants, $F_E\left(I_{tot,E}\right)$ and $F_I\left(I_{tot,I}\right)$ are the current-frequency functions ($r_E = F_E$ and $r_I = F_I$ in a steady-state), with $I_{tot,E}$ and $I_{tot,I}$ denoting the total currents to E and I populations, respectively. Neural activities are driven by recurrent synaptic inputs as well as external stimuli represented by $I_{E,ext}$ and $I_{I,ext}$ from outside of the network. The synaptic variables are, in turn, driven by neural activity (see Chapter 2):

$$ds_E / dt = \alpha_E r_E (1 - s_E) - s_E / \tau_{s,E} \tag{3.12}$$

$$ds_I / dt = \alpha_I r_I (1 - s_I) - s_I / \tau_{s,I}. \tag{3.13}$$

The four coupled equations (3.10–3.13) constitute a self-contained description for the E/I network. Two situations lead to further simplifications; both are based on a mathematical method called the "fast-slow variable analysis" (Rinzel 1987; Wang and Rinzel 1995). The idea is illustrated with a two-variable dynamical system, in which one (x) is much faster than the other (y). Let us write a generic equation for such a system:

$$\tau_x \frac{dx}{dt} = F(x, y) \tag{3.14}$$

$$\tau_y \frac{dy}{dt} = G(x, y) \tag{3.15}$$

where τ_x is much shorter than τ_y. We can rescale time by $t' = t / \tau_y$, so that the equations become

$$\frac{dx}{dt'} = F(x, y) / \epsilon \tag{3.16}$$

$$\frac{dy}{dt'} = G(x, y) \tag{3.17}$$

where $\epsilon = \tau_x / \tau_y$, which is small. This implies that given an instantaneous value of y, dx / dt is very large, until x quickly settles at its "quasi-steady-state" given by $F(x, y) \simeq 0$. In other words, x is "enslaved" by the slower variable y and approximately given by $F(x, y) = 0$. The latter, in principle, can be solved to yield x as a function of y, $x = f(y)$. Replacing x by $f(y)$, we obtain a lower-dimensional equation $\tau_y dy / dt = G(f(y), y)$ that is "self-contained." This analysis remains applicable to higher dimensional dynamical systems, if a subset of variables are much slower than the others.

Consider the first case where synapses are much faster than firing rates. As we have seen in Chapter 2, given a synaptic time constant τ_S, for firing rates roughly smaller than $1/\tau_S$ we can approximate $1 - s \simeq 1$. For AMPA receptor-mediated excitation, $\tau_{AMPA} = 0.5 - 3$ ms, $1/\tau_{AMPA} = 300 - 2000$ Hz; for $GABA_A$ receptor-mediated inhibition, $\tau_{GABA_A} = 4 - 10$ ms, $1/\tau_{GABA_A} = 100 - 250$ Hz. In these two cases, this is a good approximation with firing rates within a typical physiological range below 100 Hz. Thus, synapses described by Eqs. (3.12–3.13) are merely leaking integrators of firing activity with synaptic time constants. Moreover, because the time constants are short, s approximately keeps trace of the instantaneous firing rate, so that $s_E \simeq \alpha_E \tau_E r_E$ and $s_I \simeq \alpha_I \tau_I r_I$. We can replace s variables by these expressions in Eqs. (3.10–3.11). Because $\alpha\tau$ is multiplied with the coupling strength J, thus can be included in a redefined J, and we can simply replace s_E and s_I by r_E and r_I, respectively, yielding two self-consistent equations for the firing rate dynamics, which are often called the Wilson-Cowan model (Wilson and Cowan 1973; Ermentrout 1998; Lin et al. 2020):

$$\tau_E \frac{dr_E}{dt} = -r_E + F_E\left(J_{EE} r_E - J_{EI} r_I + I_{E,ext}\right) \tag{3.18}$$

$$\tau_I \frac{dr_I}{dt} = -r_I + F_I\left(J_{IE} r_E - J_{II} r_I + I_{I,ext}\right) \tag{3.19}$$

A second situation corresponds to the case where the time constants for the firing rate equations are very short. As a matter of fact, the meaning of τ_E and τ_I is not obvious. One might think that they are the membrane time constants of excitatory and inhibitory neurons, which would be incorrect because the firing rate and membrane potential of a neuron are not the same. Furthermore, rate models are supposed to describe neural populations, and the time constant of a neural population could be very different from that of a single cell. In a balanced network with instantaneous synaptic connections, the time constant of neural response to an external stimulus is vanishingly small ($\sim 1/\sqrt{K}$ where K is the average number of synapses per neuron) (van Vreeswijk and Sompolinsky 1998). In this case, we approximate the firing rates by their steady-states $r_E = F_E(I_{tot,E})$ and $r_I = F_I(I_{tot,I})$, yielding self-consistent equations for the synaptic drive dynamics:

$$ds_E / dt = \alpha_E F_E\left(J_{EE} s_E - J_{EI} s_I + I_{E,ext}\right)(1 - s_E) - s_E / \tau_{s,E} \tag{3.20}$$

$$ds_I / dt = \alpha_I F_I\left(J_{IE} s_E - J_{II} s_I + I_{I,ext}\right)(1 - s_I) - s_I / \tau_{s,I}. \tag{3.21}$$

Note that when saturation is ignored, $1 - s \simeq 1$, the same dynamical variables are often denoted as currents instead of gating variables (Treves 1993). However, saturation cannot be neglected with slow synapses, such as NMDA receptor-mediated excitation or $GABA_B$ receptor-mediated inhibition.

Rate models are much easier to investigate than spiking network models. However, when a phenomenon is found in a rate model, it is not guaranteed that the same can be realized by a spiking network model. For instance, in a rate model one may choose a simple formula for the input-output function F, such as a logistic function $F(I) = r_{max} / (1 + \exp(-\beta(I - I_{th})))$. The resulting network behavior may turn out to be quite different in a network with the input-output function measured from real neurons. How are rate models related to biologically more realistic models of spiking neurons? Ideally, rate models can be *derived* from spiking network models so that one can go back and forth between the two. The mathematical method for such derivations is called the "mean-field theory" (Sompolinsky et al. 1988; Amit 1989; Renart et al. 2003; Gerstner et al. 2014). The basic idea is simple: for a homogeneous neural population, the population firing rate is that of a "typical" neuron driven by a large number of inputs, which can be described as an averaged (mean) field. The actual mathematical derivation, however, can be difficult and depend on the details of the spiking network model under consideration. As expected, it requires more elaborate analysis if single neurons are described by a Hodgkin-Huxley type model than a leaky integrate-and-fire model, or if synaptic interactions are described by a conductance-based model rather than a current-based model (Chapter 2). Mean-field derivation of a rate version of a recurrent network of spiking neurons connected by conductance-based synapses, step-by-step, is covered elsewhere (Brunel and Wang 2001; Renart et al. 2003; Wong and Wang 2006).

Following we will discuss examples of rate models using Eqs. (3.18–3.19), which illustrate various dynamical behaviors exhibited by simple E/I networks.

3.3.2 Neural Integrator

Let us start with a rate model where the neuronal input-output relationship $r = F(I)$ is threshold-linear: the firing rate is zero when the total input current is below a threshold I_{th}; above I_{th} it increases linearly with the input, $F(r) = \beta(I - I_{th})$. Since the slope β is multiplied with the connection strength J and external input current, its value can be set to 1 without losing generality. Thus, $F(I) = [I - I_{th}]_+$, where $[x]_+ = x$ for $x > 0$ and 0 otherwise. For firing rates above the threshold, the rate equations are linear.

Consider first a single inhibitory neural population obeying the linear equation:

$$\tau_I \frac{dr_I}{dt} = -r_I - J_{II} r_I + I_{I,ext} \tag{3.22}$$

The steady-state of this system is $r_I = I_{I,ext} / (1 + J_{II})$, in agreement with the intuition that mutual inhibition reduces the activity of an inhibitory neural population. When a constant external input is presented, activity in this model exponentially approaches the steady-state with time constant τ_{II} given by:

$$\tau_{II} = \tau_I / (1 + J_{II}) \tag{3.23}$$

For example, the time it takes to reach the steady-state within 5% is roughly $t = 3\tau_{II}$ ($\exp(-t / \tau_{II}) = \exp(-3) \simeq 0.05$). Importantly, mutual inhibition shortens the effective time constant of an inhibitory network, speeding up its dynamical responses.

By contrast, consider an excitatory neural population endowed with mutual excitation:

$$\tau_E \frac{dr_E}{dt} = -r_E + J_{EE} r_E + I_{E,ext} \tag{3.24}$$

In the presence of recurrent excitation, this linear network has a steady-state $r_E = I_{E,ext} / (1 - J_{EE})$, which exists if $J_{EE} < 1$. Mutual excitation amplifies the overall activity level by a factor of $1/(1-J_{EE})$. The time constant of the system is given by:

$$\tau_{EE} = \frac{\tau_E}{|(1-J_{EE})|} \tag{3.25}$$

Note that τ_{EE} is independent of $I_{E,ext}$. Its value cannot be deduced from the time it takes for the firing rate to reach a fixed level, which is shorter with a larger input intensity $I_{E,ext}$. Instead, it determines the time it takes to reach the steady-state. Endowed with recurrent excitation, the system integrates an input $I_{E,ext}$ for a longer time than τ_E. Moreover, after the input is turned off, the system decays back to zero, and hence the memory about the input is lost, with the same time constant. Recurrent excitation prolongs the time constant, because more activity leads to an increase of excitatory drive, enabling further growth of activity in the larger J_{EE}, the longer the integration time.

Perfect integration is achieved when the positive feedback cancels the intrinsic leak, $J_{EE}=1$ (**Figure 3.8a**). In this case:

$$\tau_E \frac{dr_E}{dt} = I_{E,ext}; \quad r_E(t) = \frac{1}{\tau_E}\int_0^t I_{E,ext}(t')dt'. \tag{3.26}$$

This balancing of feedback and leak implies that Eq. (3.24) requires fine-tuning in order to achieve perfect integration. The tuning criterion can be viewed graphically by plotting the leak term (r_E) and the feedback term ($J_{EE}r_E$) in a single graph (**Figure 3.8a**, top row). When the two terms are equal, the network performs perfect integration. When feedback is less than leak, the stored total in the integrator decays away at a rate (time derivative) proportional to the difference between the leak and feedback. In the absence of input, this leads to exponential decay on a timescale τ_{EE} (leaky integrator, **Figure 3.8b**). When J_{EE} is larger than 1, and feedback is greater than leak, there is no steady-state, and firing activity exponentially increases with a time constant given by the same formula (unstable integrator, **Figure 3.8c**).

In a useful conceptual description, the network's firing rate can be represented by the location of a block along a surface, with the movement of this block representing the dynamics of the network (Goldman et al. 2009). External inputs slide the block with a velocity proportional to the size of the input. For a flat and level surface (**Figure 3.8a**, bottom row), corresponding to a well-tuned network, the position of the block thus represents the integral of the external input. In the absence of input, the block stays in place, corresponding to successful maintenance of the accumulated input. For a mistuned network, the slope of the surface is given by the degree of mistuning, i.e., by the difference between the feedback and leak. In this case, in the absence of input, the position of the block drifts with a velocity proportional to the mistuning. For the leaky network, the mistuning brings the activity back to the origin (**Figure 3.8b**, bottom row), whereas for the unstable network, the mistuning leads to unbounded increase in activity (**Figure 3.8c**, bottom row).

Although this model is highly simplified, it captures the essence of the positive feedback mechanism used by more biologically detailed network models. Because of the strong coupling between neurons that mediates the positive feedback in such networks, only certain activity patterns can be stably maintained internally in the absence of external inputs. Generally, a "line attractor" can be realized in the high-dimensional space of multiple neural groups (Seung 1996); the distance along this line (analogous to the distance of the block along the surface in **Figure 3.8**) represents the accumulated total of inputs being stored by the integrator.

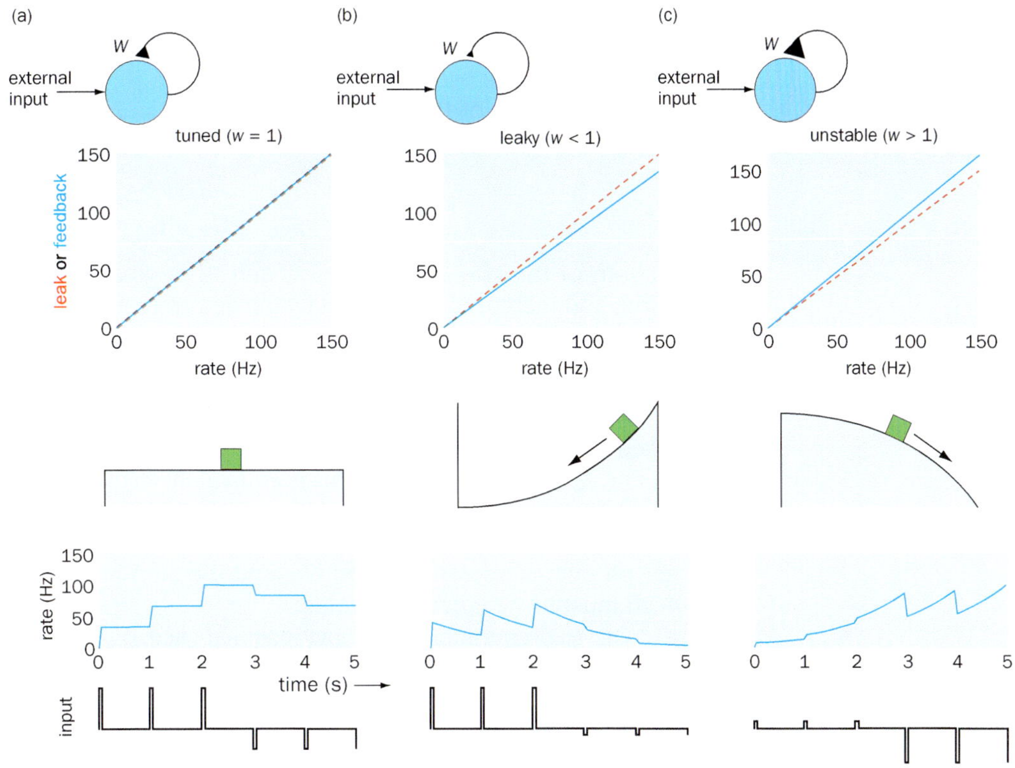

Figure 3.8 Linear network with recurrent excitation. Top: Simple population model with positive feedback onto itself carried by a synaptic weight. (a) When properly tuned, the feedback at any firing rate offsets the intrinsic leak that pulls the firing rate back towards zero; red: Leak, blue: Feedback. Middle: The rate encodes the integrated total of external inputs, conceptually represented by the location of a block along a surface. External inputs move the block along the surface. Bottom: Accumulation of pulsed excitatory or inhibitory inputs into sustained firing rates. (b) Same as (a) except that the synaptic weight is less than 1. Firing rate drift is proportional to the difference between the feedback and leak. Middle: The model firing rate behavior can be represented by the location of a block on a curved surface, where the curvature is proportional to the firing rate drift in the absence of input. Bottom: Decrease in the synaptic weight leads to exponential decay of firing rates in the absence of input. (c) A model in which feedback is too large (> 1), leading to exponential growth of firing rates. Reproduced from Goldman et al. (2009). [Goldman, M., A. Compte, and X.-J. Wang (2009). Neural integrators: Recurrent mechanisms and models. In L. Squire, T. Albright, F. Bloom, F. Gage, and N. Spitzer (Eds.), Encyclopedia of Neuroscience, pp. 165{178. MacMillan Reference Ltd.]

3.3.3 Inhibition-Stabilization and Balanced Amplification

Recurrent excitation is abundant in cortical circuits. Evidence suggests that in the hippocampus (Tsodyks et al. 1997) and neocortex (Douglas et al. 1995; Douglas and Martin 2004), recurrent excitatory dynamics is unstable (corresponding to J_{EE} larger than 1), and the local network is stabilized only because of strong feedback inhibition. Such a neural circuit is called an inhibition-stabilized network (ISN) (Ozeki et al. 2009). Consider a linear E/I network:

$$\tau_E \frac{dr_E}{dt} = -r_E + J_{EE} r_E - J_{EI} r_I + I_{E,ext}, \tag{3.27}$$

$$\tau_I \frac{dr_I}{dt} = -r_I + J_{IE} r_E - J_{II} r_I + I_{I,ext}. \tag{3.28}$$

The behavior of this system depends on both the strengths of recurrent connections and the excitation and inhibition time constants. If inhibition is much faster than excitation, so r_I can be approximated by the steady-state of Eq. (3.28), $r_I \simeq \left(I_{I,ext} + J_{IE} r_E\right) / \left(1 + J_{II}\right)$. Inserting this expression into Eq. (3.27), one can see that inhibition provides a negative feedback to r_E with the strength given by $J_{EI} J_{IE} / \left(1 + J_{II}\right)$, which counterbalances the positive feedback with the strength J_{EE}. Note that J_{II} reduces the negative feedback efficacy because it diminishes r_I for a given r_E. On the other hand, as we will see next, when recurrent connections are strong, faster excitation in interplay with slower inhibition can destabilize a steady-state, leading to network oscillations.

The stability of the steady-state of Eqs. (3.27–3.28) is given by the eigenvalues of the matrix:

$$M = \begin{bmatrix} (J_{EE}-1)/\tau_E & -J_{EI}/\tau_E \\ J_{IE}/\tau_I & -(J_{II}+1)/\tau_I \end{bmatrix}$$

The sum of the two eigenvalues λ_1 and λ_2 is given by the trace of this matrix, and their product is given by the determinant (Strogatz 2016):

$$\lambda_1 + \lambda_2 = \frac{(J_{EE}-1)}{\tau_E} - \frac{(J_{II}+1)}{\tau_I}; \lambda_1\lambda_2 = \frac{(1-J_{EE})(J_{II}+1)}{\tau_E\tau_I} + \frac{J_{EI}J_{IE}}{\tau_E\tau_I}. \quad (3.29)$$

In the case where the values of λ_1 and λ_2 are real, the steady-state is stable if λ_1 and λ_2 are both negative. This is equivalent to the requirement that [a] their sum is negative (so that each must be negative), and [b] their product is positive (so that they are of the same sign). When λ_1 and λ_2 are complex, their product is necessarily positive, and the sum is twice their real part, which must be negative for stability.

The two conditions are obviously satisfied if J_{EE} is smaller than 1. When $J_{EE} > 1$, the first term in the product of λ_1 and λ_2 is negative, hence the condition (a) requires that $J_{EI}J_{IE}$ be sufficiently large. In other words, the E-to-I-to-E loop must be strong enough, which is essential for realizing network stabilization by feedback inhibition in spite of runaway excitation (Tsodyks et al. 1997). Concerning the condition (b), J_{II} must be sufficiently large to ensure that the sum of λ_1 and λ_2 is negative. It may be surprising that the inhibitory-to-inhibitory connection is required for stability; the reason is that J_{II} shortens the effective time constant of synaptic inhibition (Eq. 3.23), which tends to stabilize a steady-state.

The ISNs display some unexpected dynamical response properties. When the inhibitory neural population receives a negative input, one would predict that its activity decreases. However, that would lead to an increase of the excitatory population activity, and the excitatory-to-inhibitory drive could more than compensate the negative external input, resulting actually in an increase of the inhibitory population activity. This paradoxical effect (Tsodyks et al. 1997) explains why, if a network operates in an ISN regime, the inhibitory and excitatory neural populations tend to fire synchronously rather than out-of-phase in response to an external oscillatory input. A rich repertoire of such paradoxical effects can be realized (Garcia Del Molino et al. 2017; Palmigiano et al. 2020) in excitatory-inhibitory networks with three types of inhibitory neurons (Chapter 2, **Figure 2.2**).

Another phenomenon is called "balanced amplification," which refers to temporal amplification of a brief input by the strong recurrent excitatory and inhibitory synaptic interactions. Murphy and Miller (2009) noted that because of the Dale's law, connectivity matrix of a neural circuit is inherently "non-normal," which implies that the eigenvectors are typically not orthogonal to each other. As a simple and extreme example, assume that $J_{EE} = J_{IE} = J_E = w$, $J_{EI} = J_{II} = J_I = -wk_I$ and $\tau_E = \tau_I = \tau$ (**Figure 3.9a**). Eqs. (3.27–3.28) can be rewritten in terms of $r_+ = r_E + r_I$ and $r_- = r_E - r_I$ as $\tau dr_- / dt = -r_- + (I_{E,ext} - I_{I,ext})$ and $\tau dr_+ / dt = -(1 + J_I - J_E)r_+ + (J_E + J_I)r_- + (I_{E,ext} + I_{I,ext})$. In other words, the circuit is equivalent to a feedforward chain from r_- to r_+ (**Figure 3.9b**). Under the condition that $J_I > J_E$, the network is stable, but displays transient amplification of response to an external input (**Figure 3.9c**). This happens because initially $r_- > 0$ ($r_E > r_I$), which provides a positive input to r_+. The latter therefore increases (both r_E and r_I grow) transiently, until the summated net input becomes negative and r_+ decays away.

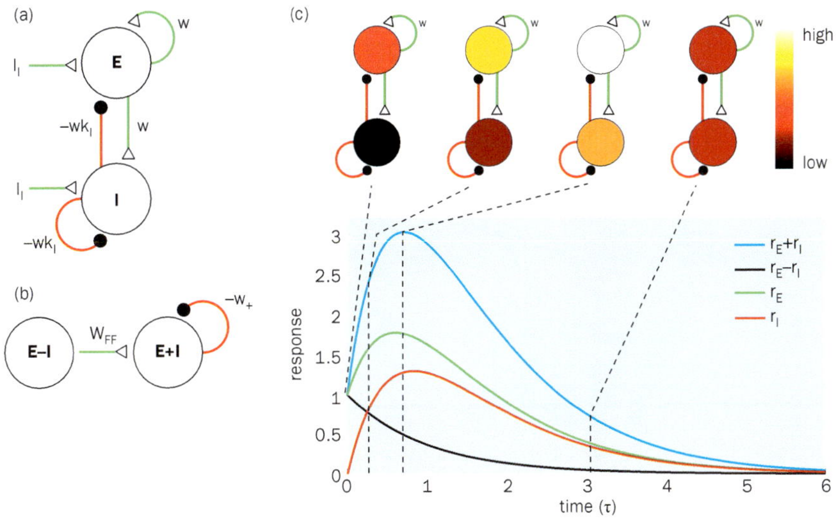

Figure 3.9 Balanced amplification of an E/I network. (a) Diagram of a balanced circuit with an excitatory and an inhibitory population. (b) The recurrent network is equivalent of a forward network from $r_{-} = r_E - r_{-}$ to $r_{+} = r_E + r_I$ with a weight $J_E + J_I$. The self-coupling of r_+ has a weight $J_I - J_E$, which is inhibitory with $J_I > J_E$. (c) Plot of the sum (blue line) and difference (black line) between activity in the excitatory (r_E, green line) and inhibitory (r_I, red line) populations in response to a pulse of input to the excitatory population at time 0 that sets $r_E = r_I(t = 0) = 1$. Diagrams above the plot represent the color-coded levels of activity in the excitatory and inhibitory populations at the time points indicated by the dashed lines. Reproduced from Murphy and Miller (2009). [Murphy, B. K., & Miller, K. D. (2009). Balanced amplification: A new mechanism of selective amplification of neural activity patterns. Neuron, 61(4), 635–648.]

3.4 COHERENT NEURAL CIRCUIT OSCILLATIONS

As mentioned earlier, network oscillations present a simple way to generate neural correlations, and the brain exhibits multiple forms of coherent rhythms. During certain stages of sleep, the entire cortex is synchronized at specific oscillation frequencies (Bremer 1958; Andersen and Andersson 1968; Steriade et al. 1993). These brain rhythms on a macroscopic scale are readily measurable by electroencephagram (EEG), intracranial electrocorticography (ECoG) or magnetoencephalogram (MEG). EEG and MEG are recordings from outside of the scalp, and they reflect averaged electrical signals from a massive number of neurons. Measurements with a much higher spatial resolution can be achieved using local field potential (LFP). LFP is a weighted spatial average of the flow of electrical currents within 1–3 mm from the electrode tip (Katzner et al. 2009; Xing et al. 2009), a spatial extent that still covers many thousands of neurons. If electrical signals from these neurons were temporally uncorrelated completely, summing over the population would yield a flat signal in time. Therefore, an oscillatory component of LFP must result from synchronization among a neural mass. Synchronous oscillation is a "collective phenomenon" of neural populations.

In contrast to large-amplitude EEG sinusoidal waves in sleep, during wakefulness active engagement in behavioral tasks is associated with distinct types of low-amplitude EEG oscillations that are intermittent in time and spatially nonuniform (Gray 1994; Buzsáki 2006; Wang 2010). Theta (4–8 Hz) rhythm is a prominent coherent oscillation observed in the hippocampus and its surrounding limbic structures during exploratory movements, e.g., when a rat navigates through a maze looking for food in a laboratory experiment (Bland 1986; Buzsáki 2002). During spatial navigation, hippocampal pyramidal neurons ("place cells") encode specific locations ("place field") of the rat during active exploration (O'Keefe and Nadel 1978). The spike timing of a hippocampal neuron systematically shifts to an earlier phase of the theta cycle as the rat moves across the cell's place field, a phenomenon called "phase precession" (O'Keefe and Recce 1993; Skaggs et al. 1996; Huxter et al. 2003) (**Figure 3.10a**).

Another common brain oscillation is called gamma or "40 Hz" rhythm, with a characteristic frequency in the range of 30–60 Hz (Adrian 1950; Freeman 1975; Buzsáki and Wang 2012). Gamma oscillations observed in spiking activity of neurons of the neocortex depend on stimulus (Gray and Singer 1989; Ray and Maunsell 2015). As a result, under a given behavioral condition, the synchronous rhythm is confined to selective neural populations rather than spatially uniform across the cortex. This is illustrated in **Figure 3.10b** from a monkey neurophysiological study of visual selective attention, a cognitive process of influencing sensory information processing by top-down control. In the monkey cortical area V4, a mid-level stage for visual information processing, LFP displays enhanced gamma

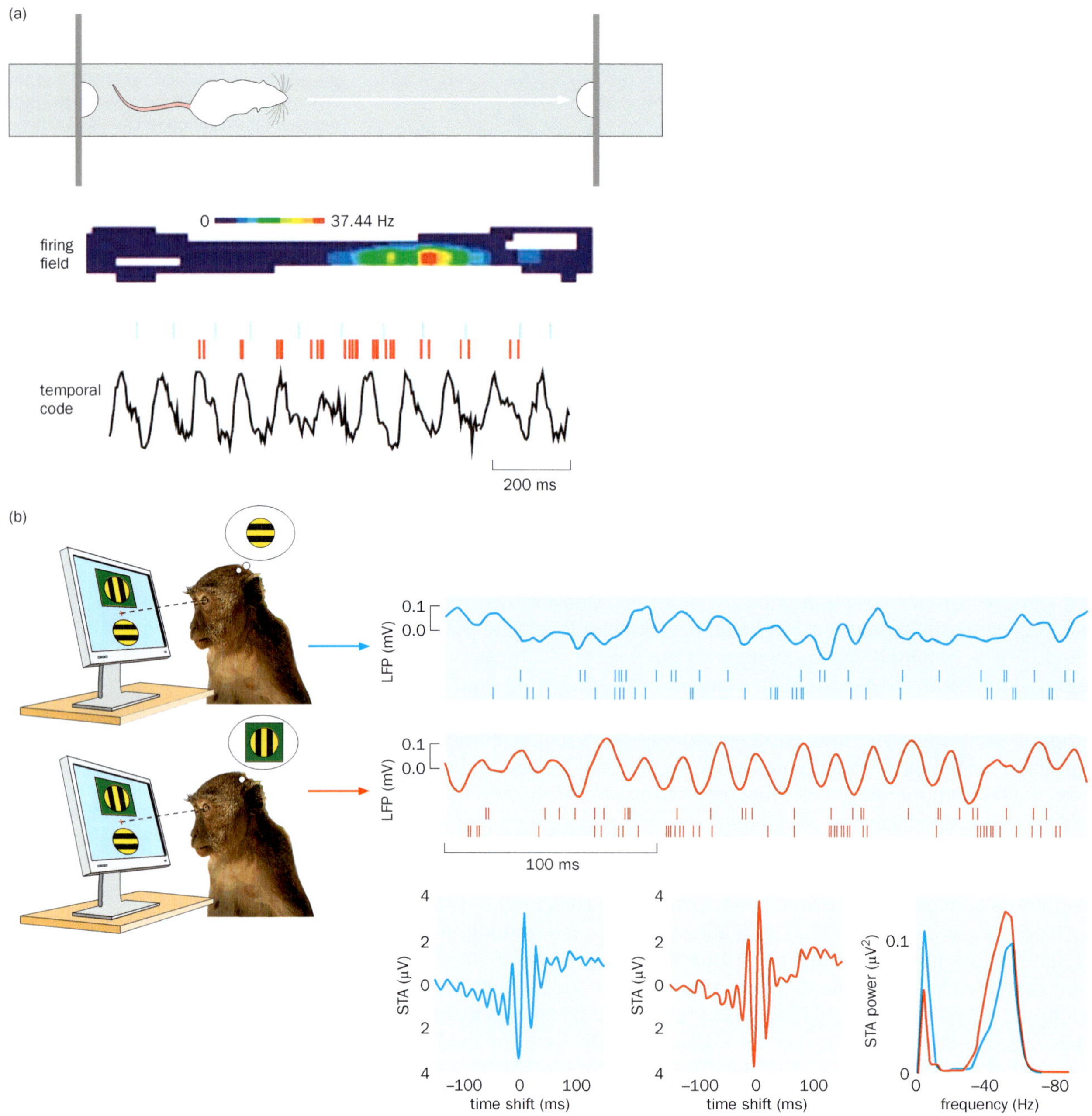

Figure 3.10 Synchronous oscillations during behaviors. (a) Theta rhythm in the hippocampus. Top: In the task, rat moves along a linear track between food rewards. Middle: Color-coded firing field of a place cell. Bottom: EEG theta rhythm and place cell firing (in red) for the same cell on a single run. Bursts of spikes move systematically to an earlier phase of the theta cycle, whereas firing rate initially rises then falls. (b) Attention induces changes in synchrony in the visual cortex. In the experiment, two visual stimuli were presented, one inside and one outside the receptive field (RF, marked by green box) of a neuron in area V4 of a behaving monkey. Red (blue): Attention directed inside (outside) the RF. Upper two panels: LFPs and spike trains. Bottom left and middle panels: Spike-triggered average (STA), the average LFP waveform around the time of a spike (0 along x-axis), computed during stimulus presentation. Bottom right panel: Power spectra of the two STAs. When attention is focused inside the RF, neural firing tends to be more in phase with the LFP at around 50 Hz and less at around 10 Hz. Reproduced from Wang (2010) with the original experimental data reported in Huxter et al. (2003) for (a) and Fries et al. (2001) for (b). [Huxter, J., Burgess, N., & O'Keefe, J. (2003). Independent rate and temporal coding in hippocampal pyramidal cells. Nature, 425(6960), 828–832.] [Fries, P., Reynolds, J. H., Rorie, A. E., & Desimone, R. (2001). Science, 291, 1560–1563.]

oscillations when attention is directed to the preferred spatial location of a recorded neural group. Spiking of neurons is more likely to be synchronized in time with the population rhythm when attention is directed toward the receptive field of the recorded neurons rather than away from it. The implication is that selective attention operates not only by changes of firing rates of sensory neurons but also through modulation of neural population synchronization.

3.4.1 Synchronization of Neural Oscillators

What are the network mechanisms for synchronization of a neural population?

As we discussed in Chapter 2, isolated neurons in response to a constant input fire spikes periodically, and in this sense they behave as nonlinear oscillators. It is thus natural to consider brain rhythms as an emerging phenomenon in a network of oscillatory neurons. Experiments have demonstrated that, whereas gamma oscillations can be synchronized over a range of spatial scales, rhythmogenesis arises locally from microcircuit mechanisms (Fisahn et al. 1998; Atallah and Scanziani 2009; Buzsáki and Wang 2012). A generic cortical microcircuit consists of two major cell types: excitatory principal neurons and inhibitory interneurons. It follows that three types of synchronization mechanisms by chemical synapses are conceivable: recurrent excitation (E-to-E) between principal neurons, mutual inhibition (I-to-I) between interneurons, and feedback inhibition through the excitatory-inhibitory (E-to-I-to-E) loop (Börgers and Kopell 2003; Wang 2010).

Intuitively, mutual excitation should be ideal for synchronizing coupled neurons. However, modeling studies have revealed that this is true only if synaptic excitation is very fast compared to the periodicity of a neural oscillation; synaptic excitation with a time constant of a few milliseconds is suitable for synchronizing a rhythm at ~ 10 Hz or lower frequencies, but not ~ 40 Hz gamma rhythm (van Vreeswijk et al. 1994; Hansel et al. 1995). On the other hand, reciprocal inhibition can bring coupled neurons to fire spikes synchronously, with zero phase difference. **Figure 3.11a** shows a model simulation of an inhibitory interneuron network, in which single neurons are type I (see Chapter 2) and synaptic coupling is mediated by $GABA_A$ receptor with a time constant of 10 ms. As can be seen in the raster plot, neurons initially fire at different times but quickly become perfectly synchronized with zero phase difference (Wang and Buzsáki 1996). At first sight, this result is counterintuitive since spiking in one neuron exerts an inhibitory effect on target neurons. However, the synaptic kinetics is not instantaneous but has an onset time and decay time. GABAergic neurons fire together, sending inhibition to each other with a delay; when the synaptic inhibitory current decays away, they can fire together anew. In this manner, mutual inhibition may indeed underlie neural synchronization. This scenario illustrates that taking into account the biophysics of synaptic transmission is essential for understanding collective network behavior, a lesson that will be demonstrated to hold for cognitive functions later in the book (Chapters 5 and 6).

It follows from this explanation that the network synchronization is optimal for oscillation periodicity roughly twice the synaptic time constant (half cycle for upswing, the other half downswing). With an experimentally measured time constant for $GABA_A$ receptor-mediated synaptic inhibition of ~ 10 ms, the optimal periodicity would be roughly 20 ms, in the gamma band. This prediction was tested in the model by varying the model parameters. Wang and Buzsaki (1996) proposed a measure of network coherence, which is 0 for complete asynchrony and 1 (maximal) for perfect synchrony. When the intensity of input current was varied, the firing frequency of neurons increased from 10 Hz to more than 100 Hz. Or, when the strength of synaptic inhibition was increased, the firing rate decreased from about 150 Hz to 20 Hz. Yet, the coherence measure displays a sharp peak and is significant only in the gamma frequency range. Therefore, $GABA_A$ receptor-mediated mutual inhibition is optimal for synchronizing gamma-band rhythm. In support of this scenario, experiments found that fast-spiking basket cells can become synchronized at 40 Hz by $GABA_A$ receptor-mediated mutual inhibition, when synaptic excitation is blocked pharmacologically (Whittington et al. 1995).

How much network connectivity is required for population synchronization? Consider a randomly connected network in which any neuron projects to another with a probability $p = K / N$, where K is the average connections per neuron. Obviously, when K is zero, synchronization is impossible without any connection. It turns out that synchronization is a nonlinear function K: there is a minimum value of K, K_c, below which the network is completely desynchronized (the coherence measure is zero), and above K_c the coherence is not zero and gradually increases with K (**Figure 3.11b**). Therefore, coherent oscillation requires a minimum connectedness, and it emerges as a qualitatively new type of collective

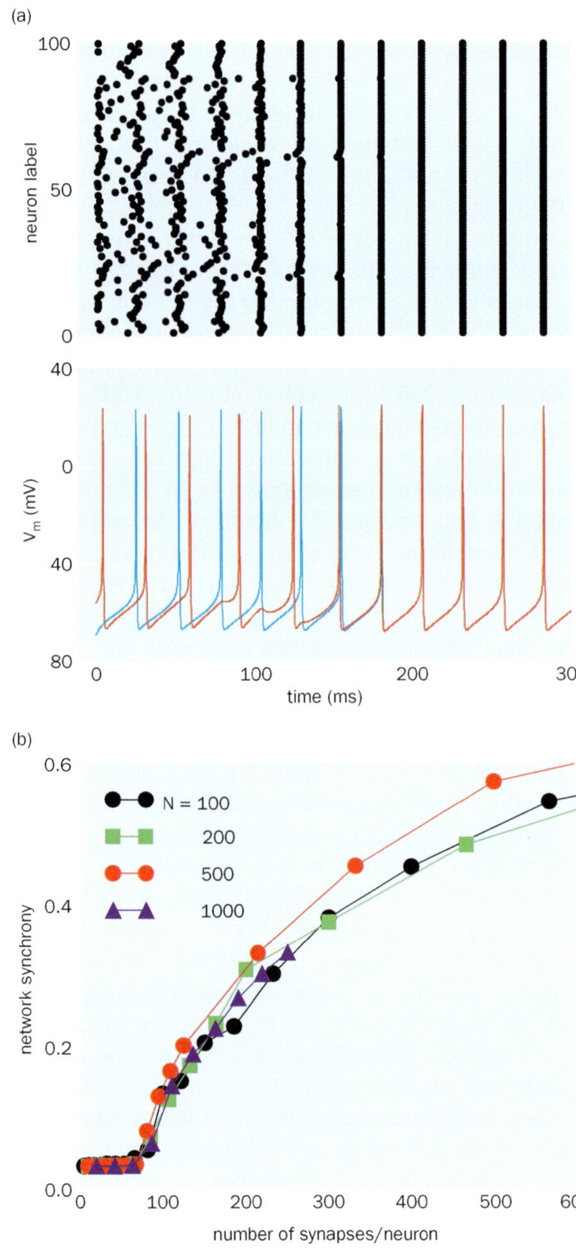

Figure 3.11 Synchronous oscillation in a randomly connected network model of GABAergic cells. (a) An example of network synchronization in a fully connected regular network. Upper panel: The rastergram of spikes (dots) of 100 neurons (ordinate) as a function of time (abscissa). Lower panel: Membrane potentials of two neurons chosen for illustration. Neurons discharge independently initially, but quickly become synchronized by mutual inhibition. (b) Network synchrony as a function of the mean number K of synaptic inputs per cell, in randomly connected networks of different sizes (N = 100, 200, 500, 1000). Note critical threshold of connectedness (each interneuron is connected to ~ 50 other neurons) for synchrony. This critical degree of connectivity is independent of the network size. Reproduced from Wang and Buzsáki (1996). [Wang, X. J., & Buzsáki, G. (1996). Gamma oscillation by synaptic inhibition in a hippocampal interneuronal network model. Journal of Neuroscience, 16(20), 6402–6413.] [Buzsáki G, Geisler C, Hinze D and Wang X-J (2004) Circuit complexity and axon wiring economy of cortical interneurons. Trends in Neurosci 27(4), 186–193.]

network behavior when K is above a critical level K_C. Remarkably, the critical value K_C is about 50, regardless of the network size N. This is shown as a function of K_{eff}, which is a theoretical quantity that converges to K with N going to infinity ($1/K_{eff} = 1/K - 1/N$) (Golomb and Hansel 2000). Thus, small and very large networks can have the same critical connectedness. The independence of K_C on the network size can be understood by random graph theory, which states that if K is increased above a critical and finite value, the network changes abruptly from a loose collection of small clusters (with no connection between any two clusters) to a system dominated by a single "giant" cluster of connected nodes with a size proportional to N, thus comparable to the entire network (Erdos and Renyi 1960; Albert and Barabási 2002). In light of the anatomical evidence that the axon arbor of a single basket cell contacts, on average, about 60 other basket neurons (Sik et al. 1995), the model suggests that the degree of connectivity among basket cells is sufficient to generate 40 Hz synchronization.

As it turns out, the value of K_C strongly depends on the biophysics of synaptic interactions and intrinsic cellular dynamics of single neurons. This is strikingly demonstrated in modeling work showing that, given similar synaptic coupling time constant, the critical connectedness K_C is ~ 50 with conductance-based Hodgkin-Huxley type I model interneurons (Wang and Buzsáki 1996), but is several thousands with LIF model neurons (Golomb and Hansel 2000). Thus, had the simpler LIP neuron model been used, one would have reached the conclusion that the connections between basket cells were vastly insufficient to generate synchronous oscillations. The finding serves as a useful reminder that biophysics of single neurons could have a major impact on the collective network dynamical behavior, and information processing (Li et al. 2020).

3.4.2 Sparsely Synchronous Rhythm

While an isolated single cortical cell behaves regularly in response to a current input, spiking activities of cortical neurons are highly irregular *in vivo* (Chapter 2, **Figure 2.5**), presumably as a result of synaptic bombardments (Destexhe et al. 2003). Hence, the framework of coupled neural oscillators is instructive but insufficient to capture the more stochastic nature of brain rhythms observed in behaving animals. If spike trains are close to a Poisson process far from being clock-like, network coherence would be weak and coexist with variable firing of its constituent neurons. At issue is a long-standing debate on whether neural representation of information uses a rate code (for which precise timing of spikes plays a negligible role) or a temporal code (for which timing and synchrony of spikes are crucial) (Abeles 1991; Shadlen and Newsome 1994; Softky 1994). To a large extent, either side of the debate tends to ignore evidence in favor of the other side's view: proponents of a temporal code often view neurons as regular oscillators and do not strive to account for high variability of neuronal spiking activity, whereas some supporters of a rate code profess not to know what to make of LPF oscillations.

Is there a way to reconcile irregular neural firing with an LFP rhythm? Let us denote by "sparsely synchronous rhythm" population dynamics in which the rhythmic modulation of neural firing probability is small; see **Figure 3.12** for hippocampal pyramidal cells during gamma oscillations. Quantitatively, suppose that the histogram is fit with the formula $r(\theta) = r_0 + r_1\cos(\theta - \theta_{peak})$, with a peak at θ_{peak}. In **Figure 3.12**, the baseline $r_0 \simeq 0.094$ and modulation $r_1 \approx 0.007$, the fraction of spikes modulated by the gamma phase is $r_1 / r_0 \simeq 8\%$.

A framework to describe such network oscillatory dynamics takes a diametrically opposite view from that of coupled oscillators. The idea is to start with a completely asynchronous network state described in the

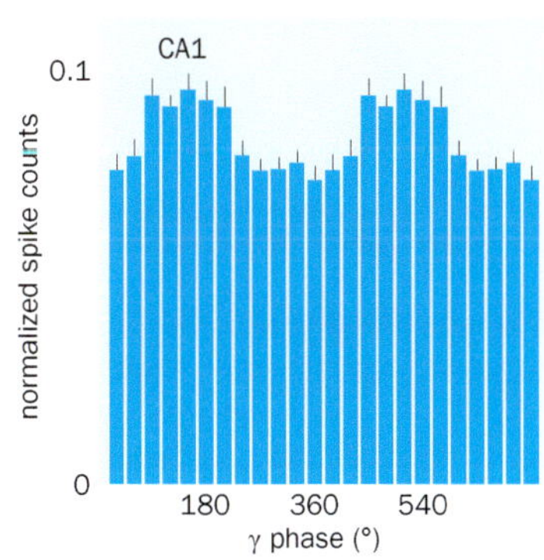

Figure 3.12 Distributions of spike times as a function of the phase of gamma rhythm for CA1 place cells in freely moving rats, shown with two oscillation periods. A small rhythmic modulation on top of a tonic baseline means that the majority of spike times are evenly distributed over all phases, whereas a small fraction of spikes have preferred phases. Reproduced from Colgin, L. L., Denninger, T., Fyhn, M., Hafting, T., Bonnevie, T., Jensen, O., ... & Moser, E. I. (2009). Frequency of gamma oscillations routes flow of information in the hippocampus. Nature, 462(7271), 353–357.

previous section, and examine how a coherent rhythm might emerge from a dynamical destabilization of the asynchronous network state (Brunel and Hakim 1999; Brunel 2000). As we discussed in Chapter 2, synaptic inhibition mediated by $GABA_A$ receptors is typically several-fold slower than glutamatergic excitation mediated by AMPA receptors. When realistic synaptic time constants are incorporated into a random network model of spiking neurons, a novel collective behavior emerges (Brunel and Hakim 1999; Brunel and Wang 2003; Geisler et al. 2005; Palmigiano et al. 2017). As shown in **Figure 3.13a**, in this regime of "sparsely synchronous rhythm," spike trains of individual neurons are highly irregular, as the network is roughly balanced between synaptic excitation and inhibition. There is also a contribution of noise into neurons, representing Poisson-like inputs from outside of the network. The population firing rate displays bursts of frequency-dependent coherence that occur intermittently in time. Most neurons participate in the generation of field oscillations, but they do not fire spikes at every oscillation cycle, and neural firing probability is weakly modulated by the population rhythm. Moreover, firing rates are highly heterogeneous in different neurons, with an exponential distribution of firing rates across the neural population.

Sparsely synchronous rhythm arises as a bifurcation phenomenon, when the asynchronous state becomes unstable: a small increase of firing activity in pyramidal neurons generates more recurrent excitation, leading to further upswing of pyramidal neural activity and thus even more mutual excitation, and the network diverges away from the asynchronous state. This destabilization happens when feedback inhibition is not fast enough to prevent reverberatory neural activity. Instead of complete asynchrony, a population rhythm emerges: reverberatory excitation also increases firing of inhibitory neurons; a slower time course feedback synaptic inhibition would grow sufficiently to counteract excitation, leading to a downswing of activity in both E and I populations. As synaptic inhibition decays over time, the network recovers and another oscillation cycle can start anew. Therefore, intuitively, population oscillations occur rather naturally when recurrent connections are strong and feedback inhibition is slow compared to feedback excitation, even with a fairly small mismatch of synaptic time constants (**Figure 3.13b–c**). In contrast to the situation where individual neurons can be treated as oscillators, there is no periodicity in single neurons. Spiking activities display a degree of variability similar to that of a Poisson process yet are modulated by the population rhythm, consistent with the experimental observations (**Figure 3.12**). The network rhythm, which may be detectable by a population measure such as LFP but not obvious at the level of spike trains of individual neurons, must be understood as an emerging phenomenon of an interacting neural circuit.

In summary, two main circuit mechanisms for synchronous oscillations are mutual inhibition of GABAergic neurons and the interplay between fast excitation and slower inhibition. When single neurons display highly irregular firing rather than behaving as regular clocks, the frequency of a network rhythm is determined by synaptic time constants as well as single neuron properties (Geisler et al. 2005). Synchrony of spiking activities can subserve multiple functional roles (Wang 2010). Perhaps somewhat counter-intuitively, weak neural oscillations are an easy way to create neural correlations to ensure highly irregular activities of neural populations, which are essential for probabilistic information processing and cognitive functions (Chapters 5 and 6).

3.4.3 At the Edge of Criticality

A brain rhythm measured by EEG, ECoG or MEG is manifest in the frequency domain by a peak in the signal's power spectrum $P(f)$. In behaving

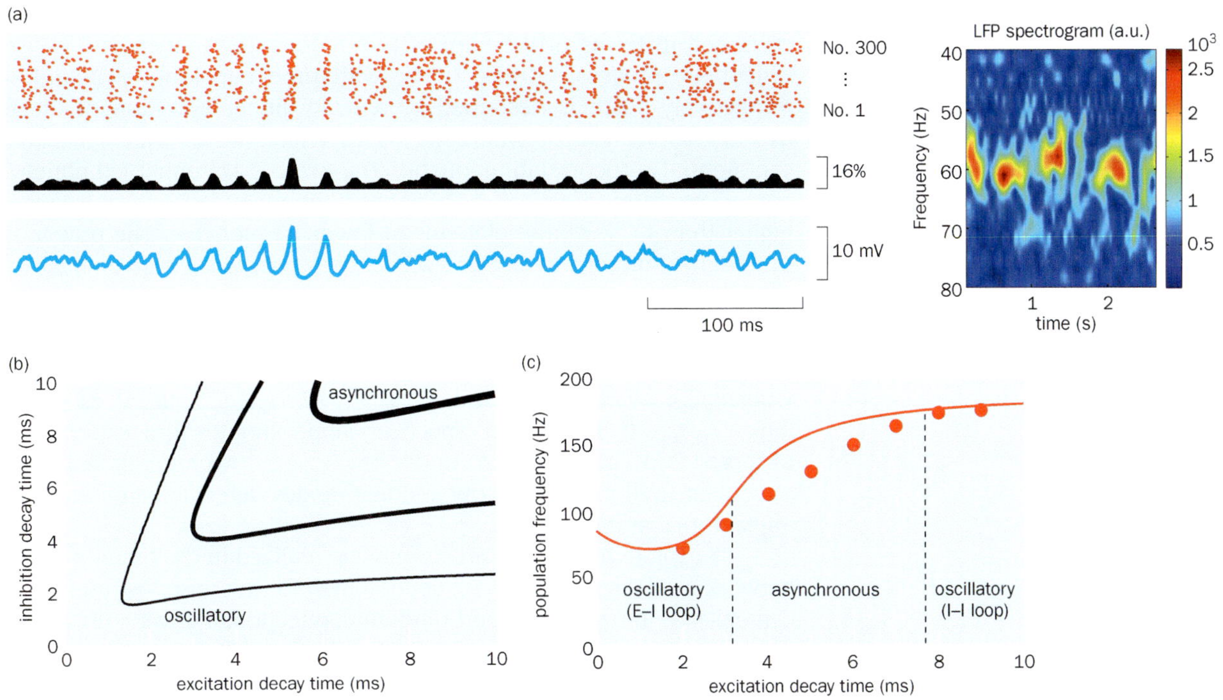

Figure 3.13 Sparsely synchronous network oscillations. (a) A network model of Wang-Buzsáki neurons. Left panel: Firing raster (red, top), multiunit activity (black, middle) and LFP (blue, bottom). Right panel: Transient gamma bursts are evident in the spectrogram of LFP activity. (b–c) Dependence of sparsely synchronized rhythm on the strength and synaptic excitation and inhibition time constants (τ_E and τ_I) in a network model of LIF neurons. (b) Network dynamical behavior as a function of τ_E and τ_I. A line separates the asynchronous state (upper right region) from the synchronous oscillation state (lower left region). The thickness of the line indicates recurrent connection strength; coherent oscillations become more prevalent with stronger recurrence of the network. (c) Population frequency as a function of τ_E, with fixed τ_I = 5 ms (from panel (b)). When τ_E is much smaller than τ_I, oscillation develops in an excitatory-inhibitory loop scenario. On the other hand, when τ_E is sufficiently large, the excitatory drive is relatively tonic, and the interneuronal network by itself generates synchronous oscillation at a higher frequency. Panel (a) is reproduced from Palmigiano et al. (2017), (b–c) from Brunel and Wang (2003). [Palmigiano, A., Geisel, T., Wolf, F., & Battaglia, D. (2017). Flexible information routing by transient synchrony. Nature Neuroscience, 20(7), 1014–1022.] [Brunel, N., & Wang, X. J. (2003). What determines the frequency of fast network oscillations with irregular neural discharges? I. Synaptic dynamics and excitation-inhibition balance. Journal of Neurophysiology, 90(1), 415–430.]

animals, when a peak is indeed present, it rides on top of a broadband component, which constitutes the majority of signal power (Buzsáki et al. 2003; Fries et al. 2008; Wang 2010). Interestingly, the broadband component commonly displays a power law, $P(f) \approx A / f^{\beta}$, where β is the power law exponent. The origins of the arrhythmic signal contributing the $1 / f^{\beta}$ component of the spectrum remain elusive (Bédard and Destexhe 2009; Roberts et al. 2015). ECoG recordings suggest that arrhythmic broadband field potential is a distinct macroscopic manifestation of the irregular firing of cortical neurons (Miller et al. 2009; He 2014).

Multiple studies using LFP and EcoG recordings have found that the power law exponent β is typically around 2 (Miller et al. 2009; Manning et al. 2009; He et al. 2010). While Bédard et al. (2006) reported slopes of 1 ($P(f) \sim 1 / f$) at low frequencies, further studies suggest that this is an artifact of finite-size effects (Korzeniowska et al. 2023). A study revealed the shape of human EcoG power spectra across a wide range of frequencies, from 0.01 Hz to higher than 100 Hz (He et al. 2010). The power spectra exhibit a distinctive shape (**Figure 3.14a**): at very low frequencies (< 0.1 Hz) and above 1 Hz, power scales approximately proportional to the inverse-square of frequency ($P(f) \sim 1 / f^2$), while power spectra in the intermediate frequency range (0.1–1 Hz) are flat. This tripartite shape was conserved across subjects, albeit with individual differences in the specific frequencies for the transitions between the decaying and flat regions of the spectrum.

The observed field power spectrum can be phenomenologically escribeed in terms of the sum of two independent stochastic Ornstein-Uhlenbeck processes (Wax 1954), with time constants τ_{fast} and τ_{slow}, respectively. Each is described by a leaky linear equation of the form $dx/dt = -x/\tau + w(t)$, where $w(t)$ is a white noise, which displays an exponential autocorrelation function (**Figure 3.14b**, upper left panel). The power spectrum (**Figure 3.14b**, upper right panel) is $1/f^2$ at frequencies higher than $1/(2\pi\tau)$ and plateaus at lower frequencies. The power spectrum of the sum of two Ornstein-Uhlenbeck processes is given by:

$$P(f) = \frac{a}{f^2 + f_{\text{slow}}^2} + \frac{b}{f^2 + f_{\text{fast}}^2}. \tag{3.30}$$

where $f_{\text{fast}} = 1/(2\pi\tau_{\text{fast}})$, and $f_{\text{slow}} = 1/(2\pi\tau_{\text{slow}})$. Thus, if τ_{slow} is very long, $f_{\text{slow}} \sim 0$, the power spectrum approximates the experimentally observed ECoG power spectrum (**Figure 3.14b**, lower panel), with the relative contributions of the slow and fast modes determined by the weighting factors a and b.

To explore how this scenario may be realized in the brain, a recurrent network model (**Figure 3.14c**, upper panel) for the observed human ECoG dynamics assumes sparse and random connections between an excitatory population and an inhibitory population of units (which could be interpreted as single neurons or neural pools) (Chaudhuri et al. 2018). The system is described by a high-dimensional version of Eqs. (3.27–3.28), as the following:

$$\tau_E \frac{dr_i^E(t)}{dt} = -r_i^E(t) + \left[\sum_{j=1}^{N_E} J^{EE} W_{ij}^{EE} r_j^E(t) - \sum_{k=1}^{N_I} J^{EI} W_{ik}^{EI} r_k^I(t) + I_{ext,i}^E(t) \right]_+ \tag{3.31}$$

$$\tau_I \frac{dr_i^I(t)}{dt} = -r_i^I(t) + \left[\sum_{j=1}^{N_E} J^{IE} W_{ij}^{IE} r_j^I(t) - \sum_{k=1}^{N_I} J^{II} W_{ik}^{II} r_k^I(t) + I_{ext,i}^I(t) \right]_+ \tag{3.32}$$

For all units to fire at nonzero rate, external inputs should be sufficiently strong so that the total current to each unit is positive, in which case the model is linear. Let us denote $\mathbf{r}$ as the vector of all excitatory and inhibitory unit firing rates, matrix A encompasses the leak and the excitatory-inhibitory synaptic connection divided by the time constant τ, and $\mathbf{I_{ext}}$ the vector of external input current. Then, we can rewrite the network equations as:

$$\frac{d\mathbf{r}}{dt} = Ar + \mathbf{I_{ext}} \tag{3.33}$$

ECoG is modeled as the sum of synaptic currents (Mazzoni et al. 2011; Wang 2010; Buzsaki et al. 2012). Under the assumption that recurrent connections rather than feedforward inputs dominate the network dynamics, one can show that ECoG(t) is dictated by the sum of eigenmodes of the matrix A. In a completely random network, each entry of matrix A is nonzero with probability p, and nonzero entries are drawn from a probability distribution. The eigenvectors are defined as vectors $\mathbf{v}_k$ that satisfy the equation $A\mathbf{v}_k = \lambda_k \mathbf{v}_k$, where λ_k is the eigenvalue corresponding to the eigenvector $\mathbf{v}_k$. There are as many eigenmodes as the number of dynamical variables $N = N_E + N_I$. The characteristic timescale of network activity corresponding to the eigenvector $\mathbf{v}_k$ is $-1/Re(\lambda_k)$. Thus, the eigenvalues tell us what timescales the network will show, and the eigenvectors tell us how this activity is distributed across nodes.

The eigenvalue λ_{slow} emerges from recurrent interactions. For the randomly connected network plus an isolated mode (Ganguli et al. 2008), the eigenvalues take a particularly simple form when the connection weight distribution is Gaussian, known from the theory of random matrices (Rajan and Abbott 2006; Tao 2011) (**Figure 3.14c**, middle panel). The slow mode corresponds to a negative eigenvalue near 0 (the red point, and $N-1$ fast modes, which are centered around $-1/\tau$ (the cloud of blue points). When the recurrent excitation balances the intrinsic decay of activity, the network is "at the edge of criticality" (similar to the one-dimensional Eq. 3.24 with $J_{EE} \simeq 1$), and λ_{slow} is very close to 0. In this case, signals reverberate in the network for a long time, giving rise to a slowly decaying autocorrelation function. On the other hand, λ_{fast} is set by the intrinsic properties of each network node and is approximately located at $-1/\tau$. There are $N-1$ such fast modes, each with time constant approximately equal to τ.

Since the slow mode emerges from inter-node interactions, it corresponds to a globally distributed pattern of network activity. This is given by the eigenvector, $\mathbf{v}_{\text{slow}}$, corresponding to the eigenvalue λ_{slow}. $\mathbf{v}_{\text{slow}}$ can be thought of as a slowly varying background state of the network that all the nodes are coupled to; in particular, for low variability in connection strengths, $\mathbf{v}_{\text{slow}}$ has approximately equal weight at each node. By contrast, the fast modes are uncorrelated with each other, and thus different nodes participate in particular fast modes to greatly varying degrees. As

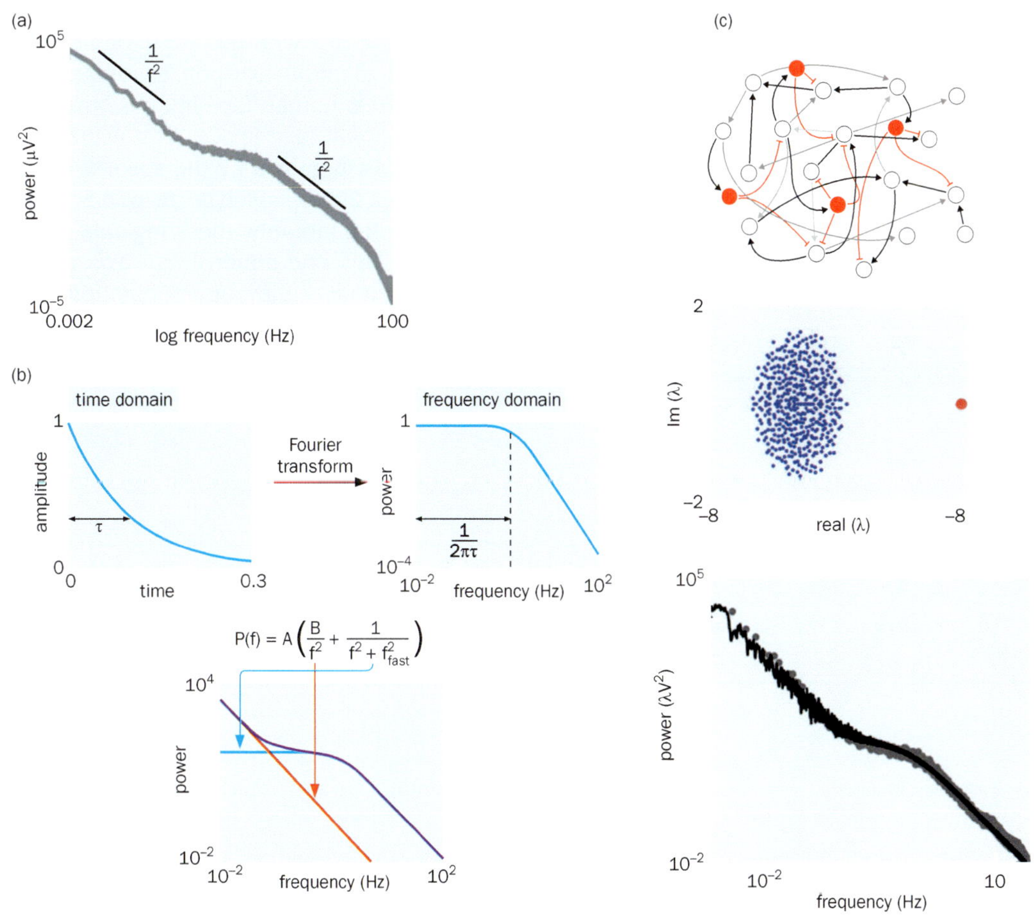

Figure 3.14 Scale-free brain dynamics. (a) Broadband power spectrum of ECoG recording from human cortex is shown in a log-log plot. Two $1/f^2$ regimes are separated by a "shoulder" (flat part) in the frequency range of 0.1–1 Hz. Power does not plateau even at frequencies as low as 0.002Hz (corresponding to a timescale of 1/0.002 = 500 seconds), indicating fluctuations across all timescales. (b) Schematic of power spectrum from Eq. (3.30) with $\lambda_{\text{slow}} \simeq 0$. (c) A random network model. Top: model scheme. Middle: distribution of eigenvalues of the neural dynamical system model, with a single eigenvalue near 0. Bottom: power spectrum from the model near criticality reproduces human's broadband power spectrum (a). Panels (a)–(c) are reproduced from Chaudhuri, R., He, B. J., & Wang, X. J. (2018). Random recurrent networks near criticality capture the broadband power distribution of human ECoG dynamics. Cerebral Cortex, 28(10), 3610–3622.

a consequence of this global distribution of the slow mode, the network model accounts for the common observation that low-frequency activity (i.e., slow timescales) shows a wider spatial correlation than does activity at high frequencies (Leopold et al. 2003; Buzsáki 2006).

3.5 NETWORK MODELS OF INFORMATION REPRESENTATION

Now that we are familiar with how to study neural networks, we can tackle the problem of how neural populations represent and process information. For instance, consider V1 where each neuron has a preferred location (receptive field) and orientation of visual stimulus, such as an elongated bar at a particular location in the visual field. Orientation is encoded by V1 neurons in the form of a bell-shaped tuning curve (Hubel and Wiesel 1968) (**Figure 3.15a**). On the other hand, neurons in the primary somatosensory cortex (S1) display a monotonic tuning for the frequency of a periodic vibrotactile stimulus (**Figure 3.15b**) (Salinas et al. 2000). In these two examples, the stimulus feature is one-dimensional. Neurons in higher sensory areas are selective for high-dimensional complex stimuli. In the temporal lobe. "face patches" are specialized in face processing (Kanwisher et al. 1997; Tsao et al. 2006). To examine coding of face-selective neurons, Chang and Tsao (2017) built a 50-dimensional space of features about a face's shape (which varies with the angle of viewing a face) and appearance (view-invariant identity). Responses of face-selective neurons were measured with 2,000 human face images, and each is mapped to a point in the 50-dimensional face space. The average stimulus that triggered a neuron to fire was identified by the spike-triggered averaging (STA) method (Schwartz et al. 2006), which corresponds to a vector called STA axis in the face space. Remarkably, the firing rate of a face neuron varies parametrically along this one-dimensional axis in the face space (**Figure 3.15c**), suggesting an algorithm for a population of face-selective neurons to encode any face.

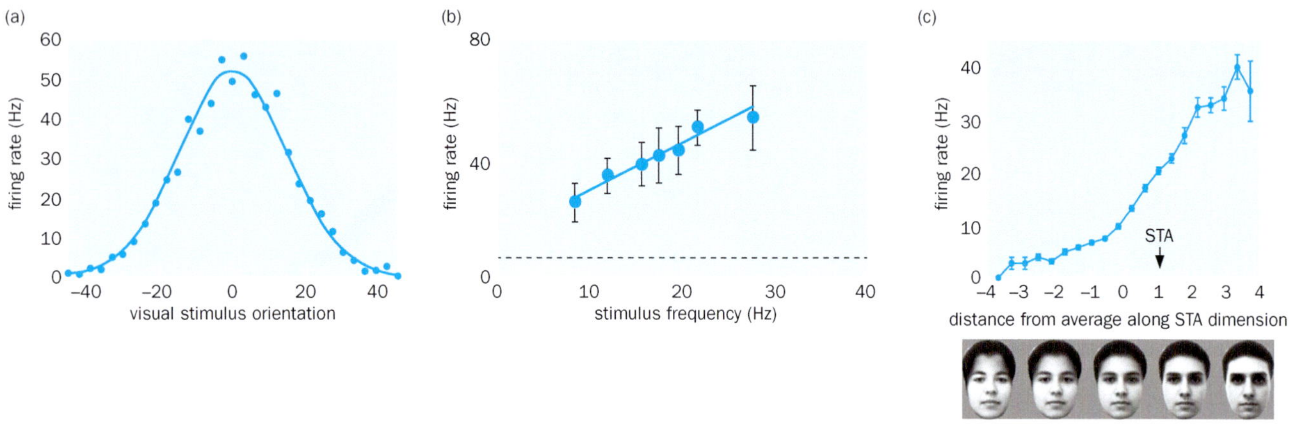

Figure 3.15 Neural coding of information. (a) Average firing rate of a cat V1 neuron plotted as a function of the orientation angle of a light bar stimulus; the curve is a fit of data (filled circles) using a Gaussian function. (b) Monotonic tuning of vibrotactile stimulus frequency in a neuron of monkey primary somatosensory cortex (S1). (c) Selectivity of a neuron in monkey interotemporal cortex for the magnitude of one of the principal components that characterize human face stimuli. Panel (a) is reproduced Henry, G. H., Dreher, B., & Bishop, P. O. (1974). Orientation specificity of cells in cat striate cortex. Journal of Neurophysiology, 37(6), 1394–1409; panel (b) is reproduced from Salinas, E., Hernandez, A., Zainos, A., & Romo, R. (2000). Periodicity and firing rate as candidate neural codes for the frequency of vibrotactile stimuli. Journal of Neuroscience, 20(14), 5503–5515; panel (c) is reproduced from [Chang, L., & Tsao, D. Y. (2017). The code for facial identity in the primate brain. Cell, 169(6), 1013–1028.

Let us consider a network with multiple neural pools that are selective for behaviorally relevant information such as a feature of simple sensory stimuli. The firing rates r_i ($i = 1, 2, ..., N$, where N is the number of neural pools) obey coupled rate equations as:

$$\tau \frac{dr_i}{dt} = -r_i + F(\sum_j w_{ij}^{rec} r_j + I_{ext,i}); \quad I_{ext,i} = \sum_k w_{ik}^{ff} r_k' \tag{3.34}$$

where r_k' are the firing rates of presynaptic neural pools, and $w_{k,i}^{ff}$ and w_{ij}^{rec} are the feedforward and recurrent connection matrices. We will first discuss simple models with one population like Eq. (3.34), with the connectivity weights as "effective" interactions that are allowed to be both positive (excitatory) and negative (inhibitory). Then we will consider models with two, excitatory and inhibitory, neural populations that obey Dale's law.

Stimulus features are often analog quantities, such as orientation or motion direction of a visual stimulus, frequency of an auditory tone, or spatial location. To model a network selective for an analog feature, it is often convenient to label neural pools by a continuous variable x. The sum of synaptic inputs then is replaced by an integral:

$$\tau \frac{dr(x)}{dt} = -r(x) + F\left(\int \left(w^{rec}(x, x') r(x') + I_{ext}(x')\right) dx'\right) \tag{3.35}$$

We shall refer to such models as continuous neural networks (CNNs), also called neural field models (Bressloff 2012).

3.5.1 Feedforward Continuous Network Model

One of the core brain processes in sensorimotor behavior is to transform sensory information into a different coordinate system appropriate for movement production. Take the example of making a rapid eye movement (saccade) to a visual target. The target input is encoded in the eye-centered retinotopic coordinates in the visual cortex. The eye movement, on the other hand, must be represented in the head-centered coordinates. The posterior parietal cortex (PPC) is believed to be an intermediate system for spatial coordinate transformation (Andersen 1989) in several sensory modalities. Single PPC cells are selective for retinotopically defined position of a visual target in the form of a bell-shaped tuning, which is multiplicatively modulated by the gaze relative to heading (**Figure 3.16a**) (Brotchie et al. 1995).

Models have been proposed for spatial transformations (Zipser and Andersen 1988; Salinas and Abbott 1995; Pouget and Sejnowski 1997b) with three layers of neural populations in a feedforward architecture ($w_{ij}^{rec} = 0$): a sensory representation layer, an intermediate layer corresponding to PPC and a motor representation layer. Pouget and Sejnowski (1997b) argued that a central idea is "basis functions," the combinations of which can approximate a large number of nonlinear functions. Specifically, neurons in PPC display spatial tuning of a visual target that can be described by a Gaussian function:

$$F_k(x) = A exp \frac{-(x - x_k)^2}{2\sigma^2} \tag{3.36}$$

where x_k is the preferred spatial location in the retinotopic coordinates of neuron k, σ is the width of the tuning curve and A is the maximum firing rate. For instance, if a network of N neurons represents a one-dimensional spatial space of size L, $x_k = kL / N$ ($k = 1, 2, ..., N$) ranging from 0 to L, with a finite resolution of $\Delta x = L / N$.

This tuning curve is modulated multiplicatively by a gain that depends on the gaze position y with respect to heading. Namely, activity of a PPC neuron depends on y, with a gain function described by a sigmoid function:

$$G_k(y) = \frac{1}{1 + \exp\left[-(y - y_k)/\lambda\right]} \tag{3.37}$$

where y_k is the midpoint of the gain for neuron k in head-centered coordinates, and λ controls the slope of the gain function.

In the model of Pouget and Sejnowski (1997b), the input layer contains two neural populations, one representing a retinotopic map and the other encoding eye position (**Figure 3.16b**). In an intermediate layer, PPC neurons instantiate the products of these functions, $r_k^{'} = F(x_k)G(y_k)$, which form "basis functions" in the sense that combinations of them can approximate a large number of functions of x and y. A desired transformation to motor command in appropriate motor coordinates can be realized by weighted projections from these neurons to a motor system in the output layer, with the synaptic input given by:

$$I_{ext}(z_i) = \sum_k w_{i,k}^{ff} r_{ik}^{'}(x_k, y_k) = \sum_k w_{i,k}^{ff} F(x_k) G(y_k) \tag{3.38}$$

where $w_{i,j}^{ff}$ is the connection weight from neuron j in the intermediate (PPC) layer to neuron i in the motor layer. Thus, output units are described by combinations of basis functions; with an appropriate choice of connection weights, a desired motor map can be realized. **Figure 3.16b** shows an example in which the gaze position with respect to heading is y, and the network transforms a target position x in the retinotopic coordinates to the head-centered coordinates, as $z = x + y$ with an appropriate choice of the weights of connections to the output units. Interestingly, various weighted combinations of the same set of basis functions can give rise to a diversity of different motor maps. Thus, in principle a single system such as PPC may underlie multiple sensorimotor transformations, such as from an auditory input to shifting gaze to the source of a sound or from a visual target to manually reaching it. This model suggests that the core function of PPC is not so much spatial representation in one particular coordinate reference framework, but to realize suitable basis functions that can be deployed to implement a diversity of desired spatial representations in downstream areas. This insight could explain behavioral symptoms associated with PPC damage such as hemineglect, a condition in which a person fails to perceive stimuli on one side of the visual field (Pouget and Sejnowski 1997a).

3.5.2 Normalization

Multiplicative gain modulation represents a general principle for neural computation; another concept is normalization (Heeger 1992; Heeger et al. 1996; Carandini and Heeger 2011). When an external stimulus drives many neurons in the network, normalization postulates that the firing rate r_i of any neuron i is given by:

$$r_i = A_0 \frac{I_i^n}{I_0^n + \sum_j (r_j / \beta)^n} \tag{3.39}$$

where n is an exponent, I_i is input to neuron i, and the parameter β in Hz/nA converts the physical unit of rate (in Hz) to current (in nA). The parameter A_0 controls the maximum firing rate, and I_0 ensures that r_i is not divided by zero in the absence of any input or activity. Normalization asserts that the overall activity of a neural population is roughly constant under different stimulus and behavioral conditions. Note that in Eq. (3.39), the firing rates are on both sides of the equation, which makes it difficult to solve r_i only as a function of inputs. For this reason, in order to fit experimental data, an alternative formula is often used in terms of the inputs:

$$r_i = A_0 \frac{I_i^n}{I_0^n + I_{sum}^n}; \quad I_{sum}^n = \sum_j I_j^n \tag{3.40}$$

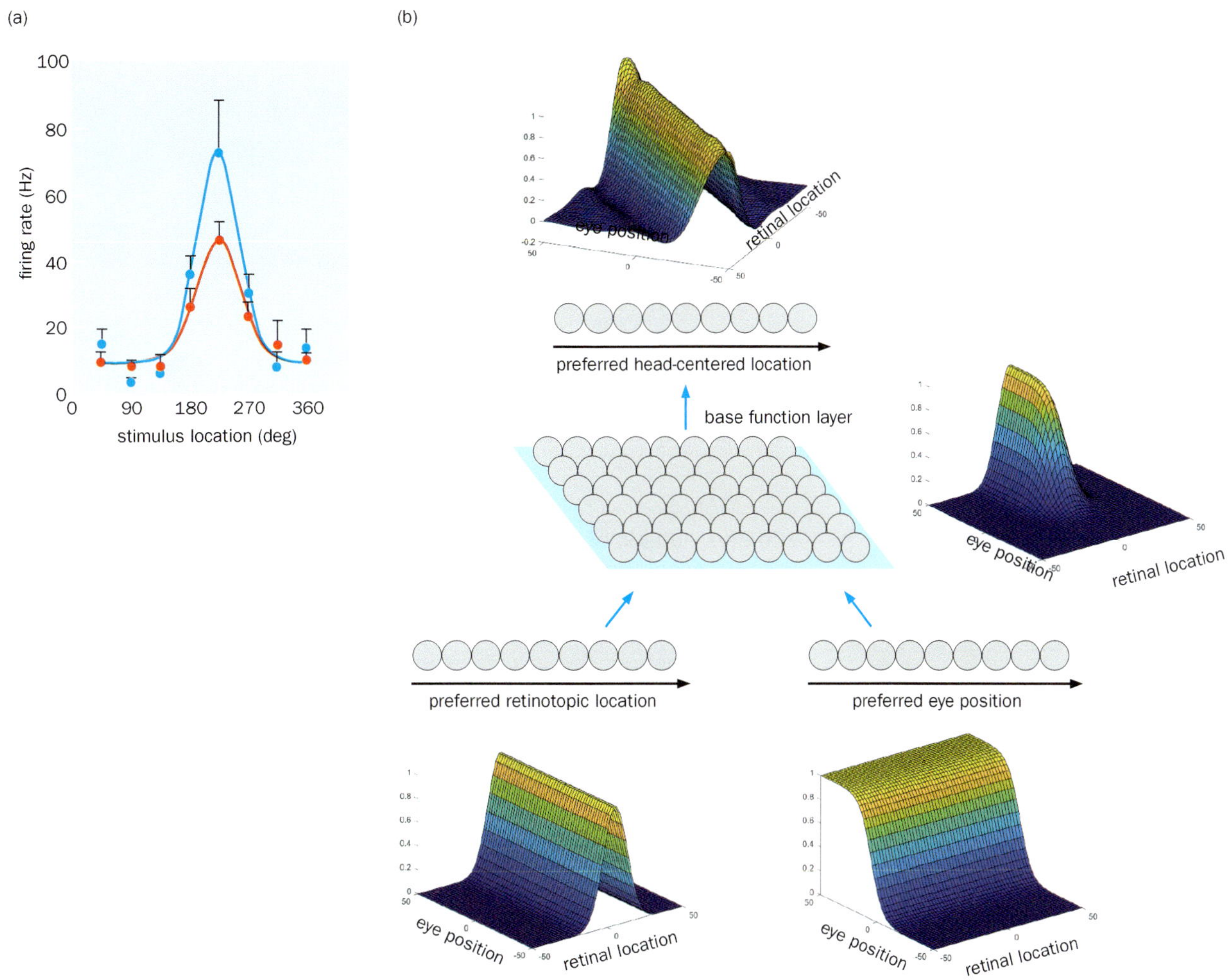

Figure 3.16 Coordinate transformation and multiplicative gain modulation. (a) Average firing rate of a neuron in the monkey posterior parietal cortex displays bell-shaped tuning as a function of the retinotopic position of a visual target. The two curves correspond to different gaze directions with respect to heading. (b) A three-layer feedforward network model of neurons. PPC is in the intermediate layer, where neurons represent basis functions as products of tuning to retinotopic position x of a visual stimulus and a gain function of the gaze direction y relative to heading. Motor neurons in the output layer combine basis functions with appropriate weights to compute $z = x + y$. Panel (a) is reproduced from Brotchie, P. R., Andersen, R. A., Snyder, L. H., & Goodman, S. J. (1995). Head position signals used by parietal neurons to encode locations of visual stimuli. Nature, 375(6528), 232–235; panel (b) is courtesy of Alex Pouget.

The normalization factor represents the sum of inputs to all neurons in the network. For instance, consider a V1 neuron with a preferred location (receptive field, RF) x and orientation θ of visual stimulus, that receives inputs from the primary visual thalamus LGN. According to the normalization model (**Figure 3.17a**):

$$r(x,\theta) = A_0 \frac{I(x,\theta)^n}{I_0^n + I_{sum}^n}; I_{sum}^n = \int I(x',\theta')^n dx'd\theta' \tag{3.41}$$

This model can explain a number of physiological observations of V1 neural responses. In response to an optimal stimulus with a neuron's preferred orientation and within its RF, the firing response increases with contrast c (which scales input intensity I_i) then saturates to r_{max} at high contrasts, because c is in both the numerator and denominator. The contrast level at which saturation occurs is controlled by the parameter I_0 (**Figure 3.17b**).

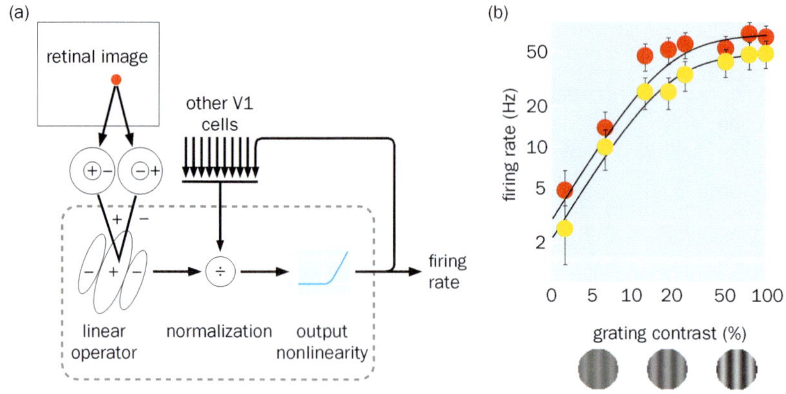

Figure 3.17 Normalization. (a) V1 model. The linear weighting of each Vl neuron is designed so that it responds selectively to intensity patterns of a particular orientation. The linear stage combines inputs from the lateral geniculate nucleus. The central excitatory subregion of the receptive field sums responses of ON-center LGN cells (which increase firing in response to a stimulus within RFs) and subtracts responses of OFF-center LGN cells (which decrease spiking activity from baseline in the presence of a stimulus in RFs) with spatially superimposed receptive fields. The flanking inhibitory subregions are obtained by the opposite arrangement of excitation and inhibition. The firing rates of all neurons in the neural population under consideration are summated to provided a divisive influence (normalization). (b) Contrast saturation. Responses as a function of the contrast for a grating stimulus having optimal orientation (red) and suboptimal orientation (yellow). Solid curves are fit of data (circles) using a normalization formula. Panel (a) reproduced from Heeger et al. (1996), (b) from Carandini and Heeger (2011). [Heeger, D. J., Simoncelli, E. P., & Movshon, J. A. (1996). Computational models of cortical visual processing. Proceedings of the National Academy of Sciences, 93(2), 623–627. Copyright (1996) National Academy of Sciences, U.S.A.] [Carandini, M., & Heeger, D. J. (2011). Normalization as a canonical neural computation. Nature Reviews Neuroscience, 13(1), 51–62.]

Moreover, V1 neurons display "surround suppression," namely, a stimulus outside of a neuron's RF reduces its response to a stimulus inside the RF, even though the surround stimulus alone does not solicit any response. This is due to an increased normalization factor in the denominator by the presence of the surround stimulus. Normalization has been found throughout the visual system, including the visual motion area MT (Heeger et al. 1996); in other words, it appears to hold from area to area along the visual hierarchy. It has also been observed in different sensory modalities and across species, as illustrated by the coding of smell by olfactory neurons in the fly (Olsen et al. 2010).

3.5.3 Recurrent Continuous Network Model

The phenomenological descriptions of multiplicative gain modulation and divisive normalization raise the question of how they are actually realized in neural circuits. Historically, an important clue came from studies of orientation selectivity in the primary visual cortex V1. As is well known, V1 neurons are tuned to the orientation of visual stimuli, and thus are capable of detecting edges in a visual scene. The underlying mechanism represents one of the most studied topics in systems neuroscience. Neurons in V1 are driven by LGN relay cells, which are selectively responsive to stimuli in a small circular part of the visual field. Given that LGN neurons have concentric RFs, how does orientation selectivity emerge in V1? The classical model of Hubel and Wiesel (1962) posits that outputs of LGN cells with neighboring RFs arranged along a particular orientation in space converge onto a neuron in V1, which thus responds maximally to an oriented stimulus that happens to be aligned spatially with the RFs of all its input LGN cells (**Figure 3.18a**). This feedforward wiring mechanism is important (Ferster and Miller 2000), even predominant in layer 4, where LGN inputs arrive in V1 (Reid and Alonso 1995). On the other hand, it is not sufficient to explain a salient property of orientation tuning of all V1 neurons. LGN neural activity increases with the contrast, so does the amount of synaptic excitation to V1 neurons. Therefore, an oriented bar with a low contrast may not produce sufficient drive for a V1 neuron to fire spikes, but with a high contrast could yield a larger excitation that exceeds the firing threshold. If so, the tuning curve of a V1 neuron is expected to widen with increased contrast. This, however, is inconsistent with the observation that the width of the orientation tuning curve does not change with the stimulus contrast (**Figure 3.18b**) (Sclar and Freeman 1982).

Ben-Yishai et al. (1995) proposed a recurrent continuous network model that offers a mechanism for contrast invariance of orientation tuning. Consider a network with neurons labeled by orientation as an analog quantity θ between 0° and 180°.

Their firing rates are described by the following equation:

$$\tau \frac{dr(\theta)}{dt} = -r(\theta) + F\left(I_{rec}(\theta) + I_{ext}(\theta) - I_{th}\right) \tag{3.42}$$

where the recurrent input current:

$$I_{rec}(\theta) = \int_0^{180} J(\theta - \theta') r(\theta') d\theta' \tag{3.43}$$

The connection weight as a function of the difference in the preferred orientations θ and θ' of a neuronal pair is often assumed to have a "Mexican hat" shape (**Figure 3.18c**, top panel), namely, neurons with similar preferred orientations excite each other, thus with positive $J(\theta - \theta')$, whereas effective interactions (through inhibitory neurons that are not explicitly included in this one-population model) are negative between neurons with dissimilar selectivity. Ben-Yishai et al. (1995) chose a simple form $J(\theta - \theta') = J_0 + J_2 \cos\left(2(\theta - \theta')\right)$, a periodic function reflecting the fact that the stimulus feature is an angle (between 0 and 180 degrees), and the network is a "ring model." A Mexican-hat connectivity is realized with a positive J_2 and a sufficiently negative value of J_0.

The external current represents the input from LGN with a degree of tuning quantified by a parameter ϵ:

$$I_{ext} = c\left(1 - \epsilon + \epsilon \cos\left(2(\theta - \theta_s)\right)\right) \tag{3.44}$$

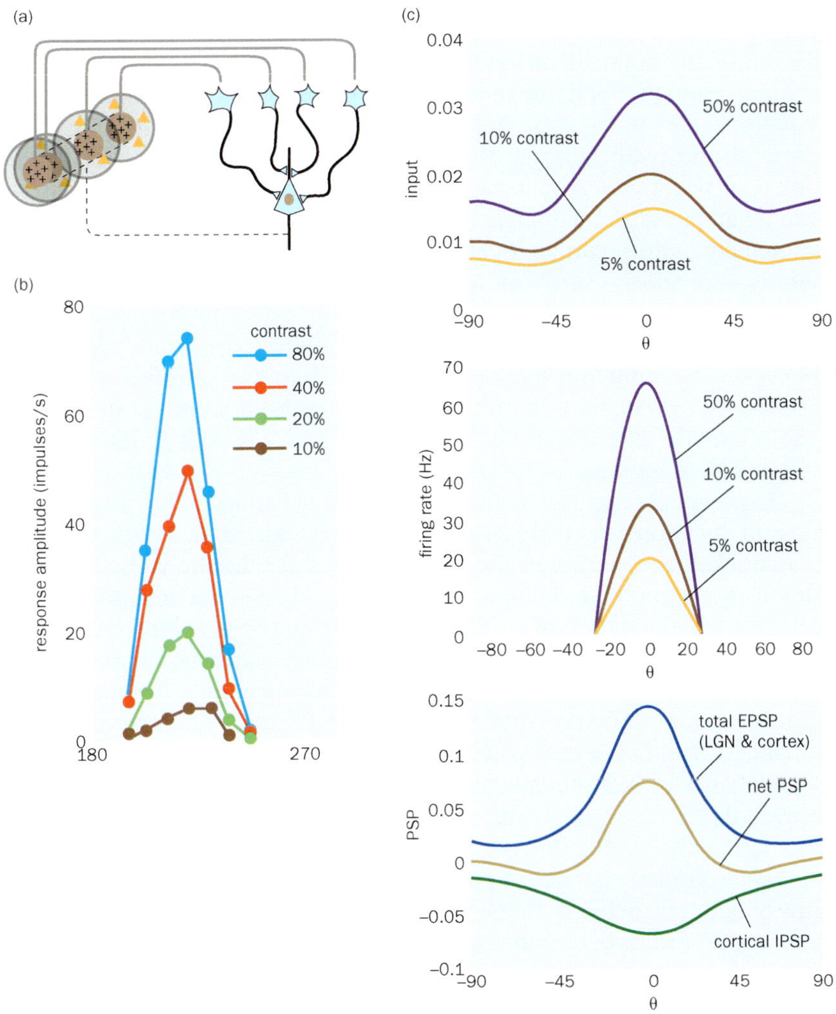

Figure 3.18 Ring model for orientation selectivity in V1. (a) The model of Hubel and Wiesel (1962) for how orientation selecitivity in V1 can be built from excitatory input from LGN relay cells (left) converging onto a V1 cell (right). (b) Orientation tuning and contrast dependence of a cell in cat V1, with four different contrasts. (c) Orientation tuning according to the cortical recurrent excitation model. Top panel: Calculated tuning curve of the peak response of the LGN afferents (relative to spontaneous levels) to a V1 cell. Middle panel: The tuning curve of the average discharge rate of a cell in response to the stimulus of upper panel, as calculated by the recurrent network model, with narrowly tuned excitatory interactions and broadly tuned inhibitory interactions. Bottom panel: Tuning of the calculated excitatory and inhibitory postsynaptic potentials (PSPs) as well as their sum, with 50% contrast. θ, stimulus orientation in degrees. Panel (a) is reproduced from Hubel, D. H., & Wiesel, T. N. (1962). Receptive fields, binocular interaction and functional architecture in the cat's visual cortex. Journal of Physiology, 160(1), panels (b)–(c) are reproduced from Sompolinsky, H., & Shapley, R. (1997). New perspectives on the mechanisms for orientation selectivity. Current Opinion in Neurobiology, 7(4), 514–522.

where c represents the contrast, and θ_S is the orientation of a stimulus. Thus, if ϵ is 1, the input is well tuned, whereas with $\epsilon = 0$ the input completely lacks any hint about orientation.

For a given stimulus θ_S, the network activity pattern is represented by $r(\theta,\theta_S)$ as a function of θ (neurons labeled by their preferred orientations). On the other hand, the tuning curve of a single neuron θ is obtained by varying the external stimulus θ_S and represented by $r(\theta,\theta_S)$ as a function of θ_S with fixed θ. In principle, the tuning curve of a neuron should be distinguished from the population activity pattern. However, in the special case of a homogeneous network model, activity only depends on the difference $\theta - \theta_S$, $r(\theta,\theta_S) = r(\theta - \theta_S)$, and the population activity pattern with a given θ_S is the same as the tuning curve of a single neuron labeled by its preferred θ.

Even though the external input to all neurons is proportional to the contrast, this model displays a tuning curve that is invariant with contrast. Namely, increased c yields higher firing rates of neurons whose preferred orientations are close to that of the stimulus, by virtue of their mutual excitation. On the other hand, more inhibition is also recruited (**Figure 3.18c**, bottom panel), so that neurons that are silent at low contrasts remain so at high contrasts, and the tuning curve width is unchanged (**Figure 3.18c**, middle panel). Hence, contrast amounts to a multiplicative gain for the Gaussian- or bell-shaped tuning curve. More recent research combined feedforward and recurrent ingredients in a more extended model of V1 (Chariker et al. 2016).

Interestingly, this model revealed a continuous family of bell-shaped network activity patterns even in the absence of any tuned input. **Figure 3.19** displays a "state diagram" of the model, on the two-dimensional parameter plane of J_0 and J_2. Three network behaviors are depicted in the absence of any tuning of an external input ($\epsilon = 0$). First, when both J_0 and J_2 are small enough, the homogeneous state corresponds to a uniform activity pattern ($r(\theta)$ is the same for all neurons). Second, if recurrent excitation is strong (with J_2 larger than 2), counter-balanced by a small (less than 1) or negative J_0, the network displays a bell-shaped activity pattern even when the external input is completely untuned. Moreover, the peak of this activity pattern can be located anywhere on the ring; in other words, there is a *continuous family* of tuned activity patterns, parametrized by the peak location. This is called a "marginal state," or a continuous attractor. Third, with sufficiently strong recurrent excitation (large positive J_2) that is unchecked by inhibition (upper right), the network is unstable, in which case the model ceases to be physiologically realistic. Contrast invariance of selectivity to external stimulus is realized typically with J_0 and J_2 in the marginal state regime.

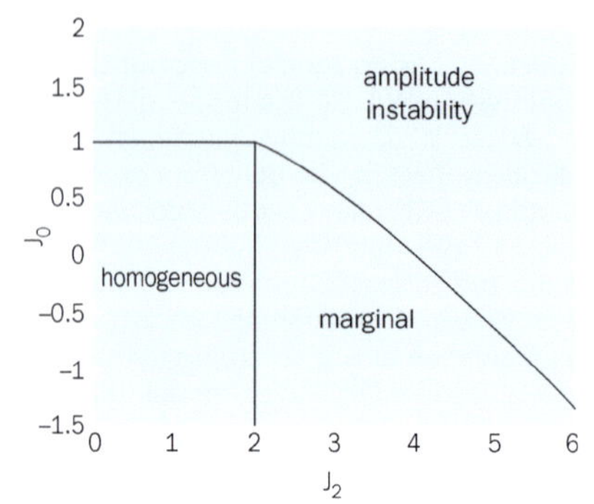

Figure 3.19 State diagram on the parameter space of J_0 and J_2 with no input tuning ($\epsilon = 0$). When J_2 is too large, the network is unstable (amplitude instability). On the other hand, when both are sufficiently small, all neurons uniformly fire at a low rate (homogeneous state). A combination of sufficiently strong recurrent excitation (large and positive J_2) and inhibition (large and negative J_0) produces a continuous family of bell-shaped activity patterns, even when the input is completely untuned (marginal states). Reproduced from Hansel and Sompolinsky (1998). [Koch, Christof, and Idan Segev, eds., Methods in Neuronal Modeling, Second Edition, Fig. 13.7 (p. 519), © 1998 Massachusetts Institute of Technology, by permission of The MIT Press.]

Experiments support a role of recurrent dynamics in V1 for orientation and direction selectivity (Hofer et al. 2011; Rossi et al. 2020), although V1 is unlikely to operate in the regime with a continuum of bell-shaped attractors (Carandini and Ringach 1997). After all, it is not desirable for V1 to have a representation of an "imagined" oriented object when it is absent in the external world. The concept of marginal states has had a broad impact in computational neuroscience. For instance, the aforementioned coordinate transformation model assumes basic functions as products of a gain factor G with a Gaussian-shaped tuning curve F. What may be a circuit mechanism for such multiplicative operation? Salinas and Abbott (1996) showed that a Ben-Yishai-type model can account for multiplicative gain modulation observed in PPC neurons.

Thus, graded changes of parameters (J_0 and J_2) can give rise to qualitatively different network behaviors (homogeneous state versus marginal state), as an example of "bifurcation" (Strogatz 2016) that takes place in a recurrent neural network. Notably, in the Ben-Yishai model with a fixed J_0 below 1, when J_2 is varied from left to right, the network displays either the homogeneous state (with J_2 smaller than 2) or the marginal state (with J_2 larger than 2), not both. By contrast, in Chapter 5, we will discuss models for working memory that require coexistence of a resting state and marginal state. Similar ring models have been developed with spiking neurons for V1 (Hansel and Sompolinsky 1998) and other neocortical and hippocampal networks.

Another population rate model realizes bell-shaped tuning by virtue of divisive normalization (Wu et al. 2008). Consider a recurrent network model for which the dynamics of synaptic input $I(x,t)$ at time t of a neuron with preferred stimulus x (such as spatial location) is described by:

$$\tau \frac{dI(x,t)}{dt} = -I(x,t) + \int J(x,x')r(x',t)dx' + I_{ext}(x,t) \tag{3.45}$$

where $J(x,x')$ is the connection weight, often chosen to be an exponentially decaying function of the difference between the preferred stimuli x and x', $J(x,x') = J_{max}\exp(-(x-x')^2 / 2\sigma^2)$, where J_{max} determines the maximal recurrent excitatory weight. Synaptic inputs evolve with a time constant τ, whereas the firing rate $r(x)$ is an instantaneous function of the inputs as:

$$r(x,t) = A_0 \frac{[I(x,t)]_+^2}{1 + k\int [I(x',t)]_+^2 dx'} \tag{3.46}$$

In the absence of external input, the model displays a continuous family of bell-shaped activity patterns (a marginal state), provided that there is sufficiently strong recurrent excitation relative to the strength of normalization k, when J_{max} is above a critical level that increases with k (more recurrent excitation is needed with stronger divisive normalization) (Wu et al. 2008). This is similar to Ben-Yishai et al. (1995), except that the effect of inhibition is captured by the parameter k of divisive normalization, rather than a subtractive term J_0.

How can divisive normalization be explained by synaptic inhibition? At the level of an isolated single neuron, consider the LIF model (Chapter 2, Eq. 3.1) with the addition of an inhibitory current, $G_I(V - V_I)$, where G_I is the inhibitory synaptic conductance and V_I its reversal potential. An inhibitory current mediated by $GABA_A$ receptors is called "shunting" because its reversal potential is close to that of a leaky current, $V_I = V_L$. When the input I is below the firing threshold, V has a stable steady-state given by $V_{ss} = V_L + I_{app} / (G_L + G_I)$, where I_{app} is an applied current and G_L is the leak conductance. Hence, the impact of a shunting inhibition on subthreshold membrane potential is *divisive*. However, when the input current I is above the firing threshold, the average membrane potential $\langle V \rangle$ during repetitive spiking is above V_I, and the inhibitory current $G_I(\langle V \rangle - V_I)$ is positive and becomes *subtractive* with respect to input current I (Holt and Koch 1997). Thus, feedforward "shunting" inhibition by itself cannot produce a divisive effect on spiking neurons.

However, normalization may arise as a collective network phenomenon. In a cortical microcircuit, inhibitory neurons receive converging inputs from excitatory neurons that are selective for different stimulus features (Harris and Mrsic-Flogel 2013). The assumption is that recruited inhibition is proportional to the integrated excitatory population activity and broadcast back to each individual excitatory neuron, leading to effective normalization of the network. This is illustrated by a biophysically realistic model of spiking neurons for MT (Ardid et al. 2007) (**Figure 3.20a**). In model stimulations, the input represents a "plaid pattern," which is a superimposition of two sine-wave gratings (**Figure 3.20b**) (Adelson and Movshon 1982). This was implemented in the model as a combination of two motion patterns with directions θ_1 and θ_2. The MT model's input reflects the outputs from neurons in V1, which display peaked responses at θ_1 and θ_2 that remain approximately unchanged when $\Delta\theta = \theta_1 - \theta_2$ is varied (**Figure 3.20c**) (Treue et al. 2000). By contrast, the peak response decreases with $\Delta\theta$ in MT neurons (Treue et al. 2000; Rust et al. 2006). This observation is reproduced by the model under the assumption that recurrent connections in MT are dominated by feedback inhibition (**Figure 3.20d**) (Ardid et al. 2007). Quantitatively, the dependence on $\Delta\theta$ is such that the summed activity (the integral of firing rates of MT cells) is roughly preserved, in line with the physiological finding from MT motion-sensitive cells.

This spiking network model demonstrates a circuit mechanism of normalization, through a collective inhibitory action that reflects the integrated activity across an excitatory neural population in the same circuit, under the condition that inhibitory neurons "pool" converging inputs from excitatory neurons.

3.6 COMPUTING WITH SPATIOTEMPORAL DYNAMICS

A bell-shaped bump attractor state can be guided by external inputs to move around in the neuronal tissue in order to accomplish neural computation. Moreover, instead of stationary activity patterns, a dynamic process, like single neuron adaptation or short-term synaptic plasticity (Chapter 2), can induce waves in a spatially extended neural network.

3.6.1 Time Integration

Neural integrators denote neurons or neural circuits capable of integrating information over time. We have seen previously an example of time integration where firing rate of a neuron varies linearly with the encoded feature (**Figure 3.8**). This "rate code" for time integration is called a "line attractor" in a network with N neurons (Seung 1996; Seung et al. 2000), exemplified by encoding an eye position in the oculomotor system (Aksay et al. 2001) or an affective state in the hypothalamus (Vinograd et al. 2024). Alternatively, in a "location code," neurons exhibit a bell-shaped Gaussian tuning curve of an encoded feature; therefore, distinct neural groups are engaged in storing different values of the analog feature. In other words, the stimulus feature is specified not by "how much active are neurons" but by "which neurons are active" in a location code. Mathematically, a perfect integrator is described by $dX / dt = I(t)$, where $I(t)$ is the input and X is the peak locus of network activity in a location code.

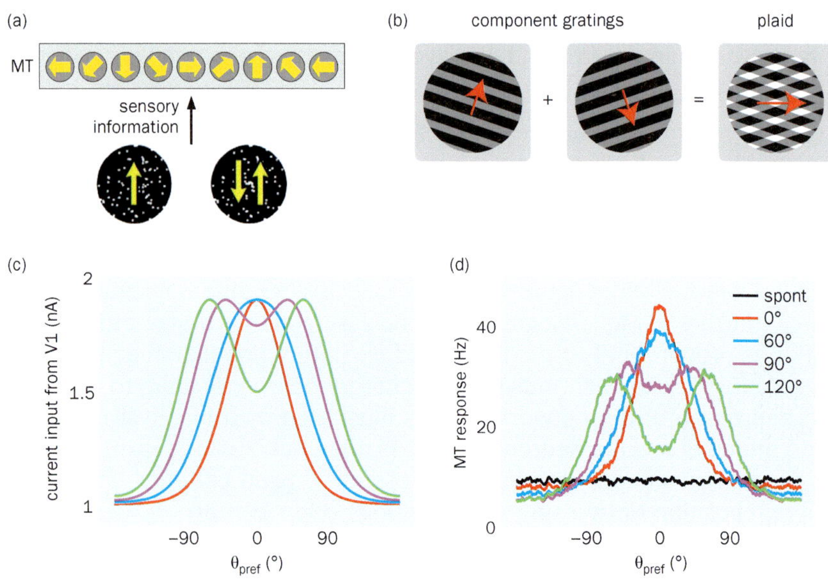

Figure 3.20 Normalization by recurrent synaptic inhibition in a model for MT, a cortical area dedicated to processing of visual motion stimuli. (a) Model scheme of a ring model where spiking neurons are selective for motion direction from 0 to 360 degress. (b) A plaid pattern of two superimposed moving gratings in the directions θ_1 and θ_2 separated by an angle $\Delta\theta$. (c) Input to MT either represents a single motion pattern (red) or two motion patterns separated by three different $\Delta\theta$ values (blue, purple, green). The peak response is constant for all four cases. (d) MT neural population activity shows a decreased peak response with increasing $\Delta\theta$, in such a way that the totality of spikes across the MT neural population remains roughly unchanged, as found experimentally. Panels (a) and (c) are reproduced from Ardid et al. (2007), (b) from Movshon et al. (2003). [Ardid, S., Wang, X. J., & Compte, A. (2007). An integrated microcircuit model of attentional processing in the neocortex. Journal of Neuroscience, 27(32), 8486–8495.] [Movshon, JA, Albright, TD, Stoner, GR, Majaj, NJ, Smith, MA (2003). Cortical responses to visual motion in alert and anesthetized monkeys. Nat Neurosci, 6, 1:3; p. 3.] [Reproduced from Ardid, S., Wang, X. J., & Compte, A. (2007). An integrated microcircuit model of attentional processing in the neocortex. Journal of Neuroscience, 27(32), 8486–8495.]

Marginal states of a ring model (Ben-Yishai et al. 1995) offers a mechanism for integration in a location code. An excellent example is the head direction (HD) system (Taube et al. 1990; Kim et al. 2019; Munn and Giocomo 2020). HD neurons are part of the spatial navigation system and signal the animal's directional heading. When an animal turns its head, an angular velocity signal carried by vestibular inputs is integrated by HD cells into a positional signal, and the latter is sustained internally when the animal keeps the head direction fixed. HD cells are selective for angular head direction according to a bell-shaped tuning curve. Interestingly, available evidence indicates that HD cells are generated in a neural circuit characterized by a paucity of local excitatory connections (Taube and Bassett 2003). How can marginal states be realized without recurrent excitation?

Figure 3.21 shows a neural integrator model in which the encoded feature is specified by the peak location of the network activity profile. Consistent with the empirical data (Taube and Bassett 2003), the model assumes that a uniform external excitation drives all neurons, which interact with each other through inhibition. This cross-inhibition mechanism without recurrent excitation gives rise to a continuous family of bell-shaped activity patterns, or direction-selective "hill of persistent activity" (Skaggs et al. 1995; Zhang 1996).

Moreover, the model surmises a shift mechanism by "rotation cells" consisting of two inhibitory cell populations (**Figure 3.21a**) (Skaggs et al. 1995; Xie et al. 2002; Song and Wang 2005). When the head direction is fixed, the inputs from the two inhibitory neural populations are balanced with each other, and the bell-shaped activity of the excitatory neural population is maintained fixed (**Figure 3.21b**, time epochs between input pulses). As the head turns, the angular velocity signal increases the firing of one inhibitory population, while it decreases the firing of the other. The resulting asymmetric inhibitory inputs induce the activity pattern in the excitatory neural population to move, at a constant speed proportional to the input amplitude (angular velocity) (**Figure 3.21b**, time epochs during input pulses). Computation performed by the model is quite close to an integral operation in the sense of Calculus. When the input intensity is doubled (second versus first input pulse), the hill of activity moves twice as fast. Moreover, the network can integrate both positive inputs (first and second pulses) and negative inputs (third pulse). In fact, with a third input half in amplitude but lasting twice as long as the second input, the hill of activity moves back to the position before the second pulse, as expected by perfect integration.

3.6.2 Spatial Navigation

The HD system provides heading information to a spatial navigation circuit in the entorhinal cortex (EC) and hippocampus (HP) (**Figure 3.22a**) (McNaughton et al. 2006; Moser et al. 2014). At any moment, a directional signal and a movement velocity signal are combined to update positional encoding of where an animal is in space. This integration process, in the absence of landmarks, is called "dead-reckoning" (Gallistel 1990). For instance, desert ants take risk to forage food in extremely hot weather, and thus must be able to stay outside as little time as possible. An ant starts from home and moves in a zigzag manner while searching for food; once food is harvested, it returns from the food location back to home in a straight line, which is mathematically equivalent to vector integration. The circuit mechanism of vector computation underlying spatial navigation has begun to be elucidated recently in small animals (Lyu et al. 2022). In mammals this dead-reckoning ability depends on a spatial navigation system in the EC-HP circuit to integrate velocity and heading information into spatial location. "Place cells" in HP are selective for the spatial location

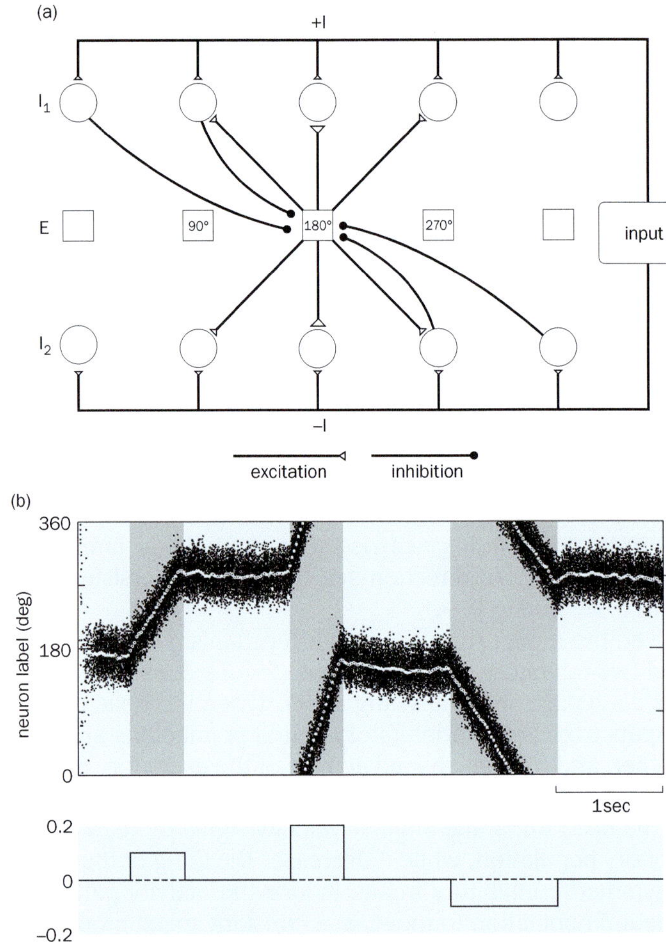

Figure 3.21 Time integration by a "hill of persistent activity." (a) Model scheme. An excitatory neural network, encoding a directional angle (0 to 360 degrees), receives inputs from two inhibitory neural populations that are balanced with each other when there is no input. A velocity signal increases input to one inhibitory population (+*I*) and decreases input to the other (–*I*), leading to a bias for the excitatory network. (b) Network firing pattern (Top) in response to a series of input signal steps (Bottom). Neurons in the network along the y-axis are labelled by their preferred directional angles, x-axis is time, each dot is a spike. The network activity pattern has the form of a bell-shaped profile, with its peak location encoding the directional angle (white line in the rastergram). In the absence of an external input, the directional information is maintained by a persistent firing pattern. An input induces the hill of activity to move at a speed proportional to the stimulus amplitude; hence, the network performs a time integral of the input. Reproduced from Goldman et al. (2009) using the model of Song and Wang (2005). [Goldman, M., A. Compte, and X.-J. Wang (2009). Neural integrators: Recurrent mechanisms and models. In L. Squire, T. Albright, F. Bloom, F. Gage, and N. Spitzer (Eds.), Encyclopedia of Neuroscience, pp. 165–178. MacMillan Reference Ltd.]

with a two-dimensional bell-shaped tuning curve of 2D space. Remarkably, upstream from HP, neurons in the superficial layer II of medial EC (MEC) display high activity selectivity at a number of spatial locations that form a regular hexagonal spatial grid (**Figure 3.22b**) (Hafting et al. 2005; Moser et al. 2014).

A continuous attractor model accounts for the formation of grid cells in medial EC as well as path integration (Fuhs and Touretzky 2006; McNaughton et al. 2006; Burak and Fiete 2009; Couey et al. 2013). A ring model can be extended to two dimensions (2D) with periodic boundary condition for each dimension, representing the geometric shape of a torus (like a doughnut). In a 2D network model (Burak and Fiete 2009), each neuron i is labeled by its preferred spatial location $\mathbf{x}_i$ and direction θ_i (for the sake of simplicity, four directions are included in the model: $\mathbf{e}_{\theta_i}$ = north, east, south and west). Connectivity from neuron j to neuron i depends on the distance between their preferred positions $\mathbf{x}_i - \mathbf{x}_j$, shifted by a small vector in the preferred direction $\mathbf{e}_{\theta_j}$ of the presynaptic neuron j. The lateral connectivity as a function of $\mathbf{x}_i - \mathbf{x}_j - l\mathbf{e}_{\theta_j}$, where l is the magnitude of the directional dependence, can be purely inhibitory (**Figure 3.22c**, left panel). This assumption is consistent with the experimental finding that stellar cells, the majority of excitatory neurons in layer II of medial EC and main source of inputs to the hippocampus, are devoid of excitatory-to-excitatory connections (Couey et al. 2013). The direction-dependent connections are balanced so that an activity pattern is stationary in the absence of an external input.

A velocity input $\mathbf{v}$, when present, is applied to all the network, with a strength proportional to the dot product $\mathbf{e}_{\theta_i} \cdot \mathbf{v}$ for each neuron i. For instance,

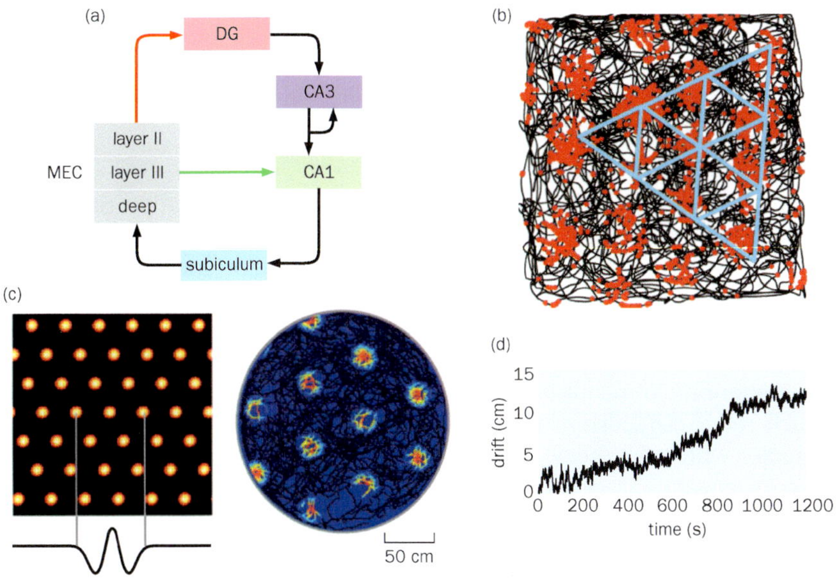

Figure 3.22 Two-dimensional continuous attractor model of grid cells in medial entorhinal cortex. (a) The grid-cell circuit in the MEC, which provides input to the hippocampus (HP), with layer II projecting to the dentate gyrus (DG) and layer III projecting to CA1. Within the HP, DG projects to CA3, which in turn projects to CA1, and CA3 has prominent recurrent connections. CA1 provides output, via the subiculum, to entorhinal cortex deep layer V. (b) Firing-rate map of a grid cell. This firing pattern does not merely reflect stimulus configurations in the external world but largely represents pattern formation processes within the entorhinal local circuit. (c) A model of grid cells. Left panel: Spatial profile of inhibitory network connectivity with the peak equal to zero. Right panel: Grid cells arise from the recurrent inhibitory circuit, with the characteristic spacing determined by the spatial footprint of recurrent inhibition. (d) In the absence of external input, the encoded spatial location drifts over time. Panel (b) is reproduced Moser, E. I., Roudi, Y., Witter, M. P., Kentros, C., Bonhoeffer, T., & Moser, M. B. (2014). Grid cells and cortical representation. Nature Reviews Neuroscience, 15(7), 466 Panels (c)–(d) are reproduced from Burak Y, Fiete IR (2009) Accurate Path Integration in Continuous Attractor Network Models of Grid Cells. PLoS Comput Biol 5(2): E1000291. https://doi.org/10.1371/journal.pcbi.1000291. Licensed under CC BY 4.0.

if velocity is in the direction Northwest, neurons with North and West as their preferred directions receive a positive input, whereas neurons that prefer South or East receive a negative input, leading the population activity pattern to move northwest bound. With a time varying $\mathbf{v}(t)$, the network's moving trajectory represents a time integral of the input, in a manner similar to **Figure 3.21**, but now as a vector integral $\int_0^t \mathbf{v}(t')\mathbf{d}t'$.

Burak and Fiete (2009) used experimentally measured trajectory $\mathbf{x}_{rat}(t)$ of a navigating rat to test the model. They computed the time derivative $\mathbf{x}_{rat}(t)$ and used it as a velocity signal $\mathbf{v}(t)$ to drive the network in a model simulation. Activity of a model neuron computed from its time course over a long trajectory lasting 20 minutes (**Figure 3.22c**, right panel) captures the triangular lattice of grid cell activity observed experimentally (**Figure 3.22b**). The period of the grid cell pattern is determined by the spatial footprint of lateral connectivity, as one would expect intuitively. The difference between the accumulated spatial change of position coded by population activity and rat's actual position (**Figure 3.22d**) shows that the total error accumulated in a trajectory of 260 meters is quite small, less than 15 cm.

How can an attractor network with recurrent inhibitory connections learn to navigate in a novel environment? In real life, animals employ landmarks: when you walk to an unfamiliar destination (without GPS), it helps if you recognize particular buildings on the way there. Campbell et al. (2018) added a sensory-driven landmark input to a continuous attractor network model, with its strength shaped by synaptic plasticity. It was found that Hebbian learning (Chapter 4) of synapses on the landmark cell enables the network to build a new spatial map in a novel environment. The model displays systematic shifts of firing locations of grid cells toward the most recently encountered landmark, a prediction that can be tested experimentally.

3.6.3 Propagating Waves

Spatial navigation illustrates how a brain function is accomplished by a neural representation updated in response to an external input. A network activity pattern can also move by itself in the absence of a stimulus, often in the form of traveling waves (Ermentrout and Kleinfeld 2001; Wang 2010; Muller et al. 2016; Kulkarni et al. 2020; Davis et al. 2020). For instance, in the model of Ben-Yishai et al. (1995), one can incorporate spike rate adaptation (SRA) of single cells (Chapter 2) in a simple manner. In Eq. (3.42) the total input now includes an adaptation current $-I_a(\theta)$, which obeys:

$$\tau_a \frac{dI_a(\theta)}{dt} = J_a r(\theta) - I_a(\theta) \tag{3.47}$$

where J_a controls the strength of SRA and τ_a is an adaptation time constant. As a result of SRA, neurons at the peak of a bump attractor become less excitable over time. In that case, a bump attractor does not disappear over time; instead, it moves with a preserved spatial profile, smoothly away from the initial peak location and throughout the network as a "traveling wave," at a constant velocity (Hansel and Sompolinsky 1998). The same phenomenon can result from short-term synaptic depression (Chapter 2), which reduces recurrent excitation, especially at the peak of the bump attractor where the presynaptic firing rates are the highest.

Propagating waves can occur without neuronal adaptation (Ermentrout 1998; Ermentrout and Kleinfeld 2001; Muller et al. 2016). In a network of excitatory and inhibitory neural populations, waves occur when, compared to excitation, inhibition is slower (destabilizing steady-state behavior, Section 3.4.2) and has a narrower or comparable spatial footprint (Huang et al. 2019). In contrast, the broader footprint of inhibition combined with narrower excitation is suitable for generating a Mexican-hat connectivity underlying stationary bell-shaped patterns (Sections 3.5.3 and 3.6.2).

Phase waves should be distinguished from a wavefront like wildfire in the prairie, in which activity is recruited successfully in time, sweeping through an initially silent network (Wang 2010). In a phase wave, neurons fire continuously and rhythmically, and the oscillation phase φ varies as a function of time and spatial location (Ermentrout and Kleinfeld 2001). For a plane wave, the phase is linear with the spatial position x and time t, like $\phi = 2\pi\left(ft - \left(f/v\right)x\right)$, where f is the oscillation frequency, v is the propagation velocity and $\lambda = v/f$ is the spatial wavelength. In two- or three-dimensional nerve tissues, more complex waves such as spiral wave can take place (Milton et al. 1993; Huang et al. 2004; Kilpatrick and Bressloff 2009). Traveling waves have been observed in a number of studies (Ermentrout and Kleinfeld 2001; Wang 2010; Bressloff 2012). Examples include MEG gamma oscillations (Ribary et al. 1991), EEG slow sleep oscillations (Massimini et al. 2004; Muller et al. 2016), evoked responses in the cortex (Arieli et al. 1995; Prechtl et al. 1997; Freeman and Barrie 2000) and hippocampal theta oscillations (Lubenov and Siapas 2009). Interestingly, the total phase shift is often $\sim \pi/2$, never larger than 2π. In other words, the observed spatial variation in peak amplitude is less than the full oscillation cycle, regardless of oscillation frequency (e.g., theta or gamma), or the physical size of the neural system or species examined (Ermentrout and Kleinfeld 2001). Modeling studies showed that a phase wave depends on the spatial extent and connectivity delay (including the axonal delay and synaptic latency). In the same model, a gradual increase in the connection delay can lead to an abrupt transition from global synchronous oscillation (when delays are short) to phase waves (when delays are long) (Crook et al. 1997).

Traveling waves are a common dynamical mode of spontaneous neural network activity (Muller et al. 2018). If so, an external stimulus is not the sole drive for activity but should be better understood as modulation of an intrinsic activity pattern (Tsodyks et al. 1999; Muller et al. 2018; Chen and Gong 2019; Davis et al. 2020). Can information representation be preserved in an activity pattern moving through a large population of neurons? Put another way, in response to a stimulus, does a wave propagate only among those cells with similar stimulus selectivity? Functional advantages and limitations of propagating waves remain to be better assessed.

3.7 RESERVOIR COMPUTING

The model of grid cells exemplifies how a recurrent network can carry out computation (path integration) underlying behavior (spatial navigation). An attractor network can continuously and swiftly change its activity pattern in the face of a stream of stochastic input (a velocity signal). On the other hand, in the absence of a time-varying input, such simple network models are not suitable for internally produced complex dynamical behavior. For instance, how does intrinsic neural population dynamics in a motor system generate a trajectory of arm movement to throw a baseball bat or sing a folklore song?

What kind of neural networks are capable of sophisticated temporal processing? One possibility is time-varying neural population attractors. Alternatively, neural computation may be carried out by *transients*, which are mathematically described as trajectories of a dynamical system outside of an attractor state. For instance, when a network has two or more stable steady-states, there must exist unstable steady-states as well. A particularly interesting type of unstable steady-states is called a "saddle." Like a saddle on a horse (**Figure 3.23a**), a saddle point is attractive in certain direction called a stable manifold, along which a trajectory converges towards it, but trajectories escape from it along another direction called an unstable manifold. Near a saddle point, dynamics slows down (Strogatz 2016), offering a mechanism for slow transients that potentially may be harnessed computationally. Indeed, in insect antennal lobe (akin to the olfactory bulb of mammals), stimuli give rise to odor- and concentration-specific spatiotemporal patterns of activity in projection neurons. Hence, odor representations appear to be realized by a successions of states, corresponding to a sequence of segments of a trajectory of neural population activity (**Figure 3.23b–c**) (Mazor and Laurent 2005; Rabinovich et al. 2008).

Models of "echo state networks" or "liquid state networks" have been proposed that emphasize slow transients (Maass et al. 2002; Jaeger et al. 2007; Lukosevicius and Jaeger 2009) in recurrent neural networks (RNNs) with random connections. These models share the characteristics of random wiring of network connections and high-dimensional dynamics, naturally giving rise to heterogeneous neural responses observed in the brain. By virtue of a wide variety of neural activity time courses, outputs from such a network (a reservoir) can be readout in different ways to accomplish a multitude of computational goals (reservoir computing).

3.7.1 State Space, Dimensionality and Manifolds

With heterogeneous activities of single neurons, coding of information at the population level can be depicted in the state space. Consider N neurons with firing rates $r_i(t,\mathbf{s}_j)$ ($i = 1,2,\ldots,N$), where $\mathbf{s}_j$ denotes stimulus, task and motor features in an experimental trial $j = 1,2,\ldots,M$. For a single trial (with fixed $\mathbf{s}_j$), in addition to the temporal course of each neuron's activity (r_i as a function of t), one can consider an N-dimensional state space in which each axis corresponds to a neuron's firing rate (**Figure 3.24a** with $N = 3$). At any time point, the state of the entire population of neurons is represented by a point in the state space, and the population dynamics is described by a trajectory parametrized by time t. Trajectories in the state space provide a general tool for the analysis of neural population coding and processing of information (**Figure 3.23c**).

The number of neurons involved in a brain function or a task is large; with new recording techniques such as Neuropixels probe (Jun et al. 2017), today it is possible to record tens of thousands of single neurons simultaneously. However, trajectories in the N-dimensional state space may be restricted to a subspace with lower dimension. In the example of

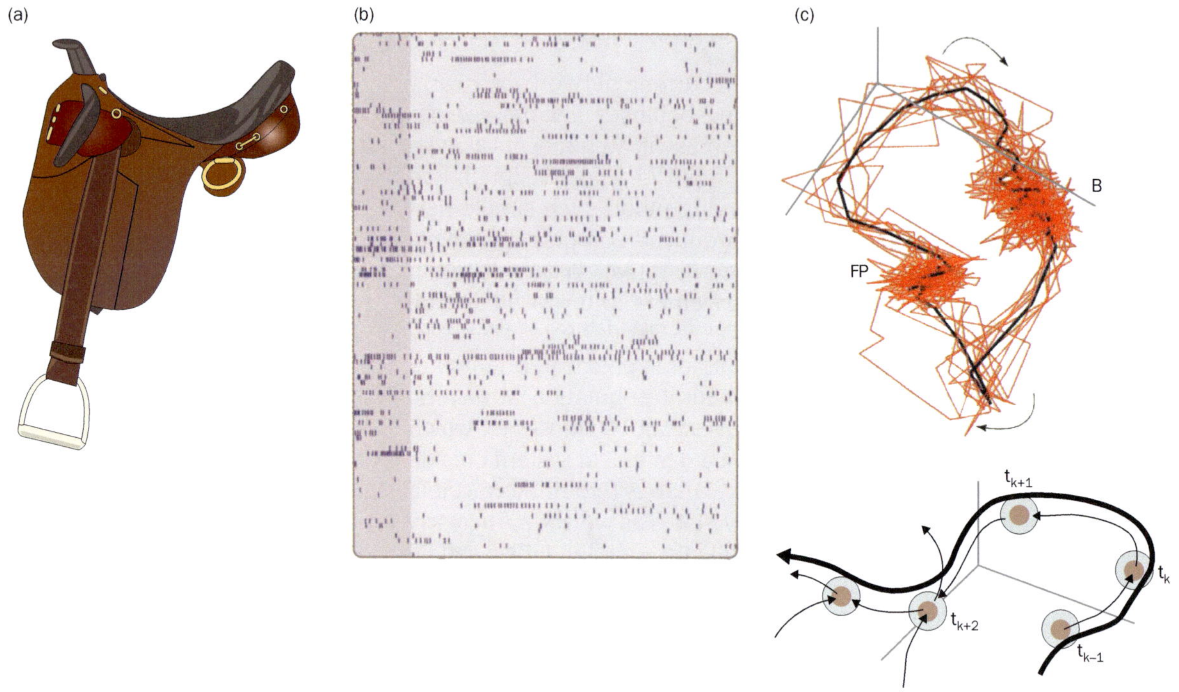

Figure 3.23 Computing by transients. (a) Schematic of a horse saddle, illustrating the mathematical concept of a saddle-type stationary state (fixed point). (b) Single-trial responses of 110 locust antennal-lobe principal neurons to one odor. (c) Upper panel: Projections of neural population trajectories on the plane, representing the succession of states visited by this neural network in response to one odor. Red lines, individual trials; black line, average of 10 trials. B, baseline state; FP, fixed point, reached after 1.5 seconds. Lower panel: Putative dynamical model of transients: A set of saddles (brown circle), sequentially connected by separatrices (thin lines). A single trajectory (thick line) connects the neighborhoods of saddles. Panel (b) is reproduced from Mazor and Laurent (2005), (c) from Rabinovich et al. (2008). [Courtesy of Gilles Laurent.]

Figure 3.24a, a trajectory in a three-dimensional state space is actually confined in a two-dimensional subspace of latent variables s_1 and s_2. In this example, the latent variables and the neural activity variables are linearly related to each other. Generally, the dimensionality of a neural representation depends on task complexity such as the number of parameters that determine the space of stimuli and motor responses (Gao and Ganguli 2015): when the behavioral attributes are low dimensional, e.g., with only two stimuli and two possible responses, the neural representation cannot be high dimensional. In a typical laboratory experiment, behavioral dimensionality is much smaller than the number of recorded neurons. This can be assessed using various methods, such as "demixed" principal component analysis (PCA) (Machens et al. 2010; Keemink and Machens 2019) and tensor component analysis (Williams et al. 2018). These methods find components that correspond to task parameters while accounting for the majority of variance of neural activities that display complex dependence on task parameters. In Figure 3.24a, the latent variable s_1 ramps up linearly, and thus is suitable for representing elapsed time, while s_2 appears to encode a sinusoidal signal.

Generally, it has been proposed that low-rank recurrent network models (with connection matrices restricted to be low rank) embody low-dimensional latent variables underlying collective dynamics and information coding of a neural population (Mastrogiuseppe and Ostojic 2018). The rank of a connection matrix is defined in the theory of linear algebra as "the dimension of the row or column of a matrix" (Larson and Falvo 2009). For instance, if a connection matrix J_{ij} is made of two vectors m_i and n_j, as $J_{ij} = m_i n_j$, then the rank is one because every column in this matrix is a multiple of the vector $\mathbf{m}$, and every row is a multiple of the vector $\mathbf{n}$.

A rank two matrix can be made of a sum of two **m** and **n** vectors, like $J_{ij} = m_i^1 n_j^1 + m_i^2 n_j^2$, etc. Since numerous tasks used in neurophysiological experiments are low-dimensional, the underlying neural latent space can potentially be well captured by a model with a low-rank connectivity (Mastrogiuseppe and Ostojic 2018; Dubreuil et al. 2022).

Such state-space analysis also applies to static data. As a point in the state space, one can plot the firing rates of N recorded neurons, averaged in a behavioral epoch such as stimulus presentation or over an entire trial. The population state varies with $\mathbf{s}_j$ as well as across trials given a fixed $\mathbf{s}_j$; the "cloud" of data points could be analyzed for dimensionality reduction. Importantly, data may be confined in a low-dimensional nonlinear manifold rather than a linear subspace, because manifold discovery requires the development of novel mathematical techniques (Gallego et al. 2017; Low et al. 2018). In a head direction system discussed in Section 3.6.1, HD neural population is expected to represent a nonlinear ring structure, parametrized by a scalar α that can be mapped on the head direction angle. To test this hypothesis, a multiple-step recipe was developed to discover robust topology of a manifold in noisy neural data recorded from neurons of the anterodorsal thalamic nucleus in awake mice (Peyrache et al. 2015). Indeed, a one-dimensional closed loop can be identified in population state data (**Figure 3.24b**), and single neural firing is tuned to the scalar parameterization of the ring manifold the same way as to the animal's head direction (**Figure 3.24c**). Interestingly, the manifold does not occupy a very low-dimensional linear subspace, and linear PCA analysis often fails

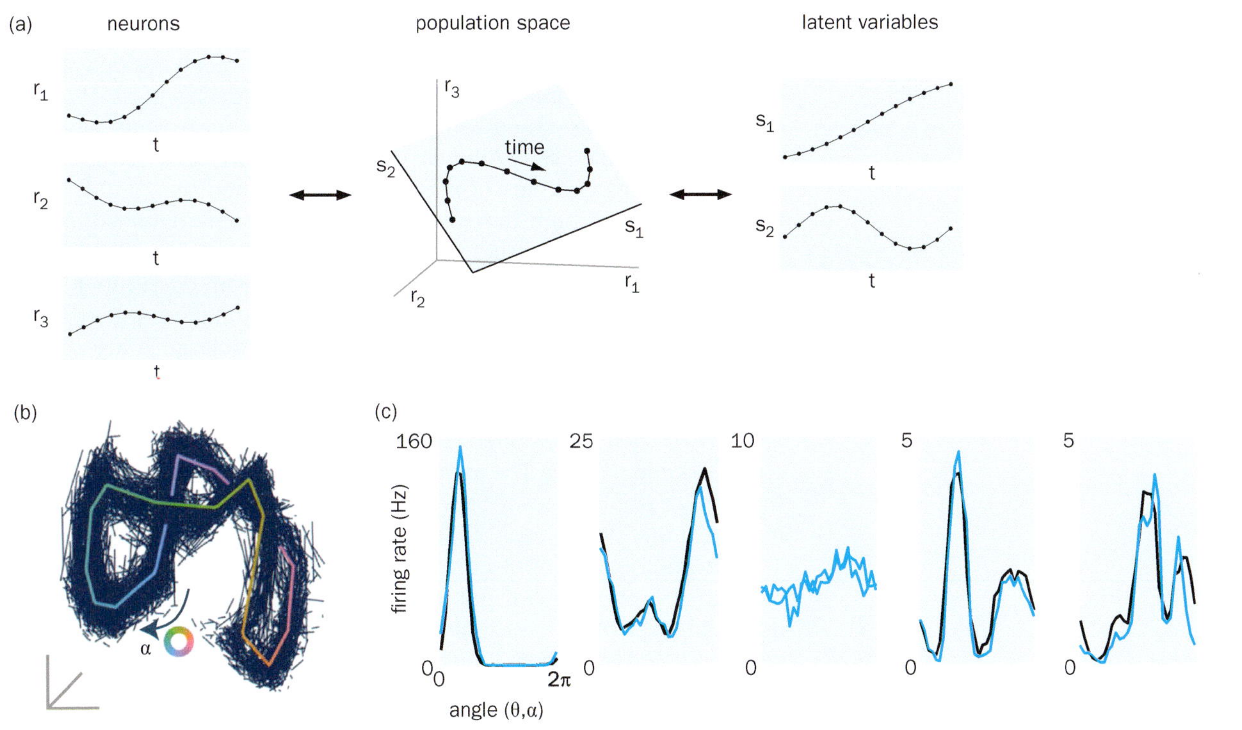

Figure 3.24 Neural representation manifolds and dimensionality reduction. (a) Conceptualization of subspace illustrated with three neurons and two latent variables. Center, the population activity (black points) lies in a plane with coordinates $[s_1, s_2]$ (shaded grey). The points trace out a trajectory over time (black curve). Left, the population activity r_1, r_2 and r_3 can be reconstructed by a combination of the latent variables with weights specified by the matrix shown. Right, the latent variables s_1 and s_2 can be obtained by a combination of the population activity with weights given by a matrix shown. (b–c) Nonlinear manifold of head direction representation. (b) Visualization of a manifold formed by neurons recorded in a behaving mouse. Colored curve is the parameterization (denoted by α) of a spline fit to the mathematically identified manifold. (c) Firing rate for five sample neurons as a function of the parametrization α (blue) of the nonlinear manifold in (b) and the experimentally measured tuning curve as a function of animal's head direction θ (black). Panel (a) is reproduced from Cunningham, J. P., & Yu, B. M. (2014). Dimensionality reduction for large-scale neural recordings. Nature Neuroscience, 17(11), 1500–1509; panels (b)–(c) are reproduced from Chaudhuri, R., Gerçek, B., Pandey, B., Peyrache, A., & Fiete, I. (2019). The intrinsic attractor manifold and population dynamics of a canonical cognitive circuit across waking and sleep. Nature Neuroscience, 22(9), 1512–1520.

to extract a ring structure demonstrating that the neural manifold is highly nonlinear (Chaudhuri et al. 2019; De and Chaudhuri 2023). The number of parameters to describe a neural representation is called the *intrinsic dimension* (here equal to one for the HD system), the *embedding dimension* of the neural manifold is typically larger than the intrinsic dimension but much smaller than N (Jazayeri and Ostojic 2021).

Computational implications of subspaces or low-dimensional manifolds are diverse. For instance, to understand population coding in a sensory system, one can differentially define a signal subspace by the set of trial-average population states as the sensory stimulus is varied, and a noise subspace, which describes the set of states visited by the population as a result of trial-to-trial variability for any given fixed stimulus (Kobak et al. 2019). Communication from a brain region to another may involve a distinct subspace of neural population activity (Semedo et al. 2019; Kohn et al. 2020; Semedo et al. 2022). Among open questions is the extent to which different subspaces are populated by distinct cell types and how signals from different subspaces are selectively readout by downstream systems. Modern genetic tools enable researchers to label distinct cell types in order to address this functional question.

3.7.2 Feedforward Random Networks

Unlike cerebral cortex in which a neuron receives thousands of inputs, other biological neural networks are characterized by sparse random connections. For instance, in an insect olfactory system, a Kenyon cell of the mushroom body randomly samples inputs from about seven projection neurons in the antennal lobe (Caron et al. 2013). In the mammalian cerebellum, inputs are coded by granule cells, each receiving only about four of the mossy fibers innervating the cerebellum (Eccles et al. 1966). These connections are assumed to be random in the Marr-Albus model of the cerebellum (Marr 1969; Albus 1971). Such sparse and random connectivity minimizes overlaps of inputs onto target neurons, and is suited for the realization of high-dimensional representations (Litwin-Kumar et al. 2017): if there is no overlap in the randomly sampled inputs, firing activities of M granule cells would be independent from each other, and their stimulus representation occupies a space of neural firing rates with a high dimensionality of M. On the other hand, if granule cells are extensively correlated due to shared inputs, their firing patterns would be confined in a small subspace with a low dimensionality. In rat cerebellum, there are about $N = 7{,}000$ mossy fibers and $M = 200{,}000$ granule cells presynaptic to each Purkinje neuron. The expansion ratio $M / N \simeq 30$. If an average $K = 4$ randomly sampled mossy fiber inputs project to a granule cell, subsets of inputs for any two granule cells are distinct with near certainty.

In addition to input connectivity, the dimensionality of representations in the granule cell population also depends on their input-output nonlinearity, which determines the fraction (the coding level f) of granule cells activated by a given input. With $f = 0.1$, the optimal dimensionality of representation in the granule neural population is $\simeq 140{,}000$, achieved with the measured $K \simeq 4$ (Litwin-Kumar et al. 2017). Such a random projection into a high-dimensional space results in separation of encoded stimuli. If stimuli are separated, readout for discrimination between them is facilitated (Ganguli and Sompolinsky 2012; Babadi and Sompolinsky 2014).

A high degree of heterogeneity typically is not taken into consideration by simple structured models, which are purposefully designed to strive for simplicity. Consider coding of movements in the primary motor cortex M1. In a classic experiment, when monkeys manually reached from a central position to one of eight target positions, single neurons in M1 were found to display spiking activity tuned to target direction (Georgopoulos et al. 1982).

The tuning curve was fit with a cosine function: $r = r_0 + r_{max}\cos(\theta - \theta_m)$, which is maximal when the moving direction θ_m is the neuron's preferred direction θ. This finding led to the idea that the motor cortex encodes target features such as direction rather than parameters of muscles at shoulder and elbow joints (Georgopoulos et al. 1986). A later study found that activities of M1 neurons are widely heterogeneous, and most of them cannot be fit well with a simple tuning curve like a cosine function but can be accounted for by a purely feedforward random network model (Lalazar et al. 2016). Moreover, a readout from M1 with trained connection weights could decode the electromyography (EMG); the heterogeneous and nonlinear response component was found to improve decoding the EMG of muscles during arm posture control.

3.7.3 Recurrent Random Networks

A feedforward model cannot describe temporal dynamics of neural activity during voluntary movement (Shenoy et al. 2011; Churchland et al. 2012). Compared to reflexive movements such as withdrawing a hand from fire, slower but more flexible voluntary movements can be planned and refined, ahead of an actual movement, to achieve a purposeful goal. An accurate mathematical description of voluntary movements is important for brain-machine interface (BMI) systems, where neural signals in the cortex are used to control a robotic arm or a computer cursor (Donoghue 2008; Andersen et al. 2010).

Preparation and execution of movements can be studied by introducing a temporal delay between the onset of a target and the go signal, akin to the time window between "get set" and "go" at the start of a 100-meter race. During the delay, the brain prepares for the impending movement; indeed, the behavioral response time between go cue and movement initiation is shorter after a longer delay, presumably thanks to more time for preparation. Interestingly, in a delayed reach task, neurons in the dorsal premotor cortex (PMd) of monkeys display time-varying activities during both delay period and movement execution (**Figure 3.25a**) (Shenoy et al. 2011; Churchland et al. 2012). The observation of heterogeneous and temporally variable motor cortical activity during preparation motivated researchers to carry out state-space analysis of neural population activity rather than single neuron tuning. Consider the time course of the firing rates $r_i(t)$ for N neurons $(i = 1, 2, \ldots, N)$ recorded in a delayed reach task. In N-dimensional state space, the system evolves over time along a trajectory parametrized by t. Kaufman et al. (2014) used a linear relationship between population activity $\mathbf{r}(t)$ and EMG (a much lower dimensional vector) $\mathbf{m}(t) = W\mathbf{r}(t)$, where W is the matrix for the linear mapping. This mapping enabled the researchers to mathematically identify an "output-null" subspace of the neural population state space, defined as the subspace of $\mathbf{r}$ such that $W\mathbf{r} = 0$. Confined in the null-subspace of the matrix W, $\mathbf{r}(\mathbf{t})$ may display considerable activity patterns, but $\mathbf{m}(t) \sim 0$ and the motor cortical activity produces virtually no EMG signal. Remarkably, neural transients in the delay period were found to be restricted within that null subspace. Therefore, neural population activity may undergo complex temporal dynamics during movement preparation without triggering an actual movement (Kaufman et al. 2014; Elsayed et al. 2016). Presumably, preparatory activity moves the system to an appropriate initial state in order to ultimately produce a desired movement (reach left or right). Notably, such insights require a consideration of network dynamics beyond single neuron analysis.

These observations inspired a recurrent network model capable of producing complex transients (Hennequin et al. 2014) (**Figure 3.25b**). The key idea is multi-dimensional extension of the inhibition stabilization

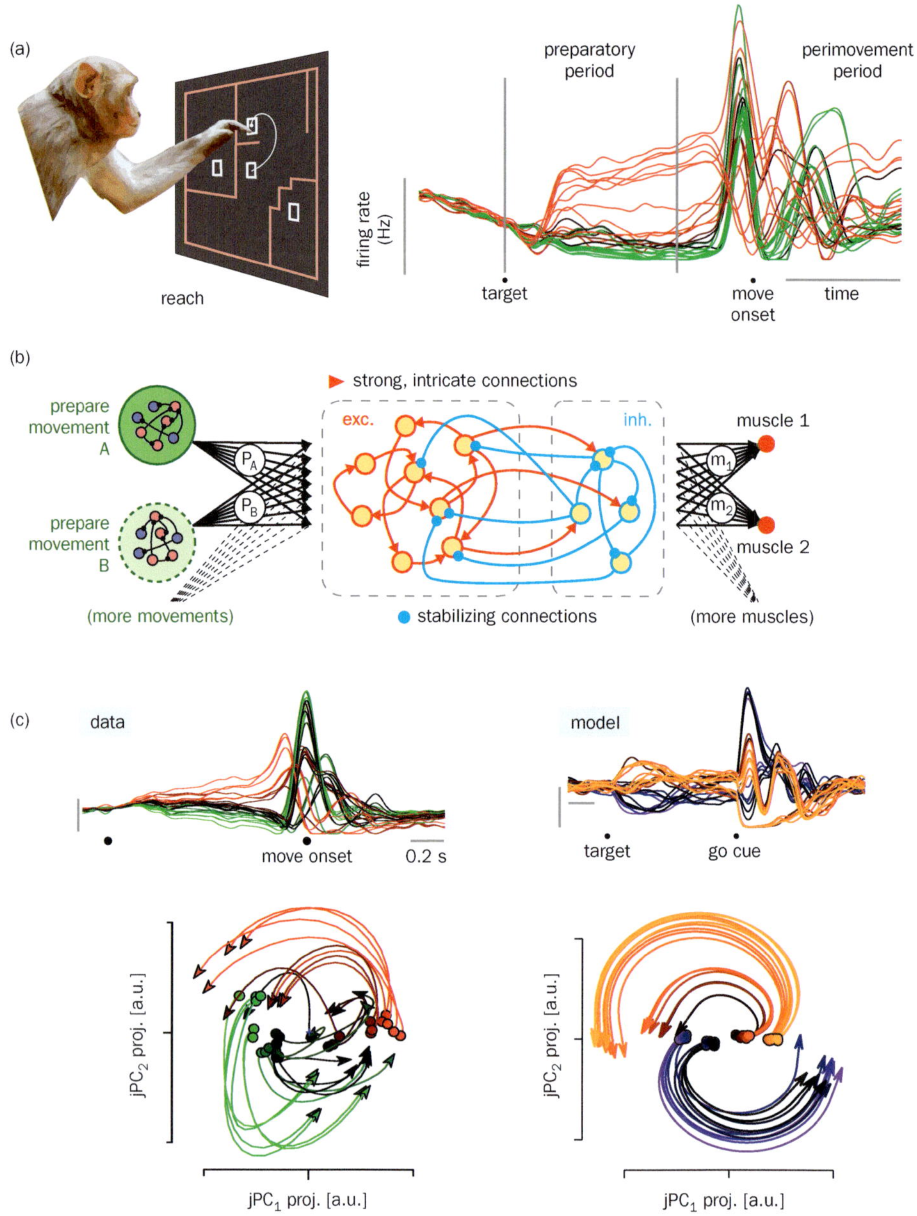

Figure 3.25 Random network model of preparatory neural activity in voluntary movement. (a) Left panel: In a monkey experiment, a target is presented followed by a delay period, during which time the subject plans a reach movement towards the target. Right panel: During the delay period, neurons in the dorsal premotor cortex (PMd) display preparatory activity; traces are colored (red to green) for different reach distances and instructed speeds. (b) Random network model of excitatory and inhibitory neurons. A population of rate units is recurrently connected, with strong and intricate excitatory pathways (red) that would normally produce chaotic activity. Stabilization is achieved through fine-tuned inhibitory feedback (blue). During a preparatory period, a control signal is applied to the model system. During movement execution, recurrent units receive an instruction cue for one of two reaches, A or B, and their outputs determine the movement execution. (c) the model (right panel) captures time course of preparatory activity (upper) and trajectory of population activity (lower) observed experimentally (left panel). Panel (a) is reproduced from Churchland, Mark M., et al. (2010). Cortical preparatory activity: Representation of movement or first cog in a dynamical machine? Neuron, 68(3), 387–400.] [Hennequin, Guillaume, Tim P. Vogels, and Wulfram Gerstner. (2014). Optimal control of transient dynamics in balanced networks supports generation of complex movements. Neuron, 82(6), 1394–1406.

network (ISN), discussed in Section 3.3. The model for motor cortex obeys Dale's law, with separate excitatory and inhibitory neurons. The excitatory connections are random and sufficiently strong to induce network instability in the absence of feedback inhibition; as a matter of fact, the system is unstable in many dimensions. The inhibitory connections are then adjusted by an algorithm to counter-balance excitation and stabilize the network.

During the preparatory period, a "control" signal is presented, possibly mimicking an input from the prefrontal cortex to motor cortex, that forces the network trajectory in a null-subspace. At the onset of the go cue, this control signal is turned off, and thereafter the network does not monotonically decay to a steady-state, even though it is dynamically stable; instead, thanks to the balanced amplification (a signature of ISN), neurons display slow transient responses, namely, large-amplitude zig-zags of activity over time (**Figure 3.25c**, right upper panel) similar to the observed motor cortical neurons (**Figure 3.25c**, left upper panel). A linear readout from these recurrent neurons remains silent during the preparatory period, then at the offset of the control signal produces a nontrivial movement trajectory, tracing a swirling pattern on a plane (**Figure 3.25c**, lower right and left panels). The generation of slow transients results from a combination of the use of a long time constant (100 ms) for the population rate dynamical equations, and sufficiently strong recurrent excitation so that some dynamical modes of the network are close to instability, which prolong the effective network time constant.

The concomitant experimental and modeling work illustrates how recurrent neural networks offer a perspective on an important brain function, namely, control of voluntary movements, according to which movement preparation and execution are understood in terms of trajectories of population dynamics in the state space. This dynamical systems view is in consonance with feedback control theory of goal-directed motor behavior (Todorov 2004).

In summary, random network models produce temporally varying dynamics and neuron-to-neuron heterogeneity, and a number of functions can potentially be realized by training synaptic weights onto a readout (motor) system. In a cortical circuit, recurrent connections are not completely random. Moreover, reproducing temporal neural signals and heterogeneity does not amount to understanding their functions. For instance, as discussed earlier, random network models capture complex dynamics of the motor cortex, which, as it turns out, encodes the timing of the impending movement during the preparatory period (Kaufman et al. 2016). Regarding heterogeneity across individual neurons, it is worth mentioning that there is a lack of information about specificity of recorded neurons in today's typical physiological experiments. Conceivably, to a large extent the observed heterogeneity results from the fact that these neurons belong to diverse types defined by their locations in different cortical layers, cell types and projection targets, and dynamics of each type could be fairly homogeneous across neurons within each type (Hirokawa et al. 2019).

3.8 SUMMARY

In this chapter, we introduced basic concepts and models of neural networks, which can be formulated in terms of either spiking neurons or population firing rates. There are different types of wiring architecture, a major distinction being feedforward versus recurrent networks. Whereas linear networks provide valuable insights, recurrent neural circuits are inherently nonlinear dynamical systems. A number of concepts have been introduced, including integrators, excitation-inhibition balance, balanced amplification, synchronous brain oscillations, normalization, tuning curve of sensory stimulus feature, contrast invariance, line attractor, state space of population activity and its subspaces, and manifold discovery. We covered networks with spatially structured and random connections, as well as slow transients of cortical neural activity during preparation and execution of voluntary movements and scale-free global brain dynamics. The richness of dynamical behaviors amply demonstrates that while the brain connectome is essential, it is insufficient to understand how a neural system works, for two reasons. First, functions arise from network dynamics that depend on not only anatomy but also neurophysiology. Second, in a recurrent neural circuit, quantitative change of a property may give rise to qualitatively different behaviors through the mathematical phenomenon of bifurcation. While we chose examples almost exclusively in the sensory and motor domains, concepts and equations introduced herein will be applicable to neural circuits underlying cognition, as we shall see in the bulk of this book.

FURTHER READING

Books and Reviews

Averbeck, B., J. Sohn, and D. Lee (2006). Activity in prefrontal cortex during dynamic selection of action sequences. *Nat. Neurosci. 9*, 276–282.

Braitenberg, V. and A. Schütz (1991). *Anatomy of the cortex.* Berlin: Springer-Verlag.

Buzsaki, G. and K. Mizuseki (2014). The log-dynamic brain: How skewed distributions affect network operations. *Nat. Rev. Neurosci. 15*, 264–278.

Cohen, M. R. and A. Kohn (2011). Measuring and interpreting neuronal correlations. *Nat. Neurosci. 14*, 811–819.

Douglas, R. J. and K. A. C. Martin (2004). Neuronal circuits of the neocortex. *Annu. Rev. Neurosci. 27*, 419–451.

Ermentrout, G. B. (1998). Neural networks as spatio-temporal pattern-forming systems. *Rep. Prog. Phys. 61*, 353–430.

Jazayeri, M. and S. Ostojic (2021). Interpreting neural computations by examining intrinsic and embedding dimensionality of neural activity. *Curr. Opin. Neurobiol.* 70, 113–120.

Lichtman, J. W., J. Livet, and J. R. Sanes (2008). A technicolour approach to the connectome. *Nat. Rev. Neurosci.* 9, 417–422.

Marder, E. and R. L. Calabrese (1996). Principles of rhythmic motor pattern generation. *Physiol. Rev. 76*, 687–717.

Shepherd, G. M. and S. Grillner (Eds.) (2018). *Handbook od Brain Microcircuits* (2nd ed.). New York: Oxford University Press.

Sporns, O. (2014). Contributions and challenges for network models in cognitive neuroscience. *Nat. Neurosci. 17*, 652–660.

Wang, X.-J. (2010). Neurophysiological and computational principles of cortical rhythms in cognition. *Physiol. Rev. 90*, 1195–1268.

Primary Research Papers

Network Dynamics

Amari, S. (1977). Dynamics of pattern formation in lateral-inhibition type neural fields. *Biol. Cybern. 27*, 77–87.

Börgers, C. and N. Kopell (2003). Synchronization in networks of excitatory and inhibitory neurons with sparse, random connectivity. *Neural Comput. 15*, 509–538.

Brunel, N. and V. Hakim (1999). Fast global oscillations in networks of integrate-and-fire neurons with low firing rates. *Neural Comput. 11*, 1621–1671.

Fisahn, A., F. G. Pike, E. H. Buhl, and O. Paulsen (1998). Cholinergic induction of network oscillations at 40hz in the hippocampus in vitro. *Nature 394*, 186–189.

Murphy, B. K. and K. D. Miller (2009). Balanced amplification: A new mechanism of selective amplification of neural activity patterns. *Neuron 61*, 635–648.

Shadlen, M. N. and W. T. Newsome (1998). The variable discharge of cortical neurons: Implications for connectivity, computation, and information coding. *J. Neurosci. 18*, 3870–3896.

van Vreeswijk, C. and H. Sompolinsky (1998). Chaotic balanced state in a model of cortical circuits. *Neural Comp. 10*, 1321–1371.

Wang, X.-J. and J. Rinzel (1992). Alternating and synchronous rhythms in reciprocally inhibitory model neurons. *Neural Comput. 4*, 84–97.

Wilson, H. R. and J. D. Cowan (1972). Excitatory and inhibitory interactions in localized populations of model neurons. *Biophys. J. 12*, 1–24.

Selectivity and Computation

Ben-Yishai, R., R. Lev Bar-Or, and H. Sompolinsky (1995). Theory of orientation tuning in visual cortex. *Proc. Natl. Acad. Sci. U.S.A. 92*, 3844–3848.

Chang, L. and D. Y. Tsao (2017). The code for facial identity in the primate brain. *Cell 169*, 1013–1028.

Hafting, T., M. Fyhn, S. Molden, M. B. Moser, and E. I. Moser (2005). Microstructure of a spatial map in the entorhinal cortex. *Nature 436*, 801–806.

Heeger, D. J., E. P. Simoncelli, and J. A. Movshon (1996). Computational models of cortical visual processing. *Proc. Natl. Acad. Sci. U.S.A. 93*, 623–627.

Hubel, D. H. and T. N. Wiesel (1962). Receptive fields, binocular interaction and functional architecture in the cat's visual cortex. *J. Physiol. Lond. 160*, 106–154.

Seung, H. S. (1996). How the brain keeps the eyes still. *Proc. Natl. Acad. Sci. U.S.A. 93*, 13339–13344.

Zhang, K. (1996). Representation of spatial orientation by the intrinsic dynamics of the head-direction cell ensembles: A theory. *J. Neurosci. 16*, 2112–2126.

Dimensionality and Subspace Analysis

Chaudhuri, R., B. Gercek, B. Pandey, A. Peyrache, and I. Fiete (2019). The intrinsic attractor manifold and population dynamics of a canonical cognitive circuit across waking and sleep. *Nat. Neurosci. 22*, 1512–1520.

Kaufman, M. T., M. M. Churchland, S. I. Ryu, and K. V. Shenoy (2014). Cortical activity in the null space: Permitting preparation without movement. *Nat. Neurosci.* 17, 440–448.

Machens, C. K., R. Romo, and C. D. Brody (2010). Functional, but not anatomical, separation of "what" and "when" in prefrontal cortex. *J. Neurosci. 30*, 350–360.

Mazor, O. and G. Laurent (2005). Transient dynamics versus fixed points in odor representations by locust antennal lobe projection neurons. *Neuron 48*, 661–673.

Semedo, J. D., A. Zandvakili, C. K. Machens, M. Y. Byron, and A. Kohn (2019). Cortical areas interact through a communication subspace. *Neuron 102*, 249–259.

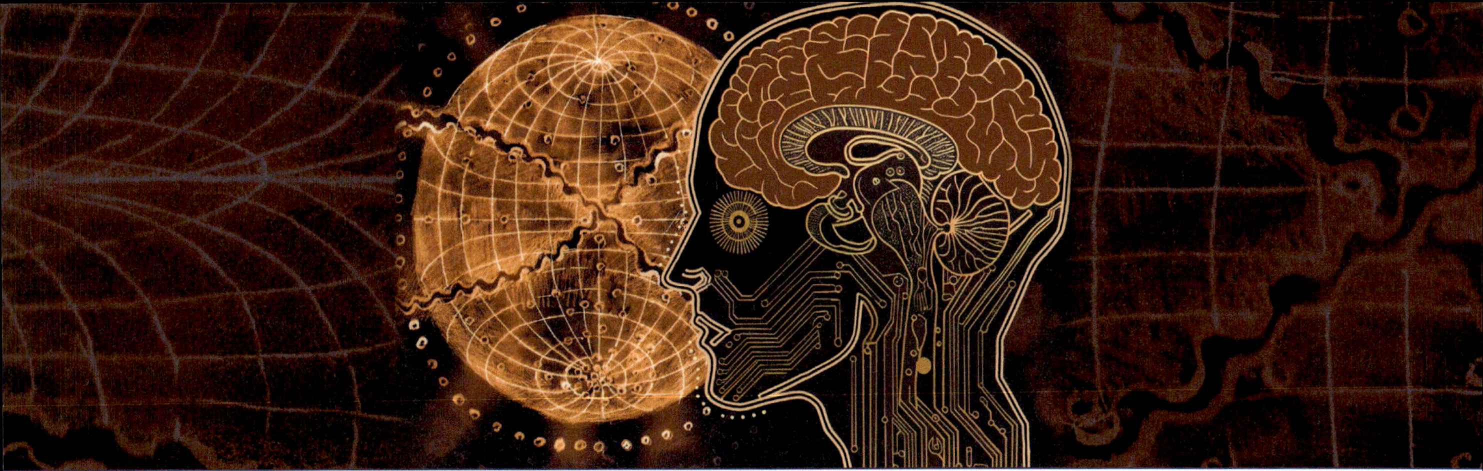

Plasticity, Learning and Memory

4

In the long silent hours, I am trampled by memories, all happening in one instant, as if my entire life were a single, unfathomable image. The child and girl I was, the woman I am, the old woman I shall be, are all water in the same rushing torrent. My memory is like a Mexican mural in which all times are simultaneous.

Isabel Allende in *Paula*

4.1 INTRODUCTION

A neural circuit is an adaptive dynamical system that constantly changes throughout a lifetime. We marvel at the brain's remarkable malleability, as we watch an infant whose vocabulary grows by the day (Bloom 2002) or attest to someone's recovery from a major stroke. This biological plasticity makes it possible for us to remember and to learn (Dehaene 2020). Scientific research on memory began in the late nineteenth century with studies of memory loss as a symptom of brain disease (Ribot 1881) and controlled psychological experiments in the laboratory (Ebbinghaus 1885). Over more than 100 years, different types of memory have been identified and their biological mechanisms investigated with careful design of task paradigms across many animal species (Mackintosh 1974; Squire 1987; Gallistel 1990; Tulving 1995; Squire and Kandel 1999; Murray et al. 2017; Magee and Grienberger 2020).

Transient short-term memory traces need to undergo an additional process of selection to reach long-term memory store (Atkinson and Shiffrin 1968). Long-term memories are divided into different classes depending on their content (**Figure 4.1**) (Squire 1987). Consciously accessible memories are usually named *declarative* or *explicit* memories. They are further divided into episodic and semantic memories. Episodic memories contain information about specific events, the time and the situation in which they occurred, such as autobiographical memories (e.g., what do you recall of your high school prom?). The medial temporal lobes, including the hippocampus and parahippocampal cortices (entorhinal and perirhinal cortices), are critical for the formation of episodic memories (Squire

DOI: 10.1201/9781003459361-5

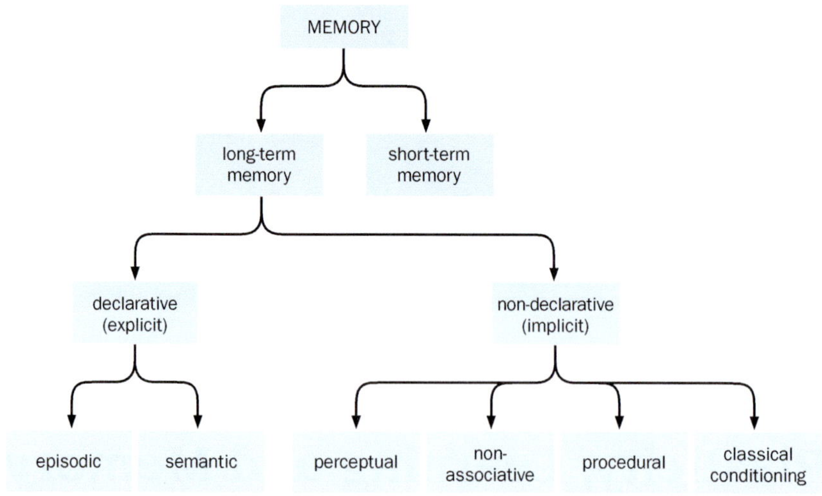

Figure 4.1 Taxonomy of memories.

1987). Semantic memories contain information about general knowledge, such as specific facts and concepts (e.g., what's the capital of Egypt?). In contrast to episodic memories, the circumstances under which they are acquired are not part of the memory. We often combine episodic and semantic memories: for example, we may remember the circumstances in which we learned about the capital of Egypt, if it happened in a memorable circumstance.

In contrast to explicit memories, implicit memories do not involve conscious recollection. They manifest themselves as any kind of influence of prior experiences on our current behavior. One typical example is perceptual priming, in which the previous exposure to a sensory stimulus affects processing of a subsequent stimulus. Non-associative memory refers to a change in response to a stimulus (such as a reflex) after previous exposures to the same stimulus. A third type of implicit memories is procedural memory, like motor skills (e.g., playing piano or skiing). Procedural memories are learned through repetition of an activity, and they are automatically retrieved and utilized. Brain regions involved in procedural memories include the basal ganglia. Finally, a fourth type of implicit memory is conditioning (Pavlov 2003), which refers to learning new association between a neutral stimulus and a physiological stimulus.

Generally, learning and memory take place in three stages: encoding, consolidation, retrieval. We remember many things only for a short period of time, such as what we ate at lunch today, while other memory traces can last for many years. A newly acquired memory needs to undergo a consolidation process for long-term storage. In addition, retrieval is a distinct process, illustrated by our daily experiences such as the trouble of recalling a person's name even though we are quite certain that the memorandum is present in our brain. Forgetting is the gradual loss of information stored in memory, which has been extensively studied experimentally. The first experiments were conducted by Hermann Ebbinghaus in the nineteenth century (Ebbinghaus 1885). He taught himself to remember long lists of nonsense syllables (consonant-vowel-consonant trigrams, such as "DIF"). After an interval ranging from tens of minutes to weeks, he then evaluated the strength of the memories by estimating the relearning time (i.e., better remembered lists could be quickly relearned). He discovered that: (1) different memories may have different memory life spans, (2) repetition makes memories last longer and (3) the rate of forgetting decreases with the passage of time. These findings have been confirmed by subsequent

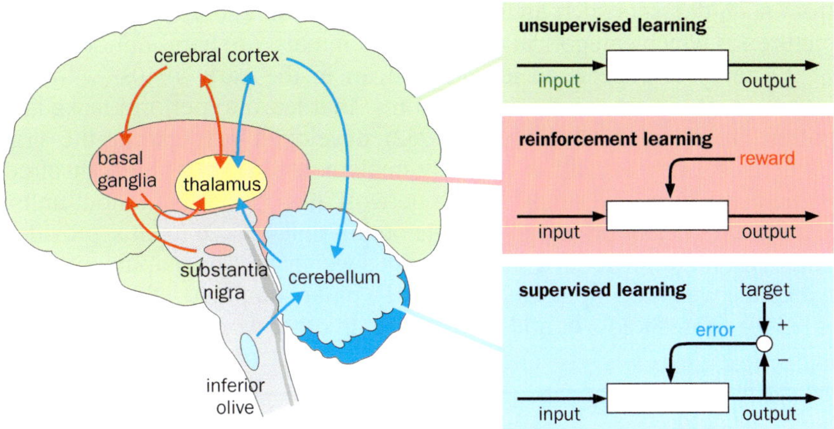

Figure 4.2 Brain structures critically involved in three types of learning processes. Supervised learning of fine movements depends on the cerebellum and aims at minimizing an error signal encoded by the input from the inferior olive. The basal ganglia and certain parts of the cortex play a central role in reinforcement learning, which is guided by the reward signal encoded in the dopaminergic input from the midbrain (including a portion of the substantia nigra). Unsupervised learning based on the statistical properties of inputs is commonplace in the cerebral cortex. Reproduced from Doya, K. (2000). Complementary roles of basal ganglia and cerebellum in learning and motor control. Current Opinion in Neurobiology, 10(6), 732–739.

experimental studies. Today, it is widely believed that the empirical forgetting functions are most accurately described by a power function, namely, a memory trace decays with time t as $1/t^{v}$ where v is a real number (Wixted and Ebbesen 1997). Such power law decay implies that there is not a single characteristic lifetime, but memory degradation occurs on all timescales.

Plasticity of the brain underlies memory. Here we shall focus on the idea of synaptic modifications that can be traced back to Santiago Ramón y Cajal, a founder of modern neuroscience (DeFelipe 2006). There are three major forms of learning processes, which involve different yet overlapping brain substrates in mammals (**Figure 4.2**). First, supervised learning denotes modifications that are dictated by explicit instruction about the correct output given each stimulus input. Second, reinforcement learning requires a reward signal about how well the system is doing in achieving a goal. Third, unsupervised learning is solely dependent on the statistics of experiencing inputs. We will first introduce supervised learning, which will be expanded in Chapter 11 in connection with machine learning algorithms, and reinforcement learning, which will be elaborated in Chapter 7 on reward-dependent decision making. The remainder of the chapter will be devoted to unsupervised learning, which has been quite extensively investigated both experimentally and computationally.

4.2 SUPERVISED LEARNING

Teaching a child to solve a problem can be done in two ways. You can observe what she just did and provide positive feedback (praise) or negative feedback (constructive criticism), similar to reinforcement learning based on reward and punishment in animal studies (see next section). Alternatively, you can instruct her on what is the correct answer explicitly. Using correct responses to guide a training process is called supervised learning. The connection weights are commonly trained by supervised learning, with correct answers provided for a training set of samples. This means that, for example, if the task is recognizing a dog in a picture, a training sample must have a dog (if present) labeled, so the system knows what is the correct answer in supervised learning. Those samples may be from such datasets such as ImageNet (Deng et al. 2009), with 15 million pictures in which categorical objects (like animals) are hand-annotated.

Theory of supervised learning began in the early 1960s, with the invention of "smart computer algorithms" that learn to perform tasks like pattern recognition. Rosenblatt (1962) developed perceptron, the first modern learning algorithm (actually implemented in machines) inspired by neuroscience. The simplest such network has a number of input units, x_i $(i = 1, 2, \ldots, N)$ that project to a single output unit y. Each input x_i provides partial information about a stimulus with a weight w_i. The output unit summates the weighted inputs and produces a binary output: 0 if the total input is below a threshold $-b$, and 1 otherwise. In other words:

$$y = H\left(\sum_i^N w_i x_i + b\right) \tag{4.1}$$

where H is the Heaviside step function. This architecture has the flavor of a neural network, and the binary output is akin to the presence or absence of action potential generation. The weights $\{w_i\}$ are modified with the goal for a perceptron to accomplish a particular task, like classification of whether a visual object is a dog. Suppose that sample animal pictures are shown to a perceptron tasked to classify dog versus non-dog. Each picture μ is represented by an array of inputs $\{x_i^\mu\}$. The correct output is known to be y_{target}^μ. The algorithm computes a quadratic error function:

$$\mathcal{L} = \sum_\mu \varepsilon_\mu^2; \quad \varepsilon_\mu^2 = \frac{1}{2}\left(y_{target}^\mu - y^\mu\right)^2; \quad y^\mu = H\left(\sum_i^N w_i x_i^\mu + b\right) \tag{4.2}$$

Learning amounts to modifying $\{w_i\}$ with the goal to minimize the error function; the system's performance is perfect when $\mathcal{L} = 0$. Importantly, this formulation can be implemented incrementally, from one sample to another, according to the following learning rule:

$$\Delta w_{i,t+1} = w_{i,t} + \alpha\left(y_{\text{target}}^\mu - y^\mu\right)x_i^\mu \tag{4.3}$$

If the t-th experienced sample is item μ. The parameter α is a learning rate that controls the speed of weight modifications, and y^μ is the actual output according to Eq. (4.1) with input $\{x_i^\mu\}$. Hence, when $y_{\text{target}}^\mu > y^k$, weight changes should be in the direction of increasing the total input. The weight w_i is potentiated ($\Delta w_i = w_{i,t+1} - w_{i,t} > 0$) if $x_i^\mu > 0$ so that its influence is enhanced, and depressed ($\Delta w_i < 0$) if $x_i^\mu < 0$ to reduce its impact on the output unit. The opposite modifications occur when $y_{\text{target}}^\mu < y^\mu$, in which case weights should be modified to reduce the total input. Step-by-step changes of the weights leads to minimizing $\mathcal{L}$, thereby optimizing the match between the desired output and the actual output.

More generally, the input-output is a smooth function, $y = F\left(\sum_i w_i x_i + b\right) \equiv F(\mathbf{w}^{\mathrm{T}} \cdot \mathbf{x} + \mathrm{b})$, where $\cdot$ represents the dot product of two vectors, $F(x)$ is differentiable (with well-defined dF/dx) and b is a bias term. Examples include sigmoid $F(x) = 1/\left(1 + \exp(-x)\right)$ and softplus $F(x) = \log\left(1 + \exp(x)\right)$. With a differentiable input-output function, learning is carried out using a gradient descent (GD) algorithm called "least mean square (LMS)" rule or Delta rule (Widrow and Hoff 1960). Consider the error function again defined by Eq. (4.2) with $\epsilon_\mu^2 = (1/2)(y_{\text{target}}^\mu - y^\mu)^2 = (1/2)\left(y_{\text{target}}^\mu - F\left(\mathbf{w}^{\mathrm{T}}\mathbf{x}^\mu + \mathrm{b}\right)\right)^2$. Eq. (4.3) becomes:

$$w_{i,t+1} = w_{i,t} - \alpha\frac{\partial\mathcal{L}(\{w_i\})}{\partial w_i} \tag{4.4}$$

$$= w_{i,t} + \alpha\delta^{\mu}x_i^{\mu} \tag{4.5}$$

with

$$\delta^{\mu} = F'\left(\mathbf{w}^{\mathrm{T}}\mathbf{x}^{\mu} + b\right)\left(y^{\mu}_{target} - F\left(\mathbf{w}^{\mathrm{T}}\mathbf{x}^{\mu} + b\right)\right) \tag{4.6}$$

This procedure is called "gradient descent": $\mathcal{L}$ is an "energy function" of the N variables $\{w_i\}$, plasticity gradually moves $\{w_i\}$ to reduce $\mathcal{L}$. When $\partial\mathcal{L}/\partial w_i$ is negative, w_i is enhanced with a rate proportional to the slope of the input-output function F if x_i is positive but is reduced if x_i is negative. The sign of change is reversed when $\partial\mathcal{L}/\partial w_i$ is positive. The quadratic form guarantees the existence of a global minimum (Minsky and Papert 1969), as illustrated by **Figure 4.3** in a simple example with $N = 1$ and two trained parameters (b and w_1).

Now consider a feedforward network with several (K) layers indexed by $k = 1, 2, \ldots, K$. Denote by r_i^k activity of unit i ($i = 1, 2, \ldots, N_k$) in k-th layer, we have:

$$r_i^k = F\left(\sum_j w_{ij}^k r_j^{k-1} + b_i\right) \tag{4.7}$$

where w_{ij}^k is the projection matrix from unit j in layer $(k-1)$ to unit i in layer k, and b_i is a bias term (e.g., representing a background input).

A loss function is defined in terms of the mismatch between the network's output and the correct target output (generally both are vectors) for a number M of training examples, indexed by $B = \{\mu_1, \cdots, \mu_M\}$:

$$\mathcal{L}_{\text{batch}} = \frac{1}{M}\sum_{\mu \in B}\mathcal{L}(\mathbf{y}^{\mu}, \mathbf{y}^{\mu}_{\text{target}}), \tag{4.8}$$

where $\mathcal{L}(\mathbf{y}^{\mu}, \mathbf{y}^{\mu}_{\text{target}})$ is typically a squared difference between $\mathbf{y}^{\mu}$ and $\mathbf{y}^{\mu}_{\text{target}}$. The gradient descent algorithm (Eqs. 4.4–4.6) is used to update connection weights for projections from one layer to another, as well as other tunable parameters such as bias current $\mathbf{b}$. Samples are often chosen randomly from a subset of available training examples, hence the term stochastic gradient descent (SGD).

In order to train a multi-layer network by supervised learning, the gradient descent algorithm uses backpropagation (Rumelhart et al. 1986a, 1986b; LeCun 1988). To compute the loss $\mathcal{L}$, the network is run in a forward pass (Eq. 4.7). Next, to efficiently compute the exact gradient, information about the loss needs to be passed backward, in the opposite direction of the forward pass, hence the name backpropagation. For the sake

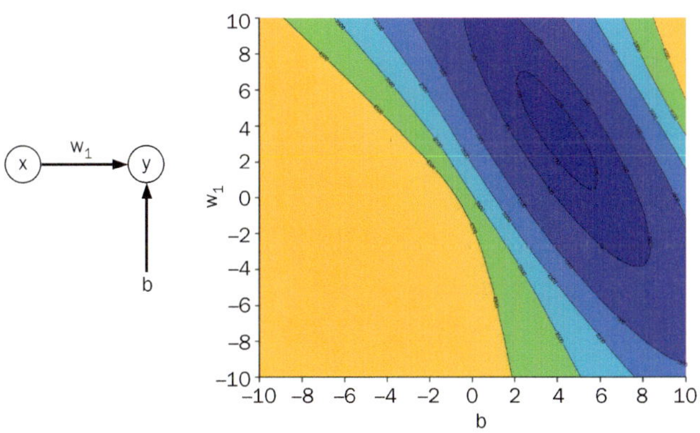

Figure 4.3 Squared error as a function of w_1 and b, exhibiting a global minimum, which is located at $b = 4$ and $w_1 = 3.5$.

of simplicity, consider a linear input-output function $\mathbf{r}^{\mathrm{k}} = \mathbf{w}^{\mathrm{k}}\mathbf{r}^{\mathrm{k}-1} + \mathbf{b}^{\mathrm{k}}$. To compute $\partial\mathcal{L}/\partial\mathbf{w}^{\mathrm{k}} = (\partial\mathcal{L}/\partial\mathbf{r}^{\mathrm{k}})(\partial\mathbf{r}^{\mathrm{k}}/\partial\mathbf{w}^{\mathrm{k}}$, we need to compute $\partial\mathbf{r}^{\mathrm{k}}/\partial\mathbf{w}^{\mathrm{k}}$. We have:

$$\frac{\partial\mathcal{L}}{\partial r_i^k} = \sum_j \frac{\partial\mathcal{L}}{\partial r_j^{k+1}} \frac{\partial r_j^{k+1}}{\partial r_i^k} = \sum_j \frac{\partial\mathcal{L}}{\partial r_j^{k+1}} w_{ji}^{k+1} = \sum_j [w^{k+1}]_{ij}^{\mathrm{T}} \frac{\partial\mathcal{L}}{\partial r_j^{k+1}}. \tag{4.9}$$

In vector notation,

$$\frac{\partial\mathcal{L}}{\partial\mathbf{r}^{\mathrm{k}}} = \left[\mathbf{w}^{\mathrm{k}+1}\right]^{\mathrm{T}} \frac{\partial\mathcal{L}}{\partial\mathbf{r}^{\mathrm{k}+1}} = \left[\mathbf{w}^{\mathrm{k}+1}\right]^{\mathrm{T}} \left[\mathbf{w}^{\mathrm{k}+2}\right]^{\mathrm{T}} \frac{\partial\mathcal{L}}{\partial\mathbf{r}^{\mathrm{k}+2}} = \cdots \tag{4.10}$$

and

$$\Delta\mathbf{w}^{\mathrm{k}} = -\alpha\frac{\partial\mathcal{L}}{\partial\mathbf{w}^{\mathrm{k}}} = -\alpha\frac{\partial\mathcal{L}}{\partial\mathbf{r}^{\mathrm{k}}}[\mathbf{r}^{\mathrm{k}-1}]^{\mathrm{T}} \tag{4.11}$$

where α is a learning rate "hyperparameter."

Therefore, starting with $\partial\mathcal{L}/\partial\mathbf{y}$, $\partial\mathcal{L}/\partial\mathbf{r}^{\mathrm{k}}$ can be recursively computed from $\partial\mathcal{L}/\partial\mathbf{r}^{\mathrm{k}+1}$, for $k = K-1,\cdots,1$. This computation flows in the opposite direction of the forward pass and is called the backward pass. Note that the algorithm would not apply to a network with a non-smooth $F(I)$ such as Heaviside function; otherwise, it works for a differentiable input-output function F that can be highly nonlinear. Generalization to networks of spiking neurons requires that spikes are smoothened so that the model equations are differentiable (Huh and Sejnowski 2017).

Backpropagation applies to multi-layer nonlinear networks with arbitrary number of units per layer. With a multi-layered deep network, weight modifications involve products of many partial derivatives. Bookkeeping in the complex backpropagation process can be cumbersome, but a most elegant way is to use Lagrangian multipliers (LeCun 1988). The supervised learning rule to networks with many layers (deep learning) is at the core of modern artificial intelligence (Chapter 11).

Does supervised learning occur in the brain (Knudsen 1994)? Perhaps the best known example is learning to execute movements (e.g., throwing a ball in a baseball game), for which arguably there are correct solutions. The cerebellum plays a central role for controlling fine movements, and its classical model by Marr (1969) and Albus (1971) bears striking similarity with an abstract supervised learning network (**Figure 4.4a**). As shown schematically of the cerebellum circuit (**Figure 4.4b**), Purkinje cells are principal neurons that receive synaptic excitation from mossy fibers via granule cells, representing sensory cues to instruct

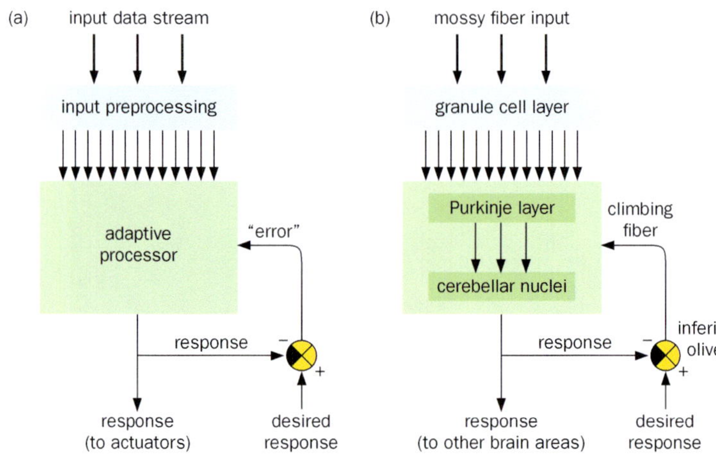

Figure 4.4 Core architecture of supervised learning networks. The essential computational frameworks for implementing supervised learning are shown for (a) an artificial neural network and (b) the cerebellar network. A preprocessing stage (cyan) transforms the input signals to create representations that are a suitable substrate for supervised learning. Those representations are then sent to an adaptive processor (green) that generates responses. Errors are computed by comparing the actual response of the network with the desired response (yellow) and are used as instructive signals to adjust the internal parameters (e.g., synaptic weights) of the adaptive processor until it learns to generate the desired response. Reproduced from Raymond, J. L., & Medina, J. F. (2018). Computational principles of supervised learning in the cerebellum. Annual Review of Neuroscience, 41, 233.

a movement. Purkinje cells are GABAergic, and thus send an inhibitory output signal to neurons in the cerebellar nucleus. The latter projects to a motor structure for the execution of a movement, as well as to the inferior olive. The inferior olive thus receives an effectively inhibitory signal from the Purkinje cells (an executed movement, X). In addition, the inferior olive receives a separate excitatory signal (a desired movement, Y), which is compared with the actual movement to compute an error signal ($Y - X$) and send it to the cerebellum via climbing fibers. This output from the climbing fiber, representing a mismatch between the desired and the actual movements, triggers a stereotypical "complex spike" (an initial prolonged large-amplitude spike, followed by a high-frequency burst of smaller-amplitude spikes) in Purkinje cells. The error signal is selective for movement parameters, such as the direction and magnitude of a saccade (Herzfeld et al. 2018). This "teaching signal," distinct from tonic spiking of Purkinje cells driven by granule cells, induces synaptic plasticity at multiple sites in the cerebellum (Ito 2000; Raymond and Medina 2018), suggesting a biological implementation of supervised learning.

4.3 REINFORCEMENT LEARNING

Reinforcement learning refers to a learning process of gathering rewards and avoiding punishments through trial and error (Sutton and Barto 2018; Daw 2016). Reward or punishment is usually the consequence of a choice or action, thus it is essential for reinforcement learning to predict or expect the outcome of a decision. Interestingly, expectation was initially studied in behaviors that do not require active response or deliberate choice. In classical conditioning pioneered by Pavlov (2003), a dog is given a piece of meat after a bell rings. After a few repetitions, the bell's sound, which is behaviorally neutral and called a conditioned stimulus (CS), triggers salivation by the dog. This conditioned response is not an action needed to obtain food; instead, it signals the expectation of food (an unconditioned stimulus, US). In another classical conditioning paradigm, a subject hears a sound (CS), followed by an electric shock (US). Over training, the sound triggers freezing, which again is not so much to avoid the shock as to signal the anticipation of it (Maren 2001). Hence, classical conditioning experiments are suitable to study "prediction learning" that does not involve choice. In this framework, prediction entails forming an association between a stimulus and a reward (or punishment).

4.3.1 The Rescorla-Wagner Rule and Reward Prediction Error

Rescorla and Wagner (1972), following earlier work by Bush and Mosteller (1951), proposed a mathematical model for learning stimulus-reward associations. Assume that a stimulus s is followed in time by a reward r_t, across trials index by t. The value of the observed stimulus V_t is updated by each observation as:

$$V_{t+1}(s) = V_t(s) + \alpha\delta_t \tag{4.12}$$

where $\delta_t = r_t - V_t(s)$, a mismatch between the actual reward and expected value called reward prediction error (RPE). The parameter α (between 0 and 1) is a learning rate controlling the magnitude of change by an experience.

Learning is error-driven: if the actual reward is larger than the expected value, δ_t is positive and $V(s)$ is increased towards r; otherwise, δ_t is negative and $V(s)$ is decreased, again towards r. A steady state is achieved when $\delta = 0$. Note that the expected value, hence the error signal, is defined for a given stimulus s, different stimuli naturally are associated with different expected values.

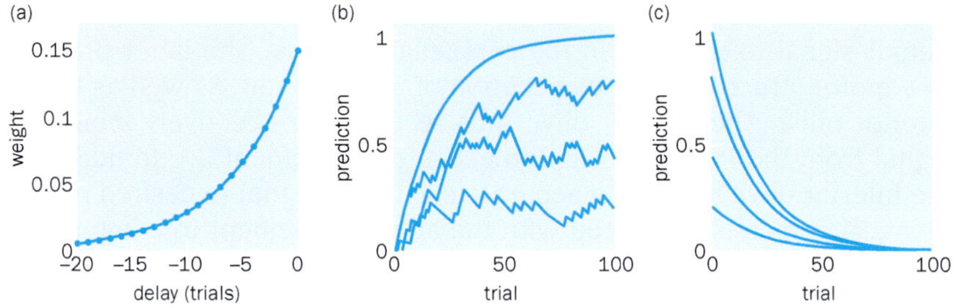

Figure 4.5 The Rescorla-Wagner model. (a) The weights on rewards received in previous trials decay exponentially into the past, with a steepness that depends on the learning rate parameter α. (b) Simulation of the Rescorla-Wagner model learning four different cues, which are reinforced (from top to bottom) 100%, 75%, 50% and 25% of the time. For the stochastically rewarded stimuli, the prediction is noisy (driven by random patterns of reward and non-reward events) around the underlying average reward. Learning curves grow to a steady-state at 1, 0.75, 0.5 and 0.25, respectively. (c) The expected value undergoes extinction of the associations learned in (b) when the stimuli are presented without reward. Reproduced from Daw, N. D. (2016). Reinforcement learning. In M. A. Arbib and J. J. Bonaiuto (Eds.), From Neuron to Cognition via Computational Neuroscience, Chapter 10, pp. 299{318. Cambridge, MA: MIT Press.

The learning rate α is related to a time constant for reward memory trace. This is readily seen by considering a situation when reward is absent in a sequence of trials. In that case, $V_{t+1} = (1-\alpha)V_t(s) = (1-\alpha)^t V_1$, where V_1 is the initial expected value. This expression is equivalent to $V_{t+1} = V_1 \exp(-t/\tau)$, with $\tau = 1/\ln(1/(1-\alpha))$. For a small learning rate $\alpha \sim 0$, τ goes to ∞; whereas for a large $\alpha \sim 1$, τ goes to zero. In general, solving Eq. (4.12) iteratively, V_t as a function of V_{t-1}, V_{t-1} in terms of V_{t-2}, etc, we obtain:

$$V_{t+1}(s) = e^{-t/\tau} V_1(s) + \alpha \sum_{k=1}^{t} e^{-(t-k)/\tau} r_k \tag{4.13}$$

where previously experienced rewards r_k are weighted by an exponentially decaying factor (**Figure 4.5a**). In other words, V dynamics represents a leaky integration of reward events with the time constant τ.

In a simple example, reward is binary, $r = 1$ with a fixed probability and 0 otherwise. Then the expected value evolves over trials to reach a steady-state that matches with reward probability (**Figure 4.5b**). If the stimuli are presented without reward, V decays to zero exponentially with the time constant τ (**Figure 4.5c**). This explains the experimental observation that the conditioned response ceases eventually when the CS is repetitively shown without reinforcer, a phenomenon called "extinction" in classical conditioning.

4.3.2 Reward Signaling by the Dopamine System

Dopamine is a major neuromodulator involved in signaling rewards in the brain, and it plays a central role in mediating rewarding effects of abused drugs (Wise and Rompre 1989). Dopamine is released from neurons in two midbrain nuclei: the ventral tegmental area (VTA) that projects to many cortical and subcortical areas including the ventral striatum, hippocampus and amygdala; and the substantia nigra pars compacta (SNc) that targets the dorsal striatum and powerfully influences movements. Spiking activity of dopamine neurons consists of a tonic baseline and phasic response, which is a brief burst of spike firing. In a classical conditioning experiment where a cue (CS) is followed by a reward after a time delay, initially dopamine neurons displayed a surge of activity at the time of reward delivery (**Figure 4.6a**, top panel). Over training, the timing of the phasic

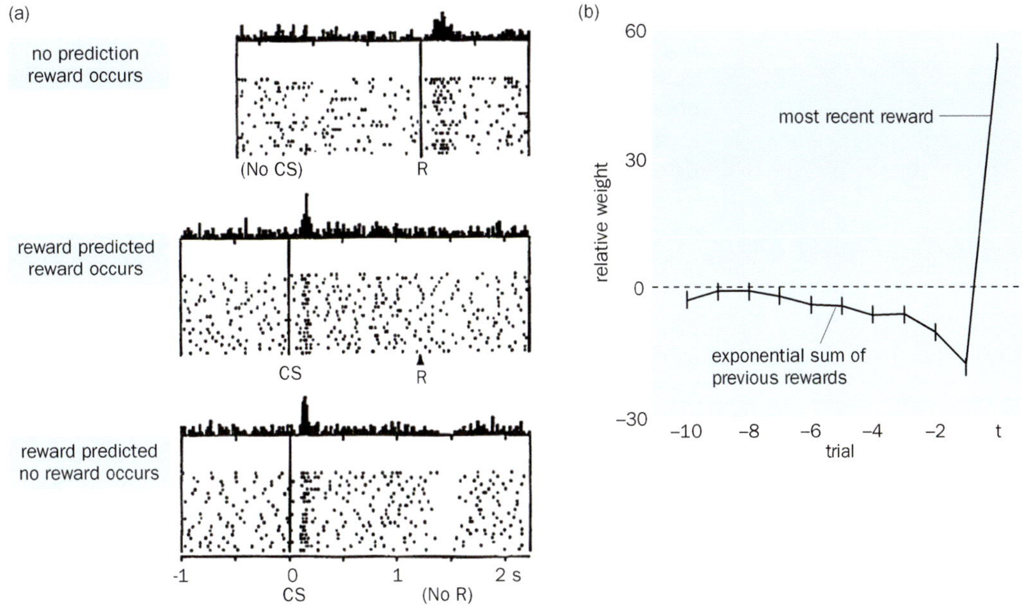

Figure 4.6 Dopamine and reinforcement learning. (a) Dopamine neurons report an error in reward prediction at the time of reward delivery (R). Top: Before learning, drop of liquid constitutes an unexpected award, thus a positive error in the prediction of reward, and phasic burst of spikes is observed. Middle: After learning, a conditioned stimulus (CS) predicts a reward, correlated with a burst of dopamine neural firing. After a delay, the reward occurs according to the prediction, at which time there is no change in dopamine neural firing. Bottom: Conditioned stimulus predicts a reward, but the reward fails to occur. Activity of the dopamine neuron is depressed exactly at the time when the reward would have occurred. Note the absence of any intervening stimuli after the conditioned stimulus, revealing an internal process of reward expectation. (b) The linear weighting function, which best relates dopamine activity to reward history. Panel (a) is reproduced from Schultz, W., Dayan, P., & Montague, P. R. (1997). A neural substrate of prediction and reward. Science, 275(5306), 1593–1599; panel (b) is reproduced from Bayer, H. M., & Glimcher, P. W. (2005). Midbrain dopamine neurons encode a quantitative reward prediction error signal. Neuron, 47(1), 129–141.

response was shifted to the CS (**Figure 4.6a**, middle panel). In trials where the reward was omitted, dopamine neurons exhibited a pause of activity (a reduced firing rate) at the time of anticipated but undelivered reward (**Figure 4.6a**, bottom panel). This behavior mirrors a reward prediction error: before training the expected value $V = 0$, hence $\delta = r$ is positive. After learning, the expected value matches the actual reward, and $\delta = 0$; but when the reward is not delivered, $\delta = -V$ is negative.

If dopamine activity indeed signals RPE, it should behave as a leaky integrator of past rewards. Specifically, according to Eq. (4.13) and $\delta_t = r_t - V_t$, dopamine neural firing rate should display a positive dependence on the reward in the current trial and negative exponential dependence on the rewards in the previous trials (with V_t given by Eq. [4.13] and $V_1 = 0$). This was confirmed by electrical recording of dopamine neurons in monkeys (**Figure 4.6b**) (Bayer and Glimcher 2005). The discovery of phasic dopamine response as a putative RPE in the brain represents a major advance thanks to an interplay between mathematical theory and experimentation (Montague et al. 1996; Schultz et al. 1997).

4.3.3 Action Valuation and Selection

In contrast to classical conditioning where reinforcers are associated with a stimulus, behaviors in which reinforcers are contingent on a response are called instrumental or operant conditioning. In instrumental conditioning, an action is required to garner a desired reward or avoid an aversive event. Experiments may be designed either as a series of discrete trials, or subjects are free to produce responses continuously in time. Probabilities of selecting different responses, response latency and reaction time are measured to assess learning and performance.

Can learning to select a desired action be guided by prediction learning described previously? Here, one must learn values associated with actions, or action values. Consider a gambling task, in which subjects choose among a set of options (e.g., four buttons: up, down, right and left), each associated with a reward probability. The value of each action a is denoted as $Q(a)$, which may be updated similarly as Eq. (4.12):

$$Q_{t+1}(a) = Q_t(a) + \alpha\left(r_t - Q_t(a)\right) \tag{4.14}$$

At any moment, given estimated action values, one is faced with the problem of a tradeoff (Daw et al. 2006). On the one hand, one should prefer to choose the action with the highest action value (exploitation of one's knowledge). However, one may have gambled just a few times, so it is uncertain that the option with the largest $Q(a)$ up to the present time is actually the best. Therefore, one should "check out" other options with smaller $Q(a)$ values (exploration of the unknown). In real life, rewards associated with action options may change over time or depending on situations; for instance, it is preferable to invest in stocks in a bull market, but a shift to safer options like bonds is warranted when the market becomes volatile. A possible "policy" π to select an action that takes into account the explore-exploit dilemma is the softmax function for the probability of choosing an action a:

$$P(a) = \frac{\exp\left(\beta Q(a)\right)}{\sum_{a'} \exp\left(\beta Q(a')\right)} \tag{4.15}$$

where the denominator is the sum over all options under consideration, the parameter β controls the extent to which actions with less than the maximal value are chosen. For instance, if there are only two options to consider, $a = A$ or B, we have:

$$P(A) = \frac{\exp\left(\beta Q(A)\right)}{\exp\left(\beta Q(A)\right) + \exp\left(\beta Q(B)\right)} = \frac{1}{1 + \exp\left(-\beta\left(Q(A) - Q(B)\right)\right)} \tag{4.16}$$

This is a familiar logistic function of the difference in action values $\Delta Q = Q(A) - Q(B)$ and $P(B) = 1 - P(A)$. When $Q(A)$ is much smaller than $Q(B)$, $P(A)$ is less than 0.5, otherwise $P(B)$ is larger than 0.5. With $\beta = 0$, $P(A) = P(B) = 0.5$, the choice is completely random, the policy is purely exploratory. With large $\beta \to \infty$, the option with a larger action value is always selected, and the policy is 100% exploitatory. Note that in principle the policy can be described by different mathematical expressions, for instance, with two options another possible expression of the decision policy is the quotient formula $P(A) = Q(A) / \left(Q(A) + Q(B)\right)$, according to which the probabilities of choosing the two options match their relative action values, $P(A) / P(B) = Q(A) / Q(B)$. Interestingly, fitting of behavioral data from a monkey experiment with both models found that animal's choice policy conforms better with the softmax formula than the quotient formula (Corrado et al. 2005). A biologically realistic neural circuit model naturally gives rise to a softmax policy rule (Soltani and Wang 2006; Wang 2008b) (see Chapters 6 and 7).

The reinforcement learning framework for action selection has two main components. The first one is an "actor" responsible for action selection. The second is a "critic" in charge of assessing how well expectation matches actual outcome and updating action values accordingly (Sutton and Barto 2018). A key brain structure for action selection and movement control is the basal ganglia (Mogenson et al. 1980; Alexander and Crutcher 1990; Hikosaka et al. 2000). The basal ganglia circuit has an input area

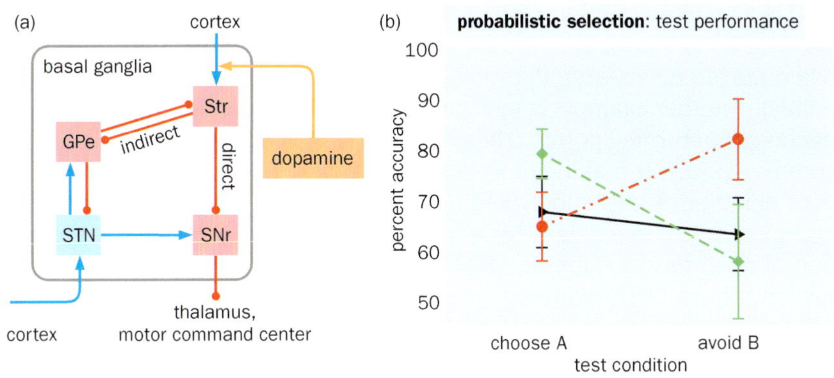

Figure 4.7 Basal ganglia for action selection under the influence of dopamine. (a) Schematic of the basal ganglia circuit. Red filled circle: Inhibitory; light blue arrow: Excitatory. (b) Reduced learning from positive feedback and enhanced learning from negative feedback in Parkinson's participants compared to healthy participants in a probabilistic choice experiment. Relative to healthy participants (black), Parkinson's patients off medication (red) perform slightly worse in responding by choosing A associated with a high probability of positive feedback, but better in not responding to B associated with a likely negative feedback. On medication (green), Parkinson's patients show the opposite trend, presumably as a result of boosted dopamine by medication. Panel (a) is reproduced from Wei, W., & Wang, X. J. (2016). Inhibitory control in the cortico-basal ganglia-thalamocortical circuit: Complex modulation and its interplay with working memory and decision-making. Neuron, 92(5), 1093.] [Frank, M. J., Seeberger, L. C., & O'Reilly, R. C. (2004). By carrot or by stick: Cognitive reinforcement learning in parkinsonism. Science, 306(5703), 1940–1943.

(the striatum [Str]) and an output area (the substantia nigra pars reticulata [SNr] and the internal segment of globus pallidus [GPi]). There are two main pathways from the input to the output of basal ganglia (**Figure 4.7a**). The direct or "Go" pathway is from Str to SNr/GPi. Because projection neurons in both Str and SNr/GPi are GABAergic, activation of this pathway reduces inhibition from the basal ganglia (disinhibition), thereby facilitating movement. In the indirect or "NoGo" pathway, the Str projects to the external segment of globus pallibus (GPe), where the principal cells are also GABAergic. The GPe connects to the subthalamic nucleus (STN), which in turn sends glutamatergic outputs to the SNr/GPi. The overall effect of the indirect pathway activation is enhancing inhibition from the basal ganglia, thereby suppressing movement.

The striatum is the brain region with the densest dopaminergic innervation, a loss of dopamine inputs in the striatum causes grave movement impairments in Parkinson's disease (DeLong 1990). Dopamine terminal forms a triad with the synapses from the cortex to the striatum (the other two components are the presynaptic terminal and the postsynaptic terminal membrane) (Freund et al. 1984; Surmeier et al. 2010), through which reward can have a potent influence on action. Specifically, plasticity modulated by reward prediction error (Wickens et al. 2003; Surmeier et al. 2010) enables the corticostriatal synapses to learn action values (the critic). As a result, neural firing of projection neurons in the striatum is correlated with action values (Samejima et al. 2005). The basal ganglia plays a central role in action selection (the actor), the outcome of which in turn updates the action values themselves. A neural network model (Frank et al. 2004) implemented reinforcement learning in the basal ganglia, taking into account the knowledge that dopamine potentiates the corticostriatal synapses of the direct pathway via D1 receptors while depressing those of the indirect pathway via D2 receptors (Surmeier et al. 2010). The opposing impact of dopamine on Go and NoGo pathways predicts a specific reinforcement deficit in Parkinson's patients. Due to reduced level of dopamine in their brains, learning from positive feedback to choose the better option A is impaired, as observed experimentally. Moreover, dopamine-modulated depression along the NoGo pathway should be less than in normal subjects, and therefore the propensity to avoid responding to the worse option B should be enhanced. This prediction was confirmed in behavioral experiments with healthy human subjects and those afflicted by Parkinson's disease (**Figure 4.7b**).

4.3.4 Temporal-Difference Learning

In simple classical conditioning experiments, behavioral trials are independent from each other. In real-life situations, however, predictions involve a sequence of events. For instance, in a basketball game, scoring a goal is a result of multiple choices between two competing teams, such as passing the ball (mostly in the forward direction, but in the home direction when judged necessary), that do not yield immediate reward (earning the points). Another example is navigating a maze, which requires a sequence of choices at turning points. These examples also illustrate two features of common decision-making processes. First, behavior unfolds in time rather than being artificially subdivided into discrete trials. Second, actions such as a pass of ball in the game or a turn in the maze are selected in expectation of rewards in the future. The temporal-difference (TD) algorithm (Sutton and Barto 2018) addresses these issues by postulating that the goal of TD learning is for an expected value V_t to predict the sum of all future rewards,

$$V_t(s) = \langle r_t + \gamma r_{t+1} + \gamma^2 r_{t+2} + \gamma^3 r_{t+3} + \ldots | s_t = s \rangle \tag{4.17}$$

where $\langle x|s \rangle$ denotes the average of x (for rewards harvested stochastically) conditioned on the current state s. States refer to an agent's observations about the environment, such as the position in a maze, sensory stimuli or the time point in a behavioral trial. In this equation, t may index real time (with a fine resolution instead of trial number), or discrete steps like turns in a maze. The parameter γ (between 0 and 1) is a discounting factor, expressing the fact that rewards more remote into the future are less desirable and should be weighted less.

The desired expected value is the total predicted future reward, so how can it be updated incrementally from a time step to another time step? Eq. (4.17) has the form of $V_t = r_t + \gamma V_{t+1}$, but this compact formula does not explicitly spell out the fact that in successful time steps, reward outcomes depend on the states that the system sequentially goes through. A decision-maker in state $s_t = s$ choses an action a_t using a state-dependent policy $\pi(s)$, which leads to the transition to another state $s_{t+1} = s'$ with probability $P(s_{t+1} | s_t, \pi(s_t))$. Taking this into consideration, we have:

$$V^{\pi}(s) = r(s) + \gamma \sum_{s'} P(s'|s,\pi(s)) V^{\pi}(s') \tag{4.18}$$

which is called the Bellman equation (Bellman 1957) with the expected value at t+1 expressed in the second term averaged over states $\{s'\}$. The TD prediction error is given by the difference between the two sides of this equation, which should converge to zero through learning. While the true values of these two quantities are not known in a decision process, an approximation can be made based on their estimates. Specifically, once the new state $s_{t+1} = s'$ is determined by an action chosen in the state $s_t = s$, define:

$$\delta_t = r_t + \gamma V^{\pi}(s_{t+1} = s') - V^{\pi}(s_t = s) \tag{4.19}$$

which is used to update V_t using Eq. (4.12) in TD learning. Learning process stops when $\delta_t = 0$ and V_t matches with the cumulative future reward.

Applying TD learning to classical conditioning, the model describes the time evolution of the expected value and RPE in a single trial. Initially, a CS has no meaning, and V_t is zero until reward r is delivered at time T, when $\delta = r - 0 = r$. If the CS is followed by r after a fixed time delay D in repeated trials, TD learning builds the expected value that predicts future rewards, and eventually we have $V_{t=T-D} = r$ at the occurrence of the CS, whereas $V_{t=T} = 0$. Phasic activity of dopamine neurons in VTA displays precisely this shift backward in time from the US to the CS (**Figure 4.6a**, top and middle panels). In another test of dopamine neurons signaling TD prediction error, distinct visual cues are associated with different reward probabilities ($p = 0, 0.25, 0.5, 0.75$ and 1). The grand average is 0.5 of these reward probabilities. In trials where the CS indicates a low $p = 0.25$ (smaller than the overall expectation of 0.5), actual reward delivery is a pleasant surprise, and dopamine neural activity should be high at the time of the US. In contrast, dopamine firing should display a dip at the time of reward in trials with a high $p = 1$ (larger than the overall expectation of 0.5). More generally, according to TD learning, after training the firing rate of dopamine neuronal phasic activity increases with the reward probability at the time of the CS, but shows the opposite dependence at the time of the US. These model predictions were confirmed in a monkey experiment (**Figure 4.8**) (Fiorillo et al. 2003).

Theory of TD learning for action selection in instrumental conditioning is called Q-learning (Watkins and Dayan 2012), where action values depend on both action a and state s. Q-learning is in principle capable of describing sequential actions like solving the problem of a maze.

While phasic firing of dopamine cells has often been studied using simple classical conditioning tasks, depending on the behavioral paradigm, dopamine neural population activity can be fairly complex, as illustrated by ramping of spike firing (Howe et al. 2013; Guru et al. 2020; Kim et al. 2020; Gao et al. 2021) and heterogeneity across individual dopamine cells (Engelhard et al. 2019). Moreover, Berridge (2007) found that dopamine neural activity signals incentive salience of actions. The perspective of incentive salience emphasizes an animal's internal state. For instance, the same food reward is highly salient as incentive to respond to it when you are hungry but less so if you are already satiated with a full stomach. In a mathematical model (Zhang et al. 2009a), dopamine is interpreted as signaling motivation or "wanting." Incentive salience is represented by $\log\kappa$, where the parameter κ reflects the bodily state. Devaluation (such as satiation) is represented by $\kappa < 1$ ($\log\kappa < 0$), while enhancement (such as appetite or sensitization) is represented by $\kappa > 1$ ($\log\kappa > 0$). The dopamine system is likely important for coding RPE, saliency and motivation in the brain (Berke 2018).

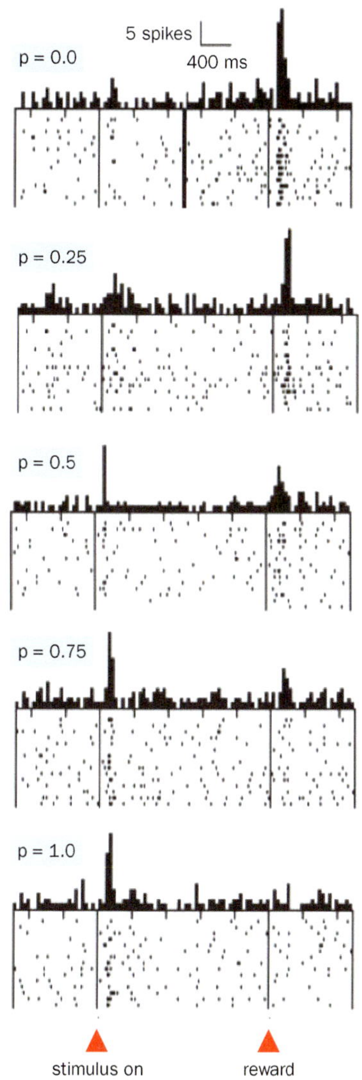

Figure 4.8 Activity in a single cell recorded in VTA of macaque monkey, in response to the conditioned stimuli and reward at various reward probabilities, increasing from top to bottom. Reproduced from Fiorillo, C. D., Tobler, P. N., & Schultz, W. (2003). Discrete coding of reward probability and uncertainty by dopamine neurons. Science, 299(5614), 1898–1902.

4.4 UNSUPERVISED LEARNING

Our brain is shaped by experience as well as genetic programming. For instance, in the primary visual cortex V1 of adult mammals, neurons display preference to inputs from one eye over the other. This ocular dominance requires visual stimulation during brain development and is absent in animals reared early in life with bilateral eye closure (Wiesel and Hubel 1965; Miller et al. 1989; Crair et al. 2001). As another example, distribution of preferred orientation in V1 cells is not uniform, but instead displays two peaks reflecting a preponderance of vertical and horizontal edges of objects in the environment (Girshick et al. 2011), which presumably would not be the case in a world devoid of vertically and horizontally oriented edges of visual objects. Modification of neural circuit merely through exposure to the environment is called unsupervised learning, and the underlying mechanism is the subject of many studies of the brain's plasticity (Kandel 2001; Holtmaat and Caroni 2016).

An influential idea for the biological basis of learning and memory was articulated by Donald Hebb (Hebb 1949):

> When an axon of cell A is near enough to excite a cell B and repeatedly or persistently takes part in firing it, some growth process or metabolic change takes place in one or both cells such that A's efficiency, as one of the cells firing B, is increased.

This Hebb rule postulates that learning takes place at the synaptic connections from one neuron to another, and memory is stored by their connection weight. Synaptic modification requires co-activation of both presynaptic and postsynaptic neurons, which is often summarized by the aphorism "neurons that fire together, wire together." This differs from short-term synaptic plasticity, which only depends on the presynaptic neuron activity (Chapter 2).

4.4.1 Hebbian Plasticity Rules

A simple mathematical equation for the Hebbian learning rule is:

$$\frac{dw}{dt} = r_{pre} r_{post} \tag{4.20}$$

where w is the synaptic weight from a presynaptic neuron with firing rate r_{pre} and a postsynaptic neuron with firing rate r_{post}. In this simplest form of plasticity, the synaptic weight w is a perfect integrator of the co-activation of the two neurons over time, without decay (forgetting). This is the simplest of a number of mathematical variants of Hebbian learning (Herz et al. 1991), including the covariance rule, where r is replaced by the difference between its current value and average value (Sejnowski 1977).

Long-term potentiation of synapses was discovered by Bliss and Lomo (1973) at a synaptic pathway in the hippocampus, a brain structure important for acquiring new memories. Subsequent research identified the crucial role of the NMDA receptor (**Figure 4.9**). By virtue of its dependence on the membrane potential (Chapter 2, **Figure 2.21a**), the NMDA receptor-mediated synaptic transmission is possible only when the postsynaptic neuron is sufficiently depolarized. On the other hand, of course the NMDA receptor-dependent synaptic transmission requires presynaptic firing. Therefore, the NMDA receptor acts as a coincidence detector of both presynaptic and postsynaptic activations. Furthermore, NMDA channel

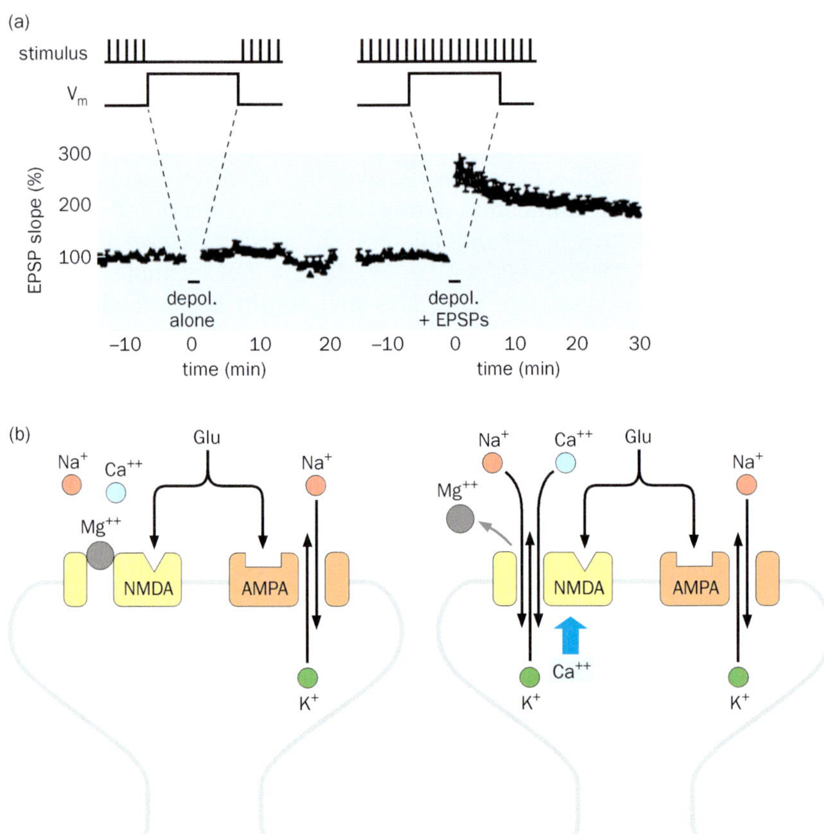

Figure 4.9 Long-term potentiation and its dependence on NMDA receptor. (a) Induction of LTP. Top: A train of stimuli and the control of a postsynaptic neuron's membrane potential (V_m). Bottom: Synaptic strength as measured by the slope of excitatory postsynaptic potential (EPSP) is unchanged by the stimulation when V_m is at rest (left). By contrast, with V_m depolarization it is increased after stimulation, and the change persists for a long time. (b) Role of NMDA. Left panel: When the postsynaptic neuron's membrane potential is near resting level, the NMDA receptor-mediated ion channel is closed due to magnesium block, and synaptic transmission is limited to AMPA receptor-mediated ion channel. Right panel: When the postsynaptic membrane potential is depolarized, the Mg^{2+} block is relieved, and thus the NMDA receptor channel opening requires both presynaptic activation (for transmitter release) and postsynaptic activation. Importantly, NMDA channel is permeable to Ca^{2+}. The intracellular calcium rise triggers a cascade of chemical processes for the induction of long-term plasticity. Reproduced from Nicoll, R. A. (2017). A brief history of long-term potentiation. Neuron, 93(2), 281–290.

opening allows Ca^{2+} ions to flow from the external medium to the cytosol, and Ca^{2+} influx has been shown to be essential for long-term plasticity induction. NMDA receptor's dependence on the postsynaptic depolarization and permeability to calcium are precisely the biological properties required for Hebbian plasticity.

Since the firing rates are positive, w can only grow according to Eq. (4.20). By contrast, synaptic efficacy of excitatory connections can either increase (long-term potentiation, LTP) or decrease (long-term depression, LTD) (Stent 1973; Malenka and Bear 2004). Generally, presynaptic activity is required (r_{pre} must be sufficiently high) for the induction of synaptic modification; LTD is observed when postsynaptic activity r_{post} is not zero but low, whereas LTP takes place when r_{post} is sufficiently high. This is captured by the Bienenstock-Cooper-Munro (BCM) rule (Bienenstock et al. 1982):

$$\frac{dw}{dt} = \phi(r_{post})r_{pre} \tag{4.21}$$

where the function $\phi(r_{post})$ is negative if r_{post} is below a threshold θ_m and positive when the activity level is above it. A simple example is $\phi(r_{post}) = r_{post}(r_{post} - \theta_m)$. This model was tested in an experiment *in vitro* where modifications were induced at a pathway called Schaffer collaterals from CA3 to CA1 in the hippocampus. By varying the frequency of "conditioning stimulation" of the connection fibers, the magnitude of the postsynaptic response was monotonically increased. The long-term change of synaptic efficacy displays a non-monotonic dependence on the stimulus frequency as predicted by the BCM model (**Figure 4.10a**). The BCM rule has several desirable properties. Unlike Eq. (4.20), the BCM rule is asymmetrical with regard to r_{pre} and r_{post}, thus the connection strength from one neuron to another generally differs from that in the opposite direction.

However, the BCM rule is not stable because of a reciprocal interplay between r_{post} and w. Consider a simple equation $dr_{post} / dt = -r_{post} / \tau + wr_{pre}$. If r_{post} is larger than θ_m, synaptic weight is potentiated, which in turn drives r_{post} to an even higher rate, and this positive feedback leads to runaway instability. Conversely, with activity r_{post} smaller than θ_m, w is depressed, leading to further reduction of r_{post}, eventually driving w to a minimum value. A solution to this problem was proposed by assuming that the threshold θ_m itself is adjustable based on the long-term average of postsynaptic neuronal firing rate $\langle r_{post} \rangle$. Specifically,

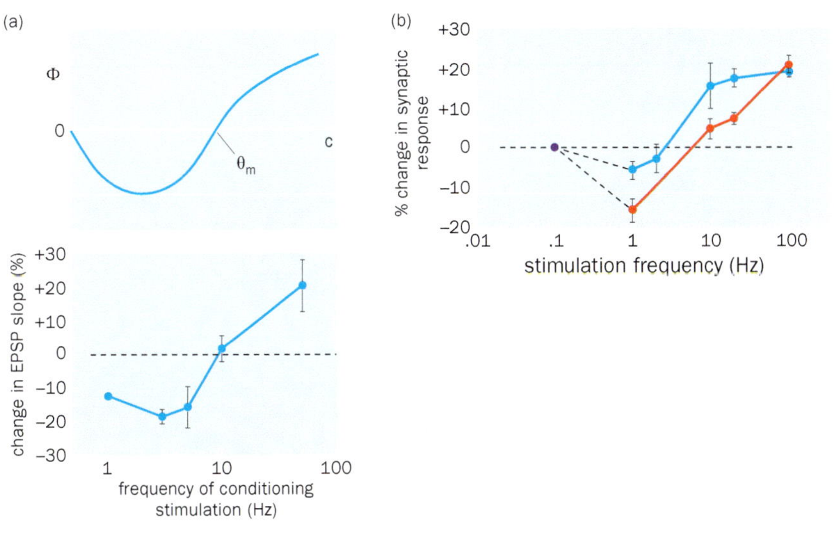

Figure 4.10 The Bienenstock-Cooper-Munro (BCM) rule for long-term potentiation (LTP) and long-term depression (LTD). (a) Comparison of the BCM theoretical dependence of synaptic plasticity on stimulation frequency (top) with physiological measurements of the strength of a synaptic pathway in the hippocampus (bottom). (b) At a synaptic connection in the primary visual cortex, the synaptic change relative to baseline is plotted as a function of the stimulation frequency, in normal rat (red circle) and after light deprivation (blue circle). The modification threshold from LTD to LTP is lower with reduced activity (after light deprivation) than in the control condition, in support of the theoretical proposal of a sliding modification threshold. Reproduced Bear, M. F. (1996). A synaptic basis for memory storage in the cerebral cortex. Proceedings of the National Academy of Sciences, 93(24), 13453–13459. Copyright (1996) National Academy of Sciences, U.S.A.; panel (b) is reproduced from A Kirkwood, M Rioult, M Bear (1996). Experience-dependent modification of synaptic plasticity in visual cortex. Nature (London), 381(6582), 526–528.

the BCM model assumes that θ_m grows nonlinearly with a long-term average of r_{post}, like $\theta_m = \langle r_{post} \rangle^2 / r_0$ with a constant r_0. Thus, after a long period with reduced activity, θ_m is lowered so that LTP becomes possible at lower r_{post}, whereas after a long period of heightened activity θ_m is raised, leading to a larger range of r_{post} for LTD. Mathematically, one can show that $\langle r_{post} \rangle = r_0$ is a stable solution of Eq. (4.21) with a "sliding modification threshold." Such "metaplasticity" was experimentally observed in the neocortex (**Figure 4.10b**) (Bear 1996). Note that there is no plasticity in either direction when the stimulus is too weak (below 0.1 Hz).

4.4.2 Pattern Formation during Brain Development

Hebbian plasticity has been applied to account for neural circuit wiring and emergence of brain functions during development. Genetic programming determines much of developmental processes before birth. Postnatally, how the brain shapes up is also under the influence of the environment. One of the best studied systems is the formation of maps in the primary visual cortex V1, such as patterns of ocular dominance (Hubel and Wiesel 1977; Shatz 1990) of strong preference to either the left (L) or right (R) eye (**Figure 4.11a**). Ocular dominance is absent in animals whose eyes are covered during the early postnatal period, demonstrating that its emergence depends on neural activity in response to visual stimulation.

Activity-dependent formation of organized structures in sensory cortical areas can be explained by Hebbian plasticity, as shown by mathematical modeling (Swindale 1980, 1996; Miller 1990; Kaschube et al. 2010). A key concept is the correlation-based learning principle. Consider ocular dominance in a two-dimensional V1 cortical sheet. Initially, V1 neurons receive randomly wired connections from neurons in the LGN, which are selective for L or R by virtue of inputs from the L or R retina. Let us denote

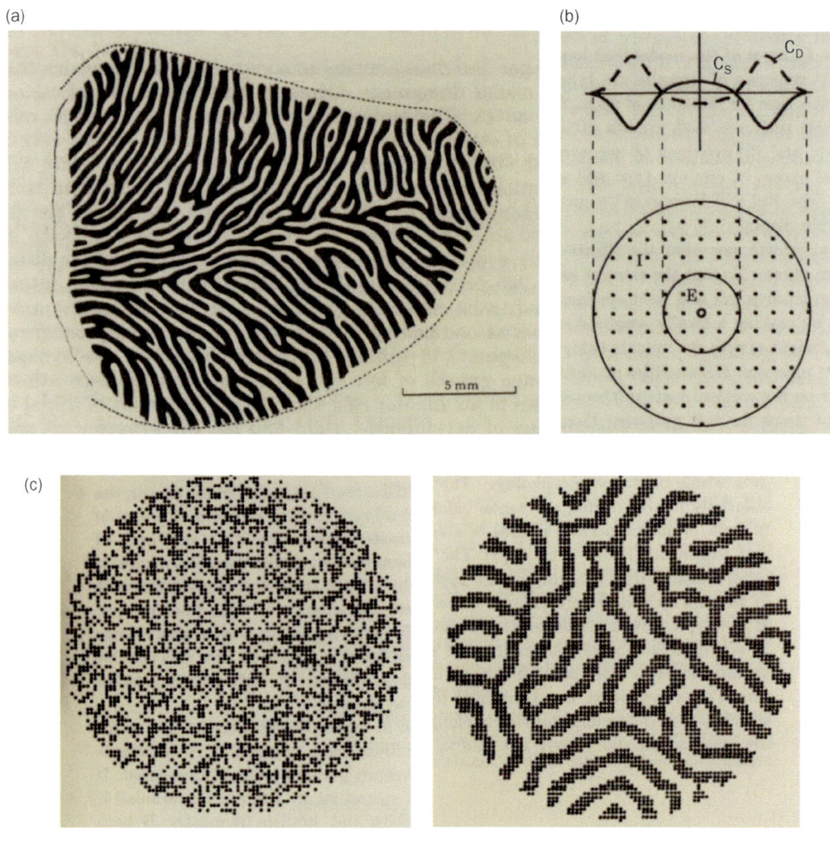

Figure 4.11 Ocular dominance in V1 and correlation-based model of its formation in V1 during development. (a) V1 cortical space displays a beautiful pattern of neurons with preference to left and right eyes (black and white, respectively). (b–c) The Swindale model. (b) Correlation of LGN neurons as a function of their spatial distance for those with the same ocular preference (C_S, solid curve) and opposite preference (C_D, dashed curve). (c) Synaptic density selective for two eyes are initially randomly distributed (left), and a strip pattern of ocular dominance (black and white for L and R, respectively) emerges through correlation-based plasticity (right). Panel (a) is reproduced from Hubel, D. H., & Wiesel, T. N. (1977). Ferrier lecture—Functional architecture of macaque monkey visual cortex. Proceedings of the Royal Society of London. Series B. Biological Sciences, 198(1130), 159; panels (b)–(c) are reproduced from Swindale, N. V. (1980). A model for the formation of ocular dominance stripes. Proceedings of the Royal Society of London. Series B. Biological Sciences, 208(1171), 243–264.

by $r^{V1}(\mathbf{x})$ the firing rate of a V1 neuron located at $\mathbf{x}$; it is driven by LGN relay cells of both L and R preferences at thalamic location $\mathbf{y}$. The LGN neurons have firing rates $r_j^{LGN}(\mathbf{y})$ and their thalamocortical synaptic weights $w_j(\mathbf{x};\mathbf{y})$ $(j = L,R)$. Assume that:

$$r^{V1}(\mathbf{x}) \sim \sum_{j=L,R} \int w_j(\mathbf{x};\mathbf{y}') r_j^{LGN}(\mathbf{y}') d\mathbf{y}' \tag{4.22}$$

According to Hebbian plasticity,

$$\frac{dw_k(\mathbf{x};\mathbf{y})}{dt} \propto r^{V1}(\mathbf{x}) r_k^{LGN}(\mathbf{y}) \sim \sum_{j=L,R} \int w_j(\mathbf{x};\mathbf{y}') r_k^{LGN}(\mathbf{y}) r_j^{LGN}(\mathbf{y}') d\mathbf{y}' \tag{4.23}$$

Replacing the product $r_k^{LGN}(\mathbf{y}) r_j^{LGN}(\mathbf{y}')$ by its temporal average (on a relatively long timescale, but short compared that of brain development), yields a correlation $C_{k,j}^{LGN}(\mathbf{y},\mathbf{y}')$ between the LGN neurons:

$$\frac{dw_k(\mathbf{x};\mathbf{y})}{dt} \sim \sum_{j=L,R} \int w_j(\mathbf{x},\mathbf{y}') C_{k,j}^{LGN}(\mathbf{y};\mathbf{y}') d\mathbf{y}' \tag{4.24}$$

Thus, synaptic plasticity is based on correlations in the input activities. Additional processes underlie cooperation and competition in plasticity (Miller 1990). For instance, within V1, a pyramidal neuron tends to excite other nearby pyramidal cells (cooperation), while exerting an effective inhibitory influence on others far away (competition), which is manifested by a Mexican-hat-shaped correlation as a function of the spatial distance between pairs of neurons (Chapter 3, Section 3.5.3). Incorporation of such interactions in the correlation functions is important for the emergence of patchy patterns in the cortex through plasticity.

Swindale (1980) proposed a simple correlation-based model for ocular dominance formation (see Miller et al. (1989) for a more elaborate model). The model describes the overall synaptic densities, $n_j(\mathbf{x})$ $(j = L,R)$, of LGN inputs selective for eye j = L and R at cortical position $\mathbf{x}$. The dynamics of plasticity obeys:

$$\frac{\partial n_L(\mathbf{x},\mathbf{t})}{\partial t} = \left(\int \left(n_L(\mathbf{y}) C_{LL}^{LGN}(\mathbf{x},\mathbf{y}) + n_R(\mathbf{y}) C_{LR}^{LGN}(\mathbf{x},\mathbf{y})\right) d\mathbf{y}\right) f(n_L); \tag{4.25}$$

$$\frac{\partial n_R(\mathbf{x},\mathbf{t})}{\partial t} = \left(\int \left(n_R(\mathbf{y}) C_{RR}^{LGN}(\mathbf{x},\mathbf{y}) + n_L(\mathbf{y}) C_{RL}^{LGN}(\mathbf{x},\mathbf{y})\right) d\mathbf{y}\right) f(n_R) \tag{4.26}$$

where $f(n)$ is a factor to ensure stability during a growth process. For instance, with a simple choice of $f(n) = n(n_{max} - n)$, synaptic weights remain bounded between 0 and n_{max}. Correlation functions of LGN neurons incorporate cooperation and competition by assuming that those driven by the same eye $(C_{LL} = C_{RR} = C_S)$ are positive if they are spatially close and negative otherwise; the opposite holds for LGN neurons driven by different eyes $(C_{LR} = C_{RL} = C_D)$ (**Figure 4.11b**).

Formation of ocular dominance pattern is then mathematically defined in terms of the dynamics of this system. Given that initially $n_L(\mathbf{x})$ and $n_R(\mathbf{x})$ are spatially random, what is their ultimate distribution in the cortical space? **Figure 4.11c** shows that, with sufficiently strong cooperation and competition, starting with a random and uniform distribution (left panel) a strip pattern of ocular dominance emerges from plasticity (right panel), similar to the experimental observation (**Figure 4.11a**).

The Swindale model and other related models illustrate how sculpting brain circuits by activity-dependent Hebbian plasticity can be mathematically framed as pattern formation (Turing 1952; Nicolis and Prigogine 1977;

Murray 1993) applied to the brain's development. Note that, whereas unsupervised learning may be adequate for describing early sensory areas, some other learning mechanism is afoot in higher cortical areas such as the prefrontal cortex (PFC). The PFC does not fully mature until the twenties in humans; executive functions mediated by the PFC continue to develop through adolescence (Goldman-Rakic 1987b; Diamond and Goldman-Rakic 1989; Huttenlocher and Dabholkar 1997; Selemon 2013), when experiences involve supervised learning and reinforcement learning and likely other hitherto unknown types of plasticity.

4.4.3 Spike-Timing Dependent Plasticity

Hebbian synaptic modification rules postulate that a presynaptic neuron takes part in causing postsynaptic firing activity; therefore, a Hebbian synapse is not only a coincidence detector but also a causality detector (Bi and Poo 2001). Specifically, consider a directed connection from a presynaptic neuron and a postsynaptic neuron, where a single spike is triggered in each of the two neurons, at time t_{pre} and t_{post}, respectively. The presynaptic neuron plays a role in triggering the spike in the postsynaptic neuron only when t_{pre} is earlier than t_{post} ($\Delta t = t_{post} - t_{pre}$ is positive), but not when Δt is negative (Gerstner et al. 1996). Long-term synaptic modification as a function of Δt indeed displays a spike-timing dependent plasticity (STDP) curve (**Figure 4.12a**). Not only the sign of synaptic modification changes with that of Δt , but also there is a relatively narrow time window (tens of milliseconds) within which synaptic plasticity is possible. STDP has been demonstrated in the hippocampus and neocortex (Markram et al. 1997; Bi and Poo 1998, 2001; Dan and Poo 2006), as well as other brain regions like the striatum (Pawlak and Kerr 2008).

In a mathematical description of STDP, the amount of synaptic modification $F(\Delta t)$ depends on Δt of paired presynaptic and postsynaptic spikes (Song et al. 2000):

$$F(\Delta t) = A_+ \exp(-\Delta t / \tau_+), \text{if } \Delta t > 0 \quad (4.27)$$

$$= -A_- \exp(\Delta t / \tau_-), \text{if } \Delta t < 0 \quad (4.28)$$

where A_+ and A_- are positive and determine the maximal changes. The time constants τ_+ and τ_- control the sizes of time window for LTP and LTD, respectively. This model mathematically describes a STDP curve like the one in **Figure 4.12a**.

Since STDP depends on the relative timing of the pre- and postsynaptic spikes, there would be no net synaptic modification if the spike discharges of a pair of neurons were uncorrelated Poisson processes (Kempter et al. 1999; Song et al. 2000). By contrast, if a neuron consistently fires right before another neuron, Hebbian STDP would create asymmetrical connections between the two. This observation led to the suggestion that STDP represents a mechanism for sequence learning in spatial navigation (Abbott and Blum 1996). As an animal moves along a linear track, location- and direction-selective place cells in the hippocampus are activated sequentially (Chapter 3, **Figure 3.10a**). If two neural pools fire one after another within tens of milliseconds, according to the STDP rule, synaptic connections from those firing earlier to those firing later should be strengthened, whereas synaptic connections in the opposite direction are weakened. An observable consequence is that the shape of the spatial tuning curve of hippocampal place cells becomes asymmetric (skewed) with experience, a prediction that was confirmed experimentally (Mehta et al. 2000).

Different synapse types display a diversity of STDP curves (Abbott and Nelson 2000), at different synapses (**Figure 4.12b**) or even at an identical

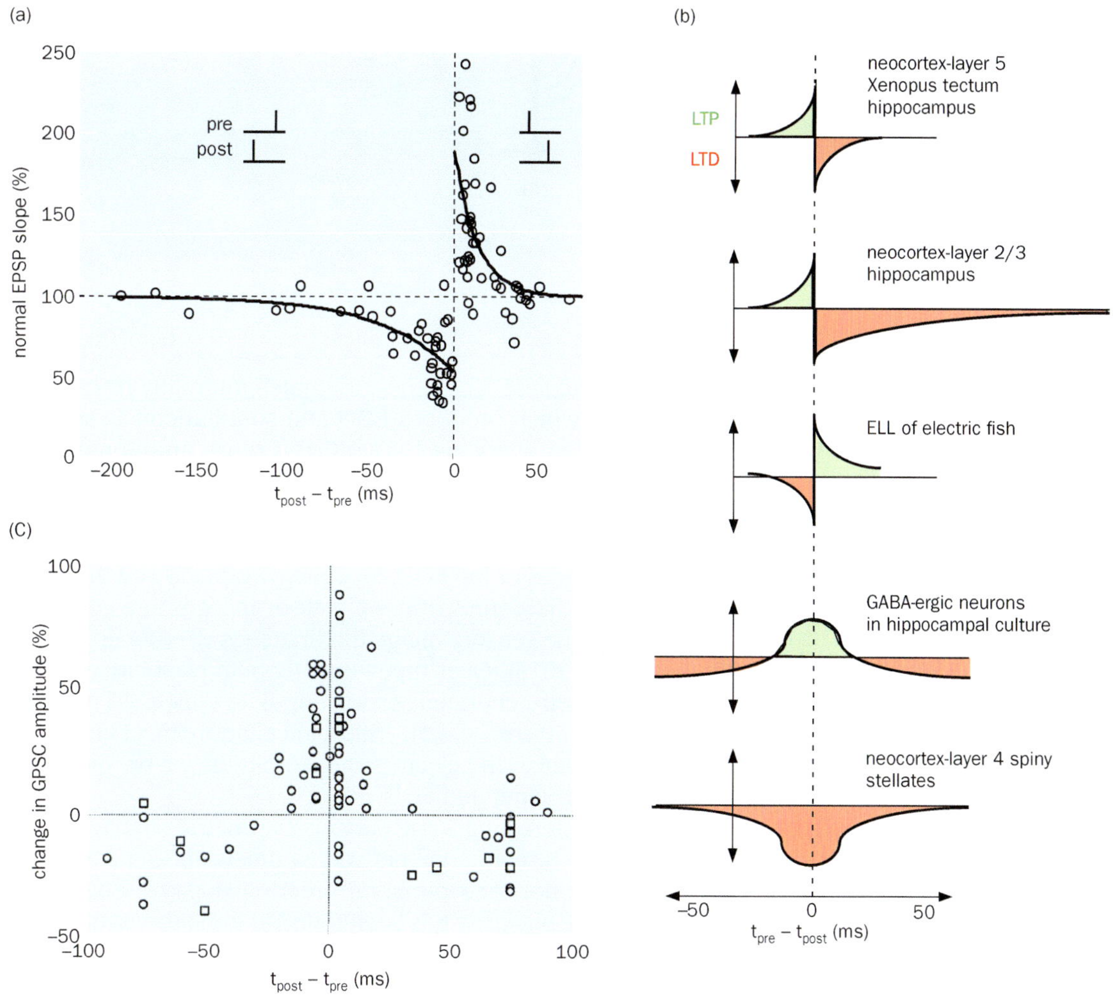

Figure 4.12 Spike-timing dependent plasticity (STDP). Modification of synaptic strength is shown as a function of time interval between a presynaptic spike and a postsynaptic spike. (a) Potentiation (respectively depression) occurs when a presynaptic spike precedes (respectively follows) the postsynaptic spike. (b) Examples of distinct STDP curves for different types of synapses. Green: LTP; Red: LTD. (c) At an inhibitory synapse in the hippocampus, the STDP curve is roughly symmetrical, with LTP for short time interval and LTD for longer time interval regardless of the relative timing of the presynaptic and postsynaptic spikes. Panel (a) is reproduced from Bi G, Poo M. (2001). Synaptic modification by correlated activity: Hebb's postulate revisited. Annu Rev Neurosci. 24(1), 139–66; panel (b) is reproduced from Abbott, L. F., & Nelson, S. B. (2000). Synaptic plasticity: Taming the beast. Nature Neuroscience, 3(11), 1178–1183; panel (c) is reproduced from Woodin, M. A., Ganguly, K., & Poo, M. M. (2003). Coincident pre- and postsynaptic activity modifies GABAergic synapses by postsynaptic changes in Cl−transporter activity. Neuron, 39(5), 807–820.

synapse under different physiological conditions (Inglebert et al. 2020). In particular, anti-Hebbian STDP (**Figure 4.12b**, third from top) has the sign flipped compared to Hebbian STDP (first from top). This was actually one of the first examples of STDP discovered experimentally, in a cerebellum-like structure of electric fish. Electric fish emit an electric organ discharge (EOD) and detect the distortion that nearby objects cause in the self-generated electric field, thereby sensing the environment. This system must solve the following computational problem: how does it "cancel out" a self-generated electric field in order to respond only to distortions for object detection?

The brain of electric fish has a cerebellum-like structure, where granule cells project to Purkinji-like neruons called medium ganglion (MG) cells with weights that undergo anti-Hebbian STDP (Bell et al. 1997; Kennedy et al. 2014). The system receives two types of inputs. One is direct sensory stimulation from electroreceptors, which are activated by the

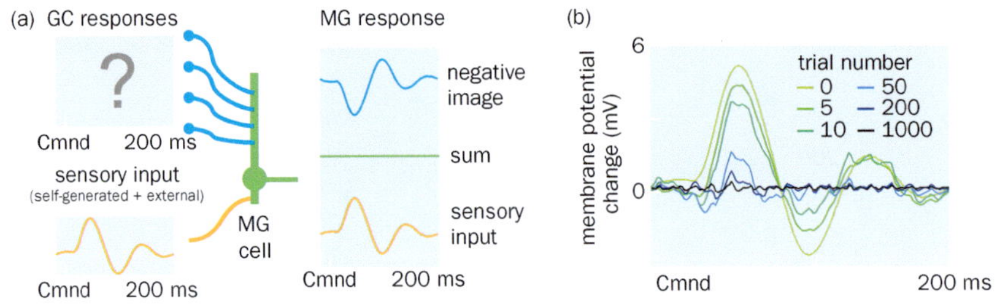

Figure 4.13 Anti-Hebbian STDP in Purkinje-like neurons. (a) Schematic of negative image formation and sensory cancellation in a medium ganglion cell (MG). Corollary discharge response in granule cells (GCs) display a wide range of response latencies. Cmnd, EOD command. (b) Cancellation of the change in membrane potential caused by sensory input locked to the EOD motor command in a model medium ganglion cell. The medium ganglion cell received 20,000 model granule cell inputs with synaptic strengths that were adjusted by anti-Hebbian STDP. Select trials show the time course of cancellation. The temporal profile of the sensory input (trial 0, yellow) was chosen to resemble the effects of the EOD on electroreceptors. Reproduced from Kennedy, A., Wayne, G., Kaifosh, P., Alviña, K., Abbott, L. F., & Sawtell, N. B. (2014). A temporal basis for predicting the sensory consequences of motor commands in an electric fish. Nature Neuroscience, 17(3), 416–422.

external electric field. The other originates internally from a motor command center, which produces EOD and simultaneously sends an "efference copy" or corollary discharge (CD) (Crapse and Sommer 2008) back to the brain. Interestingly, although EOD command is brief in time, CD induces firing of granule cells with a wide range of temporal latencies (up to about 200 ms) (Kennedy et al. 2014), covering the timespan of the sensory consequences of the EOD. Modeling work showed that anti-Hebbian STDP reduces the connection weights of the granule cells to MG cells, selectively for the sensory image that matches CD (Roberts and Bell 2000; Kennedy et al. 2014). Over time, anti-Hebbian plasticity creates a "negative image" of self-generated sensory input, to which MG neurons become unresponsive (**Figure 4.13**). This cancellation mechanism allows the electric fish to detect the electric signal emitted by a prey without interference by the self-generated EOD.

Another interesting STDP curve is symmetric so that LTP or LTD only depends on the time interval between a presynaptic spike and a postsynaptic spike but not their temporal ordering (**Figure 4.12b**, fourth from top). Such STDP plasticity has been found at inhibitory synapses on excitatory neurons (Woodin et al. 2003; Froemke 2015) (**Figure 4.12c**). Plasticity of inhibition, implemented in network models, has been shown to be crucial for the maintenance of a precise excitation-inhibition balance in the face of modifications of excitatory synaptic strengths (Vogels et al. 2011; Froemke 2015; Barron et al. 2017; Hennequin et al. 2017).

More elaborated models than Eqs. (4.27–4.28) have been developed (Pfister and Gerstner 2006; Clopath et al. 2010) to better account for the experimental observations on STDP, including the dependence of plasticity on neuronal firing rates (Sjöström et al. 2001).

4.4.4 A Calcium-Based Plasticity Model

Recall that influx of calcium through NMDA channels in the postsynaptic neuron is necessary for LTP induction (Malenka and Nicoll 1999). The entry of calcium then induces a cascade of intracellular chemical reactions leading to the induction of long-term plasticity (Kandel 2001). Based on these observations, models have been developed for calcium-based synaptic plasticity (Shouval et al. 2002). Graupner and Brunel (2012) proposed a plasticity model based on the intracellular calcium level $[Ca^{2+}]$ for STDP. In the model, a synaptic efficacy variable w can be either in a Down state ($w = 0$) or an Up state ($w = 1$) in the absence of spiking activity, an assumption that is supported by some experimental evidence (Petersen et al. 1998; O'Connor et al. 2005). This is described by a dynamical equation like $dw/dt = -w(1-w)(w_* - w)$, for which $w = 0$ and $w = 1$ are two stable states (steady-state attractors) separated by an unstable steady-state $w = w_*$ (between 0 and 1). A presynaptic spike leads to an increase of $[Ca^{2+}]$ with size C_{pre}, and a postsynaptic spike induces an increase of $[Ca^{2+}]$ with size C_{post}. With no $[Ca^{2+}]$ influx $[Ca^{2+}]$ decays exponentially with a time constant τ_{Ca}. The model assumes that LTD is induced if $[Ca^{2+}]$ exceeds a threshold θ_d, whereas LTP takes place if $[Ca^{2+}]$ exceeds a threshold θ_p. Therefore, the synaptic efficacy obeys the following equation:

$$\tau_w \frac{dw}{dt} = -w(1-w)(w_* - w) + \gamma_p(1-w)H([Ca^{2+}] - \theta_p) - \gamma_d w H([Ca^{2+}] - \theta_d) + noise(t) \tag{4.29}$$

where γ_p and γ_d control the size of LTP and LTD, respectively; $H(x)$ is the Heaviside function: $H(x) = 0$ if $x < 0$, 1 otherwise. Stochasticities of transmitter release and postsynaptic channel opening are expressed by $noise(t)$, which is present during spiking activity only when $[Ca^{2+}]$ is above the smaller value of θ_d and θ_p.

Activity-triggered calcium influx reaches a different level relative to θ_d and θ_p depending on spiking activity pattern (**Figure 4.14a**). The resulting change in w can lead to a transition from $w = 0$ to $w = 1$ (LTP) or vice versa (LTD); the model dependence on precise timing and number of spikes is supported by physiological measurements in layer 2/3 pyramidal neurons of rat somatosensory cortex (Nevian and Sakmann 2006) (**Figure 4.14b**). **Figure 4.14c** shows two stochastic samples of w dynamics starting in the Down state. During stimulation, w stochastically evolves to an intermediate level, with a probability to cross w^* and eventually settle at $w = 1$ (black but not grey trace). The model is supported by some experimental evidence. At an excitatory-to-excitatory synaptic connection in the hippocampus or neocortex, C_{post} is trigged by $[Ca^{2+}]$ influx through a high-threshold calcium ion channel: action potentials generated near the soma propagate forward along the axon as well as backward into the dendrite, and the latter is called spike backpropagation (Stuart et al. 1997; Stuart et al. 2016). Backpropagated spike depolarizes dendritic membrane, leading to the opening of the high-threshold calcium channels.

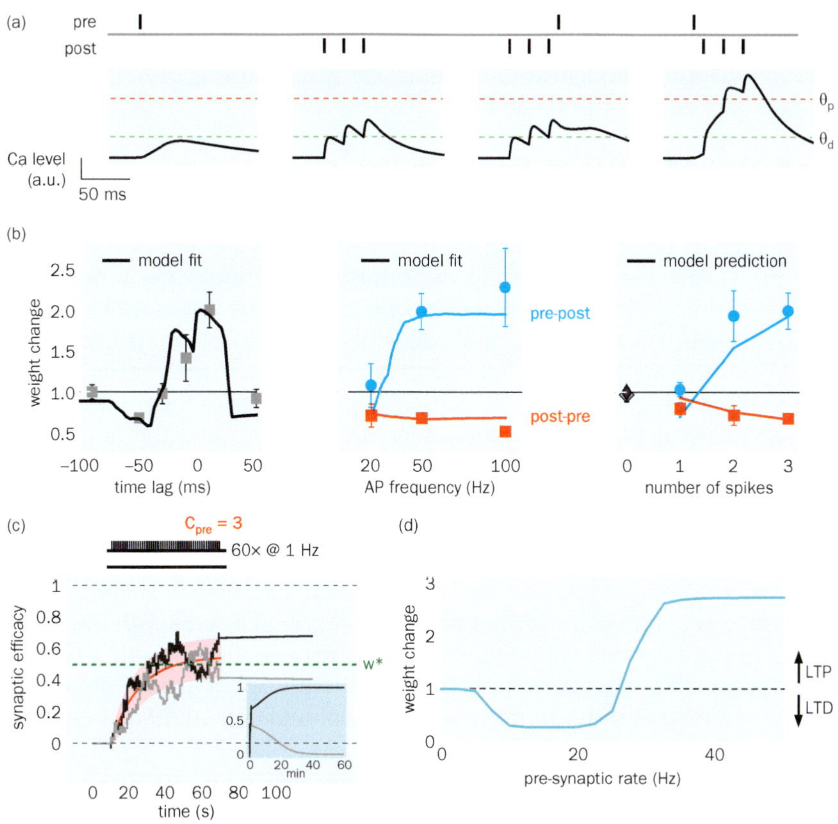

Figure 4.14 A calcium-based model of synaptic plasticity. (a) Illustration of $[Ca^{2+}]$ induced by different postsynaptic spiking patterns: No synaptic change (left), LTD (middle two), LTP (right). (b) Comparison of the model with physiological measurements (Nevian and Sakmann 2006). A presynaptic spike is paired with multiple postsynaptic action potentials (AP). Left and middle panels: Model fitting with data. Right panel: The model generalizes to predict data not used to fit the model. (c) With $w = 0$ initially, a train of stimuli produces $[Ca^{2+}]$ influx, which increases w stochastically to a level either below (grey trace) or above (black trace) w_* (green horizontal dashed line). Only the latter but not the former leads to a switch for w from the Down state to the Up state over minutes (insert). (d) The model produces the BCM rule, here as a function of presynaptic firing rate while the postsynaptic rate is fixed at 10 Hz. Panels (a–b) and (d) are reproduced from Yang, G. R., Murray, J. D., & Wang, X. J. (2016). A dendritic disinhibitory circuit mechanism for pathway-specific gating. Nature Communications, 7(1), 1–14. Licensed under CC BY 4.0; panel (c) is reproduced from Graupner, M., & Brunel, N. (2012). Calcium-based plasticity model explains sensitivity of synaptic changes to spike pattern, rate, and dendritic location. Proceedings of the National Academy of Sciences, 109(10), 3991–3996.

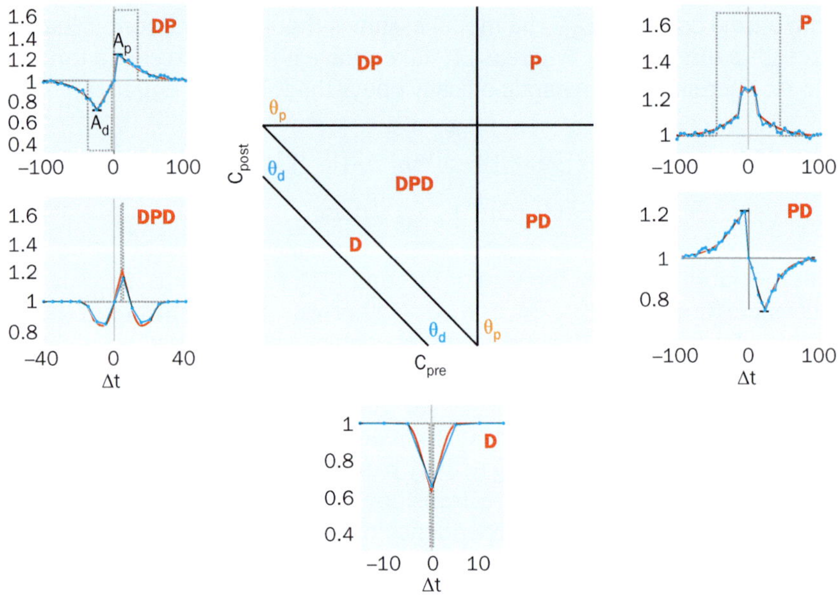

Figure 4.15 The Graupner-Brunel model exhibits various STDP curves on the plane of C_{pre} and C_{pos}, labeled according to LTD or LTP for negative, near zero and positive $\Delta t = t_{post} - t_{pre}$. DP (Hebbian): LTD for $\Delta t < 0$ and LTP for $\Delta t > 0$; PD (anti-Hebbian): The opposite; D: Only LTD; P: Only LTP; DPD: LTP for small Δt and LTD for both positive and negative larger Δt. Reproduced from Graupner and Brunel (2012). [Graupner, M., & Brunel, N. (2012). Calcium-based plasticity model explains sensitivity of synaptic changes to spike pattern, rate, and dendritic location. Proceedings of the National Academy of Sciences, 109(10), 3991–3996.] [Modified from Graupner, M., & Brunel, N. (2012). Calcium-based plasticity model explains sensitivity of synaptic changes to spike pattern, rate, and dendritic location. Proceedings of the National Academy of Sciences, 109(10), 3991–3996.] [Courtesy of Nicolas Brunel.]

Such calcium influx can be larger than that mediated by the NMDA channels, $C_{post} > C_{pre}$ (Sabatini et al. 2002). Furthermore, with presynaptic and postsynaptic spike trains, the model captures the salient dependence of plastic changes on the firing rate (Sjöström et al. 2001) and gives rise to the BCM rule (**Figure 4.14d**).

Interestingly, a plethora of STDP curves are realized with different sets of values for C_{pre}, C_{post}, θ_d and θ_p (**Figure 4.15**), including the symmetric regime DPD akin to **Figure 4.12c** and anti-Hebbian STDP (regime PD). Note that, strictly speaking, the model is not Hebbian because sufficient $[Ca^{2+}]$ influx due to either presynaptic or postsynaptic firing alone could in principle induce plasticity. However, in a typical biologically realistic situation, both presynaptic neuron and its postsynaptic target are active. The Graupner-Brunel model captures concisely various forms of STDP as well as rate dependence of LTD and LTP.

4.4.5 Molecular Basis of Memories

What is the biological basis of bistability at a synapse? The complex molecular machinery underlying synaptic plasticity involves cascades of biochemical processes. For excitatory-to-excitatory connections, a key molecule is the calcium/calmodulin-dependent protein kinase II (CaMKII) (**Figure 4.16a**). CaMKII in the postsynaptic density (PSD) is a holoenzyme consisting of two hexagonal rings, each with six subunits (Lisman et al. 2012). These subunits are phosphorylated (activated) in the presence of increased intracellular calcium. Their state affects the strength and the number of AMPA receptors, hence the strength of the synaptic connection. Modeling (Lisman 1985; Lisman and Zhabotinsky 2001; Miller et al. 2005) showed that the interplay of two mechanisms can realize bistability of CaMKII proteins (**Figure 4.16b**). Crucially, activated CaMKII subunits phosphorylate other subunits, providing an autocatalytic positive feedback. Therefore, the rate (speed of change in time) of phosphorylation increases with the number of phosphorylated subunits. Since the chemical reaction rate depends on the available reactant concentrations, the phosphorylation rate eventually decreases as the number of unphosphorylated subunits becomes scarce (**Figure 4.16b**, green curve).

This molecular positive feedback is counteracted by another enzyme (a phosphatase, named PP1) that inactivates CaMKII subunits. When

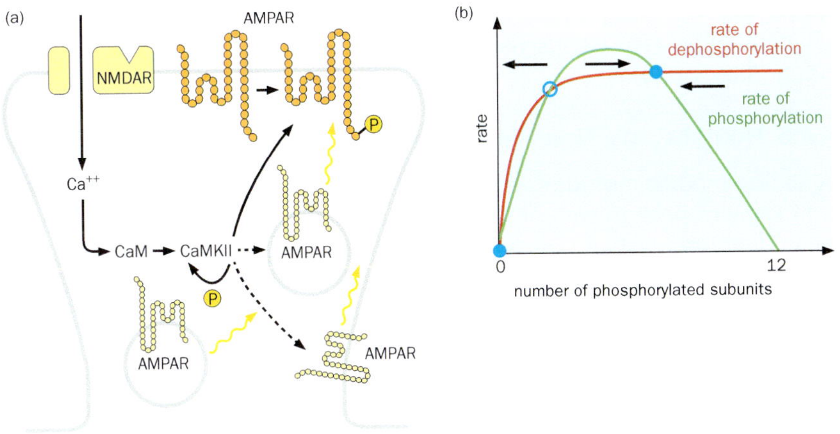

Figure 4.16 Bistability in synaptic plasticity. (a) Main molecular players in LTP induction. (b) Schematic figure indicating how bistability arises from the dependence of phosphorylation (green) and dephosphorylation (red) rates on the number of subunits phosphorylated. The number of phosphorylated subunits increases (decreases, respectively) when the green curve is above (below, respectively) the red curve. It is constant at the intersections (blue circles). Stable states are at the left and right intersections of the two curves. The middle crossing is unstable. Panel (a) is reproduced from (Nicoll 2017); panel b is modified from Miller P, Zhabotinsky AM, Lisman JE, Wang X-J (2005) The Stability of a Stochastic CaMKII Switch: Dependence on the Number of Enzyme Molecules and Protein Turnover. PLoS Biol 3(4): e107. https://doi.org/10.1371/journal.pbio.0030107. Licensed under CC BY 4.0. https://creativecommons.org/licenses/by/4.0/.

activated CaMKII subunits grow, the rate at which PP1 inactivates them increases rapidly, and then it saturates when the majority of PP1 molecules are bound to CaMKII (**Figure 4.16b**, red curve) (Miller et al. 2005). Schematically, the dynamics of phosphorylated CaMKII $([pCaMKII])$ obeys the equation $d[pCaMKII]/dt = \nu_p([pCaMKII]) - \nu_{dp}([pCaMKII])$ where $\nu_p([pCaMKII])$ and $\nu_{dp}([pCaMKII])$ are the reaction rates of phosphorylation and dephosphorylation, respectively (**Figure 4.16b**, green and red). This equation has two stable states (filled circles, the left and right intersection points of the green and red curves); in the Down state none of the subunits is active, whereas in the Up state a large proportion is active. They are separated by a third steady-state (open circle), which is unstable: a slight deviation from it triggers further changes towards either the Down or Up state. As all CaMKII molecules behave similarly, the synaptic strength is bistable.

One of the problems of this mechanism as a memory device is molecular turnover (Crick 1984): CaMKII proteins are gradually destroyed and replaced by newly synthesized unphosphorylated proteins, roughly once per day (Ehlers 2003). How can molecular memories be maintained for a long time in spite of protein turnover? Miller et al. (2005) considered a stochastic model with a small population of a dozen CaMKII molecules (that are present and immobilized in a dendritic spine). It was found that the effects of protein turnover are greatly reduced by the interplay between phosphorylation and dephosphorylation. Indeed, when CaMKII molecules are activated, the phosphatase spends a significant amount of time bonded to the existing activated CaMKII molecules, so consequently it is less available to inactivate the new molecules from the turnover. In other words, the new molecules are likely to acquire the state of the existing activated ones, preserving the state of the molecular memory that is stored at the level of a population of CaMKII molecules, in spite of ongoing protein turnover.

CaMKII is only one example of protein kinases involved in the induction of long-term synaptic potentiation. Two other kinases known to be involved in LTP are cAMP-dependent protein kinase A (PKA) and mitogen-activated protein kinase (MAPK). Modeling showed that PKA- and MAPK-dependent positive feedback loops can also lead to bistability of a synapse (Bhalla and Iyengar 1999; Smolen et al. 2006); the mathematical principle is similar to that depicted in **Figure 4.16**. Activation of these kinases is essential for the induction of plasticity but not long-term storage. Kinases translocated into the nucleus phosphorylate CREB, which regulates DNA to RNA transcription, leading to protein synthesis necessary for permanent change of a synaptic connection (Kandel 2001; Smolen et al. 2019);

this process involves a protein kinase C (PKC) isoform, protein kinase Mζ, which is expressed exclusively in neural tissue and enriched in the forebrain (Sacktor 2011).

4.4.6 Homeostasis, Non-Hebbian and Non-Synaptic Plasticity

A sliding modification threshold in the BCM model represents one of several known mechanisms of homeostatic regulation that work in concert with Hebbian plasticity in the brain. Originally proposed by Claude Bernard to maintain "la fixité du millieu intérieur (the constancy of the internal environment)" (Bernard 1865), homeostasis is a central theme of physiology (Cannon 1929). In a neural circuit, homeostasis regulates cellular excitability and synapses to preserve a function in spite of changes in the environment (Liu et al. 1998; Turrigiano et al. 1998; Marder and Prinz 2002; Davis 2006). One form of neuronal homeostatic regulation is multiplicative scaling of all synapses onto a neuron (**Figure 4.17a**), which was found to restore the long-term firing rate of cortical neurons after an acute change of the cortical tissue's environment, such as a drug that blocks spiking activity (Turrigiano et al. 1998).

Multiplicative synaptic scaling to maintain a stable long-term firing rate can be mathematically described by assuming that all synaptic weights w_i (i indexes synapses) onto a neuron are scaled by a common factor g, which obeys a simple equation (Renart et al. 2003; Toyoizumi et al. 2014):

$$\tau_g dg / dt = -g\left(r - r_{target}\right) \tag{4.30}$$

where r is the neuron's firing rate and r_{target} is its long-term average. Hence, if potentiation of w_i leads to a substantial increase in the neuronal firing rate r above r_{target}, then g is down-regulated to restore the original long-term average activity level. Conversely, when synaptic depression significantly reduces neuronal activity, g is up-regulated to recover a stable average firing rate (van Rossum et al. 2000). In this model, both the timescale for the averaged activity level r_{target} and the time constant for homeostatic dynamics τ_g are on the order of hours, allowing for a neuron to fire at different rates selectively during behavior on shorter timescales. Homeostatic regulation has been implemented in various models of neural circuits (Renart et al. 2003; Clopath et al. 2010; Litwin-Kumar and Doiron 2014).

Other forms of unsupervised plasticity also exist. For instance, new place field of hippocampal neurons (Chapter 3, **Figure 3.10a**) can be formed in a single trial through "behavioral timescale synaptic plasticity," which refers to the phenomenon of potentiation of input that arrives seconds before postsynaptic firing (Bittner et al. 2017). In addition to synaptic modifications, long-term changes also occur intrinsically in single neurons in an activity-dependent manner (Titley et al. 2017). In contrast to homeostasis, intrinsic plasticity induced by neural activity typically causes

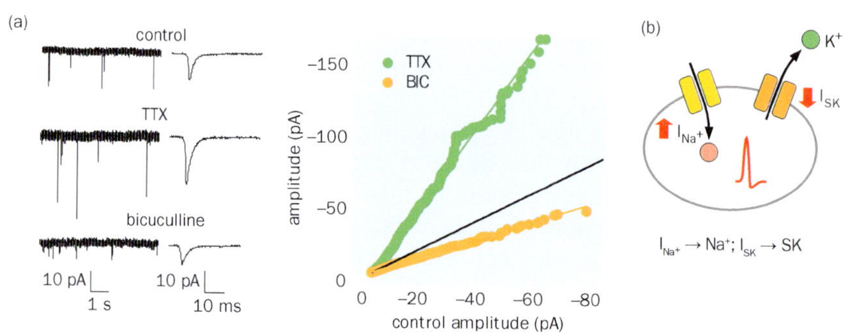

Figure 4.17 Homeostatic regulation and non-synaptic plasticity. (a) Multiplicative synaptic scaling. Left panel: Quantal amplitudes of miniature EPSCs recorded from cortical pyramidal neurons in control cultures became larger after activity was blocked with the sodium channel blocker tetrototoxin (TTX) and smaller after activity was enhanced by blocking inhibition with bicuculline (BIC) for two days. Right panel: Spontaneous EPSPs (control amplitude) are rank ordered and plotted against amplitudes recorded in other cortical cultures where activity blockade scales up mEPSC amplitude (green), whereas activity enhancement scales it down (orange). The plots are well fit by straight lines, indicating that in both cases the scaling is multiplicative. (b) Intrinsic neuronal plasticity. In the example shown here, intrinsic plasticity-related downregulation of calcium-activated, small-conductance SK-type K+ channels reduces the afterhypopolarization (AHP) following depolarizing events. As a consequence, EPSPs will be enhanced and prolonged, and the spike threshold will be reached more easily. Panel (a left) is reproduced from Turrigiano, G. G., Leslie, K. R., Desai, N. S., Rutherford, L. C., & Nelson, S. B. (1998). Activity-dependent scaling of quantal amplitude in neocortical neurons. Nature, 391(6670), 892–896; panel (a right) is reproduced from Abbott, L. F., & Nelson, S. B. (2000). Synaptic plasticity: Taming the beast. Nature Neuroscience, 3(11), 1178–1183; panel (b) is reproduced from Titley, H. K., Brunel, N., & Hansel, C. (2017). Toward a neurocentric view of learning. Neuron, 95(1), 19–32.

a further increase rather than decrease in spike firing. For instance, as schematically shown in **Figure 4.17b**, activity can reduce an outward potassium conductance, resulting in an increased neuronal excitability and more efficient drive by synaptic inputs. Interestingly, whereas Hebbian plasticity induction requires long repetitive stimulation over minutes, intrinsic plasticity may result from activation for as briefly as 3 seconds. Through long-lasting modification of firing threshold, intrinsic plasticity determines the degree of participation of a neuron in a cell assembly that defines a memory engram (Tonegawa et al. 2015; Josselyn et al. 2015). Both behavioral timescale synaptic plasticity (Bittner et al. 2017; Magee and Grienberger 2020; Milstein et al. 2021; Grienberger and Magee 2022) and neurocentric plasticity (Titley et al. 2017) could be suitable for rapid learning, but currently little is known about the brain mechanism of learning from a single event ("one-shot learning").

4.5 STORAGE CAPACITY AND MEMORY RETRIEVAL

One key question about memory storage is its capacity, defined as the maximum number of memory items that can be stored and retrievable in a neural network (Marr 1971). To address this question theoretically, consider to-be-remembered items as random and uncorrelated patterns, in the spirit of the studies of Ebbinghaus using nonsense memoranda (Ebbinghaus 1885). Of course, this does not correspond to real life where memories are correlated; however, statistically independent memory items are mathematically easier to analyze, and insights thus gained are valuable. The question we will try to answer is how does the memory capacity scale with the total number of synapses N_{syn}?

4.5.1 Ideal Observer Analysis of Memory Capacity

Let us assume that memory items occur in a sequence, where each is stored after a single exposure as a random and uncorrelated pattern of neural activity (one-shot learning) (Fusi and Wang 2016). Specifically, a synapse w_{ij}, connecting neuron j to neuron i, is modified according to:

$$w_{ij} \to w_{ij} + \Delta w_{ij}^{\mu} \tag{4.31}$$

following the storage of the μ-th memory. For example, a simple Hebbian rule is (Hopfield 1982):

$$\Delta w_{ij}^{\mu} = \xi_j^{\mu} \xi_i^{\mu} \tag{4.32}$$

where ξ_i^{μ} is the activity imposed on neuron i when memory μ is stored; $\xi_i^{\mu} = \pm 1$ depending on whether the neuron is active or inactive (ignoring Dale's law). In what follows, we will assume for simplicity that Δw_{ij} are random uncorrelated stochastic variables, which are ± 1 with equal probability (potentiation and depression are balanced).

We now choose a generic memory (item μ), estimate its initial strength immediately after encoding, and evaluate its trace during the process of storage of other memories. The chosen item μ is not special, merely a memorandum that we track. As a consequence, the dynamics of the strength of other memories would be the same. The strength of the memory trace M^{μ} can be estimated by computing the correlation between the synaptic matrix w_{ij} and the μ-th memory Δw_{ij}^{μ}:

$$M^{\mu} = \frac{1}{N_{syn}} \sum_{i,j} w_{ij} \Delta w_{ij}^{\mu} - \left(\frac{1}{N_{syn}} \sum_{i,j} w_{ij} \right) \left(\frac{1}{N_{syn}} \sum_{i,j} \Delta w_{ij}^{\mu} \right) \tag{4.33}$$

where N_{syn} is the total number of synapses, and the sum extends over all pre- and postsynaptic neurons. The subtracted term is the product of the average synaptic weight and the average synaptic modification. It expresses the similarity between w_{ij} for all memory items and Δw_{ij}^{μ} for item μ. The number of terms in each sum is N_{syn}. The memory trace is strong when M^{μ} is large and weak when $M^{\mu} \simeq 0$. In the case that we are considering, the average value of Δw_{ij}^{μ} is zero, hence we can eliminate the subtracted term and rewrite M^{μ} as:

$$M^{\mu} = \frac{1}{N_{syn}} \sum_{i,j} w_{ij} \Delta w_{ij}^{\mu} \tag{4.34}$$

Assume that we now track the first memory that is stored ($\mu = 1$). We start from a tabula rasa synaptic matrix ($w_{ij} = 0$) and then store memory 1: $w_{ij} = \Delta w_{ij}^{\mu}$. The memory strength immediately after its storage is given by:

$$M^{1} = \frac{1}{N_{syn}} \sum_{i,j} (\Delta w_{ij}^{1})^{2} \tag{4.35}$$

with $\Delta w_{ij} = \pm 1$, $M^{1} = 1$. We now store the other uncorrelated memories. As they are random, the synapses basically perform a random walk, moving up and down with the same probability. The value of the synaptic strength after the storage of p memories is:

$$w_{ij} = \sum_{\mu=1}^{p} \Delta w_{ij}^{\mu} \tag{4.36}$$

The signal $\mathcal{S}$ of each memory μ is the average of M^{μ} which, for memory 1 is equal to:

$$\mathcal{S} = \langle M^{1} \rangle = \frac{1}{N_{syn}} \left\langle \sum_{i,j} \sum_{\mu=1}^{p} \Delta w_{ij}^{\mu} \Delta w_{ij}^{1} \right\rangle \tag{4.37}$$

where the angular brackets $\langle \ldots \rangle$ denote the operation of averaging over all realizations of random memories (i.e., the average over all possible sequences of synaptic modifications, each weighted by its probability of occurrence). One can show from this formula and with $\Delta w_{ij} = \pm 1$ that $\mathcal{S}$ remains constant with an increasing number of items p (Amit and Fusi 1994).

However, the noise due to the interference with other memories constantly increases as new memories are stored. Indeed, the variance of M^{1} is given by:

$$\langle (M^{1} - \langle M^{1} \rangle)^{2} \rangle = \frac{1}{N_{syn}^{2}} \left\langle \left(\sum_{i,j} \sum_{\mu=2}^{p} \Delta w_{ij}^{1} \Delta w_{ij}^{\mu} \right)^{2} \right\rangle \tag{4.38}$$

This is equal to the variance of the sum of $\sim N_{syn} p$ independent ± 1 random variables, divided by N_{syn}^{2}. Since the variance is proportional to $N_{syn} p$, the memory noise size given by the square root of the variance of the memory trace is:

$$\mathcal{N} \sim \sqrt{\frac{p N_{syn}}{N_{syn}^{2}}} = \sqrt{\frac{p}{N_{syn}}} \tag{4.39}$$

The memory signal and the memory noise are plotted in **Figure 4.18a**. The exact number of memories that can be retrieved depends on the details of the architecture of the network and on the neural dynamics. However, the scaling properties do not usually depend on these details, and they can

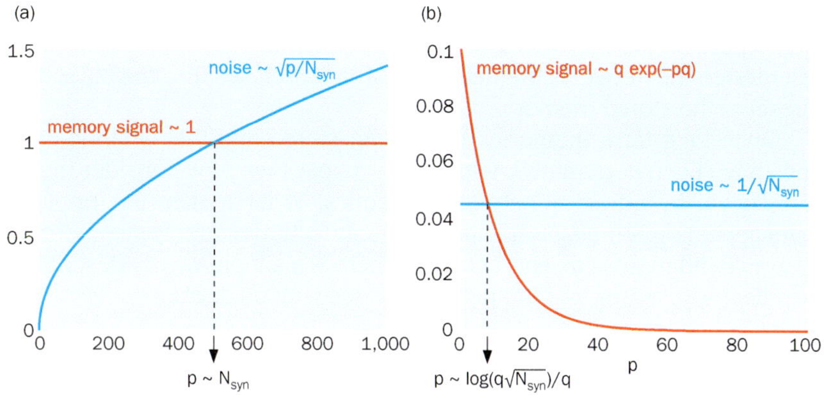

Figure 4.18 Memory signal $\mathcal{S}$ and memory noise $\mathcal{N}$ as a function of the number of stored items p. (a) In the case of unbounded synapses, the memory signal is constant, and noise increases with p. The two curves cross at $p_{max} \sim N_{syn}$. (b) In the case of bounded binary synapses, the situation is drastically different because the noise is now approximately constant, and the signal decays exponentially with p. Memory capacity is given by the crossing of the two curves, $p_{max} \sim \log\left(q\sqrt{N_{syn}}\right)/q$. Reproduced from Fusi, S. and X.-J. Wang (2016). Short-term, long-term, and working memory. In M. A. Arbib and J. J. Bonaiuto (Eds.), From Neuron to Cognition via Computational Neuroscience, Chapter 11, pp. 319{344. Cambridge, MA: MIT Press.

be determined by choosing an arbitrary threshold for the $\mathcal{S}/\mathcal{N}$ ratio. For example, if we decide that memories are retrievable as long as $\mathcal{S}/\mathcal{N} > 1$, then the memory capacity is determined by $\mathcal{S}/\mathcal{N} = 1$, yielding:

$$p_{max} \sim N_{syn} \tag{4.40}$$

which is enormous ($\sim 10^{15}$ in our cerebral cortex). However, this ideal observer analysis does not consider any plausible implementation of a memory network in the brain for which, as we shall see below, memory capacity scales with the number N of neurons because the number of independent synapses scales like N for neural population behavior of a fully connected network.

4.5.2 Hopfield Model of Associative Memory

Unlike the above ideal observer analysis where memoranda are considered in terms of "disembodied" synaptic weights, in the brain memories are represented and retrieved by collective neural populations. Memory retrieval is associative in the sense that one small cue could lead to the return of a related memory, as memorably (no pun intended) told by Marcel Proust (1913): how the taste of a little madeleine cake unleashed in the mind of his protagonist a chain of long-buried childhood memories, from his aunt soaking pieces of madeleine in her tea on Sundays, to the old grey house in Combray, the gardens, the town's streets and squares, and so forth.

In the early 1980s, inspired by then nascent theoretical advances in studying complex behavior of glasses by Giorgio Parisi (1979), for which he shared the 2021 Nobel Prize in Physics, Hopfield (1982) proposed a recurrent, attractor network model for associative memory. Consider N neurons that are either active ($s_i = +1$) or inactive ($s_i = -1$) ($i = 1, 2, \ldots, N$). Their dynamics is described by the following equation:

$$s_i(t+1) = \text{sign}\left(\sum_{j=1}^{N} w_{ij} s_j(t) + I_i\right) \tag{4.41}$$

where I_i is the external input, and w_{ij} is the synaptic strength of recurrent connections. Suppose that p memory items ($\mu = 1, 2, \ldots, p$) are to be stored as attractor states, and each one is a random pattern $\xi_i^\mu = \pm 1$. Different memories are uncorrelated, as in the examples discussed earlier. Using a Hebbian prescription,

$$w_{ij} = \frac{1}{N} \sum_{\mu} \xi_i^\mu \xi_j^\mu \tag{4.42}$$

for $i \neq j$, $w_{ij} = 0$ for $i = j$. These are examples of long-term memories as they are permanently stored in the synaptic connections. Each of these memories can be reactivated by stimulating the network. For example, when a partial cue, correlated to one of the previously stored patterns, is presented to the network, the network naturally relaxes into the state that represents the stored memory.

Once long-term memories have been stored and the synaptic couplings have been determined, with fixed synapses we now consider memory retrieval by attractor network dynamics. What makes the Hopfield model appealing is the fact that, because connection matrix is symmetrical ($w_{ij} = w_{ji}$) and there is no self-coupling, there is a simple way of describing the collective dynamics of the network. Indeed, let us define:

$$E = -\frac{1}{2}\sum_{i \neq j} w_{ij} s_i s_j \tag{4.43}$$

This "energy" function E decreases monotonically in time in the absence of external inputs ($I_i = 0$), when the network is updated asynchronously (one randomly selected neuron at a time). If neuron k is updated from t to $t+1$, since:

$$E(t) = -s_k(t)\sum_j w_{kj} s_j(t) - \frac{1}{2}\sum_{i \neq j \neq k} w_{ij} s_i s_j \tag{4.44}$$

we have

$$E(t+1) - E(t) = -(s_k(t+1) - s_k(t))\sum_{j \neq k} w_{kj} s_j(t). \tag{4.45}$$

If neuron k does not change its state, then the energy function remains the same $s_k(t+1) = s_k(t)$. If it flips from -1 to $+1$, then the input to neuron k, $\sum_{j \neq k} w_{kj} s_j(t)$, must be positive; if it flips in the opposite direction, $\sum_{j \neq k} w_{kj} s_j(t)$ must be negative. In either case, $E(t+1) - E(t) < 0$. Therefore, the network evolves over time, until it reaches a minimum of the energy function. There are many minima in a Hopfield network, corresponding to stored memory items. Each energy minimum is an attractor state, in the sense that the network perturbed slightly away will converge back to it, just like a ball rolling down the valley of a mountainous landscape (**Figure 4.19**).

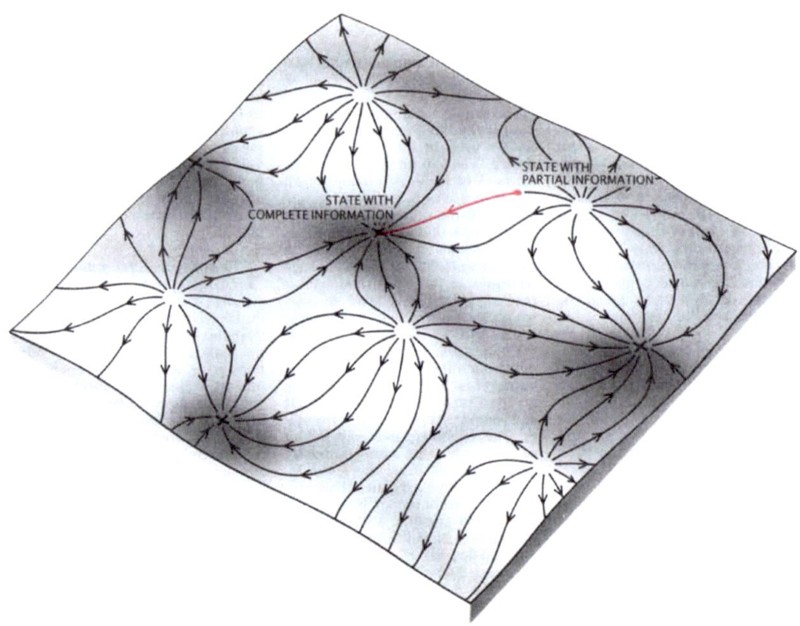

Figure 4.19 Schematics of a Hopfield network model of associative memory. Computational energy function is depicted as a landscape of hills and valleys, plotted against the neural activity states (on the XY plane). The synaptic connections and other properties of the circuit determine its contours. The circuit computes by following a path that decreases the computational energy until the path reaches the bottom of a valley, which represents a stable state of the system (an attractor). In an associative memory circuit, the valleys correspond to memories that are stored as associated sets of information (the neural activities). If the circuit is cued to start out with approximate or incomplete information, it follows a path downhill to the nearest valley (red), which contains the complete information. Reproduced from Tank and Hopfield (1987).

The attractor network suggests a mathematically rigorous formulation of content-addressable associative memory, which can be retrieved by partial cues that lead to full memory recollection: a cue brings the memory network within the basin of attraction for a memory item, and the intrinsic network dynamics evolves to the attractor state, thereby completing memory recall.

With increasing p, different memory items start to interfere with each other, and noise induces confusion during memory retrieval. What's the maximal value of p for which all the previously stored items can be reliably retrieved (Brunel 2016; Chaudhuri and Fiete 2016)? Analysis using mathematical techniques developed in theoretical physics (Amit et al. 1985; Amit 1989) revealed that as more and more memoranda are stored, there is a critical value p_{max} of p above which the signal-to-noise ratio becomes too small for the memoranda to be reliably retrievable. In the Hopfield model, each of N neurons is connected with every other neuron:

$$p_{max} = \alpha N \text{ with } \alpha \simeq 0.14$$

This formula explicitly states the dependence of long-term memory storage size on the number of neurons, which varies considerably across the animal kingdom from 302 in *C. elegans* to 86 billion in humans. Higher precisions of memories, information coding generally, can be achieved with more neurons in a brain.

Interestingly, when the critical capacity is surpassed, there is an abrupt transition to a regime in which none of the memories, not even the most recently stored, can be retrieved. This is called blackout catastrophe (Amit 1989).

The Hopfield model can be readily extended to neurons described by graded activities rather than binary units (Hopfield 1984). However, if the connection matrix w_{ij} is not symmetric, as is generally the case in biological neural circuits, an energy function no longer exists, and the mathematical analysis of the attractor landscape becomes more complicated (Derrida et al. 1987). Furthermore, as we discussed in Chapter 3, with Dale's law and realistic synaptic biophysics, attractors are no longer guaranteed to be fixed points, and more interesting spatiotemporal patterns for memory representations become possible.

Retrieval from memory storage appears to involve a complex neural process under executive control by the prefrontal cortex (Anderson and Green 2001; Yadav et al. 2022), which is expected to play a key role in determining the actual capacity of long-term memory storage estimated in behavioral studies (Anderson 1983; Landauer 1986). The underlying neural circuit mechanism remains to be elucidated in future research.

4.5.3 Plasticity-Stability Dilemma

The prior analysis uses synaptic weights that are real numbers; that is, w_{ij} values vary with the number of stored memories without bound. This is clearly unrealistic, because all variables characterizing biological synapses must vary in a limited range. Let us now consider synapses with two, the minimal number, of states for a memory device (Section 4.4.4). Basically, we assume that the upper and lower bounds, which are always present in any realistic model, coincide with only two stable states. Modification takes place with probability $q < 1$ (Tsodyks 1990; Amit and Fusi 1994). More specifically, we consider again the case of random and uncorrelated memories, and assume that each synapse is either potentiated or depressed with equal probability ($\Delta w_{ij} = \pm 1$).

Using a similar analysis as that for unbounded synapses, one can show that with binary synapses the memory signal $\mathcal{S} \sim q\exp(-qp)$, whereas the memory noise $\mathcal{N} \sim 1/\sqrt{N_{syn}}$, independent of p (Amit and Fusi 1994; Fusi and Wang 2016) (**Figure 4.18b**). Hence, the signal-to-noise ratio is:

$$\mathcal{S} / \mathcal{N} = q\sqrt{N_{syn}}exp(-qp) \tag{4.46}$$

In this case, initially $p = 0$ and $\mathcal{S} / \mathcal{N}(0) \sim q\sqrt{N_{syn}}$; it decays with the number of stored items p. The memory capacity p_{max} is determined by $\mathcal{S} \sim \mathcal{N}$, yielding:

$$p_{max} \sim log\left(q\sqrt{N_{syn}}\right) / q \tag{4.47}$$

Learning a memorandum involves probabilistic LTP from the Down to Up transition at relevant synapses, and forgetting entails LTD in the opposite transition, resulting in a tradeoff. When learning rate q is large in favor of plasticity, memories are quickly overwritten at binary synapses. With $q \sim 1$, $p_{max} \sim log\left(N_{syn}\right)$ is quite limited; it is still only several dozen even if $N_{syn} = 10^{15}$. On the other hand, when learning rate q is small, learnt memory traces can last for a long time, and the memory capacity is large. With $q \sim 0$, $p_{max} \sim \sqrt{N_{syn}}$, which is many millions. However, the stored information content of any item, given by the initial signal-to-noise ratio $\mathcal{S} / \mathcal{N}(0)$ vanishes with increasing q according to Eq. (4.46).

One possible solution to the stability-plasticity dilemma is to incorporate complexity of a synapse as a chemical system with many interactive molecules at the postsynaptic density (Kim and Sheng 2004). In a cascade model of single synapses (Fusi et al. 2005), there are numerous internal states (indexed by $i = 1, 2, .., n$) with transitions characterized by different timescales. The synaptic strength is either a weak or strong state at the top of a hierarchy, while the other states are "hidden" and not observable physiologically (**Figure 4.20a**). Transitions between states are stochastic with probabilities given by q_i and p_i. Fusi et al. (2005) assumed that the transition rates decrease exponentially with the level i in the cascade, for instance $q_i = p_i \sim (1/2)^i$, so that the deeper the state a synapse is in, the slower are transitions. Because of the presence of multiple timescales, one can have a large initial $\mathcal{S} / \mathcal{N}$ ratio, as well as slow decay of a memory trace. As shown in **Figure 4.20b**, forgetting exhibits a power law as a function of p, akin to behavioral observations (Wixted and Ebbesen 1997), up to a cutoff point that is longer with a larger n. Therefore, a mixture of fast and slow learning rates offers a possible compromise for the stability-plasticity tradeoff.

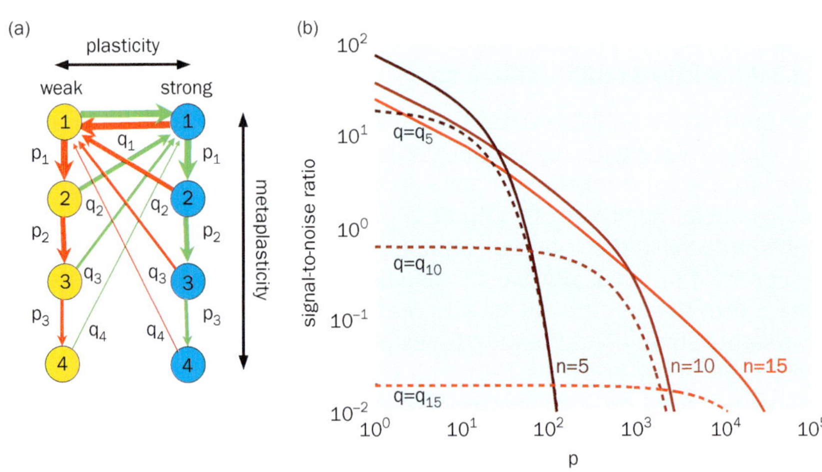

Figure 4.20 A cascade model of synaptic plasticity. (a) Model scheme. Synaptic strength is either weak (yellow) or strong (blue). Associated with each of these strengths is a cascade of n states ($n = 4$ in this example). (b) Signal-to-noise ratio as a function of the number of memory items p with a different number of internal states (solid curves). The decay initially exhibits a power law (linear in the log-log plot), up to a cutoff where it changes to an exponential decay. Increasing n expands the range over which the power law holds. By comparison, memory decay is shown in the binary model without internal states (dashed curves), with the transition probability set to the minimum transition probability in the $n = 5$, 10 and 15 cascade models (brown, purple and red, respectively). For the binary synapse model, decreasing q prolongs memory lifespan but reduces the initial signal size. Reproduced from Fusi, S., Drew, P. J., & Abbott, L. F. (2005). Cascade models of synaptically stored memories. Neuron, 45(4), 599–611.

4.6 MEMORY CONSOLIDATION

In general, memories are initially labile to rapid degradation, lest they are consolidated and become resistant to disruptions. Consolidation is believed to involve synaptic tagging, a molecular mechanism for ensuring that changes of DNA transcription in the nucleus lead to modifications specifically only at those synapses on a dendritic tree, where induction of plasticity has taken place (Frey and Morris 1997; Morris 2006; Barrett et al. 2009; Clopath 2012). Consolidation occurs not only in molecular processes at individual synapses, but also through systems consolidation, which is a reorganization of the memory trace in and across different brain systems. These mechanisms protect memories from forgetting over time and the interference or overwriting caused by other memories (Frankland and Bontempi 2005). Computational modeling suggests that plasticity of inhibition plays a key role in memory consolidation (Tomé et al. 2024).

Recall later in time after memory induction, such as review of learnt materials at school, reinforces a memory trace. In animal experiments, this is implemented by repetition of stimuli across trials. There are different ways to schedule repetition, and some are better than others. Massed training refers to a protocol with short or no intervals between stimuli, akin to cramming before an exam at school. In spaced training, by contrast, stimuli are separated by long time intervals. Ebbinghaus (1885) already realized that "with any considerable number of repetitions a suitable distribution of them over a space of time is decidedly more advantageous than the massing of them at a single time." Spaced training has been investigated extensively in human psychology (Cepeda et al. 2006). Can empirical observations of spaced learning be understood in terms of brain mechanisms? This topic represents an example of how neuroscience is inspired by learning behavior, and in turn may provide new insights for improving teaching and education (Carew and Magsamen 2010; Dehaene 2020).

Memory consolidation of different types can be investigated in laboratory experiments using a variety of species ranging from sea slugs to mammals. Insights thus gained are likely to be relevant to humans because key molecular mechanisms are conserved to a substantial degree (Kandel 2001; Alberini 2009). Zhang et al. (2011) built a computational model of memory consolidation that consists of interactions between two kinases involved in memory formation at a synapse, cAMP-dependent PKA and MAPK (also called ERK for extracellular signal-regulated kinase). Their signaling pathways, described as chemical reactions, are endowed with positive feedback loops (Smolen et al. 2006). The model describes a transient surge of the "inducer" by the activation of PKA and MAPK/ERK. The magnitude of the inducer signal presumably determines activation of transcription factors like CREB in the nucleus, which is required for protein synthesis underlying persistency of a memory trace for years. It was found (**Figure 4.21**) that a protocol with irregular time intervals between stimuli is superior to that with periodic stimuli because it produces higher inducer activity. This model prediction was tested and received support experimentally, demonstrating that computational modeling can be used to design better learning protocols for enhanced learning. Hypothetically, if there is a nonlinear threshold, for instance, the inducer must be above 1.5 μM in order to activate CREB in **Figure 4.21c**, then the standard protocol and massed protocol would not lead to memory consolidation, but the protocol with irregular intervals would. The detailed molecular processes of memory consolidation remain to be elucidated.

A memory trace may be consolidated also through internal rehearsal in the absence of external stimulation, a neural process that could, for instance, replay a one-shot experience multiple times in the brain. Hebb

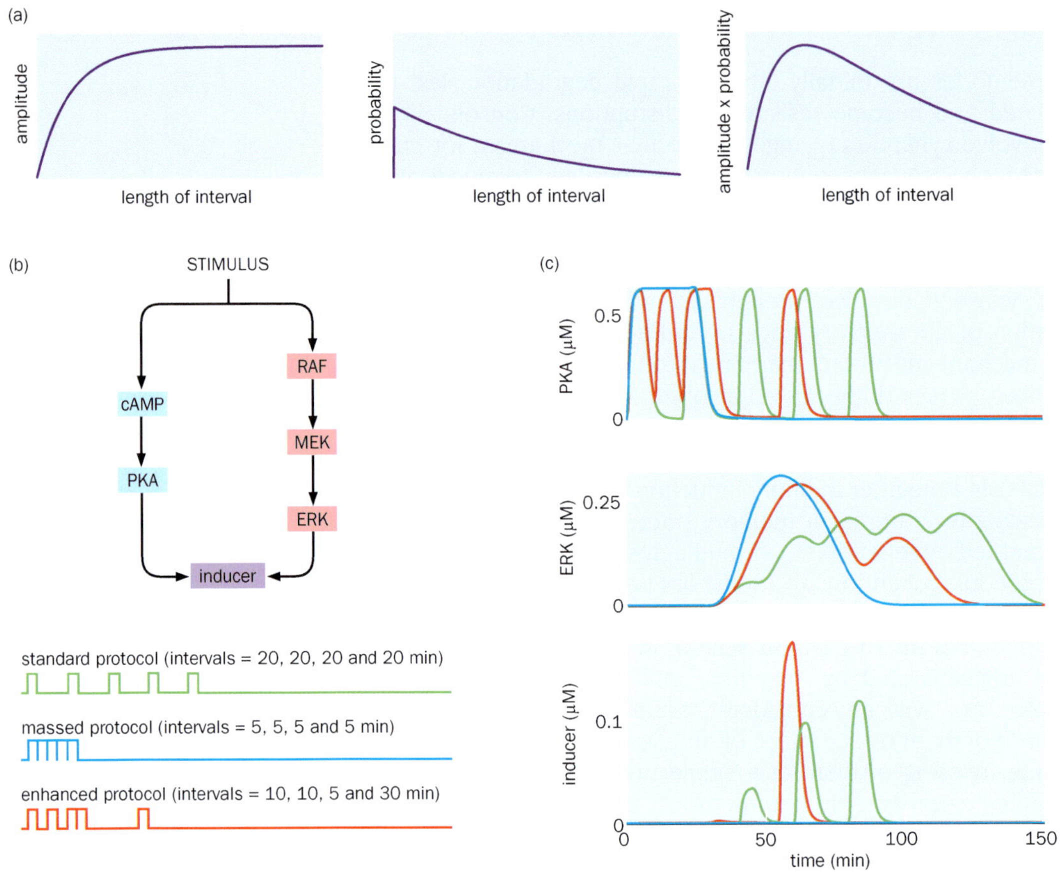

Figure 4.21 Spaced learning. (a) Benefits of spaced learning as postulated by Landauer (1969). Left panel: Potentiation increases with time interval between stimuli. Middle panel: The probability of reactivating a previously learnt memory trace declines over time. Right panel: An optimum interval corresponds to the maximum of the product of the two curves from the left and middle panels. (b) A model for spaced learning. Upper panel: Model scheme (Zhang et al. 2011) describes a chemical cascade involving two key kinases, cAMP-dependent PKA and MARK/ERK. PKA and MARK/ERK interaction (the "inducer" variable) phosphorylates transcription factors to induce LTP. Lower panel: Three samples of stimulus protocols. The standard protocol (green) is commonly used in studies of LTP. (c): Model simulated time courses of the activated inducer in response to the standard protocol (green), the massed protocol (blue) and the protocol with irregular stimulus intervals (red). The irregular-interval protocol produces the largest peak value of the inducer. Reproduced from Smolen, P., Zhang, Y., & Byrne, J. H. (2016). The right time to learn: Mechanisms and optimization of spaced learning. Nature Reviews Neuroscience, 17(2), 77–88.

(1949) posited that the underlying biological process is reverberation in a neural network, because reverberatory dynamics between neurons involved in an engram continually induce synaptic plasticity. Such reverberation is present during sleep in the hippocampus (Wilson and McNaughton 1994), a core structure for learning and memory (Andersen et al. 2007; Eichenbaum 2017). Awake immobility and non-dreaming phases of sleep are associated with sharp wave ripples in the local field potential (LFP) (Buzsáki 1986; Buzsáki et al. 1992). During sharp wave ripples, pyramidal cells fire episodically transient bursts of synchronous oscillations at about 200 Hz, which Buzsaki (1989) proposed to be a candidate manifestation of Hebbian reverberation. As described in Chapter 3, a pyramidal neuron in the CA1 region of the hippocampus exhibits increased firing activity around a selective spatial location, and different neurons have different place fields (**Figure 4.22a**). Hence, when an animal runs from left to right in a linear track, these neurons are activated sequentially, starting with the neuron at the bottom and ending with the neuron at the top. After runs, during non-rapid eye movement sleep, ripples are associated with sequential spike firing of CA1 neurons, in the same order as their activities during navigation (**Figure 4.22b**) (Lee and Wilson 2002).

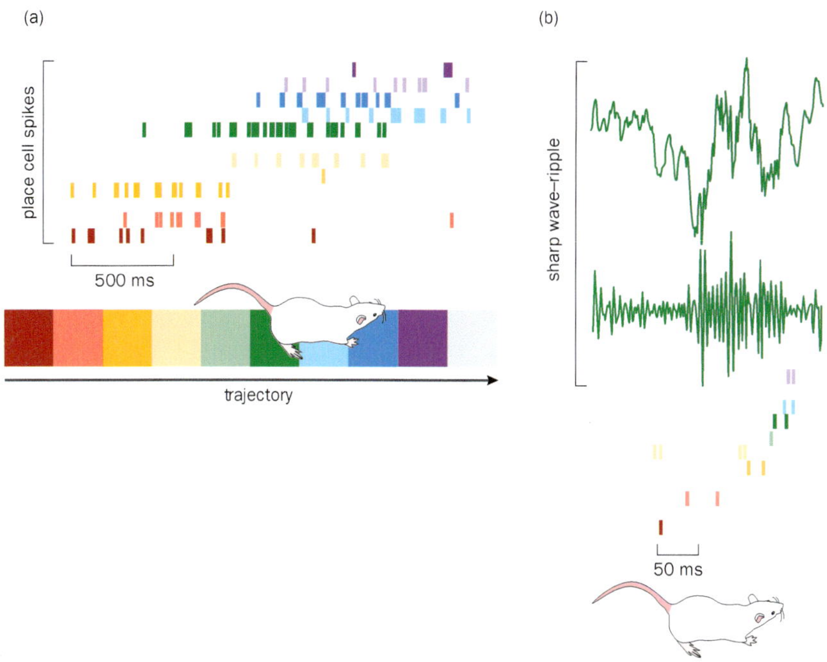

Figure 4.22 Replay during sharp wave-ripples. (a) Successively activated place cells (top) show sequential spike firing as a rodent passes through the cells' place field in a particular trajectory on a linear track (bottom). Time bar: 500 ms. (b) An example of a sharp wave ripple (top) recorded during subsequent rest at the end of the linear track; a bandpass-filtered (150–300 Hz) version of the sharp wave ripple is shown immediately below the raw recording. Spikes from the place cell ensemble are shown to reactivate during the sharp wave ripple in the same order as in exploration but on a faster timescale. Time bar: 50 ms. Reproduced from Colgin, L. L. (2016). Rhythms of the hippocampal network. Nature Reviews Neuroscience, 17(4), 239–249.

Similar replay of sequences in the hippocampal neurons also takes place in rodents during REM sleep (Louie and Wilson 2001) and awake states (Nadasdy et al. 1999; Foster and Wilson 2006; Carr et al. 2011; Foster 2017), as well as in humans (Vaz et al. 2020). These replays occur at a compressed time, about 20-fold faster than sequential activation during navigation. A recent study used optogenetic tools to artificially prolong spontaneously occurring sharp wave ripples, and found improved learning of a maze navigation task compared to controls without this manipulation, demonstrating a link between replay in the hippocampus and the strength of a memory trace (Fernandez-Ruiz et al. 2019), which is replicated in a computational model (Malerba and Bazhenov 2019). Replay has also been demonstrated in the human hippocampus (Liu et al. 2019; Kurth-Nelson et al. 2023).

While the hippocampus plays a crucial role in episodic memory formation, it is not required for remembering of past events. The famous patient HM (a.k.a. Henry Molaison), with his hippocampus and surrounding temporal lobe structures surgically removed to treat epilepsy, could recall remote memories from his childhood but was not able to form novel memories about people he just met (Scoville and Milner 1957; Squire 2009). Surprisingly, his intellectual and perceptual functions, his remote memories (i.e., memories of events occurred before the surgery) and his working memory were intact. Moreover, the medial temporal lobe may not be required for the acquisition of certain types of memories, as HM was still able to acquire new motor skills (i.e., he could form new procedural memories).

McClelland et al. (1995) proposed a theory of memory that depends on an interplay between the hippocampus and neocortex. They posited that synaptic plasticity is rapid in the hippocampus, and induced memory traces are transferred to the neocortex where synaptic modifications are slow yet essential for memory consolidation. Thus, the ability of HM to recall remote memories can be explained by his intact neocortex.

Storing a memory involves several mechanisms that start with a reorganization of the information to be stored, and end with long-term synaptic modifications. Most of the new memories are likely to bear some "resemblance" to previously stored ones. This is the typical situation in the case of semantic memories, which contain information about categories of memory items and functional relationships between them. For this type of memory, we can utilize our previous knowledge about the memoranda so that we can store only the information about the relationships between them (Rogers and McClelland 2006). In other words, we can take advantage of the correlations between the new memory and the previously stored ones that encode the relevant memoranda. Modeling this process of reorganization is of fundamental importance (Battaglia and Pennartz 2011; Fusi and Wang 2016).

4.7 SUMMARY

In this chapter, we introduced three main types of learning. Basic concepts include gradient descent for supervised learning, reward prediction error for reinforcement learning, and unsupervised learning through experienced statistics. Unsupervised learning has been the most extensively studied neurobiologically, on topics ranging from Hebbian plasticity, homeostatic regulation and memory consolidation to pattern formation during brain's development. Correspondingly, there is a large literature on Hebbian learning models. More recent years have seen dramatic advances in experimental and theoretical studies of reinforcement learning; "three-factor" Hebbian plasticity modulated by dopamine will be shown to account for adaptive decision making (Chapter 7). Supervised learning, in combination with reinforcement learning and unsupervised learning called "self-supervised" learning, has become a leitmotif in machine learning and artificial intelligence (Chapter 11).

FURTHER READING

Books and Reviews

Bi, G. and M. Poo (2001). Synaptic modification by correlated activity: Hebb's postulate revisited. *Annu. Rev. Neurosci. 24*, 139–166.

Clopath, C. (2012). Synaptic consolidation: An approach to long-term learning. *Cogn. Neurodyn. 6*, 251–257.

Dehaene, S. (2020). *How We Learn: Why Brains Learn Better Than Any Machine . . . for Now*. New York: Viking Press.

Fusi, S. and X.-J. Wang (2016). Short-term, long-term, and working memory. In M. A. Arbib and J. J. Bonaiuto (Eds.), *From Neuron to Cognition via Computational Neuroscience*, Chapter 11, pp. 319–344. Cambridge, MA: MIT Press.

Gallistel, C. R. (1990). *The Organization of Learning (Learning, Development, and Conceptual Change)*. Cambridge, MA: MIT Press.

Gerstner, W., W. M. Kistler, R. Naud, and L. Paninski (2014). *Neuronal Dynamics: From Single Neurons to Networks and Models of Cognition*. Cambridge, Britain: Cambridge University Press.

Goodfellow, I., Y. Bengio, and A. Courville (2016). *Deep Learning*. Cambridge, MA: MIT Press.

Hebb, D. O. (1949). *Organization of Behavior*. New York: Wiley.

Herz, J. A., A. S. Krogh, and R. G. Palmer (1991). *Introduction to the Theory of Neural Computation*. Boston: Addison-Wesley.

Kandel, E. R. (2001). The molecular biology of memory storage: A dialogue between genes and synapses. *Science 294*, 1030–1038.

Rosenblatt, F. (1962). *Principles of Neurodynamics*. New York: Spartan.

Smolen, P., D. A. Baxter, and J. H. Byrne (2019). How can memories last for days, years, or a lifetime? Proposed mechanisms for maintaining synaptic potentiation and memory. *Learn. Mem. 26*, 133–150.

Squire, L. R. (1987). *Memory and Brain*. New York: Oxford University Press.

Sutton, R. S. and A. G. Barto (2018). *Reinforcement Learning: An Introduction* (2nd ed.). Cambridge, MA: MIT Press.

Primary Research Papers

Synaptic Plasticity and Learning

Bienenstock, E., L. Cooper, and P. Munro (1982). Theory for the development of neuron selectivity: Orientation specificity and binocular interaction in visual cortex. *J. Neurosci. 2*, 32–48.

Bittner, K. C., A. D. Milstein, C. Grienberger, S. Romani, and J. C. Magee (2017). Behavioral time scale synaptic plasticity underlies ca1 place fields. *Science 357*, 1033–1036.

Graupner, M. and N. Brunel (2012). Calcium-based plasticity model explains sensitivity of synaptic changes to spike pattern, rate, and dendritic location. *Proc. Natl. Acad. Sci. U.S.A. 109*, 3991–3996.

Marr, D. (1971). Simple memory: A theory for archicortex. *Philos. Trans. R. Soc. Lond. Biol. Sci. 262*, 23–81.

Miller, K. D., B. Chapman, and M. P. Stryker (1989). Visual responses in adult cat visual cortex depend on n-methyl-d-aspartate receptors. *Proc. Natl. Acad. Sci. U.S.A. 86*, 5183–5187.

Rescorla, R. and A. Wagner (1972). A theory of Pavlovian conditioning: Variations in the effectiveness of reinforcement and nonreinforcement. In A. H. Black and W. F. Prokasy (Eds.), *Classical Conditioning*: Vol. 2. Current Research and Theory, Chapter 9, pp. 64–99. New York: Appleton-Century-Crofts.

Shouval, H. Z., M. F. Bear, and L. N. Cooper (2002). A unified model of nmda receptordependent bidirectional synaptic plasticity. *Proc. Natl Acad. Sci. 99*, 10831–10836.

Song, S., K. D. Miller, and L. F. Abbott (2000). Competitive oncomi learning through spike-time-dependent synaptic plasticity. *Nat. Neurosci. 3*, 919–926.

Vogels, T. P., H. Sprekeler, F. Zenke, C. Clopath, and W. Gerstner (2011). Inhibitory plasticity balances excitation and inhibition in sensory pathways and memory networks. *Science 334*, 1569–1573.

Turrigiano, G. G., K. R. Leslie, N. S. Desai, L. C. Rutherford, and S. B. Nelson (1998). Activity-dependent scaling of quantal amplitude in neocortical neurons. *Nature 391*, 892–896.

Memory

Bhalla, U. S. and R. Iyengar (1999). Emergent properties of networks of biological signaling pathways. *Science 283*, 381–387.

Foster, D. J. and M. A. Wilson (2006). Reverse replay of behavioural sequences in hippocampal place cells during the awake state. *Nature 440*, 680–683.

Fusi, S., P. J. Drew, and L. F. Abbott (2005). Cascade models of synaptically stored memories. *Neuron 45*, 599–611.

Hopfield, J. J. (1982). Neural networks and physical systems with emergent collective computational abilities. *Proc. Natl. Acad. Sci. U.S.A. 79*, 2554–2558.

Louie, K. and M. A. Wilson (2001). Temporally structured replay of awake hippocampal ensemble activity during rapid eye movement sleep. *Neuron 29*, 145–156.

Miller, P., A. M. Zhabotinsky, J. E. Lisman, and X.-J. Wang (2005). The stability of a stochastic CaMKII switch: Dependence on the number of enzyme molecules and protein turnover. *PloS. Biol. 3*, e107.

Schultz, W., P. Dayan, and P. R. Montague (1997). A neural substrate of prediction and reward. *Science 275*, 1593–1599.

Part II

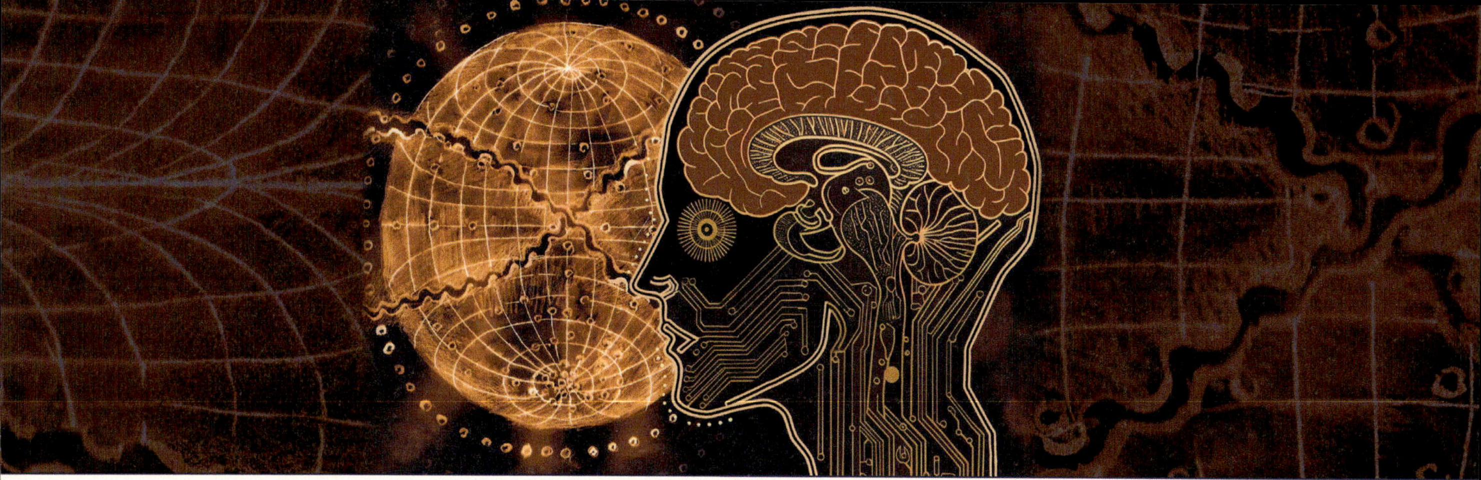

Working Memory

5

His [Isaac Newton's] peculiar gift was the power of holding continuously in his mind a purely mental problem until he had seen straight through it.

John Maynard Keynes in *Newton, the Man*

5.1 INTRODUCTION

In the previous three chapters, we introduced basic ideas and models for single neurons and synapses, networks and plasticity. With the knowledge thus acquired in theory and computational modeling, we are now in a position to tackle cognitive processes intermediate between sensation and movement. Our starting point is the observation that cognition and flexible behaviors must be freed from the immediacy of sensory stimulation. Imagine a situation where the brain could not retain or act on a sensory stimulus after it disappears—"out of sight, out of mind" (Goldman-Rakic 1995). In that case, our repertoire of sensorimotor behaviors would be reduced to quick reactions to currently present stimuli, such as jumping backward upon seeing a snake. Reasoning, thinking and planning would seem impossible. The mental function for holding and processing information in the absence of direct external stimulation is called working memory. Given its crucial role in all internal mental processes, working memory can be called "the hydrogen atom of cognition."

Archaeological evidence suggests that working memory function is markedly enhanced in *Homo sapiens* compared to Neanderthals (Wynn and Coolidge 2008, 2010). This evolutional leap appears to be crucial for the emergence of modern human cognition and intelligence: making sophisticated tools and artworks rely on holding a template in the mind through a creative process; language requires temporal integration of words as symbols in a grammatically correct way, etc. (Wynn and Coolidge 2011; Balter 2010). Indeed, working memory capacity is correlated with measures of fluid intelligence, which is our ability to solve novel problems. Behavioral studies, using working memory tasks and a battery of tests for assessing fluid intelligence, have consistently found a statistical correlation of about

DOI: 10.1201/9781003459361-7

0.4–0.5 between working memory capacity and intelligence test performance across individuals (**Figure 5.1**) (Conway et al. 2003; Johnson et al. 2013). A study reported that a child's working memory ability at 5 years of age is even a better predictor than IQ of academic success at 11 years of age (Alloway and Alloway 2010).

In psychology, the concept of working memory was first introduced by Milner et al. (1960), then developed by Baddeley in the 1970s. In Baddeley's formulation (**Figure 5.2**) (Baddeley and Hitch 1974; Baddeley 2012), working memory is postulated as a multi-component system of active short-term memory with a limited capacity in the service of cognition. The system includes three modules: (1) a visuospatial module serves as a "mental sketchpad" at the interface with the vision and spatial systems; (2) a phonological module, for verbal working memory and subvocal rehearsal, interacts with language; and (3) a central executive allocates attentional resources.

Importantly, working memory is distinct from short-term memory, although the two terms are often used interchangeably in the literature. Short-term memory generally denotes any memory trace that is not long lasting. For instance, your brain can recall where you parked your car today, but this memory is passive and fades away over a period of time,

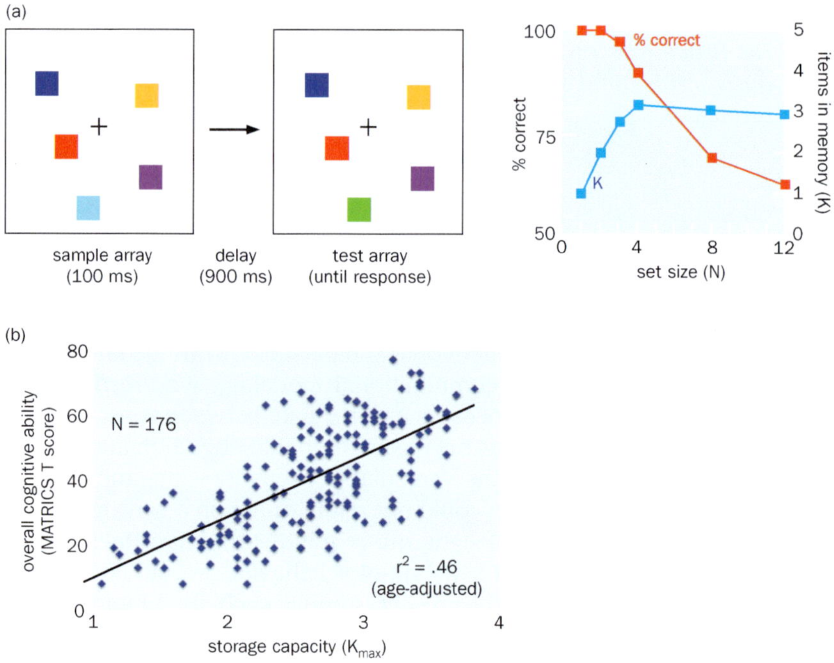

Figure 5.1 Working memory capacity and fluid intelligence. (a) A change detection task. A brief sample array of N items is followed by a delay and then a test array. The latter is either identical to the sample array or differs in a feature (color or location) of one of the objects, and the observer indicates whether a change is present. In this example, the color of the square at the bottom is changed from light blue to green. Right panel: Accuracy (% correct, red) and estimated number K of items stored in working memory (blue), based on a simple "slot model" of working memory. The capacity K_{max} is the plateau level of K for sufficiently large N. (b) Scatter plot of the relationship between storage capacity K_{max} measured in a change localization task (similar to (a) but the subject must indicate a change of location rather than color of a displayed item) versus a measure of broad cognitive function using performance test of fluid intelligence called the MATRICS Consensus Cognitive Battery. Panel (a) is reproduced from Luck and Vogel (2013), (b) from Luck and Vogel (2013) with the original data from Johnson et al. (2013). [Luck, S. J., & Vogel, E. K. (2013). Visual working memory capacity: From psychophysics and neurobiology to individual differences. Trends in Cognitive Sciences, 17(8), 391–400.] [Data from Johnson, M. K., McMahon, R. P., Robinson, B. M., Harvey, A. N., Hahn, B., Leonard, C. J., ... & Gold, J. M. (2013). The relationship between working memory capacity and broad measures of cognitive ability in healthy adults and people with schizophrenia. Neuropsychology, 27(2), 220.]

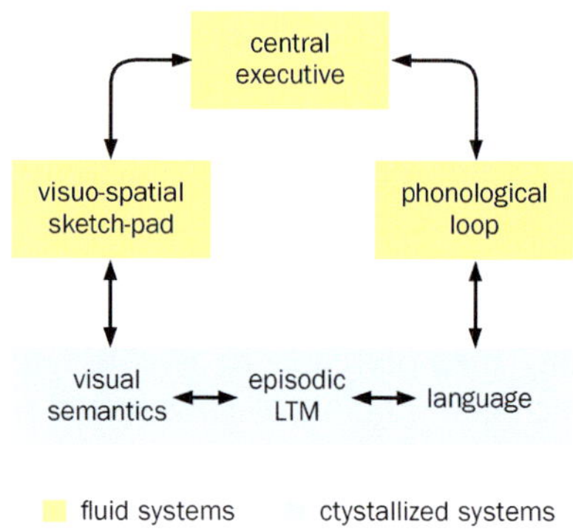

Figure 5.2 Baddeley's schematic of working memory. A working memory system consists of three subsystems: A control executive, a visuospatial sketchpad for temporal storage of visual or spatial information, a phonological loop for rehearsal. Each of the latter two modules interacts with its corresponding part in long-term memory: Visual semantics and language, respectively, which may also involve episodic memory. They are under the control of a central executive. Adopted from Baddeley (2012). [Baddeley, A. (2012). Working memory: Theories, models, and controversies. Annual Review of Psychology, 63, 1–29.]

which is ill defined. By contrast, working memeory is active ("online") and involves not only storage but also manipulation of information (Baddeley and Hitch 1974; Oberauer 2009). Unlike passive short-term memory, working memory has limited capacity, because the brain can reliably maintain only a few items concurrently (**Figure 5.1**) (Miller 1956; Cowan 2001, 2010). The executive module serves two functions: first, limited working memory resources need to be allocated to behaviorally relevant information; second, a mechanism is required to filter out continuous flow of external stimulation so that internal memory maintenance is not disrupted by interference from irrelevant stimuli.

As we shall see next, working memory representation is internally sustained in the brain in the form of persistent neural firing of spikes. Understanding the neural circuit mechanism of mnemonic activity is thus of fundamental interest and the focus of this chapter. At the present time, it is not clear what simple species such as *Drosophila* flies may be endowed with working memory. In mammals, working memory critically depends on the prefrontal cortex (PFC), which is involved in key cognitive processes (Goldman-Rakic 1987a; Miller and Cohen 2001; Fuster 1989). Working memory studies using simple task paradigms opened the door for neural circuit dissection of the PFC, as a crucial step towards elucidating the biological basis and computational principles of cognition.

5.2 NEURAL REPRESENTATION OF WORKING MEMORY

5.2.1 Delay-Dependent Task and Self-Sustained Mnemonic Activity

Working memory deployed in daily life can be rigorously investigated experimentally. For instance, when you are looking for a friend in a crowded room at a party, your brain retrieves her face from long-term memory and actively maintains it in working memory. This internal representation guides visual search while your eyes scan the crowd, until you find your friend when her image matches the template of her face inside your brain. Such a process can be studied in controlled experiments using a delayed match-to-sample (DMS) task paradigm: a sample stimulus is briefly shown and must be held in the mind for a delay period when the stimulus is absent. Then, a second stimulus is shown, and the brain must decide whether it matches with the sample based on its internally maintained representation.

Thus, the key to probing working memory in a laboratory experiment is to introduce a brief temporal delay between a sensory stimulus and a motor response, during which time the sensory information must be actively held in the mind so it can be used to guide the response later (**Figure 5.3**). Delayed response tasks, developed more than a hundred years ago by Hunter (1913), transcend simple stimulus-response reflexes and engage working memory. Studies in humans with brain damage and selective brain lesions with nonhuman animals have shown that the PFC plays an essential role in working memory (Jacobsen 1936; Szczepanski and Knight 2014).

In neurophysiological studies in which subjects perform a delayed response task, single neurons in certain parts of the brain display elevated electrical activity that is selective for sensory stimuli and persists throughout the mnemonic delay. In a pioneering study of this kind, using a task paradigm similar to **Figure 5.3**, "memory cells" were discovered in monkey PFC that displayed delay period activity for seconds (**Figure 5.4**) (Fuster and Alexander 1971; Kubota and Niki 1971). Remarkably, when the delay period is doubled in duration, mnemonic spiking activity also prolonged twice as long accordingly, demonstrating that persistent neural activity was self-sustained for as long as the information needed to be maintained, until a memory-guided behavioral response was produced.

Figure 5.3 Delayed response task. A stimulus is shown briefly (food at one of the two locations, left or right), followed by a delay period typically of a few seconds. At the end of the delay, the subject is required to produce a correct memory-guided response for reward collection. Reproduced from (Goldman-Rakic 1992). [Modified from Goldman-Rakic, P. S. (1992). Working memory and the mind. Scientific American, 267(3), 110–117.]

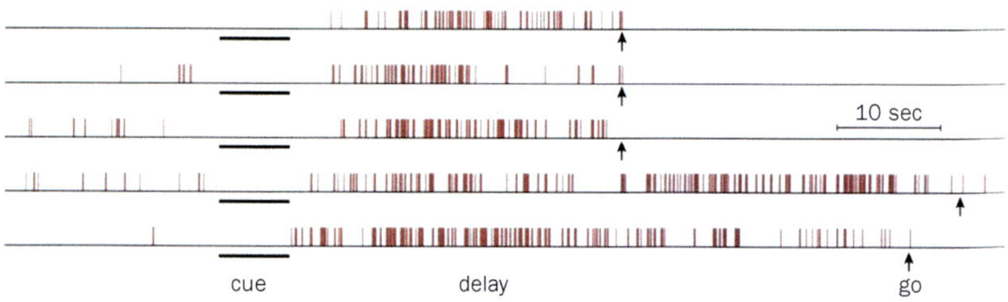

Figure 5.4 Persistent activity during working memory. In a delayed response task, shown for five trials, a single neuron recorded from monkey's PFC displays sustained activity after stimulus offset, through the delay period that lasts tens of seconds, until the go signal (upward arrow) initiates a memory-guided response and the sensory information is no longer needed. Reproduced from Fuster and Alexander (1971). [Fuster, J. M., & Alexander, G. E. (1971). Neuron activity related to short-term memory. Science, 173(3997), 652–654.]

In a delayed response task shown in **Figure 5.3**, interpretation of neural activity observed during the delay may be ambiguous because it could be, say, associated with the subject's body posture or motor preparation (Prut and Fetz 1999) rather than internal memory storage during the delay. Quantitative control of behavior is crucial. An excellent example is the oculomotor delayed response (ODR) paradigm (Wurtz 1969; Hikosaka and Wurtz 1983; Funahashi et al. 1989; Wang 2005), where sensory inputs are visual, motor responses are rapid eye movements (saccades) and eye position is constantly tracked by an eye coil. Specifically, a subject's eyes are required to fixate at the center spot of a computer monitor when a peripheral visual stimulus is briefly flashed, followed by a delay period of a few seconds. The disappearance of the fixation light spot signals the end of the delay. At that moment, the subject must make an accurate saccadic eye movement to the location where the cue was shown before the delay period in order to collect a liquid reward. During the response period, there is nothing on the computer monitor and the room is dark, so the subject must rely on working memory to perform the task.

Such delay-dependent tasks offered a concrete way to probe the neural basis of internal information coding and maintenance in the absence of external stimulation by single-neuron physiology, bridging the psychological concept of working memory and the brain. Functional brain imaging of human participants demonstrated that the PFC is especially activated in working memory dependent tasks (D'Esposito et al. 1995; Courtney et al. 1998; D'Esposito 2007), shown in **Figure 5.5** for the ODR task, thereby enabling a comparison between humans and nonhuman animals in a cognitive domain.

5.2.2 Three Types of Neuronal Working Memory Coding

Self-sustained activity during a delay period has been amply documented by single-neuron physiology in behaving animals (Wang 2001; Constantinidis

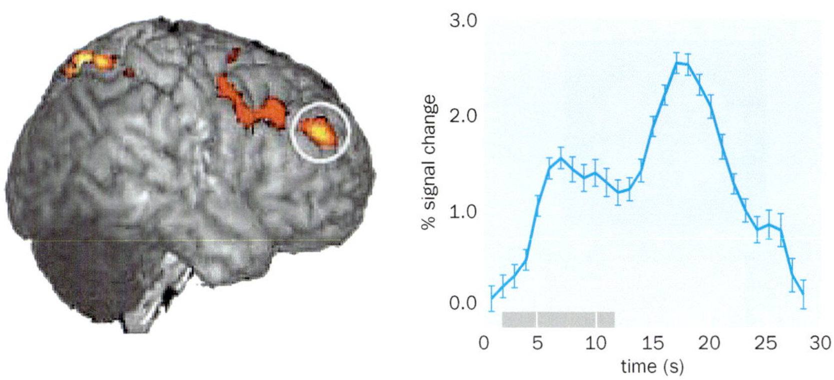

Figure 5.5 Delay period activity measured by fMRI in the human brain. Using a spatial working memory task, the lateral PFC (Brodman area 46, white circle) showed significant delay-period activity (left). Average fMRI signal persists throughout the delay, seconds after the stimulus cue has disappeared. (The grey bar represents the length of the delay interval.) Reproduced from D'Esposito (2007). [D'Esposito, M. (2007). From cognitive to neural models of working memory. Philosophical Transactions of the Royal Society B: Biological Sciences, 362(1481), 761–772.]

and Wang 2004; Pasternak and Greenlee 2005) and brain imaging (Curtis and D'Esposito 2003; Jonides et al. 2008). At the outset, a clarification is in order about the meaning of the term "self-sustained activity" or "persistent activity," which we will use interchangeably. It generally refers to a neural population state that is self-sustained and conveys behaviorally relevant information such as sensory stimulation or task rule. Here, the subjective "persistent" should be understood as in contrast to the absence of neural firing during a delay period, or a transient that eventually decays away with a sufficiently long delay. As such, it does not imply that neural activity must be a steady-state with a constant firing rate; instead, the name covers a wide range of dynamical formats of internal representation, including ramping time course (Brody et al. 2003; Fuster 1989) or stochastic oscillations (Pesaran et al. 2002; Lundqvist et al. 2016).

To encode specific information, self-sustained activity must display stimulus selectivity. At the single cell level, representation of information in working memory is in the form of tuning curves of delay activity. In sensory areas, a tuning curve quantitatively describes the response of a neuron as a function of a stimulus attribute, such as the orientation of a line stimulus in vision or the frequency of a somatosensory stimulus (Chapter 3, **Figure 3.15**). Here, we are concerned with working memory encoding, not neural responses to stimulus presentation. According to the type of the sensory stimulus for storage, one can distinguish three types of tuning curves for working memory representation:

Discrete working memory representation. Digits and words commonly used in human studies of working memory are discrete items. In nonhuman animal experiments, high-dimensional stimuli like visual objects or odors can also be considered as discrete items. In a DMS experiment (**Figure 5.6a**), a monkey subject was shown two stimuli (fractal pictures) separated by a delay period and judged whether the two are the same (match). The behavioral response depends on the memory of the first (sample) stimulus to be compared with the second (test) stimulus. Neurons of the inferotemporal cortex (ITC) display elevated activity during the delay period, which is selective for one (left) but not other (middle and right) objects (Miyashita and Chang 1988). Such tasks engage a working memory circuit where the stored information is one of a collection of discrete items.

Categorization offers another example of discrete items. In a delayed match-to-category (DMC) task, monkeys learned to classify an analog feature (motion direction, from 0 to 360 degrees) of visual moving stimuli into two arbitrarily defined categories (e.g., 45 to 225 degrees for category 1, 225 to 45 degrees for category 2). The subject performed the task by judging whether the sample and test stimuli belong to the same category. Neurons in the lateral intraparietal area (LIP) of the posterior parietal cortex (PPC) (Freedman and Assad 2006) and PFC (Sarma et al. 2016) display a

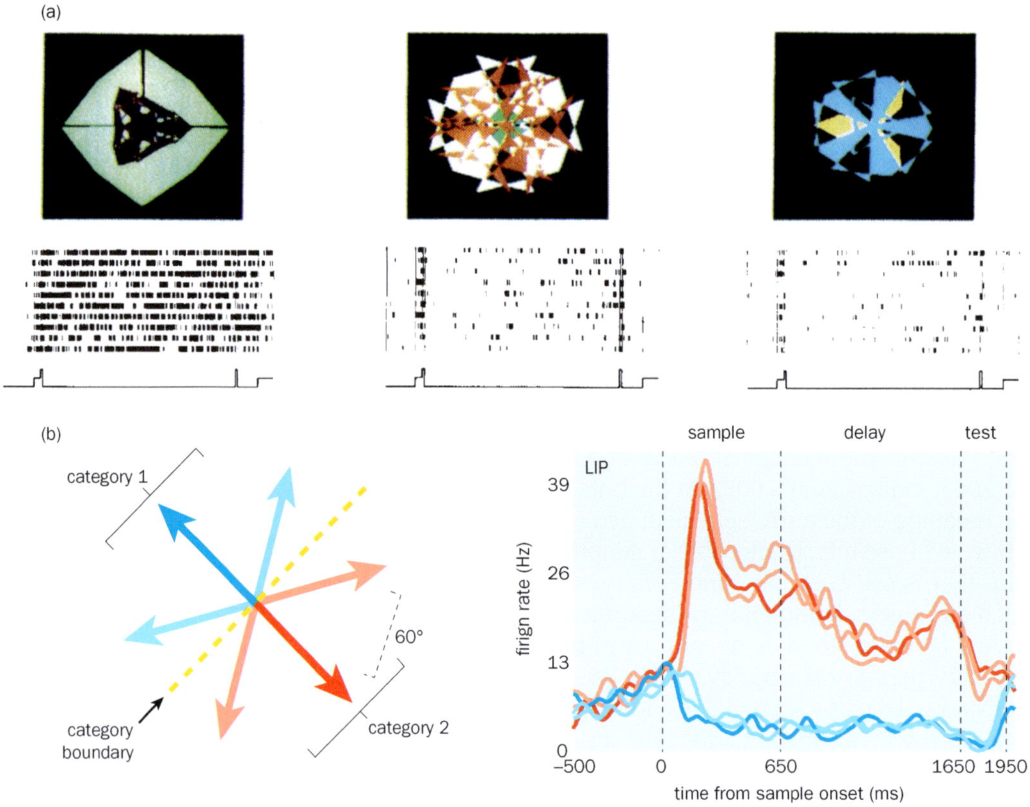

Figure 5.6 Object or category-selective persistent activity. (a) In a monkey experiment using a delayed match-to-sample task, a neuron from the intertemporal cortex displayed elevated activity in the delay period selective for one of three fractal visual stimuli. (b) In an experiment where monkeys learned to classify the direction of a visual moving pattern into two categories (1, blue; 2, red), a neuron from the PPC displays a binary activity pattern depending on the category of motion direction during the delay period. Note that two directions across the category boundary (one blue, the other red) are separated by the same angle as two adjacent directions (both blue or red) inside the same category. Panel (a) is reproduced from Miyashita and Chang (1988), (b) from Freedman and Assad (2016). [Miyashita, Y., & Chang, H. S. (1988). Neuronal correlate of pictorial short-term memory in the primate temporal cortex. Yasushi Miyashita. Nature, 331(6151), 68–70. [Swaminithan S.K., & Freedman D.J., (2012). Preferential encoding of visual categories in parietal cortex compared with prefrontal cortex. Nature Neuroscience, 15(2),315–320.]

binary pattern of firing activity depending on the categorical membership of the sample stimulus (Figure 5.6b).

Bell-shaped working memory representation. Figure 5.7 illustrates an ODR experiment, where the stored information is spatial location, which is an analog quantity. Neurons in the PPC (Gnadt and Andersen 1988) and PFC (Funahashi et al. 1989) display persistent delay activity that is spatially selective, with a cell's "memory field" characterized by a smooth tuning curve. When spatial cues are directional angles with fixed eccentricity, the bell-shaped curve is peaked at a preferred spatial cue. Peaked encoding of spatial working memory also holds when visual cues are not restricted to a fixed eccentricity, but can be at any location in the two-dimensional screen (Rainer et al. 1998). Therefore, spatial working memory representation employs a location code, where different neurons have different memory fields, the entire network represents a spatial map in working memory. Working memory coding of other stimulus features like stimulus orientation or color also displays bell-shaped tuning curves.

Parametric working memory representation. Figure 5.8 shows a vibrotactile delayed discrimination (VDD) task, where monkeys were trained to compare and discriminate the frequencies f_1 and f_2 of two somatosensory stimuli separated by a delay period. Here, the encoded information is the cue frequency, again an analog quantity. Prefrontal cells show persistent activity that varies monotonically with the cue frequency (Romo et al.

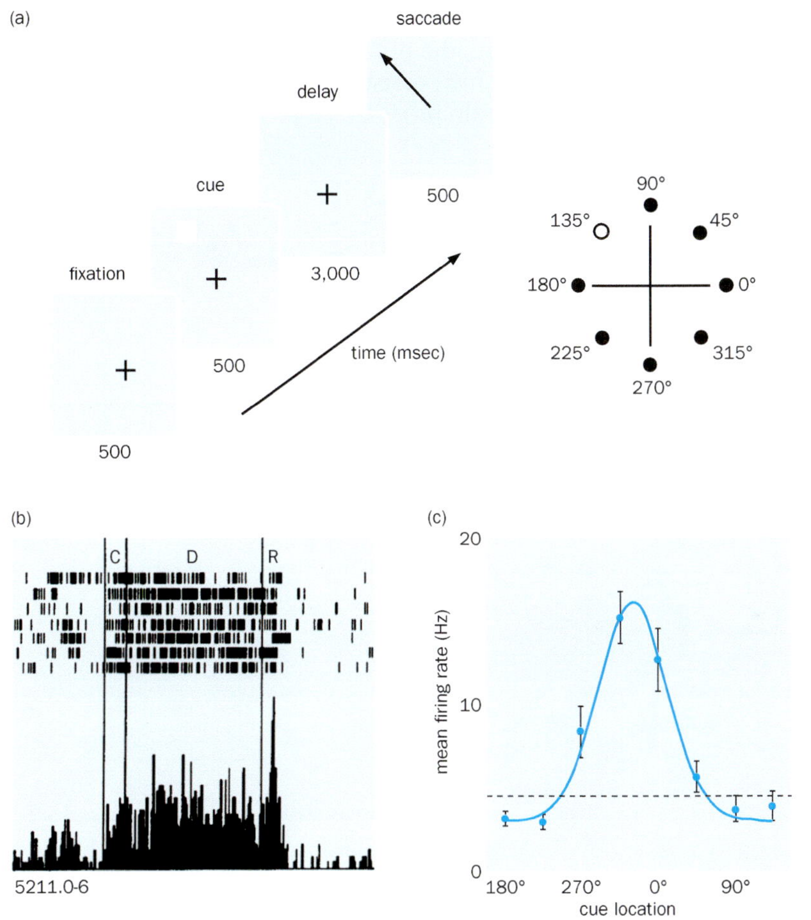

Figure 5.7 Spatial working memory. (a) In an oculomotor delayed response (ODR) task, a trial starts with a fixation period when the subject's eyes remain on a cross at the center of a computer monitor. A visual cue is briefly flashed (0.5 seconds), followed by a delay period of 3 seconds. The disappearance of the fixation cross signals the end of the delay, at which time point the subject must generate a rapid eye movement to the location of the remembered visual cue in order to collect a reward. Insert on the right: Eight cues vary across trials, with a fixed eccentricity. (b) A single neuron in the PFC displayed spontaneous activity during the fixation period and an elevated persistent activity for a preferred cue during the delay period. (c) The trial-averageed firing rate of persistent activity displays a bell-shaped, or Gaussian, tuning curve for the cue direction. Note that delay period activity is lower than spontaneous activity (dashed line) for nonpreferred cues (e.g., null direction, 180° away from the preferred cue). Reproduced from Funahashi et al. (1989). [Funahashi, S., Bruce, C. J., & Goldman-Rakic, P. S. (1989). Mnemonic coding of visual space in the monkey's dorsolateral prefrontal cortex. Journal of Neurophysiology, 61(2), 331–349.] [Modified from Funahashi, S., Bruce, C. J., & Goldman-Rakic, P. S. (1989). Mnemonic coding of visual space in the monkey's dorsolateral prefrontal cortex. Journal of Neurophysiology, 61(2), 331–349.]

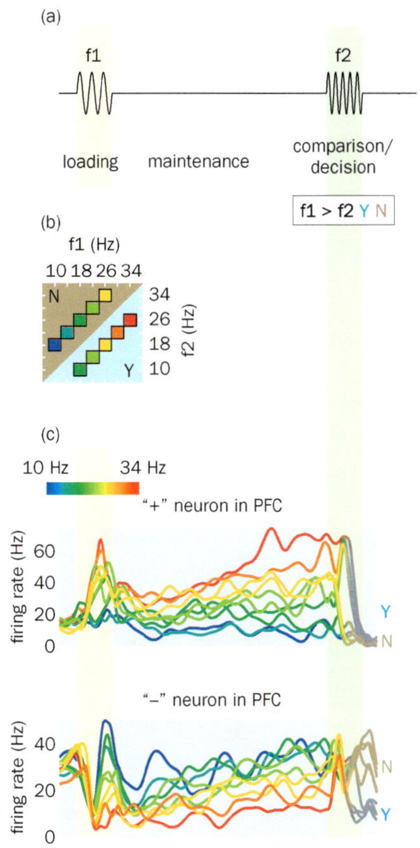

Figure 5.8 Parametric working memory. (a) In a vibrotactile delayed discrimination (VDD) task, a somatosensory stimulus at frequency f_1 is followed by a delay period of 6 seconds, and then another stimulus f_2 is shown and the subject must judge whether f_2 is higher or lower than f_1. (b) Various sample (x-axis) and test (y-axis) frequencies used in the monkey experiment. (c) Persistent activity of two single PFC cells varied with the stimulus feature f_1 indicated by color. Tuning of delay period activity is monotonic, either increasing (top) or decreasing (bottom) with f_1. Reproduced from Romo et al. (1999). [Romo, R., Brody, C. D., Hernández, A., & Lemus, L. (1999). Neuronal correlates of parametric working memory in the prefrontal cortex. Nature, 399(6735), 470–473.]

1999). Presumably, there is a continuous range of persistent activity rates, which vary in proportion to the stimulus parameter (the vibration frequency). Therefore, the stimulus is encoded by a rate code rather than a location code.

Working memory of more abstract information than sensory stimuli has also been found in the form of self-sustained persistent neural activity. In a monkey experiment (Wallis et al. 2001), a memory-guided motor response was contingent on one of two behavioral rules. Namely, a lever press was required when a sample and a test (separated by a delay period) were the same, if rule 1 DMS is currently in play; conversely, the motor response was required when the test did not match the sample according to rule 2 DNMS (delayed nonmatch-to-sample). In different blocks of trials, the rules were signaled by different cues. Single neurons in the PFC showed persistent activity that was not selective for sensory stimuli but for the behavioral rule, regardless of how the rule was signaled (**Figure 5.9**). Therefore, the brain mechanisms for the generation and maintenance of mnemonic persistent activity are of broad interest for understanding internal brain processes underlying flexible behavior.

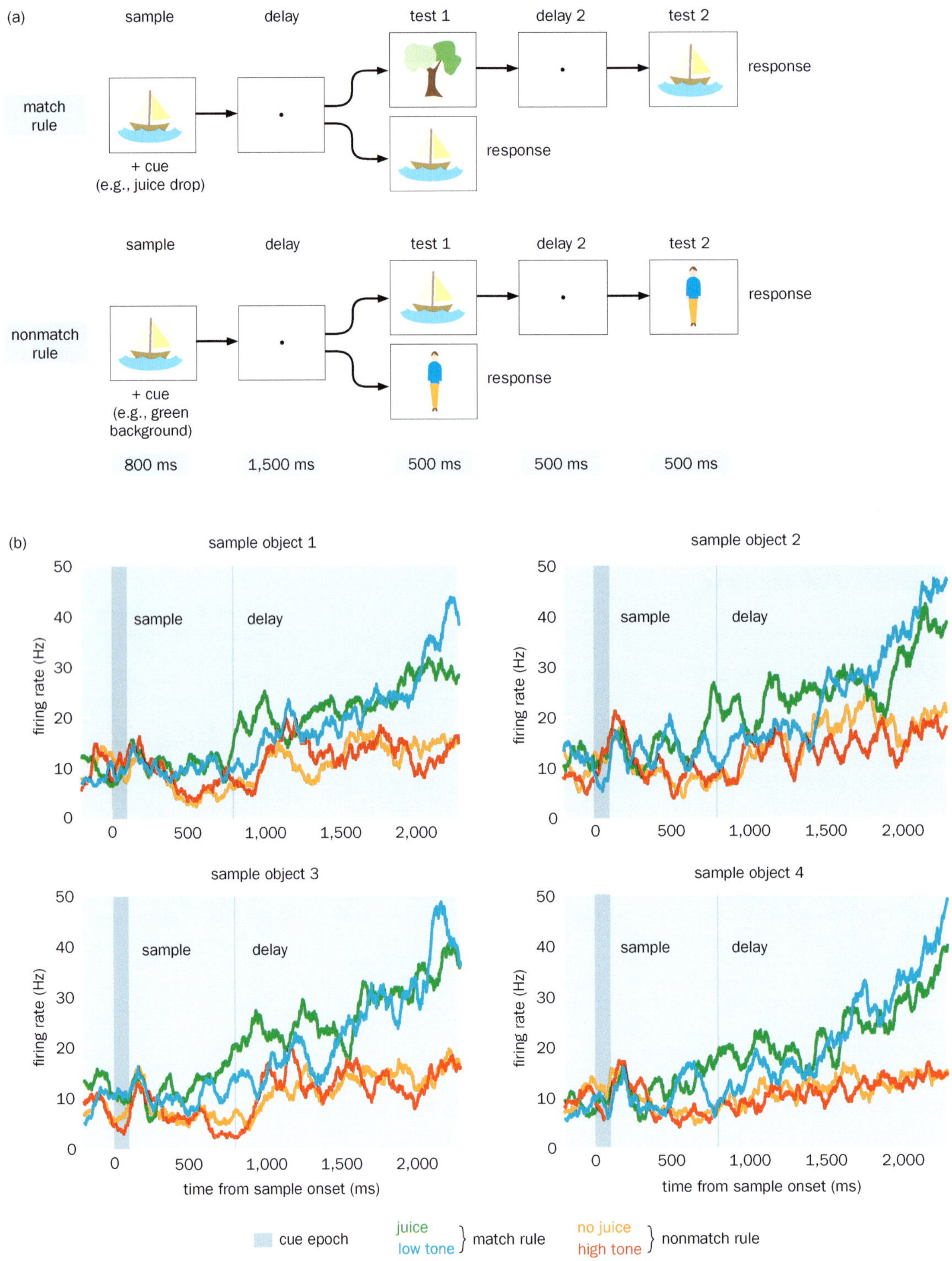

Figure 5.9 Persistent neural activity encoding a behavioral rule. (a) In a monkey experiment, a delayed response was required when sample and test objects were the same in trials when the match rule should be applied, but when they were different in trials with the nonmatch rule currently in play. (b) A single PFC neuron showed delay period activity that is not sensitive to different sample stimuli (four subpanels). By contrast, it is selectively higher for the match rule (green and blue) than for the nonmatch rule (red and orange), regardless of the different ways the rule was instructed. Reproduced from Wallis et al. (2001). [Wallis, J. D., Anderson, K. C., & Miller, E. K. (2001). Single neurons in prefrontal cortex encode abstract rules. Nature, 411(6840), 953–956.]

5.2.3 Feedback Mechanisms of Persistent Activity

The persistence time (up to tens of seconds) of sustained firing activity during working memory is orders of magnitude longer than the biophysical time constants (tens of milliseconds) of fast electrical signals in neurons and synapses (Chapter 2). For this reason, persistent activity is believed to be generated by feedback dynamics, or reverberation, in a neural circuit (Lorente de Nó 1933; Hebb 1949; Amit 1995; Goldman-Rakic 1995; Camperi and Wang 1997; Wang 2001, 2021; Arnsten et al. 2010). The central idea is that in response to a brief stimulus, recurrently connected excitatory cells in a neural assembly fire at an increased rate, and the resulting mutual synaptic interactions could sustain elevated activity after the external input is withdrawn, with the proviso that intrinsic connections are sufficiently strong. To test this hypothesis, a key issue is to identify the minimum anatomical substrate for reverberation. In the brain, several scenarios are conceivable.

Cortico-subcortical loop. Persistent activity can arise from a large neural network that involves subcortical systems, through reciprocal excitation between the cortex and thalamus, or via the cortico-basal ganglia-thalamo-cortical loop (**Figure 5.10a**). There are two loops in this circuit. The first is the reciprocal connection between the thalamus and cortex. The second engages the basal ganglia intermediate between the cortex and thalamus. Consistent with this scenario, neurons in thalamic nuclei that connect with the PFC show mnemonic activity during the delay period of a working memory task (Fuster and Alexander 1973; Funahashi 2013; Guo et al. 2017). Elevated delay period activity was also present in the striatum (Hikosaka et al. 1989), whereas output neurons in the SNr/GPi inhibited by the striatum showed suppressed firing during a delay period (Hikosaka and Wurtz 1983).

Reciprocal loop between cortical areas. It is also possible that persistent activity is maintained by reciprocal loops between cortical areas,

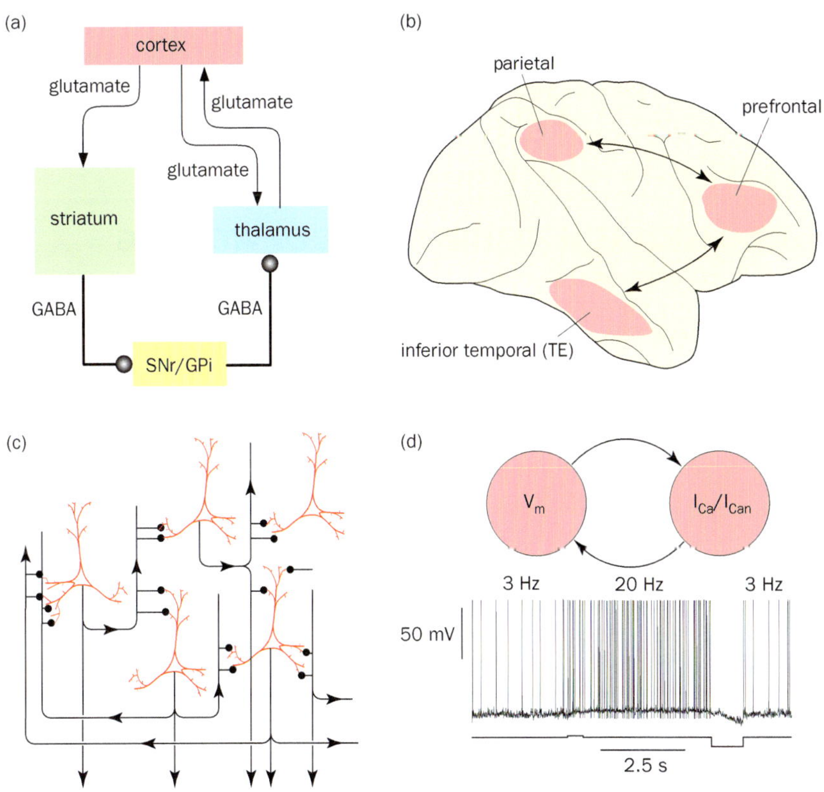

Figure 5.10 Various scenarios for the anatomical substrate of excitatory reverberation in the cortex. (a) A closed thalamo-cortical loop and/or cortico-basal ganglia-thalamic-cortical circuit. Arrow: Excitation; circle: Inhibition. In the latter case, cortical excitation of the striatum leads to an inhibition of the output neurons in SNr/GPi and thus a disinhibition of thalamic neurons, which in turn send increased excitation back to the cortex. GPi, internal segment of the globus pallidus; SNr, substantia nigra pars reticulata. (b) Reciprocal interactions between two cortical areas, prefrontal and posterior cortices (respectively, prefrontal and inferotemporal cortices) for spatial (resp. object) visual memory. (c) Excitatory recurrent collaterals within a local circuit. (d) Intrinsic regenerative dynamics of single neurons. Positive feedback between membrane depolarization/spike discharges and active inward currents (voltage-gated I_{Ca} and Ca^{2+} gated I_{Can}) can produce persistent activity that outlasts a transient input current pulse. Reproduced from Wang (2001). [Wang, X. J. (2001). Synaptic reverberation underlying mnemonic persistent activity. Trends in Neurosciences, 24(8), 455–463.]

such as between the PFC and the PPC for spatial working memory (Gnadt and Andersen 1988; Funahashi et al. 1989; Chafee and Goldman-Rakic 1998), or between the PFC and the inferotemporal cortex (ITC) for object working memory (Miyashita and Chang 1988; Miller et al. 1996; Kaminski et al. 2017) (**Figure 5.10b**).

Recurrent local cortical network. The third scenario is that persistent activity can be produced locally within a cortical area (**Figure 5.10c**). In particular, Goldman-Rakic (1995) proposed a columnarly organized cortical network model for the PFC, where persistent activity arises from reverberatory excitation, and stimulus selectivity is formed by recurrent inhibition. This proposal is consistent with the anatomical and electrophysiological data that show extensive horizontal excitatory connections in the prefrontal cortex, especially in layers 2/3 (Levitt et al. 1993; Kritzer and Goldman-Rakic 1995; González-Burgos et al. 2000). Neurophysiological recording from monkeys (Bastos et al. 2018) and high-resolution functional MRI imaging from humans (Finn et al. 2019) support the notion that working memory representation is preferentially active in superficial layers 2/3.

Intrinsic cellular bistability. Finally, feedback excitation could originate from regenerative membrane dynamics of single neurons (**Figure 5.10d**). Voltage- and calcium-gated ion channels could in principle generate bistability between a resting state and an active state sustained by a "plateau potential" (Marder et al. 1996; Camperi and Wang 1998; Major and Tank 2004; Zylberberg and Strowbridge 2017). For instance, membrane depolarization during action potentials activates a high-threshold calcium current (I_{Ca}), and the resulting calcium influx activates a calcium-activated cation current (I_{Can}), which in turn sustains membrane depolarization, thereby providing a positive feedback (Haj-Dahmane and Andrade 1998; Egorov et al. 2002; Fransén et al. 2006). The single-cell scenario predicts that in neurons recorded intracellularly from behaving animals, current injection should be able to switch on and off a persistent activity pattern.

Here, we will focus on local circuit mechanisms (which have been extensively studied experimentally and theoretically), with contributions from single neuron properties, different cell types, synaptic dynamics and short-term plasticity. In Chapter 9 we will discuss multi-regional circuit mechanisms of working memory distributed in a large-scale brain system.

5.3 ATTRACTOR NETWORK MODEL OF WORKING MEMORY

5.3.1 A Simple Rate Model

The idea of reverberation is made precise in theoretical work where persistent activity is described as "dynamical attractors" (Amari 1977; Amit 1995; Wang 2001, 2021; Brunel 2005). As discussed in Chapters 2 and 3, the mathematical term "attractor" simply means any stable state of a dynamical system, such as a neural network. It is worth emphasizing that attractors are defined differently when conditions change. For instance, the Hodgkin-Huxley model shows a steady-state at rest but an oscillation with a large input current (Chapter 2, **Figure 2.8**). For a memory system like the Hopfield model (Chapter 4, **Figure 4.19**), multiple attractors in the absence of external input are by definition self-sustained. Moreover, a memory state of persistent activity needs to coexist with a background state, so a transient input can switch the system back and forth between the resting state and a memory state.

To gain intuition for such network bistability, let us start with the consideration of a simple model of excitatory neurons described by a single population rate r_E that obeys the following rate equation (Chapter 3):

$$\tau_E \frac{dr_E}{dt} = -r_E + F(J_{EE} r_E + I_{\text{ext}}) \tag{5.1}$$

where F is the neuronal input-output function, J_{EE} is the strength of recurrent excitation and I_{ext} is an external input. A steady-state obeys $dr_E / dt = 0$, or $r_E = F(I)$ with $I = J_{EE} r_E + I_{\text{ext}}$. Assume that $F(I) = 0$ for I below a threshold $I_c > 0$, then $F(J_{EE} r_E + I_{ext})$ is zero for $J_{EE} r_E + I_{\text{ext}} < I_c$ or $r_E < (I_c - I_{\text{ext}}) / J_{EE}$. Above the threshold, $F(I)$ grows with r_E and eventually saturates for sufficiently large r_E.

What are self-sustained steady-states with $I_{\text{ext}} = 0$? The algebraic equation $r_E = F(J_{EE} r_E)$ may not be readily solvable analytically, since $F(I)$ itself depends on the firing rate r_E. However, if we plot the left-hand side and right-hand side of the equation on the same graph, any intersection of the two (that is, any intersection of $F(J_{EE} r_E)$ with the diagonal line) is a steady-state (**Figure 5.11a–c**). When recurrent excitation is weak, the only steady-state is $r_1 = 0$ (**Figure 5.11a**). This is obviously true with $J_{EE} = 0$ (for which $F(I)$ is independent of r_E; $I = J_{EE} r_E = 0$ is below I_c) and remains so for small J_{EE}. As J_{EE} is increased to a critical value J_c, $F(J_{EE} r_E)$ grows high enough to "touch on" (tangent to) the diagonal line (**Figure 5.11b**). When J_{EE} is above J_c, $F(J_{EE} r_E)$ intersects with the diagonal line twice more, corresponding to two additional steady-states r_2 and r_3 (**Figure 5.11c**).

As shown in **Figure 5.11c**, two of the three steady-states, at $r_1 = 0$ and $r_3 \sim 150$ Hz, are stable: the system that deviates slightly from each would converge back to it over time. This can be seen as follows. For the steady-state r_1, if the rate r_E happens to be slightly larger than 0 (obviously it cannot be below 0), F is below the diagonal line, $-r_E + F$ (thus dr_E / dt) is negative, and the firing rate decreases over time towards the steady-state according to Eq. (5.1). Similarly, for the steady-state r_3, if a perturbation

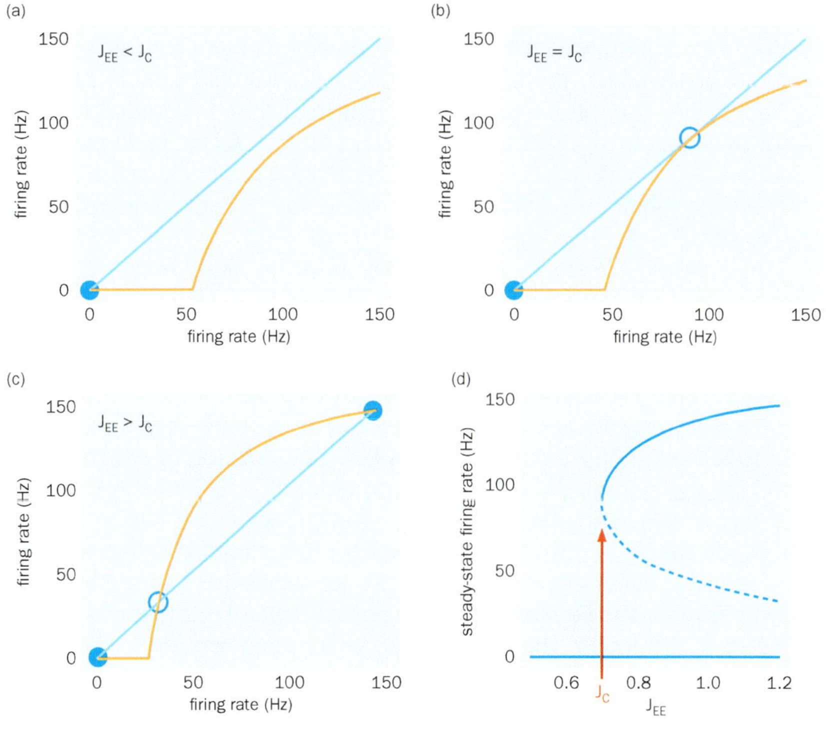

Figure 5.11 A simple one-dimensional recurrent excitatory neural network. (a–c) Graphic solution with J_{FF} smaller than, equal to and larger than J_c, respectively. Orange curve: $F(J_{EE} r)$; blue curve: r (diagonal line). An intersection of them defines a steady-state. Filled circle: Stable attractor, open circle: Unstable steady-state. (d) Steady-state(s) (y-axis) as a function of J_{EE}, showing a bifurcation at J_c (red arrow) for the emergence of bistability and persistent activity.

brings r_E to slightly below it, F is larger than r_E and the activity would increase over time. The opposite happens if r_E is above it, and in both cases the system will converge back to r_3.

The steady-state indicated by an open circle ($r_2 \sim 20$ Hz) is unstable: the same reasoning above tells us that when r_E deviates slightly from it, the system would diverge away from that steady-state. Thus, the fact that a steady-state exists does not guarantee that it is stable, and an unstable steady-state is not directly observable empirically. Nevertheless, its impact is observable: if the system is initially at rest ($r_E = r_1$), and a transient input is used to switch the system to r_3, then the unstable r_2 determines how much increase in the firing rate is needed to make the transition to the persistent activity state.

Figure 5.11d shows the bifurcation diagram, where self-sustained steady-states are plotted as a functionof J_{EE}. At the value J_c a pair of new steady-states, one is stable whereas the other is unstable, emerge through what is mathematically called a "saddle-node bifurcation" (Strogatz 2016). Note that a constant (rather than brief transient) positive external input I_{ext} would shift the bifurcation point to the left, as a background excitation reduces the need for recurrent excitation to sustain an elevated firing state. On the other hand, if I_{ext} is increased above I_c, then the resting state disappears and bistability is lost. The general insight is that, instead of being fixed, a network's "attractor landscape" can be readily modified by a sustained external input, which in the brain could correspond to an executive control signal to the working memory system.

5.3.2 Network Model of Stimulus-Selective Persistent Activity

In the prior simple example, neurons are silent in the "background state," which is inconsistent with physiological observations. It turns out to be challenging to realize nonzero spontaneous activity in an attractor network model of working memory, because strong recurrent excitation could destabilize it (Amit and Brunel 1997). Futhermore, the persistent activity has an unrealistically high firing rate (compared to typically a few tens of Hz observed in delay period activity, see **Figures 5.6–5.9**). This is due to "runaway excitation" (Douglas et al. 1995): increased firing produces more recurrent excitation, which in turn drives the firing to an even higher level. In this simple model, the positive feedback is curtailed only at high firing rates when the refractory period of neuronal firing becomes a limiting factor, reflected by the plateauing of $F(I)$ at large I. As we shall see, paying attention to biological realism in computational modeling and finding mechanistic solutions to achieve better model versus data comparison can lead to surprising revelations.

In a working memory system, there must be many memory states, each representing a stimulus item. The resting state and stimulus-selective memory states are represented as multiple attractors in a state space of neural population activity (**Figure 5.12**, left). For an analogy, consider an unusually hilly golf course with many valleys (shallow areas among hills) (**Figure 5.12**, right). The bottom (attractor) of a valley (basin of attraction) is "attractive" in the sense that a ball (the position of which corresponds to the state of the neural system) naturally rolls down towards it. This way, a sufficiently large transient input (a swipe of a ball to the air) can switch the system from rest (one valley) to a stimulus-selective mnemonic state (a different valley), and the system stays there after the stimulus withdrawal; such a state is robust against small perturbations (a tiny swing of a ball moves the ball slightly, which then returns to the same spot). Another brief but potent signal can switch the system back to the resting state, thereby erasing a memory trace. Unlike a golf course, however, attractors in a neural system do not have to be steady-states but can

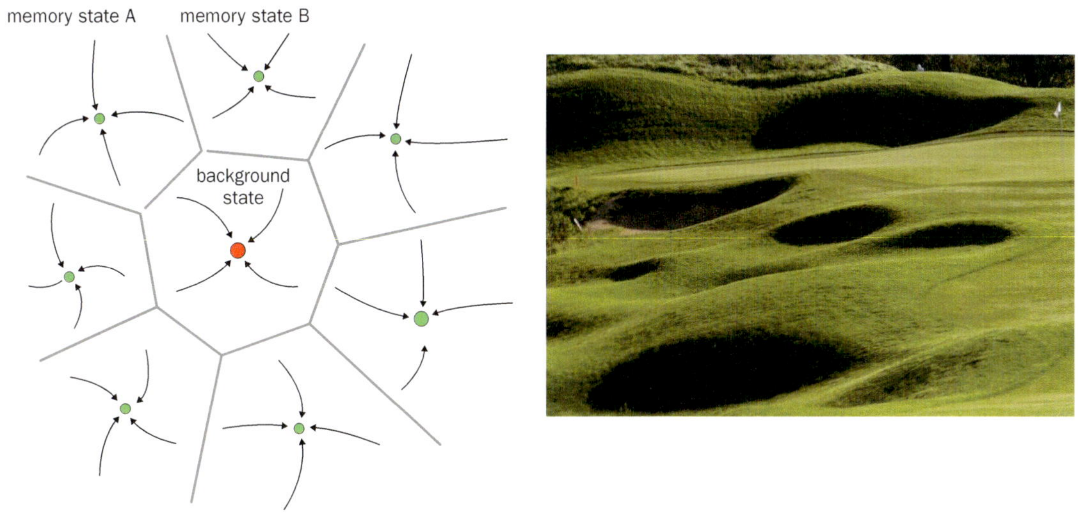

Figure 5.12 Schematics of an object working memory model. Left panel: Discrete attractor network in the state space is depicted here on a two-dimensional plane with the axes corresponding to the firing rates of two neurons. In a real network with N neurons, the state space is N-dimensional. A number of memory states (green) coexist with each other and with a background state (red). The synaptic connections and other properties of the circuit, as well as external inputs, determine the contours (the basins of attraction, within which the system follows a trajectory toward the corresponding attractor state). Right panel: A golf course analogy. Panel (a) is reproduced from Brunel (2005).

exhibit complex spatiotemporal patterns such as stochastic network oscillations (see following discussion). Furthermore, as aforementioned, the landscape of multiple attractors is readily modifiable by a sustained input, which enables executive control of working memory.

Let us build a realistic recurrent circuit model of spiking neurons for working memory. Since this model is foundational, we will dive into some details about the methodology of biologically based modeling. Consider a network with N_E excitatory pyramidal cells and N_I inhibitory interneurons ($N_E + N_I = N$). Suppose that single cells are described by the leaky integrate-and-fire model (Chapter 2). The membrane potential of each neuron is described by:

$$\begin{aligned} C_m \frac{dV_i^E}{dt} = & -\sum_{j=1}^{N_E} w_{ij}^{FF}\left(G_{EE,AMPA}s_j^{AMPA} + G_{EE,NMDA}f\left(V_i^E\right)s_j^{NMDA}\right) \\ & \left(V_i^E - V_E\right) - \sum_{k=1}^{N_I} G_{EI} w_{ik}^{EI} s_k^I \left(V_i^E - V_I\right) - G_L\left(V_i^E - V_L\right) + I_{ext,i}^E(t) \end{aligned} \tag{5.2}$$

$$\begin{aligned} C_m \frac{dV_i^I}{dt} = & -\sum_{j=1}^{N_E} w_{ij}^{IE}\left(G_{IE,AMPA}s_j^{AMPA} + G_{IE,NMDA}f\left(V_i^I\right)s_j^{NMDA}\right) \\ & \left(V_i^I - V_E\right) - \sum_{k=1}^{N_I} G_{II} w_{ik}^{II} s_k^I \left(V_i^I - V_I\right) - G_L\left(V_i^I - V_L\right) + I_{ext,i}^I(t) \end{aligned} \tag{5.3}$$

where $i = 1, 2, \ldots$ up to N_E for pyramidal neurons and N_I for interneurons. Synaptic excitation is mediated by both AMPA and NMDA receptors, and inhibition is assumed to be primarily mediated by $GABA_A$ receptors. The dimensionless weight matrices w_{ij}^{EE}, w_{ij}^{EI}, w_{ij}^{IE} and w_{ij}^{II} describe the four types of connections. G values control the absolute connection strengths. The last term of each equation describes an external current that includes a background Poisson input (from the rest of the brain) and time-dependent sensory stimuli.

How many equations does such a model have? The answer depends on how the synaptic gating variables are described. The gating variable for the *AMPA* receptor-mediated excitation and that of $GABA_A$-mediated inhibition follow one dynamical equation each (Chapter 2), and the *NMDA* gating variable obeys two equations in order to capture independent rise and decay times (Wang 1999b). Therefore, the total number of equations is four times N_E plus two times N_I. For instance, if $N = 2000$, $N_E = 1600$ and $N_I = 400$ ($N_E / N_I = 4$), the model has 7,200 equations. The number of dynamical equations is larger when more biophysical processes are incorporated, such as the Hodgkin-Huxley-like modeling of single cells and short-term synaptic plasticity.

5.3.3 How Many Parameters Does This Model Have?

One may be skeptical about biologically realistic models because in principle they have a huge number of parameters, which reminds us of John von Neumann's quip on overfitting: "with four parameters I can fit an elephant, and with five I can make him wiggle his trunk." So it is instructive to discuss how model parameters can be constrained by neurobiology, theoretical considerations and aesthetic simplicity.

For object working memory, let us assume a simple network architecture that consists of a number p of discrete neural pools or assemblies (**Figure 5.13a**) (Amit and Brunel 1997; Brunel and Wang 2001). Each neural assembly encodes a preferred stimulus, with fN_E neurons (where f is the fraction of the entire excitatory neural population for representing a

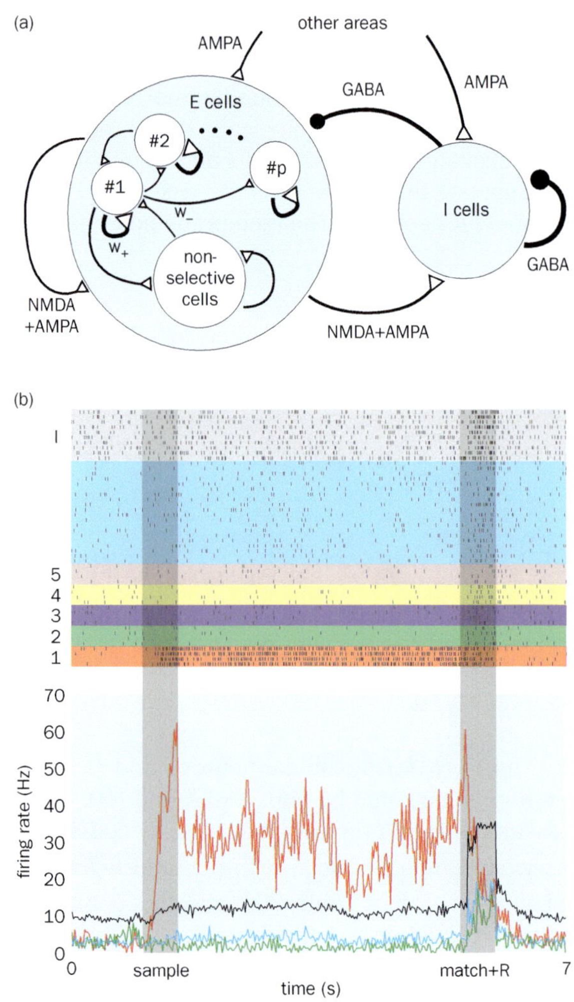

Figure 5.13 An object working memory model of spiking neurons. (a) Model scheme with p selective excitatory neural pools, nonselective excitatory neurons, and a population of inhibitory neurons. Synaptic excitation is mediated by AMPA and NMDA receptors, whereas synaptic inhibition is mediated by $GABA_A$ receptors. (b) A single-trial simulation of the model with five selective neural pools. Top: Spike trains of single neurons (red, green, purple, yellow and pink: Selective excitatory neural pools; light blue: Nonselective excitatory cells; grey: Inhibitory neurons). Bottom: Population firing rates for the five neural assemblies and for the inhibitory neural population (black). A transient input activates one of the selective neural pools (red), and the resulting reverberation generates self-sustained persistent activity at ~25 Hz during a delay period. Reproduced from Brunel and Wang (2001). [Brunel, N., & Wang, X. J. (2001). Effects of neuromodulation in a cortical network model of object working memory dominated by recurrent inhibition. Journal of Computational Neuroscience, 11(1), 63–85.]

stimulus item, $fp < 1$). The remaining excitatory population of $N_E(1 - fp)$ neurons are nonselective. Two cells within each neural assembly tend to fire together, and thus have a stronger connection $w_{ij}^{EE} = w_+$. By contrast, two cells belonging to separate assemblies are selective to different stimuli, and thus tend to be active at different times and have a weaker connection $w_{ij}^{EE} = w_-$. In this way, the connectivity matrix w_{ij}^{EE} is reduced to two parameters, w_+ and w_-.

The value of w_+ is varied to test the central hypothesis that sufficiently strong reverberation (quantified by a sufficiently large w_+) gives rise to stimulus-selective persistent activity. As w_+ is increased, too strong recurrent excitation could destabilize the resting state, but the firing rate of neurons in the resting state can be kept roughly constant if w_- is decreased concomitantly. More specifically, by choosing $w_- = 1 - f(w_+ - 1)/(1 - f)$, the overall recurrent excitatory synaptic drive in the spontaneous state remains approximately the same as w_+ is varied (Amit and Brunel 1997). This eliminates w_- as an independent parameter.

The other three connection matrices w_{ij}^{IE}, w_{ij}^{EI} and w_{ij}^{II} are concerned with inputs and outputs of inhibitory neurons. A simplified assumption is often made that synaptic inhibition is nonselective, thus w_{ij}^{EI} and w_{ij}^{II} are constant, equal to 1 without loss of generality (since the strengths of the I-to-E and I-to-I connections are determined by the maximum conductances G^{EI} and G^{II}). Note that this could be the case even if the E-to-I connection weights w_{ij}^{IE} are properly structured so that inhibitory neurons are selective for stimuli. The simplest situation is when the inhibitory neural population receives converging inputs from all excitatory neurons with equal weight, then w_{ij}^{IE} again can be set to 1 with the strength given by the maximum conductance G^{IE}. Any increased activity of a neural assembly would recruit more inhibition to all excitatory cells, and such shared negative feedback is a simple way to implement competition among selective neural assemblies.

There are six maximum synaptic conductances ($G_{EE,AMPA}$, $G_{EE,NMDA}$, $G_{IE,AMPA}$, $G_{IE,NMDA}$, G_{EI}, G_{II}). Excitation and inhibition balance (Chapter 3) requires that G_{EI} (respectively G_{II}) is fixed given $G_{EE,AMPA}$ and $G_{EE,NMDA}$ (respectively $G_{IE,AMPA}$ and $G_{IE,NMDA}$). The remaining four parameters express the strengths of synaptic excitation into pyramidal cells and interneurons, mediated by the AMPA and NMDA receptors, in addition to the fifth parameter w_+ for selective increase of recurrent excitation within each neural pool. Finally, neurons may receive some background inputs with a strength and rate as additional parameters. All in all, there are half a dozen tunable parameters for this biologically realistic neural circuit model, which is not drastically different from abstract mathematical models commonly used in cognitive psychology. These parameters are adjusted in order for the model to reproduce behavioral and neurophysiological observations.

5.3.4 Emergence of Self-Sustained Activity from a Bifurcation

With sufficiently strong recurrent excitation, the model displays stimulus-selective persistent activity. **Figure 5.13b** shows a model simulation of object working memory.

The network is initially in a resting state in which all excitatory neurons fire uniformly at a few Hz. A stimulus triggers a strong transient response in a neural pool (red) selective for this stimulus. The activated neurons send recruited excitation to each other via recurrent connections. This internal excitation is large enough to sustain elevated activity, which persists after the stimulus offset during a delay period. The firing rate of persistent activity is about 25 Hz, while neurons selective for other stimuli remain at low levels of activity. At the end of the delay period, a nonselective excitatory input is presented in the response period, leading to an increase of all excitatory neurons, which in turn drive inhibitory neurons to fire at a high rate. The recruited feedback inhibition switches the network from the elevated persistent activity state to the spontaneously active resting state. How a working memory trace is erased in the brain remains poorly examined with behaving animals. Presumably, in a delayed response task, when a movement is triggered, a copy of a motor command called "corollary discharge" is sent to the cortex (Crapse and Sommer 2008), which could cause a working memory circuit to switch from a memory state back to the baseline resting state.

Stimulus items are represented by different neural pools and stored in their persistent activity; therefore, there are p distinct attractor states that coexist with each other and with the resting state in a single network. Persistent activity cannot be sustained if the recurrent excitation between neurons within a neural subpopulation, quantified by the parameter w_+, is not strong enough. Obviously, there would be no persistent activity when neurons do not interact with each other ($w_+ = 0$). As w_+ is increased gradually, above a critical threshold (w_+^1), self-sustained persistent activity appears as an all-or-none phenomenon (**Figure 5.14a**). Below the critical threshold, only the spontaneous state exists. Above the threshold, the spontaneous activity state is still dynamically stable to small perturbations, because at low firing rates excitation is effectively counteracted by feedback inhibition (**Figure 5.14b**). However, if a stimulus generates a transient high activity in a neural pool, now recurrent reverberation is sufficiently powerful to drive this group of cells to "escape" from the spontaneous state. A higher firing activity leads to an even larger recurrent synaptic excitation, which becomes sufficient to sustain a persistent active state after the stimulus is withdrawn. As a result, a stable attractor of persistent activity with an elevated firing rate is realized, which coexists with the stable spontaneous state (**Figure 5.14b**).

On the other hand, if w_+ is too large, exceeding another threshold value w_+^2, the resting state is no longer stable, because any increase in neural activity would be amplified by powerful recurrent excitation so

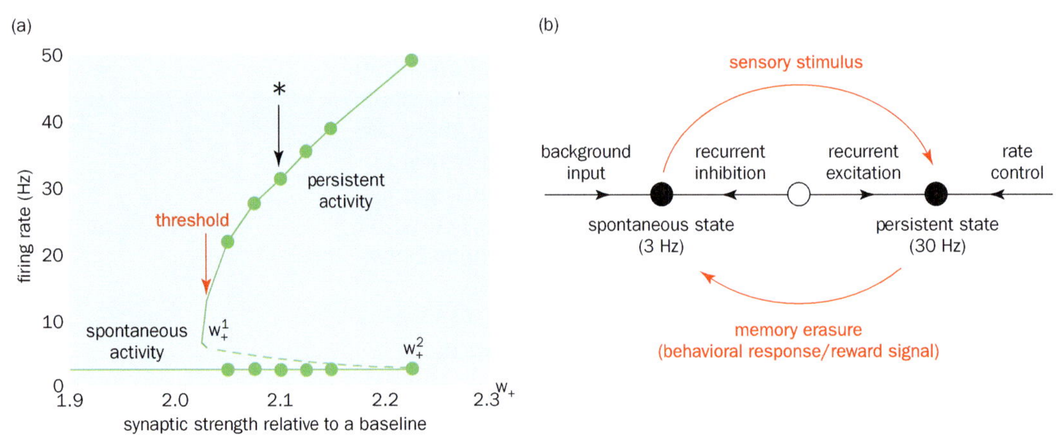

Figure 5.14 Bifurcation in an object working memory circuit model. (a) Dependence of self-sustained network activity on the strength of recurrent synaptic connectivity. Solid lines: Spontaneous state and persistent memory state; dashed line: Unstable states. There is a critical threshold of synaptic strength, above which persistent activity appears abruptly as an all-or-none phenomenon. Asterisk: The example in Figure 5.13. (b) Illustration of the attractor dynamics. An attractor is a neural firing state that is stable: When a small and brief input perturbs the network to a lower or higher activity level, there is a restoring force to bring the network back to the attractor state. In this case, the spontaneous state is stabilized from below by background inputs and from above by feedback synaptic inhibition. The persistent activity state is stabilized from below by excitatory reverberation and from above by various negative feedback control mechanisms. A transient sensory stimulus can switch the network from the spontaneous state to the memory state. During behavioral response the network is turned off, thereby erasing the memory. Reproduced from Wang (2001) with original data from Brunel and Wang (2001). [Wang, X. J. (2001). Synaptic reverberation underlying mnemonic persistent activity. Trends in Neurosciences, 24(8), 455–463.]

much that the network cannot return back to the resting state after a transient (even small) perturbation. This regime ($w_+ > w_+^2$) is not suitable for the network to operate as a working memory system, which requires the coexistence of the resting state and memory states.

Here, it is worth reiterating a point made in Chapter 3, namely bifurcations do not occur in linear dynamical systems. Note that the equations for our neural circuit model are indeed nonlinear, containing such terms as the product of a gating variable and a voltage. Thus, persistent activity emerges when w_+ exceeds the bifurcation point w_+^1. This is a powerful idea, since it enables us to conceptualize a working memory local circuit (like in the PFC) as qualitatively similar to a sensory circuit like V1, with a shared canonical architecture (Douglas and Martin 2004). A graded difference in the amount of recurrent connections could account for the emergence of working memory capability.

Rigorous experimental tests of the attractor network model of working memory became possible only recently thanks to the advances of experimental tools. In a mouse experiment, subjects learn to associate one of two sensory cues to left and right licking responses. The two sensory stimuli may be somatosensory (far and near objects that touch whiskers) or they may be auditory (high and low tones) (**Figure 5.15a**). Before the response is allowed to take place, there is a short delay period. Single neurons in the premotor area called the anterior lateral motor (ALM) cortex display elevated firing activity encoding delayed response. A series of experiments, in close interplay with computational modeling, have led to a wealth of information about the underlying neural circuit mechanism. First, the researchers used the optogenetic technique that enabled exciting or inhibiting, with millisecond precision, a genetically labeled neural population. Optogenetic inactivation systematically done across the cortex demonstrated that ALM is the crucial node for maintaining short-term memory in this task (Guo et al. 2014). Second, if persistent activity is a single-cell phenomenon rather than maintained by synaptic reverberation, current injection into a cell should be able to turn off ongoing persistent activity. This was found not to be the case. Using intracellular recording in behaving animals, injection of a brief negative current could not turn off delay period activity in a recorded cell, in support of a network but not a single-neuron mechanism (Inagaki et al. 2019). Third, despite optogenetic perturbations that transiently alter the time course of the ALM neural

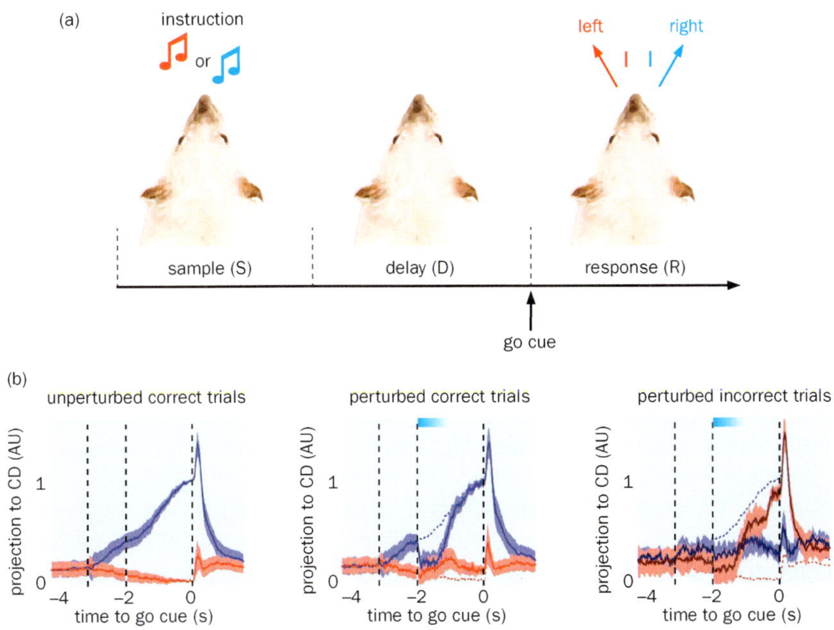

Figure 5.15 Mnemonic activity in mouse performing a delayed response task. (a) One of two sensory cues is presented briefly, which can be a high or low tone in an auditory task or near or far location of an object on the whisker in a somatosensory task. The two stimuli are mapped to left and right licking responses, shown in red and blue, respectively. A correct motor response after a delay of more than one second yields a reward. (b) Population activity from ~ 10 simultaneously recorded ALM neurons, projected in the one-dimensional coding direction (CD) optimized for mnemonic representation. Left: Control when the choice is correct (right, blue). Middle: Optogenetic inactivation at the start of the delay period suppresses right-selective neural activity, which recovers and the ultimate choice is correct. Right: Same as middle but this time optogenetic manipulation induces an incorrect response (left, red). The ALM decoding still predicts the erroneous movement direction, demonstrating its correlate with behavior performance. Panel (a) is reproduced from Svoboda and Li (2018), (b) from Inagaki et al. (2019). [Svoboda, K., & Li, N. (2018). Neural mechanisms of movement planning: Motor cortex and beyond. Current Opinion in Neurobiology, 49, 33–41.] [Inagaki, H. K., Fontolan, L., Romani, S., & Svoboda, K. (2019). Discrete attractor dynamics underlies persistent activity in the frontal cortex. Nature, 566(7743), 212–217.]

firing, the trajectory of population activity converges to one of two fixed endpoints in the state space, in support of discrete attractor models (**Figure 5.15b**) (Li et al. 2016; Inagaki et al. 2019). Fourth, brief optogenetic inactivation during the delay period revealed that robustness of memory storage against perturbations depends on reciprocal interactions between the two hemispheres (Li et al. 2016). Interestingly, thalamo-cortical connections are also important for the maintenance of delay period activity in the ALM (Guo et al. 2017). The steps outlined here demonstrate how a brain mechanism of delay period activity in behaving animals can be elucidated in details. Results from neural data analysis and experimental manipulations combined with modeling lend further support to the attractor network paradigm (Piet et al. 2017; Finkelstein et al. 2021).

In this task, since the sensorimotor transformation presumably occurs during external stimulation, persistent firing in the premotor area ALM actually encodes preparation for the impending movement rather than sensory working memory. The same point has been raised for the classic oculomotor delayed response (ODR) task in a monkey experiment where the spatial location of a briefly shown visual cue guides a delayed saccadic response (**Figure 5.7**) (Funahashi et al. 1989). To address this confound, Funahashi et al. (1993) trained monkeys to carry out a more sophisticated task in which the delayed response is either saccade to the visual cue (standard ODR) or diametrically opposite from it (an anti-saccade variant). Which of the two should be carried out in a given trial is signaled by the shape of the fixation cue. It was found that if a neuron in the dorsolateral PFC displays elevated persistent activity for a direction (say, right) in the standard ODR, then in the delayed anti-saccade its high activity is selective for the visual stimulus in that same direction (right) when the motor response direction is left, demonstrating that DLPFC persistent firing indeed represents sensory working memory. In other tasks, like the DMS or DMC (**Figure 5.6**) or VVD (**Figure 5.8**), behavior requires internal representation of the sample stimulus rather than motor preparation because the correct response is unknown and thus cannot be prepared until the test stimulus is presented at the end of a delay period.

5.3.5 Inverted U-Shape of Dopamine Dependence

Working memory function is modulated by the neuromodulator dopamine. Pioneering work by Goldman-Rakic and her colleagues revealed that pharmacological depletion of dopamine in the PFC impairs performance in delayed response tasks as severely as lesion of the PFC (Brozoski et al. 1979). Whereas striatum is rich in both D2 and D1 receptors, the cerebral cortex expresses much higher levels of D1 than D2 receptors (Lidow et al. 1998; Muly et al. 1998). When D1 receptor activation is manipulated using a drug in a dosage-dependent manner, it was found that its modulatory effect is non-monotonic: working memory performance is reduced in a dose-dependent fashion either by a dopamine D1 receptor antagonist or agonist (Sawaguchi and Goldman-Rakic 1991; Zahrt et al. 1997; Arnsten 1998; Cools and D'Esposito 2011). Too much or too little dopamine D1 modulation is detrimental to working memory function (**Figure 5.16a**).

To assess the corresponding changes of neural activity in PFC of behaving monkeys, iontophoretic technique was applied to inject a controlled amount of drug through the same electrode used to record a single unit in behaving animals. Because iontophoresing drug affects a small number of neurons, it does not change a cortical area in its entirety nor the animal's behavior. In a monkey experiment using the ODR task, a moderate dose of a drug that blocks D1 receptors enhanced stimulus-selective persistent activity in the dorsolateral PFC, whereas a higher dose led to reduced persistent activity firing (Williams and Goldman-Rakic 1995)

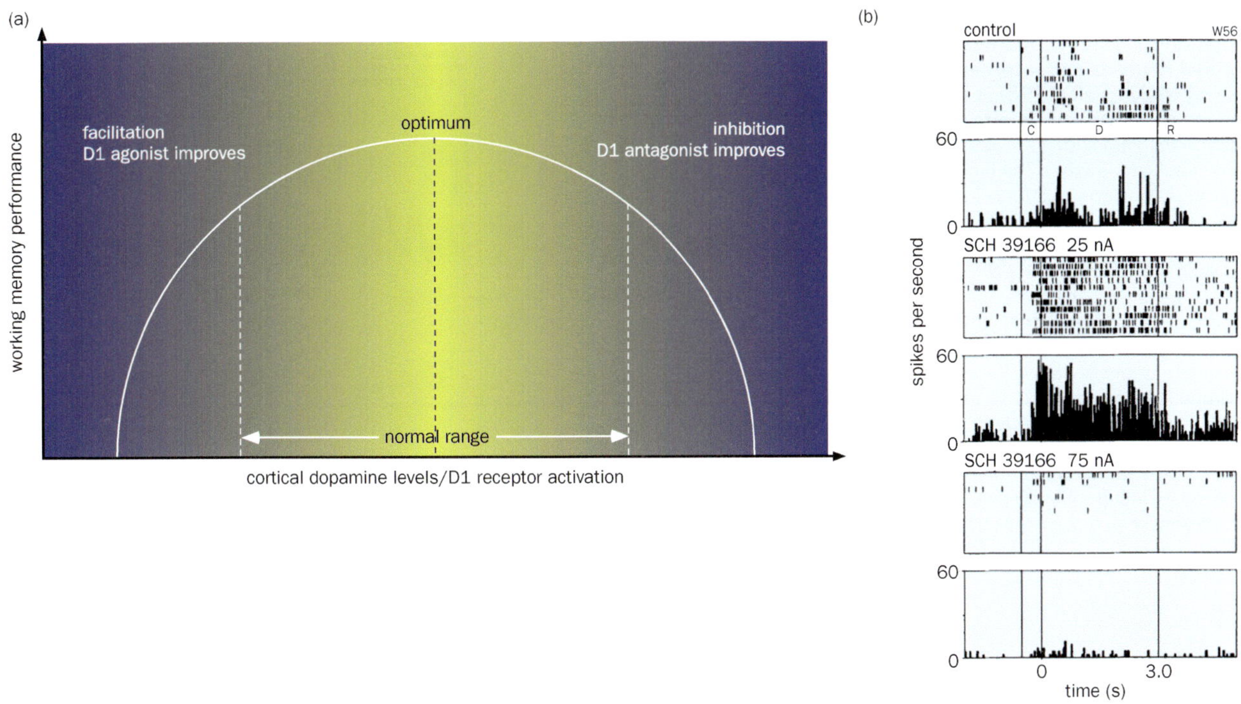

Figure 5.16 Dopamine D1 modulation of working memory performance. (a) Inverted-U dose dependence. (b) Single-unit activity of a PFC neuron recorded from a monkey performing the ODR task, showing spike trains and time course of trial-averaged firing rate with the preferred stimulus. Iontophoretic application of D1 receptor agonist SCH39166 increased persistent firing rate at a low dose (25nA) (middle panel) but reduced delay period activity at a high dose (75nA) (bottom panel) compared to the control (top panel). Panel (a) is reproduced from Lidow et al. (1998), (b) from Williams and Goldman-Rakic (1995). [Lidow, M. S., Williams, G. V., & Goldman-Rakic, P. S. (1998). The cerebral cortex: A case for a common site of action of antipsychotics. Trends in Pharmacological Sciences, 19(4), 136–140.] [Williams, G., Goldman-Rakic, P. Modulation of memory fields by dopamine Dl receptors in prefrontal cortex. Nature 376, 572–575 (1995). https://doi.org/10.1038/376572a0]

(**Figure 5.16b**), revealing an inverted U-shape of dosage dependence of D1 modulation at the neuronal level in the monkey PFC. In contrast to drugs for D1 receptors, those for D2 receptors did not significantly affect PFC mnemonic delay period activity but modulated neural firing at the time of motor response (Wang et al. 2004).

Using an object working memory model, Brunel and Wang (2001) examined how the inverted U-shape of the dopamine dependence of working memory can be explained by differential sensitivities of D1 receptor modulation along the excitatory-excitatory pathway versus the excitatory-inhibitory pathway (Muly et al. 1998; Seamans et al. 2000; Seamans and Yang 2004). Consequently, at intermediate modulation levels, recurrent synaptic excitations in pyramidal cells are selectively enhanced, giving a window of opportunity for delay activity to be enhanced. At higher activation levels, D1 modulation on pyramidal cells saturates, a further increase in dopamine modulation now results in an increased inhibition mediated by interneurons, and delay activity decreases or even disappears. **Figure 5.17** shows that this scenario indeed gives rise to an inverted U-shaped curve for persistent activity. Spontaneous activity is less affected than delay activity but also shows a peak at moderate activation levels. The stable steady-state with elevated persistent activity (solid portion) and the unstable steady-state (dashed portion) form a closed curve called "isola" in the theory of dynamical systems theory (Guckenheimer and Holmes 1983). More recent experimental results revealed that D1 receptor activation does not only modulate synaptic conductances but also, at high levels, enhances intrinsic potassium channels in pyramidal neurons,

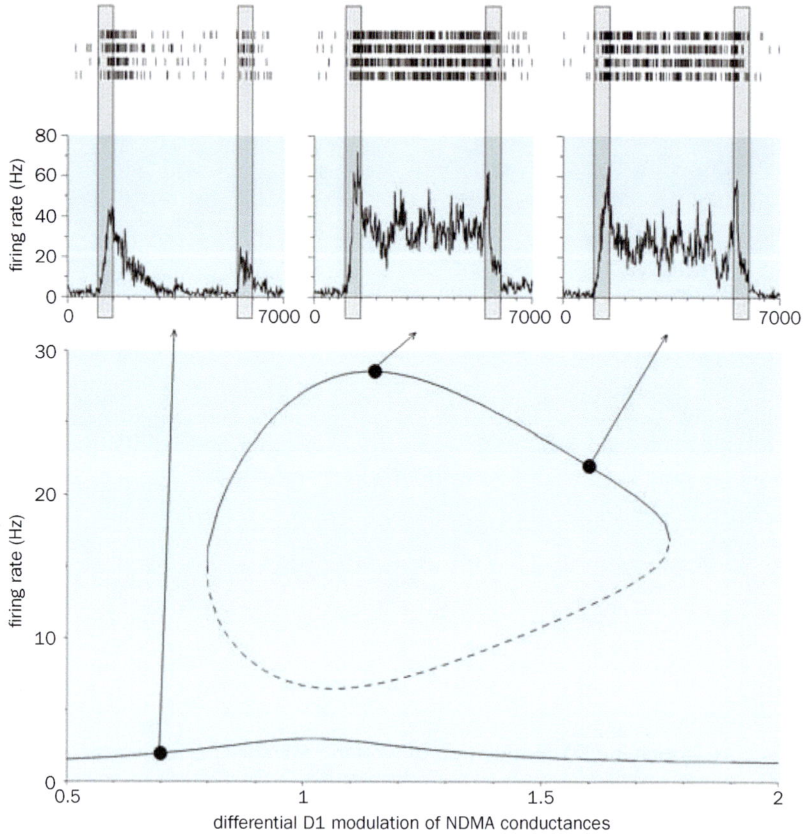

Figure 5.17 Differential dopamine D1 modulation of NMDA conductances in pyramidal neurons and interneurons following Muly et al. (1998) in an object working memory model. The bifurcation diagram shows an "isola" where persistent activity (y-axis) is the highest in an intermediate range of D1 modulation (x-axis). Solid: Stable; dashed: Unstable. Three simulations at different levels of D1 modulation indicated by filled circles are shown in the upper panels, demonstrating that too high (right panel) or too low (left panel) D1 activation would be suboptimal to working memory behavior. Reproduced from Brunel and Wang (2001). [Brunel, N., & Wang, X. J. (2001). Effects of neuromodulation in a cortical network model of object working memory dominated by recurrent inhibition. Journal of Computational Neuroscience, 11(1), 63–85.]

leading to decreased persistent activity when dopamine signaling is excessive (Arnsten et al. 2021).

5.4 CONTINUOUS ATTRACTOR MODEL FOR SPATIAL WORKING MEMORY

5.4.1 A Model of the Oculomotor Delayed Response Task

Visuospatial working memory in the ODR task requires neural representation of a spatial location as an analog quantity (**Figure 5.7**) (Gnadt and Andersen 1988; Funahashi et al. 1989; Chafee and Goldman-Rakic 1998). For neural spatial representation, a common network model architecture assumes that the strength of E-to-E connection between two excitatory neurons i and j decreases with the difference between their preferred stimuli x_i and x_j (**Figure 5.18a**),

$$w_{ij}^{EE} = w_- + (w_+ - w_-)\exp\left(-(x_i - x_j)^2 / 2\sigma_{EE}^2\right) \tag{5.4}$$

where the parameter σ_{EE} is a characteristic spatial length, or the spatial "footprint" of connectivity. Thus, $w_{ij}^{EE} = w_+$ for $i = j$ and w_- when $x_i - x_j$ is large compared to σ_{EE}. Here, the connection matrix w_{ij}^{EE} is determined by three parameters (w_+, w_- and σ_{EE}). For the ODR experiment, the spatial cues are typically limited to a directional angle θ (varying from 0 to 360 degrees) with a fixed eccentricity. In a "ring model," neurons are labeled by their preferred directions θ. Strictly speaking, the connection w_{ij}^{EE} should be a periodic function of θ, such as the von Mises function:

$$vonMises(\theta_i - \theta_j) = \exp\left(\cos(\theta_i - \theta_j) / \sigma_{EE}^2\right) \tag{5.5}$$

However, as long as the connections are sufficiently selective with $\sigma_{EE} \ll 360^o$, w_{ij}^{EE} is not negligible only with small $\theta_i - \theta_j$, and using Eq. (5.4) is adequate.

Figure 5.18b shows a model simulation of the delayed oculomotor task (Compte et al. 2000). Initially, the network is in a resting state in which all cells fire spontaneously at low rates. A transient cue stimulus (in this case at $\theta_S = 180$ degrees) drives a subpopulation of cells to fire at high rates. As a result, they send recruited excitation to each other via intrinsic connections. This internal excitation is large enough to sustain elevated activity, so that the firing pattern persists after the stimulus is withdrawn. Synaptic inhibition ensures that the activity does not spread to the rest of the network. The persistent activity has a bell-shaped pattern ("bump attractor"), where the firing rate pattern $r(\theta - \theta_{pop})$ is a function of the difference between the preferred direction θ of each neuron and the peak location θ_{pop}. At the end of the delay, the cue information can be retrieved by reading out the peak location of the persistent activity pattern; performance is satisfactory as long as θ_{pop} remains sufficiently close to the initial cue θ_S. In the model, the network is reset back to the spontaneous state by a simple input.

In another trial, a cue can be presented at any location, leading to the generation of a bump attractor of the same shape but peaked at a different angle. When a single neuron with a preferred cue θ is monitored across trials, its firing rate as a function of the stimulus θ_S can be computed in the model and compared with single-unit data from neurophysiological experiments (Funahashi et al. 1989; Constantinidis and Goldman-Rakic 2002).

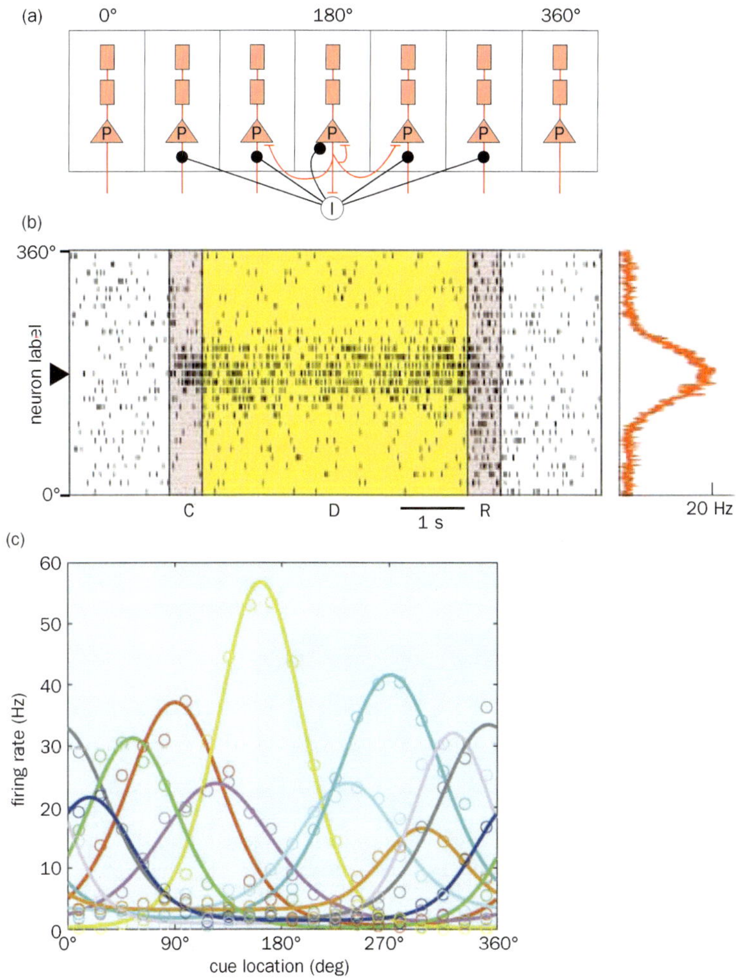

Figure 5.18 Working memory maintained by a tuned network activity pattern (a bump attractor). (a) Model architecture. Excitatory pyramidal cells are labeled by their preferred cues (0° to 360°). Pyramidal cells with similar preferred cues are connected through local E-to-E connections. Interneurons receive inputs from excitatory cells and send feedback inhibition by broad projections. (b) Spatiotemporal raster of the pyramidal cell population in a model simulation of the DOR task. Abscissa: Time; ordinate: Pyramidal neurons (labeled by their preferred cues). C: Cue period, D: Delay period, R: Response period. The elevated and localized neural activity is triggered by a brief cue stimulus at 180°, and persists during the delay period. On the right is shown the spatial profile, where the average firing rate during the delay period is plotted versus the preferred cue of the neuron. (c) Examples of tuning curves from ten neurons in a network model endowed with homeostatic synaptic scaling and in the presence of cell-to-cell heterogeneities. Panel (b) is reproduced from Wang (2001) with data from Compte et al. (2000), (c) from Renart et al. (2003). [Wang, X. J. (2001). Synaptic reverberation underlying mnemonic persistent activity. Trends in Neurosciences, 24(8), 455–463.] [Renart, A., Song, P., & Wang, X. J. (2003). Robust spatial working memory through homeostatic synaptic scaling in heterogeneous cortical networks. Neuron, 38(3), 473–485.]

There is a *continuous family* of bump attractors, suitable for working memory of an analogue stimulus feature (spatial location). Such a continuous family of bell-shaped attractor states is realized in the model with the assumption of network homogeneity, so that no part in the network is special, and a bump attractor can be peaked anywhere (Ben-Yishai et al. 1995; Camperi and Wang 1998; Compte et al. 2000). In the biological brain, however, there are various types of heterogeneities. When neuronal variations were incorporated into a ring model of spatial working memory, it was found (Renart et al. 2003) that a continuous family of bump attractors can be approximately recovered with the addition of a homeostatic mechanism of synaptic scaling (Chapter 4, Eq. 4.30) that regulates the long-term average firing rate in spite of neuron-to-neuron variations of single cell properties and synaptic inputs. In the presence of heterogeneity, different neurons exhibit a diversity of tuning curves (**Figure 5.18c**) as observed experimentally. Other activity-dependent mechanisms have also been proposed for the robustness of the continuous family of bump attractors in the presence of heterogeneities (Itskov et al. 2011; Hansel and Mato 2013; Pereira and Wang 2015; Seeholzer et al. 2019).

5.4.2 Stochastic Gamma Oscillations during Delay Period Activity

Persistent activity may exhibit temporal fluctuations with a synchronous oscillatory component. As shown in **Figure 5.19** (upper panel), population firing varies up and down in time, leading to "activity bursts" in the spatiotemporal pattern. Zooming into a short time window reveals a stochastic rhythmicity (**Figure 5.19**, lower left panel). A local field potential (LFP) computed in terms of a sum of excitatory synaptic inputs shows stochastic oscillations, and its power spectrum has a peak at around 40 Hz (**Figure 5.19**, lower right panel). Gamma (40 Hz) oscillations naturally arise in this network as a result of dynamical interplay between fast recurrent excitation mediated by the AMPA receptors and slower feedback inhibition mediated by the $GABA_A$ receptors (Chapter 3, **Figure 3.13**) (Brunel and Wang 2003; Börgers and Kopell 2003; Wang 2010). Fast mutual excitation between pyramidal neurons brings their firing rates higher and higher (producing an upswing of oscillation), until the delayed inhibition is

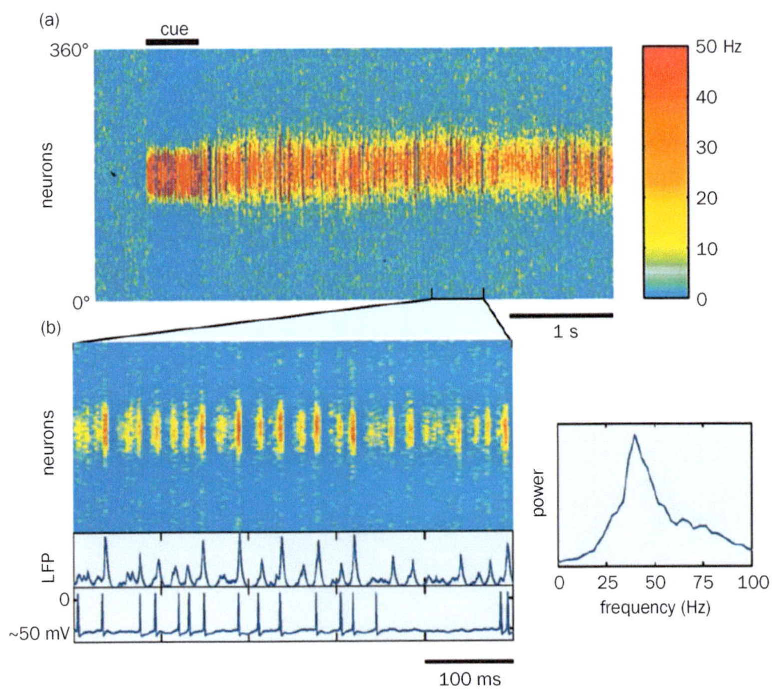

Figure 5.19 Stochastic gamma oscillations during self-sustained persistent activity in a working memory circuit model. (a) Spatiotemporal firing pattern, with neurons labeled by their preferred cues from 0 to 360 degrees along the y-axis. Firing rates are coded by color. (b) Left: Zoom-in of (a) in a 500-ms time window, showing network oscillations, the local field potential and the membrane potential of a single neuron. Note the silent epoch of more than 100 ms. Right: The power spectrum of the local field potential shows a large peak at ~40 Hz. Reproduced from Compte et al. (2000). [Modified from Compte, A., Brunel, N., Goldman-Rakic, P. S., & Wang, X. J. (2000). Synaptic mechanisms and network dynamics underlying spatial working memory in a cortical network model. Cereb Cortex. 2000 Sep;10(9):910–23. doi: 10.1093/cercor/10.9.910. PMID: 10982751]

recruited sufficiently to decrease the neural activity (a downswing). When the activity of pyramidal cells becomes very low (near zero, as seen in the population rate in **Figure 5.19**, lower left panel), interneurons lose the excitatory drive and synaptic inhibition decays away, and the oscillatory cycle restarts anew. The AMPA receptor-mediated excitation and $GABA_A$-mediated inhibition have time constants of 2 ms and 10 ms, respectively, and the addition of these two time constants roughly determine the duration of half a cycle; therefore, the periodicity is about 25 ms and the rhythmic frequency is ~ 40 Hz.

The model shows clearly that, contrary to a common misperception, the attractor of persistent activity does not need to be a "steady-state," but can display rather complex spatiotemporal patterns. Fast synchronous oscillations have been observed in mnemonic persistent activity recorded in the PPC (Pesaran et al. 2002) and the PFC (Lundqvist et al. 2016) of monkeys performing working memory tasks. The model prediction that a working memory circuit naturally displays gamma oscillations readily explains enhanced coherent oscillation in this frequency range by selective attention in sensory areas, as we will show in Chapter 8.

5.4.3 Drifts of Neural Representation across the Delay

The directional cue stored in working memory can be read out by the peak location of the bump attractor. A simple way to quantify the decoding is the population vector (Georgopoulos et al. 1986):

$$\theta_{pop}(t) = arg\left(\sum_{j=1}^{N} r_j^E(t)\exp(i\theta_j(t))\right) \tag{5.6}$$

where $\exp(i\theta_j(t)) = \cos(\theta_j(t)) + i\sin(\theta_j(t))$ are vectors on the complex plane, which are added with weights equal to firing rate of neuron j, $r_j^E(t)$, and arg denotes the angle of the summed vector.

The firing rates of pyramidal cells $r_j^E(t)$, and therefore the population vector, can be computed with a sliding time window, so that $\theta_{pop}(t)$ traces over time the directional cue stored in working memory by the population persistent activity pattern. It turns out that $\theta_{pop}(t)$ does not stay stationary but displays random drifts, and the drift varies from trial to trial even when the same cue is presented across trials. In fact, the time courses of $\theta_{pop}(t)$ in single trials all start with the cue stimulus θ_S, but diverge from each other over time (**Figure 5.20a–b**). Their average across trials is θ_S. However, behavioral performance is determined in single trials. The variance of these traces, $\sigma^2(\theta_{pop})$, increases roughly linearly with time (Camperi and Wang 1998; Compte et al. 2000; Wu et al. 2008). Functionally, this predicts that for longer decay, the saccadic responses become less accurate. This is indeed the case in the monkey experiment, where the endpoints of delayed saccade are more scattered with a longer delay (**Figure 5.20c**) (Funahashi et al. 1989). The same has been reported in human studies using the DOR task (White et al. 1994).

In the model, the random drifts reflect a basic feature of the model, namely, it is endowed with a continuous family of bump attractors, peaked at any location along the ring. During the delay, there is no tuned external input to prevent a bump attractor from drifting away from its initial location (determined by the stimulus cue) because of stochastic neural circuit dynamics. In the absence of a bias, the variance is expected to grow linearly with time (Renart et al. 2003), as predicted by the theory of random walk (Chandrasekhar 1943).

Random drifts of the bump attractor model are consistent with the observed decrease of performance as a function of the delay duration, but

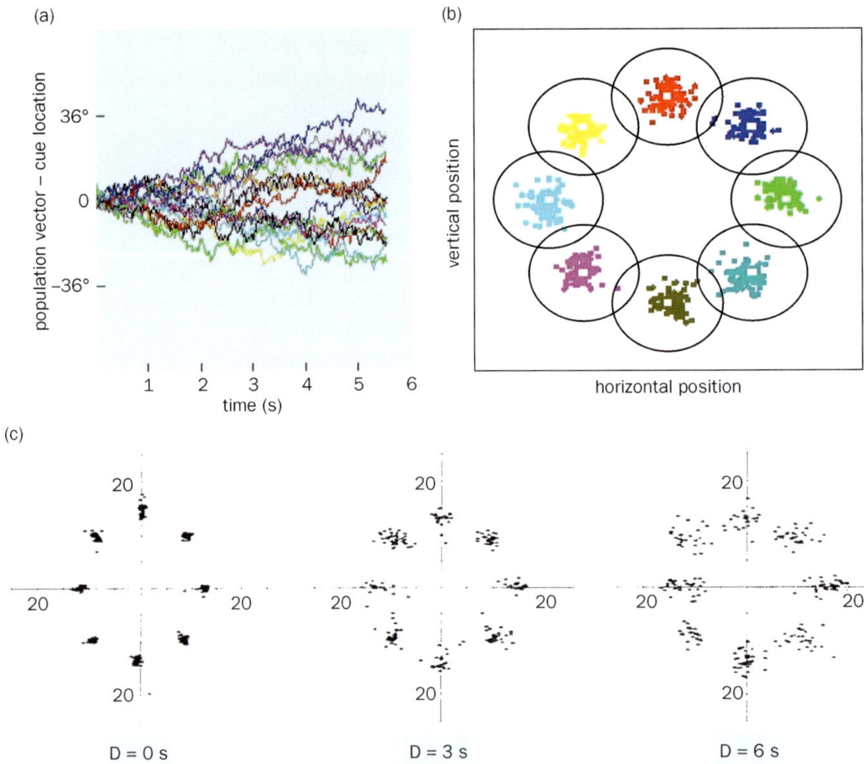

Figure 5.20 Random drifts of cue representation in working memory. (a) The temporal dynamics of the population vector of persistent activity pattern across a delay period of 5 seconds in a ring model. Colors correspond to different single trials. (b) Endpoints of memory-guided saccades in model simulations. (c) The distribution on the screen of saccadic positions in a monkey experiment in the visually guided saccade task (left), the ODR task with a delay of 3 seconds (middle) and 6 seconds (right). Panel (a) is reproduced from Compte et al. (2000), (b) is obtained by Alfonso Renart using the model of Compte et al. (2000), (c) from Funahashi et al. (1989). [Modified from Compte, A., Brunel, N., Goldman-Rakic, P. S., & Wang, X. J. (2000). Synaptic mechanisms and network dynamics underlying spatial working memory in a cortical network model. Cereb Cortex. 2000 Sep;10(9):910–23. doi: 10.1093/cercor/10.9.910. PMID: 10982751] [Funahashi, S., Bruce, C. J., & Goldman-Rakic, P. S. (1989). Mnemonic coding of visual space in the monkey's dorsolateral prefrontal cortex. Journal of Neurophysiology, 61(2), 331–349.]

do they actually happen in a working memory circuit? Wimmer et al. (2014) devised a clever way to tackle this question and relate causally behavior with random drifts of memory representation. They analyzed single-neuron activity recorded from monkeys performing an ODR task and divided trials into two groups. One consists of those trials in which the behavioral report deviates from θ_S (say 45°) clockwise (CW) so that at the end of the delay θ_{pop} is smaller than θ_S. The other group includes the remaining trials in which the errors are counterclockwise (CCW), and the final decoded θ_{pop} is larger than θ_S. The bump attractor model predicts that, due to random drifts of the population activity pattern, the preferred direction θ of a single recorded neuron defined by its tuning curve is slightly different in the two subsets of CW and CCW trials, with the preferred direction at the end of the delay larger in the CCW than CW subsets of trials. This prediction was confirmed (Wimmer et al. 2014), providing clear evidence to date for the continuous bump attractor network model of spatial working memory.

5.4.4 Resistance against Distractors

As mentioned at the beginning of the chapter, a key requirement for memory maintenance is the brain's ability to filter out distractors: while behaviorally relevant information is actively held in the mind, irrelevant sensory stimuli should be denied entrance to the working memory system. In delayed response experiments using intervening stimuli (distractors), mnemonic activity has been shown to be easily disrupted by distractors in

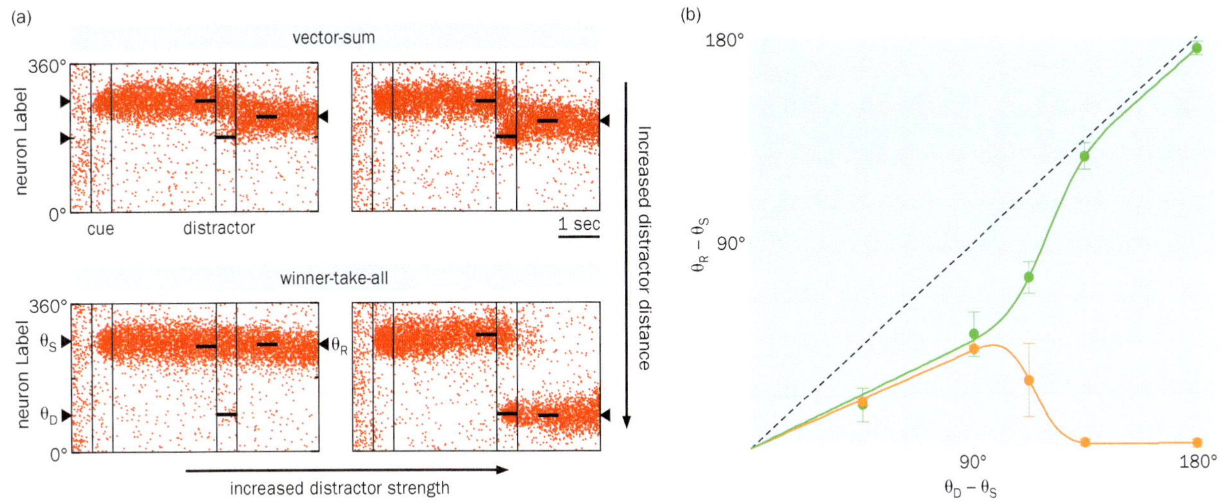

Figure 5.21 Resistance against distractors. (a) In the spatial working memory model, the initial cue θ_S triggers a persistent activity. During the delay, a second cue θ_D (distractor) is shown briefly. When the distractor is close to the initial stimulus (upper row), the network performs a vector sum (arrow on the right). On the other hand, when the distractor is far away from the initial stimulus (lower row), the network operates in a winner-take-all regime, depending on the strength of the distractor (left column versus right column). (b) Dependence of network distraction on the distance between the cue. The "distracted" distance ($\theta_R - \theta_S$) is plotted versus distraction distance ($\theta_D - \theta_S$) for several distractor cues. The dashed line indicates perfect distraction (as in (a), lower right) while points on the x-axis show absence of distraction (as in (a), lower left). Green: The network model displays vector sum with small $\theta_D - \theta_S$ and the distractor winning with large $\theta_D - \theta_S$. Orange: Distractibility is reduced when the strengths of both recurrent excitation and inhibition are enhanced (the distractor has no impact when $\theta_D - \theta_S$ is large). Panel (a) is reproduced from Wang (2006), (b) from Compte et al. (2000) (green). [Wang, X. (2006). A microcircuit model of prefrontal functions: Yin and yang of reverberatory neurodynamics in cognition (pp. 92–127). In J. Risberg & J. Grafman (Eds.), The Frontal Lobes: Development, Function and Pathology © Cambridge University Press 2006, reproduced with permission.] [Modified from Compte A, Brunel N, Goldman-Rakic PS, Wang XJ. Synaptic mechanisms and network dynamics underlying spatial working memory in a cortical network model. Cereb Cortex. 2000 Sep;10(9):910–23. doi: 10.1093/cercor/10.9.910. PMID: 10982751]

ITC neurons, while the information of the behaviorally relevant cue was preserved in PFC neurons (Miller et al. 1996). Similarly, delay period activity is sensitive to distractors in the PPC but not in the PFC (Constantinidis and Wang 2004; Suzuki and Gottlieb 2013).

In the spatial working memory model, active neurons by virtue of elevated persistent activity recruit inhibition, which project back to the rest of the network. Consequently, those excitatory cells not encoding the initial cue are less excitable during the delay than when they are in the resting state, hence less responsive to distracting stimuli presented during the delay. The impact of a distractor (θ_D) depends on its strength and the distance to the stimulus cue θ_S. In **Figure 5.21a**, θ_R denotes the population vector of the encoded cue at the end of the delay for behavioral report. The network's response to a distractor typically shows a vector sum if θ_S is close to θ_D (upper row), namely, θ_R is roughly halfway between the two, and winner-take-all if θ_S is far from θ_D (lower row), so that either θ_R still encodes the behaviorally relevant initial cue θ_S or the representation is completely switched to the distractor θ_D. This is quantified in **Figure 5.21b**, where the change in the encoded cue $\theta_R - \theta_S$ is plotted against $\theta_D - \theta_S$, for a fixed strength of distractor stimulus. Moreover, strengthening recurrent excitation and inhibition (orange) reduces the network's distractibility (Compte etal. 2000; Durstewitz etal. 2000; Brunel and Wang 2001).

5.5 LINE ATTRACTORS: PARAMETRIC WORKING MEMORY

As shown in **Figure 5.8**, another form of working memory for analog quantities is monotonic tuning of persistent activity as a function of the stimulus feature. Recall that the meaning of a tuning curve for delay period activity is profoundly different from that for responses to sensory inputs. Conventionally, the stimulus selectivity of neuronal firing during stimulus presentation is quantified by a tuning curve. For example, the higher is the input intensity, the larger is the neural response (Chapter 3, **Figure 3.15b–c**). By contrast, in a working memory task, the mnemonic neural activity is measured after the transient stimulus is withdrawn. If a working memory network exhibits a family of delay period activity that is monotonically tuned to a stimulus feature, then this entire family of mnemonic activities with different firing rates must be all realizable under exactly the same condition (during the delay when external inputs are absent).

Like **Figure 5.12**, consider a neural network of which the activity is schematically shown in a two-dimensional state space, and imagine that there is an "energy function" for which the minima define attractor states. Such a system is capable of parametric persistent activity, if the minima form a one-dimensional manifold (**Figure 5.22**), akin to a flat part of a golf course where the ball can stay at any of a continuum of positions. This is called a "line attractor" (Seung 1996) because, starting at an arbitrary initial state on the plane, the neural population evolves to converge onto it over time. The line attractor could encode parametrically a stimulus feature in working memory, with monotonic tuning curves of individual neurons.

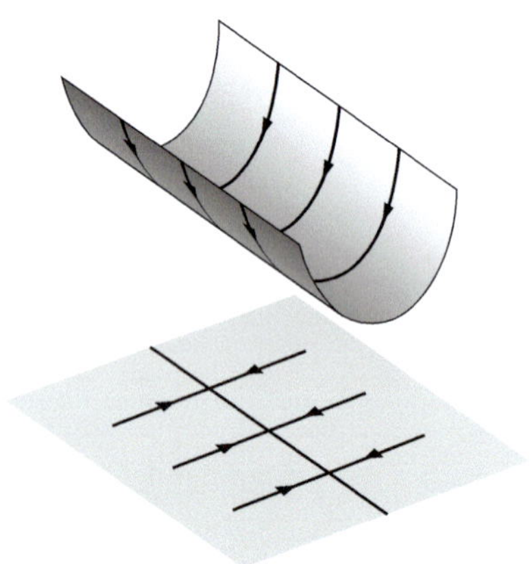

Figure 5.22 Energy landscape of a line attractor in a state space. The coordinates of the state space are the firing rates of neurons in a network. Only two of the dimensions of state space are depicted in the plane, with the height of the energy landscape as the third dimension. The dynamical trajectories in state space correspond to trajectories down the energy landscape. All trajectories in the state space of the memory network flow toward a line attractor (line without arrow). The corresponding trajectories on the energy landscape flow down the walls of a trough, toward a line of minima at the bottom. Reproduced from Seung (1996). [Seung, H. S. (1996). How the brain keeps the eyes still. Proceedings of the National Academy of Sciences, 93(23), 13339–13344. Copyright (1996) National Academy of Sciences, U.S.A.]

As discussed in Chapter 3 (Section 3.3.2), synaptic feedback needs to be adjusted in order to create such monotonically tuned neural integrators in models. The feedback must be finely tuned, so that if the firing rate of a neuron is altered by a transient input, then the resulting change in synaptic feedback to the neuron is exactly the amount required to maintain the new firing rate. Any mistuning of the feedback results in an exponential decay or growth away from the desired memory state, with a persistence time proportional to the synaptic time constant divided by the fractional error in the tuning (Chapter 3, **Figure 3.8**). Whether the fine-tuning is necessary or biologically plausible remains an open question. An approximate line attractor can be realized without the fine-tuning requirement (Koulakov et al. 2002; Miller et al. 2003b; Goldman et al. 2003). The idea is to combine many robust, bistable units that can be dendritic branches, neurons or neural groups. If they have a range of input thresholds, different input strengths (corresponding to a parametrically varying stimulus) are required to switch an increasingly larger number of bistable units from the DOWN to the UP state. The summation of neural units' outputs could be indistinguishable from a continuous quantity that encodes the stimulus feature.

Another idea (Lim and Goldman 2013) stems from the observation that, under certain conditions, feedback inhibition may approximately act as a negative feedback proportional to the time derivative of excitatory neural firing (dr_E / dt). Consider a recurrent network model of an excitatory neural population (r_E) and an inhibitory neural population (r_I). As discussed in Chapter3 (Section 3.3.3), if the dynamics of inhibition is fast, r_I keeps track of r_E in time and provides a negative feedback to r_E with the strength given by $J_{IE}J_{EI} / (1+J_{II})$. Lim and Goldman (2013) showed that this recurrent inhibition acts as a derivative negative feedback, provided that (1) the recurrent connections are strong (all four J values are large); (2) a similar excitation-inhibition (E/I) balance in the E and I cells ($J_{EE} / J_{EI} \sim J_{IE} / J_{II}$); and (3) positive feedback is slower than negative feedback. We then have:

$$\tau_E \frac{dr_E}{dt} \simeq -r_E + J_{EE}r_E - w_{der}\frac{dr_E}{dt} + I_{ext,E}(t). \tag{5.7}$$

or

$$\left(\tau_E + w_{der}\right)\frac{dr_E}{dt} \simeq -r_E + J_{EE}r_E + I_{ext,E}(t). \tag{5.8}$$

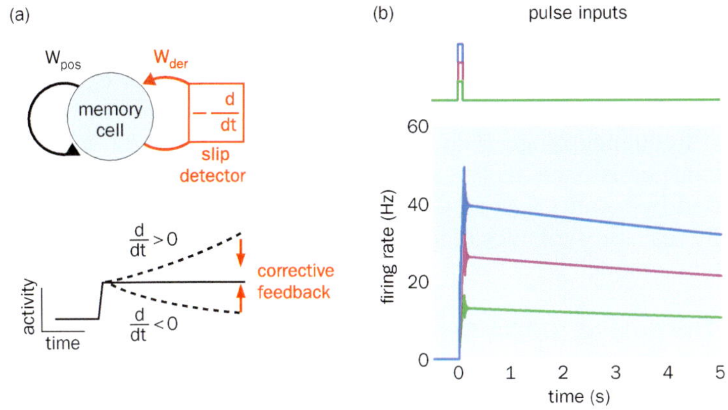

Figure 5.23 Approximate line attractor in a balanced excitatory-inhibitory recurrent network model. (a) Upper plot: Schematic excitatory neural population with a positive feedback and a negative derivative feedback. Lower plot: The negative derivative feedback term is negative (respectively positive) when r_E increases (respectively decreases) with time, in either case leading to restoration back to a desired steady-state. (b) In response to a brief input pulse, the excitatory neural population reaches a steady-state proportional to the input amplitude and roughly maintains a graded steady-state over a few seconds. Reproduced from Lim and Goldman (2013). [Lim, S., & Goldman, M. S. (2013). Balanced cortical microcircuitry for maintaining information in working memory. Nature Neuroscience, 16(9), 1306–1314.]

where w_{der} is proportional to $J_{IE}J_{EI}/(1+J_{II})$.

The effective time constant of this dynamics is $(\tau_E + w_{der})/(1-J_{EE})$, which is prolonged in comparison with τ_E not only by recurrent excitation (J_{EE}) but also by the derivative negative feedback (w_{der}). Whereas the positive feedback counteracts the leak in proportion to the firing rate r_E, the derivative negative feedback slows down the memory decay in a way that mathematically is equivalent to a viscous drag force of a fluid (Lim and Goldman 2013). It extends the network time constant of memory decay. **Figure 5.23** shows that this model displays parametric persistent activity that decays slowly over seconds, demonstrating that feedback inhibition faster than feedback excitation contributes to the maintenance of graded persistent activity. The assumed E/I balance and slow recurrent excitation generally hold true for working memory network models, which we will discuss next.

5.6 YIN AND YANG OF NEURONAL REVERBERATION

5.6.1 The Excitation-Inhibition Balance

Strong positive feedback is required for a working memory circuit, but it could lead to instability unless it is dynamically counteracted by a negative feedback. In a resting state, why does spiking activity not grow in spite of recurrent excitation? The answer is a tight balance between excitation and inhibition (Chapter 3, Section 3.2.2) (van Vreeswijk and Sompolinsky 1996; Amit and Brunel 1997). In fact, spontaneous firing is driven by random background external inputs; feedback inhibition is slightly larger than excitation, hence the overall recurrent input to a neuron is inhibitory in the resting state (**Figure 5.24a**). In a memory state in which stronger reverberatory excitation is recruited to sustain an elevated firing rate, the net recurrent input becomes excitatory (**Figure 5.24b**). Interestingly, recurrent synaptic inhibition increases proportionally with excitation. On the other hand, when the stimulus is not preferred for a neuron, which is thus not engaged in memory representation, during the delay period the net recurrent input is inhibitory and slightly larger than in the resting state (**Figure 5.24c**, brown compared to dashed box).

This excitation-inhibition balance, approximately preserved regardless of changing neuronal firing rates, represents a general characteristic of cortical network dynamics (Chapter 3, Sections 3.2.2 and 3.3.3). As a result of the balancing act of recurrent excitation and inhibition, the total synaptic current to any neuron is close to the firing threshold under all conditions (**Figure 5.24**, purple), and spike discharges are to a large degree driven by noisy synaptic fluctuations. This explains the highly irregular spiking patterns of persistent activity in the model (**Figure 5.18b**), as

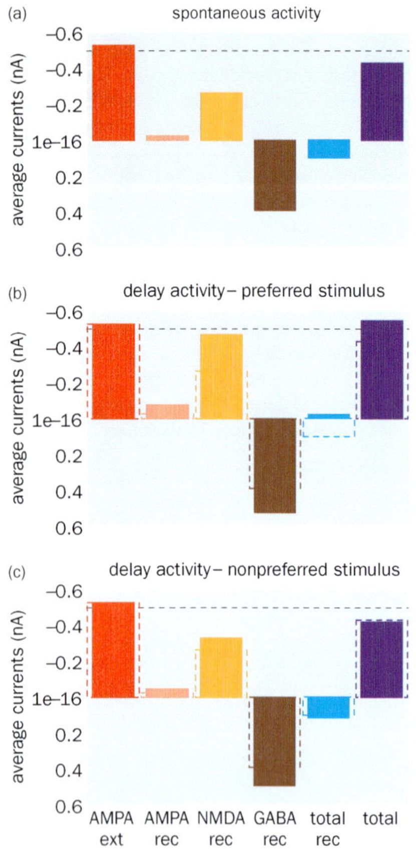

Figure 5.24 Balanced excitation and inhibition in a spiking network model of object working memory. Various components of synaptic current in a single cell during spontaneous activity (a), during delay activity following presentation of a preferred stimulus (b), and during delay activity following presentation of a nonpreferred stimulus (c). The dotted horizontal line indicates the total current needed to reach a deterministic firing threshold. In (b–c), the dotted boxes indicate the value of the corresponding component during spontaneous activity, to show the differences between delay and spontaneous activity. Background external inputs are suprathreshold. Recurrent circuit is dominated by inhibition (brown) over excitation (orange) in the spontaneous state, so that the net recurrent synaptic current is hyperpolarizing (blue). During delay activity, both recurrent excitation and inhibition are larger and dynamically balance each other, in such a way that the overall synaptic excitation becomes slightly larger following a preferred stimulus (leading to persistent activity at an elevated rate) than after a nonpreferred stimulus. Reproduced from Brunel and Wang (2001). [Brunel, N., & Wang, X. J. (2001). Effects of neuromodulation in a cortical network model of object working memory dominated by recurrent inhibition. Journal of Computational Neuroscience, 11(1), 63–85.]

observed in prefrontal neurons recorded from behaving monkeys (Chapter 2, **Figure 2.5b**) (Compte et al. 2003; Joelving et al. 2007). Moreover, as we have discussed previously, the interplay between fast excitation and slower inhibition underlies coherent network oscillations in the gamma (40 Hz) frequency range (**Figure 5.19**) (Compte et al. 2000). Therefore, stochastic persistent activity coexisting with a synchronous rhythm can be explained by balanced excitatory-inhibitory dynamics, provided that biologically realistic dynamics of synaptic transmissions is taken into consideration in a model.

5.6.2 The Role of NMDA Receptors

Fast synaptic excitation mediated by the AMPA receptors is important for explaining coherent γ oscillations. However, if reverberation is predominantly dependent on the AMPA receptors, persistent activity is often disrupted in the middle of a delay period, thereby the memory is lost (Wang 1999b). This is shown in **Figure 5.25** for the object-working memory network described earlier. Recall that in this network, $GABA_AR$-mediated inhibition has a time constant of 10 ms, and AMPAR- and NMDAR-mediated excitation have a time constant of 2 ms and 100 ms, respectively. With $\tau_{GABA_A} / \tau_{AMPA} = 5$, AMPA-mediated feedback excitation is much faster than $GABA_A$ receptor-mediated feedback inhibition. As the relative contributions of AMPA and NMDA receptors at excitatory synapses are increased, the excitation gradually becomes faster than inhibition. With a low AMPA/NMDA ratio (**Figure 5.25a–b**), the average activity is fairly constant in time and shows only small amplitude fluctuations, which do not destabilize the persistent state. As the ratio is progressively increased, the amplitude of activity fluctuations grows, and the power spectrum of the average activity shows a clear peak around 40 Hz (**Figure 5.25c** and **e**).

This is the same phenomenon described in Chapter 3 (Section 3.4.2). When the AMPA/NMDA ratio is further increased, the oscillation amplitude becomes too large. During a time epoch when the firing rate is nearly zero, the AMPA-mediated synaptic drive decays away and NMDAR-mediated excitation is too weak to bridge the temporal gap between activity bursts. As a result, the network reverts to the resting state, and the network's working memory behavior disappears (**Figure 5.25d**).

This destabilization problem is present if negative feedback is instantiated by spike-frequency adaptation or short-term synaptic depression (Chapter 2, Sections 2.4.2 and 2.7.1). Such instability does not occur, when the mismatch of time constants for positive and negative feedback processes is reversed. This is the case with slow reverberation substantially mediated by NMDA receptors (Wang 1999b). The general lesson is independent of any specific mechanism: in a powerfully recurrent circuit, active states coexisting with a resting state require that positive feedback be sufficiently slow to either produce sustained firing or bridge the temporal gaps of silence when the population undergoes stochastic oscillations with substantial up and down swings. Today, a time constant on the order of 50–100 ms is commonly assumed for synaptic excitation, even when the NMDA receptors are not explicitly mentioned, in strongly recurrent neural network models (Barak et al. 2013; Laje and Buonomano 2013; Lim and Goldman 2014; Hennequin et al. 2014; Masse et al. 2019). It is a requirement for the negative derivative mechanism (**Figure 5.23**).

A specific prediction from the modeling work, then, is that in a working memory microcircuit, if persistent activity is primarily sustained by synaptic reverberation, local E-to-E synapses should have a sufficiently high NMDA/AMPA ratio (**Figure 5.26**). An NMDA receptor is composed of an obligatory NR1 subbunit and two NR2 subunits, predominantly NR2A or/and NR2B in cortical excitatory neurons. Interestingly, in humans the mRNA levels for genes coding major NMDA receptor subunits NR1, NR2A

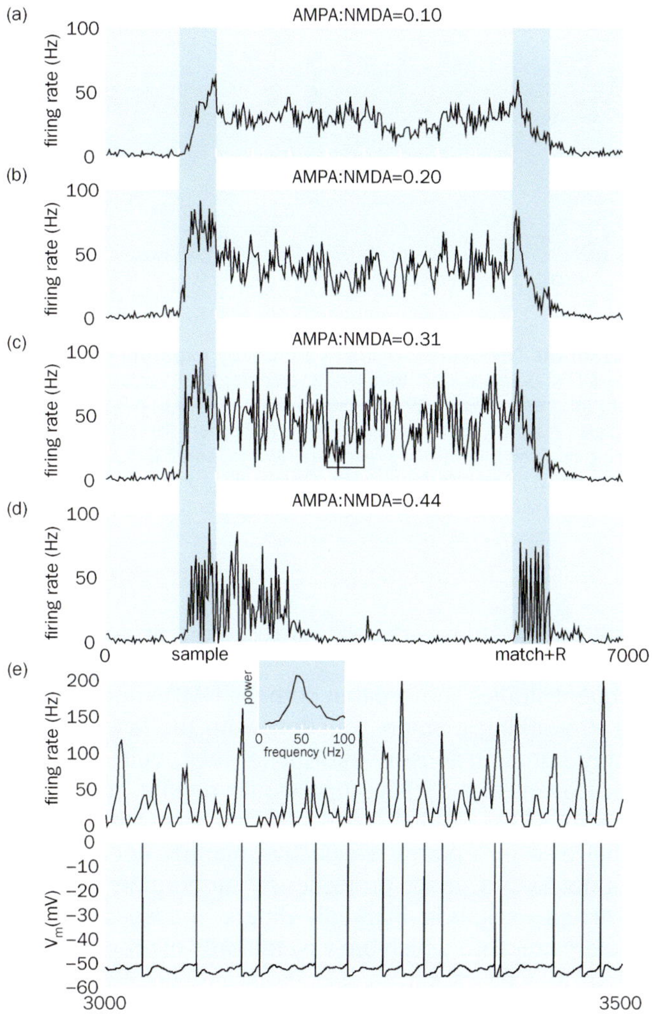

Figure 5.25 Stability of persistent activity as a function of the AMPA/NMDA ratio at local recurrent excitatory synapses. (a–d) Temporal course of the average firing rate across a sub-population of selective cells in the network of Figure 5.13 after transient excitatory input, for different levels of the AMPA/NMDA ratio. As the ratio is increased, oscillations of a progressively larger amplitude develop, which eventually destabilize the persistent activity state. (e) Snapshot of the activity of the network in (c) between 3 and 3.5 seconds. Top: Average network activity. Bottom: Intracellular voltage trace of a single neuron. Inset: Power spectrum of the average activity of the network, showing a peak in the gamma (~ 40 Hz) frequency range. Persistent activity is stable even in the presence of synchronous oscillations (d, e). Reproduced from Renart et al. (2003) using the model of Brunel and Wang (2001). [Copyright 2003. From Mean-field theory of recurrent cortical networks: Working memory circuits with irregularly spiking neurons. In J. Feng (Ed.), Computational Neuroscience: A Comprehensive Approach by Renart, A., N. Brunel, and X.-J. Wang. Reproduced by permission of Taylor & Francis Group, LLC, a division of Informa PLC.]

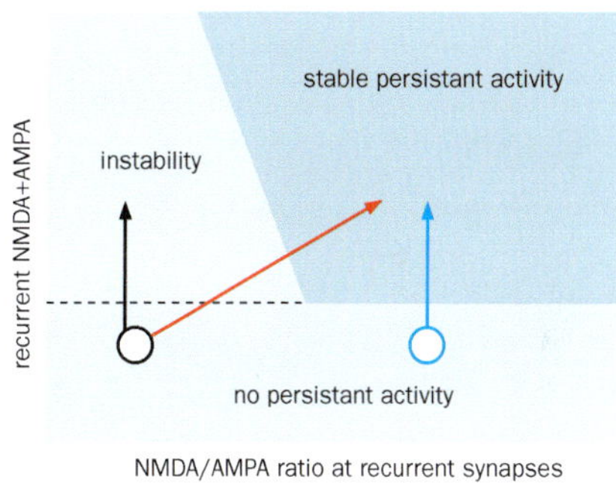

Figure 5.26 Role of NMDA receptors in a working memory circuit. Schematically shown is the dependence of stable persistent activity on both sufficiently strong recurrence (y-axis) and large NMDA/AMPA ratio at local excitatory synapses (x-axis). Starting at a point with a relatively large NMDA/AMPA ratio (open blue circle), strengthening the excitatory recurrence alone can lead to the emergence of persistent activity (blue arrow). On the other hand, starting at another point at a low NMDA/AMPA ratio (black open circle), an enhancement of recurrence can lead to network dynamical instability (black arrow), in which case the NMDA/AMPA ratio needs to be increased concomitantly (red arrow). Reproduced from Wang (2006). [Wang, X. (2006). A microcircuit model of prefrontal functions: Yin and yang of reverberatory neurodynamics in cognition (pp. 92–127). In J. Risberg & J. Grafman (Eds.), The Frontal Lobes: Development, Function and Pathology © Cambridge University Press 2006, reproduced with permission.]

and NR2B are substantially higher in the PFC compared to V1 (Scherzer et al. 1998). Note that what matters for the model prediction is the NMDA/AMPA ratio at intrinsic excitatory synapses within a microcircuit. The overall NMDA/AMPA ratio for a single neuron could still be low, if its excitatory inputs from other sources (such as external inputs from sensory areas) are predominantly mediated by the AMPA receptors. Finally, local recurrent excitation is mediated by a mixture of the NMDA and AMPA receptors, the latter are fast and important for the generation of gamma oscillations (Figure 5.19).

The role of the NMDA receptors in maintaining persistent activity has been directly tested in a monkey experiment (Wang et al. 2013). While animals perform an ODR task of spatial working memory, using ionophoresis a drug was injected to selectively block the NMDA receptors in a small number of neurons in close spatial vicinity of the recording electrode. The reasoning is that, if persistent activity of a recorded PFC single cell is sustained by synaptic excitation mediated by the NMDA receptors, blocking the latter should reduce or abolish its elevated persistent activity during the delay period. This was found to indeed be the case (Figure 5.27a). To quantify the effect, a tuning index was defined as the difference in the delay period firing rate between the trials with preferred (r_{pref}) and nonpreferred ($r_{nonpref}$) cues, divided by their sum, $TI = (r_{pref} - r_{nonpref})/(r_{pref} + r_{nonpref})$. This tuning index is greatly reduced with the NMDA receptor antagonist application compared to control (Figure 5.27b), providing empirical

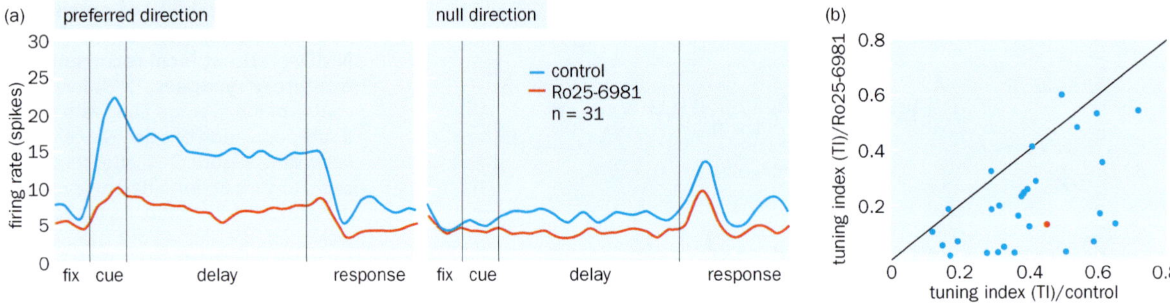

Figure 5.27 Dependence of delay period persistent activity on the NMDA receptors in a monkey experiment using the ODR task. (a) Average response showing the mean firing patterns of 31 dlPFC delay cells for their preferred (left panel) versus nonpreferred directions (right panel) under control condition (blue) and after iontophoresis of Ro25-6981, a selective antagonist of NR2B-containing NMDA receptors (red). Ro25-6981 markedly decreased task-related firing, especially for the neurons' preferred direction. (b) Iontophoresis of Ro25-6981 significantly weakened spatial selectivity by reducing tuning index. Each blue dot is a recorded neuron, and the red dot corresponds to the population average. Reproduced from Wang et al. (2013). [Wang, M., Yang, Y., Wang, C. J., Gamo, N. J., Jin, L. E., Mazer, J. A., ... & Arnsten, A. F. (2013). NMDA receptors subserve persistent neuronal firing during working memory in dorsolateral prefrontal cortex. Neuron, 77(4), 736–749.]

support for a critical role of the NMDA receptors in a working memory circuit of the primate brain.

Subsequent studies showed that both the NMDA and AMPA receptors contributed to working memory function, with the fast AMPA receptors predominantly signaling sensory information (van Vugt et al. 2020; Yang et al. 2021). However, a decisive test would require new techniques to manipulate NMDARs not only in particular cell types (pyramidal or inhibitory cells) but also in a pathway-selective manner (local recurrent connections but not long-distance projections). Furthermore, NMDAR- versus AMPAR-mediated excitation markedly differs in association areas from primary sensory areas (to which the vast majority of previous studies have been devoted) as well as across species (Wang 2020). It would thus be desirable to further rigorously test the model prediction in primate prefrontal cortex (González-Burgos et al. 2000).

It is worth noting that the model does not assume that local recurrent synapses are devoid of an AMPAR component. As a matter of fact, the precise NMDA/AMPA ratio required for a working memory microcircuit depends on the details of network biophysics and connectivity, such as the interplay of the NMDA current with slow intrinsic currents of single cells and synaptic inhibition mediated by the $GABA_B$ receptors (Tegnér et al. 2002; Papoutsi et al. 2013; Sanders et al. 2013). For instance, the time constant of a synaptic current depends on the subunit composition of its receptors. If the $GABA_A$ receptor-mediated inhibition is faster (Bartos et al. 2007), so that $\tau_{GABA_A} / \tau_{AMPA}$ is smaller in a working memory circuit, instability due to the time constant mismatch would be less severe, and the required NMDA/AMPA ratio would be lower (Tegnér et al. 2002). Moreover, mutual inhibition between GABAergic cells effectively shortens the time constant of synaptic inhibition (Chapter 3, Eq. 5.20), thereby contributing to network stability (Hansel and Mato 2003).

5.6.3 The Importance of Being Slow But Not Too Slow

We have shown that positive feedback should not be too fast compared to negative feedback, when both are strong in a working memory circuit. This principle applies to other mechanisms, and slow biophysical processes are numerous in the brain. A modeling study (Pereira and Wang 2015) considered three slow biophysical mechanisms that are prominently present in the prefrontal cortex (**Figure 5.28a**). First, an inward current like calcium-activated cation current I_{Can} in single cells opens during activity and in turn depolarizes neurons (Haj-Dahmane and Andrade 1998; Egorov et al. 2002). Second, short-term facilitation (STF) of the E-to-E synapses (Hempel et al. 2000; Wang et al. 2006) provides another positive-feedback process (Hempel et al. 2000; Mongillo et al. 2008; Wu and Zenke 2021). Third, depolarization-induced suppression of inhibition (DSI) (Wilson and Nicoll 2001; Freund et al. 2003; Carter and Wang 2007) produces disinhibition, thereby contributing to reverberating excitation.

Each of these mechanisms can play a significant role in maintaining persistent activity and is not incompatible with the attractor model of working memory representation. For instance, when I_{Can} is added in single cells, less NMDA/AMPA ratio would be necessary (Tegnér et al. 2002). A network endowed with STF (Mongillo et al. 2008) could display brief bursts of spikes interspersed with long periods of silence (**Figure 5.28b**). This repetitive pattern is self-sustained and thus mathematically described as an attractor state.

Incorporating slow positive feedbacks to the spiking network model for spatial working memory discussed prior (Compte et al. 2000), it was found that these processes enhance the robustness of memory accuracy encoding, by reducing the noise-induced random drifts of the bump attractor (**Figure 5.20**) and enhancing resistance against distractors. This is shown in **Figure 5.28c** in the case of an intrinsic I_{Can}, where the variance (due to random drifts of the bump state) of the decoding (population vector

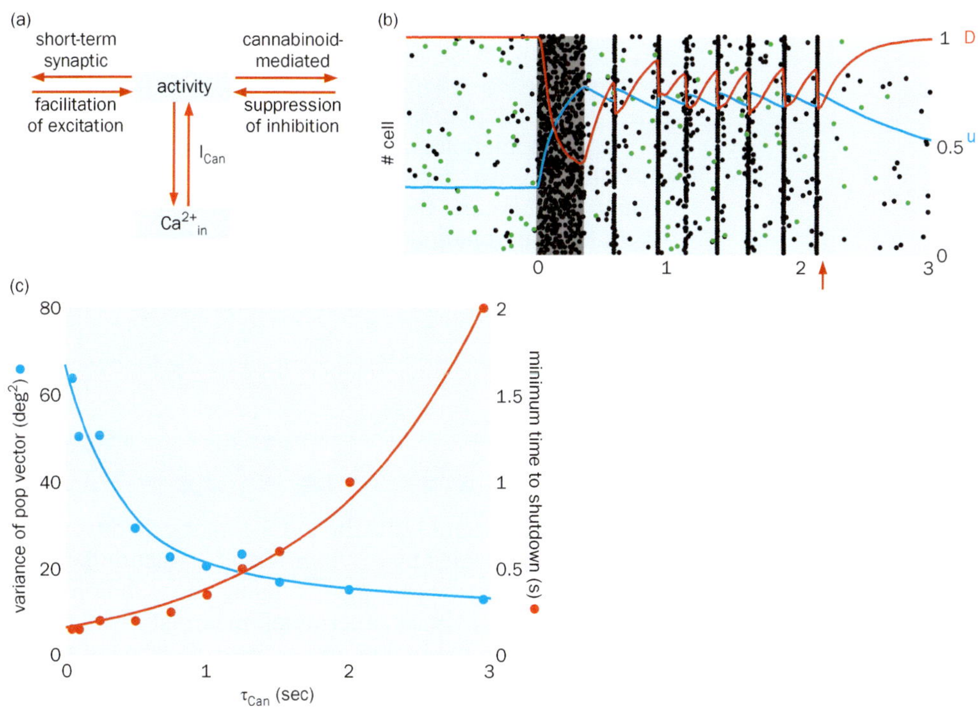

Figure 5.28 Tradeoff for contributions of slow positive feedbacks to persistent activity. (a) Three biological processes with a time constant of 0.5 to a few seconds: Calcium-activated inward cation current I_{Can} in single cells, short term facilitation (STF) of recurrent excitatory synapses and cannabinoid-mediated depolarization-induced suppression of inhibition (DSI). (b) In a recurrent network model endowed with short-term facilitation and depression, a brief stimulus (time 0) triggers repetitive bursts of spikes interspersed with long silent periods as a self-sustained but time-varying attractor state. Red and blue: Short-term depression and facilitation variable, respectively. Red arrow: The end of a delay period. (c) When the time constant for I_{Can} is increased, the cross-trial variance of the population vector $\left(\sigma^2\left(\theta_{pop}\right)\right)$ is reduced, thus working memory decoding becomes more accurate (blue circle), but the time duration of a signal required to erase the memory trace at the end of a delay increases (red circle) to be too long to meet behavioral demands. Panel (b) is reproduced from Mongillo et al. (2008) and (c) from Pereira and Wang (2015). [Mongillo, G., Barak, O., & Tsodyks, M. (2008). Synaptic theory of working memory. Science, 319(5869), 1543–1546.] [Pereira, J., & Wang, X. J. (2015). A tradeoff between accuracy and flexibility in a working memory circuit endowed with slow feedback mechanisms. Cerebral Cortex, 25(10), 3586–3601.]

θ_{pop}) at the end of a delay period is plotted against the I_{Can} time constant (blue circle). However, when a progressively slower process dominates network reverberation, it becomes increasingly more difficult to erase a memory trace (**Figure 5.28c**, red circle). Since working memory requires that its representation be switchable on and off by brief (hundreds of milliseconds) stimuli, there is a tradeoff between behavioral accuracy and flexibility with regard to slow underlying mechanisms. This finding is general and applicable to other specific mechanisms, including interactions between neurons and astrocytes (De Pittà and Brunel 2022). An optimality is achieved when the NMDA-receptor-dependent recurrent excitation is the workhorse for the maintenance of persistent activity, whereas additional but relatively weak slow synaptic or cellular processes enhance the robustness of mnemonic function.

5.6.4 Cannabinoid Modulation and Cross-Trial Serial Effect

The DSI is particularly interesting because it is mediated by cannabinoid, a brain's endogenous chemical and the psychoactive ingredient of marijuana. The basic biology of cannabinoid signaling in the cortex is well understood: cannabinoid is released by excitatory pyramidal cells in an activity-dependent way, and it reduces inhibition mediated by a subtype of interneurons that express cholecystokinin (CCK) (Wilson and Nicoll 2002). Concretely, first, calcium influx during spiking activity induces the release of endogenous cannabinoid from pyramidal neurons. Second, cannabinoid leaves the postsynaptic neuron and serves as a messenger for retrograde signaling at the presynaptic site of an inhibitory connection from GABAergic CCK-positive neurons. Third, cannabinoid binding with its presynaptic receptor lowers the release probability of GABA transmitter, thereby reducing synaptic inhibition of the same pyramidal neurons in an activity-dependent manner.

A phenomenological model of DSI (Carter and Wang 2007) captures quantitatively two most salient observations: the dependence on intracellular calcium and the slow decay time course (Heinbockel et al. 2005). Based on experimental measurements of DSI as the percent reduction in inhibitory event size, the model assumes that the inhibitory synaptic conductance G_{EI} on a pyramidal cell with preferred θ is multiplied by a factor $1 - DSI(\theta)$.

The dynamics of *DSI* is described by the following equation:

$$\frac{dDSI(\theta)}{dt} = -\frac{DSI(\theta)}{\tau_{DSI}} + \beta_{DSI}\left[Ca(\theta)^{2+}\right]^{n}\left(DSI_{max} - DSI(\theta)\right) \tag{5.9}$$

where $[Ca(\theta)^{2+}]$ represents the intracellular calcium concentration in each pyramidal cell labeled by θ. When $[Ca(\theta)^{2+}]$ accumulates, $DSI(\theta)$ increases with a rate controlled by β_{DSI}, causing disinhibition. $DSI(\theta)$ is bounded above at DSI_{max}, which determines maximal disinhibition, biophysically corresponding to the maximum number of synapses that are cannabinoid sensitive multiplied by the maximal reduction in release probability at each synapse due to DSI. When the pyramidal cell ceases to be active, $DSI(\theta)$ decays back to 0 with a time constant τ_{DSI}.

The intracellular calcium influx is triggered by spikes and obeys a first-order kinetics (Chapter 2, Eq. 2.24) (Liu and Wang 2001) with parameters calibrated by experimental measurements (Helmchen et al. 1996). Solving the steady-state of Eq. (5.9) with a fixed $[Ca^{2+}]$ level, one obtains:

$$\mathrm{DSI}_{ss} = \frac{\mathrm{DSI}_{max}[Ca^{2+}]^{n}}{K_{50}^{n} + [Ca^{2+}]^{n}} \tag{5.10}$$

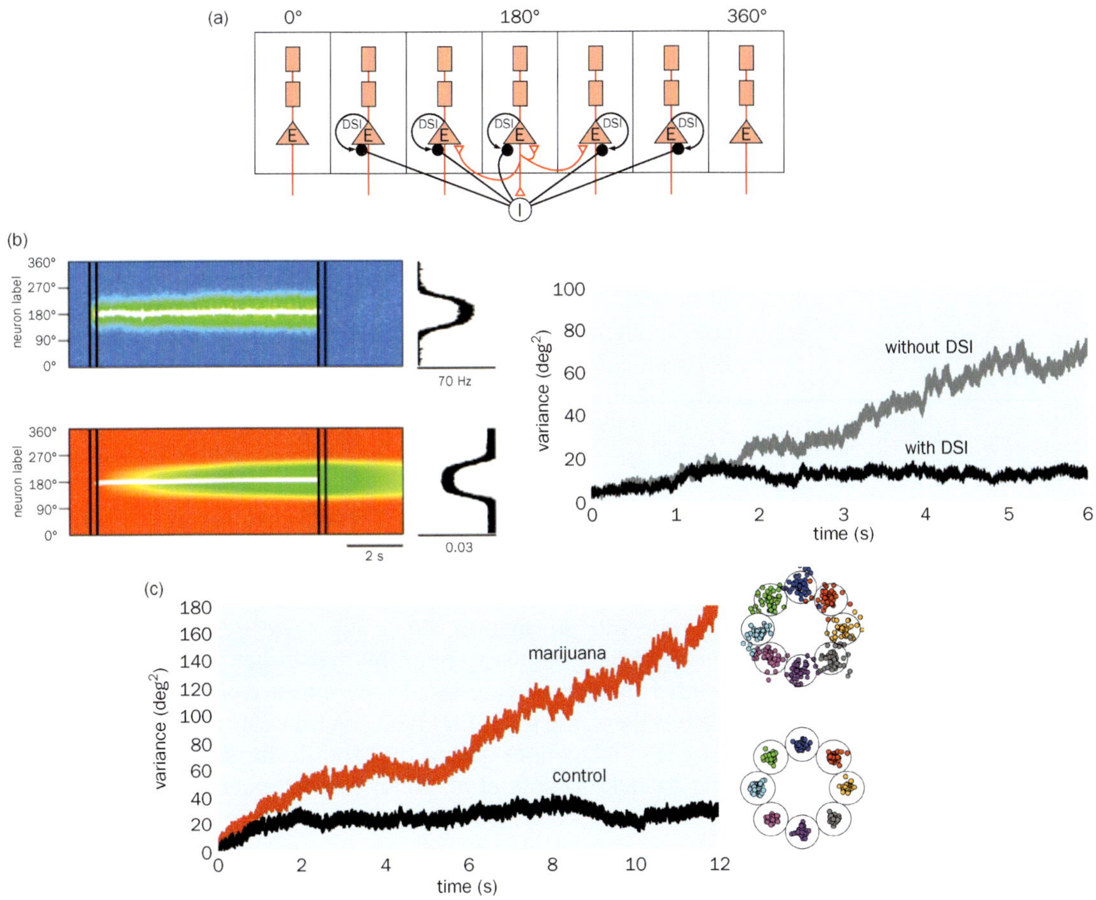

Figure 5.29 Network model of spatial working memory endowed with cannabinoid-mediated disinihibition (DSI). (a) Schematic network architecture. (b) Left: Spatiotemporal neural activity (top) and DSI signal (bottom). Right: DSI enhances the performance by reducing random drifts of persistent activity computed by $\sigma^2(\theta_{pop})$. Without DSI, the variance of the population vector grows in time (grey line). DSI creates a privileged locus to stabilize a mnemonic activity pattern, so that the variance plateaus (black line). (c) Model simulation of marijuana effect on working memory function. Variance of the remembered location during delay period with (red) and without (black) a global agonist. In the presence of an exogenous agonist, disinhibition is activity-independent, so the accuracy of memory-guided saccades deteriorates. Scatter plots on the right: Endpoints of simulated memory-guided saccades color coded by the cue location with (top) versus without (bottom) agonist. Reproduced from Carter and Wang (2007). [Carter, E., & Wang, X. J. (2007). Cannabinoid-mediated disinhibition and working memory: dynamical interplay of multiple feedback mechanisms in a continuous attractor model of prefrontal cortex. Cerebral Cortex, 17(suppl_1), i16–i26.]

where $K_{50}^n = 1/(\beta_{DSI}\tau_{DSI})$. The Hill coefficient $n = 1.36$, $K_{50}^n = 3.6$ μM and $\tau_D = 16.7$ seconds are fixed by fitting with experimental data (Wang and Zucker 2001). They in turn determine the value of $\beta_D = 1/\left(K_{50}^n \tau_D\right) = 0.0166(\mu Ms)^{-1}$. This way, the model parameters have been well constrained by experimental measurements.

As shown in **Figure 5.29**, endogenous DSI reduces random drifts of a bump attractor, because the activity-dependent reduction of synaptic inhibition is selective for those active neurons involved in working memory representation, thereby creating a favorable bias precisely where the bump attractor is located in the network. Other slow processes such as short-term synaptic facilitation can also produce similar reduction of drifts, thereby enhancing working memory (Itskov et al. 2011; Hansel and Mato 2013; Pereira and Wang 2015; Seeholzer et al. 2019). By contrast, an exogenous drug like marijuana is not activity dependent and uniformly reduces synaptic inhibition, which leads to an increase of random drifts. This explains the observation that, in human subjects, intake of marijuana decreases the accuracy of memory-guided saccades in humans performing the ODR task (Ploner et al. 2002).

Interestingly, a process like DSI characterized by a time constant of several to ten seconds does not decay away in typical inter-trial intervals (a few seconds) of a laboratory experiment. Consequently, in a given trial, a bump attractor should move towards the site of the bump in the previous trial because of the lingering DSI around that location. This finding led Carter and Wang (2007) to predict that performance of spatial working memory is biased by the previous target location, and the bias size increases with the similarity between the previous and current cues. Both predictions have been confirmed experimentally (Papadimitriou et al. 2015; Bliss et al. 2017). Such a history dependence can be induced by slow processes other than DSI, like short-term synaptic facilitation (Kilpatrick 2018; Barbosa et al. 2020), provided that the process's intrinsic time constant is longer than inter-trial intervals. The similarity-dependent cross-trial effects are likely to be commonplace in perception, attention and working memory, but the functional benefits or detrimental bias remain to be clarified.

5.6.5 Disinhibition Motif by Three Subtypes of Inhibitory Cells

As discussed previously, a cardinal requirement for working memory function is for the brain to ignore behaviorally irrelevant distractors while holding relevant information. Since inputs to pyramidal cells are located in the dendrites, inhibitory neurons that target pyramidal dendrites are in a unique position to gate input flow into a working memory circuit. Those are not the same inhibitory neurons that control spiking output of pyramidal cells. To explore gating by dendrite-targeting inhibitory neurons, the ring model (Compte et al. 2000) was extended to incorporate three types of GABAergic cells for robust working memory against distractors (**Figure 5.30a**) (Wang et al. 2004).

In the cortex, there is a wide diversity of GABAergic interneurons, with regards to their morphology, electrophysiology, chemical markers, synaptic connections and short-term plasticity, molecular characteristics (Chapter 2, **Figure 2.2**) (Freund and Buzsáki 1996; DeFelipe 1997; Kawaguchi 1997; Cauli et al. 1997; Somogyi et al. 1998; Markram et al. 2004; Buzsáki et al. 2004). Three largely non-overlapping subclasses of inhibitory cells can be defined functionally by their synaptic targets, labeled by distinct biomarkers. The first one is perisoma-targeting cells that express calcium-binding protein parvalbumin (PV) and control spiking output of pyramidal cells. Interneurons of the second type express somatostatin (SST) or calbindin (CB) and target dendrites of pyramidal cells, and therefore are in a position to control synaptic inputs into excitatory neurons. The third type is interneuron-targeting cells; they contain vasoactive intestinal peptide (VIP) or calcium-binding protein calretinin (CR), and preferentially target SST/CB interneurons cells. Interestingly, the distributions of PV, SST/CB and VIP/CR interneurons are quite different in PFC compared to V1. In primary visual cortex, PV-containing interneurons (including fast spiking basket cells) are prevalent; the proportions are about 48% (PV), 29% (CB) and 20% (CR), respectively (Kooijmans et al. 2020). By contrast, in the monkey prefrontal cortex, the proportions are about 25% (PV), 25% (CB) and 45% (CR), respectively (Condé et al. 1994; Gabbott and Bacon 1996) (**Figure 5.30b**).

The circuit model of Wang et al. (2004) suggests how these different interneuron types may work together in the prefrontal cortex (**Figure 5.31**). The model assumes that while PV interneurons project widely and sculpt the tuning of the network activity pattern, SST/CB and VIP/CR interneurons act locally as a disinhibitory motif for selective input gating. Note that the three interneuron types in the model should be more appropriately interpreted according to their synaptic targets, rather than calcium-binding protein expressions. For example, PV cells display a variety of axonal arbors, among which the large basket cells are likely candidates

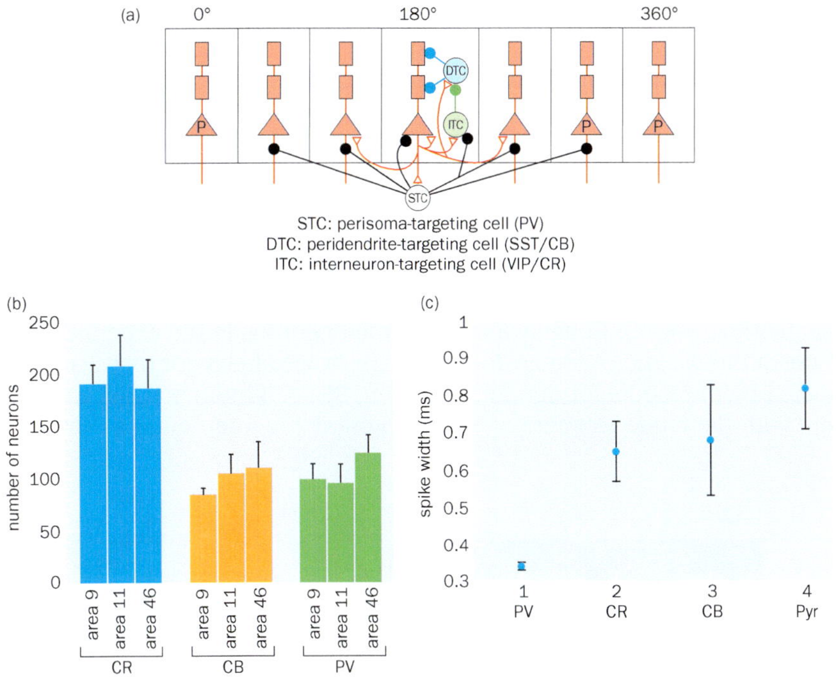

Figure 5.30 A spatial working memory model with three subclasses of GABAergic interneurons. (a) The model scheme is similar to Figure 5.18a, except with three inhibitory cell subclasses. Besides the perisoma-targeting (parvalbumin-containing, PV), broadly projecting, neurons to pyramidal (Pyr) cells, a second class of interneurons express somatostatin (SST) or calbindin (CB) and target the dendrites of Pyr neurons with a narrow projection footprint, whereas the third class of VIP or calretinin (CR)-containing interneurons preferentially project to SST/CB cells. (b) Proportional distribution of PV, CB and CR expressing GABAergic cells in three subregions of the monkey prefrontal cortex. (c) Half-peak spike width for pyramidal neurons and three types of interneurons of macaque monkey prefrontal cortex. Panel (a) is modified from (Wang et al. 2004), (b–c) are from Wang (2013) with original data for (b) from Condé et al. (1994), for (c) from Zaitsev et al. (2005) and Povysheva et al. (2006). [Wang, X. J., Tegnér, J., Constantinidis, C., & Goldman-Rakic, P. S. (2004). Division of labor among distinct subtypes of inhibitory neurons in a cortical microcircuit of working memory. Proceedings of the National Academy of Sciences, 101(5), 1368–1373. Copyright (2004) National Academy of Sciences, U.S.A.] [Wang, X.-J. (2013). The prefrontal cortex as a quintessential 'cognitive-type' neural circuit: Working memory and decision making. In D. T. Stuss and R. T. Knight (Eds.), Principles of Frontal Lobe Function (Second ed.)., pp. 226–248. New York: Oxford University Press. Reproduced with permission of the Licensor through PLSclear.]

for the widely projecting perisoma-targeting cells (Krimer and Goldman-Rakic 2001; Kisvarday et al. 2003). SST/CB interneurons show a high degree of heterogeneity, but some of them (such as Martinotti cells) are known to act locally and preferentially target dendritic spines and shafts of pyramidal cells (Gulyás et al. 1996; DeFelipe 1997; Somogyi et al. 1998; Silberberg and Markram 2007; Urban-Ciecko and Barth 2016). Finally, a subset of VIP/CR cells avoid pyramidal cells at least in the same cortical layer and preferentially target SST/CB interneurons (Gulyás et al. 1996; Meskenaite 1997; DeFelipe et al. 1999; Gonchar and Burkhalter 1999; Lee et al. 2013; Jiang et al. 2015).

Figure 5.31a–b shows a model simulation of spatial working memory. When a group of pyramidal cells are excited by a transient extrinsic input, they excite each other through interconnections. In addition, activated VIP/CR interneurons suppress SST/CB interneurons at that location, leading to reduced inhibition (disinhibition) of the dendrites of the same pyramidal cells. The concerted actions of recurrent excitation and VIP/CR interneuron-mediated disinhibition generate self-sustained persistent activity in these neurons, with the width of tuning shaped by synaptic inhibition from PV interneurons. The contribution by disinhibition added to recurrent excitation suggested that area-to-area variation of the distribution of the three interneuron types should be correlated with each area's prevalence of delay period activity, a model prediction supported by monkey experiments (Torres-Gomez et al. 2020).

In this model and that of Compte et al. (2000), fast spiking (FS) PV interneurons have a broader spatial tuning curve than do pyramidal cells. In behaving monkeys, it has been difficult to separate different cell types of recorded neurons. However, based on the fact that action potentials are narrower in FS interneurons than regular-spiking (RS) pyramidal cells (McCormick et al. 1985), tuning of delay period activity was found to be indeed broader for FS cells than RS cells in the PFC of monkeys performing the ODR task (Constantinidis and Goldman-Rakic 2002). The spike widths of SST/CB and VIP/CR interneurons are intermediate between the PV interneurons and pyramidal cells (Figure 5.30c); they are more difficult to identify by spike width alone in neurophysiological studies of behaving animals.

Moreover, during a delay period, SST/CB interneurons on the flanks of the bump attractor show enhanced firing because of long-range excitation from pyramidal cells with elevated activity, leading to surround inhibition (Adesnik et al. 2012). Therefore, pyramidal cells outside of the bump become less sensitive to external inputs, ensuring that working memory storage is not vulnerable to behaviorally irrelevant distractors. The model predicts that a small fraction of (putative SST/CB) PFC neurons recorded from a behaving monkey should show inverted tuning of mnemonic delay period activity, i.e., a reduced firing relative to spontaneous activity selectively for some sensory cues. This prediction was confirmed in data analysis (**Figure 5.31c**). Roughly 5% of recorded neurons in the experiment of Constantinidis and Goldman-Rakic (2002) showed behavior that was predicted by the model for dendrite-targeting SST/CB interneurons, consistent with the crude estimate of SST/CB-containing interneurons. Neurons

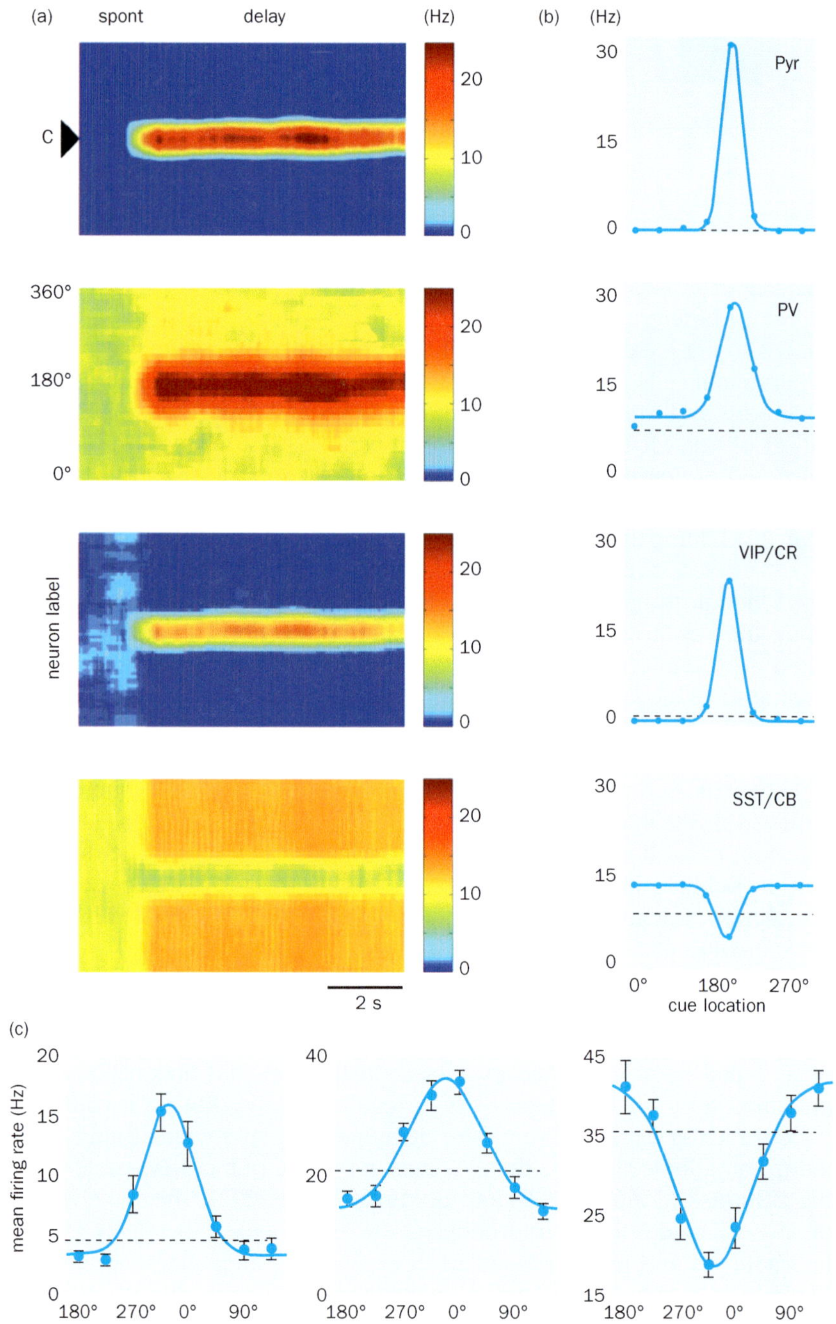

Figure 5.31 A spatial working memory model and comparison with recorded PFC neurons. (a) Bell-shaped persistent activity pattern for the pyramidal cells and the three (PV, SST/CB and VIP/CR) inhibitory neuron populations, with firing rates coded by color. (b) Neuronal tuning curves (solid lines) during the delay period in the model simulations. Dashed lines: Spontaneous firing rate during the resting state. (c) Three kinds of recorded tuning curves in monkey dorsolateral prefrontal cortex during a spatial working task, with the same conventions as in (b). Solid line, the best Gaussian fit; dashed horizontal line, average firing rate in the resting state during fixation. An example of the inverted tuning curve is shown (right), with a high baseline activity, strong reduced delay period activity for some cues, and slightly increased delay period activity for other cues. Consistent with slice physiology (Figure 5.30c), the spike width is the shortest for putative PV cells, the longest for putative pyramidal cells, and intermediate between the two for putative SST/CB interneurons. Reproduced from Wang et al. (2004). [Wang, X. J., Tegnér, J., Constantinidis, C., & Goldman-Rakic, P. S. (2004). Division of labor among distinct subtypes of inhibitory neurons in a cortical microcircuit of working memory. Proceedings of the National Academy of Sciences, 101(5), 1368–1373. Copyright (2004) National Academy of Sciences, U.S.A.]

with inverted tuning curves of delay period activity were also found in monkey's posterior parietal cortex during the ODR task (Zhou et al. 2012), suggesting that the disinhibitory motif is widespread.

How does a network's resistance against distractors depend on the relative inhibitions from the SST/CB and PV populations to the pyramidal cells? In model simulations, a distractor stimulus was introduced during the delay period at a location different from the initial cue (Compte et al. 2000; Brunel and Wang 2001). The network's distractibility is quantified by the intensity threshold of a distracting stimulus above which it is powerful enough to distract the network from the stored initial cue (**Figure 5.32a**). As expected, dendritic inhibition mediated by SST/CB cells is more efficient than perisomatic inhibition mediated by PV cells in filtering out distractor stimuli. When dendritic inhibition is insignificant, and stimulus is sufficiently strong (**Figure 5.21**), a bump state can be initiated by a brief stimulus, but a distractor of a strength comparable to that of the initial cue is able to disrupt the working memory (**Figure 5.32a**). Increasing the relative inhibition from the SST/CB population increases dramatically the threshold above which a distractor can disrupt the memory of the initial cue, and therefore the network's ability to filter out distractors is greatly enhanced (**Figure 5.32b**). SST/CB cells enhance resistance against distractors, because their inputs target the same dendritic sites as the external stimuli (such as a distractor). Furthermore, away from the bump's peak activity, SST/CB interneurons show higher firing rates during the delay period than they do in the spontaneous state (**Figure 5.31**), providing stronger dendritic inhibition to those pyramidal cells not storing the initial cue. Therefore, a working memory network endowed with synaptic inhibition is more robust against distractors when compared with a conventional two-population model.

To experimentally test the theoretically proposed disinhibitory motif by Wang et al. (2004), it is necessary to record selectively each of the three subclasses of inhibitory neurons in physiological experiments. This has become possible with the methodological advances of genetic cell labeling and optogenetic manipulation (Zemelman et al. 2002; Boyden et al. 2005), mostly for mouse up to the present day. Numerous studies have lent supporting evidence across multiple cortical areas, including primary auditory cortex (Letzkus et al. 2011), somatosensory cortex (Lee et al. 2013), visual cortex (Fu et al. 2014; Zhang et al. 2014) and frontal cortex (Pi et al. 2013) (reviewed in Kepecs and Fishell 2014, Tremblay et al. 2016 and Cardin 2018), establishing that this disinhibitory motif represents a general characteristic of the neocortical organization.

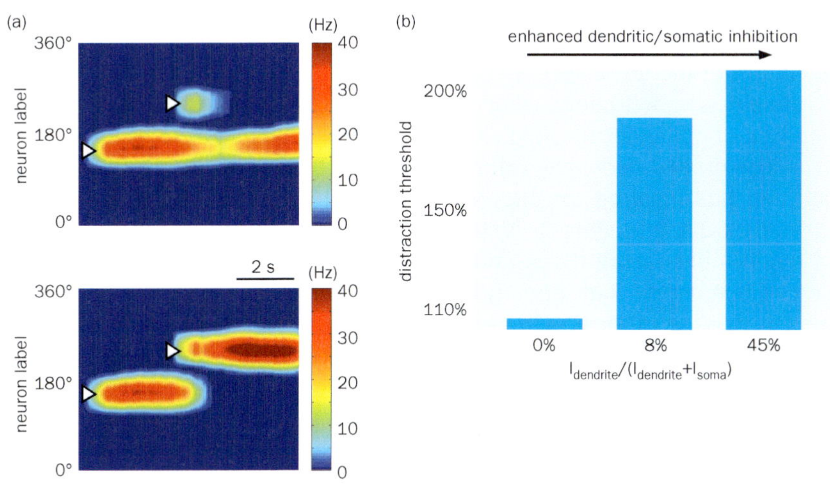

Figure 5.32 Robustness against distracting stimuli is enhanced by a higher ratio of the dendritic/somatic inhibition. (a) Spatiotemporal activity pattern in a simulated delayed response task. A transient cue elicits a spatially localized persistent activity pattern, which is resistant to a weak distractor (upper panel) but not when the distractor is stronger than a distraction threshold (lower panel). (b) Increased resistance against distractors with a larger dendritic/somatic inhibition ratio. Reproduced from Wang et al. (2004). [Wang, X. J., Tegnér, J., Constantinidis, C., & Goldman-Rakic, P. S. (2004). Division of labor among distinct subtypes of inhibitory neurons in a cortical microcircuit of working memory. Proceedings of the National Academy of Sciences, 101(5), 1368–1373. Copyright (2004) National Academy of Sciences, U.S.A.]

5.7 LIMITED WORKING MEMORY CAPACITY

A key characteristic of working memory is its limited capacity: the number of items of information we can actively hold in our mind concurrently is small, estimates range between four and seven items in humans (Miller 1956; Cowan 2001). Working memory is facilitated when items have some relational structure (Bor et al. 2003; Mathy and Feldman 2012; Amalric et al. 2017). In a simple and intuitive description of the limited capacity, the working memory system has a fixed number of discrete slots (the "discrete-slot" model) (Cowan 2001). For instance, in a visual change detection task where either there is no change or one item is changed (**Figure 5.1a**), the discrete-slot model posits that for a set size N, at most K_{max} items can be stored in working memory. Any of the remaining $(N - K_{max})$ items is reported as a result of random guess. Thus, when $N > K_{max}$, if the changed item happens to be stored in working memory (with probability K_{max} / N), the subject asserts change detection. Otherwise, if the changed item is not remembered (with probability $(N - K_{max}) / N$), the subject still says yes with a rate ρ by random guess that there was a change. The total hit rate for a correct detection is (Pashler 1988):

$$Hit\,rate = K_{max} / N + (1 - K_{max} / N)\rho \tag{5.11}$$

By fitting this simple formula with accuracy data as a function of set size N, the capacity limit K_{max} can be extracted (**Figure 5.1a**, blue in the right panel).

The discrete-slot model has been challenged using a different task paradigm. In a delayed estimation task (Zhang and Luck 2008), a number of colored squares were flashed on the screen, followed by a brief delay. Then, one of the items was cued, and the subject had to report the color of the cued square by clicking on a color wheel (as an analog quantity on a continuous ring of colors) (**Figure 5.33a**). Note that in contrast to a change detection task (Luck and Vogel 1997) where the response is binary (the presence or absence of a change) and performance is described by accuracy (**Figure 5.1a**), here the response is an estimation of a remembered continuous quantity. This estimate varies from trial to trial, and its deviation from the actual color quantifies the precision of memory coding.

With a given set size, the across-trial distribution of estimated color relative to the true color of a probed item can be fit with a combination of a von Mises function (for one of a few remembered items) and a uniform distribution corresponding to random guesses for the remaining items (**Figure 5.33b**). Note, however, the presumably remembered item is recalled with a variability quantified by the standard deviation of the von Mises function, which is larger with more items ($N = 6$ versus 3). Similar reduced precision has been reported in a separate study, defined by the standard deviation of the fitted von Mises distribution, as a function of the set size (**Figure 5.33c**, left panel) (Bays et al. 2009). Raw standard deviation independent of model fitting yields a similar trend (**Figure 5.33c**, right panel).

The graded decrease, rather than a sharp drop, of precision with the set size has inspired an alternative "shared-resource" model in which, instead of discrete slots, a finite resource is distributed among all items in the stimulus array (Bays and Husain 2008; Ma et al. 2014). Model fitting of behavioral data does not easily yield conclusions in favor of one model over the other (Cowan and Rouder 2009; Van den Berg et al. 2012; Bays 2018). Whereas the discrete-slot model is intuitively appealing, the shared resource model seems more plausible to describe graded change of memory precision (Ma et al. 2014; Koyluoglu et al. 2017). However, complex versions of the shared resource model are needed to account for a human's apparent random guessing. Van den Berg et al. (2012) assumed

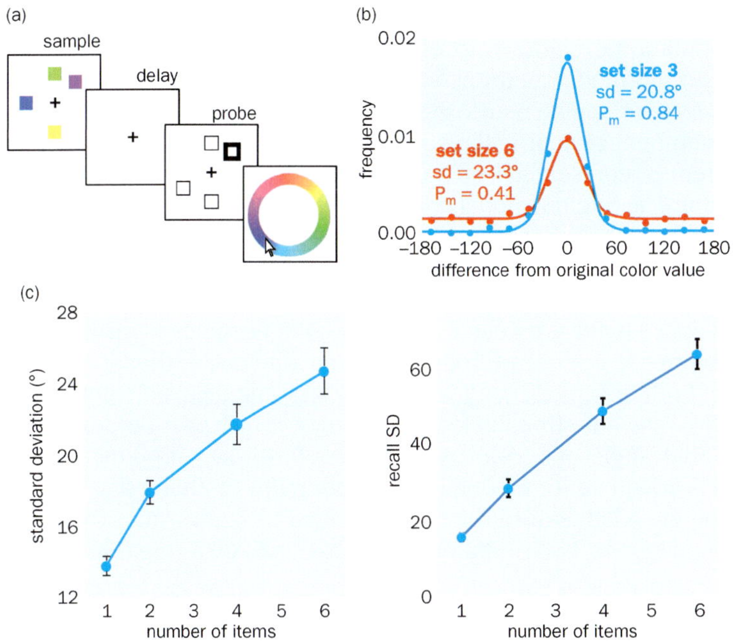

Figure 5.33 Behavioral performance in a delayed color estimation task. (a) Task paradigm. Observers must report the color in memory of the item cued at a probed location (thick box) by selecting a position on a color wheel (cursor). (b) The distribution of estimated color at the end of a delay for three items (blue) and six items (red). Data are fit with a mixture (a sum) of a von Mises function and a uniform function. A nonzero flat line is notable for six items but not three items, and the standard deviation of the von Mises function is larger for six items compared to three items. (c) Similar trend observed in another study using the same task. Left panel: Standard deviation of the von Mises component of the response distribution; Right panel: Full standard deviation of the raw data. Panel (a) is reproduced from Ma et al. (2014), (b) from Zhang and Luck (2008), (c) from Bays et al. (2009) (left) and Ma et al. (2014) (right). [Ma, W. J., Husain, M., & Bays, P. M. (2014). Changing concepts of working memory. Nature neuroscience, 17(3), 347–356.] [Zhang, W., & Luck, S. J. (2008). Discrete fixed-resolution representations in visual working memory. Nature, 453(7192), 233–235.] [Bays, P. M., Catalao, R. F., & Husain, M. (2009). The precision of visual working memory is set by allocation of a shared resource. Journal of Vision, 9(10), 7–7.] [Ma, W. J., Husain, M., & Bays, P. M. (2014). Changing concepts of working memory. Nature Neuroscience, 17(3), 347–356.]

an infinite mixture of von Mises functions, each with a different width σ and a probability $p(\sigma)$. Therefore, the probabilistic estimation is given by $p(\theta)=\int p(\sigma)\text{vonMises}(\sigma,\theta)d\sigma$. This model fits well with the trend of decreased performance with set size, but it does not reproduce big jumps in accuracy in change detection (Van den Berg et al. 2012). It is also unclear what the probability $p(\sigma)$ corresponds to biologically in the brain.

One class of computational models for working memory capacity rely on network oscillations. The idea is that, at any moment, winner-take-all competition ensures that only one of the stimulus-selective neural groups is active, but they take turns to fire in time so that the entire network may represent a small number of items serially. For instance, gamma (~ 40 Hz) oscillations ride on top of the theta rhythm at ~ 8 Hz (Stumpf 1965; Soltesz and Deschênes 1993; Bragin et al. 1995; Buzsáki et al. 2003; Jensen and Colgin 2007; Wang 2010), and a theta cycle can be fit with about 4 gamma cycles reliably (each lasting about 30 ms, allowing for jitters of spike times) (Lisman and Idiart 1995; Fiebig et al. 2020). If selective neural groups fire in different gamma cycles within a theta cycle, only about four can be represented in the network. Repetition of brief bursts controlled by short-term facilitation and depression offers another example of the same idea but with a periodicity of about one second, so each item's presentation can be separated by longer time intervals (Mi et al. 2017). This "time-slot" model is interesting because it takes advantage of collective neural dynamics, and it actually is consistent with both the "discrete-slot" model and the "shared-resource" model, with the shared resource being the finite time period of a coherent oscillation. However, this model is challenged by the observations that in recorded neurons of behaving animals, delay period spiking is not dominated by brief bursts of spikes interspersed with silent epochs (Compte et al. 2003; Joelving et al. 2007; Tang et al. 2019; Li et al. 2021a), and that during a population rhythm, the majority of spikes of cortical neurons do not have phasic information (Chapter 3, Section 3.4.2).

In alternative models, several neural pools selective for different items can be active concurrently (Edin et al. 2009; Wei et al. 2012). This scenario is supported by human studies showing that fMRI BOLD signals in certain cortical areas increase with the set size in a working memory task (Todd and Marois 2004; Xu and Chun 2006). The ring model of spiking neurons for working memory (Compte et al. 2000) (**Figure 5.18**) offers a computational platform to elucidate the two seemingly contradictory views. In the model, neurons are stimulus-selective, yet items are represented by overlapping neural groups. On the one hand, the ring represents a finite amount of neuronal resource; on the other hand, working memory of an item is represented by a bump attractor, which has the flavor of a discrete slot. The original model operates in a strict winner-take-all regime, in which only a single bump can be sustained, which explains why an initial cue or a later distractor can be remembered, but not both (**Figure 5.21**). However, when the excitatory-to-excitatory connection is sufficiently narrow in space (with a smaller footprint σ_{EE}), it becomes possible for two or more bumps to coexist without interfering with each other.

In both the color change detection task and delayed color estimation task, the relevant stimulus feature is color, which can be described as a directional angle of a color wheel. Therefore, the one-dimensional ring model is adequate for modeling the experiment of Zhang and Luck (2008) with the encoded feature now reinterpreted as color. An array of items is presented by one, two or multiple peaked inputs on a ring (**Figure 5.34a**),

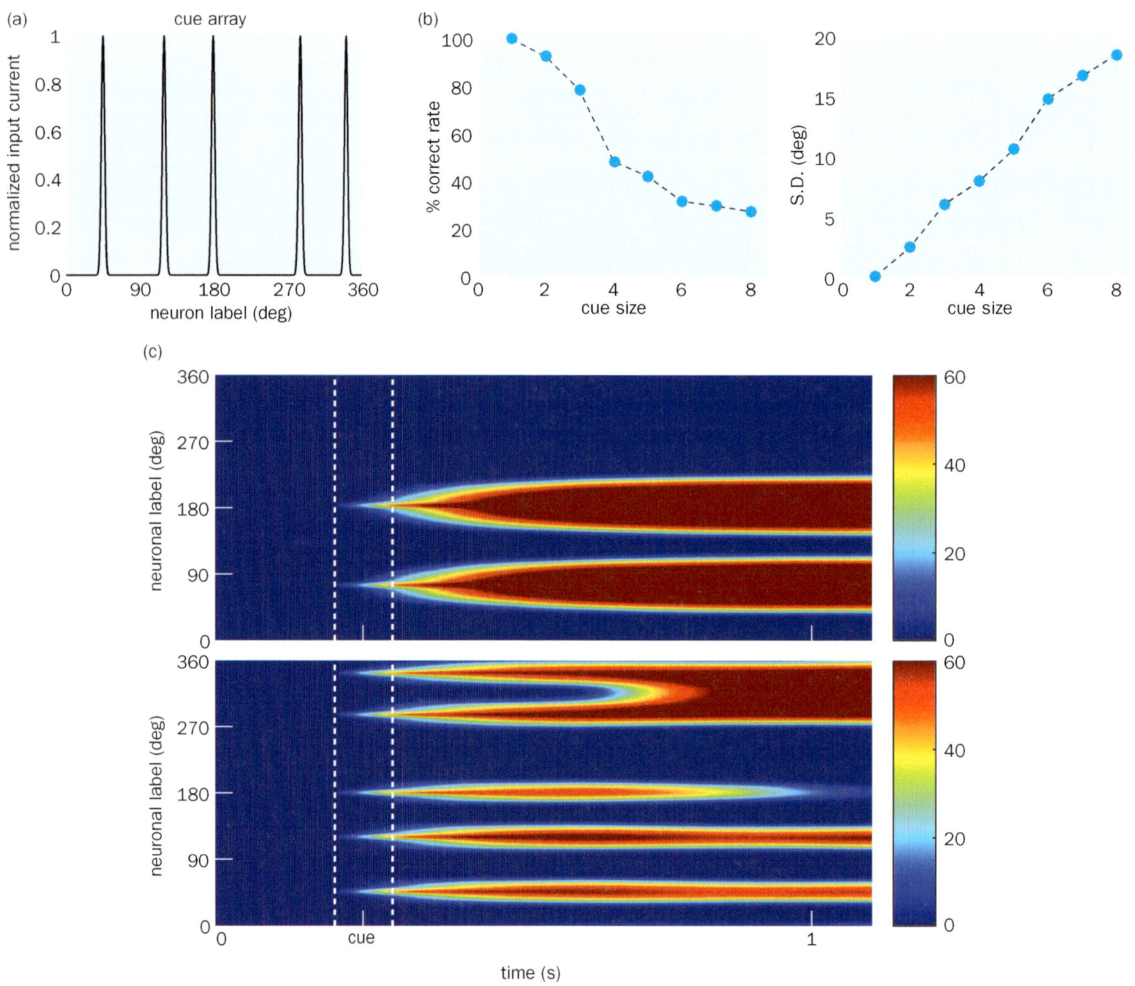

Figure 5.34 A biologically based ring of spiking neurons reconciles the discrete-slot and the shared-resource views. (a) A sample cue array of five randomly distributed angles. (b) Accuracy decreases steeply when the set size is increased above three to four (left), whereas the standard deviation of estimation increases (right) with the set size in a graded manner. (c) Network activity pattern is shown (x-axis: Time; y-axis: Neurons labeled by their preferred angle), firing rate is indicated by color. Two cues produce a stable pattern with two bumps of equal width (top panel). By contrast, with five cues, two initially distinct bumps merge, and a third disappears. Reproduced from Wei et al. (2012). [Courtesy of Ziqiang Wei.]

triggering a persistent spatiotemporal pattern. At the end of a delay period, color for each item is estimated using a simple readout method.

The model of Wei et al. (2012) reproduces both stepwise decrease of accuracy with increasing set size (**Figure 5.34b**, left panel) in favor of the discrete-slot model, and linear graded increase of standard deviation of memory-guided estimation (**Figure 5.34b**, right panel) at the core of the shared-resource model. To understand the inner working of this model, **Figure 5.34c** shows the network's spatiotemporal pattern in two sample trials. Two cues trigger a persistent activity pattern with two distinct bumps (top panel). By contrast, with five cues, bumps become sufficiently close that they interact dynamically with each other (bottom panel). Two spatially close bumps may excite each other, leading to their merging into a single wider bump (leading to less accurate readout). Or, the recruited overall inhibition is sufficiently large to stochastically destroy a bump, resulting in its disappearance so that the item's memory is lost. In this model, items are encoded by distinct bumps. At the same time, items are distributed among a limited neuronal ring. Merging (due to recurrent excitation between nearby bumps) and fade-out (due to shared feedback inhibition) are two dynamical phenomena underlying limited capacity. Therefore, this model shows that both the discrete-slot and shared-resource views can be realized by a single neural circuit. This ring model can be readily generalized to the actual experimental paradigm (**Figure 5.33**), by the addition of a two-dimensional space where colored stimuli are presented at different locations.

While the model sheds insights into how the discrete-slot and shared-resource views may be reconciled, it only considers a local circuit. By contrast, limited working memory capacity likely results from interacting PPC and PFC, as well as other cortical areas (Pessoa et al. 2002; Vogel and Machizawa 2004; Constantinidis and Klingberg 2016; Bouchacourt and Buschman 2019). We will discuss working memory distributed across multiple interacting regions in Chapter 9.

5.8 DYNAMICAL NATURE OF MNEMONIC REPRESENTATION

5.8.1 Dynamical Coding and Heterogenous Delay Activity

Elevated delay period activity compatible with working memory representation that is stable over time has been widely reported in the literature of macaque monkey neurophysiology (see above, Constantinidis et al. (2018) and Wang (2021) for reviews). In a recent experiment, cells in the PFC were wirelessly recorded from freely moving marmosets when they performed a delayed-match-to-location task, the findings support a stable working memory representation common to macaque and marmoset monkeys (Wong et al. 2022).

At the same time, increasing attention has been drawn to the observation that, in a working memory task, many single neurons in a cortical area tend to display temporal variations during the delay period (Batuev et al. 1979; Baeg et al. 2003; Bolkan et al. 2017; Lundqvist et al. 2018; Zhu et al. 2020). A relatively small number (5–10%) of recorded neurons show tonic persistent activity, whereas others display time-varying patterns: some ramp up while others ramp down their firing rates in time during the delay (Fuster 1989). Note that 5–10% is not negligible, since even a small (1 mm by 1 mm) piece of neocortex has about half a million of neurons. A brain region is engaged in many tasks, thus the number of cells activated in a single task could be a small but significant fraction of the entire population. A more puzzling finding is that stimulus selectivity may be detectable only in a portion of the delay period, and each cell shows statistically significant selectivity at different times (Zaksas and Pasternak 2006). **Figure 5.35a** shows a quantification of time-varying stimulus selectivity in an experiment with monkeys performing a delayed match-to-sample task where the to-be-remembered feature is the direction of a visual moving pattern. Neurons in a part of the PFC that receive direct inputs from MT (the visual area dedicated to process visual motion) were recorded; each neuron has a preferred direction θ_{pref} for which its delay period activity is the highest among tested motion directions, and its diametrically opposite direction is the null- or nonpreferred-direction $\theta_{\text{nonpreferred}}$. At any time point, reliable difference in an individual cell's firing rate between θ_{pref} and $\theta_{\text{nonpreferred}}$ is shown. Recorded neurons collectively portray a seemingly irregular pattern of time-varying coding of the sample stimulus.

Instead of looking at single cells, one at a time, coding/decoding should be considered collectively at the level of a neural population. One method to quantify whether working memory representation is stationary or time-varying is to train a classifier to decode the stored information from recorded neurons (Meyers et al. 2008; Crowe et al. 2010). Note that until recently typically individual neurons have been recorded one at a time, so pooling them together ignores their interactions in a single trial. This shortcoming notwithstanding, a classifier determines weights w_i for recorded neurons with firing rates $r_i(t)$ $(i = 1, 2, \ldots, N)$, to best decode a stimulus from $\sum_i w_i r_i(t)$. It is a measure of the information available in the neural signals, but does not speak to whether or how that information may be read out by a downstream neuronal circuit in the brain. For instance, in an experiment (Sigala et al. 2008), monkeys learned association between pairs of visual images: cues A, B and C were paired with targets D, E, F, respectively. Another image (G) was neutral and not paired with any image. In a trial, a cue (say, B) was shown, followed by a sequence of stimuli (for example, F, G, E), and the subject needed to produce a motor response when the target pair-associated with the cue (namely E) appeared. Note that a stimulus (e.g., F) could be the correct target in a different trial (if the initial cue was C). Serially presented stimuli were separated by a delay period of 1 second. **Figure 5.35b** shows the "cross-temporal classification matrix" (Stokes et al. 2013). A classifier was trained by optimizing the weights at time t, which was then used to decode the remembered stimulus at time t'. In this case, during the cue presentation, reliable decoding (red to orange color) is confined near the diagonal line, which means that the classifier trained at a particular time cannot decode the trial type at a different time. On the other hand, during the delay period following the initial cue, good decoding fills a square, demonstrating that working memory representation is quite stable over time.

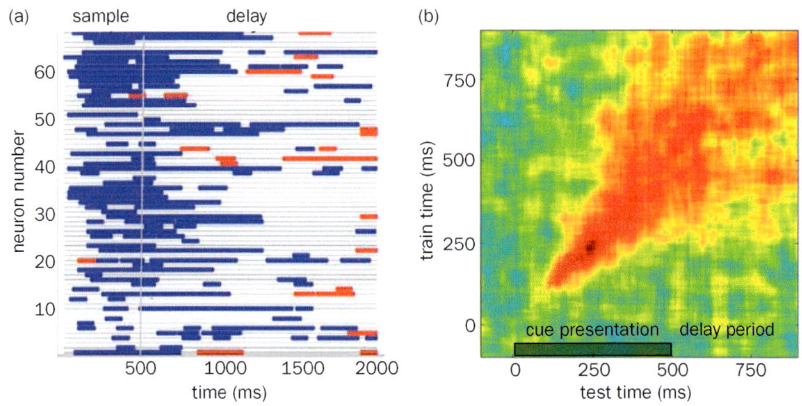

Figure 5.35 Analysis of information coding by delay period activity. (a) Time course of directional signals in 67 individual PFC neurons recorded from monkeys performing a delayed match-to-sample task with visual motion stimuli. Each horizontal line corresponds to a single neuron through the sample and delay periods. Blue: Activity is higher for the preferred direction than null direction; red: The opposite. In the last third of the delay, directional activity is rare regardless of sign. (b) Cross-time classification matrix of recorded neurons for a remembered item. Classifiers are trained to discriminate trial type at time points indicated by the y-axis, and tested at times along the x-axis, during the cue and the delay period. Panel (a) is reproduced from Zaksas and Pasternak (2006), (b) from Stokes et al. (2013). [Zaksas, D., & Pasternak, T. (2006). Directional signals in the prefrontal cortex and in area MT during a working memory for visual motion task. Journal of Neuroscience, 26(45), 11726–11742.] [Stokes, M. G., M. Kusunoki, N. Sigala, H. Nili, D. Gaffan, and J. Duncan (2013). Dynamic coding for cognitive control in prefrontal cortex. Neuron 78(2), 364–375.]

Studies using cross-temporal classification analysis yielded various cross-temporal classification matrices (King and Dehaene 2014; Meyers 2018; Kaminski and Rutishauser 2019). In general, working memory representation is stable over time in tasks that mostly involve memory maintenance, but time-varying when information processing and manipulation are required during the delay period (Meyers et al. 2008; Crowe et al. 2010; Meyers et al. 2012). Sometimes a code is stable in a time window and then evolves into time-varying in another time window. This is the case in a monkey experiment where a brief spatial cue was followed by a delay period; then a second cue indicated the amount of reward associated with that position, followed by another delay. The cross-temporal classification analysis yielded a mixture of stable code in the first delay period and dynamical code in the second delay period (Cavanagh et al. 2018).

Theoretical considerations suggest that temporal variations of neural activity can be compatible with a stable working memory representation during a delay period (Druckmann and Chklovskii 2012). To test this possibility, Murray et al. (2017) performed a "demixed" principal component analysis (PCA) (Machens et al. 2010; Kobak et al. 2016; Keemink and Machens 2019) of PFC neural trajectories using data from the ODR (Funahashi et al. 1989) and the VDD (Romo et al. 1999) monkey experiments. In the analysis, population state corresponds to a point in the N-dimensional space in which each axis corresponds to an individual neuron's firing rate $r_i(t)$ $(i = 1,2,\dots,N)$. Its temporal evolution traces a trajectory in the activity state space. It was found that single neurons display various temporal patterns in their delay period activities (**Figure 5.36a**). However, population coding of a stimulus stored in working memory is stable within a subspace where working memory coding is stationary, in spite of considerable temporal changes in the orthogonal subspace (**Figure 5.36b**). This finding demonstrates that temporal variations of delay period neural activities are not necessarily incompatible with a stable working memory representation over time.

Schematically, one can contrast stable versus dynamical coding of working memory in terms of whether the stored information remains in a fixed neural assembly (**Figure 5.37**). In some cases, firing activity in a delay period clearly moves among different neural groups. In an experiment where mice learned to associate a visual stimulus (A or B) with a delayed response (turn left or right), delay period activity of neurons (monitored by calcium imaging) in the posterior parietal cortex was transient rather than tonic, and each cell's activity peaked at a different time. Sorted by their peaked times, recorded neurons exhibited a pattern of sequential activation (**Figure 5.38a**) (Harvey et al. 2012). This and another experiment (Schmitt et al. 2017) suggest sequential activation of neural groups, each firing briefly at different times during a delay (**Figure 5.38b**) (Goldman 2009; Rajan et al. 2016). Clearly, such a delay period activity pattern is incompatible with a stationary code. However, since PPC was involved in spatial information representation and the animals underwent virtual navigation in Harvey et al. (2012), the observed sequential activation may be related to virtual movement in space (Krumin et al. 2018). In the mouse experiment of **Figure 5.15**, peak times of spiking activity of recorded neurons are inconsistent between two randomly divided subsets of trials, hence not in support of a sequential activation underlying delay period information coding in that experiment (Li et al. 2016).

If working memory is indeed represented by a sequence of transiently active neurons, the stored information must be read out from different neural groups at different times. In that case, a downstream neuron would need to constantly change its input weights over time for decoding, unless it receives converging inputs from all mnemonic cells (**Figure 5.38b**). In that case, the downstream neuron would display stationary persistent

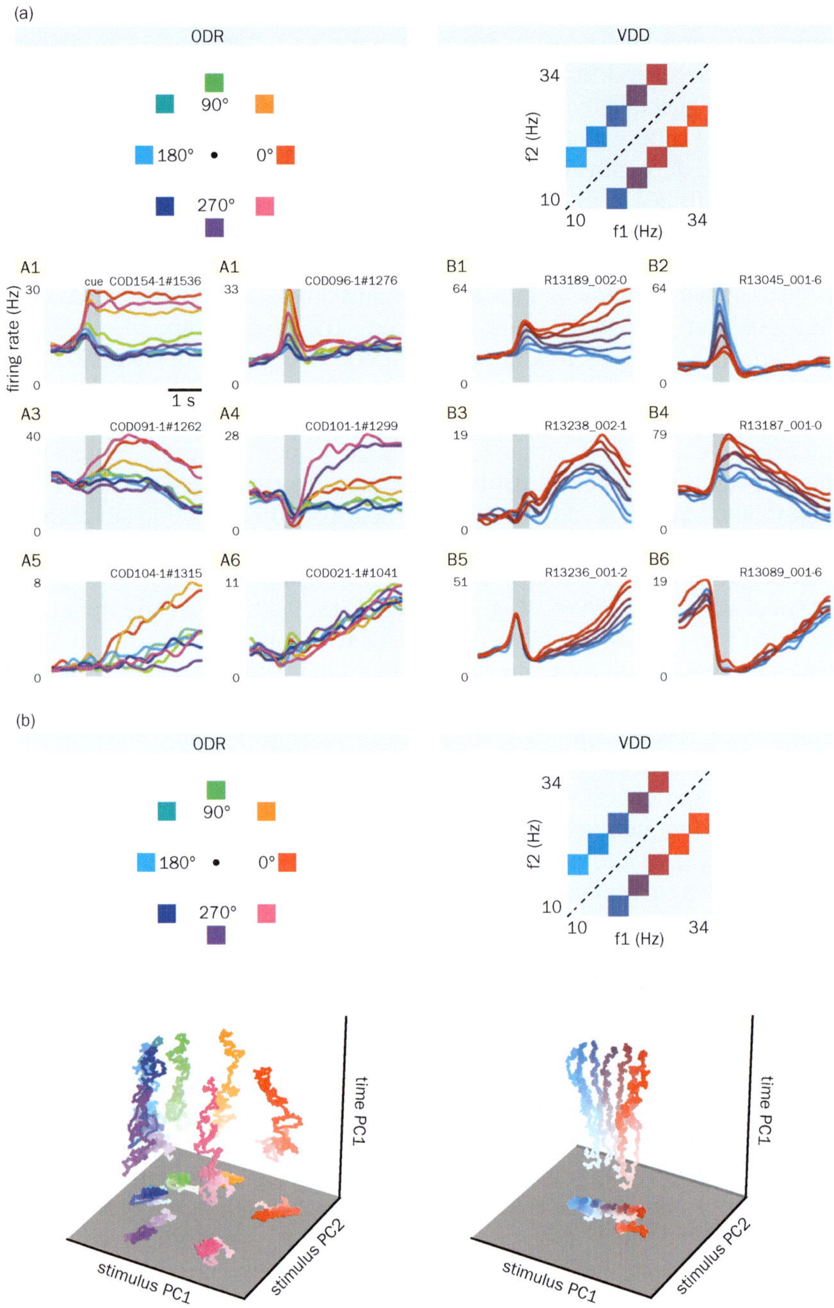

Figure 5.36 Coexistence of stable working memory coding and temporal dynamics of delay period neural population activity. (a) Six individual neurons are shown for each of two monkey experiments using ODR (left) and VDD (right) tasks. Different colors correspond to different stimuli. Grey vertical line: Sensory stimulus presentation. In both cases, there are considerable temporal changes of delay period activity and cell-to-cell heterogeneities. (b) PFC analysis of population activity reveals that coding is stable (traces for different colors are distinct) in a subspace of the population activity state space (PC1 and PC2), whereas temporal changes are confined in the orthogonal subspace (Time PC 1). Reproduced from Murray et al. (2017). [Murray, J. D., Bernacchia, A., Roy, N. A., Constantinidis, C., Romo, R., & Wang, X. J. (2017). Stable population coding for working memory coexists with heterogeneous neural dynamics in prefrontal cortex. Proceedings of the National Academy of Sciences, 114(2), 394–399.]

activity (Goldman 2009), and the computational benefit of such a scheme in comparison with a stationary code in the first place remains to be better understood.

Certain types of temporal variations of delay activity are suitable to serve specific functions. For instance, a demixed version of PCA analysis of PFC neural population data in the monkey VDD task revealed that the largest component is a quasi-linear ramping function of time (Machens et al. 2010). This is in line with an earlier study (Brody et al. 2003) showing that in the VDD experiment, ramping activity had a slope (speed of change over time) that was halved when the delay duration was doubled (6 seconds versus 3 seconds), suggesting that ramping activity reflects anticipated timing of the memory-guided response. Corroborative evidence was also reported in the mouse delayed response task (Li et al. 2016), in which ALM neurons showed ramping activity when the delay duration was fixed,

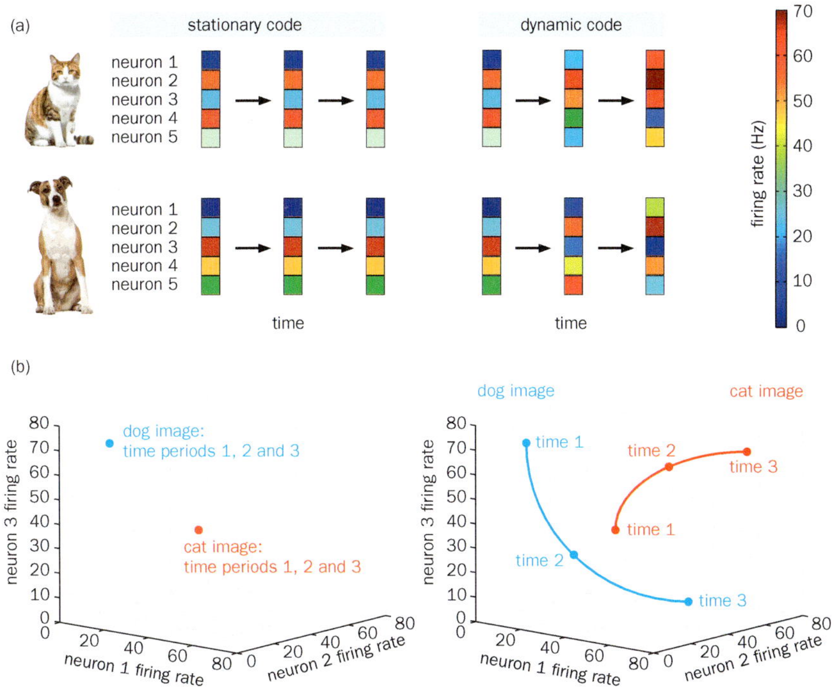

Figure 5.37 Illustrations of stationary and dynamic codes. (a) A population vector of five neurons responds to an image of a cat (top) or a dog (bottom), and the color of the square indicates the firing rate of each neuron. For a stationary code (left), the pattern of neural activity of the five neurons is the same across time, and these patterns differ depending on whether an image of a cat or a dog was shown. For a dynamic code (right), a different pattern of neural activity is present at each point in time, and the sequence of patterns also differs when a cat or a dog was shown. (b) Geometric illustration of how a population of three neurons responds to the images of a cat (red) and a dog (blue). Each axis shows the firing rates of a neuron, and thus the population response is a point. For a stationary code (left), the response to each image is a single point that does not change with time. For a dynamic code (right), the population response creates a trajectory in the state space. Because these trajectories do not cross, it is possible to distinguish between cat and dog images at all points in time. Reproduced from Meyers (2018). [Meyers, E. M. (2018). Dynamic population coding and its relationship to working memory. Journal of Neurophysiology, 120(5), 2260–2268.]

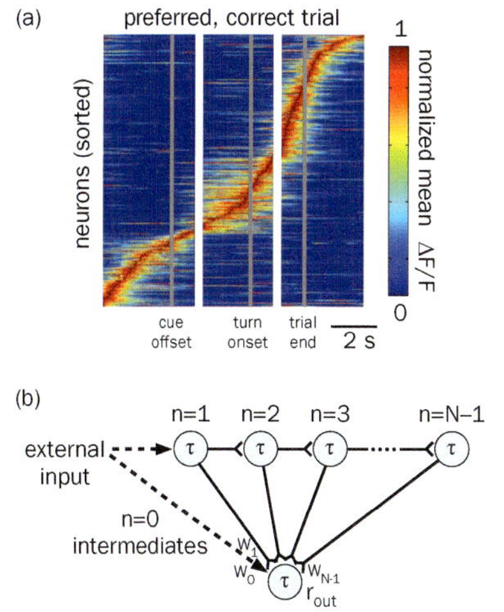

Figure 5.38 Sequential activation of neurons in a delayed response task. (a) In a delayed response task, calcium imagining was used to monitor activity of 51 choice-specific, left-preferring cells (one cell per row) in the posterior parietal cortex of a behaving mouse. Traces were normalized to the peak of each cell's mean ΔF/F trace on preferred trials and sorted by the peak time. (b) Schematics of a model where activity propagates along a chain of neurons. A downstream neuron receives converging inputs from all sequentially activated neurons, and therefore can read out the stored information at any time. Panel (a) is reproduced from Harvey et al. (2012), (b) from Goldman (2009). [Harvey, C. D., Coen, P., & Tank, D. W. (2012). Choice-specific sequences in parietal cortex during a virtual-navigation decision task. Nature, 484(7392), 62–68.] [Goldman, M. S. (2009). Memory without feedback in a neural network. Neuron, 61(4), 621–634.]

but tonic persistent activity in trials where the delay duration varied probabilistically and therefore was not predictable (Inagaki et al. 2019). Functionally, if there is a need for tonic firing neurons, why not to have them in the first place? Other temporal changes require different explanations, some of which may be related to uncontrolled factors in an experiment. As a matter of fact, in a recent study, mouse subjects were monitored by video camera, and it was found that neural signals measured by wide-field calcium imaging and neurophysiology were predominantly accounted for by movements (e.g., those of whiskers and hindlimbs) not required for behavior (when the response was to lick one of two water spouts) (Musall et al. 2019).

Separated from temporal variations of a cell's firing, delay period activity also varies considerably from cell to cell (e.g., Figure 5.36a). By design, models strive for simplicity to optimize interpretations of network behaviors observed in model simulations, and they usually do not account for the experimentally observed heterogeneities, with a few exceptions (Renart et al. 2003; Barak et al. 2013; Chaisangmongkon et al. 2017; Masse et al. 2019). Heterogeneity could arise because recording is done among several subtypes of neurons; recall (Chapter 3, Section 3.7.3) that when a subtype is identified by genetic tools, individual cells within it may be actually homogeneous (Hirokawa et al. 2019). On the other hand, heterogeneity across neurons may be essential for serving flexible cognitive functions (Rigotti et al. 2010; Rigotti et al. 2013) (see Chapter 8).

5.8.2 Self-Sustained or Decaying Transient?

Attractor network models for mnemonic activity have been mostly taken as equivalent to the coexistence of multiple stable steady-states (Kaminski and Rutishauser 2019; Masse et al. 2020). If so, the aforementioned time variations of delay period activity would appear to contradict the attractor theoretical framework. However, this is a misconception, which to some degree results from the mere fact that mathematical models are the easiest to describe and analyze if attractors are steady-states or fixed points. For instance, the Hopfield model can be solved thanks to the simplifying assumptions, under which attractors are fixed points of the recurrent network (Hopfield 1982; Amit et al. 1985). Deduction from a biologically based spiking neural network model to a population rate model of working memory is feasible through the mean-field method for fixed points (Amit and Brunel 1997; Wang 1999b; Brunel and Wang 2001; Renart et al. 2003; Wong and Wang 2006). Of course, attractor states do not have to be steady-states, as exemplified by stochastic oscillations of delay period activity (Section 5.4.2, **Figures 5.19** and **5.25**) (Compte et al. 2000; Renart et al. 2003; Mi et al. 2017; Fiebig et al. 2020). Chaotic attractors have also been proposed for persistent activity (Barbieri and Brunel 2008; Barak et al. 2013). In principle, an attractor of a dynamical system may display complex spatiotemporal patterns.

The real question is whether information-specific memory states correspond to distinct attractor states. If so, it can be properly described mathematically as a dynamical system with multi-stable states, no matter how complex their temporal variations may be. Alternatively, a network may have only a single attractor (the resting state) and delay period activity is genuinely transient: a to-be-remembered brief stimulus perturbs the system to another internal state, from which it returns back to the resting state after the input offset during a delay period. The return trajectory may be slow, governed by some biophysical processes, but eventually elevated activity should disappear if the delay period is sufficiently long.

The "activity-silent state" model posits that a memorandum can be encoded by "hidden" variables rather than at the level of neuronal spiking activity (Stokes 2015). A plausible biological substrate is synaptic short-term facilitation (STF). Currently, there is a debate about the "persistent activity" model versus "activity-silent state" model of delay period activity (Constantinidis et al. 2018; Lundqvist et al. 2018; Wang 2021). The question at the center of this debate can be sharpened by more precise definitions and appropriate word choices. Does a stochastic oscillation in which single cells stop firing for tens (**Figure 5.19**) or hundreds of milliseconds (**Figure 5.28b**) still qualify as "persistent"? Perhaps, a better descriptor is "mnemonic internal state"? Regardless, these are still internally self-sustained dynamical states mathematically described as attractors. Confusion can be avoided if we distinguish internal states (self-sustained attractor versus decaying transient) from their underlying circuit mechanisms. Thus, the attractor network model does not imply that STF is absent; on the contrary, STF may well contribute to the maintenance of persistent activity, which can be steady-state (Hempel et al. (2000); Pereira and Wang (2015)) or repetition of brief population bursts (Mongillo et al. 2008; Mi et al. 2017), both are attractor states. This point is general and holds also for models that do not incorporate STF. As we have shown in Section 5.3 (**Figure 5.14a**), if the recurrent connection strengths are above a threshold level, memory states are self-sustained; otherwise, they would decay away over time. In the latter case, because the time constant of the NMDA current is much longer than that governing spiking activity itself, there is a time window when the firing rate is back to the spontaneous state while the "hidden" NMDA synaptic variable remains substantial. This represents another example of the activity-silent state scenario but in the absence of STF. Recent monkey revealed intermittent mnemonic population activity in single trials where elevated firing is interspersed with silent periods lasting ~100 ms (Panichello et al. 2024). Note that the NMDA receptor model (Wang 1999b) and the model endowed with STF (Hempel et al. 2000; Mongillo et al. 2008) are both "synaptic theories" of delay period activity sustained by circuit reverberation. The distinction between attractors of self-sustained memory states versus decaying transients is defined mathematically and separately from the underlying cellular and synaptic hardwares.

Mongillo et al. (2008) showed that, with a suitable choice of model parameters, burst of spikes no longer repetitively occurs over time, lest it is reactivated by an external stimulus. In that case, short-term synaptic plasticity (STF) could still maintain a memory trace even when self-sustained neural spiking dies out. In other words, the activity-silent state model assumes that a dynamical variable of STF, not observable by spiking activity, could mediate short-term memory. Results from neurophysiologically testing this idea are not clearcut, partly because interpretations are not straightforward for different kinds of measurements, ranging from single-neuron physiology, EEG/MEG to fMRI BOLD signal. For instance, in monkey experiments, LFP displays brief bursts of enhanced coherence at γ frequency band (~ 40 Hz), which was interpreted as inconsistent with the sustained activity model (Lundqvist et al. 2016). However, this is a relatively modest effect, gamma power is smaller by a fraction outside these bursts. Moreover, delay period activity by no means only takes place during these burst episodes (see also Miller et al. 1996; Rainer et al. 1998; Wallis et al. 2001; Bastos et al. 2018). These data can potentially be accounted for by the theory of sparsely synchronous oscillations, where episodic bursts of network coherence coexists with sustained firing of single cells (Chapter 3, Section 3.4.2 and **Figure 3.13**) and temporally enhance information conveyed by spikes (Lundqvist et al. 2016; Palmigiano et al. 2017).

It is worth noting that the prior discussion is limited to a local circuit model. The assumed absence of any input to it is unlikely to hold in real life, when there are always external stimulation from the environment and internal processes (e.g., intruding thoughts) in other brain regions that project to the local area under consideration. A mixture of activity-silent state and episodic spikes caused by inputs from the rest of the brain represents an interesting possibility that a local network model is not suitable to investigate. Multi-regional modeling (Chapter 9) in interplay with new experiments in the future will shed insights into such a scenario.

Nevertheless, the activity-silent scenario has a specific prediction. If a brief stimulus activates one of the neural assemblies in a network, thereby inducing STF at their inter-connections, then a later global signal (a "pinging" of the entire network) would "reawaken" selectively that particular neural assembly because its hidden state is differentially primed by STF (Mongillo et al. 2008). This prediction has been tested in human experiments. In one study, a subject was shown two sample stimuli (a face and a word), followed by a delay period when a post-cue instructed which of the two would be probed (say, word). Then a test (the same or a different word) was shown, and the participant responded match or nonmatch. The trial continued with a second delay when another post-cue instructed whether face or word would be probed next (which might be face or word), a final test stimulus was shown and the subject responded match or nonmatch (Rose et al. 2016). Multivoxel patterns from the BOLD signal or EEG signal were used to decode each of the items in the initial sample set. It was found that category (face or word) decoding decayed to baseline. However, each post-cue "reawakened" significant decoding of the corresponding stimulus category, in support of the idea that information remained in some hidden state not detectable. Moreover, transcranial magnetic stimulation (TMS) reactivated representation of the latest cued category, consistent with the

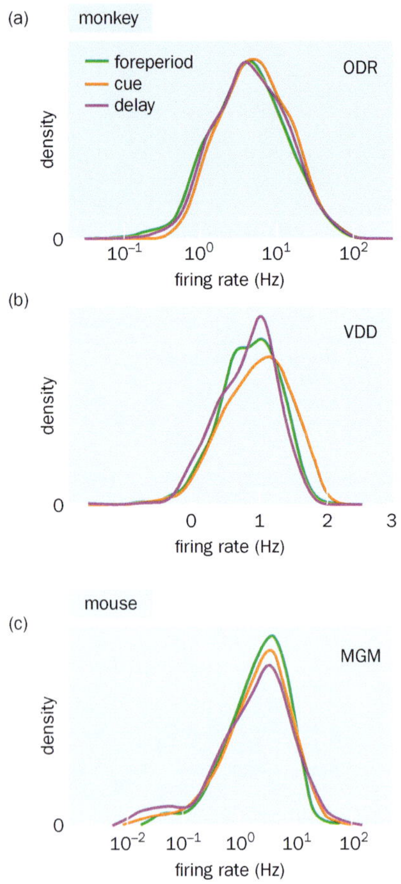

Figure 5.39 Distribution of mean firing rates across recorded neurons in three delayed response tasks. (a–b) Firing rate distributions of PFC neurons in behaving monkeys, plotted with logarithmic scale along the x-axis and linear scale along the y-axis for the oculomotor delayed response (ODR) (a) and vibrotactile delayed discrimination (VDD) (b) experiment, respectively. The observed distributions of firing rates are approximately log-normal. Interestingly, when compared across task epochs (fixation foreperiod, cue presentation, mnemonic delay), the overall distribution does not change substantially. In particular, the distribution during working memory is essentially the same as the foreperiod. (c) Firing rate distributions of ALM neurons from mice performing a memory-guided movement (MGM) task. Panels (a–b) are reproduced from Murray et al. (2017), (c) from data provided by Nuo Li (Li et al. 2016). [Murray, J. D., Bernacchia, A., Roy, N. A., Constantinidis, C., Romo, R., & Wang, X. J. (2017). Stable population coding for working memory coexists with heterogeneous neural dynamics in prefrontal cortex. Proceedings of the National Academy of Sciences, 114(2), 394–399.] [Courtesy of Nuo Li.]

model prediction about pinging a short-term memory system (Mongillo et al. 2008). Similar findings were reported in other experiments with pinging by nonspecific visual stimulation (Wolff et al. 2017) or transcranial magnetic stimulation (TMS) (Barbosa et al. 2020).

The two experiments (Rose et al. 2016; Wolff et al. 2017) were designed on the idea that a stored item can be "in" (if cued) or "out of" (when uncued) the focus of internal attention (Myers et al. 2017). The observations suggest that at a given time an item at the center of attention is represented by persistent activity, whereas information about another item encoded in a hidden variable is not lost but can be reactivated when it becomes a priority. But these studies did not distinguish active working memory from passive short-term memory traces; the latter could be realized by the activity-silent state scenario, possibly involving the hippocampus (Beukers et al. 2021). Recall that slow memory traces are difficult to be switched on and off by brief inputs (**Figure 5.28c**), a cardinal requirement for working memory function. Furthermore, modeling work showed that the activity-silent state mechanism has difficulty to filter out distractors that are presented after behaviorally relevant stimulation (Froudist-Walsh et al. 2021), because the synaptic memory trace triggered earlier in the past has decayed more than a recently shown stimulus, and thus is the strongest for the latest shown stimulus.

One argument for the activity-silent state model is that spikes are costly (Attwell and Laughlin 2001); therefore, realizing a memory trace without spike firing would save energy (Stokes 2015). If so, in the monkey ODR or VDD experiments where persistent activity is commonly observed in the PFC, neuronal spike firing rates during the delay period should be greater than during the baseline state of fixation ("foreperiod") at the start of a trial. This is not true; surprisingly, the distribution of firing rates across the recorded PFC neurons is roughly log-normal and the same across behavioral epochs for both experiments (**Figure 5.39a–b**). This is also the case for delay period activities in the mouse experiment of Li et al. (2016) (**Figure 5.39c**). What happens is that, in a given trial, neurons selective for an encoded stimulus have elevated spiking activity, whereas others reduce their firing related to the baseline. The totality of spikes is preserved. Therefore, the attractor network model for persistent activity cannot be discounted, and the activity-silent state model is not favored, on the ground of metabolic energy consumption in the brain.

A functional perspective on this topic distinguishes short-term memory (STM) only for internal maintenance of information from working memory that also involves manipulation of information (Cowan 2008; Oberauer 2009; Masse et al. 2020). Trübutschek et al. (2019) investigated this distinction in a human study where subjects were asked to hold in mind a visual stimulus at one of 24 spatial locations on a ring with the same eccentricity. The stimulus was so brief (17 ms) and faint that it was perceived in some trials but not in others. At the end of a trial, subjects rated the subjective visibility of the visual stimulus. Furthermore, during a delay period of 3 seconds, one of three cues indicated task type in the current trial, namely, whether the eventual response is pointing to the stimulus location, 120 degrees rotated clockwise or 120 degrees rotated counterclockwise from it. Interestingly, regardless of whether the stimulus was perceptually visible (with or without report), target decoding by MEG was significant after the trial type cue was shown, indicating that mental manipulation of information in working memory requires delay-period activity.

5.8.3 Persistent Activity Is Required for Manipulation of Information in Working Memory

How can one test computationally the hypothesis that maintenance and manipulation of information during a delay period have different demands and differentially engage persistent activity? In recent years, tools from machine learning have been used to train recurrent neural networks (RNNs) to perform tasks (Chapter 11) (Yang and Wang 2020). Masse et al. (2019) reasoned that an RNN is initially a "blank slate," where connection weights are not specific and the network is incapable of any function. If a to-be-learnt task involves a mnemonic delay, this approach does not make an *a priori* assumption as to whether an RNN will solve the problem by virtue of a persistent activity pattern or an activity-silent state. Therefore, it offers an opportunity to investigate which of the two scenarios emerges from task needs.

In the model (**Figure 5.40a**), an input layer signals spatial location and direction of motion stimuli, and an output layer generates a delayed

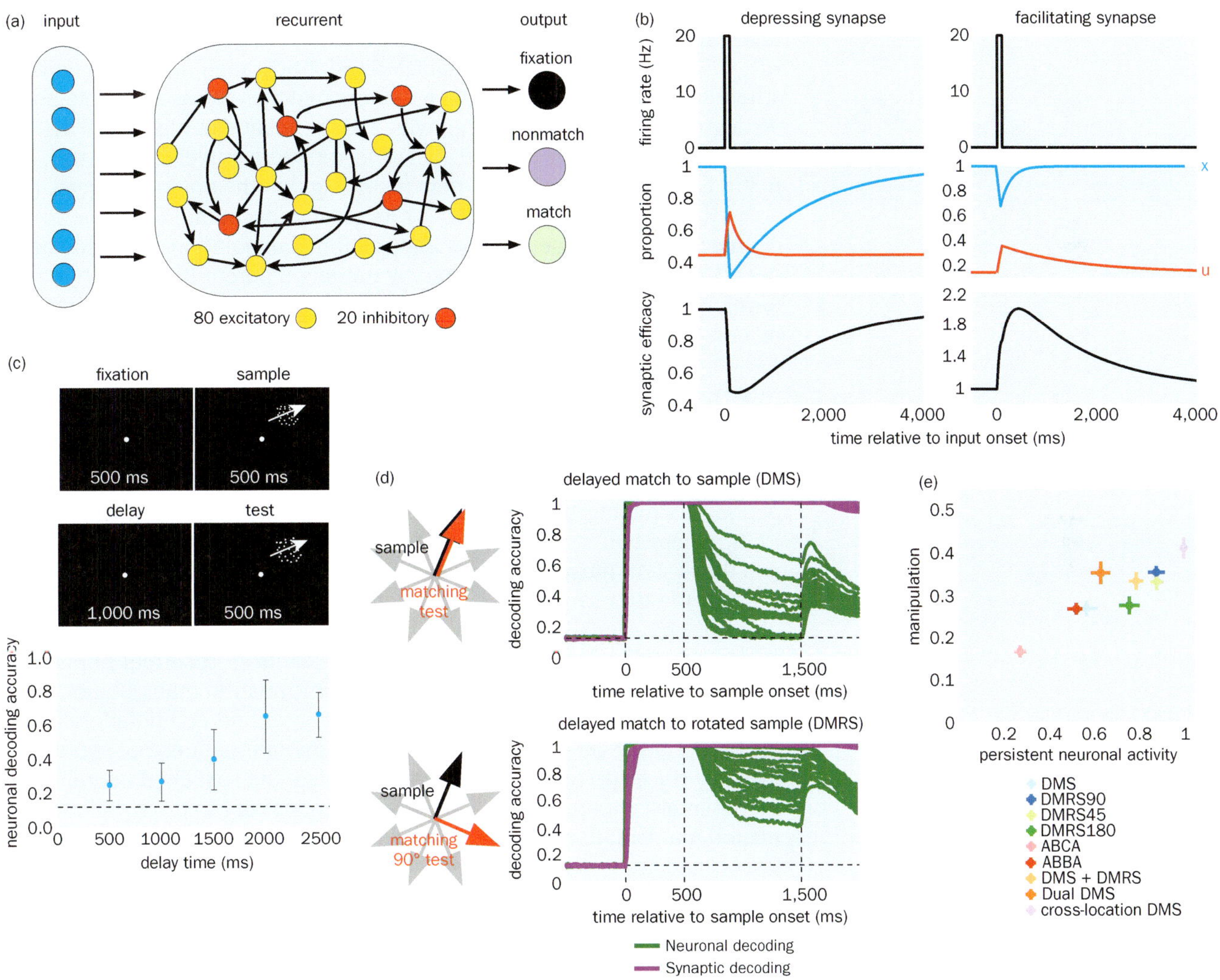

Figure 5.40 A recurrent neural network trained by machine learning to perform working memory tasks. (a) Model scheme. (b) Short-term facilitation (red) and short-term depression (blue) in response to a pulse input. (c) After the model is trained to perform a delayed match-to-sample (DMS) task (upper plot), decoding accuracy from the recurrent population activity is poor with short delay duration, but gradually increases when delay becomes longer than the biological time constants of STF (lower plot). (d) Top panel: Delay period activity (green) is weak in the DMS task, but sample decoding accuracy is high from synaptic efficacy (magenta). Bottom panel: In the delayed match-to-rotated-sample (DMRS) task, information manipulation is required and sample decoding is accurate from high delay period activity. (e) Scatterplot shows the level of persistent neuronal activity, measured as the neuronal decoding accuracy during the last 100 ms of the delay (x-axis), versus the level of manipulation (y-axis) across nine different tasks (indicated by colored crosses). Adapted from Masse et al. (2019). [Masse, N. Y., Yang, G. R., Song, H. F., Wang, X. J., & Freedman, D. J. (2019). Circuit mechanisms for the maintenance and manipulation of information in working memory. Nature Neuroscience, 22(7), 1159–1167.] [Courtesy of Nick Masse.]

response. The recurrent network between the input and output layers is wired with connections endowed with STF. Some are dominated by short-term depression (**Figure 5.40b**, left) while others by short-term facilitation (**Figure 5.40b**, right). The synaptic efficacy is the product of the depression factor and facilitation factor. After the model is successfully trained to perform a motion-direction delayed DMS task (**Figure 5.40c**), the sample is decoded either by recurrent neural population activity or by activity-silent synaptic efficacy. When the delay period is short, STF can keep a memory trace of the sample in which case activities are not necessary. Indeed, a trained RNN found the solution with chance-level performance of decoding by activity of recurrent units, but decoding by activity increases with gradually prolonged delay duration. This is because when the delay period is long compared to the biological time constants of STF, the network can no longer find a solution by an activity-silent state scenario, and persistent activity emerges from training through experience.

What happens if an RNN is trained to perform a working memory task where information must be manipulated during the delay period? **Figure 5.40d** shows a conventional DMS task (top panel) and a variant of it, where the response was whether the test matches the sample rotated by 90 degrees (bottom panel). Even with a short delay, decoding from activity is considerable in the delayed match-to-rotated-sample (DMRS) task but not in the standard DMS task, demonstrating that the amount of persistent activity, and hence the accuracy of its sample decoding, depends on the degree of behavioral demand for information manipulation during the delay period. This conclusion was tested by training different RNNs, each to perform one of nine tasks for which the degree of required information manipulation was quantified. These tasks are (1) DMS; (2–4) DMRS by 45, 90 or 180 degrees; (5) DMS where distractors are present during the delay period (ABCA, with A as sample and matched test, B and C as distractors); (6) ABBA where the correct response should be at the item matching the initial sample (A) but not a repeated distractor (B); (7) DMS+DMRS where one of two rule cues is shown after stimulus presentation and during the delay period, instructing the system to perform either DMS or DMRS; (8) dual DMS, where two sample stimuli are shown at different spatial locations, and the system must produce sequentially match/nonmatch responses for each of the two sample-test pairs defined by their common spatial location; (9) cross-location DMS, where the sample and test are at different spatial locations. Generally, decoding accuracy from recurrent population activity increases with the task demand of information manipulation (**Figure 5.40e**). Taken together, these experimental and computational findings support the general principle, in line with our intuition, that short-term memory traces do not always require elevated and sustained neural activity, but internal information processing is carried out by spikes. Since information manipulation is an integral part of working memory at the cognitive level (Baddeley 2012), persistent activity is essential for working memory.

5.9 SUMMARY

This chapter covers working memory, with a focus on stimulus-selective self-sustained activity as a neural substrate for active memory storage. Stable states of a dynamical system may be steady-states or complex spatiotemporal patterns. A working memory system is conceptualized as a neural circuit endowed with multiple attractor internal states for encoding different memory items that coexist with a baseline state. This way, a transient input can switch the system from rest to a stimulus-selective mnemonic state that is self-sustained in the absence of external stimulation, and another signal can switch the system back to rest, erasing the memory trace. The workhorse for working memory maintenance is positive feedback, which depends on the recurrent synaptic excitation but may be contributed also by single neuronal and synaptic dynamical properties. Excitatory reverberation must be slow and dependent on the NMDA receptors at local recurrent synapses between excitatory cells in a working memory circuit. Additional slow cellular and synaptic processes could produce sufficient positive feedback to maintain active working memory representation. Slow synaptic processes such as NMDA receptor-mediated excitation and short-term facilitation are desirable to ensure network stability and explain delay period neural activity that is not tonic in time but exhibits epochs of silence. As we will see in Chapter 6 next, slow reverberation turns out to be a plausible mechanism for decision-making computations as well. The NMDA receptors are also important for long-term plasticity and learning (Chapter 4), and working memory representations are shaped by learning (Miller et al. 2022). On the other hand, Hebb (1949) already speculated that reverberatory dynamics associated with working memory reinforce and consolidate long-term memory traces. The interplay between working memory and long-term memory remains under-explored.

Working memory function is marked by two salient characteristics. First, it cannot function properly without the brain's ability to filter out behaviorally irrelevant distractors. We proposed that a disinhibitory motif composed of three subtypes of inhibitory neurons plays a crucial role in gating out distractors. Secondly, unlike long-term memory and passive short-term memory, working memory can hold simultaneously only a small number of items. A biologically based model offers a way to understand this limited capacity and reconciles the seemingly contrasting discrete-slot versus shared-resource models in cognitive neuroscience.

An alternative to the attractor model posits that information is stored in hidden variables, not manifest at the level of spike firing. Pure "activity-silent" memory traces may be sufficient for passive short-term memory but not active working memory for several reasons: (1) it works only when the delay period is short compared to the time constant of the underlying biological process; (2) slow synaptic or cellular processes cannot be switched on and off by brief inputs required for working memory; (3) a hidden memory trace does not filter out distractors; (4) there is no small limited capacity for passive short-term memory; and (5) internal information manipulation in the brain requires spiking activity. Even in simple delayed response tasks, it is not known at what time point the sensory-to-motor transformation occurs. If it happens during the delay period, it represents one type of manipulation of information internally in the brain, and spiking activity would be required for this purpose (Takeda and Funahashi 2002).

In real-life situations, the brain normally does not stay in a fixed internal state for long, which could account for temporal variations of observed neural activity. When a task involves several stages, internal coding is expected to be non-stationary because the stored content changes over time. For example, in an "N-back" task, subjects are shown a sequence of stimuli, such as letters or numbers, and required to respond when the current item is the same as the one presented *N* items previously (Kirchner 1958; Owen et al. 2005). This task requires that the working memory system keeps loading in and out items in order to represent the last *N* stimuli, how mechanistically the content of working memory is dynamically updated according to task demands is unknown. Thinking involves a sequential internal process that cannot be accounted for by a simple attractor. Generally, the biological mechanisms and functions of temporal variations of self-sustained activity and heterogeneity across neurons remain a widely open topic for future research.

FURTHER READING

Books and Reviews

Amit, D. J. (1995). The Hebbian paradigm reintegrated: Local reverberations as internal representations. *Behav. Brain Sci. 18*, 617–626.

Arnsten, A. F., C. D. Paspalas, N. J. Gamo, Y. Yang, and M. Wang (2010). Dynamic network connectivity: A new form of neuroplasticity. *Trends Cogn. Sci. 14*, 365–375.

Baddeley, A. (2012). Working memory: Theories, models, and controversies. *Annu. Rev. Psychol. 63*, 1–29.

D'Esposito, M. (2007). From cognitive to neural models of working memory. *Philos. Trans. R. Soc. Lond. B. Biol. Sci. 362*, 761–772.

Goldman-Rakic, P. S. (1995). Cellular basis of working memory. *Neuron 14*, 477–485.

Luck, S. J. and E. K. Vogel (2013). Visual working memory capacity: From psychophysics and neurobiology to individual differences. *Trends Cogn. Sci. (Regul. Ed.) 17*, 391–400.

Ma, W. J., M. Husain, and P. M. Bays (2014). Changing concepts of working memory. *Nat. Neurosci. 17*, 347–356.

Miller, E. K. and J. D. Cohen (2001). An integrative theory of prefrontal cortex function. *Annu. Rev. Neurosci. 24*, 167–202.

Wang, X.-J. (2001). Synaptic reverberation underlying mnemonic persistent activity. *Trends Neurosci. 24*, 455–463.

Wang, X.-J. (2021). 50 years of mnemonic persistent activity: Quo vadis? *Trends Neurosci. 44*, 888–902.

Primary Research Papers

Delay Period Persistent Activity

Bastos, A. M., R. Loonis, S. Kornblith, M. Lundqvist, and E. K. Miller (2018). Laminar recordings in frontal cortex suggest distinct layers for maintenance and control of working memory. *Proc. Natl. Acad. Sci. 115*, 1117–1122.

Funahashi, S., C. J. Bruce, and P. S. Goldman-Rakic (1989). Mnemonic coding of visual space in the monkey's dorsolateral prefrontal cortex. *J. Neurophysiol. 61*, 331–349.

Fuster, J. M. and G. Alexander (1971). Neuron activity related to short-term memory. *Science 173*, 652–654.

Romo, R., C. D. Brody, A. Hernández, and L. Lemus (1999). Neuronal correlates of parametric working memory in the prefrontal cortex. *Nature 399*, 470–474.

Recurrent (Attractor) Network Mechanism

Amari, S. (1977). Dynamics of pattern formation in lateral-inhibition type neural fields. *Biol. Cybern. 27*, 77–87.

Amit, D. J. and N. Brunel (1997). Model of global spontaneous activity and local structured activity during delay periods in the cerebral cortex. *Cereb. Cortex. 7*, 237–252.

Brunel, N. and X.-J. Wang (2001). Effects of neuromodulation in a cortical network model of object working memory dominated by recurrent inhibition. *J. Comput. Neurosci. 11*, 63–85.

Compte, A., N. Brunel, P. S. Goldman-Rakic, and X.-J. Wang (2000). Synaptic mechanisms and network dynamics underlying spatial working memory in a cortical network model. *Cereb. Cortex. 10*, 910–923.

Inagaki, H. K., L. Fontolan, S. Romani, and K. Svoboda (2019). Discrete attractor dynamics underlies persistent activity in the frontal cortex. *Nature 566*, 212–217.

Seung, H. S. (1996). How the brain keeps the eyes still. *Proc. Natl. Acad. Sci. U.S.A. 93*, 13339–13344.

Wang, X.-J. (1999). Synaptic basis of cortical persistent activity: The importance of NMDA receptors to working memory. *J. Neurosci. 19*, 9587–9603.

Wimmer, K., D. Q. Nykamp, C. Constantinidis, and A. Compte (2014). Bump attractor dynamics in prefrontal cortex explains behavioral precision in spatial working memory. *Nat. Neurosci. 17*, 431–439.

Dynamical Nature of Working Memory Representation

Masse, N. Y., Yang, G. R., Song, H. F., Wang, X.-J., and Freedman, D. J. (2019). Circuit mechanisms for the maintenance and manipulation of information in working memory. *Nat. Neurosci. 22*, 1159–1167.

Mongillo, G., O. Barak, and M. Tsodyks (2008). Synaptic theory of working memory. *Science 319*, 1543–1546.

Murray, J. D., A. Bernacchia, N. A. Roy, C. Constantinidis, R. Romo, and X.-J. Wang (2017). Stable population coding for working memory coexists with heterogeneous neural dynamics in prefrontal cortex. *Proc. Natl. Acad. Sci. U.S.A. 114*, 394–399.

Rose, N. S., J. J. LaRocque, A. C. Riggall, O. Gosseries, M. J. Starrett, E. E. Meyering, and B. R. Postle (2016). Reactivation of latent working memories with transcranial magnetic stimulation. *Science 354*, 1136–1139.

Stokes, M. G., M. Kusunoki, N. Sigala, H. Nili, D. Gaffan, and J. Duncan (2013). Dynamic coding for cognitive control in prefrontal cortex. *Neuron 78*, 364–375.

Limited Working Memory Capacity

Wei, Z., X. J. Wang, and D. H. Wang (2012). From distributed resources to limited slots in multiple-item working memory: A spiking network model with normalization. *J. Neurosci. 32*, 11228–11240.

Xu, Y. and M. M. Chun (2006). Dissociable neural mechanisms supporting visual shortterm memory for objects. *Nature 440*, 91–95.

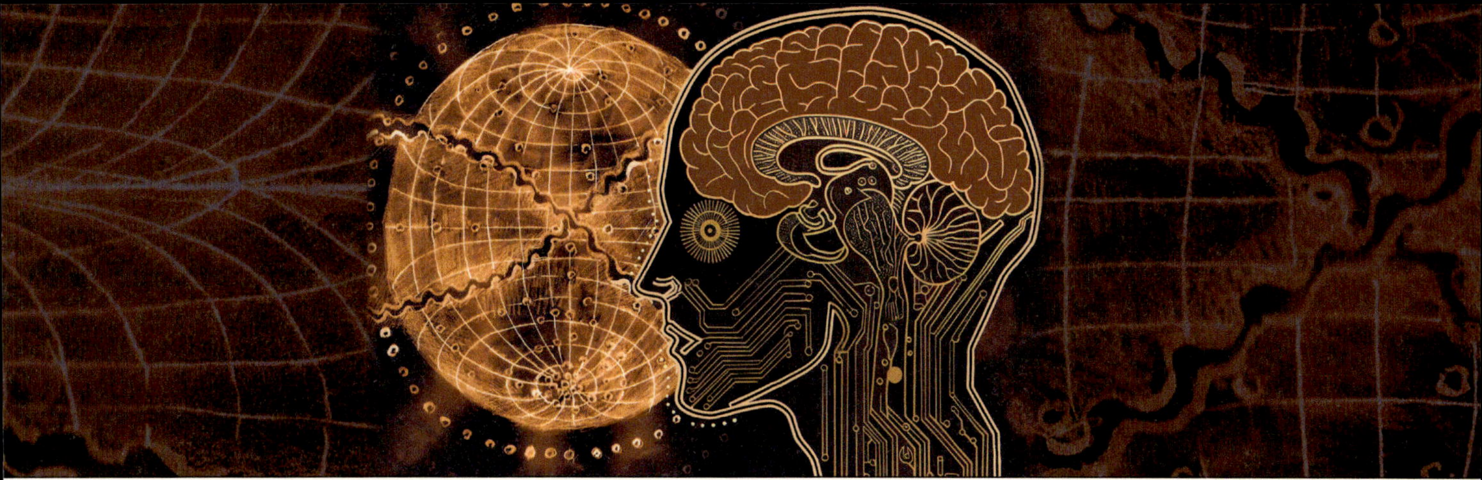

Decision Making

6

Strictly speaking, we do not make decisions, decisions make us.

José Saramago in *All the Names*

6.1 INTRODUCTION

One's life is shaped by a few momentous choices, such as selecting a school or a career. More generally, decision making pervades all aspects of flexible behavior; every day we go through a slew of decisions, often without being really aware of them. Consider Christmas shopping if you live near Yale University. Should you drive 20 minutes to Milford shopping mall or 90 minutes to New York City? Once there, which stores do you enter, what items catch your eyes and how do you reach a purchase decision? At the end of the day, you want to have dinner at a restaurant, so what kind of cuisine would you like this evening? Sampling a menu, what do you select for appetizer, entrée and dessert? Driving home in a snowstorm, you missed an exit, what should you do now? And so on.

How do we, i.e., our brains, make decisions? Decision making is a cognitive process of choosing an opinion or an action among a set of two or more alternatives. A hallmark of hard decisions is deliberate consideration of accumulated information about the pros and cons over time (Etter 1974). This process is manifested by the fact that we usually take a longer time to make a more difficult decision (Luce 1986). Deliberation is based on assessment of the expected outcomes from choice alternatives, in the face of incomplete information and risk of potential undesirable or unintended outcome from a choice. Indeed, one can say that the essence of great leadership is to have the foresight in making a right decision in the face of enormous uncertainty, illustrated by historical figures such as Ernest Shackleton, who made a number of risky choices in his *Endurance* expedition to Antarctica. Decision making is much more than sensation and movement, offering a window on cognition (Shadlen and Kiani 2013).

Behavioral studies of decision making have two traditions. A decision may be perceptual about the sensory world. Imagine a situation of

DOI: 10.1201/9781003459361-8

driving in the night in a blizzard. Through the view of a snow-splashed windshield, you need to stay safe by detecting cars around you, determining the color of a traffic light, etc. Detection and discrimination are perceptual decisions in the realm of psychophysics (Green and Swets 1966; Vickers 1979; Luce 1986). The second type of choices, like sampling a menu in a restaurant, are based on subjective values that vary from individual to individual. Valuation-based choice behavior is the theme of foraging theory, economics and game theory (von Neumann and Morgenstern 1944; Stephens and Krebs 1986; Luce 2005). Psychology of human decision making studies judgments that are based on logical reasoning as well as heuristics and influenced by biases (Kahneman et al. 1982; Gigerenzer and Gaissmaier 2011).

Pioneering neurobiological research on decision making started with recording from neurons in sensory areas (Mountcastle et al. 1969; Newsome et al. 1989; Mountcastle et al. 1990; Britten et al. 1992; Parker and Newsome 1998). The field took off around the turn of the twenty-first century, thanks to a confluence of animal physiology, human brain imaging, theory and neural circuit modeling. Single-neuron recordings from candidate cortical areas for subjective decisions, downstream from sensory areas, of behaving monkeys were carried out in a variety of decision tasks, including perceptual discrimination (Shadlen and Newsome 1996, 2001; Romo and Salinas 2001; Roitman and Shadlen 2002; Romo et al. 2004), action selection (Hanes and Schall 1996; Schall 2001; Schall 2004; Cisek and Kalaska 2005; Scherberger and Andersen 2007), economic choice behavior (Platt and Glimcher 1999; Sugrue et al. 2004; Sugrue et al. 2005; Padoa-Schioppa and Assad 2006) and competitive games (Glimcher 2003; Barraclough et al. 2004; Dorris and Glimcher 2004; Lee 2008). These experiments and those using behaving rodents in more recent years (see later section) have uncovered neural signals at the single-cell level that are correlated with specific aspects of decision computation (Schall 2001; Gold and Shadlen 2007; Wang 2008; Cisek and Kalaska 2010; Lee and Seo 2016; Padoa-Schioppa and Conen 2017; Najafi and Churchland 2018; O'Connell et al. 2018; Schall 2019).

In the brain a decision is made by the collective dynamics of a neural circuit. How are the observed neural signals generated? What are the properties of a local cortical area (e.g., in the prefrontal cortex or posterior parietal cortex) that enable it to subserve decision computations, in contrast to early processing in primary sensory areas? How can one establish the chain of causal links from molecules, circuits to decision behavior? In this chapter, we will discuss how the brain makes perceptual decisions, going back and forth between experimentation and theory. Value-based economic choices will be covered in Chapter 7. We will first introduce abstract mathematical models, then focus on a biologically realistic neural network model needed to probe brain mechanisms across levels. As we shall see, this line of research has led to the concept of "cognitive-type" neural circuits underlying decision making as well as working memory.

6.2 MATHEMATICAL MODELS OF DECISION MAKING

6.2.1 Signal Detection Theory

A simple kind of perceptual decision is detection, for which you are tasked to answer if a signal (such as a visual target or an auditory tone) is present or absent. There are four conditions (**Figure 6.1a**): when the signal is shown, you either see it (a hit) or not (a miss); when it is not there, your answer is either no (correct rejection) or yes (false alarm). For example, in a laboratory experiment, the luminance of a visual target can be varied parametrically, so that the brightest target is readily detectable (100% hit rate) and the dimmest

(a)

decision / stimulus	probability	
	correct	error
present	hit	miss
absent	correct rejection	false alarm

(b)

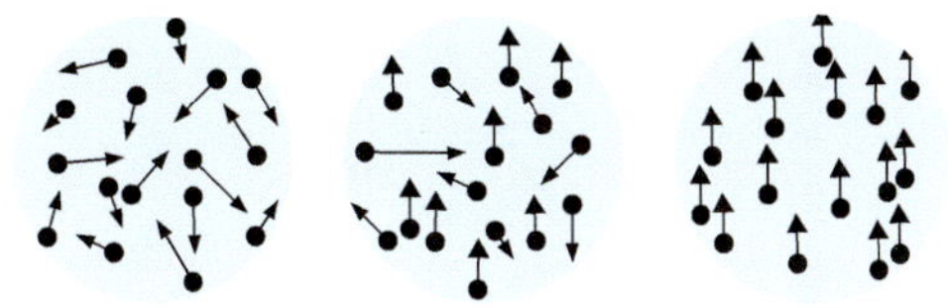

(c)

decision / stimulus	probability	
	correct	error
A	β	1–β
B	1–α	α

Figure 6.1 Signal detection theory. (a) In a detection task, when a signal is shown, the subject may report yes (hit) or no (miss); in the absence of the signal, the report may be yes (false alarm) or no (correct rejection). (b) A random dot pattern in which a fraction c' of dots move together in one of two possible directions (e.g., upward or downward), the task difficulty can be parametrically varied by changing the motion coherence, also called motion strength (c'= 0, 50% and 100% from left to right). (c) In a two-alternative forced choice (TAFC) task, when the stimulus is A, the decision may be A (correct) with probability β or B (error) with probability $1-\beta$; conversely, when the stimulus is B, the decision may be A (error) with probability α and B (correct) with probability $1-\alpha$.

target is invisible (100% miss rate). The probabilities for the four answers are expected to change gradually with the luminance level.

Another common perceptual problem is binary discrimination, in which one must assess whether a stimulus is A or B. This is a form of two-alternative forced choice (TAFC) between two options. Consider a well-known example of random dot motion (RDM) direction discrimination task. A visual random dot pattern is presented, in which a fraction (c', called coherence or motion strength) of dots move in one of two possible directions (A or B) (**Figure 6.1b**). When the coherent motion is in direction A, the answer is expected to be A with probability 1 if $c' = 100\%$; A or B with equal probability of 0.5 if $c' = 0\%$. By varying c' between 0 and 1, the task difficulty is parametrically changed and behavioral performance can be assessed. Again, there are four conditions (**Figure 6.1c**): the subjective judgment can be A (correct) with probability β and B (error) with probability $1-\beta$ when the evidence is in favor of A; when evidence is in favor of B, the probabilities of reporting A (error) and B (correct) are denoted as α and $1-\alpha$, respectively.

In signal detection theory (Green and Swets 1966), a quantity x about the decision process under consideration is measured in an experiment. The likelihood of observing x given that the signal s is A or B is represented by conditional probability $p_A(x) = p(x \mid s = A)$ or $p_B(x) = p(x \mid s = B)$, respectively (**Figure 6.2a**). In a detection task, the measured signal x is larger on average in the presence (A) than absence (B) of the stimulus, $p_A(x)$ is on the right side of $p_B(x)$ (which is merely caused by noise). The question is: given a measured value for x, can one decode the behavioral report (A = yes or B = no)? A simple decoding rule assumes a threshold z, the report is A if $x \geq z$ and B otherwise. Let us define:

$$\alpha(z) = P_B(x \geq z \mid B) = \int_z^{\infty} p_B(x)dx; \tag{6.1}$$

$$\beta(z) = P_A(x \geq z \mid A) = \int_z^{\infty} p_A(x)dx \tag{6.2}$$

Ideally, we would like to find a value of z that maximizes the hit rate β and minimizes the false alarm rate α. If $p_A(x)$ and $p_B(x)$ are well separated without any overlap, one can readily choose a value of z in the middle, above $p_B(x)$ and below $p_A(x)$. Consequently, $\alpha = 0$ and $\beta = 1$. However, the situation is more difficult when the two probability distributions overlap strongly. With a small enough z, both distributions are on the right side of z along the x-axis, thus $\alpha = \beta = 1$; on the other hand, for a sufficiently large z, both distributions are on the left side and $\alpha = \beta = 0$. A graphic plot $\beta(z)$ against $\alpha(z)$ (both between 0 and 1) is called receiver operating characteristic (ROC) curve (**Figure 6.2b**). Each point on the ROC curve corresponds to a z value; varying z traces the entire curve.

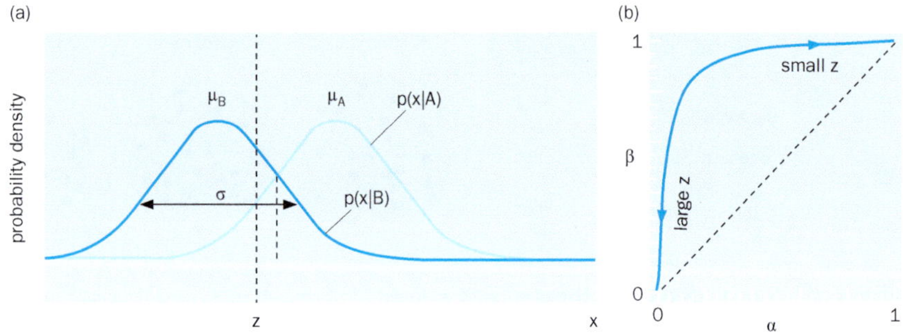

Figure 6.2 Receiver operating characteristic (ROC) analysis. (a) Probability densities $p_A(x) = p(x \mid s = A)$ and $p_B(x) = p(x \mid s = B)$, with the mean indicated by μ_A and μ_B, respectively, and the standard variation σ. The two probabilities are equal when x is at the value indicated by the dashed vertical line. A threshold is indicated by z. (b) β plotted against α forms an ROC curve parameterized by z.

This approach was applied to a monkey physiological experiment in which monkeys performed the RDM direction discrimination 2AFC task (Newsome et al. 1989; Britten et al. 1992). A single unit was recorded at a time from the visual medial temporal (MT) cortical area, known to be specialized in processing visual motion stimuli (Albright et al. 1984; Newsome and Paré 1988; Salzman et al. 1990). An MT neuron is selective for motion direction, with its maximal response at a preferred direction θ_{pref}. The direction of a shown motion stimulus was either θ_{pref} (A) or diametrically opposite from it called the null or antipreferred direction, $\theta_{\text{null}} = \theta_{\text{pref}} + 180°$ (B). Note that MT cells respond quickly at the onset of a stimulus, show spike rate adaptation. Monkeys were required to report by a saccadic eye movement to one of the two targets in a screen indicating A or B (separated by 180 degrees), at the end of a 2-second-long stimulus presentation. The activity of an MT neuron was measured as the total number of spikes divided by the stimulus duration of 2 seconds, and that firing rate varies from trial to trial stochastically. **Figure 6.3** displays firing patterns of a single MT neuron that increases (respectively decreases) with motion strength c' when the stimulus direction is preferred (left) (anti-preferred [right], respectively).

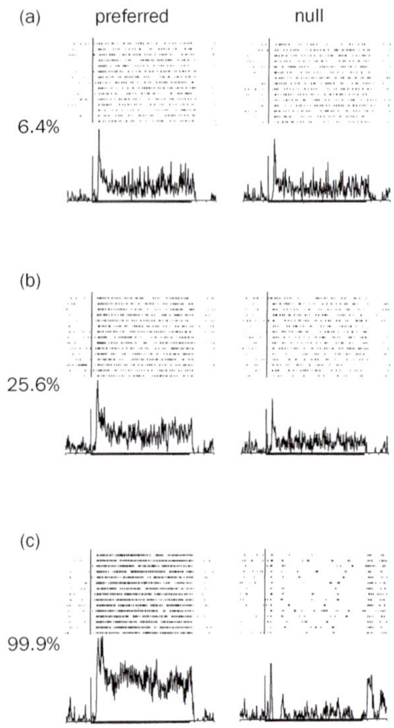

Figure 6.3 MT neural activity in response to a random dot motion stimulus with three levels of motion strength c′. Spiking activity increases with c' for preferred direction (left) and decreases with c' for null direction (right). In each panel, a post-stimulus time histogram (PSTH) is shown below spike trains in single trials. The vertical bar at the onset of the stimulus indicates 100 spikes per second, and stimulus duration is 2 seconds. Reproduced from Britten et al. (1993). [Britten, K. H., Shadlen, M. N., Newsome, W. T., & Movshon, J. A. (1993). Responses of neurons in macaque MT to stochastic motion signals. Visual Neuroscience, 10(6), 1157–1169.]

Figure 6.4a shows three histograms of an MT neuron's firing rate when the net motion direction is the preferred (open bar) or null (filled bar) direction. As expected, when the coherence is high (c' = 12.8%), the two histograms are well separated; on the other hand, at low coherence (c' = 0.8%), the two histograms are virtually indistinguishable from each other because in this case the random dot pattern is essentially the same whether the tiny net motion is A or B. For an intermediate c' = 3.2%, the two histograms are distinguishable albeit with a significant overlap. Britten et al. (1992) interpreted **Figure 6.4a** as if they recorded simultaneously from two MT neurons with the opposite preferred directions, A and B, respectively. In that hypothetical situation, the two firing rate histograms were viewed as the distributions of firing rates r_A and r_B of the two neurons, sampled across trials when the net motion was always A. Using neural firing distributions of **Figure 6.4a**, the ROC curve computed with this interpretation is shown with five coherence levels in **Figure 6.4b**.

Could the animal's report be decoded by the simple rule: choose A if $r_A \geq r_B$? For a 2AFC task, the optimal solution is to use the integral of ROC (for all possible threshold z values), or the area under the ROC curve, which is between 0 and 1, as the probability to report A (Green and Swets 1966). In signal detection theory, one of the two firing rates serves as threshold $z = r_B$; thus, the rule becomes: report A if $r_A \geq z = r_B$; B otherwise. Consequently, the overall probability of reporting A is the integral over z of $P_A(r_A \geq z) = \beta(z)$ weighted by $p_B(r_B = z) = -d\alpha / dz$. Recall that α decreases with increasing z monotonically, so one can use a change of variable from z to α and $\beta(z)$ can be replaced by $\beta(\alpha)$, leading to:

$$P(\text{report A}) = \int_0^{\infty} p_B(z) P_A(r_A \geq z) dz = \int_0^1 \beta(\alpha) d\alpha \tag{6.3}$$

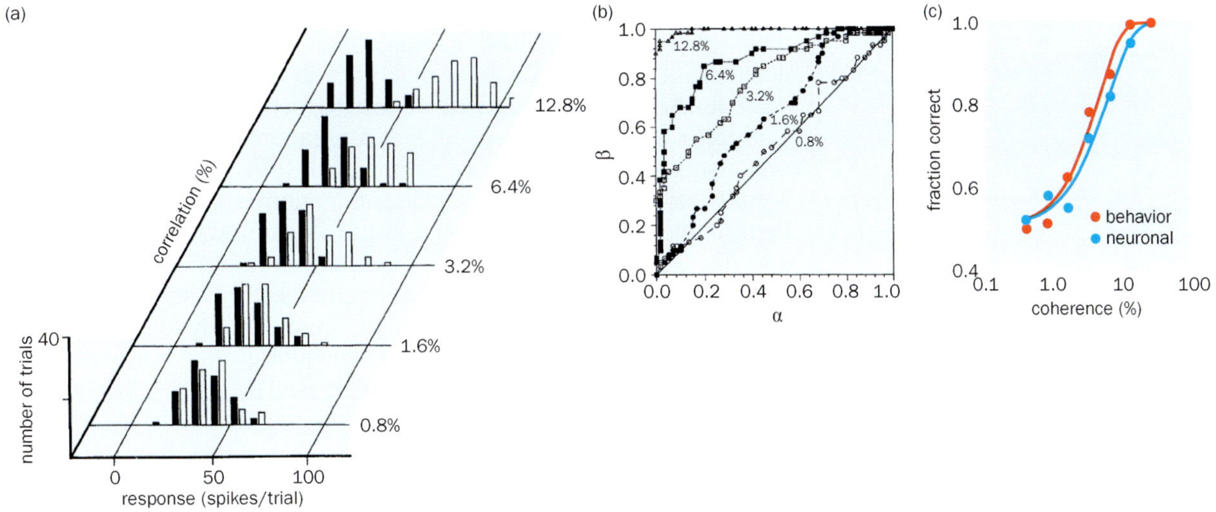

Figure 6.4 MT neural activity in a random dot visual motion discrimination task and neurometric function. (a) Histograms of spike count of a single MT neuron when the stimulus direction is preferred (open bar) or null (filled bar), for five different coherence levels. The two histograms are well separated at high c' (12.8%), while almost indistinguishable at low c' (0.8%). (b) ROC curve is near the diagonal line for c' = 0.8% and close to $\beta = 1$ for c' = 12.8%. (c) Neurometric function of a single MT neuron (blue circle), defined by the integral of ROC curve as a function of c', is shown together with monkey's behavioral psychometric function (red circle). Reproduced from Britten et al. (1992). [Britten, K. H., Shadlen, M. N., Newsome, W. T., & Movshon, J. A. (1992). The analysis of visual motion: A comparison of neuronal and psychophysical performance. Journal of Neuroscience, 12(12), 4745–4765. Copyright 1992 Society for Neuroscience.]

which is precisely the integral of, or the area under, the ROC curve. As can be seen in **Figure 6.4b**, this ideal observer's report is near chance level ~0.5 at a low coherence of 0.8%, and close to 1.0 at a high coherence of 12.8%. $P(\text{report A})$ as a function of c' from such a neuronal ideal observer is called a "neurometric" function, which was computed for each of the recorded MT neurons. Surprisingly, many MT cells produced neurometric functions comparable to the behavioral psychometric function (**Figure 6.4c**).

Subsequent work clarified the relationship between MT neural activity and an animal's discrimination performance. Sensory stimulus is noisy, and the signal-to-noise ($\mathcal{S} / \mathcal{N}$) ratio of neural representation can be improved by pooling (averaging over) more neurons. However, if activities of neurons are correlated with each other, noise cannot be averaged out by summing over more and more neurons, and the $\mathcal{S} / \mathcal{N}$ ratio stopped increasing with the number of neurons when the latter is large enough. It was estimated that in MT, the $\mathcal{S} / \mathcal{N}$ ratio plateaus with about 100 MT cells (Zohary et al. 1994; Shadlen et al. 1996). Moreover, using a fixed duration of stimulus representation, it was not clear whether the animal's decision was based on the MT neural representation through the entire 2-second time window. This issue was addressed in later experiments using a reaction time version of the task when the animal can produce a decision any time during stimulus viewing, and the time it takes to reach a choice is measured by reaction time (RT).

6.2.2 Drift Diffusion Model

The ROC signal detection analysis does not account for the reaction time, the time it takes to make a decision. Reaction time or response time (RT) represents an important measure of how difficult a decision is and how confident we are about a choice; RT is longer when decision is harder and we are less confident (Donders 1969; Posner 1978; Luce 1986; Meyer et al. 1988). Although we are capable of producing rapid responses, rushed decisions may yield ill effects. There is often a tradeoff between speed and accuracy: performance improves with slower response times (Wickelgren 1977).

Although the concept of time integration is appealing, it is not obvious what types of choice behavior engage such accumulation process (a characteristic of deliberate decision making) and on what timescale (Uchida et al. 2006). Selection among a set of possible actions is a form of choice that can occur quickly, when speed is at a premium. Simple choices in some organisms can take as little as 100 ms (Real 1991; Ewert 1997). In rodents, a study reported that olfactory discrimination is fast (~ 300 ms) in rats. Performance varied from chance level to near 100% correct, as the task difficulty was varied by adjusting the relative proportions of two odorants in a binary mixture, but the RTs were changed only slightly (by ~ 30 ms) (Uchida and Mainen 2003). In contrast, another study found that, in mice, olfactory discrimination performance was high (~ 95 % correct) regardless of discrimination difficulty, while RTs increased by ~ 80 ms from the easiest to the hardest (Abraham et al. 2004), suggesting that in these tasks rodents exhibit speed-accuracy tradeoff on a timescale of less than 100 ms. In primate studies, even simple saccades to a suddenly shown visual target take a couple of hundreds of milliseconds (Carpenter 1981; Schall and Hanes 1993; Salinas et al. 2010); more generally, RTs typically range from hundreds of milliseconds to about 1 second in simple perceptual tasks (Luce 1986; Usher and McClelland 2001).

Dynamical models are needed to account for reaction times and its variability. Indeed, even the reaction time of an automatic ballistic eye movement to a bright visual stimulus varies significantly from saccade to saccade. A mathematical model to describe this RT variability assumes an activity variable X that starts to increase from a baseline value X_0 at the onset of the visual target, increases linearly over time with a slope μ, and a response is triggered when X reaches a preset threshold level θ. This model is called LATER (linear accumulation to threshold with ergodic rate) (Carpenter 1981). The RT is then given by $(\theta - X_0)/\mu$. RT variability results from trial-to-trial variations of the ramping slope μ, which are often assumed to obey a Gaussian distribution (Reddi and Carpenter 2000). Hanes and Schall (1996) performed a neurophysiological experiment where monkeys made a saccade to a suddenly appearing visual target while activity of single neurons from the front eye field (FEF, a key saccade generation cortical area in the frontal lobe) was recorded. From trial to trial, the behavioral RT varied, and so did the slope of ramping activity of FEF neurons. Interestingly, FEF neural firing rate reached an approximately fixed level at the time when saccade was initiated, regardless of the RT (**Figure 6.5**), demonstrating the ramping-to-threshold phenomenon in the brain.

The LATER model is suitable to describe a Go process but not a decision process with more than one option. More generally, the intuitive idea that a decision is reached when accumulation of information about alternative choices reaches a particular threshold has been formulated mathematically as sequential sampling models. For 2AFC tasks, a popular implementation is called the drift diffusion model (DDM) (Laming 1968; Ratcliff 1978; Link 1992; Busemeyer and Townsend 1993; Ratcliff et al. 2016; Forstmann et al. 2016). In this model, a "decision variable" X represents the difference between the respective amounts of accumulated information about two alternatives A and B, say X_A and X_B, $X = X_A - X_B$. The dynamics of X is given by the drift diffusion equation,

$$\frac{dX}{dt} = \mu + \sigma w(t) \tag{6.4}$$

where the drift rate μ represents the bias in favor of one of the two choices, and is zero if there is no net bias. For instance, in a random-dot motion direction discrimination task, $\mu = kc'$ is proportional to the motion coherence c' with k as a measure of sensitivity. The second term $w(t)$ is a white noise of zero mean and unit variance. The paramter σ determines the noise size and has the physical unit of $X/\sqrt{t}$.

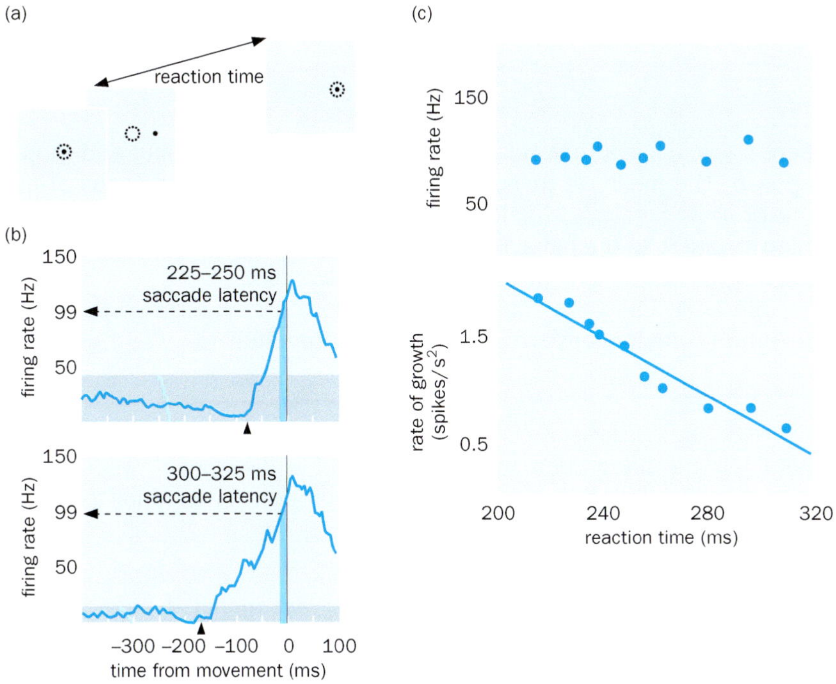

Figure 6.5 Ramping-to-threshold neural dynamics in a Go task. (a) In a saccade task, a monkey subject fixated on a spot at the center of the screen and makes a saccade to a target that suddenly appeared while simultaneously the center spot disappeared. (b) Firing rate from a single neuron in the frontal eye field (FEF) aligned at the saccade onset time, in trials with saccade reaction times between 225 and 250 ms (top) and 300–325 ms (bottom). Time zero marks saccade onset; the firing rate at that moment is similar in the short- and long-RT trials. (c) FEF neural activity ramps up with a different slope and is faster with shorter reaction times (bottom) but reaches roughly the same maximum indicative of a threshold around the time of movement initiation (top). [Hanes, D. P., & Schall, J. D. (1996). Neural control of voluntary movement initiation. Science, 274(5286), 427–430.]

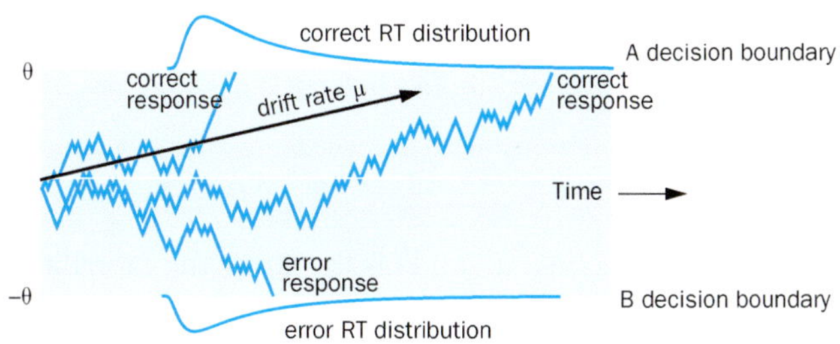

Figure 6.6 Drift diffusion model. A decision variable X undergoes a random walk in continuous time biased by the difference in the evidence in favor of one of two choice options. In a single trial, the process of information accumulation is terminated when X reaches a preset threshold $+\theta$ or $-\theta$, both the decision (A or B) and reaction time are read out. Reproduced from Ratcliff and McKoon (2008). [Ratcliff, R., & McKoon, G. (2008). The diffusion decision model: Theory and data for two-choice decision tasks. Neural Computation, 20(4), 873–922.]

Unlike the LATER model, DDM assumes that momentary sensory evidence fluctuates stochastically in time, and the signal-to-noise ratio increases by integration over time. The integration process is terminated and the reaction time is read out, whenever $X(t)$ reaches a positive threshold θ (choice A) or a negative threshold $-\theta$ (choice B). If the drift rate μ is positive, then choice A is correct, whereas choice B is an error (**Figure 6.6**). Thus, DDM accounts for both choice and RT in single trials.

The drift diffusion model has been widely applied to fit behavioral data of perceptual and memory experiments. This model is a continuous time version of the sequential probability ratio test, which is a procedure of hypothesis testing (A or B?) by temporal accumulation of randomly sampled inputs to a criterion. It is the optimal procedure for making binary choices under uncertainty, in the sense that, for a fixed error rate, it minimizes the mean decision time among all statistical tests; in other words, there is no better algorithm that can achieve the same performance within a shorter time window (Wald 1948; Bogacz et al. 2006).

The accuracy or psychometric function of the DDM model can be calculated analytically (Shadlen et al. 2007):

$$P(A) = \frac{1}{1 + \exp\left(-2\mu\theta / \sigma^2\right)} \tag{6.5}$$

and the mean reaction time called the chronometric function is given by

$$\langle RT \rangle = \frac{\theta}{\mu} tanh\left(\frac{\mu\theta}{\sigma^2}\right) + t_{\text{nondecision}} \tag{6.6}$$

where $t_{nondecision}$ is a time related to latencies of sensory input and motor response. The mean RT increases with decreasing μ and plateaus at the value of $\theta^2 / \sigma^2 + t_{\text{nondecision}}$ for $\mu = 0$. As shown in **Figure 6.7**, these formulae fit well with the behavioral data from a reaction time version of the RDM discrimination task (Roitman and Shadlen 2002).

The DDM can be modified to incorporate a leak term so that time integration is no longer perfect but becomes forgetful:

$$\frac{dX}{dt} = -\frac{X}{\tau} + \mu + \sigma w(t) \tag{6.7}$$

where τ is the integration time. When input μ disappears, information decays exponentially with time τ; perfect integration corresponds to $\tau = \infty$.

6.2.3 Race Models

In contrast to the DDM, in a two-dimensional race model, dynamical variables X_A and X_B accumulate evidence separately for the two alternatives in perceptual decision making or action selection (Usher and McClelland 2001; Bogacz et al. 2006; Boucher et al. 2007). Furthermore, instead of being independent from each other, the two racers may compete through mutual inhibition:

$$\frac{dX_A}{dt} = -\frac{X_A}{\tau} - JX_B + \mu_A + w_A(t); \tag{6.8}$$

$$\frac{dX_B}{dt} = -\frac{X_B}{\tau} - JX_A + \mu_B + w_B(t) \tag{6.9}$$

where τ is the leaky time constant and J is the strength of inhibitory interaction, μ_A and μ_B are mean external inputs, and $w_A(t)$ and $w_B(t)$ are white noise terms. In this model the decision is determined by which one of the two decision variables reaches a threshold θ first before the other does, and the time it takes for threshold crossing is read out as reaction time. An independent race model corresponds to $J = 0$ and is called a "linear ballistic accumulator model" when there is no noise (Brown and Heathcote 2008).

Race models resemble neural population rate models and potentially account for leakage and synaptic interactions between competing neural

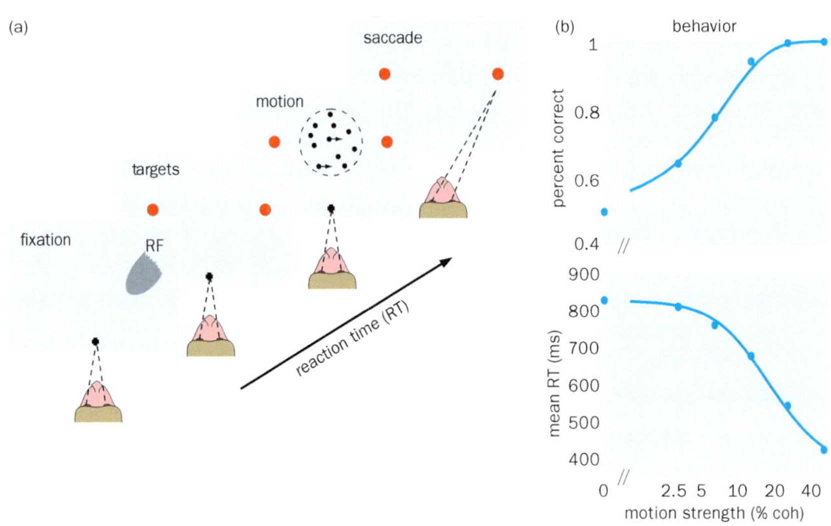

Figure 6.7 Behavioral performance of a monkey in a TAFC task. (a) A reaction time version of the RDM direction discrimination task. A monkey subject views a patch of dynamic random dots, and the decision is indicated by an eye movement to one of two peripheral targets. (b) Psychometric function (top) and chronometric function (bottom) are shown with data (dots) fitted by the DDM (Eqs. [6.5–6.6]) (solid curves). Reproduced from Gold and Shadlen (2007). [Gold, J. I., & Shadlen, M. N. (2007). The neural basis of decision making. Annual Review of Neuroscience, 30(1), 535–574.]

pools selective for different choice options. A race model is readily extendable to decisions with more than two choice options by adding an arbitrary number of accumulators.

6.2.4 Bayesian Modeling

A probabilistic formulation of decision making is explicitly based on probability distributions (Berger 1980; Pouget et al. 2003; Doya et al. 2011; Ma 2019). Here, a subjective decision is called a belief or hypothesis h (e.g., that the direction of a RDM pattern is A). Consider an empirical observation as evidence e, and the conditional probability that e is observed if a hypothesis h holds true, $p(e \mid h)$, is called likelihood function. The probability of the hypothesis $p(h)$ is called "prior," expressing pre-existing knowledge about h. Bayes' rule (Bayes 1763) asserts that:

$$p(h \mid e) = \frac{p(e \mid h)p(h)}{p(e)} \tag{6.10}$$

where $p(e)$ describes the statistics of observing a value e.

The conditional probability $p(h \mid e)$ is called posterior, a measure of belief in the hypothesis given the prior and the observed evidence. This equation results from the calculus of conditional probabilities, namely, the joint probability $p(h,e) = p(h)p(e \mid h) = p(e)p(h \mid e)$. Bayes' rule adheres to common-sense reasoning (Howie 2002). It states that the following assertions are equivalent: (a) evidence supports the hypothesis if the probability of h being true increases by knowing e $\left(p(h \mid e) > p(h)\right)$; (b) observing e is more likely if h holds true than its overall probability $\left(p(e \mid h) > p(e)\right)$. Evidence increases the posterior about a hypothesis in proportion to the likelihood $p(e \mid h)$. So if the hypothesis is incompatible with e, $p(e \mid h) = 0$, then the belief is refuted and $p(h \mid e) = 0$. If the evidence deterministically follows from the hypothesis, $p(e \mid h) = 1$, the observation provides maximum support, $p(h \mid e) = p(h) / p(e)$, or $p(h,e) = p(h)$.

Bayesian inference provides a recipe for combining prior knowledge with new observations. For instance, with two independent pieces of evidence $e = \{e_1, e_2\}$, $p(e_1, e_2 \mid h) = p(e_1 \mid h)p(e_2 \mid h)$, and the posterior $p(h \mid e_1, e_2)$ is updated by observations through multiplication of probabilities:

$$p(h \mid e_1, e_2) = \frac{p(e_1, e_2 \mid h)p(h)}{p(e)} = \frac{p(e_1 \mid h)p(e_2 \mid h)p(h)}{p(e)} \tag{6.11}$$

The Bayesian inference approach (Eqs. 6.10–6.11) has been applied widely to perception (Knill and Richards 1996; Doya et al. 2011). Consider subjective estimation of an analogue stimulus feature s, where a key question is how to incorporate uncertainty about s in veridical judgment (Ernst and Banks 2002; Knill and Pouget 2004). The problem is not trivial in multisensory integration that involves estimation based on two independent cues. For instance, a bird hidden behind leaves may be located using a combination of a visual cue (its partially invisible body) and an auditory cue (its chirping). When two cues are uncertain, each from a Gaussian distribution with a mean μ_1 or μ_2 and variance σ_2 or σ_2, a Bayesian decoder is optimal in the sense that its estimate has the smallest variance possible (Cochran 1937; Kalman and Bucy 1961). The average estimate is given by:

$$\mu = w_1\mu_1 + w_2\mu_2; w_1 = \frac{\sigma_2^2}{\sigma_1^2 + \sigma_2^2}; w_2 = \frac{\sigma_1^2}{\sigma_1^2 + \sigma_2^2} \tag{6.12}$$

and the variance σ^2 of the estimate is given by:

$$\frac{1}{\sigma^2} = \frac{1}{\sigma_1^2} + \frac{1}{\sigma_2^2} = \frac{\sigma_1^2 + \sigma_2^2}{\sigma_1^2\sigma_2^2} \tag{6.13}$$

According to Eq. (6.12) and in agreement with our intuition, the Bayesian estimate μ is a weighted sum of μ_1 and μ_2. When σ_1 is near zero, $\mu \simeq \mu_1$; and when $\sigma_1 \simeq \sigma_2$, $\mu \simeq (\mu_1 + \mu_2)/2$. The variance σ of the estimate is smaller than both σ_1 and σ_2 (Eq. 5.13), and the estimation accuracy is increased by integrating pieces of evidence.

Figure 6.8a shows an example where location is informed by a visual cue (1) and an auditory cue (2). When you look straight at the sound source, the visual cue has a low uncertainty (σ_1^2 is small), thus the Bayesian estimate is close to that of the visual cue alone (top panel). On the other hand, if you orient your gaze away from the sound source, visual uncertainty is high (σ_1^2 is large), and the Bayesian estimate is moved to the middle between the two estimates by independent cues (lower panel).

In the sensory cortex, a location or motion direction is represented by neurons with bell-shaped tuning curves, say, $f_i(s) = f_{\max} \exp\left(-(s - s_i)^2 / (2\sigma_f^2)\right)$, where s_i is the preferred stimulus for neuron i and σ_f is the width of the tuning curve. Now take the view of a population of N neurons at a given moment, with spike counts in a time window denoted as a vector $\mathbf{r}(s) = \{r_1(s), r_2(s), \ldots, r_N(s)\}$. The probabilistic population coding (Pouget et al. 2003; Ma et al. 2006) posits that variability of neural spike firing encodes uncertainty. The idea is that, in response to a stimulus s, $\mathbf{r}(s)$ represents a sample in a single trial, from a probability

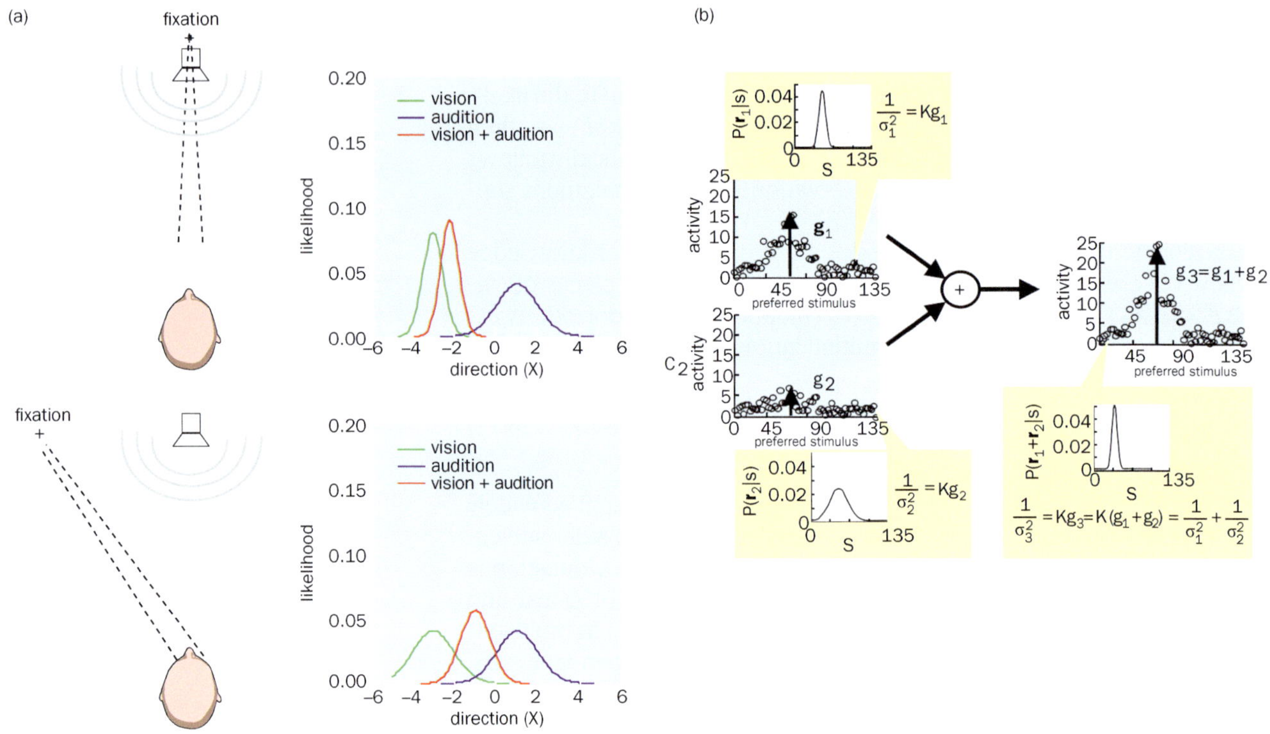

Figure 6.8 **(a) Auditory and visual cues provide conflicting information about the direction of a target.** The conflict is apparent in the difference in means of the likelihood functions associated with each cue, although the functions overlap. Such conflicts are always present, owing to noise in the sensory systems. To integrate visual and auditory information optimally, a multimodal area must take into account the uncertainty associated with each cue. Upper panel: When the vision cue (green) is more reliable than the auditory cue (purple), the peak of the posterior distribution (red) is shifted toward the direction suggested by the vision cue. Lower panel: When the reliabilities of the cues are more similar, for example, when the stimulus is in the far periphery, the peak is shifted toward the direction suggested by the auditory cue. (b) Inference with probabilistic population codes for Gaussian probability distributions and Poisson variability. Panel (a) is reproduced from Knill and Pouget (2004), (b) from Ma et al. (2006). [Knill, D. C., & Pouget, A. (2004). The Bayesian brain: The role of uncertainty in neural coding and computation. Trends in Neurosciences, 27(12), 712–719.] [Ma, W. J., Beck, J. M., Latham, P. E., & Pouget, A. (2006). Bayesian inference with probabilistic population codes. Nature Neuroscience, 9(11), 1432–1438.]

distribution $p(\mathbf{r} \mid s)$ conditioned on s. The likelihood $p(\mathbf{r} \mid s)$ should be understood as a function of s with a given sample $\mathbf{r}$. It is not a probability density of s; its integral over s is not 1. Bayes' rule (Eq. 5.10) is used to compute the posterior:

$$p(s \mid \mathbf{r}) = \frac{p(s)p(\mathbf{r} \mid s)}{p(\mathbf{r})} \tag{6.14}$$

which is a probability density with $\int p(s \mid \mathbf{r})ds = 1$. For instance, if neural spiking follows a Poisson process, the likelihood is:

$$p(\mathbf{r} \mid s) = \prod_i \frac{e^{-f_i(s)} f_i(s)^{r_i}}{r_i!} \tag{6.15}$$

where $f_i(s)$ is the tuning function of the average firing rate of neuron indexed i. One can show that, with large N, the likelihood $p(\mathbf{r} \mid s)$ is a Gaussian (Seung and Sompolinsky 1993; Ma 2010):

$$p(\mathbf{r} \mid s) \simeq \exp\left(-\frac{1}{2\sigma_f^2}\sum_{i=1}^{N}(s - s_i)^2 r_i\right) \tag{6.16}$$

Given the posterior, the best vertical judgment is given by the maximum-likelihood estimation (MLE) of s, the value s_{MLE} for which $dp(\mathbf{r} \mid s)/ds = 0$ and $d^2p(\mathbf{r} \mid s)/ds^2 < 0$. It can be readily shown that $s_{\text{MLE}}(\mathbf{r}) = \sum_i r_i s_i / \left(\sum_i r_i\right)$, which is the same as the population vector decoder (Georgopoulos et al. 1982). The variance of this estimate is $\sigma_f^2 / \left(\sum_i r_i\right) \simeq \sigma_f^2 / \left(\sum_i f(s - s_i)\right)$, which increases with σ_f and decreases with the peak (f_{max}) of the tuning curve as well as the number of neurons (Ma 2010). The denominator is sometimes called a gain factor g (Ma et al. 2006); increasing the gain reduces the variance of the posterior distribution.

Ma et al. (2006) applied this idea to multisensory integration of two cues, c_1 and c_2, related to the same variable s (**Figure 6.8b**). Each of them encodes a probability distribution with a variance inversely proportional to the peaks, g_1 and g_2, of the population codes (K is a constant depending on the width of the tuning curve and the number of neurons). In the model, a neuron in the multisensory network (right) receives inputs from a single cell from each of the two sensory streams (left), and spikes are added arithmetically. With this simplistic setting, the output population activity encodes a probability distribution with a variance inversely proportional to the gain g, which is the peak of the population activity pattern in the multisensory integration circuit. Because the gain of this code is $g_1 + g_2$, and g_1 and g_2 are inversely proportional to σ_1^2 and σ_2^2, respectively, the inverse variance of the output population code is the sum of the inverse variances associated with c_1 and c_2. This is precisely the variance expected from an optimal Bayesian inference (Eq. 6.13). In other words, taking the sum of two population codes is equivalent to taking the product of their encoded distributions.

The Bayesian inference theory is elegant, and it highlights key concepts like prior. Mathematically, a powerful idea is that the posterior as a probability distribution can be updated by one or several sample(s) of evidence, and this framework provides an algorithm for such updating. Regarding Bayesian optimality, whether statistical optimality is the norm in perceptual decision making in human behaviors remains a matter of debate (Rahnev and Denison 2018) for three reasons. First, achieving optimality may be hindered by constraints, e.g., due to the limited resources, and by insufficient statistical learning about the sensory world. Second, the prior $p(s)$ is often ill defined; while it can be specified in a laboratory experiment, usually it is not quantified and thus unconstrained in real-life situations (Jones and Love 2011). Third, people are notoriously bad with Bayesian computations using conditional probabilities and often neglect prior (Eddy 1982; Gigerenzer and Hoffrage 1995).

Relating to the brain, we would like the Bayesian theory to be applicable to nonhuman animals, presumably without symbolic mathematics. An open question is how a Bayesian decoder can be realized by a neural circuit (Gold and Shadlen 2001; Jazayeri and Movshon 2006; Zhang et al. 2022), downstream from stimulus representation in a neural circuit where vertical judgment is produced (Adelson and Movshon 1982; Nichols and Newsome 2002; Liu and Wang 2008). Moreover, exact Bayesian calculations are costly. In the case of aforementioned time integration, a sequence of pieces of evidence $e_i; i = 1,2,\ldots,T$ are accumulated over a large number of time steps, and the posterior entails computing a product of $\sim T$ probabilities (Beck et al. 2008). Such large multiplication of probability distributions is computationally prohibitive (Shi et al. 2010; Sanborn and Chater 2016). Finally, a Bayesian decoder uses the posterior to produce a statistical estimate of s, but the underlying neural process is unspecified; whether that actually happens, in a single trial, and how it would happen neurobiologically remain unknown.

It is worth keeping in mind that any stochastic process can in principle be mathematically framed in two different ways. Consider decision making described by the DDM. Eq. (6.4) is a stochastic differential equation, called the Langevin equation for random Brownian motion in Physics (Chandrasekhar 1943), which generates a stochastic sample path $X(t)$ in a single trial. Equivalently, the process can be described by the Fokker-Planck equation for the probability density $p(X,t)$ of the random variable $X(t)$ (Gardiner 1985; Kampen 2007). The performance and reaction time of the DDM can be efficiently calculated using the Fokker-Planck equation (Ratcliff 1978; Shadlen et al. 2007). It does not follow, however, that in a single trial a Brownian particle somehow represents a probability distribution. This is also true for spatiotemporal dynamics: a decision is made in a single trial by a stochastic neural population activity pattern that unfolds in space and time, whereas their statistics across trials can be described by probability distributions. Therefore, both approaches are useful, but we should distinguish what a system does stochastically from how it may be described mathematically. Bayesian computations in the brain do not necessarily require explicit neural representation of probability distributions in real time by neural population activity, at least for nonhuman animals with the proviso that symbolic mathematics be excluded.

6.3 NEURAL CIRCUIT MECHANISM OF DECISION MAKING

6.3.1 Neural Correlates

How do neural populations in the brain underlie decision making? Electrical activity of a "decision neuron" should signal the subject's choice rather than the physical stimulus, which are opposite from each other in an error trial of a TAFC task (stimulus is A, decision is B). Moreover, a neural correlate of a deliberate decision should reflect temporal accumulation of evidence in favor or against alternative propositions. Electrophysiological studies with behaving monkeys have identified brain regions that exhibit decision-related neural activities.

The vibrotactile delayed discrimination (VDD) paradigm (Romo et al. 1999) discussed in Chapter 5 represents a decision task in the somatosensory domain. Recall that in that experiment, monkeys report a decision based on the comparison of two mechanical vibration frequencies f_1 and f_2 applied sequentially to the fingertips, separated in time by a delay of 3–6 seconds (**Figure 6.9a**). Trained monkey subjects displayed a psychometric function with near-chance performance when f_1 and f_2 are similar and high accuracy when they are significantly different (**Figure 6.9b**).

The behavioral response (signaling whether f_1 is perceived as larger or smaller than f_2) requires the animal to hold in working memory the frequency of the first stimulus across the delay period. It was found that neurons in the premotor cortex (PMC), part of the prefrontal cortex, showed persistent activity during the delay, with the firing rate of memory activity increasing (a "plus cell") or decreasing (a "minus cell") monotonically with the stimulus frequency (Chapter 5, **Figure 5.8**). During comparison decision, neural activity became binary: a "plus neuron" ("+" neurons) showed high firing in trials when the monkey's choice was $f_1 > f_2$, and low firing in trials when the monkey's choice was $f_1 < f_2$ (**Figure 6.9c**). A "minus neuron" ("–" neuron) showed the opposite trend. In this task, working memory precedes decision making, and the same circuit is engaged in both processes. In a modified version of the task, the decision report is postponed a few seconds after the comparison period, and PMC neurons were found to retain monkey's categorical choice in the form of persistent activity across the second delay (Lemus et al. 2007), reinforcing the point that the same circuit is involved in both working memory and decision making. A systematic exploration during the same task across a number of cortical areas

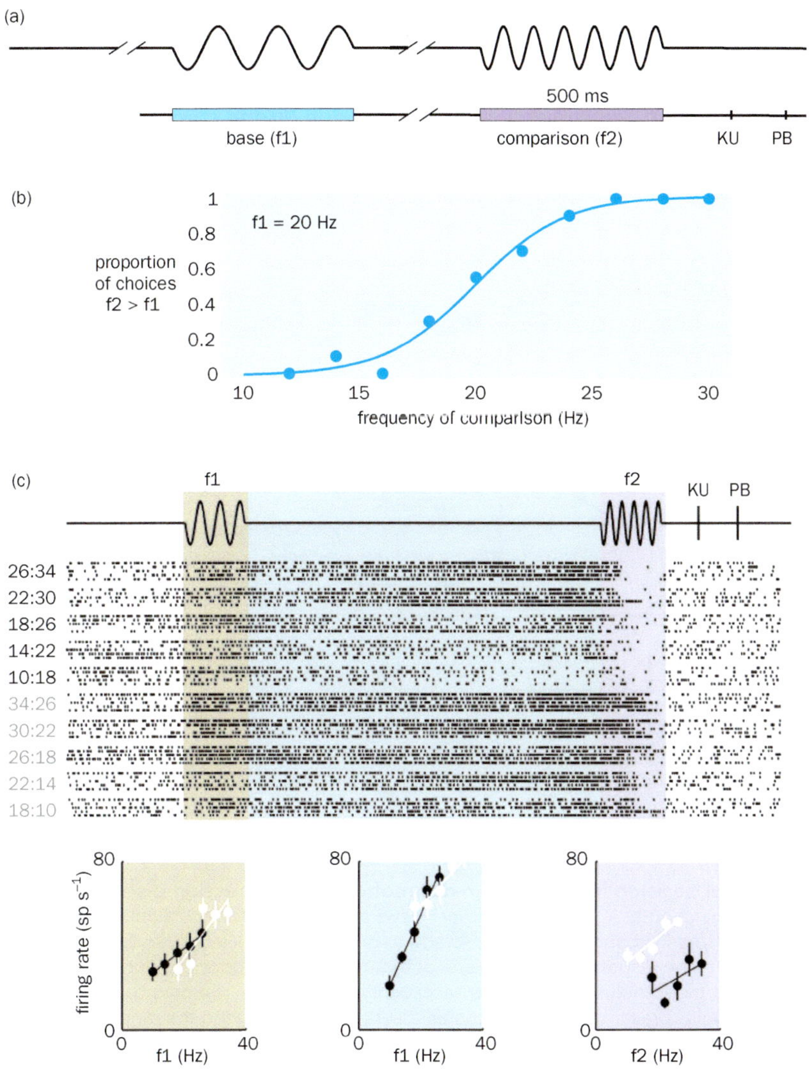

Figure 6.9 Neural correlates of a decision about vibrotactile frequency. (a) Vibrotactile delayed discrimination task. A sinusoidal tactile stimulus at base frequency f_1 is delivered to a finger. After a delay period, a comparison stimulus is delivered at frequency f_2, when the monkey must decide whether $f_2 > f_1$, a decision it indicates by releasing a key (KU) and pressing one of two buttons (PB) with its free hand. (b) Psychometric function. The task is difficult when the base (20 Hz) and comparison frequencies are similar ($f_2 \simeq f_1$). (c) Response of a neuron in the premotor cortex. This neuron carries information about the base frequency during the delay period (blue). In the comparison epoch (purple), the neuron is more active when $f_2 < f_1$ (white) than otherwise (black). Note that for all trials, the difference between the base and comparison frequencies is fixed to 8 Hz across trial types. Reproduced from Gold and Shadlen (2007) with original data from Romo et al. (2004). [Gold, J. I., & Shadlen, M. N. (2007). The neural basis of decision making. Annual Review of Neuroscience, 30(1), 535–574. Original data comes from Romo, R., Hernández, A., & Zainos, A. (2004). Neuronal correlates of a perceptual decision in ventral premotor cortex. Neuron, 41(1), 165–173.]

(primary and secondary somatosensory areas, premotor and lateral prefrontal areas) yielded a rich picture of neural dynamics in these areas as the process unfolds in time (Hernández et al. 2002; Romo et al. 2004; Romo and Rossi-Pool 2020).

In an RDM task, microstimulation studies have shown that while direction-sensititve neurons in the area MT encode the motion stimulus (Newsome et al. 1989; Britten et al. 1992; Britten et al. 1993; Britten et al. 1996), the decision process itself occurs downstream of MT, in the posterior parietal cortex and prefrontal cortex. Shadlen and Newsome (1996) found that activity of neurons in the lateral intraparietal area (LIP) was correlated with monkey's perceptual choice in both correct and error trials. Decision-related activity has also been observed in the prefrontal cortex in monkey (Kim and Shadlen 1999) and human studies (Heekeren et al. 2004; Heekeren et al. 2008). In a fixed-duration version of the RDM direction discrimination task, the monkey was required to maintain fixation and view a RDM stimulus for one second, followed by a delay period, then the monkey signaled its choice by a saccadic eye movement. In contrast to the RT version of the task (Roitman and Shadlen 2002), where the interpretation of neural activity may be confounded because perceptual decision and saccadic response occur concurrently, now spike firing of recorded neurons correlated with decision (during motion-viewing) and that with motor response (after the delay) can be temporally dissociated. Neurons in the lateral prefrontal cortex (Kim and Shadlen 1999) and LIP (Shadlen and Newsome 1996, 2001; Roitman and Shadlen 2002) showed similar activity pattern: their activity reflected the monkey's choice but, while it was a graded function of the motion strength c' during motion-viewing, spiking activity during the delay became insensitive to c' (**Figure 6.10**). The implication was that the subject made the decision during stimulus viewing and maintained the binary choice across the delay to guide a later behavioral

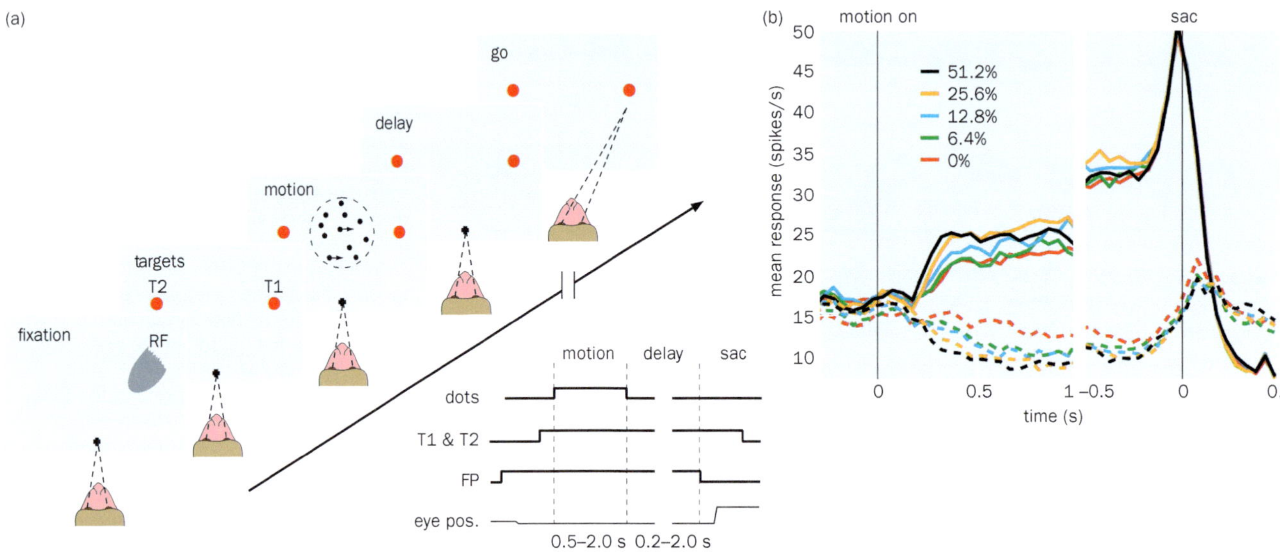

Figure 6.10 Neural dynamics correlated with perceptual decision in the random dot motion (RDM) direction discrimination task. (a) A fixed-duration version of the RDM direction discrimination task. A monkey subject views a centrally presented stimulus, followed by a delayed period of variable durations across trials. The decision was indicated by a saccade to one of two choice targets. (b) Averaged neural activity of LIP in trials when the decision was the target in the response field of the recorded neuron (solid) or opposite from it (dashed), for five coherence levels. The activity time course is aligned by the stimulus onset (left) or saccade initiation (right). The time course and steady-state of neural activity are dependent on c' during stimulus presentation but becomes essentially binary during the mnemonic delay period prior to the saccadic response. Reproduced from Shadlen and Newsome (2001). [Shadlen, M. N., & Newsome, W. T. (2001). Neural basis of a perceptual decision in the parietal cortex (area LIP) of the rhesus monkey. Journal of Neurophysiology, 86(4), 1916–1936.]

response. These results suggest that decision making and working memory can be subserved by the same cortical mechanism, presumably residing in the parieto-frontal circuit.

Rodent studies of perceptual decision making, with rigorous task design and behavioral control as in primate studies, include TAFC paradigms with olfactory (Uchida and Mainen 2003), somatosensory (Li et al. 2015), visual (Harvey et al. 2012), auditory (Brunton et al. 2013; Pardo-Vazquez et al. 2019), and visual and auditory multisensory (Najafi et al. 2020) stimuli. For instance, in an auditory task (Brunton et al. 2013; Erlich et al. 2015; Hanks et al. 2015), random clicks with Poisson-generated inter-click intervals were presented from left and right speakers, and a subject was required to decide which of the two click trains had a higher rate. Behavioral analysis and single-neuron recordings showed that animals summated click events over time, separately for those coming from the left versus right side. Neural correlates of this process were found in the posterior parietal cortex and a frontal cortical area called the frontal orienting field (FOF) (Hanks et al. 2015), as well as subcortical structures (Kopec et al. 2015).

In summary, substantial evidence from monkey and rodent single-neuron physiology suggests that the parietal and frontal cortices form a core brain system for temporal accumulation of data and choice formation in perceptual judgments. By contrast, upstream sensory areas encode momentary evidence based on actual physical stimulus. Nevertheless, spiking activity of sensory neurons commonly displays weak trial-to-trial correlation with an animal's choice, which is quantified by "choice probability" (Britten et al. 1996; Shadlen and Newsome 1996; Parker and Krug 2003). In the RDM task, choice probability is computed as the ROC area like in **Figure 6.2a–b**, except that the two histograms of MT firing rates are defined by the animal's choices rather than by stimulus directions. The choice probability varies across MT neurons, with a mean of ~0.55, slightly above the chance level of 0.5 (which would imply the absence of correlation with behavioral choice). Hence, sensory neurons co-vary weakly trial-by-trial with animal's choice. One interpretation of this result is that in a single trial, a particular fluctuating activity pattern in sensory neurons affects downstream decision neurons, hence the animal's choice, in a feedforward and causal manner. If so, the amount of stochasticity in choice behavior should be smaller when the exact same random dot pattern is repeated across trials, which was found not to be the case in monkey experiments. A more plausible interpretation is that choice probability results from top-down signaling from decision neurons (Wimmer et al. 2015), consistent with the observation that it grows to above chance level slowly over time (Nienborg and Cumming 2009).

6.3.2 A Recurrent Neural Circuit Model

How are decision computations instantiated in a cortical circuit? A clue came with the observation that decision-related neural activity has been reported in cortical areas that typically exhibit mnemonic persistent activity during working memory maintenance. It has been proposed (Wang 2002b) that this is not a mere coincidence, but suggests a common circuit mechanism underlying decision making and working memory. It is not obvious why a circuit capable of mnemonic persistent activity can underlie a decision process; a working memory system dominated by rapid switching between multiple attractors would be inadequate. As we discussed in Chapter 5, biologically realistic modeling revealed that positive feedbacks underlying working memory must be slow. Therefore, a modified view is to conceptualize persistent activity as a form of time integral of a brief input (**Figure 6.11**, left panel). Can the same slow reverberation subserve accumulation of information in decision making (**Figure 6.11**, right panel)?

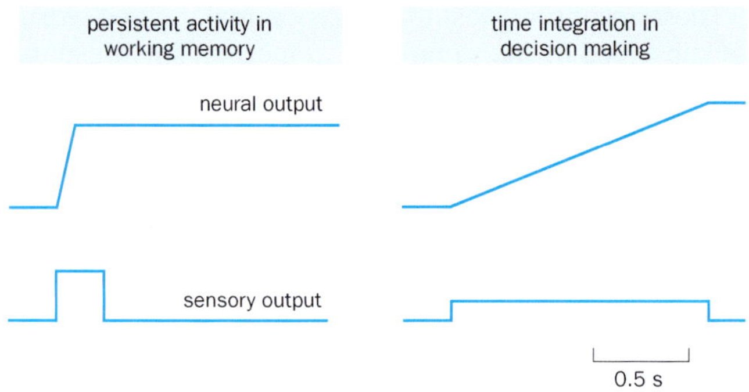

Figure 6.11 Time integration as a unifying concept for working memory and decision making. Persistent activity that outlasts a transient input (left panel) or linear ramping during stimulus presentation (right panel) can be viewed as an integral of the stimulus over time, in the sense of calculus. [Wang, X. (2006). A microcircuit model of prefrontal functions: Ying and yang of reverberatory neurodynamics in cognition (pp. 92–127). In J. Risberg & J. Grafman (Eds.), The Frontal Lobes: Development, Function and Pathology © Cambridge University Press 2006, reproduced with permission.]

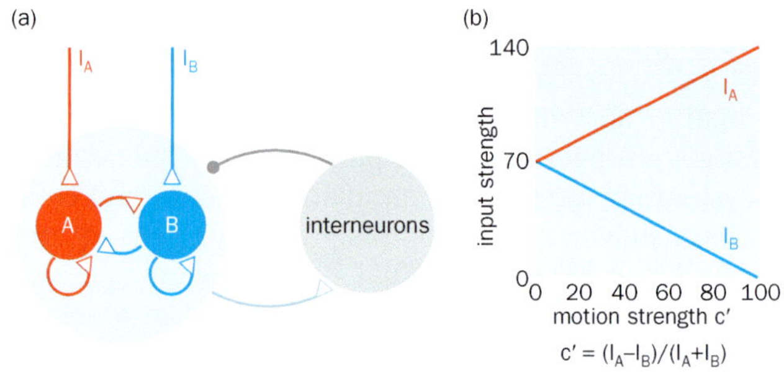

Figure 6.12 A recurrent circuit model of spiking neurons for perceptual decision. (a) Schematic depiction of the model. There are two pyramidal cell groups (A and B), and each is selective to one of the two stimuli (e.g., motion to the right or left). Within each pyramidal neural group, there are strong recurrent excitatory connections; the two neural groups compete through feedback inhibition from interneurons. (b) Inputs I_A and I_B represent momentary sensory evidence. They are sampled from Gaussian distributions with means μ_A and μ_B, respectively, and a standard deviation σ. At every 50 ms, the two stimuli are independently re-sampled using their Gaussian distributions, so that the inputs vary stochastically in time. If $\sigma = 0$, the two inputs would be constant in time. The means μ_A and μ_B represent outputs from MT shown in Figure 6.3. If a stimulus is in favor of A, μ_A increases and μ_B decreases with c' linearly. [Wang, X. J. (2002b). Probabilistic decision making by slow reverberation in cortical circuits. Neuron, 36(5), 955–968.]

To address this question, Wang (2002b) proposed a biophysically based model originally developed for working memory and endowed with the NMDA receptor-mediated slow reverberation (Wang 1999b), for the RDM discrimination experiment. Recall that excitatory-to-excitatory synaptic connections increase a network's operation timescale. A simple estimate of the network's time constant is given by $\tau_{network} = \tau_{syn} / (|1 - w_{rec}|)$ (where τ_{syn} is the synaptic time constant, w_{rec} is the strength of recurrent excitation) (Chapter 3, Eq. 3.25). For instance, if $\tau_{syn} = 100$ ms for the NMDA receptor-mediated synapses (Hestrin et al. 1990), and $w_{rec} = 0.9$, then $\tau_{network} = 1$ second. Thus, the network displays transient dynamics of ramping activity on a timescale of up to 1 second, and this ability critically depends on the NMDA receptors (Wang 2002b; Wong and Wang 2006). Note that this mechanism relies on the slow kinetics of the NMDA receptor-mediated channel and emphasizes the importance of NMDA receptors for online cognition.

Figure 6.12 illustrates such a recurrent neural circuit model. In a two-pool version of the model, subpopulations of spiking neurons are selective for two choice alternatives (e.g., A = left motion, B = right motion). Within each pyramidal neural group, strong recurrent excitatory connections can sustain persistent activity triggered by a transient preferred stimulus. The two neural pools compete through feedback inhibition from interneurons. Conflicting sensory inputs are fed into both neural pools in the circuit, with the motion strength c' implemented as the relative difference in the inputs (**Figure 6.12b**). In particular, when $c' = 100\%$ in favor of A, I_A is large and I_B is zero; on the other hand, $I_A = I_B$ at $c' = 0\%$. The inputs correspond to spiking output from neurons on the sensory coding area MT, which display precisely such dependence on c' for preferred (red) and nonpreferred (blue) stimulus directions (Britten et al. 1993). The perceptual choice is decided based on which of the two neural populations wins a winner-take-all competition, which results from both reverberatory excitation and shared feedback inhibition.

Figure 6.13 shows model behavior in three single trials with c' = 0%, 12.8% and 51.2%. The model captures important features of activity of

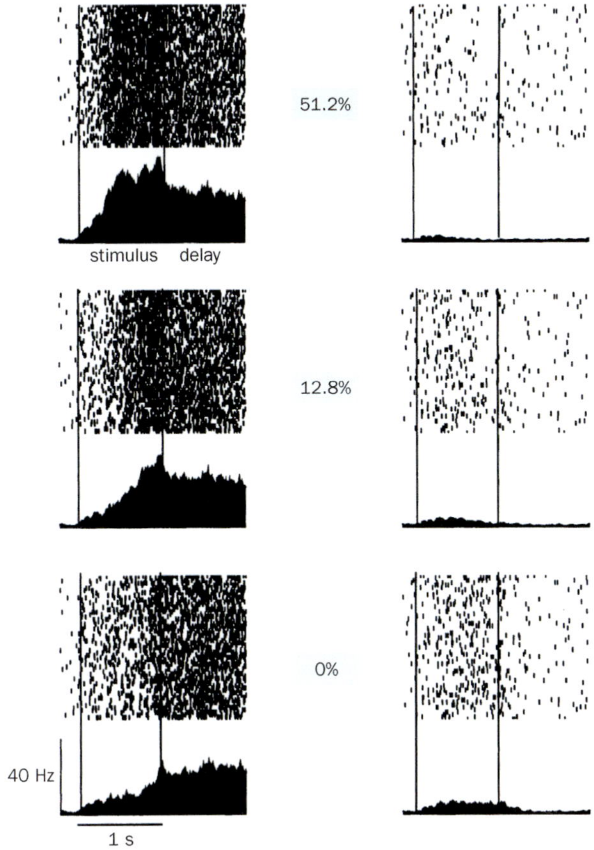

Figure 6.13 A recurrent circuit model of spiking neurons reproduces salient characteristics of decision-correlated neural activity in behaving monkeys. Left panels: Neurons in group A; right panels: Neurons in group B. Three trials are displayed with the signal's coherence $c' = 0\%$ (bottom), 12.8% (middle) and 51.2% (top). In all three cases, the attractor A wins the competition; therefore, the network's choice is said to be A (correct decision for $c' > 0\%$). Similar to the neural activity in LIP, there is a slow time course of activity in group A, with the ramping slope increasing with the signal strength. Moreover, even when the coherence is zero, the firing patterns of the two neural groups diverge dramatically over time during the stimulation, leading to a categorical (binary) decision formed by the network. Finally, the persistent activity in group A during the mnemonic delay period, with a level independent of the stimulus strength, stores working memory of the decision. [Wang, X. J. (2002b). Probabilistic decision making by slow reverberation in cortical circuits. Neuron, 36(5), 955–968.]

LIP cells recorded from behaving monkeys. First, neural activity is primarily correlated with the decision choice (even when the motion strength is zero). Second, the neural decision process proceeds in two steps: sensory data are first integrated over time in a graded fashion, followed by a winner-take-all competition leading to a binary choice. Third, after the stimulus is withdrawn, the network stores the decision choice in working memory, in the form of persistent activity that is insensitive to c'.

Note that with $c' = 0\%$, the input is identical to the two neural groups and the system is completely symmetrical, yet strong stochastic circuit dynamics produces a winner and a loser (**Figure 6.14**). At the stimulus onset, the firing rates (r_A and r_B) of the two neural populations initially ramp up together for hundreds of milliseconds, before diverging from each other when one increases (by virtue of recurrent excitation within that neural pool), while the other declines. The phenomenon of asymmetric states emerging from a symmetric system is called "spontaneous symmetry breaking" (Anderson 1963; Nambu 2009).

In addition to time courses $r_A(t)$ and $r_B(t)$ (**Figure 6.14a**), the network dynamics can also be shown in the two-dimensional state space where $r_A(t)$ is plotted against $r_B(t)$, which is shown for six single trials in **Figure 6.14b**. In this plot, the network is initially at a resting state where r_A and r_B are both low and symmetrical. In response to an external stimulus, the system traces a trajectory parameterized by time t. With $c' = 0\%$, the inputs are the same in average, the two neural populations first increase their firing activities along the diagonal line corresponding to the initial phase of ramping activity when $r_A(t)$ and $r_B(t)$ are indistinguishable (**Figure 6.14a**). The winner-take-all dynamics leads to a divergence of $r_A(t)$ and $r_B(t)$, and the trajectory moves towards one of the two attractor states where $r_A(t)$ is high and $r_B(t)$ is low or vice versa, and the choice is random across trials.

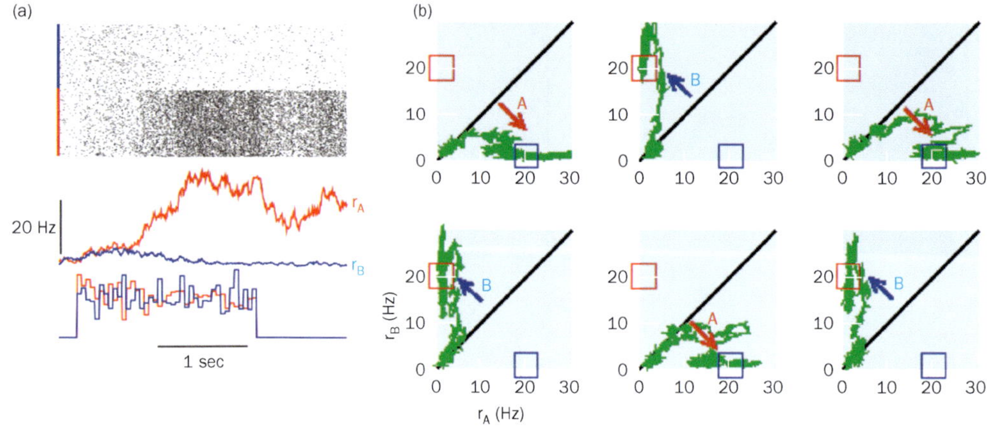

Figure 6.14 Stochastic attractor dynamics in the state space. (a) Neural population firing rates of two neural pools are shown as a function of time in a single trial with $c' = 0$. (b) The two population firing rates r_A and r_B are plotted against each other with zero coherence for six sample trials. Even though the model network is symmetrical and inputs have identical means, the strongly recurrent dynamics gives rise to symmetry breaking spontaneously in the presence of noise, leading to a binary choice A or B in single trials.

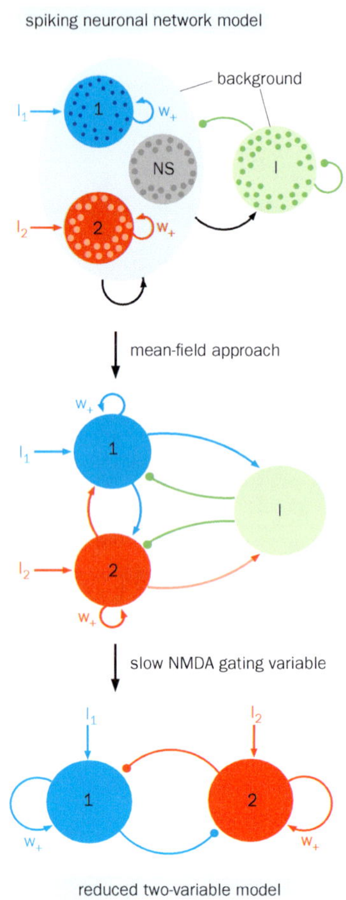

Figure 6.15 Schematic of mean-field reduction from a network of thousands of spiking neurons (top) to an eight-variable (middle) and then two-variable (bottom) population rate model. Reproduced from Wong and Wang (2006). [Wong KF, Wang XJ. A recurrent network mechanism of time integration in perceptual decisions. J Neurosci. 2006 Jan 25;26(4):1314–28. doi: 10.1523/JNEUROSCI.3733-05.2006. PMID: 16436619; PMCID: PMC6674568]

Cross-trial statistics for choice behavior can be described by a probability of 0.5, like tossing of a fair coin at the level of mathematical description. Importantly, this holds true even when external input is completely deterministic, showing that stochasticity is dominated by intrinsic decision circuit dynamics (Wang 2002b), in line with the experimental finding that the psychometric function is virtually identical regardless of whether the sequence of random dots varies across trials. In single trials, a decision is made by high variability of neural population firing (Figure 6.13), without the need for an explicit representation of a probability nor a literal implementation of coin-tossing as a neural process.

6.3.3 State-Space Trajectories of Population Dynamics

The spiking neural circuit model has many interacting single neurons described by thousands of dynamical equations (Figure 6.15, top panel). To shed insights into its inner workings, mean-field theory can be applied to reduce the model into a simpler population rate model (Renart et al. 2003). As shown in Figures 6.13 and 6.14, neurons within a selective pool are dedicated to the same computation, and their activities can be averaged out to yield a population rate. The mean-field calculation produces firing rates for three distinct populations: the two selective neural pools (r_A and r_B) and one inhibitory neural pool (r_I); their inputs are described by the synaptic conductances driven by presynaptic neural activities (Chapter 3, Eqs. 3.12–13). For each of the excitatory neural pools, there are two synaptic variables s_{AMPA} and s_{NMDA} mediated by the fast AMPA and slow NMDA receptors; for the inhibitory neural pool, there is one synaptic variable s_{GABA} mediated by $GABA_A$ receptors. All together, there are eight dynamical variables for the firing rates and synaptic drives (Figure 6.15, middle panel).

Moreover, we are mostly interested in the slow dynamics, which is dominated by the NMDAR-mediated recurrent excitation. All of the other dynamical variables for the firing rates, AMPAR-mediated excitation and $GABA_A$R-mediated inhibition are relatively fast, which can be eliminated using the "fast-slow variable analysis" (Chapter 3, Eqs. 3.14–17) (Rinzel 1987; Wang and Rinzel 1995). Applying this approach to the eight-variable population rate model, we obtain a simple two-variable dynamical system (Figure 6.15, bottom panel) describing the slow gating variables of the NMDA receptor-mediated recurrent synaptic excitation in a self-contained form,

$$\frac{ds_A}{dt} = F(I_A)\gamma(1-s_A) - s_A / \tau_{\text{NMDA}} \quad (6.17)$$

$$\frac{ds_B}{dt} = F(I_B)\gamma(1-s_B) - s_B / \tau_{\text{NMDA}} \quad (6.18)$$

where $F(I)$ is the input-output function of an LIF neuron (see Chapter 2) and the input current I includes both external stimulus and recurrent drive that depend on s_A and s_B. For instance, the recurrent input to the neural pool A is given by $J_{EE}s_A - J_I(s_A + s_B)$, where J_{EE} is the strength of recurrent excitation between neurons within the pool selective for A, and J_I measures the effective inhibition mediated by interneurons whose activity is driven by summated outputs from the two excitatory neural pools. Ditto for the neural pool B. This two-variable model reproduces salient behaviors of the spiking network model (Wong and Wang 2006), such as the psychometric function (**Figure 6.16**), representing a suitable reduced version of the biologically based spiking network model from (Wang 2002b).

Recurrent dynamics underlying decision making in single trials can be better described in the two-dimensional (r_A, r_B) state space of the reduced model. In the absence of noise, one can calculate the steady-states and their stabilities. As shown in **Figure 6.17a**, without external input the network displays three stable steady-states (attractors), and each has a "basin of attraction" indicated by a colored region of the state space. The system starting with any initial condition in a basin of attraction evolves to converge onto the corresponding attractor state. The first attractor corresponds to a resting state of spontaneous activity where both r_A and r_B are low, and its basin of attraction is around the diagonal line. The two other attractor states corresponding to stimulus-selective persistent activity where either r_A is high and r_B is low, or vice versa.

When a random dot motion stimulus is shown at a low coherence c' = 6.4% (**Figure 6.17b**), the attractor landscape is changed. In this example, the resting state attractor disappears. The sensory evidence is in favor of the choice A, so the attractor A has a larger basin of attraction (orange) than that of the attractor B (brown). The system is initially in the spontaneous state, which falls in the basin of attraction A, and evolves toward the decision state A in a correct trial (blue). However, at low c' the bias is not strong, and noise can induce the system's trajectory to traverse the boundary of the two attraction basins, in which case the system eventually evolves to the decision state B in an error trial (red). At increasingly higher coherence (**Figure 6.17c**), the basin of attraction for the attractor A gradually enlarges. As a result, it becomes less likely for the network's trajectory to cross the boundary by noise, and the error rate decreases. For sufficiently high coherence (**Figure 6.17d**), the attractor B corresponding to the erroneous choice disappears, and the performance is perfect. Therefore, the state-space analysis hammers home the point that the system is not rigid but flexible in response to external signals. Attractor states can be created or destroyed by inputs.

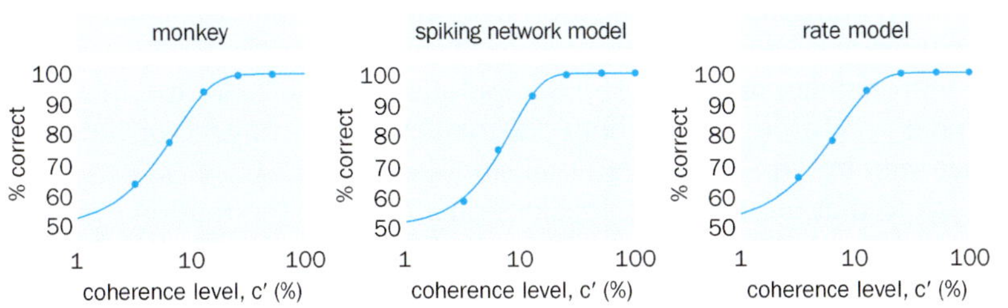

Figure 6.16 Psychometric function of the population rate model. The reduced model (right) captures the behaviors of the spiking network model (middle) and a monkey (left). Reproduced from Wong and Wang (2006). [Wong KF, Wang XJ. A recurrent network mechanism of time integration in perceptual decisions. J Neurosci. 2006 Jan 25;26(4):1314–28. doi: 10.1523/JNEUROSCI.3733-05.2006. PMID: 16436619; PMCID: PMC6674568]

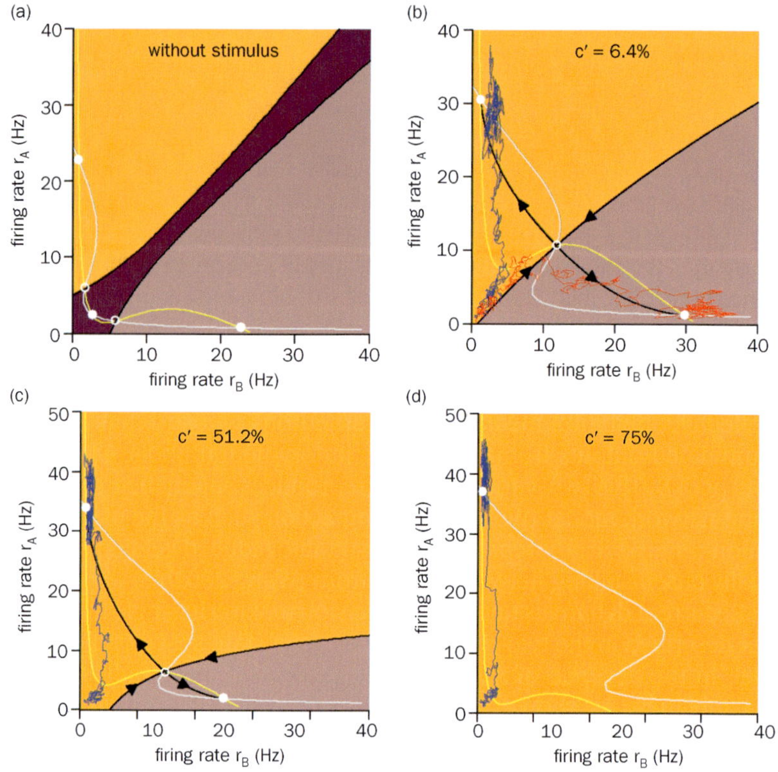

Figure 6.17 Basin of attraction of the neural population rate model. The population dynamics of the mean-field population rate model is displayed in the state space of firing rates r_A and r_B, without external input (a) and in the presence of a motion stimulus at three different c' values (b–d). The attractor landscape sensitively depends on the input. A stable steady-state (attractor) is shown in the filled white circle with its basin of attraction indicated by a colored region, and an unstable steady-state (a saddle) is shown by an open white circle. In (b–c), the arrowed black curves correspond to a separatix, and their intersection defines a saddle fixed point. The boundary of the two attraction basins is given by the stable manifold of the saddle point. [Wang, X. J. (2008). Decision making in recurrent neuronal circuits. Neuron, 60(2), 215–234.] [Wong KF, Wang XJ. A recurrent network mechanism of time integration in perceptual decisions. J Neurosci. 2006 Jan 25;26(4):1314–28. doi: 10.1523/JNEUROSCI.3733-05.2006. PMID: 16436619; PMCID: PMC6674568]

The framework of state-space trajectories offers a new look at time integration in decision making. In the drift diffusion model, the signal and noise are integrated as $\mathcal{S} \sim \mu t$ and $\mathcal{N} \sim \sigma\sqrt{t}$, respectively, thus the signal-to-noise ratio $\mathcal{S} / \mathcal{N}$ grows over time by virtue of integration. By contrast, in a recurrent neural circuit, input currents depend on neural firing itself, and thus noise is part of the signal, and time-domain analysis of the signal-to-noise ratio is not straightforward. Instead, consider the state space of population activity from a dynamical systems perspective. At any given time, define signal as the distance $d(t)$ between the current network state (given by the two neural pool firing rates r_A and r_B) and the boundary that separates the basins of attraction of the two choice attractors. Noise can be quantified by the trial-to-trial standard deviation $(\sigma(t))$ of the network state. The signal-to-noise ratio is $d(t) / \sigma(t)$. At the onset of a RDM stimulus, the initial signal $d(t=0)$ depends on the motion strength c'; it is zero if c'=0, but it is positive for nonzero c' because the network already starts inside the basin of the correct choice attractor. However, the network remains "undecided" as long as $d(t) / \sigma(t)$ is small. As the network evolves further into one of the basins of attraction, both $d(t)$ and $\sigma(t)$ may increase over time, but the ratio $d(t) / \sigma(t)$ grows; therefore, it becomes increasingly unlikely that noise can "bring back" the network to the boundary and transition to the other, alternative, attraction basin. In this sense, one may say that a categorical choice is reached when $d(t)$ becomes much larger than $\sigma(t)$, even though the network may be still far away from the actual attractor state. According to this population dynamics approach to signal-to-noise ratio, there is no need to separately treat external noisy stimulus and internal neuronal stochasticity. The computational benefit of time integration is understood through the network dynamics in an attractor landscape, rather than in terms of the external stimulus alone.

Note that the network does not directly "see" the motion coherence c', only the RDM stimulus, but the decision space landscape is sensitive to the mean input that reflects c'. At a higher c' (51.2% vs 6.4%), the network starts out deeper in the territory of the correct choice attractor, d is larger at the stimulus onset, and the time point of a categorical choice is earlier. With a sufficiently large c', the performance is 100% because, as soon as the stimulus is presented, $d(t)/\sigma(t)$ is already so large that switching to an alternative attractor (an error) is impossible. One can say that the system has already "made up its mind" at the stimulus onset. Note that each state-based depiction corresponds to a (sufficiently long) stationary input. It does not mean that the decision is irreversible; a change in the external input (e.g., reversing the direction of the motion stimulus) can radically alter the attractor landscape, leading to a "change of mind" (Wang 2002b) (see Section 6.7).

The analysis of neural dynamical trajectories can be extended to high-dimensional state space. This dynamical systems approach is increasingly needed as simultaneous monitoring of activities of many single neurons become possible, for instance, using calcium imaging (Helmchen and Denk 2005) or neuropixel electrode recording (Jun et al. 2017). Harvey et al. (2012) recorded ~65 single neurons in the parietal cortex by calcium imaging from behaving mice. In a virtual navigation decision task, animals were shown a cue indicating, after a delay period, which of two turns (left or right) yielded a reward on a given trial (**Figure 6.18a**). Neurons displayed sequential activation, where individual cells showed high activity only in brief periods of time during the delay period (Chapter 5, **Figure 5.38a**), in contrast to persistent activity sustained for a long time. Interestingly, dynamical trajectory in the recorded population's state space is well separated for two alternative choices (**Figure 6.18b**). Moreover, in error trials, the animal's choice decoded from the population activity either started in the wrong direction (**Figure 6.18c**, trials shown in black) or began along the path of the correct choice but later veered to be aligned with the path of the erroneous choice (**Figure 6.18c**, trials shown in grey).

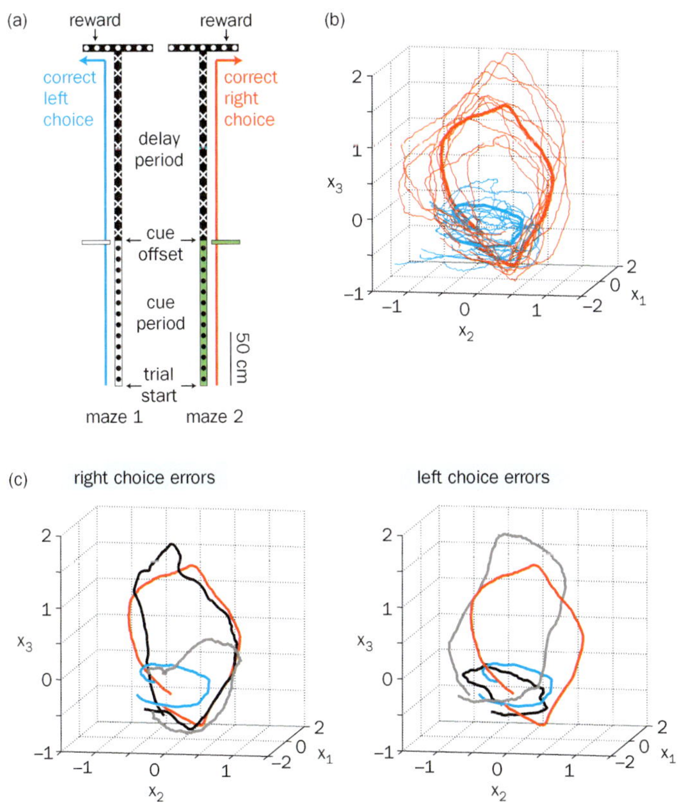

Figure 6.18 Population dynamics of ~65 cells recorded from the posterior parietal cortex in mice performing a virtual navigation decision task. (a) Task design. A cue is followed by a delay period, at the end of which the animal must make a correct turn (left or right) to collect a reward, based on the cue stored in working memory. (b) Trajectories of the population activity, in the state space of three principal components, are choice specific (red: Right choice trials, blue: Left choice trials). (c) Sample trial trajectories in error trials. Two individual trial trajectories (grey and black) are shown together with the mean trajectories for correct right (red) and left (blue) choice trials, for each of the two error types. Reproduced from Harvey et al. (2012). [Harvey, C. D., Coen, P., & Tank, D. W. (2012). Choice-specific sequences in parietal cortex during a virtual-navigation decision task. Nature, 484(7392), 62–68.]

Simultaneous electrical recording from multiple single neurons showed that the population dynamics framework also applies to movement selection and preparation (Churchland et al. 2010; Churchland et al. 2012). In a mouse experiment already discussed previously (Chapter 5, **Figure 5.15b**), this approach was deployed to define a "coding direction" in the state space of neural population activity during motor preparation (Li et al. 2016; Svoboda and Li 2018; Inagaki et al. 2019). Coding direction is basically the orthogonal vector from the boundary of attraction basins (**Figure 6.17**), which proved to be useful in interpreting the results of optogenetic perturbations (Li et al. 2016) and identifying the attractor dynamics (Inagaki et al. 2019). However, the separatrix in **Figure 6.17** is a nonlinear manifold, and manifold discovery from neurophysiological data (Gallego et al. 2017; Low et al. 2018; Chaudhuri et al. 2019) would lead to a refined definition of coding direction in the future.

In summary, the recurrent neural circuit model demonstrates that a single network can subserve different functions, such as decision making during stimulus presentation followed by active memory of the choice across a delay in the fixed-duration RDM task.

6.4 TERMINATION RULE FOR A DECISION PROCESS

6.4.1 Ramping-to-Threshold in the Brain

In the reaction time version of an RDM discrimination task (Roitman and Shadlen 2002; Huk and Shadlen 2005), the subject's reaction time (RT) increased by ~ 400 ms from the easiest (with $c' = 100\%$) to the hardest (with $c' = 3.2\%$) (**Figure 6.7b**). LIP cells exhibited characteristic firing time courses that reflected the monkey's response time and perceptual choice (**Figure 6.19a**). From the onset of a random dot motion stimulus until the time the monkey produced a choice response by a rapid saccadic eye movement, spike activity of LIP neurons selective for a particular saccadic target increased for hundreds of milliseconds. The ramping slope was larger with a higher c' (when sensory information has a higher quality). Furthermore, it was observed that at the time of saccade onset, the firing rate of LIP neurons (selective for a choice) appeared to reach a fixed level that was independent of c' and the response time (**Figure 6.19a–b**), suggestive of a ramping-to-threshold process, although threshold-crossing could be causally related to either a perceptual decision or a motor response.

Human studies on the physiological correlates of RTs in perceptual discrimination tasks began in the 1980s, with the development of event related potential (ERP) measurements (Gratton et al. 1988). Interestingly, the P300 component of ERP in the parietal cortical area displayed a time-to-peak that increased with the difficulty of stimulus discriminability and RT (Kutas et al. 1977; McCarthy and Donchin 1981). More recent EEG studies confirmed that ERP signal from the central parietal cortex is correlated with perceptual decision making (O'Connell et al. 2012; Kelly and O'Connell 2013; van Vugt et al. 2019). The centro-parietal positivity amplitude built up over time with a slope that increases with motion coherence, revealing a ramping-to-threshold process in the human brain (**Figure 6.20**) (Kelly and O'Connell 2013).

In neurophysiological experiments where one cell is recorded at a time, ramping activity is usually reported as trial-averaged neural activity. This leaves the question open as to whether a spike train in a single trial indeed displays a quasi-linear ramp of firing rate. Alternatively, neurons could actually undergo a sudden jump from one rate to another at a particular moment in a trial, and the jumping time varies from trial to trial in such a way that the trial average shows a smooth ramp

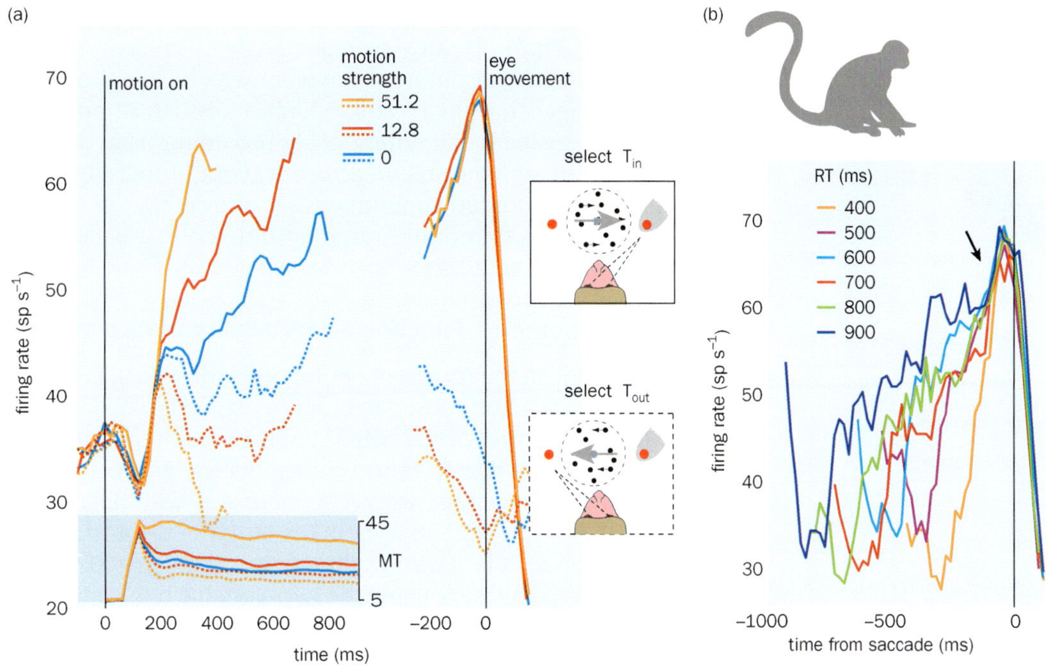

Figure 6.19 Firing activity of LIP neurons during decision formation in time. (a) Average firing rate from 54 LIP neurons is shown for three levels of task difficulty. Responses are grouped by motion strength and direction of choice, as indicated (T_{in}, T_{out}: Trials where the choice was towards or away from the response field of a recorded neuron, respectively). Left: The responses are aligned to onset of random dot motion. Insert: Average responses from direction selective neurons in area MT to motion in the preferred and antipreferred directions. After a short transient, MT responds at a nearly constant rate. Right: Replotted to be aligned to the eye movement, LIP neural activity ramps up to a fixed rate level at the time of saccadic response, regardless of the motion strength. (b) Responses grouped by RT. Only T_{in} choices are shown, revealing that ramping slope decreases with longer RTs. Reproduced from Gold and Shadlen (2007), with original data from Roitman and Shadlen (2002) and Britten et al. (1992). [Gold, J. I., & Shadlen, M. N. (2007). The neural basis of decision making. Annual Review of Neuroscience, 30(1), 535–574. Original data comes from Roitman, J. D., & Shadlen, M. N. (2002). Response of neurons in the lateral intraparietal area during a combined visual discrimination reaction time task. Journal of Neuroscience, 22(21), 9475–9489 and Britten, K. H., M. N. Shadlen, W. T. Newsome, and J. A. Movshon (1992). The analysis of visual motion: A comparison of neuronal and psychophysical performance. J. Neurosci. 12, 4745–4765.]

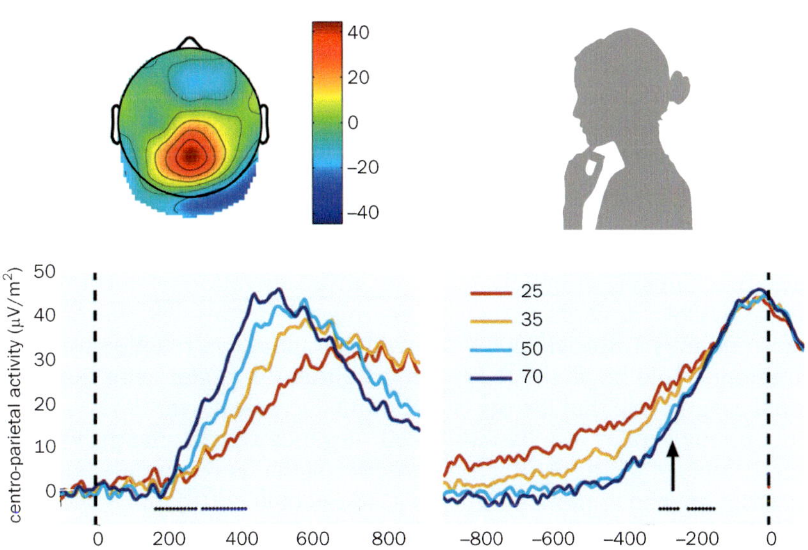

Figure 6.20 Accumulate-to-threshold signals in humans performing an RDM task. The centro-parietal positivity aligned to stimulus onset displays ramping with a slope that increases with motion coherence (left), and reaches a fixed threshold at the time of response execution (right). Color bar: Motion coherence c' = 25%, 35%, 50%, 70%. Reproduced from Kelly and O'Connell (2013). [Kelly, S. P., & O'Connell, R. G. (2013). Internal and external influences on the rate of sensory evidence accumulation in the human brain. Journal of Neuroscience, 33(50), 19434–19441.]

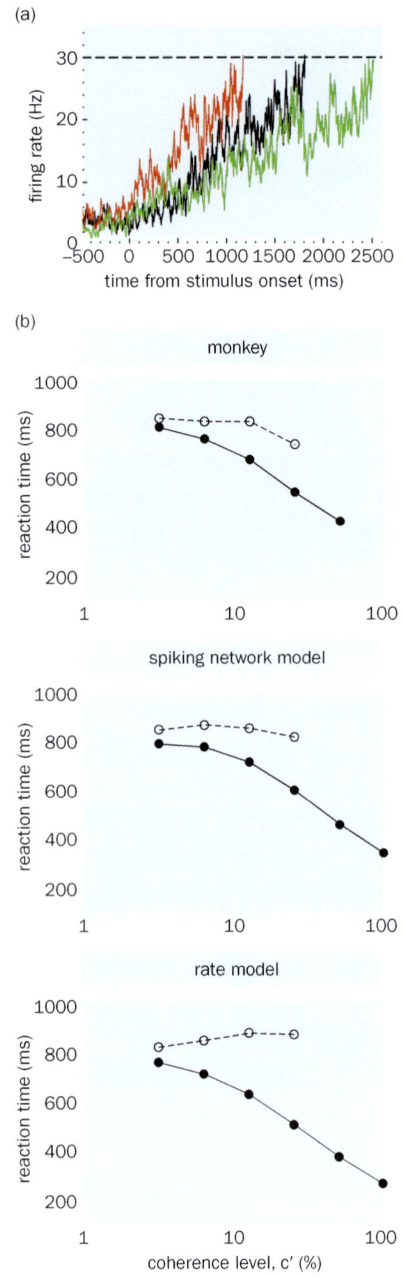

Figure 6.21 Chronometric function of the population rate model. (a) Ramping activity in three sample trials of the spiking network model; the time to cross a preset threshold level (horizontal dashed line) is read out as the decision time. (b) Mean RT as a function of c' is shown for monkey (top), spiking network model (middle) and the two-variable reduced model (bottom). Filled circle: Correct trials; open circle: Error trials. Reproduced from Wong and Wang (2006). [Wong KF, Wang XJ. A recurrent network mechanism of time integration in perceptual decisions. J Neurosci. 2006 Jan 25;26(4):1314–28. doi: 10.1523/JNEUROSCI.3733-05.2006. PMID: 16436619; PMCID: PMC6674568]

(Okamoto et al. 2007; Latimer et al. 2015). Either neurons in a population jump their firing rates synchronously at the same time, or neurons jump sequentially at different times, in which case the population average in a single trial might still exhibit a smooth ramp. These possibilities can be tested by simultaneously recording from a large population of neurons in single trials. In a recent work, hundreds of single cells in the LIP were recorded simultaneously of monkeys performing the RDM task, and it was found that ramping time course occurs in single trials (Steinemann et al. 2022).

6.4.2 Chronometric Function and Scale Invariance of Reaction Times

Like the DDM, in the recurrent neural network model, an appropriate firing rate threshold can be chosen in order to extract a reaction time (RT), as the time it takes for one of the neural pools to reach the threshold before the other neural pool does (**Figure 6.21a**). RT varies stochastically from trial to trial, the trial-average RT, $\langle RT \rangle$, increases with decreasing coherence c' (**Figure 6.21b**). Note that the RT is larger in error trials than in correct trials for monkey's behavior (Roitman and Shadlen 2002), the spiking network model (Wang 2002b) and the two-variable reduced rate model (Wong and Wang 2006). Dynamical systems analysis (**Figure 6.17b**) revealed that in such an error trial, the network's trajectory moves close to a saddle (Chapter 3, **Figure 3.23a**), whose separatrix defines the boundary of the two attraction basins. Dynamics near a saddle point slows down, which explains longer reaction times when the response is erroneous.

Figure 6.22a shows both the mean and standard deviation of RT as a function of c' from the monkey experiment (top) and spiking network model (bottom). Notably, the standard deviation increases concomitantly with the mean. Indeed, the model predicts that one is a linear function of the other, which is confirmed by monkey's behavioral data (**Figure 6.22b**). In other words, the variability of RT is proportional to the mean.

The histograms of RTs are considerably different for different c' values in the model (**Figure 6.23a**, top), but become virtually identical when replotted as a function of $RT/\langle RT \rangle$ (**Figure 6.23a**, bottom) (Wang 2006b). This model prediction is supported by monkey's behavioral data (**Figure 6.23b**). Therefore, the entire distribution of RTs is approximately "scale invariant" in the sense that the statistics of RTs normalized by its mean remain the same across task conditions.

The drift diffusion model produces reaction time histograms with long tails (reflecting unusually long RTs) (Luce 1986; Ditterich 2006), inconsistent with the monkey experiment. Unlike the DDM, in the recurrent neural circuit model, the ramping slope is not exactly constant but eventually increases (**Figure 6.21a**). This is because the positive feedback mechanism ultimately leads to an acceleration of the ramping neural activity towards a decision bound, preventing excessively long decision times. Ditterich (2006) showed that monkey's reaction time distributions could be well fitted by an accumulator model, with the additional assumption that the decision bound decreased over time, which is functionally similar to a hypothesized "urgency signal" that grows over time (Hanks et al. 2007). Such an urgency signal is commonly added to the DDM applied to monkey studies (Carland et al. 2019), but evidence is mixed (Hawkins et al. 2015). In the recurrent circuit model, the desired effect can be accomplished by a temporally increasing ramping slope, which naturally occurs by virtue of synaptic reverberation without additional assumptions. Noteworthy is that, the DDM with the addition of an urgency signal has a half dozen adjustable parameters to fit data, which are comparable to those of the spiking network model (Chapter 5, Section 5.3.3).

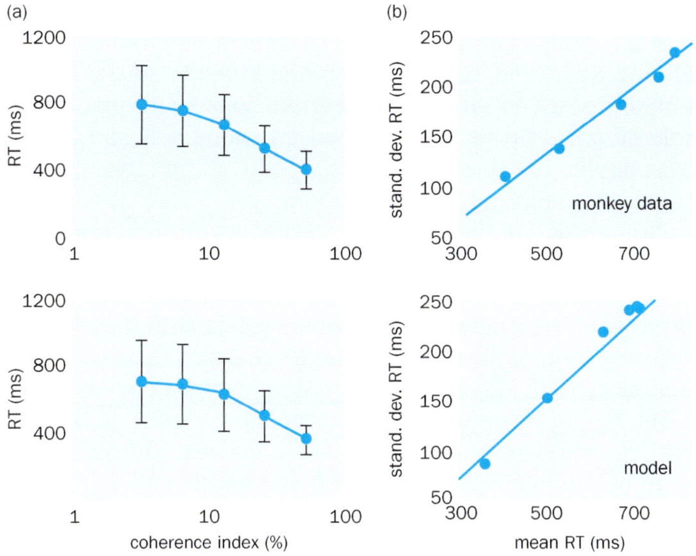

Figure 6.22 The variability is proportional to the mean of reaction times. (a) Standard deviation of RT is shown on the chronometric function from monkey's behavior (top) (Roitman and Shadlen 2002) and spiking network model (bottom) (Wang 2002b). (b) The standard deviation is a linear function of the mean of RT.

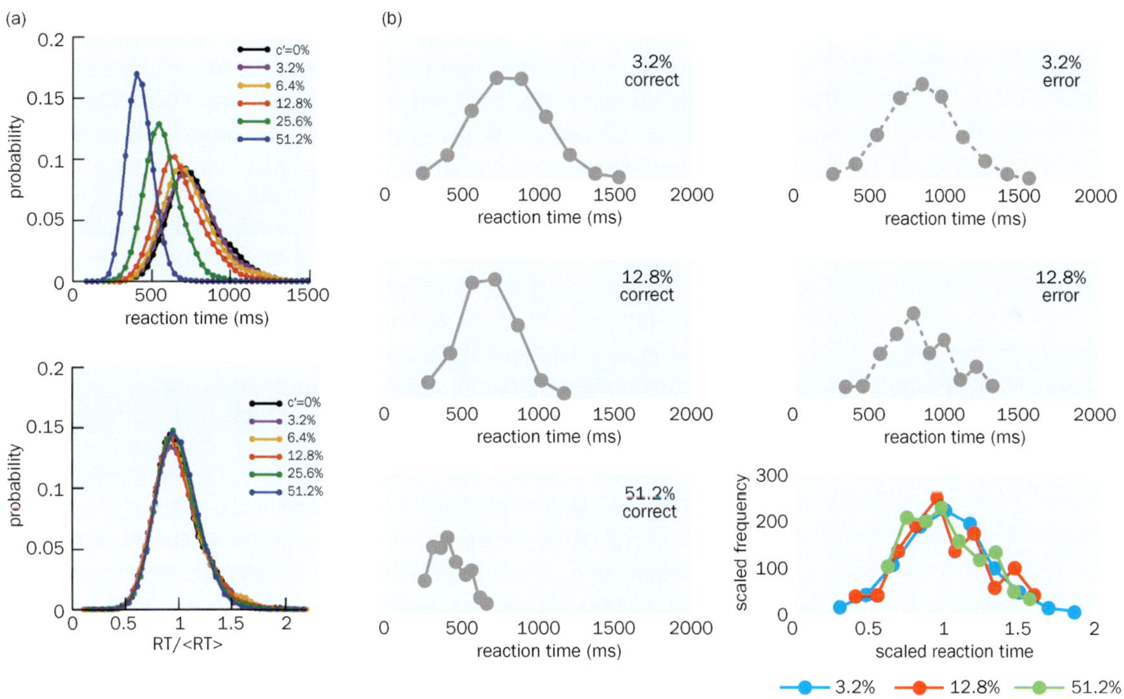

Figure 6.23 Scale invariance of RT distributions. (a) In the spiking network model, the RT distribution is wider and shifted to larger RT values with lower coherence (top), but becomes virtually identical when replotted against RT divided by its mean (bottom). (b) Observed monkey RT histograms in correct and error trials (grey), combined and replotted for three coherence levels (colored), display scale invariance.

6.4.3 The Biological Substrate of a Decision Threshold

How do we know precisely when a graded accumulation process ends and a categorical decision is formed? Although decision threshold is a critical behavioral measure in reaction time tasks, it should not be equated to a general decision rule that terminates an accumulation process. This is clearly demonstrated in the fixed-duration version of the RDM task in which the viewing time is controlled externally, and the subject is required to refrain from making an overt response until either at the stimulus offset (Britten et al. 1992; Kiani et al. 2008) or after a mnemonic delay period (Shadlen and Newsome 2001; Roitman and Shadlen 2002). The subjects, and LIP neurons, do not appear to integrate sensory information through the whole stimulus presentation period (provided it is sufficiently long). One possibility is that threshold crossing may still be the rule for terminating time integration of evidence (Kiani et al. 2008). If so, in the fixation duration version of the RDM task, neural firing rates should reach the same fixed level and stay there regardless of the motion coherence. This prediction, however, is contradicted by the observation that firing activity LIP neurons reaches a plateau level, which varies with the motion coherence (**Figure 6.10**). At the neural circuit level, a simple explanation is that the system eventually reaches a steady-state, which depends on c' (Wang 2002b). The neural integration process runs its course without triggering an overt behavioral response, presumably because downstream movement neurons are inhibited by virtue of an external cue (e.g., the fixation signal) and/or internally generated control signals (see Chapter 8).

The state-space framework puts the emphasis on population dynamics rather than single neurons. According to this population dynamics view, the notion of a decision threshold is not really fundamental, because the decision is made as long as the system has gone into the basin of a choice attractor state far enough, even though the neural firing rate may not yet have attained a pre-set threshold level. Algorithmically and biologically, the reaction time should be interpreted as the time at which neural signals from a decision circuit are read out by a downstream motor system. Consistent with this perspective is the common observation that, in a visual search task, behavioral reaction time co-varies with the time it takes for the firing rates of competing neurons (selective for different actions) to begin to significantly diverge from each other in areas involved in saccade generation, such as the FEF (Sato et al. 2001; Bichot and Schall 2002; Woodman et al. 2008; Cohen et al. 2010; Heitz et al. 2010), the SC (McPeek and Keller 2002) and the LIP (Ipata et al. 2006; Thomas and Paré 2007; Balan et al. 2008).

Hence, to understand the biological meaning of a decision threshold, one needs to consider a motor command system engaged in the generation of a behavioral output. A well-established example is an escape decision triggered by crossing a synaptic threshold in the motor system responsible for behavioral responses to threatening stimuli (Evans et al. 2018). In the case of primate oculomotor tasks, a natural candidate is the superior colliculus (SC) located in the midbrain. The SC plays a powerful role in selecting, preparing and initiating saccadic eye movements (Sparks 2002; Hall and Moschovakis 2003). In the SC, superficial neurons receive direct inputs from the retina, providing a mechanism for rapid reflexive visuomotor responses. In the intermediate layer, movement neurons are also retinotopically organized, but their synaptic drives originate from the cortical FEF and LIP. They are selective for saccade amplitude and direction, and therefore form a map for saccades. These movement neurons display virtually no spontaneous activity and fire a stereotypical burst of spikes immediately before a saccade is initiated (**Figure 6.24a**) (Wurtz and Goldberg 1971; Munoz and Fecteau 2002). We shall focus on the SC as a saccade command center, while keeping in mind that the FEF (Bruce and Goldberg 1985; Hanes and Schall 1996), cerebellum and other brain areas jointly form a multi-regional network for controlling the eye movements.

In a multiple-module model (Lo and Wang 2006), excitatory accumulator neurons in the cortex project to movement neurons in the SC (**Figure 6.24b**). This model also includes a direct pathway in the basal ganglia, which is known to play a major role in controlling voluntary movements (Hikosaka et al. 2000). As a neural pool in the cortex ramps up in time, so do their synaptic inputs to the corresponding pool of SC movement

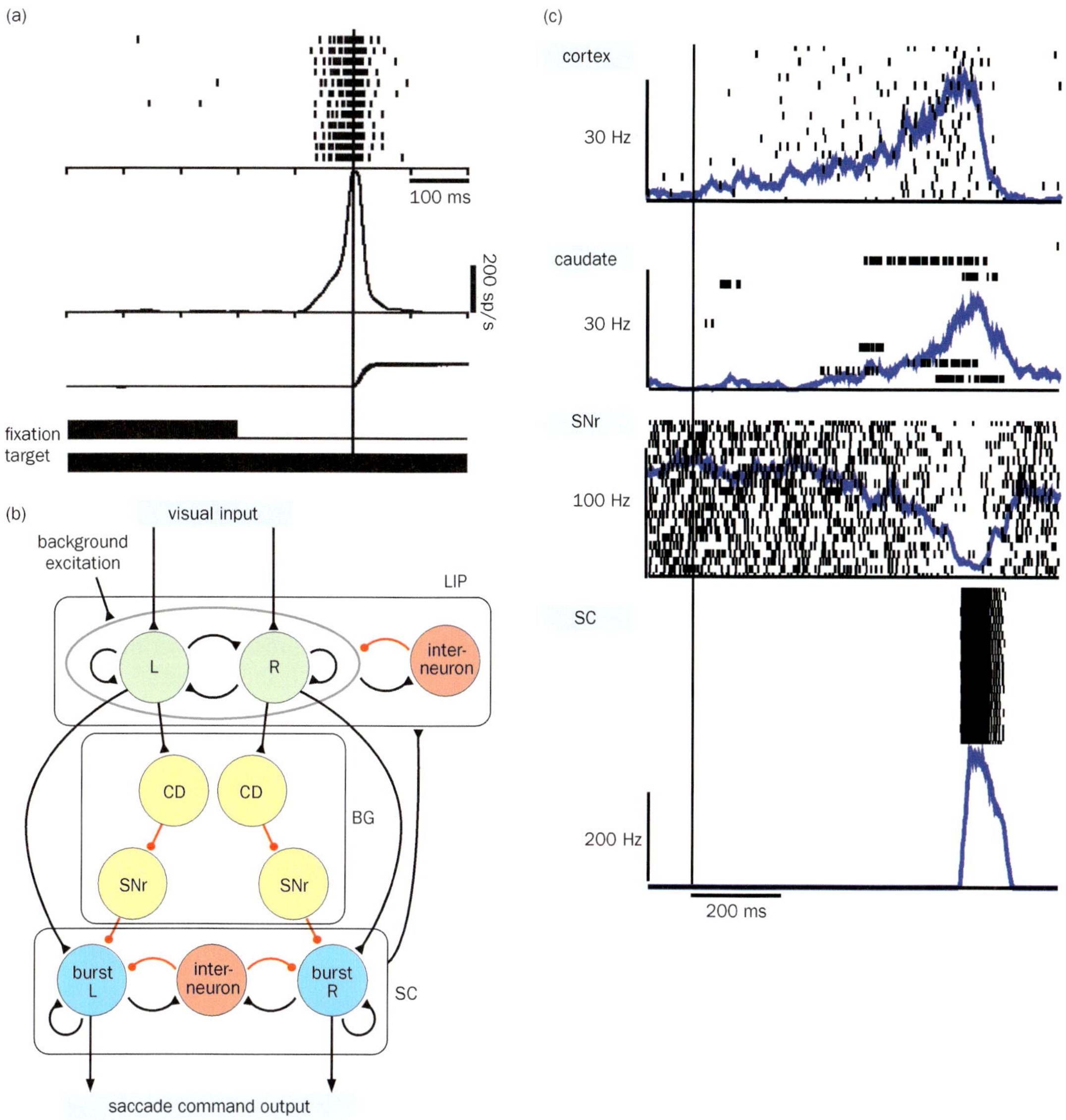

Figure 6.24 A cortico-basal ganglia circuit model for decision threshold. (a) In a monkey experiment, a visual target is shown, but saccade (vertical line) to it was withheld during the fixation period. A movement neuron in the superior colliculus (SC) is silent until it fires a brief burst at the time of saccade. Top: Spike raster, middle: PTSH, bottom: Eye position, which displays a sudden change at saccade onset. (b) Schematic model of spiking neurons with a cortical module (Cx) as in Figure 6.12, a premotor module exemplified by SC for saccadic eye movement, and a basal ganglia module. (c) A single-trial model simulation, showing ramping activity in cortical neurons, resulting in gradual increase of cortical excitatory inputs in neurons of the caudate nucleus (CD), which in turn leads to increased inhibition into the substantia nigra pars reticulata (SNr). There is a strong nonlinearilty in the SC, where neurons remain below the firing threshold until the total synaptic drive exceeds a threshold due to a confluence of increased cortical excitation and decreased inhibition from basal ganglia, leading to a brief burst of firing at more than 200 Hz. Panel (a) is reproduced from Munoz and Wurtz (1995), (b–c) from Lo and Wang (2006). [Munoz, Douglas P., et al. (2000) On your mark, get set: Brainstem circuitry underlying saccadic initiation. Canadian Journal of Physiology and Pharmacology 78(11), 934–944.] [Wang, Xiao-Jing, (2008). Decision making in recurrent neuronal circuits, Neuron 60(2), 215–234.]

neurons, as well as neurons in the input layer caudate (CD) of the basal ganglia. The CD cells in turn inhibit neurons in the substantia nigra reticulata (SNr), which is the output from the basal ganglia, thereby disinhibiting the SC. The SC model circuit is endowed with a switch-like nonlinearity. When the total drive, combining cortical excitation and release of inhibition from the basal ganglia, exceeds a threshold level, an all-or-none burst of spikes is triggered in the SC movement cells, signaling a particular (A or B) motor output. In this scenario, a decision threshold (as a bound of firing rate of cortical decision neurons) is instantiated by a hard threshold of synaptic input for triggering a special event in downstream motor neurons. **Figure 6.24c** shows a sample trial of such a model simulation of the RDM direction discrimination experiment.

The rate of ramping activity fluctuates from trial to trial, as a result of stochastic firing dynamics in the cortical module, and is inversely related to the decision time (as defined by the time when a burst is triggered in the SC) on a trial-by-trial basis (**Figure 6.25**). Moreover, when the task is more difficult (with a lower motion coherence), ramping activity is slower, leading to longer reaction times. However, the threshold of cortical firing activity that is read out by the downstream motion system has the same mean and narrow distribution, regardless of the ramping speed or reaction times. Therefore, the variability of reaction times is mostly attributed to the irregular ramping of neural activity itself. This model reproduced the monkey's behavioral performance and reaction times quantitatively (Lo and Wang 2006).

Consistent with the model, a recent experiment with monkeys performing the RDM reaction time task found that burst firing in SC is correlated with the termination of a deliberative decision (Stine et al. 2023). In another study not with mammals but with larval zebrafish, Bahl and Engert (2020) used RDM stimuli and examined the induced direction of animal's swimming as a readout of subjective direction discrimination. Whole-brain imaging revealed a neural circuit with three neural clusters: an evidence accumulator cluster excites while a "dynamical threshold" cluster inhibits, and a motor command cluster is responsible for swimming output. This is roughly similar to the Cx-BG-SC circuit, with BG playing the role of dynamically controlling the decision threshold.

While here we focus on response execution, FEF and SC are also involved in other aspects of oculomotor decision and response selection (Basso et al. 2021). In addition to movement neurons, SC "buildup" or "prelude" neurons display ramping activity (Munoz and Wurtz 1995), with a

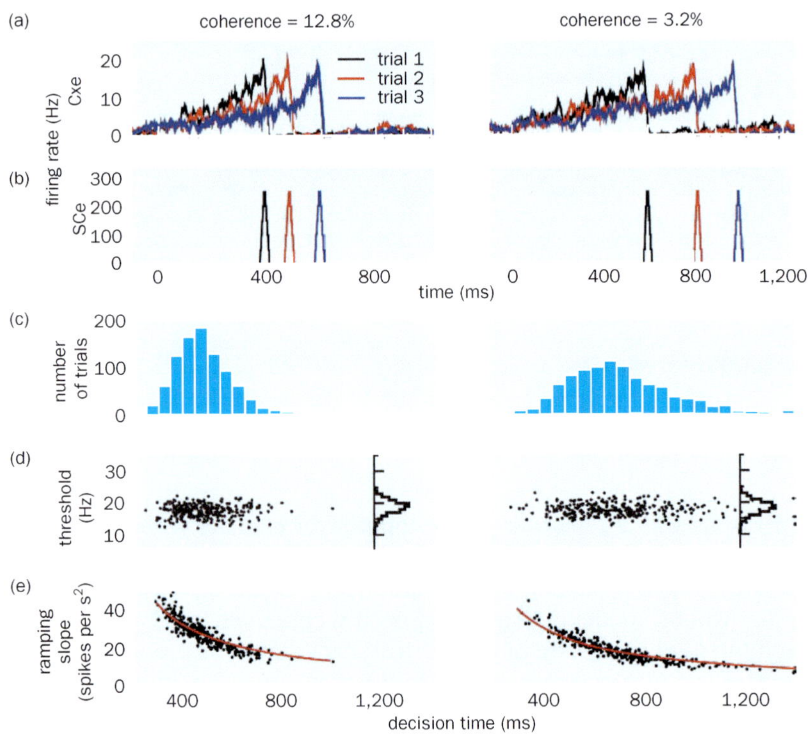

Figure 6.25 Invariance of the threshold across trials, regardless of decision times and coherence levels (c' = 12.8%, left column; 3.2%, right column). (a–b) Three single trials. Firing rate traces of a selective excitatory neural population in the cortex (Cxe) (a) indicate that Cx always triggers a burst in a saccade-selective excitatory population in the SC (SCe) (b) when the firing rate of Cxe reaches about the same level. (c) The distribution of decision times. (d) Threshold level (defined as the Cx firing rate that triggers an SC burst) as a function of the decision time (each data point corresponds to an individual trial). Insert on the right: Distribution of threshold values. The threshold is independent of the decision time, even though the latter varies considerably with the motion coherence as shown in (c). (e) The ramping slope of Cx firing rate is inversely related to the decision time. Reproduced from (Lo and Wang 2006). [Lo, C. C., & Wang, X. J. (2006). Cortico-basal ganglia circuit mechanism for a decision threshold in reaction time tasks. Nature Neuroscience, 9(7), 956–963.]

slope correlated with the motion coherence in the RDM task (Horwitz and Newsome 2001). These prelude neurons have been shown to bias decision threshold (Crapse et al. 2018). Importantly, the SC projects back to the cortex through an intermediate stage of mediodorsal thamalus (Sommer and Wurtz 2002); therefore, a more complete picture requires an understanding of this closed loop circuit. Finally, the actual trigger for saccade is downstream from SC, in the brainstem (Moschovakis et al. 1996; Munoz and Fecteau 2002), where additional control mechanisms set the ultimate threshold for a behavioral response. In summary, the main insight from computational modeling is that a decision threshold is defined in the motor system, providing a neurobiological explanation of "setting a bound" for spiking activity in neural accumulators.

6.4.4 Speed-Accuracy Tradeoff

The ability to dynamically adjust speed versus accuracy is a salient feature of decision making (Wickelgren 1977; Gold and Shadlen 2002; Wang 2008b; Heitz and Schall 2012): if "to get it right" is the priority, we slow down to gather more information and strive for a better performance. On the other hand, if time is at a premium (e.g., when detecting a predator), we make a quick judgment at the potential cost of less accuracy. Adjustment can often be made quickly, "on the fly," either based on task instruction, a changing environment or a controlled perturbation inside the brain (Edwards 1965; Wickelgren 1977; Luce 1986; Palmer et al. 2005; Forstmann et al. 2008; Peixoto et al. 2021).

Speed-accuracy tradeoff (SAT) can be realized by adjusting a decision threshold of an integrator, such as that described by the drift diffusion model (Bogacz et al. 2010). In this framework, under speed emphasis the decision threshold is decreased, so it can be reached faster to trigger a response, whereas accuracy emphasis is instantiated by an increase of the decision threshold in order to integrate more information before a decision is made. Equivalently, the baseline of the decision variable may be adjusted instead of the decision threshold, since what matters is the distance that the integrator travels from its initial state to the decision threshold.

Can a decision threshold be adaptively tuned in the Cx-BG-SC circuit? In the neural circuit model, if we define the decision threshold as the cortical firing needed to induce a burst response in the downstream SC neurons, then one would expect that this threshold could be adjusted by plastic changes in the cortico-collicular pathway. The synaptic drive is the product of the connection weight and presynaptic firing rate; with an enhanced cortico-collicular strength, the same level of cortical drive to the SC could be achieved with less firing of cortical neurons. Interestingly, this is not the case when the system is gated by the basal ganglia (**Figure 6.26a–b**). The reason is that neurons in the SNr normally fire tonically at a high rate (**Figure 6.24b**) and provide a sustained inhibition to SC movement neurons (Hikosaka et al. 2000). Ramping activity in the cortex activates CD neurons, which in turn suppresses the activity in the SNr, and the resulting release from inhibition is required for SC neurons to produce a burst output. The highly nonlinear disinhibition mechanism implies that the decision threshold is much more readily adjustable by tuning the synaptic strength of cortico-striatal pathways than by changes of the cortico-collicular synaptic strength (Lo and Wang 2006). This conclusion still holds when the indirect pathway of the basal ganglia is added, with the proviso that the direct (Go) pathway is balanced with the indirect (Stop) pathway (Wei et al. 2015).

This theoretical finding is particularly appealing in light of the fact that cortico-striatal synapses represent a prominent target of innervations by dopamine neurons. Given that long-term potentiation and depression of the cortico-striatal synapses depend on the dopamine signals (Reynolds et al. 2001; Shen

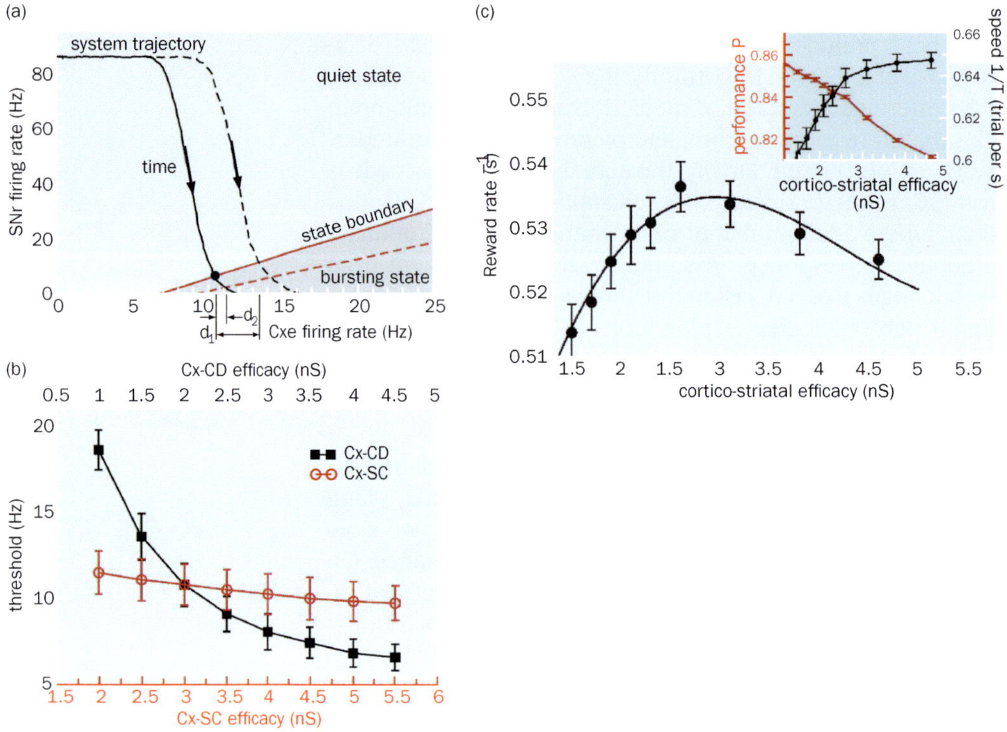

Figure 6.26 Speed-accuracy tradeoff by changing the cortico-striatal connection strength. (a) SNr and Cx firing rates are plotted against each other, showing two SC regimes: Quiet state (light region) and burst firing state (dark region). Two trajectories parametrized by time (arrow) are shown, starting at the upper left conner with high SNr and low Cx firing rates. The decision threshold (black circle) is determined by crossing the state boundary. The position of the state boundary can be shifted by increasing the efficacy of Cx-SC synapses (red dashed to solid curve), and the position of the dynamical trajectory can be shifted by increasing the Cx-CD synaptic efficacy (black dashed to solid curve). The decision threshold is more efficiently adjusted by synaptic modification of the Cx-CD than the Cx-SC pathway (d_1 is larger than d_2). (b) Quantification of the dependence of the decision threshold on the efficacy of the Cx-CD (black) and Cx-SC synapses. (c) A maximum reward rate is achieved by tuning the Cx-CD efficacy. The reward rate R is defined as R = P/T, where P is the average performance and T is the average trial time duration (see insert for P and 1/T). Reproduced from Lo and Wang (2006). [Lo, C. C., & Wang, X. J. (2006). Cortico–basal ganglia circuit mechanism for a decision threshold in reaction time tasks. Nature Neuroscience, 9(7), 956–963.]

et al. 2008), dopamine-dependent plasticity of cortico-striatal synapses represents a candidate neural locus for adaptive tuning of a decision threshold in the brain. The overall reward rate (the average rewards per trial) exhibits an inverted U-shaped dependence of the cortico-striatal synaptic strength (**Figure 6.26c**), with an optimality in the tradeoff between fast and error-prone versus time-consuming but more accurate decisions.

Human studies with functional brain imaging and computational modeling have identified several prefrontal cortical areas, striatum and subthalamic nucleus that are implicated in SAT (Frank et al. 2007; Forstmann et al. 2008; Furman and Wang 2008). Furthermore, single-neuron physiological studies of SAT have revealed changes in the baseline activity and ramping slope of neural activity, in addition to the decision threshold (Heitz and Schall 2012; Hanks et al. 2014; Thura and Cisek 2016). As noted earlier, an increase in the baseline is functionally equivalent to a decrease of the decision threshold. What might an adjustment of ramping slope be good for?

In contrast to slow plasticity of cortico-striatal connection weight, rapid SAT may be required whenever warranted by behavioral demands. This can be accomplished by a control signal into a decision circuit that changes the ramping slope. In **Figure 6.27a**, a balanced excitatory-inhibitory (E/I) input is added to the recurrent network model (Wang 2002b). Assume that without such a "control" signal, the network tends to ramp up quickly in favor of speeding decisions. An E/I balanced external input results in a divisive gain modulation of single neurons (Chance et al. 2002). The net effect is to reduce ramping slope, thereby prolonging temporal integration in decision making (Lo et al. 2015). When integration is enhanced, performance is improved (**Figure 6.27b**). Surprisingly, with a significant control input, the distribution of reaction times becomes skewed with a long tail, implying a higher probability for long RTs than expected by a symmetrical Gaussian distribution (**Figure 6.27c**). With the balanced E/I input, the *modus operandi* of the network changed qualitatively. The spontaneous state of the network does not vanish upon the stimulus presentation, unlike the case without the control input (**Figure 6.17**). This is schematically shown in **Figure 6.27d**, where a dynamical system displays three attractor states as minima of an "energy potential" as a function of the difference in the firing rates of the two competing neural pools ($r_A - r_B$). The spontaneous state corresponds to $r_A - r_B = 0$, stochastic fluctuations of neural activity eventually brings the network out of its basin of attraction towards one of the two choice attractors (with large r_A or r_B, respectively). The model predicts that, with accuracy emphasis realized by a balanced control signal, the ramping rate of neural integrator

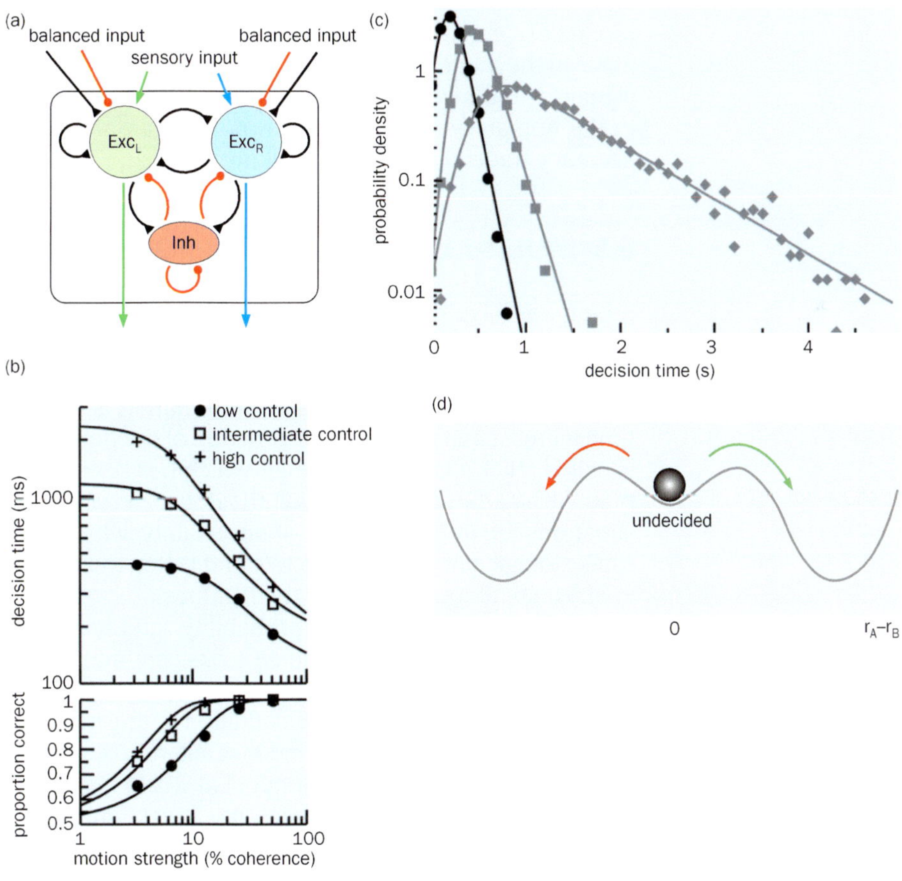

Figure 6.27 Balanced E/I input leads to a transition from the ramping mode to the jumping mode of operation and prolongs RTs. (a) Recurrent neural network model for TAFC perceptual decision in the presence of an additional excitatory-inhibitory balanced synaptic input for top-down control. (b) Larger balanced input increases accuracy and reaction time. (c) The addition of an excitatory-inhibitory balanced input produces a long exponential tail in the RT distribution. (d) Excessive reaction times is caused by noise-induced switching from a weakly stable resting state to one of the choice attractor states. Adopted from Lo et al. (2015). [Lo, C. C., Wang, C. T., & Wang, X. J. (2015). Speed-accuracy tradeoff by a control signal with balanced excitation and inhibition. Journal of Neurophysiology, 114(1), 650–661.] [Modified from Lo, C. C., Wang, C. T., & Wang, X. J. (2015). Speed-accuracy tradeoff by a control signal with balanced excitation and inhibition. Journal of Neurophysiology, 114(1), 650–661.]

and the baseline neural activity are lower, which is consistent with studies in which single neurons in the lateral intraparietal area and the frontal eye field of behaving monkeys displayed a reduction in both the ramping rate and in the baseline activity, when the emphasis was shifted from speed to accuracy (Heitz and Schall 2012; Hanks et al. 2014).

This "jumping mode," denoting escape from a resting state by noise as part of the decision process, is in contrast to the "ramping mode" (**Figure 6.17**), demonstrating that the same model network can exhibit different scenarios and operate in different regimes (Deco et al. 2007; Deco et al. 2009; Gigante et al. 2009; Miller and Katz 2010; Wang 2012). The model in the jumping mode has been suggested to account better than the ramping mode for neurophysiological data from a taste discrimination task by rats, where simultaneously recorded neurons exhibited discrete transitions between quasi-stable states over the time course of animal's decision process (Miller and Katz 2010). The distribution of transition times across trials is exponential when switching is induced by noise, as predicted by the Kramers' escape rate theory for crossing a potential barrier (Chandrasekhar 1943; Gardiner 1985; Kampen 2007).

Skewed RT distributions are not found in the RDM monkey experiment (**Figure 6.23**), but commonplace in human behavior (Ratcliff 1978; Luce 1986; Ratcliff and Rouder 1998; Usher and McClelland 2001; Sigman and Dehaene 2005). As a matter of fact, it is virtually impossible to find reports of symmetrical RT distributions in the human psychology literature. Since in the model a long-tail of RT distributions involves top-down control signaling, the neuronal machinery of executive control may be enhanced in the human brain compared to nonhuman animals, for flexible speed-accuracy tradeoff in decision making.

6.5 MULTI-ALTERNATIVE DECISIONS

So far, we have only discussed choices between two options. How does the brain make decisions with more than two alternatives? At the behavioral level, more options under consideration lead to slower reaction times. Under the condition that sensory information is not ambiguous, reaction time increases as a logarithmic function of the number of alternatives. This is called "Hick's law" (Hick 1952), which has been reproduced with a race model (Usher and McClelland 2001; Bogacz et al. 2007). More interesting situations are in the presence of uncertainty, which has been studied neurophysiologically with three (Niwa and Ditterich 2008) or four (Churchland et al. 2008) alternatives using the RDM task.

Whereas the sequential probability ratio test (Wald 1948) is optimal for two-alternative forced choice tasks, optimal tests for three or more options are not known. A neural network model can be readily generalized to multi-alternatives by the addition of any number of selective neural pools (**Figure 6.28a**) (Usher and McClelland 2001; Brunel and Wang 2001; Albantakis and Deco 2009; Kriener et al. 2020). As shown in **Figure 6.28b**, the decision behavior of a three-choice version of the attractor network model (Wong and Wang 2006) can be well described by a softmax function:

$$P(1) = exp(\beta I_1) / \left(exp(\beta I_1) + exp(\beta I_2) + exp(\beta I_3)\right) \tag{6.19}$$

where I_1, I_2 and I_3 are the strengths of evidence for the three options, and β is a parameter that quantifies the amount of stochasticity. This model fits well with behavioral performance data from a three-choice RDM direction discrimination experiment (**Figure 6.28b**) (Niwa and Ditterich 2008).

When the alternatives are values of an analog feature, such as directions of a moving dot pattern, decision difficulty depends on the similarities among choice options. This can be captured by a continuous recurrent network model rather than a model with discrete neural pools. This is shown in **Figure 6.29a** (Furman and Wang 2008) using the ring model for working memory of Compte et al. (2000), appropriate since choice options can be any direction from 0 to 360 degrees. In a sample trial model simulation with four options and $c' = 0\%$. As in a monkey experiment, four targets are first presented at equal distances (adjacent targets are separated by $\Delta\theta = 90°$), which are necessary to indicate discrete directional choices in a continuous network. The corresponding neural populations display increased activities in response to target inputs. Then, a random dot motion stimulus is shown, which triggers competition eventually leading to a winner (in this case at $\theta = 315°$), even when the motion coherence is zero.

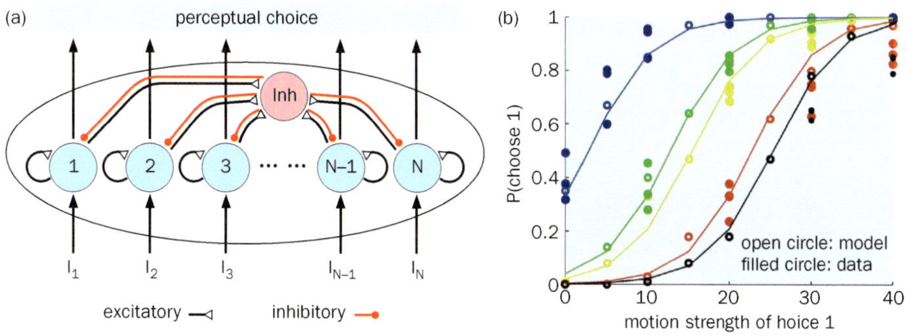

Figure 6.28 Three-alternative perceptual decision. (a) Schematic of the recurrent neural network model from Wong and Wang (2006) generalized to multiple discrete neural pools. (b) Model (open circles) captures behavioral performance (filled circle) from Niwa and Ditterich (2008). Model behaviors are well fitted with a softmax function Eq. (6.19) (smooth curves). Different colors correspond to five pairs of motion strengths of the directions 2 and 3. Reproduced from Wang (2012) with original data obtained by Nathaniel J. Smith. [Wang, X. J. (2012). Neural dynamics and circuit mechanisms of decision-making. Current Opinion in Neurobiology, 22(6), 1039–1046. Original data comes from Nathaniel J. Smith.]

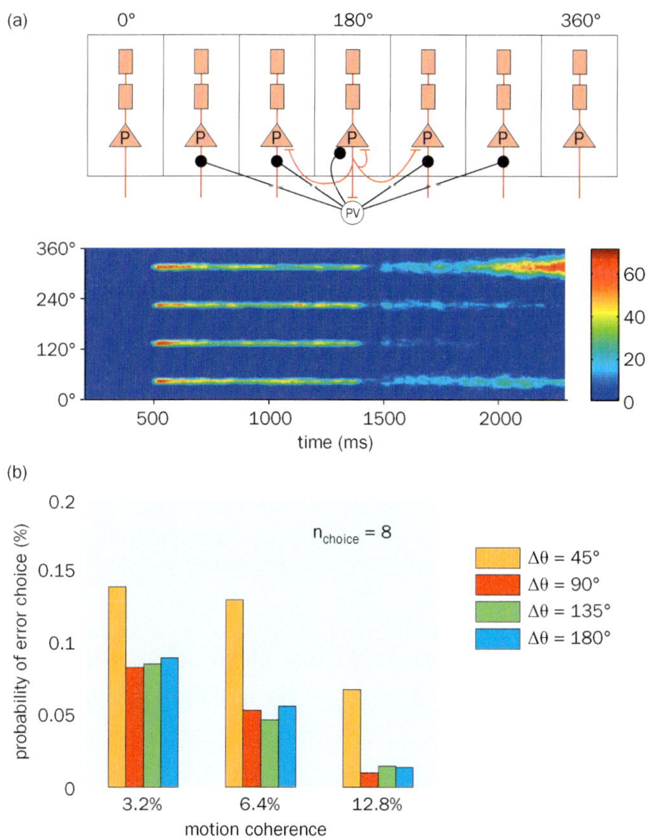

Figure 6.29 Multi-alternative decision making in a continuous recurrent network model. (a) The continuous recurrent network model (upper) is simulated in a trial with four options (lower). Four target inputs are first shown to indicate choice alternatives, producing four bumps in the spatiotemporal activity pattern. This is followed by the presentation of an RDM stimulus when recurrent network dynamics gives rise to a winning bump by attractor dynamics. (b) With eight options, the error rate is the highest for the choice that is the closest to the correct one (orange), shown with three low to intermediate c' values, demonstrating the similarity effect. Reproduced from Furman and Wang (2008). [Furman, M., & Wang, X. J. (2008). Similarity effect and optimal control of multiple-choice decision making. Neuron, 60(6), 1153–1168.]

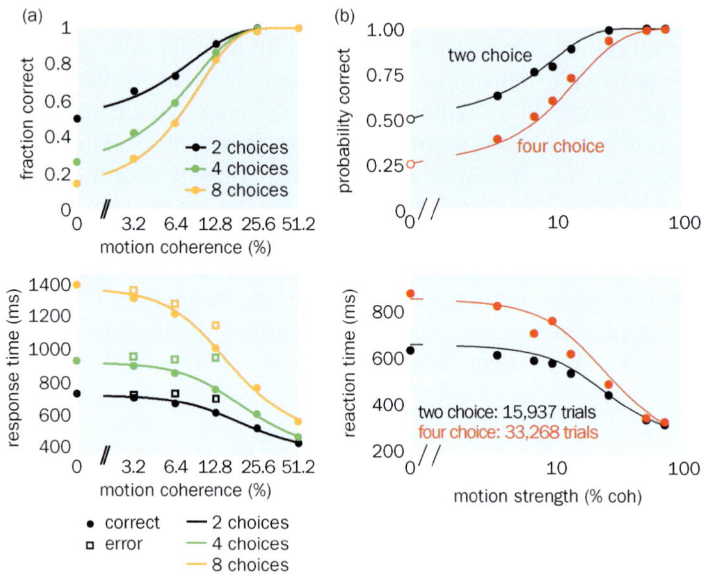

Figure 6.30 Multi-alternative decision behavior. (a) Psychometric function and chronometric function of the continuous recurrent network model with two, four, and eight alternatives. (b) Psychometric function and chronometric function of the monkey experiment with two and four alternatives. Panel (a) is reproduced from Furman and Wang (2008), (b) from Churchland et al. (2008). [Furman, M., & Wang, X. J. (2008). Similarity effect and optimal control of multiple-choice decision making. Neuron, 60(6), 1153–1168.] [Churchland, A. K., Kiani, R., & Shadlen, M. N. (2008). Decision-making with multiple alternatives. Nature Neuroscience, 11(6), 693–702.]

The model reproduces the salient behavioral characteristics of the monkey experiment (Churchland et al. 2008); namely, the accuracy decreases and reaction time increases with an increasing number of choice options (**Figure 6.30**). The traditional one-dimensional DDM cannot be obviously extended to multiple choices (McMillen and Holmes 2006; Niwa and Ditterich 2008; Churchland et al. 2008; Krajbich and Rangel 2011). Race models are readily generalizable (Usher and McClelland 2001; Bogacz et al. 2007), but they typically do not capture the effects of similarity among choice options. Another merit of a continuous network model is its ability to account for both categorical decision and veridical judgment that requires continuous report rather than a discrete choice (Liu and Wang 2008; Smith 2016).

When the choice targets are close to each other, interactions between neural pools selective for them through lateral connections become important in determining the network dynamics. Does this similarity effect have observable consequences at the behavioral level? Indeed, for low and intermediate coherence levels (between 3.2% and 12.8%), the probability of an erroneous choice in the eight-choice task depended on the location of the selected target (**Figure 6.29b**). Specifically, the probability of making an erroneous choice to a target adjacent to the correct one was higher than it was for the other targets. This spatial pattern of errors provides a testable prediction that is specific to the continuous model.

6.6 DIVERSE TYPES OF PERCEPTUAL DECISIONS

Insofar as decisions refer to making a judgment under uncertainty, many perceptual tasks represent diverse forms of decision making. Can they be described in terms of shared key computational steps such as accumulation of evidence and categorical choice? Can models with shared core features account for their underlying neural processes and circuit mechanisms? Here we discuss several examples illustrating how problem solving in perception can be cast in a common computational framework.

6.6.1 Detection

Detecting a stimulus represents a simple task that involves a transformation from sensory stimulation to subjective perception (Romo and

Rossi-Pool 2020). As discussed at the beginning of this chapter, a stimulus near the psychophysical threshold of detection may be perceived (hit) or not (missed). Moreover, the brain sometimes reports its presence even when sensory organs receive no physical input (false alarm). What is the neurobiological basis of this all-or-none perceptual decision?

In an experiment with monkeys performing a detection task, neurons in the frontal cortex were found to display higher activity in hit than miss trials and in false alarm than correct rejection trials (Thompson and Schall 1999). In another monkey experiment, a vibration stimulus is shown on a fingertip, in some but not in other trials. After a delay period of 3 seconds, subjects reported "yes" or "no" by pushing one of two buttons (Figure 6.31a). The fraction of hit trials displayed a threshold of the stimulus amplitude for detection (Figure 6.31b). Neurons in the medial premotor cortex (MPC), a frontal area involved in somatosensory decision making (Hernández et al. 2002), displayed spiking activity correlated with subjective perception. Namely, with a near-threshold stimulus, MPC neural activity was substantially higher in hit trials than in miss trials (Figure 6.31c, left panel), even when the physical input was identical and neurons in the primary somatosensory area S1 showed no difference in their firing activities. In the absence of a stimulus, MPC neurons exhibited a slightly higher overall firing rate in false alarm trials than in correct rejection trials (Figure 6.31c, right panel). In hit trials, in contrast to graded dependence of S1 neural firing rate, MPC neural firing rate displayed a stepwise dependence on the stimulus amplitude (Figure 6.31d) that mirrors monkey's detection performance (Figure 6.31b). Similar detection psychometric functions were obtained with premotor cortex microstimulation instead of external mechanical stimulation, suggesting that the stimulated frontal site is causally related to the decision behavior.

Subjective detection signals are also present in the posterior parietal cortex. While monkeys performed a task of detecting the presence of a visual motion stimulus, neurons in the ventral intraparietal (VIP) area of

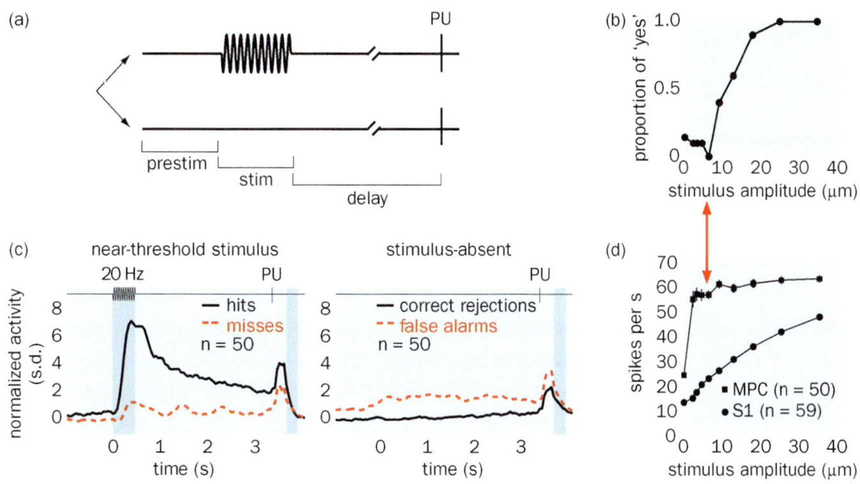

Figure 6.31 Subjective perception and its neural correlate in a detection task performed by monkeys. (a) A somatosensory stimulus at 20 Hz was shown for 0.5 seconds in a fraction of trials. After a delay period of 3 seconds, at the time of a probe moved up (PU), subjects reported the presence or absence of the sensory input. (b) Hit rate as a function of stimulus amplitude. (c) Average spiking activity recorded from single neurons in the medial premotor cortex in four trial types: Hit (solid black) and miss (dashed red) in near-threshold trials (left), false alarm (dashed red) and correct rejection (solid black) in stimulus-absent trials (right). (d) Firing rate of MCP and S1 neurons in hit trials as a function of the stimulus amplitude. While S1 activity varies gradually, MPC neural firing shows a binary pattern, where the onset of its high firing coincides with the stimulus amplitude threshold for behavioral report (double arrow). Reproduced from de Lafuente and Romo (2005). [de Lafuente, V., & Romo, R. (2005). Neuronal correlates of subjective sensory experience. Nature Neuroscience, 8(12), 1698–1703.]

the posterior parietal cortex exhibited ramping activity that was correlated with the subjective judgment (a higher activity in hit and false alarm trials than in miss and correct rejection trials) and response time (ranging from 400 to 700 ms) (Cook and Maunsell 2002). In VIP but not MT, neuronal activity was significantly larger than baseline in false alarm trials, suggesting that the subjective judgment was computed downstream of MT. This study thus supports the notion that the posterior parietal cortex (PPC) is part of the brain system underlying accumulation of information and subjective judgment.

The attractor network model (Wang 2002b; Wong and Wang 2006) was adopted to this detection task, where the two competing neural pools represent "yes" and "no," respectively (Deco et al. 2007). The "no" neural pool signaled the absence of a stimulus as the default choice. A selective input drives the "yes" neural pool, and a probabilistic decision (yes or no) results from stochastic neural circuit dynamics as described prior. The model was also used to account for findings in a mouse experiment on auditory detection (Zagha et al. 2015), suggesting that variations of the recurrent circuit model framework can account for detection. The source of input drive that leads to false alarm remains unknown. Presumably, the likelihood of a stimulus is internally estimated, which is higher with a larger fraction of trials with stimulus shown. In the brain, the "no" response could be encoded and signaled by neurons that suppressed an over movement (Hanes and Schall 1996; Zagha et al. 2015) (see Chapter 8).

6.6.2 Comparison and Discrimination

Recurrent circuit attractor models have also been developed for the vibrotactile delayed discrimination experiment where subjects make a decision about which of the two vibrotactile stimuli separated by a delay period has a higher frequency (is f_1 larger than f_2?) (Romo et al. 2002; Romo et al. 2004) (Chapter 5, Section 5.2.2). In the VDD task, parametric working memory of an analog quantity (f_1) precedes a two-choice decision process ($f_1 > f_2$ or $f_1 < f_2$). This is in contrast to the fixed-duration version of the RDM task, where decision precedes a delayed period when a binary choice must be stored in working memory. Both working memory and decision making are required for behavior, but the type of information coding and the temporal order are quite different in the two tasks.

Machens et al. (2005) proposed a network model that could perform discrimination computation ($f_1 > f_2$ or $f_1 < f_2$) during the comparison period (**Figure 6.32a**), provided that a switching mechanism is included. The network model receives inputs from a somatosensory area, where positive ("+") and negative ("−") neurons show monotonic increase and decrease, respectively, of spiking activity with f_1 during sample presentation. The "+" (respectively "−") sensory neurons activate "+" ("−") prefrontal neurons, which show mnemonic activity during the delay period that is positively (negatively) proportional to f_1 (Chapter 5, **Figure 5.8**). During f_2 presentation, however, the projections are reversed ("+" sensory neurons project to "−" prefrontal neurons, and vice versa), thereby enabling $f_1 > f_2$ versus $f_1 < f_2$ decision by virtue of two competing attractor states. A phase plane analysis showed how the attractor landscape is modified by external signals for each of the task epochs (cue loading, memory maintenance, comparison), and hence the same circuit can subserve both working memory and decision computation (Machens et al. 2005).

This model leaves open the question as to the brain mechanism of the hypothesized switching of synaptic connection from sensory to memory neurons in the middle of a trial. In an alternative scenario, an integral feedback instantiates comparison computation in a neural network with two ("+" and "−") interacting subnetworks (Miller and Wang 2006). As shown in **Figure 6.32b**, the model assumes a loop between "+" type comparison neurons and memory neurons. The comparison neural pool receives sensory stimulus, displays a transient response proportional to input frequency, and excites the memory neural pool that acts as a perfect integrator (Chapter 3, Eq. 3.26). The latter hence fires persistently at a level parametrically tuned to f_1 during the delay period. The feedback projection from the memory neurons to the comparison neurons is inhibitory. As a result, when f_2 is shown, the comparison neurons are under the influence of both internal inhibition proportional to f_1 (from the memory neurons) and external excitation proportional to f_2 (from the sensory coding neurons); therefore, they fire at high rates when $f_2 - f_1 > 0$ or $f_1 < f_2$. Another subnetwork composed of the "−" type comparison and memory neurons signals $f_1 > f_2$. The two subnetworks compete with each other, leading to a binary decision (Miller and Wang 2006). Like Machens et al. (2005), this model offers a computational explanation for the need of both "+" and "−" neuron types, but it does not assume a switching of synaptic projections from sensory to memory circuits that occurs between f_1 and f_2 presentations. The model predicts positive trial-by-trial correlation between neural pairs of the same (both are either "+" or "−" type) tuning type, and negative correlation between neural pairs of opposite tuning (one is "+" while the other is "−" type). This prediction is confirmed by neural data recorded from the prefrontal cortex. On the other hand, the biological basis of the hypothesized negative integral feedback remains untested. Although the model is cast in terms of comparison neurons and memory neurons, **Figure 6.32b** could be interpreted differently, in terms of one neural population (instead of two, comparison and memory neurons) with the feedback instantiated by some intracellular process in single cells. For instance, calcium dynamics could temporally integrate inputs

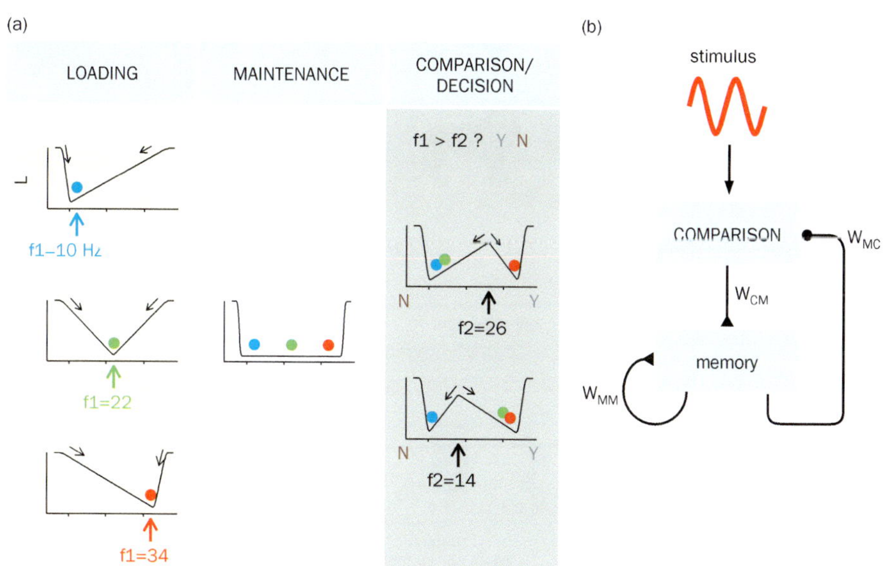

Figure 6.32 Models for delayed discrimination. (a) A dynamical algorithm for two-stimulus interval discrimination. Abscissa: The state variable (e.g., the difference in the firing rates of the plus and minus neurons; ordinate: A computational energy function (with minima corresponding to stable attractor states). During the loading period, the first stimulus creates a unique attractor state encoding the f_1 value. The energy landscape is flat during the delay period, so the memory of f_1 is maintained in the form of parametric persistent activity. During the comparison period, the system is reconfigured, and the second stimulus f_2 in interplay with the internal memory state gives rise to a categorical decision $f_1 > f_2$ or $f_1 < f_2$. (b) An alternative circuit model of discrimination. The comparison neurons are quiescent unless excited by external input. The memory neurons are integrators of the activity of the comparison neurons (W_{CM}) and in turn send feedback inhibition (W_{MC}), thereby providing integral feedback control. Panel (a) is reproduced from Machens et al. (2005), (b) from Miller and Wang (2006). [Machens, C. K., Romo, R., & Brody, C. D. (2005). Flexible control of mutual inhibition: A neural model of two-interval discrimination. Science, 307(5712), 1121–1124.] [Miller, P., & Wang, X. J. Inhibitory control by an integral feedback signal in prefrontal cortex: A model of discrimination between sequential stimuli. Proceedings of the National Academy of Sciences, 103(1), 201–206. Copyright (2006) National Academy of Sciences, U.S.A.]

(Loewenstein and Sompolinsky 2003), and accumulated intracellular calcium could provide a negative feedback to the neuronal excitability. Negative integral feedback has been shown in cellular biochemical networks (Barkai and Leibler 1997; Yi et al. 2000); whether it indeed exists in neurons and plays a role in cognition remains to be explored in the future.

6.6.3 Pattern Match Decisions

Delayed match-to-sample (DMS) is a commonly used task that involves both working memory and decision making (Chapter 5, **Figure 5.6a**) (Fuster and Jervey 1981; Miyashita and Chang 1988; Miller et al. 1993, 1996; Constantinidis and Steinmetz 1996; Pasternak and Greenlee 2005; Sereno and Amador 2006; Sigala et al. 2008; Liu et al. 2014; Wu et al. 2020). In a DMS task, a sample (object A) is first presented, followed by a delay period. Then, one test stimulus or a sequence of several test stimuli are shown, and a behavioral response is required if the current test stimulus is the same (match) as the sample (**Figure 6.33a**). If the comparison is about a one-dimensional feature such as motion direction, then one possible algorithm is to compute a difference signal $X - Y$ and decide whether it is zero, where X and Y are the sample stored in working memory and the test stimulus, respectively. However, in general, the objects to be compared with can be in a high dimensional space (like images in **Figure 6.33a** or odorants), and the match/nonmatch decision should be based on a comparison between patterns rather than scalar numbers.

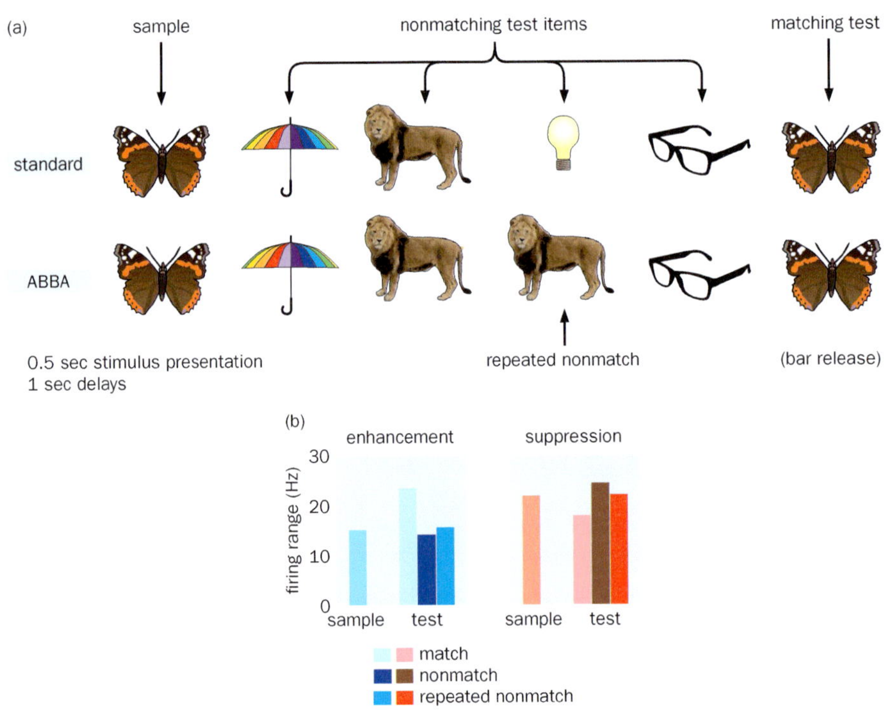

Figure 6.33 A visual delayed match-to-sample experiment with behaving monkeys. (a) Top: In a standard DMS task, a sample (A = butterfly) is followed by several distinct test objects, and the match must be detected when the same object as the sample is presented again. Bottom: In an "ABBA" task, a test object (B = lion) is also repeated but should be ignored, and the correct response is matching the test with the initial sample A. (b) Averaged firing activities of PFC neurons under four different conditions, during the presentation of a test stimulus. Match enhancement (ME) neurons (left) show an increased activity with matching test item than nonmatch but not repeated nonmatch item; match suppression (MS) neurons (right) exhibit the opposite trend. Hatched bar: Response to the sample. Reproduced from Miller et al. (1996). [Miller, E. K., C. A. Erickson, and R. Desimone (1996). Neural mechanisms of visual working memory in prefrontal cortex of the macaque. Journal of Neuroscience 16(16), 5154–5167.] [Engel, T. A., & Wang, X. J. (2011). Same or different? A neural circuit mechanism of similarity-based pattern match decision making. Journal of Neuroscience, 31(19), 6982–6996. With the original data from Miller, Earl K., Cynthia A. Erickson, and Robert Desimone (1996). Neural mechanisms of visual working memory in prefrontal cortex of the macaque. Journal of Neuroscience 16(16), 5154–5167.]

Evidence from physiological studies with behaving monkeys (Miller et al. 1993, 1996; Rawley and Constantinidis 2010; Hussar and Pasternak 2013) and human brain imaging (Duncan et al. 2009) pointed to two candidate neural processes involved in match versus nonmatch computation. One is referred to as repetition suppression, a passive reduction of neural response to any stimulus repetition regardless of behavioral relevance (**Figure 6.33b**, right). Repetition suppression is the predominant neural signal observed in standard DMS tasks when the matching test is the only stimulus that is repeated within a trial (Miller et al. 1993; Zaksas and Pasternak 2006). The other is referred to as match enhancement (**Figure 6.33b**, left), an active mechanism that is engaged whenever feature matching is not sufficient to perform a task, as for example when nonmatch (object B) can also be repetitive (e.g., ABBA) and irrelevant repetitions of nonmatch stimuli (BB) should be ignored. Neurophysiological recordings in the prefrontal and parietal cortices revealed two populations of neurons whose selectivity for visual stimuli is modulated by match/nonmatch context in complementary ways: match enhancement (ME) cells show higher firing response for a match than nonmatch to the sample, whereas match suppression (MS) cells exhibit the opposite trend.

Based on these observations, a neural circuit model was proposed (**Figure 6.34a–b**). For the sake of concreteness, consider a DMS task where stimuli are visual moving patterns with its direction as the relevant feature (Zaksas and Pasternak 2006). The core of the model is a comparison module that receives bottom-up directional information (encoded in MT) and a top-down directional signal from working memory in the form of a bump attractor (Chapter 5, **Figure 5.18**). Three assumptions are made about comparison neurons. First, they show activity-dependent spike rate adaptation (Chapter 2, Section 2.4.2). Second, they are assumed to be heterogeneous in terms of top-down influence. Some do not receive inputs from the working memory, and their response to a repeated stimulus is reduced due to adaptation, so therefore they display match suppression (**Figure 6.34c**, right). For others that are targeted by the working memory circuit, the top-down excitation more than compensates for adaptation effect so that their response to a repeated stimulus is larger (match enhancement) (**Figure 6.34c**, left). Third, a homeostatic regulation mechanism (Chapter 4, Section 4.4.6) leads to the same overall excitation for ME and MS neurons; consequently, recurrent excitation is smaller among ME cells than MS cells, because the former but not latter receives an additional excitation from the working memory circuit. Consequently, when the test does not match the sample, the ME neurons selective for the test feature do not receive excitation from working memory, and because of comparatively strong recurrent connections, the MS cells selective for the test stimulus display larger responses than the ME ones. The model (**Figure 6.34c**) reproduces the salient physiological findings (**Figure 6.33b**).

MS and ME neurons in the comparison circuit project to a decision circuit with two competing neural pools (Wong and Wang 2006), which are selective for "match" and "nonmatch" choices. The weights of connections undergo reward-dependent synaptic plasticity (Chapter 4, Section 4.3). These weights are initially random, so the performance is at chance level. When a choice results in a reward, synaptic weights are enhanced by an amount proportional to the presynaptic firing rate. Thus, if the choice is match and rewarding, connections to match decision neurons from ME neurons are potentiated more than those from MS neurons because the former have a higher activity level. In a trial, if the choice is nonmatch, MS neurons fire at a higher rate than ME cells; if the response is correct and thus rewarding, connections to nonmatch decision neurons from MS neurons are potentiated more than those from ME neurons. Eventually, connections from the comparison circuit to the decision circuit are self-organized

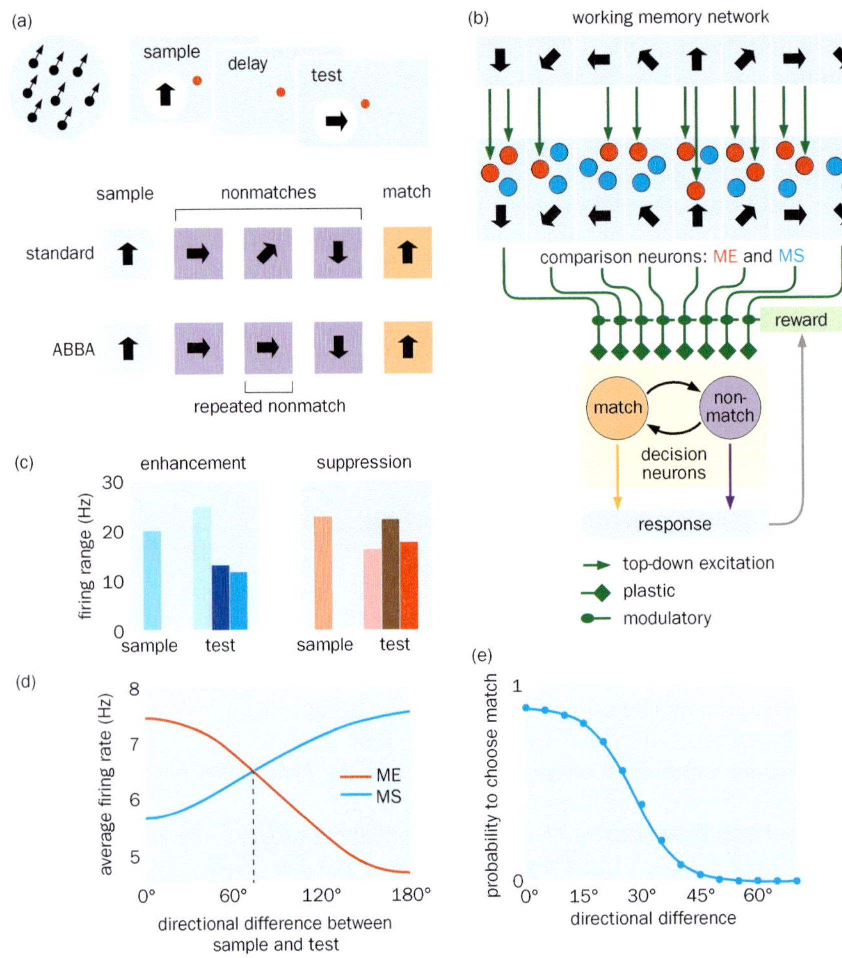

Figure 6.34 A neural circuit model for pattern match decisions. (a) A delayed match-to-sample task where the stimulus feature is direction of a visual motion pattern. (b) Schematic of a multi-module network model. Comparison neurons receive two types of inputs representing sensory stimulus and its working memory, and they project to a two-pool decision circuit (Figure 6.12). (c) Average neural responses in the model to a preferred stimulus appearing as a sample, match, nonmatch and repeated nonmatch, comparable to single-neuron activity data (Figure 6.33b). (d) Similarity tuning. Average population firing rate for the ME (red line) and MS (blue line) neurons as a function of directional difference between the sample and test. The ME and MS populations are parametrically tuned to the sample-test similarity in complementary ways. (e) Psychometric function is defined by the probability (the fraction of trials) for choosing "match" as a function of the difference between the sample and test stimuli. Reproduced from Engel and Wang (2011). [Engel, T. A., & Wang, X. J. (2011). Same or different? A neural circuit mechanism of similarity-based pattern match decision making. Journal of Neuroscience, 31(19), 6982–6996. https://www.jneurosci.org/content/31/19/6982]

through learning, with ME cells tending to target the "match" decision neural pool while MS cells prefer the "nonmatch" decision neural pool. Let us denote I_{ME}^{match} the total input currents from the ME neural population to the match decision pool, which is roughly the averaged ME neural firing weighted by a connection strength, and $I_{MS}^{nonmatch}$ the total input currents from the MS neural population to the nonmatch decision pool.

During test presentation, the average firing rates of ME and MS neurons depend on the similarity between the test and sample stimuli parametrically (**Figure 6.34d**). Namely, when the two are similar, activity is higher in ME cells than MS cells because of top-down excitation, and hence the overall current is larger to the "match" decision pool than the "nonmatch" one. Consequently, the probability of choosing match is high (**Figure 6.34e**). As the sample-test similarity decreases, the difference in the ME and MS firing activities decreases and then reverses the sign (**Figure 6.34c**), leading to "nonmatch" as the predominant choice. The probability of choosing "match" is a sigmoid function of $I_{ME}^{match} - I_{MS}^{nonmatch}$ (**Figure 6.34e**), just like **Figure 6.16**. The engagement of the working memory circuit is especially important in the ABBA task, where persistent activity stores the initial sample A, and ME neurons selective for A (not those selective for B) are enhanced, because without this mechanism, repetition of B could not be differentiated from repetition of A.

The psychometric threshold in **Figure 6.34e** (the sample-test difference at which probability of choosing match becomes non-zero) is about 45°, in agreement with the threshold reported in monkey DMS paradigms (Zaksas and Pasternak 2006). In other studies, a much smaller behavioral threshold of 1°–2° was reported using fine discrimination paradigms (Hol and Treue 2001; Purushothaman and Bradley 2005). The two task types, however, are different in at least two respects. First, fine motion discrimination tasks often do not include a mnemonic delay, and working memory representation is noisy (Li et al. 2021b) and may deteriorate over time (Chapter 5, Section 5.4.3), thereby rendering the DMS more difficult. Second, for fine discrimination the sample typically has a fixed reference direction (e.g., upward), and the task is to judge whether a subtle deviation in a test direction is clockwise or counterclockwise relative to this reference (Purushothaman and Bradley 2005). When the model was applied to the fine discrimination task with a fixed sample, synaptic plasticity reflects the statistics of shown stimuli, and the model system learns to perform with a psychometric threshold of 1°–2°, demonstrating that the model endowed with plasticity can account for different types of pattern-match decisions (Engel and Wang 2011). Note that plastic changes here take place upstream from MT, and learning does not transfer to a novel direction (e.g., after learning with vertical direction as reference, testing with the horizontal reference direction yields poor performance). This is in line with the idea that plasticity occurs by "re-weighting" connections from sensory coding neurons to decision neurons, in the literature of perceptual learning (Law and Gold 2008; Dosher and Lu 2009).

Other types of decision making have been described using multi-attractor models. For instance, binocular rivalry refers to the phenomenon of perceptual bistability where a visual stimulus is subjectively seen as one of two images, and the two perceptions alternate spontaneously in irregular time intervals (Breese 1899; Levelt 1965; Blake 1989; Leopold and Logothetis 1996). A neural circuit model typically assumes two coexisting attractor states corresponding to the neural populations selective for the alternative decisions (percepts). In one of the attractor states, single neuron adaptation (Chapter 2, Section 2.4.2) or synaptic depression (Chapter 2, Section 2.7.1) of the recurrent excitatory drive gradually reduces the firing of the dominant neural population, eventually leading to a transition to the other attractor state, which in turn becomes dominant. Random noise can also induce switching between the two states; both adaptation and noise are commonly incorporated into models of bistable perception (Wilson 2003; Moreno-Bote et al. 2007; Theodoni et al. 2011; Cao et al. 2021).

In summary, variations of a core neural circuit model (Wang 2002b) can account for diverse types of perceptual decisions, suggesting a general mechanism for both working memory and decision making in cognitive-type neural circuits.

6.7 CONFIDENCE AND CHANGES OF MIND

In addition to accuracy and reaction time, confidence is the third major component of decision making (Grimaldi et al. 2015; Meyniel et al. 2015; Mamassian 2016; Pouget et al. 2016). In daily life, confidence judgment about a choice influences subsequent decisions and actions (van den Berg et al. 2016). Confidence assessment is generally important for problem-solving and learning; for example, progression in studies at school relies on how confident the learner is about knowing the previously acquired materials. Therefore, confidence estimation has been an important subject of research in cognitive and developmental psychology (Flavell 1979; Vickers 1979). Confidence is related to uncertainty; usually we are less confident when we are faced with a higher degree of uncertainty, such as making investment choices in a volatile stock market. However, confidence is inherently subjective and represents an individual's trait (Ais et al. 2016; Navajas et al. 2017); given the same stock market environment, confidence varies considerably from one investor to another.

Such a subjectively defined measure of confidence can be reported, say, on a scale between 0 and 1, verbally in human studies (Vickers 1979). Nonverbal report is also possible, for instance, by pointing to a position between 0 and 1 of a visually presented bar (Kiani et al. 2014a). Other behavioral paradigms are better suited for studies using nonhuman animals (Smith et al. 2014). A carefully designed protocol to probe confidence is crucial to avoid confounds as a consequence of confidence being correlated with reaction time and accuracy. In one paradigm, in addition to choice alternatives, an opt-out option is offered, which may be selected by a subject due to a lack of confidence. In another paradigm, a two-stage wagering task requires a subject to first make a choice among two or more alternatives, and then gamble over two options. In the second stage, selecting one of the two options yields a large reward but only when the first decision is correct, whereas the outcome of selecting the other option is a guaranteed small reward regardless of the first decision. One is more likely to select the risky option in the second stage if confidence is higher about the decision in the first stage. Therefore, confidence can be measured by the probability of choosing the high-award gambling option (Shields et al. 2005; Kornell et al. 2007; Middlebrooks and Sommer 2012).

In these experiments, behavioral data may be fit with mathematical models in which confidence is quantified. In a race model (Eqs. 6.8–6.9), confidence about a decision can be read out as the difference between the decision variables $X_A - X_B$, properly normalized, at the time one of them (say, A) has first reached the decision threshold. This measure makes intuitive sense: when X_A hits the threshold, internally estimated confidence should be lower if X_B is closer to X_A (Vickers 1979). A similar measure can be defined for the drift diffusion model (Moreno-Bote 2010; Drugowitsch et al. 2012).

To probe the brain mechanism of confidence judgment, Kiani and Shadlen (2009) carried out a monkey physiological experiment using the RDM direction discrimination task, with the addition of an "opt-out" manipulation. Usually, a subject views the motion stimulus for a fixed duration; after a delay period, the decision was indicated by a saccade to one of two diametrically opposed targets T_A and T_B, and a large reward was delivered with a correct decision. In a fraction of trials, during the delay period a "sure" target T_S (90° away from both T_A and T_B) was shown, in which case the subject could avoid completing the discrimination task by saccading to T_S and collecting a sure, albeit smaller, reward as a result. In this setting, it was observed that the probability of selecting T_S, P_{sure}, in a trial is higher when certainty is reduced with a lower motion coherence c' or a briefer stimulus presentation. Hence, P_{sure} reflects quantitatively a lack of confidence. Concomitantly, activities of neurons in LIP correlated with decision making itself also signal confidence. Specifically, a neuron with preferred choice A displayed a large difference in its activity depending on the animal's choice A versus B when the sure target was present but not chosen. By contrast, its activity is almost the same regardless of the actual motion direction in trials when the sure target was selected.

The neurobiological interpretation of this finding is that the difference in the activities of two neural populations selective for A and B correlates with confidence at the behavioral level. Wei and Wang (2015) investigated this hypothesis using the same continuous attractor network model of spiking neurons as Furman and Wang (2008), with an appropriate setting of external inputs (**Figure 6.35a**). The parametric directional selectivity of model neurons accurately captures the relationships between T_A and T_B separated by 180° as well as T_S orthogonal to both. **Figure 6.35b** shows the spatiotemporal pattern of network activity in two sample trials where the sure target is presented during the delay. In one trial (top), the neural pool selective for the correct choice T_A wins the competition, and T_S did not alter the network's decision. Under the same condition, in the other trial (bottom), stochastic population dynamics does not yield a winner between

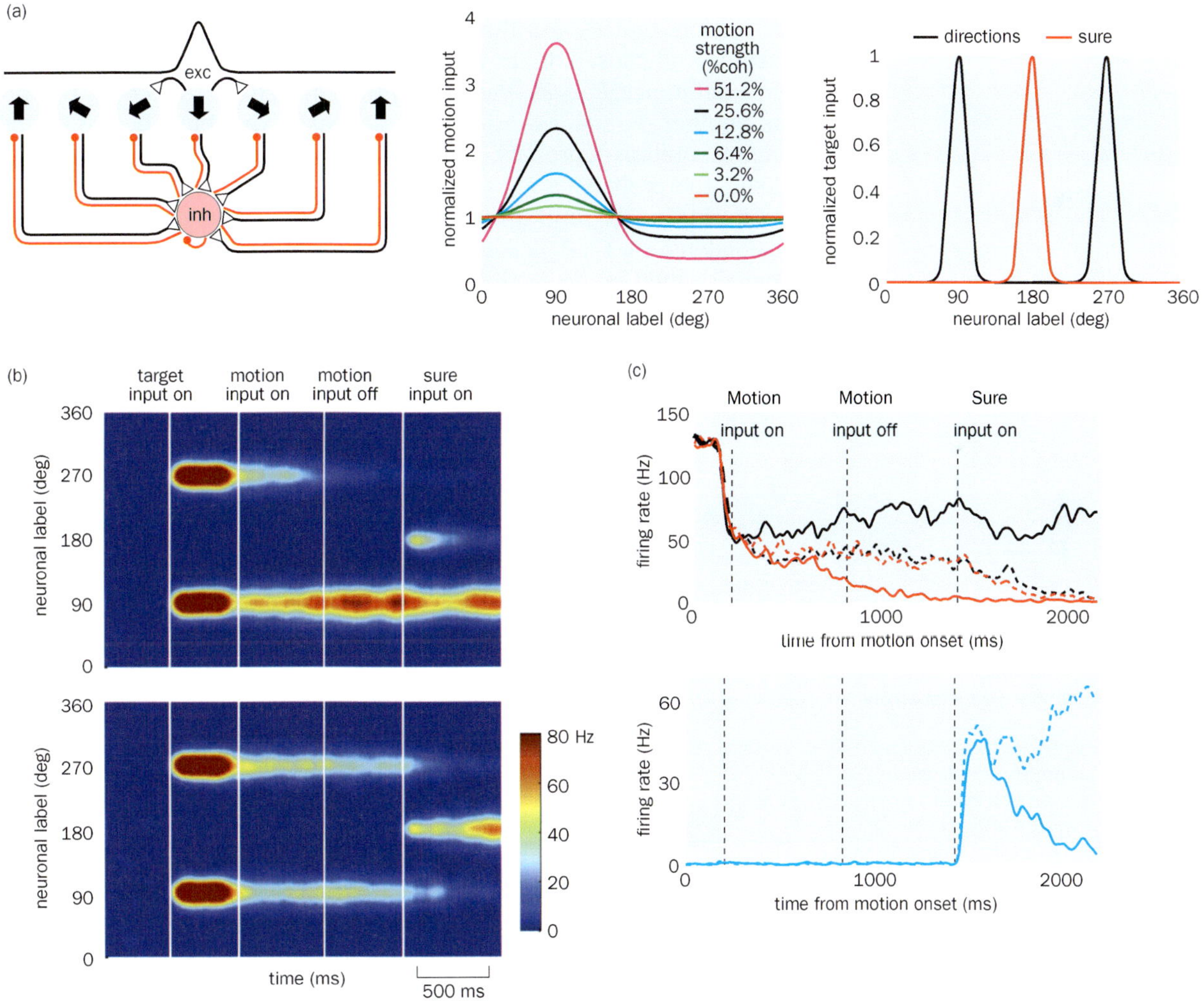

Figure 6.35 Confidence estimation by recurrent neural circuit dynamics. (a) Left: A continuous recurrent network model of spiking neurons. Middle: Dots stimulus is presented at one of two directions T_A and T_B, with a coherence c' ranging from 0% to 100%. Right: In trials when a sure target T_S is presented, it is orthogonal to T_A and T_B, as in the monkey experiment of Kiani and Shadlen (2009). (b) Spatiotemporal network activity pattern of the model in two sample trials. X-axis: Time; Y-axis: Neurons labeled by their preferred directions, and their firing rates are coded by color. Two target inputs, around $T_A = 90°$ and $T_B = 270°$, generate high activity of neural populations R_A and R_B. When the dots stimulus is presented at T_A, R_A wins the competition in the first trial (upper panel), but R_A and R_B remain similar in the second trial (lower panel). Upon presentation of a sure target, the neural population selective for $T_S = 180°$ prevails, leading the network's ultimate decision in the second but not first trial. (c) R_A, R_B and R_S as a function of time from the top trial (solid) and the bottom trial (dashed) of (b). Reproduced from Wei and Wang (2015). [Wei, Z., & Wang, X. J. (2015). Confidence estimation as a stochastic process in a neurodynamical system of decision making. Journal of Neurophysiology, 114(1), 99–113.]

T_A and T_B; neurons selective for T_S eventually become dominant, and the network's choice is the sure target. Similar to the neurophysiological findings, the firing rates R_A and R_B of neural populations selective for T_A (black) and T_B (red) are widely separated in the trial when T_S does not impact the network's decision, whereas they are indistinguishable in the trial when T_S is selected (**Figure 6.35c**, top).

The response of the neural population selective for T_S, during sure target presentation, decays away in the first trial, while it grows over time in the second trial (**Figure 6.35c**, bottom). Therefore, this network model reproduces salient neurophysiological observations of Kiani and Shadlen (2009). At the behavioral level, the probability of opt-out, P_{sure} in a single decreases with increased motion coherence (**Figure 6.36a**, left); the accuracy is higher for trials when T_S is presented but not selected than when T_S is absent (**Figure 6.36a**, right), as in the monkey experiment.

Furthermore, the model captures the dependence of the opt-out probability on the motion coherence c' and stimulus presentation duration for a fixed c' (**Figure 6.36b**), observed in monkey's behavior (**Figure 6.36c**).

Note that P_{sure} is the fraction of trials when the sure target is presented and chosen. In the continuous recurrent network model, the absolute difference $|R_A - R_B|$ varies over time in a single trial and from one trial to another (**Figure 6.35b–c**) as a stochastic variable. There is no explicit representation of probability or confidence in neurons. As a cross-trial statistic, the probability P_{sure} was found to be a nonlinear decreasing function of $|R_A - R_B|$, and a measure of choice confidence can be defined as Confidendce $= 1 - P_{\text{sure}}$ between 0 and 1. This reading of confidence, internally defined by network dynamics, increases with motion coherence c' (**Figure 6.37a**). However, this confidence estimation generally deviates from the behavioral accuracy (**Figure 6.37b**): unexpectedly, the model displays the "hard-easy effect" observed in human behavioral studies

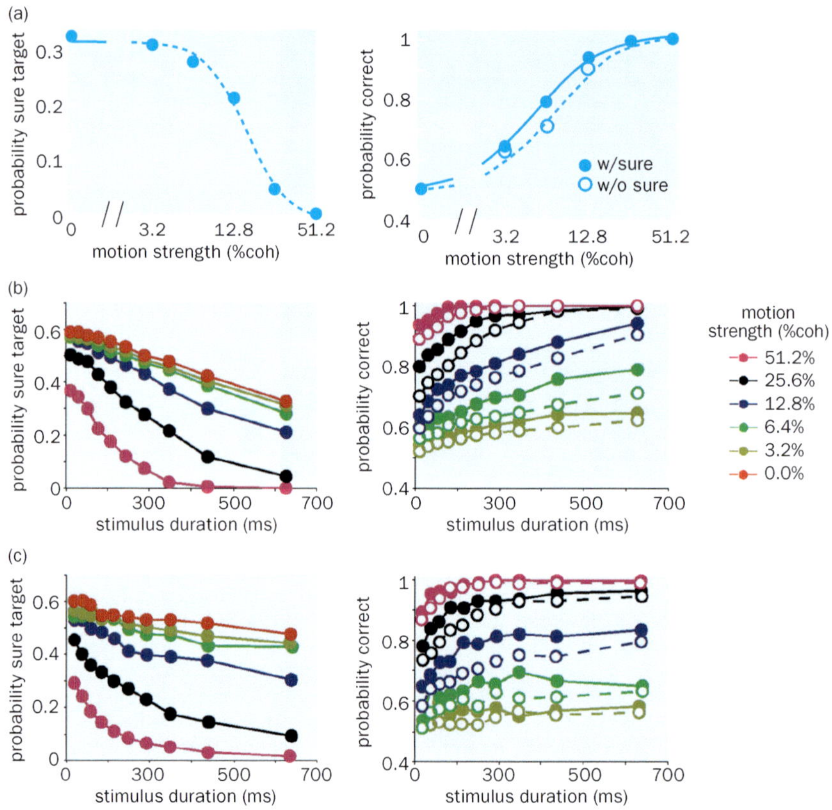

Figure 6.36 Confidence and performance. (a) Model performance (at a fixed stimulus duration). Left: Probability P_{sure} of choosing T_S decreases as an function of the motion strength. Right: Compared to trials where T_S is not shown (dashed curve), accuracy is improved in trials when T_S was shown but waived (solid curve). (b) In the model, P_{sure} decreases with motion strength and stimulus duration (left panel); accuracy is higher in trials where T_S was shown but waived (solid lines, filled circles) than in those trials where T_S was not shown (dashed lines, open circles) (right panel). (c) Behavioral data from the monkey experiment. Panels (a–b) are reproduced from Wei and Wang (2015), (c) from Kiani and Shadlen (2009). [Wei, Z., & Wang, X. J. (2015). Confidence estimation as a stochastic process in a neurodynamical system of decision making. Journal of Neurophysiology, 114(1), 99–113.] [Kiani, R., & Shadlen, M. N. (2009). Representation of confidence associated with a decision by neurons in the parietal cortex. Science, 324(5928), 759–764.]

(Gigerenzer et al. 1991). Namely, confidence is higher than accuracy (overconfidence) at low c'; in particular, it is about 0.7 rather than 0.5 at $c' = 0$, which has been proposed to hold generally (Hangya et al. 2016); on the other hand, confidence is lower than accuracy (underconfidence) at high c', except for $c' = 1$, when both are 1. Furthermore, confidence reading decreases with reaction time in a reaction time version of task simulation, as commonly reported experimentally (Vickers 1979; Kiani et al. 2014a).

The dependence of this confidence measure on c' shows an opposite trend for correct and error trials (**Figure 6.37c**). In error trials when evidence is in favor of A but the winner is B, R_B is large. The input to neural pool A, hence its firing rate R_A, increases with c', and therefore the difference $|R_A - R_B|$ (and internally estimated confidence) decreases with c'. We all have experiences where we are less confident about a judgment in the face of of strong evidence supporting a contrary alternative; the model suggests a neural mechanism of this behaviorally measurable trait.

What would be a neural circuit mechanism for computing an absolute difference like $|R_A - R_B|$? Single neurons in the orbitofrontal cortex (OFC) (Kepecs et al. 2008) and in a part of the thalamus called pulvinar (Komura et al. 2013) display confidence correlates akin to $|R_A - R_B|$. In a decision experiment performed by monkeys (**Figure 6.38a**, left panel), the firing rate of a neuron in the pulvinar increased with stimulus strength in correct trials; by contrast, the dependence was the opposite in error trials (**Figure 6.38a**, right panel), mirroring the confidence estimation at the behavioral level (**Figure 6.38a**, middle panel). Computational modeling (Jaramillo et al. 2019) proposed a neural circuit mechanism for computing $|R_A - R_B|$ in the pulvinar (**Figure 6.38b**). In the model, pulvinar neurons receive converging excitatory inputs from two selective neural populations in the cortex; cortical circuit also projects to GABAergic neurons in the thalamic reticular nucleus (TRN), which in turn send inhibition to the principal (projection) excitatory neurons in the pulvinar.

Crucially, the model incorporates the experimental finding that direct cortical projection to principal neurons in the thalamus displays short-term synaptic facilitation, while TRN-mediated inhibition shows short-term synaptic depression (Crandall et al. 2015). As we learnt in Chapter 2 (Section 2.7), a synaptic connection dominated by short-term depression operates the best at relatively low presynaptic rates. By contrast, a synaptic connection endowed with strong short-term facilitation is weak at low presynaptic rates but becomes effective at high rates. Therefore, high firing of a winning cortical neural pool (say, A) generates a net excitation $\approx R_A$ via direct cortical projection to the pulvinar principal neurons. By contrast, a

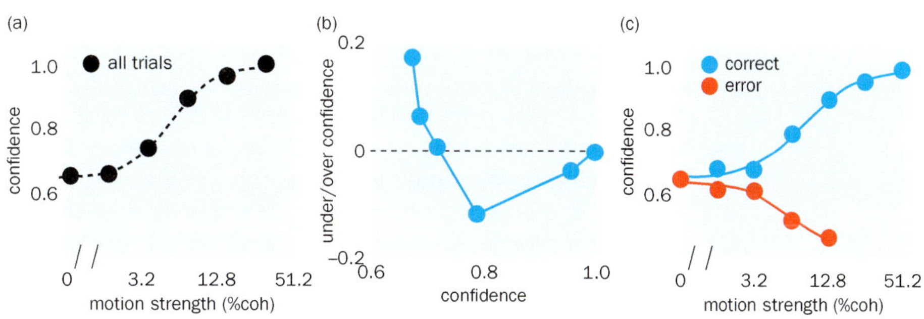

Figure 6.37 Hard-easy effect of confidence estimation in a spiking network model of decision making. (a) Choice confidence as a function of coherence c'. (b) Confidence is higher than accuracy (overconfidence) at low c'; the opposite is true at high c' and accuracy (underconfidence) except for 100%. Y-axis: Confidence minus accuracy. (c) Confidence increases with c' in correct trials, whereas it decreases with c' in error trials. Reproduced from Wei and Wang (2015). [Wei, Z., & Wang, X. J. (2015). Confidence estimation as a stochastic process in a neurodynamical system of decision making. Journal of Neurophysiology, 114(1), 99–113.]

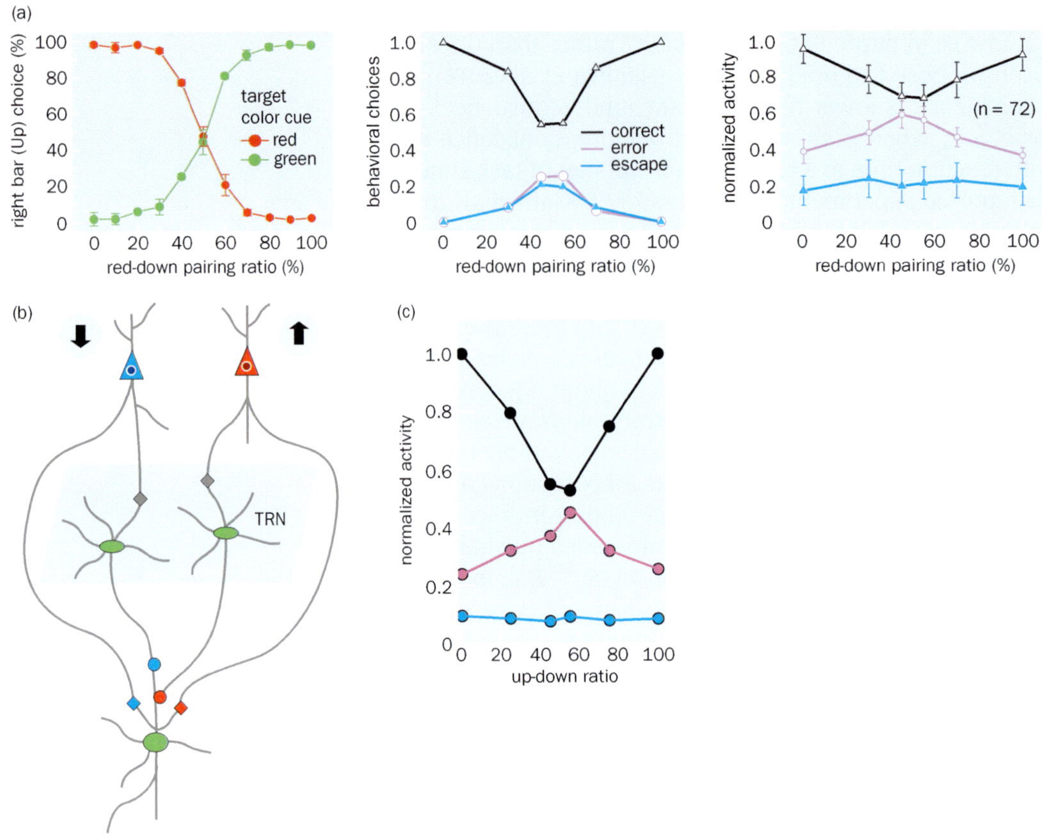

Figure 6.38 A pulvinar model of confidence estimation. (a) A monkey experiment. In a discrimination task, a fraction X% of random dots in moving pattern are red, and the net direction is either Up or Down, the remaining (100-X)% are green with the opposite net direction from the red dots. Left panel: The probability of choosing Up as a function of X%. Middle panel: The Accuracy (black) increases, while error rate (light purple) decreases, with the stimulus strength. When opt-out was offered, monkeys chose that option while abandoning the discrimination with a probability that is maximal at X = 50% and decays both for smaller and larger red-green pairing ratio (light blue). Right panel: Neurons in the pulvinar display firing rates that mirror the behavioral correct and error rates (black and light purple) but are low and independent of X in opt-out trials (light blue). (b) A schematic model of the cortico-pulvinar network. (c) The model reproduces the pulvinar neural representation, thereby offering a circuit mechanism for computing an absolute difference for confidence reading. Panel (a) is reproduced from Komura et al. (2013), (b–c) from Jaramillo et al. (2019). [Komura, Y., Nikkuni, A., Hirashima, N., Uetake, T., & Miyamoto, A. (2013). Responses of pulvinar neurons reflect a subject's confidence in visual categorization. Nature Neuroscience, 16(6), 749–755.] [Jaramillo, J., Mejias, J. F., & Wang, X. J. (2019). Engagement of pulvino-cortical feedforward and feedback pathways in cognitive computations. Neuron, 101(2), 321–336.]

losing cortical neural pool (B) at a low firing rate is not effective for mediating direct cortical excitation to the principal pulvinar neurons, but through RTN exerts a net inhibitory influence proportional to R_B. Consequently, the pulvinar neural activity represents $R_A - R_B$. Conversely, in the case where the cortical neural population B wins the competition, firing of pulvinar neurons represents $R_B - R_A$. The model reproduces salient physiological observations from pulvinar neurons (Komura et al. 2013) (**Figure 6.38c**). Beyond confidence estimation, this work proposes a biologically based mechanism for computing the absolute value of a difference signal.

In a decision process, confidence may impact the course of a deliberation process on the fly; one observable phenomenon of this influence is referred to as "changes of mind" (Kiani and Shadlen 2009). A change of mind occurs when empirical evidence varies over time, and the brain may tentatively make a choice but is uncertain about it and eventually settles on a different choice. This is difficult to assess when the response is a ballistic saccade but more readily measured with slower manual movement. In a human study of the RDM direction discrimination (Resulaj et al. 2009), a decision was indicated by a hand movement to either a leftward or rightward target (**Figure 6.39a**). The moving dot visual stimulus was extinguished as soon as the movement was initiated, so no sensory information was available during the movement. In some trials, subjects generated a hand trajectory that first was toward one target but then changed the path and eventually ended at the other target (**Figure 6.39b**). Changes of mind occur more frequently from an incorrect to correct decision than in the reverse direction (**Figure 6.39c**), so overall performance is better based on the decision at termination than in the early phase of hand movement (**Figure 6.39d**). Motion energy analysis (Adelson and Bergen 1985) revealed that early stimulus influenced, but late stimulus fluctuations made no impact, on the decision across all trials (**Figure 6.39e**, blue). By contrast, in the subset of trials where a change of mind took place, late fluctuating sensory evidence from random dots, ~ 100 ms prior to motion onset, supported the final rather than initial decision (**Figure 6.39e**, red). The interpretation is that changes of mind may result from a reversal of the sign of momentary evidence in the middle of a decision process (Wang 2002b).

In psychology, confidence is considered in the realm of metacognition, a term that denotes self-knowledge of cognitive processes, such as assessment of one's own decisions, memory, etc. (Flavell 1979; Terrace and Son 2009; Mamassian 2016). In humans, evidence from functional brain imaging (Fleming et al. 2012; Bang and Fleming 2018) and subjects with brain damage (Fleming et al. 2014) suggest that the medial prefrontal cortex is selectively engaged in confidence signaling, presumably in interplay with other brain

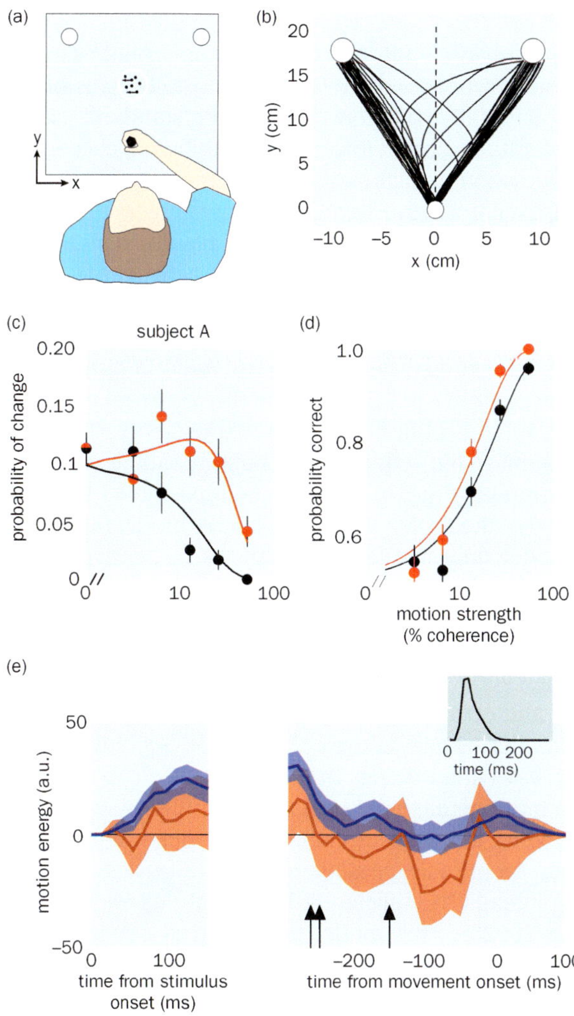

Figure 6.39 Changes of mind. (a) In a human study, subjects view an RDM pattern and indicate the decision A or B by a manual movement to the left or right. (b) In some trials, hand trajectories started to move in one target but changed to the other one in mid-course, betraying a change of mind. (c) Probability of changes of mind as a function of coherence c' from an erroneous to a correct response (red) or vice versa (black). (d) The probability of a correct decision based on hand movement at initiation (black) is lower than at termination (red). (e) Influence of motion energy fluctuations on initial and final decisions. Data are shown for all the trials (blue) and the subset of trials with a change of mind (red), aligned at stimulus onset (left) and movement onset (right). Reproduced from Resulaj et al. (2009). [Resulaj, A., Kiani, R., Wolpert, D. M., & Shadlen, M. N. (2009). Changes of mind in decision-making. Nature, 461(7261), 263–266.]

regions, including the pulvinar. The relationship between changes of mind and neural dynamics was analyzed in the continuous recurrent network model of decision making (Wei and Wang 2015). In the model, a form of changes of mind occurs when a decision is determined by attractor dynamics between neural populations selective for target A or B, but later the network switches to the sure target as the final decision. With a fixed motion coherence, an initial decision is determined when one of the two competing neural populations crosses a population firing threshold (50 Hz) prior to the onset of a sure target. However, the neural population selective for T_{sure} can still prevail and determine the final decision in trials when the choice confidence estimated by $|R_A - R_B|$ is low, as expected intuitively. The probability for this change of mind is roughly constant at low and intermediate c', and then decreases at high c', similar to human's behavior (**Figure 6.39c**). Experimentally, neural decoding of changes of mind in real time has been examined with simultaneous recording from a population of neurons in the dorsolateral prefrontal cortex. While monkeys performed an RDM task, the neuronally decoded decision occasionally reversed the signal in a single trial, and the probability of such a reversal is higher at lower c' (Kiani et al. 2014b). Therefore, even such elusive concepts as confidence and changes of mind can be explained by sampling stochastic neural population dynamics in single trials.

6.8 DUALITY OF COGNITIVE-TYPE NEURAL CIRCUITS

We have discussed a biologically based model of local neural circuits capable of both decision making and working memory. This class of neural networks, which we shall call the "cognitive-type" (Wang 2013), shares a lot of commonalities with sensory circuits and motor systems, built by the same types of single neurons and synapses as well as anatomically plausible network connectivity. Two quantitative differences, however, stand out. First, the strength of recurrent synaptic excitation (balanced by inhibition) exceeds a critical level to induce a bifurcation, which explains the emergence of multiple attractor states. Second, excitatory reverberation is slow and dependent on the NMDA receptors. The combination of these features results in a characteristic duality: on the one hand, slow transients (ramping activity) are suitable for accumulation of information over time; on the other hand, attractor dynamics is required for categorical choice as well as for working memory storage.

As a study of contrasts, let us compare the drift diffusion model with a long tradition in cognitive psychology and the recurrent neural circuit model of Wang (2002b) that is built on modern neuroscience. The DDM is normative and equivalent to the sequential probability ratio test optimal for binary choices in continuous time. Is the neural circuit model merely an approximate realization of the DDM? If a neural circuit model faithfully implements a normative model, they should produce the same behavior, which is not the case for the DDM and the recurrent neural circuit model. First, the recurrent circuit model produces longer response times in error trials than in correct trials (Wong and Wang 2006), consistent with the monkey experiment (Roitman and Shadlen 2002) (**Figure 6.21**). By contrast, the DDM yields the same RTs in correct and error trials, unless additional assumptions are included (Ratcliff and Rouder 1998; Mazurek et al. 2003). Second, the DDM does not have a steady-state and predicts that performance can potentially improve indefinitely with an increasing duration of stimulus viewing, e.g., by raising the decision bound. In the recurrent circuit model, ramping activity eventually stops when an attractor state is reached (**Figure 6.13**). Consequently, performance plateaus at sufficiently long stimulus-viewing times (**Figure 6.40a**) (Wang 2002b). This model prediction is confirmed by a monkey experiment (Kiani et al. 2008) (**Figure 6.40b**). Third, the attractor network model has been shown to be able to subtract negative signals as well as add

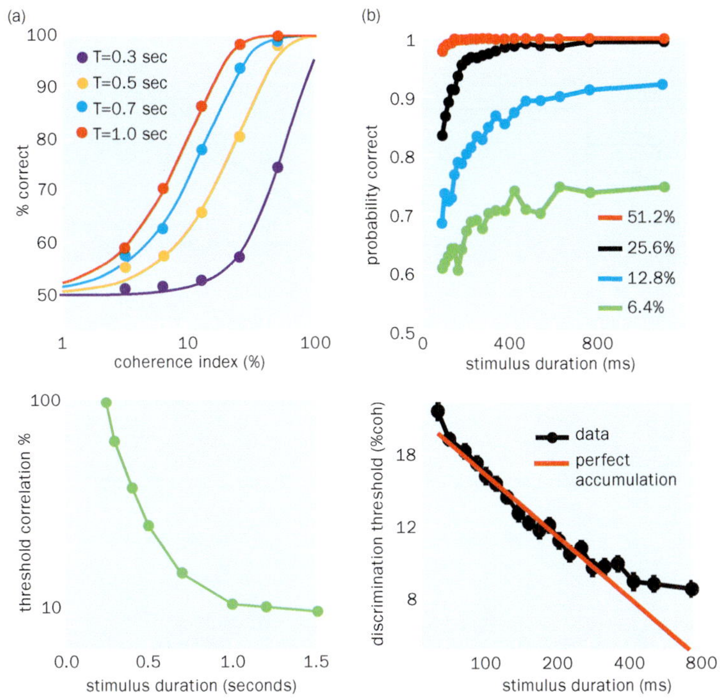

Figure 6.40 Dependence of the decision performance on the duration of stimulus presentation. (a) Model performance. Top: Psychometric function is shifted to the left with longer stimuli. Bottom: Coherence threshold (defined by the coherence level required to achieve the accuracy of 82% correct choices) decreases with the stimulus duration and plateaus for stimulus presentation longer than 0.5–1 seconds. (b) Monkey's behavior performance increases with longer stimulus presentation, plateauing at around the same time duration as in the model. Panel (a) is reproduced from Wang (2002b), (b) from Kiani et al. (2008). [Wang, X.-J. (2002b). Probabilistic decision making by slow reverberation in cortical circuits. Neuron, 36(5), 955–968.] [Kiani, R., Hanks, T. D., & Shadlen, M. N. (2008). Bounded integration in parietal cortex underlies decisions even when viewing duration is dictated by the environment. Journal of Neuroscience, 28(12), 3017–3029.]

positive evidence about choice alternatives, but the influence of newly arriving inputs diminishes over time, as the network converges toward one of the attractor states representing the alternative choices (Wang 2002b). This prediction is also confirmed by the monkey experiment, which showed that the impact of a brief motion pulse in addition to the random-dot stimulus was greater with an earlier onset time (Huk and Shadlen 2005; Wong et al. 2007) (**Figure 6.41**). This violation of time-shift invariance cannot be accounted for by the inclusion of a leak. In fact, in contrast to the recurrent circuit model, a leaky integrator model (Eq. 6.7) actually predicts the opposite: that later rather than earlier signals influence more behavior, because an earlier pulse is leaked out and gradually "forgotten," and therefore does not significantly affect the decision that occurs much later (Wong et al. 2007). Fourth, as shown by Berlemont et al. (2020), the neural circuit model but not the DDM accounts for the observations that confidence estimation is influenced by early but not late sensory evidence (Zylberberg et al. 2012) and that lower confidence in a trial induces "more caution" in the subsequent trial with slower and more accurate decisions (Desender et al. 2019).

The two models also have distinct predictions about the decision variable dynamics. Interestingly, in contrast to the one-dimensional diffusion model, a state-space analysis of neural population activity (**Figure 6.17**) showed that the dynamics of the attractor neural network is inherently two-dimensional (Wong and Wang 2006). This is consistent with the finding that, in the RDM experiment, LIP neuronal activity displays a biphasic time course, with neurons selective for the two opposite targets first ramping up together before diverging away from each other (Roitman and Shadlen 2002; Huk and Shadlen 2005). The same type of behavior was also observed in a free-choice task (Scherberger and Andersen 2007). Under some assumptions, a network model for two choices has been approximately reduced to a one-dimensional system (with the dynamical variable representing $r_A - r_B$), which nevertheless is a nonlinear equation endowed with attractor dynamics (Roxin and Ledberg 2008). Moreover, the DDM predicts that at zero motion strength, the overall trial-averaged neural activity trace should be flat in time. In contrast, the neural circuit model predicts a nonzero time course of neural activity on average (**Figure 6.14**), which has been observed in LIP

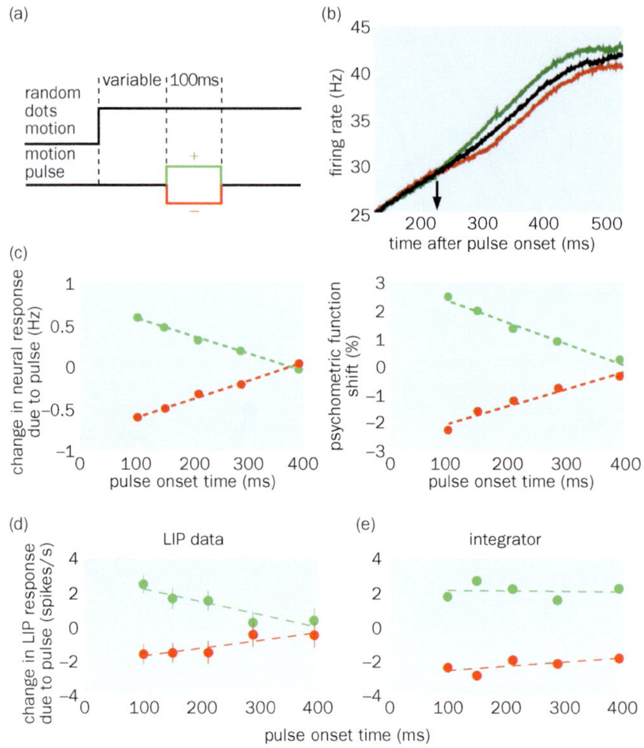

Figure 6.41 Model testing with a pulse perturbation. (a) A brief input pulse is added in the same direction as the dots stimulus (green) or in the opposite direction (red). (b) Change of firing rate (red or green) relative to control (black) persists after a brief pulse of 100 ms, with onset time indicated by an arrow. (c) In the model, the impact of a pulse is substantial on both neural activity (left) and performance (right) when the pulse is presented early, and is negligible with late pulse onset. (d) The time dependence is observed in the neural activity of LIP neurons in the monkey experiment. (e) The DDM, by contrast, predicts that the impact of a pulse is insensitive to onset time. Panels (a–c) are reproduced from Wong et al. (2007), (d–e) from Huk and Shadlen (2005). [Wong, K. F., A. C. Huk, M. N. Shadlen, and X.-J. Wang (2007). Neural circuit dynamics underlying accumulation of time-varying evidence during perceptual decision-making. Frontiers in Comput. Neurosci. 1:6, doi: 10.3389/neuro.10/006.2007. Licensed under CC BY 4.0.] [Huk, A. C., & Shadlen, M. N. (2005). Neural activity in macaque parietal cortex reflects temporal integration of visual motion signals during perceptual decision making. Journal of Neuroscience, 25(45), 10420–10436.]

neurons (Churchland et al. 2007). Moreover, the scale-invariance of reaction time statistics suggests that noise (from irregular spiking firing) co-varies in time with the mean neural activity in a recurrent decision circuit, in contrast to the DDM, for which noise is constant in time.

The DDM assumes that a difference signal, about the conflicting inputs, is computed at the input level (Ratcliff 1978; Mazurek et al. 2003). This idea has been taken seriously in a human fMRI experiment, in which the task was to discriminate whether an ambiguous image is a face or a house. The BOLD signal in the dorsolateral prefrontal cortex was found to covary with the difference signal between the face- and house-selective regions in the ventral temporal cortex (Heekeren et al. 2004). However, a significant correlation from a regression analysis does not necessarily prove that a difference signal is explicitly computed in a brain region. One should not interpret the DDM model literally in neuronal terms; a differential input is not needed for another mathematical model such as the race model (Eqs. 6.8–6.9). In the recurrent circuit model, competition between neural pools selective for choice alternatives is instantiated by lateral synaptic inhibition (Wang 2002b; Wong and Wang 2006). This feedback mechanism, not the feedforward subtraction, is supported by the observation that microstimulation of one neural pool in LIP not only sped up the decisions in its preferred direction, but also slowed down the decisions in the antipreferred direction (Hanks et al. 2006).

In principle, one can be skeptical about the merit of such a comparison, on the ground that a neural circuit model could have a lot more parameters than a normative model. As a matter of fact, a neural circuit model is constrained not only by behavior but also by neurobiology. The two types of models may actually have a similar number (a half dozen) of adjustable parameters: on the one hand, an extended DDM (Drugowitsch et al. 2012) is characterized by the stimulus sensitivity, decision threshold, noise size, two parameters for a time-dependent urgency signal, and possibly parameters about trial-to-trial variability of ramping rate (Ratcliff and McKoon 2008) and decision threshold (Desender et al. 2019); on the other hand, a reduced neural network model (Wong and Wang 2006) is characterized by the strengths of recurrent excitation and inhibition, the time constant of the NMDA receptor-mediated excitation, noise level, decision threshold and external stimulation. Furthermore, the model comparison under consideration is entirely different from the traditional model comparisons in psychology, where differences are small and the exact number of parameters for each model should be taken into account, e.g., using the Akaike information criterion. As is clear from this comparison, all of the differences between the two models are about *qualitative* behaviors, not small statistical differences. Finally, the recurrent neural network model displays more than one operating regime (ramping mode versus jumping mode) (Wang 2012), potentially suitable for different behavioral demands.

6.9 SUMMARY

In this chapter, we covered three main components of decision processes: accuracy, response time and confidence. Over the last two decades, neurophysiological experiments have made remarkable advances in uncovering neural dynamics associated with various aspects of perceptual decision making. In interplay with behavioral and neurophysiological studies, we discussed linear dynamical systems models, probabilistic Bayesian models and nonlinear neural circuit models of decision making. We focused on our biologically based recurrent network model, characterized by a duality of slow transients for information integration and attractor dynamics that underlies winner-take-all leading to a choice. These findings form the basis of the concept of "cognitive-type" neural circuits capable of both decision making and working memory, which constitutes a theoretical cornerstone for understanding the prefrontal cortex (Wang 2013).

The recurrent neural circuit model accounts for a host of experimental observations and has yielded several unexpected predictions. It represents a framework for cross-level understanding that bridges multiple levels from proteins (e.g., NMDA receptors) to neural circuit reverberatory dynamics and quantitative animal performance. The model can and should be extended to incorporate diverse subtypes of inhibitory neurons (Chapter 5, Section 5.6.4) to explain selectivity of inhibitory neurons in a decision process (Najafi et al. 2020). In the same vein, the model for pattern match decisions incorporated single neuron rate adaptation, synaptic reward-dependent plasticity and homeostatic regulation in a network with working memory and decision-making modules; and short-term synaptic plasticity is crucial in the model for computing an absolute difference of two neural population signals in confidence estimation. Brick by brick, it is possible to build a framework in which building blocks of biophysical mechanisms and basic cognitive processes are assembled to account for increasingly more complex cognitive functions across spatiotemporal scales.

Future work will explore strategies employed in decision making, which may even vary over time in a single individual (Ashwood et al. 2022). In particular, our discussion has been limited to situations where evidence is simultaneously considered for choice alternatives. In real life, however, evidence of the pros and cons may be present at different time points separated by temporal gaps (Kiani et al. 2013). For instance, while shopping for a coat, generally we do not evaluate all available options at once. Instead, we often assess individual items sequentially, reconsider previously examined items, maybe compare particular pairs of items, until a choice is made. In the laboratory experiments, it was found that subjects switch their gaze and pay attention back and forth between choice alternatives (Krajbich et al. 2010; Hunt et al. 2018; Kaanders et al. 2021). Moreover, deliberation in choice behavior takes a longer time when required computation is more complex (Hong and Stauffer 2023). Neural network models remain to be developed to account for these temporally elaborated processes of decision making.

FURTHER READING

Books and Reviews

Cisek, P. and J. F. Kalaska (2010). Neural mechanisms for interacting with a world full of action choices. *Annu. Rev. Neurosci. 33*, 269–298.

Gigerenzer, G. and W. Gaissmaier (2011). Heuristic decision making. *Annu. Rev. Psychol. 62*, 451–482.

Gold, J. I. and M. N. Shadlen (2007). The neural basis of decision making. *Annu. Rev. Neurosci. 30*, 535–574.

Heekeren, H. R., S. Marrett, P. A. Bandettini, and L. G. Ungerleider (2004). A general mechanism for perceptual decision-making in the human brain. *Nature 431*, 859–862.

Luce, R. D. (1986). *Response Time: Their Role in Inferring Elementary Mental Organization*. New York: Oxford University Press.

Ma, W. J. (2019). Bayesian decision models: A primer. *Neuron* 104, 164–175.

O'Connell, R. G., M. N. Shadlen, K. Wong-Lin, and S. P. Kelly (2018). Bridging neural and computational viewpoints on perceptual decision-making. *Trends Neurosci. 41*, 838–852.

Parker, A. J. and W. T. Newsome (1998). Sense and the single neuron: Probing the physiology of perception. *Ann. Rev. Neurosci. 21*, 227–277.

Usher, M. and J. McClelland (2001). On the time course of perceptual choice: The leaky competing accumulator model. *Psychol. Rev. 108*, 550–592.

Vickers, D. (1979). *Decision Processes in Visual Perception*. London: Academic Press.

Wang, X.-J. (2008). Decision making in recurrent neuronal circuits. *Neuron 60*, 215–234.

Primary Research Papers

Brain Basis of Perceptual Decision

Britten, K. H., M. N. Shadlen, W. T. Newsome, and J. A. Movshon (1992). The analysis of visual motion: A comparison of neuronal and psychophysical performance. *J. Neurosci. 12*, 4745–4765.

de Lafuente, V. and R. Romo (2005). Neuronal correlates of subjective sensory experience. *Nat. Neurosci. 8*, 1698–1703.

Harvey, C. D., P. Coen, and D. W. Tank (2012). Choice-specific sequences in parietal cortex during a virtual-navigation decision task. *Nature 484*, 62–68.

Krajbich, I., C. Armel, and A. Rangel (2010). Visual fixations and the computation and comparison of value in simple choice. *Nat. Neurosci. 13*, 1292–1298.

Leopold, D. A. and N. K. Logothetis (1996). Activity changes in early visual cortex reflect monkeys' percepts during binocular rivalry. *Nature 379*, 549–553.

Roitman, J. D. and M. N. Shadlen (2002). Response of neurons in the lateral intraparietal area during a combined visual discrimination reaction time task. *J. Neurosci. 22*, 9475–9489.

Mathematical Models of Perceptual Decision

Bogacz, R., E. Brown, J. Moehlis, P. Holmes, and J. D. Cohen (2006). The physics of optimal decision making: A formal analysis of models of performance in two-alternative forced-choice tasks. *Psychol. Rev. 113*, 700–765.

Ratcliff, R. (1978). A theory of memory retrieval. *Psychol. Rev. 85*, 59–108.

Biologically Based Neural Circuit Models

Deco, G., M. P´erez-Sanagust´ın, V. de Lafuente, and R. Romo (2007). Perceptual detection as a dynamical bistability phenomenon: A neurocomputational correlate of sensation. *Proc. Natl. Acad. Sci. U.S.A.* 104, 20073–20077.

Lo, C. C. and X.-J. Wang (2006). Cortico-basal ganglia circuit mechanism for a decision threshold in reaction time tasks. *Nat. Neurosci. 9*, 956–963.

Machens, C. K., R. Romo, and C. D. Brody (2005). Flexible control of mutual inhibition: A neural model of two-interval discrimination. *Science 18*, 1121–1124.

Wang, X.-J. (2002). Probabilistic decision making by slow reverberation in cortical circuits. *Neuron 36*, 955–968.

Wong, K. F. and X.-J. Wang (2006). A recurrent network mechanism of time integration in perceptual decisions. *J. Neurosci. 26*, 1314–1328.

Confidence

Kiani, R. and M. N. Shadlen (2009). Representation of confidence associated with a decision by neurons in the parietal cortex. *Science 324*, 759–764.

Wei, Z. and X.-J. Wang (2015). Confidence estimation as a stochastic process in a neurodynamical system of decision making. *J. Neurophysiol. 114*, 99–113.

Value-Based Economic Choice

7

How do you know whether something is really what you want to do or just some kind of nerve impulse in the brain? Some minor little activity takes place somewhere in this unimportant place in one of the brain hemispheres and suddenly I want to go to Montana or I don't want to go to Montana.

Don DeLillo in *White Noise*

7.1 INTRODUCTION

In the previous chapter, we discussed decision making in the perceptual domains, where we strive for an objectively correct judgment. By contrast, value-based decisions, or economic choices, are subjective. For instance, selection of dishes at a restaurant is not only determined by available options but also influenced by what you like (beef or tofu, spicy or curry sauce, etc.), which varies from individual to individual. No one is alike; each has distinct preferences determined by factors such as memories of our past experience (did you love spicy tofu?) and internal physiological needs (how hungry are you?).

As schematically shown in **Figure 7.1**, economic choice behavior involves several processes. At the core is valuation, which combines sensory stimuli and internal factors into estimated values for different offered options. These values will be called "expected values," denoting the subjective sense of reward or punishment as outcome by choosing one of available alternatives. The next step is to make a choice informed by expected values. Note that here we are concerned with value-based decisions, which require an action to yield an outcome, unlike Pavlovian classical conditioning, where reward is independent of voluntary behavioral response (Chapter 4). A choice may be made first among offers in the space of goods rather than actions. For example, you can choose to have steak for dinner first, then decide to either cook it yourself at home or go out and have it at a steakhouse. Hence, the third step is to transform a choice into a motor act. The outcome of an action may or may not meet your expectation, producing a feedback signal to your brain. When you are pleasantly

DOI: 10.1201/9781003459361-9

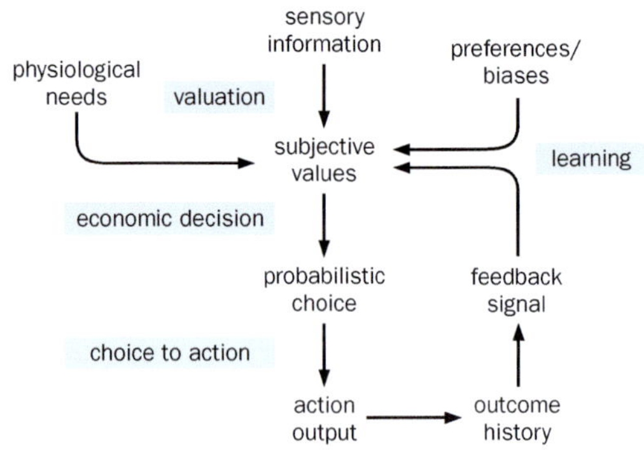

Figure 7.1 Basic steps and processes in value-based decision making. Valuation, economic choice, transformation from choice to action and reinforcement learning are four main processes of adaptive economic choice behavior.

surprised by what you get, you are more likely to choose the same option again in the future; conversely, a disappointment (e.g., filet mignon you just had was not as good as you thought) or even a punishment (you got sick afterwards) would reduce the likelihood that you will make the same choice. Reinforcement learning based on choice outcomes is the fourth key process in adaptive choice behavior.

The topic of economic choice covers a wide range of behaviors. Most fundamental to animals is foraging for survival in a complex environment where weather changes unpredictably, individuals compete or cooperate in gathering food, dangerous predators lurk around. The field of studies of foraging is called behavioral ecology (Stephens and Krebs 1986). Another rich field of studies of individual choices with a long history is microeconomics (Frank 2003), where subjective value is called utility function. von Neumann and Morgenstern (1944) formalized an axiomatic expected utility theory as the foundation of classical economics. Both foraging theory and expected utility theory assume that choice behavior is rational in the sense of achieving optimality or maximal rewarding outcomes contingent on minimal costs and aversive outcomes.

In contrast to a static description of utility optimization, adapting decision making in order to increase rewards (even when failing to achieve maximization) is a dynamical process. The latter has long been the subject of learning in laboratory studies of choice behavior under controlled conditions (Mackintosh 1974; Gallistel 1990), with reinforcement learning (Chapter 4, Section 4.3) as a central theoretical underpinning. Hence, value-based decision making involves at least three fields: behavioral ecology, economics and reinforcement learning. In addition, choice behavior cannot be dissociated from emotions. Our mood, stress, fear and joy strongly influence our choice behavior. There is an increasing recognition that the brain systems for value-based economic choice and affects are intertwined (Lempert and Phelps 2014), and a substantial literature exists on the brain circuit of emotions (LeDoux 2000; Dalgleish 2004; Barrett 2017; Adolphs and Anderson 2018).

In this chapter, we will introduce basic ideas from economics and foraging theory, and discuss what is known about the brain mechanism for subjective valuation. Our focus will be on the neural circuit model of decision making discussed in Chapter 6 endowed with reward-dependent plasticity. We will discuss how this model and its extensions can account for behavioral and neurophysiological observations from experiments using several tasks, ranging from foraging to probabilistic inference to competitive games. Finally, we will touch on cooperation in social decision making.

7.2 NEUROECONOMICS AND FORAGING THEORY

Valuation, essential for determining preference in economic choice, is a complex undertaking. For instance, you love books as well as Belgian chocolates, but how do you measure and compare their worth to you?

Classical microeconomics does not directly address this question; instead, it assesses people's preference through their choices, referred to as the revealed preference. Let us denote subjective value or "utility" by $V(x_i)$ as the benefit or satisfaction from a choice i (with an amount of x_i) minus its cost (e.g., money spent). Utility increases with the amount of good, hence $V(x_i)$ has a positive first derivative ($dV/dx_i > 0$). Moreover, an increment Δx_i from a larger x_i yields a smaller increase of satisfaction: the added joy from obtaining one more chocolate bar is less if you already have 20 bars than it is if you have none. This is called "diminishing return," and mathematically it means that the second derivative of $V(x_i)$ is negative ($d^2V/dx_i^2 < 0$). In other words, the curve for gain is concave (**Figure 7.2a** for $x_i > 0$). If each option is associated with a probability p_i, as in a lottery, then the expected utility theory postulates that individuals choose options among available alternatives that collectively maximize the average $\sum_i p_i V(x_i)$. Here, utility is revealed only by preference in actual choice.

It is possible to describe a set of choices as maximizing a utility function, with a small set of plausible axioms including transitivity: if you prefer X to Y and Y to Z, then you will choose X over Z (von Neumann and Morgenstern 1944). However, a host of phenomena in human behavior appear to be inconsistent with the assumed rationality (Kahneman and Tversky 1979; Tversky and Kahneman 1981).

Prospect theory (Kahneman et al. 1982) was developed to account for some of the observed suboptimal behavioral biases, as deviations from expected utility theory. First, values can be positive (gain) or negative (loss); people tend to weigh loss more than gain. This aversion is expressed by assuming that $V(x_i)$ has a "kink" at zero, with the slope larger on the left for loss than on the right for gain (**Figure 7.2a**). Second, gain or loss is not absolute, but relative to a reference point r, hence utility is rewritten as $V(x_i - r)$. This reference point may change, for instance, consuming a chocolate bar may actually be loss rather than gain if you are on a strict diet. Third, given probabilities for different options, people tend to overweight small p_i and underweight large p_i. Hence, the expected utility $V(x_i - r)$ is multiplied by a nonlinear function $\pi(p_i)$ of p_i (**Figure 7.2b**); $\pi(p_i) > p_i$ for small p_i, $\pi(p_i) < p_i$ for large p_i. Not only humans but also nonhuman primates (Lakshminarayanan et al. 2011; Yamada et al. 2013) and rodents (Constantinople et al. 2019) display those traits.

In a choice task, two goods are presented and subjects choose one of them. For instance, back to the example of Belgian chocolate bars (A) versus books (B). Do you prefer a book to a Belgian chocolate bar? If so, how about x chocolate bars (good A) versus y books (good B)? For a utility function $V(x, y)$, an indifference curve (IC) is defined by combinations of x and y that yield a fixed $V(x, y) = \text{constant}$. **Figure 7.2c** shows a plane where two goods x and y are plotted against each other. A fixed total utility (e.g. monetary budget) defines an IC (**Figure 7.2c**, thick black curve), along which various quantities of two goods are equivalent. Writing an IC as $y = f(x)$, it should be a decreasing function ($dy / dx = f'(x) < 0$) because for a fixed budget an increase in x implies a decrease in y (compare (x_2, y_2) with (x_1, y_1)). Furthermore, the slope is steep at small x and large y, reflecting the intuition that starting with a large amount of y and little x, you'd be willing to give up a lot of y in exchange for a unit increase in x; conversely, the slope is shallow at large x and small y. Mathematically, this is guaranteed with $dy^2 / dx^2 = f''(x) > 0$; in other words, IC is usually convex. Using standard calculus, it can be shown that, when $dy / dx = f'(x) < 0$, the second derivative has the same sign for $y = f(x)$ and $x = f^{-1}(y)$; therefore, convexity of an IC holds regardless of

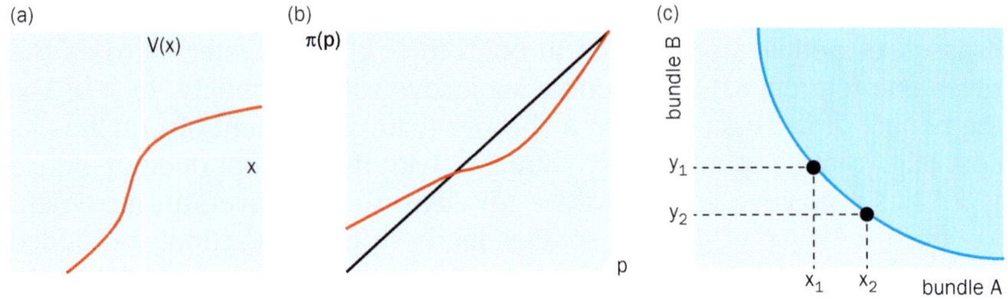

Figure 7.2 Economic concepts for utility functions. (a) Utility is a concave function of gain (positive *x*), whereas the dependence is convex for loss (negative *x*). (b) Subjective probability as a function of actual probability according to the prospect theory. (c) A convex indifference curve on the plane of two goods.

which of the two ways to plot the relationship between two goods. In Section 7.4, we will encounter the possibility of a concave IC.

Given goods x and y, a choice is made according to a decision criterion. The simplest is a step function: choose y if $V(y) > V(x)$ (above IC); x otherwise (below IC). However, when the two values are similar, close to an indifference curve, decision making is not deterministic and choices present trial-to-trial variability. This is described by a logistic function, namely, the probability of choosing x is $\text{Prob}(x) = 1/(1+\exp(-\beta(V(x)-V(y)))$, with β controlling the sensitivity to small $(V(x)-V(y))$ values. In other words, utilities are computed deterministically, then a die is thrown with probability equal to the softmax; the source of randomness in choice behavior is left unaddressed.

In addition to a choice's type, amount and probability, a fourth important factor to consider is temporal delay before the good can be collected. Decision makers behave as though immediately consumable goods are more valuable than those available after a delay, implying that utility function is "discounted" by a function $D(t)$ of the waiting time t, $V(x,t) = V(x)D(t)$, where $D(t)$ is a decreasing function (Frederick et al. 2002). Potential reasons for temporal discounting include immediate physiological needs, risk of losing a promised good during a waiting time and limited self-control of impulses. As shown by the famous "marshmallow test," a 4-year-old child's "willpower" to postpone immediate gratification represents a predictor of her/his success in adolescence (Mischel et al. 1989). If the discounting factor is an exponential function, $D(t) = \gamma^t$ (with $\gamma < 1$), an individual's choice is consistent across time: if one chooses x_A after a delay t_A over y_B after a delay t_B, $V(x_A)D(t_A) > V(y_B)D(t_B)$; the inequality remains true if both t_A and t_B are shifted by the same amount t. This is shown in **Figure 7.3a** (upper row) with $V(x_A) < V(y_B)$ and $t_A < t_B$, $V(x_A)D(t)$ and $V(y_B)D(t+(t_B-t_A))$ do not cross for all t.

However, many species, including pigeons (Ainslie 1974), rats (Richards et al. 1997), monkeys (Louie and Glimcher 2010) and humans (Rachlin et al. 1991), exhibit "hyperbolic discounting" (Mazur 1987):

$$D(t) = \frac{1}{1+kt} \tag{7.1}$$

where the parameter k is a positive discounting rate. Hyperbolic discounting decays more steeply at short t but more slowly at long t than exponential discounting as the delay gets larger (compare **Figure 7.3a**, upper and lower left plots). It implies that preference may reverse, as evidenced by an intersection of $V(x_A)D(t)$ and $V(y_B)D(t+(t_B-t_A))$ (**Figure 7.3a**, lower row). Reversal occurs in favor of the more proximate reward if both rewards are closer to the present time. For instance, one may prefer \$60 in 15 days over \$55 in 13 days (two more days of delay are not so significant compared to two weeks), but \$55 today over \$60 the day after tomorrow (immediate reward is only 10% less without having to wait for two days). Preference reversal is common in humans and nonhuman animals (Laibson 1997; Frederick et al. 2002). Interestingly, the discounting rate progressively

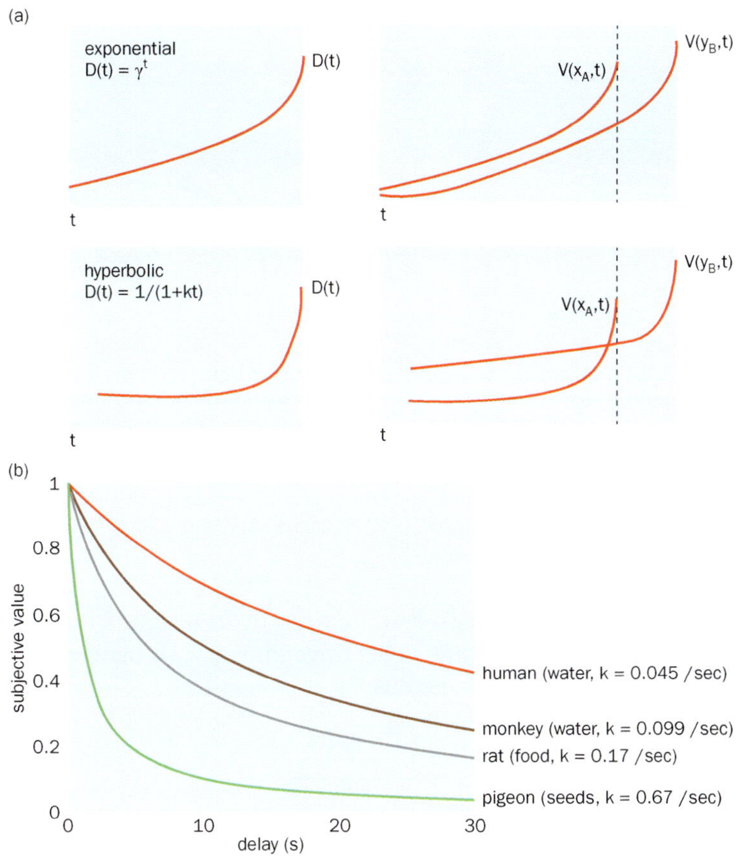

Figure 7.3 Temporal discounting.
(a) Exponential discounting function with a rate $\gamma < 1$ (upper left), for which preference (A over B) remains the same over time (upper right). By contrast, hyperbolic discounting function with a rate k (lower left) predicts preference reversal (lower right). (b) Hyperbolic discounting in four different species with measured discounting rate k values. Panel (b) was made by Kenway Louie with data from Ainslie (1974), Richards et al. (1997), Louie and Glimcher (2010) and unpublished data with water rewards in a human study. [Courtesy of Kenway Louie.]

decreases from pigeons, rats, monkeys to humans (**Figure 7.3b**), suggesting that an animal's ability for self-control is bettered through biological evolution. Note that it is not clear how hyperbolic discounting can be incorporated into the temporal-difference model of reinforcement learning (Chapter 4, Section 4.3.4), since the sum of discounted future rewards now is not guaranteed to remain finite.

Economic ideas briefly discussed here are powerful by virtue of their generality for describing choice behavior. At the same time, constraints under which agents operate in the world are often not taken into consideration in normative economic theories. Behavioral ecology offers a different perspective on choice behavior that emphasizes natural settings in real life, like humans in the prehistoric era and nonhuman animals in general (Kolling et al. 2012; Pearson et al. 2014). A cornerstone of foraging theory is the Marginal Value Theorem (MVT) (Charnov et al. 1976). Consider a famous example of passerine birds called starlings. Every day, a busy starling parent makes hundreds of trips from its nest to a feeding site, where it gathers larvae called leatherjackets, then returns home to feed its young (Kacelnik 1984; Stephens and Krebs 1986). There is usually not enough food, so there is no diminishing return with an increasing number of leatherjackets, but another form of diminishing return is caused by physical constraints. A starling probes leatherjackets with its beak, which also must hold collected goods, so the search becomes less efficient when a starling carries more load of leatherjackets in its beak. This is described by a concave yield $y(t)$ (load of leatherjackets) as a function of search time t (**Figure 7.4a**, black curve), $dy/dt > 0$ and $d^2y/dt^2 < 0$.

Hence, there is an optimal tradeoff between travel time and foraging time: gathering as much food as possible on each trip to save time of traveling between home and a feeding site, but maximal load would mean long search time spent on feeding site and reduced efficiency. The question is: what is the optimal search time t^* at a feeding site? The total

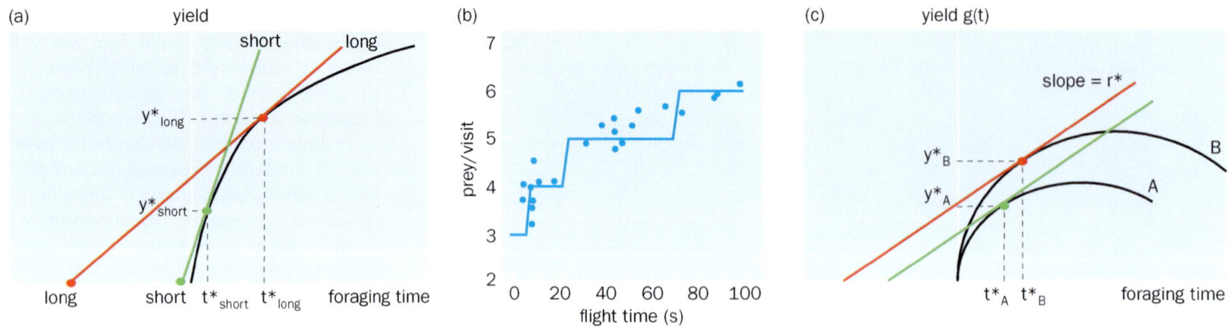

Figure 7.4 Marginal Value Theorem of foraging theory. (a) The problem of maximizing efficiency in trips between home and a feeding site has a solution according to the MVT, which asserts that the optimal search time t^* is given by the intersection of the yield function (black curve) and a tangent line with the slope equal to the rate of return. The intercept at $y = 0$ corresponds to the travel time (red: Long t_{travel}, green: Short t_{travel}). (b) The MVT is confirmed in an experiment with starlings, showing larger load of prey per visit with longer t_{travel}. Solid line: Prediction, dots: Empirical observation. (c) In foraging between two (here) or more patches of food, the MVT predicts optimum search time on each patch, corresponding to the value t^*, where the slope of the yield (marginal yield rate) is the same for all patches (A and B) and equal to the grand average yield rate for the entire habitat. Panel (b) is reproduced from Kacelnik (1984). [Kacelnik, A. (1984). Central place foraging in starlings (Sturnus vulgaris). I. Patch residence time. Journal of Animal Ecology, 53(1), 283–299.]

time for each trip is search time t plus travel time (t_{travel}) between feeding site and nest. The rate of gain is thus:

$$r(t) = \frac{y(t)}{t_{travel} + t} \tag{7.2}$$

This rate has a maximum r^* at t^*, given by $dr/dt = 0$ and $d^2r/d^2t < 0$. Since $dr/dt = \left((dy/dt)(t_{travel} + t) - y(t)\right)/(t_{travel} + t)^2$, it is zero when:

$$\frac{dy}{dt}\left(t^*\right) = \frac{y\left(t^*\right)}{t_{travel} + t^*} = r\left(t^*\right) \equiv r^* \tag{7.3}$$

In other words, there is a unique straight line with the slope $r\left(t^*\right) = r^*$ that is tangent with the function $y(t)$ at $t = t^*$. The corresponding search time is the optimum t^*. The straight line is given by $y = r^*\left(t + t_{travel}\right)$, and the intercept at $y = 0$ is $-t_{travel}$ (**Figure 7.4a**). That $r^* = r\left(t^*\right)$ is indeed a maximum can be proven analytically, $d^2r/dt^2 = \left(d^2y/dt^2\right)/(t_{travel} + t)^2 < 0$ evaluated at t^*. This is an example of the MVT. According to this theorem, if a feeding site is farther away from home, t_{travel} is longer (red compared to green in **Figure 7.4a**), and another line tangent to $y(t)$ predicts a longer search time (t^*_{long} versus t^*_{short}) and larger load per trip (y^*_{long} versus y^*_{short}). This prediction was verified in a laboratory experiment with starlings where the distance between nest and feeding site was varied in a controlled manner (**Figure 7.4b**) (Kacelnik 1984).

Charnov et al. (1976) proved the MVT in the case of foraging among a number N of "patches" of food in a natural habitat. If an animal stays on a patch, available food decreases over time, but moving to another patch entails travel time. What's the optimal time spent on each of N patches? Denote by $y_i(t_i)$ the yield as a concave function of staying time and by P_i the proportion of time spent on the visited patches ($i = 1, 2, \ldots, N$). The rate of yield per unit time is:

$$r(t_1, t_2, \ldots t_N) = \frac{\sum_i P_i y_i(t_i) - a}{\sum_i P_i t_i + b} \tag{7.4}$$

where constants a (expressing energy cost while searching) and b (proportional to travel time between patches) are not functions of t_i. Analytical calculation similar to the above derivation (Eq. 7.3) shows that there is a unique optimum strategy for patch leaving. Namely, a maximal r^* is achieved with $(t_1^*, t_2^*, \ldots, t_N^*)$, which are determined by $\partial y_i / \partial t_i = r^*$ for every i. Thus, the MVT asserts that the animal should leave whatever patch it is currently in, when the marginal yield rate $(\partial y_i / \partial t_i)$ in the patch, which decreases with t, drops to the average yield rate r^* of *the entire habitat* (Charnov et al. 1976). Comparing any two patches (A and B) with different yield functions (**Figure 7.4c**), their respective optimal foraging times (t_A^* and t_B^*) are given by the intersections of two straight lines with the same slope r^* that are tangent to $y_A(t_A)$ and $y_B(t_B)$, respectively.

Foraging theory has broad appeal in describing and predicting choice behavior. However, in natural behavior, what often matters for our well-being is to learn in a stochastic environment, by trial and error through feedbacks (**Figure 7.1**), which is the realm of reinforcement theory discussed in Chapter 4 (Section 4.3).

Economics, behavioral ecology and psychology offer concepts and tools to define quantities (such as utility) and computations (like temporal discounting) to describe value-based choice behavior. With the appropriate task designs in physiological experiments with behaving animals, starting at the turn of this century, neurophysiology with behaving monkeys has demonstrated value correlates of single cell firing in various brain regions, including the posterior parietal cortex (PPC) (Platt and Glimcher 1999); Sugrue et al. 2004), the dorsolateral prefrontal cortex (DLPFC) (Barraclough et al. 2004), the orbitofrontal cortex (Padoa-Schioppa and Assad 2006) and the basal ganglia (Ding and Hikosaka 2006). There are also numerous experiments using reward-dependent choice tasks with humans (Odoherty 2004; Rangel et al. 2008; Rushworth et al. 2012) and rodents (Schoenbaum et al. 2009; Cox and Witten 2019). Simultaneously, economists baffled by the idiosyncrasies of people's choice behavior became interested in understanding how the brain actually makes decisions, at the interface between neuroscience and behavioral economics (Camerer et al. 2005). The integration of economics and neuroscience produced a fruitful cross-disciplinary field called neuroeconomics (Glimcher 2003; Rangel et al. 2008). It is a broad field that covers topics ranging from simple conditioning to our sense of fairness and altruism in social decision making.

7.3 NEURAL CIRCUIT MECHANISM FOR VALUE-BASED CHOICE

7.3.1 Dopamine and Synaptic Plasticity

In Chapter 4 (Section 4.3), we introduced basic ideas of reinforcement learning (RL) and its biological underpinning in the brain. A learning signal at the core of RL is phasic firing of dopamine neurons in the midbrain correlated with reward prediction error (RPE) (Schultz et al. 1997; Schultz 1998). How such an RPE is computed mechanistically in a neural circuit is still under experimental investigation, involving the ventral tegmental area (Cohen et al. 2012), nucleus accumbens (Parker et al. 2022), ventral pallidum (Ottenheimer et al. 2020), as well as the neocortex (Takahashi et al. 2011; Neubert et al. 2015).

Action of dopamine in downstream neurons depends on the D1 and D2 receptor subtypes. Unlike fast actions of activating ionotropic receptors for glutamate or GABA, dopamine engages complex intracellular biochemical processes through G-protein-coupled receptors, the activation of which triggers cAMP-dependent intracellular signaling (Tritsch and Sabatini 2012). Therefore, influence of dopamine release is slow, and its physiological targets are multiple, including synaptic transmission and intrinsic ion channels (Seamans and Yang 2004). Generally, the end effect mediated by a D1-class receptor is depolarizing, whereas that mediated by a D2-class receptor is hyperpolarizing a postsynaptic neuron.

Here our goal is to explore neural circuit models of value-based economic decisions. For that purpose, we will limit ourselves to dopamine modulation of Hebbian synaptic plasticity. An anatomical basis for this modulation is a triad arrangement, with dopamine input co-localizing with a synapse (**Figure 7.5a**), which was found at the excitatory synapses on prefrontal neurons and cortico-striatal synapses (Freund et al. 1984; Goldman-Rakic 1995; Surmeier et al. 2010; Tritsch and Sabatini 2012; Bazzari and Parri 2019). Mathematically, a Hebbian learning rule for modification of a synaptic weight w (Chapter 4, Section 4.4.1) can be extended to a three-factor dependence on the dopamine input $DA(t)$, as well as the presynaptic and postsynaptic activities (r_{pre} and r_{post}):

$$\frac{dw}{dt} = F\left(r_{pre}, r_{post}, w, DA\right). \tag{7.5}$$

Various forms of three-factor learning rules have been developed (Wang 2014; Brea and Gerstner 2016; Kuśmierz et al. 2017; Dayan and Nakahara 2018); for instance, the modulatory signal may be the reward-prediction error or reward itself. Also, reinforcement learning may be non-Hebbian and does not require co-activation of presynaptic and postsynaptic neurons.

Interestingly, in *in vitro* studies, dopamine has been found to reverse the sign of plasticity (from depression to potentiation) at cortico-striatal synapses (Reynolds et al. 2001) and synapses on prefrontal neurons (Matsuda et al. 2006; Xu and Yao 2010). This finding has been refined with the use of stimulation protocols that induce spike-timing dependent plasticity (STDP) (Bi and Poo 2001; Dan and Poo 2006). As discussed in Chapter 4 (Section 4.4.3), STDP refers to the fact that Hebbian synaptic modification depends on the relative timing of presynaptic and postsynaptic spikes. At cortico-striatal synapses, in the presence of dopamine D_1 receptors, positive timing (presynaptic firing before postsynaptic firing) leads to potentiation (**Figure 7.5b**, red), and negative timing does not induce synaptic change (**Figure 7.5b**, black). When D_1 receptors are blocked pharmacologically, however, negative spike timing yields depression (**Figure 7.5b**, blue) (Shen et al. 2008). In hippocampal neurons, bath application of dopamine enlarged the temporal window for potentiation with positive spike timing and converted depression to potentiation with negative spike timing (**Figure 7.5c**) (Zhang et al. 2009b). These experimental results support the notion that dopamine activation (presumably mediated by D_1 receptors) can reverse the signal of synaptic modification. Other neuromodulators such as noradrenaline (Seol et al. 2007) can also alter synaptic modification or reverse its sign (reviewed in Pawlak et al. 2010).

In computational neuroscience, various reward-dependent learning rules have been proposed, where plasticity is gated by either reward or reward prediction error (RPE) (Loewenstein and Seung 2006; Pfeiffer et al. 2010; Legenstein et al. 2010; Izhikevich 2007; Frémaux et al. 2010). It is presently unclear whether the existing models are fundamentally different or they are essentially similar in different mathematical forms. For instance, it has been shown that the covariance of neural activity and reward is a common denominator of several reward-dependent learning rules (Loewenstein and Seung 2006; Frémaux et al. 2010). Furthermore, a major open issue is the so-called eligibility trace linking an action and its reward outcome that are temporally separated (Sutton and Barto 2018; Dayan and Abbott 2001; Izhikevich 2007). The biological substrate of such eligibility trace remains unknown.

Similar to perceptual decision making, there are computational models for value-based choice with varying degrees of abstraction, including

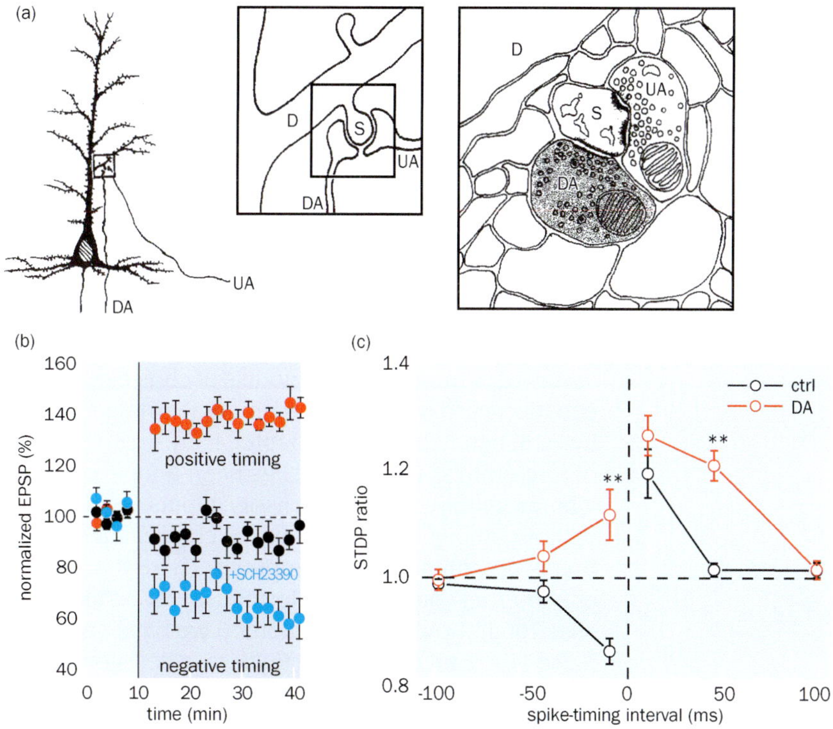

Figure 7.5 Dopamine modulation of synaptic plasticity. (a) Triad arrangement at a cortical synapse. Left: Afferents labeled with a dopamine (DA)-specific antibody together with an unidentified axon (UA). Middle: Enlargement from the left panel. Right: Ultrastructural features of the axospinous synapses; the dopamine terminal forms a symmetrical synapse; the unidentified profile forms an asymmetrical (excitatory) synapse. D: Dendrite, S: Spine. (b) Dopamine gates the sign of plasticity for cortical synapses on D1 receptor-expressing medium spiny neurons in the striatum. Positive spike timing produces potentiation (red), whereas negative timing does not induce plastic changes (black). D1 receptor antagonist SCH23390 unmasked depression with negative timing (blue). (c) Dopamine alters the STDP window in hippocampal neurons. Black: Control; red: Bath application of dopamine. With negative spike timing, dopamine reverses the sign of modification from depression in control to potentiation. Panel (a) is reproduced from Goldman-Rakic (1995), (b) from Surmeier et al. (2010) and (c) from Zhang et al. (2009b). [Goldman-Rakic, P. S. (1995). Cellular basis of working memory. Neuron, 14(3), 477–485.] [Surmeier, D. J., Shen, W., Day, M., Gertler, T., Chan, S., Tian, X., & Plotkin, J. L. (2010). The role of dopamine in modulating the structure and function of striatal circuits. Progress in Brain Research, 183, 148–167.] [Zhang, J. C., Lau, P. M., & Bi, G. Q. (2009). Gain in sensitivity and loss in temporal contrast of STDP by dopaminergic modulation at hippocampal synapses. Proceedings of the National Academy of Sciences, 106(31), 13028–13033.]

Bayesian learning (Behrens et al. 2007), reinforcement learning (Dayan and Abbott 2001; Daw 2016) (Chapter 4, Section 4.3) and neural circuit models (Wang 2014). Here we shall focus on a recurrent neural circuit model (Wang 2002b; Wong and Wang 2006) extended to incorporate reward-dependent synaptic plasticity in a simple way.

7.3.2 A Decision-Making Network Model Endowed with Reward-Dependent Learning

In a recurrent neural circuit model of decision making (Chapter 6), decisions are made by stochastic neural dynamics of selective neural pools, say A and B, in any given trial. Across many trials, the probability of choosing A (i.e., the fraction of trials when the neural pool selective for option A wins the competition through attractor dynamics) defines the psychometric function (Chapter 6, **Figure 6.16**). Since the motion coherence is expressed by the difference in the inputs to A and B, the choice probability (the psychometric function) is, more generally, a function of the difference in the strengths (c_A and c_B) of inputs to the two competing neural pools. This function can be well fit by a logistic function (Soltani and Wang 2006):

$$P_A = 1/(1 + \exp(-\beta(c_A - c_B))) \tag{7.6}$$

where β (sometimes called inverse temperature) expresses the degree of stochasticity of choice behavior. Decision making is deterministic in the limiting case of $\beta = \infty$, $P_A = 0$ when $c_A < c_B$; 1 otherwise. At the other extreme, with $\beta = 0$, the system behaves completely randomly like tossing a fair coin, with $P_A = P_B = 0.5$ regardless of c_A and c_B. Interestingly, P_A is not sensitive to the overall strength of inputs, remaining essentially the same when the sum $c_A + c_B$ is varied (**Figure 7.6**). On the other hand, increases in the input drive speed up neural dynamics, leading to shorter reaction times (Soltani and Wang 2006). These two characteristics were recently confirmed in a value-based decision experiment with mouse (Bari et al. 2019).

As we have seen previously, a softmax decision criterion is widely assumed in more abstract models of choice behavior; indeed, it is the same equation used in the reinforcement learning model for fitting monkey (and human) behavioral data (Chapter 4, Eq. 4.15). Is it a coincidence? The origin of randomness in behavior has been a topic of interesting debate. In most laboratory experiments, sensory inputs used to cue rewards are salient without ambiguity in value-based choice tasks. In economic theory, valuation is deterministically computed. There thus appears a need for a different, internal source of indeterminism for economic choice behavior (Glimcher 2005). However, in Chapter 6 we have shown that the main source of noise is irregular neural activity of a decision circuit itself (Wang 2002b). Indeed, even in the perceptual RDM task, decisions by monkeys were as stochastic as in the control case when the exact sequence of random dots was repeated across trials. Since the recurrent neural circuit model of decision making (Chapter 6) captures stochastic choices in single trials, can it be extended to economic choices, provided that reward-dependent learning is incorporated? Suppose that input synaptic connections c_A and c_B are plastic, synaptic modifications will alter the network's future decision behavior. For instance, in a trial the choice (say, A) results in a positive outcome, which triggers dopamine release and leads to a potentiation of c_A. Then the value of $c_A - c_B$ is larger, thus the system is "moved" to the right in **Figure 7.6** (lower panel), and the probability of choosing A is increased in the next trial.

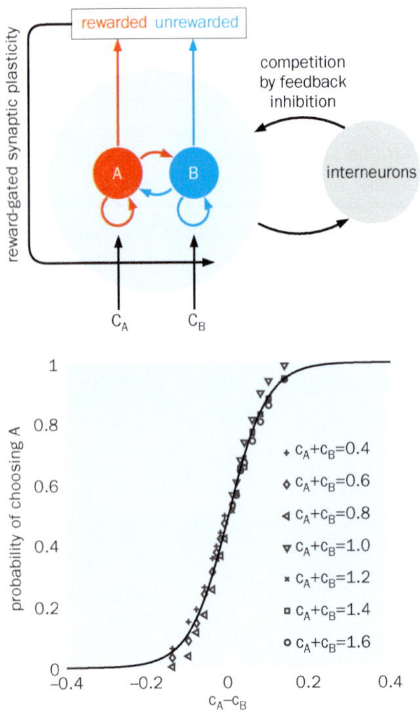

Figure 7.6 A recurrent neural circuit model (Wang 2002b) with reward-dependent plasticity at input pathways. The probability of choosing option A versus B in a single trial is a softmax function of the input strengths c_A and c_B, independent of the overall strength $c_A + c_B$. Lower panel is reproduced from Soltani, A., & Wang, X. J. (2006). A biophysically based neural model of matching law behavior: Melioration by stochastic synapses. Journal of Neuroscience, 26(14), 3731–374. Upper panel is reproduced from Wang, X. J. (2002). Probablistic decision making by slow reverberation in cortical circuits. Neuron, 36(5), 955–968.

A simple way to realize reinforcement learning is to use binary synapses (Amit and Fusi 1994; Fusi 2002) that undergo a Hebbian learning (Chapter 4) rule modulated by a dopamine-mediated reward signal. Specifically, synapses between two neurons are assumed to have two (Down and Up) states, and c_A (respectively c_B) is the fraction of synapses from an input pathway to a decision neuron in the pool A (respectively B) that are in the Up state. Presynaptic neurons represent choice options, and therefore they fire at high rates for both decision neural pools. Postsynaptically, only the winning pool fires at a high rate, whereas the losing pool fires at a low rate for which a condition of Hebbian plasticity is not fulfilled. Thus, the synaptic weight onto the winner undergoes plasticity, while that onto the loser remains unchanged in a behavioral trial.

Based on the experimental evidence (**Figure 7.5**), let us assume that synapses of inputs to decision neurons of the winning pool are potentiated only if the choice is rewarded (because dopamine is high) and depressed otherwise (Soltani and Wang 2006; Soltani et al. 2006; Fusi et al. 2007). If the choice is rewarded, then c_A is potentiated with a change Δc_A according to:

$$\Delta c_A = q_+ (1 - c_A) \tag{7.7}$$

if no reward is delivered, c_A is depressed according to:

$$\Delta c_A = -q_- c_A \tag{7.8}$$

where q_+ and q_- are the learning rates with values between 0 and 1. Synaptic variables are also guaranteed to remain between 0 and 1, because Δc_A is negative if c_A is above 1, and positive if c_A is below 0. It is worth noting that here, we used reward for modulating plasticity rather than reward prediction error (RPE), which ensures that learning stops when the error signal becomes zero (Chapter 4).

This does not make a substantial difference, since the synaptic variables are bounded, and thus termination of plasticity is guaranteed. The formulation is reasonable given the diverse signals transmitted by dopamine neurons (Berridge 2007; Engelhard et al. 2019).

As a result of synaptic modifications, the input strengths for the competing neural groups of the decision network vary from trial to trial, leading to adaptive dynamics of choice behavior. The learning rates q_+ and q_- determine the time constants with which the system keeps the memory trace for past reward and nonreward outcomes. When A is chosen, yielding reward in consecutive trials indexed by $i = 1, 2, .., n$,

$$c_A(n) = c_A(n-1) + q_+\left(1 - c_A(n-1)\right) = q_+ + (1-q_+)c_A(n-1) \tag{7.9}$$

$$= \ldots = 1 + \left(c_A(0) - 1\right)(1-q_+)^n \tag{7.10}$$

where $c_A(0)$ is the initial value of c_A. With a large n, c_A converges to 1. Since $(1-q_+)^n = \exp\left(n\ln\left(1-q_+\right)\right)$, the linear dynamics has a time constant of $\tau_+ = -1/\ln\left(1-q_+\right)$, which is shorter with a larger q_+. On the other hand, when there is no reward in consecutive trials:

$$c_A(n) = (1-q_-)c_A(n-1) = (1-q_-)^2 c_A(n-2) \tag{7.11}$$

$$= \ldots = (1-q_-)^n c_A(0) = \exp(-n/\tau_-)c_A(0) \tag{7.12}$$

and c_A decays to 0 exponentially with a time constant $\tau_- = -1/\ln\left(1-q_-\right)$.

The neural circuit model of decision making (Wang 2002b) endowed with this simple three-factor synaptic plasticity is a general computational framework, not designed for any particular task. Now we apply the model to specific value-based decision paradigms.

7.3.3 Computation of Returns by Synapses: Matching Law through Melioration

In a classic foraging task, a subject makes a choice between two target options (A and B), each associated with a reward probability (p_A and p_B). The optimal behavior is obviously to always select the one with a higher probability, say A if $p_A > p_B$. In a naturalistic setting, however, this does not necessarily yield maximal rewards because of constraints. One is that when a reward becomes available, it stays there for some time. In a laboratory experiment, this translates into the "baiting condition": if a reward is assigned to a target, it stays there until it is harvested. Under the baiting condition, it is desirable to choose the option B from time to time, because the probability that no reward has been delivered on B in n consecutive trials is $(1-p_B)^n$, which decreases exponentially with n. Once delivered, the reward probability becomes 1 for B until it is chosen. In that case, the optimal strategy is to occasionally choose B, but not to repeat it in the next trial when its probability is reset back to p_B, whereas the probability of reward is higher for option A. This behavior has been observed experimentally (Lau and Glimcher 2005). Another constraint is to impose a cost for every switch between A and B, reflecting efforts of traveling between two food patches in foraging. This is called changeover delay penalty, so if the subject switches from one target to the other, reward is suspended until it chooses the new target for the second time.

Under the baiting condition and changeover delay, the behavior of animals, ranging from pigeons, rodents and monkeys to humans, display "matching law," namely a subject allocates choices in a proportion that matches the relative reinforcement obtained on these choices (Herrnstein 1961, 1990; Mazur 1981; Rachlin and Laibson 1997; Kubanek 2017).

To investigate the brain mechanism of matching law, physiological experiments were carried out with monkeys (Sugrue et al. 2004; Lau and Glimcher 2007) and rodents (Bari et al. 2019; Hattori et al. 2019). In a monkey experiment, subjects chose between a red target (A) and a green target (B), each associated with a reward schedule (**Figure 7.7a**). In a block of trials, reward probabilities p_A and p_B were fixed. Running through trials indexed by $t = 1, 2, 3\ldots$, the cumulative number of times of choosing option $i = A$ or B is denoted by $N_i(t)$. The ratio of reinforcement is computed as the ratio of cumulative numbers $M_i(t)$ of harvested rewards on target $i = A$ or B. Matching law states that $N_A / N_B = M_A / M_B$. Plotting the cumulative numbers of trials with red and green choices against each other, the slope is N_A / N_B, which should match the reward ratio M_A / M_B. This is shown in **Figure 7.7b**, where the reward probabilities varied in six different blocks of trials. Matching is mostly good but not perfect (the slopes for blue and black are not identical). When the reward ratio (black) is high in favor of one option (3:1 or 6:1 for red versus green target), the choice ratio (blue) is smaller, an effect called undermatching, which is commonly observed (Herrnstein 1990).

Let us apply our model to this experiment. With a large number N of trials, $N_A + N_B = N$. The choice probabilities are estimated to be $P_i = N_i / N$.

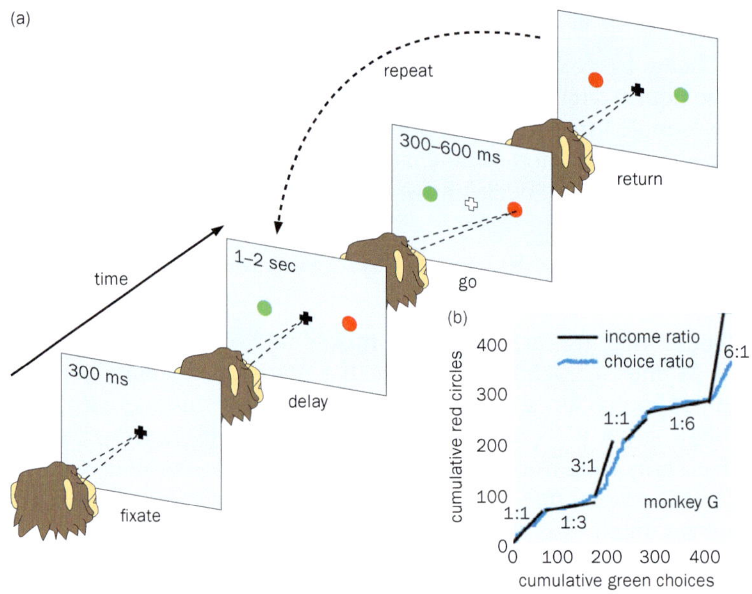

Figure 7.7 Matching behavior in monkeys. (a) The sequence of events of an oculomotor foraging task: (i) To begin a run of trials, the animal must fixate the central cross. (ii) Saccade targets appear in opposite hemifields while the animal maintains fixation. (iii) Dimming of the fixation cross cues a saccadic response and hold. (iv) Reward is delivered at the time of the response, if at all. Overall maximum reward rate is set at 0.15 rewards per second. Relative reward rates changed in blocks without warning. (b) Dynamic matching behavior. Representative behavior of a monkey during a single session. Continuous blue curve shows cumulative choices of the red and green targets. Black lines show average ratio of rewards per trial for the two choices (red:green) within each block (here, 1:1, 1:3, 3:1, 1:1, 1:6, and 6:1). Matching predicts that the blue and black curves are parallel. Reproduced from Sugrue et al. (2004). [Sugrue, L. P., Corrado, G. S., & Newsome, W. T. (2004). Matching behavior and the representation of value in the parietal cortex. Science, 304(5678), 1782–1787.]

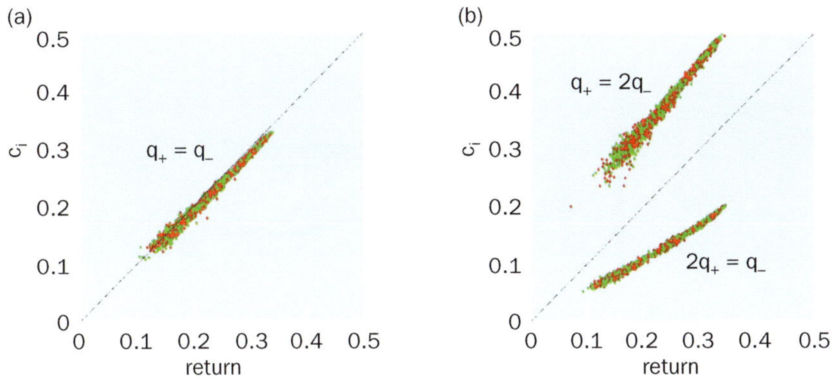

Figure 7.8 Plastic synapses approximately compute the return from each choice or a linear function of it. Block-averaged synaptic strengths are plotted against the obtained returns in the same block (dots indicate individual trial blocks in model stimulation). Two colors represent two different choices (red for A and green for B). (a) The average synaptic strength is equal to the return from each choice, when the two learning rates are equal ($q_+ = q_- = 0.06$). (b) If the potentiation rate is greater than the depression rate ($q_+ = 0.06, q_- = 0.03$), the average synaptic strength is larger than the return (above the diagonal line), whereas if the depression rate is greater than the potentiation rate ($q_+ = 0.03, q_- = 0.06$), then the synaptic strength is smaller than the return (below the diagonal line). Reproduced from Soltani and Wang (2006). [Soltani, A., & Wang, X. J. (2006). A biophysically based neural model of matching law behavior: Melioration by stochastic synapses. Journal of Neuroscience, 26(14), 3731–3744.]

If the total numbers of reward are M_A and M_B, the reward per trial defines income $I_i = M_i / N$. The amount of reward per choice is called return, given by $R_i = M_i / N_i = I_i / P_i$. According to Eqs. (7.7–7.8), in N trials the synapse strength c_i is potentiated M_i times and depressed $N_i - M_i$ times, the cumulative change is:

$$\Delta c_i = q_+ (1 - c_i) M_i - q_- c_i (N_i - M_i). \tag{7.13}$$

The steady-state is given by $\Delta c_i = 0$. Dividing by N, we have $q_+ (1 - c_i^{ss}) I_i - q_- c_i^{ss} (P_i - I_i) = 0$. This yields:

$$c_i^{ss} = \frac{q_+ I_i}{(q_+ - q_-) I_i + q_- P_i} = \frac{q_+ R_i}{(q_+ - q_-) R_i + q_-} \tag{7.14}$$

If $q_+ = q_-$, we have $c_i = R_i$. Even when the two learning rates are not the same but $(q_+ - q_-) R_i \ll q_-$, c_i values still display a linear dependence on R_i (**Figure 7.8**). Therefore, through learning the synaptic strengths, c_A and c_B compute returns R_A and R_B (the amount of reward per choice) rather than income (the amount of reward per trial) for the alternative options A and B.

The network of spiking neurons produces a choice by stochastic dynamics in a trial. The fraction of times with one of the two options (say, A) chosen is given by the softmax function (**Figure 7.6**). Because neural firing is driven by inputs, firing rates of model neurons naturally depend on the values of synaptic strengths c_A and c_B. This is shown in **Figure 7.9a** with four sets of baiting probabilities. Thus, the model reproduces the observation from Sugrue et al. (2004) that single neurons involved in decision making are modulated by relative values of targets A and B (**Figure 7.9b**). The general insight is that value modulation of neural firing does not necessarily mean that valuation is computed at the neural level. Instead, values can be estimated at synapses through reward-based plasticity. Value-coding synapses may not even be in a brain region where neural activity is monitored and correlated with choices, but reflect inputs from a distinct valuation system (see Section 7.4).

As the reward probability ratio p_A / p_B varies from one block of trials to the next block in model simulations, the choice behavior changes adaptively (**Figure 7.10a**). The input synaptic strengths (c_A and c_B) display a considerable variability due to stochastic network behavior. However, on average the choice probability ratio is close to that of reward rates for the two targets (black line). The model's performance is close to the matching behavior (**Figure 7.10b**).

Matching is achieved dynamically, from trial to trial, through a *melioration process* (Herrnstein 1990; Herrnstein and Prelec 1991). The model has no prescription for global optimization, and it does not have perfect memory of past reward events. Indeed, the system is forgetful: if there is

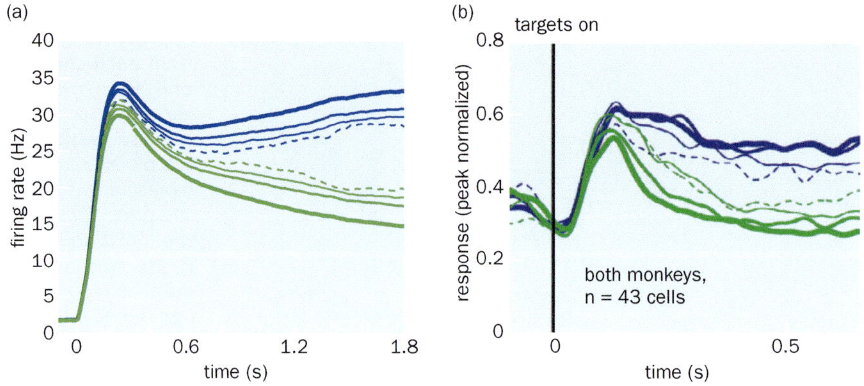

Figure 7.9 Modulation of neural firing activity by values. (a) Graded activity of model neurons as a function of the input synaptic strengths, which encode the values (returns) of choice options in a matching task. The activity of decision neurons shows a graded pattern, if single-trial firing rates are sorted and averaged according to the network's choice and the difference between synaptic strengths. Activity is aligned by the onset of two targets, and it is shown separately for the choices corresponding to the neurons' preferred (red) or non-preferred (blue) target. In addition, trials are subdivided into four groups according to the difference between the strength of synapses to the two competing neural populations: $c_A - c_B = -0.16$ to -0.05 (dashed), -0.05 to 0 (thin), 0 to 0.05 (medium), 0.05 to 0.16 (thick). (b) Neural firing activity of LIP neurons in a monkey experiment display modulation by fractional income, which is related to return by $I_i = R_i P_i$. Different traces correspond to averages in subsets of trials for the recorded neuron's preferred (purple) or non-preferred (green) target. Panel (a) is reproduced from Soltani and Wang (2006), (b) from Sugrue et al. (2004). [Soltani, A., & Wang, X. J. (2006). A biophysically based neural model of matching law behavior: Melioration by stochastic synapses. Journal of Neuroscience, 26(14), 3731–3744.] [Sugrue, L. P., Corrado, G. S., & Newsome, W. T. (2004). Matching behavior and the representation of value in the parietal cortex. Science, 304(5678), 1782–1787.]

Figure 7.10 Stochastic choice behavior in the foraging task. (a) In a model simulation, baiting probability ratios [1:1, 1:3, 3:1, 1:1, 1:6, 6:1] are used in a sequence (black), each block with 200 trials. The synaptic strengths, c_A (red) and c_B (green), and the choice probability (blue) are plotted as a function of time (trial number). (b) Cumulative choice ratio as a function of cumulative reward ratio from (a), which is comparable to the monkey's behavior (Figure 7.7b). (c) Normalized choice-triggered-average (CTA) of rewards as a function of previous trial number. Reproduced from Soltani and Wang (2006). [Soltani, A., & Wang, X. J. (2006). A biophysically based neural model of matching law behavior: Melioration by stochastic synapses. Journal of Neuroscience, 26(14), 3731–3744.] [Courtesy of Ali Soltani.]

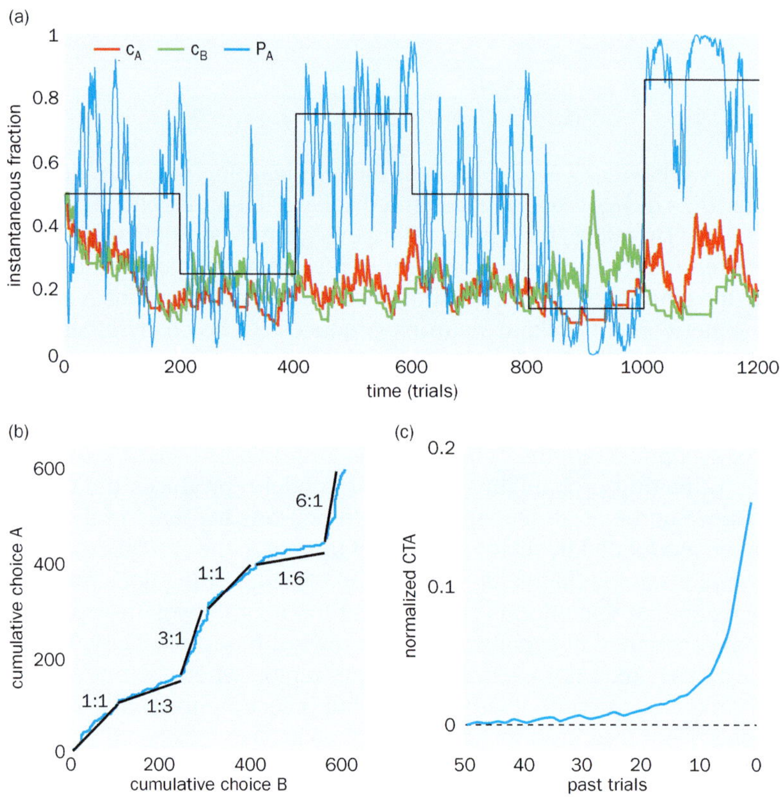

no reward delivered in n consecutive trials, c_i decays to zero (Eq. 7.12). Thus, choice behavior is guided only by information about rewards locally in time. One way to quantify how choice in a trial is influenced by past experience is to compute the "choice-triggered-average (CTA) of rewards" (Corrado et al. 2005), which measures how choice in the current trial is influenced by the harvested rewards in the past $t = 1, 2, 3, \ldots$ trials. Because synapses are potentiated or weakened stochastically over time, they behave like a leaky integrator of past choice outcomes. Hence, CTA decays with increasing t (**Figure 7.10c**), as observed in the monkey experiment (Corrado et al. 2005). One can extract a time constant τ_{memory} from this decay curve fitted with either a single exponential function or a sum of two exponential functions. In the latter case, CTA = $w_1\exp(-t/\tau_1)+w_2\exp(-t/\tau_2)$ with $w_1+w_2=1$, $\tau_{memory}=w_1\tau_1+w_2\tau_2$. In the model τ_{memory} is controlled by the learning rates; model reproduction of the experimentally observed CTA (Corrado et al. 2005) constrains the values of q_+ and q_-. In the example of **Figure 7.10c**, $\tau_{memory}\sim 7.5$ trials.

Matching is not perfect, however. Specifically, the relative probability of choosing the more rewarding option is slightly smaller than the relative reward rate ("undermatching") (**Figure 7.11a**), which was also found in monkey's behavior (**Figure 7.11b**) (Sugrue et al. 2004; Lau and Glimcher 2005). A model analysis explained this finding, revealing that undermatching is a natural consequence of stochasticity in neural activity and leaking integration of past rewards (Soltani and Wang 2006). In **Figure 7.12a** are shown returns R_A (red curve) and R_B (green curve) computed for each assumed value of P_A (x-axis), with $p_A/p_B=3$. R_A decreases, and R_B increases, with P_A. Knowing R_A and R_B, c_A and c_B are computed using Eq. (7.14), and P_A is given from Eq. (7.6). The predicted P_A is plotted against the assumed P_A (black curve). The two must agree with each other; hence, the actual choice probability is determined by the intersection of the black curve with the diagonal line (self-consistency), corresponding to $P_A\simeq 0.7$ in this example.

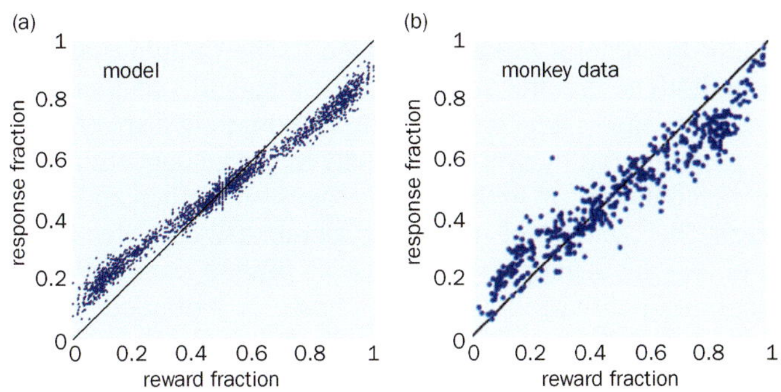

Figure 7.11 Undermatching (a) and by a monkey (b). Perfect matching would correspond to the diagonal line. Panel (a) is reproduced from Soltani and Wang (2006), (b) from Sugrue et al. (2004). [Soltani, A., & Wang, X. J. (2006). A biophysically based neural model of matching law behavior: Melioration by stochastic synapses. Journal of Neuroscience, 26(14), 3731–3744.] [Courtesy of Bill Newsome.]

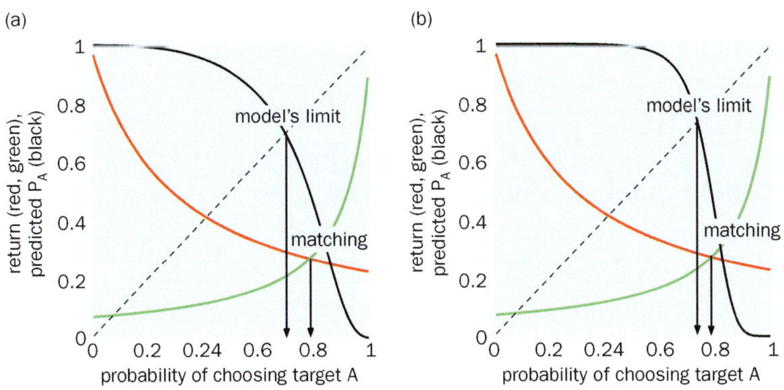

Figure 7.12 Mechanism of melioration and the limit for achieving perfect matching. (a) With $p_A/p_B=3$ and overall baiting rate of 0.3, returns R_A and R_B are plotted for target A = red and B = green as a function of assumed choice probability for A (x-axis). For given R_A and R_B (hence c_A and c_B) values, the calculated choice probability for A is shown as the black curve with $\beta=10$. The actual P_A is determined by self-consistency, i.e., the intersection of the black curve with the diagonal line. Its value falls short of the prediction of matching (corresponding to the intersection of the red and green curves), a phenomenon called undermatching. (b) With the level of stochasticity reduced by half, $\beta=20$, the black curve is steeper and P_A is closer to the prediction of matching (compare panels (a) and (b)). Reproduced from Soltani and Wang (2006). [Soltani, A., & Wang, X. J. (2006). A biophysically based neural model of matching law behavior: Melioration by stochastic synapses. Journal of Neuroscience, 26(14), 3731–3744.]

Perfect matching $N_A / N_B = M_A / M_B$ would imply that the returns $R_A = M_A / N_A$ and $R_B = M_B / N_B$ are equal, corresponding to the intersection between the red and green curves, which would predict $P_A \simeq 0.8$, above the actual choice probability (undermatching). It can be shown that this is caused by stochastic neural dynamics underlying choice behavior in single trials (β is finite); when β is increased by a factor of two (reflecting reduced stochasticity), P_A is now closer to the value predicted by perfect matching (**Figure 7.12b**). Matching becomes perfect in the absence of any noise, which is unrealistic and undesirable inasmuch as stochastic neural population dynamics underlie the indeterminacy of decision making itself. In the brain, the amount of stochasticity (measured by β in the model) could be attributed to multiple sources. Learning on multiple timescales offers a complementary factor to explain undermatching (Iigaya et al. 2019).

In summary, a general model of decision making (Wang 2002b) endowed with a simple reward-modulated Hebbian plasticity rule (Eqs. 7.7–7.8) is introduced. Recent experiments questioned whether reinforcement learning in the brain is indeed realized by synaptic plasticity or, alternatively, in the form of time integration of reward signals across trials (Bari et al. 2019). Pereira-Obilinovic et al. (2024) compared the two contrasting scenarios, with a line attractor (Chapter 5, Section 5.5) implementing time integration. They found that the contrasting models yield distinct and experimentally testable predictions. In particular, brief perturbations during an inter-trial interval would disrupt estimated values according to the integration mechanism but not the synaptic mechanism.

When the general model is applied to the foraging task, it accounts for salient, both behavioral and neurophysiological, observations from foraging experiments in which animals approximately display matching law. Values are computed by synapses locally in time, suggesting a biologically realistic mechanism for melioration process hypothesized at the behavioral level. The concept of melioration emphasizes a dynamical process instead of a static description. Because of limited memory and stochastic dynamics, rationality of value-based choice is "bounded" by constraints, and the system follows a "satisficing" strategy to achieve "good enough" outcomes in decision making (Simon 1997; Gigerenzer et al. 2000). This example illustrates how a biologically based neural circuit model offers an understanding across levels bridging stochastic neural population dynamics with economic choice theory.

7.4 VALUATION

7.4.1 Computation of Common Currency

Contrary to the common saying, "you cannot compare apples and oranges," the brain needs a common currency for comparing different goods. The orbitofrontal cortex (OFC) and ventromedial prefrontal cortex (vmPFC) are especially important for valuation in decision making (Wallis 2007; Padoa-Schioppa 2013; Rudebeck and Murray 2014; Neubert et al. 2015; Rich and Wallis 2016; Hiser and Koenigs 2018; Banerjee et al. 2020; Enel et al. 2020; Klein-Flügge et al. 2022). Anatomically, vmPFC includes Brodmann areas 25 and 32; the OFC includes areas 11,12 and 13; and area 14 and polar area 10 are often included in both. Humans with damage in these cortical regions were found to be oblivious to the future consequences of their choices (Harlow 1868; Bechara et al. 2000). Functional MRI (Knutson et al. 2005; Rushworth and Behrens 2008) and magnetoencephalography (MEG) (Hunt et al. 2012) in normal human participants support the importance of vmPFC for valuation of choice options. More refined information was collected from single-neuron recording in OFC and vmPFC of primates (Thorpe et al. 1983; Roesch and Olson 2005;

Padoa-Schioppa and Assad 2006) and rodents (Schoenbaum et al. 1998; Schoenbaum et al. 2009).

Padoa-Schioppa and Assad (2006) trained monkeys to choose between two types of reward (A = grape juice, B = peppermint tea), each with a different amount. For instance, in a trial, one yellow square and three blue squares are shown, associated with one drop of grape juice ($n_A = 1$) and three drops of peppermint tea ($n_B = 3$), respectively (**Figure 7.13a**). The left and right locations of two sensory cues are randomly assigned in each trial; subjects indicated the decision by a saccade to the chosen cue. A monkey prefers A (with a subjective value v_A of a unit quantity) over B (with a unitary subjective value v_B; $v_B < v_A$), but how to choose between $v_A n_A$ and $v_B n_B$? By parametrically varying the relative amounts n_A and n_B across trials, the fraction of times a monkey chooses A versus B produces a sigmoid function of the relative amounts n_B / n_A (**Figure 7.13b**). A curve fitting is used to identify the indifference point, the relative values $(n_B / n_A)^*$ (equal to 4.1 in the example of **Figure 7.13b**) at which the choice is purely random. From there one may deduce $v_A = (n_B / n_A)^* v_B$. Note that the conclusion is the same if the softmax is expressed in terms of the difference $v_A n_A - v_B n_B$ (instead of n_B / n_A), for which the indifference point ($v_A n_A^* = v_B n_B^*$) corresponds to $v_A / v_B = (n_B / n_A)^*$.

Single neurons from the OFC Brodmann area 13 display activity that broadly falls into three types. Subclasses of "offer value" encoding neurons are selective for one of the two options, with the firing rate monotonically increasing with the offer amount of their preferred good (B for the example cell of **Figure 7.13c**). By contrast, the second type of "chosen value" neurons exhibits a

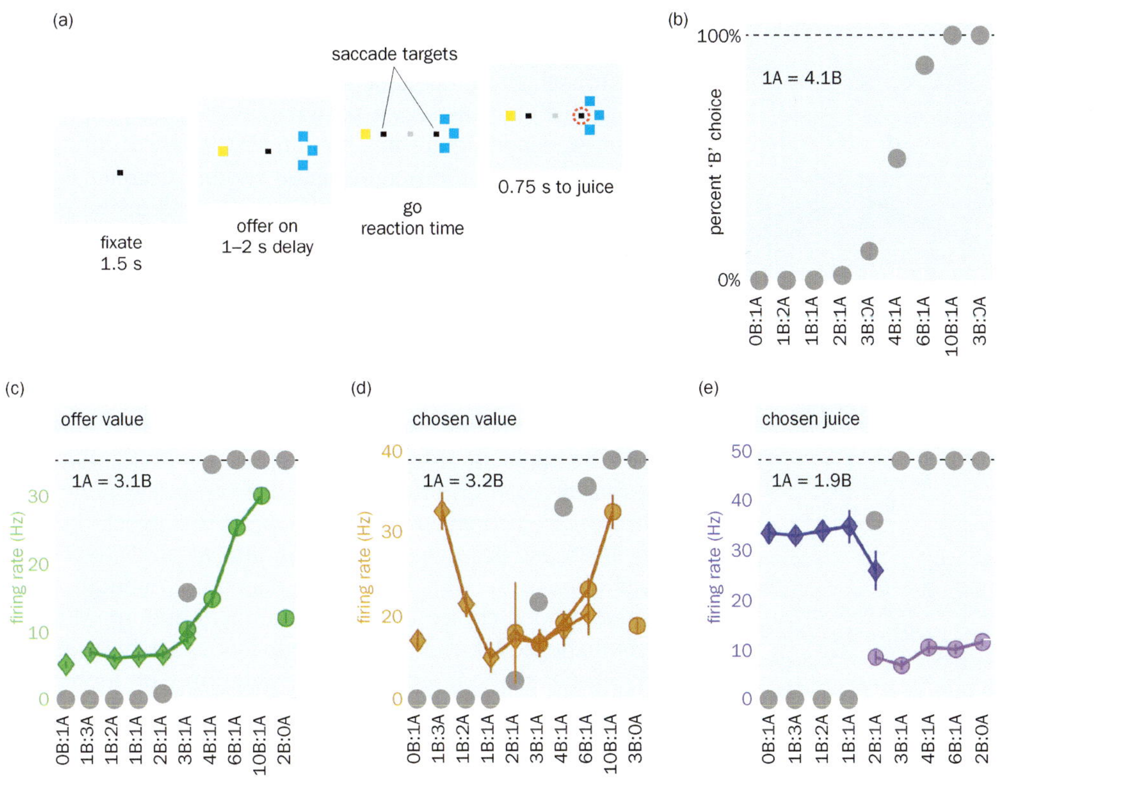

Figure 7.13 Three types of OFC neurons for computation of common currency. (a) Task design. In each session, monkeys chose between two juices (labeled A and B, with A preferred) offered in variable amounts. Offers were represented by colored squares on a computer monitor, and monkeys indicated their choice with a saccade. (b) The proportion of trials where B is chosen (P_B) as a function of the ratio n_B / n_A, where n_i is the quantity of juice i = A or B offered to the animal. (c–e) The activity (average firing rate) of three neurons that encode the offer value (c), chosen value (d) and chosen juice (e); grey circles: P_B. In (b–e), on the upper left corner is indicated the indifference point, which varies across sessions when the three shown neurons were recorded. Reproduced from Padoa-Schioppa (2013). [Padoa-Schioppa, C. (2013). Neuronal origins of choice variability in economic decisions. Neuron, 80(5), 1322–1336.]

U-shaped dependence on n_B / n_A; they represent common currency (with $(n_B / n_A)^*$ as the conversion rate for the two goods) (**Figure 7.13d**). For instance, if $(n_A / n_B)^* \simeq 3$, the firing rate is the same when the offers are 1B:2A or 4B:1A since both amounts to 7B in a common currency. The third type of "chosen juice" neurons are also selective for A or B, but are insensitive to the amount; their firing rate is approximately binary, high with choice A and low with choice B (or vice versa) (**Figure 7.13e**). The neural firing of chosen juice neurons is independent of the spatial location (left or right) of the chosen option, indicating that the decision is made in the space of goods rather than actions. The fact that different neurons in OFC encode the input and the output of the decision suggests that economic decisions may be generated within a neural circuit formed by these groups of cells (Padoa-Schioppa 2011).

The binary spiking activity pattern by chosen juice neurons is reminiscent of attractor dynamics underlying a categorical decision (Chapter 6). In a computational work (Rustichini and Padoa-Schioppa 2015), the monkey experiment of Padoa-Schioppa and Assad (2006) was simulated using a version of the model from Wong and Wang (2006), under the assumption that inputs to the two competing neural populations A and B reflect the firing rates of "offer value" neurons. It was found that the model decision neurons capture firing patterns of experimentally observed chosen juice neurons. Moreover, in the model, the inhibitory neural pool receives converging inputs from the two excitatory selective neural populations; therefore, its firing rate increases with both and behaves as that of chosen value neurons in the OFC. Of course, any neural population driven by summated activities r_A and r_B would show the same trend and could well be excitatory. The specific cell type corresponding to chosen value cells can be experimentally identified by genetic labelling in the future. This is now possible with the demonstration that the same task can be performed by mouse (Kuwabara et al. 2020), for which optogenetic methods are readily available. The model of Wang (2002b) has also been shown to successfully capture the time course of value-selective signals in a human MEG study (Hunt et al. 2012).

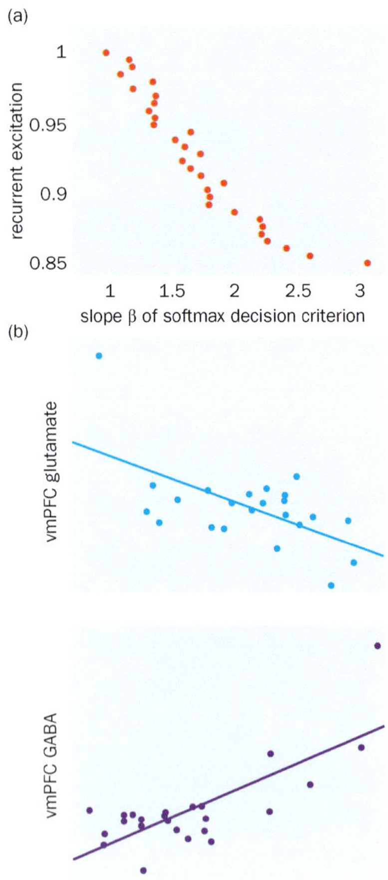

Figure 7.14 Performance of a value-based decision task depends on the strengths of recurrent synaptic excitation and inhibition in a decision neural circuit. (a) An increase in excitation relative to inhibition in the model of Wang (2002b) leads to a smaller softmax slope (less accuracy in difficult trials). (b) In a human experiment, subjects with low levels of vmPFC glutamate (upper plot) and high levels of vmPFC GABA (lower plot) are associated with good performance (large softmax slope). Reproduced from Jocham et al. (2012). [Courtesy of Laurence Hunt and permission from Jocham, G., Hunt, L. T., Near, J., & Behrens, T. E. (2012). A mechanism for value-guided choice based on the excitation-inhibition balance in prefrontal cortex. Nature Neuroscience, 15(7), 960–961.] [Jocham, G., Hunt, L. T., Near, J., & Behrens, T. E. (2012). A mechanism for value-guided choice based on the excitation-inhibition balance in prefrontal cortex. Nature Neuroscience, 15(7), 960–961.]

According to the neural circuit model (Chapter 6), strong reverberatory excitation (large parameter w_+) speeds up neural firing and thus reduces time integration, so a rushed decision could result in reduced performance in difficult trials. This is reflected in a smaller β of the softmax decision criterion (**Figure 7.14a**). If so, a lower glutamate level (correlated with a smaller w_+ value) in the vmPFC should lead to a larger β in the choice behavior. A higher GABA level, on the other hand, should signal stronger winner-take-all competition, a larger β value and more deterministic choice behavior. Measurements of glutamate and GABA in human vmPFC by magnetic resonance spectroscopy found that, indeed, the slope β of the decision criterion computed from fitting behavioral data with a softmax varied considerably from subject to subject; the β value increased with GABA and decreased with glutamate, as predicted by the neural circuit model (**Figure 7.14b**). This result links our computational understanding of value-guided choice to a neurochemical underpinning, and suggests a mechanistic explanation for interindividual variability in choice behavior.

How does the brain carry out valuation with more than two options? The neural circuit model with three neural pools (Chapter 6, Section 6.5) has been shown to account for behavioral and neural signals from human subjects performing a value-based trinary choice task, with inputs assumed to reflect option values (Chau et al. 2014). As shown in **Figure 6.28b** of Chapter 6, the decision behavior of a three-choice version of the attractor network model can be well described by a softmax function:

$$P_A = \frac{\exp(\beta V_A)}{\exp(\beta V_A) + \exp(\beta V_B) + \exp(\beta V_C)} \tag{7.15}$$

where V_A, V_B and V_C are the inputs representing strengths of evidence for the three. If a value-coding circuit exhibits divisive normalization (Carandini and Heeger 2011) (Chapter 3, Section 3.5.2), then value signals should be divided by the sum of value signals (Louie et al. 2011; Soltani et al. 2012; Louie et al. 2014):

$$\widetilde{V_i} = \frac{V_i + V_0}{\sigma + \sum_k V_k} \tag{7.16}$$

where σ and V_0 are constants; i = A, B or C. When $\widetilde{V_i}$ instead of V_i are used in Eq. (7.15), it is clear that the impact of normalization is rescaling of β to be $\beta / (\sigma + \sum_k V_k)$. Whether the brain actually utilizes normalized values remains a matter of debate (Gluth et al. 2020).

7.4.2 Cost and Regret

In decision making, an option among available alternatives is associated with not only a payoff and its probability but also a cost due to efforts for collecting it. For instance, in an experiment, monkeys made a choice between two out of a collection of visual images, each is associated with either the amount of reward, its probability, or effort (the number of lever presses required to obtain the chosen reward). It was found that spiking activities of different neurons in OFC, as well as ACC and lateral PFC, are correlated with combinations of the three factors (Kennerley et al. 2009). OFC, ACC and LPFC areas are interconnected with each other (Averbeck and Seo 2008), and their relative roles in estimating payoff, probability and cost remain to be determined. Also open is the question of how the three factors are combined mechanistically in a neural circuit.

Furthermore, adaptive behavior depends on monitoring a choice made compared with unselected options. If the choice is not the best among available alternatives, one might feel regret. Tendency to minimize future regret produces a bias in one's behavior, which is not incorporated in standard choice theory. As remarked by D. E. Bell (1982):

> After making a decision under uncertainty, a person may discover, on learning the relevant outcomes, that another alternative would have been preferable. This knowledge may impart a sense of loss, or regret. The decision maker who is prepared to tradeoff financial return in order to avoid regret will exhibit some of the behavioral paradoxes of decision theory.

There is evidence that regret depends on the OFC. For instance, in a gambling task, human subjects were asked to make a choice between two alternatives, and then both the obtained outcome (gain or loss) from the chosen option and unobtained outcome from the unselected option were revealed. A trial ended with rating emotional feeling from being sad to happy on a graded scale (–50 to +50) (Camille et al. 2004). In normal participants, knowledge about an unobtained reward shifts emotional rating negatively regardless of the obtained outcome, which was used as a measure of regret. It was found that subjects with the OFC lesion did not report regret, suggesting that this introspective process depends on the OFC. Nonhuman animals appear to also show regret. In a physiological experiment (Steiner and Redish 2014), rats were placed at the center of an arena with four arms; each was associated with a different type of food (banana, cherry, chocolate and unflavored) and a waiting time that was sampled from a range between 1 to 30 seconds. A rat went through a sequence of choice options. Once it has entered an arm, it could decide to stay (and wait until the delay ended and a reward was harvested) or skip to the next option. It was found that rats displayed signs of regret: after skipping a short-delay option and encountering a long-delay one, animals often looked backwards toward the previous option. This "pause-and-look" phenomenon is correlated with selective activity of a subpopulation of neurons in the OFC. These findings suggest that adaptive value-based decisions rely on both the estimation of future outcomes and the process of regretting previous poor choices; the underlying circuit computation is poorly understood.

7.4.3 Predictive Valuation

Foraging exhibiting matching law described prior is an example of instrumental or operant conditioning, where association between an action and its outcome is formed based on long-run expected value of each action. The situation is more complex when a sequence of actions is required in order to achieve a goal, exemplified by navigation in a maze where one must devise a policy to guide multiple turning decisions. Importantly, one can deploy knowledge about the maze (e.g., a spatial map of the maze's layout) to choose flexibly different routes, in contrast to habitually following a single fixed path. Learning that is solely based on reinforcing successful actions but eschews representation of the world is called "model-free" RL; by contrast, "model-based" RL uses internal knowledge of the task and environment in reward-dependent learning (Dolan and Dayan 2013; Balleine 2019).

In a task paradigm especially designed to dissociate model-free versus model-based RL (Daw et al. 2011), an agent first chooses one of two options, A1 and A2; selecting A1 leads to second-stage state B with a high probability of 70% and state C with a low probability of 30% (**Figure 7.15a**). The opposite holds if A2 is chosen. In state B (respectively C), another choice must be made (B1 or B2) (respectively C1 or C2), which produced an outcome. Consider a trial in which choosing A1 leads to state C with a low probability of 30% and a second choice yields reward. Model-free RL predicts that this experience would increase the likelihood of choosing A1 in the future (**Figure 7.15b**). On the other hand, model-based RL

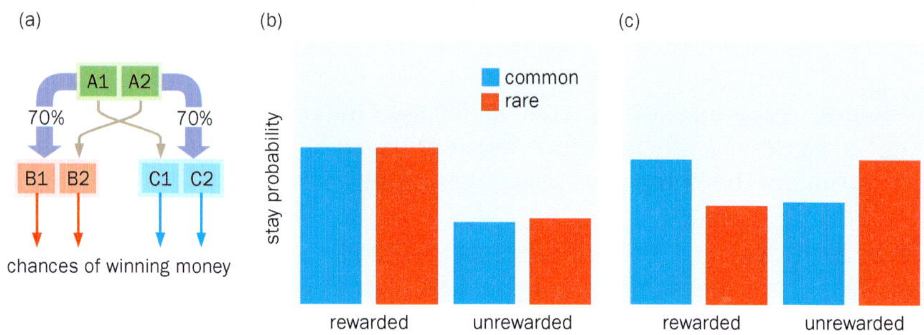

Figure 7.15 Sequential task dissociating model-based from model-free learning. (a) A two-step decision-making task, in which each of two options (A1, A2) at a start state leads preferentially to one of two subsequent states (A1 to B, A2 to C), where choices (B1 versus B2 or C1 versus C2) are rewarded stochastically with money. (b and c) Model-free and model-based RL can be distinguished by the pattern of staying versus switching of a top-level choice following bottom-level winnings. (b) A model-free learner using temporal-difference algorithm tends to repeat a rewarded action without regard to whether the reward occurred after a common transition (blue, e.g., A1 to B) or a rare one (red). (c) A model-based learner evaluates likely consequences of top-level actions, so that reward following a rare transition (e.g., A1 to C) actually increases the value of the unchosen option (A2) and thus predicts switching. Human subjects exhibited a mixture of both effects. Reproduced from Daw et al. (2011). [Daw, N. D., S. J. Gershman, B. Seymour, P. Dayan, and R. J. Dolan (2011). Model-based influences on humans' choices and striatal prediction errors. Neuron 69(6), 1204–1215.]

incorporates the knowledge about the task's transition structure, which dictates that A2 rather than A1 is more likely to lead a transition to state C, and hence produce the same reward as experienced. Therefore, the expected value for the unchosen option A2 should be increased and that for the chosen one A1 should be reduced (**Figure 7.15c**). In other words, RL should be informed by the knowledge of the world. Human subjects performing this task display choice behavior intermediate between those predicted by the model-free and model-based RL (Daw et al. 2011).

Brain imaging studies (Doll et al. 2012; Dolan and Dayan 2013) and single-neuron physiology (Deserno et al. 2015; Sadacca et al. 2016) found that brain regions, such as the dopamine system, which is important for model-free RL, are also engaged in model-based RL. Perhaps a better distinction is automated habit versus effortful action selection (Graybiel 2008; Kahneman 2011; Balleine 2019), because the brain system constantly deals with a tradeoff between the two. The introduction of a predictive state in the valuation process enhances flexibility in RL (Dayan 1993; Russek et al. 2017; Dayan and Nakahara 2018; Collins and Cockburn 2020). Arbitration between fast habit and slower deliberate choice involves executive control, which we will discuss in Chapter 8.

7.4.4 Multi-Attribute Choice

So far we have discussed valuation of two (A and B) goods with varying amounts (n_A and n_B). In real life, each option among alternatives has a bundle of two or more attributes or components. This is illustrated by the example at the beginning of the chapter: choosing between beef curry and spicy tofu, each dish is characterized by nutrition (beef or tofu) and flavor (curry or spicy) and other attributes. As another example, consider choosing between books and chocolate bars: do you prefer option A (with $n_{A,1}$ books and $n_{A,2}$ chocolate bars) or option B (with $n_{B,1}$ books and $n_{B,2}$ chocolate bars)? Consideration of multi-component bundles is necessary for trading goods on the market, a major preoccupation of microeconomics (Samuelson 1937; Arrow 1959). In general, a bundle contains two or more types of goods with varying quantities. For instance, in a laboratory experiment, monkeys are given two options to choose from: A consists of $n_{A,1}$ amount of juice type 1 (blackcurrant) and $n_{A,2}$ amount of juice type 2 (strawberry); B consists of $n_{B,1}$ amount of juice type 1 and $n_{B,2}$ amount of juice type 2 (Pastor-Bernier et al. 2017).

Conventional economics theory posits that an individual should first compute the integrated values of two components 1 and 2, $V_A\left(n_{A,1}, n_{A,2}\right)$ and $V_B\left(n_{B,1}, n_{B,2}\right)$, then make a decision based on the integrated values V_A versus V_B. In this "common currency" scenario, values from different components or attributes are combined by a weighted summation. However, our brains can process different types of value information in parallel as well as serially along a hierarchy (Rushworth et al. 2012; Hunt and Hayden 2017; Yoo and Hayden 2018; Farashahi et al. 2019). Back to the example of selecting dishes at a restaurant: you may make a decision merely based on a particular attribute (e.g., you love hot pepper, so go for the dish with red chili regardless of its other attributes). In the "hierarchical decision" scenario, two choices are initially made in parallel, one for each component: $V_{A,1}\left(n_{A,1}\right)$ versus $V_{B,1}\left(n_{B,1}\right)$; $V_{A,2}\left(n_{A,2}\right)$ versus $V_{B,2}\left(n_{B,2}\right)$. These separate, attribute-specific, categorical decisions may be the same or disagree with each other (one in favor of A, the other in favor of B). In either case, they inform a final choice (A versus B).

The OFC is crucial for normal multi-attribute decision making (Fellows 2006). Neurophysiological recording found a population of OFC neurons to represent "common currency" (Pastor-Bernier et al. 2019), in the sense that the firing rate of a neuron is constant along an indifference curve (revealed by choice behavior) in the two-dimensional plane of two bundles (**Figure 7.2c**). These neurons represent chosen value, which is invariant on an indifferent curve, in multi-attribute decision making. Other OFC cells coded option components separately, rather than integrating the values of bundle components, which potentially could be used to compute preference (A versus B) within each attribute (1 or 2). These results show that the OFC circuit is endowed with neural representations suitable to subserve both the common currency scenario and the hierarchical decision scenario.

Pettine et al. (2021) extended the recurrent neural circuit model of decision making (Chapter 6), with inputs representing values, to differentiate these two scenarios for multi-attribute decision making. A linear network (**Figure 7.16a**, left panel) assumes the common currency scenario, where weighted sums of the attribute inputs are computed for each bundle, and used as converging inputs to a circuit that produces the ultimate choice. In a hierarchical network (**Figure 7.16a**, right panel), value-cuing inputs were first transformed in an intermediate layer with two parallel, attribute-specific modules. Each of these modules is modeled as in Wong and Wang (2006). To denote the transformation from a value input signal by an intermediate layer area, its output will be referred to as T, $T_{(A,1)}$ for option A and attribute 1, etc. These transformed signals were then fed into a choice circuit that determined the ultimate decision on a given trial. This framework represents the simplest form of a network with parallel processing streams, where transformations are performed on attribute signals passing through attribute-specific modules prior to the final decision.

The behavior of a transformation module, as a nonlinear dynamical system, depends on the strengths of recurrent excitation (controlled by a parameter J_+) and inhibition (J_-). In this setting, the linear summation network becomes a special case of the hierarchical network, when recurrent connections are absent ($J_+ = J_- = 0$) and the neuronal input-output function is linear. At the other extreme, with sufficiently strong J_+ and J_-, each intermediate module can behave as an attractor circuit producing attribute-specific categorical choice. A multi-attribute decision was simulated with a fixed bundle B (inputs $I_{(B,1)}$ and $I_{(B,2)}$ were maintained constant), whereas inputs for bundle A ($I_{(A,1)}$ and $I_{(A,2)}$) were varied. As before, an ultimate decision from the choice circuit implements a softmax function of the relative difference in its inputs from the intermediate layer. An indifference curve (IC) on the plane of $I_{(A,1)}$ and $I_{(A,2)}$ was defined by equal probabilities of choosing A and B, $P_A = P_B$. Interestingly, the model shows three different regimes (**Figure 7.16b**) depending on whether the IC is linear (I), convex (II) or concave (III). These three regimes are mapped onto distinct regions of the J_+ and J_- space (**Figure 7.16c**). Regime I corresponds to a narrowly balanced excitation and inhibition, and srtong excitation produces regime II. Regime III requires both sufficiently strong excitation and inhibition, which give rise to within-attribute binary decisions. By configuring the levels of excitation and inhibition in the intermediate layer, the hierarchical network can adopt different decision making *modus operandi*.

As mentioned in Section 7.2, a convex IC is expected by common sense, whereas a concave IC contradicts the principle of diminishing marginal utility. The latter, however, assumes perfect information about values. In the model, it turns out that optimal operating regime depends on the amount of uncertainty in the environment, which in the model can be parametrically varied by turning a knob for the amount of noise in the external inputs. When uncertainty is low, the overall collected reward per unit time is higher in regime I or II than in regime III, and thus the common-currency scenario, using a linear integration of attribute values, is the best strategy. By contrast, in the presence of a high degree of uncertainty, the hierarchical decision scenario (regime III) yields optimality. This is because

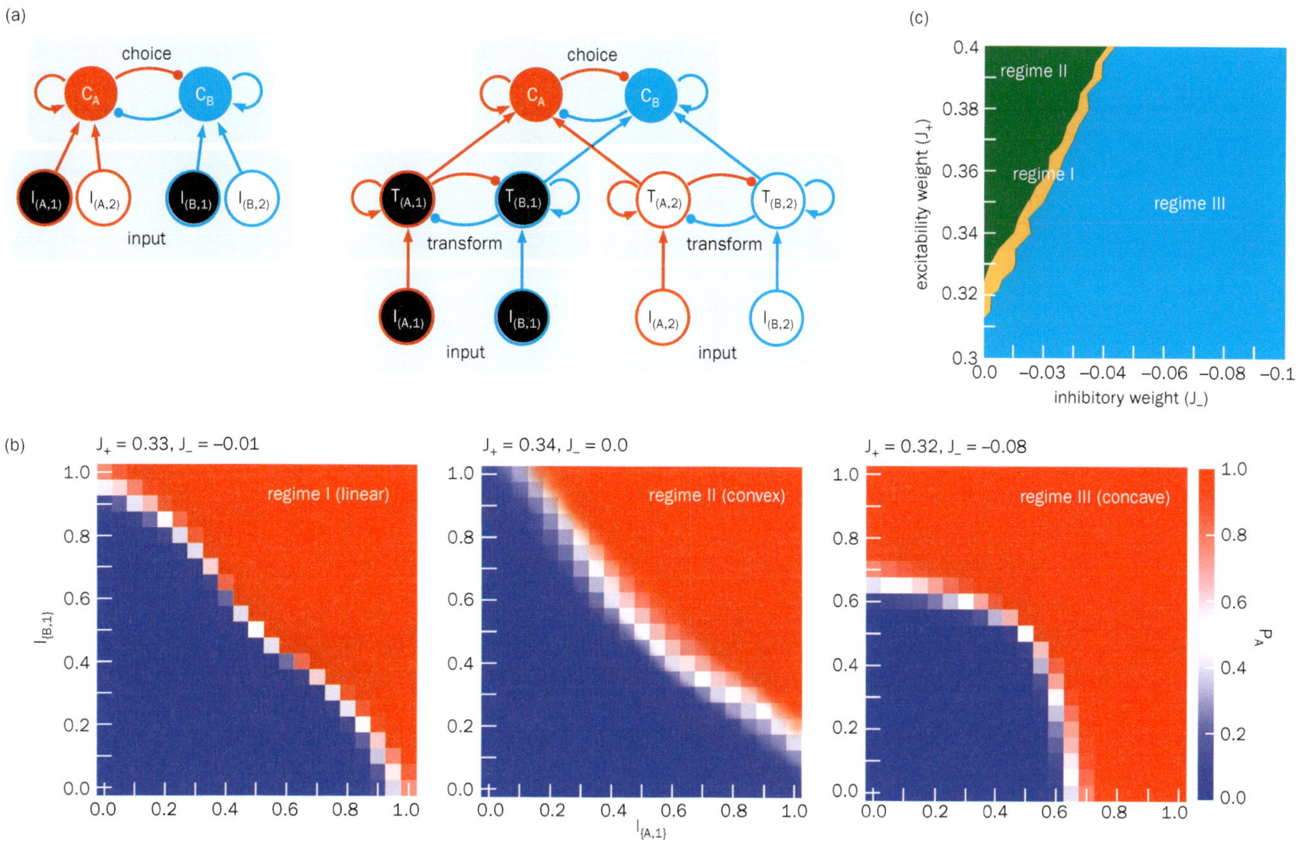

Figure 7.16 A hierarchical decision model for multi-attribute choices. (a) Network schematics. Left: In a linear network, the choice circuit directly receives summated attribute signals weighted by projection strengths. Right panel: A hierarchical network includes an additional intermediate layer that performs a transformation on the attribute signals prior to their passing to the choice circuit. Arrow: Excitation; circle: Inhibition; red: Choice A; cyan: Choice B; filled black: Attribute 1; white: Attribute 2. (b) Decision regimes I, II and III with linear, convex and concave indifference curves, respectively. (c) The space of E/I synaptic weights of an intermediate module is partitioned according to the decision regimes of their indifference curves. Forest green: Regime II; tan: Regime I; sky blue: Regime III. Reproduced from Pettine et al. (2021). [Pettine, W. W., Louie, K., Murray, J. D., & Wang, X. J. (2021). Excitatory-inhibitory tone shapes decision strategies in a hierarchical neural network model of multi-attribute choice. PLoS Computational Biology, 17(3), e1008791. Licensed under CC BY 4.0., https://creativecommons.org/licenses/by/4.0/]

categorical decision within each attribute represents a robust approach to computation with low signal-to-noise ratio. This model prediction is supported by an experiment where the amount of noise in the external inputs informing choice options was manipulated, in a multi-attribute decision task (Farashahi et al. 2019). Furthermore, this model suggests that it is desirable to switch between the two scenarios, depending on the amount of uncertainty in the environment; behavioral assessment of such flexibility and its underlying brain mechanism represent an open topic for future research.

7.5 PROBABILISTIC REASONING

Can our neural circuit model serve as a basis for understanding the brain mechanisms of more complex decision processes? One class of decision problems involves probabilistic inference. For example, given a number of meteorological measurements, such as temperature, air pressure, humidity and wind speed, what are the odds that it will rain or shine today? Other inference problems are of this type, such as assessing the odds of having cancer based on genomic testing, X-ray exam, biopsy, etc. Bayesian theory offers a specific recipe to compute odds through an integration of prior and likelihoods. In Chapter 6 (Section 6.2.4), we discussed Bayesian theory in the domain of perceptual decisions. Here we will extend the discussion to probabilistic inference through outcome-based learning.

In a probabilistic inference paradigm called "weather prediction task," several cues are given, each associated with the likelihood odds that one of the two propositions (A = rain or B = shine) is true (Gluck and Bower 1988; Gluck et al. 2002). Likelihood log odds is defined as the logarithm of the likelihood ratio $\log(\text{LR})_i$, $\log(p(S_i \mid A) / p(S_i \mid B))$, where S_i are informative cues indexed by i. Probabilities $p(A)$ and $p(B)$ are priors, which in a Bayesian model should be combined with likelihoods to determine posteriors. Specifically, according to a generalization of Eq. (6.11) in Chapter 6, we have:

$$p(A \mid \{S_i\}) = p(A)\prod_i \frac{p(S_i \mid A)}{p(S_i)}. \tag{7.17}$$

Similarly for $p(B \mid \{S_i\})$. Therefore, posterior log odds is given by:

$$\log\frac{p(A \mid \{S_i\})}{p(B \mid \{S_i\})} = \log\frac{p(A)}{p(B)} + \sum_i \log\frac{p(S_i \mid A)}{p(S_i \mid B)}. \tag{7.18}$$

In the special case where priors are equal, $\log\left(p(A)/p(B)\right) = 0$ and log posterior odds is an integration of log likelihood ratio (logLR) values.

In a monkey experiment (Yang and Shadlen 2007), reward probabilities assigned to two targets (A = red, B = green) are informed by visual cues (**Figure 7.17a**), and subjects make a choice by a saccade to one of the targets. In every trial, four cues ($S_i, i = 1,2,3,4$) are randomly sampled from ten shapes, and each is assigned a weight of evidence (WOE) defined by $w_i = \text{logLR}_i = \log(p(S_i \mid A) / p(S_i \mid B))$. Reward is delivered if target A is chosen with a probability $p_A = p(A \mid \{S_i\}) = 10^{\sum_i w_i} / (1 + 10^{\sum_i w_i})$, $p_B = p(B \mid \{S_i\}) = 1 - p_A$, and the two reward probabilities depend on cues and vary from trial to trial. If the priors are equal, as in Yang and Shadlen (2007), the sum of logLRs yields log posterior odds, which should be used to make a choice (Gold and Shadlen 2001).

The number of possible combinations of four non-overlapping shapes out of ten is given by "10 choose 4," equal to $10!/4! = 210$. It is unlikely, if not impossible, to memorize such a large number of specific four-shape patterns. A subject can learn the probabilities and use Eq. (7.18) to compute log posterior odds, but the probabilities are not explicitly given in the monkey experiment. Remarkably, the choice behavior of a well-trained monkey, aggregated across trials, showed that the probability of choosing A is a softmax function of the log posterior odds in favor of A (**Figure 7.17b**). Presumably, the probabilistic information conferred by each shape can be learned by virtue of experienced frequencies of different shape patterns and outcomes of an agent's choices. Moreover, single neurons recorded from the posterior parietal cortical area LIP engaged in saccade selection display firing activities modulated by logLRs (Yang and Shadlen 2007). This is most clearly shown by a spike train of a neuron in a single trial (**Figure 7.18**), in a variant of the task where four shapes were shown sequentially in time rather than simultaneously (Kira et al. 2015), demonstrating an accrual process for probabilistic reasoning at the single cell level.

How can such a quantity as summated log likelihood ratios or log posterior odds be actually computed in the brain? **Figure 7.19a** shows schematically a model with three layers for the weather prediction task (Soltani and Wang 2010). The first layer contains sensory neurons that are selective for visual cues (shapes). They project to the second layer, where neurons learn to encode reward values of the two alternative responses (action values) through plastic synapses that undergo the same reward-dependent

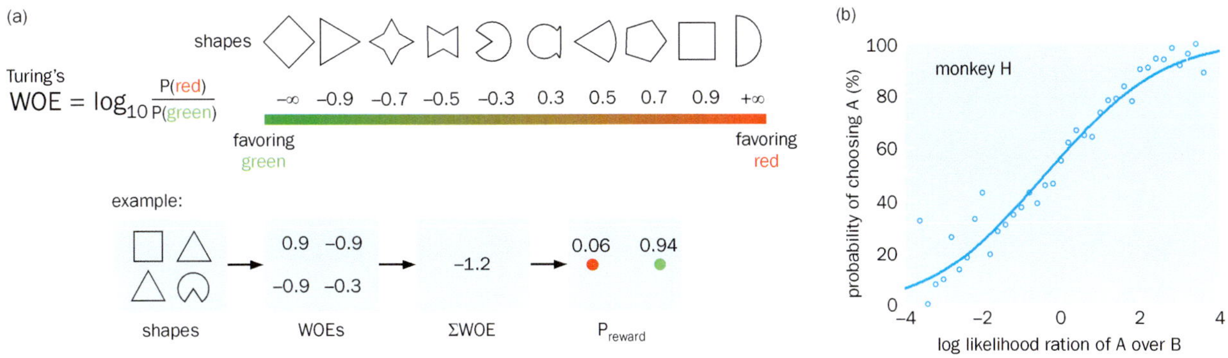

Figure 7.17 Weather forecasting in a monkey experiment. (a) Task design. Four shapes were presented on the computer monitor. After a brief delay period, the monkey made an eye movement (saccade) to either the red or green choice target. During neural recording, one of the choice targets was in the response field of the neuron. The shapes were selected randomly in each trial from a larger set of ten (top). The reward was determined probabilistically by summing the weights of evidence (WOEs) associated with the four shapes. (b) Performance. The fraction of choices for A is plotted as a function of the summated logLRs conferred by the four shapes in favor of A. Reproduced from Yang and Shadlen (2007). [Yang, T., & Shadlen, M. N. (2007). Probabilistic reasoning by neurons. Nature, 447(7148), 1075–1080.]

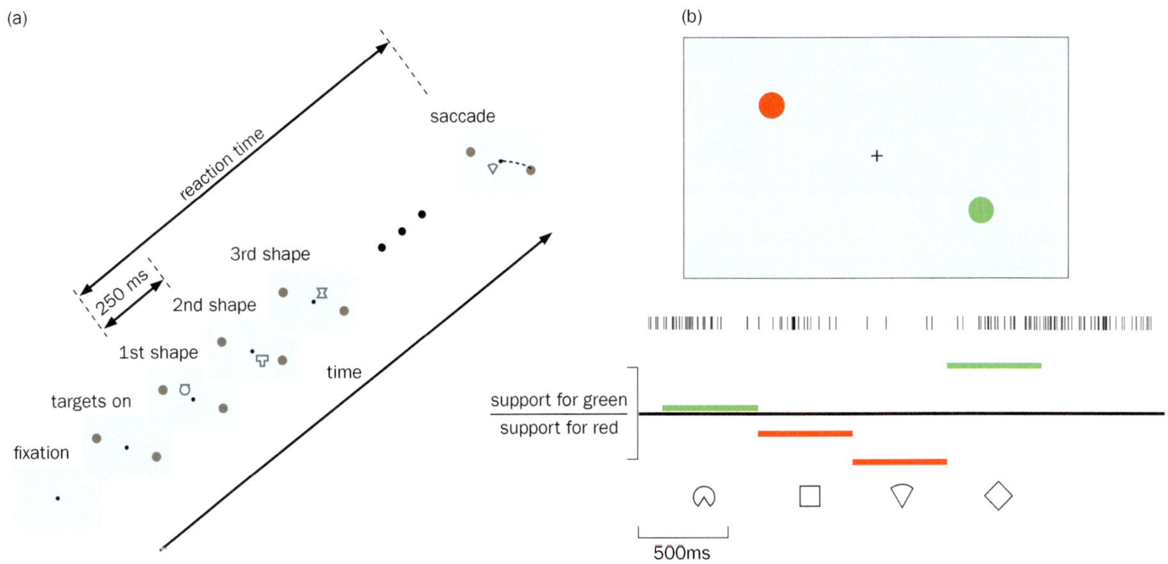

Figure 7.18 Neural integration of sequential samples in probabilistic inference. (a) In a variant of the weather prediction task, four shapes were presented one at a time, in a series. One of two targets (here B = green) is in the response field of a recorded neuron in LIP of behaving monkey, and the other is diametrically opposite to it (A = red). LogLR values associated with different shapes are similar to those in **Figure 7.17**. (b) Spike train in a single trial displays a stochastic pattern with rate that increases with shapes in favor of target B (green) and decreases with shapes in favor of A (red), as expected by time integration. Provided by Michael Shadlen with data from Kira et al. (2015). [Courtesy of Michael Shadlen with data and permission from Kira, S., Yang, T., & Shadlen, M. N. (2015). A neural implementation of Wald's sequential probability ratio test. Neuron, 85(4), 861–873.] [Courtesy of Michael Shadlen.]

Hebbian modifications as Eqs. (7.7–7.8). The convergence from cue-selective neurons enables action value-coding neurons to combine information from different cues. The third layer is a decision circuit with two competing neural pools that are selective for choice ("preferred target") A and B, respectively.

Consider a simple situation where one cue (S_i) alone is presented in a trial, which determines the reward probability for each of the two alternative responses, $p(A \mid S_i)$ and $p(B \mid S_i) = 1 - p(A \mid S_i)$. Since reward assignment is independent of choice selection, the strength c_{iA} for synapses selective for cue S_i and choice A is potentiated with a probability equal to the product of three probabilities: the probability $p(S_i)$ that cue S_i is presented, the probability $P_A(S_i)$ that A is selected when cue S_i is presented, and the

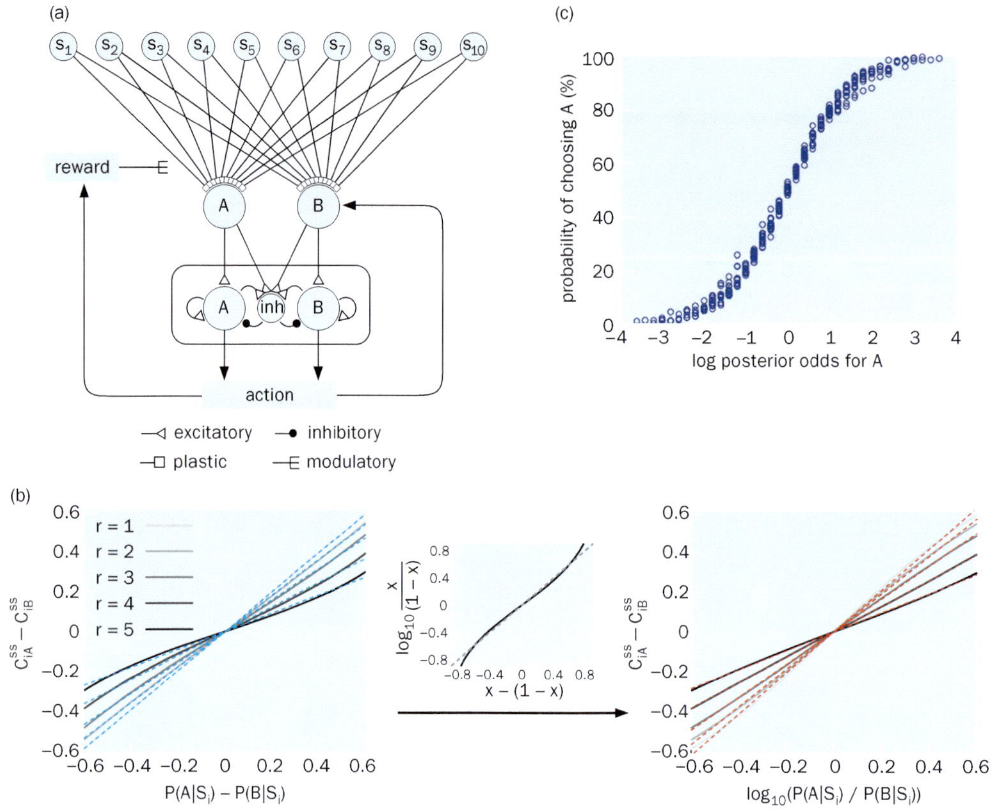

Figure 7.19 Probabilistic inference in a decision-making circuit model endowed with reinforcement learning. (a) Schematic of the three-layer model for a weather prediction task. Combination of cues is accomplished through convergence of cue-selective neurons onto action value-encoding neurons, which in turn project to a decision-making circuit. Depending on the reward schedule, a chosen action may be rewarded or not. The presence (respectively absence) of a reward signal leads to potentiation (respectively depression) of plastic synapses. (b) Difference in the steady-state of the synaptic strengths is a softmax function of the difference in posteriors (left panel) and log posterior odds (right panel), for different values of the learning rate ratio r. Dashed lines show linear fits for the values of posterior between 0.2 and 0.8 ($x = p(A \mid S_i)$ and $p(B \mid S_i) = 1 - p(A \mid S_i) = 1 - x$). (c) Choice behavior of the model in the weather prediction task. Probability of choosing A as a function of the sensory evidence favoring this option, defined by the sum of logLRs of four cues. Reproduced from Soltani and Wang (2010). [Soltani, A., & Wang, X. J. (2010). Synaptic computation underlying probabilistic inference. Nature Neuroscience, 13(1), 112–119.]

probability $p(A \mid S_i)$ that choice A yields reward given cue S_i is presented; $p(S_i)P_A(S_i)p(A \mid S_i)$. The probability of depression for the same set of synapses is $p(S_i)P_A(S_i)(1 - p(A \mid S_i))$. Through ongoing learning, the synaptic strength for each set of plastic synapses eventually reaches a steady-state value, computed by setting the overall change equal to zero:

$$\Delta c_{i,A} = q_+ (1 - c_{i,A}) p(S_i) P_A(S_i) p(A \mid S_i) - q_- c_{i,A} p(S_i) P_A(S_i)(1 - p(A \mid S_i)) = 0 \tag{7.19}$$

Solving this algebraic equation yields the steady-state of the synaptic strength:

$$c_{i,A}^{SS} = \frac{r p(A \mid S_i)}{1 + (r - 1) p(A \mid S_i)} \tag{7.20}$$

where r is the learning rate ratio ($r \equiv q_+ / q_-$). Therefore, the steady-state is independent of the choice behavior (i.e., $P_A(S_i)$).

In the special case of equal potentiation and depression rates ($r = 1$), the steady-state of the synaptic strength is equal to the posterior probability, $c_{i,A}^{SS} = p(A \mid S_i)$. Through experience across a sequence of trials, synaptic strengths learn to encode posteriors. In general, when the learning rates are not equal, the synaptic strength is a nonlinear monotonic function of the posterior probability.

Recall that the decision circuit generates choices in such a way that the probability of selecting A is a softmax function of the differential input, $c_A - c_B = p(A \mid S) - p(B \mid S) = 2p(A \mid S) - 1$ (with $p(B \mid S) = 1 - p(A \mid S)$). Mathematically, $x - (1 - x) \simeq \log(x / (1 - x))$, if $0.2 \leq x \leq 0.8$. It follows that, for the intermediate range of posteriors where the model's choice behavior is stochastic, the difference in the synaptic strengths is linearly proportional to log posterior odds (**Figure 7.19b**). For smaller or larger values of posteriors, the choice behavior is deterministic (the probability of choosing A is close to 0 or 1). As a result, decision making is based on log posterior odds, as required by probabilistic inference.

When several cues are presented to inform a decision, logLRs for the presented stimuli are readily added by virtue of convergence of cue-encoding neurons to decision neurons (**Figure 7.19a**). Therefore, a decision circuit endowed with such synapses makes choices on the basis of the summed log posterior odds and performs near-optimal cue combination. This model thus reproduces the behavioral performance of the monkey experiment (**Figure 7.19c**).

Furthermore, because decision neurons are driven by input synapses, their firing activity is naturally modulated by the accrual of logLR values, in the same way as LIP single-neuron recorded from behaving monkeys (**Figure 7.20**). Like the case of a foraging task (Soltani and Wang 2006), this result again demonstrates that through plasticity synapses encode values (here, reward probabilities), which are manifested at the level of firing activities of downstream neurons. In the brain, a neural circuit such as the OFC encodes values; another circuit transforms the chosen option to an action (Cisek and Kalaska 2010; Cai and Padoa-Schioppa 2014; Yim et al. 2019), such as the LIP and DLPFC for saccades, where neural activity is correlated with the input synaptic strength.

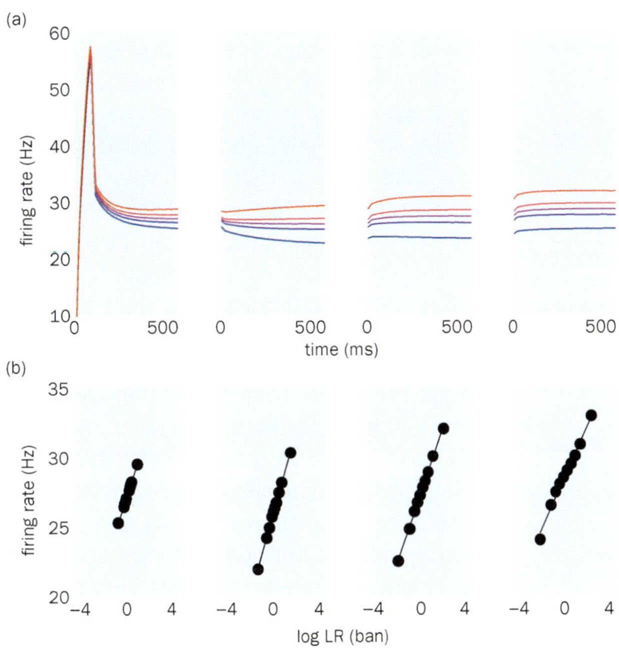

Figure 7.20 Neural population activity is parametrically correlated with the logLR. (a) Effect of the logLR on the firing rate of decision neurons. Population activity in each epoch is aligned on the onset of each shape stimulus presentation, and the average over many trials is computed for five quintiles of logLR in that epoch (plotted in different colors; more red means larger logLR favoring the alternative A). The logLR in each epoch is equal to the sum of the logLR of shapes that are presented before and during that epoch. (b) Average population firing rate as a function of logLR for four epochs. Average firing rate during the last 250 ms is computed by grouping logLR (in base 10, or a unit called "ban") into ten equal bins in each epoch. These neural activity patterns from the model are similar to those observed experimentally in the posterior parietal cortex of monkeys performing the weather prediction task (Yang and Shadlen 2007). Reproduced from Soltani and Wang (2010). [Soltani, A., & Wang, X. J. (2010). Synaptic computation underlying probabilistic inference. Nature Neuroscience, 13(1), 112–119.]

When the choice alternatives have unequal priors, the model (Soltani and Wang 2010) predicts deviations from the Bayes decision rule that are akin to an effect called "base-rate neglect" commonly observed in human studies; namely, people tend to be primarily informed by the likelihoods while underweighting prior (base-rate) in estimating the posteriors (Kahneman et al. 1982; Gluck and Bower 1988; Soltani et al. 2016; Yang and Wu 2020). Therefore, the recurrent neural circuit model with appropriate extensions accounts for inference in certain forms of probabilistic problem solving.

7.6 SOCIAL DECISION MAKING

A broad class of decisions engage social interactions among a group of individuals. Social decisions are often studied using interactive games, pioneered by von Neumann and Morgenstern (1944). Both zero-sum competitive games and games that probe cooperation between agents have been investigated in neuroscience, so we shall discuss an example of each.

7.6.1 Random Choice Behavior in Matching Pennies Game

Social decisions are not made by a single individual; instead, choice outcomes are determined jointly by interacting agents. Game theory represents a dominant framework in studies of social decision making for several reasons (Seo et al. 2011). First, people display complex behavior even in simple games, offering opportunities to investigate rigorously such processes as strategies, cooperation and fairness in a social setting. Second, there is a long history of mathematical modeling of games (Camerer 2003), perhaps best illustrated by the Nash equilibrium theorem for zero-sum competitive games (Nash et al. 1950), which provides theoretical guidance for describing optimality and deviations from it. Third, those games can be learnt by a nonhuman animal playing either with another animal or with a computer in experimental laboratories, enabling researchers to carry out brain imaging or physiological recording and explore the underlying brain mechanisms.

A well-known game is "rock-paper-scissors" played in the schoolyard. A similar game but with two instead of three options is called matching pennies: you decide to show a penny head or tail, and I do the same; if the two pennies match (head-head, tail-tail), you win a point and I lose a point; the converse holds when there is no match (head-tail, tail-head). This game is zero-sum: your gain is my loss, and vice versa. When two rational players participate in the matching pennies game, each player should choose the two targets with equal probabilities and independently across successive trials. Any other strategy could be exploited by the opponent; for example, if you choose head with a probability higher than 0.5, I'll choose tail more frequently and increase my gain and your loss.

This task was used in a physiological experiment where monkeys played with a computer (Barraclough et al. 2004). In the experiment, the strategy of the computer opponent (programmed by the experimenter) was systematically manipulated to determine how the animal's choice behavior was affected by the strategy of its opponent. Initially, the computer opponent chose the two targets with the same probabilities while ignoring animal's choices. In this case, a monkey's expected payoff was fixed regardless of its behavior, and it displayed the easiest strategy of always selecting the same target. Later in the experiment, the computer algorithm kept track of animal's past choices and outcomes, with the payoff matrix given by **Figure 7.21a**, the animal's behavior became very close to be random, as predicted by the Nash equilibrium.

A model of two recurrent neural circuits has been applied to a matching pennies game (Soltani et al. 2006). The idea is that each circuit simulates a decision maker or an agent, and the two interact with each other merely though synaptic plasticity determined by the agents' choices and outcomes according to a payoff matrix. This class of models can thus be used to simulate monkey experiments using game-theoretic tasks (Barraclough et al. 2004; Dorris and Glimcher 2004). The model reproduces salient behavioral observations (Soltani et al. 2006). If the opponent is not interactive, the model decision behavior is idiosyncratic and might, for instance, choose one of the targets exclusively. When the opponent is fully interactive according to the rules of matching pennies, the model behavior becomes quasi-random. This is shown in **Figure 7.21b** with ten different initial values for the synaptic variables c_A and c_B. Different c_A and c_B values yield different initial probability P_A of choosing target A as a softmax function of $(c_A - c_B)$. Competitive interaction with the opponent, however, quickly equalizes the synaptic variables (**Figure 7.21b**), and the choice probability becomes very close to 0.5 (**Figure 7.21c**), regardless of the initial state of the system. For instance, if initially the system chooses target A more frequently because c_A is larger than c_B, it would be exploited by the opponent, and the unrewarded outcomes from choosing A induce depression of c_A of the synapses to the neural pool A, so that the difference $c_A - c_B$ decreases over time, and the system gradually chooses B more frequently so the behavior eventually becomes random.

Although the model can reproduce monkey behavior, different model parameters are required for different computer algorithms from ignoring animal's behavior to fully taking in consideration of it. How can these model parameters be tuned adaptively, as the opponent's algorithm is changed? To address this question, Soltani et al. (2006) incorporated a meta-learning rule proposed by Schweighofer and Doya (2003) that maximizes long-term rewards. They found that the enhanced model captures very slow changes of monkey's behavior, as the opponent's algorithm changes from day to day of animal's learning experience.

A general insight is that a decision circuit produces random choice behavior, not necessarily because the system has a prescribed "random number generator" but because the trial-to-trial choice and learning dynamics force the decision agent to play randomly. This is well demonstrated in the model, because the same model produces either stereotypical responses or random responses, depending on the behavior of its opponent. The model decision maker thus does not have a goal to play randomly, but simply tries to garner rewards at its best, dynamically from trial to trial, given the environment and other decision agents involved in the game. This conclusion is consistent with behavioral studies and models, emphasizing the critical importance of feedback in the production of quasi-random behavior (Rapoport and Budescu 1992; Camerer 2003). Moreover, the model suggests that irregular neural firing and the stochastic nature of synaptic learning contribute to the generation of random choice behavior, which can be desirable in interactive decision tasks. This model sheds insights into neural processes in the brain underlying the randomness observed at the psychological level (Glimcher 2005; Wang 2008b).

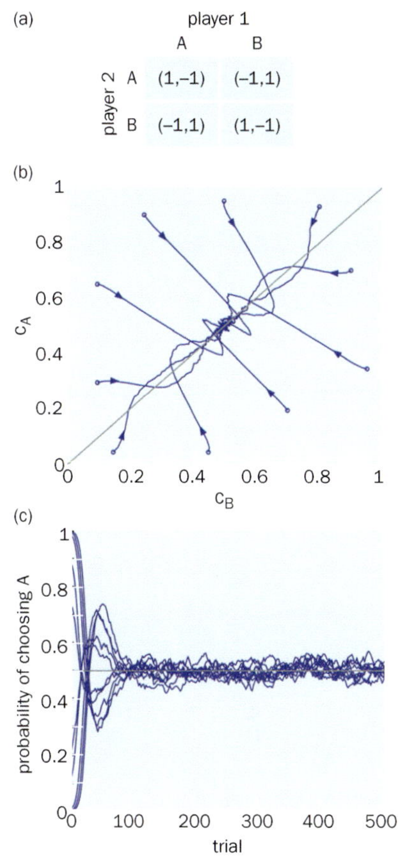

Figure 7.21 Model simulation of random choice behavior during the matching pennies game. (a) Payoff matrix for the matching pennies game. Each box corresponds to a particular combination of two player's choices, and the outcomes for players 1 and 2 are given by the first and second number, respectively. (b) The synaptic strengths c_A and c_B plotted against each other show the time evolution of adaptive learning. (c) The corresponding choice probability for A, which is a softmax function of $c_A - c_B$. Regardless of the initial state of the neural circuit (different c_A and c_B values), the network quickly evolves towards the Nash equilibrium state of random mixed strategy, with $c_A \approx c_B$ (the diagonal line in (b)), and the choice probability becomes chance level (0.5 in (c)). Reproduced from Wang (2014) using the model of Soltani et al. (2006). [Wang, X.-J. (2014). Neuronal circuit computation of choice. In P. Glimcher and E. Fehr (Eds.), Neuroeconomics (Second ed.)., pp. 435–453. Academic Press.]

7.6.2 Volatility and Reinforcement Learning on Multiple Timescales

When a monkey performed the matching pennies game, its behavior in a trial was influenced by what happened in the past few trials. Correlatively, single neurons in the LIP, ACC and DLPFC displayed firing activity modulated by the events (own choice, opponent's choice, outcome) in the last three trials (**Figure 7.22**).

What is the time constant for memory trace of experienced rewards? Recall that, in the case of the Rescorla-Wagner learning rule, the parameter α (between 0 and 1) is a learning rate, and the model behaves as a leaky integrator of reward events with an integration time of $\tau = -1/\ln(1-\alpha)$

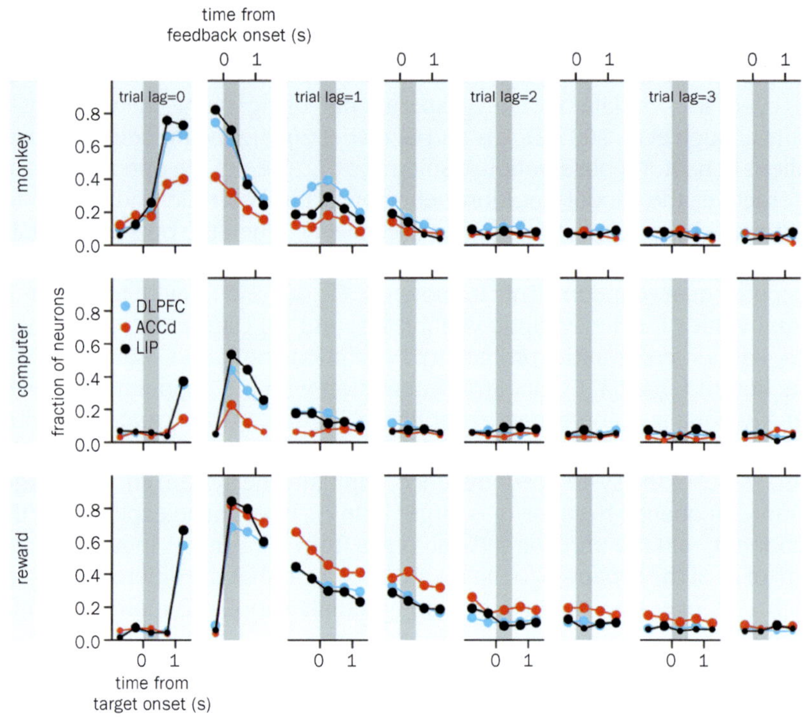

Figure 7.22 Modulation of neural activity by previous experience across trials, monitored from monkeys during a matching pennies game with the computer. Time course of signals related to the animal's choice (top), choice of the computer opponent (middle), and choice outcome (bottom) in three different cortical areas (DLPFC, dorsolateral prefrontal cortex; ACCd, dorsal anterior cingulate cortex; LIP, lateral intraparietal cortex). Data in each row represent the fraction of neurons significantly modulating their activity according to choices (top and middle) or outcome (bottom) in the current (trial lag = 0) or previous (trial lag = 1 to 3) trials. Grey background corresponds to the delay period (left) and feedback period (right) in each trial. Reproduced from Seo and Lee (2008). [Courtesy of Daeyeol Lee and permission from Curtis CE, Lee D. (2010) Beyond working memory: The role of persistent activity in decision making. Trends in Cognitive Sciences. 14(5):216–22.]

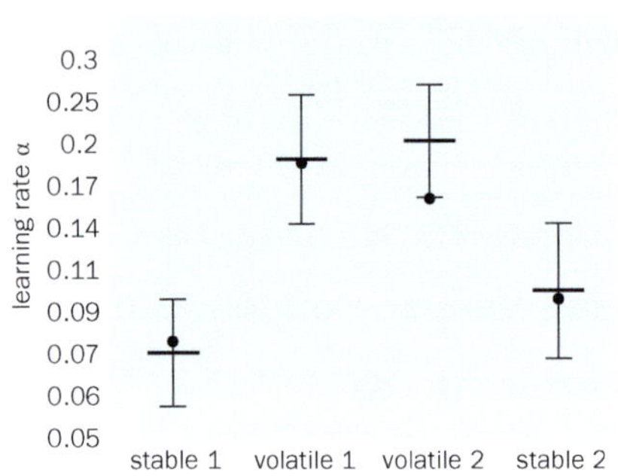

Figure 7.23 Change of learning rate α (in unit of one over trial number) with varying volatility. Average learning rates during the stable and volatile phases (stable, volatile, volatile, stable, respectively) in a human experiment. Reproduced from Behrens et al. (2007). [Behrens, T., Woolrich, M., Walton, M. et al. (2007). Learning the value of information in an uncertain world. Nature Neuroscience 10(9), 1214–1221.]

(Chapter 4, Section 4.3.1). This time constant is short when α is large (zero with $\alpha = 1$) and long with a small α (∞ with $\alpha = 0$). Intuitively, the learning rate α should be dynamically adjustable: if the environment is stochastic but stable, then it is desirable to deploy a long integration time in order to learn about and exploit the statistics of the environment, whereas if the environment is highly uncertain, one should use a short time constant and high learning rate to explore different options quickly. In a human experiment (Behrens et al. 2007), the volatility of reward delivery statistics was varied between blocks of trials. It was found that the learning rate of human subjects, estimated by fitting behavioral data with a mathematical model, was higher when the environment was more unpredictable (Figure 7.23).

Can such a time constant for integrating past reward events be extracted from single cells in decision making? Bernacchia et al. (2011) analyzed how rewards in previous trials affect the firing activity of neurons in the DLPFC, ACCd and LIP from monkeys performing the matching pennies task. For each recorded single cell, modulation of firing rate by past rewards was computed: relative to the overall trial average, what is the change of neural firing rate conditioned on that a reward was collected in the previous trial, two trials before, etc.? The initial modulation is denoted as the memory magnitude A, which may be positive or negative.

The modulation size decreases with more remote reward events, and the decay time course is fit with an exponential or a weighted sum of two

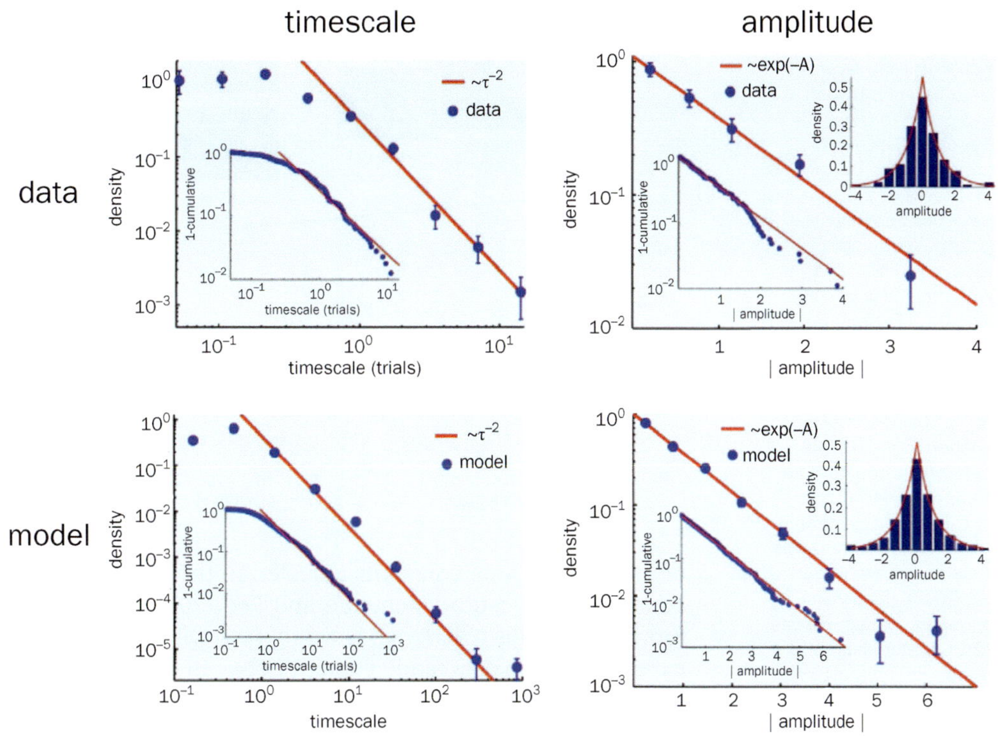

Figure 7.24 Power law of time constants for reward memory trace. Upper panels: The histograms of time constants (left) and memory amplitudes (right) extracted from single neurons in the prefrontal and parietal cortices in monkeys performing a decision task. Insert: 1-Cumulative histograms. The time constant distribution exhibits a power law tail (within a limited range of τ values), and the distribution of the memory amplitude is exponential. Lower panels: The histograms of time constants (left) and memory amplitudes (right) in a high-dimensional and recurrent linear neural network model. Reproduced from Bernacchia et al. (2011).

exponentials as a function of previous trial numbers, from which a time constant τ was extracted. Surprisingly, it was found that different neurons showed widely different τ values. The histogram of reward memory trace time constants over 800 individual neurons displayed a power law like $\sim 1/\tau^2$ (**Figure 7.24**, upper left plot), whereas the histogram of the memory trace amplitude behaved as $\exp(-A)$ (upper right plot). The power law tail of the time constant distribution means that there is an excessive probability for very long time constants compared to a Gaussian or an exponential distribution.

The implications are twofold: first, a "reservoir" of time constants are heterogeneously distributed in a population of single neurons; second, the reward memory system operates in a high-dimensional space. Indeed, a simple linear neural network model in an infinite dimensional space, under the condition of being "at the edge of criticality" (Chapter 3, Section 3.4.3), reproduces the same power law distribution of memory time constants and the exponential distribution of memory trace amplitudes (**Figure 7.24**, lower panels). This finding suggests that reinforcement learning modeling should be extended to more than one dimension, in order to allow for reward-dependent learning over many timescales. In principle, short or long time constants (high or low learning rates) from this reservoir can be selectively readout, flexibly, depending on the degree of volatility of the environment.

In the matching pennies game, trials are uncorrelated, and therefore history does not matter and one should merely play random. Yet, the animals made their decisions largely on the basis of their reward history in the past few trials (Seo and Lee 2008). Perhaps in the persistent search for an appropriate strategy, they sampled different timescales across experimental sessions. In tasks that do have a temporal structure, is there a selection

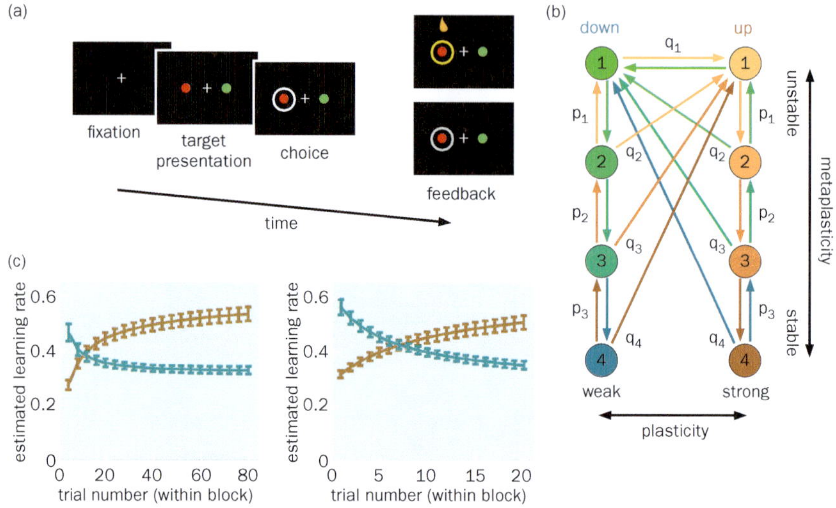

Figure 7.25 Probabilistic reversal learning task and reward uncertainty. (a) Timeline of the probabilistic reversal learning task. Subjects select between two options (red and green targets); the reward is assigned probabilistically to one of the two targets, and the better target flips between blocks of trials. (b) Schematic of meta-plastic synapses. Meta-plastic synapses have multiple internal states associated with each of the two levels of synaptic efficacy: Weak (Down) and strong (Up). (c) Experimental evidence for meta-plasticity revealed by time-dependent, choice-specific learning rates upon a transition from a volatile to a stable environment (Left) and a transition in the opposite direction (right). Brown: The average estimate of learning rate for trials when reward was assigned to the better target; blue: The average estimate of learning rate for trials when reward was assigned to the worse target. Reproduced from Farashahi et al. (2017). [Farashahi, S., Donahue, C. H., Khorsand, P., Seo, H., Lee, D., & Soltani, A. (2017). Metaplasticity as a neural substrate for adaptive learning and choice under uncertainty. Neuron, 94(2), 401–414.]

of appropriate time constants adjusted to the environmental volatility? To address this question, Donahue and Lee (2015) trained monkeys to perform a probabilistic reversal learning task with two volatility levels. In the experiment, subjects made choices between two targets, each associated with a reward probability (e.g., red target: 80%, green target: 20%) so that only one of the choices is assigned a reward in a given trial. After a number L of trials, the two probabilities reversed (red target: 20%, green target: 80%) (**Figure 7.25a**). The environment was either stable with $L = 80$ or volatile with $L = 20$. Farashahi et al. (2017) developed a reinforcement learning model for this probabilistic reversal learning task, by extending the learning rule (Eqs. 7.7–7.8) to the cascade synapse model (Fusi et al. 2005) (Chapter 4, **Figure 4.20**). **Figure 7.25b** shows a synapse model with four internal states, which describes the input to one of two decision neural pools (selective for A = red and B = green targets). When A wins and yields a reward, its input synapses undergo potentiation, through those transitions between internal meta-states indicated by brown arrows, ultimately to the Up state. Conversely, if a choice of A does not result in reward, those transitions in favor of the Down state indicated by green arrows take place, and the synapses to neural pool A are depressed.

The introduction of meta-states endows reward-dependent plasticity with multiple time constants and a natural mechanism for adjustments to reward statistics in the environment. When the model was applied to the probabilistic reversal learning task, overall learning rates can be estimated, which depend on the relative occupancies of the internal states of the synapses. The estimated overall learning rates for the two targets dynamically adjust when the volatility in the environment suddenly changes. Namely, after a block transition, the estimated overall learning rate increased over time in trials where reward was assigned to the better option, whereas the overall learning rate decreased in trials when reward was assigned to the worse option. The same trend was found in monkeys' behavior (**Figure 7.25c**), revealing that learning rates are not fixed but vary in a choice-specific and time-dependent manner.

7.6.3 Cooperation

Many social decisions are not zero-sum games but benefit from cooperation among decision makers. Cooperation is crucial for collective decision making and depends on predicting the intentions of others, as illustrated by the famous prisoner's dilemma (Rapoport and Chammah 1970; Axelrod 1984; Poundstone 1992). In the classic version of the game (**Figure 7.26a**), two suspects (A and B) are arrested and in solitary confinement. The police

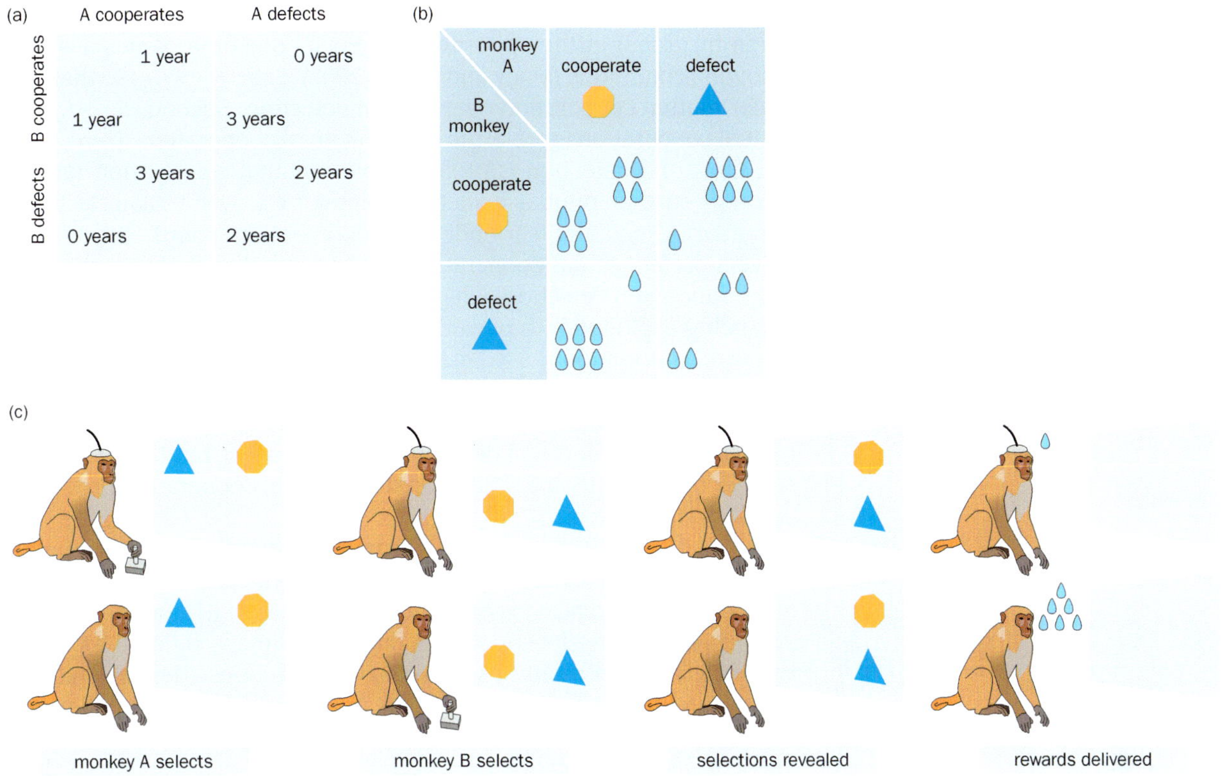

Figure 7.26 The prisoner's dilemma. (a) In the classic version of the game, in an entry the upper right and lower left numbers are the punishment outcome (number of years in prison) for players A and B, respectively. (b) A reward version of the task using a monkey experiment, the amounts of juice reward are shown for both players. (c) Sequence of events in a behavioral trial. Panels (b–c) are reproduced from Haroush and Williams (2015). [Haroush, K., & Williams, Z. M. (2015). Neuronal prediction of opponent's behavior during cooperative social interchange in primates. Cell, 160(6), 1233–1245.]

do not have sufficient evidence, so they plan to sentence the pair to a year in prison on a lesser charge in the absence of a confession (when A and B cooperate with each other). Each is offered the option to confess (defect). If one of them confesses but the other does not, the former is set free, whereas the latter gets a heavier punishment of three years in prison. If both defect, however, they will be convicted of the principal charge and receive a two-year sentence. What should one do? Without the opportunity to communicate with the other, A could reason that the punishment is smaller if one chooses to defect than cooperate, regardless of B's decision (0 versus 1 year if B cooperates; 2 versus 3 years if B defects). The trouble is, B will come to the same conclusion, in which case both will get two years. This is worse than if both cooperate, in which case each gets only one year in jail. Hence, at the core of such a social dilemma, decisions that make sense to each individual result in an outcome in which everyone suffers.

Variants of the prisoner's dilemma games involve rewards rather than punishment, in which case the goal is to maximize the numerical payoff; and iterative prisoner's dilemma (iPD) games entail repetitions of such a game between the same players. In an iPD game, one would expect that players should cooperate all the time. However, one is prone to temptations: if one player cooperates consistently, the other might become greedy and defect in order to collect the best outcome. One might also fear to behave as a sucker who keeps cooperating even when the other defects. How to deal with such social dilemmas in general?

Let us denote by R_C and R_D a player's outcomes in the cases of mutual cooperation and defection, respectively; R_S (S for sucker) the payoff for cooperation by the player and defection by the other, and R_G (G for temptation by greed) in the opposite case. Social dilemma games assume the following conditions (Macy and Flache 2002). First, $R_C > R_D$, so that players prefer mutual cooperation over mutual defection. Second, $R_C > R_S$, hence mutual cooperation is preferred over unilateral cooperation. Third, $R_G > R_C$, i.e., unilateral defection is preferred over mutual cooperation (greed); or $R_D > R_S$, so that mutual defection is preferred over unilateral cooperation (fear). Greed and fear undermine cooperation. Fourth, $2R_C > R_G + R_S$, players prefer mutual cooperation over an equal probability of unilateral cooperation and defection, which means that an even chance of exploiting and being exploited by a player is not as good as mutual cooperation. For instance, a commonly used strategy in iPD is called Tit-for-Tat (Axelrod 1984): a player starts with cooperation, then choses the same action taken by the other player in the previous move. This could lead to alternations between cooperation and defection by each player, but this would yield a smaller overall outcome than consistent mutual cooperation provided that condition four is met.

Figure 7.26b shows the payoff matrix that satisfies all four conditions: $R_C = 4$, $R_D = 2$, $R_S = 1$ and $R_G = 6$. This game was played by two monkeys in a neurophysiological experiment (**Figure 7.26c**) (Haroush and Williams 2015). Like the classical prisoner's dilemma game, a rational monkey, not knowing the other monkey's intention, should choose defect, which yields a larger reward. However, both defecting would yield a smaller reward (two drops of juice each) than both cooperating (four drops of juice). On the other hand, temptation derives from the greed of collecting six drops of juice, and hence there is incentive to defect in case the other's intention to cooperate is assessed to be high. The key then for a player in an iPD game is to estimate the likelihood that the other player will cooperate.

Behavioral analysis found that monkeys did not cooperate consistently; overall, a monkey cooperates in ~35% trials (**Figure 7.27a**, right blue bar). The most salient trait of interactions is that the probability of cooperation is substantially enhanced, to 60%, conditioned on mutual cooperation in the previous trial (**Figure 7.27a**, left blue bar). This enhancement of

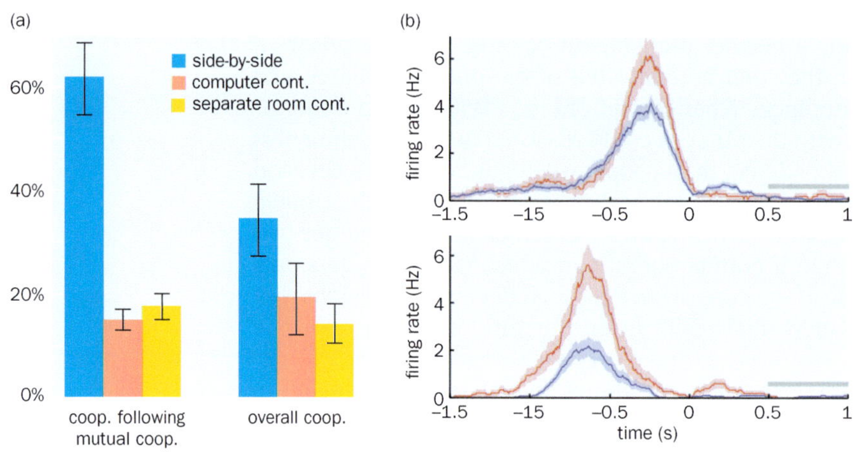

Figure 7.27 Salient behavior and neural signaling of other's intention to cooperate in a monkey experiment. (a) Monkeys cooperate significantly more playing together in the same room (blue), vs. playing in separate rooms (yellow), or playing against a computer opponent (orange). Notably, the probability of cooperation is larger following a trial when both cooperated (left versus right blue bar).
(b) Activity of neurons in the dorsal anterior cingulate cortex correlates with self or other upcoming decision of cooperating (red) or defecting (blue). Top: An example neuron showing a higher firing rate when the monkey's own decision was to cooperate than defect prior to the onset of monkey's response (t = 0). Bottom: Another neuron showing a higher firing rate when the other's upcoming decision was to cooperate than defect. Reproduced from Haroush and Williams (2015). [Courtesy of Keren Haroush.] [Courtesy of Keren Haroush and permission from Haroush, K., & Williams, Z. M. (2015). Neuronal prediction of opponent's behavior during cooperative social interchange in primates. Cell, 160(6), 1233–1245.]

conditional cooperation was absent if the two monkeys were in different rooms (yellow) or a monkey played the same game with a computer rather than a co-species (orange), suggesting that social interactions were important for monkey's behavior.

Ablation (Rudebeck et al. 2006) and brain imaging (Behrens et al. 2008) have shown that the dorsal part of the anterior cingulate cortex (ACCd) plays a key role in signaling social interest in others. Neurophysiological recording revealed single neurons in ACCd whose firing activity is higher when a player was about to cooperate than defect (**Figure 7.27b**, upper panel), whereas the firing time course is the same regardless of the other player's action. Other neurons exhibited the opposite trend: their firing rate is higher when the other player was going to cooperate than defect (**Figure 7.27b**, lower panel), and insensitive to its own imminent decision. Crucially, the peaked responses occur before the onset of the subject's response, and therefore can in principle be used to causally inform monkey's decision.

The model for the matching pennies game discussed earlier can be extended to the iPD game. Namely, the model for each player now has two modules, for self and other player, respectively. The inputs to the "self player" module are updated by its own decisions and outcomes, whereas the "other player" module are driven by inputs that are updated by the other player's decisions and outcomes. When the payoff matrix is changed from **Figure 7.21a** to **Figure 7.26b**, the model can simulate the iPD game, reproducing the observed increased conditional probability of cooperation after mutual cooperation takes place (Panagiota Theodoni, unpublished results). Note that in the prisoner's dilemma game, a player does not directly observe the choice of the other player and can only infer its outcome indirectly by her own payoff, the underlying brain mechanism remains to be elucidated.

7.7 SUMMARY

In this chapter, we covered value-based decisions. In the spirit of searching for cross-level understanding of cognition, we showed how the neural circuit model discussed in Chapter 6, endowed with a simple reward-dependent plasticity rule, can account for various economic choice behavior. First, applied to a stochastic foraging task, synapses learn to compute returns (the reward amount per choice on each option) through experience, and the model exhibits matching law as well as slight deviations (undermatching) observed experimentally across species. Second, in simulations of a weather prediction task, plastic synapses now encode posteriors used for probabilistic inference that approximates Bayesian theory. Third, two such models play a competitive game according to a payoff matrix, leading to random choice behavior. These studies yielded some unexpected findings. In particular, the brain appears to possess a reservoir of time constants for reward memories that obey a power law, suggesting that the reinforcement learning theory should be extended to a high-dimensional space for flexible adjustment to a change in volatility. Taken together, the emerging picture is that perceptual decision making and value-based choice are accounted for by a unified model. We emphasized the importance of going beyond statistic descriptions; instead, choice behavior is better understood in terms of dynamical adaptation across a few trials, which offers an explanation of behavior according to Herbert Simon's satisficing principle (Simon 1997; Todd and Gigerenzer 2012).

This model emphasizes that stochasticity of choices is inherently explained by irregular neural population dynamics in single trials, suggesting that there is no need for an explicit neural representation of probabilities, nor invoking "flip a coin" for the generation of randomness in our decisions. Free will in our choices may be understood in terms of stochastic neural dynamics in interplay with the world underlying self-initiated and adaptive behavior (Heisenberg 2009). This model framework can be extended to other aspects of decision making like confidence (Lak et al. 2020) and hierarchical behavior (Botvinick et al. 2009). Cross-level investigations of the brain mechanisms and computational modeling of social cognition remain a widely open area of neuroscience (Wittmann et al. 2018; Lockwood et al. 2020; Testard et al. 2024).

FURTHER READING

Books and Reviews

Camerer, C., G. Loewenstein, and D. Prelec (2005). Neuroeconomics: How neuroscience can inform economics. *J. Econ. Lit. 43*, 9–64.

Glimcher, P. W. (2003). *Decisions, Uncertainty, and the Brain: The Science of Neuroeconomics*. Cambridge, MA: MIT Press.

Hunt, L. T. and B. Y. Hayden (2017). A distributed, hierarchical and recurrent framework for reward-based choice. *Nat. Rev. Neurosci. 18*, 172–182.

Kahneman, D. and A. Tversky (1979). Prospect theory: An analysis of decision under risk. *Econometrica 47*, 263–292.

Lempert, K. M. and E. A. Phelps (2014). Neuroeconomics of emotion and decision making. In P. Glimcher and E. Fehr (Eds.), *Neuroeconomics* (2nd ed.), pp. 219–236. Cambridge, MA: MIT Press.

Padoa-Schioppa, C. (2011). Neurobiology of economic choice: A good-based model. *Annu. Rev. Neurosci. 34*, 333–359.

Rushworth, M. F., N. Kolling, J. Sallet, and R. B. Mars (2012). Valuation and decisionmaking in frontal cortex: One or many serial or parallel systems? *Curr. Opin. Neurobiol. 22*, 946–955.

Schoenbaum, G., M. R. Roesch, T. A. Stalnaker, and Y. K. Takahashi (2009). A new perspective on the role of the orbitofrontal cortex in adaptive behaviour. *Nat. Rev. Neurosci. 10*, 885–892.

Seamans, J. and C. Yang (2004). The principal features and mechanisms of dopamine modulation in the prefrontal cortex. *Prog. Neurobiol. 74*, 1–58.

Surmeier, D. J., W. Shen, M. Day, T. Gertler, S. Chan, X. Tian, and J. L. Plotkin (2010). The role of dopamine in modulating the structure and function of striatal circuits. *Prog. Brain Res. 183*, 149–167.

Wallis, J. D. (2007). Orbitofrontal cortex and its contribution to decision-making. *Annu. Rev. Neurosci. 30*, 31–56.

Wang, X.-J. (2014). Neuronal circuit computation of choice. In P. Glimcher and E. Fehr (Eds.), *Neuroeconomics* (2nd ed.), pp. 435–453. Cambridge, MA: MIT Press.

Primary Research Papers

Behavioral Laws

Mazur, J. (1987). An adjusting procedure for studying delayed reinforcement. In M. L. Commons, J. E. Mazur, J. A. Nevin, and H. Rachlin (Eds.), *Quantitative Analysis of Behavior: The Effects of Delay and Intervening Events on Reinforcement*, pp. 55–73. Hillsdale, New Jersey: Lawrence Erlbaum Associates, Publishers.

Rachlin, H. and D. I. Laibson (Eds) (1997). Richard J. Herrnstein, *The Matching Law: Papers in Psychology and Economics*. Cambridge, MA: Harvard University Press.

Valuation and Economic Choice

Bari, B. A., C. D. Grossman, E. E. Lubin, A. E. Rajagopalan, J. I. Cressy, and J. Y. Cohen (2019). Stable representations of decision variables for flexible behavior. *Neuron 103*, 922–933.

Behrens, T., M. Woolrich, M. Walton, and M. Rushworth (2007). Learning the value of information in an uncertain world. *Nat. Neurosci. 10*, 1214–1221.

Cohen, J. Y., S. Haesler, L. Vong, B. B. Lowell, and N. Uchida (2012). Neuron-typespecific signals for reward and punishment in the ventral tegmental area. *Nature 482*, 85–88.

Padoa-Schioppa, C. and J. Assad (2006). Neurons in the orbitofrontal cortex encode economic value. *Nature 441*, 223–226.

Pereira-Obilinovic, U., H. Hou, K. Svoboda, and X.-J. Wang (2024). Brain mechanism of foraging: Reward-dependent synaptic plasticity or neural integration of values? *Proc. Natl. Acad. Sci. 121*, e2318521121.

Platt, M. L. and P. W. Glimcher (1999). Neural correlates of decision variables in parietal cortex. *Nature. 400*, 233–238.

Rich, E. L. and J. D. Wallis (2016). Decoding subjective decisions from orbitofrontal cortex. *Nat. Neurosci. 19*, 973–980.

Sugrue, L. P., G. C. Corrado, and W. T. Newsome (2004). Matching behavior and representation of value in parietal cortex. *Science 304*, 1782–1787.

Neural Circuit Model With Reward Dependent Plasticity

Bernacchia, A., H. Seo, D. Lee, and X.-J. Wang (2011). A reservoir of time constants for memory traces in cortical neurons. *Nat. Neurosci. 14*, 366–372.

Frémaux, N., H. Sprekeler, and W. Gerstner (2010). Functional requirements for rewardmodulated spike-timing-dependent plasticity. *J. Neurosci. 30*, 13326–13337.

Loewenstein, Y. and H. Seung (2006). Operant matching is a generic outcome of synaptic plasticity based on the covariance between reward and neural activity. *Proc. Natl. Acad. Sci. U.S.A. 103*, 15224–15229.

Louie, K., L. E. Grattan, and P. W. Glimcher (2011). Reward value-based gain control: Divisive normalization in parietal cortex. *J. Neurosci. 31*, 10627–10639.

Soltani, A. and X.-J. Wang (2006). A biophysically based neural model of matching law behavior: Melioration by stochastic synapses. *J. Neurosci. 26*, 3731–3744.

Probabilistic Inference

Soltani, A. and X.-J. Wang (2010). Synaptic computation underlying probabilistic inference. *Nat. Neurosci. 13*, 112–119.

Yang, T. and M. Shadlen (2007). Probabilistic reasoning by neurons. *Nature 447*, 1075–1080.

Games

Barraclough, D. J., M. L. Conroy, and D. Lee (2004). Prefrontal cortex and decision making in a mixed-strategy game. *Nat. Neurosci. 7*, 404–410.

Haroush, K. and Z. M. Williams (2015). Neuronal prediction of opponents behavior during cooperative social interchange in primates. *Cell 160*, 1233–1245.

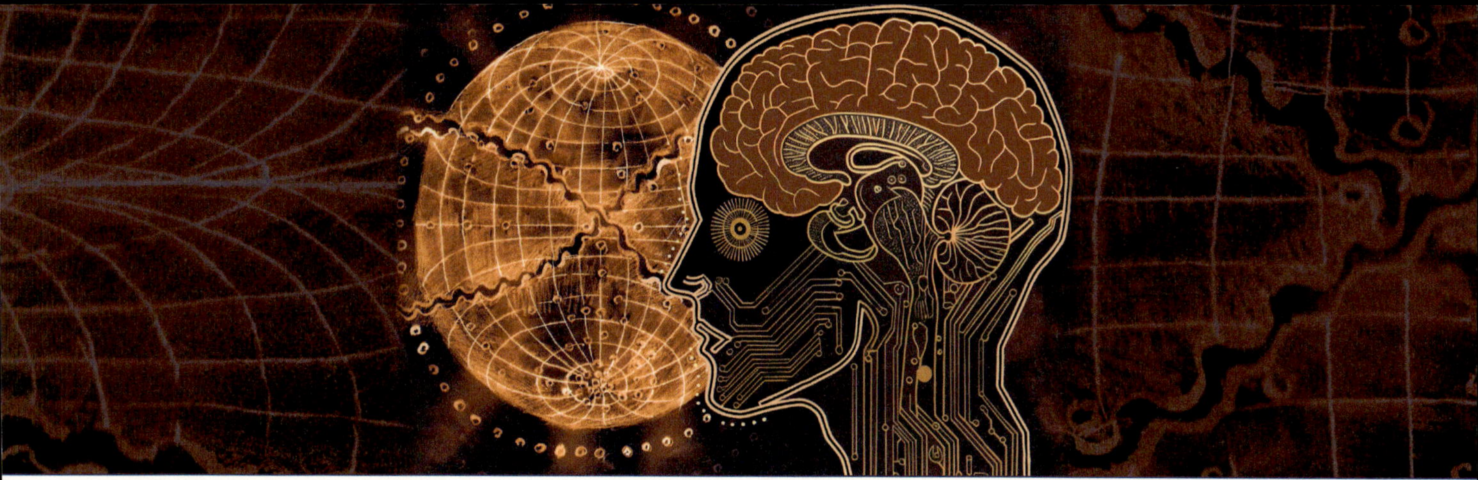

Executive Function

8

Attention . . . is the taking possession by the mind, in clear and vivid form, of one out of what seem several simultaneously possible objects or trains of thought. Focalization, concentration of consciousness are of its essence. It implies withdrawal from some things in order to deal effectively with others, and is a condition which has a real opposite in the confused, dazed, scatterbrained state.

William James in *The Principles of Psychology*

8.1 INTRODUCTION

Imagine that, you go on a trip and you have forgotten to bring your automatic toothbrush. You take an old-fashioned manual toothbrush provided in the hotel room. While brushing your teeth, your mind starts to meander; after a long awhile, you might suddenly ask yourself: this has been going on for more than the usual two minutes, why has the toothbrush not stopped? The anecdote illustrates the fact that we carry out routine activities "on autopilot," without being aware of them. Moreover, our brain has a way to estimate a time interval (two minutes), which helps us to monitor behavior. Repeated practice in a consistent environment results in automaticity (Logan 1988). Automated habits simplify everyday life and save mental effort but, if unchecked, they may lead us astray. For instance, if you live in New York City, while crossing a street you look carefully first left and then right, to make sure no car is coming. When you visit London, however, this habit could put you in danger because a car could come from the right at the moment you start at a crosswalk. "Attention to action" denotes executive control of behavior when prepotent tendency needs to be suppressed or an action plan must be altered (Norman and Shallice 1986). Response inhibition plays an important role in dealing with two or more courses of action, and the most prepotent one may not be the best for you. Descartes wrote: "if anger makes the hand rise in order to strike, the will can ordinarily restrain it; if fear incites the legs to flee, the will can stop them" (Descartes 1989). Hence, a major component of attention to action is the brain's ability to override actions that are driven by habit or emotional outburst. Deficits

DOI: 10.1201/9781003459361-10

of this function cause impulsivity (Bari and Robbins 2013), manifested by people who are prone to aggression and violence (Sapolsky 2017).

At the other end of the sensorimotor behavioral spectrum, sensory stimuli arriving upon our sense organs compete for allocation of our attention (Klingberg 2009). Attention shift may be automatic or reflexive, caused exogenously by salient stimuli: an object may "pop-out" because it has a different shape or color from other objects in a cluttered visual scene. By contrast, endogenous attention is an internal process in the brain that biases sensory information processing in order to achieve a goal (Chun et al. 2011). Consider the example of visual search mentioned at the beginning of Chapter 5: to find your friend in a crowded party, your eyes constantly look for cues (e.g., body silhouette) that are associated with your friend while ignoring many other visual objects irrelevant to your task. Guided visual search engages endogenous selective attention as a form of executive control. Biases also exist in endogenous attention, exemplified by the pattern of our eyes' fixation as we freely view an image (Yarbus 1967): our gaze tends to linger on eyes and mouth much more than the rest of a face (**Figure 8.1**), reflecting our priorities as well as past experience. It takes volition to overcome such biases. In the Stroop task (Stroop 1935; MacLeod 1991), words naming colors are shown in color, for example "yellow" in green, "blue" in red, etc. Asked to identify the color of a word (e.g., green for the word "yellow"), we tend to err by reading out the word (yellow) and our response time is longer with a correct answer (green). Selective attention is crucial for us in order to focus on a relevant feature (color of a word) while filtering out an irrelevant feature (word itself). As a matter of fact, we are often oblivious of what come into our senses that we are not paying attention to.

What is the brain mechanism of executive control, both in the motor and sensory domains? In 1876, David Ferrier wrote:

> In proportion to the development of the faculty of attention are the intellectual and reflective powers manifested. This is in accordance with the anatomical development of the frontal lobes of the brain, and we have various experimental and pathological data for localizing in these the centres of inhibition, the physiological substrata of this psychological faculty.

Among empirical evidence supporting Ferrier's proposal was the famous case of Phineas Gage, who exhibited unordinary profanity and compulsivity after a railroad accident damaged his frontal lobe (Harlow 1868). Neuroscientific research has documented that the prefrontal cortex (PFC) is

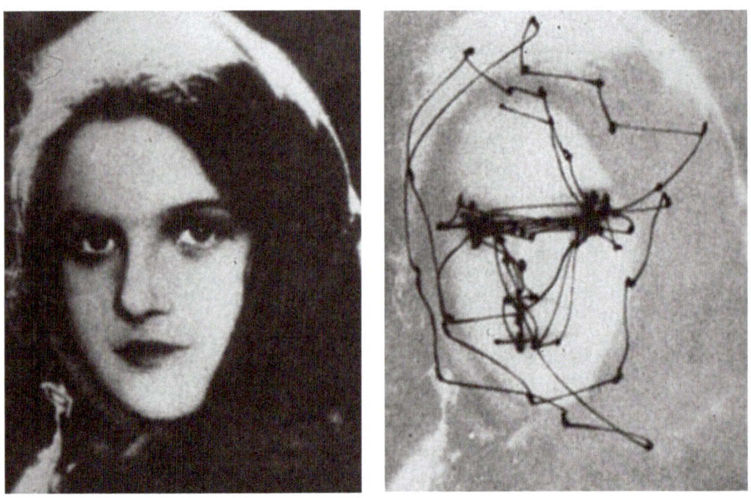

Figure 8.1 Eye fixation and saccade pattern while viewing a face with no instruction for 1 minute. Reproduced from Yarbus (1967). [Yarbus, A. L. (1967). Eye Movements and Vision. New York: Plenum Press. (B.Haigh, Trans. The original book was published in Russian in 1965).]

Figure 8.2 Wisconsin Card Sorting Test. Participants sort cards based on one of three features: Color, shape or number. In this example, the card with two blue crosses should be put in pile 2, 3 or 4 if the rule is number, shape or color, respectively. The relevant feature changes unpredictably, after a transition must figure out the new rule by trial-and-error.

essential for executive control of motor responses (Aron 2007; Bari and Robbins 2013; Diamond 2013; Apšvalka et al. 2022). The frontal lobe is also crucial for executive control in the sensory domain (Desimone and Duncan 1995; Awh and Jonides 2001; Corbetta and Shulman 2002; Soto et al. 2008; Gazzaley and Nobre 2012). An early and commonly practiced assessment of frontal damage uses the Wisconsin Card Sorting Test (WCST) (Milner 1963). In the WCST, a participant is given a deck of cards, and there are colored shapes on each. The task is to sort cards into piles depending on a rule about the relevant feature, which is not explicitly instructed. In the example of **Figure 8.2**, the card with two blue crosses should be put in pile 2, 3 or 4, if the relevant feature is number, shape or color, respectively. From time to time, the rule changes without warning, and the participant must figure out the new rule by feedback (correct or wrong) after each response. Therefore, one feature of visual stimuli (cards) is relevant for behavioral goal; attention must be selected and sustained to guide task performance. Subjects with frontal damage typically have difficulty with switching from one rule to another and persevere in committing errors in spite of repeated negative feedbacks.

Executive control requires decision making (selection of a goal) and working memory (internal maintenance of a focus). In this chapter, we will discuss how "cognitive-type" neural circuits capable of working memory and decision making (Chapters 5 and 6) subserve volitional control of the motor and sensory systems.

8.2 RESPONSE INHIBITION

A simple behavioral paradigm for probing response inhibition is the "stop-signal" task (Lappin and Eriksen 1966; Logan and Cowan 1984; Verbruggen et al. 2019). In a visual version of the stop-signal task, a behavioral trial starts with gaze fixation. In a "go trial," a salient target appears in the periphery of vision at the same time the fixation cue disappears, and the subject makes a saccade to the target as a prepotent response (**Figure 8.3a**) In a small number of "stop-signal" trials randomly interleaved with go trials, after the saccade target is shown, a stop signal "countermands" the imminent saccadic response, in which case the subject is required to cancel the planned action (**Figure 8.3b**). The stop signal can be a reappearance of the fixation cue or from another sensory modality such as an auditory tone. Obviously, one's ability to cancel an automatic reaction (signal-inhibit) crucially depends on the time interval between the target stimulus and stop signal, which is called stop-signal delay (SSD). When SSD is too long, saccade would have already been produced before the

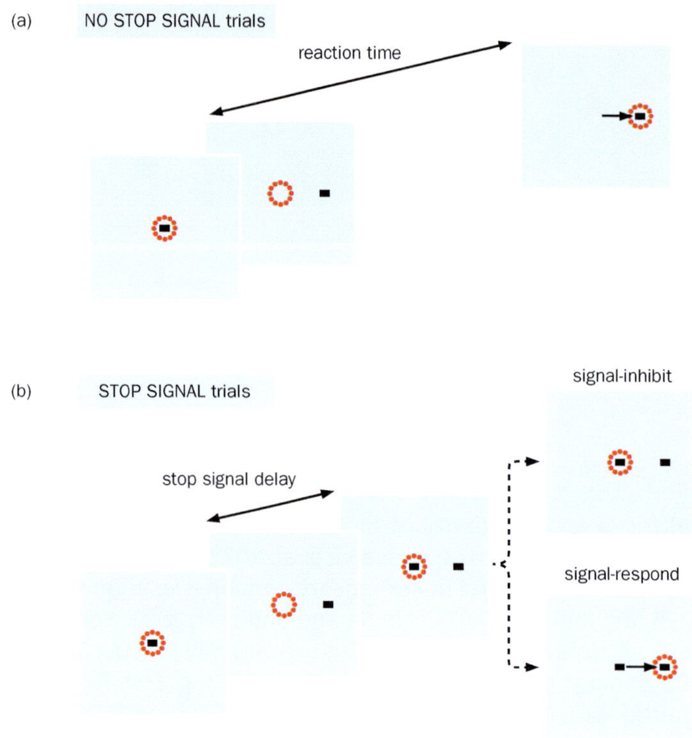

Figure 8.3 Stop-signal task paradigm. (a) In go trials, a subject makes a saccade to a suddenly appearing visual target. (a) In stop-signal trials, a stop signal is presented (the reappearance of the fixation cue), in which case the planned response should be cancelled. Reproduced from Boucher et al. (2007). [Boucher, L., Palmeri, T. J., Logan, G. D., & Schall, J. D. (2007). Inhibitory control in mind and brain: An interactive race model of countermanding saccades. Psychological Review, 114(2), 376–397.]

stop signal occurs, thus the probability of signal-respond (saccade in spite of a stop signal) is 1. That probability decreases with reduced SSD, offering a controlled manipulation for assessing inhibitory control. The fraction of trials where a stop signal is shown can also be titrated; if the stop signal is more frequent, then one has a higher expectation for a stop signal in a given trial and may engage more response inhibition. In this way, the brain mechanism of monitoring behavioral response across trials can be studied as well.

8.2.1 Race Model and Neurophysiology of a Stop-Signal Task

In an independent race model for the stop-signal task (Logan and Cowan 1984), a go process is activated by the go stimulus, whereas a stop process is triggered at the onset of the stop signal. The two ramp up linearly with time (as perfect integrators) to reach a threshold, and whichever gets there first determines whether a response is ultimately withheld (a signal-inhibit trial) or not (a signal-respond trial) (**Figure 8.4a**). As expected by intuition, the probability of signal-respond increases with SSD (**Figure 8.4b**). How can one explain this performance curve by an internal stop process?

In a signal-respond trial, there is no visible sign of an internal stop process; in a signal-inhibit trial, nothing is observed except the absence of a response. Therefore, properties that characterize the stop process can only be inferred indirectly from behavior. Logan and Cowan (1984) assumed that the stop process takes a fixed amount of time from the onset of a stop signal to finish, called stop-signal reaction time (SSRT). In a race model (**Figure 8.4a**), if the STOP process has a ramping slope β and the threshold is θ, then SSRT $= \theta / \beta$. The method to estimate SSRT is shown in **Figure 8.4c**. Consider the distribution of reaction times in go trials. For stop-signal trials, the combined time of SSD and SSRT divides the distribution into two parts: those with RTs below SSD+SSRT (cyan) correspond to signal-respond trials in which responses are too fast to be cancelled, and the remaining trials (purple) correspond to signal-inhibit trials. Remember

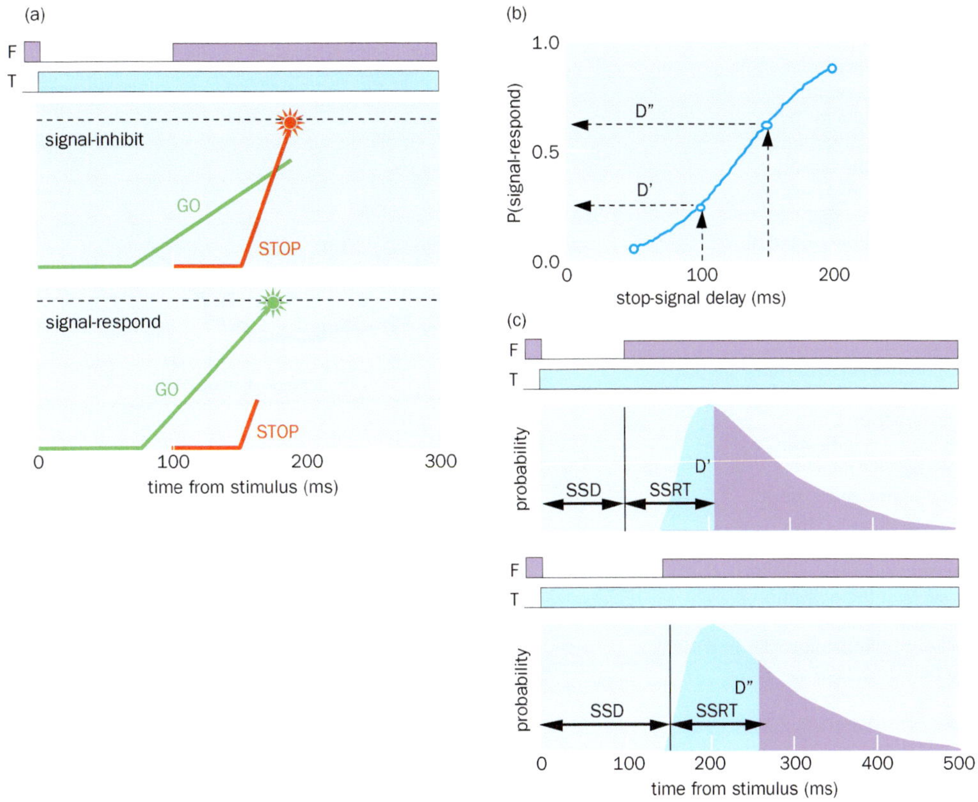

Figure 8.4 Race model for the stop-signal task. (a) Race between a go process and a stop process. When the stop process reaches a threshold before the go process, the response is cancelled in a signal-inhibit trial (top); otherwise, the response is executed in a signal-respond trial (bottom). (b) The probability of signal-respond increases with the stop-signal delay (SSD). (c) The reaction time distribution is shown for go trials. The race model assumes that the time it takes for the stop process from stop-signal onset to reach the threshold is a fixed quantity, called stop-signal reaction time (SSRT). The integral of the blue-shaded portion with go RTs shorter than the sum of SSD and SSRT corresponds to responses that are too fast to be cancelled, thus this fraction of trials predicts the probability of signal-respond in (b). The blue area is larger and the probability of signal-respond is higher with a longer SSD (bottom plot compared to top plot). Reproduced from Boucher et al. (2007). [Boucher, L., Palmeri, T. J., Logan, G. D., & Schall, J. D. (2007). Inhibitory control in mind and brain: An interactive race model of countermanding saccades. Psychological Review, 114(2), 376–397.]

that SSRT is not directly observable. The race model maps the probability of signal-respond (b) onto the fraction of RT distribution in go trials [integrated cyan area in (c)], thereby estimating the covert variable SSRT, which is independent of SSD.

Neurophysiological studies reported larger signals from various frontal areas in stop-signal trials than in go trials. Evoked response potential around the inferior frontal cortex displays an N200 (negativity at 200 ms) component in signal-inhibit trials, presumably correlated with efforts to suppress an automatic response; the amplitude of this signal is substantially reduced in children diagnosed with ADHD, which is characterized by a propensity for impulsivity (lack of response inhibition) (Pliszka et al. 2000). For the visual stop-signal task of **Figure 8.3**, the frontal eye field (FEF) is of special interest because of its potent role in controlling saccades. Single-neuron recording revealed two populations of FEF neurons in macaque monkeys (Hanes and Schall 1996; Hanes et al. 1998). Movement neurons, also called gaze-shifting neurons, already shown in Chapter 6 (**Figure 6.5**), are characterized by ramping-to-threshold in go trials (**Figure 8.5a**). In stop-signal trials, they display different patterns in signal-respond versus signal-inhibit trials (**Figure 8.5b**, top panel). A second neural population, called gaze-holding or fixation neurons, display a high activity during fixation, and their activity decays away starting with the go

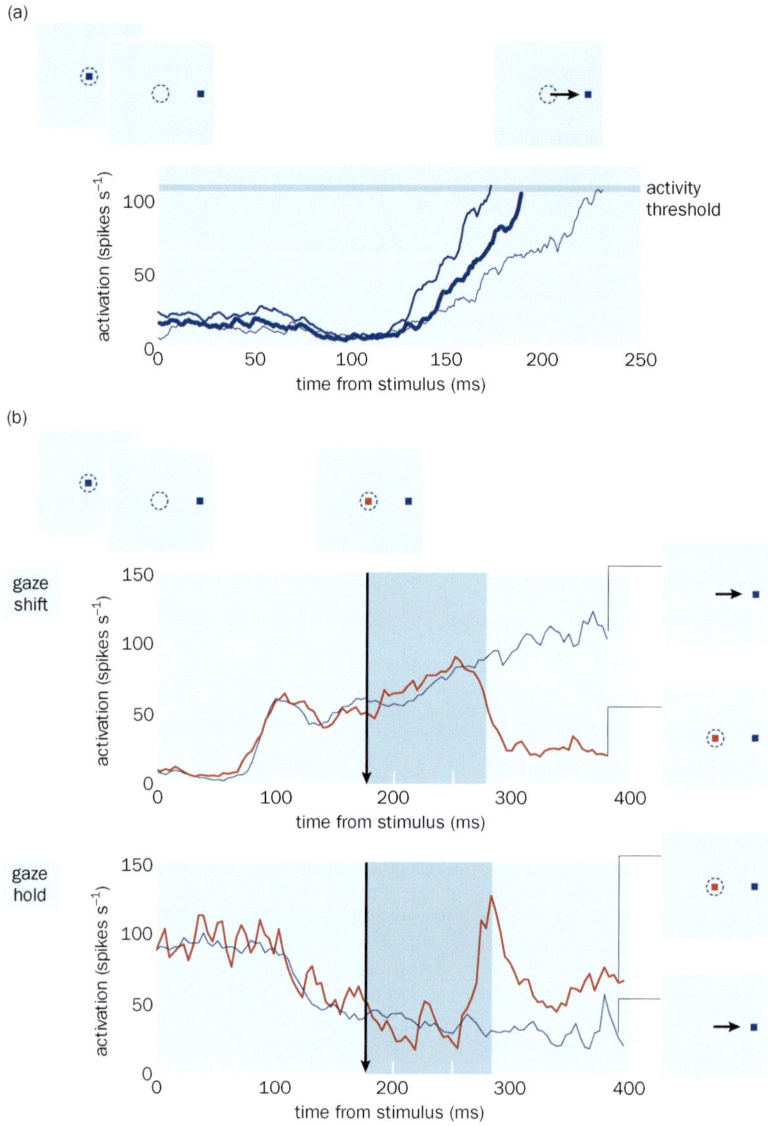

Figure 8.5 Neural control of a purposeful eye movement. (a) Schematics where a response is produced when the activity of movement neurons reaches a fixed threshold. Variability in the time of initiation of the action originates in the variable time taken for the activity to grow to the threshold. The thickness of the line is proportional to the probability of occurrence. Insets show the sequence of visual events. (b) Executive control through cancelling a planned movement. Plots compare the activity of a gaze-shifting neuron (top) and a gaze-holding neuron (bottom) in the FEF of monkeys performing a stop-signal task. Red trace: A signal-inhibit trial; blue trace: A signal-respond trial. The onset of the stop signal is indicated by the solid vertical arrow. The time needed to cancel the planned movement is indicated by the dark blue area. The rapid change in activation of both types of neurons immediately before the movement cancellation time shows that the activation of these neurons is sufficient to control whether an action will occur. Reproduced from Schall (2001). [Schall, J. D. (2001). Neural basis of deciding, choosing and acting. Nature Reviews Neuroscience, 2(1), 33–42.]

stimulus onset (Figure 8.5b, bottom panel). In a signal-inhibit trial, firing activity of gaze-holding neurons surges, whereas gaze-shift neurons decay away after the stop-signal onset (Figure 8.5b, red traces), in contrast to a signal-respond trial where gaze-holding neurons do not ramp up while spike firing of gaze-shifting neurons follows the same temporal course as in go trials (Figure 8.5b, blue traces). These observations differ from the mathematical race model in that the neuronal stop process starts at a high activity level. Moreover, the go and stop processes are not independent from each other; instead, they are mutually inhibitory and modeled by an interactive race model (Boucher et al. 2007; Schall et al. 2017).

8.2.2 A Neural Circuit Model of Countermanding

Consider the decision-making network model (Wang 2002b) discussed in Chapter 6, adapted to simulate an FEF network (Figure 8.6a). Here, selective neural pools encode movement or gaze-shifting (MOV^L and MOV^R) neurons, say, left versus right saccade. A third excitatory pool is added to represent fixation or gaze-holding neurons. All three excitatory neural populations project to an inhibitory neural population, which provides feedback inhibition underlying competition. Furthermore, the fixation

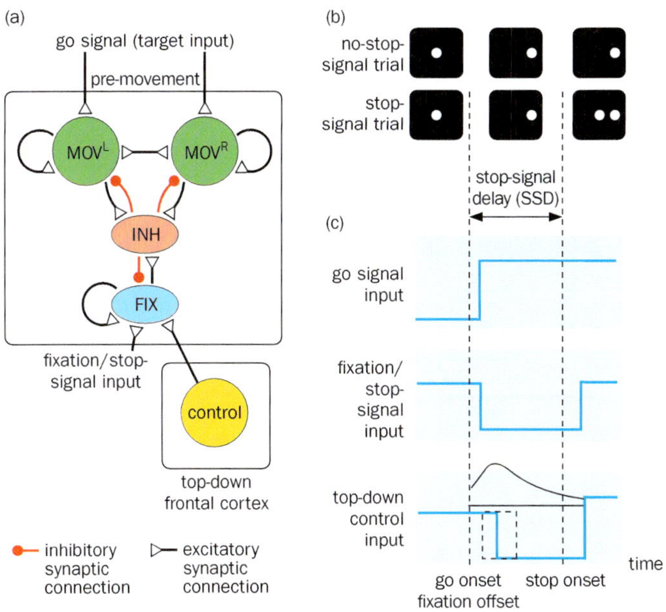

Figure 8.6 A spiking neural network model of inhibitory control for the countermanding task. (a) The schematic model architecture. The pre-movement module consists of two populations of excitatory movement neurons (MOV^L and MOV^R), a population of inhibitory interneurons (INH) and a population of excitatory fixation neurons (FIX). (b) The countermanding task paradigm. In stop-signal trials, after a variable delay following the go signal onset, the central signal is turned back on, serving as the stop signal. The two types of trials are randomly interleaved. (c) Time courses of the inputs in stop-signal trials. The left and right dashed lines mark the go and stop-signal onsets, respectively. After the fixation signal offset, the top-down control remains for a variable interval (holding period) with a probability distribution (insert). The holding period mimics a subject's endogenous tendency to delay the saccade response. Reproduced from Lo et al. (2009). [Lo, C. C., Boucher, L., Paré, M., Schall, J. D., & Wang, X. J. (2009). Proactive inhibitory control and attractor dynamics in countermanding action: A spiking neural circuit model. Journal of Neuroscience, 29(28), 9059–9071. https://www.jneurosci.org/content/29/28/9059.]

neural pool is under the influence of a top-down signaling from other frontal areas. This internal control signal maintains the high tonic activity of fixation neurons.

The model simulation follows the same protocol as in the monkey experiment (Figure 8.6b). The go stimulus sends an input to one of the movement neural pools, and the stop signal activates the fixation pool (Figure 8.6c). The control signal to the fixation pool is internal; upon the go stimulus onset, it decays with a widely variable latency (Figure 8.6c, bottom panel with the distribution of latency shown in the insert). As a result, when there is a temporal gap between the offset of a fixation cue offset and the appearance of a go target, the fixation neural activity drops down but remains significant during the temporal gap (Figure 8.7), reproducing a firing pattern of FEF gaze-holding neurons (Hanes et al. 1998).

The model of spiking neurons fits well with monkey's performance data. The probability of signal-respond increases roughly linearly with SSD (Figure 8.8a). Figure 8.8b shows the cumulative reaction time distributions in go trials without a stop signal and in signal-respond trials with three different SSD values. For a shorter SSD, response is produced in spite of the stop signal in a smaller fraction of trials corresponding to fast RTs.

SSRT can be deduced, both for the monkey data and the model, as follows (Figure 8.9a and b). The cumulative probability of reaction times in stop-respond trials is virtually indistinguishable from that in go (no stop-signal) trials (green) in a range of short RTs, then plateaus at a level that is the signal-respond probability shown in Figure 8.8a. A horizontal line at this probability (blue dashed line) intersects with the cumulative RT probability of go trials (black) at a particular RT value. The difference between this time (red shaded vertical line) and SSD (red solid vertical line) defines SSRT. The estimated SSRT values from the model and monkey data are similar. Thus, there is a principled method to estimate SSRT, in spite of the strong synaptic interactions between movement neurons (the go process) and fixation neurons (the stop process).

The trial-averaged neural spike firing in the model (Figure 8.10) reproduces physiological observations (Figure 8.5b). In stop-signal trials, activity of movement neurons increases and that of fixation neurons decreases, after the go target onset. The activity time course diverges between signal-respond and signal-inhibit trials. When fixation neural

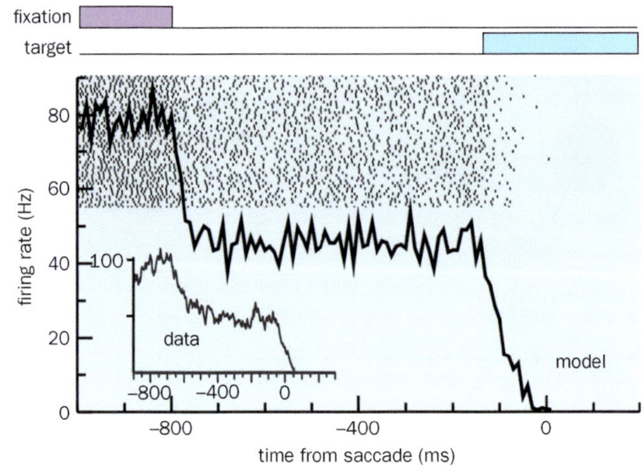

Figure 8.7 Simulated top-down modulation in the absence of sensory input. Spike rastergram (black dots) on top of the average firing rate (black curve) (200 trials) from a FIX neuron in a gap saccade task. Each row in the rastergram represents the spike train from one trial. In the task, a subject has to hold saccade after the offset of the fixation signal until the onset of the target signal, which appears after a gap period. In the proposed model, although the fixation signal is off, the internal top-down control remains during the gap period until the onset of the target, resulting in a lower but persistent activity in fixation neurons during the gap period. The simulated activity resembles the observed activity data of fixation neurons of a behaving monkey, as shown in the insert (Hanes et al. 1998). Reproduced from Lo et al. (2009). [Lo, C. C., Boucher, L., Paré, M., Schall, J. D., & Wang, X. J. (2009). Proactive inhibitory control and attractor dynamics in countermanding action: A spiking neural circuit model. Journal of Neuroscience, 29(28), 9059–9071. https://www.jneurosci.org/content/29/28/9059.]

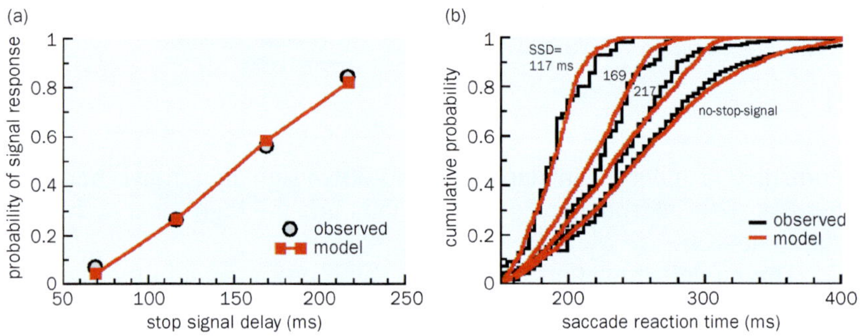

Figure 8.8 The spiking network model quantitatively fits the behavioral data from a rhesus monkey (Boucher et al. 2007). (a) Inhibition function shows the probability of (non-canceled) signal-respond in stop-signal trials as a function of SSD. (b) Cumulative reaction time distributions of go trials (right most) and signal-respond trials for different stop-signal delays (from left to right: SSD = 117, 169 and 217 ms). Reproduced from Lo et al. (2009). [Lo, C. C., Boucher, L., Paré, M., Schall, J. D., & Wang, X. J. (2009). Proactive inhibitory control and attractor dynamics in countermanding action: A spiking neural circuit model. Journal of Neuroscience, 29(28), 9059–9071. https://www.jneurosci.org/content/29/28/9059.]

activity displays a surge in response to the stop signal, whereas movement neurons decay back to baseline (thick solid curve), a response is successfully cancelled. On the other hand, with exactly the same external events of go and stop-signal presentations, movement neurons may keep increasing to a threshold level, whereas fixation neurons fail to build up (thick dashed curve), leading to a signal-respond trial. In the model, the antagonistic behavior of movement and fixation neurons is a result of effective mutual inhibition through a shared inhibitory neural population. The temporal trajectories of the two neural populations are virtually indistinguishable between signal-respond trials and go (thin solid curve) trials, in agreement with the physiological experiment.

As in Chapter 6 (**Figure 8.14b**). where we examined the trajectory in a state space of neural population activity, it is instructive to plot the population firing rates r_{MOV} against r_{FIX} of the movement and fixation neurons in a two-dimensional state space (**Figure 8.11**). The red and black curves are the nullclines for the dynamical variables r_{MOV} and r_{FIX}, respectively; intersection of the two defines a steady-state that can be stable (filled circle) or unstable (grey circle). The network is initially in a stable STOP state with high r_{FIX} and low r_{MOV} (**Figure 8.11b**, phase i, upper left). The onset of a go stimulus creates a new GO attractor state with high r_{MOV} and low r_{FIX}, which coexists with the STOP state (**Figure 8.11b**, phase ii, lower left) thanks to the internal control signal, although r_{FIX} is smaller because of the withdrawal of the external fixation cue. When the internal control signal decays away, the STOP state disappears and the system moves towards the GO state. The upper right panel of **Figure 8.11b** (phase iii) shows the phase portrait during this temporal epoch from the end of the top-down control signal to the start of a stop signal. Also shown are eight single-trial trajectories, on which open circles correspond to the end time point of this epoch. When a stop signal is introduced, ushering the final epoch of a behavioral trial (**Figure 8.11b**, phase iv, lower right), there is again a bistability between a Go attractor and a STOP attractor. Their basins of attraction are indicated by yellow and grey, respectively, separated by the

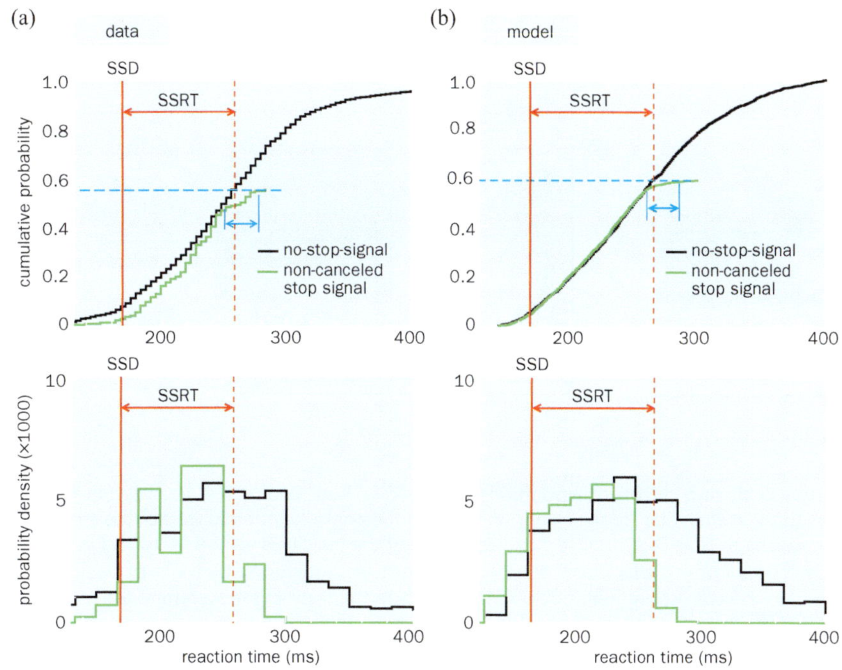

Figure 8.9 Reaction time distributions from a monkey experiment (a) and a neural circuit model (b). To demonstrate how saccade reaction times are matched between go (no-stop-signal) trials and signal-respond (non-canceled stop-signal) trials, we plotted the cumulative reaction time distributions for go trials (black curves) and for stop-respond trials (SSD = 169 ms, green curves) for monkey data (a, top panel) and model (b, top panel). The corresponding probability density functions are shown in the bottom panels. For easy comparison between the two types of trials, the curves of signal-respond trials are rescaled vertically by the probability of signal-respond (blue dashed lines, 56% for the data and 58% for the model). The result shows that for both the model and the data, the reaction times of signal-respond trials roughly correspond to those of go trials for RTs shorter than SSRT, estimated by the intersections between the blue dashed line and the back curve (SSRT = 90 ms for the data and 95 ms for the model). Blue arrows indicate the ranges of STOP process variability. Reproduced from Lo et al. (2009). [Lo, C. C., Boucher, L., Paré, M., Schall, J. D., & Wang, X. J. (2009). Proactive inhibitory control and attractor dynamics in countermanding action: A spiking neural circuit model. Journal of Neuroscience, 29(28), 9059–9071. https://www.jneurosci.org/content/29/28/9059.]

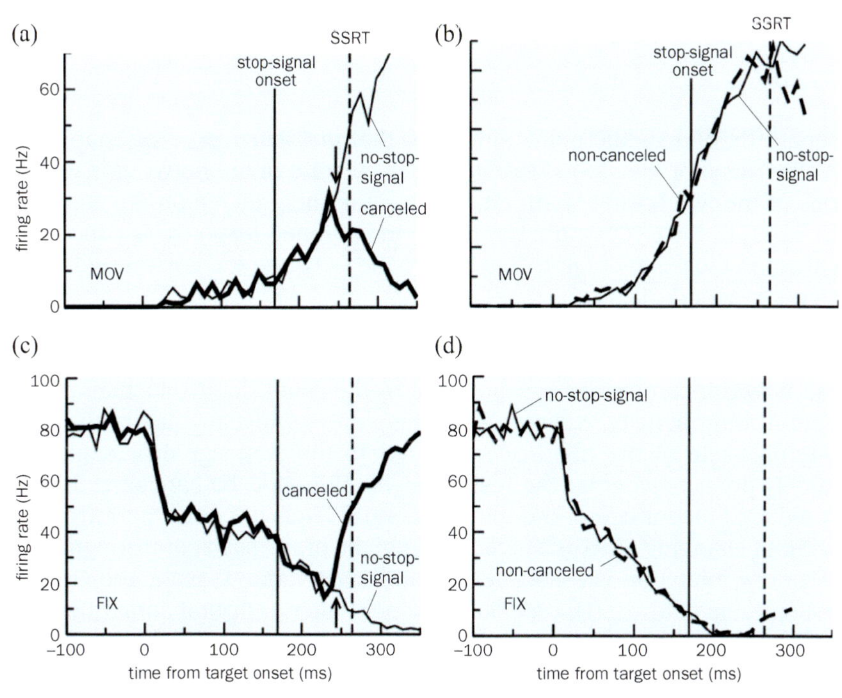

Figure 8.10 Model neural activity patterns in stop-signal trials (SSD = 169 ms) and go trials. (a–b) Mean firing rates of signal-stop (thick solid curve), signal-respond (thick dashed curve) and latency-matched go (thin solid curve) trials from representative MOV neurons in the model for signal-inhibit (a) and signal-respond (b) trials, respectively. (c–d) Same as (a–b), respectively, for FIX neurons. Solid vertical line: Onset of the stop signal; dashed vertical line: Estimated stop-signal response time (SSRT). Arrow indicates the cancel times. Reproduced from Lo et al. (2009). [Lo, C. C., Boucher, L., Paré, M., Schall, J. D., & Wang, X. J. (2009). Proactive inhibitory control and attractor dynamics in countermanding action: A spiking neural circuit model. Journal of Neuroscience, 29(28), 9059–9071. https://www.jneurosci.org/content/29/28/9059.]

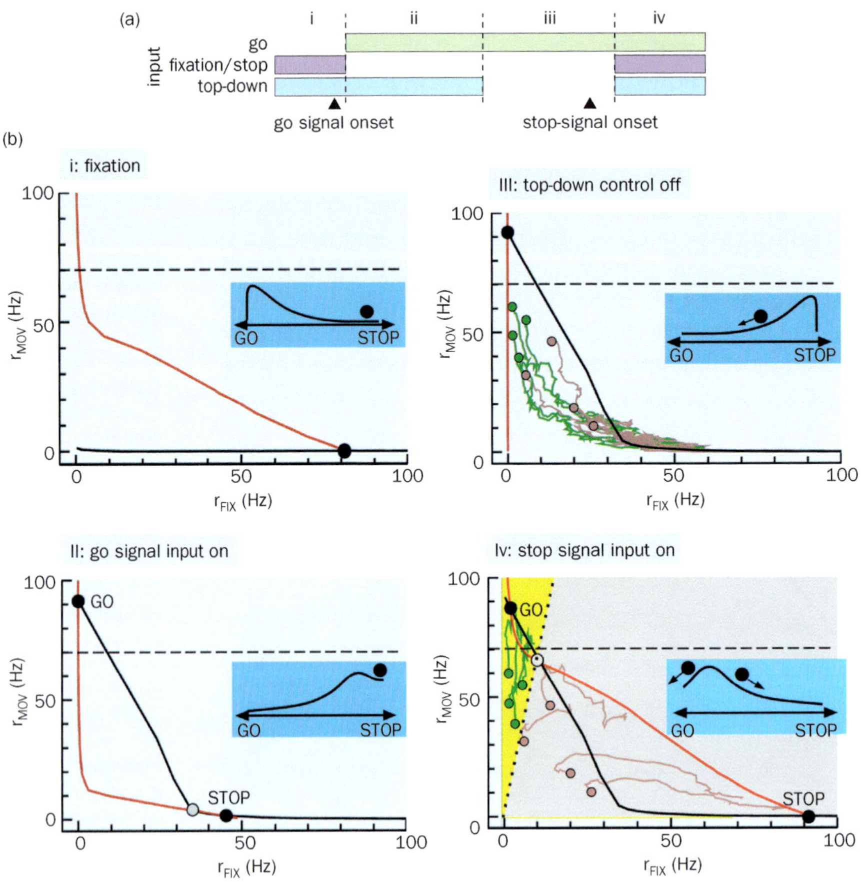

Figure 8.11 Attractor dynamics of the model network during different epochs of the task. (a) Four epochs (i, ii, iii and iv) in a stop-signal trial defined by time windows separated by three events: Start of the go input, offset of the control input and start of the stop input (SSD = 169 ms). (b) Phase plane plot for the four epochs of task. Black and red curves represent the nullclines for the MOV and FIX neurons, respectively. The two nullclines are defined by $dr_{MOV}/dt = 0$ (black) and $dr_{FIX}/dt = 0$ (red). In phases iii and iv (right panels) are shown eight sample trajectories ($r_{MOV}(t)$ versus $r_{FIX}(t)$) in the corresponding epochs. Depending on the state of the system at the moment when the stop input starts (marked by open circles), the network may continue converging into the GO attractor and trigger a response (green), or may turn back into the STOP attractor and cancels the response (brown). Horizontal dashed line: The threshold for the saccade response. Insets: Schematic plots illustrating how the one-dimensional "energy landscape of the action space" changes with stimulus inputs. Reproduced from Lo et al. (2009). [Lo, C. C., Boucher, L., Paré, M., Schall, J. D., & Wang, X. J. (2009). Proactive inhibitory control and attractor dynamics in countermanding action: A spiking neural circuit model. Journal of Neuroscience, 29(28), 9059–9071. https://www.jneurosci.org/content/29/28/9059.]

separatrix of a saddle point, similar to that in **Figure 6.17** of Chapter 6. At the onset of the stop signal, the internal state of the network is within one of the two basins of attraction. Those in the grey region are destined to travel to the STOP attractor, hence the planned response is cancelled. Others in the yellow region converge to the GO attractor, corresponding to signal-respond trials.

In principle, the starting point does not predict the behavior with complete certainty, because if the starting point is close to the separatrix, the network can cross from one side to the other due to fluctuations of neural firing activity. Nevertheless, this analysis reveals clearly that the internal state of the neural circuit prior to the stop-signal presentation largely predicts whether the response will ultimately be cancelled or not. A reduced magnitude of the internal control leads to poorer performance without changing SSRT (Lo et al. 2009). In the brain, such control signal for response inhibition is associated with PFC activation (Sasaki and Gemba 1986; Rubia et al. 2001), especially the right inferior frontal cortex (Broadmann area 44/45) (Aron 2007). In summary, a version of the attractor network model can be quantitatively applied to response inhibition. Unlike a race model with zero baseline of the stop process, suitable for capturing reactive inhibition triggered by a stop stimulus, the neural circuit model distinguishes proactive inhibition (controlled by the high baseline activity of the STOP process) from reactive inhibition (inhibition of the GO process by the STOP process in response to a stop signal). SSRT measures the system's ability to reactively cancel a planned action, whereas internal top-down signaling represents a distinct form of proactive executive control.

8.2.3 Role of Basal Ganglia in "Holding the Horse"

Movement and fixation neurons exist in the superior colliculus (SC) as well as FEF (Munoz and Wurtz 1993, 1995; Paré and Hanes 2003). Indeed, response inhibition in general, stop-signal task in particular, engages multiple brain regions (Li 2020). The basal ganglia (BG) is of special interest because of its central role in controlling voluntary movements (Hikosaka et al. 2000). The BG is already present in vertebrates such as reptiles that do not have a neocortex (Grillner and Robertson 2016). Recall that in mammals the cortex projects to the striatum (Str), the input structure of the BG, where the majority of neurons are GABAergic medium spiny neurons (MSNs). The substantia nigra pars reticulata (SNr) and the internal segment of globus pallidus (GPi) are BG output stations with GABAergic neurons. SNr/GPi project to the SC, as well as thalamic nuclei (TH), which in turn connect to the frontal areas, including premotor and motor cortices. From the input to the output, there are three major BG pathways (**Figure 8.12a**) (Alexander and Crutcher 1990; Hikosaka et al. 2000). The direct pathway from the Str to the SNr/GPi is also called the Go pathway, because cortical activation of the Str suppresses projection neurons in the SNr/GPi, thereby releasing the SC and TH from tonic inhibition by the basal ganglia (**Figure 6.24** in Chapter 6). In the indirect pathway, the Str projects to the external segment of the globus pallidus (GPe), which also consists of GABAergic principal neurons projecting to the SNr/GPi. Therefore, through a chain of three GABAergic synapses, the net impact onto the SC and TH is enhanced inhibition, and the indirect pathway is also called the NoGo pathway. MSNs in the Str that mediate the direct and indirect pathways are distinct cell populations that are modulated by dopamine in an opposite way. Those

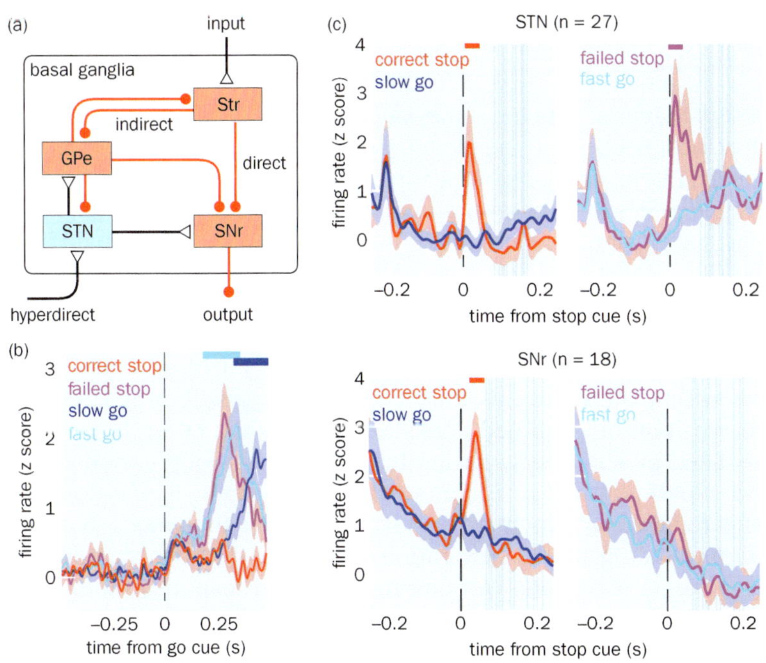

Figure 8.12 The STN in stop-signal task. (a) Simplified scheme of neural circuitry in the basal ganglia during the stop-signal task. Projections from striatum (Str) and STN converge on SNr, which provides tonic inhibition of motor output. GPe, globus pallidus (the external segment). (b) Mean firing rate z-score for recorded striatal neurons. Cyan bar at the top indicates times with significantly different firing rates on fast versus slow Go trials, and blue bar indicates the same for slow go versus signal-inhibit trials. (c) A stop cue increases firing in STN (top) before SNr (bottom). Firing rate time courses for the neuronal subpopulations that distinguish stop-signal trials (red) from go trials (blue). Horizontal colored bars at the top of each panel indicate times with significantly different firing rates in go versus stop-signal trials. Vertical light lines show SSRTs deduced separately for each experimental session. Reproduced from Schmidt et al. (2013). [Wei, W., & Wang, X. J. (2016). Inhibitory control in the cortico-basal ganglia-thalamocortical circuit: Complex modulation and its interplay with working memory and decision-making. Neuron, 92(5), 1093.] [Schmidt, R., Leventhal, D. K., Mallet, N., Chen, F., & Berke, J. D. (2013). Canceling actions involves a race between basal ganglia pathways. Nature Neuroscience, 16(8), 1118–1124.]

driving the direct pathway express D1 receptor and are depolarized by dopamine, whereas those at the origin of the indirect pathway express D2 receptor and are hyperpolarized by dopamine (Kravitz et al. 2010). Experiments (Cui et al. 2013; Markowitz et al. 2018) and computational modeling (Wei et al. 2015) found that a balance between the direct and indirect pathways must be maintained for proper basal ganglia function in cognition.

The third major BG route is called the hyperdirect pathway, with the subthalamic nucleus (STN) at its core (**Figure 8.12a**). The STN is part of the BG with glutamatergic principal neurons; they receive inhibitory inputs from the GPe and send excitatory output to the SNr/GPi, as well as back to the GPe. Importantly, the medial frontal cortex provides rapid and potent excitation to the STN. Cortical stimulation produced spiking activity with the shortest latency in the STN ($\simeq$ 5–6 ms) compared to the SNr/GPi and GPe (Nambu et al. 2000). This and other lines of evidence suggest that cortical activation of the STN instantiates a mechanism for a surge of spike firing in the BG output, thereby "holding your horse" (Frank et al. 2007; Isoda and Hikosaka 2007) to suppress an imminent response so that an alternative action has a chance to be produced instead. If so, the STN should play a significant role in a stop-signal task.

Schmidt et al. (2013) trained rats to perform a stop-signal task with two different auditory stimuli signaling go and stop, and the behavioral response was turning nose to a food port. Single-neuron recording revealed that Str neurons show early ramping activity in signal-respond trials but not in signal-inhibit trials, and the activity is indistinguishable between slow go and failed stop (signal-respond) trials (**Figure 8.12b**). STN neurons display a burst of spike firing upon the stop-signal onset with minimal latency in stop-signal trials (**Figure 8.12c**, upper panel). Neurons in SNr also show a transient increase of activity with a longer time-to-peak than the STN, but only in stop-inhibit trials, not in signal-respond trials (**Figure 8.12c**, lower panel). Taken together, these observations support the notion that the stop process (mediated in part by the STN) is stereotypical for all stop-signal trials, whereas the variability of the go process determines whether a planned response is ultimately cancelled.

Wei and Wang (2016) extended the spiking neuron model of Lo and Wang (2006) to include all three pathways of the basal ganglia as well as the thalamus. The model system has three feedback loops: the intra-cortical recurrent connections as in the standard decision-making model (Wang 2002b); the thalamocortical loop; and the cortico-basal ganglia-thalamocortical loop. The model reproduces salient findings from Schmidt et al. (2013), as shown in **Figure 8.13a**, and SSRT can be extracted from the model's behavioral data the same way as in **Figure 8.9**, in spite of multiple feedback connections in the model. Unexpectedly, the model shows that a surge of STN not only enhances firing in the SNr but also curtails ramping activity in MSNs through excitation of the GPe, which in turn inhibits the Str, via a newly found feedback pathway mediated by the arkypallidal cells in the GPe (Mallet et al. 2016; Aristieta et al. 2021). When the strength of this connection is reduced in the model, the probability of non-cancelled response in stop-signal trials is increased and SSRT is longer.

This study investigated the scenario that intrinsic recurrent excitation within the cortex module does not have sufficient strength for the generation of sustained persistent activity; the latter, however, can be realized by sufficiently strong thalamocortical loop connections, as found experimentally in rodents (Guo et al. 2017). It was found that SSRT is reduced with an increased strength of the thalamocortical reverberation, leading to the generation of self-sustained persistent activity (**Figure 8.13b**); thus working memory and response inhibition are correlated with each other. This model finding is interesting in light of empirical evidence that the brain's substrates for working memory and

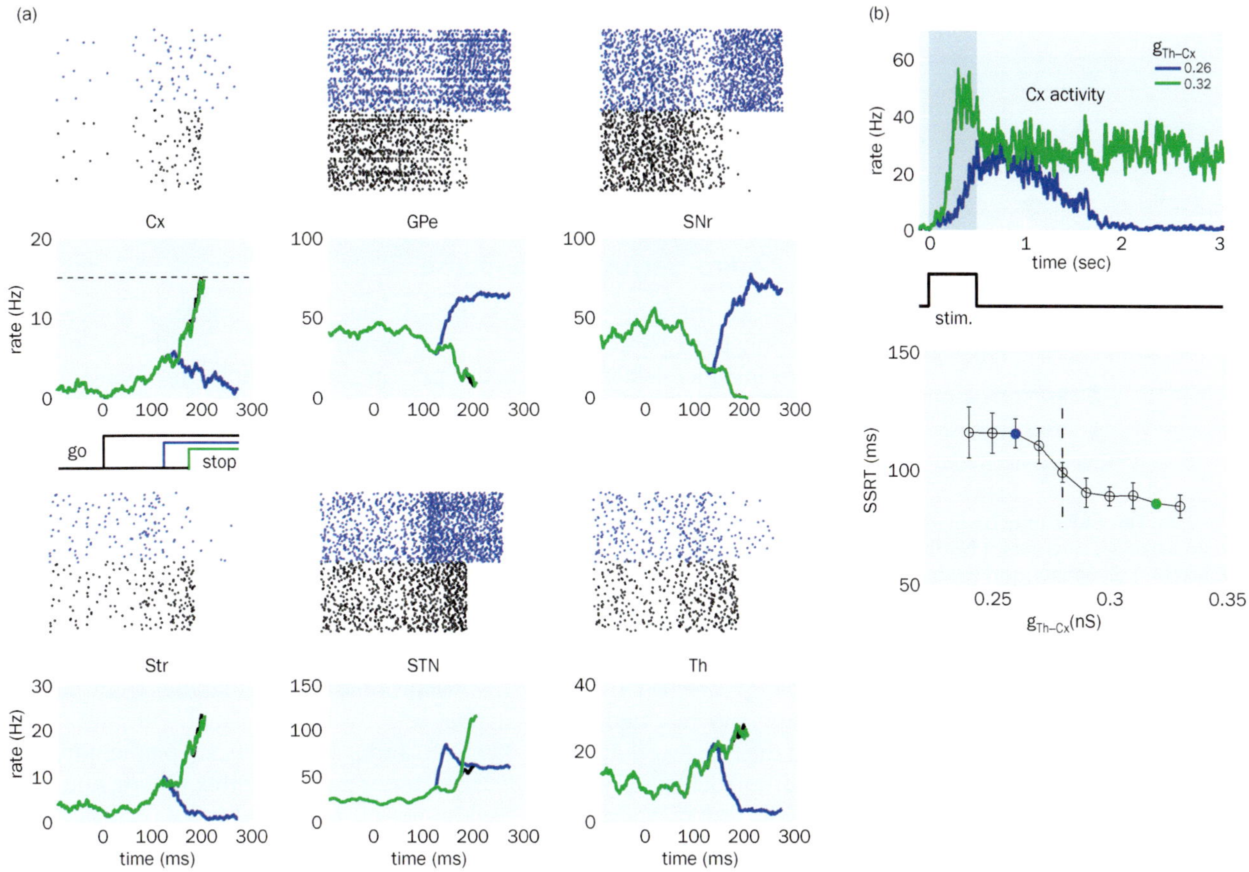

Figure 8.13 Spiking neuron model of a cortico-basal ganglia-thalamocortical loop. (a) In a model simulation of a stop-signal task, spike raster plots are shown for a go trial (black) and a signal-inhibit trial (blue) in each area. Population firing rates in each area are shown for signal-inhibit (blue) and signal-respond (green) trials, with SSD of 120 ms (blue) and 170 ms (green), respectively. Also shown is a go trial (black), which is essentially invisible because network dynamics is virtually the same as in the signal-respond trial (green). (b) Upper panel: Cortical activity after stimulus offset is persistent with sufficiently strong thalamocortical connection. Lower panel: SSRT displays a step decrease as the thalamocortical connection strength is increased above a threshold level (vertical dotted line) for the generation of self-sustained persistent activity. The blue and green dots correspond to the connection values used in model simulations of the upper panel (blue and green traces, respectively). Reproduced from Wei and Wang (2016). [Wei, W., & Wang, X. J. (2016). Inhibitory control in the cortico-basal ganglia-thalamocortical circuit: Complex modulation and its interplay with working memory and decision-making. Neuron, 92(5), 1093.]

response control are shared (Machizawa and Driver 2011; Gazzaley and Nobre 2012), even though they are not identical (Schecklmann et al. 2013). It could explain why subjects with attention deficit hyperactivity disorder (ADHD) exhibit both working memory deficits and impulsivity (Barkley 1997; Castellanos and Tannock 2002).

Since this system also underlies decision making, as discussed in Chapter 6, the same model was used to simulate an experiment where the RDM discrimination task was combined with a stop signal in a small fraction of trials (Middlebrooks and Schall 2014). Interestingly, when the motion coherence was varied in stop-signal trials, the SSRT is independent of the task difficulty, as found on human and monkey behaviors. Another monkey experiment combined a value-based choice task with a stop-signal task, and found that neural populations in the orbitofrontal cortex were correlated with value encoding and response inhibition in orthogonal subspaces (Balasubramani et al. 2020). Functionally independent signals for decision and executive control may be realized by the fact that the latter but not the former relies on fixation (NoGo) neurons in the motor cortical areas as well as STN-mediated response inhibition in the basal ganglia. Therefore, a single network model captures three

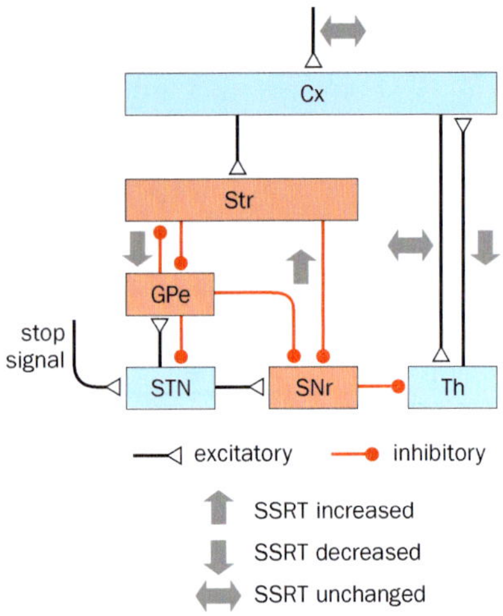

Figure 8.14 The SSRT may be increased (wide up arrows), decreased (wide down arrows) or unchanged (wide horizontal arrows) by an increase of the connection strength from one area to another area. The SSRT is independent of the task difficulty which is varied by changing the input to the Cx. Reproduced from Wei and Wang (2016). [Wei, W., & Wang, X. J. (2016). Inhibitory control in the cortico-basal ganglia-thalamocortical circuit: Complex modulation and its interplay with working memory and decision-making. Neuron, 92(5), 1093.]

basic cognitive processes: working memory, decision making and inhibitory control.

Whereas SSRT can be extracted from behavioral data as proposed by the independent race model (Logan and Cowan 1984), it is modifiable in the recurrent system by varying the strengths of distinct neural circuit pathways (**Figure 8.14**). These modeling results suggest multiple ways to modulate response inhibition at the service of behavioral flexibility.

8.2.4 Pro- versus Anti-Response

Besides the stop-signal task, pro- versus anti-response is another commonly used paradigm for response inhibition (Munoz and Everling 2004). For instance, when an attractive person suddenly appears, your gaze naturally shifts to that person (pro-saccade). Under certain contexts, however, this prepotent response may be inappropriate socially, in which case you should instead opt to look in the opposite direction (anti-saccade). The pro- versus anti-saccade (**Figure 8.15a**) (Hallett 1978) has been widely used in clinical studies of impaired response inhibition, which showed that subjects with ADHD, obsessive-compulsive disorder and schizophrenia make more mistakes in the anti-saccade task compared to healthy subjects (Everling and Fischer 1998; Damilou et al. 2016). Similar tasks include body orientation towards (pro-response) or away from (anti-response) a visual stimulus by rats (Duan et al. 2015). Neurophysiological studies with behaving monkeys reported neural signals selectively associated with anti-saccade in the FEF (Everling et al. 1999; Sato and Schall 2003), supplementary motor cortex (Schlag-Rey et al. 1997), posterior parietal cortex (Zhang and Barash 2000), superior colliculus (Everling et al. 1999) and basal ganglia (Watanabe and Munoz 2009). In other words, a largely overlapping brain system is involved in the stop-signal and anti-saccade tasks.

Figure 8.15b shows activity patterns of a movement neuron (MOV) and a fixation neuron (FIX) in each hemisphere of the FEF in monkeys performing a pro- and anti-saccade task. FEF MOVs are selective for

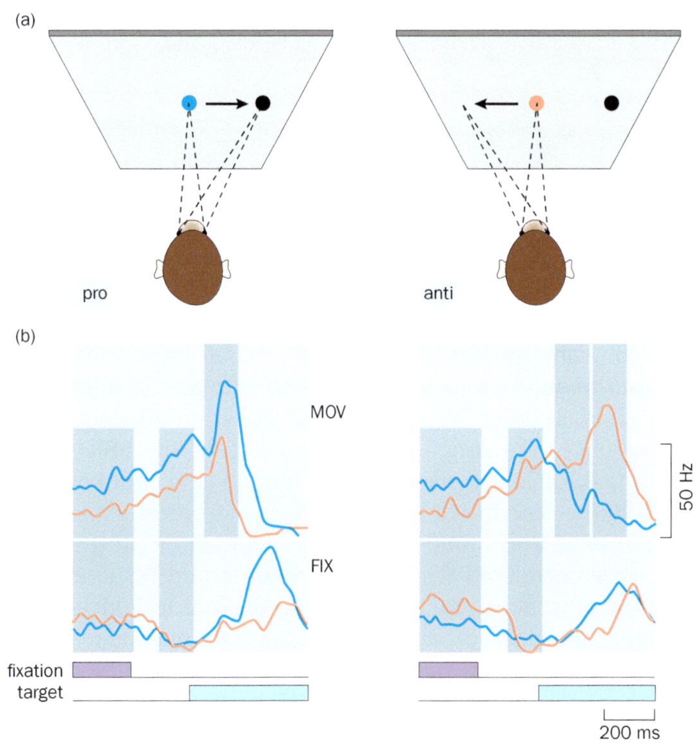

Figure 8.15 Neural activity in monkey frontal eye field (FEF) in pro- versus anti-saccade. (a) In a pro- and anti-saccade task, the color (blue or orange) of a central fixation marker instructs the subject to generate either a pro-saccade (left) or an anti-saccade (right). (b) In this version of the task, target is shown after a gap in time of the fixation offset. Activity patterns in response to a right target in FEF neurons selective for right (left panel) and left (right panel) saccade. MOV: Movement neuron; FIX: Fixation neuron. Cyan: Pro-saccade trials; orange: Anti-saccade trials. Traces are aligned on target appearance. In an anti-saccade trial, movement neurons on the left panel must be inhibited and those on the right panel activated to drive the leftward anti-saccade. Before target appearance, fixation neurons are more active on anti-saccade trials, while movement neurons are more active on pro-saccade trials. Reproduced from Munoz and Everling (2004). [Munoz, D. P., & Everling, S. (2004). Look away: The anti-saccade task and the voluntary control of eye movement. Nature Reviews Neuroscience, 5(3), 218–228.]

contralateral saccades. Therefore, in a pro-saccade trial (blue), a right stimulus produces a burst of firing in the left MOV and a reduced activity in the right MOV. In a correct anti-saccade trial (orange), where the ultimate response is leftward, the opposite occurs. There is still a brief increase of activity at target onset in the left MOV, but the right MOV exhibits a robust ramping activity to a higher level, leading to the actual motor response. Notably, prior to the target stimulus in anti-saccade compared with pro-saccade trial, FIXs show a higher activity during the holding period, mirroring a lower anticipatory activity in MOVs, which presumably reflects an enhanced internal control signal in anti-saccade trials.

Computational modeling of pro- and anti-saccade tasks typically assumes a network with two (pro- and anti-) mappings from a sensory target to a motor response (Salinas 2004; Heinzle et al. 2007; Wiecki and Frank 2013), and vector inversion (from pro- to anti-saccade) is learnt as an arbitrary sensorimotor association (Passingham 1995; Wise and Murray 2000; Fusi et al. 2007). Saccade is a prepotent response. In anti-saccade trials, there is a competition between a fast automatic response pathway and a slower voluntary response pathway. A top-down control signal influences which one wins the competition, conceptually similar to a model of the Stroop task (Cohen et al. 1990). An error in an anti-saccade trial (a saccade toward rather than away from the target) is attributed to a failure to suppress the automatic response, and therefore is inevitably associated with a short reaction time. However, experiments with the anti-saccade task have shown diversity in erroneous responses; there are fast errors associated with express saccade, but slow errors with reaction times comparable to those of correct anti-saccade were also observed (Everling et al. 1999).

Figure 8.16a depicts a spiking neural network for the pro- versus anti-saccade behavior. In the model, a sensory stimulus sends inputs to an action selection module and to a remapping module. The remapping module is a decision network (Wang 2002b) driven by a direct input (A to A, B to B) and an inverted input (A to B, B to A), and its output projects to the action selection module, which is under the influence of both the direct module and the remapping module. It is described by a model similar to that of the stop-signal task with MOVs and FIXs (Lo et al. 2009) and ultimately responsible for saccade generation. An internal control signal (see the prior section) is required in anti-saccade trials to both strengthen the input to the remapping module and enhance fixation neurons in the action selection module. This control signal is assumed to fluctuate across a range of levels on a trial-by-trial basis.

The model reproduces salient physiological observations in pro- and anti-saccade tasks (**Figure 8.16b–c**). As in the monkey experiment, the model makes two types of erroneous responses in anti-saccade: "slow error" and "fast error." In some anti-saccade trials, stochastic control is weak and cannot efficiently suppress the response of the saccade neurons (Sac_L) to the left target. Therefore, a quick pro-saccade is triggered even before decision neurons (Dec_R and Dec_L) in the remapping circuit reach a high firing level (**Figure 8.16c**, left). In other anti-saccade trials, the system exhibits a strong gaze-holding control and successfully suppresses the direct pathway. However, the stochastic dynamics of the remapping circuit produces a higher activity of Dec_L neurons than Dec_R, leading an erroneous response (black versus grey in **Figure 8.16c**, right). The activity time courses of neural activity in the three modules are similar to those of a correct anti-saccade trial (**Figure 8.16b**), except that the wrong response (black) rather than the correct (grey) response is generated. The reaction times in this type of error trials are comparable to those of correct anti-saccade trials because in both cases the system goes through the neural competition in the decision layer. The probability of a fast or slow error is

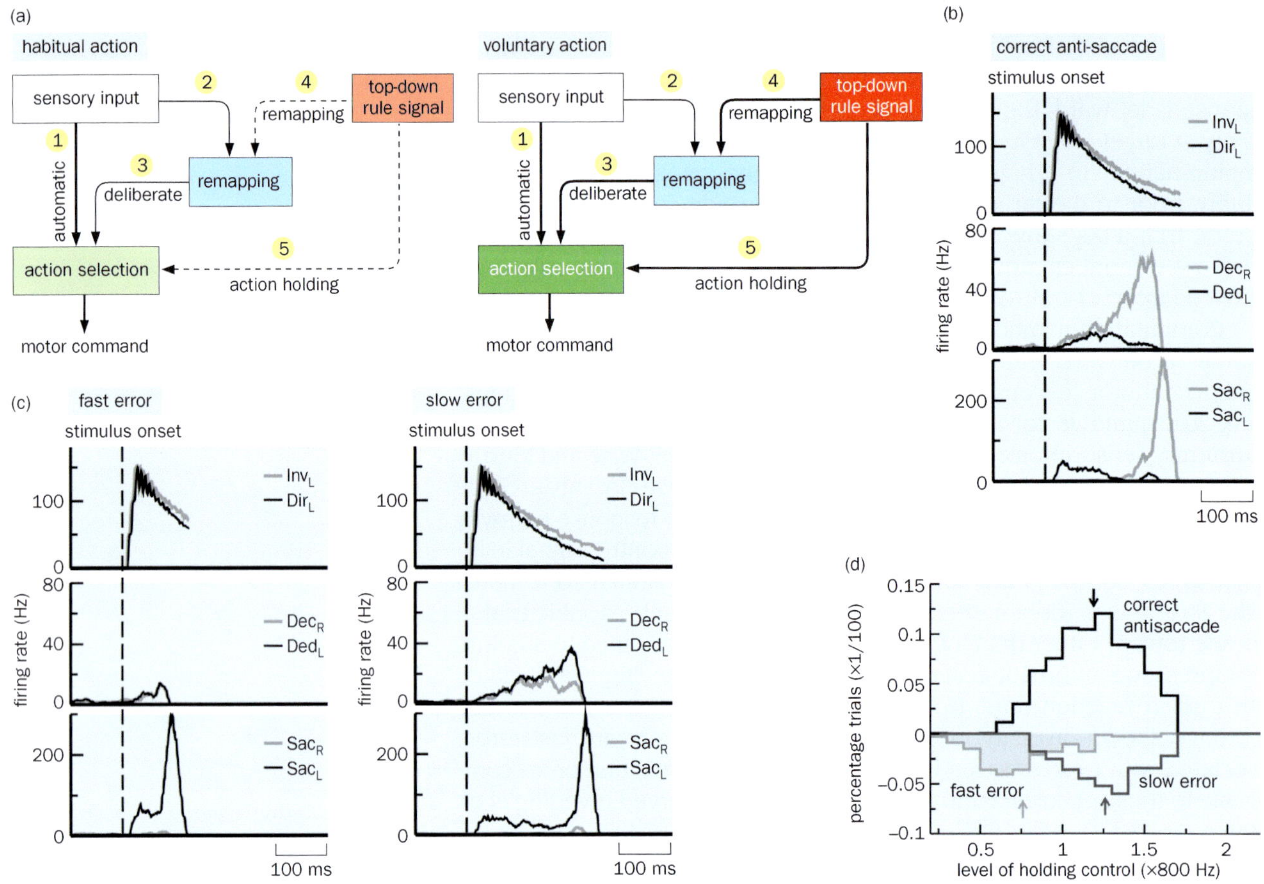

Figure 8.16 A spiking neural network model of pro- and anti-saccade. (a) The model consists of an action selection and a remapping module. (b–c) Three anti-saccade trials with a left target, where direct and inverted inputs are similar (top panels). (b) In a trial leading to a correct response, in the remapping module (middle) the neural pool selective for right (grey) wins and that selective for left (black) loses the competition, resulting in correct anti-saccade produced by the action selection module (bottom). (c) Same as (b) in two anti-saccade trials, where the response is a fast or slow error. A fast error (left) happens when the direct module prevails and the remapping module does not have a chance to act, whereas a slow error (right) is caused by a wrong outcome of deliberation in the decision circuit of the remapping module. (d) The probabilities of a correct response (black), a fast error (light grey) and a slow error (dark grey) as a function of the strength of the internal executive control. Arrows indicate the mean of the corresponding distributions. Fast errors are associated with weaker control levels. Reproduced from Lo and Wang (2016). [Lo, C. C., & Wang, X. J. (2016). Conflict resolution as near-threshold decision-making: A spiking neural circuit model with two-stage competition for antisaccadic task. PLoS Computational Biology, 12(8), e1005081. Licensed under CC BY 4.0.]

correlated with the strength of internal control in a given trial, as shown in **Figure 8.16d**.

In the model, a successful anti-saccade requires that voluntary response (driven by top-down control) wins two competitions: the slower remapping pathway prevails over the direct pathway, and a deliberative decision circuit produces a correct winner. These explain two types of erroneous pro-saccades made in anti-saccade trials. Slow errors from the decision module happens only when the relative difference in the inputs to the two competing neural pools of the remapping module is zero or very small (Wang 2002b) (**Figure 6.16** of Chapter 6). This finding reveals a common mechanism for near-threshold decision-making in perceptual decisions and conflict resolution in executive control of voluntary actions.

8.3 TIMING

Timing is important generally for automated responses as well as controlled behaviors. Anticipation of timing of a sensory event or a motor response requires an internal process in the brain that can start at an arbitrary time and stop after an appropriate time interval. Behavioral studies of interval timing have shown a phenomenon called scalar timing, namely, time estimation is variable with a standard deviation proportional to the mean (Weber's law) (Gibbon 1977; Gibbon et al. 1997). Scalar timing has been observed across a wide range of timescales in different animal species (Buhusi and Meck 2005). Various mathematical models have been proposed for scalar timing (Merchant et al. 2013; Paton and Buonomano 2018; Issa et al. 2020). In one class of models (Treisman 1963; Killeen and Fetterman 1988; Miall 1989), elapsed time is represented by accumulation of input pulses, which can be stochastic or oscillatory. Another class is ramping-to-threshold models, which must satisfy certain conditions to yield scale-invariance of response times (Simen et al. 2011) (Chapter 6, Section 6.4.2). In such models, if the time interval is doubled, the ramping rate should be halved. This prediction is supported experimentally. For instance, in a rat experiment where a cue predicts a reward after a time delay, neurons in the posterior thalamus display linear ramping to a fixed level over time, with a smaller slope when the delay is longer (Komura et al. 2001). In the monkey delayed vibrotactile discrimination experiment, neurons in the PFC showed ramping activity with a shallower slope when the mnemonic delay was prolonged from 3 seconds to 6 seconds (Brody et al. 2003). Furthermore, human fMRI (Hinton and Meck 2004) and lesion in rats (Meck 2006) demonstrated that the striatum plays a major role in interval timing.

Neural activity patterns correlated with interval timing may also be more complex than simple ramping, better described by trajectories in a state space of neural population activity (Buonomano and Laje 2010; Remington et al. 2018; Goudar and Buonomano 2018). In a mouse experiment (Mello et al. 2015), subjects pressed a lever in order to collect a reward, but only when the action takes place after a fixed interval (FI) since the previous reward delivery. FI was 12, 24, 36, 48 or 60 seconds in different blocks of trials. Firing activity of individual neurons in the striatum exhibited a bump-shaped time course, with an onset latency and peak time that increased with FI. When shown as a function of time divided by FI, the population activity pattern was remarkably similar, demonstrating that the neural population dynamics can compress or stretch in time by virtue of a scaling.

In a monkey experiment, a cue instructed the subjects to make a motor response either after 800 ms or 1500 ms; a second cue indicated whether the response ought to be a saccade or a hand button press (Wang et al. 2018a). This cued task was designed such that timing needed to be flexibly varied on a trial-by-trial basis instead of being fixed in a block of trials. This flexibility on the fly is characteristic of controlled behavior and requires, metaphorically, a "stopwatch" in the brain, which can start and end at will. The appearance of a brief target signaled the "Set" signal, and the subject generated an appropriate response after the estimated time interval. Monkeys performed well on the task, with a higher variability of response time in long trials than short trials, as expected by Weber's law. Neurons were recorded in the medial frontal cortex (MFC), including the supplementary motor area, caudate and thalamic nuclei that project to the MFC. As shown in **Figure 8.17a**, time courses of spike firing of MFC cells are categorically different in the two types of trials and display wide heterogeneity across individual neurons. In the state space defined by PCAs of MFC neural population, trajectories are distinct in long versus short trials, but appear to start and end at the same points (**Figure 8.17b**, left panel). A new dimensionality reduction analysis, designed to maximize scaling (instead of maximizing variance by PCA) identified scaling components (SCs) that were ranked by the degree of scaling in the data. Scaling was found to be nearly perfect in a low-dimensional subspace of

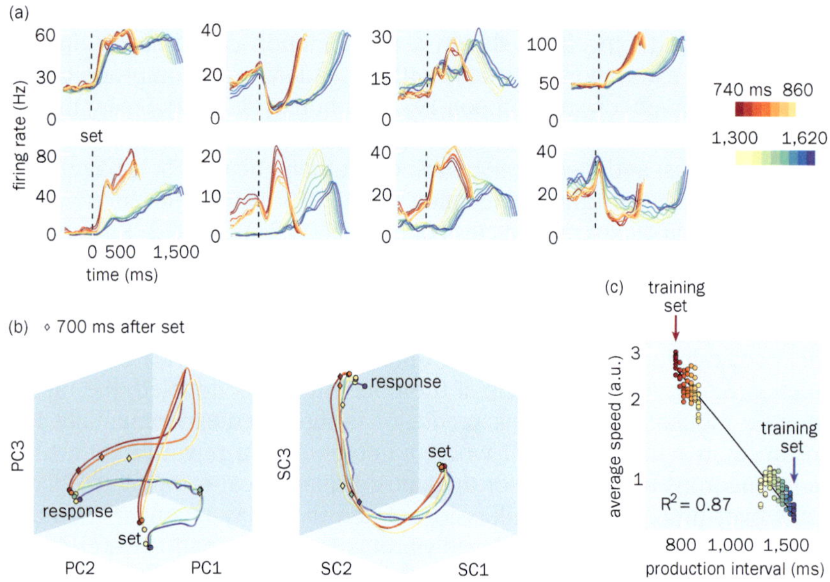

Figure 8.17 MFC neural activity recorded from monkeys performing a timing task. (a) Activity temporal profiles of eight example neurons in blocks of trials when the report was hand button press (top row) or saccade (bottom row). Different colors correspond to binned production interval (Tp) values in short trials (near orange) and long trials (greenish). (b) Left panel: Population activity projected onto the first three principal components (PCs) from the start (Set) of time estimation to the time of button press (Response). Activity profiles associated with different produced intervals are plotted in different colors. Right panel: Population activity sorted according to produced interval (Tp) bins and projected onto the first three scaling components (SCs). In this subspace, the trajectories overlap. (c) The speed of neural trajectory within the scaling subspace spanned by the first three SCs predicted average Tp values across bins. It is smaller with a longer time interval. Reproduced from Wang et al. (2018a). [Wang, J., Narain, D., Hosseini, E. A., & Jazayeri, M. (2018). Flexible timing by temporal scaling of cortical responses. Nature Neuroscience, 21(1), 102–110.]

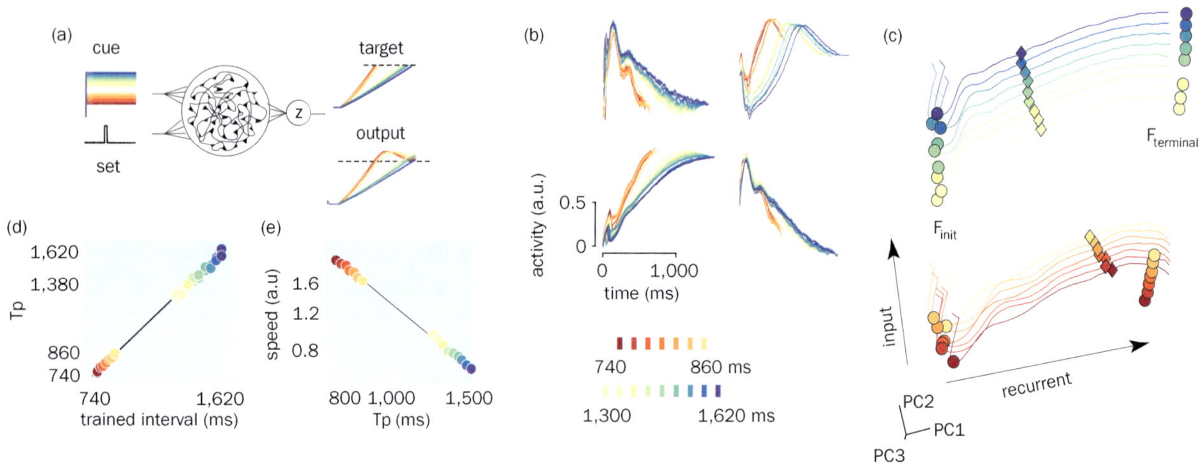

Figure 8.18 Trained recurrent neural network (RNN) for the timing task of Figure 8.17. (a) A model receives an input (Cue) whose strength depends on the desired interval (different colors) and a transient Set pulse that initiates the timing interval. The model produces a response when its output (Z) reaches a fixed threshold (dashed horizontal line). Network was trained to produce a linear ramp at its output (right plots: Top: Target output; bottom: Actual output after training). (b) Response time courses of four randomly selected recurrent units of the model, aligned to the time of Set. (c) Network activity projected onto the first three principal components (PCs) across all trials. Different traces correspond to trials with different durations (red for shortest to blue for longest). In the recurrent subspace, trajectories traverse the same path at different speeds. (d) After training, the network accurately produced the intervals according to the presented Cue input. Reproduced from Wang et al. (2018a.). [Wang, J., Narain, D., Hosseini, E. A., & Jazayeri, M. (2018). Flexible timing by temporal scaling of cortical responses. Nature Neuroscience, 21(1), 102–110.]

SCs (**Figure 8.17b**, right panel), with the average speed that is higher in short trials than long trials (**Figure 8.17c**). Neural activity patterns in the caudate displayed similar time scaling as MFC, as well as neurons in the thalamus, albeit to a less degree.

A recurrent neural network (RNN) model was built for this timing task (**Figure 8.18a**). Using supervised learning (Chapter 4, Section 4.2), the network was trained to produce an output that ramps linearly with time.

The ramping slope takes one of two values for short versus long trials, respectively, such that on average the output reaches a fixed threshold at the desired production interval (800 ms or 1500 ms). After training, recurrent units display distinct time courses in short versus long trials (**Figure 8.18b**), and the trajectory in the state space displays scaling of time in a particular direction (**Figure 8.18c**) that naturally emerges from learning. After a rotation of the state space, it becomes clear that in the recurrent subspace, trajectories traverse the same path at different speeds in short and long trials. The interval estimation is accurate (**Figure 8.18d**, left), and the speed of network dynamics is smaller in long trials than short ones (**Figure 8.18d**, right). Thus, this model reproduces salient observations from the monkey experiment.

Among open issues are: (1) Is there a "master" timing system for information processing and behavioral control across sensory and motor modalities or, alternatively, is timing intrinsic in individual subnetworks or modules dedicated to different functions (Hinton and Meck 2004; Meck 2006; Jin et al. 2009; Merchant et al. 2013; Emmons et al. 2017; Issa et al. 2020)? (2) What is the biological mechanism of slow neural transients on the timescale of more than one second? (3) Computationally, if the desired output is linear ramping, what are the advantages of using complex trajectories of neural population activity in the state space? (4) What circuit dynamics underlies Weber's law (Chapter 6, Section 6.4.2)?

8.4 SELECTIVE ATTENTION

Executive control of sensory information processing refers to the brain's attention to a part of the external world in order to achieve a behavioral goal, such as finding your friend (visual search) or focusing on a single conversation amid chatters (auditory cocktail party effect) at a crowded reception. Here we will discuss such endogenous attention, which should be distinguished from exogenous attention like pop-out of a red square among black ones. Selective attention is most vividly demonstrated by inattentional blindness, the phenomenon of not seeing things as if they did not exist because they are not at the focus of our attention. In a famous study (Simons and Chabris 1999), a subject was shown a video in which three people wearing white passed a basketball among themselves, and three wearing black did the same with another basketball; the two groups moved around and were interspersed in space. The subject focused attention on counting the number of passes in the group wearing white. At the end of the video clip, the participant gave the answer about the number of passes, and then was asked: "Did you see the gorilla?" Many participants reported no. Knowing now that there was a "gorilla" (selective attention is directed to it), watching the clip again they were surprised to see, halfway through the video, a big person in a black gorilla costume thumbing their chest right in front of the camera. The "invisible gorilla" makes it evident how little we notice in our environment outside of our attention focus.

Selective attention is not a bug but essential for our organization of behavior, as articulated by William James in the quote at the start of this chapter (James 1890). Cognitive and neuroscientific studies of selective attention range from visual search, to sustained attention over time, to spatial neglect (Nobre and Kastner 2014). Mathematical models of selective attention began to be built in the 1990s in behavioral psychology and increasingly in neuroscience (Bundesen 1990; Wolfe 1994; Olshausen et al. 1993; Itti and Koch 2001; Bundesen et al. 2005). Here we shall focus on neural network modeling of selective attention.

8.4.1 Biased Competition and Multiplicative Gain Modulation

As illustrated by visual search and inattentional blindness, vision represents a sensory modality that is especially suited for studying selective attention. Indeed, starting with the pioneering work of Moran and Desimone (1985), neurophysiological research with behaving monkeys has been an active area of neurobiology of cognition (Desimone and Duncan 1995; Bisley and Goldberg 2003; Reynolds and Chelazzi 2004; Maunsell 2015; Moore and Zirnsak 2017). Selective attention may be overt by orienting the gaze on a particular location or an object, and the brain system for overt attention is thus connected with the motor system of heading and saccades. Attention may also be covert when the item of interest is in the periphery of vision but the focus of the mind. Many experiments on attentional influence of perception use covert attention paradigms, which presumably are dissociated from the motor system.

In a typical experiment, monkeys performed a discrimination on a target stimulus (the behavioral goal) at a particular location (the focus of spatial attention). A neuron is recorded from visual areas V1, V2, V4 or MT. The receptive field (RF) of the neuron is determined, as well as a preferred stimulus (triggering a high response, r_{high}) and a nonpreferred one (leading to a low response, r_{low}). It was found (Moran and Desimone 1985; Reynolds et al. 1999) that when both stimuli are placed inside the RF, the neural response was intermediated between r_{high} and r_{low}, suggesting a competition between the two. When the subject was required to draw attention to the preferred stimulus, neural firing is enhanced to close to r_{high}, as if the nonpreferred stimulus was not present, reminiscent of inattentional blindness at the behavioral level. Attention effect on neural activity is quite robust in extrastriate visual areas such as V4 in the ventral pathway and MT in the dorsal pathway.

Figure 8.19 shows an example of V4 neural activity modulated by spatial attention. In a monkey experiment (Chelazzi et al. 2001), the task was a version of delayed match-to-sample (Chapter 6): a sample was either a preferred (face) or nonpreferred (house) stimulus of the recorded neuron, and this sample should be the focus of attention that need be maintained across a delay period. Then the same stimulus reappeared in the RF, sometimes accompanied by the other, and the correct response was to make a saccade to the target that matches the sample. The neural firing patterns under the four conditions show an early phase (blue horizontal bar) where the peak was the highest for the preferred face and lowest for the nonpreferred house, and intermediate when both were presented. When the two stimuli were shown together, the neural firing time course diverges remarkably in the second later phase (red bar), depending on whether the attention was directed to the preferred stimulus or not. In the former, the firing rate ramps back to the same high level as the case of face alone, whereas in the latter it decays to the similar low level as the case of house alone. The fast response is driven by feedforward input, whereas the slower second phase of activity likely reflects a top-down signaling with a longer latency than external stimulation.

These observations led to the proposal that visual attention influences sensory information processing through biased competition (Desimone and Duncan 1995; Reynolds et al. 2001): stimuli compete with each other through mutual inhibition in a sensory circuit, and top-down attention biases in favor of one at the focus of attention at the expense of the others.

Attention is also engaged even when a single stimulus is presented; in which case there is no competition. For instance, many V4 neurons show Gaussian-shaped tuning of oriented visual stimuli (gratings) with maximal firing at a preferred orientation. When attention is directed to the RF of a

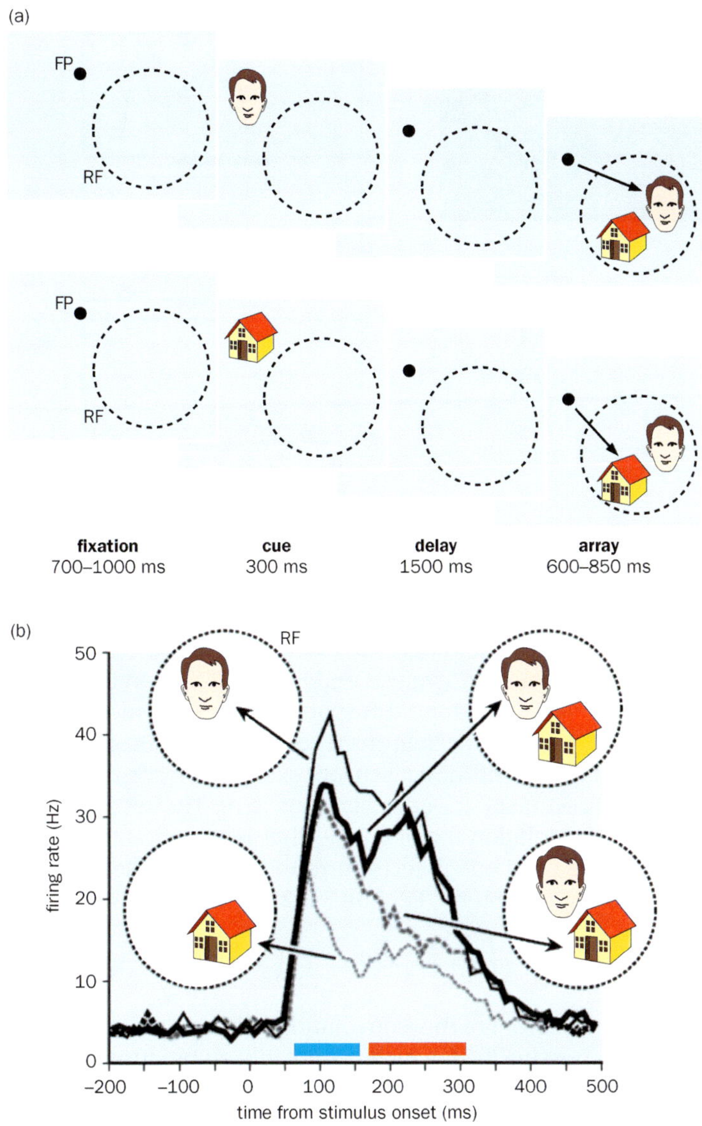

Figure 8.19 Biased competition by selective attention. (a) In a spatial attention task, a sample stimulus appears near the fovea (a face above or a house below). After a delay period of 1.5 seconds, both face and house are shown in the receptive field (RF) of a recorded neuron (dotted circle), and the monkey is required to make a saccade to the one that matches with the sample. (b) Trial-averaged firing activity of neurons in V4 under four conditions: A preferred (face) or nonpreferred (house) is shown, or both are shown with attention (instructed by the sample) to either the preferred or nonpreferred object. Time zero is the onset of the test stimulus. Reproduced from Chelazzi et al. (2001). [Reynolds, J. H., & Chelazzi, L. (2004). Attentional modulation of visual processing. Annual review of neuroscience, 27(1), 611–647, with the original data from Chelazzi, L., Miller, E. K., Duncan, J., & Desimone, R. (2001). Responses of neurons in macaque area V4 during memory-guided visual search. Cerebral Cortex, 11(8), 761–772.]

recorded neuron, its tuning curve is scaled multiplicatively by a gain factor larger than 1 compared to the case in the absence of attention (McAdams and Maunsell 1999). Directional tuning of MT neurons also displays multiplicative gain modulation by attention (**Figure 8.20a**) (Treue and Maunsell 1999). In another experiment, the contrast c of an oriented stimulus was varied, and V4 neural activity was enhanced by spatial attention as if the contrast dependence curve (e.g., Chapter 3, **Figure 3.17b**) was shifted to the left: the firing rate is enhanced by covert attention at an intermediate contrast level (**Figure 8.20b**, middle plot with $c = 10\%$) but not at a very low (bottom plot with $c = 5\%$) or high (top plot with $c = 80\%$) contrast. In this sense, attention acts as an extra contrast from within the brain, and thereby the saliency of a stimulus item is highlighted (Reynolds and Chelazzi 2004). Recall that neural tuning curves in the visual cortex display contrast invariance (Chapter 3, **Figure 3.18b**), hence contrast gain is similar to response gain. However, they are not equivalent, because contrast gain implies a shift of the contrast dependence curve and predicts no change of response at a high contrast, which appears to contradict pure response gain (always multiplying a tuning curve by a factor larger than 1).

In addition to multiplicative gain, visual cortical neurons display a shift of receptive field towards the attention focus (Connor et al. 1996;

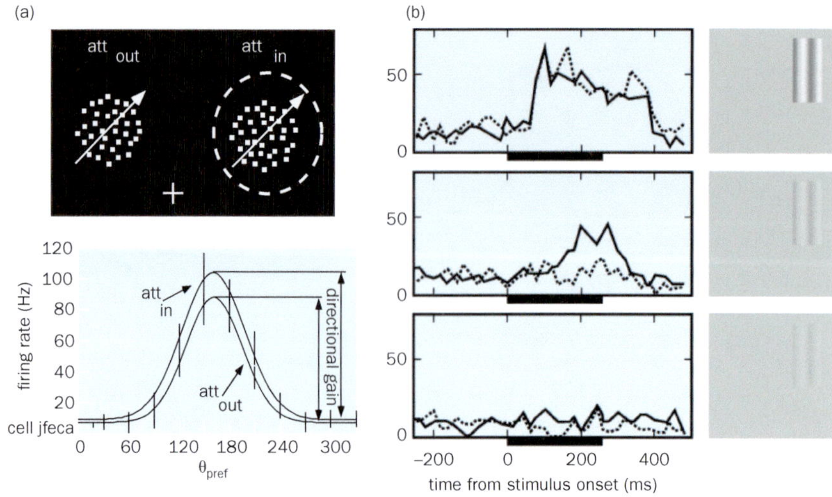

Figure 8.20 Attention and contrast gain modulation of neural response. (a) Top: Two moving patterns are shown, one inside and the other outside of the receptive field of an MT neuron. Bottom: Directional tuning curve of firing response is multiplicatively enhanced by attention in the RF versus outside of it. (b) Responses of a V4 neuron as a function of attention and stimulus contrast. The contrast of the stimulus in the receptive field varied from 5% (bottom panel) to 10% (middle panel) to 80% (upper panel). On any given trial, attention was directed to either the location of the stimulus inside the receptive field (solid line) or a location far away from the receptive field (dotted line). The animal's task was to detect a target grating at the attended location. Attention reduced the threshold level of contrast required to elicit a response (middle panel, solid curve is larger than dotted curve) without causing a measurable change in response at saturation contrast (upper panel, solid and dotted curves are similar). Panel (a) is reproduced from Treue and Martínez Trujillo (1999), (b) from Reynolds and Chelazzi (2004). [Treue, S., & Trujillo, J. C. M. (1999). Feature-based attention influences motion processing gain in macaque visual cortex. Nature, 399(6736), 575–579.] [Reynolds, J. H., & Chelazzi, L. (2004). Attentional modulation of visual processing. Annual Review of Neuroscience, 27(1), 611–647.]

Connor et al. 1997; Compte and Wang 2006), resulting in an increased neuronal resource to represent information selected by attention.

A phenomenological model has been proposed to capture these diverse attentional effects (Reynolds and Heeger 2009). In the model, neurons have receptive fields centered at spatial location x and are selective for a feature θ (e.g., orientation). Neurons receive three types of inputs (**Figure 8.21a**). The first represents direct sensory excitation $I(x,\theta,c)$, which is proportional to the contrast c for each stimulus item. The second component is an attention modulation factor $A(x,\theta)$ that is assumed to be multiplied with $I(x,\theta,c)$. The third one reflects surround inhibition described by divisive normalization (Chapter 3, Section 3.5.2):

$$S(x,\theta,c) = \int\int h(x-x',\theta-\theta')A(x',\theta')I(x',\theta',c)\,dx'd\theta' \tag{8.1}$$

where $h(x,\theta)$ is a kernel of this convolution, with an integral equal to 1. It describes a suppressive field, which results in a summation of inputs over a large population of neurons with different RFs and preferred feature values. According to divisive normalization, the firing rate is given by:

$$r(x,\theta,c) = \frac{A(x,\theta)I(x,\theta,c)}{S_0 + S(x,\theta,c)} \tag{8.2}$$

where S_0 is a constant so that the firing rate is well defined in the absence of an input ($S(x,\theta,c)=0$). The functions $I(x,\theta,c)$, $A(x,\theta)$ and $h(x,\theta)$ all have a Gaussian shape characterized by a peak value and a width parameter. Varying the stimulus θ_S and x_S provides a different input $I(x,\theta,c)$ to the neuron with the RF entered at x and the preferred feature θ. The peak value γ of $A(x,\theta)$ is larger than 1, expressing enhancement by attention. Without attention, $A(x,\theta)=1$.

With appropriated choice of parameter values, this model describes a number of experimental observations. For instance, multiplicative response gain directly results from the model construction (**Figure 8.21b**). Less obviously, the aforementioned seeming contradiction between findings of contrast gain versus response gain can be explained by the size of attention field relative to that of a stimulus. Contrast gain can be derived analytically with this model. Let us assume that the stimulus is the preferred ($x_S = x$ and $\theta_S = \theta$). In the control case without attention, it can be shown that neural firing response to a preferred and small stimulus is $r \sim c/(c+S_0$ (normalized so that its maximum is 1), which plateaus for high contrast ($c \gg S_0$). When attention is present with a broad focus compared to the stimulus size, the model neural response becomes $r \sim \gamma c/(\gamma c + S_0)$ with

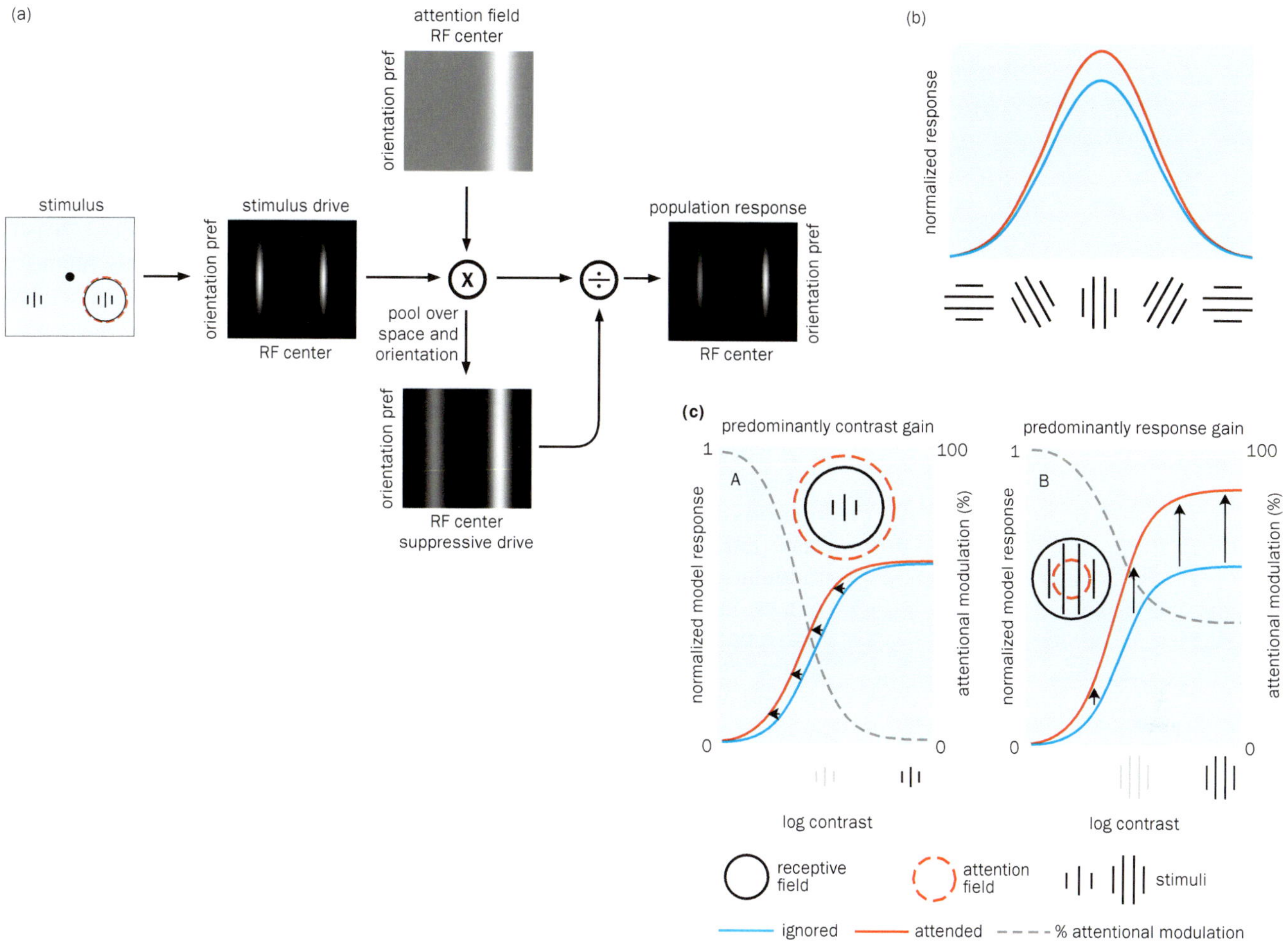

Figure 8.21 A phenomenological model of attention modulation. (a) Model scheme, where neurons are labeled by their receptive field center x (abscissa) and preferred stimulus feature θ (ordinate). Direct sensory excitation is multiplied by attention modulation. Their product is used to compute a suppressive field for divisive normalization. (b) Multiplicative gain of tuning curve of a single neuron by attention (red) compared to control (blue). (c) Dependence on the relative sizes of the stimulus and the attention field. Responses (left y-axis label) as a function of contrast when the stimulus in the receptive field was attended (red) or not (blue). Attentional modulation is indicated by the dashed grey curve, which quantifies the percentage (right y-axis label) increase in the responses when the stimulus within the neuron's receptive field was attended versus not. Left panel: Contrast gain for small stimulus size and large attention field. Right panel: Response gain for larger stimulus size and smaller attention field. Reproduced from Reynolds and Heeger (2009). [Reynolds, J. H., & Heeger, D. J. (2009). The normalization model of attention. Neuron, 61(2), 168–185.]

$\gamma > 1$. In other words, the attention effect is equivalent to an increased contrast from c to γc (**Figure 8.21c**, left panel). The firing rate at high contrast remains unchanged ($\simeq 1$).

On the other hand, a large stimulus also activates neurons in the surround, resulting in an additional term βc in the denominator, where β (between 0 and 1) represents surround inhibition. A small attention focus does not affect neurons in the surround, and therefore the firing rate modulated by attention becomes $r \sim \gamma c / (\gamma c + \beta c + S_0)$. In this case, not only is there a shift of contrast dependence to the left, but the plateau level at large c (given by $\gamma / (\gamma + \beta)$) is also increased from that in the absence of attention $(1/(1+\beta))$. Under this condition, the attention impact is a mixture of contrast gain and response gain (**Figure 8.21c**, right panel), as observed experimentally (Williford and Maunsell 2006).

8.4.2 An Integrative Circuit Model of Selective Attention

Functional MRI and neurophysiological experiments identified an "attention system" with the parieto-frontal cortical circuit at its core (Desimone and Duncan 1995; Kastner and Ungerleider 2000; Awh and Jonides 2001; Corbetta and Shulman 2002; Moore 2006; Soto et al. 2008; Gazzaley and Nobre 2012; Petersen and Posner 2012; Moore and Zirnsak 2017; Panichello and Buschman 2021). Since an attentional focus must be internally sustained, it not surprising that attention and working memory share a brain substrate, at least partly. This was directly demonstrated in monkey experiments where inactivation of a prefrontal area abolished attention effects in an extrastriate visual area (Moore and Armstrong 2003; Bichot et al. 2019). Can synaptic interactions between a sensory area and a working memory area account for diverse observed attentional effects and provide a unified explanation for both gain modulation and biased competition?

To address this question, Ardid et al. (2007) investigated a biophysically based model of spiking neurons with a reciprocally connected loop between a sensory area and an executive control area. Specifically, consider a sensory module consisting of MT neurons selective for motion direction (Albright et al. 1984; Born and Bradley 2005) and a prefrontal area where neurons display direction-selective persistent activity in a delayed response task (Zaksas and Pasternak 2006). **Figure 8.20a** showed an example of modulation of MT neurons by spatial attention. Related monkey experiments also examined feature-based attention where the attention focus was a particular motion direction (e.g., upward motion), and neurophysiological evidence suggests that feature-based and location-based attention depend on similar mechanisms (Maunsell and Treue 2006; Maunsell 2015). A simple model setup, then, is a reciprocal loop between two continuous "ring" networks representing MT and PFC, respectively, where in each neurons are selective for directional angle of a motion stimulus.

In the model of Ardid et al. (2007), the PFC module was adopted from the model of spatial working memory (Compte et al. 2000) (Chapter 5, Section 5.4.1), with the stimulus feature now reinterpreted as motion direction and neurons labeled by their preferred angles $\theta_{\text{pref}}^{PFC}$. The MT module is also a continuous network (a ring), where neurons are selective for direction θ_{pref}, and recurrent excitation is sufficiently weak that no persistent activity can be self-sustained (Zaksas and Pasternak 2006; Mendoza-Halliday et al. 2014). As a matter of fact, intrinsic connections within MT are dominated by feedback inhibition, giving rise to normalization (Chapter 3, **Figure 3.20**). What happens when the two local circuits are put together? **Figure 8.22** shows a loop model where the two modules are reciprocally connected, where strengths of inter-areal connections decrease with the difference in the preferred direction of pre- and post-synaptic neurons ($\theta_{\text{pref}}^{PFC} - \theta_{\text{pref}}$).

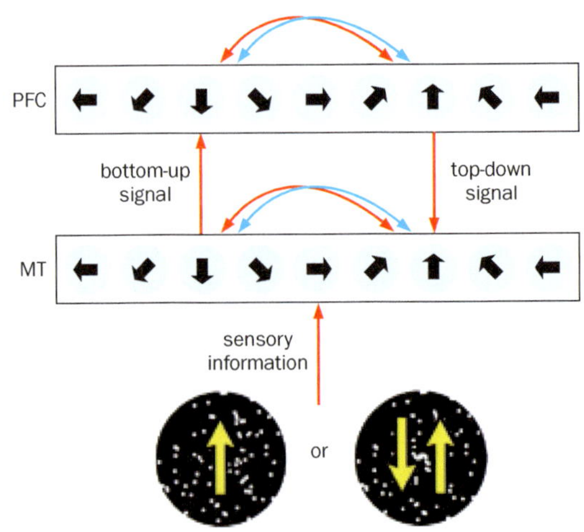

Figure 8.22 A loop model of spiking neurons for selective attention. The circuit consists of two continuous (ring) networks, representing the middle temporal (MT) cortex and the prefrontal cortex (PFC), respectively. WM: Working memory. Each of the two neural populations is selective for a stimulus feature (direction of a visual motion pattern). The MT module conveys sensory information to the PFC, which in turn mediates top-down selective attention. Reproduced from Ardid et al. (2007). [Ardid, S., Wang, X. J., & Compte, A. (2007). An integrated microcircuit model of attentional processing in the neocortex. Journal of Neuroscience, 27(32), 8486–8495. Licensed under CC BY 4.0., https://www.jneurosci.org/content/jneuro/27/32/8486.full.pdf]

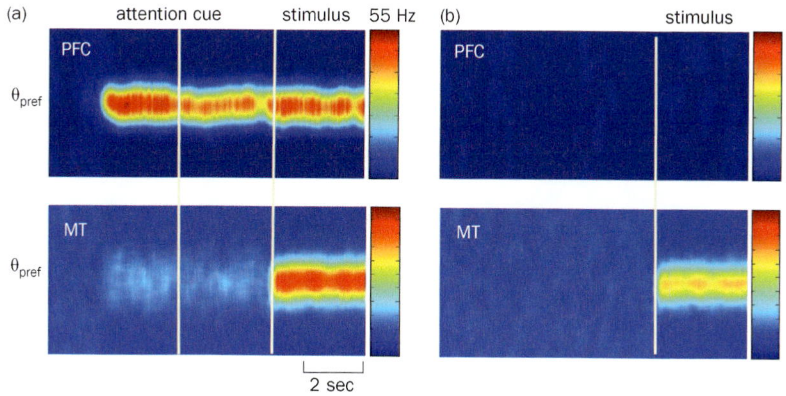

Figure 8.23 Single-trial model simulation with attention (a) or without it (b) when the attended feature θ_A is the same as the stimulus θ_S. In each of the four panels, neurons in a population are shown along the y-axis by their preferred direction θ_{pref}, and their firing activities unfold as a function of time (along the x-axis). Firing rate is indicated by color. An attentional cue triggers persistent activity in a group of neurons in the PFC that displays gamma oscillations (upper panel in (a)), leading to enhanced spontaneous activity in the corresponding group of MT neurons with preferred directions near θ_A. The spiking responses of these neurons to a visual motion stimulus are stronger with attention (lower panel in [a]) compared to the case without attention (lower panel in [b]). Reproduced from Ardid et al. (2007). [Ardid, S., Wang, X. J., & Compte, A. (2007). An integrated microcircuit model of attentional processing in the neocortex. Journal of Neuroscience, 27(32), 8486–8495. Licensed under CC BY 4.0., https://www.jneurosci.org/content/jneuro/27/32/8486.full.pdf]

In a simulation of an attention experiment by this simple model (**Figure 8.23a**), an attentional cue (θ_A) is first presented, which triggers a bump attractor in the PFC module (Chapter 5, **Figure 5.18**). During a brief delay, the spiking activity of a subset of MT neurons (with preferred direction θ_{pref} near θ_A) is slightly enhanced by top-down sustained input from the PFC, consistent with the observation that spontaneous activity in extrastriate areas is increased by attention (Luck and Vogel 1997; Kastner et al. 1999). Then a motion stimulus θ_S is presented to the MT module in the form of a Gaussian-shaped input, in this case peaked at $\theta_S = \theta_A$, triggering a population response. As expected, sensory response by MT is enhanced by attention (**Figure 8.23a**), compared with another simulation in the absence of an attention cue (**Figure 8.23b**). Note that the PFC module exhibits stochastic oscillations in the gamma (~ 40 Hz) frequency range (Chapter 5, **Figure 5.19**), leading to enhanced rhythmic coherence in MT, which will be discussed in the next section.

Let us denote by $r(\theta, \theta_S, \theta_A)$ the population activity pattern of MT neurons labeled by their preferred direction θ_{pref} in response to a stimulus θ_S and under the influence of attention directed to θ_A (when present). Population activity in a single trial (**Figure 8.23**) is $r(\theta_{pref}, \theta_S, \theta_A)$ as a function of θ_{pref} with fixed θ_A and θ_A. By contrast, in a monkey experiment where neurons are recorded one at a time, the tuning curve of a neuron is computed by probing a given cell with a fixed preferred stimulus θ_{pref} while varying stimulus direction θ_S. **Figure 8.24** (left panel) shows tuning curves of a model neuron without attentional input (blue), in the presence of attention focused on its preferred direction (green, $\theta_A = \theta_{pref}$) or null direction (red, $\theta_A = \theta_{pref} + 180^o$). Multiplying with a constant factor larger than one for the red curve and smaller than one for the green curve, all three tuning functions become indistinguishable (**Figure 8.24**, right panel). The reproduction of multiplicative gain was not prescribed by model design. Analysis revealed that neural firing is a power law function of the sum of the inputs from bottom-up stimulation and top-down attention, which underlies multiplicative gain, as shown by Murphy and Miller (2003).

Hence, attention effect on a given neuron labeled by θ_{pref} is enhancing if θ_A coincides with its preferred feature θ_{pref}, but suppressive when θ_A is in its null direction. How does the attention modulation depend on the difference $\theta_{pref} - \theta_A$ parametrically? Martinez-Trujillo and Treue (2004) designed an experiment where the attended feature θ_A and the stimulus feature θ_S were the same and co-varied from trial to trial (**Figure 8.25a**). **Figure 8.25b** shows the neural firing rate of an MT neuron as a function of the co-varied directional angle $\theta_A = \theta_S$. The ratio of firing rates with attention (blue line) and without it (red line) is larger than 1 near the neuron's preferred feature θ_{pref} and less than one near its null direction (**Figure 8.25c**). This observation forms the empirical basis

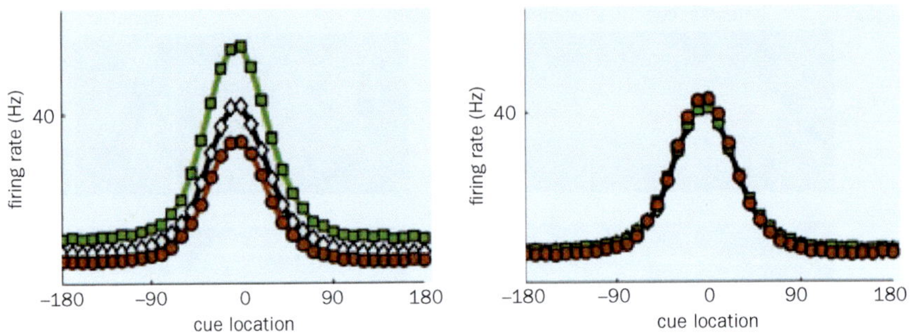

Figure 8.24 Multiplicative response gain by attention. (a) Tuning curve of an MT neuron as a function of stimulus direction θ_S, when attention is focused onto the preferred direction (green) or diametrically opposed (red), or is not presented (blue). Right panel: Rescaling of tuning curves (factors, 0.73 for green, 1.28 for red) reveals multiplicative scaling. Reproduced from Ardid et al. (2007). [Ardid, S., Wang, X. J., & Compte, A. (2007). An integrated microcircuit model of attentional processing in the neocortex. Journal of Neuroscience, 27(32), 8486–8495. Licensed under CC BY 4.0., https://www.jneurosci.org/content/jneuro/27/32/8486.full.pdf]

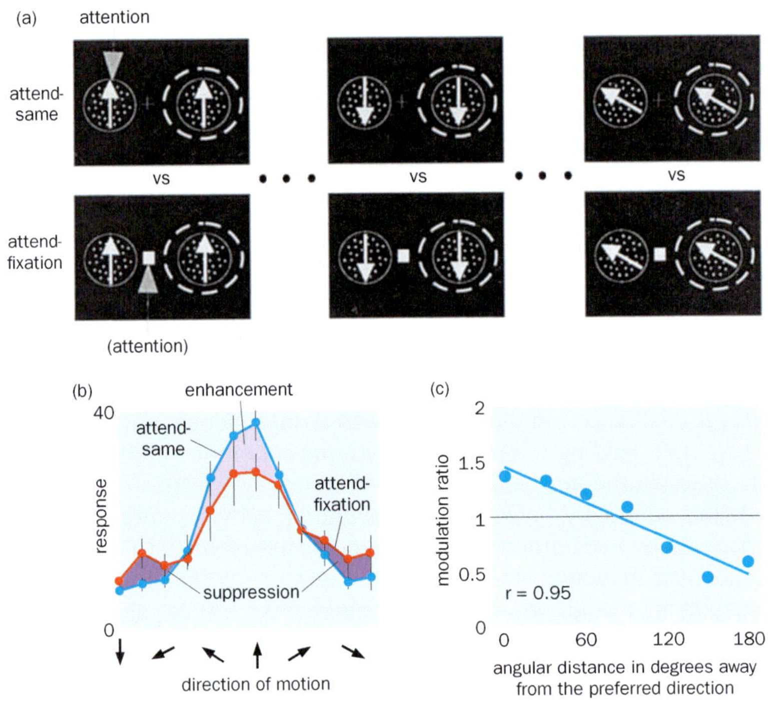

Figure 8.25 Feature-similarity gain principle revealed in a monkey experiment. (a) The top panels represent the conditions in which the monkey attended to the same motion direction presented inside the receptive field (dotted circle) (attend-same). The bottom panels represent the condition without attention, in which the animals attended to a color square on the fixation point (attend-fixation). (b) The average firing rate of an MT neuron (ordinate) was plotted as a function of the direction of the pattern (abscissa), in the attend-same (blue) and attend-fixation (red) conditions. The pink shaded area between the two curves represents the region of response enhancement, and the purple shaded area represents the area of response suppression. (c) Modulation ratios between the responses in the two conditions shown in (b). The abscissa represents the angular distance between the attended direction and MT cell's preferred direction, and the ordinate represents the modulation ratio. Reproduced from Martinez-Trujillo and Treue (2004). [Martinez-Trujillo, J. C., & Treue, S. (2004). Feature-based attention increases the selectivity of population responses in primate visual cortex. Current Biology, 14(9), 744–751.]

of the proposal of a feature-similarity gain principle, which posits that the effect of selective attention on a sensory neuron is solely determined by the difference between its preferred feature at the single cell level and the attended feature at the behavioral level (Treue and Maunsell 1999; Martinez-Trujillo and Treue 2004). The feature-similarity gain principle applies to spatial attention as well as feature attention (Treue and Maunsell 1999).

It turns out that the model naturally gives rise to the feature-similarity gain principle (**Figure 8.26**). Note that $r\left(\theta_{\text{pref}}, \theta_S, \theta_A\right)$ actually depends on the differences between angle variables, so it can be rewritten as $r\left(\theta_{\text{pref}} - \theta_S, \theta_{\text{pref}} - \theta_A, \theta_A - \theta_S\right)$. When $\theta_A = \theta_S$ is varied together, the third argument is zero. Moreover, since the other two variables are $\theta_{\text{pref}} - \theta_A$ and $\theta_{\text{pref}} - \theta_S$, watching a single neuron with a fixed preferred θ while varying $\theta_A = \theta_S$ across trials is equivalent to probing all the neurons in MT with a fixed $\theta_A = \theta_S$ in a single trial. In other words, the peculiar stimulus protocol of Martinez-Trujillo and Treue (2004) amounted to effectively monitor an entire MT neural population in a behavioral trial while actually recording from a single cell across trials.

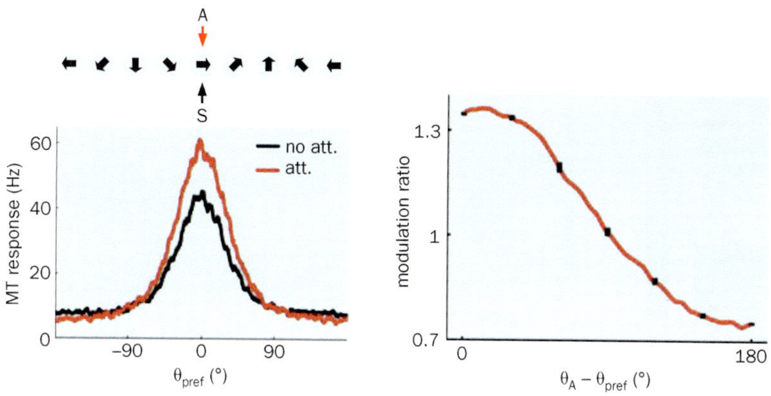

Figure 8.26 Feature-similarity gain principle emerges from the biophysically based model of spiking neurons. Left panel: MT population activity pattern as a function of θ_{pref} in a single trial with (red) or without (black) attention on θ_A. Right panel: The ratio of the firing rates under the two conditions (red over black from the left panel) accounts for the experimentally observed feature-similarity attention modulation curve. Reproduced from Ardid et al. (2007). [Ardid, S., Wang, X. J., & Compte, A. (2007). An integrated microcircuit model of attentional processing in the neocortex. Journal of Neuroscience, 27(32), 8486–8495. Licensed under CC BY 4.0., https://www.jneurosci.org/content/jneuro/27/32/8486.full.pdf]

The multiplicative gain by attention requires that the modulation factor does not depend on S. If we denote the neuronal tuning curve without attention by $r_0\left(\theta_{pref} - \theta_S\right)$, then:

$$r\left(\theta_{pref} - \theta_S, \theta_{pref} - \theta_A\right) = G\left(\theta_{pref} - \theta_A\right) r_0\left(\theta_{pref} - \theta_S\right) \tag{8.3}$$

This equation can be used to fit model simulation data with $G\left(\theta_{pref} - \theta_A\right) = 1 + g_0 \cos\left(\theta_{pref} - \theta_A\right)$ and $g_0 \simeq 0.3$ (Ardid et al. 2007) (**Figure 8.26**). Thus, $G = 1 + g_0$, larger than 1 with $\theta_{pref} = \theta_A$; and $G = 1 - g_0$, smaller than 1 with $\theta_{pref} = \theta_A + 180^\circ$. Importantly, behavior is determined by neural population activity pattern in a single trial. According to the feature-similarity gain principle, attention selectively enhances a group of sensory neurons while suppressing others (rather than uniformly enhances all neurons as in multiplicative gain), which is functionally desirable.

Attention modulation solely determined by $\theta_{pref} - \theta_A$ implies that it is independent of θ_S. If so, the same Eq. (8.3) should predict MT neural activity responses to more complex motion stimuli. This prediction was tested by simulating the model with two moving stimuli in superimposition (**Figure 8.27**) in a plaid pattern (Adelson and Movshon 1982) (Chapter 3, **Figure 3.20b**). Indeed, the population activity pattern with attention (green trace) is well predicted (dotted trace) by that without attention (grey trace) multiplied by $G\left(\theta_{pref} - \theta_A\right)$ for each neuron. Remarkably, a simple principle can explain attention modulation of representations of complex sensory stimuli.

Without changing parameters, the same model also reproduces biased competition (**Figure 8.28**). MT neural response to a single stimulus is the highest when it is the preferred direction (up, red) and lowest when it is the null direction (down, black). Simultaneous presentation of the two stimuli (**Figure 8.27**) produces an intermediate firing rate, due to normalization realized by recurrent inhibition in the MT circuit. The response is enhanced by attention (blue oval) to the preferred stimulus and reduced if the attended feature is the nonpreferred. Therefore, multiplicative gain and bias competition is explained by a single circuit.

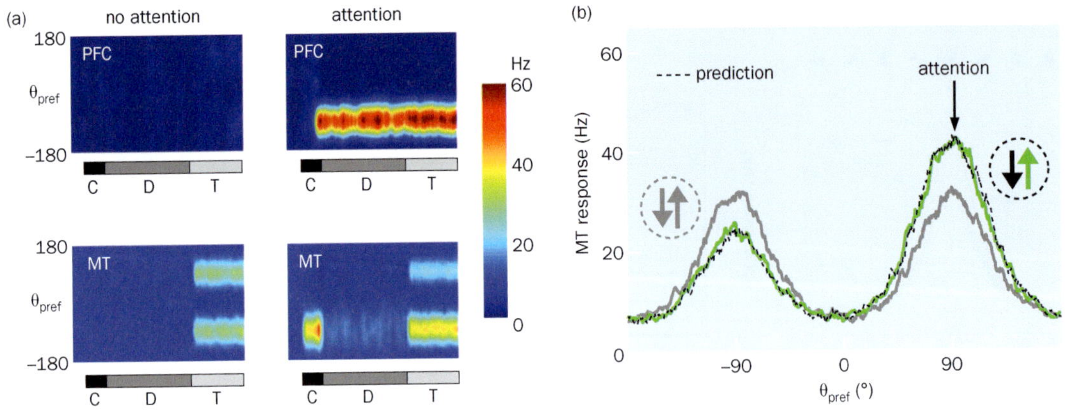

Figure 8.27 Model simulation with two superimposed moving patterns in the directions 90° and –90°, separated by 180°. (a) Activity patterns without attention (left) and with attention directed at $\theta_A = 90^\circ$ (right). (a) MT population firing pattern without attention (grey) and with attention (green). Activity is enhanced near the attention focus ($\theta_{pref} \simeq 90^\circ$) while suppressed away from it ($\theta \simeq -90^\circ$). The grey curve multiplied by the feature-similarity gain function $G(\theta_{pref} - \theta_A)$ yields the black dotted curve, which matches perfectly with the green curve. Reproduced from Ardid et al. (2007). [Ardid, S., Wang, X. J., & Compte, A. (2007). An integrated microcircuit model of attentional processing in the neocortex. Journal of Neuroscience, 27(32), 8486–8495. Licensed under CC BY 4.0., https://www.jneurosci.org/content/jneuro/27/32/8486.full.pdf]

The model of Ardid et al. (2007) can be extended to a higher-dimensional space where neurons are selective for both spatial location and stimulus feature. Such a model would be suitable to explore the dependence on the stimulus size of the interplay between contrast and attention (Section 8.4.1), and simulate visual search in a cluttered visual scene (Bichot et al. 2005; Bichot et al. 2019).

8.4.3 Attention Modulation of Network Synchrony and Noise Correlation

In Chapter 3 (Section 3.4), we introduced brain rhythms with an example from Fries et al. (2001) of enhanced neural population coherence at gamma frequency by selective attention. Network synchronization is measured by LFP, and its power spectrum displays a peak in case a genuine oscillation is present. Based on LFP and spike-field coherence measurements, sinusoidal waves have been proposed to underlie gating by selective attention: imagine a source area *A* projecting to two target areas *B* and *C*, where all three areas show sinusoidal waves at a similar frequency. If the wave in the source area *A* is synchronized at zero phase with *B* and anti-phase with *C*, spikes at a wave's peak of the source area arrive in area *B* when it is maximally excited and in area *C* where excitability is at a trough; consequently, communication is "open" to area *B* and "closed" for area *C* (Fries 2005). However, on the face value of it, this picture is too simplistic, because neural firing is far from being clock-like; instead, spike trains of single neurons are nearly Poissonian (Chapter 2, **Figure 2.5**), and only a small fraction of spikes are timed to specific phases of an LFP oscillation (Chapter 3, **Figure 3.12**).

The role of network synchronization in the mammalian cortex during behavior is better appraised by reconciling weak network oscillations and highly stochastic neural spike discharges (Chapter 3, Section 3.4.2). Furthermore, sometimes in an experiment the LFP power spectrum does not have a peak, implying the absence of a rhythm, but spike-field coherence and spike-to-spike coherence can still show a peak at certain frequency. Such an example is shown in **Figure 8.29a** from a selective visual attention experiment with monkeys (Fries et al. 2008). Note that the two

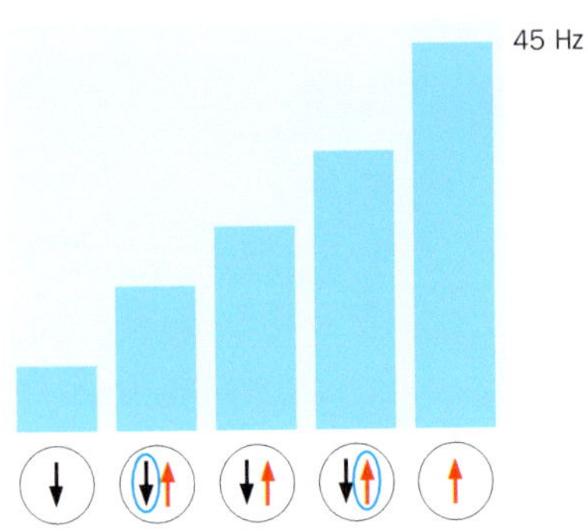

Figure 8.28 Biased competition from the spiking network model, with firing rate of a single MT neuron shown in five different stimulus conditions. Blue oval: Attended feature. Reproduced from Ardid et al. (2007). [Ardid, S., Wang, X. J., & Compte, A. (2007). An integrated microcircuit model of attentional processing in the neocortex. Journal of Neuroscience, 27(32), 8486–8495. Licensed under CC BY 4.0., https://www.jneurosci.org/content/jneuro/27/32/8486.full.pdf]

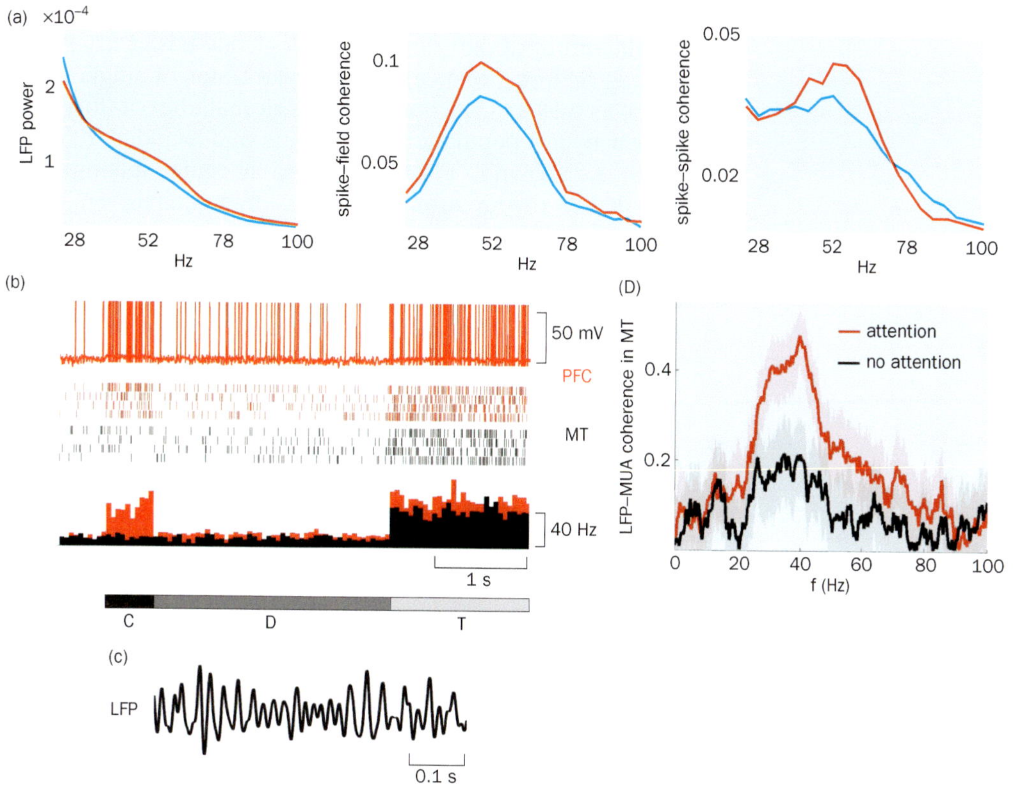

Figure 8.29 Enhancement of coherence by attention. (a) Coherence in V4 during sustained stimulation from behaving monkeys in an attention task. Red: Attended condition; blue: Non-attended condition. Left: LFP power spectrum from area V4. Middle: Spike-to-field coherence (between multiunit activity and LFP). Right: Spike-to-spike coherence between two simultaneously recorded multiunit spike trains. (b–d) Weak synchrony at gamma frequency modulated by attention in the MT-PFC loop model. In a single-trial model simulation, highly irregular neuronal spike trains (b) coexist with a stochastic gamma oscillation measured by population LFP (c). (d) LFP-spike coherence is enhanced at gamma frequency by attention (red) relative to the condition without attention (black). Panel (a) is reproduced from Fries et al. (2008), (b–d) from Ardid et al. (2010). [Fries, P., Womelsdorf, T., Oostenveld, R., & Desimone, R. (2008). The effects of visual stimulation and selective visual attention on rhythmic neuronal synchronization in macaque area V4. Journal of Neuroscience, 28(18), 4823–4835. https://www.jneurosci.org/content/28/18/4823] [Ardid, S., Wang, X. J., & Compte, A. (2007). An integrated microcircuit model of attentional processing in the neocortex. Journal of Neuroscience, 27(32), 8486–8495. Licensed under CC BY 4.0., https://www.jneurosci.org/content/jneuro/27/32/8486.full.pdf] [Ardid, S., Wang, X. J., Gomez-Cabrero, D., & Compte, A. (2010). Reconciling coherent oscillation with modulation of irregular spiking activity in selective attention: Gamma-range synchronization between sensory and executive cortical areas. Journal of Neuroscience, 30(8), 2856–2870. Licensed under CC BY 4.0. https://www.jneurosci.org/content/jneuro/30/8/2856.full.pdf]

coherences are defined between 0 and 1, the measured values are small (below 0.1) and the change by selective attention is only about 20%, e.g., from 0.08 to 0.1. Spike-field coherence and spike-to-spike coherence of around 10% and 10–20% attention effect are typical across experiments, highlighting the fact that network coherence is weak and its attentional modulation is significant but relatively small in the extrastriate cortex (Buschman and Kastner 2015).

As noted in Chapter 5 (**Figure 5.19**) and **Figure 8.23**, a working memory circuit naturally displays stochastic gamma oscillations due to strong recurrent synaptic reverberation required for the maintenance of persistent activity (Compte et al. 2000). This sparsely synchronous rhythm coexists with irregular spike firing of individual neurons, and the rhythmic frequency is determined by an interplay of fast AMPA receptor-mediated excitation and slower $GABA_A$-mediated inhibition (Brunel and Wang 2003; Börgers and Kopell 2003; Wang 2010). Recall that neural population coherence at ~40 Hz during a mnemonic delay was reported experimentally (Pesaran et al. 2002; Wang 2010; Lundqvist et al. 2016). Although we focus on the cortex here, subcortical structures such as the pulvinar also play a role in the enhancement of neural population synchrony by selective attention (Saalmann et al. 2012). In the biologically based MT-PFC loop model of spiking neurons, one can probe spike trains of single units and multiple units, as well as LFP computed by summated excitatory synaptic currents over a neural population (Logothetis and Wandell 2004; Mazzoni et al. 2009). As shown in **Figure 8.29b-c**, irregular spiking coexists with stochastic LFP rhythm around 40 Hz. Gamma coherence in the MT network is enhanced by attention automatically as a result of the engagement of top-down signaling from the PFC network endowed with strong reverberation (**Figure 8.29d**) (Ardid et al. 2010). Hence, the observed increase of spike-field coherence at gamma frequency in sensory neurons, under attended condition, supports the idea that attentional bias originates in a working memory circuit. Note that the rhythmic frequency depends on the precise parameter values in the model; for instance, coherent oscillations can be in the beta frequency range with slight changes of model parameters. The model also reproduces the experimental observation that Fanor factor of spike trains in the extrastriate cortex is reduced by attention (Mitchell et al. 2007).

To assess whether network coherence has an impact on attentional modulation, one should find a way to eliminate enhancement of synchrony in the MT network by top-down signaling. This was achieved in the model by the introduction of sufficiently large heterogeneity of latencies for the PFC-to-MT synaptic projections, so that top-down inputs became totally asynchronous between MT neurons. It was found that compared to the asynchronous regime, in the control case enhanced network coherence by top-down further increased by about 20% the attentional gain, which is typically 10–30%, as we have seen earlier (Ardid et al. 2010). Such slight enhancement of attentional effect by synchrony may be amplified by recurrent circuit dynamics locally, as we shall see in the next Section 8.5.

As we discussed in Chapter 3, weak coherence is a manifestation of correlations between neurons. Interestingly, studies consistently found that noise correlation (Chapter 3, Section 3.2.3) in visual neurons is reduced by selective attention (Cohen and Maunsell 2009; Maunsell 2015). While coherence is slightly increased in the gamma frequency range (**Figure 8.29a**), there is a concomitant decrease of coherence at low frequencies by attention (Chapter 3, **Figure 3.10b**) (Fries et al. 2001; Fries et al. 2008; Gregoriou et al. 2009; Buffalo et al. 2011), and noise correlation is computed with relatively large time windows (hence, low frequencies). Interestingly, a modeling study suggests that to account for the observed reduction of noise correlations, top-down attentional signaling enhances inhibition in a sensory area (Huang et al. 2019). This conclusion is not necessarily inconsistent with excitatory gain for a subgroup of neurons in a sensory area. As is clear in the feature-similarity gain curve (**Figure 8.25c**, **Figure 8.26** [right panel]), some neurons are excited, while others are suppressed under a given attention condition. The net effect would be inhibitory if the averaged gain factor over a neural population is less than one, a prediction yet to be tested experimentally (Yoo et al. 2022).

8.5 TASK SWITCHING

A logical next step in our discussion on attention is to consider task switching (Allport et al. 1994; Grange and Houghton 2014). In this internet era, we are bombarded by stimuli and a pyramid of choices; cognitive control is of paramount importance to our behavior that is not solely driven by stimulation of our senses but organized to achieve desired goals (Klingberg 2009). In laboratory experiments, task switching is commonly studied with more than one (typically two) tasks, and which one of them needs to be carried out in a given trial is cued by a rule signal. For instance, the color of fixation point instructs a participant that the rule currently in play is pro-saccade (rule 1) or anti-saccade (rule 2) to a visual target (**Figure 8.15a**), and the rule cue changes randomly from trial to trial. In a variant of task design like the WCST, a rule is not cued and remains the same for a block of trials, at some point suddenly switches to another rule in another block.

Figure 8.30a shows a task switching paradigm in human studies (Rogers and Monsell 1995; Meiran 2000; Brown et al. 2007). In each trial, a task cue instructs a subject whether the response should be based on rule 1 = letter (vowel, left button press (L); consonant, right button press (R)) or rule 2 = number (odd, L; even, R). After a time delay (rule cue and target stimulus interval, CTI), a stimulus array with a letter and a number is shown, and the subject should ignore the irrelevant of the two and produce a response correctly based on the relevant item (the letter according to rule 1; the number otherwise). Studies using such a task robustly found that response time (RT) and error rate (ER) increase in trials following a task switch. Switch cost splits into a component that decreases with increased CTI that presumably facilitates preparation for implementing the new sensorimotor mapping and a residual component that remains for long CTIs (**Figure 8.30b**). Switch cost is believed to reflect the time course of cognitive control dedicated to task switching (Rogers and Monsell 1995; Meiran 2000; Altmann and Gray 2008).

A second robust observation is task-response interaction: on task repeat trials (when the rule is the same as in the previous trial), RT is shorter if the same response as the previous trial (RR or LL) is selected than if it is different (RL or LR); by contrast, the opposite trend holds on switch trials (when the rule just changed) (**Figure 8.30c**) (Rogers and Monsell 1995; Meiran 2000). The third salient observation is congruency effect: RT is longer and ER is higher when the stimulus is incongruent than when it is congruent, defined by whether the mapped behavioral response is different or the same according to the two rules (Rogers and Monsell 1995; Meiran 2000; Brown et al. 2007). For instance, the example of **Figure 8.30a** represents an incongruent trial, where the stimulus array is X9, the response is R with the letter rule (X is a consonant), but L with the number rule (9 is an odd number). A trial with stimulus array A7 would be a congruent trial, because the response is L regardless of the rule currently in play. In the latter case, the rule does not matter, and thus the task is expected to be easier and RT is shorter than in the former case.

Task set denotes a collection of component processes required to accomplish a task. Neuroimaging and neurophysiological experiments

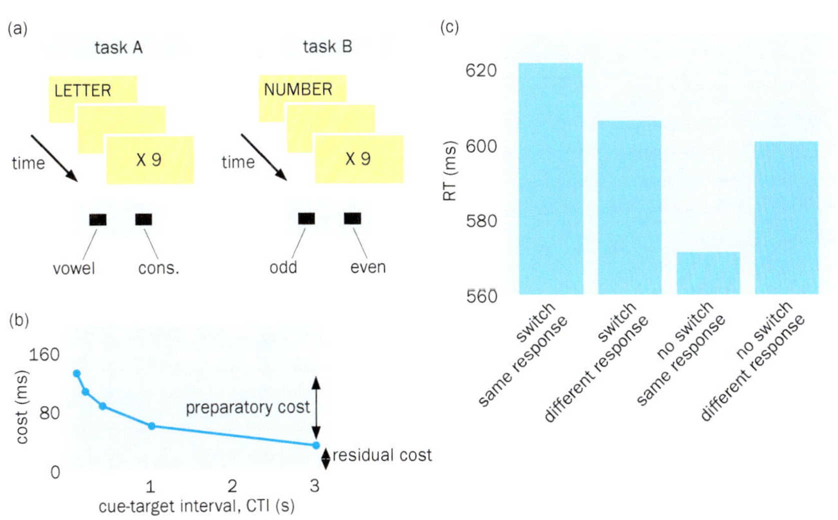

Figure 8.30 Task switching and switch-induced behavioral effects. (a) A task switching paradigm adapted from Rogers and Monsell (1995). Each trial begins with the cue "Letter" or "Number" appearing briefly. After a time interval during which the screen is blank, a letter and a number appear side-by-side. If the rule cue is Letter, a subject must respond differentially for vowels and consonants. If the cue is Number, the response is based on whether the number is odd or even. (b) Increased reaction time after a rule switch. (c) Task-response interaction manifested by RTs in the four cases of task and response combinations. Panel (a) is reproduced from Brown et al. (2007), (b-c) from Meiran (2000). [Modified from Nachshon Meiran, Ziv Chorev, and Ayelet Sapir, (2000). Component Processes in Task Switching, Cognitive Psychology 41(3), 211–253 Doi:10.1006/cogp.2000.0736.] [Modified from Ardid, S., & Wang, X. J. (2013). A tweaking principle for executive control: Neuronal circuit mechanism for rule-based task switching and conflict resolution. Journal of Neuroscience, 33(50), 19504–19517.]

have shown that prefrontal subregions are implicated in distinct components of a task set (Sakai 2008; Stoet and Snyder 2009), in particular the dorsolateral PFC (DLPFC) for representation of a task rule (Wallis et al. 2001; Everling and DeSouza 2005; Bunge and Wallis 2007; Buckley et al. 2009). In task switching, such a rule signal needs to remap the same stimulus to a different response in a task-dependent manner, which is called "network reconfiguration." Network reconfiguration could be readily realized if the rule signal is sufficiently powerful. For instance, in a model for the Stroop task (Section 8.1), the word pathway is more prepotent and stronger than the color pathway alone. A sensory input of the relevant feature (color, not word) is combined with a rule signal that is as strong or stronger than the stimulus, so that the overall input about color overcomes the strength of the word pathway (Cohen et al. 1990). However, a task rule modulates activity of sensory neurons only by a modest amount, similarly to attention modulation (**Figure 8.25**). This was the case in a monkey experiment (Mirabella et al. 2007) shown in **Figure 8.31a**. In this study, the stimulus is a colored oriented bar, and the shape of fixation instructs which of the two color- and orientation-based tasks needs to be performed in a given trial. After a time interval between the cue and target stimulus (CTI), monkeys were trained to classify the relevant feature into one of two categories (say, A or B for color; C or D for orientation) and produce a mapped response (A or C, leftward turn of a response lever; B or D, rightward turn). Some stimuli are congruent, whereas others are incongruent, defined by whether the response is depending on the rule currently in play (**Figure 8.31b**). Monkeys exhibited longer RTs in incongruent trials, similar to human behavior. Spiking activity of neurons in V4 showed significant but modest modulation (about 10%) by the task rule about 100–150 ms after the stimulus onset; this latency suggests that task-dependent signaling originates from a top-down pathway.

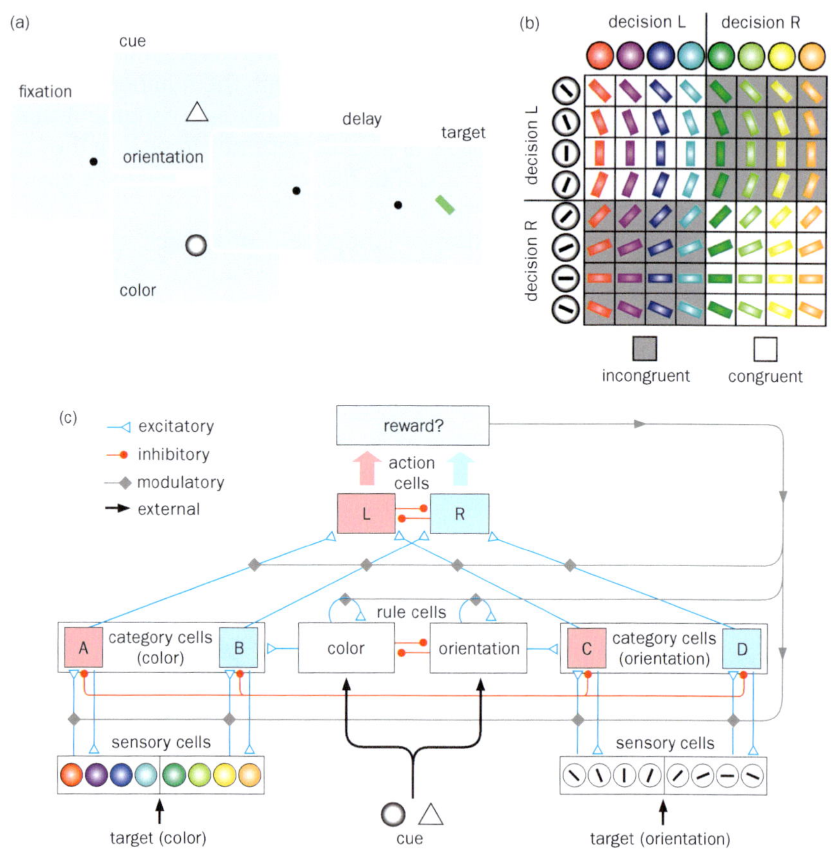

Figure 8.31 Task switching paradigm and its biologically based network model. (a) Task design. The shape of fixation point indicates relevant feature in a given task: Triangle, orientation; circle, color. After a short delay, a colored oriented bar is shown, and a subject is required to categorize the relevant feature, which is an analog quantity, into one of two classes learnt through experience. (b) Color and orientation each can be described as a ring, which is arbitrarily divided into two classes, one mapped onto Right response, the other to Left response. Shaded: Incongruent stimuli. (c) Model architecture with four interacting neural circuits that instantiate distinct cognitive processes: Rule encoding, sensory processing, feature categorization and action selection. Connection pathways modulated by reward-dependent plasticity are indicted by a grey diamond. Panels (a–b) are reproduced from Mirabella et al. (2007), (c) from Ardid and Wang (2013). [Ardid, S., & Wang, X. J. (2013). A tweaking principle for executive control: Neuronal circuit mechanism for rule-based task switching and conflict resolution. Journal of Neuroscience, 33(50), 19504–19517. https://www.jneurosci.org/content/33/50/19504]

Ardid and Wang (2013) built a biologically based model for task switching focusing on task rule-guided sensorimotor mapping (**Figure 8.31c**). The model consists of four interacting neural circuits that reify distinct cognitive processes: rule encoding and maintenance, sensory processing, feature categorization and action selection. The three non-sensory (rule, category, action) modules are modeled as in Wang (2002b) and Wong and Wang (2006), where the two selective neural pools represent two rules, categories and motor responses, respectively. A rule neural pool maintains the task context in the form of persistent activity (Wallis et al. 2001). A transition to the other rule neural pool occurs in response to a transient cue. Sensory cells independently process color and orientation features, based on the observation that feature preference for color and orientation is segregated in distinct bands of V4 neurons (Tanigawa et al. 2010). Moreover, both orientation and color can be represented by a circular variable (a ring). They receive top-down modulation from rule and category cells. Category cells group feature values into classes; they project to action cells that compete with each other and select a motor response. Connection strengths are plastic through reward-dependent plasticity, as described in Chapter 7 (Section 7.3.2), at recurrent synapses in the rule circuit, at feedforward synapses from sensory to category cells and from category to action cells.

The model reproduces the observed switch cost (**Figure 8.32a–b**). In the model, the preparatory component of the switch cost results from attractor dynamics in the rule circuit. The rule is actively maintained from one trial to the next by the activity of a rule-selective neural pool, in agreement with the observation of sustained rule-selective activity in PFC across trials (Mansouri et al. 2006; Durstewitz et al. 2010; Yamada et al. 2010). On a switch trial, rule updating requires a transition between the two rule-selective attractor states, and its time course then gives rise to the preparatory cost, which decreases with increased CTIs. In a switch trial, ER is maximum without CTI because the system is guided by the wrong rule attractor, and ER decreases with longer time for preparation similarly to the RT. The residual component is the asymptote of the switch cost for long

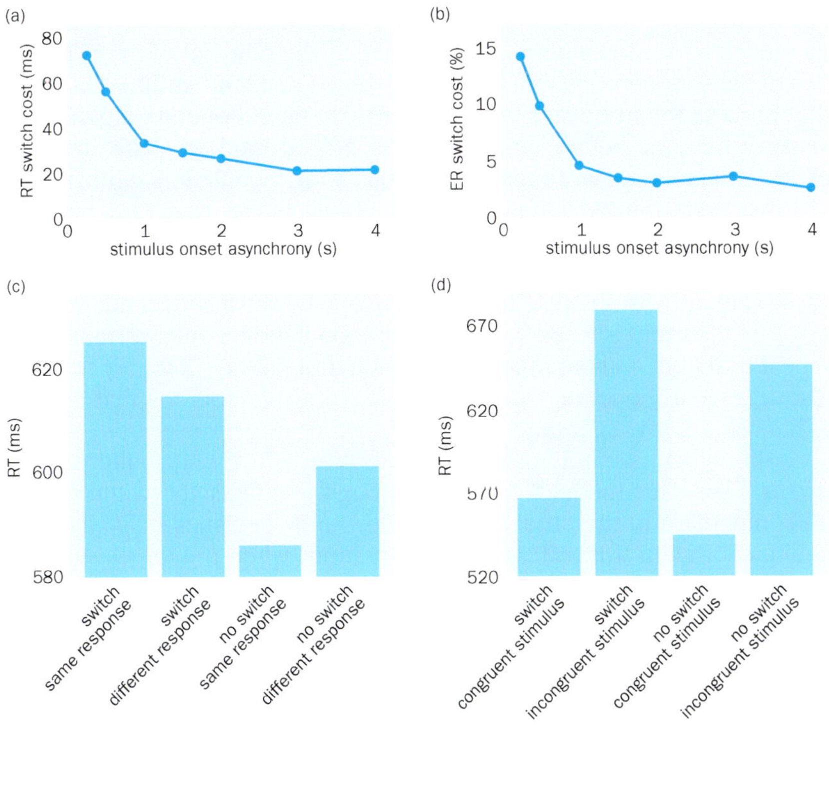

Figure 8.32 The model captures salient behavioral manifestations of task switching. (a) Reaction time (RT) switch cost. The cost is computed as the difference in RT between switch and repeat trials. The difference is displayed as a function of time interval between cue and target stimulus (CTI). The total switch cost consists of two components: A preparatory component that decays with CTI and a residual component that represents the asymptotic level for long CTIs. (b) Error rate (ER) switch cost. (c) Task-response interaction. RT is shorter for rule repeat trials but longer for rule switch trials in response repetition. (d) Task-congruency interaction, with longer RTs on incongruent compared to congruent condition. Reproduced from Ardid and Wang (2013). [Ardid, S., & Wang, X. J. (2013). A tweaking principle for executive control: Neuronal circuit mechanism for rule-based task switching and conflict resolution. Journal of Neuroscience, 33(50), 19504–19517. https://www.jneurosci.org/content/33/50/19504]

CTIs, which cannot be avoided even after a long time for preparation. In the model, neural interactions are modulated by reward-dependent synaptic plasticity: on correct trials, activated pathways are facilitated through synaptic potentiation and inactivated pathways get depressed through synaptic depression. On repeat trials this reinforcement leads to behavioral melioration, but on switch trials the depressed pathway becomes relevant and competes with the facilitated pathway, leading to the residual switch cost that remains for arbitrarily long CTIs (**Figure 8.32a–b**). It was noted in the model that a difference in the strength of the learning rate could underlie across species (human versus monkey) differences in the residual switch cost. Monkeys learn more slowly than humans and present lower residual switch costs (Stoet and Snyder 2007). A lower plasticity rate implies slower learning and, according to our model, reduces the magnitude of the residual switch costs. Plasticity through experience provides a biological mechanism for task set priming at the behavioral level (Allport and Wylie 1999).

The model captures the aforementioned intriguing effect commonly observed in humans (Rogers and Monsell 1995; Meiran 2000; Altmann 2011): selecting the same behavioral response as in the previous trial shortens RT under rule repeat but lengthens RT under rule switch (**Figure 8.32c**). This interaction effect is explained in the model by a combination of two counteracting factors. On the one hand, a same response under rule repeat requires activation of a category-to-action pathway that has been facilitated through reward-dependent plasticity, which shortens RT. On the other hand, response repetition leads to intrinsic adaptation in action cells, which makes RT longer. The combination of the two effects explains the task-response interaction, with the assumption that the benefit from potentiation overcomes the effect of adaptation in rule repeat and same response trials. A third salient behavioral signature is the congruency effect (**Figure 8.32d**). In the model, this effect is explained by the fact that the activated category pools for the two features cooperate on congruent stimuli (A for color and C for orientation dictating the leftward response; or B for color and D for orientation, leading to the rightward response), leading to faster responses, so that the role of rule bias is less important.

The neural circuit model also accounts for observed neural activities during task switching. In particular, modulation by task rule can be quantified by a "modulation ratio," defined as the ratio between the activities of sensory neurons that process the cued relevant feature and those that process the irrelevant target feature. The attentional modulation for sensory neurons before target onset is indeed modest, ~10% increase of spontaneous activity that corresponds to a modulation ratio of ~1.1 (**Figure 8.33a**). Interestingly, the model predicts that reward-dependent plasticity in the rule circuit affects this modulation, due to the activation of rule pools that have been previously reinforced with task repetition or weakened with task switch, respectively, by reward-dependent plasticity in recurrent synapses. Thus, this modulation is higher for a rule repeat trial (red) compared to a rule switch trial (black). Category neurons also receive a moderate top-down input from rule neurons that slightly enhances the low spontaneous activity before target onset from ~2.4 to ~3.2 Hz, which is barely distinguishable from activity fluctuations (**Figure 8.33b**). This modulation is not enough to trigger the winner-take-all competition in the attractor network. It only becomes fully visible in conjunction with the bottom-up sensory input when the target stimulus is presented, and the attention modulation ratio grows rapidly over time (**Figure 8.33b**). By contrast, without amplification by attractor dynamics, a rule signal must be much stronger than sensory stimulation, inconsistent with experimental observations (Ardid and Wang 2013). Attentional modulation to both sensory and category neurons is modest, yet sufficient

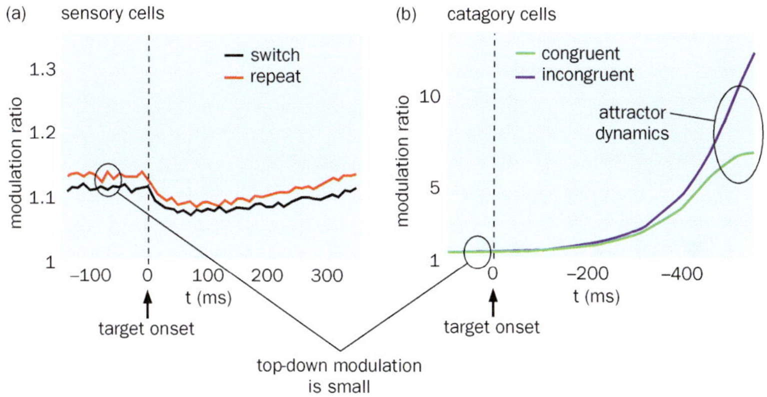

Figure 8.33 Task switching by weak top-down modulation. (a) Modulation ratio time course in feature-selective neurons. The modulation ratio is defined as the ratio of the activity of neurons selective for the relevant target feature and that of neurons selective for the irrelevant feature. The modulation ratio deviates from 1 (by ~10%) before the target onset (CTI), which is explained by weak top-down inputs originated from the rule circuit. The rule bias is slightly but significantly higher after task repeat (red) compared to switch (black). (b) Modulation ratio in category-selective neurons. The relevant category pool receives a small top-down input from rule neurons that barely enhances its activity before target onset. This weak modulation is not enough to trigger winner-take-all competition in the category circuit. However, in conjunction with the bottom-up sensory input during target presentation, it acts to bias the winner-take-all competition. Once the attractor dynamics is triggered, the modulation ratio ramps up very fast. Reproduced from Ardid and Wang (2013). [Ardid, S., & Wang, X. J. (2013). A tweaking principle for executive control: Neuronal circuit mechanism for rule-based task switching and conflict resolution. Journal of Neuroscience, 33(50), 19504–19517. https://www.jneurosci.org/content/33/50/19504]

to properly bias the winner-take-all competition in the category circuit during target presentation.

Conceptually, these experimental and modeling results give rise to the tweaking hypothesis for executive control: a weak rule bias, amplified by strong reverberatory neural population dynamics, can drastically reconfigure sensorimotor circuits.

8.6 BEHAVIORAL FLEXIBILITY AND MIXED SELECTIVITY

While the prior model is suitable for task switching guided by a rule cue in each trial, it cannot describe task switching by block design, in which case there is no explicit rule cure. Consider a simplified version of the WCST (**Figure 8.2**) used in a monkey experiment (Mansouri et al. 2006; Mansouri et al. 2009) (**Figure 8.34**). A sample stimulus (a colored shape, e.g., blue square) is first shown, then three test stimuli are added and a subject selects the one that matches the sample either in color or shape, depending on the task rule (color or shape) currently in play (Mansouri et al. 2006). When the rule, say color, remains the same across a block of trials, it is maintained internally by persistent activity of "color rule cells" across trials, but switched off when the rule has changed suddenly without warning. Feedback signal is the same for two transitions from rule 1 to rule 2 and vice versa, switching is based on an external negative feedback (e.g., no reward) for the transition from color to shape or in the opposite direction.

Figure 8.35a shows a two-by-two matrix in which each row corresponds to the internal state encoding behavioral rule (COLOR or SHAPE), high (H) or low (L) activity denotes current firing rate of color-selective neurons. Each column corresponds to the external feedback at the end of the current trial (Error or Reward). In the matrix, H (high, blue) or L (low, red) is their desired activity in the subsequent trial. When task rule is COLOR, color rule-selective neurons show a high firing rate, thus recurrent excitation among them is high (first row). It should stay H when the response yields reward, but transition to L with a negative feedback. Conversely, when task rule is SHAPE, activity of these neurons is low (second row), and they should remain L with a positive feedback but switch to H with a negative feedback. In an error trial, the transition is from H to L when the context has been COLOR and just changed to SHAPE, but from L to H in the opposite case. These two transitions cannot be accomplished by an identical feedback signal. Mathematically, this is the problem of nonlinear separability of the Boolean operation of exclusive OR (XOR).

The key to solving this problem is to introduce neurons that show conditional responses, for instance, their firing is selectively high for a particular stimulus only when rule 1 but not rule 2 is currently in play. This

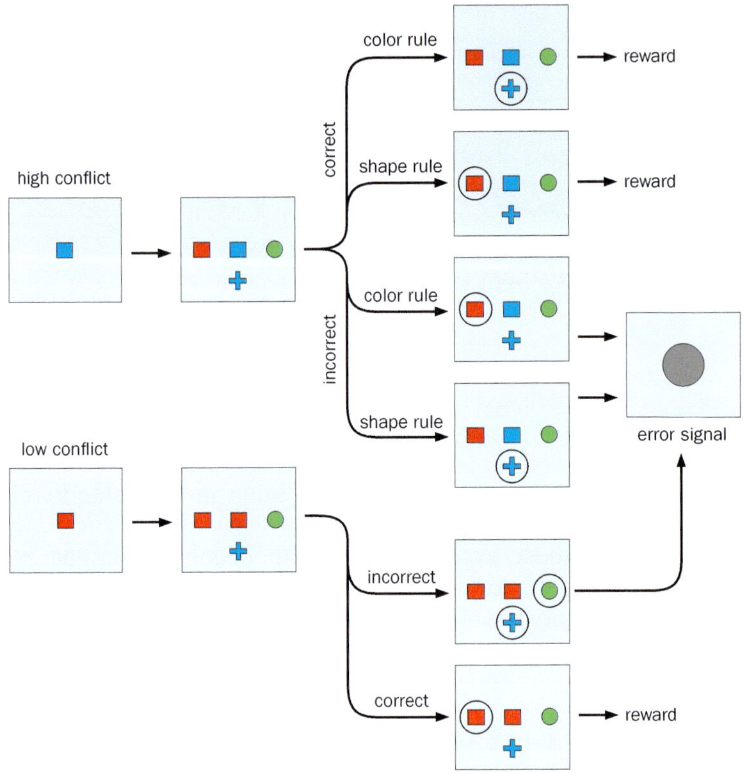

Figure 8.34 An analogue of WCST used in a monkey experiment. In each trial, the sample appeared first, then three test items were added when the monkey was required to touch the test item that matched the sample in the relevant dimension (either color or shape). The relevant feature was not cued and changed without notice whenever performance reached 85% correct in 20 consecutive trials. Trials with two different levels of conflict, high and low, were randomly mixed. In the incongruent (high-conflict) condition, the sample matched one test item only in color and another only in shape. Therefore, the monkey had to resolve the competition between two potential responses. In the congruent (low-conflict) condition, the sample matched one test item in both color and shape, and it did not match the other two test items in either color or shape; thus, there was no conflict between matching rules. Circle indicates the test item selected by the monkey. Reproduced from Mansouri et al. (2009). [Mansouri, F. A., Tanaka, K., & Buckley, M. J. (2009). Conflict-induced behavioural adjustment: A clue to the executive functions of the prefrontal cortex. Nature Reviews Neuroscience, 10(2), 141–152.]

(a)

stimulus	error	reward
state=COLOR H activity	L	H
state=SHAPE H activity	H	L

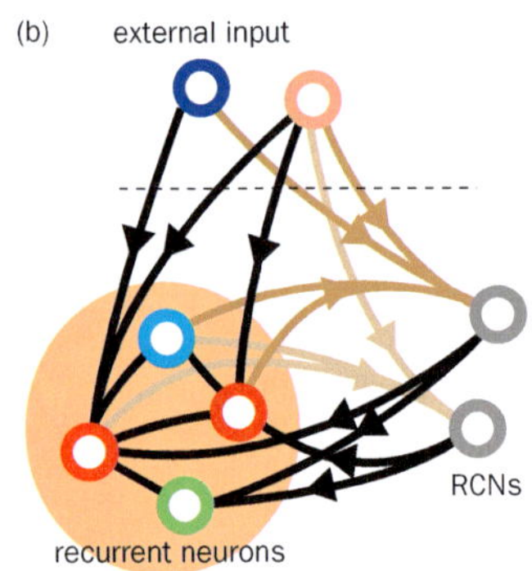

Figure 8.35 Mixed selectivity for behavioral flexibility. (a) Exclusive OR (XOR) computation by a cell that encodes the rule Color in a simple variant of the Wisconsin Card Sorting Test (WCST). (b) Neural network model architecture: A large reservoir of randomly connected neurons (RCNs) are connected both to a standard decision network of recurrent neurons and external neurons encoding sensory and feedback inputs, by random weights (brown). Each RCN projects back to the recurrent network by means of plastic synapses (black). Not all connections are shown for the sake of clarity. Reproduced from Rigotti et al. (2010). [Wang, X.-J. (2013). The prefrontal cortex as a quintessential 'cognitive-type' neural circuit: Working memory and decision making. In D. T. Stuss and R. T. Knight (Eds.), Principles of Frontal Lobe Function (Second ed.), pp. 226–248. New York: Oxford University Press. Reproduced with permission of the Licensor through PLSclear.] [Rigotti, M., Rubin, D. B. D., Wang, X. J., & Fusi, S. (2010). Internal representation of task rules by recurrent dynamics: The importance of the diversity of neural responses. Frontiers in Computational Neuroscience, 4, 24.]

reasoning led Rigotti et al. (2010) to introduce, theoretically, the concept of "mixed selectivity" for neurons with activities that depend on more than one behavioral attribute. A simple way to obtain mixed selectivity is to have a large number of randomly connected neurons (RCNs) that receive random connections from multiple sources (**Figure 8.35b**). By virtue of random connections, RCNs are naturally sensitive to a combination of synaptic inputs from external stimuli as well as rule-coding neurons (e.g., the COLOR rule is currently in play *and* the network receives a negative feedback signal), and such mixed selectivity is exactly what is needed to solve the task. This model introduced a key idea for describing context- or rule-dependent tasks (Rigotti et al. 2010).

Figure 8.36a shows an implementation of such a network model for the simplified WCST experiment of Mansouri et al. (2006). Notable is the high degree of temporal variability of single-cell firing activity and heterogeneity across neurons (**Figure 8.36b**), as observed in PFC neurons. Mixed selectivity is a salient yet puzzling characteristic of frontal cortical neurons recorded from behaving animals (Asaad et al. 1998; Sigala et al. 2008; Lapish et al. 2008; Parthasarathy et al. 2017; Dang et al. 2022). Our model suggests that mixed selectivity is computationally desirable, as it allows the network to encode a large number of combinations of behaviorally relevant information, which are critically important for enabling the PFC to subserve rule-dependent flexible behavior.

The idea of representations in a high-dimensional space has been proposed to be crucial for brain functions, in particular "untangling" in object recognition (DiCarlo and Cox 2007). Nonlinear mixed selectivity, defined by nonlinear dependence on two or more behavioral-relevant attributes (sensory stimuli, rule or context), empowers high-dimensional neural representations. This is illustrated with four hypothetical neurons (**Figure 8.37a**). The upper left neuron is selective for behavioral attribute y but not x, so its firing response is fixed along a vertical line (a rate contour). Similarly, the lower left neuron is selective for attribute x but

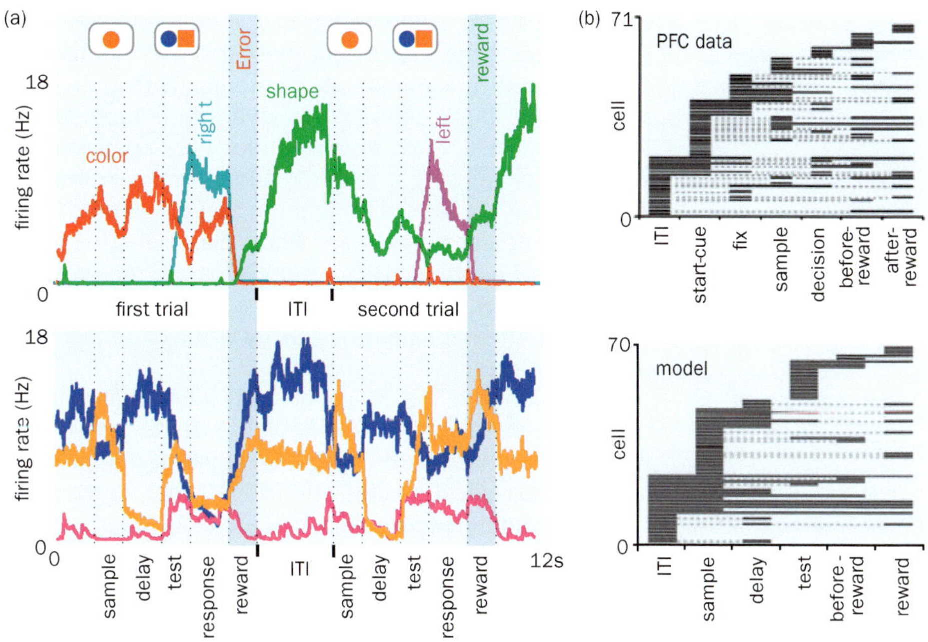

Figure 8.36 Decision-making network model with a reservoir of randomly connected neurons (RCNs) for rule-based behavior. (a) Simulation of a WCST after a rule shift (first vertical strip). In a model simulation activity, as a function of time is shown for two sample rule-selective neurons (top panel) and three RCNs (bottom panel). Key events of the two trials are indicated above the traces. With a rule switch, color rule-selective neuron is switched off (red), whereas shape-selective neuron is turned on (green), and it remains high after the end of the second trial without rule shift (second vertical strip). (b) Rule selectivity pattern is heterogeneous over time and across neurons. Lower panel: Rule selectivity for 70 simulated cells in the model: For every trial epoch (x-axis), a black bar indicates that the neuron had a significantly different activity in shape and in color rule blocks. The neurons are sorted according to the first behavioral epoch of a trial in which they show rule selectivity. Upper panel: Rule selectivity for spiking activity of single units recorded in prefrontal cortex of monkeys performing an analog of the WCST (Mansouri et al. 2006). Reproduced from Wang (2013), with original data from Rigotti et al. (2010). [Rigotti, M., Rubin, D. B. D., Wang, X. J., & Fusi, S. (2010). Internal representation of task rules by recurrent dynamics: The importance of the diversity of neural responses. Frontiers in computational neuroscience, 4, 24.] [Wang, X.-J. (2013). The prefrontal cortex as a quintessential 'cognitive-type' neural circuit: Working memory and decision making. In D. T. Stuss and R. T. Knight (Eds.), Principles of Frontal Lobe Function (Second ed.)., pp. 226–248. New York: Oxford University Press. Reproduced with permission of the Licensor through PLSclear.] [Mansouri, F. A., Matsumoto, K., & Tanaka, K. (2006). Prefrontal cell activities related to monkeys' success and failure in adapting to rule changes in a Wisconsin Card Sorting Test analog. Journal of Neuroscience, 26, 2745–2756. Licensed under CC BY 4.0.]

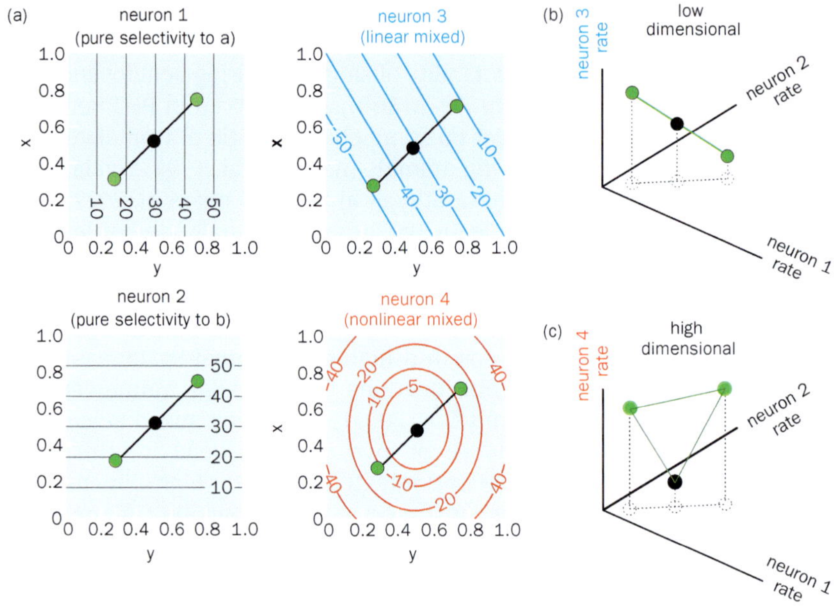

Figure 8.37 Nonlinear mixed selectivity and high dimensionality of neural representation. (a) The responses of four hypothetical neurons to two stimulus features *x* and *y*. Upper and lower left pure selectivity for *y* or *x*, respectively. Upper right: Linear combination of *x* and *y*. Lower right: Nonlinear mixed selectivity with neural response. The two green and one dark circles indicate the responses to three sensory stimuli. Numbers on a rate contour (line or circle) indicate a constant firing rate. (b) The responses of the pure and linear mixed selectivity neurons from (a) in the space of activity patterns (the axes correspond to neural firing rates) elicited by the three stimuli lie on a line, spanning a low-dimensional space, and it is impossible for a linear classifier to decode the darker central state differently from the two green states. (c) As in (b) but with the third neuron being nonlinear mixed selective. The representations of the stimuli lie on a plane, no longer confined on a line. This higher dimensionality enables a linear classifier to separate the two green states from the dark state by an appropriately chosen plane (in general, by a hyperplane). Reproduced from Rigotti et al. (2013). [Rigotti, M., Barak, O., Warden, M. R., Wang, X. J., Daw, N. D., Miller, E. K., & Fusi, S. (2013). The importance of mixed selectivity in complex cognitive tasks. Nature, 497(7451), 585–590.]

not *y*, and its rate contours are horizontal lines. The upper right neuron encodes a linear combination of *x* and *y* with rate contours as slanted lines, whereas the lower right neuron shows nonlinear mixed selectivity characterized by circular rate contours. With pure selectivity for single attributes or linear combination of attributions, neural representation is low-dimensional. Such representations are commonly difficult or impossible to decode using a simple readout. In the example of **Figure 8.37a**, suppose that the task is to map the two green dots and the one dark dot into two different responses. A linear classifier corresponding to a plane in the space of neural activity cannot separate the green versus dark states (**Figure 8.37b**). By contrast, using a neuron with nonlinear mixed selectivity (red), this becomes possible by virtue of an increased dimensionality of neural representation (**Figure 8.37c**).

Furthermore, theoretical analysis found that representations for rule- or context-dependent tasks should be dense, and optimality is achieved when ~30% of RCNs are engaged in any task (Rigotti et al. 2010). Interestingly, physiological studies often found that ~30% of randomly sampled neurons from the prefrontal cortex are involved in whatever task a subject is engaged in (see, for example, Cromer et al. 2010), leading to the proposal that neural coding in the PFC is extraordinarily adaptive (Duncan 2001). Dense representations in the PFC are in sharp contrast to early sensory areas such as V1, where coding is sparse, with a small fraction of neurons selective for a particular stimulus (Olshausen and Field 1996; Vinje and Gallant 2000; Haider et al. 2010). Another consequence of mixed selectivity is a high degree of heterogeneity of neural responses (**Figure 8.36b**), commonly observed in PFC neurons, which appear to defy simple quantification of their selectivity. Therefore, this modeling work, in particular the introduction of the concept of mixed selectivity therein, offers a simple explanation for complex neural activities observed in cognitive-type cortical circuits.

To test mixed selectivity in the PFC, Rigotti et al. (2013) analyzed single-neuron data from an experiment (Warden and Miller 2010) in which monkeys performed two tasks in different blocks of trials (**Figure 8.38a**).

In the experiment, each trial began with eye fixation. A first sample object (one of four pictures) was shown, followed by a brief delay (the one-object delay), then a second sample object (different from the first sample object), then another delay (the two-object delay) (**Figure 8.38a**, top). In the Recognition task (**Figure 8.38a**, middle), the sample sequence was followed by a test sequence, which was either a match to the sample sequence, in which case the monkeys were required to release the bar, or a nonmatch, in which case the monkeys were required to hold the bar until a matching sequence appeared. In the Recall task (**Figure 8.38a**, bottom), the sample sequence was followed by an array of three objects that included the two sample objects. Monkeys were required to make a sequence of saccades in the correct order to the two sample objects. In this task, there are $c = 24$ different conditions with two contexts, four first samples and three second samples (different from the first sample in a given trial). A binary classifier separates a subset of conditions from the others; therefore, there are at most 2^c binary classifications. How many classifiers are actually implementable depends on the neural representation, which is optimally flexible if it enables as many classifiers as possible.

Single neurons in the dorsolateral PFC (Brodmann area 46) displayed a wide range of activity patterns. Some showed pure selectivity or linear mixed selectivity, for example, a large response to one of the four stimuli regardless of the task (Recognition versus Recall) rule. Others were characterized by nonlinear mixed selectivity, i.e., their firing rates are drastically different depending on the task rule, or nonlinearly depend on a combination of the first and second sample stimuli. The number N_c of binary classifications implementable by PFC neural activity patterns can be estimated as a function of the number N of recorded cells. The number N_c of implementable binary classifications grows exponentially with the dimensionality d of neural representation (Vapnik 2013; Fusi et al. 2016), d can be estimated through the expression $d = \log_2 N_c$, when the number of neurons is sufficiently large.

Figure 8.38b shows d as a function of the number N of neurons; for N along the x-axis larger than the number of recorded cells, the performance was computed by introducing additional resampled neurons. Notably, the estimated dimensionality is nearly maximal (solid curve), but is much smaller in the absence of the subset of neurons showing nonlinear mixed selectivity (grey curve), in support of high-dimensional representation in the PFC and the hypothesis that mixed selectivity enhances flexibility, as illustrated by examples of Figure 8.37. The behavioral impact was examined in the Recall task (with $c = 12$), and it was found that the estimated dimensionality was substantially lower in error trials (grey) than in correct trials (solid) (**Figure 8.38c**), which held true even after linear mixed selectivity neurons were removed. In conclusion, this analysis of PFC neurons in behaving monkeys (Rigotti et al. 2013) lends support to the theoretically proposed mixed selectivity (Rigotti et al. 2010) as a general concept for understanding neural dynamics of the PFC and associated areas underlying executive control of flexible behavior.

It is worth noting that random connectivity is merely a convenient way to give rise to mixed selectivity, but not a prerequisite for it; dimensionality of neural population activity varies from task to task (Dubreuil et al. 2022). Moreover, linear classifier analysis probes information available in all neurons recorded in an experiment; what information is read out depends on specific cell classes that project to diverse downstream brain regions (Harris and Shepherd 2015) and their connectivity, as well as dynamical gating of information flow (see Chapter 9).

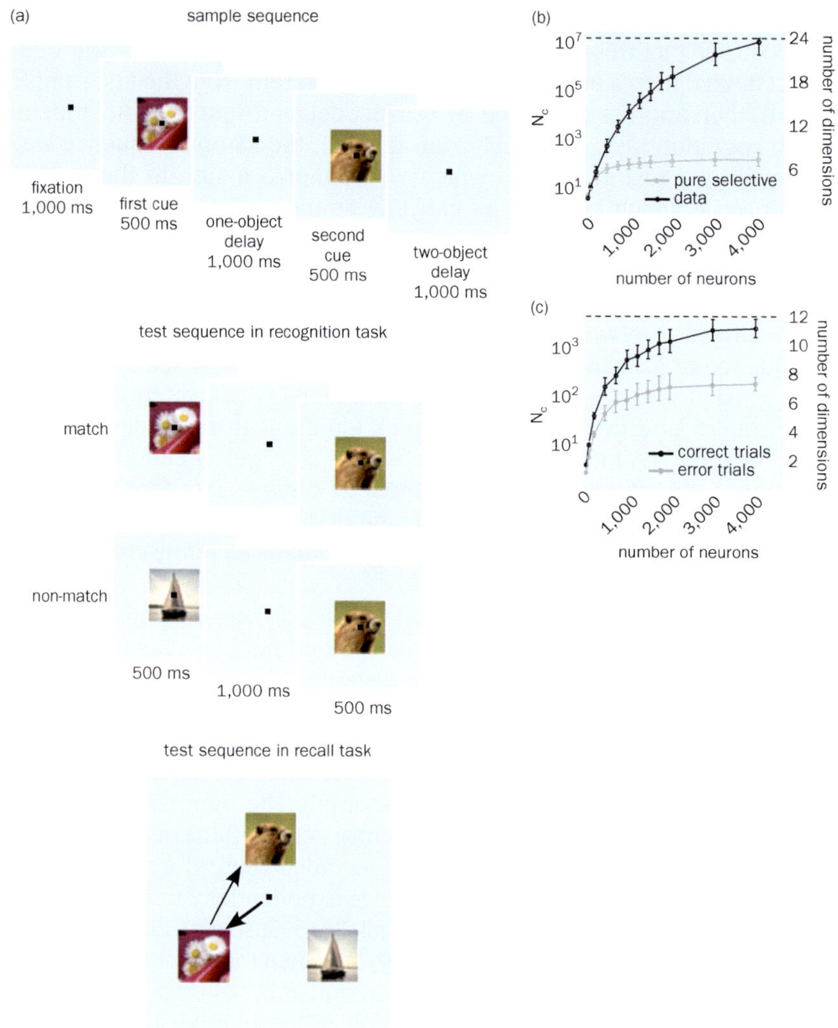

Figure 8.38 PFC neurons from monkey display nonlinear mixed selectivity. (a) Monkey experiment task. (b) Number of implementable binary classifications, N_c (ordinate left, on logarithmic scale), and dimensionality (ordinate right, linear scale) for varying number of neurons of the population readout by a linear classifier. The advantage of the recorded neural representations over the pure selectivity neurons is huge. For the recorded neural data (black line). N_c reaches 2^{24}, the value that corresponds to the maximal dimensionality $d = 24$, given by all possible combinations of cue 1 object, cue 2 object and two (Recognition and Recall) task types (solid line). On the other hand, representations based on pure selectivity (grey line) generate fewer than eight dimensions. (c) The dimensionality of the neural representations predicts animal behavior. For the correct trials, the number of implementable classifications corresponds to a dimensionality that is close to maximal ($d = 12$, dashed line, for Recognition task). In the error trials, the dimensionality drops significantly. Reproduced from Rigotti et al. (2013). [Rigotti, M., Barak, O., Warden, M. R., Wang, X. J., Daw, N. D., Miller, E. K., & Fusi, S. (2013). The importance of mixed selectivity in complex cognitive tasks. Nature, 497(7451), 585–590.]

8.7 SUMMARY

In this chapter, we covered executive control for sensory information processing and motor responses. We have seen that the same attractor network model as that of perceptual decision making can be extended to response inhibition, with the addition of a mechanism for suppressing automated habitual response in the cortex and the basal ganglia. Presumably, a similar mechanism underlies suppression of unwanted thoughts (Anderson and Green 2001; Anderson et al. 2004). On the other hand, two local circuit models built for different purposes, a working memory module and a sensory module, endowed with normalization put together in a reciprocal loop, can account for selective attention. Furthermore, a model composed of circuit modules dedicated to various cognitive components and endowed with reward-dependent plasticity explains a complex set of behavioral observations in task switching. Biologically realistic neural circuit modeling bridges neural dynamics and brain functions.

We can extract four computational themes shared by various forms of executive control. First, some internal signal must be selected depending on the behavioral goal and maintained in the form of persistent activity, which may represent behavioral rule, attentional focus or the level of executive control for automated habits. Thus, the attractor model of working memory and decision making (Chapters 5 and 6) serves as a core principle. Second, control signals may be relatively weak but amplified by local reverberatory dynamics in the sensorimotor circuits, according to the tweaking hypothesis for network reconfiguration. Third, behavioral inhibition may result from either action selection through winner-take-all competition (Chapters 6 and 7) or suppression of prepotent responses, or a combination of the two. They involve different neural circuit substrates; the latter depends on the subthalamic nucleus as well as specific excitatory cell types dedicated to movement suppression in the cortex, which are the recipients of executive command signals. Fourth, the notion of mixed selectivity helps explain highly complex neural signals observed in the prefrontal cortex underlying behavioral flexibility.

Looking into the future, beyond single-cell considerations, the state-space perspective of neural population activity is likely to shed light on how attention sculpts representational structure (Chapman and Störmer 2024) as well as control of actions. Moreover, while brain substrates for working memory and attention have much in common, they engage distinct subsets of cortical areas (Mendoza-Halliday et al. 2024), which can be investigated using a multi-regional cortex model. Another underexplored aspect of executive control is how the brain monitors poor performance and adjusts control accordingly. Behavioral and physiological experiments have shown that the medial frontal cortex is crucial for detecting a behavioral error (Gehring et al. 1993; Ridderinkhof et al. 2004; Mansouri et al. 2007; Johnston et al. 2007; Ullsperger et al. 2014; Sajad et al. 2019; Kawai et al. 2019; Domenech et al. 2020; Fu et al. 2023). While mathematical models for error computation have been proposed (Alexander and Brown 2011), the underlying circuit mechanism remains unknown. Since errors usually are correlated with enhanced conflict, whether conflict or error representation better describes the function of the medial frontal cortex remains a topic of active research (Botvinick et al. 2004; Shenhav et al. 2013). Also poorly understood is the brain mechanism of deploying such a monitoring signal to dynamically adjust the level of executive control. Another widely open question concerns the distributed brain network of cognitive control (Badre 2021). These are a few examples of how the philosophy of understanding "building blocks" of cognition can be extended in multiple directions, in the development of a theory capable of accounting for increasingly complex mental processes.

FURTHER READING

Books and Reviews

Bari, A. and T. W. Robbins (2013). Inhibition and impulsivity: Behavioral and neural basis of response control. *Prog. Neurobiol. 108*, 44–79.

Buschman, T. J. and S. Kastner (2015). From behavior to neural dynamics: An integrated theory of attention. *Neuron 88*, 127–144.

Chun, M. M., J. D. Golomb, and N. B. Turk-Browne (2011). A taxonomy of external and internal attention. *Annu. Rev. Psychol. 62*, 73–101.

Desimone, R. and J. Duncan (1995). Neural mechanisms of selective visual attention. *Annu. Rev. Neurosci. 18*, 193–222.

Diamond, A. (2013). Executive functions. *Annu. Rev. Psychol. 64*, 135–168.

Hikosaka, O., Y. Takikawa, and R. Kawagoe (2000). Role of the basal ganglia in the control of purposive saccadic eye movements. *Physiol. Rev. 80*, 953–978.

Mansouri, F. A., K. Tanaka, and M. J. Buckley (2009). Conflict-induced behavioural adjustment: A clue to the executive functions of the prefrontal cortex. *Nat. Rev. Neurosci. 10*, 141–152.

Maunsell, J. H. R. (2015). Neuronal mechanisms of visual attention. *Annu. Rev. Vis. Sci. 1*, 373–391.

Munoz, D. P. and S. Everling (2004). Look away: The anti-saccade task and the voluntary control of eye movement. *Nat. Rev. Neurosci. 5*, 218–228.

Nobre, A. C. and S. Kastner (Eds.) (2014). *The Oxford Handbook of Attention*. Oxford: Oxford University Press.

Reynolds, J. H. and D. J. Heeger (2009). The normalization model of attention. *Neuron 61*, 168–185.

Schall, J. D. (2001). Neural basis of deciding, choosing and acting. *Nat. Neurosci. 2*, 33–42

Primary Research Papers

Selective Attention

Ardid, S. and X.-J. Wang (2013). A tweaking principle for executive control: Neuronal circuit mechanism for rule-based task switching and conflict resolution. *J. Neurosci. 33*, 19504–19517.

Ardid, S., X.-J. Wang, and A. Compte (2007). An integrated microcircuit model of attentional processing in the neocortex. *J. Neurosci. 27*, 8486–8495.

Ardid, S., X.-J. Wang, D. Gomez-Cabrero, and A. Compte (2010). Reconciling coherent oscillation with modulation of irregular spiking activity in selective attention: Gammarange synchronization between sensory and executive cortical areas. *J. Neurosci. 30*, 2856–2870.

Bundesen, C., T. Habekost, and S. Kyllingsbæk (2005). A neural theory of visual attention: Bridging cognition and neurophysiology. *Psychol. Rev. 112*, 291–328.

Chelazzi, L., E. K. Miller, J. Duncan, and R. Desimone (2001). Responses of neurons in macaque area V4 during memory-guided visual search. *Cereb. Cortex 11*, 761–772.

Fries, P., J. Reynolds, A. Rorie, and R. Desimone (2001). Modulation of oscillatory neuronal synchronization by selective visual attention. *Science 291*, 1560–1563.

Martinez-Trujillo, J. and S. Treue (2004). Feature-based attention increases the selectivity of population responses in primate visual cortex. *Curr. Biol. 14*, 744–751.

McAdams, C. J. and J. H. R. Maunsell (1999). Effects of attention on orientation-tuning functions of single neurons in macaque cortical area V4. *J. Neurosci. 19*, 431–441.

Moore, T. and K. M. Armstrong (2003). Selective gating of visual signals by microstimulation of frontal cortex. *Nature 421*, 370–373.

Moran, J. and R. Desimone (1985). Selective attention gates visual processing in the extrastriate cortex. *Science 229*, 782–784.

Executive Control of Action

Boucher, L., T. Palmeri, G. Logan, and J. Schall (2007). Inhibitory control in mind and brain: An interactive race model of countermanding saccades. *Psychol. Rev. 114*, 376–397.

Frank, M. J., J. Samanta, A. A. Moustafa, and S. J. Sherman (2007). Hold your horses: Impulsivity, deep brain stimulation, and medication in parkinsonism. *Science 318*, 1309–1312.

Hanes, D. P., W. F. Patterson, and J. D. Schall (1998). Role of frontal eye fields in countermanding saccades: Visual, movement, and fixation activity. *J. Neurophysiol. 79*, 817–834.

Isoda, M. and O. Hikosaka (2007). Switching from automatic to controlled action by monkey medial frontal cortex. *Nat. Neurosci. 10*, 240–248.

Lo, C. C., L. Boucher, M. Pare, J. D. Schall, and X.-J. Wang (2009). Proactive inhibitory control and attractor dynamics in countermanding action: A spiking neural circuit model. *J. Neurosci. 29*, 9059–9071.

Lo, C. C. and X.-J. Wang (2016). Conflict resolution as near-threshold decision-making: A spiking neural circuit model with two-stage competition for antisaccadic task. *PLoS. Comput. Biol. 12*, e1005081.

Logan, G. D. and W. B. Cowan (1984). On the ability to inhibit thought and action: A theory of an act of control. *Psych. Rev. 91*, 295–327.

Milner, B. (1963). Effects of different brain lesions on card sorting. *Arch. Neurol. 9*, 90–100.

Rigotti, M., O. Barak, M. R. Warden, X. J. Wang, N. D. Daw, E. K. Miller, and S. Fusi (2013). The importance of mixed selectivity in complex cognitive tasks. *Nature 497*, 585–590.

Rigotti, M., D. B. Rubin, X.-J. Wang, and S. Fusi (2010). Internal representation of task rules by recurrent dynamics: The importance of the diversity of neural responses. *Front. Comput. Neurosci.* 4, 24.

Schmidt, R., D. K. Leventhal, N. Mallet, F. Chen, and J. D. Berke (2013). Canceling actions involves a race between basal ganglia pathways. *Nat. Neurosci. 16*, 1118–1124.

Wei, W. and X. -J. Wang (2016). Inhibitory control in the basal ganglia-thalamocortical loop: Complex regulation and interplay with memory and decision processes. *Neuron 92*, 1093–1105.

Part III

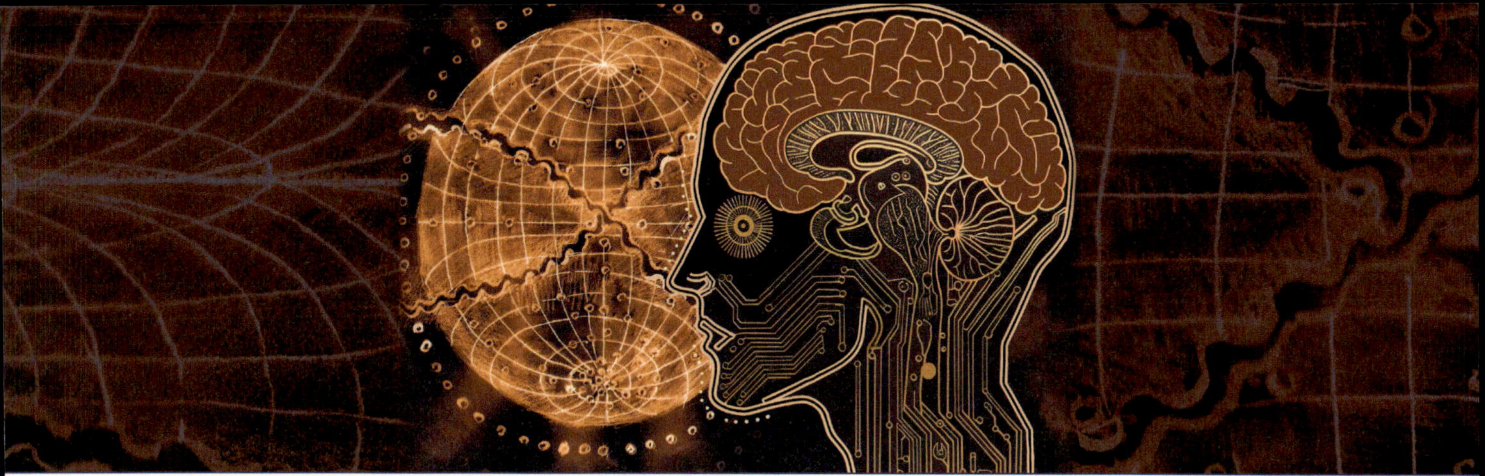

Large-Scale Multi-Regional Brain

9

Faculty psychology is impressed by such prima facie differences as between, say, sensation and perception, volition and cognition, learning and remembering, or language and thought.

J. A. Fodor in *The Modularity of Mind*

9.1 INTRODUCTION

Modules refers to subsystems, each of which is able to accomplish a specific function with autonomy, although its operation may be altered in the absence of interactions with the rest of the intact whole system (Shallice 1988). Fodor argued that mental processes are modularly organized; to some extent the brain also subdivides into modules, exemplified by a small "patch" in the inferotemporal cortex that is specialized for perception of faces (Desimone et al. 1984; Kanwisher et al. 1997; Tsao et al. 2006; Tsao and Livingstone 2008). In the early nineteenth century, phrenology promoted by Franz Joseph Gall claimed that mental traits could be assessed by measuring bumps of various parts of one's head (**Figure 9.1**). While phrenology was non-scientific and misleading, historically it helped spread the idea of localization of distinct functions in the brain. In particular, it influenced Paul Broca in his discovery of a part of the frontal lobe dedicated to speech production. In the brain, low-level sensory systems, language and other cognitive faculties are subserved by different neural subsystems, which nevertheless depend on each other in a complicated interconnected network. The relatively independent modules do not seem to be readily reconciled with the holistic perspective of emergent abilities (Bechtel 2008).

In the previous chapters, we covered various types of cognitive functions and executive control; by and large, theories and models were developed for local circuits. This is mainly because theory is driven by empirical data, and until quite recently neurophysiology with behaving animals in an experiment has been mostly limited to recording from one brain region at a time, be it the primary visual cortex (V1), prefrontal cortex (PFC) or hippocampus. Presently, there is a dearth of established theory, models or

DOI: 10.1201/9781003459361-12

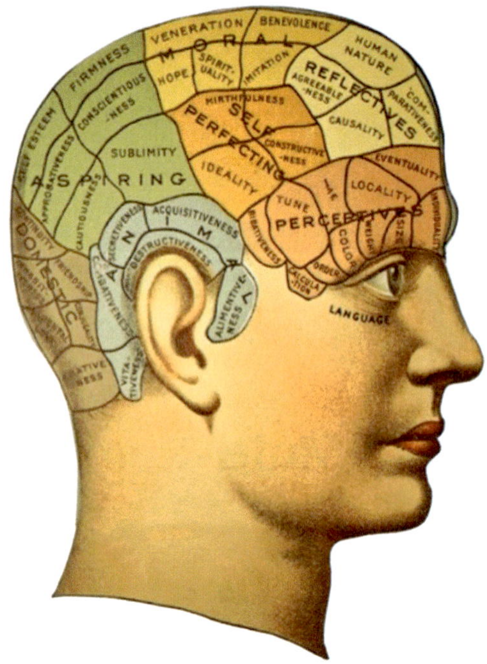

Figure 9.1 Phrenology posits that a mental faculty is instantiated by an identifiable region of the brain, and its prominence can be measured by the size of the corresponding part of a person's skull.

general principles for distributed brain processes underlying behavior. A cognitive function such as working memory may depend on a local area, but it is still possible that its interactions with other brain regions are important. For instance, in an oculomotor delayed response (ODR) task, location-selective persistent neural activity during a mnemonic delay period has been observed in the posterior parietal cortex (Gnadt and Andersen 1988), prefrontal cortex (Funahashi et al. 1989), thalamus (Funahashi et al. 2004) and basal ganglia (Hikosaka and Wurtz 1983) of behaving monkeys. How can we understand the distributed nature of working memory (Christophel et al. 2017)? Similarly, a simple decision engages multiple brain regions (Siegel et al. 2015; Pinto et al. 2019; Gallero-Salas et al. 2021). Neural signals correlated with decision in a direction discrimination task (Chapter 6) have been observed in the posterior parietal cortex (Shadlen and Newsome 2001; Roitman and Shadlen 2002; Hanks et al. 2015), prefrontal cortex (Kim and Shadlen 1999; Kiani et al. 2014b; Hanks et al. 2015), superior colliculus (Horwitz et al. 2004; Kim and Basso 2008; Kopec et al. 2015) and basal ganglia (Ding and Gold 2013; Yartsev et al. 2018). Moreover, flexible behavior and cognition depend on cooperation of various cognitive processes, each of which engages a subnetwork of the brain. Therefore, a major challenge today is to develop theory and models of multi-regional large-scale brain systems (Eliasmith et al. 2012; Wang 2022).

Rigorous research on large-scale complex brain circuits is presently taking off thanks to technical advances in the two main pillars of neuroscience: anatomy and physiology. Characterization of large-scale cortical connectivity provides a necessary structural foundation. More than 30 years ago, Felleman and Van Essen (1991) published a landmark paper where they collated data from existing literature to propose a hierarchy model of the macaque monkey cortex (**Figure 9.2**). Since then, novel

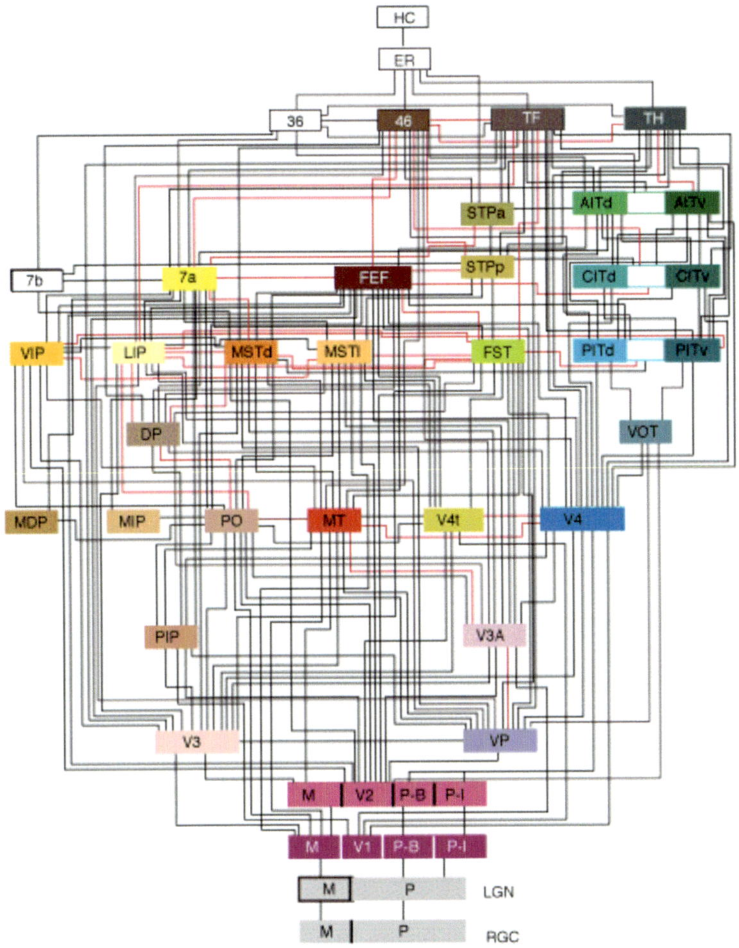

Figure 9.2 Felleman–van Essen cortical hierarchy (from bottom to top). The circuit wiring diagram was based on connectivity data from 32 visual areas and a few areas outside of the visual system; connection weights are binary, and the directionality of connections is not indicated. Reproduced from Felleman and Van Essen (1991). [Felleman, D. J., & Van Essen, D. C. (1991). Distributed hierarchical processing in the primate cerebral cortex. Cerebral Cortex, 1(1), 1–47.]

approaches and technologies have ushered in the field of brain connectomics (Sporns et al. 2005; Seung 2012; Swanson and Lichtman 2016; Wang and Kennedy 2016; Wang et al. 2020; Abbott et al. 2020), where quantitative connectivity analysis is carried out at microscopic, mesoscopic and macroscopic scales (Hagmann et al. 2008; Bohland et al. 2009; Glasser and Van Essen 2011; Helmstaedter et al. 2013; Markov et al. 2014a; Zheng et al. 2018; Harris et al. 2019; Pereira et al. 2019; Turner-Evans et al. 2020; Scheffer et al. 2020; Xu et al. 2021; Foster et al. 2021; MICrONS Consortium et al. 2021; Wildenberg et al. 2021; Xu et al. 2022). At the same time, simultaneous recording (Dotson et al. 2017) or calcium imaging from many neurons (Grienberger and Konnerth 2012; Sofroniew et al. 2016; Musall et al. 2019) in multiple brain regions of behaving animals has become feasible. Progress is accelerating with the invention of neuropixels, the new-generation electrodes that record the activity of hundreds of neurons in several brain regions (Jun et al. 2017; Stringer et al. 2019; Steinmetz et al. 2021; Paulk et al. 2022; Trautmann et al. 2023; Chen et al. 2023b). Both anatomical and physiological studies are aided by transcriptome, which provides genetic tools for classification of cell types and cell-type specific connectivity analysis (Luo et al. 2008; Zeisel et al. 2015; Tasic et al. 2018; Zhu et al. 2018; Hodge et al. 2019; Krienen et al. 2020; Chen et al. 2023a). Challenges are daunting as new statistical methods are needed to analyze massive neural data and lack of theory hinders efforts to understand complex multi-regional brain systems (Urai et al. 2021). Tackling the new challenge, in this chapter we will discuss recent research on connectivity, global neural dynamics and distributed cognition in large-scale brain circuitry.

9.2 CORTEX-WIDE CONNECTIVITY

9.2.1 Connectome

At the turn of the twenty-first century, new tools were developed for the analyses of the brain's connectivity. On the one hand, quantitative microscopic analysis came to the fore thanks to the invention of serial block-face scanning electron microscopy (EM) (Denk and Horstmann 2004) for three-dimensional reconstruction of single neurons and their connections with a spatial resolution of tens of nanometers. Today, nanoscale EM-level analysis is in various stages of development for a number of systems, including the complete wiring diagrams of brains in zebrafish (Hildebrand et al. 2017) and *Drosophila* fly (Scheffer et al. 2020; Dorkenwald et al. 2022), the mouse retina (Helmstaedter et al. 2013) and a cubic-millimeter chunk of the mouse neocortex with 75,000 neurons and 500 million connections (MICrONS Consortium et al. 2021). On the other hand, on a macroscopic scale, the advances of magnetic resonance imaging (MRI) diffusion-weighted tractography made it possible to probabilistically measure, non-invasively, connection fibers in the human brain (**Figure 9.3a**) (Mori et al. 1999; Catani et al. 2002; Hagmann et al. 2008; Johansen-Berg and Rushworth 2009; Glasser and Van Essen 2011).

The first decade of this century also saw the start of mathematical analysis of the brain's connectivity. Felleman and Van Essen (1991) inspired efforts that, 10 years later, led to an inter-areal cortical connectivity matrix, the Collation of Connectivity data on the Macaque brain (CoCoMac) (Stephan et al. 2001). Graph theory that was pioneered by Erdos and Renyi (1960) and has matured in recent decades (Albert and Barabási 2002; Newman 2010; Barabási 2016) began to be applied to quantify the brain's connectivity in search of its organizational principles (Sporns et al. 2005;

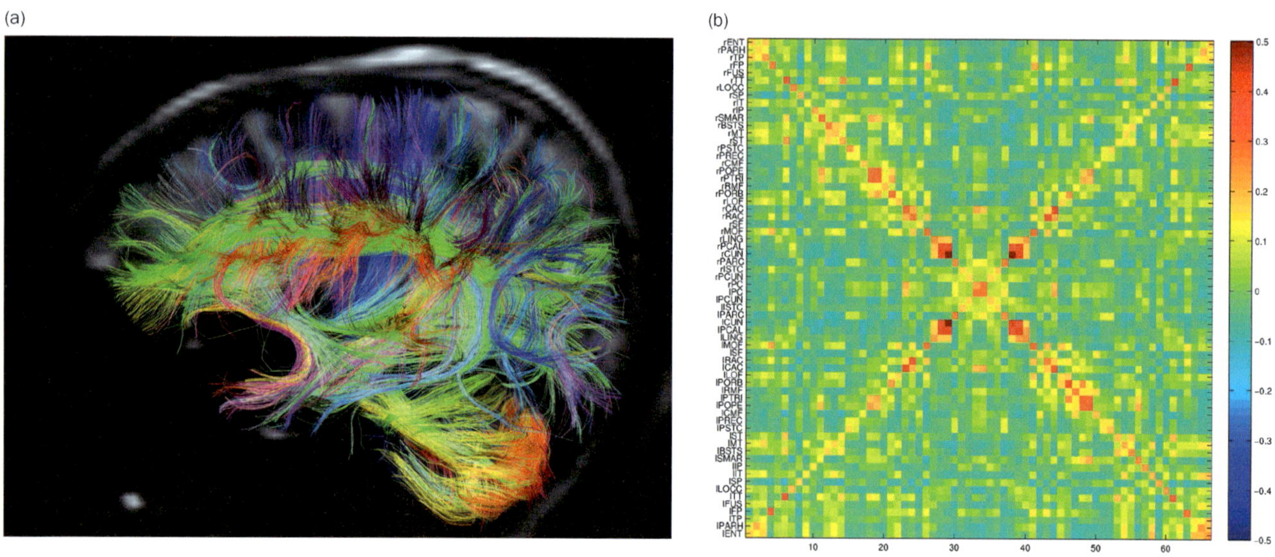

Figure 9.3 Structural and functional connectivity of the human cortex. (a) A sagittal slice of diffusion tensor imaging data, where coloring indicates the orientation of measured axonal tracts in the white matter. (b) Functional connectivity matrix computed from fMRI BOLD signals (five human subjects) in the resting state, where 33 cortical regions in each hemisphere are labeled by their name abbreviations along the y-axis, totaling 66 (along the x-axis). Pairwise correlations are indicated by color; high correlations are concentrated around the diagonal line, as well as the line orthogonal to it corresponding to the same regions on the left and right hemispheres. Panel (a) is reproduced from www.humanconnectomeproject.org, (b) from Deco and Jirsa (2012) using data from Hagmann et al. (2008). [Deco, G., & Jirsa, V. K. (2012). Ongoing cortical activity at rest: Criticality, multistability, and ghost attractors. Journal of Neuroscience, 32(10), 3366–3375. Data from Hagmann, P., Cammoun, L., Gigandet, X., Meuli, R., Honey, C. J., Wedeen, V. J., & Sporns, O. (2008). Mapping the structural core of human cerebral cortex. PLoS Biology, 6(7), e159.]

Bullmore and Sporns 2009; Sporns 2009). By then, human brain imaging had gone mainstream in cognitive science (Posner and Raichle 1994; Raichle 2009). Functional MRI (fMRI) (Ogawa et al. 1990; Logothetis and Wandell 2004), enabling researchers to non-invasively monitor the blood-oxygen-level-dependent (BOLD) signal that co-varies with neural activity, has become an important tool in brain connectomics (Van Essen 2013). fMRI has been widely used to measure large-scale cortical functional connectivity (FC). In a given brain state, for instance "at rest" when a subject is in an fMRI scanner but does not perform any task, BOLD signals can be measured and processed to yield $\{x_i;\ i=1,2,\ldots,N\}$ for N parcellated cortical areas. These signals fluctuate stochastically, and a covariance matrix measures how correlated are the variations of pairs of areas i and j in their activities (x_i and x_j):

$$\mathrm{FC}_{i,j} = \frac{\left\langle (x_i - \langle x_i \rangle)(x_j - \langle x_j \rangle) \right\rangle}{\sqrt{\left\langle (x_i - \langle x_i \rangle)^2 \right\rangle \left\langle (x_j - \langle x_j \rangle)^2 \right\rangle}} \tag{9.1}$$

where $\langle \cdot \rangle$ denotes the average. The functional connectivity is symmetrical ($\mathrm{FC}_{i,j} = \mathrm{FC}_{j,i}$) with values between −1 and 1. **Figure 9.3b** shows an example from human fMRI measured in the resting state (Hagmann et al. 2008; Deco et al. 2012).

Two areas more strongly connected anatomically are naturally expected to show a larger FC; indeed, many studies have been devoted to relating functional connectivity and structural connectivity (Honey et al. 2007; Hagmann et al. 2008; Smith et al. 2009; Greicius et al. 2009; Glasser et al. 2016; Breakspear 2017). Functional connectivity can also be measured with a better spatial resolution using focal stimulation (Hu et al. 2020). Electroencephalography (EEG), intracranial electrocorticography (ECoG) and magnetoencephalography (MEG) offer a better temporal resolution but poorer spatial resolution than fMRI (Bressler and Menon 2010).

Diffusion tractography does not yield directional information about structural connectivity: estimated fibers may be from area A to area B, or vice versa. Functional connectivity defined by correlation of BOLD signals also does not have directionality. In both cases, the resulting connection matrix is symmetrical. Also, due to limited signal-to-noise ratio (Donahue et al. 2016), analysis of diffusion tractography often focuses on an all-or-none binary matrix (**Figure 9.4**). Thus, cortical areas are nodes connected by links, corresponding to vertices and edges, respectively, on a mathematically defined graph. Using graph theory, a number of network characteristics can be identified: degree of a node is the number of its links; hubs are nodes with an unusually large number of connections; clustering

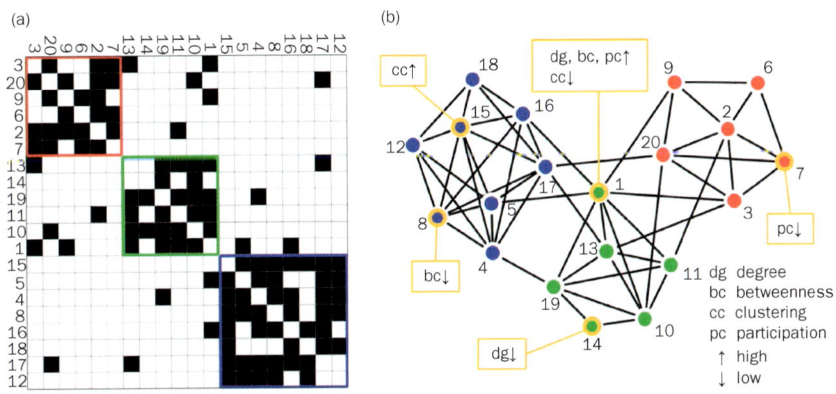

Figure 9.4 Network concepts. (a) A connection matrix summarizing binary pairwise relations among 20 nodes. These relations express the presence (black) or absence (white) of a symmetric connection. The nodes are ordered according to an optimal modularity partition. Three network modules are shown (red, green, blue). (b) Graph diagram of the network summarized in (a) with module assignments indicated by node color. Various nodes are highlighted, according to various network measures. For example, node 1 stands out because of its high degree (number of connections), betweenness centrality (placement on many of the network's short paths), and participation coefficient (connections broadly distributed across network modules). Node 1 also has low clustering, as most of its topological neighbors are mutually unconnected. Other nodes with high clustering (node 15), low betweenness (node 8), low degree (node 14) and low participation coefficient (node 7) are indicated. Reproduced from Sporns (2014). [Sporns, O. (2014). Contributions and challenges for network models in cognitive neuroscience. Nature Neuroscience 17, 652–660.]

coefficient measures the degree to which nodes are clustered together (nodes that are connected with a given node also are connected with each other); etc. While network science offers a mathematical tool for analyzing the brain's connectome (Sporns 2009; Deco et al. 2015; Fornito et al. 2016; Bassett and Sporns 2017), this framework needs to be dramatically extended in order to meet the needs of neuroscience, as it will become clear later.

Structural data provide crucial information for large-scale cortical modeling (Honey et al. 2007; Izhikevich and Edelman 2008; Deco et al. 2012; Eliasmith 2013; Breakspear 2017). For example, Deco et al. (2012) used inter-areal connectivity matrix obtained from diffusion spectrum imaging (Hagmann et al. 2008) to build a dynamical model for the resting state of the human cortex; each local area was modeled as a network of spiking neurons (Brunel and Wang 2001; Wang 2002b) or by population rate equations (Wong and Wang 2006). The Balloon model for hemodynamic response (Friston et al. 2003) converted neural activity into BOLD signal. The model-generated FC matrix was compared with measured FC, the correlation between the two was optimized by varying a model parameter that globally scaled all inter-areal connections, and the maximal correlation achieved by this model was around 0.5. This and other early models assumed that each area is described by the same model, in line with the graph theory, which treats nodes as being identical.

9.2.2 Directed and Weighted Inter-Areal Cortical Connections

Classical anatomical methods use a retrograde or anterograde tracer, sometimes both in a study (Zingg et al. 2014). Injected into a brain region, an anterograde tracer travels to downstream target areas, whereas a retrograde tracer travels backward and labels neurons in upstream source areas. Up to recently, anatomical studies were essentially qualitative with results shown in images. While it may be true that a picture is worth a thousand words, systematic measurements translated into quantitative numbers are essential for discovering general principles of large-scale cortical organization.

Regarding macroscopic cortico-cortical connectivity, a notable advance was the publication of a quantitative matrix of macaque cortical connectivity from the group of Henry Kennedy (Markov et al. 2014a). In that study of a subset of cortical areas, a retrograde tracer was injected into one of 29 parcellated areas (**Figure 9.5a**). Denote by $N_{i,j}$ the number of labeled neurons from source area j to injected area i, the relative input weight of area j, among all source areas, was quantified by the fraction of labeled neurons (FLNs), $FLN_{i,j} = N_{i,j} / \sum_k N_{i,k}$ (defined between 0 and 1) (**Figure 9.5b**). Interestingly, by analyzing more than 1,600 connection pathways, it was found that this measure of connection weights spans five orders of magnitudes (**Figure 9.5c**); some inter-areal inputs are strong while many others are quite weak but robustly present. The FLN statistics obeys a log-normal distribution.

With 29 areas analyzed, the total number of their potential interconnections is $29 \times 28 = 812$. Markov et al. (2014) found 536 directed connections, about 65% of all possible connections, thus inter-areal connectivity is quite dense in macaque cortex. Marmoset monkey cortex displays a comparable density of inter-areal connectivity, as well as five orders of magnitudes of FLN values (Theodoni et al. 2022). In mouse cortex, inter-areal connection weights measured using an anterograde tracer span three orders of magnitudes (Oh et al. 2014), and the overall connection density is even higher than in monkeys (Gămănuţ et al. 2018). For all three species, the inter-areal connection weights exhibit a lognormal distribution.

These new datasets are in contrast to previously published cortical connectivity matrices, which are sparse due to lack of sufficient information. An early analysis of a macaque connectivity matrix that included only 15% of all possible projections suggested that the mammalian cortex exhibits small-world topology (Sporns and Zwi 2004; Bassett et al. 2006). Small-world networks are known as "six degrees of separation" in the general public (Watts 2004) and were made well known in scientific research by the model of Watts and Strogatz (1998). In a small-world network (such as a social network of people), the majority of links are locally clustered (most of your friends are near you), whereas a few long-range connections (a few acquaintances live far away) ensure that a message from a node can reach any other node through a surprisingly small number (about six) of connections. In the cortex, most connections are local in a parcellated area, thus highly clustered. However, the concept of small-world network for the macroscopic cortico-cortical network assumes sparse connectivity, whereas inter-areal connections are actually dense, as shown by recent data. One can argue that with weighted connections, the network would become sparser when weak connections are discarded, and small-world properties may apply to a backbone of the cortex with strong connections. This reasoning is unsatisfying because presently potential functions of weak connections measured by FLNs remain unknown and could be significant for some purposes.

Buzsáki et al. (2004) proposed that on a fine spatial scale, neuron-to-neuron connectivity is sparse and may follow small-world network attributes, which would be desirable for efficient communication while optimizing wiring costs. Indeed, wires in the small volume of a cortex must be minimized as much as possible (Chklovskii et al. 2002). In *Drosophila* flies, an electron microscopy–based connectome of the whole brain showed that the neuron-to-neuron connection matrix is sparse with the graph density of 0.016% (Dorkenwald et al. 2024). A human brain has connection fibers with a total length comparable to the distance between the earth and the moon. With 86 billion neurons and 10^{15} connections, in a randomly wired brain the average probability that two arbitrarily chosen neurons are connected is $\sim 10^{-7}$, exceedingly small. If 60% of all neuron-to-neuron connections were present, the human brain would have 10^{21} connections, and our head would need to be a million $(10^{21}/10^{15})$-fold bigger.

In summary, macroscopically inter-areal connectivity is dense and follows a log-normal distribution of connection weights; by contrast, at the neuronal level, synaptic connections are sparse and highly clustered, needed for the economy of wiring cost as well as a tradeoff between modularity and integration of brain functions.

9.2.3 Exponential Distance Rule

Another finding from the same experiment was that $FLN_{i,j}$ decreases exponentially with the wiring distance d_{ij} between a pair of areas i and j (**Figure 9.5d**), as $\exp(-\lambda d_{ij})$ where λ is the inverse of a characteristic spatial scale (Ercsey-Ravasz et al. 2013). For macaque cortex, $1/\lambda_{\text{macaque}} \approx 5.7$ mm. Therefore, a cortical network is not only a topological graph made of nodes connected by links (**Figure 9.4**), but the spatial relationship between nodes (areas) must also be taken into account in a network description.

This exponential distance rule (EDR) holds for marmoset monkeys as well as mouse. It was previously shown that the decay rate of the EDR is about $\lambda \simeq 0.78$ mm^{-1} in the mouse (Horvát et al. 2016), and the decay rate in the marmoset shows an intermediate value between macaque and mouse, $\lambda \simeq 0.3$ mm^{-1} (Theodoni et al. 2022). The three data points obtained using the same tract tracing method are plotted as a function of the grey matter volume (GMV) in **Figure 9.6**, suggesting a power law of the EDR, $\lambda \sim GMV^{-2/9}$.

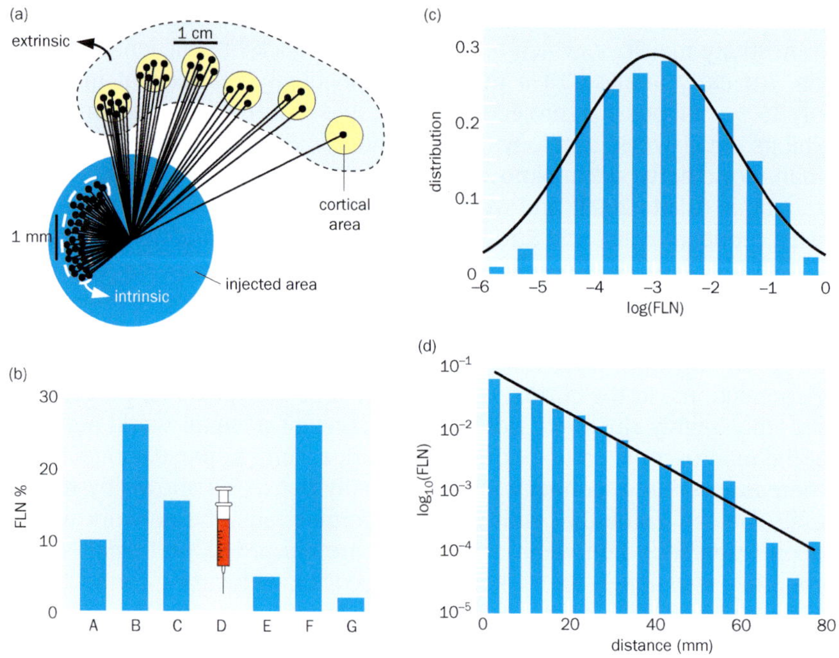

Figure 9.5 Inter-areal connections of macaque monkey cortex. (a) A retrograde tracer is injected into a (target) area, and neurons in (source) areas that project to it are labeled. (b) Relative weight of connection between a source area and the target area is quantified by the fraction of labeled neurons (FLN), which is the number of labeled neurons in the source area divided by the total number of labeled neurons. (c) FLNs span five order of magnitudes and are fit by a log-normal distribution. (d) FLN between a pair of cortical areas is an exponential function of their distance (black line), with the decay rate $\lambda \simeq 0.175$ mm^{-1}. Reproduced from Wang and Kennedy (2016), with original data from Markov et al. (2014) and Ercsey-Ravasz et al. (2013). [Wang, X. J., & Kennedy, H. (2016). Brain structure and dynamics across scales: In search of rules. Current Opinion in Neurobiology, 37, 92–98.] [Markov, N. T., Vezoli, J., Chameau, P., Falchier, A., Quilodran, R., Huissoud, C., ... & Kennedy, H. (2014). Anatomy of hierarchy: Feedforward and feedback pathways in macaque visual cortex. Journal of Comparative Neurology, 522(1), 225–259.]

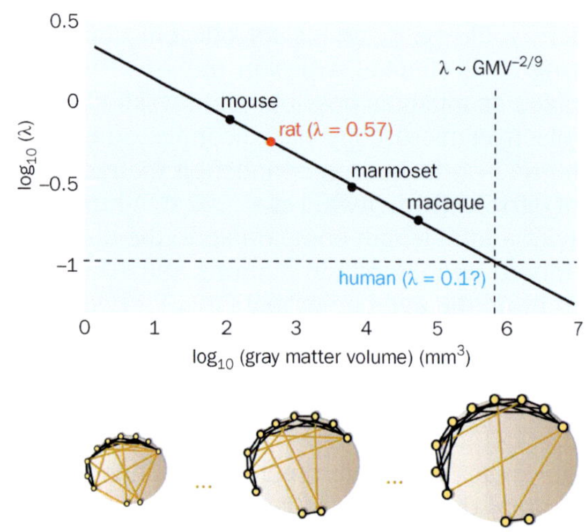

Figure 9.6 Cortical-connectivity spatial length as a function of brain size. Top: Shown is the base 10 logarithm of the decay rate λ of the exponential distance rule (EDR) of the mouse, marmoset and macaque, as a function of the grey matter volume (GMV). The plot is a linear fit on these three points with a slope of $-2/9$. The red square is the measured value of the decay rate in the rat, and the intersection of the dotted lines is the extrapolation of the decay rate in the human, given the rat and human GMV values. Bottom: A schematic representation of the decrease of long-range connections, showing that the larger the brain, the fewer long-range connections. Reproduced from Theodoni et al. (2022). [Theodoni, P., Majka, P., Reser, D. H., Wójcik, D. K., Rosa, M. G., & Wang, X. J. (2022). Structural attributes and principles of the neocortical connectome in the marmoset monkey. Cerebral Cortex, 32(1), 15–28. Licensed under CC BY 4.0.]

Given the GMV of the rat, this power law predicts its EDR decay rate to be $\simeq 0.57\ \text{mm}^{-1}$, which is validated by fitting the EDR to the rat connectome (Noori et al. 2017). Using this power law with the human GMV, the decay rate of the projection lengths in the human cortical connectome is expected to be about $\simeq 0.1\ \text{mm}^{-1}$ (**Figure 9.6**, dotted lines), or a characteristic spatial length of 10 mm. Currently available estimates are not precise due to the limitations of the presently available method and the likelihood that inter-areal connectivity exhibits several spatial scales. Nevertheless, they provide a quantification of the large-scale cortical organization.

The volume of white matter containing long-distance cortico-cortical connection fibers increases superlinearly with that of grey matter across many species, according to an allometric scaling relationship $\text{WMV} \sim \text{GMV}^{4/3}$ (WMV: white matter volume) (Zhang and Sejnowski 2000). Therefore,

$$\lambda \sim \text{GMV}^{-2/9} \sim \text{WMV}^{-1/6} \tag{9.2}$$

In other words, the characteristic spatial length $1/\lambda$ of the axonal projections does not scale with linear dimension $(WMV)^{1/3}$ of white matter, but its square root $\sqrt{(WMV)^{1/3}}$. The sublinear dependence of the characteristic spatial length on the linear dimension of the white matter suggests that the inter-areal connections become more spatially restricted in a larger cortex, shown schematically in **Figure 9.6** (bottom), which presumably is desirable for increasing the brain's modularity. The square-root allometric scaling law of the EDR remains to be explained in future research.

9.2.4 A Generative Model of Spatially Embedded Neocortex

While certain attributes of a network's connections can be defined in terms of a topological graph, a fuller description of a network should take into consideration the spatial relationship among nodes (Barthélemy 2011; Song and Wang 2014). Spurred by recently available data, Song et al. (2014) proposed a new class of spatially embedded random networks for the inter-areal cortical system. The starting point was the observation that the overall distribution of inter-areal wiring distances is well described by the distribution of distances in a spheroid (i.e., a rugby ball). Assume a model cortex in a spheroid with N parcellated areas, their "centers" at positions $\mathbf{R}_i$, $i = 1,\ldots,N$, are assigned randomly (**Figure 9.7a**, with $N = 7$). Next, cortical area i is then defined as all points in the spheroid closer to $\mathbf{R}_i$ than to any other center, i.e., the model parcellation is mathematically a Voronoi partition (Okabe et al. 2000) of the spheroid (**Figure 9.7a**). Although the true cortical sheet is nearly two-dimensional, this three-dimensional Euclidean manifold provides a useful approximation to its convoluted geometry with respect to axonal pathways.

Axonal connections are generated one by one. Starting with a uniformly sampled point at $\mathbf{s}$ in any source area (**Figure 9.7a**), a projection is determined by its direction and length. In the brain, the growth of axons is guided by a variety of attractive and repulsive molecular cues (Price et al. 2006; Stiles and Jernigan 2010). Inspired by developmental neurobiology, in the model each target area is assumed to exert an attractive "force" that decays with the distance as $(\mathbf{R}_i - \mathbf{s}) / |\mathbf{R}_i - \mathbf{s}|^{\alpha+1}$, where $|\cdot|$ indicates the vector's magnitude and α controls the distance dependence of the cue. The trajectory of each axon is determined by the sum of these forces (**Figure 9.7a**); the overall effect on an axon originating at $\mathbf{s}$ is given by the vector:

$$\mathbf{L}(\mathbf{s}) = \sum_{i=1}^{N} \frac{\mathbf{R}_i - \mathbf{s}}{|\mathbf{R}_i - \mathbf{s}|^{\alpha+1}} \tag{9.3}$$

The direction of the axon's growth is given by the unit vector $\mathbf{L}/|\mathbf{L}|$ computed from the sum over forces, while the length of the axon, ℓ, is drawn from an exponential distribution with decay rate parameter λ (**Figure 9.7a**). The axon terminates at position $\mathbf{s} + \ell\mathbf{L}/|\mathbf{L}|$, with the constraint that axons are limited to within the bounds of the spheroid. A consequence of the combination of Eq. (9.3) and the EDR is that most axons terminate in the same area they originate from, consistent with the experimental observation. After many projections are formed this way, inter-areal connection weight can be determined as the sum of all directed projections from one area to another (**Figure 9.7a**).

The generation of axons as described earlier is not dynamic: both the direction and length of an axon are calculated once, and the forces that determine the direction of axon growth are evaluated only at the initial position of the axon. This static description retains the advantage of being computationally simple; in future work, a dynamical growth model can be developed that incorporates a more realistic cortical geometry.

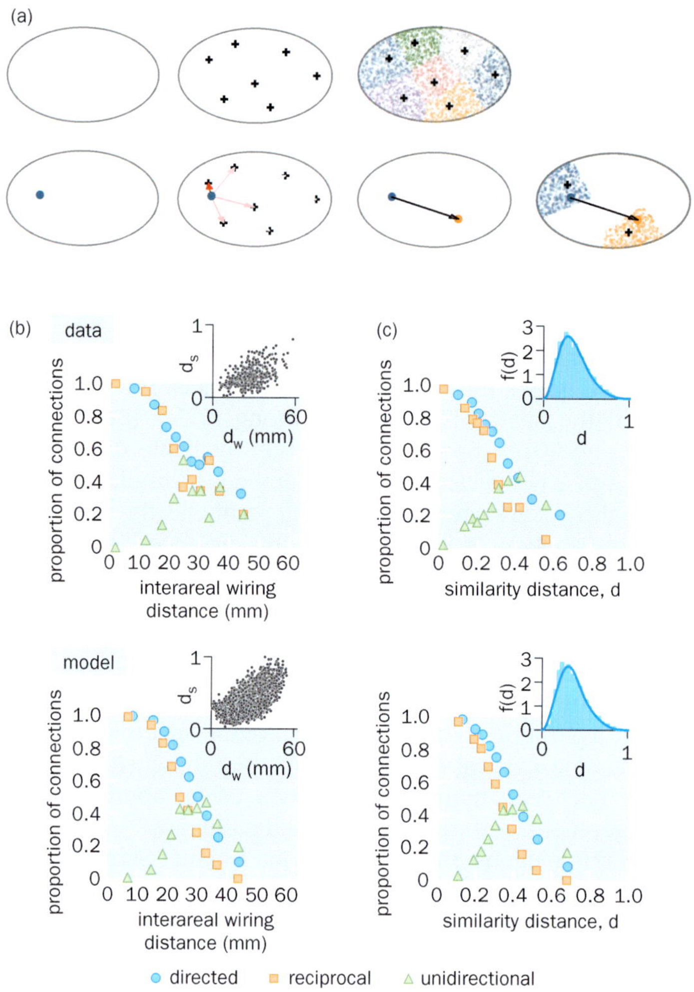

Figure 9.7 A spatially embedded random network model of a large-scale cortical system. (a) Model construction. (b) Proportion of directed (light blue), reciprocal (orange) and unidirectional (green) connections as a function of inter-areal (center-to-center) wiring distance in macaque monkey data (top) and model (bottom). (c) Same as (b) but as a function of output similarity distance. Inset: Distribution of similarity distances. Reproduced from Song et al. (2014). [Song, H. F., Kennedy, H., & Wang, X. J. (2014). Spatial embedding of structural similarity in the cerebral cortex. Proceedings of the National Academy of Sciences, 111(46), 16580–16585.]

With $N = 91$ areas as in Markov et al. (2014) and appropriate chosen values of model parameters (in particular, $\alpha = 2.5$ in Eq. 9.3), a realization of the model reproduces the macaque data quite well. The probability or fraction of directed (for example, from area A to area B), reciprocal (from A to B and from B to A) and unidirectional (from A to B but not from B to A) connections as a function of the wiring distance is similar between the monkey data and the model (**Figure 9.7b**). Another way to view the same data was as a function of similarity of a pair of areas. Song et al. (2014) introduced a measure of similarity of two areas based on their output patterns. The idea is that two nodes are functionally similar (the similarity distance equals to 0) if their projection patterns overlap strongly, e.g., they project to virtually an identical set of areas. Conversely, they are dissimilar if they project to entirely non-overlapping sets of areas (the similarity distance equals to 1). The proportion of connections as a function of this similarity measure from macaque data is also captured by the model (**Figure 9.7c**). This conclusion holds when the similarity is defined in terms of shared inputs rather than outputs. Notable is the fact that not all connections are reciprocal; unidirectional connections are more common for area pairs separated by relatively large distances.

Furthermore, The model displays a comparable span of connection weights as the macaque cortex (**Figure 9.8a**). Interestingly, other network properties not assumed in the model design turn out to be also supported by the monkey data, such as rank-ordered in-degree and out-degree sequences (**Figure 9.8b**) and the distribution of triad motifs (graphs made of three areas) (**Figure 9.8c**). Taken together, Song et al. (2014) offers a

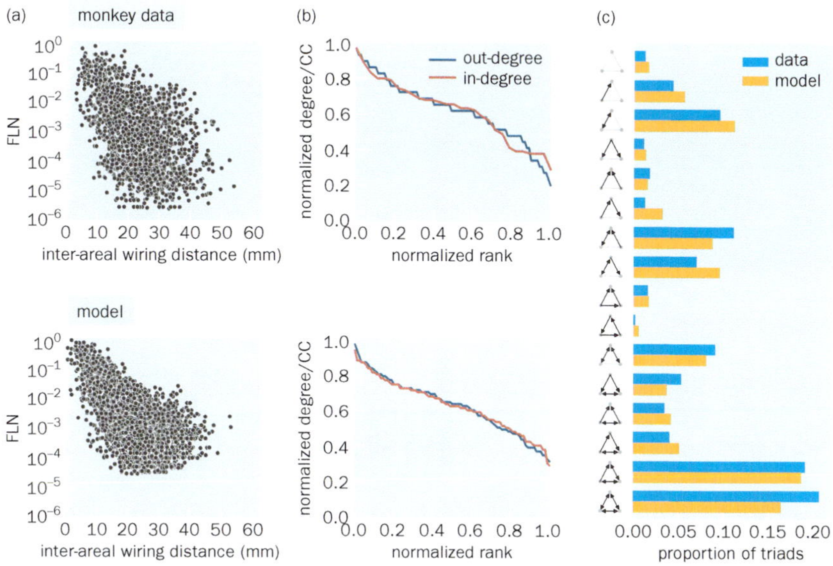

Figure 9.8 The spatially embedded model reproduces salient characteristics of the macaque cortex: (a) five-fold span of connection weights, (b) rank-ordered sequences of in-degree and out-degree and (c) distribution of triad motifs. Reproduced from Song et al. (2014). [Song, H. F., Kennedy, H., & Wang, X. J. (2014). Spatial embedding of structural similarity in the cerebral cortex. Proceedings of the National Academy of Sciences, 111(46), 16580–16585.]

spatially embedded generative model that captures observed cortical connectivity properties.

9.2.5 Cortical Hierarchy

The neocortex is organized hierarchically. As shown in **Figure 9.2**, starting with a primary sensory area like V1, information is processed and sent to higher areas through multiple steps. This hierarchy is based on the notion that a connection from one area to another is either feedforward (from a lower to a higher position) or feedback (from a higher to a lower position). In the primate, generally a feedforward projection, say from visual area V1 to V2, originates in the superficial layers 2/3, whereas a feedback projection (from V2 to V1) is mediated by neurons in the deep layers below layer 4. A parametric measure called supragranular labeled neurons (SLNs) is defined as the fraction of labeled neurons that are located in the superficial layers 2/3 of a given source area (Markov et al. 2014b; Vezoli et al. 2021a) (**Figure 9.9**, insert), with values between 0 and 1. Thus, if SLN is 1, the connection is purely feedforward and is 0 for a pure feedback projection. Importantly, SLN parametrically quantifies the degree of "feedforwardness" or "feedbackness." Using a statistical model of all measured pairwise connections, a consistent hierarchy can be defined where each area is assigned a hierarchy position, normalized between 0 and 1 (**Figure 9.9**). For instance, a qualitative description asserts that V2 is higher than V1, V4 is higher than V2, and TEO is higher than V4 along the visual hierarchy. Quantitatively, V2, V4 and TEO are now assigned hierarchical positions of 0.17, 0.42 and 0.71, respectively, with V1 at the starting position 0.

A similar hierarchy has been reported for marmoset monkeys (Theodoni et al. 2022). In mouse, cortical hierarchy is also present but quite shallow (Gămănuţ et al. 2018; Harris et al. 2019), suggesting a less hierarchical form of cortical organization in rodents than in primates.

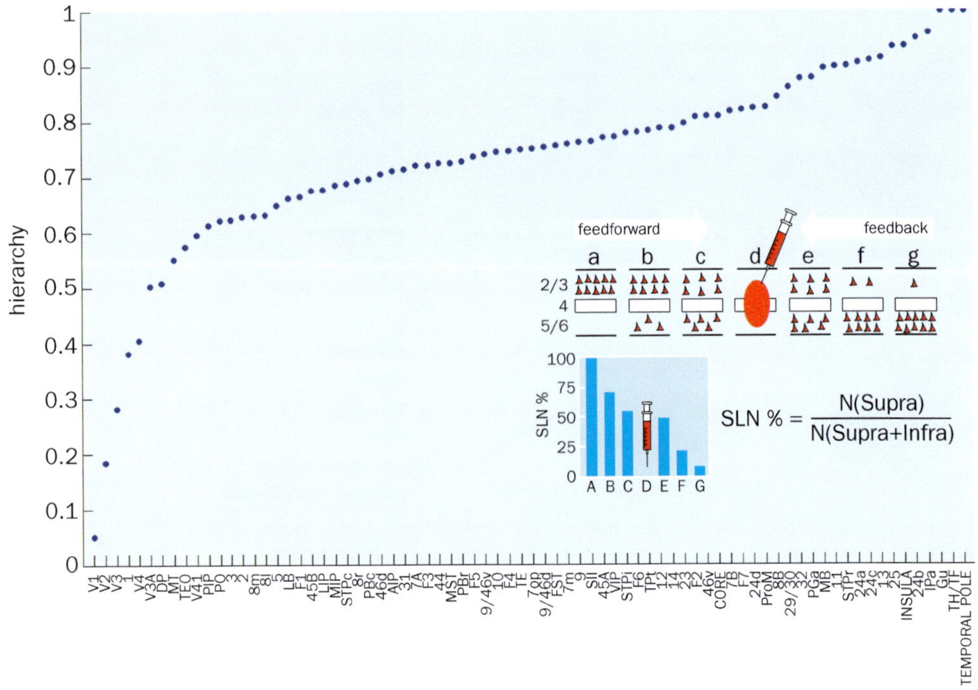

Figure 9.9 **Quantification of a cortical hierarchy, where each area (x-axis) is assigned a particular position (y-axis), with V1 at the bottom.** The hierarchy is steep in early and intermediate visual areas (V1, V2, V4, MT), but becomes shallow for association areas. Reproduced from Markov et al. (2014). [Markov, N. T., Vezoli, J., Chameau, P., Falchier, A., Quilodran, R., Huissoud, C., . . . & Kennedy, H. (2014). Anatomy of hierarchy: Feedforward and feedback pathways in macaque visual cortex. Journal of Comparative Neurology, 522(1), 225–259.] [Courtesy of Henry Kennedy.]

9.3 MACROSCOPIC GRADIENTS

9.3.1 Heterogeneous Variations of a Canonical Circuit

The concept of canonical circuits has been a cornerstone of neuroscience ever since the discovery of columns in the 1950s and 1960s. According to this view (Mountcastle 1997), a column of neurons is the basic unit of cortical organization. Its horizontal spatial extent varies little (300–600 μm in diameter), even between species whose brain volumes vary over 100,000-fold in volume (**Figure 9.6**). This expansion of cortical volumes corresponds to an increased number of columns (copies of a canonical unit) across species (Rakic 2009). Decades of *in vitro* neurophysiological studies of cortical circuits have been largely done using slices of primary sensory areas, often with the implicit assumption that results thus obtained remain valid for all cortical areas. A limited number of studies have revealed marked differences between V1 and an associated area like the PFC (Hempel et al. 2000; Wang et al. 2006; Wang et al. 2008), as well as across rodent and primate species (Wang et al. 2015; Boldog et al. 2018; Gonzelez-Burgos et al. 2019; Goriounova et al. 2018; Gidon et al. 2020; Campagnola et al. 2022), but these differences are not systematically documented and tend to be under-appreciated. As it was put by Douglas and Martin (2012): "Our view is that the rapid evolutionary expansion of neocortex has been made possible by building an 'isocortex' – a structure that uses repeats of the same basic local circuits throughout a single [cortical] sheet."

The idea of a canonical local circuit is reflected in the applications of graph theory to connectome, where cortical areas as "nodes" are mathematically indistinguishable from each other. Similarly, in studies of functional connectivity, such as those using functional MRI, areas are typically treated as identical. Functional connectivity measured by a covariance matrix of activity of pairs of areas is then interpreted in terms of their inter-areal structural connections (Glasser et al. 2016; Eickhoff et al. 2018). Areal differences are by and large ignored in current graph theoretical analysis of the brain connectome (Sporns 2014; Breakspear 2017).

From this perspective, how can one explain different functional capabilities of such disparate areas as V1 and PFC? Differential functions of various cortical areas could emerge from their relative proximities to sensory peripheries, their input and output connections, and synaptic plasticity. Take, for instance, the primate visual system, which is organized in a hierarchy: external stimuli arrive in the retina, which projects to the thalamus en route to V1, whose output propagates to V2, which in turn connects to MT and V4, etc. (**Figure 9.2**). Along the hierarchy, progressive information representation and processing are accomplished by connections from one area to the next, and the connection patterns are determined during development and sculpted by plasticity. In the visual ventral pathway, neurons gradually acquire larger receptive fields and selectivity for more abstract stimulus features, ultimately leading to size- and position-invariant object recognition. Beyond the visual areas, some areas like the orbitofrontal cortex may receive prominent inputs from the amygdala and consequently are involved in coding and processing of emotions, etc. In this picture, areas differ from each other exclusively in terms of inputs and outputs.

In contrast to this view of distinct areas being different only by their inputs and outputs, it is well known that spatial heterogeneity is a salient anatomical characteristic of mammalian cortex. Neuron density, pyramidal cell size, myelin content in the grey matter, cortical thickness and laminar differentiation all vary across the cortical mantle (von Economo 1929; Sanides 1972; Pandya et al. 1988; Amunts and Zilles 2015; Goulas et al. 2018). Starting with Brodmann, von Economo, C and O Vogt at the dawn of the twentieth century, these variations of cytoarchitecture and myeloarchitecture have been measured and utilized as an anatomical basis of parcellating the cortex into discrete areas and defining cortical hierarchy (Amunts and Zilles 2015; Barbas 2015); in connectomic analysis, microscale cellular variations were taken as correlates of macroscale inter-areal long-range connections of different areas (Scholtens et al. 2014). Besides the use of biological variations as markers, their functional implications across the cortical mantle have eluded much attention until recently.

9.3.2 Macroscopic Gradients of Synaptic Excitation

How can area-to-area heterogeneities inform theory and computational modeling? Our starting point is to consider biological properties that determine synaptic excitation and inhibition, which are important for understanding network dynamics and functions. In Chapters 5 and 6, we introduced "cognitive-type" cortical circuits characterized by more powerful excitatory reverberation than in early sensory areas, especially in the superficial layers 2/3. What are the anatomical substrates of this difference? In pyramidal neurons, excitatory synapses are located on small protrusions of pyramidal dendrites called spines, one synapse per spine; therefore, the number of spines can be considered a proxy of the strength of synaptic excitation per pyramidal cell, even though the physiological strength of an individual synapse may vary considerably from spine to spine. Elston (2007) carried out extensive analysis of spine count in the basal dendrite, where local recurrent excitatory synapses make preferential contacts, of layer 2/3 pyramidal neurons. Remarkably, in macaque monkeys, the spine count per pyramidal cell displays a strong positive correlation with the hierarchical position of cortical areas derived from SLN measurements (**Figure 9.10a**) (Chaudhuri et al. 2015). In particular, a pyramidal cell in prefrontal areas has about 10 times more spines than in

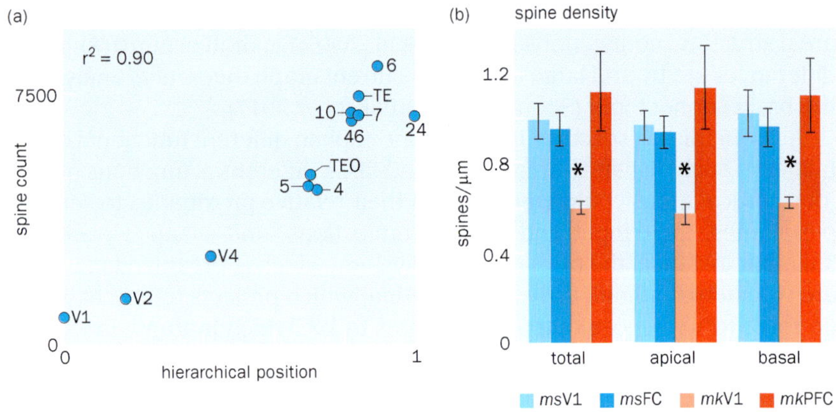

Figure 9.10 Macroscopic gradients of spine count per pyramidal neuron in the primate cortex. (a) The number of spines on the basal dendrites of layer 3 pyramidal cells in an area is strongly correlated with the area's hierarchical position determined by layer-dependent projections. (b) Comparative graphs showing the density per unit dendritic length of spines on the total, apical and basal dendritic trees in the four groups of neurons, showing a substantial difference between V1 (pink) and PFC (red) of macaque monkey but not between V1 (light blue) and frontal cortex (blue) of mouse. Panel (a) is reproduced from Wang (2020), with data from Elston (2007) and Chaudhuri et al. (2015), (b) from Gilman et al. (2017). [Wang, X. J. (2020). Macroscopic gradients of synaptic excitation and inhibition in the neocortex. Nature Reviews Neuroscience, 21(3), 169–178. Data from Elston, G. N. (2007). Specialization of the neocortical pyramidal cell during primate evolution and Chaudhuri, R., Knoblauch, K., Gariel, M. A., Kennedy, H., & Wang, X. J. (2015). A large-scale circuit mechanism for hierarchical dynamical processing in the primate cortex. Neuron, 88(2), 419–431.] [Gilman, J. P., Medalla, M., & Luebke, J. I. (2017). Area-specific features of pyramidal neurons—a comparative study in mouse and rhesus monkey. Cerebral Cortex, 27(3), 2078–2094.]

V1 (Elston 2000). A graded increase of spine count per pyramidal neuron has also been found in marmoset monkeys (Theodoni et al. 2022). By contrast, in mouse, total spine count per pyramidal cell appears to be uniform across the cortex (**Figure 9.10b**) (Ballesteros-Yanez et al. 2010; Gilman et al. 2017), suggesting that the macroscopic gradient of spine counts is a distinct feature of the primate cortex.

Furthermore, synaptic excitation is mediated by the AMPA and NMDA receptors. Computational modeling predicted that NMDA receptor-dependent slow reverberation provides a circuit mechanism for working memory (Wang 1999b) and decision-making computations (Wang 2002b) (Chapters 5 and 6). Is there also a macroscopic gradient of NMDA receptor signaling along the cortical hierarchy? Relevant evidence is emergent for human, monkey and mouse. As brain-wide transcriptome data are becoming available, one approach is to examine expressions of genes that encode NMDA receptor subunits. Generally, genes that encode receptors and other proteins of importance for synaptic excitation and inhibition are heterogeneously distributed in the cortex. In humans there are more than 2,000 brain-specific genes, so are the area-to-area variations of gene expressions random heterogeneities?

To address this question, it is necessary to first deal with the hurdle that, presently, hierarchy as defined anatomically by tract tracing analysis is not available for human cortex. In human brain imaging, it was noted that MRI-based T1-weighted versus T2-weighted (T1w/T2w) ratio is high in V1 and low in PFC (Glasser et al. 2016). Burt et al. (2018) showed that, in macaque monkeys, the T1w/T2w ratio is correlated negatively (Spearman coefficient of −0.76), with the hierarchical position defined independently in terms of layer-dependent connections (Markov et al. 2014b), in support of T1w/T2w ratio as a non-invasive index of cortical hierarchy in the reversed direction. A principal component analysis (PCA) of RNA microarray data of the human cortex (Hawrylycz et al. 2015) revealed that genetic heterogeneities across the cortex are not random; instead, gene expression patterns are low-dimensional, with the first principal component that is strongly correlated with the T1w/T2w axis and captures close to 30% of genetic data variance, more than twice the second principal component (Burt et al. 2018).

Genes involved in synaptic transmission display macroscopic gradients along the T1w/T2w axis. In particular, the gene GRIN2B (**Figure 9.11a**) encoding the NR2B NMDA subunit decreases with the T1w/T2w value, and hence increases with the hierarchy. NMDA receptors are heterotetramers, with the NR1 as an obligatory subunit and two other subunits. Synapses where the NMDA receptors are dominated by the NR2B subunits are especially plastic; this is because the synaptic ion channels mediated by the NR2B-containing NMDA receptors are longer lasting

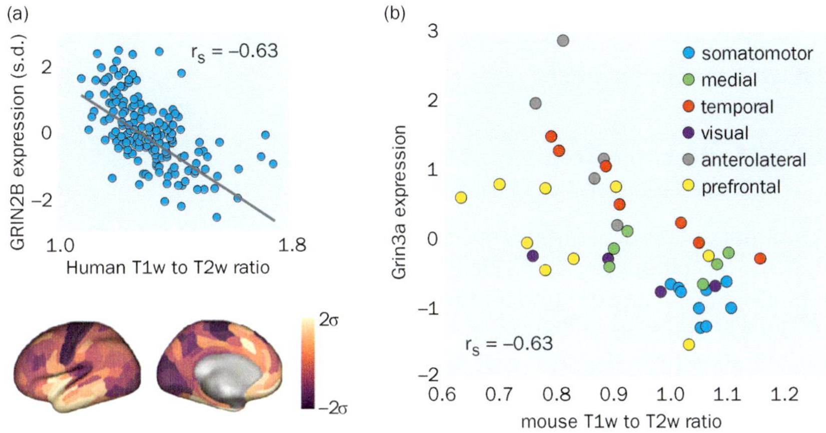

Figure 9.11 Macroscopic gradients of gene expression level for NMDA receptor subunits. (a) In human cortex, GRIN2B, which codes the NMDA receptor subunit NR2B, exhibits a negative correlation with T1w/T2w ratio (Spearman correlation is −0.63, $p<10^{-5}$). (b) In mouse cortex, Grin3A, which encodes the NMDA receptor NR3A subunit, as a function of T1w/T2w ratio (Spearman correlation is −0.63; $p < 5\times10^{-4}$). Reproduced from Wang (2020), with original data from Burt et al. (2018) for (a) and Fulcher et al. (2019) for (b). [Burt, J. B., Demirtaş, M., Eckner, W. J., Navejar, N. M., Ji, J. L., Martin, W. J., ... & Murray, J. D. (2018). Hierarchy of transcriptomic specialization across human cortex captured by structural neuroimaging topography. Nature Neuroscience, 21(9), 1251–1259.] [Fulcher, B. D., Murray, J. D., Zerbi, V., & Wang, X. J. (2019). Multimodal gradients across mouse cortex. Proceedings of the National Academy of Sciences, 116(10), 4689–4695.]

than those mediated by the NMDA receptors that contain the NR2A but not NR2B subunits, allowing more calcium influx of critical importance for Hebbian plasticity (Chapter 4, Section 4.4). The NR2B-dependent slow synaptic excitation is also desirable for reverberatory dynamics underlying working memory (Chapter 5, **Figure 5.27**) and decision making. In the primary visual cortex, a switch occurs early in the development, starting near the time of eyelid opening, from NR2B to NR2A dominance of NMDA receptors (Quinlan et al. 1999), marking a transition from high to low plasticity in V1. It is unknown when such a transition takes place in the PFC during brain development given that the PFC matures much later than V1 into adolescence. Interestingly, both the genes that encode the NR1 and NR2A subunits increase with the T1w/T2w, and thus decrease along the cortical hierarchy (Burt et al. 2018). These results are consistent with the converging physiological evidence that the NR2B-containing NMDR receptors mediate more prominent slow reverberation in PFC areas than in primary sensory areas in adult mammals (Wang et al. 2008; Wang et al. 2013).

Similar analysis of genetic data among cortical areas ranked along the T1w/T2w ratio led to a different picture for the mouse cortex. Using *in situ* hybridization transcriptome data (Lein et al. 2007), several macroscopic gradients were identified (Fulcher et al. 2019). Unlike the primates, in the mouse cortex expression levels of both NR2B and NR2A coding genes are positively correlated with the T1w/T2w ratio. By contrast, a negative correlation of NR3A coding gene (**Figure 9.11b**) suggests that it increases along the cortical hierarchy in mouse. NR3A-containing NMDA receptors are mostly found perisynaptically, its role in NMDA signaling could be quite different from those of NR2A and NR2B subunits (Pérez-Otaño et al. 2016), and its newly reported macroscopic gradient should be explored in future studies. To rigorously link with functions, it is necessary to establish that expression level of a gene correlates strongly with physiological strength of its encoded protein. For intrinsic ion channels, mRNA levels can be compared with physiologically measured channel conductances (Schulz et al. 2007). Quantitative labeling of receptors that are more directly linked to physiology than genes (Zilles and Palomero-Gallagher 2017; Froudist-Walsh et al. 2023) provides a valuable intermediate step. For synaptic receptors, there is an added complexity. Genes encode proteins for the whole neuron, whereas receptor prevalence is pathway specific, which has functional implications. Consider a hypothetical scenario where the amount of the NR2B subunit expression increases at local recurrent excitatory synapses but decreases at a subset of long-range pathways, from primary sensory areas to association areas. The overall NR2B expression per neuron could decrease along the mouse cortical hierarchy, compatible with the genomic data analysis. Novel experimental tools are required for assessing macroscopic gradients of synaptic excitation in a cell-type and pathway-specific manner.

9.3.3 Macroscopic Gradient of Input- versus Output-Controlling Inhibition

A hallmark of cortical organization is the balance between synaptic excitation and inhibition (Chapter 3, Section 3.2.2). Given that densities of various inhibitory cell types (Freund and Buzsáki 1996; DeFelipe 2002; Markram et al. 2004; Paul et al. 2017) are heterogeneous in the cortex, does synaptic inhibition also display a macroscopic gradient? In a disinhibitory motif (Chapter 2, **Figure 2.2**; Chapter 5, Section 5.6.5) (Wang et al. 2004; Kepecs and Fishell 2014; Tremblay et al. 2016; Cardin 2018), PV-positive interneurons target the perisomatic region of pyramidal cells and control their spiking output, whereas SST/CB-positive interneurons target pyramidal dendrite and control synaptic input flow. The third interneuron subpopulation expressing VIP/CR preferentially project to SST/CB interneurons (**Figure 9.12a**). A comprehensive cell count analysis of GABAergic cells in the mouse brain revealed that the ratio of input-controlling SST cells and output-controlling PV cells varies considerably across cortical areas (Kim et al. 2017). When plotted by rank order, it became clear that the SST/PV ratio is generally low in early sensory areas and motor areas, but is high in association including frontal areas (**Figure 9.12b**), revealing a macroscopic gradient of synaptic inhibition in the mouse cortex. Notably, PV cells are twice as much as SST/CB cells in V1, but the latter is more numerous than the former by fourfold in frontal areas. Macaque monkey cortex displays the same characteristic trend (Condé et al. 1994; Kondo et al. 1999; Kooijmans et al. 2020; Torres-Gomez et al. 2020), although systematic data for the entire cortex is currently not available. Burt et al. (2018), using an entirely different methodology, found that the expressions of genes coding for PV, CB and CR all display strong correlations with the T1w/T2w ratio in the human cortex (**Figure 9.12c**), with the PV-encoding gene expression level decreasing, while those for CB and CR increasing, with the hierarchy.

Functional implications of these macroscopic gradients remain to be elucidated for a range of processes that critically depend on inhibition, such as stimulus selectivity (Somers et al. 1995; Callaway 2004) or synchronous oscillations (Buzsáki 2006; Wang 2010). Since higher areas receive more converging inputs than do lower primary sensory areas, the increased input-controlling SST/CB interneurons versus output-controlling PV cells may be especially important for gating inputs into pyramidal dendrite flexibly according to behavioral demands (see Section 9.5.3). The identification of macroscopic gradients of synaptic inhibition represents an important clue for extending our understanding of inhibitory neurons from local circuits towards multi-regional large-scale cortical systems.

Macroscopic gradients can be extended to more than one dimension. As a matter of fact, defining a single one-dimensional hierarchy tends to be vision-centric and does not fairly take into consideration different sensory modalities. In addition, motor areas are not readily placed in a linear framework from sensory to association areas. Decades ago, Mesulam (2000) proposed a two-dimensional diagram of cortical organization, with the radial direction along the hierarchy and the polar direction covering different sensory modalities and motor domains. This view was recently confirmed, using a sophisticated analysis called the diffusion map (Coifman et al. 2005) of inter-areal functional correlations of the human cortex (Margulies et al. 2016; Huntenburg et al. 2017), shown in **Figure 9.13a** according to the seven-network parcellation (Yeo et al. 2011). A compatible

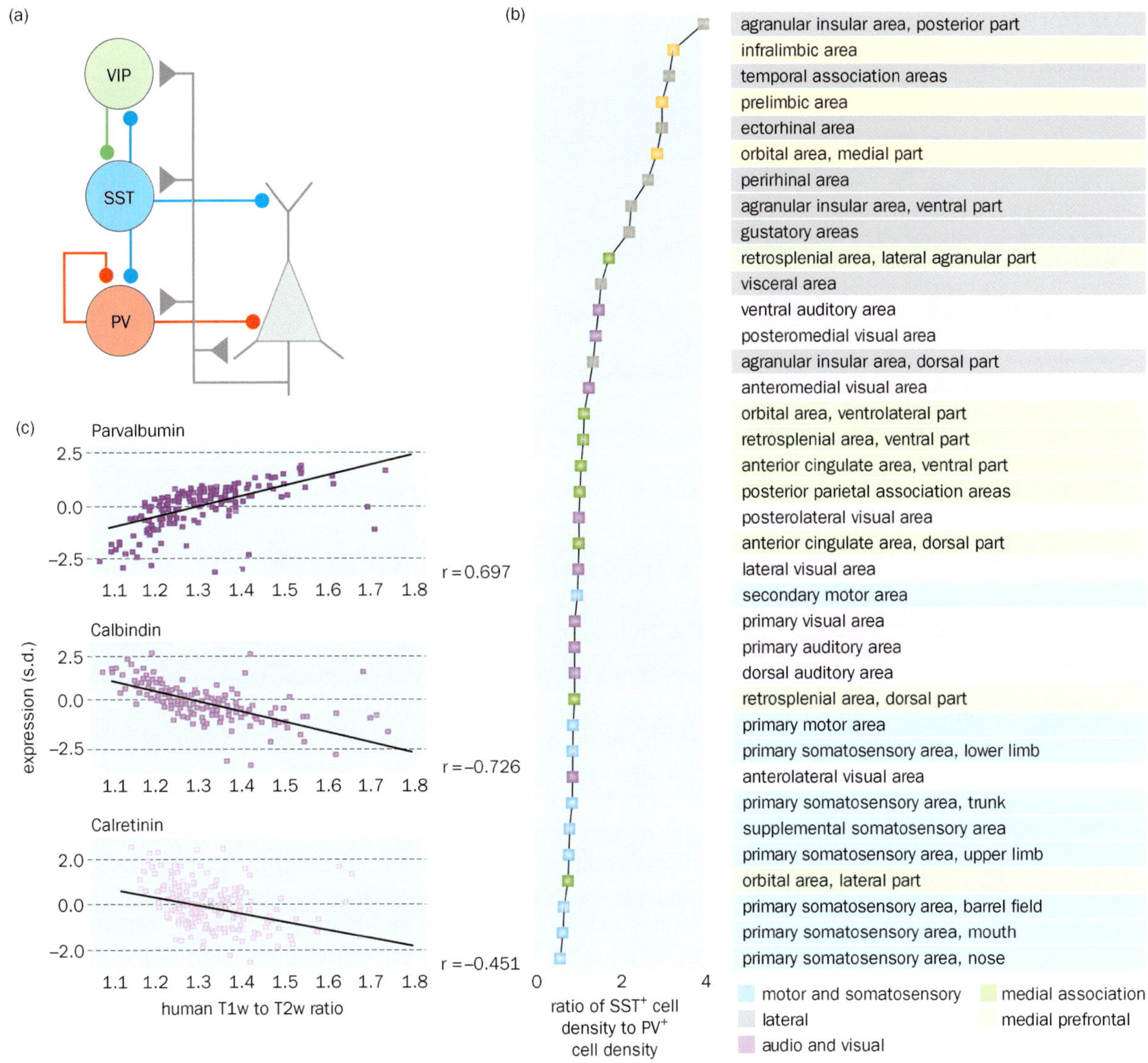

Figure 9.12 Macroscopic gradients of synaptic inhibition. (a) A disinhibitory circuit realized by PV, SST and VIP neurons. (b) The SST/PV cell density ratio plotted for areas ordered by the SST/PV ratio value of mouse cortex. PV neurons are abundant in primary sensory areas, while lateral (grey) and medial (yellow) frontal areas are dominated by SST neurons. The colors code which of the five types (indicated at the bottom) of cortical subnetwork that each area belongs to. (c) Genes coding for parvalbumin, calbindin and calretinin exhibit macroscopic gradients in the human cortex. Reproduced from Wang (2020), with original data from Kim et al. (2017) for (b) and Burt et al. (2018) for (c). [Wang, X.-J. & Yang, G. R., (2018), A disinhibitory circuit motif and flexible information routing in the brain. Current Opinion in Neurobiology 49, 75–83.] [Wang, X. J. (2020). Macroscopic gradients of synaptic excitation and inhibition in the neocortex. Nature Reviews Neuroscience, 21(3), 169–178. Original data from Burt, J. B., Demirtaş, M., Eckner, W. J., Navejar, N. M., Ji, J. L., Martin, W. J., ... & Murray, J. D. (2018). Hierarchy of transcriptomic specialization across human cortex captured by structural neuroimaging topography. Nature Neuroscience, 21(9), 1251–1259.]

two-dimensional organization of the macaque monkey cortex, in the polar coordinates, is shown in Figure 9.13b, with the radial direction defined by the hierarchy and the angular distance defined by the inverse of inter-areal connection strength (Chaudhuri et al. 2015).

In summary, macroscopic gradients represent an emerging principle of large-scale cortical organization (Wang 2020; Bernhardt et al. 2022). Quantified low-dimensional gradients make it possible to incorporate variations of synaptic excitation and inhibition in a computational model of the large-scale multi-regional cortex. Next, we will discuss dynamical and functional implications of these macroscopic gradients across the cortex.

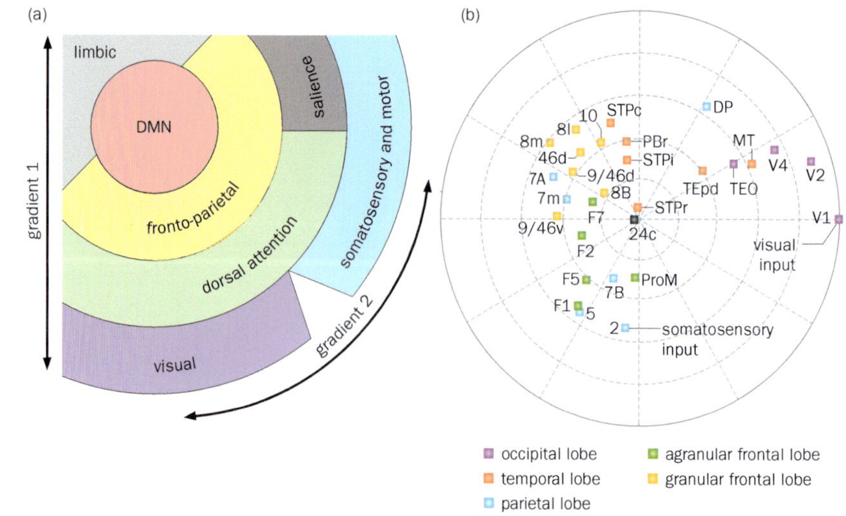

Figure 9.13 Two-dimensional gradients. (a) Spatial relationships of the human cortical subnetworks in the resting state. dmn, default-mode network; fronto-parietal: Multiple demand network; dorsal attn, dorsal attention network; sal, salience (ventral attention) network; somato/mot, somatosensory/motor network. (b) Two-dimensional plot of areas of the macaque monkey cortex determined by long-range connectivity and hierarchy. The radial distance of an area from the edge corresponds to its hierarchical position, while the angular distance between two areas is inversely related to their connection strength. Areas are colored by cortical lobes. Panel (a) is reproduced from Margulies et al. (2016), (b) from Wang (2020), with original data from Chaudhuri et al. (2015). [Margulies, D. S., Ghosh, S. S., Goulas, A., Falkiewicz, M., Huntenburg, J. M., Langs, G., ... & Smallwood, J. (2016). Situating the default-mode network along a principal gradient of macroscale cortical organization. Proceedings of the National Academy of Sciences, 113(44), 12574–12579.] [Wang, X. J. (2020). Macroscopic gradients of synaptic excitation and inhibition in the neocortex. Nature Reviews Neuroscience, 21(3), 169–178. Original data from Chaudhuri, R., Knoblauch, K., Gariel, M. A., Kennedy, H., & Wang, X. J. (2015). A large-scale circuit mechanism for hierarchical dynamical processing in the primate cortex. Neuron, 88(2), 419–431.

9.4 A HIERARCHY OF TIMESCALES

9.4.1 A Dynamical Model of Multi-Regional Monkey Cortex

Using FLNs (Markov et al. 2014a) and SLNs (Markov et al. 2014b), Chaudhuri et al. (2015) developed a dynamical model of a subnet of the macaque cortex. The inter-areal connections are given by a directed- and weighted-connectivity matrix, so how should each area be modeled? The answer depends on the scientific aim of a study.

Let us first consider a simple model to explore the specific idea of a canonical circuit valid for all cortical areas from V1 through PFC, but there is a macroscopic gradient of synaptic excitation. Mathematically, a canonical architecture is described by an identical excitatory-inhibitory recurrent network model for each local area, and a gradient is incorporated as a systematic variation of model parameters along the hierarchy.

To be concrete, take a previously published model for the macaque V1 (Binzegger et al. 2009):

$$
\begin{aligned}
\tau_E \frac{dr_E}{dt} &= -r_E + \beta_E \left[I_E\right]_+ \\
\tau_I \frac{dr_I}{dt} &= -r_I + \beta_I \left[I_I\right]_+ .
\end{aligned}
\tag{9.4}
$$

r_E is the firing rate of the excitatory population, with intrinsic time constant τ_E and input current I_E. Its f-I curve has a slope β_E, $\left[I_E\right]_+ = I_E$ if $I_E > 0$; 0 otherwise. The inhibitory population with rate r_I has corresponding parameters τ_I, I_I and β_I. The model parameter values were calibrated from neurophysiological measurements. In particular, the time constants τ_E and τ_I are about 10 milliseconds and the same for all areas.

In a multi-regional network, the input currents to an area indexed by i have a component from within the area (i.e., local input) and another that comes from the other areas:

$$
\begin{aligned}
I_E^i &= \left(1+\eta h_i\right)\left(w_{EE} r_E^i + I_{lr,E}^i\right) - w_{EI} r_I^i + I_{ext,E}^i \\
I_I^i &= \left(1+\eta h_i\right)\left(w_{IE} r_E^i + I_{lr,I}^i\right) - w_{II} r_I^i + I_{ext,I}^i .
\end{aligned}
\tag{9.5}
$$

where h_i is the hierarchical position of area i defined by its SLN value according to **Figure 9.9**. The parameters w_{EE} and w_{EI} are local couplings from the excitatory and inhibitory populations, respectively, and $I_{lr,E}^i$ is the long-range input to the excitatory population. $I_{ext,E}^i$ is external input, which may include sensory stimuli and background noise in a resting state. The corresponding parameters w_{IE}, w_{II}, $I^i_{lr,I}$ and $I^i_{ext,I}$ are for the inhibitory population. A macroscopic gradient of synaptic excitation is introduced by assuming spine count (**Figure 9.10a**) as a proxy of synaptic input strength. Specifically, the strengths of both local and long-range excitatory inputs to an area are scaled by its position in the hierarchy, h_i. Note that h_i is normalized between 0 and 1, and η is a parameter that controls the gradient slope.

Long-range inputs are excitatory, targeting both excitatory and inhibitory cells:

$$I^i_{lr,E} = \mu_{EE} \sum_{j=1}^{N} \text{FLN}_{ij} r^j_E; \quad I^i_{lr,I} = \mu_{IE} \sum_{j=1}^{N} \text{FLN}_{ij} r^j_E. \tag{9.6}$$

Here j ranges over all areas. The parameters μ_{EE} and μ_{IE} are parameters that control the strengths of long-range input to the excitatory and inhibitory populations, respectively.

How many parameters are there in this model? With a fixed local area model, and measured FLNs as well as the hierarchy, key adjustable parameters are three: η that controls the steepness of the macroscopic gradient, μ_{EE} and μ_{IE} that determine the absolute strengths of long-range connections. Therefore, based on experimental data and with a careful design, large-scale network modeling of macaque cortex can be well constrained with only a few parameters.

The model can be used to simulate a resting state where each area receives a random input (white noise). The resulting activities fluctuate stochastically in time, which is clearly much faster in V1 than DLPFC area 9/46d (**Figure 9.14a**). To quantify, the activity from each area i produces

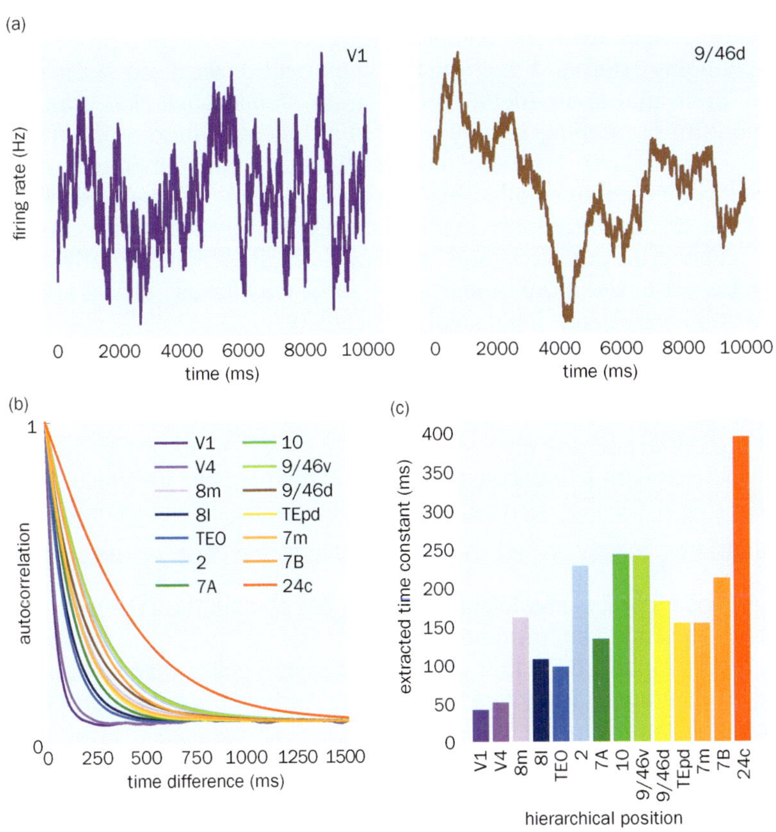

Figure 9.14 A hierarchy of time constants. (a) Stochastic activities of V1 (left) and area 9/46d (right) in a simulation where all areas are driven by random noise in a simulated resting state. (b) Autocorrelation of activity for 14 different areas. (c) Dominant time constant extracted from (b) for each area is plotted as a function of its anatomically defined hierarchical position. Provided by Aldo Battista using the model from Chaudhuri et al. (2015). [Chaudhuri, R., Knoblauch, K., Gariel, M. A., Kennedy, H., & Wang, X. J. (2015). A large-scale circuit mechanism for hierarchical dynamical processing in the primate cortex. Neuron, 88(2), 419–431.]

a time series, say $r_E^i(t)$, of which autocorrelation $AC_i(\tau)$ can be computed. $AC_i(\tau)$ measures how two values separated by a time τ ($r_E^i(t)$ and $r_E^i(t+\tau)$) are correlated with each other. This correlation function decays with τ (**Figure 9.14b**). A measured $AC_i(\tau)$ can be fit with an exponential function like $\exp(-\tau/\tau_i)$, from which a time constant τ_i can be extracted. In some cases AC_i is fit better with a sum of two exponentials like $w_1\exp(-\tau/\tau_{i1}) + w_2\exp(-\tau/\tau_{i2})$, and a dominant time constant is given by $\tau_i = w_1\tau_{i1} + w_2\tau_{i2}$ (with $w_1 + w_2 = 1$).

Figure 9.14c shows τ_i as a function of the hierarchical position h_i for 14 sample areas. The time constants span a wide range. Generally, this time constant increases with h_i, revealing a hierarchy of time constants along the anatomically defined hierarchy. The dependence is not monotonically, however. For example, area 8m (light purple) is part of the frontal eye field (FEF). It is at a relatively low hierarchical position, but belongs to the frontal lobe and receives inputs from other prefrontal areas with slow dynamics, which explains FEF's long time constant. Note that in a multi-regional cortex model, the range of time constants depends on the global strength of synaptic excitation, which prolongs the longest time constant similarly as in a local circuit (Chapter 3, Section 3.3).

In the model, when all long-range connections are deleted, areas are isolated from each other, and the time constant of a local circuit depends on its synaptic excitation and inhibition (Chapter 3, Section 3.3.3). When only the feedback projections (all connections with SLN below 0.5) are deleted, the system is feedforward and the range of time constants shrinks considerably. Finally, in the original model but with FLNs randomly shuffled while preserving the same set of FLN values, the range of time constants τ_i also becomes quite narrow, demonstrating the importance of experimentally measured inter-areal connectome.

9.4.2 A Spatial Localization Measure

The hierarchy of time constants is functionally desirable. On the one hand, primary sensory areas should operate quickly to encode and process rapidly changing external stimuli. On the other hand, as we discussed in Chapter 6, prefrontal and parietal cortical areas should have slow transients in the form of ramping activity required for accumulation of information in decision making. The wide range of time constants emerges from the model, which requires both a macroscopic gradient of synaptic excitation and complex feedback loops in the multi-regional cortical system. Assuming a number of time constants in this system, it is highly nontrivial why they are not mixed more or less in the same way for all cortical areas, in spite of the very dense inter-areal connectivity.

In Chaudhuri et al. (2015), when E and I neural populations are active with positive firing rates for all areas, the model is linear. For a linear dynamical system of N variables, one can compute its eigenvalues λ_k and the corresponding eigenvectors $\mathbf{v}_k$, $k = 1, 2, .., N$. Each eigenvalue λ_k is associated with a time constant given by the negative inverse of its real part, $\tau_k = -1/Re(\lambda_k)$ (Strogatz 2016). Let us normalize each eigenvector so that $|\mathbf{v}_k|^2 = \sum_j |v_{k,j}|^2 = 1$. As illustrated in **Figure 9.15a**, the dynamics of any node (cortical area) is given by a linear combination of eigenmodes with weights that vary from node to node.

From condensed matter physics, a measure called inverse participation ratio (IPR) was introduced to describe Anderson localization (Anderson 1958). For an eigenvector $\mathbf{v}_k$, IPR is:

$$IPR(\mathbf{v}_k) = \sum_j^N |v_{k,j}|^4 \tag{9.7}$$

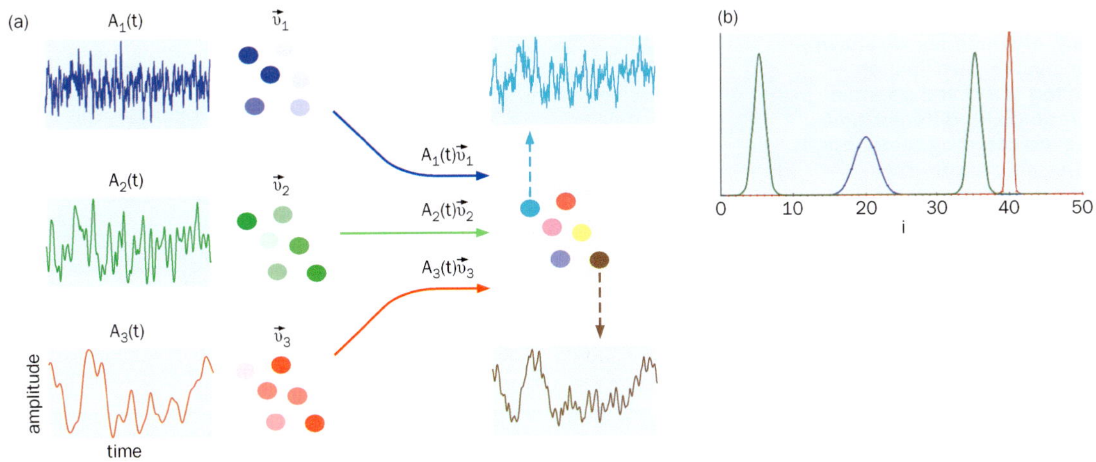

Figure 9.15 Definition and examples of a spatial localization measure for eigenmodes. (a) Schematics of a linear dynamical system with eigenmodes $\mathbf{v}_k$ ($k = 1,2,3,\ldots$) indicated by blue, green and red, etc. The dynamics of a node is determined by a weighted sum $\sum_k A_k(t)\mathbf{v}_k$, with a different set of A_k for each node i. Thus, the aqua blue node is dominated by the fast blue mode, the brown node by the slow red mode, etc. (b) Eigenvectors for three eigenmodes (red, blue and green) plotted in a network with 50 nodes indexed by i. Panel (a) is reproduced from Chaudhuri et al. (2014), (b) unpublished data by Luis Carlos Garcia del Molino and Xiao-Jing Wang. [Chaudhuri, R., Bernacchia, A., & Wang, X. J. (2014). A diversity of localized timescales in network activity. Elife, 3, e01239.]

For instance, if $v_{k,j}$ is 1 only for one node and 0 elsewhere, this eigenmode is perfectly localized and $IPR = 1$. At the other extreme, if $v_{k,j} = 1/\sqrt{N}$ uniformly, the eigenmode is maximally delocalized, $IPR = 1/N$ and is negligible with large N. The IPR quantifies the fraction of nodes that participate in an eigenmode, but not the degree of spatial proximity of participating nodes. Let us introduce another, spatial localization (SL) measure:

$$SL(\mathbf{v}_k) = \sum_{i,j}^{N} |v_{k,i}|^4 |v_{k,j}|^4 \exp(-d_{ij}/d_0) \tag{9.8}$$

where d_{ij} is the distance between any pair of nodes i and j, and d_0 is the average distance between all pairs. For a perfectly localized eigenmode, we have $SL = 1$; for a maximally delocalized eigenmode, $SL = \langle \exp(-d_{ij}/d_0) \rangle / N^2$, which is nearly zero with large N. Mathematically, $SL(\mathbf{v}_k) \le IPR^2(\mathbf{v}_k)$. We can define a spatial localization ratio $SLR(\mathbf{v}_k) = SL(\mathbf{v}_k)/IPR^2(\mathbf{v}_k)$, which is between 0 and 1. **Figure 9.15b** shows three examples: the red eigenvector is strongly localized in space, with both IPR and SLR close to 1; the blue eigenvector is less localized with $IPR = 0.032$, but participating nodes are spatially clustered and $SLR = 0.95$; the green eigenvector has the same IPR as the blue one, but its participating nodes are not spatially together and $SLR = 0.58$.

The IPR and SLR measures can be computed for any eigenvectors. Applying to the large-scale model of the monkey cortex (Chaudhuri et al. 2015), the IPR and SLR values for each of its eigenmodes are shown in **Figure 9.16** (lower middle plot), with color indicating the associated time constant (fast in blue color, slow in red color). Interestingly, all eigenmodes are relatively localized in space with SLR above 0.5. Some involve a fairly large number of areas (examples on lower left and right), while others are quite localized spatially (upper middle and right). The two eigenmodes in the upper and lower right plots have the same IPR but different SLR, demonstrating the utility of SLR for quantifying spatial localization.

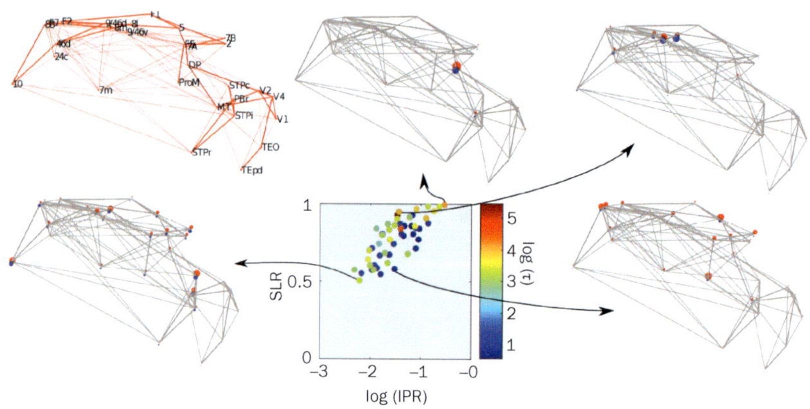

Figure 9.16 In a two-dimensional plot (lower center), eigenmodes are shown, each with an inverse participation ratio (IPR) (in log scale) and a spatial localization ratio (SLR); different time constants are indicated by color. Four sample eigenvectors visualize various degrees of spatial localization in the macaque cortical network. Unpublished data by Luis Carlos Garcia del Molino and Xiao-Jing Wang.

These results raise new questions for both theory and experimentation. On the computational side, mathematical analysis (Chaudhuri et al. 2014; Li and Wang 2022) identified specific conditions under which a cortical circuit realizes spatial localization of time constants. A macroscopic gradient of biological properties is important for generating a broad range of time constants. Intuitively, recurrent excitation prolongs the effective time constant of a neural circuit, because a positive feedback enables neural activity to increase for a longer time than in the absence of reverberation (Seung 1996; Wang 2001). Thus, heterogeneity of synaptic excitation as well as inhibition creates a broad range of time constants. Furthermore, each long-range projection pathway must be balanced, in the sense that the long-range excitatory-to-excitatory projection is counteracted by the local inhibitory-to-excitatory one (Vogels and Abbott 2009), which is aided by a larger gain of the input-output relationship for inhibitory neurons than for excitatory neurons. Consequently, the net inter-areal input is weak relative to local recurrent dynamics. This is consistent with the observation in the tract tracing experiments that the majority (80%) of cortical connections are part of the local circuit and originate within the injected area itself; long-range inputs from other cortical areas are relatively weak compared to local synaptic connections (Markov et al. 2014a).

9.4.3 Experimental Observations of Timescale Hierarchy

A hierarchy of temporal response windows or timescales is supported by several lines of empirical evidence. In close interplay with the computational modeling (Chaudhuri et al. 2015), Murray et al. (2014b) analyzed single-neuron data recorded from macaque monkeys performing different tasks in five laboratories. For a meaningful comparison across laboratories, the analysis considered neural spiking activities during a foreperiod when monkeys held their gaze on a fixation cue on a computer screen, prior to the start of a behavioral trial. The resting state is close to a linear system (Nozari et al. 2020), thus modeling it represents an appropriate way to test the model prediction of a hierarchy of time constants qualitatively. Autocorrelation of spike trains in this baseline condition was calculated and averaged across neurons from each recorded cortical area. The resulting decay time constant of autocorrelation is shown in **Figure 9.17a** for five areas ordered along their anatomically defined hierarchical positions (Markov et al. 2014b). In support of the model, the characteristic time constant of activity fluctuations display a systematical increase along the hierarchy. This finding was supported by other single-cell physiological studies with monkeys (Ogawa and Komatsu 2010;

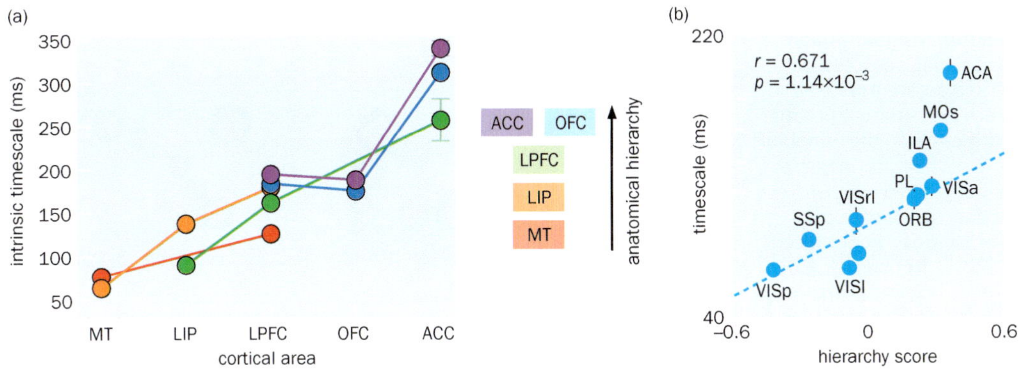

Figure 9.17 Hierarchy of time constants derived from single-neuron recording. (a) Time constant extracted from autocorrelation function of neuronal spiking fluctuations during baseline of eye fixation in five cortical areas of macaque monkeys, as a function of an anatomically defined hierarchical position. (b) Time constant of neurons during spontaneous activity increases along the cortical hierarchy in mouse. Panel (a) is reproduced from Murray et al. (2014b), (b) is provided by Daeyeol Lee using original data from Song et al. (2023). [Murray, J. D., Bernacchia, A., Freedman, D. J., Romo, R., Wallis, J. D., Cai, X., ... & Wang, X. J. (2014). A hierarchy of intrinsic timescales across primate cortex. Nature Neuroscience, 17(12), 1661–1663.] [Courtesy of Daeyeol Lee.]

Cavanagh et al. 2018; Wasmuht et al. 2018; Spitmaan et al. 2020). Supporting evidence was reported for mouse (**Figure 9.17b**) (Runyan et al. 2017; Siegle et al. 2021; Song et al. 2023). A number of studies have also demonstrated a hierarchy of timescales in the human cortex (Hasson et al. 2008; Honey et al. 2012; Lerner et al. 2011; Stephens et al. 2013; Shafiei et al. 2020; Gao et al. 2020; Pang et al. 2023).

It is well known that the size of a single neuron's spatial receptive field (RF) is progressively increased along the hierarchy. Spatial integration can be achieved through convergence of projections from neurons with smaller RFs onto neurons in an upstream area. The model of Chaudhuri et al. (2015) offers a mechanism for a hierarchy of response temporal windows (Honey et al. 2007; Hasson et al. 2008; Kiebel et al. 2008), in parallel with the enlargement of response spatial fields in the cortical hierarchy. Thus, along the cortical hierarchy, neural circuits are endowed with increased capability of integration both in space and in time.

A hierarchy of timescales is a new general principle of the neocortical organization of functional importance. First, it can be deployed for representation of temporal context of stimuli from the environment (Hasson et al. 2008; Kiebel et al. 2008; Baldassano et al. 2017). Comprehension depends on temporal integration of stimuli, for instance, following the storyline when watching a movie (Hasson et al. 2008) or listening to a narrative story (Lerner et al. 2011). Second, since prefrontal and posterior parietal cortices with slow dynamics mediate selective attention, their top-down signaling can instantiate proactive attention (Chapter 8) on multiple timescales (Nobre and Stokes 2019). Third, speech is an example of information processing on multiple timescales (Poeppel 2003; Hickok and Poeppel 2007): phonemes and words processed on a subsecond timescale are integrated into sentences and phrases over seconds, which in turn are comprehended as an integrated narrative in minutes. Human experiments using ECoG (Honey et al. 2012) and fMRI pattern analysis (Raut et al. 2020; Chien and Honey 2020) revealed that the underlying processes are carried out in different language-related cortical areas (**Figure 9.18**). Presumably perception and appreciation of music (Zatorre et al. 2002; Levitin 2006) also rely on a hierarchy of time constants in the brain.

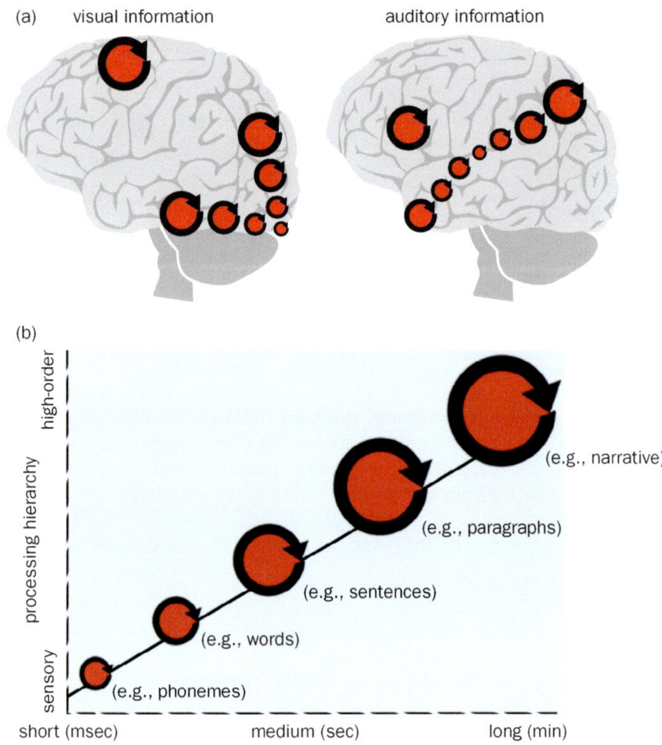

Figure 9.18 Hierarchy of time constants in cortical areas underlying language processing. (a) A schematic process hierarchy for visual and auditory systems. (b) A hierarchy of speech processing timescale (operationalized by measuring the temporal receptive window) in each region increases in a topographically organized manner, from milliseconds in early sensory areas up to minutes in higher-order areas. Reproduced from Hasson et al. (2015). [Hasson, U., Chen, J., & Honey, C. J. (2015). Hierarchical process memory: Memory as an integral component of information processing. Trends in Cognitive Sciences, 19(6), 304–313.]

9.5 FUNCTIONAL CONNECTIVITY AND INTER-AREAL COMMUNICATION

9.5.1 Functional Connectivity in a Resting State

The existence of macroscopic gradients implies that cortical areas are not the same, in contrast to commonly practiced graph theoretic analysis of functional connectivities. Intuitively, one expects that functional connectivity, be it measured by fMRI, MEG or ECoG, would resemble more closely anatomical connectivity if "nodes" were identical; therefore, the global dynamics would be predominantly determined by the interactions among nodes. This was confirmed in simulations of our multi-regional macaque cortex model (Chaudhuri et al. 2015), where all areas were driven by noise, and co-variance of activity of pairs of areas defined FC. The correlation between the simulated FC and the anatomical connectivity (FLNs) is about $r^2 = 0.53$ (**Figure 9.19**, right panel). To assess what happens without such a gradient, in another model simulation $(1 + \eta h_i)$ in Eq. (9.5) was replaced by its average over i, so that the macroscopic gradient is removed and isolated areas become identical. Notably, the functional connectivity is dramatically altered in the absence of the macroscopic gradient (**Figure 9.19**, left panel). In that case, the correlation between the functional and anatomical connectivities is higher ($r^2 = 0.83$). The difference is largely due to reverberatory dynamics in association areas that have a large impact on the global neurodynamical pattern in the control case.

It follows from this finding that long-range connections alone cannot predict global brain activity patterns. A recent study of human cortex showed that correlation between functional connectivity and structural connectivity (measured by diffusion tensor imaging, DTI (Iturria-Medina et al. 2007)) gradually decreases from unimodal sensory areas to transmodal or association areas (Vázquez-Rodríguez et al. 2019). This is consistent with the model behavior where functional connectivity is not affected with deletion of the macroscopic gradient in early sensory areas (**Figure 9.19**, upper left corner of the matrix), but the opposite is true in association areas (**Figure 9.19**), lower right corner of the matrix). Therefore, Chaudhuri et al. (2015) predicted that functional connectivity analyses revised to take into account a heterogeneous distribution of properties in a large-scale cortex, notably in the form of macroscopic gradients, would better relate functional and structural connectivities.

Demirtaş et al. (2019) tested this prediction by comparing a computational model of the human cortex with functional imaging measurements from more than 300 healthy subjects. In this model, the inter-areal connectivity was based on the structural MRI data of the Human Connectome Project (Glasser et al. 2016). As in a previous work (Deco et al. 2012; Deco et al. 2014), dynamics of each local area is described by a population rate model adopted from Wong and Wang (2006) (**Figure 9.20a**). The procedure of comparison between the model and functional connectivity data are illustrated in **Figure 9.20b**: the model is built with inter-areal structural connectivity data measured by DTI, and the model is simulated dynamically and neural activity is converted into BOLD signal using the Balloon model (Friston et al. 2003), yielding model-simulated functional connectivity (Model Generated Data). This is compared with the experimentally measured functional connectivity in the resting state (empirical data), and the fitting is optimized by adjusting model parameters.

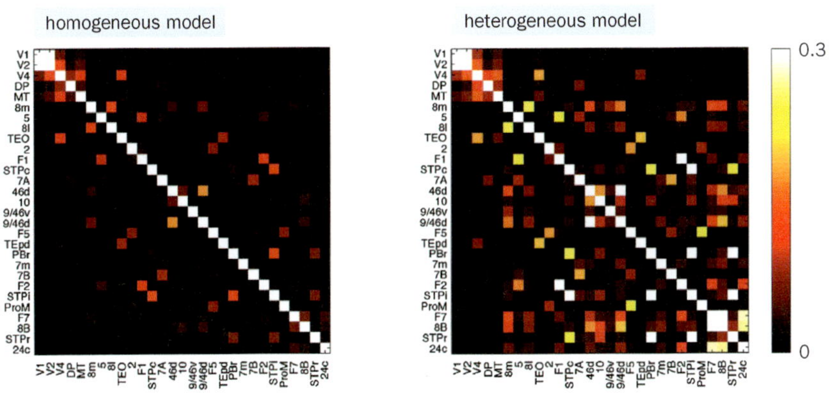

Figure 9.19 Functional connectivity of the macaque cortex model (right) and when the macroscopic gradient was removed from the model (left). A gradient of synaptic excitation enhances functional connectivity specially for association areas with slow time constants. Reproduced from Chaudhuri et al. (2015). [Chaudhuri, R., Knoblauch, K., Gariel, M. A., Kennedy, H., & Wang, X. J. (2015). A large-scale circuit mechanism for hierarchical dynamical processing in the primate cortex. Neuron, 88(2), 419–431.]

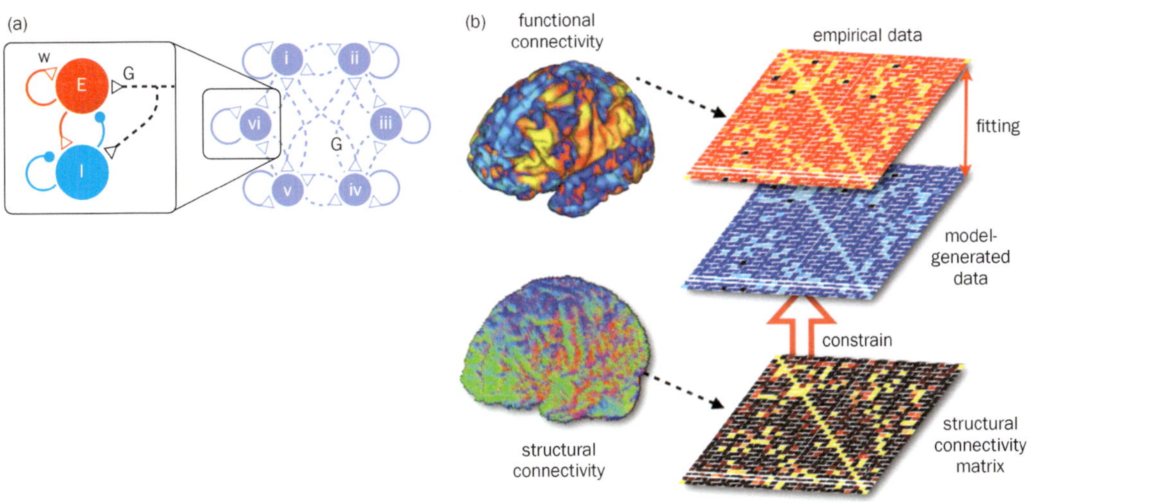

Figure 9.20 Schematic of a large-scale model of the human cortex. (a) Each of the local circuits (from i to iv in this example) has an excitatory neural population and an inhibitory neural population; long-range connections target both. (b) Structural connectivity measured by diffusion MRI is used to build a dynamical model, and the model-simulated functional connectivity is compared with the measured one by BOLD signal from humans in the resting state. Figure provided by John Murray. [Courtesy of John Murray.]

A global brain connectivity (GBC_i) for each parcellated area i is defined as the average of its functional connectivities with all the other cortical areas (**Figure 9.21a**). Two cases were compared in the model: either areas are identical, thus differing from each other only by inter-areal connectivity, or macroscopic gradients are introduced (**Figure 9.21b**), in which case the strengths of local recurrent excitation increase linearly along the human cortical hierarchy defined approximately by decreased T1w/T2w ratio. **Figure 9.21c–d** show model simulated global brain connectivity values (one for each area) plotted against those measured using human fMRI in the resting state. With areas differing only in their input and output patterns, the correlation (r) between the computational model and data is about 0.48, comparable to a previous study (Deco and Jirsa 2012). When a linear gradient of strength for local synaptic excitation is introduced along the T1w/T2w axis, the correlation is significantly higher (~ 0.74), in support of Chaudhuri et al. (2015). In a separate work combining modeling and fMRI measurements in the resting state of more than 1000 subjects, it was shown that the inclusion of macroscopic gradients also improved model fitting with human data of dynamical fluctuations on multiple timescales in addition to static functional connectivity measures (Kong et al. 2021). These works highlight the importance of taking into consideration of macroscopic gradients in network studies of large-scale brain dynamics.

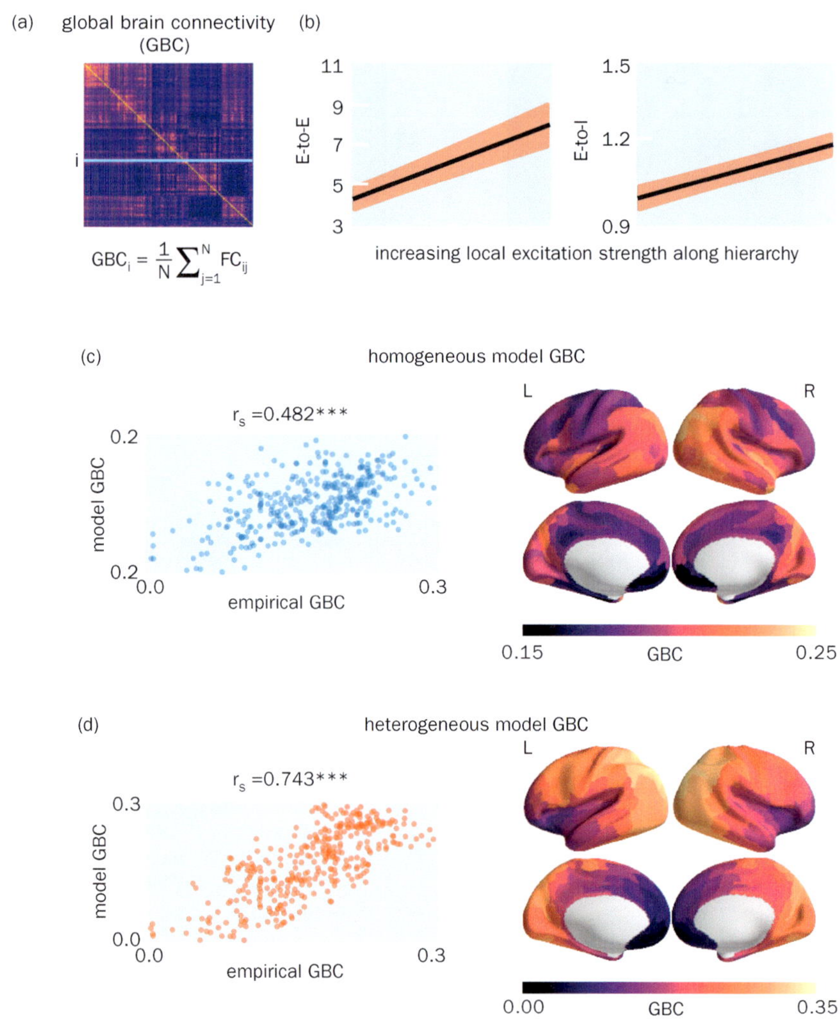

Figure 9.21 Incorporation of a hierarchy of local synaptic excitation improves model fitting with experimentally measured functional connectivity. (a) Definition of global brain connectivity (GBC) of each area from functional connectivity matrix. (b) Macroscopic gradient of synaptic excitation strength as a function of cortical hierarchy defined in humans by the T1w/T2w ratio. (c) Model-simulated GBC versus the measured GBC using fMRI in humans, in the absence of a macroscopic gradient so that isolated areas are identical in the model. (d) Same as (c) except with macroscopic gradients shown in (b). Reproduced from Demirtaş et al. (2019). [Demirtaş, M., Burt, J. B., Helmer, M., Ji, J. L., Adkinson, B. D., Glasser, M. F., ... & Murray, J. D. (2019). Hierarchical heterogeneity across human cortex shapes large-scale neural dynamics. Neuron, 101(6), 1181–1194.]

9.5.2 Layer-Dependent Feedforward and Feedback Processes

Feedforward information processing across multiple areas in a "deep network" has been extensively studied in neuroscience, especially for the visual system (Serre et al. 2007a; Yamins et al. 2014), and influenced recent developments in artificial intelligence (LeCun et al. 2015). However, the cortex is endowed with an abundance of feedback projections (Felleman and Van Essen 1991; Gilbert and Li 2013; Harris and Mrsic-Flogel 2013; Harris and Shepherd 2015; Kreiman and Serre 2020). They play important roles such as top-down selective attention (Chapter 8). Feedback projections are also essential for the framework of predictive coding (Clark 2013). The central idea of predictive coding is that the brain is not merely a passive receiver of sensory stimulation but constantly, and actively, generates inference about the external world, such as predicting what is going to happen next while watching a movie (Lee et al. 2021). Predictive coding involves a comparison between an expectation signal mediated by top-down projection and sensory stimulus from bottom-up projection (Mumford 1992; Ullman 1995; Rao and Ballard 1999; Summerfield et al. 2006; Bastos et al. 2012; O'Reilly et al. 2021).

As we know, feedforward and feedback projections are layer-dependent (**Figure 9.9**). To examine the interplay between bottom-up and top-down processes in a multi-regional model, feedforward and feedback projections should be wired in a layer-dependent manner. Mejias et al. (2016) extended the model of Chaudhuri et al. (2015) to incorporate a laminar structure that is simple yet anatomically and physiologically constrained. The extended model is developed in three steps. First, each cortical area now consists of two layers, a superficial layer (including layer 4) and a deep layer (below layer 4); this is minimal but adequate for implementing inter-areal connections in a layer-dependent fashion. The synaptic connections between the two layers are informed by neurophysiology. Studies using techniques such as intracellular recording of connected pairs (Thomson et al. 2002) or a combination of recording and laser scanning photostimulation (Xu et al. 2016) found that the connection from the superficial layer to the deep layer is dominated by pyramid-to-pyramid excitation, while the connection in the reversed direction is primarily from pyramidal cells in the deep layer to inhibitory neurons in the superficial layer (**Figure 9.22a–b**). Other studies showed that in some cortical areas the deep-to-superficial connection also has substantial excitatory-to-excitatory projections (Petreanu et al. 2009; DeNardo et al. 2015).

Second, numerous reports have documented that the superficial layer generates fast gamma oscillations at around 40 Hz (Chapter 3), while the deep layer shows slower synchronous oscillations such as alpha rhythm around 10 Hz (Buffalo et al. 2011) or beta rhythm around 15–25 Hz (Bastos et al. 2018) (for a review, see Wang 2010). In the model, each of the two layers is a recurrent network of excitatory and inhibitory neurons, which can produce a rhythm with frequency controlled by the time constant of feedback inhibition (Chapter 3, Section 3.4.2). This is realized in the model of Mejias et al. (2016), with slower synaptic time constants in the deep layer than in the superficial layer, so that the former shows a rhythm at 10 Hz, whereas the latter oscillates at 40 Hz (**Figure 9.22c**). Because of the synaptic interactions between the two layers, gamma oscillations in the superficial layer are modulated by alpha rhythm (**Figure 9.22d**), as observed experimentally (Spaak et al. 2012).

Third, for the inter-areal connections, a feedforward projection now originates from the superficial layer in a source area and targets the superficial layer in a target area, and a feedback projection starts with the deep layer of a source area, according to neuroanatomy. Less known are layer- and cell-type-specific targets of a feedback projection. This question cannot

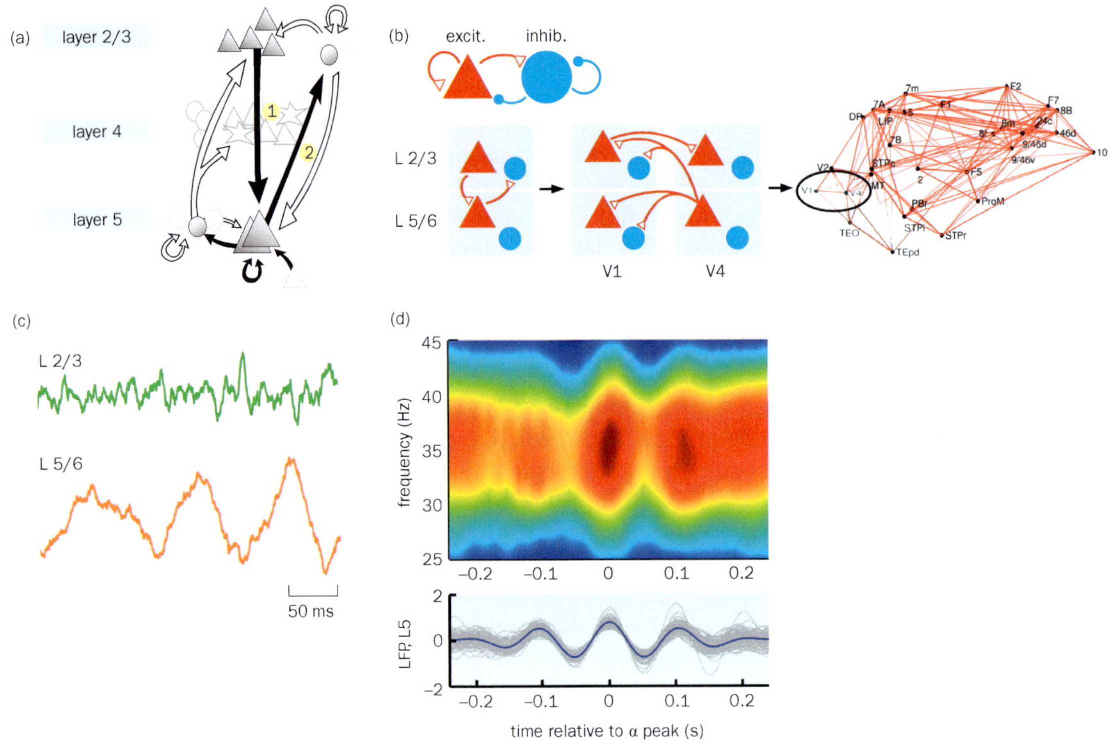

Figure 9.22 A multi-regional model of the macaque monkey cortex endowed with a laminar structure. (a) Summary of intra-laminar connectivity based on a physiological study of intracellular recording from pairs of neurons in rat somatosensory cortex. (b) Model scheme shows the four levels considered: A within-layer local microcircuit consisting of an excitatory (red) and an inhibitory (blue) population (upper left), a laminar circuit with two laminar modules (corresponding to superficial and deep layers, lower left), an inter-areal circuit with laminar-specific projections (lower middle), and a large-scale network of cortical areas based on macaque anatomical connectivity (right). Each level is anatomically constrained. Only the connections at each level not shown at a lower level are plotted, for clarity. (c) The superficial layer and deep layer display gamma (upper panel) and alpha (lower panel) oscillations, respectively. (d) The periodogram of the superficial layer shows gamma rhythm modulated by alpha rhythm (top), whereas the deep layer is dominated by alpha rhythmicity (bottom). Panel (a) is reproduced from Thomson et al. (2002), (b–d) from Mejias et al. (2016). [Thomson, A. M., Bannister, A. P., Mercer, A., & Morris, O. T. (2002). Target and temporal pattern selection at neocortical synapses. Philosophical Transactions of the Royal Society of London. Series B: Biological Sciences, 357(1428), 1781–1791.] [Mejias et al., Sci. Adv.2016;2:e1601335. © The Authors, some rights reserved; exclusive licensee AAAS. Distributed under a CC BY-NC 4.0 license http://creativecommons.org/licenses/by-nc/4.0/] [Modified from Mejias et al., Sci. Adv.2016;2:e1601335. © The Authors, some rights reserved; exclusive licensee AAAS. Distributed under a CC BY-NC 4.0 license http://creativecommons.org/licenses/by-nc/4.0/]

be answered by retrograde track tracing, which does not ascertain layers of target cells. Nor can the information be obtained using anterograde tracing of axonal fibers because, for example, labeled terminals in a superficial layer may still make synaptic contacts with deep layer neurons, which have dendrites in superficial layers. In the model, for a feedback pathway, the weights of projections into the superficial and deep layers of a target area are adjusted as free parameters. Carefully calibrated by experimental literature, again only a few parameters are unconstrained and freely tunable in this multi-regional cortex model with a laminar structure.

This model reproduces salient observations from a monkey experiment on feedforward versus feedback signaling. van Kerkoerle et al. (2014) stimulated V1 while recording from V4 using a laminar probe in macaque monkeys, or vice versa, and used a frequency-dependent Granger causality measure to assess communication between V1 and V4 separately in each of the two directions. Granger causality is a linear regression analysis measure to determine whether one time series $x(t)$ forecasts another time series $y(t)$ (Granger 1969); frequency-dependent Granger causality is an extension where $x(t)$ and $y(t)$ are first filtered at different frequencies, and then Granger causality is computed as a function of the frequency

(Barnett and Seth 2014). It was found that a feedforward signal from V1 to V4 selectively enhances Granger causality in the gamma frequency range, whereas a feedback signal from V4 to V1 differentially increases Granger causality in the alpha frequency range. Similar results were found in another experiment (Bastos et al. 2015). These observations were reproduced by the model (**Figure 9.23a**), provided that the feedback projection primarily targets excitatory neurons in the deep layer. Whether the latter condition is necessary remains to be further studied both in models and experimentally.

Based on this general trend, Bastos et al. (2015) proposed a measure to deduce cortical hierarchy from frequency-dependent Granger causality analysis (**Figure 9.23b**). Using the same analysis as in the monkey experiment, the model captures fairly well the hierarchy deduced from Granger causality analysis (**Figure 9.23c**). This procedure is solely based on physiological measurements, and in the monkey experiment it was found that the hierarchy thus deduced correlates strongly with the anatomically deduced hierarchy shown in **Figure 9.9**. This success led the same research group to deduce cortical hierarchy based on non-invasive MEG measurements of the human cortex (**Figure 9.23d**) (Michalareas et al. 2016). Note that in humans the cortical hierarchy has not been firmly established by purely anatomical means, and therefore a physiology-based approach is especially valuable.

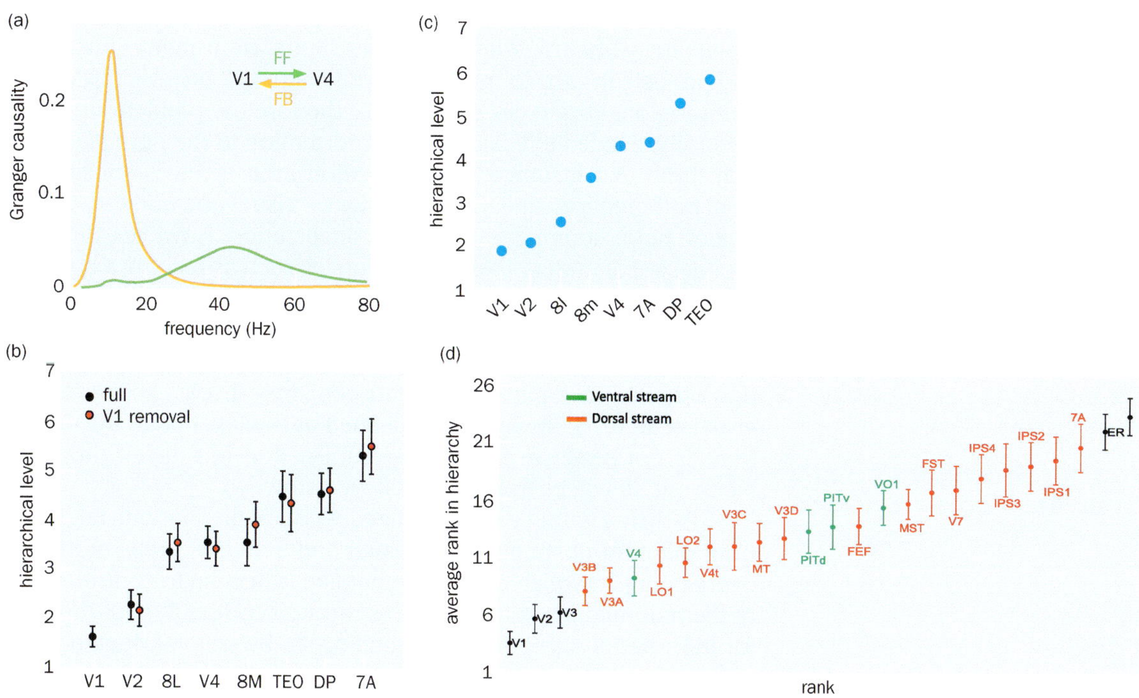

Figure 9.23 Cortical hierarchy deduced from frequency-dependent Granger causality. (a) Granger causality as a function of frequency for feedforward signaling from V1 to V4 (green) and feedback (orange) in the model. (b) Cortical hierarchy deduced from the frequency-dependent Granger causality measure in a monkey experiment. (c) Cortical hierarchy in the model of **Figure 9.22**, computed using the same method as the monkey experiment. (d) Cortical hierarchy deduced from frequency-dependent Granger causality analysis of MEG data in the human cortex. Panels (a) and (c) are reproduced from Mejias et al. (2016), (b) from Bastos et al. (2015) and (d) from Michalareas et al. (2016). [Reproduced from Mejias et al., Sci. Adv.2016;2:e1601335. © The Authors, some rights reserved; exclusive licensee AAAS. Distributed under a CC BY-NC 4.0 license http://creativecommons.org/licenses/by-nc/4.0/] [Bastos, A. M., Vezoli, J., Bosman, C. A., Schoffelen, J. M., Oostenveld, R., Dowdall, J. R., ... & Fries, P. (2015). Visual areas exert feedforward and feedback influences through distinct frequency channels. Neuron, 85(2), 390–401.] [Michalareas, G., Vezoli, J., Van Pelt, S., Schoffelen, J. M., Kennedy, H., & Fries, P. (2016). Alpha-beta and gamma rhythms subserve feedback and feedforward influences among human visual cortical areas. Neuron, 89(2), 384–397.]

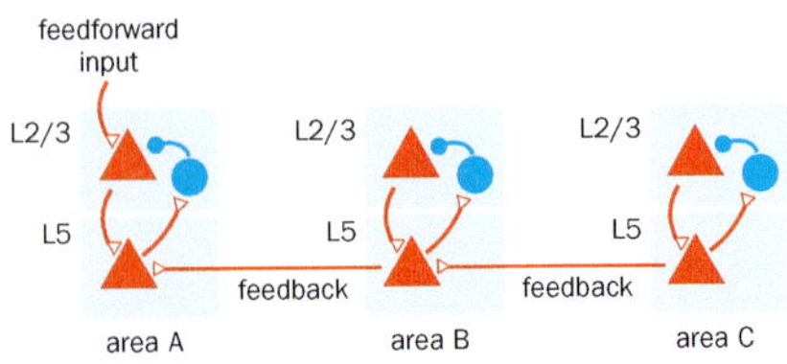

Figure 9.24 Schematic of a circuit substrate for predictive coding. Under the assumption that the dominant effect of feedback projection is excitation of infragranular pyramidal cells, which in turn project to supragranular inhibitory neurons, a top-down prediction signal Y changes the sign and is inhibitory for infragranular excitatory neurons to be compared with excitation X from feedforward stimulation, thereby implementing a comparison mechanism. This process may apply to a chain of cortical areas in a counterstream of the hierarchy. [Wang, Xiao-Jing. (2022). Theory of the Multiregional Neocortex: Large-Scale Neural Dynamics and Distributed Cognition. Annual Review of Neuroscience 45:533–560.]

The laminar model of Mejias et al. (2016) suggests a circuit substrate for predictive coding, as schematically shown in **Figure 9.24**. Consider an area receiving a feedforward signal X onto pyramidal neurons in the superficial layer, such as external stimulation propagating along the hierarchy. A prediction signal Y from a higher area excites pyramidal neurons in its deep layer, which in turn project to supragranular inhibitory neurons, thereby changing the sign from Y to $-Y$, and can be compared with X in pyramidal neurons of the superficial layer. Such a process can occur as an inverse cascade in a deep cortical system, from area C to area B, area B to area A, etc. Consistent with this scenario, there is experimental evidence that top-down prediction reduces sensory responses (Eliades and Wang 2008; Fiser et al. 2016; Keller and Mrsic-Flogel 2018), and alpha/beta coherence has an inhibitory effect (Hagan and Pesaran 2022). It is important, however, to note that response to a stimulus X and prediction Y are not numbers, but represented by neural population activity. The comparison thus is between two spatiotemporal firing patterns, and an alternative mechanism to **Figure 9.24** was discussed in the context of pattern match decisions (Chapter 6, Section 6.6.3). Much future work is needed to dissect the brain mechanisms of predictive coding (Keller and Mrsic-Flogel 2018).

9.5.3 Gating of Inter-Areal Communication

One widely open question is how inter-areal transmission of spikes is gated flexibly based on behavioral demands (Olshausen et al. 1993; Abbott 2006; Vogels and Abbott 2009; Ardid and Wang 2013; Wang and Yang 2018). We have already encountered examples of gating in Chapter 8, in the context of selective attention and task switching. Given the dense inter-areal connectivity, a cortical area receives inputs from many "upstream" areas and projects in turn to numerous "downstream" areas. Consideration of a multi-regional brain circuit brings to the fore the problem of gating, more generally, needed to flexibly route information to the right place and at the right time. For instance, an association area receives converging inputs from both auditory and visual pathways, visual and auditory sensory integration helps comprehension in a conversation. However, this integration is undesirable under other circumstances such as reading a book in a noisy cafe, when your brain should "gate in" visual inputs and "gate out" auditory ones to language areas. We shall call this form of routing of information "pathway gating" in a multi-regional brain system. Unlike selective attention oriented to a particular location or object (e.g., your friend's face in visual search), pathway gating defined at the level of area-to-area projections is not selective for specific information in a given projection (e.g., block out all voices from the auditory pathway).

Pathway gating can be realized if neural populations receiving inputs from different upstream (e.g., visual and auditory) areas are largely non-overlapping, in which case a mechanism is needed to control whether one of them dominates over the others through competition or they cooperate for information integration, depending on behavioral demands. An alternative or complementary mechanism takes advantage of the complex dendritic tree of single pyramidal neurons consisting of two dozen relatively independent units (Polsky et al. 2004; Poirazi and Papoutsi 2020). Yang et al. (2016) investigated this scenario with a model under the assumption that quasi-independent dendritic compartments are targeted by disinhibitory motifs (Chapter 2, **Figure 2.2**) involving subsets of SST and VIP cells (**Figure 9.25a**). The model incorporates several experimental observations. First, dendritic membrane potential exhibits plateau potential (Schiller et al. 2000; Larkum et al. 2009), and its response to an excitatory input is highly nonlinear and can be effectively controlled by dendritic inhibition (**Figure 9.25b**). Second, a pyramidal neuron receives projections

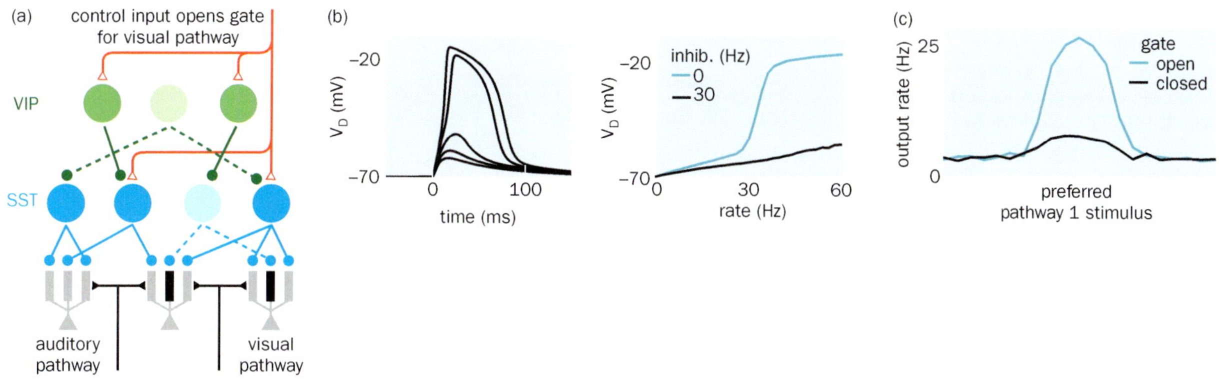

Figure 9.25 Pathway gating by dendritic inhibition and disinhibition. (a) A schematic illustration where dendritic segments are differentially targeted by different SST-positive interneurons. In this example, visual input is gated in, because a control signal excites VIP-positive and SST-positive cells that lead to disinhibition of dendritic branches (black) targeted by the visual pathway (light blue), where inhibition of those dendritic branches (grey) targeted by the auditory pathway (dark blue) is enhanced. (b) Dendritic nonlinearity in the form of a plateau potential that is triggered when a graded input exceeds a threshold level (left), and this nonlinear response is suppressed with dendritic inhibition (right). (c) Neural response as a function of a stimulus feature is much higher with disinhibition (light blue) than when the gate is closed (black). Panels (b–c) are reproduced from Yang et al. (2016). [Yang GR, Murray J and Wang X-J (2016) A dendritic disinhibitory circuit mechanism for pathway-specific gating. Nature Communication, 7:12815. Doi: 10.1038/ncomms12815. Licensed under CC BY 4.0.]

from upstream areas with synapses that are clustered in dendrite branches (Yang et al. 2014; Kastellakis et al. 2015). Third, VIP interneurons are a major recipient of top-down signaling (Wall et al. 2016). Consider a behavior requiring that vision is gated in and audition is gated out, and an executive control signal activates a subset of VIP cells, which in turn inhibit their targeted SST cells. As a result, those dendrite branches receiving visual inputs are disinhibited selectively (**Figure 9.25c**). This mechanism is effective even when connections are random, as long as projections from SST interneurons to pyramidal dendrites are sufficiently sparse (Yang et al. 2016b). In this scenario, behavioral context or rule is hypothesized to involve executive control signaling onto selective sets of SST and VIP neurons, a prediction that can be tested experimentally.

Several other candidate mechanisms for gating of inter-areal communication have been proposed. For illustration in **Figure 9.26**, a cortical area receives inputs from two source areas and sends its output to two downstream areas. Input gating assumes that information in a source is broadcast widely to all of its target areas, which may or may not "listen to" it. This may be gated by the aforementioned disinhibitory motif and modulated by rhythmic synchronization, as discussed in Chapter 8 (Section 8.4.3) (Fries 2005; Wang 2010; Buschman and Kastner 2015; Schneider et al. 2021). On the other hand, gating can take place at the output end: if neural populations in a source area projecting to different downstream areas are distinct, they may be controlled by different groups of chandelier cells, which synapse on the axon initial segment of pyramidal cells (Somogyi 1977; DeFelipe et al. 2013; Lu et al. 2017). Gating may also involve subcortical structures; in particular, information flow may be contingent on suppression of inhibition from the basal ganglia onto the thalamus. This mechanism has been proposed to gate access of selective information to the working memory system (Frank et al. 2001) and decision making (Wei and Wang 2016).

Mathematically, area-to-area communication has been described in terms of subspaces of neural population activity (Semedo et al. 2019; Kohn et al. 2020). To illustrate the idea, consider a toy example of a valuation area with three selective neural pools that encode values V_A, V_B and

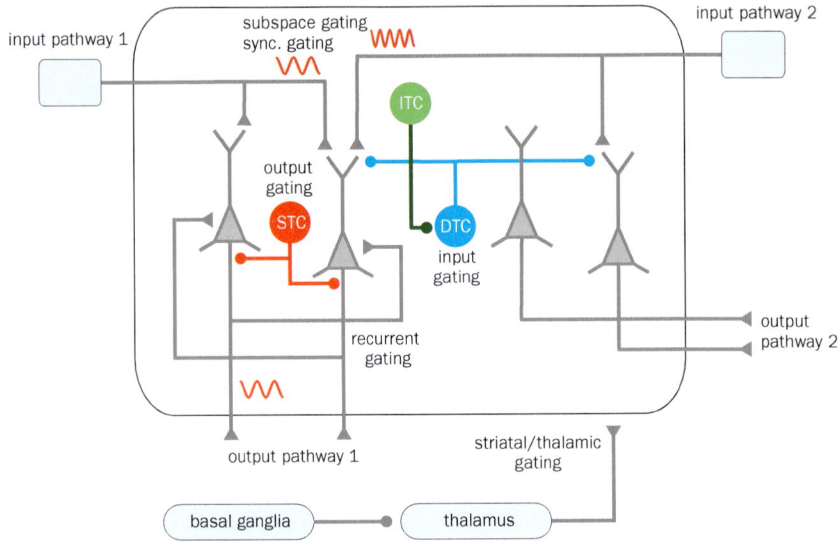

Figure 9.26 Various mechanisms for information gating in the brain. Communication between two areas may selectively channel presynaptic population activity in a particular subspace of neural population activity in a source area, or/and depend on the degree of temporal synchrony of neural activity between the source and target areas. Pathway gating can be achieved by dendrite-targeting interneurons that selectively control inputs to pyramidal dendrites. Recurrent gating mechanism involves selective integration of inputs based on context-dependent dynamics of the network. Output gating is instantiated interneurons that specifically target the initial segment of the axon in pyramidal neurons, projecting to one pathway but not others. Gating may also involve subcortical structures, especially basal ganglia and thalamus. Reproduced from Wang and Yang (2018). [Wang, X. J., & Yang, G. R. (2018). A disinhibitory circuit motif and flexible information routing in the brain. Current Opinion in Neurobiology, 49, 75–83.]

V_C for choice options A, B and C, respectively. The population activity is represented in a three-dimensional state space (with axes corresponding to outputs x_A, x_B, x_C which, for the sake of simplicity are not restricted to be positive). The population state is given by the vector $\mathbf{x} = (x_A, x_B, x_C)$. Suppose that it is desirable to readout difference signals $y_1 = V_A - V_B$, $y_2 = V_A - V_C$ and $y_3 = V_B - V_C$ to inform decision making in a downstream area. Such difference signaling can be achieved with a rank-2 connectivity matrix as:

$$W = \begin{pmatrix} 1 & -1 & 0 \\ 1 & 0 & -1 \\ 0 & 1 & -1 \end{pmatrix}, \tag{9.9}$$

so that the downstream system is driven by neural activity in the two-dimensional subspace $x_A + x_B + x_C = 0$, determined by the row vectors of W ($\mathbf{v}_1 = (1,-1,0)$, $\mathbf{v}_2 = (1,0,-1)$ and $\mathbf{v}_3 = (0,1,-1)$). The projection of $\mathbf{x}$ onto this subspace is communicated to the downstream area. If $\mathbf{x}$ happens to be orthogonal to that subspace, which corresponds to $x_A = x_B = x_C$ (the null-space of the matrix W), nothing will be transmitted to the target area.

The same source area could, in principle, communicate to another target area where the total sum of values ($z = V_A + V_B + V_C$) is encoded, by a rank-1 connectivity matrix with all the elements equal to 1. The communication subspace in the source area is a line orthogonal to the subspace for communicating to the value-difference encoding area.

This hypothetical example makes it clear that a communication subspace is determined by the connectivity matrix. When sender and receiver neurons are simultaneously recorded from a pair of connected areas, such as visual V1 and V2 areas, one can identify a communication subspace in each area within which neural population activity patterns of the two areas are maximally correlated (Semedo et al. 2022). Here, "connectivity" is defined functionally. One must not be content with a phenomenological description without probing the corresponding neural circuit mechanisms. For example, computing difference signals requires a set of target inhibitory neurons with selectivity (e.g., changing V_A to $-V_A$ for computing a value difference), but how this might take place biologically in inter-areal communication is unknown.

Recurrent dynamics within a local area may also gate particular computational processes rather than input flow. For instance, in a trained recurrent network model of context-dependent decision making, in a given context one of two stimulus features is relevant and should be integrated over time, and a selection vector can guide temporal accumulation of evidence only for the relevant feature but not the irrelevant one, even though both inputs are received by a decision circuit (Mante et al. 2013). This work demonstrates that a context or a task rule can modify a population activity with respect to the communication subspace defined by area-to-area connectivity, thereby altering the kind of information that needs to be communicated to fulfill behavioral demands. Moreover, reverberatory dynamics can amplify weak control signals to achieve task-dependent reconfiguration of sensorimotor systems, according to the tweaking hypothesis of executive control (Chapter 8, Section 8.5).

Although candidate mechanisms are numerous, experimental tests to validate or falsify them are scarce. This situation is likely to change in the near future, with our newly acquired ability to record from multiple brain regions in animals performing cognitive tasks. Progress in this direction would shed light on the mystery of gated communication in a large-scale brain system underlying behavioral flexibility.

9.6 DISTRIBUTED WORKING MEMORY

As we discussed in Chapter 5, working memory represents an "atom" of cognition. While this may be an apt metaphor, working memory engages multiple brain regions. Self-sustained neural persistent activity during working memory of a delay-dependent task has been found in multiple brain regions, and often areas appear in co-activation (Christophel et al. 2017; Leavitt et al. 2017; Dotson et al. 2018; Sreenivasan and D'Esposito 2019). This does not necessarily imply that working memory representation lacks specificity (**Figure 9.27a**); recent studies reported that widespread activation in the cortex is correlated with bodily movements, including microbehavior during a delay period (Musall et al. 2019; Schaffer et al. 2023; Wang et al. 2023b), presumably reflecting proprioceptive signals (Proske and Gandevia 2012) or/and corollary discharges from motor command centers (Sommer and Wurtz 2008; Zagha et al. 2022). A systematic survey of the existing literature found that working memory representation is distributed yet selectively involves a subset of cortical areas in the macaque cortex (**Figure 9.27b**).

Let us tackle the "distributed coding versus modularity" dichotomy using multi-regional cortical network modeling. In psychology, modularity denotes an organization of the mind into distinct component capabilities (Fodor 1983); in neuroscience, it refers to functional specialization of brain areas (Kanwisher 2010). We propose that modularity is understood in terms of a selective subset of cortical areas, which are not necessarily spatially congruent, engaged in a distinct brain function. This definition is compatible with distributed neural representation and processing across multiple brain regions, but in contrast to the absence of modularity manifest by merely graded variations of engagement across the entire cortical mantle.

9.6.1 The Parieto-Frontal Loop

While most biologically based models describe local circuits, a few works explored two interacting areas (Edin et al. 2009; Murray et al. 2017), particularly inspired by the experimental observation that the PFC and PPC often are engaged in the same working memory tasks (Gnadt and Andersen 1988; Funahashi et al. 1989; Chafee and Goldman-Rakic 1998, 2000; Courtney et al. 1998; Curtis and D'Esposito 2003; Constantinidis and Wang 2004). **Figure 9.28a** shows schematically a parieto-frontal loop model, with each of the two areas modeled as in Wong and Wang (2006). Recall

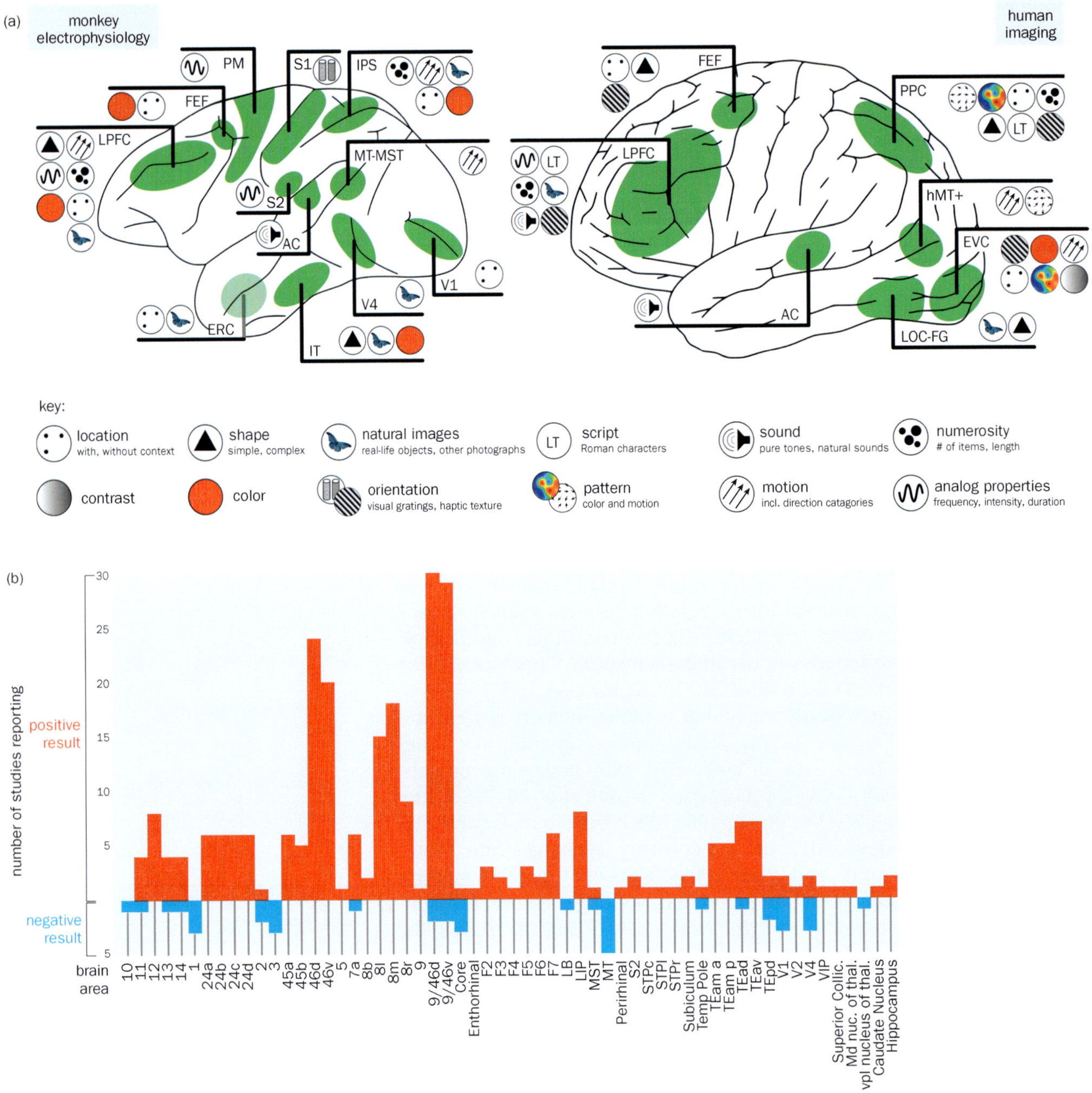

Figure 9.27 Distributed working memory. (a) Content-specific delay period activity during working memory in the macaque (left) and human (right) brains. Icons indicate persistent stimulus-selective activity for each stimulus type indicated by the icon at the respective locations. AC, auditory cortex; ERC, entorhinal cortex; EVC, early visual cortex; FEF, frontal eye fields; FG, fusiform gyrus; hMT+, human analog to MT/MST; IPS, intraparietal sulcus; IT, inferior temporal cortex; LOC, lateral occipital complex; lPFC, lateral prefrontal cortex; PM, premotor cortex; PPC, posterior parietal cortex. (b) Positive and negative reports of working memory-related activity in different brain regions from more than 90 monkey physiological experiments. Panel (a) is reproduced from Christophel et al. (2017), (b) from Leavitt et al. (2017). [Christophel, T. B., Klink, P. C., Spitzer, B., Roelfsema, P. R., & Haynes, J. D. (2017). The distributed nature of working memory. Trends in Cognitive Sciences, 21(2), 111–124.] [Leavitt, M. L., Mendoza-Halliday, D., & Martinez-Trujillo, J. C. (2017). Sustained activity encoding working memories: Not fully distributed. Trends in Neurosciences, 40(6), 328–346.]

that, in an isolated area when the strength of local recurrent excitation, denoted by a parameter J_{EE}, is varied gradually beyond a threshold value J_{th}, a family of self-sustained persistent activity patterns, each encoding a stimulus item, suddenly emerges (Chapter 5, **Figure 5.14a**) (Brunel and Wang 2001; Wang 2001). There are three ways to produce persistent activity in both areas (**Figure 9.28b**). First, each in isolation has sufficient local

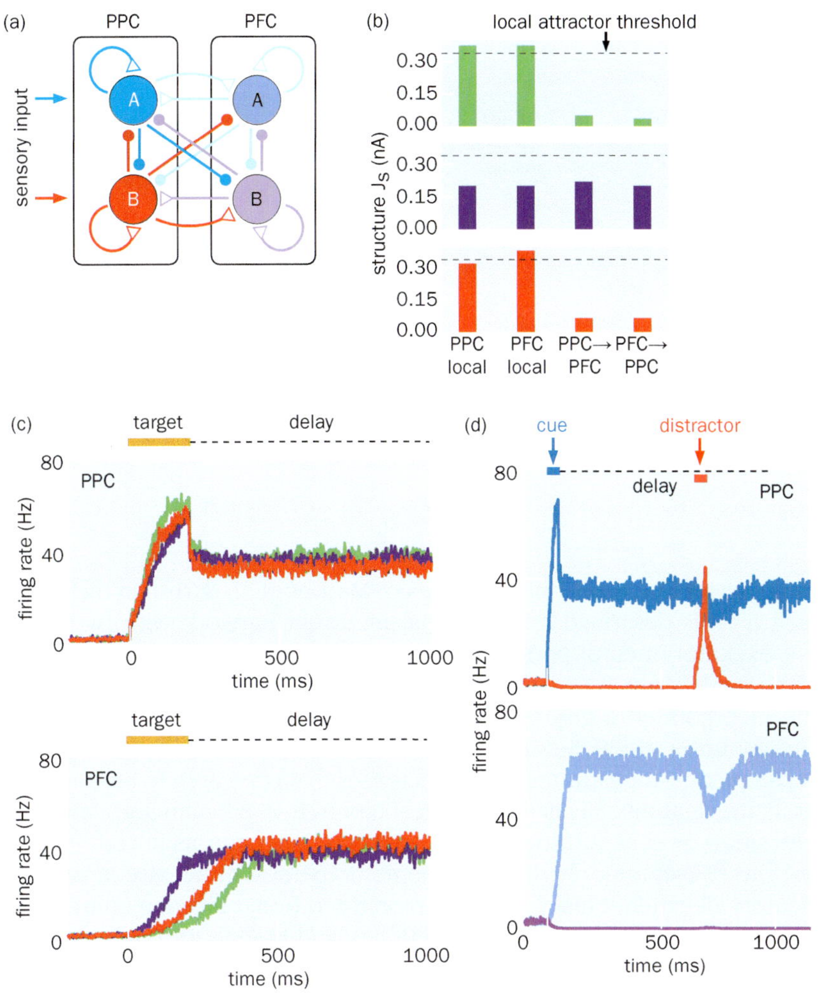

Figure 9.28 Working memory in a parieto-frontal loop model. (a) Model scheme with two selective excitatory neural pools in each of the areas. (b) Three scenarios for the generation of persistent activity, defined by whether the strength of an isolated area exceeds the threshold J_{th} for the emergence of stimulus-selective mnemonic states. (c) Persistent activity is similar in the three scenarios, therefore insufficient to differentiate them. (d) The model reproduces the experimental observation that the PPC is sensitively responsive to a distractor in the middle of a delay period, but the PFC is resistant and thus preserves the memory of behaviorally relevant information. Colors are as in panel (a) for four selective neural pools. Reproduced from Murray et al. (2017). [Murray, J. D., Jaramillo, J., & Wang, X. J. (2017). Working memory and decision-making in a frontoparietal circuit model. Journal of Neuroscience, 37(50), 12167–12186.]

recurrent excitation to maintain mnemonic firing trigged by a brief external stimulus (**Figure 9.28b–c**, green). Second, one of the two (say, the PFC) is endowed with sufficient reverberation, but not the other (the PPC), which nevertheless displays persistent firing as a result of sustained PFC input (**Figure 9.28b–c**, red). Finally, recurrent excitation is insufficient in each area by itself, and persistent activity depends on the inter-areal reciprocal loop (**Figure 9.28b–c**, purple). The persistent activity pattern may be virtually indistinguishable in the three cases.

The demand of filtering out distractors for proper working memory function provides an additional consideration to pin down a particular mechanism. Neurophysiological experiments showed that delay period activity is readily disrupted by a distractor in the PPC but not PFC (Katsuki and Constantinidis 2012; Suzuki and Gottlieb 2013). This differential function makes sense, for the brain to simultaneously maintain behaviorally relevant information while being sensitive to potentially useful new information from the environment. The model reproduces this finding (**Figure 9.28d**) under the condition that the PFC has a stronger recurrent excitation strength than that of the PPC. If a two-area loop already has three possible scenarios, then how many possibilities are there for a cortical system with many interacting parcellated areas?

9.6.2 Distributed Mnemonic Activity in the Cortex

Mejias and Wang (2022) developed a large-scale model of distributed working memory, using the directed and weighted inter-areal connectivity

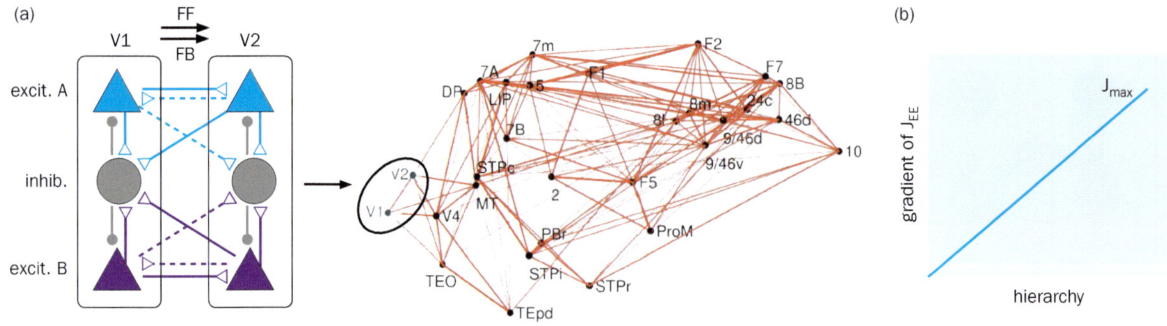

Figure 9.29 A model of distributed working memory representation in a large-scale monkey cortex model endowed with a macroscopic gradient of synaptic excitation. (a) Model schema with two selective excitatory neural pools and one inhibitory neural pool in each local area. (b) Linear gradient of the strength of local recurrent excitation J_{EE}. Reproduced from Mejias and Wang (2022). [Mejias, J. F., & Wang, X. J. (2022). Mechanisms of distributed working memory in a large-scale network of macaque neocortex. Elife, 11, e72136. Licensed under CC BY 4.0.]

database for macaque monkey cortex of Markov et al. (2014). In the model, each area is described by a minimal recurrent network with two selective excitatory neural pools and an inhibitory neural pool (Wang 2002b; Wong and Wang 2006) (**Figure 9.29a**). The large-scale circuit modeling incorporated a gradient of synaptic excitation in the primate cortex (Chaudhuri et al. 2015; Mejias et al. 2016; Burt et al. 2018; Wang 2020). Consequently, the parameter J_{EE} increases from one area to another area along the anatomically defined cortical hierarchy (Felleman and Van Essen 1991; Markov et al. 2014b) (**Figure 9.29b**). While neurons in certain areas like the PFC commonly display mnemonic persistent activity in working memory-dependent tasks, they are monitored from behaving animals and not necessarily maintained locally. Mejias and Wang (2022) explored different scenarios. For instance, if the maximum of J_{EE}, J_{Max}, is below J_{th}, none of the cortical areas in isolation is capable of persistent activity. Consequently, the emergence of persistent activity observed in the interconnected cortical system necessarily depends on long-distance connection loops between cortical areas, extending the concept of synaptic reverberation (Lorente de Nó 1933; Amit 1995; Goldman-Rakic 1995; Wang 2001) to a large-scale multi-regional brain system.

Figure 9.30a–b shows model simulation of a visual delayed response task under this condition. Notably, responses to an input during stimulus presentation engage posterior parts of the cortex, whereas persistent activity during the delay period displays a spatial pattern involving frontal, parietal and temporal areas. Persistent activity level of each area plotted as a function of its hierarchical position exhibits a gap in the firing rate that separates the areas with and those without mnemonic activity (**Figure 9.30c**). This is reminiscent of a bifurcation (**Figure 5.14a** of Chapter 5), but occurring in space rather than as a function of a parameter. One could say that the parameter is mapped onto the cortical space, by virtue of a macroscopic gradient of synaptic excitation. However, such an interpretation would be straightforward if areas were disconnected from each other (with J_{EE} varying along the hierarchy), and J_{Max} was larger than J_{th} so that the J_{EE} value is above J_{th} for some areas. Neither of these conditions holds for **Figure 9.30**, where the distributed working memory representation critically depends on inter-areal interactions in a highly recurrent whole system. Yet, a sharp transition is spatially localized, which is remarkable given that inter-areal cortical interactions are dense and involve long-range connections.

It is also possible to develop a multi-regional model of the mouse cortex using connectomic data of inter-areal connectivity (Oh et al. 2014;

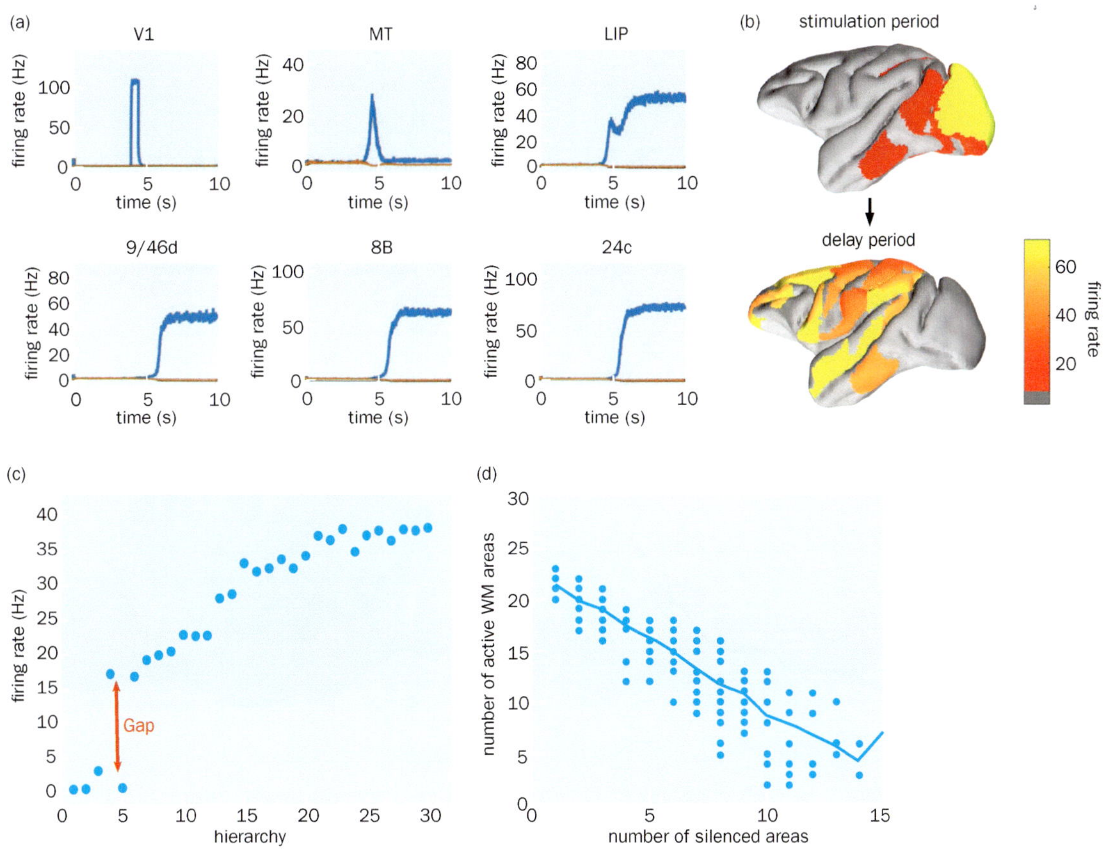

Figure 9.30 Bifurcation in space leading to distributed working memory representation in a connectome-based large-scale monkey cortex model. (a) In model simulation of a visual delayed response task, activities of the two excitatory neural populations are shown for six sample areas. Blue: Preferred to the shown stimulus; brown: Nonpreferred to it. None of the isolated areas is capable of generating persistent activity. (b) Activity is confined to the posterior part of the cortex during stimulus presentation. By contrast, it is distributed in the frontal, parietal and temporal areas during the delay period after the stimulus withdrawal. Firing rate is shown in color. (c) Mnemonic firing rate of the selective neural pool in each area during the delay period is plotted as a function of its hierarchical position. Those areas displaying persistent activity are separated from those that do not, by a gap in the firing rate. (d) Persistent activity is robust with lesioning of an increasing number of randomly sampled areas. Adapted from Mejias and Wang (2022). [Modified from Mejias, J. F., & Wang, X. J. (2022). Mechanisms of distributed working memory in a large-scale network of macaque neocortex. Elife, 11, e72136. Licensed under CC BY 4.0.] [Mejias, J. F., & Wang, X. J. (2022). Mechanisms of distributed working memory in a large-scale network of macaque neocortex. Elife, 11, e72136. Licensed under CC BY 4.0.]

Harris et al. 2019). Such an effort is timely with the rapid advances of genomic mapping, neurophysiological recording and calcium imaging, optogenetic manipulation for causal analysis of the mouse brain. Interestingly, spine count per pyramidal cells seems to be the same between primary visual cortex and frontal cortex (Gilman et al. 2017); whether there is a macroscopic gradient of synaptic excitation in the mouse cortex is presently unclear. Ding et al. (2024) built a whole mouse cortex model that assumed only a gradient of synaptic inhibition mediated by the parvalbumin (PV) positive interneurons. Since PV cells control spiking output of pyramidal neurons, they play a predominant role in determining the excitability and stability of local areas. It is thus expected that areas with less PV cells show weaker local feedback inhibition and are more likely to display elevated persistent activity during working memory.

Figure 9.31a shows PV neural density measured experimentally (Kim et al. 2017) (see also **Figure 9.12b**) that are expressed as the fraction of all neurons and rank-ordered across cortical areas. PV density is generally high in somatomotor and visual areas and low in anterolateral and

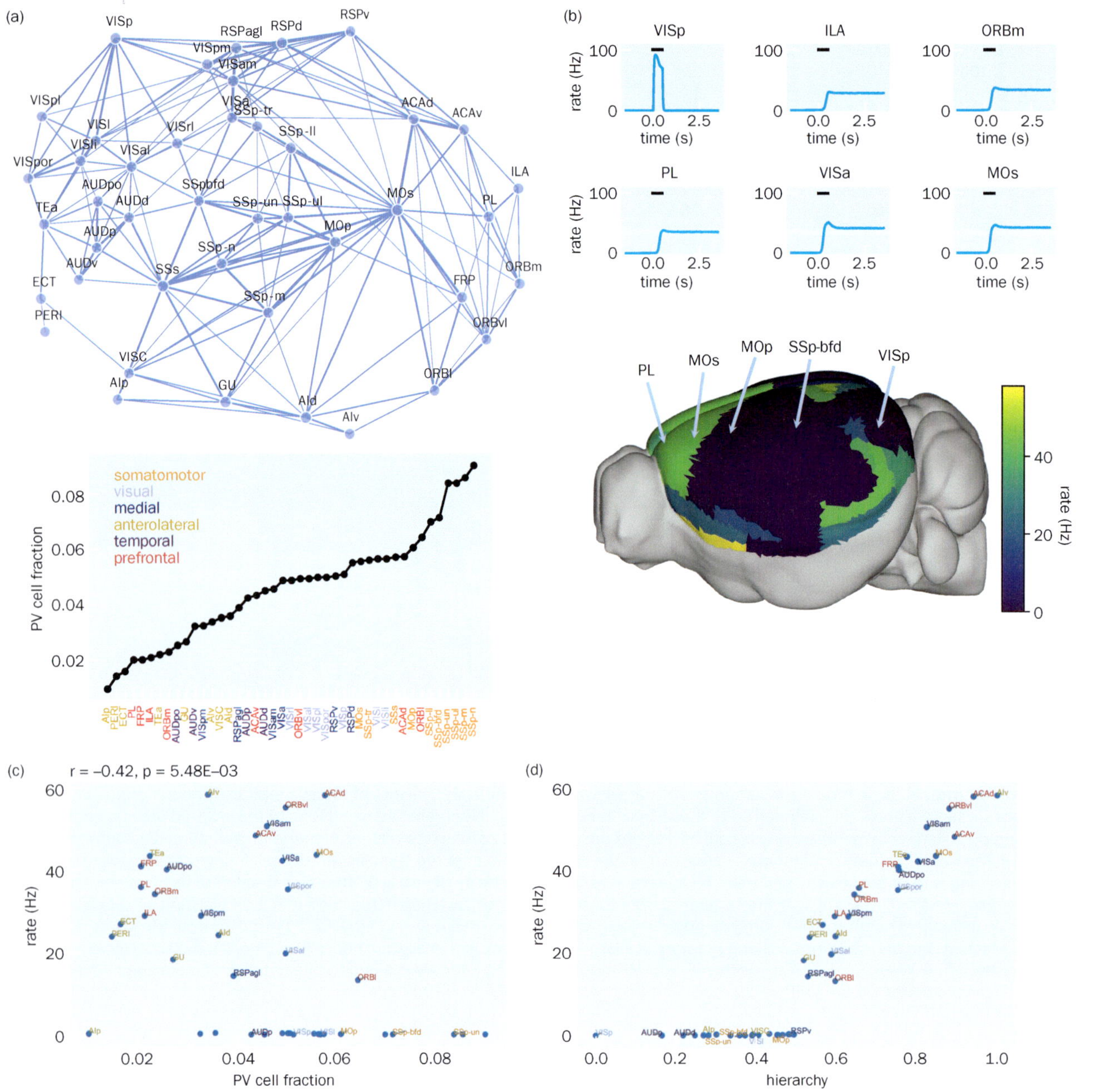

Figure 9.31 Distributed working memory in a connectome-based whole mouse cortex model. (a) Upper panel: Connectivity graph; lower panel: Parvalbumin positive cells as the fraction of all neurons ranked ordered across the mouse cortex. (b) A model simulation of delayed visual response task. Top: Time course of neural activity in sample areas; bottom: Self-sustained delay period activity map. (c) Delay period activity firing rate as a function of PV density showing a negative correlation. (d) Delay period activity firing rate as a function of the mouse cortical hierarchy defined anatomically. Reproduced from Ding et al. (2024). [Ding XY, Froudist-Walsh S, Jaramillo J, Jiang JJ, Wang X-J (2022) Predicting distributed working memory activity in a large-scale mouse brain: The importance of cell-type specific connectome. BioRxiv doi: https://www.biorxiv.org/content/10.1101/2022.12.05.519094v4]

frontal areas. In a model-simulated delayed visual response task, in contrast to transient response of V1 to a brief input, other areas display self-sustained persistent activity (Figure 9.31b). In this example, none of the isolated areas is capable of maintaining persistent activity, and the distributed working memory emerges from long-distance area-to-area connection loops. The delay period activity in a given area is negatively correlated with its PV density (Figure 9.31c) but also depends on the statistics of its inter-areal connections as well as the relative strengths of long-range inputs targeting excitatory versus inhibitory neurons (Ding et al. 2024). This finding highlights the need for connectomic data to be cell-type specific in order to predict the global cortex behavior. When mnemonic firing rate is plotted against the anatomically defined mouse hierarchy, a modularly organized pattern of distributed working memory is apparent (**Figure 9.31d**), similar to that of macaque monkey cortex (**Figure 9.30c**).

The model provides a platform to investigate distributed working memory, in particular the question of whether sensory areas are critically involved in working memory storage (Xu et al. 2020). There are reports that the content of working memory can be decoded from fMRI signals in early sensory areas (Harrison and Tong 2009; Curtis and Sprague 2021). In the model, those areas showing elevated persistent activity send synaptic inputs to those that do not. Since BOLD measurements are correlated with synaptic currents that include excitation and inhibition (Logothetis and Wandell 2004; Mazzoni et al. 2010), our model proposes that decoding from BOLD signals can be explained in terms of top-down synaptic currents rather than neural spike firing in sensory areas like V1, while highlighting the distinction between a subnetwork at the core of working memory and other brain regions as recipients of synaptic mnemonic signals.

To further analyze self-sustained collective dynamics of the cortex, individual areas are randomly sampled for lesioning in model simulations. As shown in **Figure 9.30d**, the persistent activity pattern did not disappear abruptly by such manipulations. Instead, the number of areas engaged in working memory representation gradually decreases with lesioning increasingly more areas, demonstrating the robustness of distributed working memory. However, certain areas are more important than others, as we shall see as follows. In principle, the model can also be used to simulate cell-type-specific transient inactivation as in optogenetic manipulation at different time points, in order to fully understand the underlying network mechanism of distributed working memory.

9.6.3 Bifurcation in Space: Emergence of Modularity

We shall call "bifurcation in space" the phenomenon of a sudden transition in a large-scale cortex underlying the emergence of functional modularity. Unlike a traditional bifurcation for which an externally controlled parameter must be fine-tuned for it to occur, bifurcation in space is robust and does not require fine-tuning parameters of the model system. Changes of network parameters, such as J_{Max} or scaling all inter-areal connection strengths by a factor G, would alter the location of cortical tissue where the transition occurs, thus precisely which areas show mnemonic persistent activity, but does not abolish the transition itself. The model predicts unusual neural fluctuations in areas at the edge of the rate gap, as a manifestation of a bifurcation (Nicolis and Prigogine 1977; Scheffer 2009). Specifically, there is a "critical slowing down" near a bifurcation, where the rate of change in a dynamical system is reduced to zero, as illustrated by the toy model of **Figure 5.11** (panel b) in Chapter 5, implying that the time constant of activity becomes infinite. For a bifurcation in space, this is expected to happen locally at a particular location rather than for the whole dynamical system.

The generative model of a spatially embedding mammalian cortex (Section 9.2.4) (Song et al. 2014) with a large number of parcellated areas enables one to "zoom in" arbitrarily close to an abrupt transition point along the cortical hierarchy (Wang et al. 2023a). **Figure 9.32** shows such a model with 1,000 areas that reproduces the basic persistent activity pattern along the cortical hierarchy (blue) as in the connectome-based cortex models. The time constant extracted from neural firing fluctuations from different areas (orange) grows near the location that separates the areas engaged in working memory representation from those that are not, from both sides. From the peak, the time constant decreases for both areas

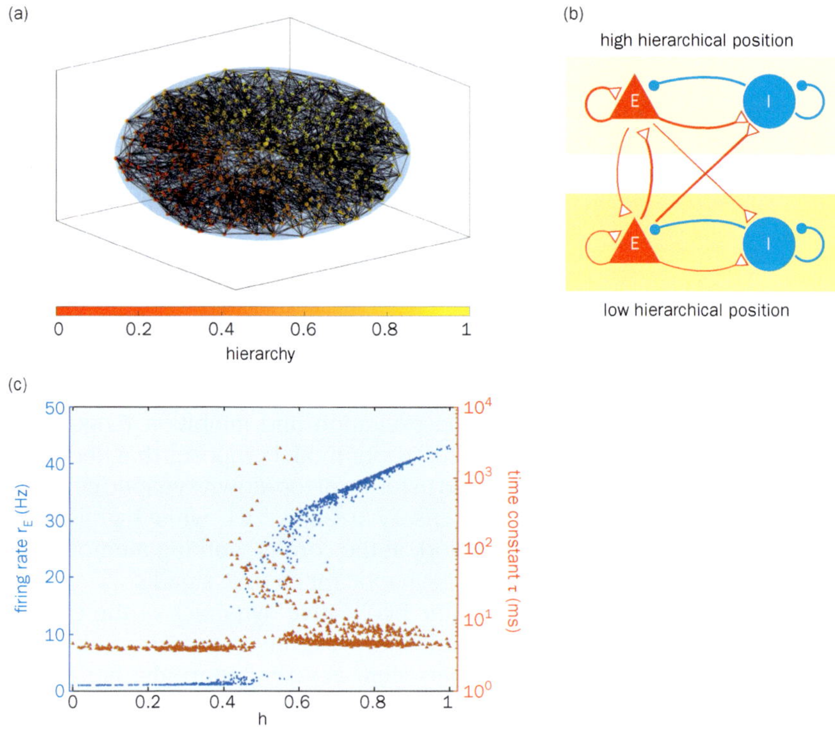

Figure 9.32 Critical slowing down as a signature of bifurcation in space.
(a) A generative model of the neocortex with 1,000 areas, constructed using Song et al. (2014). Color indicates hierarchy normalized between 0 and 1, which roughly but not exactly increases from left to right.
(b) Schematic local circuit and inter-areal connections between two areas; the thickness indicates that excitations are stronger in a higher area than a lower area, implementing a macroscopic gradient.
(c) A self-sustained persistent activity pattern is shown along the hierarchy, with the firing rates in blue. The dominant time constant in each area (in log scale, red) displays a peak around the transition. Reproduced from (Wang et al. 2023a). [Wang X-J, Jiang J, and Pereira-Obilinovic U (2023). Bifurcation in space: Emergence of function modularity in the neocortex. bioRxiv https://doi.org/10.1101/2023.06.04.543639]

with weaker and stronger recurrent excitation following a macroscopic gradient along the hierarchy.

This finding is consistent with a roughly increasing pattern of time constants along the hierarchy in the resting state (Section 9.4). For a nonlinear neural circuit, time constants depend on the internal state of the system, and their spatial pattern is different for the spontaneous state and a mnemonic representation. The model prediction of an inverted V-shaped pattern of time constants during working memory was replicated in the connectome-based cortex models of macaque monkey and mouse, providing an experimentally testable prediction as a specific signature of bifurcation in space. In particular, for working memory of visual motion information, the work in Mendoza-Halliday et al. (2014) suggests MST as a candidate area close to a criticality. Furthermore, since working memory and decision making are believed to share a common cortical substrate (Wang 2008b), the inverted V-shaped timescale profile is likely to hold during a decision process, a proposal in line with the existing evidence that neurons in the posterior parietal cortex display longer integration times underlying accumulation of information than both sensory areas and the prefrontal cortex, located at lower and higher hierarchical positions, respectively (Brody and Hanks 2016; Murray et al. 2017).

9.6.4 A Diversity of Spatially Distributed Persistent States

A multi-regional cortex can display a large number of distributed persistent activity patterns (attractors); six examples are shown in **Figure 9.33**. Here, distinct attractor states are defined in terms of their spatial patterns, which do not depend on the number of information-selective excitatory neural pools per area. For instance, in a two-area network, there are at most three spatial pattern attractors (with one of the two, or both, showing persistent activity), independent of the number of selective neural pools in each area. Using a numerical approach, many attractors are identified. Some engage the temporal lobe, while others do not; a few involve the frontal polar area

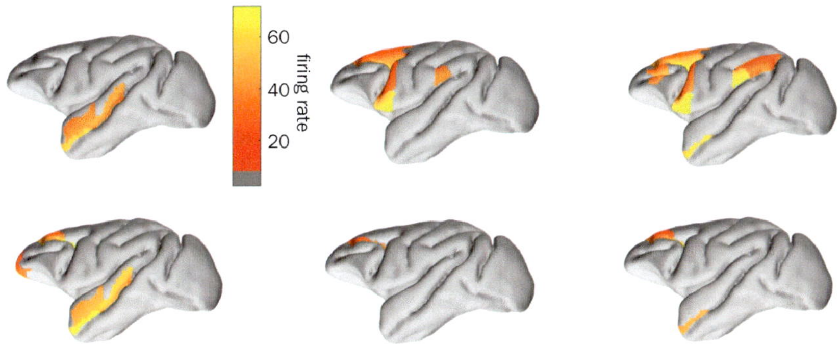

Figure 9.33 Six examples of self-sustained persistent activity patterns, each engaging a subset of areas. Reproduced from Mejias and Wang (2022). [Mejias, J. F., & Wang, X. J. (2022). Mechanisms of distributed working memory in a large-scale network of macaque neocortex. Elife, 11, e72136. Licensed under CC BY 4.0.]

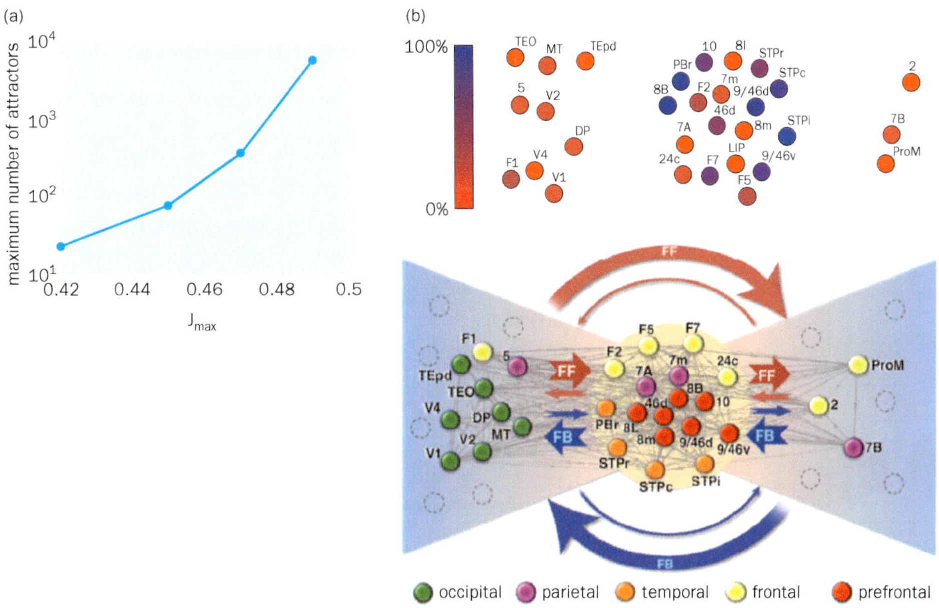

Figure 9.34 Multi-regional attractor states (distributed persistent activity patterns). (a) The number of distinct attractor states increases with the maximal local synaptic excitation strength J_{max}. (b) Lesioning a cortical area leads to reduction of the number of persistent activity patterns to a varying degree (0%, red; 100%, blue), and areas that matter the most (top, areas in blue) correspond to those that constitute the anatomically defined cortical core (bottom). Reproduced from Mejias and Wang (2022), with the lower panel of (b) from Markov et al. (2013). [Mejias, J. F., & Wang, X. J. (2022). Mechanisms of distributed working memory in a large-scale network of macaque neocortex. Elife, 11, e72136. Licensed under CC BY 4.0.] [Markov, N. T., Ercsey-Ravasz, M., Van Essen, D. C., Knoblauch, K., Toroczkai, Z., & Kennedy, H. (2013). Cortical high-density counterstream architectures. Science, 342(6158), 1238406.]

10, etc. Each of these distributed working memory attractors engages a subset of cortical areas, and frontal areas are commonly involved.

Interestingly, some spatially distributed attractor states are not accessible by stimulation of a primary sensory area, suggesting that they serve internal processes other than working memory of external stimulation. Instead, they potentially can serve various internally driven cognitive processes. For example, one spatially distributed attractor stores sensory information, another maintains a behavioral rule that guides sensorimotor mapping, etc. After all, mental processes such as thinking and reasoning as well as emotional feelings are largely freed from immediate sensory stimuli, which can be realized by various self-sustained attractor patterns of distributed brain activity. Each of these is modularly organized in the sense of selectively engaging a subset of areas but not the others, mathematically described by its own bifurcation in space. In other words, there are many bifurcations in space in a given large-scale cortical system.

As one would expect, the number of distributed persistent activity patterns increases with the maximal strength of local recurrent excitation J_{max} (**Figure 9.34a**). A systematic evaluation of the effect of cortical inactivation revealed that silencing most areas has only limited consequences on the total number of distributed attractors. The exceptions are several temporal and prefrontal areas for which the overall impact is large, and sometimes more than half of the initially available attractors are lost when a single such area is inactivated. The impact of the inactivation on the number of available attractors increases with the location of the silenced area in the hierarchy. In addition to this hierarchical dependence, the temporal and prefrontal areas that had the strongest impact are part of the core of the macaque cortex identified in anatomical (Markov et al. 2013) and physiological studies (Vezoli et al. 2021b) (**Figure 9.34b**). Overall, silencing areas at the center of the bowtie (the core) has a much greater impact than areas on the sides of the bowtie (the periphery).

9.6.5 Macroscopic Gradient of Dopamine Modulation

As discussed in Chapter 5 (Section 5.3.5), dopamine has a major impact on working memory function; either too little or too much D1 modulation could alter delay period activity and impair working memory performance (Sawaguchi and Goldman-Rakic 1991; Arnsten et al. 1994; Williams and Goldman-Rakic 1995; Vijayraghavan et al. 2007). *In vitro* studies of slices of prefrontal cortex found various intricate effects of dopamine on interactions between cell types in the prefrontal cortex (Seamans and Yang 2004) that potentially underlie this inverted U-shaped relationship between stimulation of dopamine D1 receptors and working memory. D1 receptor agonists increased the slow NMDA component of synaptic excitation (Seamans et al. 2000), which is critical for maintaining persistent activity (Wang et al. 2013). Dopamine also modulates synaptic inhibition (Gao et al. 2003) and affects excitability of pyramidal cells by virtue of modulating potassium channels that contribute to neuronal adaptation (Arnsten et al. 2019).

While it is known that dopamine signaling is heterogeneous in different parts of the cortex (Berger et al. 1991), there has been little quantification of how it differs throughout the cortex. Using the technique of receptor autoradiography to quantify the density of receptors across brain areas (Zilles and Palomero-Gallagher 2017), it was found that dopamine D1 receptor density per neuron varies in the form of a systematic gradient across the cortex (Froudist-Walsh et al. 2021). D1 receptor density increases along the cortical hierarchy, such that early sensory areas have relatively few D1 receptors per neuron, while prefrontal and lateral parietal areas have a large amount of D1 receptors (**Figure 9.35**). This finding showed that the concept of macroscopic gradients in the cortex (Wang 2020) also holds for neuromodulation.

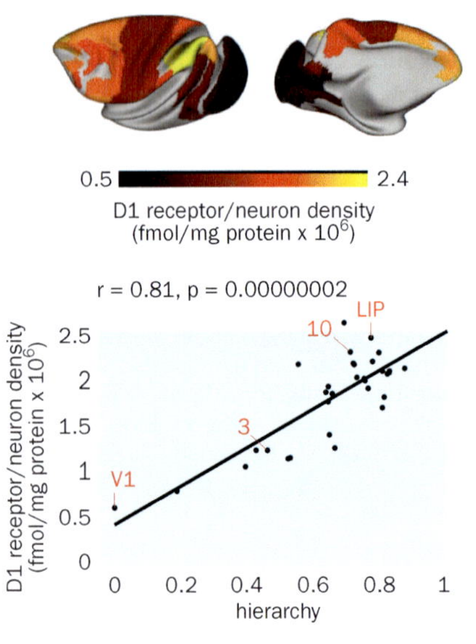

Figure 9.35 Dopamine D1 modulation of a multi-regional cortical system. Upper panel: The dopamine D1 receptor density per neuron visualized on the macaque cortical surface, showing a relatively low D1 receptor density in sensory areas like V1 (dark), while prefrontal and lateral parietal areas (orange and yellow) have a high density. Lower panel: The D1 receptor density increases along the anatomically defined cortical hierarchy. Reproduced from Froudist-Walsh et al. (2021). [Froudist-Walsh, S., Bliss, D. P., Ding, X., Rapan, L., Niu, M., Knoblauch, K., . . . & Wang, X. J. (2021). A dopamine gradient controls access to distributed working memory in the large-scale monkey cortex. Neuron, 109(21), 3500–3520. Licensed under CC BY 4.0.]

In an extension of the macaque cortex model of Mejias and Wang (2022) to incorporate a gradient of dopamine modulation (Froudist-Walsh et al. 2021), the hypothesis was examined that dopamine differentially enhances dendritic inhibition and reduces somatic inhibition (Gao et al. 2003), in line with the anatomical evidence that D1 receptor expression is higher in dendrite-targeting calbindin or somatostatin-positive GABAergic cells than in perisoma-targeting parvalbumin-positive ones (Mueller et al. 2020). For that purpose, the local area model is extended to include dendrite-targeting SST/CB and soma-targeting PV inhibitory cells, as well as SST/CB-targeting VIP/CR interneurons (**Figure 9.36a**).

In model simulations of a delayed visual response task, a behaviorally relevant stimulus (target, red) is first shown, which must be remembered to produce a memory-guided response at the end of the trial. Following a delay, a distractor stimulus (purple) is presented, which should be ignored for task performance. A second delay period precedes the end of the trial. The activity map depicts spatial distribution of neural firing for

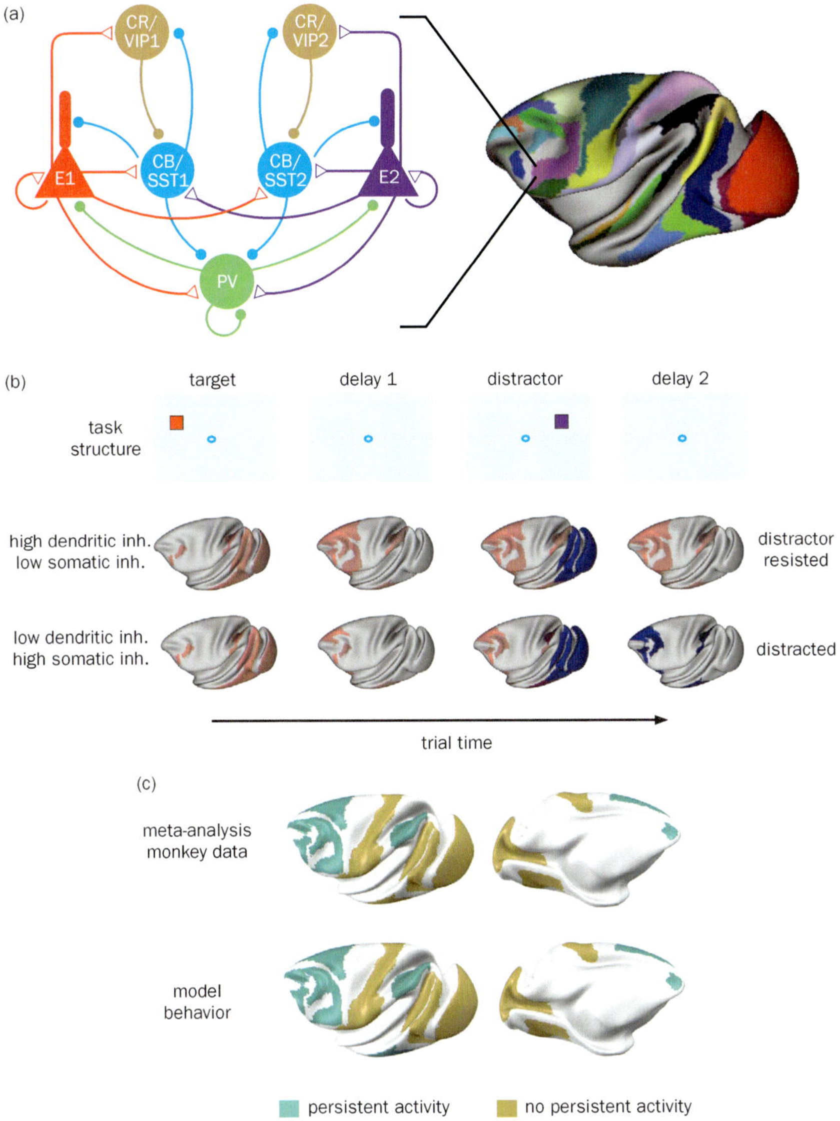

Figure 9.36 An extended model of the macaque cortex with diverse types of inhibitory neurons in each local area. (a) Model scheme. (b) Simulations of a working memory task. With a high dendritic-somatic inhibition ratio (upper panel), target-selective activity (red) is maintained in the second delay in spite of the distractor that briefly causes activity in the distractor-selective neural population (purple) of visual cortical areas. With a low dendritic-somatic inhibition ratio, persistent activity becomes selective for the distractor in the second delay (lower panel). (c) Comparison between the model and meta-analysis of persistent activity by Leavitt et al. (2017). Reproduced from Froudist-Walsh et al. (2021). [Froudist-Walsh, S., Bliss, D. P., Ding, X., Rapan, L., Niu, M., Knoblauch, K., . . . & Wang, X. J. (2021). A dopamine gradient controls access to distributed working memory in the large-scale monkey cortex. Neuron, 109(21), 3500–3520. Licensed under CC BY 4.0.] [Froudist-Walsh, S., Bliss, D. P., Ding, X., Rapan, L., Niu, M., Knoblauch, K., ... & Wang, X. J. (2021). A dopamine gradient controls access to distributed working memory in the large-scale monkey cortex. Neuron, 109(21), 3500–3520. Licensed under CC BY 4.0. / modified from original.]

two selective neural populations in each cortical area, red for target and blue for distractor (**Figure 9.36b**). When dopamine enhances dendritic inhibition and reduces somatic inhibition, the target leads to activity in the target-selective neural population (red) that is self-sustained after the brief input is withdrawn, during the first delay period. The distractor only produces transient activity in the distractor-selective neural population (purple) in the visual areas, but does not gain access to most of the frontal and parietal cortices, and behaviorally relevant memory is preserved (**Figure 9.36b**, upper panel). When a reduced D1 modulation leads to higher somatic inhibition and lower dendritic inhibition, working memory was easily distracted, and persistent activity in the second delay becomes selective for the distractor (purple) rather than the target (**Figure 9.36b**, lower panel), in which case the working memory system malfunctions. Notably, the model captures the meta-analysis results of more than 90 experiments (**Figure 9.27b**) (Leavitt et al. 2017) on mnemonic persistent activity in various cortical areas of the macaque cortex (**Figure 9.36c**).

Based on empirical evidence (Seamans et al. 2000; Gao et al. 2003; Arnsten et al. 2019), in the model different targets of D1 receptor have differential sensitivities: low-level dopamine release enhances the NMDA

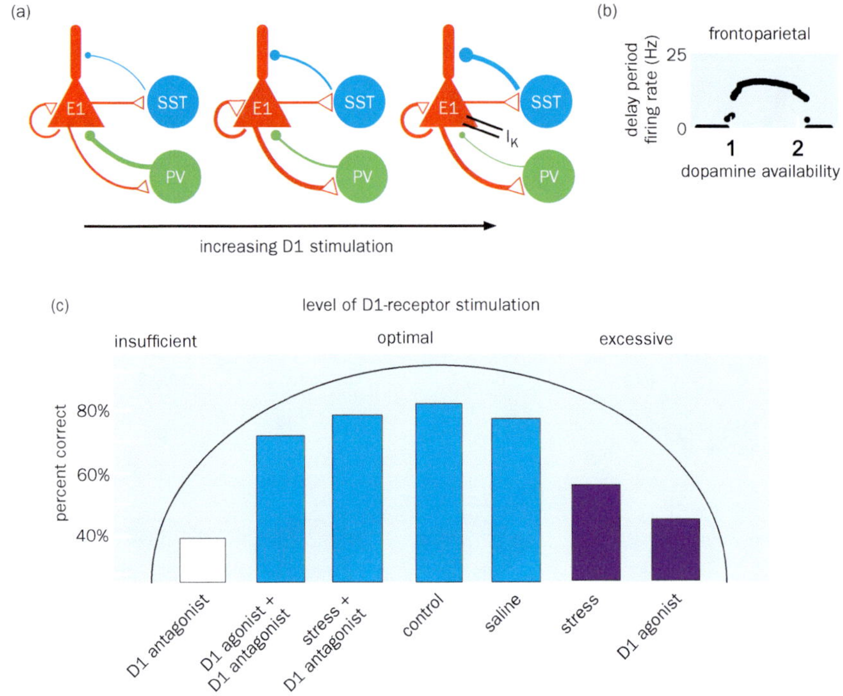

Figure 9.37 Dopamine modulation of working memory in a multi-regional cortical system. (a) An increase in dopamine release strengthens recurrent excitation and dendritic inhibition (thicker lines in the middle and right plots compared with the left plot) while reducing somatic inhibition (thinner lines); a potassium current is activated at high dopamine level (right plot). (b) Inverted-U dependence on dopamine of persistent activity in the parietal and prefrontal areas in the large-scale macaque cortex model. (c) Influence of D1 receptor stimulation on spatial working memory performance, based on a summary of rodent and monkey experiments. Panels (a–b) are reproduced from Froudist-Walsh et al. (2021), (c) from Arnsten (1998). [Froudist-Walsh, S., Bliss, D. P., Ding, X., Rapan, L., Niu, M., Knoblauch, K., ... & Wang, X. J. (2021). A dopamine gradient controls access to distributed working memory in the large-scale monkey cortex. Neuron, 109(21), 3500–3520. Licensed under CC BY 4.0. / modified from original.] [Froudist-Walsh, S., Bliss, D. P., Ding, X., Rapan, L., Niu, M., Knoblauch, K., ... & Wang, X. J. (2021). A dopamine gradient controls access to distributed working memory in the large-scale monkey cortex. Neuron, 109(21), 3500–3520. Licensed under CC BY 4.0.] [Arnsten, A. F. (1998). Catecholamine modulation of prefrontal cortical cognitive function. Trends in Cognitive Sciences, 2(11), 436–447.]

receptor-mediated excitation and dendritic inhibition, whereas high-level dopamine release has a net suppressive effect because it increases a potassium current (**Figure 9.37a**). Consequently, the model exhibits an inverted U-shaped dependence on the dopamine modulation of persistent activity in the parietal and prefrontal cortices (**Figure 9.37b**), similar to the experimental observations (**Figure 9.37c**), but now from the perspective of a large-scale cortical system. Optimality in the intermediate dopamine level can be understood by the preferential targeting of long-range connections to the distant dendrites, while local connections synapse the perisomatic region of pyramidal cells (Petreanu et al. 2009). With high D1 receptor density, once a stimulus-selective persistent activity has been engaged, strong inhibition to distal dendrites prevents distractor information in the sensory areas from entering the frontal and parietal areas, while leaving excitatory reverberation relatively unaffected to maintain the information about the behaviorally relevant target stimulus.

Furthermore, this work also sheds new insights into the activity-silent state model of working memory (Stokes 2015; Rose et al. 2016) (Chapter 5, Section 5.8.2). By incorporating short-term synaptic plasticity, the model implements the activity-silent state scenario in which a memory trace is encoded at synapses in the absence of persistent activity (Mongillo et al. 2008). Interestingly, activity-silent state memory trace can be realized by short-term facilitation of long-range connections from sensory areas, rather than at recurrent synapses within a local area as proposed originally, demonstrating the utility of a large-scale cortex model for findings that are not feasible with local circuit models. However, if memory traces are in the synaptic states, a non-selective "ping" would always activate the strongest neural population selective for the most recent stimulus, in which case working memory is vulnerable to distraction. Enhanced dopamine modulation can induce a transition, in the same model, from the activity-silent state regime (unable to filter out distractors) to the persistent activity regime (which is robust against distraction). These predictions can be tested in future experiments.

Taken together, recent work illustrates how a cross-level understanding can be achieved that links receptors and cell types to recurrent circuit dynamics and to a core cognitive function distributed across a multi-regional cortex.

9.7 DISTRIBUTED DECISION MAKING

Signal propagation in a cortical hierarchy has been studied under the assumption that areas were identical, and projections were feedforward with the same statistics for all area-to-area connections, as we discussed in Chapter 3 (Section 3.2.1) (Perkel and Bullock 1968; Shadlen and Newsome 1998; Diesmann et al. 1999; van Rossum et al. 2000; Kumar et al. 2010). Using the connectome-based multi-regional model of macaque cortex, Joglekar et al. (2018) revisited the problem of signal propagation in a network with an abundance of feedback loops and a high degree of connection weight heterogeneity. Intuitively, one expects that propagation can be boosted by increasing the strength of feedforward excitation. However, model simulations found (Joglekar et al. 2018) that such a change can readily lead to network instability because excitatory reverberation could cause runaway activity (Douglas et al. 1995). A potential solution is the balanced amplification (Chapter 3, Section 3.3.3), initially proposed for local networks (Murphy and Miller 2009) and now generalized to the inter-areal network (**Figure 9.38a**). When the strengths of inter-areal excitatory-to-excitatory connection (μ_{EE}) and local inhibitory-to-excitatory connection (w_{EI}) are enhanced simultaneously, inhibition permits

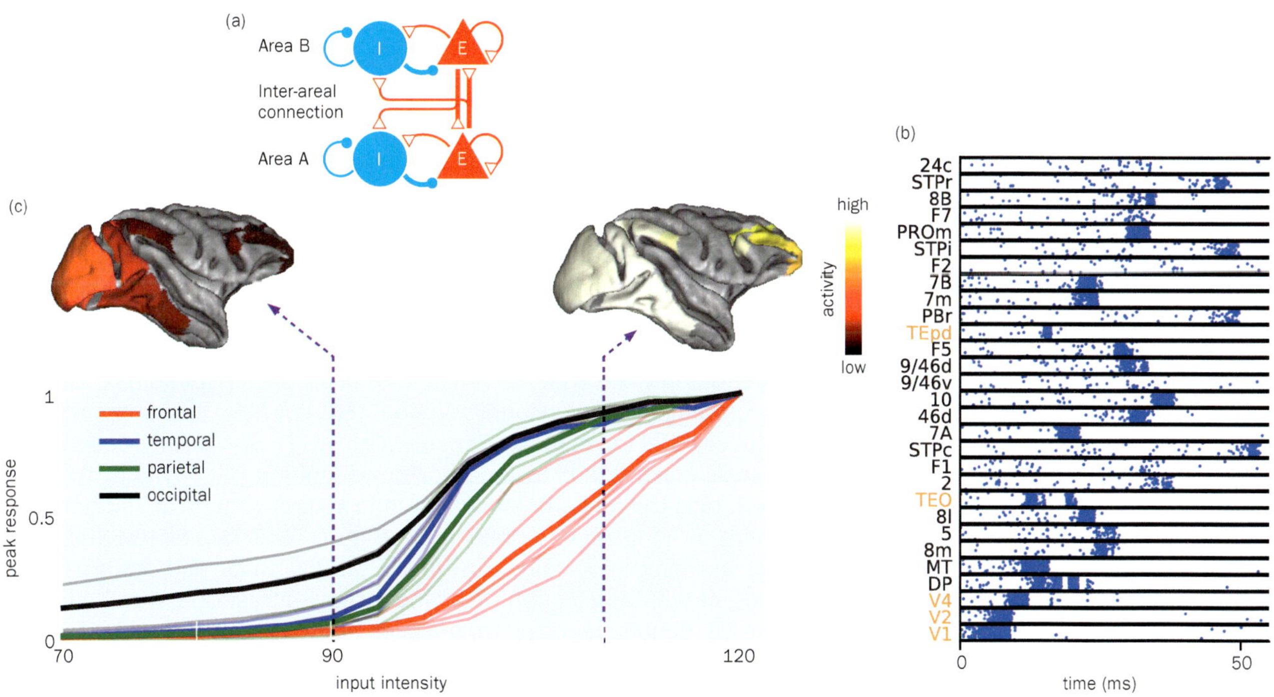

Figure 9.38 Signal propagation in a large-scale monkey cortex model of spiking neurons. (a) Generalization of balanced amplification from local circuit to inter-areal communication. The strengths of inter-areal excitatory-to-excitatory connection and local inhibitory-to-excitatory connections are enhanced (thick lines). (b) Propagation of a burst of spikes across a number of areas but not others, in response to a brief input to V1. The areas along the ventral stream showing strong response activity are indicated with orange labels. (c) Peak response as a function of the stimulus amplitude. The maximum firing rate is normalized to 1 to facilitate a comparison between areas grouped according to the four cortical lobes. Reproduced from Joglekar et al. (2018). [Joglekar, M. R., Mejias, J. F., Yang, G. R., & Wang, X. J. (2018). Inter-areal balanced amplification enhances signal propagation in a large-scale circuit model of the primate cortex. Neuron, 98(1), 222–234.]

a time window for a rigorous surge of response to inputs from upstream areas, while ensuring dynamical stability of the entire large-scale recurrent network.

Figure 9.38b displays a macaque cortex model with spiking neurons where synaptic transmission latency is implemented as a function of wiring distance with a constant axonal conduction velocity of 3.5 m/s (Swadlow 1990). In response to a brief visual input, V1 and other cortical areas fire a burst of spikes with various onset latencies. A few areas like TEO show more than one burst, as a manifestation of a complex pattern of communication across the highly recurrent cortical system. The dynamic pattern varies from trial to trial, and on average those cortical areas in ventral visual areas (indicated by orange) show more robust responses. It turns out that the connection strengths measured by FLN values are stronger for those areas along the ventral stream than the average, leading to effective signal propagation along the ventral pathway.

Figure 9.38c displays the peak response averaged over areas in each of the four cortical lobes, as a function of the stimulus intensity. Interestingly, the curve for the prefrontal areas (red) appears to show a threshold (around input strength equal to 95), below which the PFC response is essentially zero. Two examples of activity map are shown: when the input strength is below the threshold, the response is mostly confined into the posterior areas (top left), whereas when the input is above the threshold, frontal areas are activated (top right). The posterior areas display higher activities (top right versus left, with firing rate coded by color) because not only the input is larger, but also the PFC is engaged and sends its widespread top-down signals to the whole system.

The observed threshold for PFC responsiveness in the model resembles the so-called "ignition phenomenon" postulated by the theory of global neuronal workspace (GNW) for consciousness (Baars 1988; Dehaene et al. 2006; Mashour et al. 2020). In philosophy of mind, individual conscious experiences called qualia, such as toothache, are differentiated from access consciousness, which refers to subjective awareness of information (Dennett 1991; Block 2005). In well-controlled neuroscientific experiments, access consciousness can be studied using simple tasks (Crick and Koch 1990; Koch 2004; Dehaene 2014). For instance, in a detection task (Chapter 6, Section 6.6.1) with a stimulus intensity near detection threshold, a subject sometimes reports the presence of the stimulus but other times reports its absence, even though the physical stimulation is identical. While activity of neurons in sensory areas reflect external inputs, the PFC neural firing is correlated with subjective awareness and judgment (Chapter 6, **Figure 6.31**). Human studies using such paradigms combined with recording of event-related potential (ERP) or MEG lend support to the GNW theory (Del Cul et al. 2007; Baria et al. 2017). Specifically, access consciousness corresponds to an all-or-none ignition of activity in the PFC, which in turn triggers a broadcast of information content in the brain via its extensive long-range connections (**Figure 9.39a**). To test the GNW model neurophysiologically, van Vugt et al. (2018) carried out a monkey experiment with a detection task where the contrast of a visual stimulus was varied. Simultaneous recording from neurons in V1, V4 and prefrontal cortex found that firing of neurons in V1 and V4 was similar whether a near-threshold stimulus was reported in a given trial, whereas activity of neurons in PFC displayed robust difference in hit (green) versus miss (red) trials (**Figure 9.39b**). When no stimulus was shown, V1 and V4 neural activity was similar in false alarm (blue) and correction rejection (black) trials. By contrast, PFC neurons ramped up their firing in false alarm but not correct rejection trials, up to a level similar to that in hit trials.

This detection experiment was computationally simulated (Klatzmann et al. 2024) using an extended connectome-based large-scale macaque

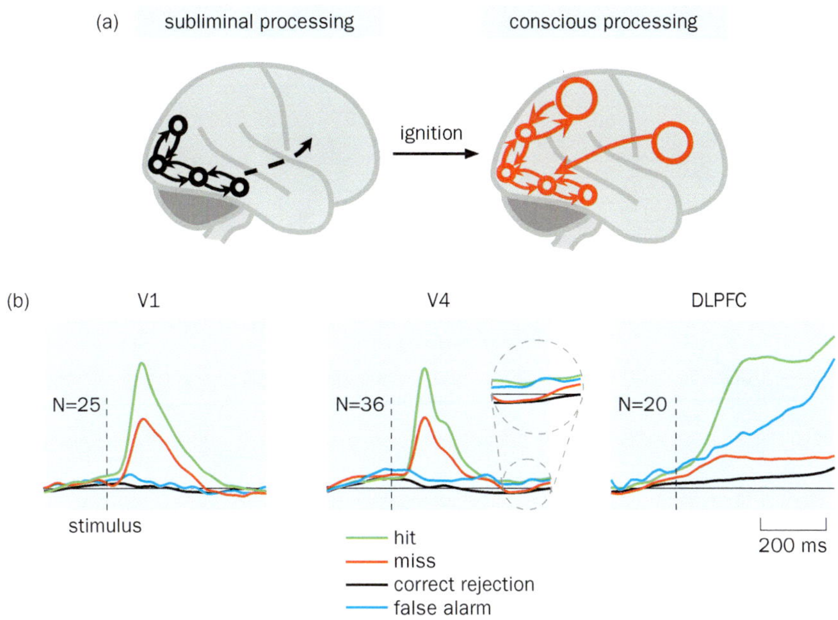

Figure 9.39 Global neuronal workspace model. (a) Schematics of the idea of ignition for access to consciousness. (b) Electrophysiological test of global workspace in awake macaque monkeys. Recordings were performed in V1, V4 and PFC while monkeys attempted to detect a stimulus of a contrast around the psychophysical threshold. Monkeys reported target presence with an eye movement, thus resulting in four trial types: Hit, miss, correct rejection and false alarm. The missed stimuli could evoke strong early transients in V1 and V4, indicating that such firing was not sufficient for a consciously reportable representation. The main difference between conscious perception (hits and false alarms) versus the absence of it (misses and correct rejections) was late, sustained activity in PFC, together with small but significant concomitant late, sustained activation in V1 and V4 (see inset in middle panel). Missed stimuli evoked only transient, decaying PFC activity. Panel (a) is kindly provided by Claire Sergent; (b) is reproduced from van Vugt et al. (2018). [Courtesy of Claire Sergent.] [Van Vugt, B., Dagnino, B., Vartak, D., Safaai, H., Panzeri, S., Dehaene, S., & Roelfsema, P. R. (2018). The threshold for conscious report: Signal loss and response bias in visual and frontal cortex. Science, 360(6388), 537–542.]

cortex model. The main difference from Mejias and Wang (2022) is that the ratio of the NMDA receptor and AMPA receptor-mediated components of synaptic excitation is small for feedforward pathways so that bottom-up signal propagation is quick; it is larger at local recurrent synapses and feedback projections underlying slow reverberatory dynamics. By varying the magnitude of an input to V1, mimicking contrast of a visual stimulus, the model captures the basic physiological observations and behavioral performance from the monkey experiment of van Vugt et al. (2018). In particular, V1 activity reflects the physical stimulation in hit trials and miss trials trials, whereas all-or-none activity in PFC is correlated with subjective decision in hit and false alarm trials (**Figure 9.40a**). In false alarms, PFC exhibits an all-or-none transition; the jumping time varies stochastically from trial to trial, leading to an averaged smooth ramping pattern (**Figure 9.40b**). The model performance as a function of the input strength (i.e., stimulus contrast) (**Figure 9.40c**) is comparable to monkeys' psychometric function (van Vugt et al. 2018).

A time-dependent analysis revealed a dynamic signature of the onset of ignition in the model, corresponding to a bifurcation in both space and time. This is similar to observations in a human experiment using an auditory detection paradigm combined with EEG recordings (Sergent et al. 2021). In the experiment, the stereotypical EEG signature of consciousness access was correlated with reports during the task, but also was observed in passive listening when no detection was required. This result shows that ignition in the cortex represents a biological marker of conscious processing independently from behavioral response (the report), an important dissociation for elucidating the role of the PFC in consciousness (Block 2019).

Whereas access consciousness can be dissociated from a variety of cognitive processes (Cohen et al. 2020; Dellert et al. 2021; Kapoor et al. 2022), its neural mechanism is shared with that of decision making (Shadlen and Kandel 2021), as demonstrated by Klatzmann et al. (2024). At the present time, we are still far from understanding the brain basis of consciousness; there is a debate on two adversarial proposals: the Global Neuronal Workspace theory with the PFC at its core versus the Integrated Information theory, which emphasizes a "hot spot" associated with consciousness in the posterior part of the cortex (Tononi et al. 2016), our model is in support of the GNW theory. Regardless, the example of detection illustrates the promise that the concept of

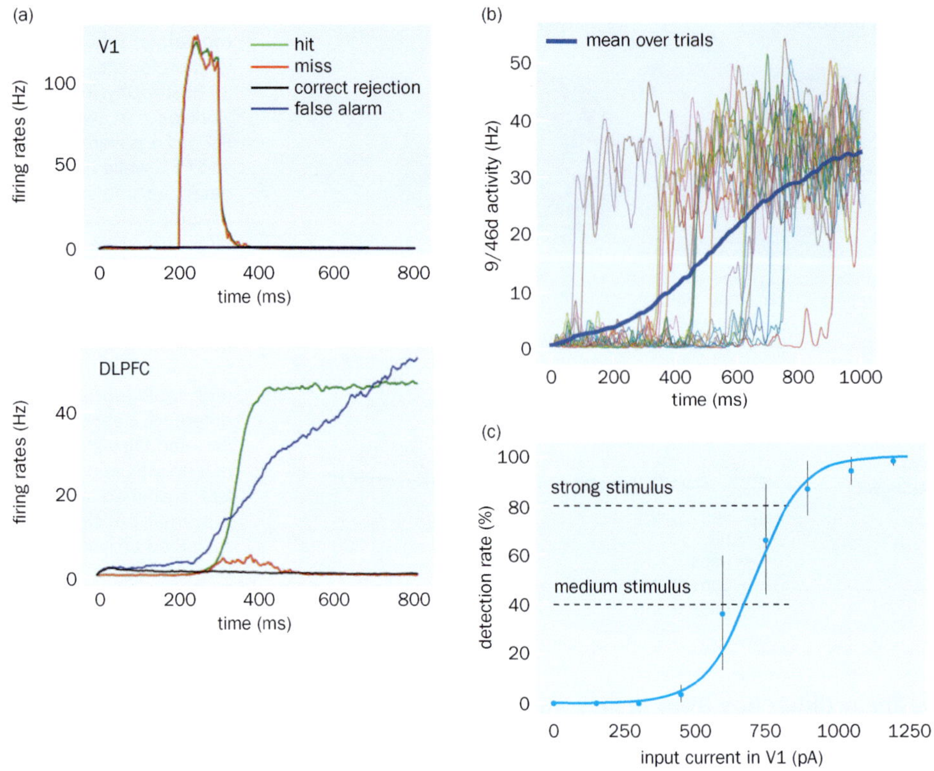

Figure 9.40 A connectome-based macaque cortex model applied to detection as a simple decision. (a) Neural population activity in V1 (top) and PFC (bottom) in four types of trials, same color code as in Figure 9.39b. (b) In false alarms, neural firing activity in the dorsolateral prefrontal cortex area 9/46d undergoes a discrete jump at a time that varies substantially across single trials, and the trial average shows a smooth ramping time course (blue curve). (c) Psychometric function of detection as a function of the input strength. Reproduced from Klatzmann et al. (2024).

a neural circuit capable of both working memory and decision making can be extended from local circuits (Wang 2013) to a large-scale multi-regional cortex.

9.8 SUMMARY

In this chapter, we covered multi-regional brain systems, starting with analysis of connectome and ending with a biologically based model of distributed working memory and decision making. Five general principles have emerged from this line of research about the global organization of the whole neocortex. First, the inter-areal connectivity weights are heterogeneous and fit with a log-normal distribution. Second, the cortical network is spatially embedded and obeys an exponential distance rule. These two characteristics are captured in a new class of random network models of the cortex. Third, heterogeneities of synaptic excitation and inhibition are not random among cortical areas; instead, they vary systematically along certain low-dimensional axes in the form of macroscopic gradients. Fourth, there is a hierarchy of temporal integration windows in parallel with a gradual enlargement of spatial receptive fields of neurons along the cortical hierarchy. Fifth, the biological macroscopic gradients combined with the mathematical notion of bifurcation gives rise to the new concept of bifurcation in space, which may hold promise for understanding cortical areas that share a canonical circuit architecture while exhibiting distinct functional capabilities such as working memory and simple decisions.

Neural processes underlying cognition are both distributed and exhibit modularity in the brain (Wang 2022). **Figure 9.33** offers concrete examples of how a module can be defined by a specific set of brain areas that may not be spatial neighbors but subserve an internal state potentially underlying a cognitive process. The alternative and unlikely hypothesis is that there is no modularity, and all areas participate in any brain function in a graded manner. A specific test to falsify one of the two scenarios is presented in the form of an inverted V-shape pattern of timescales during working memory.

The global ignition phenomenon associated with consciousness access represents merely one example of a wide variety of distributed brain processes that can now be rigorously studied experimentally and theoretically. Future research on distributed cognitive processes, including perceptual decisions (Siegel et al. 2015; Steinmetz et al. 2019; Wilming et al. 2020; International Brain Lab et al. 2023), value-based choices (Vickery et al. 2011; Redish 2013; Hattori et al. 2019; Botvinick et al. 2020; Shin et al. 2021) and memory storage (Roy et al. 2022; Lavi et al. 2023), will spur advances in the nascent neuroscience of multi-regional large-scale brain systems.

FURTHER READING

Books and Reviews

Breakspear, M. (2017). Dynamic models of large-scale brain activity. *Nat. Neurosci. 20*, 340–352.

Christophel, T. B., P. C. Klink, B. Spitzer, P. R. Roelfsema, and J. D. Haynes (2017). The distributed nature of working memory. *Trends Cogn. Sci. 21*, 111–124.

Felleman, D. J. and D. C. Van Essen (1991). Distributed hierarchical processing in the primate cerebral cortex. *Cereb. Cortex 1*, 1–47.

Fodor, J. A. (1983). *The Modularity of Mind: An Essay on Faculty Psychology*. Cambridge, MA: MIT Press.

Fornito, A., A. Zalesky, and E. Bullmore (2016). *Fundamentals of Brain Network Analysis*. London: Academic Press.

Kanwisher, N. (2010). Functional specificity in the human brain: A window into the functional architecture of the mind. *Proc. Natl Acad. Sci. 107*, 11163–11170.

Leavitt, M. L., D. Mendoza-Halliday, and J. C. Martinez-Trujillo (2017). Sustained activity encoding working memories: Not fully distributed. *Trends Neurosci. 40*, 328–346.

Markov, N. T., M. Ercsey-Ravasz, D. C. Van Essen, K. Knoblauch, Z. Toroczkai, and H. Kennedy (2013). Cortical high-density counterstream architectures. *Science 342*, 1238406.

Seung, H. S. (2012). *Connectome: How the Brain's Wiring Makes Who We Are*. Boston, MA: Houghton Mifflin Harcourt.

Sporns, O. (2014). Contributions and challenges for network models in cognitive neuroscience. *Nat. Neurosci. 17*, 652–660.

Wang, X.-J. (2020). Macroscopic gradients of synaptic excitation and inhibition in the neocortex. *Nat. Rev. Neurosci. 21*, 169–178.

Wang, X.-J. (2022). Theory of the multiregional neocortex: Large-scale neural dynamics and distributed cognition. *Ann. Rev. Neurosci. 45*, 533–560.

Primary Research Papers

Connectome

Dorkenwald, S., Matsliah, A., Sterling, A. R., Schlegel, P., Yu, S. C., McKellar, C. E., . . . & Murthy, M. (2024). Neuronal wiring diagram of an adult brain. Nature, 634, 124–138.

Ercsey-Ravasz, M., N. T. Markov, C. Lamy, D. C. Van Essen, K. Knoblauch, Z. Toroczkai, and H. Kennedy (2013). A predictive network model of cerebral cortical connectivity based on a distance rule. *Neuron 80*, 184–197.

Hagmann, P., L. Cammoun, X. Gigandet, R. Meuli, C. Honey, V. Wedeen, and O. Sporns (2008). Mapping the structural core of human cerebral cortex. *PLoS. Biol. 6*, e159.

Markov, N. T., M. M. Ercsey-Ravasz, A. R. Ribeiro Gomes, C. Lamy, L. Magrou, J. Vezoli, P. Misery, A. Falchier, R. Quilodran, M. A. Gariel, J. Sallet, R. Gamanut, C. Huissoud, S. Clavagnier, P. Giroud, D. Sappey-Marinier, P. Barone, C. Dehay, Z. Toroczkai, K. Knoblauch, D. C. Van Essen, and H. Kennedy (2014). A weighted and directed interareal connectivity matrix for macaque cerebral cortex. *Cereb. Cortex 24*, 17–36.

Oh, S. W., J. A. Harris, L. Ng, B. Winslow, N. Cain, S. Mihalas, Q. Wang, C. Lau, L. Kuan, A. M. Henry, M. T. Mortrud, B. Ouellette, T. N. Nguyen, S. A. Sorensen, C. R. Slaughterbeck, W. Wakeman, Y. Li, D. Feng, A. Ho, E. Nicholas, K. E. Hirokawa, P. Bohn, K. M. Joines, H. Peng, M. J. Hawrylycz, J. W. Phillips, J. G. Hohmann, P. Wohnoutka, C. R. Gerfen, C. Koch, A. Bernard, C. Dang, A. R. Jones, and H. Zeng (2014). A mesoscale connectome of the mouse brain. *Nature 508*, 207–214.

Song, H. F., H. Kennedy, and X.-J. Wang (2014). Spatial embedding of similarity structure in the cerebral cortex. *Proc. Natl. Acad. Sci. U.S.A. 111*, 16580–16585.

Theodoni, P., P. Majka, D. H. Reser, D. K. Wojcik, M. G. Rosa, and X.-J. Wang (2022). Structural attributes and principles of the neocortical connectome in the marmoset monkey. *Cereb. Cortex 32*, 15–28.

Macroscopic Gradients

Chen, A., Y. Sun, Y. Lei, C. Li, S. Liao, J. Meng, Y. Bai, Z. Liu, Z. Liang, Z. Zhu, et al. (2023). Single-cell spatial transcriptome reveals cell-type organization in the macaque cortex. *Cell 186*, 3726–3743.

Burt, J. B., M. Demirta, W. J. Eckner, N. M. Navejar, J. L. Ji, W. J. Martin, A. Bernacchia, A. Anticevic, and J. D. Murray (2018). Hierarchy of transcriptomic specialization across human cortex captured by structural neuroimaging topography. *Nat. Neurosci. 21*, 1251–1259.

Froudist-Walsh, S., T. Xu, M. Niu, L. Rapan, L. Zhao, D. S. Margulies, K. Zilles, X.-J. Wang, and N. Palomero-Gallagher (2023). Gradients of neurotransmitter receptor expression in the macaque cortex. *Nat. Neurosci. 26*, 1281–1294.

Margulies, D. S., S. S. Ghosh, A. Goulas, M. Falkiewicz, J. M. Huntenburg, G. Langs, G. Bezgin, S. B. Eickhoff, F. X. Castellanos, M. Petrides, E. Jefferies, and J. Smallwood (2016). Situating the default-mode network along a principal gradient of macroscale cortical organization. *Proc. Natl. Acad. Sci. U.S.A. 113*, 12574–12579.

Distributed Dynamics and Function

Chaudhuri, R., K. Knoblauch, M. A. Gariel, H. Kennedy, and X.-J. Wang (2015). A largescale circuit mechanism for hierarchical dynamical processing in the primate cortex. *Neuron* 88, 419–431.

Deco, G. and V. K. Jirsa (2012). Ongoing cortical activity at rest: Criticality, multistability, and ghost attractors. *J. Neurosci. 32*, 3366–3375.

Demirtaş, M., J. B. Burt, M. Helmer, J. L. Ji, B. D. Adkinson, M. F. Glasser, D. C. Van Essen, S. N. Sotiropoulos, A. Anticevic, and J. D. Murray (2019). Hierarchical heterogeneity across human cortex shapes large-scale neural dynamics. *Neuron 101*, 1181–1194.

Ding, X., S. Froudist-Walsh, J. Jaramillo, J. Jiang, and X.-J. Wang (2024). Predicting distributed working memory activity in a large-scale mouse brain: The importance of the cell type-specific connectome. *eLife 12*, e85442.

Eliasmith, C., T. C. Stewart, X. Choo, T. Bekolay, T. DeWolf, Y. Tang, C. Tang, and D. Rasmussen (2012). A large-scale model of the functioning brain. *Science 338*, 1202–1205.

Froudist-Walsh, S., D. P. Bliss, X. Ding, L. Rapan, M. Niu, K. Knoblauch, K. Zilles, H. Kennedy, N. Palomero-Gallagher, and X.-J. Wang (2021). A dopamine gradient controls access to distributed working memory in the large-scale monkey cortex. *Neuron 109*, 3500–3520.

Honey, C. J., R. Kötter, M. Breakspear, and O. Sporns (2007). Network structure of cerebral cortex shapes functional connectivity on multiple time scales. *Proc. Natl. Acad. Sci. 104*, 10240–10245.

Mejias, J. F. and X.-J. Wang (2022). Mechanisms of distributed working memory in a large-scale model of the macaque neocortex. *eLife 11*, e72136.

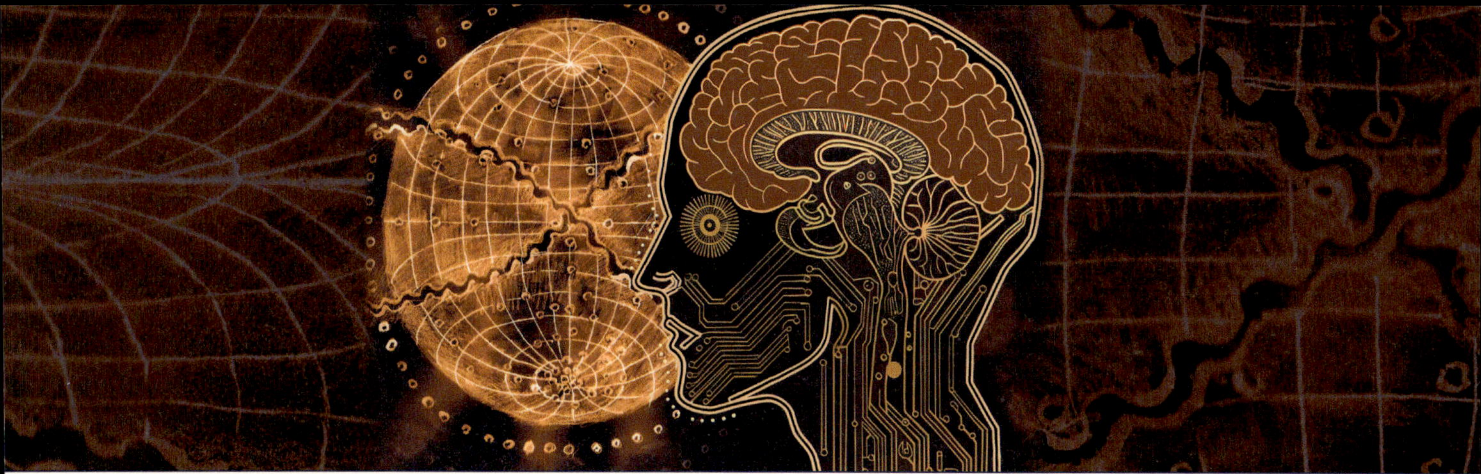

Computational Psychiatry

10

The subject of the following course of lectures will be the Science of Psychiatry, which, as its name implies, is that of treating mental disease . . . from the medical point of view, it is disturbances in the physical foundations of mental life which should occupy most of our attention.

E. Kraepelin in *Lectures on Clinical Psychiatry*

10.1 INTRODUCTION

These opening words were written by Emil Kraepelin (1904), a founding father of psychiatry as a medical discipline. He recognized the complex nature of mental disorders and started diagnosis classification by differentiating manic depression from dementia praecox, and the latter was renamed as schizophrenia by Eugene Bleuler. Biological studies of mental disorders had to await the arrival of modern neuroscience (Kandel 1979). One hundred years after Kraepelin published his lectures, Eric Kandel (2005) declared: "It is becoming increasingly apparent that a biological approach to psychiatry will enable us to reach a deeper understanding of human behavior." Notable are the future tense he used and his prescience; since the turn of the twenty-first century, Kraepelin's vision of understanding mental illness through science has begun to take shape. The stakes are high, given that mental illness puts an enormous burden on society (Kessler et al. 2008). Because psychiatric diseases usually strike at a young age and resist treatment, they account for a large proportion (20–30%) of years lived with disability (Whiteford et al. 2015; Vigo et al. 2016).

A closely related field is neurology. The central role of the frontal lobe in both fields was recognized with the famous case of Phineas Gage, a foreman working on the construction of a railroad in Vermont in the mid-nineteenth century. A terrible accident happened on September 13, 1865. A 6-kg tamping iron (32 mm in diameter, 1.1 m long) blasted through Gage's face in an upward direction, into his frontal lobe and then out of the top of the skull. Because of the angle of the tamping iron's thrust, the accident damaged greatly the ventromedial prefrontal cortex

DOI: 10.1201/9781003459361-13

(vmPFC). Gage survived the accident but was no longer his old self; in particular, he now made bad choices against his own interests (Harlow 1868), suggesting that the prefrontal cortex is critical for decision making and control of behavior.

The main hindrance in advancing the field has been a lack of understanding of the brain mechanisms of behavior and cognition. The situation has changed greatly over the last two decades, as discussed in the previous chapters. In particular, the prefrontal cortex plays a key role in many psychiatric disorders, such as schizophrenia (Lewis et al. 2005; Barch and Smith 2008; Insel 2010), addiction (Goldstein and Volkow 2011) and autism (Courchesne et al. 2011). With advances in our studies of the neural mechanisms and general computational principles of basic cognitive functions, are we in a position to meet the challenge of elucidating what goes wrong in the brain that causes functional and behavioral abnormalities?

For instance, addiction affects a large population regardless of socioeconomic class. People with addiction continue to choose actions (e.g., taking cocaine) in spite of devastating consequences, thus the basic problem is abnormal decision making (Volkow and Li 2004; Redish et al. 2008). In Chapters 6 and 7, we discussed the neurobiology of decision making and reinforcement learning. Can the newly acquired knowledge lead to understanding and treatment of drug abuse? Take another example of excessive distractibility in people diagnosed with schizophrenia (Park and Holzman 1992; Goldman-Rakic 1994; Barch 2005; Anticevic et al. 2011), which makes it difficult to hold onto a job and go about the business of a normal life. As discussed in Chapters 5, 8 and 9, we now possess a wealth of information on the brain mechanisms of working memory and filtering out distraction by selective attention, can these insights be translated into pharmacotherapy and non-drug treatments (such as brain stimulation) of working memory and attention deficits? Progress in answering this question requires a two-way interaction between basic and clinical research, exemplified by the neurophysiological elucidation of how methylphenidate, a commonly prescribed drug for ADHD, affects selective attention, discussed in Chapter 8 (Ni et al. 2022).

Drug design that targets a particular receptor type requires knowledge about its differential expressions in specific cell classes and distinct parts of the brain. Consider a drug for the dopamine system. In the monkey cortex, excitatory pyramidal cells, output-controlling PV and input-controlling SST interneurons have different mixtures of dopamine D1 and D2 receptors with varying binding affinities for a dopamine drug (Muller et al. 2019); and dopamine receptor density displays a macroscopic gradient across the cortical mantle (Froudist-Walsh et al. 2021). For this reason, understanding cell types, their receptors and distributions in the brain (Chapter 9, Section 9.3) (Wang 2020) will help design effective therapeutic drugs that potentially target a particular cell type or specific brain regions.

The emerging field that integrates computational, cognitive, neurobiological and clinical approaches to mental illness is called computational psychiatry (Montague et al. 2012; Wang and Krystal 2014; Redish and Gordon 2016; Seriès 2020; Browning et al. 2020; Huys et al. 2021; Hitchcock et al. 2022). The computational approach has also begun to be developed in neurology, exemplified by models for sensory dysfunction (Parr 2019), dementia related to Alzheimer's disease (Wong-Lin et al. 2020) and movement disorders (Parr et al. 2021). This chapter aims at familiarizing the reader with this new way of thinking about cognitive and behavioral disorders in terms of neural circuits bridging molecules and brain functions, in order to establish a solid biological foundation of psychiatry as well as neurology.

10.2 MENTAL DISORDER CLASSIFICATION VERSUS DIMENSIONAL PSYCHIATRY

Currently, psychiatrists assess patients guided by the *Diagnostic and Statistical Manual for Mental Disorders*, Fifth Edition (DSM-5) (American Psychiatric Association et al. 2013), which is based on checklists of subjective symptoms. For instance, a person is diagnosed with depression if at least five of the following nine symptoms are present for two weeks:

- Depressed mood or irritable most of the day, nearly every day, as indicated by either subjective report (e.g., feels sad or empty) or observation made by others (e.g., appears tearful)
- Decreased interest or pleasure in most activities, most of each day
- Significant weight change (5%) or change in appetite
- Change in sleep: Insomnia or hypersomnia
- Change in activity: Psychomotor agitation or retardation
- Fatigue or loss of energy
- Guilt/worthlessness: Feelings of worthlessness or excessive or inappropriate guilt
- Concentration: Diminished ability to think or concentrate, or more indecisiveness
- Suicidality: Thoughts of death or suicide, or has suicide plan

At least one symptom must be one of the first two, low mood or anhedonia (inability to feel pleasure). Thus, two people may have only one of five symptoms in common, yet could be classified with the same illness and prescribed with the same medication. Moreover, "for two weeks" seems quite arbitrary. Roiser (2012) likened the current state of psychiatry to a situation where the doctor attempts to diagnose a patient with the sole symptom of cough, which

> might commonly co-occur with runny nose, sore throat, fatigue, etc. A cough has many potential causes: viral or bacterial infection in the lung; asthma; or even cancer. But these are conditions that are defined mechanistically, not according to symptoms. Most importantly they require radically different treatments: antibiotics for bacterial infections; steroids for asthma; chemotherapy, radiotherapy or even surgery for lung cancer.

Behavioral symptoms across all disorders described in the DSM form a highly connected "psychopathology network," where a link between a pair of two symptoms (e.g., "often easily distracted by extraneous stimuli" and "disorganized speech") when they are present in the same disorder (e.g., schizophrenia) (**Figure 10.1**) (Borsboom and Cramer 2013). A given symptom may be present in more than one disorder (e.g., distractibility is also a characteristic of attention-deficit/hyperactivity disorder, ADHD). Enormous clinical heterogeneity causes low inter-rater reliability, i.e., inconsistency of assessment by different raters on the basis of the same body of evidence, for many common psychiatric diagnoses (Freedman et al. 2013). How can we build a solid biological foundation of psychiatry based on etiology (causes of a disease) and pathophysiology (abnormal physiological processes associated with a disease) (Adam 2013)?

Against this background, it has been argued that an alternative to DSM classification, called the dimensional approach, is to focus on specific dysfunctions that may be present across diagnostic categories (Krueger 1999). For instance, compulsivity describes excessive repetition of an

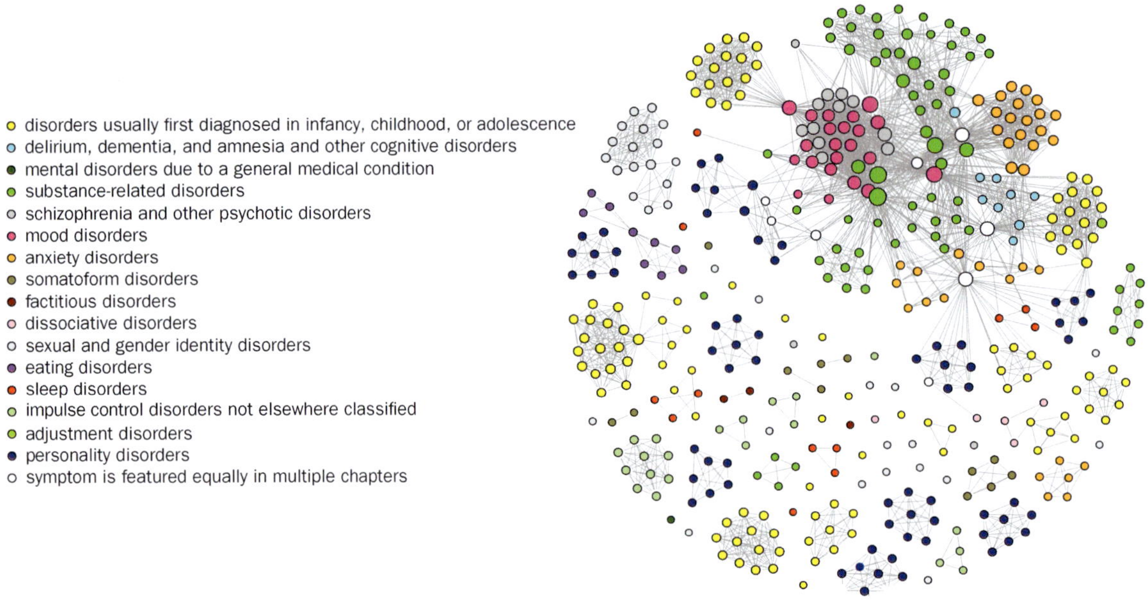

Figure 10.1 Network of symptoms among psychiatric disorders in the DSM-IV symptom space. Symptoms are represented as nodes and connected by edges whenever they are present in the same diagnostic disorder category. The color of a node represents the DSM-IV chapter in which they occur most often. Reproduced from Borsboom and Cramer (2013). [Borsboom, D., & Cramer, A. O. (2013). Network analysis: An integrative approach to the structure of psychopathology. Annual Review of Clinical Psychology, 9(1), 91–121.]

action without a beneficiary outcome. It is characteristic of obsessive-compulsive disorder (OCD) (Graybiel and Rauch 2000) and a main drive of addictive behavior (Everitt and Robbins 2005). The dimensional research strives to uncover transdiagnostic abnormalities of brain circuits that causally explain compulsivity, and breakthroughs would provide a biological basis for development of medication to treat people with problematic compulsivity in OCD, addiction, and other conditions. This line of thinking is at the core of the Research Domain Criteria (RDoC) advocated by the National Institute of Mental Health (Insel et al. 2010; Insel 2014; Clark et al. 2017), as part of a new approach to mental diseases (Wang and Krystal 2014; Krystal and State 2014; Clementz et al. 2016, Kyzar and Denfield 2023).

The key is rigorous understanding of the neural mechanisms across molecular and circuit levels and plastic changes in the brain as a result of the interplay with the external environment. Starting with the Human Genome Project, knowledge of brain genomics has opened up the possibility of uncovering specific genetic vulnerabilities of psychiatric disorders of an individual. Studies have revealed gene variations associated with psychiatric disorders, such as schizophrenia risk gene DISC1 (Brandon et al. 2009). Typically, however, the predisposition to psychiatric illness is polygenetic (a single disease involves multiple genes) and pleiotropic (a single gene is implicated in several disorders); genetic analysis revealed strong overlap of gene expression patterns for certain disorders but not others (**Figure 10.2**) (Ripke et al. 2014; Gandal et al. 2018; Schork et al. 2019).

Linking genes to behavior may be facilitated by considering quantitative traits intermediate between risk genotypes and the psychiatric phenotypes, called endophenotypes (**Figure 10.3**) (Gottesman and Gould 2003; Meyer-Lindenberg and Weinberger 2006). An endophenotype should be heritable, measurable with psychometric properties, stable over time, and related to the disorder and its symptoms in the general population. Moreover, endophenotypes may be further subdivided into sub-endophenotypes.

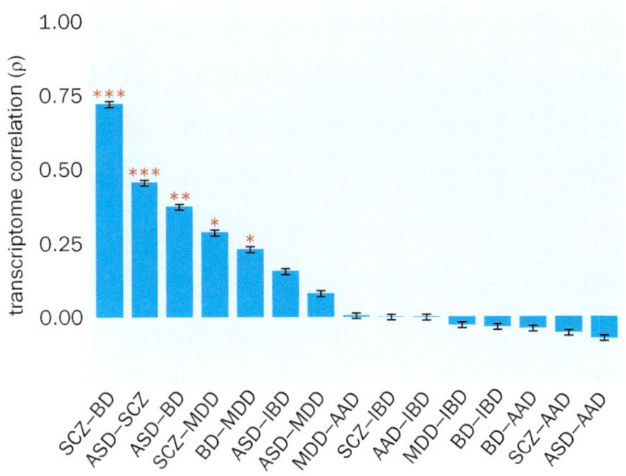

Figure 10.2 Correlation of gene expressions between psychiatric disorders. Correlation of gene expression patterns between pairs of disorders, revealed by microarray data from 700 postmortem brains of individuals with schizophrenia (SCZ), bipolar disorder (BD), autism spectrum disorder (ASD), major depressive disorder (MDD), alcoholism (AAD) and inflammatory bowel disease (IBD) for comparison. Strong correlations exist between certain pairs of diseases, such as schizophrenia and bipolar disorder, or between either of the two and autism; by contrast, there is no significant correlation of gene expressions between AAD and the three psychiatric disorders. Reproduced from Gandal et al. (2018). [Gandal, M. J., Haney, J. R., Parikshak, N. N., Leppa, V., Ramaswami, G., Hartl, C., ... & Geschwind, D. H. (2018). Shared molecular neuropathology across major psychiatric disorders parallels polygenic overlap. Science, 359(6376), 693–697.]

For instance, a behavioral endophenotype is impulsivity, which denotes the tendency to act prematurely without deliberate consideration or foresight. Impulsivity is behaviorally different from compulsivity, and they are associated with distinct anatomical abnormalities in the prefrontal cortex (Ziegler et al. 2019). Like compulsivity, impulsivity cuts across a range of diagnostic categories (**Figure 10.4**) (Robbins et al. 2012), including substance dependence (Bickel and Marsch 2001) and ADHD (Patros et al. 2016). Impulsivity in turn subdivides into a set of psychological processes (Evenden 1999), which themselves are candidate endophenotypes. For example, one could show impulsive behavior in intertemporal choice because of an aversion to delayed gratification (Lempert et al. 2019), measured by the temporal discounting function (Chapter 7, **Figure 7.3**) at the behavioral level, and quantified biologically in terms of abnormal frontal executive control (McClure et al. 2004) and serotonin signaling (Schweighofer et al. 2008). Another distinct sub-endophenotype is impulsive response due to motor disinhibition or timing impairment (Chapter 8, Section 8.2), quantified by the stop-signal response time at the behavioral level and deficits in its underlying brain circuitry (Bari and Robbins 2013). While this dimensional approach should not supplant the prevailing psychiatric diagnostic schema, it has powerfully stimulated psychiatry research.

To accurately identify endophenotypes, it is beneficial to assess what people do (choices and actions) rather than what they say (in questionnaires), and search for neurophysiological biomarkers. By using behavioral paradigms that are designed to probe a specific cognitive domain, one can parametrically measure the abnormalities of a particular function that are shared by multiple mental disorders. Preferably, those carefully designed tasks should be doable by both human subjects and nonhuman animals, thereby enabling neural circuit dissection for productive translational research (Carter et al. 2008; Insel 2014).

The challenge of a rational psychiatry cannot be met purely through experimentation for two reasons. First, big data of all kinds require the development of novel statistical tools to aid discoveries, for that purpose modern tools of machine learning are increasingly deployed in computational psychiatry (Brodersen et al. 2014; Chekroud et al. 2016; Stephan et al. 2017; Frässle et al. 2020). Second, between genes and behavior, there is a neural circuit in interaction with the world, which exhibits complex positive and negative feedback loops. A close interplay between experimental research, theory and modeling is necessary to achieve understanding that would be impossible with intuition alone. Theories are poised to play an increasingly substantial role in psychiatry.

By incorporating computational modeling within translational research programs, it may be possible to develop more specific hypotheses

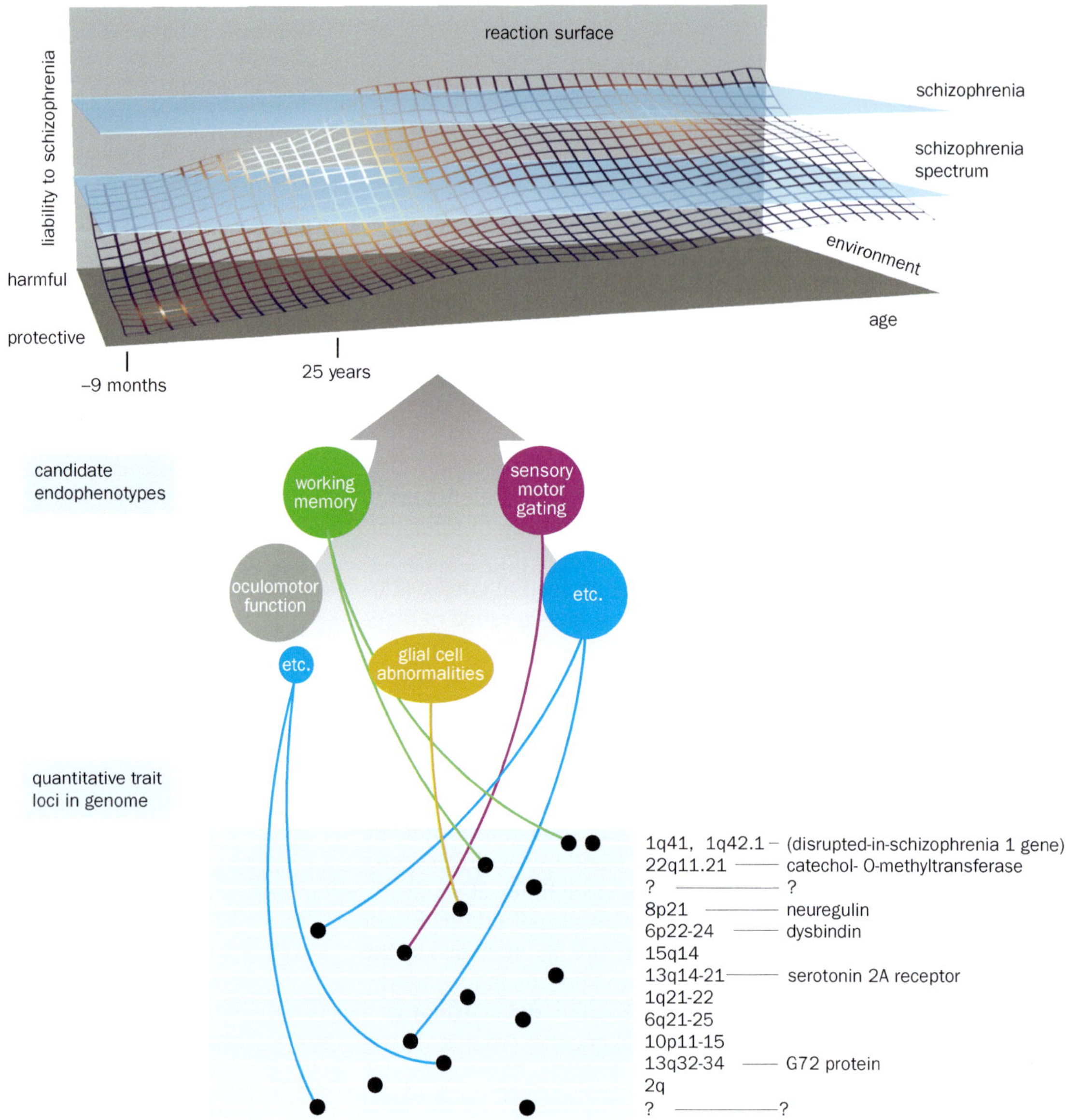

Figure 10.3 Putative endophenotypes implicated in a biological systems approach to schizophrenia research. Endophenotypes as schizophrenia discriminators involve sensorimotor gating, oculomotor function, working memory, and glial cell abnormalities. Many more gene loci, genes and candidate endophenotypes remain to be discovered (represented by question marks). Top panel is a three-dimensional schematic plot depicting the dependence of liability (z-axis) for schizophrenia on brain development age (x-axis) and environment (ranging from protective to harmful) (y-axis). Reproduced from Gottesman and Gould (2003). [Reprinted with permission from the American Journal of Psychiatry, 160:4, Figure 2, pages 636–645 (Copyright © 2003). American Psychiatric Association. All Rights Reserved.]

related to circuit dysfunction in nonhuman animal model systems and psychiatric disorders (Wang and Krystal 2014; Paulus et al. 2016; Stephan et al. 2016; Friston et al. 2017). Computational models of mental illness broadly fall into two categories. Algorithmic models from psychology and computer science are enormously useful for fitting behavioral data and relating model parameters to neural computations (Montague et al. 2012; Huys et al. 2016). On the other hand, biophysically informed computational modeling, constrained by the biophysical properties of synaptic signaling mechanisms and connection properties of neural circuits, is needed for a cross-level mechanistic understanding of brain functions and mental disorders (Wang and Krystal 2014; Murray et al. 2017). We will discuss how these two approaches are developed and applied in psychiatry.

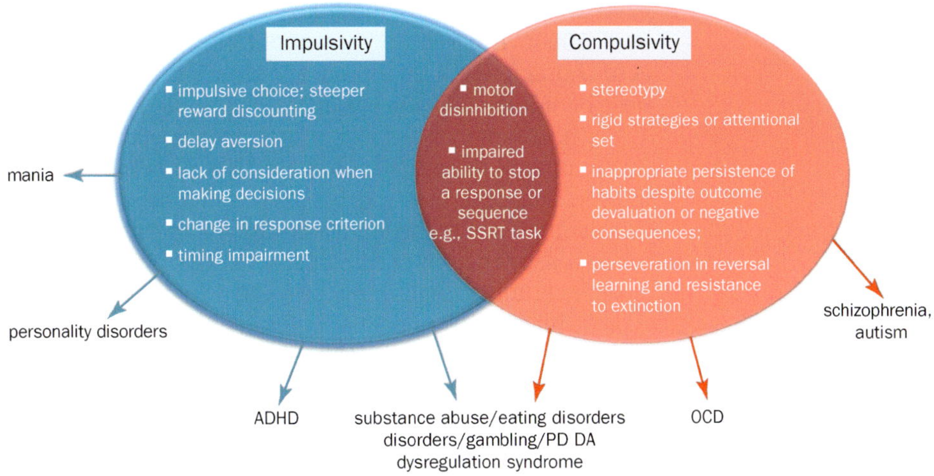

Figure 10.4 Impulsivity and compulsivity. The diagram describes possible psychological component mechanisms underlying impulsivity and compulsivity. These different measures likely do not intercorrelate well, which would argue against a unitary construct for either impulsivity or compulsivity. Both impulsivity and compulsivity involve motor/response disinhibition, but at different stages of the response process and on different timescales (with impulsivity manifest by quick and inappropriate reactions and compulsivity characterized by undesirable repetition of an act over a prolonged time). Reproduced from Robbins et al. (2012). [Robbins, T. W., Gillan, C. M., Smith, D. G., de Wit, S., & Ersche, K. D. (2012). Neurocognitive endophenotypes of impulsivity and compulsivity: Towards dimensional psychiatry. Trends in Cognitive Sciences, 16(1), 81–91.]

10.3 REINFORCEMENT LEARNING MODELS OF BEHAVIORAL DISORDERS

Computational psychiatry combines sophisticated behavioral paradigms with modeling to probe specific dysfunctions. A particularly fruitful area is applications of reinforcement learning theory (Chapters 4 and 7).

10.3.1 Task Design and Behavioral Quantification

To illustrate this approach, consider the hypothesis that a dysfunction shared among multiple disorders is a shift from flexible behavior to automatized responses such as compulsive repetition of unnecessary actions. One way to test this idea is to assess performance of a two-stage choice task (Daw et al. 2011) that probes the interplay between a model-free system and a model-based system (see Chapter 7, **Figure 7.15**). In a behavioral experiment (Gillan et al. 2016), a large number of participants, using Amazon's Mechanical Turk online, performed such a task as well as answered questionnaires for assessment of seven psychiatric diagnostic categories. A bias measure can be defined in such a way that it is negative if behavior is more dominated by the model-free (habit) system than in healthy subjects. It was found that different disorders display a varying degree of abnormal reliance on the model-free system relative to the control group (**Figure 10.5**). This example shows that differences with the control subjects are graded and transdiagnostic, a theme that will recur throughout this chapter. It is worth noting that the dichotomy of model-free and model-based systems has strong theoretical appeal, but there is presently no solid evidence for separate cognitive processes or neural circuits for the postulated model-free and model-based systems (Collins and Cockburn 2020). A more sophisticated perspective, such as that of multiple decision-making systems (Redish 2013), remains to be fully explored.

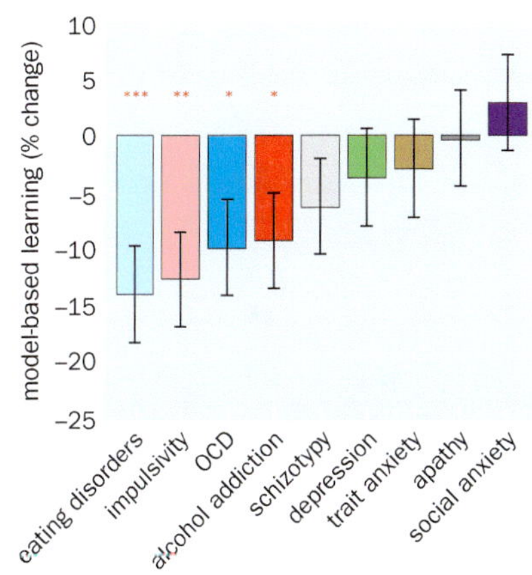

Figure 10.5 Deficits in goal-oriented behavior associated with mental disorders. Performance of a two-step reinforcement learning task was assessed for psychiatric disorders; a negative percentage change relative control indicates bias in favor of the model-free system at the expense of the model-based system. OCD and impulsivity, but not depression and anxiety, were associated with deficits in goal-directed reinforcement learning. This deficit generalized to symptoms associated with other disorders similarly characterized by a loss of control over behavior, such as alcohol addiction and eating disorders. Reproduced from Gillan et al. (2016). [Gillan, C. M., Kosinski, M., Whelan, R., Phelps, E. A., & Daw, N. D. (2016). Characterizing a psychiatric symptom dimension related to deficits in goal-directed control. Elife, 5, e11305. Licensed under CC BY 4.0.]

10.3.2 Mood and Depression

At least 4% of the United States population is affected by major depressive disorder (MDD) each year, and 15% of the population is expected to suffer from MDD in their lifetime (Blazer et al. 1994; Kessler et al. 2003; Alonso et al. 2004). The first step of computational research on a mental problem is to characterize it quantitatively. How can a feeling of happiness or sadness be numerically measured?

The moment-based approach to "objective happiness" (Kahneman 2000) aims at assessing our sense of well-being that varies continuously in time. Rutledge et al. (2014) proposed a simple measure of "happiness" that fluctuates over time in a laboratory experiment. In their study, subjects made a sequence of choices between two options: one yielded a reward r_0 with certainty (certain reward, CR) while the other was a gamble (with a probability p for a positive outcome r_1 and $1-p$ for a negative outcome r_2) leading to an expected value $EV(j) = pr_1 + (1-p)r_2$. After every two to three choice trials, subjects were asked to report "How happy are you right now" by moving a cursor along a line (**Figure 10.6a**). This mood rating was correlated with the expected value and reward prediction error (RPE), which is the difference between the actual reward and EV (see Chapter 4, Section 4.3.1) in the previous trials. Specifically, the empirically observed mood rating was compared against a mathematical measure defined as:

$$\text{Happiness}(t) = w_0 + \sum_{j=1}^{t} \gamma^{t-j}\left[w_1 CR(j) + w_2 EV(j) + w_3 RPE(j)\right] \quad (10.1)$$

where, in trial $t-j$ (jth previous trial from the present time t), if the certain option is selected, certain reward $CR(j)$ is r_0, $EV(j)$ and $RPE(j)$ are set to zero; when the gamble is chosen instead, $CR(j)$ is zero, the expected value $EV(j)$ and reward prediction error $RPE(j)$ are used in the formula. The factor γ (between 0 and 1) is the forgetting rate that discounts progressively remote events back into the past. As discussed in Chapters 4 and 7, at the heart of RL theory is estimating values of choice options, using an RPE signal mediated by phasic spike activity of dopamine neurons. RPE is not the only factor that determines one's emotional well-being, because reward receipt itself often leads to a positive mood swing even when it is expected. Therefore, although EV and RPE are correlated, it is reasonable to have them as separate regressors, equivalent to treating expected value and actual reward as independent variables.

As a person went through ups (wins) and downs (losses) in the task, the internal mood state was associated with a discounted integration of received outcomes and surprises over the choice history. Interestingly, this formula of Eq. (10.1) could fit fairly well with the subjective rating of momentary happiness over time during task performance ($r^2 = 0.79$ for the subject shown in **Figure 10.6b**, and 0.47 across subjects). After the model was validated, the researchers applied Eq. (10.1) to subjects diagnosed with MDD while they performed the same task (Rutledge et al. 2017). Notably, in addition to laboratory experiments, these studies also used smartphone technology to enlist more than 1,800 participants. It was found that MDD subjects displayed normal association between RPE and happiness (the value of w_3 is indistinguishable from those of the control group). Brain imaging in the laboratory confirmed that RPE signals in the ventral striatum (the part of the striatum associated with the limbic system), presumably reflecting dopamine inputs, are the same for MDD and control participants. The values of w_1 and w_2 were also not statistically different between the control and MDD subjects. By contrast, a significant difference between the two subject groups was found in the value of the baseline mood parameter w_0. As shown in **Figure 10.6c**, MDD symptom severity correlated with the baseline happiness parameter in laboratory and smartphone participants. Thus, a simple mathematical model revealed that major depression is associated with a reduced baseline around which

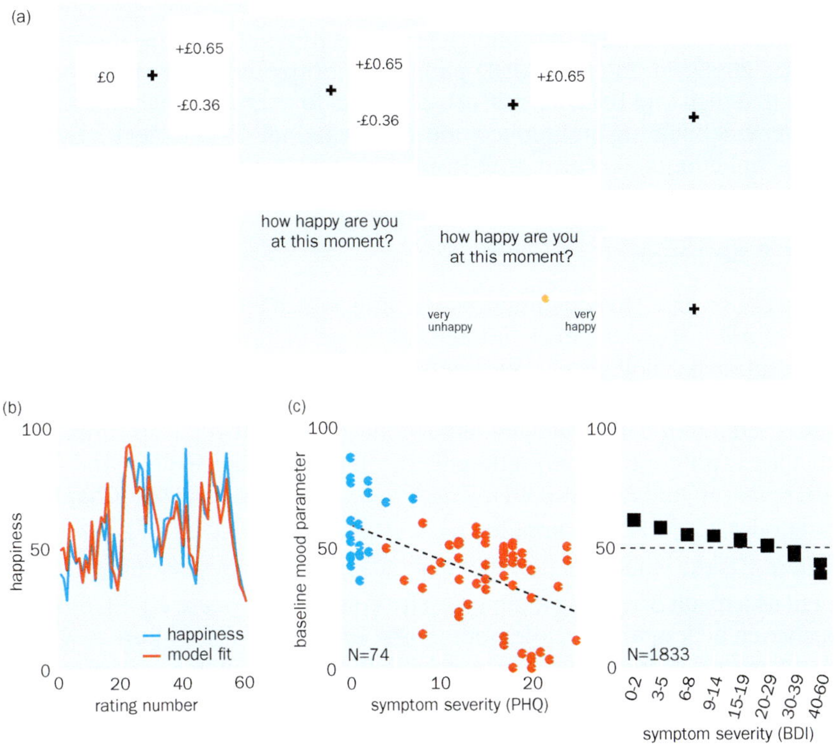

Figure 10.6 Happiness ratings with healthy and depressed subjects.
(a) Experimental design. In each trial, subjects chose between an option with certain reward and a gamble. The outcome of a chosen gamble was revealed later. Every two to three trials, subjects were asked to indicate "How happy are you at this moment?" by using a button press to move a cursor. (b) Happiness ratings in an example subject ($r^2 = 0.79$ between rating data and model fitting). (c) Left panel: Baseline mood parameter value was negatively correlated with symptom severity (Patient Health Questionnaire [PHQ], $r = -0.54$; $P < 10^{-6}$); Right panel: Participants (N = 1,833) played a similar risky decision task on their smartphones and also completed the Beck Depression Inventory second edition (BDI-II) questionnaire. Baseline mood parameter value was negatively correlated with depression symptom severity ($r = -0.30$; $P < 10^{-39}$). Reproduced from Rutledge et al. (2017). [Courtesy of Robb Rutledge and permission from Rutledge, R. B., Skandali, N., Dayan, P., & Dolan, R. J. (2014). A computational and neural model of momentary subjective well-being. Proceedings of the National Academy of Sciences, 111(33), 12252–12257.] [Courtesy of Robb Rutledge.]

momentary mood fluctuates, but the emotional impact of RPE is intact compared with normal subjects.

This observation is consistent with another computational study of MDD (Huys et al. 2013), where an RL model was fit with behavioral data from depressed subjects performing a probabilistic reward task (Pizzagalli et al. 2005). Recall that according to the Rescorla-Wagner learning rule (Chapter 4, Eq. 4.12), the RPE $\delta_t = r_t - V_t$, where r_t is the actual reward and V_t is the expected reward, at time t. The expected values for choice options are updated by $\alpha\delta_t$, with the rate α controlling the speed of learning. The authors introduced an additional "sensitivity" parameter ρ, so that the RPE becomes $\rho r_t - V_t$. A meta-analysis of six studies has been carried out by fitting behavioral data with an RL model. It was found that the level of anhedonia was associated with a significantly reduced reward sensitivity (ρ value is smaller than 1), but no change in the learning rate α. Note that $V_{t+1} = V_t + \alpha(\rho r_t - V_t)$ can be rewritten as $\tilde{V}_{t+1} = \tilde{V}_t + \alpha\left(r_t - \tilde{V}_t\right)$ with $\tilde{V}_t = V_t / \rho$, which becomes the conventional Rescorla-Wagner learning rule. In other words, an alternative interpretation of the reward sensitivity parameter ρ is a rescaling of the value function. Presumably, MDD patients display intact RPE-dependent learning, but the valuation at the time of making a choice is blunted compared to that of normal subjects ($V_t = \rho\tilde{V}_t < \tilde{V}_t$).

These results illustrate how quantitative modeling can assist in dissecting potential processes associated with a psychiatric disorder like depression. The work reveals, quantified by the parameter ρ, an attenuated ("pessimistic") estimation of reward expectations, consistent with the anhedonia and lack of motivation found in patients with depression. Moreover, inasmuch as RPE remains unchanged in MDD, deficits of the dopamine system observed in MDD (Nestler and Carlezon 2006) are likely to be linked to tonic rather than phasic activity (Grace 1991) of dopamine neurons, which is thought to signal incentive salience and effort (Berridge 2007; Niv et al. 2007). This would be in line with the finding that a baseline "happiness" is reduced in proportion with the severity of MDD (Rutledge et al. 2017). If so, attention is warranted towards elucidating brain circuit mechanisms that control tonic activity of dopaminergic and other neuromodulatory, especially serotonergic, systems (Huys et al. 2015; Husain and Roiser 2018). In addition, MDD patients prioritize negative over positive events (Pulcu and Browning 2017); concerns for negative outcomes could become debilitating and reduce one's motivational drive to seek rewards. Thus, anhedonia may also arise from more sensitivity to punishment than reward (Chen et al. 2015).

10.3.3 Addiction

Drug abuse refers to consumption of a drug, which may be legal (e.g., alcohol) or not (e.g., cocaine), in amounts or intake patterns that are harmful to the user. Neurobiologically, drug abuse involves maladaptive changes in the brain's reward circuit (Kauer and Malenka 2007; Volkow and Morales 2015), and therefore it can be viewed as RL gone awry. Redish (2004) formalized this idea using a temporal-difference RL (TDRL) model (Sutton and Barto 2018; Dayan and Abbott 2001). Recall that, according to TDRL (Chapter 4, Section 4.3.4), RPE is computed as $\delta(s(t)) = r(s(t)) + \gamma V(s(t+1)) - V(s(t))$, where V is the value function that depends on the state s, r is reward delivered as a result of making a transition from state $s(t)$ to state $s(t+1)$ through a chosen action, and γ is a temporal discounting factor for future rewards. TDRL learns to predict reward by driving RPE to zero. It was proposed that addiction involves the same RL system as in the normal brain with RPE signaled by dopamine neurons, but drug-induced positive prediction errors could produce unbounded increases in the expected value of drug receipt. Specifically, an addictive drug is assumed to produce a transient increase in dopamine release independent of the change in value; thus the new RPE function δ^D becomes:

$$\delta^D(t) = max\{\delta(t) + D, D\} \tag{10.2}$$

where D expresses a dopamine surge in response to an addictive drug. Thus, according to $V(t+1) = V(t) + \alpha\delta^D(t)$, even with δ approaching zero, the value keeps increasing through repetitive drug intake without bound.

Several phenomena of drug addiction can be accounted for by an actor-critic model (Chapter 4, Section 4.3.3), where the critic updates values based on the modified RPE (Eq. 10.2) and the actor is responsible for action selection (e.g., using a softmax criterion, Eq. 4.16 in Chapter 4). First, when faced with a choice between a drug and a nondrug food, the probability of selecting a drug initially depends on the difference between the current values of drug receipt and reward size of food, but this dependence diminishes over time because eventually the *RPE* vanishes for food but not drug. Second, the probability of choosing a reward option depends on the cost for obtaining it less sensitively for drug than food. Third, phasic activity of dopamine neurons is expected to display two peaks: one at the time of sensory cue predicting the reward, and the other when drug is delivered even though δ is zero, and the latter would be absent with nondrug awards. Double-peaked dopamine signals have been observed in the nucleus accumbens (a major recipient of dopaminergic innervation) of rats learning a cue-cocaine association (Aragona et al. 2009).

Such a quantitative model is concrete enough to be testable by new experiments. Redish's model predicts that a behavioral trait called blocking does not occur when drugs are used as unconditioned reinforcers. Blocking refers to the observation that after a subject learns to associate a stimulus A with a reward, pairing A with another stimulus B should not lead to learning to associate B with the reward (Kamin 1967). If drugs lead to unlimited value increase, blocking should not be observed. Behavioral experiments using cocaine as an unconditioned stimulus showed that this is not the case, i.e., blocking does occur (Panlilio et al. 2007). One possibility is that blocking is absent in drug abuse only for a specific form of reinforcement learning model. Indeed, blocking is present in a modified model in which the expected reward V was computed by a weighted average over past reward events (Dezfouli et al. 2009).

A crucial issue concerning addiction is to identify the brain mechanism that explains the differences between controlled drug use and drug abuse. Under what conditions does an initial use lead to an escalation of intake that is characteristic of drug abuse? Generally, restricted drug access produces a stable pattern of moderate consumption, whereas with continuous drug access, consumption behavior changes to a binge-like pattern. For instance, in a self-administration setting, rats were given the opportunity to produce a behavior, e.g., a lever press, that triggers an intravenous injection of a dose of cocaine (Ahmed et al. 1998). In this paradigm, drug intake is a direct result of the agent's action, thus drug-seeking behavior can be rigorously measured. When cocaine access was limited to one hour per session (short access [ShA] condition), drug intake per session was found to remain moderate and stable over sessions (days). In contrast, with a longer access of six hours per session (long access (LgA) condition), drug intake per session gradually increased over a dozen of sessions, before it reached a plateau level. This finding suggests the existence of an endogenous hedonic "setpoint," which is increased from the baseline in the ShA condition to a higher plateau in the LgA condition by escalated drug use (Koob and Le Moal 1997; Ahmed et al. 1998).

Keramati et al. (2017) proposed an RL model that incorporates homeostasis (Bernard 1865; Cannon 1929). The central idea can be illustrated with a one-dimensional system where h denotes the level of internal hedonic state (such as blood sugar level), but the model can be readily extended to multiple physiological variables. Motivational drive $d(h_t)$ at time t is defined as a convex function, like a parabola, of the current state h_t, with a minimum that defines a homeostatic setpoint h^* (**Figure 10.7a**, left). Assuming that a certain action changes the internal state, the reward value of this outcome can be computed. Mathematically, an action yields an outcome equal to a change in the internal state $h_t \rightarrow h_{t+1}$, and its rewarding value is determined by the resulting change in the drive $r_t = d_t - d_{t+1}$ (such as the quantity of glucose received). A reward ($r_t > 0$) amounts to a decrease of the drive function. In other words, a reward moves the internal state closer to, and a punishment shifts it away from, the homeostatic setpoint h^*, respectively. The quantity r_t is the reward signal used to compute the reward prediction error in an actor-critic reinforcement learning model (**Figure 10.7a**, right).

To model the self-administration experiment (Ahmed et al. 1998), it is assumed that the homeostatic setpoint h^* starts with a default value h_0^*, and is increased upon each drug infusion by an amount μK proportional to drug dose K. In the absence of drug intake, h^* recovers back to normal, with a slow adaptation time constant $\tau_{recovery}$. Thus, the dynamical homeostatic equation is:

$$\frac{dh^*(t)}{dt} = -\frac{\left(h^* - h_0^*\right)}{\tau_{recovery}} + \mu K \Sigma_i \delta\left(t - t_i^{\text{infusion}}\right) \tag{10.3}$$

where t_i^{infusion} are drug infusion times.

With appropriate values for the model parameters (in particular, $\tau_{recovery} \simeq 4$ hours), the model captures the salient experimental observations (**Figure 10.7b–c**). Under the short access (ShA) condition (the

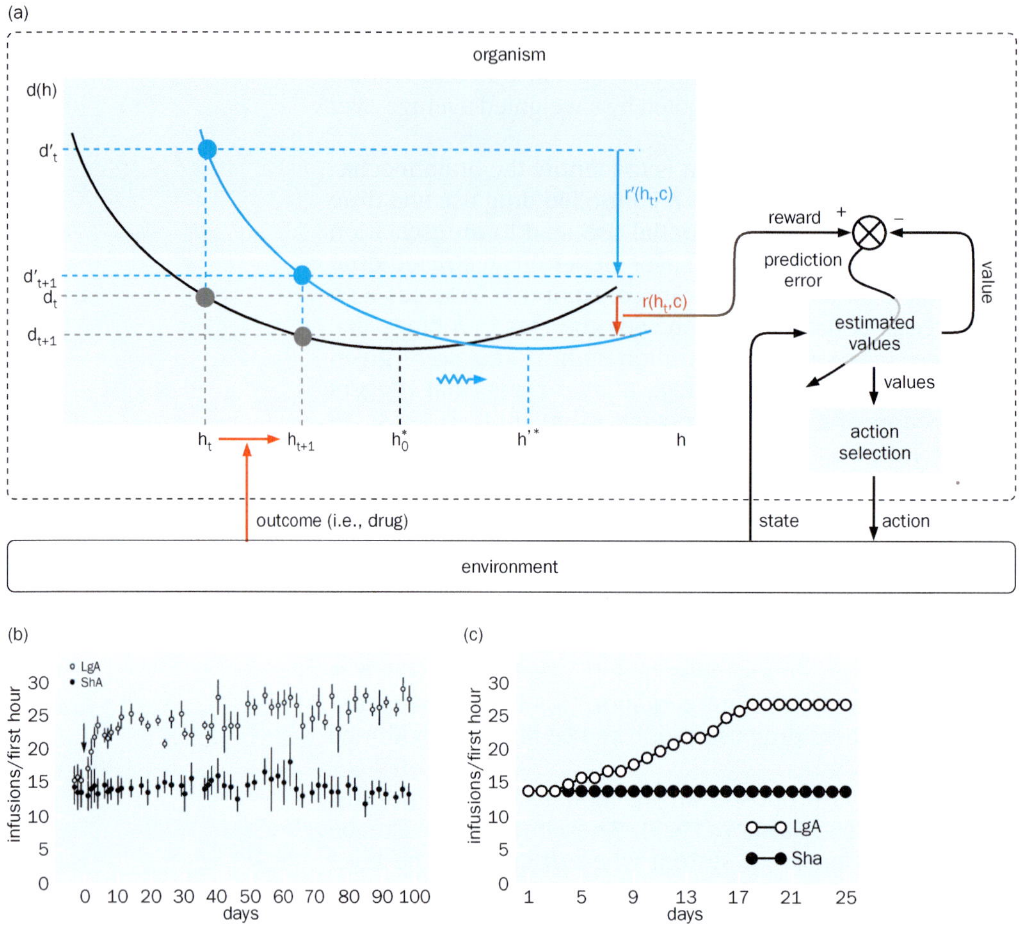

Figure 10.7 Homeostatic reinforcement learning model of drug addiction. (a) Model scheme. The organism's internal state at time *t* is represented by *h(t)*. Given a desired internal state, h^*, known as the setpoint, the organism's drive *d(h)* is defined in terms of the distance between the current internal state and the setpoint. Seeking rewards is equivalent to the objective of reaching physiological stability. Red: Effect of a volitional drug use, which results in the internal state to move from h_t to h_{t+1}, causing a drive-reduction reward $r(h)$. If the drug intake continues for a long time, the drive function shifts from the grey curve to the cyan curve, and the setpoint shifts from $h_0{}^*$ to $h'{}^*$ (squiggly cyan arrow). (b) Effect of drug availability on first-hour cocaine self-injections in a rat experiment. The arrow indicates the start of the escalation phase, during which rats had access to cocaine (250 μ g/injection) for short access (ShA) or long access (LgA). (c) The model of (a) reproduces the experiment observation (b) that drug use remains moderate and stable under the ShA condition, in contrast to the LgA condition, which leads to increased drug intake across days before saturating at a plateau level. Adapted from Keramati et al. (2017). [Courtesy of Boris Gutkin.] [Courtesy of Boris Gutkin with original data and permission from Keramati, M., Durand, A., Girardeau, P., Gutkin, B., & Ahmed, S. H. (2017). Cocaine addiction as a homeostatic reinforcement learning disorder. Psychological Review, 124(2), 130.]

sequence $\{t_i^{\text{infusion}}, i = 1,2,...\}$ is limited to 1 hour), the increment of h^* over 1 hour is small, and the setpoint has sufficient time to recover to h_0^* for the remaining 23 hours of the day; thus drug intake remains moderate and stable across sessions. On the other hand, under the long access (LgA) condition (the sequence $\{t_i^{\text{infusion}}, i = 1,2,...\}$ continues for 6 hours), the increment of h^* is sufficiently large, and the remaining 18 hours are too short for h^* to decay back to h_0^*, so that recovery is incomplete, leading to a buildup of the setpoint over days. Eventually, however, an elevated steady state h'^* is reached when the intake is balanced by the recovery, which would explain the observed plateauing of drug intake (**Figure 10.7b–c**). According to this model, the threshold required to make a transition from moderate to excessive drug intake is determined by the tradeoff between cumulative shift of the homeostatic setpoint and recovery speed. This homeostatic reinforcement learning model accounts for a number of phenomena observed with drug-seeking behavior, including a surge of drug intake at the start of a session and elevated dose dependence curve associated with excessive drug consumption. The latter results in the same dose of drug to become more rewarding than before escalation (r' (cyan) is larger than r (red) in **Figure 10.7a**), which could explain compulsivity (reduced sensitivity of drug to punishments associated with drug-taking behavior). In this view, tolerance is interpreted by the fact that after escalation the same dose of drug as before hardly gets close to the upshifted setpoint. So in order to reach the elevated setpoint h'^*, much higher doses are required. The model explains why the drug intake rate is the highest at the start of each session when presumably the difference between the current hedonic state and the setpoint is the largest, and predicts that the entire dose-response curve is scaled upward without significant horizontal shift after escalation, both were indeed observed (Ahmed et al. 1998).

Conceptually as well as neurobiologically, this model represents a first step towards linking reinforcement to internal physiological processes regulated by allostatics (McEwen 1998; McEwen and Wingfield 2003), which refers to changes in order to achieve stability in the face of challenges, here drug intake escalation. An allostatic model does not assume a fixed setpoint but a dynamical one, as is the case under the long access (LgA) condition when the shifted setpoint becomes the "new norm" for the organism. This perspective may potentially shed light into drug craving, a curse of addiction Heinz (2017). Craving becomes increasingly more intense over time during abstinence, similar to hunger that gets worse without eating for a longer time. Presumably, lack of drug consumption causes progressive drift of the internal state further and further away from the homeostatic setpoint, and the augmented drive eventually leads to drug seeking at all costs, resulting in relapse. Computational considerations of these effects require a nonlinear model going beyond Eq. (10.3). The hypothetical drift of the internal state away from a setpoint has yet to be observed physiologically, possibly involving the nucleus accumbens, a brain center for reward processes and motivation, where long-term drug abuse produces abnormal long-term plasticity (Kauer and Malenka 2007) and D2 receptor signaling (Volkow et al. 1993; Volkow and Morales 2015).

In summary, we discussed reinforcement learning models of depression and addiction, which illustrate the utilities of computational modeling: it can shed specific insights into a psychiatric disorder, such as deficient mood baseline but not reinforcement learning rate associated with depression; and it helps formulate a quantitative framework for testing a hypothesis, like allostatic dysregulation of natural reward processing underlying drug abuse.

10.4 DEFICITS OF EXECUTIVE CONTROL

10.4.1 Loss of Control in Addiction and Depression

Models in the prior section deal with "low-level" sensorimotor and reward processes, but the core of mental illness involves executive control of behavior. For instance, addiction is viewed as the endpoint of transitions from voluntary drug use to loss of control over drug intake (Volkow and Li 2004). It is desirable to describe this transition in a way that can be rigorously investigated in systems neuroscience. In one formulation (Everitt and Robbins 2005, 2016), volitional drug use is a goal-oriented behavior in order to garner a hedonic outcome, and thus represents an instrumental or operant conditioning (action-outcome learning). In drug abuse, environmental cues that have been paired with a drug (conditioned stimuli) trigger repeated actions necessary for drug consumption. The process through which a Pavlovian, conditioned stimulus becomes influential on instrumental behaviors is a phenomenon called Pavlovian-instrumental transfer. When deeply instated, drug-seeking responses become a habit (Graybiel 2008), compulsive and out of control in spite of aversive consequences.

What are the brain mechanisms underlying such loss of control? At the level of the subcortical system, the basal ganglia is a structure critically implicated in inhibitory control of prepotent and habitual behaviors. As we explained in Chapter 8 (Section 8.2.3), in the dorsal striatum, a direct pathway facilitates motor responses, whereas indirect and hyperdirect pathways serve as a brake whenever an automatic response should be stopped. These two counteracting systems must be balanced, as demonstrated in a computational model of the cortico-basal ganglia-thalamocortical loop (Wei and Wang 2016). The required balance would explain, for instance, how an impaired hyperdirect pathway via the subthalamic nucleus leads to reduced ability to suppress premature actions (Frank 2006; Hikosaka and Isoda 2010; Schmidt et al. 2013).

At the level of the prefrontal cortex, two subnetworks form an opponent system: on the one hand, a ventromedial "affective" subnetwork for rewards and motivation including the ventral anterior cingulate cortex (vACC) and medial orbitofrontal cortex (mOFC); and on the other hand, a dorsolateral "executive" subnetwork for cognitive control including the dorsal ACC, DLPFC and inferior frontal cortex (IFC, BA 44). Flexible behavior depends on a balance between the two; drug addiction is characterized by dominance of the former and weakening of the latter, although the interaction between the two is not unidirectional, and the executive control system is under the strong influence of the emotional evaluation system. Furthermore, projection from the PFC to the dopamine system is also impaired in addiction (Dalley et al. 2011).

Understanding the neural circuit substrates for volitional control is equally applicable to other psychiatric disorders such as depression. Indeed, in MDD individuals, persistent negative mood is associated with hypoactivity of DLPFC and hyperactivity of vACC, which is why deep brain stimulation of vACC or subgenual cingulate white matter represents a candidate treatment for a subset of MDD patients, presumably by reducing vACC activity (Mayberg et al. 2005). Interestingly, the vACC has a smaller volume in MDD subjects, yet its activation is consistently correlated with acute sadness. This apparent paradox can be explained by the observation that reduced vACC volume reflects a smaller number of glial cells responsible for high-affinity glutamate transporters; weaker glutamate uptake in the synaptic cleft causes an enhancement of glutamate-mediated activity in the vACC observed in MDD (Choudary et al. 2005).

The vACC and the DLPFC appear to be antagonistic with each other; increased activation in one is concomitant with decreased activity in the other (Drevets and Raichle 1998). Ramirez-Mahaluf et al. (2017) developed a model for this opponency system, where synaptic interaction between the affective (vACC) and executive (DLPFC) modules is mutually inhibitory through long-range excitatory-to-inhibitory connections. The reduced glutamate uptake is instantiated as a modest increase of the AMPA receptor signaling in the vACC but not DLPFC. The model was used to simulate behavior where trials for emotional processing of a sad face and trials for cognitive processing in a working memory task were randomly interleaved. Enhanced excitation in the vACC leads to predominance of the affective module at the expense of the executive module. Consequently, the system "gets stuck" and becomes unable to switch from emotional to cognitive trials (**Figure 10.8**). This work illustrates how an attractor network model can describe a tradeoff between stability of any attractor state and flexibility for switching between attractor states; excessive stability impairs flexibility and vice versa (see Section 10.5).

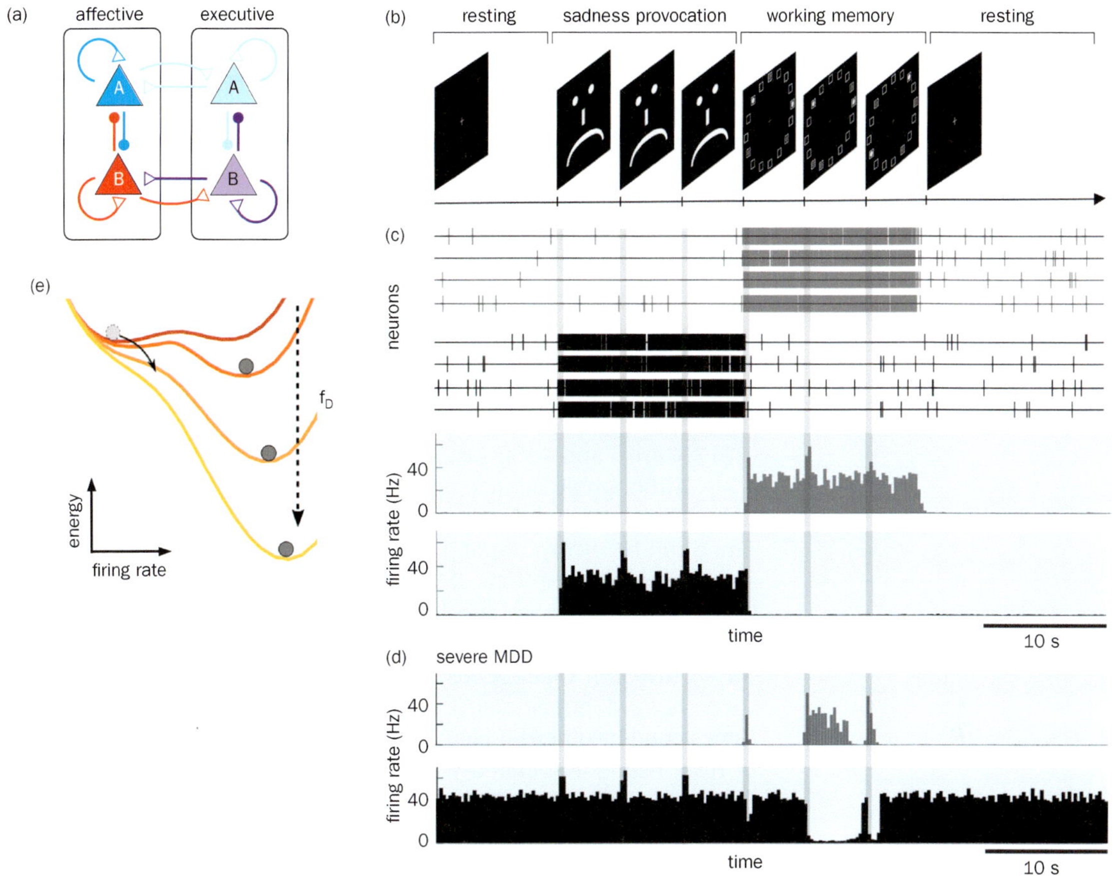

Figure 10.8 Predominance of the affective module at the expense of the executive module in depression. (a) The computational model consists of an affective circuit and an executive circuit that interact with each other through effective reciprocal inhibition. (b) A series of tasks alternating between viewing a sad face and working memory. (c) Model behavior in the control condition. The affective module (black) is active during viewing of a sad face, and the executive module (grey) is active during working memory. (d) MDD was simulated with modified glutamatergic transmission selectively in the affective module, resulting in a predominant attractor state corresponding to high activity of the affective module even in two out of three cognitive task trials. (e) Schematic attractor landscape with two attractors, corresponding to the executive (left) and affective (right) minima of an energy function. The parameter f_D denotes increased glutamatergic signaling in the affective module, leading to an exceedingly deep basin of attraction for its predominance (right attractor). Reproduced form Ramirez-Mahaluf et al. (2017). [Courtesy of Albert Compte and permission from Ramirez-Mahaluf, J. P., Roxin, A., Mayberg, H. S., & Compte, A. (2017). A computational model of major depression: The role of glutamate dysfunction on cingulo-frontal network dynamics. Cerebral Cortex, 27(1), 660–679.]

10.4.2 Negative Bias in Anxiety and Obsessive-Compulsive Disorder

Studies of mood disorders need to take into account both positive and negative emotional valences (Panksepp 2004). Arguably, acute sadness associated with MDD points to an abnormally high negative valence. This is also true for clinical anxiety, and there is a high degree of comorbidity of depression and anxiety (Craske and Stein 2016).

Anxiety denotes a feeling of apprehension or worry about something. It is a fairly stable personality trait and highly subjective; what causes strong distress to one person may hardly affect another's emotion (Ledoux 2015). Can computational models be developed and applied to such a seemingly elusive affective trait as abnormal anxiety? One important step is to identify a well-defined behavior as manifestation of anxiety that can be studied in controlled experiments. A commonly used such paradigm is avoidance conditioning, in which subjects learn to avert action in order to reduce the likelihood of an aversive outcome (Mobbs and Kim 2015; LeDoux et al. 2017; Cain 2019). Debilitating anxiety is characterized by excessive avoidance. For instance, if you are overly anxious about attending social parties, you may routinely choose not to go even when your worries are not justified. Avoidance is generally enhanced with heightened anxiety. In a go/nogo task, a subject learns to produce a response (go) when a sensory cue A is shown, and withhold the response (nogo) if a different cue B is presented. When go trials have a much higher probability than nogo trials, the go response is prepotent and the ability to suppress it in a nogo trial provides an index for response inhibition (Bari and Robbins 2013). In a variant of the go/nogo task, each trial is either in a safe condition or a threat condition (when a shock may be administrated). It was found that when anxiety is heightened with a threat of shock, the rate of success to inhibit a go response was increased. Subjects with clinical anxiety displayed increased errors in go trials (incorrectly withholding the response in go trials) in both safe and threat conditions, suggesting an excessive increase of response inhibition compared to healthy subjects (Grillon et al. 2017; Carlisi and Robinson 2018).

Although avoidance appears to be a form of response suppression, it actually is better understood as a goal-directed action (Cain 2019). Avoidance can be incorporated in a reinforcement learning model as one of the learnt response options (Maia 2010). However, reward and punishment are not symmetrical, and learning what not to do is fundamentally different from learning what to do. Guitart-Masip et al. (2012) developed a task paradigm that probes action and valence with a two (go/nogo) by two (reward/punishment) design (Guitart-Masip et al. 2012), with trial types cued by different visual images (**Figure 10.9a**). In a "go to win" trial, a response is associated with a reward, whereas in a "go to avoid" trial, action is required to prevent a punishment; on the other hand, in a "nogo to win" trial, response must be avoided in order to collect a reward and, finally, in a "nogo to avoid" trial, no response entails absence of a punishment. As expected intuitively, behavior is asymmetrical. People performed the best with an active response (go) and the worst when withholding a response (nogo) was required in order to obtain a reward (**Figure 10.9b**). The behavioral data were fitted well by a reinforcement learning model that included in the action value function an "avoidance bias," which represented a model-free Pavlovian state in favor of nogo. When participants included individuals self-defined as suffering from anxiety disorder, model fitting with performance data shows a unimodal distribution of model parameter values, rather than a bimodal distribution that would readily differentiate healthy subjects and others with anxiety disorder (Mkrtchian et al. 2017). Moreover, this experiment assessed behavioral change under stress, which increased anxiety, by including a threat condition under which an unpleasant but not painful shock was administrated in a fraction of trials. It was found that the avoidance bias was larger in the anxiety group compared to heathy controls. Even the avoidance bias value for the anxiety group under a safe condition is significantly larger than that for the healthy group under a stressful threat condition (**Figure 10.9c**). Again, this study shows how computational modeling in concert with sophisticated behavioral experiments can quantitatively reveal characteristics of a psychiatric disorder.

Unusually high tendency for avoidance is also characteristic of obsessive-compulsive disorder (OCD): repetition of an action (e.g., washing hands) regardless of whether it serves a useful purpose is caused by the urge to avoid undesirable or fearful outcomes (getting germs and becoming sick). Experiments using an avoidance task design showed that OCD subjects displayed enhanced avoidance (Gillan et al. 2014). Similarly, OCD subjects display an increased propensity, compared with control subjects, to use the model-free system relative to a model-based system in the two-stage decision task (Voon et al. 2015). Recent research has begun to elucidate abnormalities associated with anxiety of the affective brain circuit,

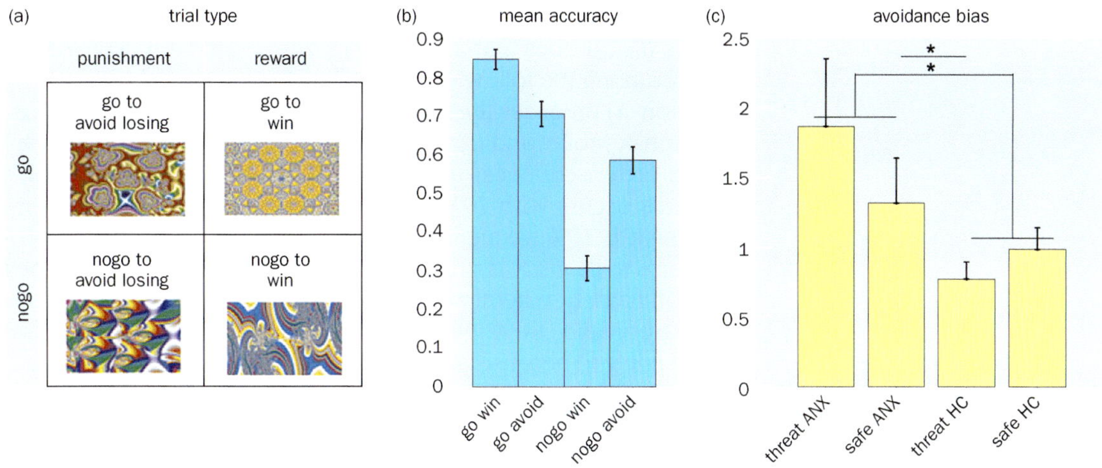

Figure 10.9 Avoidance in anxiety. (a) Four trial types probing the two axes of behavioral control: Valence (reward versus punishment) and action (go versus nogo). (b) Accuracy for different trial types, showing the highest error rate when inhibition of response (nogo) is required for winning a reward. (c) Compared to healthy control (HC), subjects with abnormal anxiety (ANX) show a larger avoidance bias in both safe and stressful threat conditions. Panel (a) is reproduced form Guitart-Masip et al. (2012), (b–c) Mkrtchian et al. (2017). [Guitart-Masip, Marc, et al. (2012). Go and no-go learning in reward and punishment: Interactions between affect and effect., Neuroimage 62(1), 154–166. Licensed under CC BY 3.0.] [Mkrtchian, A., Aylward, J., Dayan, P., Roiser, J. P., & Robinson, O. J. (2017). Modeling avoidance in mood and anxiety disorders using reinforcement learning. Biological Psychiatry, 82(7), 532–539. Licensed under CC BY 4.0.]

which involves interactions between the medial frontal cortex, amygdala and ventral striatum (Calhoon and Tye 2015; Tovote et al. 2015; Carlisi and Robinson 2018; Diehl et al. 2020). OCD also involves impairments of a cortico-basal ganglia circuitry that includes the orbitofrontal cortex, the anterior cingulate cortex and the dorsal striatum (Saxena et al. 1998; Graybiel and Rauch 2000).

Thus, anxiety, OCD and addiction all display a weakening of cognitive control necessary for goal-directed behavior. The differences among them lie in the extent to which positive versus negative valence systems are faulty. Based on newly acquired knowledge (Roberts 2020; Coley et al. 2021), cortico-basal ganglia thalamocortical loop models (Frank et al. 2004; Wei and Wang 2016) should be extended to incorporate multiple loops (Alexander et al. 1986; Alexander and Crutcher 1990; Haber et al. 2000; Cox and Witten 2019) for elucidating neural circuit deficits that lead to a variety of emotional abnormalities.

10.4.3 Reactive versus Proactive Control in Schizophrenia

As we mentioned in Chapter 8 (Section 8.2), there is a distinction between reactive versus proactive control. Reactive control is a "late correction" mechanism recruited only in a just-in-time manner when a warning signal is detected, while proactive control entails an early selection of what event (e.g., stop signal) demands attention and maintenance in a sustained manner prior to event occurrence (Aron 2011; Braver 2012). For instance, in the stop-signal task, when the probability for a stop signal to be presented in a trial is higher, the efficacy of the stop process should be upregulated by a control signal to accommodate the increased need for withholding a response (Lo et al. 2009; Wei and Wang 2016; Balasubramani et al. 2020). This control signal is proactive and should be sustained across trials.

Reactive and proactive controls can be differentially examined in a task paradigm called "AX continuous performance task (AX-CPT)" (Cohen et al. 1999). In this task, participants are instructed to make a response to a cue-probe sequence, in which the letter A appears as the cue, followed by the letter X as the probe when a response is required. The remaining letters of the alphabet serve as invalid cues (i.e., cues that are not As) and probes (i.e., probes that are not Xs) (**Figure 10.10a**). Non-AX are divided evenly among the following three trial types: (a) In BX trials, an invalid cue (i.e., non-A) precedes the probe X; (b) in AY trials, a valid cue A is followed by a non-X probe; and (c) in BY trials, an invalid cue is followed by a non-X probe. Letter sequences are presented in pseudo-random order, such that AX trials occurs with 70% frequency and each of the three non-AX trials occurs with 10% frequency. Because of the preponderance of AX trials, response to probe X is prepotent. Performance in AY trials probes reactive control, namely, whether one can successfully suppress a response anticipated by a valid cue A. By contrast, performance in BX trials assesses proactive control, because a non-A cue should proactively engage control to withhold automatic response to the valid probe X presented later in time.

The AX-CPT probes context-dependent stimulus-response mapping. As we discussed in Chapter 8, context representation depends on the prefrontal cortex (Passingham 1995). Abnormalities of the PFC are associated with cognitive deficits in schizophrenia, which are more prognostic than psychosis. Impaired executive control was examined in schizophrenic patients using the AX-CPT (MacDonald et al. 2005). Functional MRI revealed that activation of the dorsolateral prefrontal cortex associated with a higher demand of control in BX trials is reduced in schizophrenia relative to control subjects (**Figure 10.10b**). Concomitantly, schizophrenic patients display a much higher error rate in BX trials but not AY trials, compared to healthy controls (**Figure 10.10c**), suggesting that proactive control is selectively impaired, whereas reactive control remains intact in schizophrenia.

This human study inspired a monkey physiological experiment using a variant of the AX-CTP task (Blackman et al. 2016). It was found that, compared to the valid A cue, neurons in the DLPFC tend to display a higher activity upon presentation of non-A cues that inhibit prepotent response X, and the differential activity persists across a delay period and through probe presentation. In a study with the same monkeys performing the same task, subjects received a low-dose ketamine injection, which is known to induce cognitive deficits associated with schizophrenia (Krystal et al. 1994). Administration of ketamine resulted in impaired performance

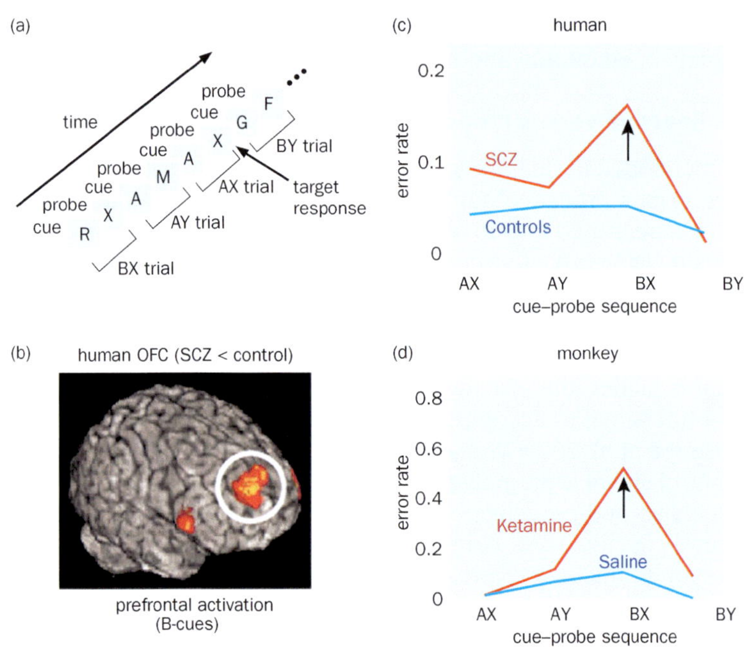

Figure 10.10 Selective deficit of proactive control in schizophrenia. (a) The AX continuous performance task. (b) BOLD signal reflecting maintenance of the B-cue (which countermands the habitual response to subsequent X-probe) in working memory is reduced in the prefrontal cortex of schizophrenia patients relative to controls performing the AX-CPT. (c) Schizophrenic patients show a higher error rate compared to controls selectively for BX trials, a differential effect that is reproduced in monkeys with administration of ketamine (d). Panel (a) is reproduced from Braver et al. (2001), (b–d) from Averbeck and Chafee (2016), with original data from MacDonald et al. (2005) and Blackman et al. (2013). [Braver, T. S., Barch, D. M., Keys, B. A., Carter, C. S., Cohen, J. D., Kaye, J. A., ... & Reed, B. R. (2001). Context processing in older adults: Evidence for a theory relating cognitive control to neurobiology in healthy aging. Journal of Experimental Psychology: General, 130(4), 746.] [Averbeck, B. B., & Chafee, M. V. (2016). Using model systems to understand errant plasticity mechanisms in psychiatric disorders. Nature Neuroscience, 19(11), 1418–1425. Original data from MacDonald III, A. W., Carter, C. S., Kerns, J. G., Ursu, S., Barch, D. M., Holmes, A. J., ... & Cohen, J. D. (2005). Specificity of prefrontal dysfunction and context processing deficits to schizophrenia in never-medicated patients with first-episode psychosis. American Journal of Psychiatry, 162(3), 475–484.]

selectively in the BX trials similar to the observation with schizophrenic individuals (**Figure 10.10d**) (Blackman et al. 2013). This finding demonstrates that impaired proactive cognitive control represents a characteristic of schizophrenia that can be examined in nonhuman experiments, opening the possibility of elucidating the underlying circuit mechanism. Conceivably, DLPFC sends an internally sustained, top-down executive signal as hypothesized in the model of inhibitory control (Chapter 8, Section 8.2.2 and Figure 8.6) (Lo et al. 2009), and its reduction leads to impaired proactive response inhibition. Neural and LFP recordings from monkeys performing the AX-CPT showed that ketamine administration reduced top-down signal from the PFC to PPC, which influences whether a behavioral response is produced correctly in a context-dependent manner (Kummerfeld et al. 2020).

10.5 NEURAL CIRCUIT MODELS OF COGNITIVE DEFICITS

Proactive control requires top-down signaling sustained over time, so could it be that its impairment is caused by abnormal working memory? Low-dose ketamine induces working memory deficits similar to that observed in schizophrenia and reduces activation of the prefrontal cortex during working memory in humans (Krystal et al. 1994; Driesen et al. 2013). To delineate circuit abnormalities in the PFC underlying cognitive deficits, it is paramount to go beyond behavioral modeling and leverage biophysically based neural circuit models. From a computational psychiatry perspective, biophysically based neural circuit modeling can address key questions that are inaccessible to abstract computational modeling (Wang and Krystal 2014; Murray and Wang 2017; Wang and Murray 2020). For instance, circuit models are well suited to mechanistically study the impacts of synapse-level alterations. One can perturb specific synaptic parameters in the model that are physiologically interpretable, thereby allowing direct implementation of hypothesized perturbations related to disease mechanisms and pharmacological manipulations. One can then characterize the impact of these perturbations on collective neural population activity and task performance, and relate them to behavioral findings from healthy and clinical populations, as well as in animal models of disease states. Biophysically based models may also inform the rational design of pharmacological therapies, by mechanistically bridging molecular, cellular, recurrent synaptic circuit and ultimately behavioral levels of analysis.

The linking proposition between a neural state and a psychological state provides a framework for detailed hypothesis generation and experimental examination of psychological models (Schall 2004). To establish linking propositions for core cognitive functions, neural circuit modeling plays a central role in identifying circuit mechanisms underlying required computations in a cognitive process, and model predictions can be tested in the laboratory. In the case of working memory, the model prediction of a critical role by the NMDA receptors (Wang 1999b) (Chapter 5, Section 5.6.2) has been supported by experiments that found focal antagonism of the NMDA receptors impaired mnemonic persistent activity in behaving animals, thereby informing the biological mechanisms at the receptor and synaptic levels (Wang 2013; Krystal et al. 2017; Yang et al. 2021). In this way, circuit models can iteratively contribute to our understanding of these links across levels of analysis and leverage them to study dysfunction in neuropsychiatric disorders.

To illustrate this biophysically based neural circuit modeling framework, we will specifically discuss the idea of altered E/I balance using cortical attractor network models of working memory (Chapter 5) and decision making (Chapter 6) applied to mental disorders.

10.5.1 Working Memory

Working memory function is a promising candidate endophenotype in clinical neuroscience. It is important to consider the component processes and features involved in overall working memory function: encoding, maintenance, robustness to distraction, precision and capacity. Ongoing work in clinical cognitive neuroscience aims at resolving how these processes are impaired in schizophrenia (Barch and Ceaser 2012; Smucny et al. 2022). Many studies have found that patients with schizophrenia exhibit deficits in encoding and in maintenance, which result in a loss of precision for working memory representation (Lee and Park 2005; Badcock et al. 2008; Starc et al. 2017; John et al. 2018; Zhao et al. 2021).

Abnormality of bistability has been proposed to underlie working memory deficits associated with schizophrenia (Tanaka 2006; Durstewitz and Seamans 2008; Rolls et al. 2008). As we discussed in Chapters 5 and 9, dopamine modulation is important for working memory, which has been described by attractor network models. The idea is that normal working memory function depends on a balance between D1 and D2 receptor activations: excessive D2 activation reduces robustness of attractor states and therefore impairs working memory maintenance and enhances distractibility; on the other hand, too much D1 receptor-mediated signaling leads to inflexibility, because in this case it becomes unusually difficult to induce transitions between attractor states. This dual-state model offers a potential explanation of the observed diversity of working memory performance in healthy human subjects and schizophrenic patients (Egan et al. 2001; Cools and D'Esposito 2011).

Dopamine modulates both synaptic excitation and inhibition (Seamans and Yang 2004). In schizophrenia, cortical microcircuit alterations are complex, with observed dysfunction in both glutamatergic excitation and GABAergic inhibition. Postmortem investigations of prefrontal cortex in schizophrenia found reductions in spines on layer-3 pyramidal cells (Glantz and Lewis 2000) compared to control subjects. Studies also have revealed multiple impairments in inhibitory interneurons (Lewis et al. 2005). Pharmacological models of schizophrenia provide complementary evidence. One such approach is to use NMDA receptor antagonists (e.g., ketamine), which transiently, safely and reversibly induce cardinal symptoms of schizophrenia in healthy subjects (Krystal et al. 2003). A leading hypothesis regarding ketamine's effects on neural function proposes a state of cortical disinhibition, potentially via preferential blockade of NMDA receptors on GABAergic interneurons (Homayoun and Moghaddam 2007; Kotermanski and Johnson 2009; Cohen et al. 2015). However, questions remain regarding the neural effects of ketamine, such as which NMDA receptor subunits and neuronal cell types may be the preferential sites of action (Khlestova et al. 2016; Zorumski et al. 2016). Existing data suggest that hypofunction of NMDA receptors can alter E/I ratio and produce cognitive deficits in two different ways: on inhibitory interneurons, which elevates E/I ratio via disinhibition, or on excitatory pyramidal neurons, which lowers E/I ratio (**Figure 10.11a**).

Murray et al. (2014a) examined the effects of altered E/I balance in a cortical circuit model of visuospatial working memory (Chapter 5, Section 5.4) (Compte et al. 2000; Carter and Wang 2007; Wei et al. 2012). Disinhibition was implemented through antagonism of NMDA receptors preferentially onto interneurons. In this model, disinhibition leads to a broadening in the neural-activity patterns in the mnemonic attractor states (**Figure 10.11b**). This neural change induced specific cognitive deficits. During maintenance, the mnemonic activity pattern undergoes random drifts

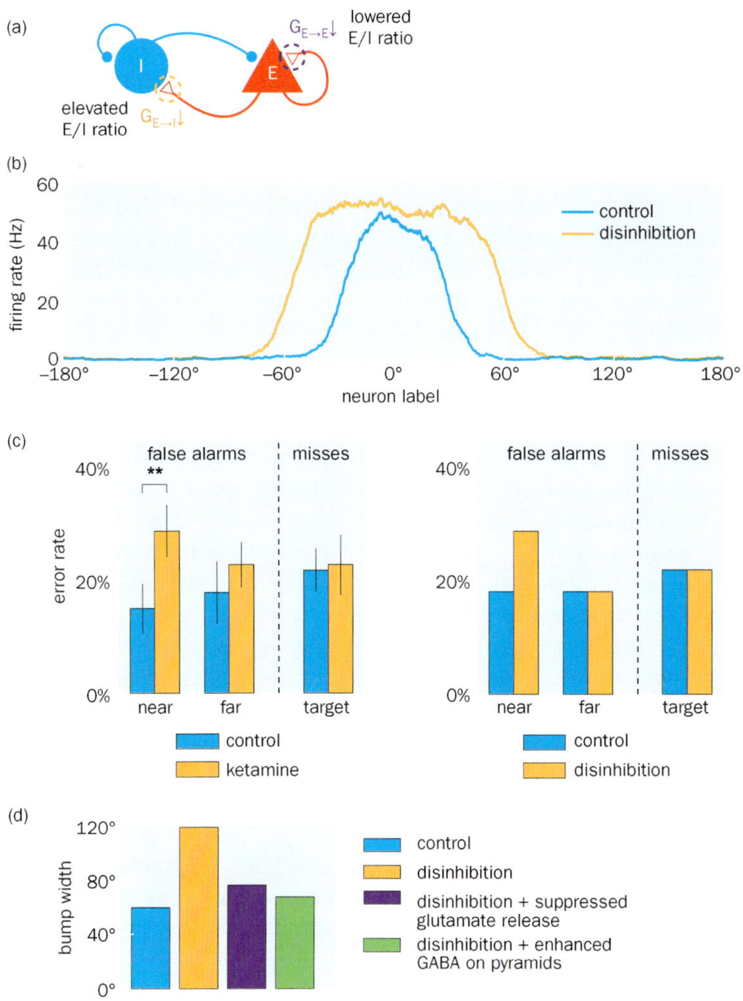

Figure 10.11 Impaired working memory in a cognitive-type neural circuit model. (a) E/I ratio was perturbed via hypofunction of NMDA receptors at two recurrent synaptic sites: On inhibitory interneurons, which elevates E/I ratio via disinhibition, or on excitatory pyramidal neurons, which lowers E/I ratio. (b) A spatial working-memory model generates a bump-shaped stimulus-selective persistent activity pattern following stimulus withdrawal. Disinhibition, mediated by NMDAR hypofunction on interneurons, broadens working-memory representations (orange) compared to control (cyan). (c) Broadening of working-memory representation was tested using behavioral data from human subjects performing a spatial working-memory task combined with ketamine infusion, a pharmacological model of schizophrenia. Consistent with broadening, ketamine induced errors specifically for near distractor probes (left), as predicted by the model (right). (d) Compensations can restore E/I balance and ameliorate behavioral deficits in the model, when disinhibition is paired with either reduced excitation (purple) or increased inhibition (green). Reproduced from Murray et al. (2014a). [Murray, J. D., Anticevic, A., Gancsos, M., Ichinose, M., Corlett, P. R., Krystal, J. H., & Wang, X. J. (2014). Linking microcircuit dysfunction to cognitive impairment: Effects of disinhibition associated with schizophrenia in a cortical working memory model. Cerebral Cortex, 24(4), 859–872, by permission of Oxford University Press.]

(Chapter 5, **Figure 5.20**), which lead to decreased precision of memory-guided responses. Disinhibition increased the rate of this drift, thereby inducing a specific deficit in mnemonic precision during working memory maintenance.

Additionally, Murray et al. (2014a) found that broadened neural representations make working memory more vulnerable to intervening distractors. In the model, a distractor is more likely to "attract" the memorandum towards it if the two representations overlap. Distractibility therefore depends on the similarity between the representations of the mnemonic target and the intervening distractor. Consistent with this model behavior, it has been found empirically that in visuospatial working memory, a distractor can attract the memory report toward its location, but only if the distractor appears within a "distractibility window" around the target location (Herwig et al. 2010). Because disinhibition broadens the mnemonic activity patterns, there is an increased range of distractors that can disrupt working memory.

To test the model prediction of broadened working memory representations under disinhibition, Murray et al. (2014a) analyzed behavior from healthy humans administered ketamine during a spatial delayed match-to-sample task (Anticevic et al. 2012). The model predicted a pattern of errors depending on whether the probe was similar to a target held in working memory. Analysis of the behavioral data guided by the model revealed a specific pattern of errors under ketamine predicted by the computational model (**Figure 10.11c**). Namely, ketamine increased the rate of errors specifically for distractors that would overlap with a broadened mnemonic representation. A similar pattern of errors has been observed in schizophrenia, with a selective increase in false alarms for near nontarget probes but not for far nontarget probes (Mayer and Park 2012). These findings consistent with disinhibition contradict the alternative of a reduced E/I ratio by insufficient recurrent excitation, which would lead to a collapse of persistent activity and induce an error pattern of misses and spatially random errors. Modeling showed that a reduced excitation or an enhanced inhibition could restore the bump attractor width (hence decoding accuracy for working memory) to the control level (**Figure 10.11d**), suggesting potential ways to remediate behavioral deficits associated with E/I ratio abnormalities associated with schizophrenia.

Applying this model to schizophrenia and more directly testing model predictions, Starc et al. (2017) designed a working memory task to be explicitly aligned with the model and with the primate electrophysiology task paradigms for which the model was developed. Such an alignment, in which a clinical study is linked to basic neurophysiology findings through a computational model, allows stronger inferences and testing of hypotheses through clinical experiments. In the working memory task of Starc et al. (2017), the memorandum was a single visuospatial location, and the response was a direct report of the remembered location, which provided a continuous measure of mnemonic coding. To test the model prediction of increased drift during working memory maintenance, the duration of the mnemonic delay was varied, to characterize how response variability increased with the duration of maintenance. To test the model prediction of increased distractibility dependent on target-distractor similarity, a set of trials included a distractor during the delay with a variable distance from the target. It was found that results largely followed model predictions, whereby patients exhibited increased variance and less working memory precision as the delay period increased relative to healthy controls. Schizophrenia patients also exhibited increased distractibility, with reports biased toward distractors at spatial locations close to the behaviorally relevant cue. This study illustrates a productive computational psychiatry approach in which predictions from biophysically based neural circuit models of cognition can be translated into experiments in clinical populations.

10.5.2 Decision Making

Perceptual and value-based decision functions are impaired in multiple psychiatric disorders (Lee 2013). Since the biologically based model is capable of both working memory and decision making (Wang 2002b, 2008b; Wong and Wang 2006), let us consider decision dysfunction in that recurrent neural circuit model. Recall (Chapters 6) that in a two-alternative forced choice task, a random-dot motion stimulus is presented, and the subject must report the net direction of motion (e.g., left vs. right). The coherence of the random-dot pattern can be parametrically varied to control the strength of perceptual evidence and thereby task difficulty. Performance is quantified by accuracy as a function of coherence, and the coherence level for achieving a certain high level of accuracy defines the discrimination threshold. Random-dot motion paradigms have been applied to clinical populations and have revealed impaired perceptual discrimination in schizophrenia, as measured by a higher discrimination threshold (Chen et al. 2004; Chen and Fetz 2005).

To investigate dysfunction in evidence accumulation, Lam et al. (2022) studied the effects of altered E/I balance in the cortical circuit model of decision making (Wang 2002b). E/I ratio was perturbed, to compare the impact of elevated vs. lowered E/I ratio, via NMDA receptor hypofunction on inhibitory vs. excitatory neurons, respectively. Interestingly, disruption of E/I balance in either direction can similarly impair decision making as assessed by psychometric performance, following an inverted-U dependence on E/I ratio. Therefore, the standard psychophysical measurements from clinical populations cannot dissociate distinct circuit-level alterations: elevated E/I ratio, lowered E/I ratio, or an upstream sensory coding deficit could lead to an identical performance change.

As discussed in Chapter 6 (Section 6.8), the attractor neural circuit model and empirical psychophysical behavior of nonhuman primates showed that, while the RDM task involves temporal accumulation of evidence, the choice is not uniformly sensitive to the stimulus value at all time points. In the "pulse" task paradigm (Huk and Shadlen 2005; Wong et al. 2007), a brief pulse of additional coherence is inserted at a variable onset time during the otherwise constant-coherence stimulus (**Figure 10.12a**). This pulse induces a shift of the psychometric function according to pulse coherence. The dependence of this shift on pulse onset timing reflects the weight of evidence at that time point on the ultimate choice (Chapter 6, **Figure 6.41**).

The pulse paradigm was able to dissociate distinct decision-making impairments under altered E/I ratio (**Figure 10.12b**). Under elevated E/I ratio, decision making is impulsive: perceptual evidence presented early in time is weighted much more than late evidence. In contrast, under lowered E/I ratio, decision making is indecisive: evidence integration and winner-take-all competition between options are weakened. These effects are qualitatively captured by modifying the drift diffusion model (Ratcliff 1978; Gold and Shadlen 2007). The standard drift diffusion model assumes perfect integration with an infinite time constant for memory (Chapter 6, Section 6.2.2). Lowered E/I ratio in the circuit model can be captured by "leaky" integration with a finite time constant so that choice is more influenced by recent information compared to control. In contrast, elevated E/I ratio can be captured by "unstable" integration, which has an intrinsic

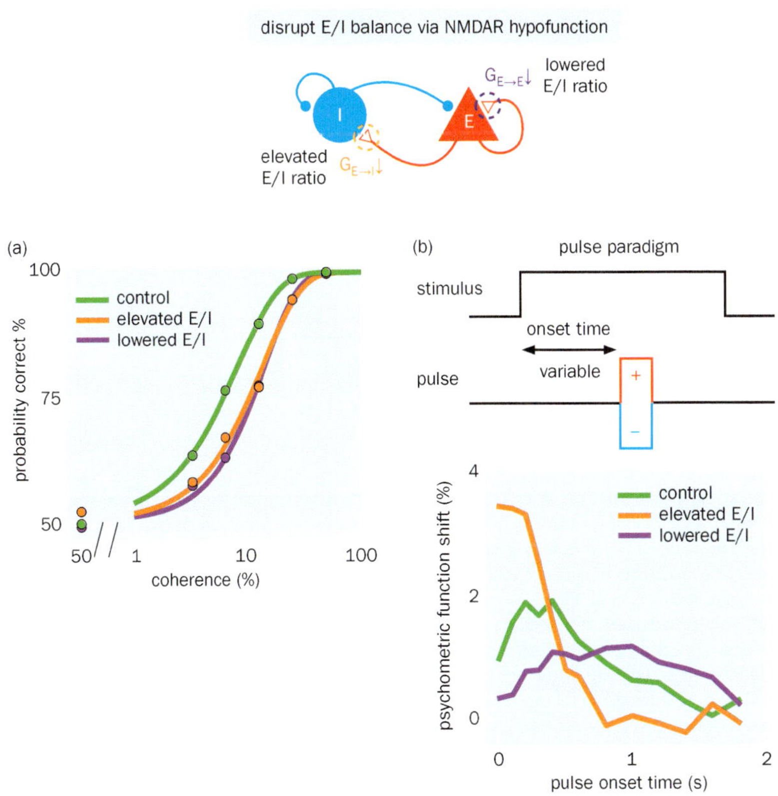

Figure 10.12 Impaired decision making in a cognitive-type neural circuit model. (a) Performance quantified by the psychometric function, i.e., the proportion of correct choices as a function of stimulus coherence. Both perturbations, elevated and lowered E/I ratio, can comparably degrade performance relative to the control circuit. (b) The time course of evidence accumulation can test dissociable behavioral predictions from elevated vs. lowered E/I ratio. Top: The pulse paradigm uses a brief pulse of additional perceptual evidence at different onset times. This pulse induces a shift in the psychometric function, which quantifies the sensitivity of the choice on evidence presented at that time point. Bottom: Shift in the psychometric function as a function of pulse onset time, for the three E/I regimes. Relative to control (green), in the elevated-E/I circuit the pulse has a stronger impact at early-onset times (orange); by contrast, the circuit with reduced E/I ratio shows a flattened profile of the shift, with greater impact at late-onset times (purple). Reproduced from Lam et al. (2022). [Lam, N. H., Borduqui, T., Hallak, J., Roque, A., Anticevic, A., Krystal, J. H., ... & Murray, J. D., & Wang, X. J. (2022). Effects of altered excitation-inhibition balance on decision making in a cortical circuit model. Journal of Neuroscience, 42(6), 1035–1053.]

tendency to diverge toward the decision threshold, leading to impulsive decision making. This study demonstrates the merit of sufficiently sophisticated task designs, in close interplay with cross-level modeling, for dissecting alternative scenarios of brain circuit malfunctions.

10.5.3 Critical Role of E/I Balance

As described previously, neural circuit models of cognitive functions can generate dissociable predictions for how distinct synaptic perturbations impact behavior under various task paradigms. Biophysically based models can also suggest what aspects of neural activity or behavior may be differentially sensitive or robust to particular manipulations by pathology, compensation or treatment. Changes in certain network parameters, or the combinations of parameters, have much stronger impact on model behavior than do changes in other parameter combinations. A "sloppy" axis in parameter space is one along which the model response is relatively insensitive to perturbations in that parameter combination, whereas a "stiff" axis is one in which the model response is highly sensitive to perturbations (Gutenkunst et al. 2007).

Murray et al. (2014) and Lam et al. (2022) assessed behavioral performance in these neural circuit models while parametrically varying the E/I ratio. This analysis characterizes the sensitivity of model function to joint perturbations of two parameters. Specifically, they explored the parameter space of reductions of NMDA receptor conductances onto both inhibitory interneurons $G_{E\to I}$ (elevating E/I ratio) and onto excitatory pyramidal neurons $G_{E\to E}$ (reducing E/I ratio) (**Figure 10.13**). For the spatial working memory model, the width of a bell-shaped persistent activity pattern determines mnemonic precision and distractibility during working memory maintenance (Chapter 5, Section 5.4); therefore, circuit function is determined by the width of the mnemonic persistent activity pattern. For the decision-making model, circuit function can be measured through discrimination sensitivity (inverse of the discrimination threshold). For both circuit models for working memory and decision making, E/I ratio was

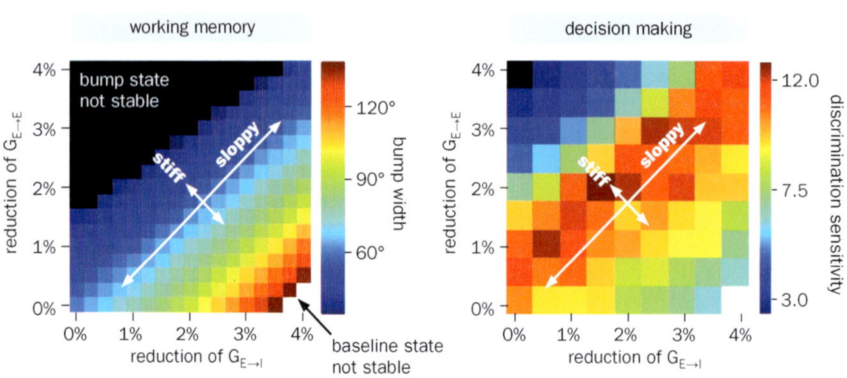

Figure 10.13 Dependence of circuit function on synaptic parameters: A critical role of excitation-inhibition (E/I) balance in both working memory and decision making. Performance of working memory and decision making as a function of two recurrent NMDAR conductance strengths from excitatory pyramidal neurons onto inhibitory interneurons ($G_{E\to I}$) and onto excitatory pyramidal neurons ($G_{E\to E}$). Reduction of $G_{E\to I}$ elevates E/I ratio, which leads to reduced precision of working memory representation and highly unstable integration, leading to impulsive decision making. In contrast, reduction of $G_{E\to E}$ lowers E/I ratio, which also degrades performance: For working memory, due to loss of mnemonic persistent activity; for decision making, due to lack of winner-take-all attractor dynamics. These findings indicate that E/I ratio is a crucial effective parameter for cognitive function in these circuits, with an inverted-U dependence on the E/I ratio along the stiff axis. Left panel is reproduced from Murray et al. (2014a) and right panel from Lam et al. (2022). [Murray, J. D., Anticevic, A., Gancsos, M., Ichinose, M., Corlett, P. R., Krystal, J. H., & Wang, X. J. (2014). Linking microcircuit dysfunction to cognitive impairment: Effects of disinhibition associated with schizophrenia in a cortical working memory model. Cerebral cortex, 24(4), 859–872, by permission of Oxford University Press.] [Lam, N. H., Borduqui, T., Hallak, J., Roque, A., Anticevic, A., Krystal, J. H., ... & Murray, J. D., & Wang, X. J. (2022). Effects of altered excitation-inhibition balance on decision making in a cortical circuit model. Journal of Neuroscience, 42(6), 1035–1053.]

found to be a key parameter for optimal network function. When $G_{E \to I}$ and $G_{E \to E}$ are varied together in a certain proportion, circuit performance is essentially unaltered, because E/I balance is maintained. Preserved E/I ratio therefore corresponds to a "sloppy" axis in this parameter space. In contrast, even moderate changes to E/I ratio along a "stiff" axis (orthogonal to the sloppy axis) have a strong impact on model function.

If E/I imbalance is either too elevated or too lowered, the circuit can lose multistability. If disinhibition is too strong (via elevated E/I ratio), then the spontaneous state is no longer stable. Conversely, if recurrent excitation is too weak (via lowered E/I ratio), then the circuit cannot support persistent activity. Collectively, these analyses reveal that E/I balance is vital for optimal cognitive performance in these cortical circuit models. This suggests that despite the complexity of synaptic alterations in a disorder such as schizophrenia, the impact on cognitive function in neural circuits may be understandable in terms of their "net effect" on effective parameters, such as E/I ratio, to which the circuit is preferentially sensitive. Moreover, synaptic inhibition plays a critical role in the generation of gamma oscillations during awake behaving conditions (Chapter 3, Section 3.4), and reduced inhibition is associated with disrupted gamma-band oscillations (Spencer et al. 2003; Vierling-Claassen et al. 2008; Volman et al. 2011; Rotaru et al. 2011), which can be measured by EEG/MEG from psychiatric patients (Uhlhaas and Singer 2010).

To address increasingly complex and detailed questions about neural circuit dysfunction, future models will need to incorporate additional knowledge from neurobiology. Notably, there are key differences between PV-expressing and SST-expressing interneurons, which differ in their synaptic connectivity and functional responses (Gonzalez-Burgos and Lewis 2008; Jiang et al. 2015; Urban-Ciecko and Barth 2016; Munoz et al. 2017; Campagnola et al. 2022). Both PV- and SST-mediated inhibitions are reduced in schizophrenia (Smucny et al. 2022). While PV cells have drawn much attention in schizophrenia research (Lewis et al. 2005), inhibition of pyramidal dendrite mediated by SST cells deserves more attention because it may play a central role in resisting distractibility (Chapter 5, Section 5.6.5 and Chapter 9, Section 9.5.3). Microcircuit models that incorporate a division of labor among interneuron classes (Wang et al. 2004; Yang et al. 2016b; Litwin-Kumar et al. 2016; Lee et al. 2017; Garcia Del Molino et al. 2017; Hertäg and Sprekeler 2019; Palmigiano et al. 2020) and large-scale cortex models (Froudist-Walsh et al. 2021) have the potential to make dissociable predictions for dysfunction in distinct cell types.

In an experiment where monkeys performed a decision task that required integration of evidence, infusion of the NMDA receptor antagonist ketamine led to behavioral impairments consistent with a reduction of recurrent excitation and E/I balance (Cavanagh et al. 2020). Another study adapted the working memory model of Compte et al. (2000) to the AX-CPT task, and found that selective impairment of proactive control in the BX trials could be reproduced with a reduced E/I ratio (Calvin and Redish 2021). These results appear to be at odds with the idea that ketamine increases the E/I ratio. In a review of attractor models of schizophrenia, Stein et al. (2021) surveyed the literature on whether schizophrenia is associated with an increase (less NMDA signaling in inhibitory neurons) or decrease (less NMDA signaling in excitatory neurons) of the E/I ratio in a working memory circuit. They concluded that currently available data are insufficient, and future examination of neural correlations would help to draw a firm conclusion.

These modeling studies offer insights into the design of cognitive tasks applied to computational psychiatry. In each model, multiple distinct cortical disruptions (e.g., elevated vs. lowered E/I ratio vs. upstream sensory coding deficit) can impair performance. Standard performance analyses in common task paradigms (e.g., error rate in working memory, or psychometric threshold in decision making) may be insufficient to resolve dissociable predictions. Fine-grained analyses of task behavior should distinguish different types of errors or deficits, rather than simply measuring overall impaired performance, which could be due to deficits in distinct cognitive sub-processes (e.g., encoding vs. maintenance for working memory) or opposing deficits in a single sub-process (e.g., leaky vs. unstable integration in decision making). Circuit modeling can provide insight into the variety of potential "failure modes" in a cognitive function, and suggest which task designs can reveal them. In turn, alignment of a task design with a circuit model allows for generation of mechanistic neurophysiological hypotheses that would not be possible with behavior-level mathematical modeling.

This perspective is applicable to other psychiatric disorders (Foss-Feig et al. 2017). Experiments have shown that the E/I ratio is increased in the cortex of rodent and nonhuman primate models of autism (Rubenstein and Merzenich 2003; Yizhar et al. 2011; Antoine et al. 2019; Zhou et al. 2019). The model explains how this biological abnormality yields impaired discrimination threshold in patients with autism spectrum disorder observed in the performance of the RDM direction discrimination task (Milne et al. 2002; Milne et al. 2009; Koldewyn et al. 2010). OCD subjects also show decision deficit in performing the RDM direction discrimination task, in its own peculiar way: compared to control subjects, their performance is intact, but reaction times are longer only when the coherence is low (Banca et al. 2015). Applying the decision-circuit model (Wong and Wang 2006), one can show that this deficit can result from a different kind of impaired E/I balance. Recall that a top-down E/I balanced input slows down ramping neural activity in a decision circuit (Chapter 6, **Figure 6.27**). Hence, abnormal executive control in OCD is hypothesized to involve an excessive top-down E/I balanced input into a perceptual decision circuit, leading to longer reaction times, especially when decision is difficult (indecisiveness) associated with OCD.

Furthermore, this computational framework serves as a starting point to probe implications of abnormal neuromodulation in psychiatric illness (Lisman et al. 2008), when molecular targets of a neuromodulator are explicitly described in a cross-level neural circuit model. Consider the serotonin (5-HT) signaling mediated by 5-HT1A, 5-HT2A and other receptor subtypes (Jacobs and Azmitia 1992; Carhart-Harris and Nutt 2017; Liu et al. 2020). In the frontal cortex, the activation of 5-HT1A receptors hyperpolarizes pyramidal neurons; by contrast, activation of 5-HT2A receptors results in neuronal increased excitability of pyramidal cells and excitatory postsynaptic currents (Celada et al. 2013). The attractor neural circuit model of working memory and decision making has been used to elucidate how 5-HT deficits affect cognitive functions (Cano-Colino et al. 2014; Wong-Lin et al. 2017). Maia and Cano-Colino (2015) simulated the effects of 5-HT1A and 5-HT2A on the model behavior and found that reduced 5-HT levels lead to abnormality observed in OCD. Specifically, a decrease in 5-HT levels makes the network more stable and difficult to get out of an attractor state (a possible neural substrate for obsession and compulsion). Simulating the effects of a 5-HT2A blocker (mimicking the partial effects of second-generation antipsychotics) or a 5-HT1A agonist was effective in reducing the OCD state, through decreasing the overall network's excitability.

10.6 DEFICITS IN MULTI-REGIONAL BRAIN SYSTEMS

So far our discussions have been largely limited to local circuit considerations; when discussing executive control, top-down signaling is treated as an "external input" to a local area under consideration. By contrast, imaging studies reveal deficient inter-areal interactions of a complex brain system in mental illness. For instance, functional connections between the anterior and posterior cortical areas as well as between the thalamus and cortex are altered in schizophrenia compared to control (Volkow et al. 1988; Bassett et al. 2008; Anticevic et al. 2014; John et al. 2018; Dong et al. 2018). These observations have led to the "dysconnectivity hypothesis" in schizophrenia (Stephan et al. 2009), and network theory has been used to quantify abnormal brain connectivity in psychiatric disorders (Fornito et al. 2015; Braun et al. 2018). Dysconnectivity denotes an atypical connectome, which can be higher or lower than the control, in contrast to disconnectivity referring to weakening or absence of connections. Newly developed multi-regional brain circuit modeling (Chapter 9) provides a promising computational platform to investigate dysconnectivity associated with mental disorders.

10.6.1 Abnormal Default-Mode Network

Research that explicitly combines behavior, brain imaging and computational modeling has begun to identify circuit mechanisms of large-scale dysconnectivity associated with schizophrenia. One study (Anticevic et al. 2012) focused on the anticorrelation between two cortical subnetworks: one is called default-mode network (DMN) of areas showing a higher activity during "resting state" (when subjects are not performing specific tasks) than during behavioral performance; the other network includes "task-activated" areas, which increase activity when the brain is cognitively engaged with the external world (**Figure 10.14**) (Raichle et al. 2001; Fox et al. 2005; Buckner 2012; Buckner and DiNicola 2019; Frankland and Greene 2020). Healthy volunteers were either "at rest" or performed a demanding working memory task while brain activities were monitored using fMRI. It was found that with ketamine administration compared to control, activation of the fronto-parietal network was attenuated. In addition, there was less deactivation of the DMN during working memory task, leading to a reduced anticorrelation between the two systems. In a large-scale human cortex model, an inter-areal connectivity matrix was based on the diffusion-weighted structural data (Hagmann et al. 2008), local areas were modeled as in Wong and Wang (2006), and inter-areal connections targeted both excitatory neurons and inhibitory neurons (**Figure 10.15a**). The diminished anticorrelation between the DMN and the task-activated network was reproduced by assuming that NMDA receptors play an important role in mediating long-distance inter-areal projections, and

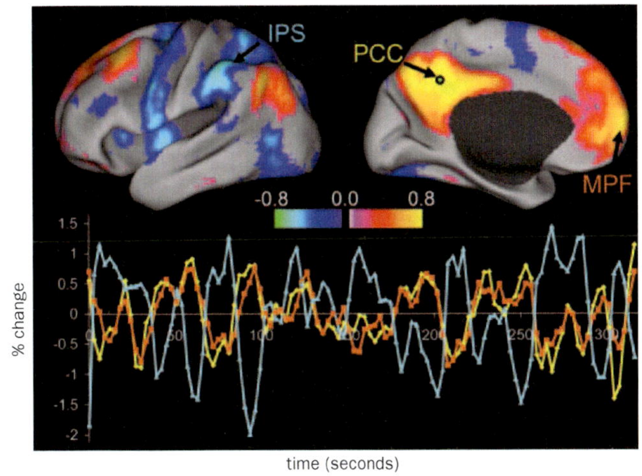

Figure 10.14 Anti-correlation of default-mode network (DMN) and task-activated network. Resting state functional connectivity reveals correlations and anticorrelations with the default-mode network. Top: Correlations between a seed region in the posterior cingulate/precuneus (PCC) and all other voxels in the brain for a single subject during resting fixation. Both correlations (positive values, warm colors) and anticorrelations (negative values, cold colors) are shown, thresholded at r = 0.3. Bottom: The time course for a single run is shown for the seed region (PCC, yellow), a region positively correlated with this seed region in the medial prefrontal cortex (MPF, orange), and a region negatively correlated with the seed region in the intraparietal sulcus (IPS, blue). Reproduced from Fox et al. (2005). [Fox, M. D., Snyder, A. Z., Vincent, J. L., Corbetta, M., Van Essen, D. C., & Raichle, M. E. (2005). The human brain is intrinsically organized into dynamic, anticorrelated functional networks. Proceedings of the National Academy of Sciences, 102(27), 9673–9678. Copyright (2005) National Academy of Sciences, U.S.A.]

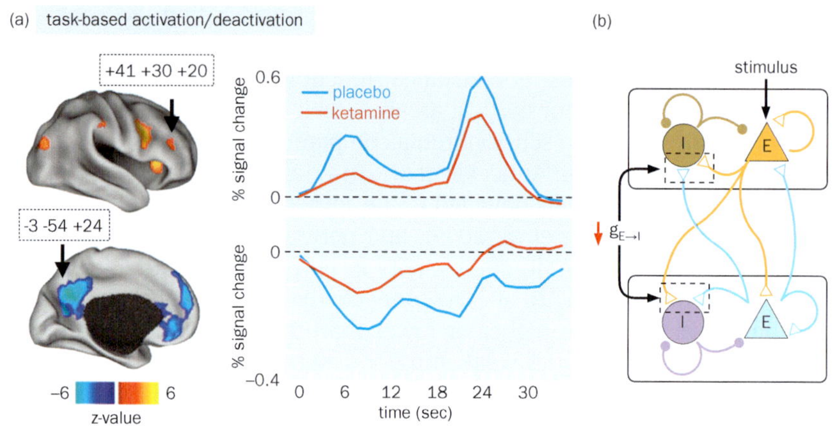

Figure 10.15 Anticorrelation of default-mode network (DMN) and task-activated network with ketamine. (a) Left column: During working memory, BOLD signal is enhanced in task-activated areas (Upper) and reduced in default-mode areas (Lower). Right column: Working memory time courses are shown for dorsolateral prefrontal cortex (part of task-activated network) and precuneus (part of DMN) for ketamine (red) and placebo (blue) (coordinates are shown in boxes in the left column). Less negative value for DMN under ketamine reflects less deactivation relative to baseline compared with the placebo condition. (b) Computational model scheme, comprised of task-activated (Upper) and task-deactivated (Lower) modules. The effects of ketamine are simulated by a reduction of NMDA conductance onto inhibitory interneurons. In this model, "disinhibition" via reduced NMDA conductance onto GABA cells results in effects similar to experimental findings using BOLD measurements under ketamine. Reproduced from Anticevic et al. (2012). [Anticevic, A., Gancsos, M., Murray, J. D., Repovs, G., Driesen, N. R., Ennis, D. J., ... & Corlett, P. R., & Wang, X.J. (2012). NMDA receptor function in large-scale anticorrelated neural systems with implications for cognition and schizophrenia. Proceedings of the National Academy of Sciences, 109(41), 16720–16725.]

their receptors on interneurons have a high affinity for ketamine. Thus, under ketamine mutual inhibition between the two macroscopic subnetworks is weakened (**Figure 10.15b**). Notably, in schizophrenic patients, the degree of cognitive deficits varied parametrically with the amount of attenuation of the DMN deactivation during working memory, suggesting that impaired task-induced DMN deactivation could obstruct goal-directed cognition.

10.6.2 Altered Macroscopic Gradients

There is a global-versus-local dilemma regarding psychiatric pathologies. Dysconnectivity is not uniform over all cortical regions and may result from local pathologies. For instance, analysis of postmortem brain tissues found that the spine density per layer-3 pyramidal neurons is reduced by 20% in the dorsolateral prefrontal cortex (area 46) of schizophrenic subjects compared to normal subjects, but it is not significantly different in V1 between the two groups of individuals (Glantz and Lewis 2000). Through interactions of the PFC with the rest of the brain, local pathologies in the PFC could produce multi-regional dysconnectivity.

On the other hand, a drug acts on its target(s) in all the brain regions. Consider the hypothesis that low-dose ketamine preferentially blocks NR2C-containing subtypes of NMDA receptors in interneurons (Khlestova et al. 2016) throughout the brain. How can localization of malfunction (in cognitive areas like PFC) be explained by a global deficit? Here, the recent finding that neurobiological properties vary quantitatively across cortical areas in the form of systematic macroscopic gradients (Wang 2020) offers a clue. The idea is that because of macroscopic gradients, a general pathology could cause notable abnormalities differentially in some areas but not in others. This hypothesis was tested in a study that combined fMRI with large-scale modeling of the human cortex (Yang et al. 2016a). In the model, parcellated cortical areas were divided into association areas and sensory areas, and inter-areal structural connections were based on diffusion tractography in humans (Hagmann et al. 2008). Moreover, recurrent excitation strength was simulated by a scaling parameter W_A for association areas and W_S for sensory areas (**Figure 10.16a**). The heterogeneity in synaptic excitation was compensated by adjusting local inhibitions strength in each area (Deco et al. 2014), so that an E/I balance was achieved, and the activity level in the spontaneous state was roughly the same (at ~ 3 Hz) in all areas. Functional connectivity between a pair of areas was defined by the covariance of their activities, and "within network connectivity" was computed by the average of functional connectivities between association areas, or between sensory areas, separately.

Using this model, Yang et al. (2016a) simulated the effect of low-dose ketamine by a reduction of excitatory drive to inhibitory neurons. When there is no heterogeneity ($W_A = W_S$), weakened inhibition results in increased functional connectivity similarly for the sensory network and association network. By contrast, under the assumption of a higher recurrent excitation in association areas than sensory areas ($W_A > W_S$), reduced E-to-I coupling produced an increase of the functional connectivity in the association network, while that in the sensory network remains essentially "normal" without noticeable change (**Figure 10.16b**). In support of this modeling result, fMRI measurements of the resting state revealed a differential increase of functional connectivity in the association areas of schizophrenic patients compared to healthy individuals, but no statistically significant difference for sensory areas between the two subject groups (**Figure 10.16c**).

This model can be expanded by incorporating more quantitatively macroscopic gradients of synaptic excitation and inhibition (Burt et al. 2018). Moreover, there is evidence from postmortem genomic analysis that macroscopic gradient of glutamatergic signaling is blunted while that of GABAergic signaling is accentuated, in schizophrenia (**Figure 10.17**) (Hoftman et al. 2018). These findings suggest that abnormalities of macroscopic gradients represent potential biomarkers for large-scale brain organization in mental illness.

Macroscopic gradients can be defined anatomically or functionally. Margulies et al. (2016) applied a method called diffusion map for analyzing structural data in high dimensions (Coifman et al. 2005) to functional

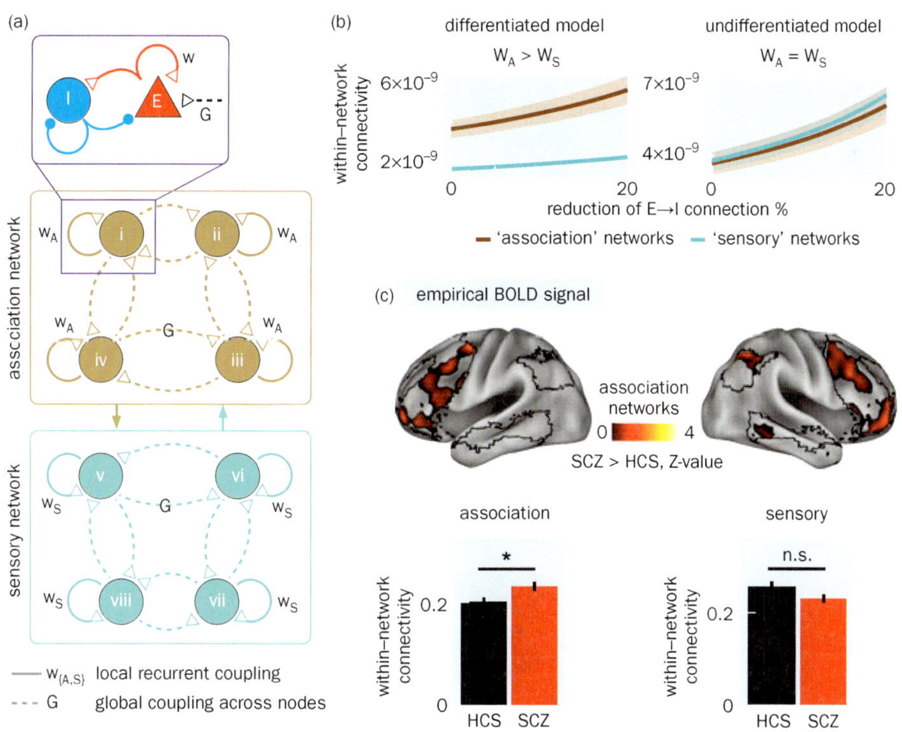

Figure 10.16 Preferential functional connectivity changes in schizophrenia. (a) Schematic of a biophysically based cortical network model with a difference in recurrent excitation for sensory (cyan) and associative (brown) areas. (b) The preferential functional effect of reducing feedback inhibition on the association versus sensory areas in the model when the networks differed in recurrent excitation (i.e., differentiated model on the left). For the undifferentiated model on the right, there was no preferential effect. (c) A fully data-driven functional connectivity analysis in SCZ, illustrating a preferential hotspot emergence largely across association cortical areas, but not sensory areas. BOLD, blood oxygen-level dependent; G, global connectivity scaling parameter; SCZ, schizophrenia; HCS, healthy comparison subjects. Adapted from Yang et al. (2016a). [Yang, G. J., Murray, J. D., Wang, X. J., Glahn, D. C., Pearlson, G. D., Repovs, G., ... & Anticevic, A. (2016). Functional hierarchy underlies preferential connectivity disturbances in schizophrenia. Proceedings of the National Academy of Sciences, 113(2), E219–E228.]

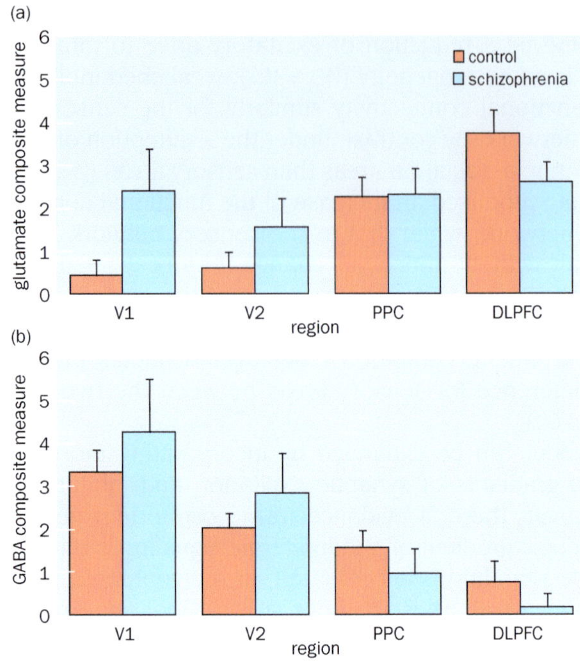

Figure 10.17 Altered macroscopic gradients of glutamatergic signaling and GABAergic signaling in schizophrenia. Effect of schizophrenia on composite measure of glutamate and GABA transcripts, respectively, in four areas of the visuospatial working memory network. (a) In unaffected comparison subjects, the composite measure of glutamate signaling shows a caudo-to-rostral increasing gradient, which is lost in schizophrenic patients. (b) By contrast, the composite measure of GABA signaling shows a decreasing gradient, which becomes steeper with schizophrenia. Reproduced from Hoftman et al. (2018). [Hoftman, G. D., Dienel, S. J., Bazmi, H. H., Zhang, Y., Chen, K., & Lewis, D. A. (2018). Altered gradients of glutamate and gamma-aminobutyric acid transcripts in the cortical visuospatial working memory network in schizophrenia. Biological Psychiatry, 83(8), 670–679.]

connectivity of the human cortex, measured by fMRI in the resting state. Functional connectivity was found to be decomposed into two dominant gradients: the first gradient roughly corresponds to cortical hierarchy from early sensory areas to transmodal association areas in the DMN, and the second gradient distinguishes different sensory modalities. The same analysis of more than 100 individuals with autism spectrum disorder revealed that compared to healthy subjects, the autistic cortex exhibits a compression of the principal gradient (**Figure 10.18**, right compared to left plots). Moreover, symptom severity in social cognition and repetitive actions was indexed by the Autism Diagnostic Observation Schedule (ADOS) (Lord et al. 2000) for autistic subjects.

A classifier was trained using supervised learning to predict ADOS from functional connectivity, yielding a significant correlation between the behavioral impairments measured by ADOS and large-scale cortical functional connectivity assessed by macroscopic gradients. The observed shifts

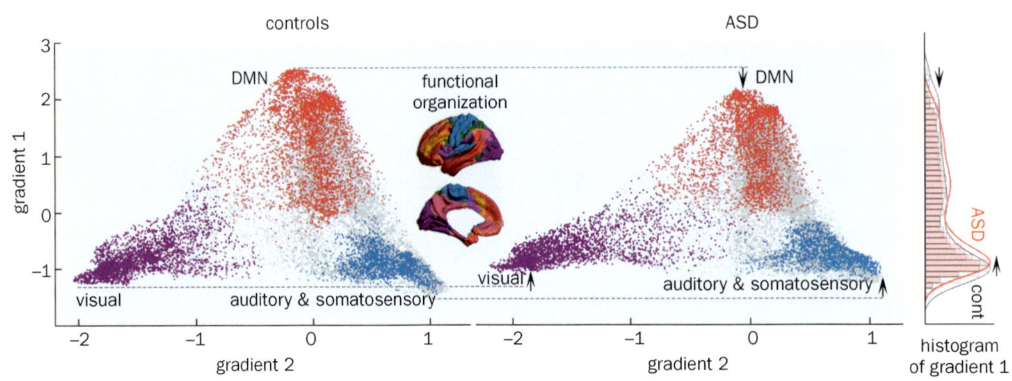

Figure 10.18 Atypical functional gradient in autism spectrum disorder (ASD). Scatterplot of the first two connectivity-embedding gradients in controls and ASD. Gradient 1 (y-axis) runs from primary sensorimotor (purple and dark turquoise) to transmodal default-mode network (DMN) (sienna). Gradient 2 (x-axis) separates somatomotor and auditory cortex from visual cortex. Right plot: Histogram of the principal gradient is shrunk (arrows) in autistic subjects (red) compared to controls (grey). Reproduced from Hong et al. (2019). [Hong, S. J., Vos de Wael, R., Bethlehem, R. A., Lariviere, S., Paquola, C., Valk, S. L., ... & Bernhardt, B. C. (2019). Atypical functional connectome hierarchy in autism. Nature Communications, 10(1), 1–13. Licensed under CC BY 4.0. / modified from original.]

of sensory and transmodal association areas to the middle of the hierarchy are interpreted functionally to cause reduced segregation between stimulus-driven information processing and internally driven cognitive processes, leading to behavioral deficits associated with autism spectrum disorder. Relating gradients of brain-wide gene expressions with brain connectivity represents a new way to bridge the genetic and functional levels (Fornito et al. 2019).

10.6.3 Deficits in Top-Down Signaling

Impairments may be differentially pronounced in certain projection pathways. As mentioned in Subsection 10.4.2, disorders ranging from anxiety to OCD to addiction all implicate abnormal executive control via top-down signaling. This is also the viewpoint taken in experimental and computational research on psychosis. Consider hallucinations, such as hearing voices, which appear to be sensations but actually only exist within the mind. Hallucination is a common symptom of schizophrenia, but healthy people also "see" or "hear" things that are not real (Honig et al. 1998). The brain constantly negotiates between internal signaling and external stimulation, as Taine (1875) observed: "External perception is an internal dream which proves to be in harmony with external things; and instead of calling hallucination a false external perception, we must call external perception a true hallucination." Neuroscientists have long postulated that hallucination (false percepts) and delusion (false beliefs) result from a deficit of prediction signaling (Frith 2005; Corlett et al. 2010; Schmack et al. 2021).

In an experiment (Powers et al. 2017), human subjects were asked to detect a 1-kHz auditory tone in white noise that occurred concurrently with a visual stimulus (a checkerboard picture). Through experience, subjects built an association between the visual and auditory stimuli and came to expect their co-occurrence. By changing the intensity of the auditory stimulus, the tone detection threshold was determined. In an initial block of training trials, the tone was presented at threshold, producing an expectation of audiovisual association. In subsequent blocks of test trials, the probability of absence of tone stimulus was gradually increased, and a subject's subjective detection threshold was measured. Participants included four groups: people with a diagnosed psychotic illness who heard voices, those with a similar illness who did not hear voices, a control group who daily heard voices but had no diagnosed illness, and controls without diagnosis or hearing voices. It was found that the probability of reporting "yes" in no-tone test trials was substantially higher in the two groups of subjects who experience hallucination in daily life than in the other two groups, regardless of the diagnosis. Presumably, conditioned hallucination is caused by weighting prior experience about audiovisual association formed in training trials too heavily so that it overrides sensory evidence (absence of a tone) in test trials. This interpretation was supported by a computational model where a system representing "perceptual beliefs" sends top-down prior signaling to influence detection in the sensory circuit. fMRI showed that conditioned hallucination engages multiple brain regions, including the anterior insula cortex, inferior frontal gyrus and superior temporal sulcus, which are involved in subjective feelings (Craig and Craig 2009), control of response (Aron 2007) and social cognition (Beauchamp et al. 2004; Hein and Knight 2008), respectively. Hallucination is also related to dopamine signaling in the striatum and cortex (Laruelle and Abi-Dargham 1999; Horga and Abi-Dargham 2019; Schmack et al. 2021); how the cortex and basal ganglia collectively produce hallucination in healthy and diseased subjects remains to be elucidated in the future.

Corlett et al. (2019) proposed a schematic circuit mechanism to explain how strong priors could produce hallucinations. Akin to **Figure 9.24** of Chapter 9, sensory input X is computed in pyramidal cells of the superficial layer, posited to be mediated by fast AMPA glutamate receptors on feedforward synapses. Prior (*Y*) is represented by a top-down input that encodes expectation and targets pyramidal cells in the deep layer. The latter in turn sends that information to inhibitory neurons in the superficial layer (*Y* is converted to $-Y$), which is compared with the sensory input *X* to yield a prediction error signal $X-Y$. Top-down prior signaling likely depends on the slow NMDA receptors as well as excitation/inhibition balance, both of which are impaired in schizophrenia, leading to an abnormally high propensity for hallucinations.

Physiologically, the discrepancy between top-down expectation with bottom-up experience is detectable as a mismatch negativity (MMN) signal in EEG or MEG (Garrido et al. 2007; Garrido et al. 2009). The amplitude of MMN is reduced in schizophrenic patients (van der Stelt and Belger 2007), which correlates with impaired NMDA signaling (Javitt et al. 1996; Javitt and Sweet 2015). Subanesthetic administration of ketamine results in reduced MMN measured in both humans and monkeys (Gil-da Costa et al. 2013). Given the robust signal and its alternation by ketamine, MMN represents a well-defined phenomenon that should be amenable for mechanistic studies using experiments (Heilbron and Chait 2018; Ross and Hamm 2020; Parras et al. 2021) and neural circuit modeling.

10.7 BIG DATA AND MODEL-AIDED DIAGNOSIS

Can modeling be directly deployed to aid diagnosis? Typically, the process of building from a behavioral experiment to a computational model follows several steps: (1) a cognitive task is strategically designed to probe a particular function (e.g., reward-related learning in decision making); (2) an appropriate computational model (e.g., reinforcement learning) is chosen to simulate the process (e.g., valuation) under consideration; and (3) model-fitting of data yields estimation of model parameters (e.g., reward sensitivity and learning rate). Many of these studies compare people deemed to be free of a psychiatric diagnosis to people who have been recruited for the presence of a specific psychiatric diagnosis according to DSM or international classification of diseases (ICD) criteria (Clark et al. 2017). Significant differences between the healthy group and patient group in some model parameters (e.g., reward sensitivity) provide the basis for characterizing the presumed "abnormality" in the patient group.

As discussed in Section 10.2, current classification of mental disorders for diagnosis lacks a solid biological basis. It is thus desirable to extend this approach to considerations where people are not separated as control versus patient populations *a priori*, and model fitting of experimental data is carried out on single subjects (Wiecki et al. 2015). This requires a fourth step, i.e., to use statistical analysis algorithms to investigate whether model parameter values extracted from single subjects are clustered into distinct clusters or groups. This step is crucial for the data-driven approach to potentially serve as a clinical tool, since diagnosis must obviously be done on an individual basis (**Figure 10.19a**).

The framework is conceptually general and applicable to a wide range of task paradigms and brain measurements. Based on the idea that the analysis of functional connectivity patterns probes pathoconnectomics associated with mental disorders (Rubinov and Bullmore 2013; Fornito et al. 2015), Brodersen et al. (2014) proposed a cross-disciplinary approach that combines behavior, brain measures (fMRI) and dynamical causal modeling (Friston et al. 2003; Stephan et al. 2007). In that study, the aforementioned four steps were followed by two more steps: (5) to assess whether clusters of subjects obtained by model-fitting are correlated with different severity

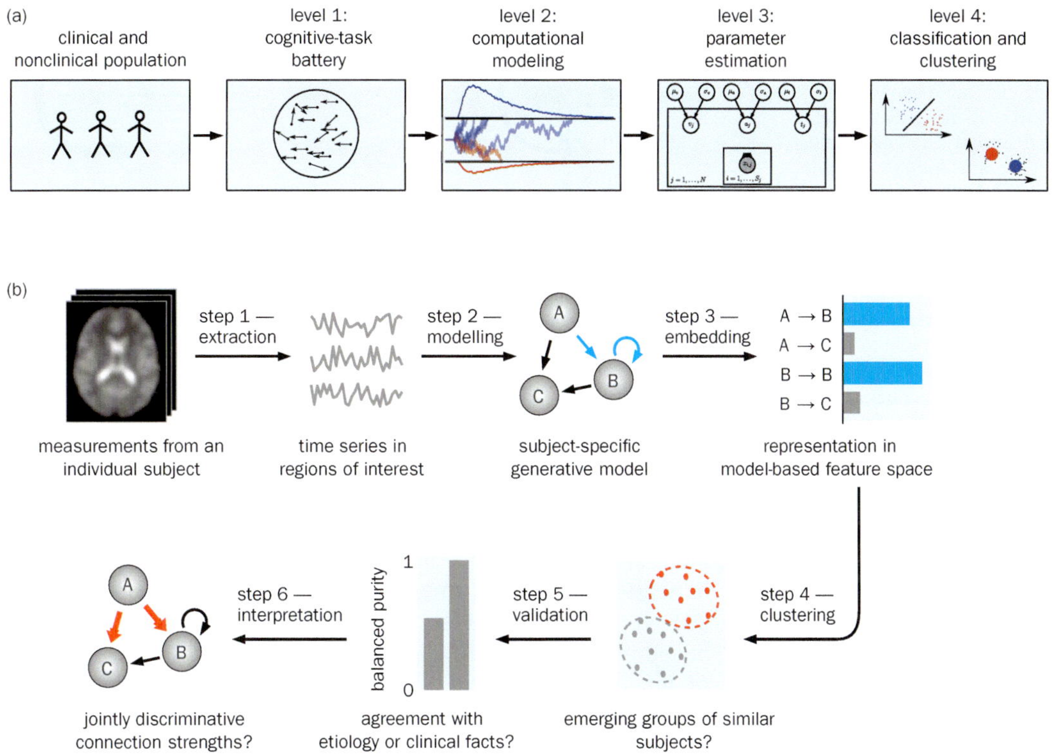

Figure 10.19 Model-aided diagnosis. (a) Methodology in four steps. First, subjects not pre-classified into clinical and nonclinical populations are tested on a battery of cognitive tasks. Second, computational models relate task performance to psychological and/or neurocognitive processes. Third, model parameters are estimated. Fourth, model training using machine learning algorithms can either uncover groups and subgroups in clinical and healthy populations or parametrically relate model parameters to clinical symptom severity. (b) Model-aided clustering of fMRI data. First, BOLD signals are extracted from a number of regions of interest. Second, subject-specific time series are used to estimate the parameters of a model. Third, subjects are embedded in a score space in which each dimension represents a specific model parameter. This space implies a similarity metric under which any two subjects can be compared. Fourth, a clustering algorithm is applied to the data. Fifth, the resulting clusters are validated against clinical variables. Sixth, a clustering solution is interpreted in terms of the underlying model circuit. Panel (a) is reproduced from Wiecki et al. (2015), (b) from Brodersen et al. (2014). [Wiecki, T. V., Poland, J., & Frank, M. J. (2015). Model-based cognitive neuroscience approaches to computational psychiatry: Clustering and classification. Clinical Psychological Science, 3(3), 378–399.] [Brodersen, K.H., Deserno, L., Schlagenhauf, F., Lin, Z., Penny, W.D., Buhmann, J.M., and Stephan, K.E. (2014). Dissecting psychiatric spectrum disorders by generative embedding. Neuroimage Clin 4, 98–111. Licensed under CC BY 4.0.]

of behavioral impairment, and (6) from the results from step 5 to uncover a possible brain substrate (visual-parietal-prefrontal circuitry connectivity) underlying quantified behavioral deficits (**Figure 10.19b**). In a working memory study of schizophrenic patients, they focused on estimates of effective connectivity among visual, parietal and prefrontal cortex, since these cortical areas are critically involved in visual working memory. An unsupervised clustering procedure operating on the individual connectivity patterns yielded three distinct patient subgroups (**Figure 10.20a**): those with greater fronto-parietal connectivity, those with weaker fronto-parietal connectivity and those with greater visuo-frontal connectivity. In step 5, they found that subjects in the three clusters displayed a different degree of negative symptom severity (**Figure 10.20b**).

In another study with 700 subjects, data analysis of fMRI resting-state connectivity in the limbic and fronto-striatal systems revealed four clusters, which were used as biomarkers to define subtypes of depression and anxiety (Drysdale et al. 2017; Tozzi et al. 2024). This line of work raises the possibility of using brain imaging data and computational modeling for data-driven diagnostic classification of mental disorders and identification of their brain circuit mechanisms (Clementz et al. 2016).

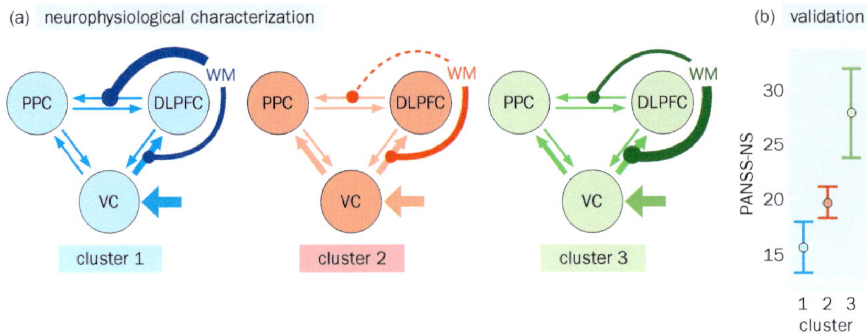

Figure 10.20 Model-based clustering of fMRI data from schizophrenic patients in a working memory (WM) task. (a) An unsupervised clustering analysis of the patient group only, using dynamical causal model parameter estimates, yield the parameter estimates for each coupling and input parameter in the model. This is displayed graphically by the thickness of the respective arrows. PPC: Posterior parietal cortex, DLPFC: Dorsolateral prefrontal cortex, VC: Visual cortex. (b) The three subgroups defined on the basis of connection strengths differ in terms of negative clinical symptoms (ability losses such as social withdrawal and lack of initiative) as operationalized by the negative symptoms (NS) subscale of the Positive and Negative Syndrome Scale (PANSS) score (Kay et al. 1987). Reproduced from Brodersen et al. (2014). [Brodersen, K. H., Deserno, L., Schlagenhauf, F., Lin, Z., Penny, W. D., Buhmann, J. M., & Stephan, K. E. (2014). Dissecting psychiatric spectrum disorders by generative embedding. NeuroImage: Clinical, 4, 98–111. Licensed under CC BY 4.0.]

This line of work raises the possibility of using brain imaging data and computational modeling for data-driven diagnostic classification of mental disorders and identification of their brain circuit mechanisms (Clementz et al. 2016). In this endeavor, it is necessary to quantify performance and brain measurement data from a large number of participants (Marek et al. 2022), which is only becoming possible now. Case studies discussed herein benefitted from fMRI imaging with more than 100 subjects (Yang et al. 2016a) or behavioral tests by nearly 2,000 individuals using a smartphone-based platform (Rutledge et al. 2017). Because of the size and complexity of data, machine learning plays an increasingly indispensable role (Bzdok and Meyer-Lindenberg 2018; Beam et al. 2021; Chen et al. 2022).

10.8 SUMMARY

In this chapter, we introduced the nascent field of computational psychiatry. A number of factors will determine the success of this framework: very large samples of subjects, judicious choices of behavioral tests, biological measurements of potential biomarkers, efficient and statistically reliable analysis methods, and innovative computational models.

In our efforts to understand the brain mechanisms of mental deficits, a future direction of research is linking genes to basic cognitive and behavioral impairments across levels. Progress in this area is crucial for providing a firm basis for a diagnostic system and mitigations for common psychiatric disorders (Krystal and State 2014). Take the familiar example of ketamine, which is an antagonist of the NMDA receptors, and NMDA dysfunction is a primary pathology of schizophrenia (Coyle et al. 2003). NMDA receptors are crucial for long-term synaptic modifications (Chapter 4), thus impaired plasticity of long-distance inter-areal connections contributes to the dysconnectivity observed in schizophrenia (Stephan et al. 2009). On the other hand, NMDA receptors are crucial for mediating recurrent synaptic excitation underlying cognitive processes such as working memory (Chapter 5), decision making (Chapters 6) and top-down signaling (Chapter 9), which is more relevant to understanding the reversible effect of a one-time ketamine infusion. Therefore, the functional consequences of deficient NMDA signaling also need to be appraised in terms of online cognitive functions in daily life (Wang 2006c), as well as synaptic plasticity resulting from experiences. Generally, consideration of a single receptor subtype, neurotransmitter or synaptic pathway is severely limited. A more sophisticated approach is needed to quantify a multi-dimensional space of abnormalities and for a cross-level understanding linking molecules to circuit dynamics and behavior.

Numerous mental disorders have a neurodevelopmental origin, and the prefrontal cortex matures much later than sensory cortical areas. In the human prefrontal cortex, the number of spines per pyramidal neuron, which are the anatomical substrates for excitatory connections on pyramidal neurons, does not reach a steady-state until 30 years old (Petanjek et al. 2011) (**Figure 10.21a**). Individuals with different psychiatric disorders exhibit distinct abnormal temporal patterns of spine numbers over a lifetime, especially during brain development in childhood and adolescence (**Figure 10.21b**). In ASD, exaggerated spine formation or incomplete pruning may occur in childhood, leading to increased spine numbers. In schizophrenia, exaggerated spine pruning during late childhood or adolescence may lead to the emergence of symptoms during these periods. In Alzheimer's disease, spines are rapidly lost in late adulthood, suggesting perturbed spine maintenance that may underlie cognitive decline.

More broadly, biological properties ranging from dopamine signaling to grey and white matters undergo temporal evolutions that are abnormal in the mentally ill brain (**Figure 10.22**). Computational modeling of brain development underlying maturation of cognitive functions and executive control is scarce (Averbeck 2022, Astle et al. 2023). A prerequisite for filling this void is the availability of relevant experimental data, for instance, by connectomic analysis at different time points during brain development (Zuo et al. 2017), or following maturation of a cognitive function like working memory or inhibitory control in children (Diamond 2013; Satterthwaite et al. 2013; Rosenberg et al. 2020). A theoretical framework for investigating the complex process of brain development over a long period of time, that is sculptured by both genetics and experience, is essential for understanding neurodevelopmental underpinnings of mental disorders.

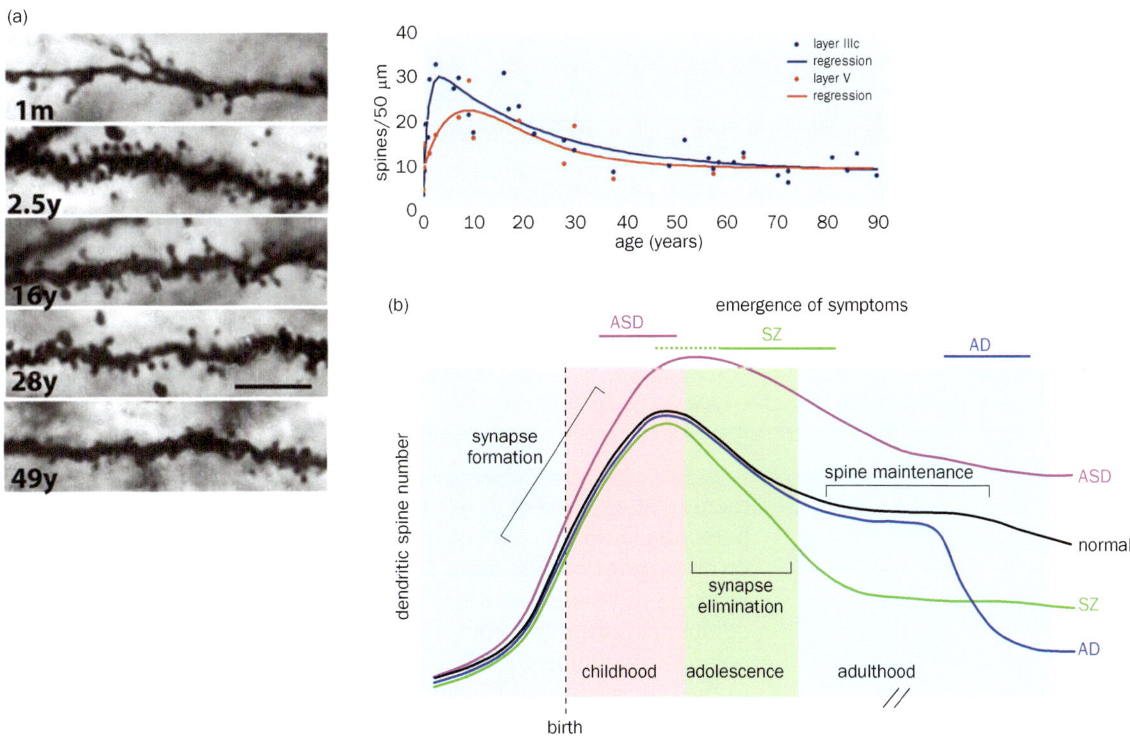

Figure 10.21 Lifetime trajectory of dendritic spine number in the brain. (a) Measured spine number for pyramidal neurons of human dorsolateral prefrontal cortex. Left panel: Example of spines on a dendritic branch from 1-month-old (top) to 49-year-old (bottom). Right panel: Spine density (number per 50 μm dendritic length) as a function of age for layer-3 (blue) and layer-5 (red) pyramidal neurons. (b) Putative trajectory in a normal subject (black), ASD (pink), schizophrenia (SZ, green) and Alzheimer's disease (AD) (blue). Bars across the top indicate the period of emergence of symptoms and diagnosis. In normal subjects, spine numbers increase before and after birth; spines are selectively eliminated during childhood and adolescence to adult levels. Panel (a) is reproduced from Petanjek et al. (2011), (b) from Penzes et al. (2011). [Petanjek, Z., Judaš, M., Šimić, G., Rašin, M. R., Uylings, H. B., Rakic, P., & Kostović, I. (2011). Extraordinary neoteny of synaptic spines in the human prefrontal cortex. Proceedings of the National Academy of Sciences, 108(32), 13281–13286.] [Penzes, P., Cahill, M. E., Jones, K. A., VanLeeuwen, J. E., & Woolfrey, K. M. (2011). Dendritic spine pathology in neuropsychiatric disorders. Nature Neuroscience, 14(3), 285–293.]

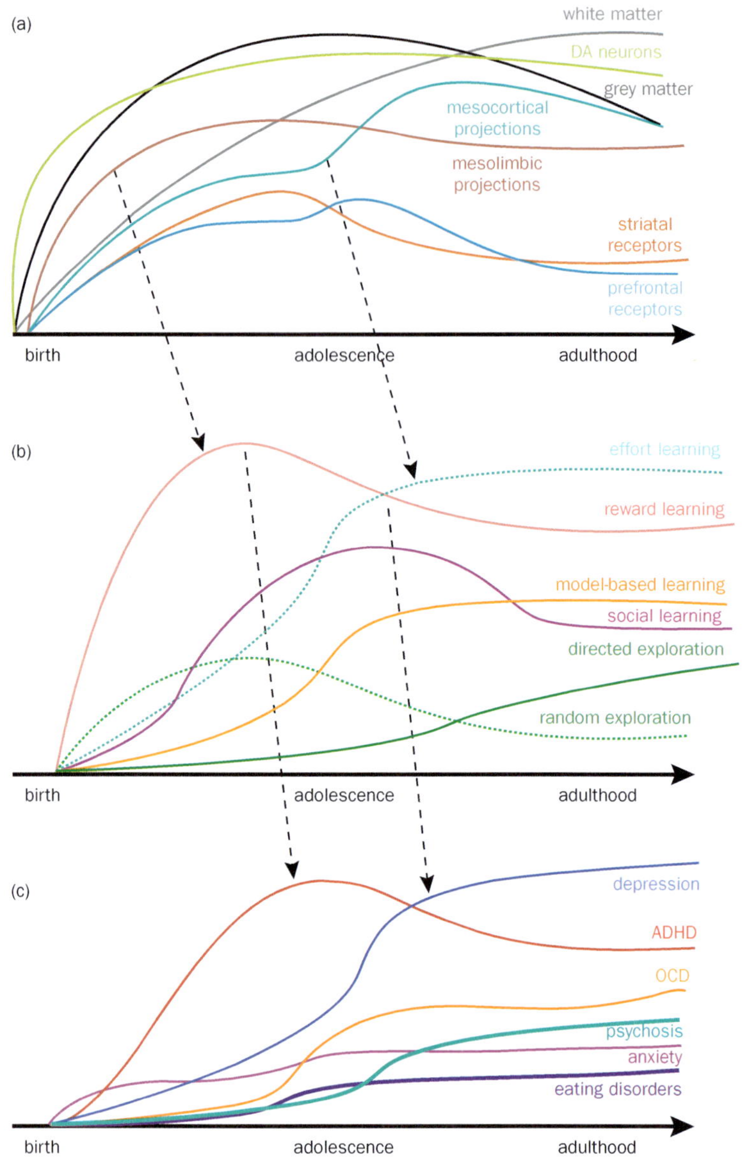

Figure 10.22 Developmental trajectories across multiple systems. A key challenge in developmental computational psychiatry is to determine how developmental trajectories across multiple systems interact. We need to understand (a) how the development of a neurotransmitter system and other biological properties (b) influences the emergence of a specific cognitive function, and (c) how an impaired unfolding of this trajectory might induce the emergence of a psychiatric disorder (even decades later in life). Different colors indicate various psychiatric disorders, but the time trajectories are not based on data and thus remain speculative. Reproduced from Hauser et al. (2019). [Hauser, T. U., Will, G. J., Dubois, M., & Dolan, R. J. (2019). Annual research review: Developmental computational psychiatry. Journal of Child Psychology and Psychiatry, 60(4), 412–426.]

Of utmost relevance to psychiatry is to offer novel therapeutic treatments (Moutoussis et al. 2018), drug discovery (Wiltschko et al. 2020) and personalized prediction of treatment outcomes (Webb et al. 2018). As a proof-of-principle example of this, Murray et al. (2014) examined in the working memory circuit model how E/I balance can be restored through compensations acting on multiple parameters; in turn, restoration of E/I balance ameliorated the associated deficits in working memory behavior in the computational model. Further development of biophysically based models will incorporate more detailed microcircuitry, and receptors will better capture pharmacological effects. Integration of biomarkers and behavior in the models will allow refinement through more direct testing with empirical data from pharmacological manipulations in animal models and humans. Cross-level neural circuit modeling has the potential to provide a method for simulating possible treatments (or compensations) at the molecular, cellular and circuit levels.

From the dimensional perspective, instead of looking for a "magic pill" for a psychiatric disorder, an alternative can be called the "cocktail strategy." Since various specific deficits (e.g., positive, negative and cognitive symptoms in schizophrenia) are caused by distinct biological abnormalities and should be treated with different medicines, a subject could be prescribed with a cocktail of pills and non-drug treatments that is designed individually. Future studies in this area hold real promise of rationally guiding treatment development in psychiatry, grounded in basic neuroscience.

What is normal versus pathological is not always clear. As asked by the French philosopher Canguilhem (1989): "Is the pathological state merely a quantitative modification of the normal state?" In other words, do psychiatric disorders represent extremes of human diversities in a continuum? There could be evolutionary advantages of being mentally out of the ordinary. As Seneca wrote: "No great genius has ever existed without some touch of madness." There is no shortage of historical personalities who were both geniuses and "madmen" in sciences (Pickover 1998), as well as in arts and poetry. Perhaps some people are biologically predisposed for extreme ends of human behavior, mental capabilities and emotional feelings.

A personal experience could lead the brain into a so-called pivotal mental state characterized by hyperplasticity in which the likelihood of major psychological change is enhanced (Brouwer and Carhart-Harris 2020). Depending on the context, the brain circuit would evolve into one of two or more possible new states, as a bifurcation (**Figure 10.23**). For example, "an experience" could be taking a drug like lysergic acid diethylamide (LSD) (Carhart-Harris et al. 2016) or psilocybin (Carhart-Harris et al. 2021) that increases serotonin 2A receptor-mediated signaling and induces hallucination. The use of the term "bifurcation" may be more than a metaphor, as Brouwer and Carhart-Harris (2020) suggested that such pivotal mental states could correspond to a near-criticality situation described by the mathematics of bifurcations for nonlinear dynamical systems (Strogatz 2016).

Up to the present time, population data appear to show unimodal statistical distributions of biological measurements, rather than multimodalities that could be readily used for categorical classifications (van Os et al. 2009; Howard et al. 2019). The "elephant in the room" is whether any distribution of such a measure will show well-defined categories or a continuum. Even if it turns out that individual variations of biological

Figure 10.23 Use of the "fork in the river analogy" to communicate the nature and effect of pivotal mental states. The boat, named "5-HT2A," represents the molecular gateway leading to the occurrence of a pivotal mental state, and the nature of that change is strongly context dependent. The fork in the road reflects a bifurcation towards a state of either exaltation (left) or desperation (right). Reproduced from Brouwer and Carhart-Harris (2020). [Brouwer, A., & Carhart-Harris, R. L. (2021). Pivotal mental states. Journal of Psychopharmacology, 35(4), 319–352. Licensed under CC BY 4.0.]

measures are merely dimensional, quantitative changes in the brain could give rise to categorically different behaviors by virtue of bifurcations. Put differently, the possibility remains that graded genetic variations, in interaction with the environmental risk factors for an individual, give rise to qualitatively different categories in the space of functional phenotypes. The missing link between molecules and behavior is the brain's neural circuit in interaction with the world, building a foundation across levels calls for computational psychiatry.

FURTHER READING

Books and Reviews

Averbeck, B. B. and M. V. Chafee (2016). Using model systems to understand errant plasticity mechanisms in psychiatric disorders. *Nat. Neurosci. 19*, 1418–1425.

Barch, D. M. and A. Ceaser (2012). Cognition in schizophrenia: Core psychological and neural mechanisms. *Trends Cogn. Sci. (Regul. Ed.) 16*, 27–34.

Fornito, A., A. Zalesky, and M. Breakspear (2015). The connectomics of brain disorders. *Nat. Rev. Neurosci. 16*, 159–172.

Insel, T. R. (2010). Rethinking schizophrenia. *Nature 468*, 187–193.

Kandel, E. R. (2005). *Psychiatry, Psychoanalysis, and the New Biology of Mind*. Washington DC: American Psychiatry Publishing, Inc.

Montague, P. R., R. J. Dolan, K. J. Friston, and P. Dayan (2012). Computational psychiatry. *Trends Cogn. Sci.* 16, 72–80.

Penzes, P., M. E. Cahill, K. A. Jones, J.-E. VanLeeuwen, and K. M. Woolfrey (2011). Dendritic spine pathology in neuropsychiatric disorders. *Nat. Neurosci. 14*, 285–293.

Redish, A. D. and J. A. Gordon (Eds.) (2016). *Computational Psychiatry: New Perspectives on Mental Illness*. Cambridge, MA: MIT Press.

Robbins, T. W., C. M. Gillan, D. G. Smith, S. de Wit, and K. D. Ersche (2012). Neurocognitive endophenotypes of impulsivity and compulsivity: Towards dimensional psychiatry. *Trends Cogn. Sci. (Regul. Ed.) 16*, 81–91.

Series, P. (Ed.) (2020). *Computational Psychiatry: A Primer*. Cambridge, MA: MIT Press.

Stephan, K. E., D. R. Bach, P. C. Fletcher, J. Flint, M. J. Frank, K. J. Friston, A. Heinz, Q. J. Huys, M. J. Owen, E. B. Binder, P. Dayan, E. C. Johnstone, A. Meyer-Lindenberg, P. R. Montague, U. Schnyder, X. J. Wang, and M. Breakspear (2016). Charting the landscape of priority problems in psychiatry, part 1: Classification and diagnosis. *Lancet. Psychiatry. 3*(1), 77–83.

Wang, X.-J. and J. H. Krystal (2014). Computational psychiatry. *Neuron 84*, 638–654.

Primary Research Papers

Bridging Basic and Clinic Research

Anticevic, A., M. Gancsos, J. D. Murray, G. Repovs, N. R. Driesen, D. J. Ennis, M. J. Niciu, P. T. Morgan, T. S. Surti, M. H. Bloch, R. Ramani, M. A. Smith, X.-J. Wang, J. H. Krystal, and P. R. Corlett (2012). NMDA receptor function in large-scale anticorrelated neural systems with implications for cognition and schizophrenia. *Proc. Natl. Acad. Sci. U.S.A. 109*, 16720–16725.

Brodersen, K. H., L. Deserno, F. Schlagenhauf, Z. Lin, W. D. Penny, J. M. Buhmann, and K. E. Stephan (2014). Dissecting psychiatric spectrum disorders by generative embedding. *Neuroimage Clin. 4*, 98–111.

Corlett, P. R., J. R. Taylor, X.-J. Wang, P. C. Fletcher, and J. H. Krystal (2010). Toward a neurobiology of delusions. *Prog. Neurobiol. 92*, 345–369.

Hoftman, G. D., S. J. Dienel, H. H. Bazmi, Y. Zhang, K. Chen, and D. A. Lewis (2018). Altered gradients of glutamate and Gamma-aminobutyric acid transcripts in the cortical visuospatial working memory network in schizophrenia. *Biol. Psychiatry*. 670–679.

Hong, S.-J., R. V. De Wael, R. A. Bethlehem, S. Lariviere, C. Paquola, S. L. Valk, M. P. Milham, A. Di Martino, D. S. Margulies, J. Smallwood, et al. (2019). Atypical functional connectome hierarchy in autism. *Nat. Comm. 10*, 1–13.

Wiecki, T. V., J. S. Poland, and M. J. Frank (2015). Model-based cognitive neuroscience approaches to computational psychiatry: Clustering and classification. *Clin. Psychol. Sci. 120*, 329–355.

Yizhar, O., L. E. Fenno, M. Prigge, F. Schneider, T. J. Davidson, D. J. Oshea, V. S. Sohal, I. Goshen, J. Finkelstein, J. T. Paz, et al. (2011). Neocortical excitation/inhibition balance in information processing and social dysfunction. *Nature 477*, 171–178.

Reinforcement Learning Models

Huys, Q. J., D. A. Pizzagalli, R. Bogdan, and P. Dayan (2013). Mapping anhedonia onto reinforcement learning: A behavioural meta-analysis. *Biol. Mood Anxiety Disord. 3*. https://doi.org/10.1186/2045-5380-3-12.

Keramati, M., A. Durand, P. Girardeau, B. Gutkin, and S. H. Ahmed (2017). Cocaine addiction as a homeostatic reinforcement learning disorder. *Psychol. Rev. 124*, 130–153.

Redish, A. D. (2004). Addiction as a computational process gone awry. *Science 306*, 1944–1947.

Neural Circuit Modeling

Durstewitz, D. and J. K. Seamans (2008). The dual-state theory of prefrontal cortex dopamine function with relevance to catechol-o-methyltransferase genotypes and schizophrenia. *Biol. Psychiatry 64*, 739–749.

Lam, N. H., T. Borduqui, J. Hallak, A. Roque, A. Anticevic, J. H. Krystal, X.-J. Wang, and J. D. Murray (2022). Effects of altered excitation-inhibition balance on decision making in a cortical circuit model. *J. Neurosci. 42*, 1035–1053.

Maia, T. V. and M. Cano-Colino (2015). The role of serotonin in orbitofrontal function and obsessive-compulsive disorder. *Clin. Psychol. Sci. 3*, 460–482.

Murray, J. D., A. Anticevic, M. Gancsos, M. Ichinose, P. R. Corlett, J. H. Krystal, and X.-J. Wang (2014). Linking microcircuit dysfunction to cognitive impairment: Effects of disinhibition associated with schizophrenia in a cortical working memory model. *Cereb. Cortex 24*, 859–872.

Ramirez-Mahaluf, J. P., A. Roxin, H. S. Mayberg, and A. Compte (2017). A Computational model of major depression: The role of glutamate dysfunction on cingulofrontal network dynamics. *Cereb. Cortex 27*, 660–679.

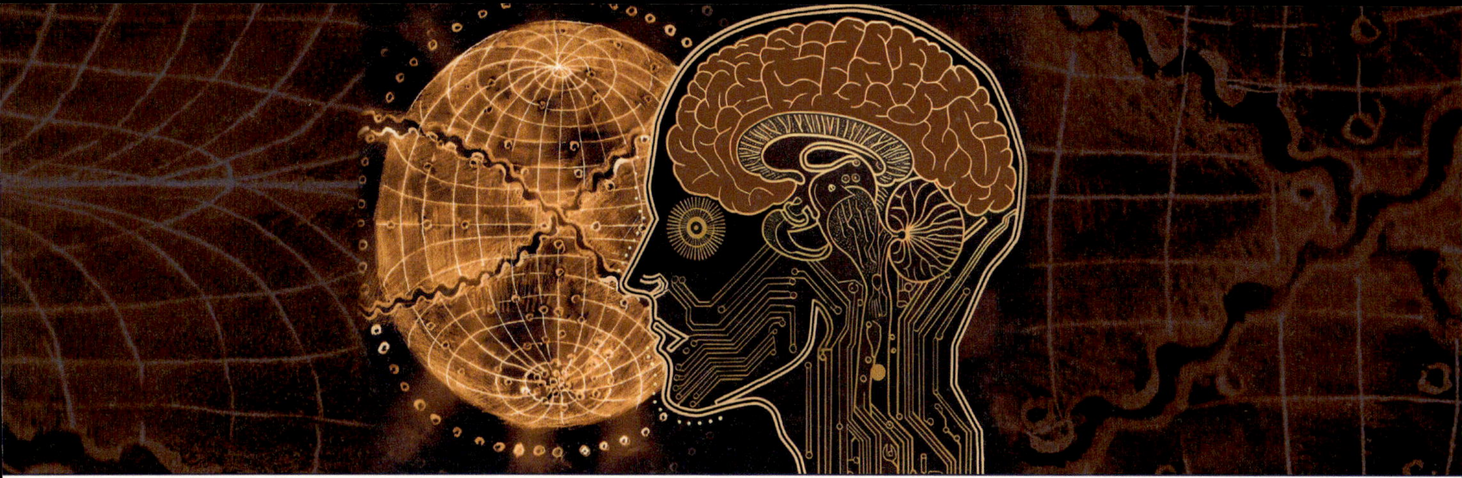

Biological and Artificial Intelligence

11

Intelligence is a very general mental capability that, among other things, involves the ability to reason, plan, solve problems, think abstractly, comprehend complex ideas, learn quickly and learn from experience. It is not merely book learning, a narrow academic skill, or test-taking smarts. Rather, it reflects a broader and deeper capability for comprehending our surroundings—"catching on," "making sense" of things, or "figuring out" what to do.

Linda A. Gottfredson

11.1 INTRODUCTION

We are living in an era where artificial intelligence (AI) pervades society, thanks to the revolutionary advances in AI in the last decade. Although machines now are superior to humans in specialized domains, according to the quoted description (Gottfredson 1997), there is still a chasm between today's smartest machines and human-level intelligence. Understanding the biological mechanism of human intelligence is considered by many to be the holy grail of neuroscience and cognitive psychology, but when asked "what is intelligence?", scientists would offer a wide range of opinions (Sternberg and Detterman 1986; Neisser et al. 1996; Legg et al. 2007; Duncan 2010a; Chollet 2019, Lee 2020, Bennett 2023). There are multiple intelligences (Gardner 2011) including emotional intelligence (Goleman 1996; Mayer et al. 1999). As a starting point, the essence of intelligence should be appraised in terms of what it is not. When we study intelligence scientifically, we are not primarily concerned with instinctive behaviors (in which a sensory stimulus automatically triggers a fixed response pattern) that are needed for survival, such as escaping from a predator or mating (Tinbergen 1951; Lorenz 1961; von Frisch 1974). Rather, our focus is to search for a scientific understanding of such higher functions as reasoning, planning and thought. In a sense, this whole book is about understanding intelligence: while cognition refers to mental processes, intelligence denotes cognitive ability of these processes as an individual's quality (Carroll 1993). In this chapter, we shall discuss how computational neuroscience, in interplay with the field

DOI: 10.1201/9781003459361-14

of artificial intelligence, tackles the challenge beyond the coverage of the previous chapters.

We shall not pretend to offer all-encompassing theories about how neural circuits achieve intelligence. Certain higher cognitive functions in humans, e.g., serial organization of a complex task like cooking a meal, have so far eluded rigorous experimentation in the laboratory. Some abilities such as language may be unique to humans (Newell 1990; Penn et al. 2008; Passingham 2008; Maclean 2016), and their brain mechanisms can be investigated using tools ranging from functional magnetic resonance imaging (fMRI) to invasive recording in patients (Quiroga et al. 2005; Dehaene 2010; Poeppel et al. 2012; Cogan et al. 2014; Cheung et al. 2016; Helfrich et al. 2018; Dichter et al. 2018; Yi et al. 2019; Minxha et al. 2020). By contrast, elemental cognition we have discussed so far can be investigated in nonhuman animal experiments, which are critical for providing clues and data to inform theory building. While keeping in mind these limitations, we will push forward our central theme of bridging the neurobiology and cognitive science, by identifying a new set of cognitive building blocks.

As a pioneer of AI argued: "The deliberate process we call reasoning is, I believe, the thinnest veneer of human thought, effective only because it is supported by this much older and much more powerful, though usually unconscious, sensorimotor knowledge" (Moravec 1988). In line with this perspective, abstraction-dependent sensorimotor tasks that do not resort to language can and have been used in experiments with human subjects and nonhuman animals. The increasing sophistication of these cognitive building blocks through biological evolution likely provides the groundwork for human thought and language. Theory can now be informed by innovative experiments such as those on the brain mechanism of abstract knowledge (Constantinescu et al. 2016; Behrens et al. 2018; Schuck and Niv 2019; Liu et al. 2019; Frankland and Greene 2020).

The field of artificial intelligence is currently confronted with a similar challenge. Deep neural networks with a purely feedforward architecture are excellent at solving input-output mapping problems, however complex they may be. Today's generative AI systems can continuously produce a string of outputs like a short text in response to a query, by virtue of using its output as new input fed into the system. When no input is provided, the whole system would become silent. In that sense, they are devoid of a "mental life". By contrast, the human brain is never silent; thinking, reasoning, planning, etc. all are in essence internal processes. Therefore, research covered in this chapter represents a frontier area at the interface with the field of artificial intelligence (Zador et al. 2023; Gleeson et al. 2023).

Contemporary studies of cognition and language have been very much influenced by the connectionism movement (Rumelhart et al. 1986; Rogers and McClelland 2014). The connectionist approach utilizes artificial neural networks as dynamical systems, sharing mathematically a kinship with the neural circuit modeling discussed in the previous chapters. Recent advances in AI rely on machine learning in deep neural networks, and thus can be viewed as neoconnectionsim (Saxe et al. 2021). Thanks to this revolution, we now have at our disposal a set of new and powerful machine learning tools that are applicable to neuroscience for data analysis and building computational models.

The exchanges between AI research and brain science are bidirectional (LeCun et al. 2015; Hassabis et al. 2017). Computer science has a long history of symbiosis with brain research (Wiener 1948; Ashby 1956; von Neumann 1958; Mindell and Mindell 2002; Pickering 2010). In the machine learning field, supervised learning rules at the core of today's AI have their roots in the theory of training perceptrons, which in turn was inspired by the brain (Rosenblatt 1962); the hierarchical architecture (Fukushima and Miyake 1982) and convolutional principle (LeCun and Bengio 1995) were closely linked to our knowledge about the primate visual system (Hubel and Wiesel 1962; Felleman and Van Essen 1991). The state-of-the-art computer vision systems share similarities with the ventral pathway of the primate visual system (**Figure 11.1**) for object perception.

Notable in **Figure 11.1** is the absence of the equivalent of the prefrontal cortex (PFC) in currently available AI systems. As is already evident in the previous chapters, the PFC and its interplay with the rest of the brain hold the key to scientific understanding of cognitive functions and executive control. In this chapter, we will discuss how, with the aid of artificial neural networks, we can develop models for higher-level cognitive functions (Yang and Wang 2020). We shall focus on abstraction in its various forms, like abstract rules that guide multi-tasking in sensorimotor behavior, concepts and reasoning. Hence, this chapter engages the interplay of the two fields from machine learning to neuroscience. Conversely, computational neuroscience of cognition should yield new ideas for the next generation of artificial intelligence, given that today's machines lack the learning abilities for abstraction and problem solving in the face of the unexpected in real-life situations. Deep neural networks need the prefrontal cortex.

11.2 DEEP FEEDFORWARD NEURAL NETWORKS

11.2.1 Basic Methods of Deep Neural Network Models

What are deep neural network models? As schematically shown in **Figure 11.1**, a multi-layer neural network has "layers" indexed by $k = 1,2,\ldots,K$, and it is called "deep" if there is more than one "hidden" layer between the system's input and output. In a feedforward architecture, there are only connections that project from one layer to higher layers in a series, but not those in the reversed direction nor local recurrent connections within a layer. The computational goal is to realize an input-output mapping $\mathbf{y} = \mathbf{F}(\mathbf{x})$. For instance, the input may be a two-dimensional visual image with many thousands of pixels, represented by a vector $\mathbf{x}$ in a very high-dimensional space. The classification task is to produce an output $\mathbf{y}$, e.g., whether the picture includes one of categorical objects like dog, cat, horse, panda, etc. In principle, such a high-dimensional function $\mathbf{F}$ can be overwhelmingly complex. Deep neural networks redefine the problem in terms of learning connection weights and using a single one-dimensional function F, which is akin to the transfer function of a neuron from an input I to an activity output $r = F(I)$.

Denote by r_i^k activity of unit i ($i = 1,2,\ldots,N_k$) in k-th layer, we have:

$$r_i^k = F\left(\sum_j J_{ij}^k r_j^{k-1} + b_i\right) \tag{11.1}$$

where J_{ij}^k is the projection matrix from unit j in layer $(k-1)$ to unit i in layer k, and b_i is a bias term (e.g., representing a background input). The choice of F can be the rectified linear unit (ReLU) $F(I) = [I]_+$ or the softplus $F(I) = \log(1+\exp(I))$ or some other simple function.

The connection weights are commonly trained by supervised learning (Chapter 4, Section 4.2), with correct answers provided for a training set of samples. This means that, for example, if the task is recognizing a dog in a picture, a training sample must have a dog (if present) labeled, so the system knows what is the correct answer in supervised learning. Those

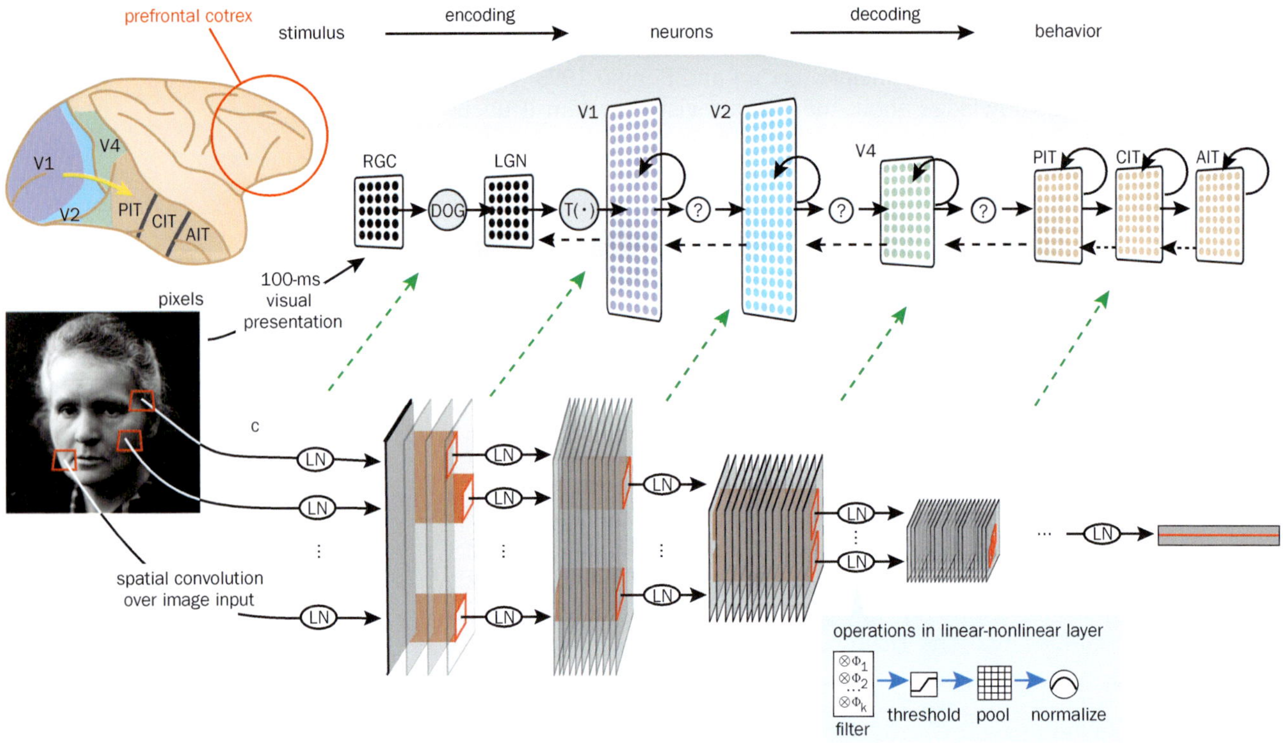

Figure 11.1 Schematic of a deep convolutional network. It roughly maps onto the ventral stream of the primate visual cortical system (V1, V2, V4; posterior, central and anterior inferotemporal cortex [PIT, CIT, AIT]) (yellow arrow), while leaving out the vast territories of the neocortex (light orange), including the prefrontal cortex (red circle and arrow) in the cortex (upper left). Reproduced from Yamins and DiCarlo (2016). [Yamins, D. L., & DiCarlo, J. J. (2016). Using goal-driven deep learning models to understand sensory cortex. Nature Neuroscience, 19(3), 356–365.]

samples may be from a dataset such as ImageNet (Deng et al. 2009) with 15 million pictures, in which categorical objects (like animals) are hand-annotated. A loss function is defined in terms of the mismatch between the output from the network's last layer $\mathbf{y}$ and the correct target output $\mathbf{y}_{\text{target}}$. The form of the cost function depends on the problem under consideration, for categorization the commonly used one is a cross entropy loss function (Goodfellow et al. 2016; Zhang and Sabuncu 2018). Here for ease of exposition, let us limit ourselves to the least mean square cost function (Chapter 4, Section 4.2). For a number M of training examples indexed by $B = \{\mu_1, \cdots, \mu_M\}$,

$$\mathcal{L}_{\text{batch}} = \frac{1}{M} \sum_{\mu \in B} \mathcal{L}(y^{\mu}, y^{\mu}_{\text{target}}), \tag{11.2}$$

where $\mathcal{L}\left(\mathbf{y}^{\mu}, \mathbf{y}^{\mu}_{\text{target}}\right)$ is typically a squared difference between $\mathbf{y}^{\mu}$ and $\mathbf{y}^{\mu}_{\text{target}}$. The gradient descent algorithm (Chapter 4, Eq. 11.4) is used to update connection weights for projections from one layer to another, as well as other tunable parameters such as bias current $\mathbf{b}$. Samples are often chosen randomly from a subset of available training examples, hence the term "stochastic gradient descent" (SGD).

In order to train a multi-layer network by supervised learning, the gradient descent algorithm uses backpropagation (Rumelhart et al. 1986; LeCun 1988). To compute the loss $\mathcal{L}$, the network is run in a forward pass (Eq. 1). Next, to efficiently compute the exact gradient, information about the loss needs to be passed backward, in the opposite direction of the forward pass, hence the name backpropagation. To illustrate the concept by a simple example, consider a linear network of three neurons in a chain, with activity denoted by input unit x, hidden unit $r = J_X x$ and output unit $y = J_r r$, with modifiable connection weights J_X and J_r. Assuming that $\mathcal{L} = \frac{1}{2}(y - y_{\text{target}})^2$, the change of weight J_r is given by $\Delta J_r = -\alpha \partial \mathcal{L} / \partial J_r = -\alpha(\partial \mathcal{L} / \partial y)(\partial y / \partial J_r) = \alpha(y_{\text{target}} - y)r$, with learning rate α as a "hyperparameter." This modification only depends on local information about the input and output units of this connection. Hence, if $y_{\text{target}} > y$, J_r should change to increase the net input and ΔJ_r has the same sign as r. Conversely, if $y_{\text{target}} < y$, J_r should change to decrease the net input, and ΔJ_r has the opposite sign from r. Learning stops when $y = y_{\text{target}}$.

The change of weight J_X desired for improving the performance must be assessed through the hidden unit r linking x and y. We have $\Delta J_X = -\alpha \partial \mathcal{L} / \partial J_X$, and the gradient

$$\frac{\partial \mathcal{L}}{\partial J_X} = \frac{\partial \mathcal{L}}{\partial r}\frac{\partial r}{\partial J_X} = \frac{\partial \mathcal{L}}{\partial y}\frac{\partial y}{\partial r}\frac{\partial r}{\partial J_X} = \left(y - y_{\text{target}}\right) J_r x \tag{11.3}$$

according to the chain rule in calculus. Therefore, modification of J_X is not local but depends on a synaptic weight downstream J_r. The error $\partial \mathcal{L} / \partial y = (y - y_{\text{target}})$ is "back-propagated" from the output side to the input side. Importantly, this algorithm provides a recipe for credit assignment in determining how much J_X and J_r should be modified for improving performance.

More generally, consider a K-layer feedforward network with a linear transfer function, $r_i^k = \sum_j J_{ij}^k r_j^{k-1} + b_i^k$, the input $x_i = r_i^0$ and the output $y_i = r_i^K$. The learning equation is $\Delta J_{ij}^k = -\alpha \partial \mathcal{L} / \partial J_{ij}^k$. To compute $\partial \mathcal{L} / \partial J_{ij}{}^k = \left(\partial \mathcal{L} / \partial r_i{}^k\right)\left(\partial r_i^k / \partial J_{ij}{}^k\right)$, we have

$$\frac{\partial \mathcal{L}}{\partial r_i^k} = \sum_j \frac{\partial \mathcal{L}}{\partial r_j^{k+1}} \frac{\partial r_j^{k+1}}{\partial r_i^k} = \sum_j \frac{\partial \mathcal{L}}{\partial r_j^{k+1}} J_{ji}^{k+1} = \cdots. \tag{11.4}$$

In vector notation,

$$\frac{\partial \mathcal{L}}{\partial \mathbf{r}^{\mathrm{k}}} = [\mathbf{J}^{\mathrm{k+1}}]^{\mathrm{T}} \frac{\partial \mathcal{L}}{\partial \mathbf{r}^{\mathrm{k+1}}} = [\mathbf{J}^{\mathrm{k+1}}]^{\mathrm{T}}[\mathbf{J}^{\mathrm{k+2}}]^{\mathrm{T}} \frac{\partial \mathcal{L}}{\partial \mathbf{r}^{\mathrm{k+2}}} = \cdots - \prod_{\mathrm{m=k+1}}^{\mathrm{K}} [\mathbf{J}^{\mathrm{m}}]^{\mathrm{T}} \frac{\partial \mathcal{L}}{\partial \mathbf{y}}. \tag{11.5}$$

Therefore, starting with $\partial \mathcal{L} / \partial \mathbf{y}, \partial \mathcal{L} / \partial \mathbf{r}^{\mathrm{k}}$ can be recursively computed from $\partial \mathcal{L} / \partial \mathbf{r}^{\mathrm{k+1}}$, for $k = K-1, \cdots, 1$. This computation flows in the opposite direction of the forward pass and is called the backward pass. Note that the algorithm relies on a differential input-output function F; its applicability is problematic when f is not smooth or for a network of binary units, e.g., with $r = F(I) = 0$ if $I < 0$ and 1 otherwise. Generalization to networks of spiking neurons requires that spikes are smoothened so that the model equations are differentiable (Huh and Sejnowski 2017).

Backpropagation applies to multi-layer nonlinear networks with an arbitrary number of units per layer. With an increasing number of layers in a deep network, weight modifications involve products of many partial derivatives. Bookkeeping in the complex backpropagation process can be cumbersome, but a most elegant way is to use Lagrangian multipliers (LeCun 1988).

Fully connected networks have several problems. First, the number of trainable weights is very large. For example, if images are given by 100×100 pixels, and inputs project to a layer with just 100 neurons, the connection matrix has one million weight elements. Insufficient training data could cause overfitting. Second, there is no build-in invariance regarding shifts in positions of inputs, such as a dog at different parts of the visual field. Third, topology of inputs is ignored; by contrast, naturalistic images display local features such as edges, etc., and thus pixels are correlated on multiple spatial scales. A proposed solution is called convolutional neural network (CNN) (LeCun and Bengio 1995; LeCun et al. 1998), built on visual neuroscience (Hubel and Wiesel 1962, 1968) and noncognition modeling of pattern recognition (Fukushima and Miyake 1982; Fukushima 1988). The key architectural idea is illustrated in **Figure 11.2a** with two layers. In this example, the first layer has four "channels." A channel computes a feature map, which uses a filter to extract a stimulus feature like color or orientation from the sensory data in the visual system. More generally, a feature map in a layer is built on applying a filter to the population output from a previous layer.

Units or neurons on a given feature map share the same input projection weights but have different "receptive fields" (squares in the image), so they can detect an elementary feature by a linear kernel (filter) at any location in the image. For instance, if the receptive field is a 10×10 square for images with 100×100 pixels (Left in **Figure 11.2a**), there are 91×91 neurons in the first layer, and they all share the same weights (Middle in **Figure 11.2a**). Thus, the number of weight parameters is reduced from $100 \times 100 \times 91 \times 91 = 82{,}810{,}000$ to $10 \times 10 = 100$. Similarly, in the next layer, neurons receive filtered inputs from the previous one, and neurons on a given channel (feature map) share a single set of weights. This way, not only the number of weights as free parameters is reduced dramatically, but also local relationships in the inputs are taken into consideration and shift invariance is automatically guaranteed by forcing the replication of weights (LeCun and Bengio 1995). Typically, there are more feature maps in each succeeding layer (Right in **Figure 11.2a**), while the spatial resolution becomes coarser, consistent with the neurophysiology of the visual cortical hierarchy.

Modeling that combines backpropagation learning algorithm and convolutional neural networks, two ideas dated to the 1980s and early 1990s, demonstrated dramatic success when it won a competition of

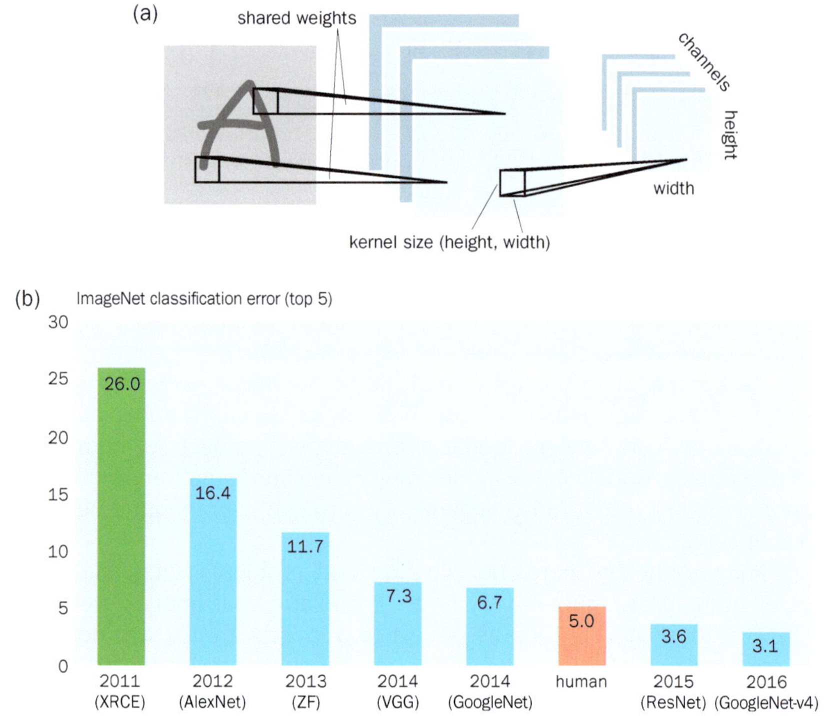

Figure 11.2 Convolutional neural network and ImageNet classification. (a) Schematics of a convolutional neural network for processing images. Each layer contains a number of channels (four in layer 1, six in layer 2). A channel or feature map consists of spatially organized units or artificial neurons, each receiving connections from neurons in the previous layer with similar spatial preferences. The spatial extent of these connections is described by the kernel size. (b) Performance of classification of labeled pictures from ImageNet by various machine learning algorithms and human subjects. The outputs are rank-ordered guesses, and top-five error rate is defined as one minus the fraction of images for which the correct answer is one of the top five guesses. Panel (a) is reproduced from Yang and Wang (2020) inspired by LeCun et al. (1998), (b) is downloaded from devopedia.org. [Yang, G. R., & Wang, X. J. (2020). Artificial neural networks for neuroscientists: A primer. Neuron, 107(6), 1048–1070.] [Devopedia.org. https://creativecommons.org/licenses/by-sa/4.0/deed.en.]

classifying objects in natural images with an error rate half of that by the next-best rival (Krizhevsky et al. 2012). The watershed event would have been impossible without a large database of pictures with labeled objects by ImageNet and the availability of fast computers powered by graphics processing units (GPUs). Since then, neural networks have become dominant in the field of computer vision, and by 2015 CCNs surpassed human performance at ImageNet classification (**Figure 11.2b**). Rapid progress has also been made in other domains, including machine translation of natural language (Sutskever et al. 2014). Deep networks captured people's imagination when AI systems from DeepMind beat the world champion of the game Go (Silver et al. 2017), and made a huge leap in solving three-dimensional folded protein structures with 90% accuracy compared to the previously held record of 40% (Jumper et al. 2021). More recently, the transformer architecture (Vaswani et al. 2017) based on an "attention" mechanism (Bahdanau et al. 2014; Xu et al. 2015) has powered generative AI systems such as ChatGPT (Radford et al. 2018).

The spectacular success of deep learning raised the epistemological question of what does it mean to "understand" the world in science. Is understanding equivalent to predicting? A deep network may predict correct outputs for samples not included in the training set, but that does not mean that the system "understands" the computation to accomplish the task in the sense of compact summary of general principles like the Newtonian laws in physics. Efforts have been devoted to elucidate the inner working of deep learning (Schwartz-Ziv and Tishby 2017; Li and Sompolinsky 2020; Papyan et al. 2020; Cao and Yamins 2024; Frank 2023). Importantly, there is an ongoing debate on the appropriate methods, beyond the Turing test (Turing 1950), for assessing whether AI systems exhibit human-like intelligence: "Better experimental methods and benchmarks will be brought about through collaborations between AI researchers and cognitive scientists who have long investigated how to do robust tests for intelligence, understanding, and other cognitive capabilities in children, animals, and other "alien" intelligences" (Mitchell 2023).

11.2.2 Deep Neural Network Modeling and the Brain

Can deep networks, initially inspired by the primate visual system, be used to understand biological information processing in return? Multi-layer models for the neural cascade of visual information processing, such as the HMAX model (Riesenhuber and Poggio 1999; Serre et al. 2007b), have focused on position- and size-invariant object recognition. DiCarlo and Cox (2007) argued that the core problem is that the axes of sensory input space in the retina are not aligned with the axes along which behaviorally relevant constructs like object shapes vary. Thus, variations of an object's shape entail complex nonlinear changes in the input space; conversely, two distinct objects such as faces of two people may be close to each other in pixelated images. Thus, high-level behaviorally relevant dimensions need to be "untangled" from natural dimensions in the input space, by virtue of a highly complex nonlinear transformation.

Deep networks offer a systematic framework for achieving complex input-output function, through a hierarchy of simpler transformations. After a CNN is trained to perform an object recognition task, how do units (artificial neurons) in their successive layers compare with neural activities of various areas in the primate ventral stream? Yamins et al. (2014) investigated this question by comparing spiking firing of neurons recorded from monkeys in response to photos, with artificial neurons of a hierarchical CNN (HCNN). More specifically, units in a given layer are linearly summated with weights optimized to best replicate recorded neural activities. **Figure 11.3a** shows an example of model prediction and actual activity recorded in the monkey inferotemporal (IT) cortex. Interestingly, model units in the last layer account for an impressive 50% variance of IT neuronal responses (**Figure 11.3b**, right panel), whereas the predictivity of neural responses in V4 is the highest in the penultimate layer (**Figure 11.3b**, left panel), which is consistent with the hierarchical relations of V4 and IT. The HCNN model displays better matching with neural responses from both V4 and IT than do other existing models.

Comparison can also be made between a model and recorded neural population data instead of single cells. Consider population activity as an N-dimensional vector $\mathbf{x}_i$ in response to a stimulus i, which can be BOLD signals from N voxels or firing rates from N neurons. Dissimilarity of population responses to two stimuli i and j is defined as $\left(1-\mathrm{Cov}\left(\mathbf{x}_i,\mathbf{x}_j\right)\right)/\sqrt{\mathrm{Var}(\mathbf{x}_i)\mathrm{Var}\left(\mathbf{x}_j\right)}$. It is 1 if $\mathbf{x}_i$ and $\mathbf{x}_j$ are completely uncorrelated, and 0 if they are identical. Using pairwise dissimilarity as a measure of distance between two vectors $\mathbf{x}_i$ and $\mathbf{x}_j$, multidimensional scaling (Shepard 1980) or another algorithm can be used to identify clusters with similar population responses grouped together. The dissimilarity matrix measured from human or monkey IT imaging exhibits clustering, e.g., clusters corresponding to animate and inanimate objects, and the animate cluster divides into face and body subclusters (Kriegeskorte et al. 2008). Similar structure was found in the highest layer of a CNN trained for object recognition (**Figure 11.4a**) (Khaligh-Razavi and Kriegeskorte 2014). Match between a model and IT data can be quantified by the Spearman's rank correlation of all elements of the two dissimilarity matrices. **Figure 11.4b** shows that match with IT measurement is quite high for CNNs but not V1-like, V2-like or HMAX models. Combining functional MRI, electrophysiology and computational modeling, Bao et al. (2020) found that IT cortex exhibits a map in a low-dimensional object space (rather than the physical space of cortical tissue). Four clusters of cells according to the first two components of their stimulus preference axes were found; two of them representing the face and body of animals, whereas the other two encode inanimate objects (separated by the degree of stubby versus spiky appearance). This organization was captured by a feedforward deep neural network model.

These results demonstrate that training deep networks has been more successful than hand-crafted models by certain criteria. However, commonly used metrics are only indirectly related to neurons; for instance, it is unclear what would be a biological interpretation of the linear readout of artificial units with optimized weights, an obligatory intermediate step in the process of comparison with biological neurons. Furthermore, equivalence between network models and brains may be overstated (Saxe et al. 2021); networks that perform much worse in object classification account for only 5% less variance of neural data than the best deep network currently available (Schrimpf et al. 2018). In addition, contributions by subcortical structures need to be explored in applying artificial network models to the brain (Suzuki et al. 2023).

Regardless, developing benchmarks for model performance comparison has become a trend of the future (Schrimpf et al. 2020). Importantly, to shed insights into how the brain works, it is essential not to be content with treating deep networks as black boxes. Instead, they serve as model systems of which the inner working should be thoroughly examined, such as the connectivity that emerges from training. Computational principles and underlying circuit mechanisms thus identified would constitute genuine understanding.

With impressive progress accomplished by feedforward deep networks, one may ask what are the roles of abundant feedback projections and local recurrent connections in the cortex. As covered in the previous

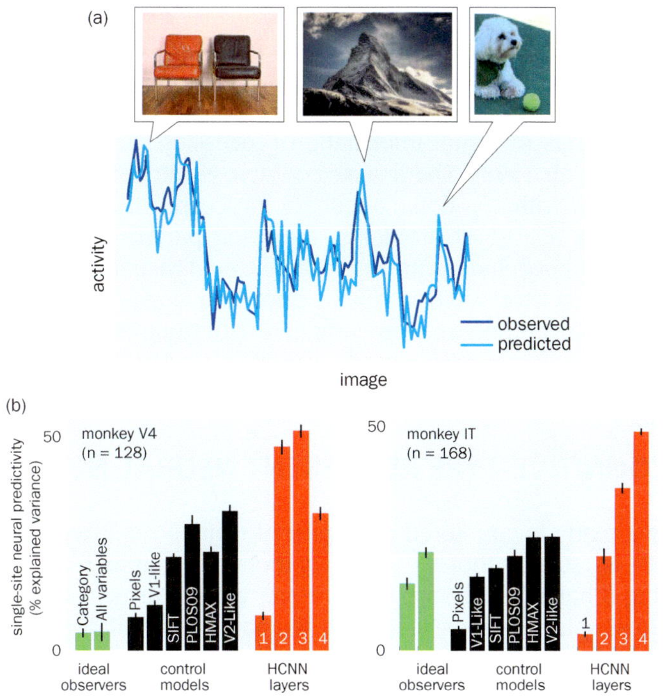

Figure 11.3 Comparison of neural activities from the monkey visual cortical areas and those predicted by linear combinations of units in computational models. (a) An example of actual neural responses (dark blue) to three images and prediction from a hierarchical CNN (HCNN) model (light blue). (b) Comparison of IT and V4 neural response predictivity for various models. Bar height shows median predictivity, taken over 128 predicted units in V4 (left panel) or 168 units in IT (right panel). The last hidden layer of the HCNN model best predicts IT responses, while the second-to-last hidden layer best predicts V4 responses. Panel (a) is reproduced from Saxe et al. (2021), with original data from Yamins et al. (2014), (b) from Yamins and DiCarlo (2016). [Saxe, A., Nelli, S., & Summerfield, C. (2021). If deep learning is the answer, what is the question? Nature Reviews Neuroscience, 22(1), 55–67. Original data from Yamins, D. L., Hong, H., Cadieu, C. F., Solomon, E. A., Seibert, D., & DiCarlo, J. J. (2014). Performance-optimized hierarchical models predict neural responses in higher visual cortex. Proceedings of the National Academy of Sciences, 111(23), 8619–8624.] [Yamins, D. L., & DiCarlo, J. J. (2016). Using goal-driven deep learning models to understand sensory cortex. Nature Neuroscience, 19(3), 356–365.]

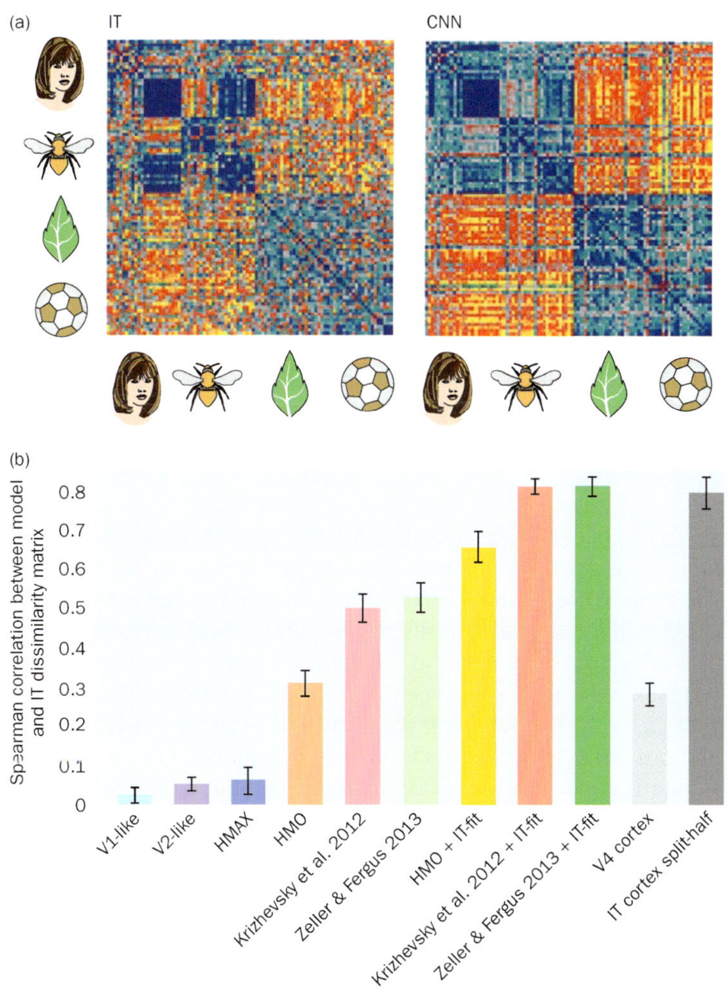

Figure 11.4 Comparison of dissimilarity matrices from neurophysiology and computational models. (a) Representational dissimilarity matrices from the inferior temporal cortex (IT) measured by human fMRI (left panel) and the final layer of a CNN (right panel). (b) Deep neural network representations more closely resemble IT than do three simpler models (V1-like, V2-like and HMAX). The similarity between each model and monkey IT neurons was measured using the Spearman rank correlation coefficient to compare representational dissimilarity matrices. Panel (a) is reproduced from Kriegeskorte et al. (2008), (b) from Kriegeskorte (2015), with original data from Cadieu et al. (2014). [Kriegeskorte, Nikolaus, et al. (2008). Matching categorical object representations in inferior temporal cortex of man and monkey. Neuron 60(6), 1126–1141.] [Cadieu, C. F., Hong, H., Yamins, D. L., Pinto, N., Ardila, D., Solomon, E. A., ... & DiCarlo, J. J. (2014). Deep neural networks rival the representation of primate IT cortex for core visual object recognition. PLoS computational biology, 10(12), e1003963. Licensed under CC BY 4.0.]

chapters, internal states and cognitive processes are inherently dependent on the feedback loops in the neural circuits. In the visual system, a purely feedforward architecture essentially computes static input-output functions; in the brain, fast object classification corresponds to the initial ~200 milliseconds of visual information processing (Thorpe et al. 1996; Schmolesky et al. 1998). The first sweep across the ventral stream hierarchy interacts with top-down feedback processes that are crucial for vision to solve a host of problems, including pattern completion from noisy or partially occluded objects in a cluttered visual scene, inference that combines sensory evidence with prior knowledge (van Bergen and Kriegeskorte 2020; Kang and Druckmann 2020; Kreiman and Serre 2020; Kar and DiCarlo 2021).

11.3 COGNITIVE-TYPE RECURRENT NEURAL NETWORKS

We have already encountered examples of trained recurrent neural networks (RNNs), such as those for working memory (Chapter 5, **Figure 5.40**) and timing (Chapter 8, **Figure 8.18**). Here, we will discuss this approach applied to simple decisions, for which temporal accumulation of information is essential.

A "vanilla" RNN is described by a simple equation like:

$$\frac{d\mathbf{r}}{dt} = -\frac{\mathbf{r}}{\tau} + \mathbf{F}\left(\mathbf{W}^{\text{rec}}\mathbf{r} + \mathbf{W}^{\text{in}}\mathbf{x} + \mathbf{b}\right). \tag{11.6}$$

where $\mathbf{r}$ is the unit activity (firing rate), $\mathbf{x}$ is the external input and $\mathbf{F}$ is the input-output transfer function. The output of an RNN is a readout given by $\mathbf{z} = \mathbf{g}\left(\mathbf{W}^{\text{out}}\mathbf{r}\right)$, with another input-output transfer function $\mathbf{g}$. As schematically shown in **Figure 11.5**, an RNN receives stimuli and produces responses; therefore, there are three (input, recurrent, output) sets of connection weights $\mathbf{W}^{\text{in}}$, $\mathbf{W}^{\text{rec}}$ and $\mathbf{W}^{\text{out}}$. Using the Euler numerical method, Eq. (11.6) is simulated according to:

$$\mathbf{r}_t = \left(1 - \Delta t / \tau\right)\mathbf{r}_{t-1} + \Delta t\mathbf{F}\left(\mathbf{W}^{\text{rec}}\mathbf{r}_{t-1} + \mathbf{W}^{\text{in}}\mathbf{x}_t + \mathbf{b}\right) \tag{11.7}$$

where Δt is the step for discretized time. A white noise input is often added in an appropriate way.

If the input weights W^{in} and recurrent weights W^{rec} are sampled from probability distributions but fixed, only output weights W^{out} are plastic. The advantage of this setting is that training can be done using the recursive least-square method (Haykin 2002; Jaeger and Haas 2004; Sussillo and Abbott 2009). The same algorithm can train recurrent connection weights when the outputs of recurrent units learn to explicitly reproduce target signals like recorded neural activities (Laje and Buonomano 2013; Rajan et al. 2016). The backpropagation method is commonly used when recurrent connection weights are also trained, but supervision only imposes target responses of output units. This is the case for training an RNN to perform a desired task, when the supervised learning is solely based on behavioral performance, without specifying the activities of recurrent units, and thus how the network actually learns to solve the problem.

Regularization is a general concept for supervised learning algorithms (Goodfellow et al. 2016). For instance, to achieve sparseness of connections, we may apply a soft constraint by adding the sum of the L_1-norms of the weights to the loss function, $\lambda^{\text{rec}}\sum_{i,j=1}^{N}\left|W_{ij}^{\text{rec}}\right|$, which encourages a fraction of weights to be near zero, whereas others are strong. The

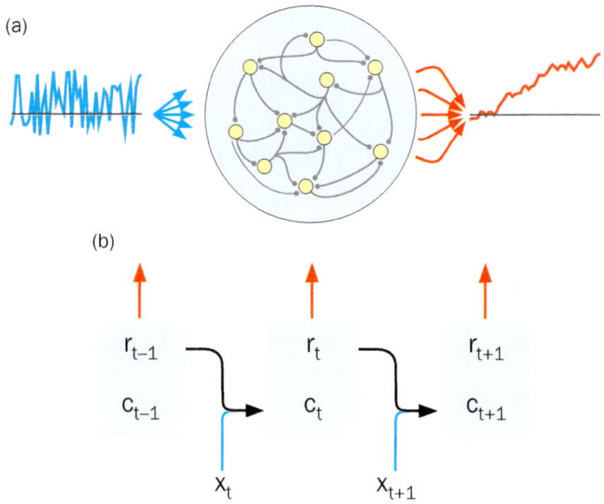

Figure 11.5 Schematic of a (a) trained recurrent neural network (RNN) and (b) training algorithm called backpropagation in time. The variable $\mathbf{c}_t$ is called cell state in machine learning, corresponding to the total input current to a unit, as a sum of an external input $\mathbf{x}_t$ and a recurrent input that depends on the activity variable $\mathbf{r}_t$. Reproduced from Yang and Wang (2020). [Mante, V., Sussillo, D., Shenoy, K.V., and Newsome, W.T. (2014). Context dependent computation by recurrent dynamics in prefrontal cortex. Nature 503(7474), 78–84.] [Yang, G. R., & Wang, X. J. (2020). Artificial neural networks for neuroscientists: A primer. Neuron, 107(6), 1048–1070.]

hyperparameter λ^{rec} controls the relative strength of this term in the loss function $\mathcal{L}$. Or, we may choose to encourage solutions with small firing rates through regularization of the L_2-norms of the firing rates (Sussillo et al. 2015). Automatic differentiation in such software as PyTorch (Paszke et al. 2019) greatly facilitates the exploration of loss functions with various constraints.

An RNN can be formally considered as a very deep neural network, in which its state at each discretized time step is a different layer and its state at the "current layer" $t+1$ is determined by that in the "previous layer" t (**Figure 11.5b**). All the "layers" share the same weights, akin to weight sharing in convolutional neural networks. Computing its gradient involves propagating information backward in time (backpropagation-through-time) (Rumelhart et al. 1986; Werbos 1990). For instance, if the time step is chosen to be $\Delta t = 10$ ms, and a simulated behavioral trial takes $T = 2$ s, then an RNN amounts to a deep network with $K = T / \Delta t = 200$ layers. Note that the neural network is actually a dynamical system in real time, and model simulation results should be validated using sufficiently small Δt values.

As one can see in Eq. (11.5), gradients entail products of weight matrices, like $\prod_{k=1}^{K} \mathbf{J}^k$. Think of $\mathbf{J}^k$ as real numbers for a moment, if they are larger than 1, the product would explode exponentially with K, whereas if they are smaller than 1, the product would be vanishingly small. The same holds true for matrices. Pascanu et al. (2013) proposed "tricks" to deal with these problems. First, gradient explosion is avoided by simply "clipping" the gradient when its norm exceeds a maximum. Second, the vanishing gradient is addressed by adding a regularization term to $\mathcal{L}$ that preserves the size of the gradients as errors backpropagated through time. Another approach is the Hessian-free algorithm of backpropagation which, instead of a gradient (first-order derivatives), uses a second-order optimization method that alleviates the vanishing gradient problem for RNNs (Martens and Sutskever 2011; Sussillo and Barak 2013; Chaisangmongkon et al. 2017).

Song et al. (2016) trained an RNN to perform a random dot motion (RDM) direction discrimination task (Roitman and Shadlen 2002). The weights are initiated randomly from a Gaussian distribution, the loss function $\mathcal{L}$ is large and the network produces nonsensical responses. Through training the network gradually learns the task while $\mathcal{L}$ decays to a small plateau level, and eventually the performance reaches a high level characterized by psychometric function (**Figure 11.6a**) comparable to the observed monkey behavior (Roitman and Shadlen 2002) (Chapter 6,

Figure 6.7). Slow time integration in the form of ramping activity emerges from training (Figure 11.6b).

In order to relate this trained RNN to the brain, it is desirable to incorporate biological properties of neural circuits. Dale's law can be imposed by separating units into excitatory and inhibitory neurons. All projections of an excitatory unit j are zero or excitatory, i.e., $W_{ij}^{\text{rec}} \geq 0$ for all i; and similarly $W_{ij}^{\text{rec}} \leq 0$ for all i receiving inputs from an inhibitory unit j (Song et al. 2016). After training, the recurrent connection matrix W^{rec} displays a structure with recurrent excitation between units of similar decision selectivity, especially when Dale's law is imposed (Figure 11.6c, right versus left panel). This wiring diagram resembles that of the structured network model (Wang 2002b; Wong and Wang 2006) discussed in Chapter 6, albeit with heterogeneity, demonstrating that training RNN can in principle be deployed as a tool to explore and uncover biologically plausible neural circuits to carry out a cognitive task.

Time integration represents an important form of dynamics underlying decision making; another is sequence generation. Training an RNN to perform a short-term memory-dependent decision (Harvey et al. 2012), Rajan et al. (2016) used recorded activities of neurons in the posterior parietal cortex (PPC) in behaving mice as the target outputs of recurrent units, resulting in sequential activation in the RNN (Figure 11.6d), as observed experimentally (Chapter 5, Figure 5.38a). For this approach to be genuinely powerful in neuroscience, systematic efforts need to be devoted to "opening the box" and probing the inner working of the network dynamics

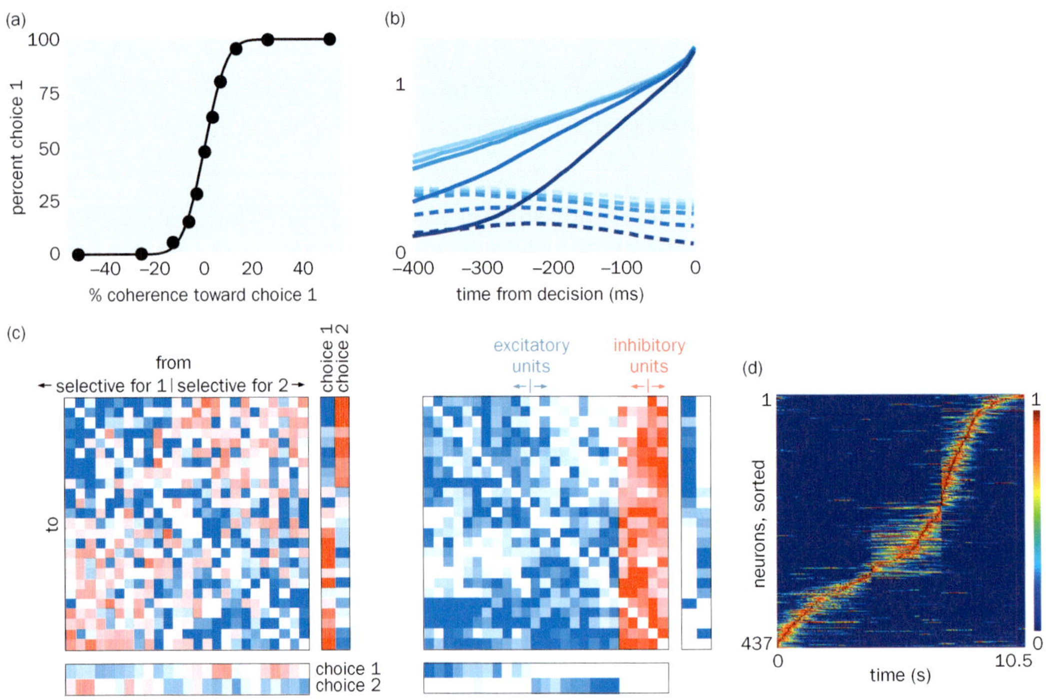

Figure 11.6 RNNs for temporal dynamics during decision making. (a–c) An RNN trained to perform the random dot direction discrimination task. (a) Performance as a function of motion coherence. (b) Single unit activity in the reaction time version of the task. Solid (dashed) lines denote preferred (nonpreferred) direction with three coherence levels. (c) Recurrent connection matrix without (left) or with (right) the Dale's law. Connections go from columns (pre-synaptic) to rows (post-synaptic). Blue: Positive weights; red: Negative weights. Matrix: Recurrent connections; right two columns: Input weights; lower two rows: Output weights. (d) An RNN trained to reproduce measured sequential activation of neurons in the mouse posterior parietal cortex in a delayed decision task (Harvey et al. 2012). Panels (a–c) are reproduced from Song et al. (2016), (d) modified from Rajan et al. (2016). [Song, H. F., Yang, G. R., & Wang, X. J. (2016). Training excitatory-inhibitory recurrent neural networks for cognitive tasks: A simple and flexible framework. PLoS computational biology, 12(2), e1004792. Licensed under CC BY 4.0.] [Rajan, K., Harvey, C. D., & Tank, D. W. (2016). Recurrent network models of sequence generation and memory. Neuron, 90(1), 128–142.]

or the structure of the wiring diagram that emerge from training (Sussillo and Barak 2013; Song et al. 2016; Yang and Wang 2020).

It is worth emphasizing that training an RNN by neurophysiological data so that recurrent units reproduce neural firing should be contrasted with training by behavior alone; in the latter case, the unit activity patterns (e.g., whether activation is sequential along a chain), thus the computational algorithm to solve the task, are not posited at the outset, but instead the solution emerges from learning.

Mante et al. (2013) applied an RNN trained to perform a context-dependent decision as in a monkey physiological experiment (**Figure 11.7a**). In that study, random dots in a moving pattern were either red or green. A rule cue instructed the subject whether color or motion direction was the relevant feature in a given trial. If the cue was a yellow square motion direction was the rule currently in play, and the classic RDM discrimination (left or right direction) should be carried out. On the other hand, if the cue was a blue cross, the behavioral context was color, in which case motion direction should be ignored, and a judgment was required as to whether there were more red or green dots. The response was indicated by a saccade from fixation to one of two targets. Note that the rule was present throughout a trial and therefore did not need to be held in working memory.

As shown in **Figure 11.7b**, in the motion context (upper row, black), monkey subjects showed a psychometric function as a function of motion coherence but were insensitive to the color coherence defined as (#green dots − #red dots)/(#green dots + #red dots). The opposite holds in the color context (lower row, blue), demonstrating excellent behavioral performance. Single-neuron activities recorded from the frontal eye field (FEF), known to be important for saccade behavior, appeared complicated for simple interpretation. An RNN model was simulated in which the network received a mixture of color and motion inputs, and was trained to perform the same task as in the monkey experiment. It was found that, in the state space of the RNN population, a line attractor along a one-dimensional axis was formed for temporal integration (Chapter 3, Section 3.3.2). A "selection vector" determines which of the two input types is integrated according to the behavioral rule. In the motion context, the selection vector is orthogonal to that of color input, and as a result a color pulse perturbed the network, but afterwards the network dynamically converged back to the same point on the line attractor. By contrast, a motion

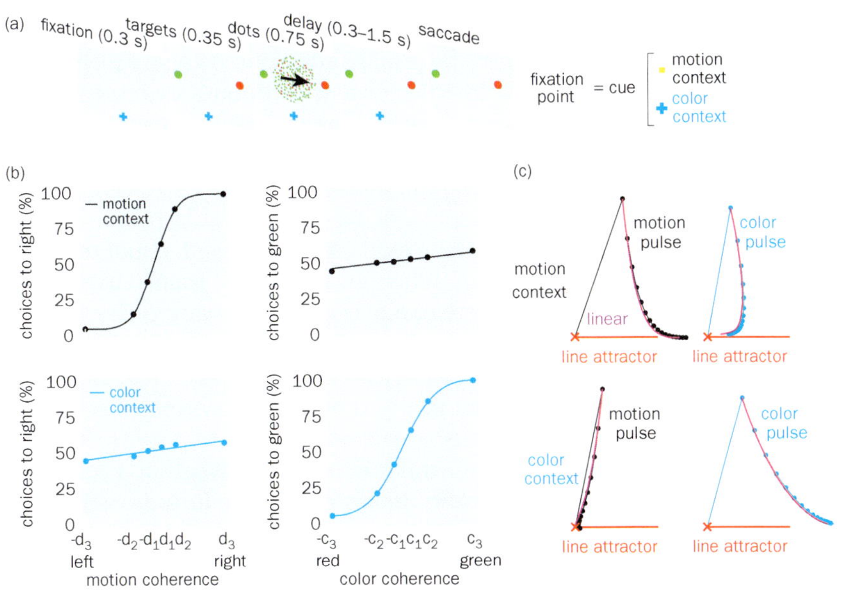

Figure 11.7 Trained RNN for context-dependent decision making. (a) Task design. Depending on the rule cue signaled by the shape (yellow square or blue cross) of the fixation point, subjects are required to either judge the net motion direction (left or right) or prevalence of color (whether there are more green or red dots). (b) Monkey's performance as a function of motion coherence and color coherence in the motion context (upper row) or color context (lower row). (c) RNN modeling revealed selection vector for gated integration mechanism. In the motion context (upper row), a pulse input that perturbs the system from the red-cross state is integrated if the stimulus was motion (left) but not color (right); the converse is true in the color context (lower panel). Reproduced from Mante et al. (2013). [Mante, V., Sussillo, D., Shenoy, K. V., & Newsome, W. T. (2013). Context-dependent computation by recurrent dynamics in prefrontal cortex. Nature, 503(7474), 78–84.]

pulse was integrated, and the network settled at a different point on the line attractor as desired computationally for accumulating evidence about motion signal over time (**Figure 11.7c**, upper row). The converse was true in the color context (**Figure 11.7c**, lower row). Analysis of FEF neural activities broadly lent support to the selection vector idea.

We have previously discussed rule-cued task switching through top-down signaling of behavioral context that weakly enhances the relevant pathway while weakening the irrelevant pathway (Mirabella et al. 2007; Ardid and Wang 2013) (Chapter 8, Section 8.5). Here, a trained RNN revealed a novel, recurrent gating mechanism that could work in concert with other gating mechanisms (Chapter 9, **Figure 9.26**). In the monkey experiment, sensory areas coding the two stimulus features, MT for motion direction and V4 for color, are likely to be modulated by the top-down context signaling, and the brain utilizes a combination of gating mechanisms. This was confirmed in a work that combined rodent experiments using a context-dependent decision-making paradigm and trained RNN; as a matter of fact, different individuals may use various mixtures of the two gating mechanisms (Pagan et al. 2022).

Supervised learning is not suitable for training an RNN to perform economic decision making based on subjective valuation of options (Chapter 7), where there is no "correct" output. Instead, reinforcement learning (RL) (Sutton and Barto 2018) (Chapter 4, Section 4.3) should be used when learning is driven by rewards. RL has been developed for training deep neural networks (Mnih et al. 2015; Botvinick et al. 2020) and recurrent neural networks (Song et al. 2017). In the model shown in **Figure 11.8a**, there are two modules for valuation and action selection, respectively. The decision module is an RNN, implementing the decision policy $\pi_\theta(a_t \mid u_{1:t})$ (θ denotes parameters like connection weights that are modifiable by learning). At each time t, the agent chooses an action a_t (e.g., choose left, L; right, R; or fixation) after observing noisy inputs $\mathbf{u}_t$ provided by the environment. The policy at time t can depend on all past and current inputs $\mathbf{u}_{1:t} = (\mathbf{u}_1, \mathbf{u}_2, \ldots, \mathbf{u}_t)$, allowing the agent to integrate sensory evidence or use working memory to perform the task. The choice yields an outcome ρ_{t+1}, which can be a positive reward according to a delivery schedule, e.g., in a foraging task, unless fixation is broken too early, in which case the outcome is negative.

The goal of the agent is to maximize the sum of expected and temporally discounted future rewards at time $t = 0$, $J(\theta) = \sum_{t=0}^{T} \exp(-t/\tau)\rho_{t+1}$, where τ is the discounting time constant. The function J is not computed on a single trial but averaged over all possible trial histories. Modification of recurrent connections is given by gradient descent, using a REINFORCE (Williams 1992; Sutton and Barto 2018) or asynchronous advantage actor-critic (Mnih et al. 2016) or proximal policy optimization (Schulman et al. 2017) algorithm. The valuation module is also an RNN, and its output v_φ is the expected discounted return with modifiable parameters φ (such as connection weights). It is trained to represent expected rewards, by minimizing the average difference between the expected value v_φ and actual reward J (both with temporally discounting). When the model was trained to perform an economic task of choosing between two types of reward offered in different amounts (Padoa-Schioppa and Assad 2006) (**Figure 11.8b**), recurrent units displayed a variety of selectivity (**Figure 11.8c**), quite similar to heterogeneous neurons observed in the orbitofrontal cortex of behaving monkeys (Chapter 7, **Figure 7.13**).

In a model trained by RL, information regarding whether a response is desirable is not contained in the inputs but rather in the association between inputs, actions and rewards. Moreover, the output is stochastic,

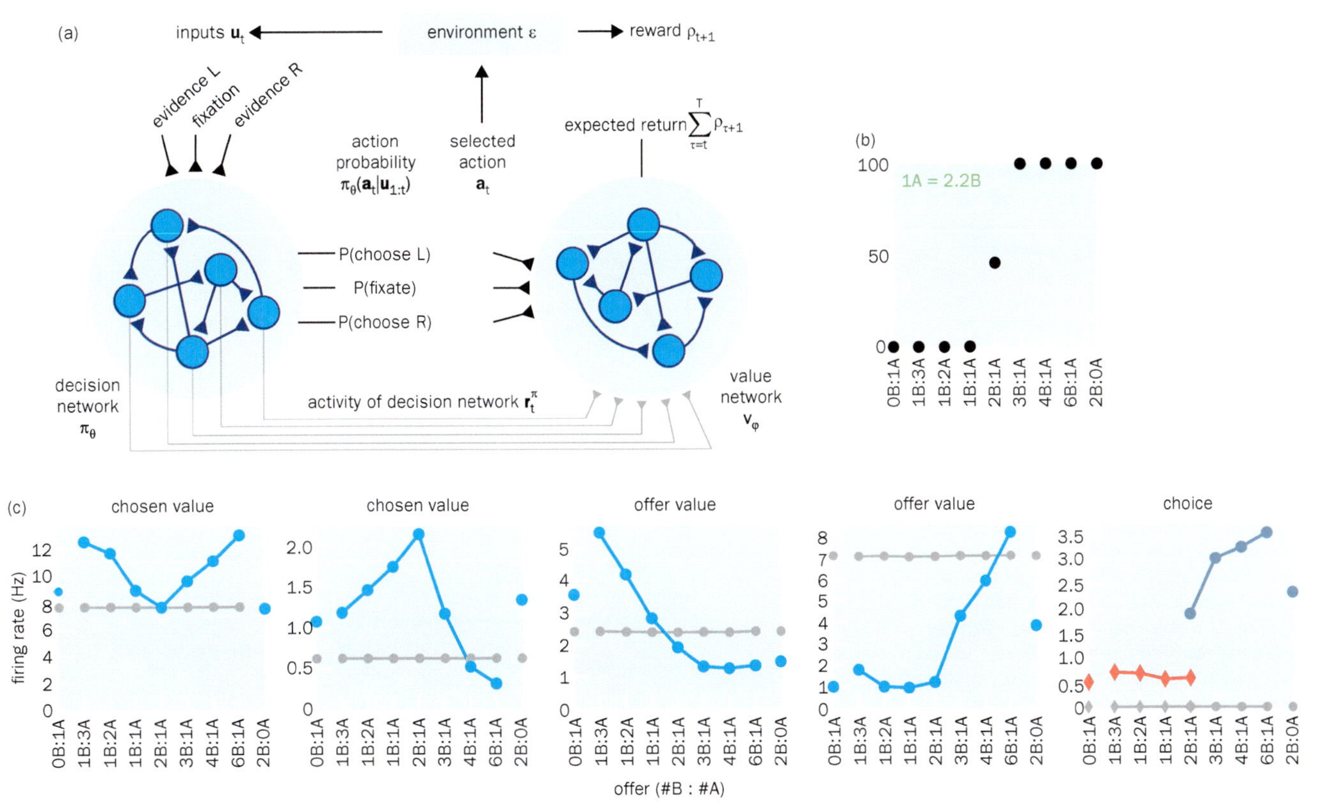

Figure 11.8 An RNN trained by reinforcement learning. (a) Model scheme. At each time t, the agent selects action $\mathbf{a}_t$ according to the output of the decision network π_θ, which can depend on all past and current inputs $\mathbf{u}_{1:t}$ provided by the environment. In response, the environment provides reward ρ_{t+1} to the agent. The value network v_φ uses the output of the decision module to predict discounted future rewards. All of the weights shown are plastic, i.e., trained by gradient descent. (b–c) Value-based economic choice task (Padoa-Schioppa and Assad 2006) performed by a trained RNN. (b) Choice pattern when the reward contingencies are indifferent for roughly one "juice" of A and two "juices" of B. (c) Mean activity of example neurons in the valuation network. Units in the value network exhibit diverse selectivity as observed in the monkey orbitofrontal cortex. For "choice" (last panel), trials were separated into choice A (red) and choice B (blue). Reproduced from Song et al. (2017). [Song, H. F., Yang, G. R., & Wang, X. J. (2017). Reward-based training of recurrent neural networks for cognitive and value-based tasks. Elife, 6, e21492. Licensed under CC BY 4.0.]

e.g., with a response $\mathbf{a}$ sampled using a policy $\pi_\theta(\mathbf{a})$ as a probability distribution. Finally, this model differs from the previously described RNNs in that it has two interacting modules. The decision module outputs selected actions and projects to the valuation module, which computes the expected values used in the reward prediction error signal $(J - v_\phi)$ during learning. It is conceivable, in principle, that weights of inputs and recurrent connections in the action selection module acquire the value information, in which case after training, the appropriate decisions can be produced without the valuation module. However, an experiment showed that disruption of the OFC impaired performance of economic choices after training (Ballesta et al. 2020). Alternatively, if the valuation module is the only source of value information to drive the action selection module, then it would always be required to generate adaptive economic choice. Development of multi-module RNNs represents a new theme of current research (Kleinman et al. 2021; Perich et al. 2021; Yang and Molano-Mazón 2021).

Training the model of Song et al. (2017) benefitted from including gated recurrent units (GRUs) (Cho et al. 2014; Chung et al. 2014). GRUs are recurrent units for which updating and leak (degree of forgetfulness) are controlled dynamically by gating variables (**Figure 11.9a**). A more general gating mechanism is long-short-term-memory (LSTM) (**Figure 11.9b**) (Hochreiter and Schmidhuber 1997; Gers et al. 2000), which also assumes

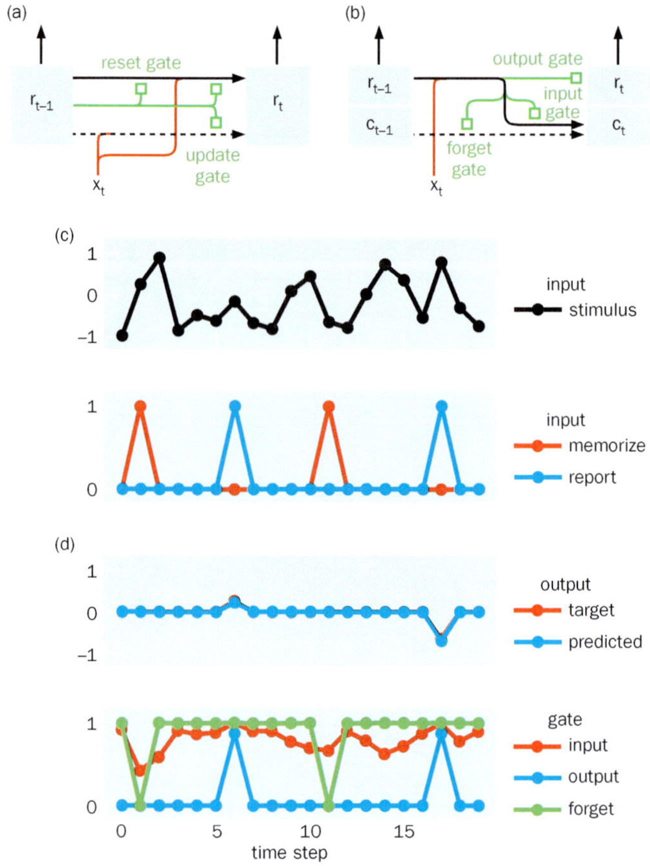

Figure 11.9 Gating in RNNs. (a) Gated recurrent units (GRUs), where x_t is the input and r_t is a neuron's activity. Two gating variables determine, at any time, the extent to which r_t is updated by inputs and influenced by its previous value. (b) Long-short-term memory (LSTM) is similar to GRU but with the addition of output gating. (c–d) An example of LSTM in a simple memory task. (c) Upper panel: The network receives a continuous stream of input stimulus $x(t)$. Lower panel: A "memorized input" (red) instructs the network to remember the current value of the stimulus, $x(t=0)=-0.2$ and $x(t=22)=-0.9$. A "report input" (blue) signals when the output should be produced, the first at $t=15$ and the second at $t=30$. (d) After training, a single-unit LSTM can perform the task almost perfectly. Upper panel: Actual output matches predicted output. Lower panel: When the memorized input is active, the network opens the input gate (red). Forgetting is absent except when the input gate is open (green). The output gate is opened (blue) when the report input is active. Reproduced from Yang and Wang (2020). [Courtesy of Robert Guangyu Yang.] [Yang, G. R., & Wang, X. J. (2020). Artificial neural networks for neuroscientists: A primer. Neuron, 107(6), 1048–1070.]

dynamical gating of output, in addition to input and leak. Gated networks are generally easier to train and more powerful than vanilla RNNs. Gating variables dynamically control information flow within these networks through multiplicative interactions. In an LSTM network, there are three types of gating variables. Input and output gates, $\mathbf{g}_t^i$ and $\mathbf{g}_t^o$, control the inputs to and outputs of the cell state $\mathbf{c}_t$, while forget gate $\mathbf{g}_t^f$ controls whether cell state $\mathbf{c}_t$ (which as we recall is the total input to $\mathbf{r}_t$) keeps its memory $\mathbf{c}_{t-1}$.

$$
\begin{aligned}
\mathbf{g}_t^f &= \sigma_g(\mathbf{W}_f\mathbf{x}_t + \mathbf{W}'_f\,\mathbf{r}_{t-1} + \mathbf{b}_f),\\
\mathbf{g}_t^i &= \sigma_g(\mathbf{W}_i\mathbf{x}_t + \mathbf{W}'_i\,\mathbf{r}_{t-1} + \mathbf{b}_i),\\
\mathbf{g}_t^o &= \sigma_g(\mathbf{W}_o\mathbf{x}_t + \mathbf{W}'_o\,\mathbf{r}_{t-1} + \mathbf{b}_o),\\
\mathbf{c}_t &= \mathbf{g}^f{}_t \odot \mathbf{c}_{t-1} + \mathbf{g}_t^i \odot \sigma_c(\mathbf{W}_c\mathbf{x}_t + \mathbf{W}'_c\,\mathbf{r}_{t-1} + \mathbf{b}_c),\\
\mathbf{r}_t &= \mathbf{g}^o{}_t \odot \mathbf{F}(\mathbf{c}_t).
\end{aligned}
\tag{11.8}
$$

Here the symbol $\odot$ denotes the element-wise (Hadamard) product of two vectors of the same length ($\mathbf{z} = \mathbf{x} \odot \mathbf{y}$ means $z_i = x_i y_i$). Gating variables are bounded between 0 and 1 by the sigmoid function σ_g, which can be viewed as a smooth differentiable approximation of a binary step function. All the weights ($\mathbf{W}$ and $\mathbf{W}'$ matrices) are trained. By introducing these gates, an LSTM can in principle keep a memory in its cell state $\mathbf{c}_t$ indefinitely by having the forget gate $\mathbf{g}_t^f = 1$ and input gate $\mathbf{g}_t^i = 0$ in some temporal epochs, as illustrated by a toy example (**Figure 11.9c–d**). In addition, the network can choose when to read out from the memory by setting its output gate $\mathbf{g}_t^o = 0$ or 1.

In Eq. (11.8), leak is described in the input $\mathbf{c}_t$, instead of the output firing rate for traditional network models (Eq. 11.7). When the forget gate

$\mathbf{g}_t^f$ is 1 (equivalent to $\tau = \infty$ in Eq. 11.7), $\mathbf{c}_t$ acts as a perfect integrator of inputs; when $\mathbf{g}_t^f$ is smaller than 1, $\mathbf{c}_t$ decays over time exponentially in the absence of input (Chapter 3, Section 3.3.2). Learning produces a time-varying $\mathbf{g}_t^f$, which entails a range of time constants used across a task's duration. It remains to be seen whether the functional power of LSTM (and GRUs) can be realized in the brain by biological neural circuits (Costa et al. 2017; Krishnamurthy et al. 2022).

In summary, trained RNN has emerged as a useful tool for discovering novel concepts (such as selection vector) and network models (e.g., for value-coding of OFC cells), which would be difficult to come up with otherwise. This methodology will help us in computationally tackling higher-level cognitive functions.

11.4 ABSTRACTION

Abstract knowledge, extracted from past experience and generalized to new situations, is at the heart of learning and intelligence. Here we will discuss two broad types of abstraction (**Figure 11.10**). The first involves categories and concepts. Using categorization of dogs as an example, Immanuel Kant wrote:

> The conception of a dog indicates a rule, according to which my imagination can delineate the figure of a four-footed animal in general, without being limited to any particular individual form which experience presents to me, or indeed to any possible image that I can represent to myself *in concreto*.
>
> (p. 103 of Kant 1900)

Categories and concepts are general schemata in the service of conceptual knowledge (Locke 1706; Rosch 1999; Murphy 2004; Rogers and McClelland 2006; Kemp and Tenenbaum 2008; Carey 2011). A second type of abstraction is called task rule or task set, which organizes serial behavior in order to achieve a goal (Harlow 1949; Bunge and Wallis 2007; Sakai 2008; Mansouri et al. 2020). For instance, to make a smoothie, one first prepares fruits (pineapple, papaya, kiwi) and vegetables (celery, avocado, cucumber), puts them into a blender and holds the ON button for an appropriate time interval until it's done. This general routine is applicable to specific situations with available ingredients at hand (e.g., the task would not be derailed simply because one ran out of avocados).

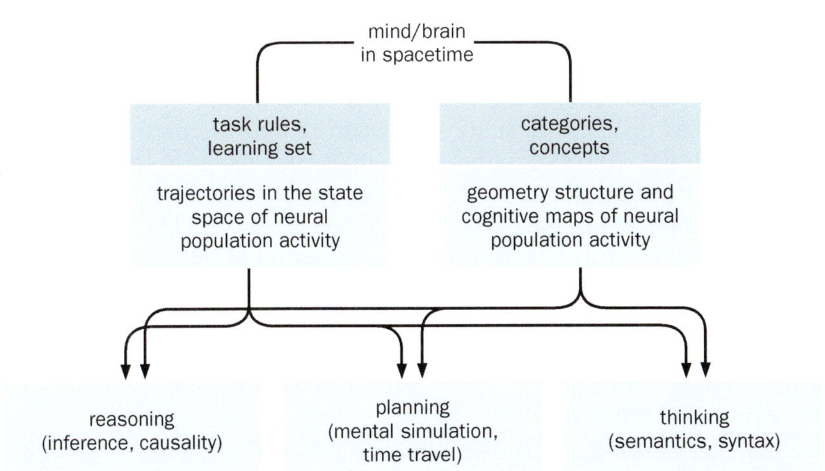

Figure 11.10 Two forms of abstraction. Right: Categories and concepts, their relationships may be organized by a cognitive map in the brain. Left: Task rules (or task sets) organize serial behavior to achieve a goal and learning sets subserve continual learning over time, they may be instantiated by trajectories of neural population activity in the state space. Combinations of these two types of abstraction cement a foundation for reasoning, planning and thinking.

Serial order of behavior is important (Lashley 1951); it would be absurd to turn on the blender before putting fruits and vegetables into it. Comparing to a smoothie, cooking for a crowd with many dishes engages a whole different level of behavioral organization in serial order. Task rules or task sets refer to abstract mental representation that must be maintained in the brain and deployed to guide the temporal sequence of behavioral epochs. This second type of abstraction also includes learning set, an abstract form of knowledge that guides continual learning processes over time.

The two types of abstraction often work in tandem. Consider playing chess (Konidaris 2019): There are an unlimited number of chess sets with different sizes and colors for chessboard and pieces, and each produces a distinct pixel-by-pixel image on the retina. The same is true, albeit to a lesser extent, for a single chess set viewed from different angles or under different lighting. Applicable to any specific chessboard, a child's brain learns to understand a chessboard configuration by forming an abstraction representation in her brain: a chessboard is an 8x8 array of positions, each of which is either empty or contains one of six piece types of one of two colors. A sequence of actions consists of legal moves according to the rules of the game (task set).

The combination of these two forms of abstraction enables complex reasoning, planning and thinking (**Figure 11.10**). Human judgments and decisions are often not determined by a particular past experience. Rather, inferential reasoning is based on our understanding of relations between concepts and unfolds over time (e.g., A is larger than B, B is larger than C, therefore A is larger than C) (Piaget 1923; McGonigle and Chalmers 1977); causality analysis deals with relationships of events across time (Pearl 2009). Semantics (meanings of and relationships between symbols) (Rogers and McClelland 2006) and syntax (sequence of symbols in accordance to a grammar) (Haegeman 2005) are defining characteristics of logic. Hence, neuroscientific elucidation of abstraction in the brain would go a long way toward understanding human intelligence.

Today's AI lacks human-like inductive reasoning about relationships of instances (as in a Sherlock Holmes or Hercule Poirot detective story). Thus, research in this area bridges between neuroscience and AI. Recent experiments in close interplay with computational modeling have began to shed insights into the brain mechanisms of abstraction and reasoning (Behrens et al. 2018).

11.4.1 Categorization

Clustering of the dissimilarity matrix of IT responses to objects, discussed in Section 11.2, does not necessarily mean that IT neurons are selective for categories. Instead, it shows that neural representations of visual information in IT are "untangled" into a suitable format for object classification (DiCarlo and Cox 2007; Hong et al. 2016). As a matter of fact, category learning relies on the interactions between IT and the prefrontal cortex (PFC) (Jiang et al. 2007; Kumaran et al. 2009), and category-selective neurons are found not in IT but the PFC as well as the posterior parietal cortex (PPC) (Freedman et al. 2001; Miller et al. 2003a; Freedman and Assad 2006; Seger and Miller 2010; Cromer et al. 2010; Sarma et al. 2016; Freedman and Assad 2016). Semantic dementia, a mental disorder that causes the loss of the ability to match words with their meanings, vividly described as "the quicksand of forgetfulness" in *One Hundred Years of Solitude* by Gabriel Garcia Márquez (Rascovsky et al. 2009), involves abnormal temporal lobe and PFC (Patterson et al. 2007).

Categories may be innate (e.g., is an approaching animal prey or predator?) or arbitrary with meaning acquired through experience

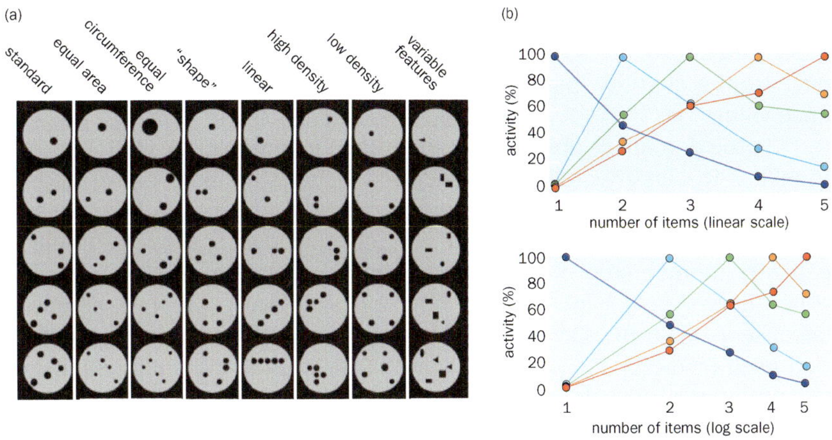

Figure 11.11 Number-selective neurons in the prefrontal cortex. (a) Stimuli used in a monkey experiment. The physical appearance of the displays varied widely for the same quantities. (b) Normalized spiking activity of neurons with peaks at different numbers of items, plotted linearly (top) or on log scale (bottom). Reproduced from Nieder et al. (2002). [Nieder, A., Freedman, D. J., & Miller, E. K. (2002). Representation of the quantity of visual items in the primate prefrontal cortex. Science, 297(5587), 1708–1711.]

(e.g., a chair is defined by its functionality as seating furniture rather than sensory attributes). Numerosity as a symbolic system is uniquely human, but the brain also has a separate non-symbolic system for estimating a quantity of discrete entities (Gallistel and Gelman 2000; Cantlon et al. 2009). This "approximate number system" is present in nonhuman animals (Brannon and Terrace 1998; Dehaene 2011; Nieder 2019) for classification of small numbers (less than 10). In a monkey experiment, Nieder et al. (2002) trained subjects to perform a numerosity-dependent decision task. A visual input consists of a number of filled circles or other shapes (**Figure 11.11a**), and a decision is based on the number (1 through 6) of items regardless of their physical appearances or spatial locations. It was found that in the dorsolateral prefrontal cortex, firing rates of neurons are selective for the number of items in the form of peaked tuning curves (**Figure 11.11b**, top panel). Unlike symmetrical spatial tuning (Chapter 5, **Figure 5.7c**), however, number tuning curves are asymmetrical, with firing rate falling off from its maximum (which defines the preferred number) more slowly with increasing than decreasing number. When plotted with number on log-scale, the tuning curves become more symmetrical (**Figure 11.11b**, bottom panel) (Nieder and Miller 2003). This is interesting because, behaviorally, comparison between two numbers ("is X larger than Y?") depends on the difference relative to their absolute values (Nieder 2019; Cheyette and Piantadosi 2020). This holds true generally for veridical judgment of magnitudes (such as stimulus intensity) as a general principle in psychology. Sensory coding of magnitude on log-scale underlies the Weber-Fechner law, which states that a just-noticeable difference in X is a constant fraction of X (Weber 1834; Fechner 1860; Pardo-Vazquez et al. 2019). To convey the idea, consider judging whether two numbers X and Y are the same or different, using a non-symbolic representation of numbers as in **Figure 11.11a** by nonhuman animals. Correct decision depends on how X and Y differ from each other. If numerosity is represented by firing rate r of number-selective neurons on log-scale, $r \sim \log X$ on a "mental number line" (Dehaene et al. 2004), then numeral judgment is now determined by the difference $r(Y) - r(X) \sim \log(Y/X)$. Let us denote just-noticeable difference by $\Delta X = Y - X$. We have $\log(Y/X) = \log((X + \Delta X)/X) \simeq \Delta X / X$ with small $\Delta X / X$. In other words, just-noticeable difference depends on the relative change. The Weber-Fechner law asserts that $\Delta X / X = c$ is independent of X, in agreement with the intuition that a discernible difference from a quantity (be it intensity, weight, etc.) is a fraction of its reference value.

In the domain of object categorization, Freedman et al. (2001) trained monkeys to classify computer-generated visual stimuli as cat or dog by an arbitrarily defined category boundary (**Figure 11.12a**). In human studies, categorization often requires subjects to judge a shown stimulus's category membership (Rosch 1978; Ashby and Maddox 2005). In such a "one-interval task," interpreting neurophysiological data is ambiguous because a single-neuron activity could be correlated with either classification itself or motor response for report. To avoid this potential confound, Freedman et al. (2001) used a delayed match-to-category (DMC) task, in which visual stimuli (sample and test) were temporally separated by a delay period, and subjects learnt to decide whether they belonged to the same category. Stimuli were morphed objects from three prototype cats and three prototype dogs. Morphing was graded, e.g., x% cat and $(1-x)$ % dog, and classifying the stimulus as cat was correct if x was larger than 50%. Importantly, a pair of stimuli might look alike but belonged to different categories, whereas quite distinct stimuli could share a same category membership. Many neurons recorded from the prefrontal cortex displayed binary responses for the two categories regardless of the amount of evidence x (**Figure 11.12b**). By contrast, spiking activity of neurons in the IT cortex encodes similarity-based object representation in a graded manner. IT neurons also show a weak signal of category with a slow onset, which likely results from its interactions with the PFC (Freedman and Miller 2008). It is noteworthy that, in artificial deep neural networks, categorization is

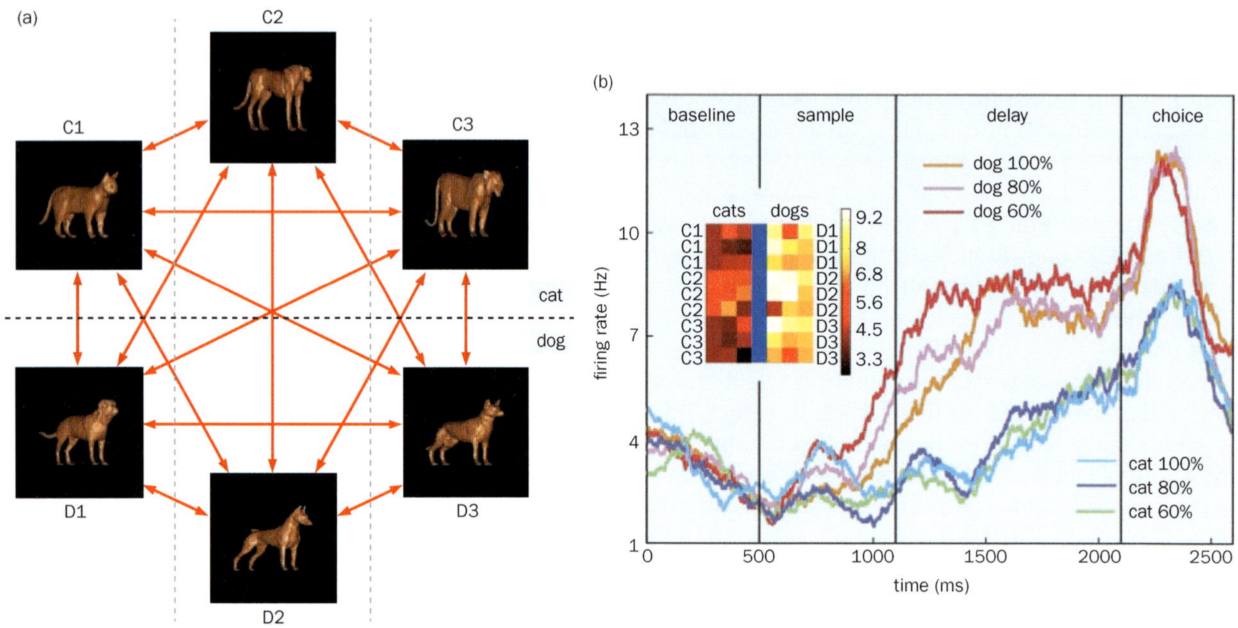

Figure 11.12 Category selectivity of single neurons in the monkey prefrontal cortex. (a) Monkeys learned to categorize randomly generated morphs from the vast number of possible blends of six prototypes. The placement of the prototypes in this diagram does not reflect their similarity. (b) The average activity of a single neuron in response to stimuli at the six morph blends. The vertical lines correspond (from left to right) to sample onset, offset and test stimulus onset. The inset shows the neuron's delay activity in response to stimuli along each of the nine between-class morph lines (see (a)). The prototypes (C1, C2, C3, D1, D2 and D3) are represented in the outermost columns; each appears in three morph lines. A color scale indicates the activity level. Reproduced from Freedman et al. (2001). [Freedman, D. J., Riesenhuber, M., Poggio, T., & Miller, E. K. (2001). Categorical representation of visual stimuli in the primate prefrontal cortex. Science, 291(5502), 312–316.]

produced by a classifier at the output end of the system, as a black box. This is where perception ends and cognition begins.

Category-selective neurons were reported in experiments where visual stimuli were simply random moving dot patterns (Freedman and Assad 2006). The motion direction θ was arbitrarily divided into two classes, category A (red) and category B (blue) in **Figure 11.13a**, with 135°–315° as the category boundary. For instance, $\theta_{sample} = 105^{o}$ is similar to $\theta_{test} = 165^{o}$, but they belong to different categories; it is very different from $\theta_{test} = 345^{o}$, yet both belong to category A. It was found that while neurons in the visual area MT represented veridically θ as an analog feature, single cells in the PFC as well as PPC signaled binary categories (Freedman and Assad 2006; Sarma et al. 2016). The meanings of categories were deduced by experience since their formation was necessary to perform the task in order to garner rewards. Engel et al. (2015) considered computationally how reinforcement learning can underlie categorization. The model had an input MT layer as a ring where neurons were selective for θ, and an output PFC layer was described by the recurrent attractor neural circuit (Wang 2002b; Wong and Wang 2006) for producing a binary choice about the category membership. Between the two, a hidden layer learnt to transform stimulus feature to category. Before learning, the connections were randomly initiated, and RL could train the network for category learning. Notably, neurons in the hidden layer displayed a mixture of feature and category selectivities, a model prediction that was confirmed by analysis of LIP neurons recorded from behaving monkeys, suggesting that PPC plays a role in feature-to-category transformation. A top-down signal from the decision neurons to the hidden layer is required so that synaptic plasticity was influenced by choice outcomes (reward versus nonreward). Computationally, this conclusion is in line with the proposed important role of top-down signaling in plasticity for learning to perform behavioral tasks (Roelfsema and van Ooyen 2005; Roelfsema and Holtmaat 2018).

In cognitive psychology, a prevalent behavioral model of categorization posits that a category is defined by its experienced individual members (exemplars). Generalization to a new item entails computing its similarity with all the retrieved past exemplars, which define a category "prototype" (their "center of gravity") (Posner and Keele 1968; Nosofsky 1986). By contrast, neurophysiology and a neural circuit model highlight the formation of category boundary. In the stimulus space, those with the largest distances from a category boundary can be considered as prototypes. For instance, if the category boundary for classifying θ is 135°–315°, $\theta = 45^{o}$ and $\theta = 225^{o}$ represent prototypes for A and B, respectively. This principle explains a higher categorization error rate for stimuli further away from a prototype (closer to a categorization boundary), because an attractor decision network is prone to errors when sensory evidence is similar for choice alternatives (Wang 2008b). Judging category membership may be ambiguous for stimuli near a classification boundary, as observed behaviorally (Rosch and Mervis 1975); for instance, a Quaker dining chair is more typical than a bean bag for the category of seating furniture. This is explained in the neural circuit model by a smooth softmax decision criterion as a function of the differential evidence for alternative categories. Furthermore, category coding neurons could provide top-down signaling in a neural population that represents stimuli, and therefore exemplar and prototype signals may overlap in certain parts of the brain (Bowman et al. 2020). Such feedback produces a bias by category in purely feature-based perceptual tasks such as delayed match-to-sample (DMS) even when categorization is irrelevant, a widely observed phenomenon called "categorical perception" (Harnard 1987; Goldstone and Hendrickson 2010), which can be modeled by expanding the model of Engel and Wang (2011) (Min et al. 2020).

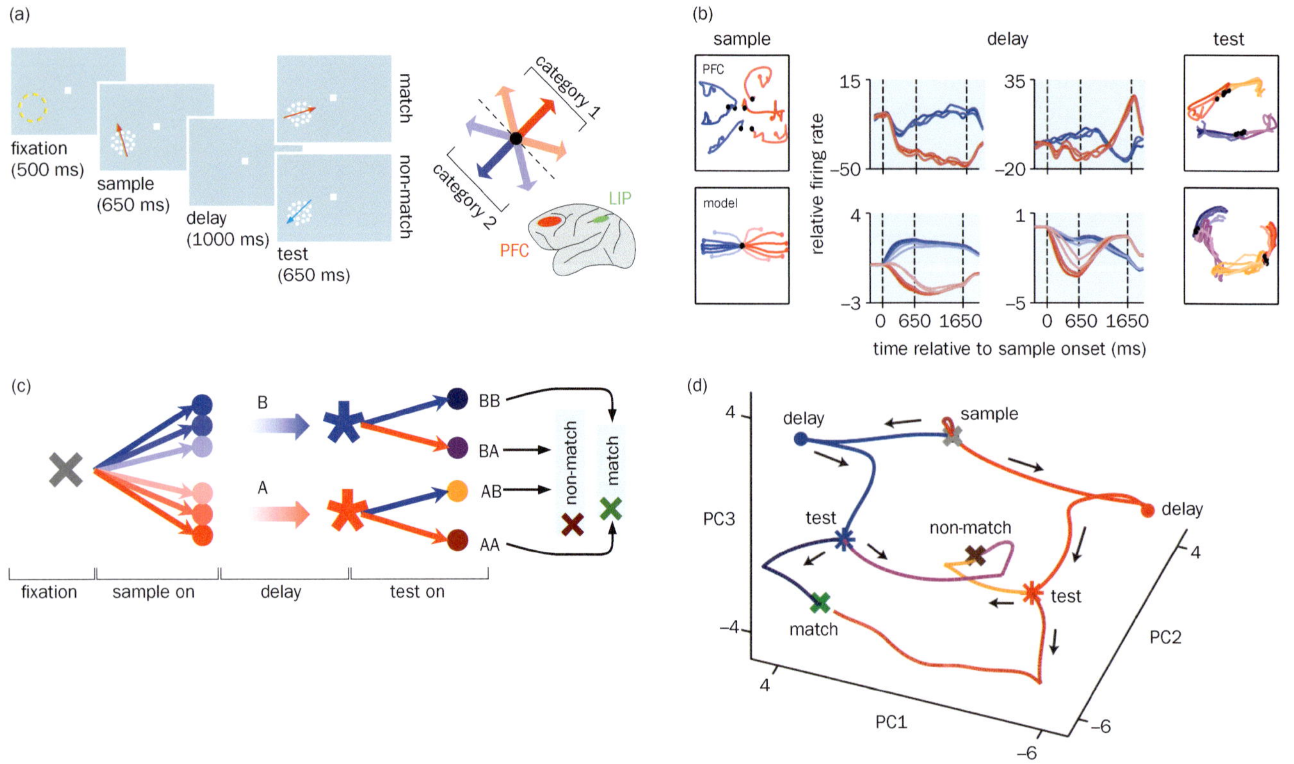

Figure 11.13 A neural circuit for category decisions. (a) Left: The task is to decide whether the test and sample stimuli (visual moving patterns), presented in time separated by a mnemonic delay period, belong to the same category. Right: The two categories are defined based on the motion direction of the stimulus (red: Category A; blue: Category B). (b) Upper row: The population trajectory of PFC neurons during sample period (left) and test period (right), with individual units showing heterogenous activity patterns during the mnemonic delay period (middle). Lower row: An RNN model captures general trends of PFC neural dynamics. (c) Four types of sequence of behavioral epochs. (d) Neural dynamics of an RNN underlying the performance of the DMC task. The final decision, match (AA or BB) or nonmatch (AB or BA), corresponds to distinct attractor states located at separate positions in the state space. Similar trajectories of population activity have been found in experimental data. Reproduced from Chaisangmongkon et al. (2017). [Chaisangmongkon, W., Swaminathan, S. K., Freedman, D. J., & Wang, X. J. (2017). Computing by robust transience: How the fronto-parietal network performs sequential, category-based decisions. Neuron, 93(6), 1504–1517.]

How would a neural circuit not only learn categorization but also carry out sequential decisions such as those in a DMC task? This question was addressed by Chaisangmongkon et al. (2017) using an RNN trained by a Hessian-free algorithm (Martens and Sutskever 2011). The sensory input was a directional angle, and the network was initiated with random connections and could not do any computation. After training to perform the DMC task, an RNN shows diverse neural activity patterns similar to neurons recorded from the PFC and PPC in monkeys doing the same task (**Figure 11.13b**). **Figure 11.13c–d** displays the RNN population trajectory for the entire time course of a behavioral trial. During the sample period, population trajectory converges over time to different parts of the state space according to the category (red or blue) of six sample stimuli. During the delay period, self-sustained activities maintaining category coding display considerable temporal variations and cell-to-cell heterogeneity. At the onset of the test stimulus, population state is indicated by the stars at two possible locations for the sample category, A or B. Depending on the category of the test (A = red or B = blue), there are four possibilities: AA (red), BB (blue), AB (orange) and BA (purple) (**Figure 11.13c**). Neural population trajectory converges to one of two attractor states for match (AA, BB) or nonmatch (AB, BA) (**Figure 11.13d**), the same dynamical behavior was found in the population activity trajectories of recorded PFC neurons. Moreover, recurrent units display a wide range of onset latencies for category selectivity of the sample stimulus, as observed experimentally. The comparison between model and experimental data is remarkable, given that the model network is initiated with random connections and all emerge from training. Furthermore, RNN connectivity developed during learning exhibits strong mutual excitation among units with the same category selectivity, and mutual inhibition between units with opposite categories. This is similar to the attractor network model of Wang (2002b) in support of the idea that categorization and perceptual choice can be instantiated by a common neural circuit mechanism (Freedman and Assad 2016).

11.4.2 Factorized Code for Abstract Knowledge

In **Figure 11.13d**, a linear classifier can easily separate category A (red) and B (blue) of the sample during stimulus presentation and delay period by a two-dimensional decision plane (e.g., PC1 = 0), insensitive to exemplars. During the decision epoch, match (brown cross) and nonmatch (green cross) can be decoded by another two-dimensional decision plane (e.g., PC2 = −4), robustly regardless of particular instances. Hence, representation of categories and that of match/nonmatch decisions are "factorized." This type of factorized code has been proposed in machine learning for efficient memory storage of abstract knowledge (Higgins et al. 2017).

In search for the neural basis of abstraction such as behavioral rule, we need a way to assess whether a neural representation is indeed in an abstract format. The answer is straightforward when there is a "pure" abstract representation, exemplified by "rule cells" in the PFC with neural firing insensitive to stimuli or sensory cue that instructs the rule (Wallis et al. 2001) (Chapter 5, **Figure 5.9**). Commonly, however, PFC neurons display nonlinear mixed selectivity (Rigotti et al. 2010; Rigotti et al. 2013; Parthasarathy et al. 2017; Dang et al. 2022), and hence the majority of them may be correlated with other behavioral attributes (stimuli, values and actions) as well as rules. In such more complicated cases, a neural population representation of abstraction can be generally assessed using multi-dimensional scaling (MDS): if the analysis reveals that the population activity states for different stimuli and actions fall into separable clusters that correspond to behavioral rules, then one can conclude that the circuit encodes rules in an abstract format. This was indeed found experimentally (Bernardi et al. 2020; Minxha et al. 2020). In a monkey experiment, subjects performed a context-dependent stimulus-response-outcome (SRO) association task (**Figure 11.14a**). In rule 1, four visual objects were mapped to two possible responses (hold or release a lever), which led to either reward or nonreward. The stimulus-response mapping needed to be correct even in a nonrewarding trial, or otherwise there would be a timeout and repetition of the same trial. After a block of trials, the mapping suddenly changed to another one (context 2), and subjects adjusted behavior accordingly. Animals learnt to quickly adapt to a change of behavioral rule (**Figure 11.14b**) through numerous transitions back and forth between the two types of trial blocks.

MDS analysis of neural activities revealed an interesting signature of abstract neural representation in the hippocampus and prefrontal cortex. Concretely, at any moment in time, one of eight trial types corresponds to a vector in the N-dimensional state space of neural population activity (N is the number of recorded cells). A neural presentation of context would link the four SRO instances for each context in the form of a cluster in the state space of neural population activity. In a block design, subjects needed to maintain internally across trials the context information, which thus should be present during the fixation period at the start of a trial in the monkey

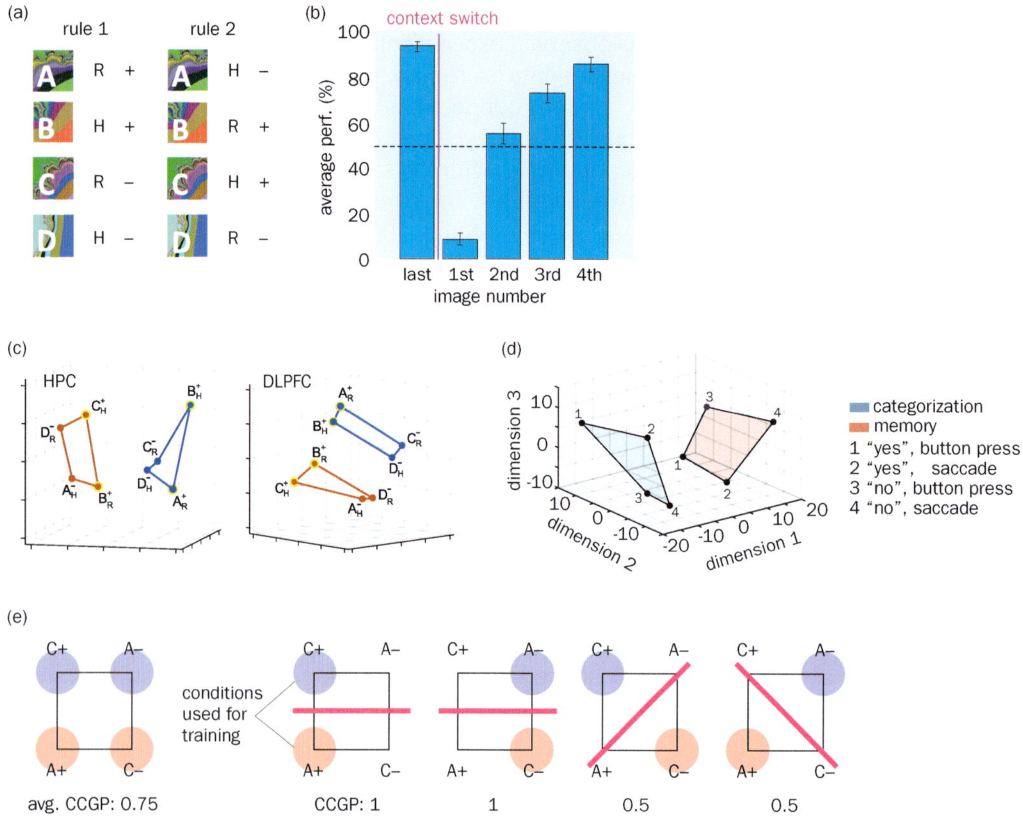

Figure 11.14 Neural representation in abstract format. (a) In a monkey experiment, subjects learn to associate four visual objects with two responses (H: Hold, R: Release a lever), which may (+) or may not (–) yield a reward. Two stimulus-response-outcome mappings alternate between blocks of trials. (b) A monkey shows a rapid switch after rule transition in performance. (c) Neural population activities from the hippocampus (left) and the dorsolateral prefrontal cortex (right) show an abstract representation of behavioral rules in the form of two parallelograms in the state space. (d) Similar findings from the MDS analysis of neural activity of the medial prefrontal cortex in a human experiment using two (memory and categorization) tasks. (e) Illustration of the cross-condition generalization performance (CCGP) measure. In this example, there are two stimuli (A and C) and two rules (purple and orange). A classifier (magenta line: Decision criterion) is trained using a group of sample conditions (shaded), then applied to the complementary group of conditions (not shaded). Panels (a–c) and (e) are reproduced from Bernardi et al. (2020), (d) from Minxha et al. (2020). [Bernardi, S., Benna, M. K., Rigotti, M., Munuera, J., Fusi, S., & Salzman, C. D. (2020). The geometry of abstraction in the hippocampus and prefrontal cortex. Cell, 183(4), 954–967.] [Minxha, J., Adolphs, R., Fusi, S., Mamelak, A. N., & Rutishauser, U. (2020). Flexible recruitment of memory-based choice representations by the human medial frontal cortex. Science, 368(6498), eaba3313.]

experiment. Figure 11.14c shows the results of an MDS analysis of neural signals prior to stimulus presentation, with two clusters in the shape of parallelograms corresponding to two rules in a low-dimensional subspace. Further analysis showed that such representations in abstract format did not rely on the contribution of neurons with selectivity for only one variable, indicating that neurons with mixed selectivity for multiple variables likely play a key role. Notably, similar parallelograms were reported in a human study with single-cell recording from the PFC (Figure 11.14d) (Minxha et al. 2020). Such factorized abstract representation is suitable for generalization: new instances of rule 1 would remain separated from those of rule 2. Furthermore, the parallelogram shape has the advantage of decoding more than one behavioral attribute. For instance, for the DLPFC representation (Figure 11.14c, right), a linear classifier can use a horizontal plane to decode contexts and a vertical plane to decode values (reward versus nonreward) that are insensitive to particular instances.

Bernardi et al. (2020) proposed a quantitative metric of the extent to which a variable (e.g., rule) is represented in an abstract format, called cross-condition generalization performance (CCGP). To calculate CCGP, trials are split into a training set and a test set. In the simple example of **Figure 11.14e** with two (A and C) stimuli under rule 1 (purple) and rule 2 (orange), there are four possible ways to choose training samples (with one instance from each rule). In each case, a classifier is trained to discriminate the two rules using a training sample, then applied to the other samples. For instance, in the first case, the classifier realizes a horizontal criterion to separate C+ from A+, which then predicts the discrimination of A– and C–, thus CCGP is 1. The same holds true in the second case. By contrast, in the third and fourth cases, the classifier produces chance-level performance for the instances not used in training, and CCGP is 0.5. The average CCGP over all instances is 0.75 in that example; it would be close to 0.5 for a random neural representation, and close to 1 if coding is in an abstract format. It was found that CCGP is high (0.75–0.95) for neural representations in the hippocampus and dorsolateral prefrontal cortex in the monkey experiment.

How is an abstract neural representation of behavioral rule formed through experience? One possibility is Hebbian plasticity based on temporal proximity of events within a given rule (Bouchacourt et al. 2020), since in a block design behavior is guided by the same rule across a number of trials, and thus instances in one context are closer to each other in time than are those in different contexts. Appropriate timescales are given by Hebbian learning rates, which can be adjusted by neuromodulators (Hasselmo and Stern 2018; Hall-McMaster et al. 2019). Bernardi et al. (2020) considered a model in which the triplet of stimulus, action and outcome in the previous trial is part of inputs to the network in the current trial. Note that the triplet is sufficient to determine the correct rule in the previous trial, which is therefore assumed to inform the network about the rule in the current trial. How such information is maintained and updated from trial to trial in the brain remains unclear.

11.4.3 Task Set

The prefrontal cortex is known to be engaged in many cognitive tasks (**Figure 11.15**), in part because it plays a critical role in encoding task set, defined as "a configuration of cognitive processes that is actively maintained for task performance" (Sakai 2008). In the brain, a neural circuit that internally sustains a task set reconfigures sensorimotor circuitry to realize behavioral flexibility (Miller and Cohen 2001, Duncan 2010a; Wang 2013; Liu and Wang 2024). What is the circuit mechanism of task set presentation?

To address this question, Yang et al. (2019) trained a single RNN to perform 20 cognitive tasks, which are commonly used in monkey physiological experiments and covered in various chapters of this book: delayed discrimination (Dly DM) (Chapter 5, Section 5.5), multi-sensory integration (MultSen DM) (Chapter 6, Section 6.2.4) with or without a mnemonic delay, delayed match-to-sample (DMS) (Chapter 6, Section 6.6.3), delayed nonmatch-to-sample (DNMS), context-dependent decision making (Ctx DM 1 and Ctx DM 2) (this chapter, Section 11.3), delayed match-to-category (DMC) (this chapter, Section 11.4.1), delayed nonmatch-to-category (DNMC), go (pro-response) or anti-response (Chapter 8, Section 8.2.4) with or without a mnemonic delay, etc.

To simulate these tasks with a single RNN, a suitable setting has two stimulus pathways (Mod 1, Mod 2), each represented by an analogue feature (**Figure 11.16a**). For some tasks such as DMC, there is one sensory stimulus (Freedman and Assad 2006). Other tasks involve both stimuli, like context-dependent decision making (Mante et al. 2013) where motion direction and color are encoded by analog quantities, or multisensory integration where two pathways may convey auditory and visual inputs. In

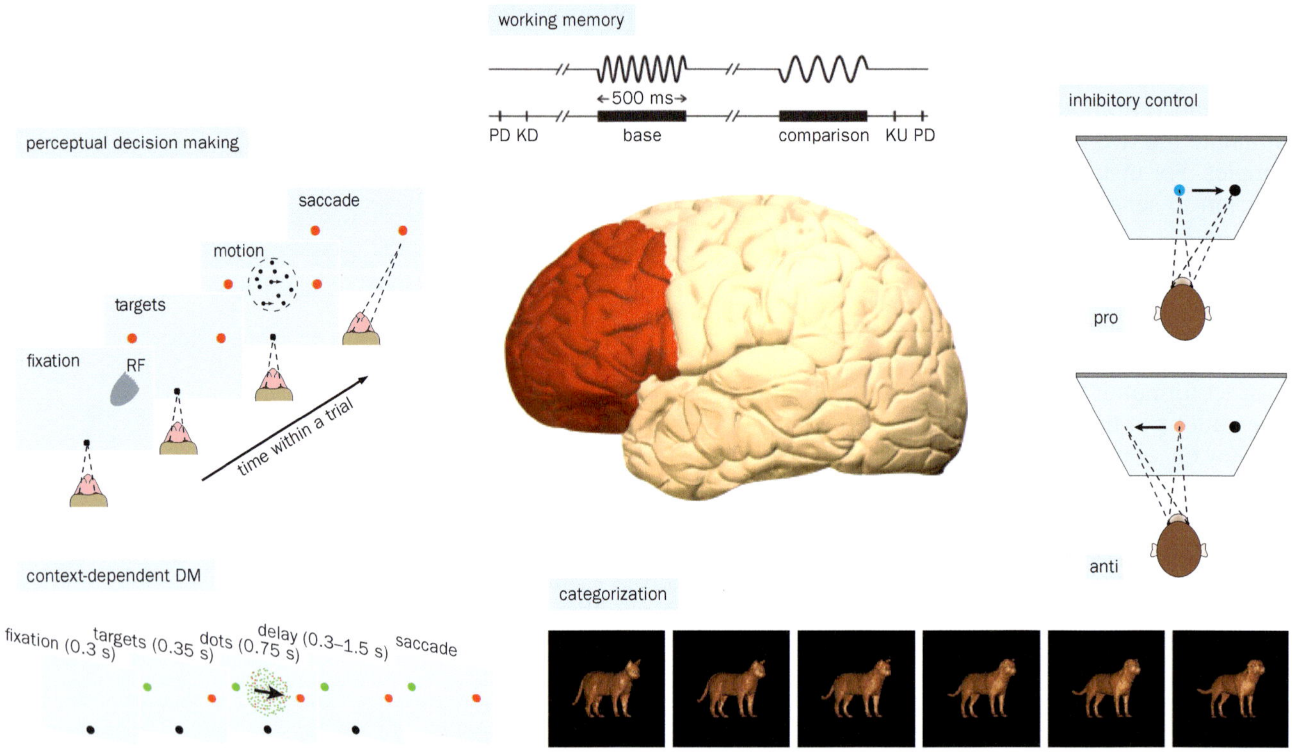

Figure 11.15 The frontal lobe (indicated in red on the human brain) plays an essential role in many cognitive functions, including working memory, perceptual judgment, context-dependent decision making (DM), inhibitory control in tasks like pro- versus anti-saccade, and categorization. [Romo, R., Brody, C. D., Hernández, A., & Lemus, L. (1999). Neuronal correlates of parametric working memory in the prefrontal cortex. Nature, 399(6735), 470–473.] [Gold, J. I., & Shadlen, M. N. (2007). The neural basis of decision making. Annual Review of Neuroscience, 30(1), 535–574.] [Mante, V., Sussillo, D., Shenoy, K.V., and Newsome, W.T. (2014). Context dependent computation by recurrent dynamics in prefrontal cortex. Nature 503(7474), 78–84.] [Munoz, D. P., & Everling, S. (2004). Look away: The anti-saccade task and the voluntary control of eye movement. Nature Reviews Neuroscience, 5(3), 218–228.] [Freedman, D. J., Riesenhuber, M., Poggio, T., & Miller, E. K. (2001). Categorical representation of visual stimuli in the primate prefrontal cortex. Science, 291(5502), 312–316.]

addition, there are 20 separate rule inputs; in a given trial, one of them is presented to dictate which task the network should perform. On the output side, there is a fixation unit and a ring for action in any specific direction. The fixation unit shows a high activity, e.g., during a mnemonic delay; except for reaction time (RT) tasks, a response should be produced only when the fixation input is withdrawn at the end of a trial.

Using a protocol where rules are randomly interleaved across trials, the model learns to accomplish all 20 tasks (**Figure 11.16b**). After learning, one can "open the box" and examine the inner working of the trained RNN. For instance, what is the internal representation of working memory? What in the network underlies better judgment by multisensory integration than that informed by a single stimulus? How is a vector inversion from a stimulus to its diametrically opposite location computed in the anti-saccade task? Importantly, all of these taks as well as categorization and decision making happen in a single RNN, which reconfigures itself according to the task rule currently in play.

The model of Yang et al. (2019) offers a platform to ask a new question: how are different task sets represented by recurrent units in the model? To quantify task representation rather than stimulus or response coding, consider the time course of a neuron's activity, which varies with stimulus condition for a task A (**Figure 11.17a**, top panel). The task variance computed over stimulus conditions and time yields a task value,

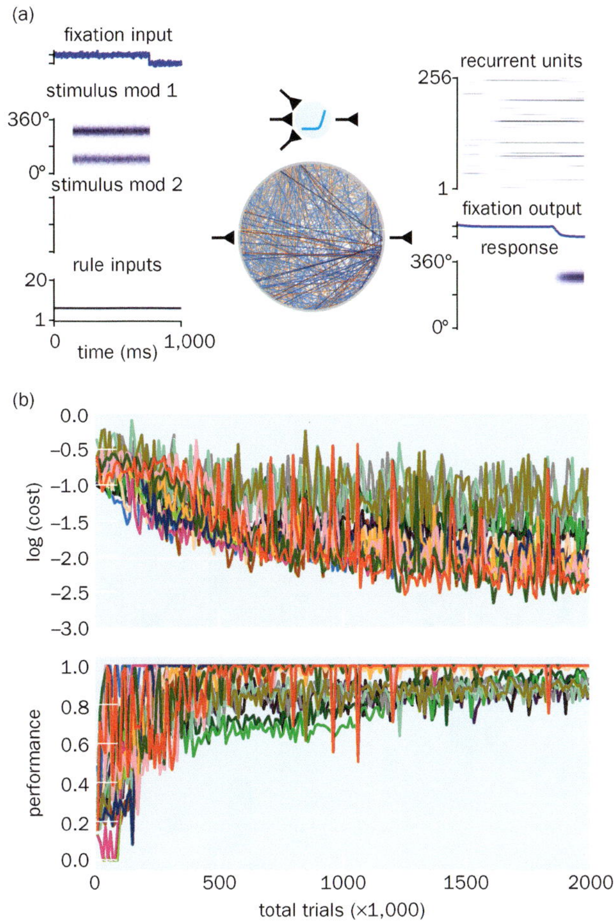

Figure 11.16 An RNN trained to perform 20 cognitive tasks. (a) Schematic of model setting. Left: In a trial, the RNN receives a rule cue, sensory stimuli and a fixation signal when the network should not produce a motor response. In this example of motion direction discrimination, the stimulus is shown in Mod 1 pathway. Right: Network dynamics with RNN units (top), fixation unit (middle), motor response unit (bottom). (b) Using randomly interleaved protocol, the network's loss function decreases (top) and performance increases (bottom) through learning for all tasks (shown in different colors). Reproduced from Yang and Wang (2020).

TV(A). This measure quantifies the amount of behaviorally relevant information a unit conveys during a task; it would be zero if the unit does not show any time- and behavioral attribute-dependent variations and thus is not engaged in a task. In a task, recurrent units with different stimulus tuning can have an identical TV value, and the latter thus genuinely probes task set representation.

By computing the task variance for all trained tasks, one can study how an individual unit is differentially selective in all of the tasks (**Figure 11.17a**, bottom panel). For comparison across units, the task variance of each unit is divided by its maximum; the normalized variance over all tasks is between 0 and 1. To analyze the patterns of normalized task variance for all active units, the TVs of each unit form a vector in the 20-dimensional space of tasks (**Figure 11.17b**). Relationships between units can be assessed using clustering algorithms applied to the distances between the task representation vectors. Units were self-organized into distinct clusters through learning (**Figure 11.17c**); those belonging to the same cluster are mainly selective in the same subset of tasks. For instance, anti-saccade primarily engages a distinct cluster #3 (purple). Cluster #8 is involved in both working memory and decision tasks, while cluster #9 participates in delayed response tasks but not much in decisions that do not involve a mnemonic delay. DMS and DNMS strongly engage distinct clusters #10 and #11, respectively. To further assess the division of labor among different clusters, each is inactivated, and their differential impacts on performances of the 20 tasks are quantified (**Figure 11.17d**): inactivating units in cluster #3 impairs only anti-response tasks but not the others, etc.

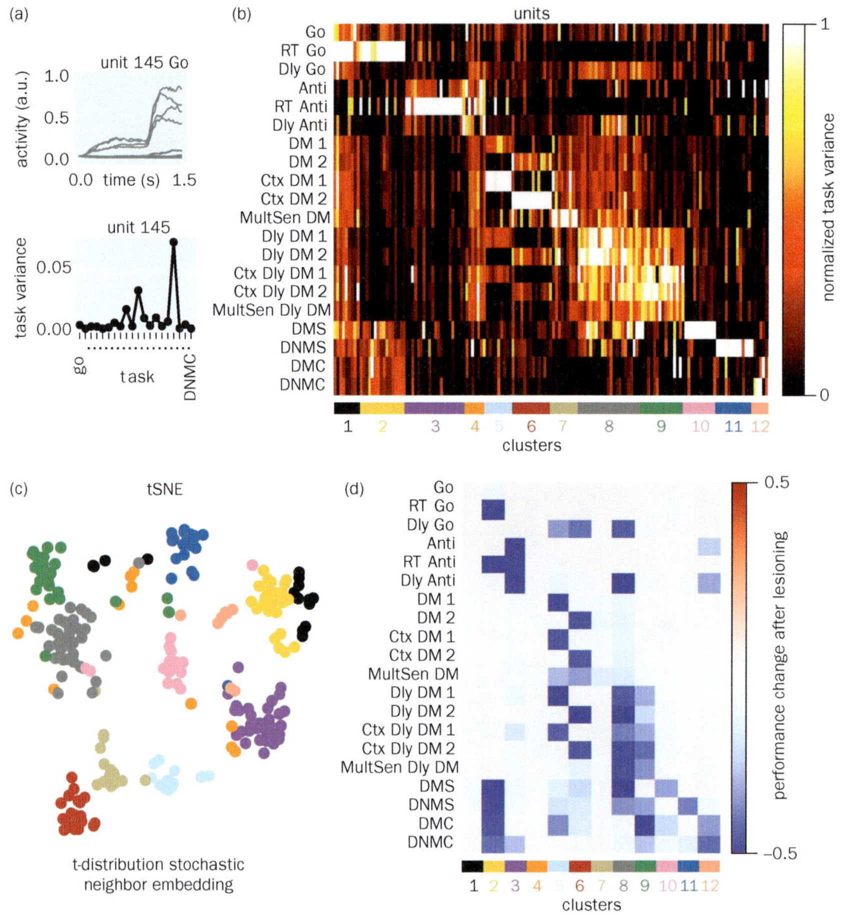

Figure 11.17 Task representations in an RNN model. (a) Time course of a unit's activity in different stimulus conditions (top), where its variance measures its participation in a given task. Each unit's task representation is given by 20 task variance values (bottom). (b) Task variances across all tasks and active units, normalized by the peak value across tasks for each unit. Units form distinct clusters identified based on normalized task variances. Each cluster is specialized for a subset of tasks, such as those that involve a mnemonic delay (Dly). A task can involve units from several clusters; for example, DMS engages clusters #1, 2, 8 and 10. Units are sorted by their cluster membership, indicated by colored lines at the bottom. (c) Visualization of the task variance map. For each unit, task variances form a vector that is embedded in the two-dimensional space using t-distributed stochastic neighbor embedding (tSNE). Units are colored according to their cluster membership. (d) Effect of inactivating each cluster on the performance of all 20 tasks. Reproduced from Yang et al. (2019). [Yang, G. R., Joglekar, M. R., Song, H. F., Newsome, W. T., & Wang, X. J. (2019). Task representations in neural networks trained to perform many cognitive tasks. Nature Neuroscience, 22(2), 297–306.]

We define a measure for the extent to which a neuron participates in two tasks A and B as the fractional task variance (FTV):

$$FTV(A,B) = \frac{TV(B) - TV(A)}{TV(A) + TV(B)} \tag{11.9}$$

which is between −1 and +1, corresponding to a neuron exclusively involved in task A or B, respectively. FTV is close to 0 in case the neuron is equally engaged in both tasks. This value is different for each unit, and thus a histogram can be computed over all recurrent units for each pair of tasks. In this RNN model, the FTV distribution has a peak at zero for some task pairs such as the two context-dependent decisions (Mante et al. 2013), meaning that some neurons show mixed selectivity. It is bimodal with peaks at −1 and +1 for other tasks, such as perceptual decision and anti-saccade, which are realized by distinct neural pools.

The FTV measure is readily applicable to single neural spiking data from neurophysiological experiments where nonhuman animals perform two or more tasks (Mansouri et al. 2006; Mirabella et al. 2007; Mante et al. 2013; Eiselt and Nieder 2013; Siegel et al. 2015; Duan et al. 2015; Okazawa et al. 2021; Panichello and Buschman 2021; Liu et al. 2021). Furthermore, analysis of neural population activity will shed insights into collective network encoding of task sets in the brain.

11.5 LEARNING-TO-LEARN

In Yang et al. (2019), training an RNN was done with task rules randomly interleaved across trials. One would expect that a better strategy is to design a curriculum in which tasks are learnt sequentially from the simplest to more complex ones like learning at school (Wang et al. 2021; Kepple et al. 2022): Go followed by Anti; perceptual decisions followed by context-dependent decisions; working memory followed by DMS and DMC. It turns out that when the model is trained using such block design, one is confronted with the so-called catastrophic forgetting phenomenon. Namely, learning more than one task or even a single task in a nonstationary environment alters connection weights, resulting in abrupt erasure of knowledge acquired through previous experience (McCloskey and Cohen 1989; French 1999; Parisi et al. 2019). In other words, new learning happens at the expense of existent memories.

A possible solution is to add a regularization term to the loss function, which preserves those weights that are important for performing previously learnt tasks (Kirkpatrick et al. 2017; Zenke et al. 2017). This is schematically shown in **Figure 11.18a**, where the loss function is plotted as a function of the modifiable parameter θ (connection weights) for task 1 (green) and task 2 (red). Using a conventional algorithm (traditional learning,

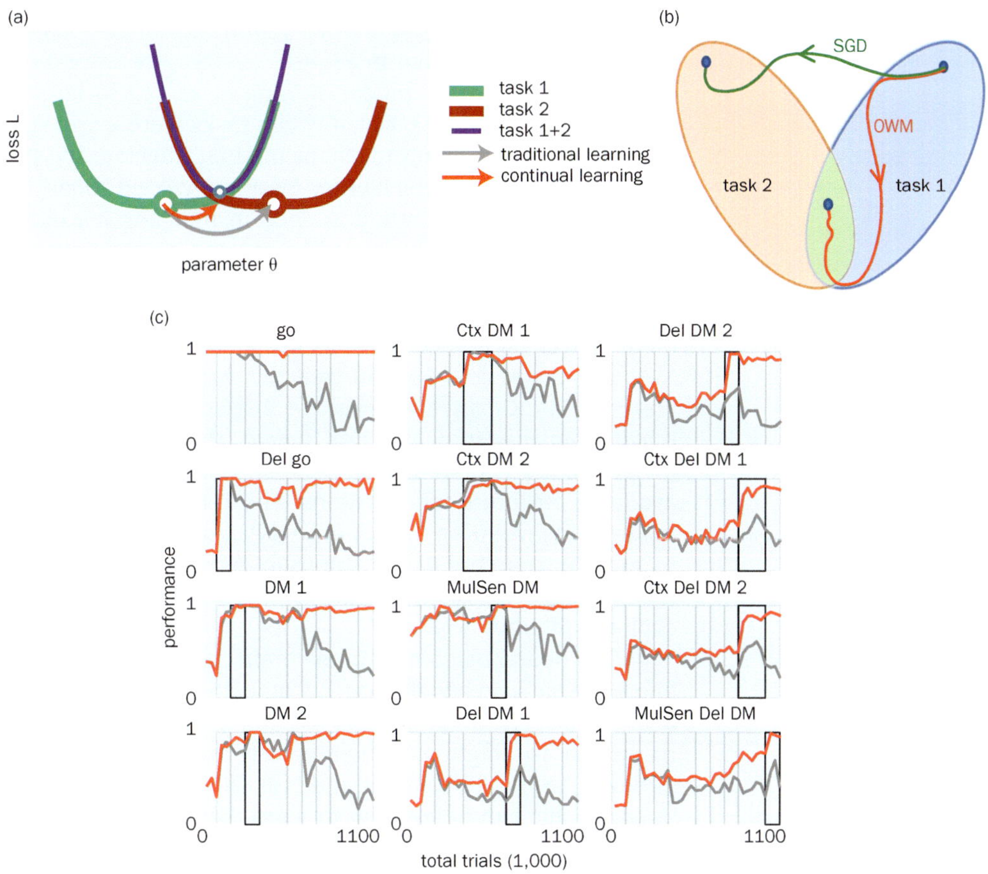

Figure 11.18 Sequential training of cognitive tasks. (a–b) Continual learning compared with traditional learning. (a) Cost function for task 2 (red) leads to the forgetting of task 1 (green). An added regularization shifts the red cost function to the purple one suitable for performance of both tasks 1 and 2. (b) Task 1 and task 2 require connection weights in their respective domains of the weight space (blue and orange, respectively). Learning task 2 by traditional stochastic gradient descent (SGD) typically brings the system to a part of the orange region detrimental for task 1, whereas orthogonal weight modification drives the system to the overlapping region (green). (c) An RNN model endowed with continual learning with block design to solve 20 cognitive problems sequentially. Example networks used traditional (grey) or continual (red) learning techniques, respectively. For each task, the black box indicates the period in which this task was trained. Panels (a) and (c) are reproduced from Yang et al. (2019), (b) from Zeng et al. (2019). [Yang, G. R., Joglekar, M. R., Song, H. F., Newsome, W. T., & Wang, X. J. (2019). Task representations in neural networks trained to perform many cognitive tasks. Nature Neuroscience, 22(2), 297–306.] [Zeng, G., Chen, Y., Cui, B., & Yu, S. (2019). Continual learning of context-dependent processing in neural networks. Nature Machine Intelligence, 1(8), 364–372.]

grey arrow), when task 2 is learnt, the network parameter is shifted to the minimum of the red loss function, which is far from the minimum of the green loss function, and consequently memory about task 1 is forgotten. By contrast, continual learning (red arrow) avoids too many changes crucial for performing task 1, by realizing a new loss function (purple) with a minimum where the red and green loss functions are both close to their smallest values. In an alternative algorithm of orthogonal weight modification (Zeng et al. 2019) (**Figure 11.18b**), the training process searches for configurations in the weight space that can accomplish task 2 (orange area), within the subspace that enables the network to accomplish task 1 (blue area). A successful search (red trajectory) stops at a location inside the overlapping subspace (green area). In comparison, the traditional solution obtained by stochastic gradient descent search (SGD) is more likely to end outside this overlapping area (green trajectory).

Concretely, one way to introduce regularization for continual learning in the RNN model is as follows. When the system learns to perform the k-th task, an extra term is added in the loss function as $\lambda_k \sum_{i,j}(W^{\text{rec},*}_{i,j} - W^{\text{rec}}_{i,j})^2$, where $\mathbf{W}^{\text{rec},*}$ is the synaptic weight matrix at the end of the last task. Consequently, $\mathbf{W}^{\text{rec}}$ undergoes modifications while minimizing its difference with $\mathbf{W}^{\text{rec},*}$, which is useful for performing the previous tasks. The hyperparameter λ_k controls the relative importance of this penalty in the overall cost function. As shown in **Figure 11.18c**, incorporation of this mathematical trick into the model of Yang et al. (2019) largely mitigated the catastrophic forgetting problem when tasks were learnt sequentially.

What are the brain mechanisms that alleviate the problem of catastrophic forgetting during lifetime learning (Wixted 2004)? First, connection weight variables currently described in machine learning are simplistic. Kaplanis et al. (2018) proposed to use the cascade synapse model (Fusi et al. 2005; Benna and Fusi 2016) (Chapter 4, **Figure 4.20**) in training an artificial neural network. Deeper internal synaptic states have increasingly slower memories, thereby keeping previously acquired knowledge over all timescales. Second, reactivation through replay (Colgin 2016; Foster 2017) (Chapter 4, **Figure 4.22**) could either consolidate memory traces (Kaplanis et al. 2020) or "freeze" those synaptic weights important for previously acquired memories, and only the remaining synapses are available for future modifications. Third, memory can be consolidated at the system level (Kaplanis et al. 2019), akin to the transfer from an initial memory trace in the hippocampus to permanent storage in the neocortex (McClelland et al. 1995; O'Reilly et al. 2014). These ideas remain to be fully explored in future research.

The neural machinery in the brain can not only alleviate catastrophic inference but also benefit from accumulated knowledge. This "learning-to-learn" ability (Thrun and Pratt 1998) makes it possible to design school curriculum, according to which previously learnt materials serve as the basis of new learning (Blakemore and Frith 2005; Dehaene 2020). To probe the brain mechanism of learning-to-learn, studies can be carried out on how animals learn to solve a series of motor problems (Braun et al. 2010) or sensorimotor mappings (Sohn et al. 2020) for which neuronal recordings and circuit dissection are feasible.

The simplest form of such learning is association between sensory stimuli and rewards (Harlow 1949). John M. Harlow described an experiment where monkeys solved a series of problems, one after another. In a given problem, monkeys were shown two visual objects, and one was associated with reward while the other was not; subjects needed to point to the former object in order to collect a reward. After the subjects solved the problem, a new problem was presented with two different visual objects, etc. Initially, finding the solution took many trials (**Figure 11.19a**,

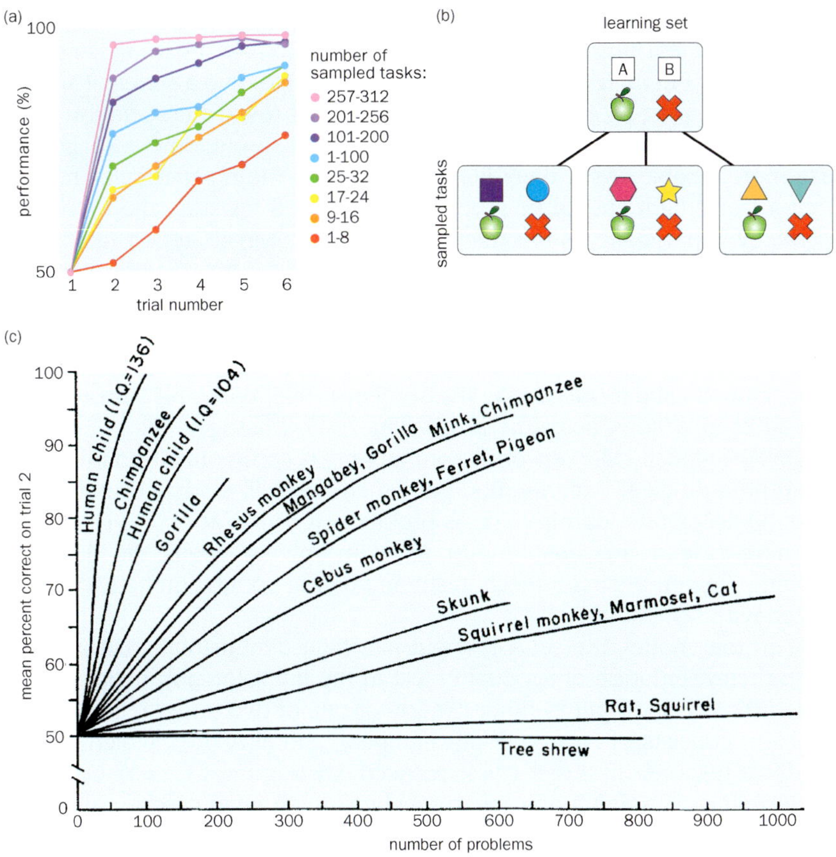

Figure 11.19 Learning-to-learn and learning set. (a) Gradual speedup leads to one-shot learning in a stimulus-reward association task where one of two visual objects yielded award, but the other did not, performed by monkeys. (b) Learning set (A, reward; B, nonreward) deployed to different pairs of sensory stimuli. (c) Meta-analysis across different species, with primates but rarely lower animals showing learning-to-learn. Panels (a–b) are reproduced from Behrens et al. (2018), with raw data from Harlow (1949), (c) from Warren (1974). [Behrens, T. E., Muller, T. H., Whittington, J. C., Mark, S., Baram, A. B., Stachenfeld, K. L., & Kurth-Nelson, Z. (2018). What is a cognitive map? Organizing knowledge for flexible behavior. Neuron, 100(2), 490–509. Raw data comes from Harlow, H. F. (1949). The formation of learning sets. Psychological Review, 56(1), 51.] [Hodos, W. (1970). Evolutionary interpretation of neutral and behavioral studies of living vertebrates. In (F. 0. Schmitt, Ed.), The Neurosciences: Second Study Program. New York: Rockefeller University Press. Pp. 26–39.]

red). Learning gradually sped up and, by the end of the experiment, the subjects exhibited one-shot learning (**Figure 11.19a**, light purple), which was the optimal solution (if one gets it wrong in the first trial, the correct answer must be the alternative option). Harlow designated "learning how to learn a kind of problem" by the term "learning set," which can be applied to any concrete problems (**Figure 11.19b**). Comparative studies suggest that such learning-to-learn ability is a prominent characteristic of primates (**Figure 11.19c**) (Warren 1974). However, caution is warranted for claiming the uniqueness of primates in learning-to-learn ability, because experiments were done with widely different tasks, so learning-to-learn may be observable in a species only when tasks resonate with its naturalistic behavior.

In machine learning, learning-to-learn is commonly considered as "meta-learning," using an algorithm to modify hyperparameters or a learning rule itself (Hochreiter et al. 2001; Li et al. 2017; Finn et al. 2017; Wang et al. 2018b). It is unknown if such a process of meta-learning exists in the brain. Psychology emphasizes the idea that learning is founded on the formation of "schema," a notion originally defined in Philosophy by Kant (1900) as "pure conception of understanding." Schema refers to a knowledge structure that reflects abstracted commonalities of multiple experiences (Piaget 1923; Bartlett 1932; Rumelhart 1980; Gilboa and Marlatte 2017) and can serve to incorporate new knowledge efficiently. The notion of schemas has been investigated using brain imaging (Schlichting et al. 2015; Sommer 2017; Takeuchi et al. 2022) and, only rarely up to now, single-neuron physiology (Tse et al. 2007; Zhou et al. 2020).

Observations of neural correlates are but the first step in our search for the brain basis of schema formation. By solving many problems of the same type and without resorting to meta-learning, can a neural circuit form a learning set that facilitates future learning? If so, what does the schema of a learning set look like neuronally? To investigate these questions, Goudar et al. (2023) trained a vanilla RNN to solve a series of sensorimotor mapping problems. Unlike the stimulus-reward association used in Harlow's original experiment, stimulus-response-outcome associations are more flexible. As a daily life example, there is no particular reason that a traffic light should be green for go and red for stop; the reverse is perfectly fine as long as the rule for stimulus (color) to action (go/nogo) mapping is understood. The association is arbitrary and acquires meaning only through learning in order to achieve a desirable outcome (to safely drive or walk in a city). This is called "conditional sensorimotor learning," which is known to critically depend on the frontal lobe in interplay with the hippocampus and basal ganglia (Passingham 1995; Wise and Murray 2000; Asaad et al. 1998; Pasupathy and Miller 2005; Boettiger and D'Esposito 2005; Fusi et al. 2007). Sensorimotor mapping is commonly used in animal neurophysiological experiments, for which there is evidence that monkeys do accelerate learning across problems (Cromer et al. 2011). Hence, learning-to-learn in sensorimotor problem solving opens the door for probing the neural circuit mechanism and neural population activity associated with schemata.

In this model, the sensory input is a high-dimensional vector mimicking representation of a visual object in the inferotemporal cortex (Bao et al. 2020). The output is either fixation or one of two target responses. A problem consists of two randomly sampled stimuli A or B, which should lead to a response to target 1 or 2, respectively (**Figure 11.20a**). Once the performance in solving a problem reaches a high level (performance criterion), the system transitions to learn a new problem with another pair of stimuli. The RNN is trained to solve many hundreds of problems in a series. As shown in **Figure 11.20b**, solving the first problem takes thousands of trials to learn, which is not surprising because the network model is initiated as a clean slate with random connection weights. Starting with problem 2, the number of trials to reach the performance criterion (N_k) to solve the k-th problem gradually decreases following a time course that can be well fit with an exponential function:

$$N_k = N_{\text{plateau}} + \left(N_2 - N_{\text{plateau}}\right)\exp\left(-(k-2)/\tau_{L2L}\right) \tag{11.10}$$

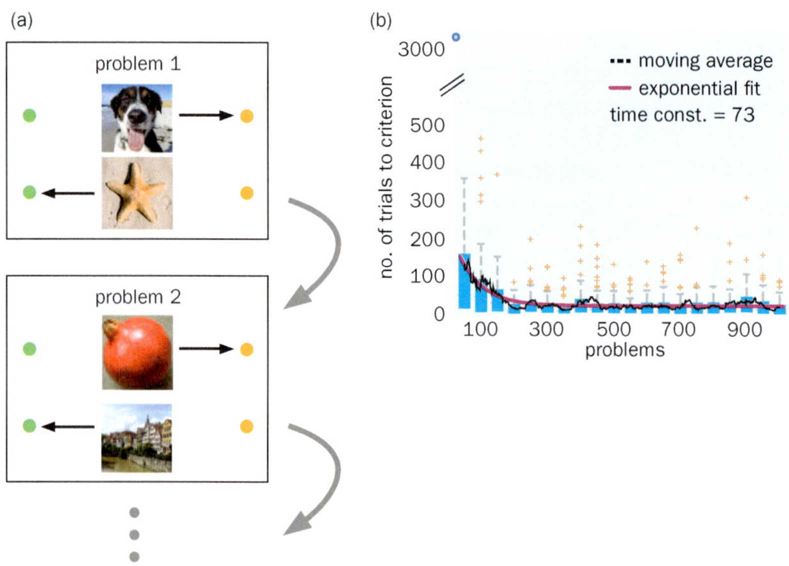

Figure 11.20 Learning-to-learn by an RNN model. (a) A problem is defined by mapping two visual stimuli onto their respective responses (left, target 1; right, target 2). Once performance on problem 1 reaches a criterion, the learning process transitions to problem 2, etc. (b) The number of trials to reach the desired performance criterion is a few thousand for the first problem, followed by a learning-to-learn time course that can be fit with an exponential function (purple curve). Reproduced from Goudar et al. (2023). [Goudar, V., Peysakhovich, B., Freedman, D. J., Buffalo, E. A., & Wang, X. J. (2022). Schema formation in a neural population subspace underlies learning-to-learn in flexible sensorimotor problem-solving. bioRxiv, 2021–09.]

where τ_{L2L} is the learning-to-learn time constant, and N_{plateau} is the steady-state number of trials required for solving a new problem. According to this formula, $N_k = N_2$ with $k = 2$, and N_{plateau} as k increases to ∞. The parameters τ_{L2L} and N_{plateau} quantify the learning-to-learn dynamics.

Since learning is realized by connection weight modifications, schema formation may entail the emergence of a low-dimensional subspace of the weight space, which is deployed to guide new learning. With 100 recurrent units in the RNN model, its initial weight matrix has 100×100 recurrent weights. Exploring in a 10,000-dimensional space takes a long time. If, through experience, a manifold of a few dimensions is formed on which exploration is confined when a new problem is presented, learning can be much accelerated. This low-dimensional manifold would be the circuit substrate of a schema. Indeed, when the system has reached the plateau of learning-to-learn, the modifications of connections are limited to a weight subspace of about four dimensions.

Even though connection weights are typically not directly accessible by neurophysiology in behaving animals, a structural manifold of weights should be reflected by a subspace of neural population activity dynamics. Concretely, trajectories of recurrent neural population activity are decomposed into a decision subspace and a stimulus subspace. The decision subspace is defined by the demixed PCA method (Machens et al. 2010; Keemink and Machens 2019), for which stimulus information is averaged out. The stimulus subspace is orthogonal to the decision subspace (**Figure 11.21a**; central). Trajectories for two decisions (targets 1 and 2) are separated in the decision subspace (blue and red in **Figure 11.21a**, left) but not in the stimulus subspace (**Figure 11.21a**, right). Crucially, a trajectory shown in **Figure 11.21a** corresponds to solving a particular problem, and thus each is from a different network with modified connection weights from one trajectory to another. These different networks share a common low-dimensional subspace, which can reasonably be referred to as the neural substrate of schema formation.

In a set of model simulations, learning to solve problem 2 takes place when the response selection is independent of the decision subspace but

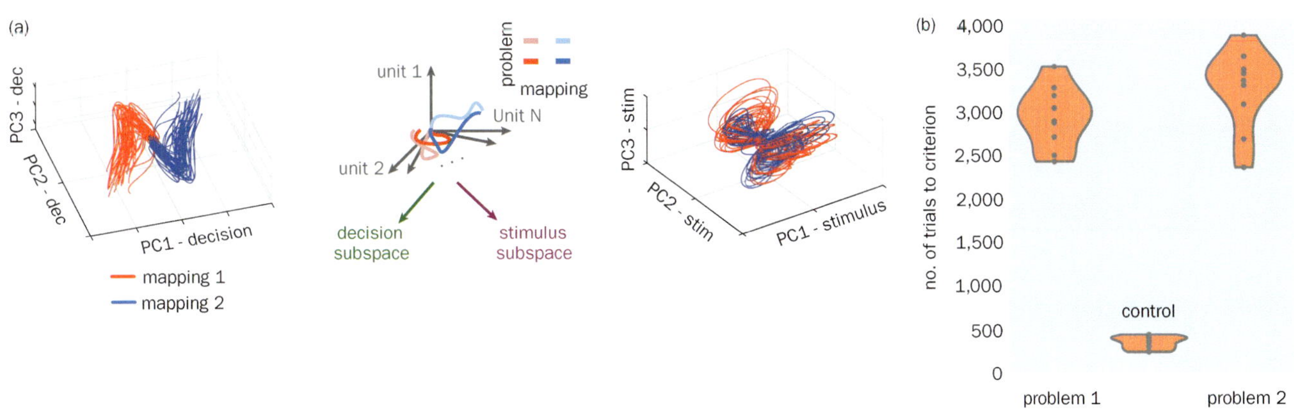

Figure 11.21 Subspace analysis of RNN dynamics during learning-to-learn. (a) Population activity of the RNN lives in a high-dimensional state space, where each axis corresponds to a neuron's firing rate (middle). The network undergoes a trajectory over time, which is different for the two mappings. Schema corresponds to the low-dimensional decision subspace (left) shared by many networks trained to solve a series of problems, and a stimulus subspace (right) is defined as the subspace orthogonal to the decision subspace. Red and blue trajectories are not discriminable in the stimulus subspace. (b) Number of trials to performance criterion for problem 2 (control) is much lower than that for problem 1. If behavioral response from the RNN is readout from the stimulus subspace, the time course of learning problem 2 reverted back to the initial situation with problem 1. Reproduced from Goudar et al. (2023). [Goudar, V., Peysakhovich, B., Freedman, D. J., Buffalo, E. A., & Wang, X. J. (2022). Schema formation in a neural population subspace underlies learning-to-learn in flexible sensorimotor problem-solving. Nature Neurosci. 26, 879–890.]

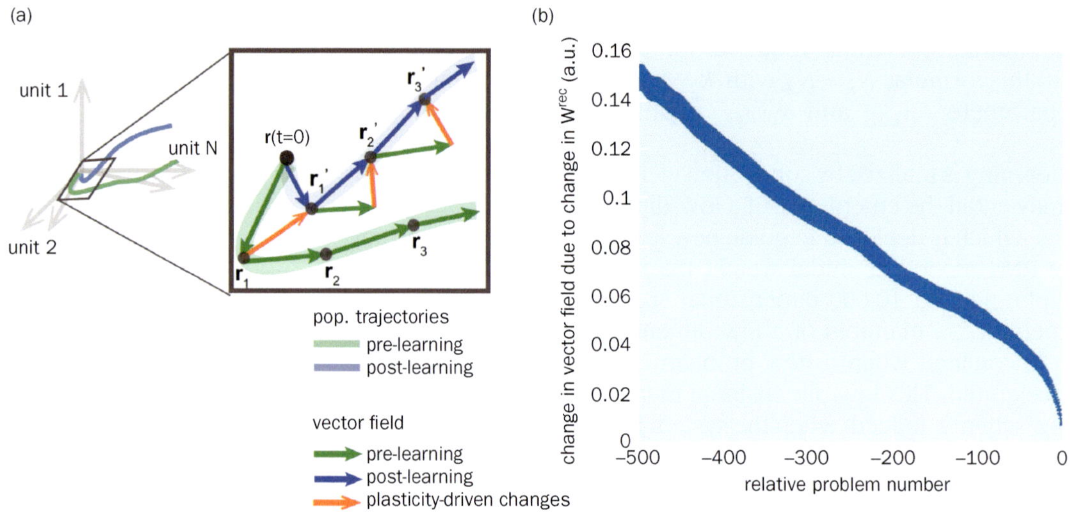

Figure 11.22 Cumulative changes of vector field underlies learning-to-learn. (a) Schematic showing that weight change is multiplied by activity to yield current change (vector field), ultimately producing a change of neural population activity trajectory pre (green) and post (blue) learning a new problem. (b) Vector field changes are additive across problems over previously learnt problems. Reproduced from Goudar et al. (2023). [Modified from Goudar, V., Peysakhovich, B., Freedman, D. J., Buffalo, E. A., & Wang, X. J. (2022). Schema formation in a neural population subspace underlies learning-to-learn in flexible sensorimotor problem-solving. Nature Neurosci. 26, 879–890.]

given by the readout from the stimulus subspace. In that case, the number of trials to reach performance criterion reverts back to that of learning problem 1 (**Figure 11.21b**), demonstrating the importance of the low-dimensional schema manifold for learning-to-learn.

Since a decision is guided by the neural population trajectory that converges to an appropriate place of the state space, what matters for learning is only indirectly linked to changes of connection weights $\Delta\mathbf{W}^{\text{rec}}$ because temporal trajectories of a recurrent network are determined by the vector field of a dynamical system. This is schematically shown in **Figure 11.22a** with a trajectory before learning (green) and after learning (blue) in discrete time steps ($t = j\Delta t$, $j = 1, 2, \ldots$). Starting at an initial population activity vector $\mathbf{r}(t=0)$ (filled circle in the state space), because of weight change (orange), at $t = \Delta t$ the new state ($\mathbf{r}_1{}'$) differs from that before learning ($\mathbf{r}_1$). The next post-learning state $\mathbf{r}_2{}'$ further diverges from that of pre-learning $\mathbf{r}_2$ for two reasons: not only because the weight is modified (orange) but also the starting state is no longer the same as in the pre-learning case, and so on, leading to two (green and blue) distinct trajectories. The changes in the vector field add up across learning experience (**Figure 11.22b**), which underlies learning-to-learn.

In summary, learning-to-learn takes place on three disparate timescales, each subserved by a distinct identifiable substrate: a decision within a trial is given by recurrent neural network dynamics; learning to solve a problem is scaffolded by a low-dimensional decision subspace of neural population dynamics; and learning-to-learn is associated with cumulative changes of the vector field across the entire experience (**Figure 11.23**). The exponential learning-to-learn time course is in contrast to the Harlow study, where monkeys displayed one-shot learning. Neurobiological studies of fast learning with rodents (Rosenberg et al. 2021) and primates (Achterberg et al. 2022) have just begun, future experiments and modeling will uncover the underlying neural circuit mechanism.

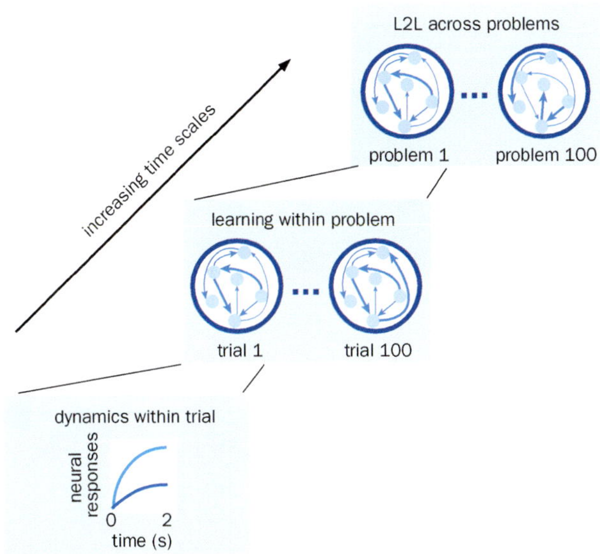

Figure 11.23 Dynamics on three timescales for decision within a trial, learning to solve a given problem across trials, and learning-to-learn (L2L) across problems, respectively. Reproduced from Goudar et al. (2023). [Goudar, V., Peysakhovich, B., Freedman, D. J., Buffalo, E. A., & Wang, X. J. (2022). Schema formation in a neural population subspace underlies learning-to-learn in flexible sensorimotor problem-solving. Nature Neurosci. 26, 879–890.]

11.6 REASONING AND FLUID INTELLIGENCE

11.6.1 Compositionality

Humans reason with language, which is compositional in nature (Pinker 1994), so that the same collection of words can be strung together to express different facts or thoughts. We learn to carry out tasks rapidly by instructions thanks to compositional coding of language. In an experiment, subjects performed a paradigm of tasks instructed by permuted rule operations (Cole et al. 2013a, 2013b). There were four logic rules (same, first, second, just one), four semantics (sweet, loud, soft, green) and four motor responses (press left index, left middle, right index and right middle finger). For instance, suppose that in a trial the instruction is "If the answer to 'is it sweet [semantic]?' is the same [logic rule] for both words, press your left index finger [motor response]"; then the correct response should be yes (left index) if inputs were the words "grape" and "sugar" as they are both sweet, but no (left middle finger, the other finger of the same hand) if input words were "parsley" and "candy" because parsley is not sweet. In a different trial, with the instruction "If the answer to 'is it sweet [semantic]?' is true for the second [logic rule] word, press your right middle finger [motor response]", the response should be yes (right middle finger) for both aforementioned pairs of words, and no (right index finger) if the words were "sugar" and "parsley". Thus $4 \times 4 \times 4$ combinations of these three components produced a number of task rules applicable to arbitrary pairs of inputs that were words. fMRI measurements identified the parieto-frontal circuit at the heart of compositional task set representation (Cole et al. 2013).

Not limited to language, let us define more generally compositionality as the principle that a whole is achieved by employing a set of parts that are combined according to certain rules. Compositionality is a defining characteristic of thought and flexible behavior (Fodor 1975; Duncan et al. 2017; Frankland and Greene 2020). Elucidation of the brain basis of

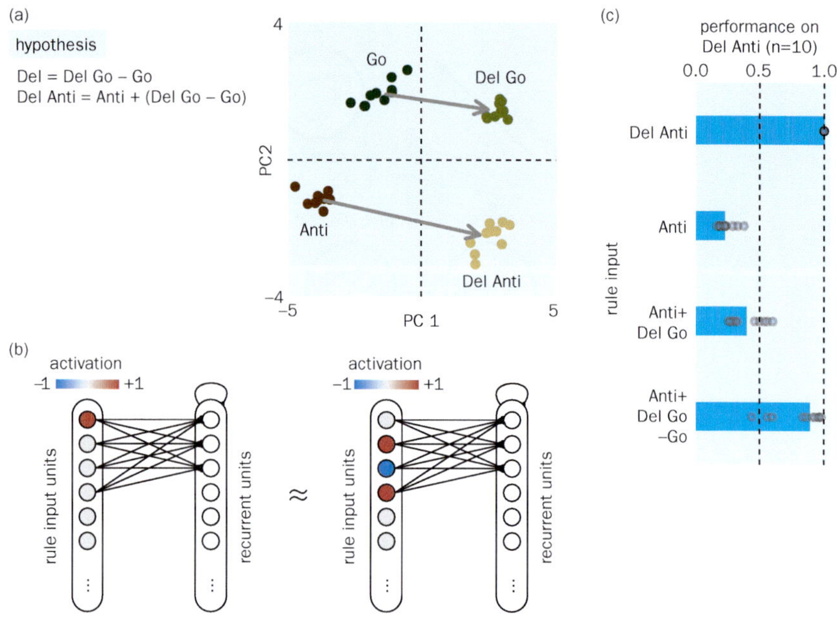

Figure 11.24 Compositionality in an RNN trained to perform 20 cognitive tasks. (a) PCA analysis of population states demonstrates compositionality between the Go, Anti-Go, Del Go and Del Anti-Go tasks. (b) During training, a task is always instructed by activation of the corresponding rule input unit (left). After training, the network can potentially perform a task by activation or deactivation of a set of rule input units meant for other tasks (right). (c) Performance of the delayed anti-saccade task instructed by four different combinations of rule signals. "Anti" rule or "Anti" plus "Del Go" rules yield poor performance, but the accuracy is the same for the "Anti" plus "Del Go" minus "Go" rules and the "Del Anti" rule. Reproduced from Yang et al. (2019). [Yang, G. R., Joglekar, M. R., Song, H. F., Newsome, W. T., & Wang, X. J. (2019). Task representations in neural networks trained to perform many cognitive tasks. Nature Neuroscience, 22(2), 297–306.]

compositionality relies on sensorimotor experiments with both human and nonhuman animals as well as related computational modeling (Smolensky 1990; Eliasmith 2013; Lake et al. 2017). Factorized coding described in Section 11.4.2 has been proposed to facilitate compositionality (Higgins et al. 2017; Behrens et al. 2018), so what neural networks are capable of abstract knowledge endowed with a compositional structure?

Interestingly, in the trained RNN of Yang et al. (2019), simple forms of compositionality emerge from training by experience without language. This is shown in Figure 11.24 for a simple example: delayed anti-saccade. In this task, a stimulus is shown, followed by a delay period (Del), at the end of which the correct response is to generate a response diametrically opposed to the initial stimulus's location (Anti). Can the Del Anti task be decomposed into a sum of "Anti" (without a delay) and "Del Go" (without vector inversion) minus "Go" (response without delay)? PCA analysis of the RNN indeed reveals compositionality (Figure 11.24a): The population activity vector from "Go" to "Del Go" and that for "Anti" to "Del Anti" are parallel with each other; therefore, summation of two ("Anti" and "Del Go") vectors and then subtraction of the "Go" vector yields the "Del Anti" vector. This is tested by using the three corresponding task rules (Figure 11.24b), which reproduces the same performance as in trials where the single "Del Anti" rule is presented to guide the network model's behavior (Figure 11.24c). The model thus exhibits compositionality, which enables combinations of building blocks such as mnemonic delay and vector inversion for anti-response to perform required tasks.

Thanks to compositionality, dynamical motifs such as attractors can be reused to subserve different tasks (Driscoll et al. 2024, Riveland and Pouget 2022).

11.6.2 Inference and Cognitive Maps

Externally oriented sensorimotor behaviors engage a "task-activated" network of cortical areas, including the attention system (Corbetta and Shulman 2002), while activity in the default-mode network (DMN) is reduced related to the brain states in the absence of obvious interactions with the world (Chapter 10, Section 10.6.1 and Figure 10.14). By contrast, inferential reasoning is largely an internally oriented mental process of reaching a conclusion from acquired knowledge, and it activates the DMN. For example, if you know that Alan is Evelyn's father, and Lucian is the son of Alan's uncle, then you can conclude that Lucian and Evelyn are first cousins once removed. This is called "transitive inference" (Piaget 1923). Inference is decision making informed by sequential sampling of evidential information retrieved from memory (Shadlen and Shohamy 2016); it enables us to make judgments or choices based on acquired knowledge applied to novel situations (Zeithamova et al. 2012). Causality analysis (Pearl 2009) is a form of inference about relationships that are diachronic, i.e., between events occurring at different times (Holland et al. 1986).

Inference depends on compositionality of concepts, so we can recombine them (e.g., father, son, aunt, cousin) to form new knowledge (the relationship between Connor and Lucian). Human fMRI studies show that the combinatory process in support of inference engages the hippocampus and the default-mode network (DMN) (Frankland and Greene 2020). The DMN includes the medial prefrontal cortex (mPFC), precuneus and posterior cingulate cortex (Raichle et al. 2001; Raichle 2015), which are less active when subjects are engaged in performing a task than they are "at rest" (when thoughts can meander freely), in an antagonistic interplay with the task-activated brain network. The same system underlies "time travel" in planning that requires consideration of possibilities, internally in the brain, beyond one's past and immediate present into the future (Suddendorf and Busby 2003; Buckner and Carroll 2007; Luhmann et al. 2008; Klein 2013; Mullally and Maguire 2014; Szpunar et al. 2014, Neupane et al. 2024).

What are the neural circuit mechanisms and computational principles of inferential reasoning? Research addressing this question can be traced back to the classic work of Tolman (1948), who observed that foraging food in a maze by rats involved inference based on acquired knowledge. For instance, when a particular path was blocked in the experiment, if behavior was totally dependent on simple stimulus-response associative learning, rats would have to explore available routes randomly before a new solution could be identified. Instead, behavior showed that rats deployed systematic knowledge about the spatial layout (spatial map) of the maze to find a shortcut in reaching the location where reward was delivered. The discovery of place cells in the hippocampus (HP) (O'Keefe and Dostrovsky 1971) (Chapter 3, **Figure 3.10a**) led to the proposal that the HP and associated areas like the entorhinal cortex (ERC) encode a spatial map, and more generally "cognitive maps" (O'Keefe and Nadel 1978). As a matter of fact, the HP is also necessary for non-spatial inference in rodents, as shown by the work of Eichenbaum (2017) and others. In an early experiment, Dusek and Eichenbaum (1997) trained rats to perform a set of two-item olfactory discrimination problems, in which the stimuli were ordered by choice preferences as $A > B$, $B > C$ and $C > D$ (premise pairs) in different trials. A probe for transitive inference could be a trial where B and D were presented, and a correct response indicated that subjects understood $B > D$. Rats performed equally well in probe trials as in trials with learnt premise pairs, at $\sim 80\%$ correct. Lesioning the ERC and perirhinal cortex, or the fornix (which connects the HP to the neocortex), reduced the performance in probe trials to chance level, whereas the accuracy remained intact with premise pairs. The findings with rodents are in consonance with the human literature, showing that the HP is a crucial component of the memory system (Scoville and Milner 1957; Squire 2009) and is involved in statistical learning that allows the brain to extract generalizable knowledge from experienced incidences (Turk-Browne et al. 2010; Schapiro et al. 2016). Notably, the HP projects to medial and orbital parts but not lateral parts of the prefrontal cortex (Aggleton et al. 2015), and therefore is well suited to work in concert with the DMN for inferential reasoning.

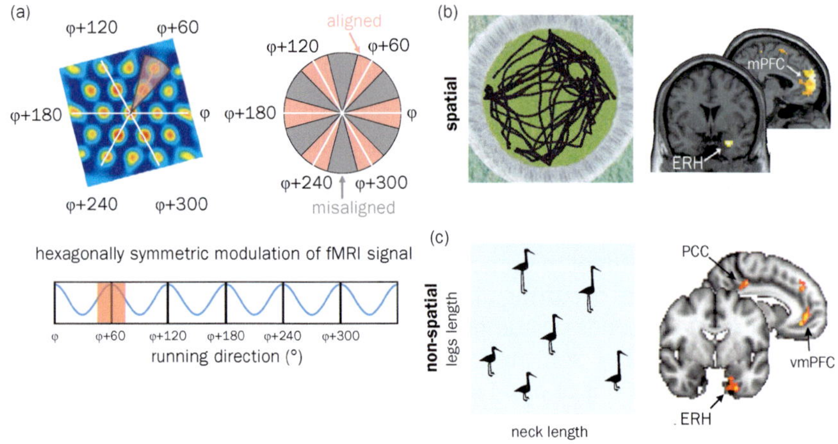

Figure 11.25 Grid-cell-like organization in the human brain. (a) Data analysis in search of grid cells with fMRI. Trajectories through a two-dimensional space can be either aligned or misaligned with the axes of a grid code (white lines denote grid axes). Signal is greater for trajectories aligned versus misaligned with the grid. As a result, the fMRI signal displays a sinusoidal modulation by movement direction (bottom). (b) Significant such modulation of the fMRI signal, providing evidence for grid cells, is observed in the entorhinal cortex (ERH) and medial prefrontal cortex (mPFC) when participants navigate through virtual reality spatial worlds. Left: An aerial view of the spatial task map. (c) The same signals are observed in ERH, ventromedial PFC and posterior cingulate cortex (PCC), when participants navigate through an abstract conceptual space defined by two continuous dimensions: The neck length and leg length of "stretchy birds," suggesting grid cells code for non-spatial dimensions. Reproduced from Behrens et al. (2018), with raw data from Doeller et al. (2010) for (a–b) and from Constantinescu et al. (2016) for (c). [Behrens, T. E., Muller, T. H., Whittington, J. C., Mark, S., Baram, A. B., Stachenfeld, K. L., & Kurth-Nelson, Z. (2018). What is a cognitive map? Organizing knowledge for flexible behavior. Neuron, 100(2), 490–509. Raw data comes from Doeller, C. F., Barry, C., & Burgess, N. (2010). Evidence for grid cells in a human memory network. Nature, 463(7281), 657–661.] [Behrens, T. E., Muller, T. H., Whittington, J. C., Mark, S., Baram, A. B., Stachenfeld, K. L., & Kurth-Nelson, Z. (2018). What is a cognitive map? Organizing knowledge for flexible behavior. Neuron, 100(2), 490–509. Raw data comes from Constantinescu, A. O., O'Reilly, J. X., & Behrens, T. E. (2016). Organizing conceptual knowledge in humans with a gridlike code. Science, 352(6292), 1464–1468.]

Soon after the discovery of grid cells in the rat medial entorhinal (mEC) cortex (Hafting et al. 2005; Moser et al. 2014) (Chapter 3, **Figure 3.22**), fMRI experiments were designed to probe grid-like neural representation of space in the human brain (Doeller et al. 2010). The periodicity and alignment of firing phase between grid cells predict that BOLD signal from the ERH would show a peak when running direction during virtual navigation is aligned with the grid axes separated by 60° and a trough with misaligned directions (**Figure 11.25a**). As a human subject undertook virtual navigation (**Figure 11.25b**, left panel), the ERC and mPFC exhibited significant modulation by the grid axes (**Figure 11.25b**, right panel), in support of a grid-like structure in a human brain network including the ERC and medial PFC. Both rodent (Aronov et al. 2017) and human (Constantinescu et al. 2016; Garvert et al. 2017; Bongioanni et al. 2021) (**Figure 11.25c**) experiments revealed grid-like neural representation of non-spatial information, suggesting that the same system supports cognitive processes beyond spatial navigation.

Grid cells offer a quantification of relationships between locations in a spatial map, or concepts mapped onto a cognitive map. Importantly, grid-like code is naturally compositional: a new relation between two concepts of a cognitive map can be inferred by vector sum from their relations with other concepts, just like navigating between two locations on a spatial map. Theoretical work revealed that grid-like encoding exhibits a large combinatory capacity (Fiete et al. 2008). Relations may also be organized according to other forms of graphs rather than on a spatial grid, such as a family tree (Schuck et al. 2016; Knudsen and Wallis 2020). These findings are opening the door to investigate inferential reasoning in neurobiological terms (Behrens et al. 2018; Morton et al. 2020; Momennejad 2020).

A model called the Tolman-Eichenbaum machine (TEM) was developed that incorporated recent findings into a theoretical framework (**Figure 11.26**) (Whittington et al. 2020; Whittington et al. 2022). Based on neuroscientific research, the TEM model consists of three modules. The lateral entorhinal cortex (lERC, red at the bottom) represents non-spatial sensory stimuli (Deshmukh and Knierim 2011), which can be, e.g., visual objects or words. The medial ERC encodes a cognitive map to define relationships of stimuli in an abstract space (blue at the top); it is modeled as an RNN trained by a machine learning algorithm, similar to previous models of grid cells (Cueva and Wei 2018; Banino et al. 2018; Sorscher et al. 2023). The HP (orange in the middle) encodes a conjunction of sensory and spatial information (Manns and Eichenbaum 2006); it is described as a Hopfield network (Hopfield 1982) (Chapter 4, Section 4.5.2) that learns quickly, by unsupervised Hebbian plasticity, through associations

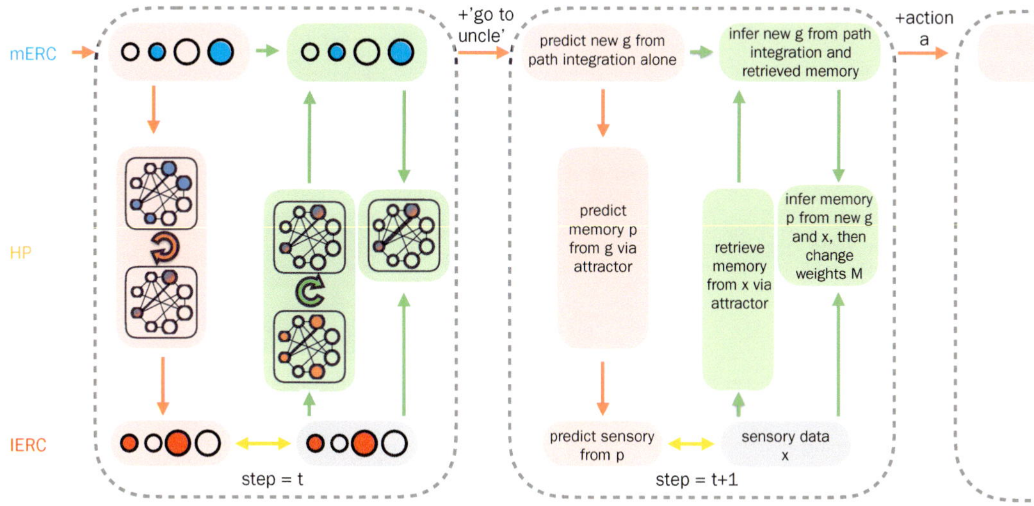

Figure 11.26 The Tolman–Eichenbaum machine of inference. Left: Model architecture (mERC: Medial entorhinal cortex; HP: Hippocampus, lERC: Lateral entorhinal cortex). Right: Key computational processes. Path integration in the mERC module provides a mechanism for presenting relationships between items (e.g., visual objects or words). Conjunctional coding of sensory information from the lERC and relational information from the mERC in the HP predicts the next sensory item **x** in an inference process. Reproduced from Whittington et al. (2020). [Whittington, J. C., Muller, T. H., Mark, S., Chen, G., Barry, C., Burgess, N., & Behrens, T. E. (2020). The Tolman-Eichenbaum machine: Unifying space and relational memory through generalization in the hippocampal formation. Cell, 183(5), 1249–1263. Licensed under CC BY 4.0.]

between locational information from the mERC and sensory information from the lERC.

The model learns from experience to perform inference in the entorhinal-hippocampal system. In the network model, an inference process operates as follows: an action (e.g., "go to uncle") provides an input to the mERC module, triggering a movement from one location to another (**g**), like in a spatial map during navigation. The combination from the new mERC **g** and present lERC input drives the HP to retrieve the conjunctive item from memory as the predicted next stimulus (**p**). The predicted next sensory input is compared with the actual non-spatial input **x** encoded in lERC (horizontal yellow arrows), yielding a mismatch error signal $\mathbf{p} - \mathbf{x}$. The latter drives modification of connection weights in order to minimize overall errors. Crucially, path integration by vector sums in the mERC underlies transitive relationships. For example, "go to uncle" must cause the same change in **g** as to "go to father" followed by "go to brother," but different from "go to brother" followed by "go to father," which should result in "father" rather than "uncle."

When items (e.g., visual objects) are used as stimuli and their relationships are organized by a large two-dimensional grid (**Figure 11.27a**, left panel), TEM learned structural representations through random walks on two-dimensional graphs. After learning, neurons in the mERC exhibits grid-cell properties (**Figure 11.27a**, top right), whereas neurons in the HP behave like place cells (**Figure 11.27a**, bottom right). This representation is compositional, so that in two different environments (green and purple in **Figure 11.27a**, left panel), relations between the same set of items are arbitrarily reassigned, and the network can use the structural knowledge of the grid to carry out inference by path integration. The relational map formed by experience may or may not exhibit a grid pattern; for example, a regular grid does not naturally represent relational knowledge of a hierarchical tree for family members.

This explains why TEM exhibits learning-to-learn when trained to solve a series of problems, such as transitive inference with chain-like

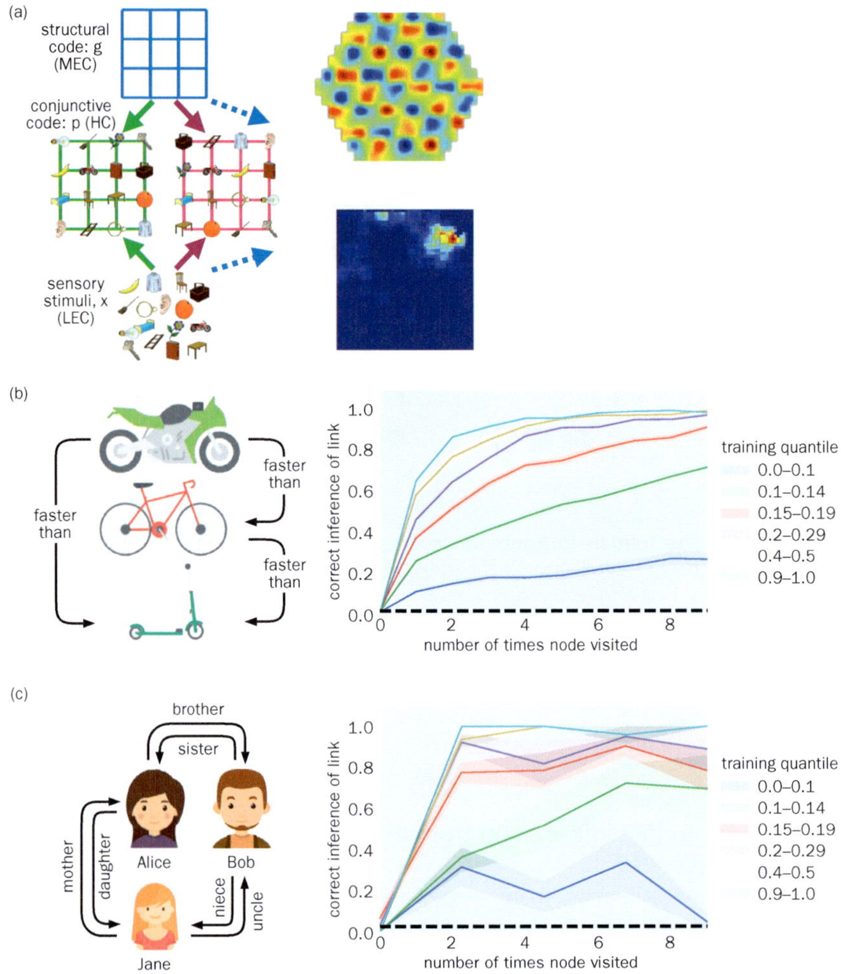

Figure 11.27 Generalization using cognitive maps. (a) Left panel: Factorization and conjunction as a principle for generalization. Separating structural codes (the transition rules of the graph) from sensory codes allows generalization over environments sharing the same structure. The conjunctive code represents the current environment in the context of this learned structure. Right panel: After learning, hexagonal grid cells in the mERC module (g, top) and unimodal place cells in the HP module (p, bottom) emerge from learning in TEM. A cell's firing rate map is obtained by allowing the agent to explore the environment and then calculating its average firing rate at each point (graph node) in the environment. (b–c) Learning-to-learn in solving transitive inference problems (b) or understanding family relations (c). Reproduced from Whittington et al. (2020). [Whittington, J. C., Muller, T. H., Mark, S., Chen, G., Barry, C., Burgess, N., & Behrens, T. E. (2020). The Tolman-Eichenbaum machine: Unifying space and relational memory through generalization in the hippocampal formation. Cell, 183(5), 1249–1263. Licensed under CC BY 4.0.]

relationships (motorbike is faster than bicycle, which is faster than scooter) (**Figure 11.27b**) or a family tree-like graph (**Figure 11.27c**). At the start of learning (blue/green traces), TEM does not yet understand the structure of the graph and has not learned how to use memories. Therefore, learning takes a lot of exposures to an environment. After visiting more environments and learning the common structure (cyan/yellow traces), TEM now understands both the rules of the graph (path integration) that guide how to store and retrieve memories, as a result the model correctly predicts an item on the second visit. Such fast learning can potentially be realized not only in two-dimensional planes but also in high-dimensional abstract spaces (Klukas et al. 2019). Studies of abstraction and inference discussed earlier underscore the importance of synergy of experimentation and computational modeling, in addition to close interactions between human cognitive neuroscience and neurophysiology with nonhuman animals. The latter is necessary in order to identify circuit mechanisms across levels.

11.6.3 Mental Programming and Intelligence

Planning in our daily life, e.g., for a party or a trip, is a form of inference through a search of possibilities in our memory system to determine the anticipated sequence of chosen actions. It is a topic of active research in both neuroscience (Hunt et al. 2021; Mattar and Lengyel 2022; Jensen et al. 2023; van Opheusden et al. 2023) and AI (Webb et al. 2023a). Planning requires "mental time travel," thinking towards the future (Mullally and Maguire 2014; Szpunar et al. 2014). More generally, thought itself is not some mysterious mental process beyond reach of neuroscientific understanding, if we consider it as sophisticated inference that involves a long chain of reasoning steps, often with information retrieved from our knowledge rather than the sensory world. Planning and thinking depend on "mental programming" to implement a sequence of events that are either externally observable actions or unseen thoughts, and the next move is informed by the previous ones and abstract rules.

Interestingly, the concept of working memory was originally introduced in psychology by necessity to account for planning of structured behavior. As explained by Milner et al. (1960, p. 65):

> When we have decided to execute some particular Plan, it is probably put into some special state or place where it can be remembered while it is being executed. Particularly if it is a transient, temporary kind of Plan that will be used today and never again, we need some special place to store it. The special place . . . maybe somewhere in the frontal lobes of the brain. . . . [W]e should like to speak of the memory we use for the execution of our Plans as a kind of quick-access working memory.

In other words, mental programming shares a core brain substrate with working memory (Chapter 5). It is thus fitting to end this chapter by circling back to the working memory system and its connection with planning, thought and general intelligence.

In research with animals performing a delayed response task, a longstanding question has been whether working memory and the associated neural signals encode retrospective information or prospective event/action (Roitblat 1980; Fuster 1989; Honig and Thompson 1982; Boettcher et al. 2021; Henderson et al. 2022). For instance, when monkeys carry out a delayed anti-saccade task, the response is diametrically opposed to the sensory target, so does the delay period activity store the target information or signal the planned response (Funahashi et al. 1993)? In a delayed match-to-sample (DMS) task, persistent neural activity after a sample stimulus could be interpreted as retrospective sensory short-term memory; alternatively, it may specify prospectively the context that determines the correct mapping of an upcoming test stimulus and a response (Wu et al. 2020). Different brain regions contribute in varying degrees to retrospective versus prospective representations; premotor as well as parietal areas are generally considered to be important for motor planning (Snyder et al. 1997; Andersen and Buneo 2002; Kaufman et al. 2014; Li et al. 2015; Berger et al. 2020). Of course, planning is more general than just for motor responses; organizing behavior or thinking can take many steps to achieve a goal. Studies of people with frontal damage (Kimberg and Farah 1993; Duncan et al. 1995; Woolgar et al. 2018) and brain imaging with healthy subjects (Duncan 2010b; Gray et al. 2003; Cole et al. 2011; Hampshire et al. 2012) have shown that the lateral prefrontal cortex plays a central role in such "mental programming."

Rainer et al. (1999) designed a monkey experiment to differentiate retrospective versus prospective information representation. In alternating blocks of trials, the task was either DMS with three visual objects (C1, C2 and C3) or delayed paired associate (DPA). In the DPA task, subjects learnt to associate three new pictures (S1, S2 and S3) shown during the sample period with paired C1, C2 and C3 as targets shown during test period, respectively, and again the sample and target were separated by a short temporal delay. Single neurons in the dorsolateral prefrontal cortex showed either retrospective coding of a sample or prospective coding of a target. For instance, a retrospective neuron showed early in the delay period high activity selective for S3 (associated with C3 presented after the delay period) in the DPA task, but little delay activity in the DMS task when the sample was C3 presented before the delay period. By contrast, a prospective neuron's activity is elevated late in the delay period, similarly when C1 was the target in the DPA task or C1 was the sample in the DMS task (in which case the subjects anticipated to choose C1 at the end of the delay). Across recorded neurons, it was found that retrospective coding is predominant at the start of the delay, while prospective coding gradually prevails towards the end of the delay, a process suitable for future planning. Note that such a dynamical shifting of neural population activity could account for, or at least contribute to, the temporal changes of representation in a working memory circuit discussed in Chapter 5 (Section 5.8).

What may be a neural circuit mechanism for planning? Ehrlich and Murray (2022) considered this question formulated in terms of a simple delay-dependent Boolean logic task, with inputs that are 0 or 1 (**Figure 11.28a**). For instance, if the rule is AND, the response is 1 only when both the sample (Cue A) and the test (Cue B) are 1, 0 otherwise. In that case, when Cue A is 0, the response would be 0 regardless of Cue B, and this case is called the contingency state 0–0. In such a situation of "task closure," the system can plan for the upcoming response before Cue B. On the other hand, if Cue A is 1, the ultimate response depends on Cue B (0 if B = 0, and 1 otherwise). The situation is similar if the rule is OR and Cue A is 0, when the response should also be the same as Cue B. Hence, different combinations of a task rule and Cue A (AND rule with A = 1, OR rule with A = 0) can lead to a common "contingency state" (called 0–1 in the last two examples, where the response should match B=0 or 1). An example of contingency state 1–0 is with XOR rule and A = 1, in which case the response should be different from Cue B. There are four types of contingency states, 0–0, 0–1, 1–0 and 1–1, shown in **Figure 11.28a** for three rules (AND, OR,

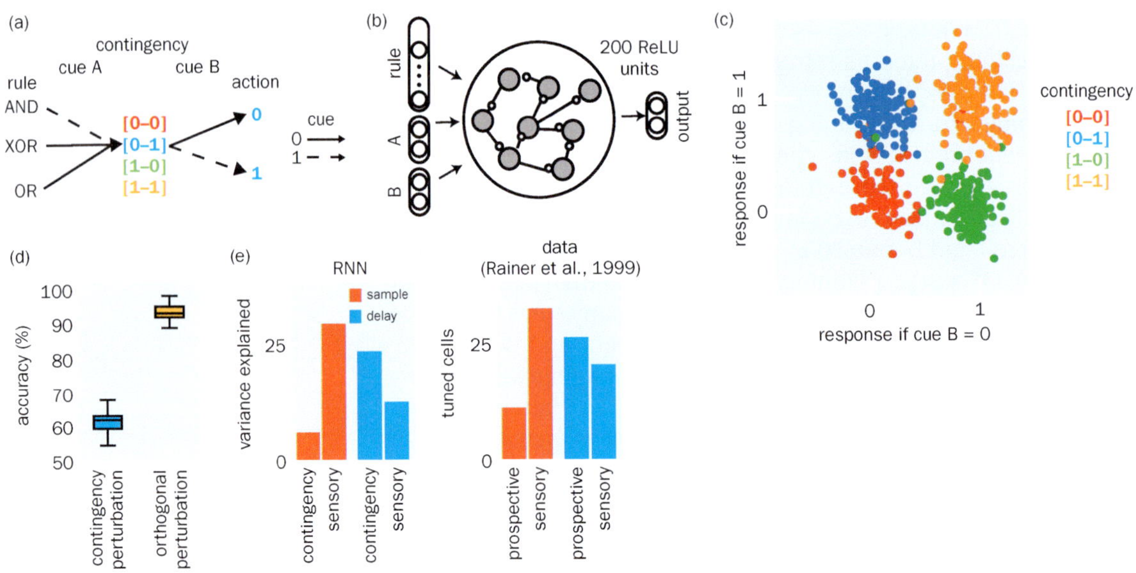

Figure 11.28 An RNN model for planning. (a) Delayed logic task using contingency representation, which maps expected future stimulus to actions, in this case from Cue B to the correct logical output. Each possible intermediate stage represents a unique Cue B-response mapping. (b) A schematic artificial RNN model trained on delayed logic problems in (a). Trained RNN reached >99% accuracy on the task. (c) A two-dimensional subspace of hidden internal state of recurrent units in the trained RNN. Each point is a trial, colored by its associated contingency. (d) A perturbation was added to the hidden state during the delay, either within the contingency subspace (c) or orthogonal to it. Perturbations in the subspace were far more effective in reducing accuracy than were matched orthogonal perturbations. (e) The RNN model showed the same phenomenon with individual unit variance in activity, moving from cue-dominated (retrospective) to contingency-dominated (prospective) as in the monkey experiment of Rainer et al. (1999). Reproduced from Ehrlich and Murray (2022). [Courtesy of John Murray.]

XOR). In the 0–0 and 1–1 states, the future response is independent of Cue B (closure), thus in principle the network can prepare for the response ahead of time; the opposite is true in the 0–1 and 1–0 states, for which future planning is not feasible.

Ehrlich and Murray (2022) trained a vanilla RNN model to perform the delayed logic task (**Figure 11.28b**). Similar to Yang et al. (2019), in addition to sensory inputs for cue A and cue B, a rule input is instructing the network about which of the three rules (OR, AND or XOR) to follow in a trial. This approach was agonistic about how the problems should be solved. Interestingly, it was found that after the network has learnt the task, the recurrent neural population activity displayed four separate clusters in a low-dimensional subspace of the state space, corresponding to the four contingencies (**Figure 11.28c**). A perturbation within the contingency subspace dramatically reduced task performance, whereas a perturbation orthogonal to it had little impact (**Figure 11.28d**), showing the functional importance of the contingency representation that emerged from learning by the RNN model. Decoding from recurrent units revealed a gradual shift in time during the delay period from retrospective memory to prospective planning in the model (**Figure 11.28e**), reproducing the monkey experimental results of Rainer et al. (1999).

While the model is fairly simple, it opens the door for further computational research on more complex planning with multiple mental and behavioral steps to achieve a goal, such as winning a Go game or proving Fermat's last theorem. One difficulty is for nonhuman animals to learn cognitively demanding multiple-step behaviors. Perhaps a more feasible approach is to train animals to perform sophisticated serial movements, designed to probe compositionality and grammar of actions in a rule-guided sequence (Goodnow and Levine 1974; Tanji and Shima 1994; Procyk et al. 2000; Averbeck et al. 2003; Mushiake et al. 2006; Lake et al. 2015). To the extent that thinking can be considered as akin to an internal motor act (Llinás 2002), insights gained from experiments and modeling using language-independent tasks should be of broad relevance for understanding planning and thought in terms of neural circuit mechanisms and computational principles.

11.6.4 Cross-Scale Brain Basis of Intelligence

As mentioned in Chapter 5, intelligent abilities associated with cognitive functions can be subdivided into two types: crystallized intelligence refers to abilities to learn and absorb information based on knowledge, such as someone's vocabulary; fluid intelligence refers to reasoning abilities for solving new problems (Cattell 1987; Duncan 2010a; Haier 2017). While crystallized intelligence is relatively stable in a lifespan, fluid intelligence generally decays with age. The two are related to each other, and of course problem solving obviously benefits from cumulative knowledge, but they are distinct most notably by the more flexible and creative use of knowledge associated with fluid intelligence. Intelligence quotient (IQ) tests assess both, which is typically measured with a battery of tests and a factor analysis of performance measures, yielding the so-called general *g* factor (Spearman 1927; Carroll 1993). For instance, Raven's matrices are used in a nonverbal test of fluid intelligence, in which subjects are shown a multiple-part geometric figure with a missing piece, and the task is to correctly select one among several given options according to a rule inferred from observing the existing pieces (**Figure 11.29**) (Raven et al. 1988). The performance of the single test using Raven's matrices varies among individuals and is highly correlated with the overall performance of fluid intelligence (Duncan 2010a). Solving Raven's matrices represents a challenge in machine learning and may require a neuro-symbolic approach

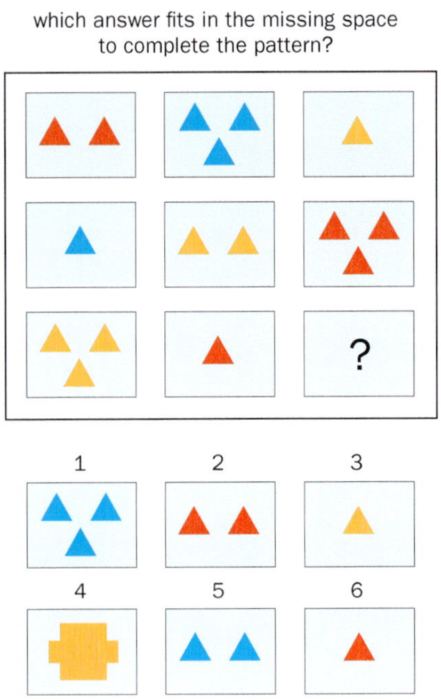

Figure 11.29 Raven's matrices as a fluid intelligence test. Given a matrix (top), one of the six pieces (bottom) should be used to complete the pattern; #5 is the correct answer (so that each column has a orange, a red and a blue shape, as well as 1, 2, 3 triangles). The completed pattern has 3 × 3 = 9 pieces, with 1, 2, 3 triangles of each color. Downloaded from www.stuartbell.co.uk/ravens.html.

(Hersche et al. 2023; Webb et al. 2023b) that combines neural network models and symbolic AI models (Marcus 2003; Mao et al. 2019; Feinman and Lake 2020; Sablé-Meyer et al. 2022).

What are the biological basis of unique human intelligence compared to nonhuman animals, and of individual differences among people? Comparative neurophysiology revealed that anatomical and physiological properties of single neurons (Goriounova et al. 2018; Beaulieu-Laroche et al. 2021; Bakken et al. 2021) and synaptic circuits (Campagnola et al. 2022) show only quantitative differences across species including humans. Mice modified with a human-specific gene SRGAP2C expression displayed substantially higher cortical connectivity than the wild-type mice and are more successful in learning a sensory discrimination task than their wild-type littermates (**Figure 11.30a**) (Schmidt et al. 2021). Single neurons of the human cortex show individual variability of intrinsic properties, such as the dendritic length, which are correlated with the Intelligence Quotient test score (**Figure 11.30b**) (Goriounova et al. 2018).

Moreover, there appear to be qualitative evolutionary changes uniquely for the primates including humans (Smaers et al. 2017), possibly as a result of punctuated equilibrium through biological evolution (Gould and Eldredge 1993). Take the example of metabotropic glutamate receptor subtype mGluR3 reviewed in Arnsten et al. (2021). In the rodent cortex, mGluR3 is predominantly located at the presynaptic terminals, and its activation reduces release probability of transmitters and thus synaptic excitation. By contrast, in the dorsolateral prefrontal cortex (DLPFC) of the primates, mGluR3 receptors are present postsynaptically on the spines, where they inhibit calcium-cyclic AMP-dependent protein kinase cAMP-PKA, resulting in a reduced opening of a potassium channel, and thereby synaptic excitation is enhanced. This idea was tested in an experiment where monkeys performed the oculomotor delayed response (ODR) task (Chapter 5, Section 5.4) (Jin et al. 2018). Based on the knowledge that glutamate carboxypeptidase II (GCP II) reduces mGluR3 signaling, a drug that blocks GCP II is expected to boost mGluR3 receptor activation. It was found that, compared to the control condition, application of a GCP II antagonist yielded increased stimulus-selective persistent firing (**Figure 11.30c**), in support of an mGluR3 contribution to working memory. The GCP II is encoded by the gene folate hydrolase 1 (FOLH1). Since working memory is correlated with intelligence (Chapter 5, **Figure 5.1**), can variations of mGluR3 signaling, at the molecular level, be correlated with individual differences in IQ? This question was addressed in a recent human study with hundreds of healthy and schizophrenia subjects, some of whom are carriers of the missense G allele (a mutation from adenine [A] to guanine [G]) of the FOLH1, which causes higher GCP II expression (Zink et al. 2020). Carriers with the mutation showed lower IQ scores (**Figure 11.30d**), in line with a beneficial role of the mGluR3 signaling in cognition. This example illustrates how neural circuit studies of core cognitive functions represent an effective, even obligatory, way to elucidate the brain basis of intelligence across levels.

On the macroscopic level, human brain imaging showed that activation of the multiple-demand (MD) system is significantly correlated with performance of fluid intelligence tests across human subjects (Duncan et al. 2000; Gray et al. 2003). The MD system, which consists of areas of the parietal and prefrontal cortices (Duncan 2010b; Ji et al. 2019; Assem et al. 2020), is domain-general; it plays a central role in executive control of externally oriented flexible behavior as well as internally oriented mental processes (Spreng et al. 2010; Duncan 2013). Interestingly, studies of people with selective brain damage found a double dissociation between the MD and language systems: fluid intelligence is largely independent of the cortical language areas, whereas performance on verbal tests of

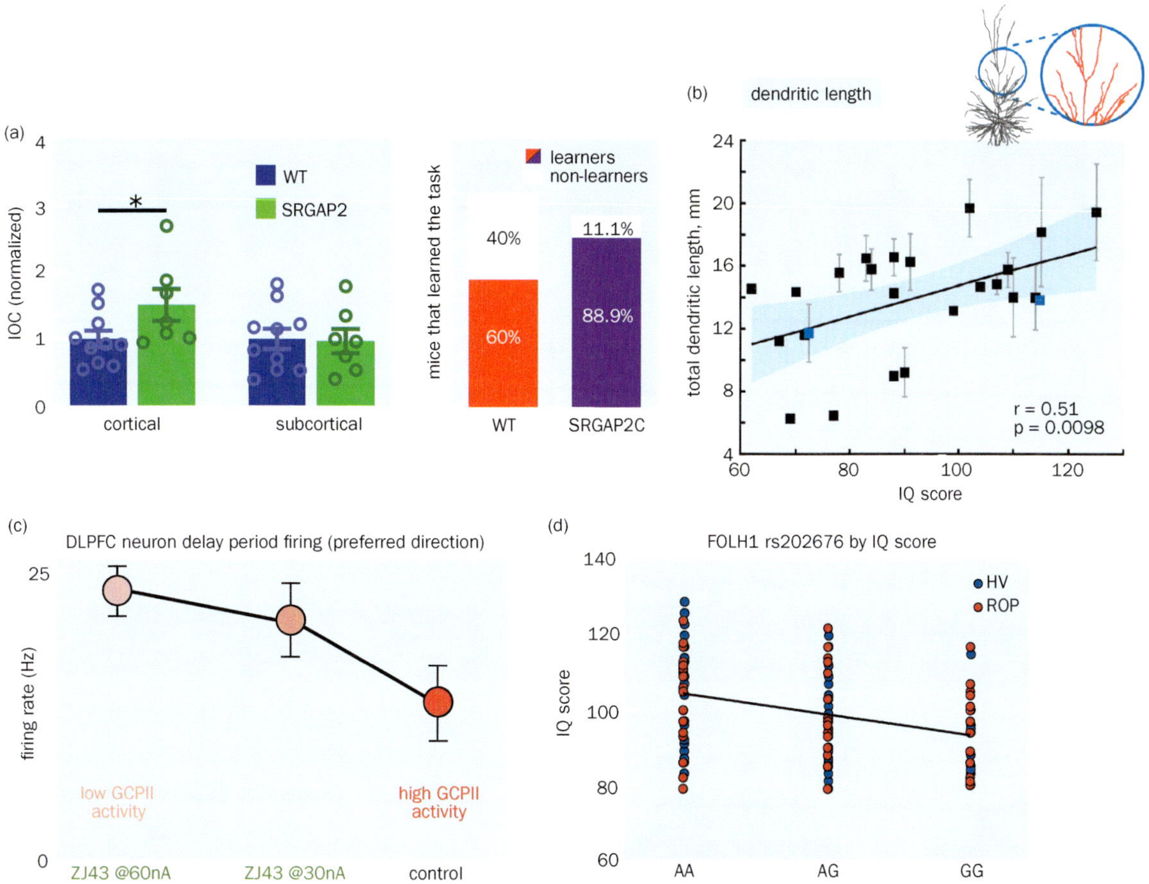

Figure 11.30 Biological properties of the human brain and their correlation with Intelligence Quotient (IQ) score. (a) Cortico-cortical connections measured by an index of connectivity (IOC) are more abundant in mice with the human-specific gene SRGAP2C (green) than the wild control (blue) in the neocortex but not in subcortical brain regions. (b) Dendritic length of layer 2/3 pyramidal cells of the human cortex is positively correlated with the IQ score. (c) Iontophoretic application of an antagonist (ZJ43) of the GCP II (which diminishes mGluR3 signaling) with two different doses progressively increased mnemonic delay period firing in the dorsolateral prefrontal cortex of monkeys during working memory. (d) In healthy volunteers (HV) and schizophrenia subjects with recent-onset psychosis (ROP), the FOLH1 alleles in favor of mGluR3 activation in the DLPFC (GG to AG to AA) are positively associated with the IQ score ($p < 0.01$) of human individuals. Panel (a) is reproduced from Schmidt et al. (2021), (b) from Goriounova et al. (2018), (c) from Arnsten and Wang (2020), with the original data from Jin et al. (2018) and (d) from Zink et al. (2020). [Schmidt, E. R., Zhao, H. T., Park, J. M., Dipoppa, M., Monsalve-Mercado, M. M., Dahan, J. B., . . . & Polleux, F. (2021). A human-specific modifier of cortical connectivity and circuit function. Nature, 599(7886), 640–644.] [Goriounova, N. A., Heyer, D. B., Wilbers, R., Verhoog, M. B., Giugliano, M., Verbist, C., . . . & Mansvelder, H. D. (2018). Large and fast human pyramidal neurons associate with intelligence. Elife, 7. Licensed under CC BY 4.0.] [Courtesy of Amy Arnsten.]

crystalized intelligence remains intact with damaged MD areas (Woolgar et al. 2018). fMRI BOLD signals of language areas do not increase significantly in numerous cognitive tasks (**Figure 11.31**), suggesting that core intelligence abilities are dissociable from the brain's language system (Fedorenko et al. 2024).

Hampshire et al. (2012) carried out a human experiment using 12 intelligence tests, while imaging the MD system (**Figure 11.32a**). Consider a number N of voxels as the "regions of interest" in the MD system, where their BOLD signals represent a vector in the N-dimensional space for each of the 12 tasks. Correlations of vector pairs (for any two tasks) produce a 12×12 matrix. The factor analysis of this co-activation matrix had two principal components that together accounted for $\sim 90\%$ variance of task-related activation of the MD system, and they were used to quantify "weighting" for each task on the two principal components (**Figure 11.32b**, red and blue bars). When factor scores were projected back

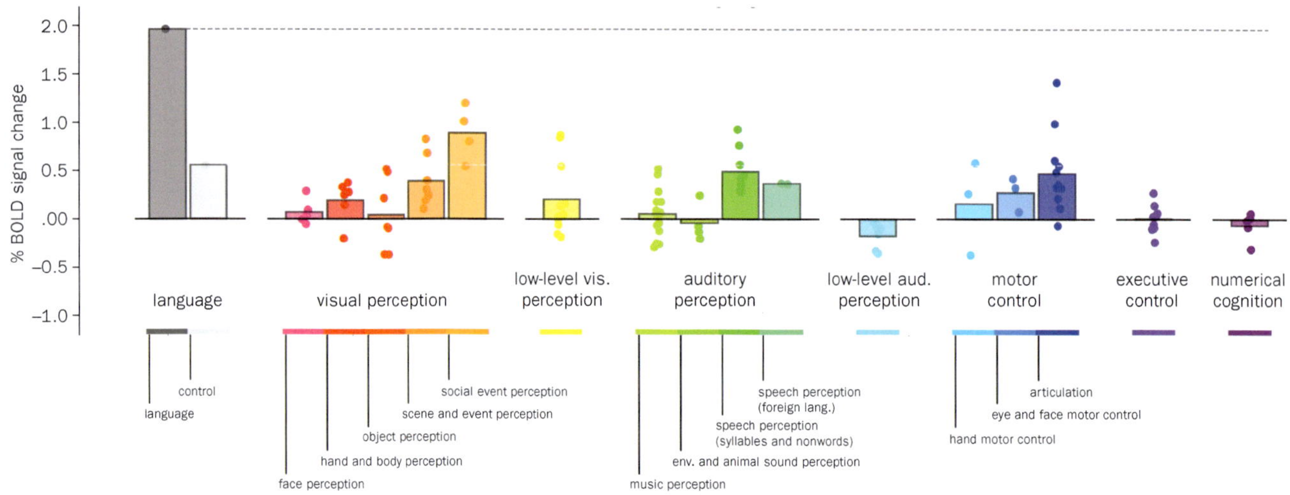

Figure 11.31 Functional selectivity of the language network. fMRI BOLD responses in the language-responsive brain areas identified with an extensively validated language localizer task based on a contrast between the processing of sentences and nonword sequences. This figure summarizes responses in the language areas to diverse non-linguistic conditions. Zero corresponds to the fixation baseline. Provided by Ev Fedorenko, with original data from several studies (Fedorenko et al. 2011; Fedorenko et al. 2012; Pritchett et al. 2018; Jouravlev et al. 2019; Ivanova et al. 2020). [Courtesy of Evelina Fedorenko.]

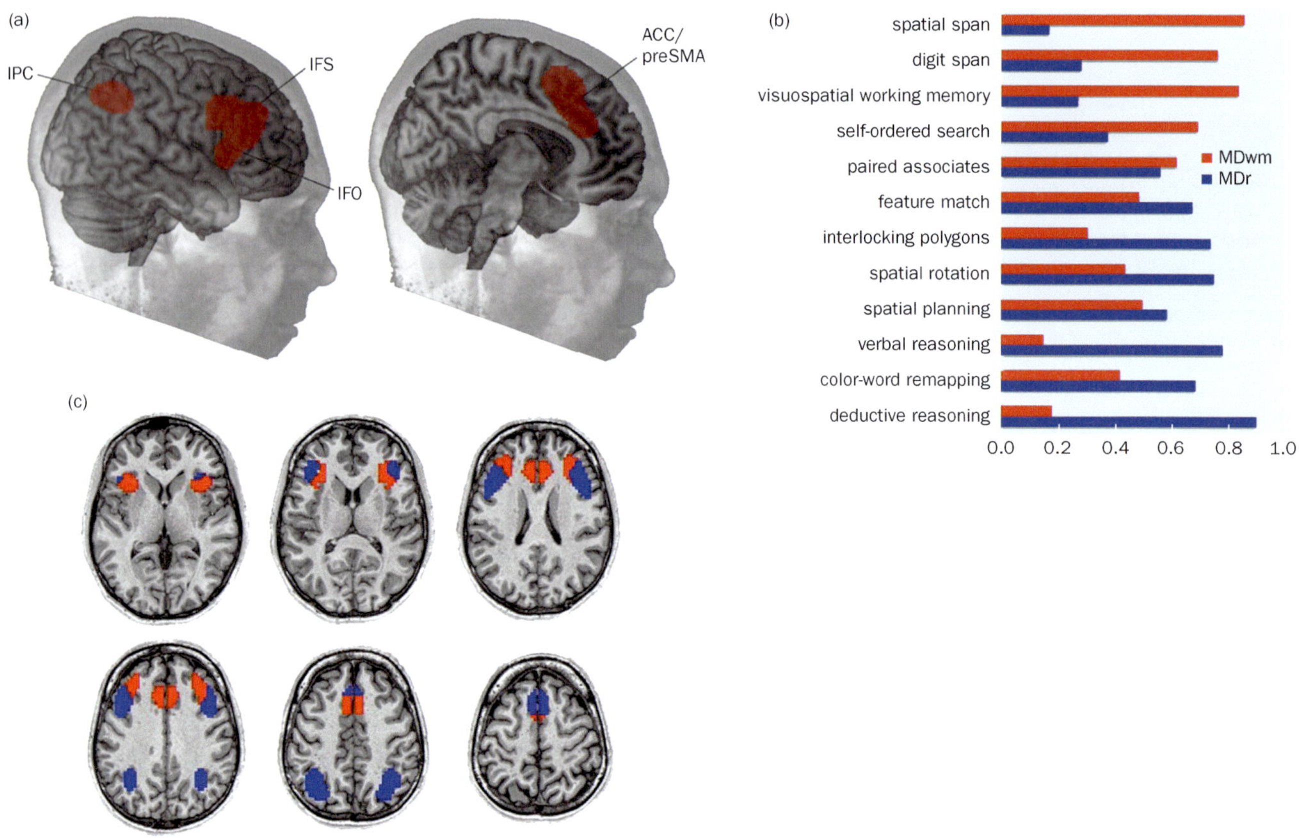

Figure 11.32 Factor analysis of functional brain imaging data from within the multiple demand (MD) cortex. (a) The MD cortex regions of interest (ROIs). (b) Principal component analysis of the average activation patterns within the MD cortical system shows two subnetworks, the MD working memory (MDwm, red) and the MD reasoning (MDr, blue), with different weighting for each of 12 tasks. (c) Component scores from the analysis of the MD task-related activations averaged across individuals. Voxels that loaded more heavily on the MDwm subnetwork are displayed in red; voxels that loaded more heavily on the MDr subnetwork are displayed in blue. Reproduced from Hampshire et al. (2012). [Hampshire, A., Highfield, R. R., Parkin, B. L., & Owen, A. M. (2012). Fractionating human intelligence. Neuron, 76(6), 1225–1237.]

onto the brain, it was found that two functional subnetworks of the multiple-demand (MD) system were selectively engaged in nonverbal working memory tasks (MDwm) and reasoning tasks (like Raven's matrices) (MDr), respectively (**Figure 11.32c**). In agreement with the previous findings, performances on these two types of fluid intelligence tasks were not strongly correlated with those of verbal tasks. These results suggest that, actually, fluid intelligence (as assessed by a *g* factor) is not a unitary entity but fractionated into working memory, reasoning and verbal subsystems.

11.7 SUMMARY

In this chapter, we covered machine learning tools applied to computational neuroscience, with a focus on ideas and methods that enable researchers to investigate the brain mechanisms of increasingly sophisticated mental processes underlying intelligence. We discussed a range of topics of interest for both brain science and AI, such as learning-to-learn, abstract concepts, compositionality, inference and planning. Under the premise that fluid intelligence is largely independent of human language, in principle its brain mechanism can be investigated using both human and nonhuman animal experiments, combined with multi-regional large-scale brain circuit modeling. Time will tell whether research using this approach can form a foundation for human cognition, including mathematics (Dehaene 2011; Nieder 2019), language (Pinker 1994; Dehaene 2010), music (Levitin 2006; Sacks 2010) and arts (Solso 1994; Zeki 2001; Fuster 2013).

What does it mean to build a computational model of the primate brain? The field of computational neuroscience has yet to reach a consensus, and a diversity of models is desirable. Here, we propose that a milestone would be the development of a neurobiologically based whole-brain model that is successful in three respects. First, it serves as a computational platform to account for experimental observations on multiple levels from molecules to cell types to neural population dynamics. Second, it provides general principles for understanding how the brain works. Third, it is capable of carrying out a battery of fluid intelligence tests, quantifiable by a *g* factor. Ongoing research suggests that this ambitious goal is reachable, perhaps in the not-too-distant future.

FURTHER READING

Books and Reviews

Ashby, W. R. (1956). *An Introduction to Cybernetics*. London: Chapman and Hall.

Behrens, T. E., T. H. Muller, J. C. Whittington, S. Mark, A. B. Baram, K. L. Stachenfeld, and Z. Kurth-Nelson (2018). What is a cognitive map? organizing knowledge for flexible behavior. *Neuron 100*, 490–509.

Carroll, J. B. (1993). *Human Cognitive Abilities: A Survey of Factor-Analytic Studies*. New York: Cambridge University Press.

Dehaene, S. (2011). *The Number Sense: How the Mind Creates Mathematics*. New York: Oxford University Press.

Duncan, J. (2010). *How Intelligence Happens*. New Haven, CT: Yale University Press.

Fedorenko, E. and R. Varley (2016). Language and thought are not the same thing: Evidence from neuroimaging and neurological patients. *Ann. N. Y. Acad. Sci. 1369*, 132–153.

Hassabis, D., D. Kumaran, C. Summerfield, and M. Botvinick (2017). Neuroscienceinspired artificial intelligence. *Neuron 95*, 245–258.

Kriegeskorte, N. (2015). Deep neural networks: A new framework for modeling biological vision and brain information processing. *Ann. Rev. Vis. Sci. 1*, 417–446.

LeCun, Y., Y. Bengio, and G. Hinton (2015). Deep learning. *Nature 521*, 436–444.

Mattar, M. G. and M. Lengyel (2022). Planning in the brain. *Neuron 110*, 914–934.

Nieder, A. (2019). *A Brain for Numbers: The Biology of the Number Instinct*. Cambridge, MA: MIT Press.

Spearman, C. (1927). *The Abilities of Man.* New York: McMillan.

Yamins, D. L. and J. J. DiCarlo (2016). Using goal-driven deep learning models to understand sensory cortex. *Nat. Neurosci. 19*, 356–365.

Yang, G. R. and X.-J. Wang (2020). Artificial neural networks for neuroscientists: A primer. *Neuron 107*, 1048–1070.

Primary Research Papers

Machine Learning and the Brain

Hochreiter, S. and J. Schmidhuber (1997). Long short-term memory. *Neural Comput. 9*, 1735–1780.

Krizhevsky, A., I. Sutskever, and G. E. Hinton (2012). Imagenet classification with deep convolutional neural networks. *Adv. Neural Inf. Process. Syst. 25*, 1097–1105.

LeCun, Y. (1988). A theoretical framework for back-propagation. In D. Touretzky, G. Hinton, and T. Sejnowski (Eds.), *Proceedings of the 1988 Connectionist Models Summer School*, pp. 21–28. Burlington, MA: Morgan Kaufmann.

Rumelhart, D. E., G. E. Hinton, and R. J. Williams (1986). Learning representations by back-propagating errors. *Nature 323*, 533–536.

Yamins, D. L., H. Hong, C. F. Cadieu, E. A. Solomon, D. Seibert, and J. J. DiCarlo (2014). Performance-optimized hierarchical models predict neural responses in higher visual cortex. *Proc. Natl. Acad. Sci. U.S.A. 111*, 8619–8624.

Training Recurrent Neural Network Models

Chaisangmongkon, W., S. K. Swaminathan, D. J. Freedman, and X. J. Wang (2017). Computing by robust transience: How the fronto-parietal network performs sequential, category-based decisions. *Neuron* 93, 1504–1517.

Ehrlich, D. B. and J. D. Murray (2022). Geometry of neural computation unifies working memory and planning. *Proc. Natl Acad. Sci. 119*, e2115610119.

Rajan, K., C. D. Harvey, and D. W. Tank (2016). Recurrent network models of sequence generation and memory. *Neuron 90*, 128–142.

Song, H. F., G. R. Yang, and X.-J. Wang (2016). Training excitatory-inhibitory recurrent neural networks for cognitive tasks: A simple and flexible framework. *PLoS Comput. Biol. 12*, e1004792.

Sussillo, D. and L. F. Abbott (2009). Generating coherent patterns of activity from chaotic neural networks. *Neuron 63*, 544–557.

Abstraction and Compositionality

Bernardi, S., M. K. Benna, M. Rigotti, J. Munuera, S. Fusi, and C. D. Salzman (2020). The geometry of abstraction in the hippocampus and prefrontal cortex. *Cell 183*, 954–967.

Cole, M. W., J. R. Reynolds, J. D. Power, G. Repovs, A. Anticevic, and T. S. Braver (2013). Multi-task connectivity reveals flexible hubs for adaptive task control. *Nat. Neurosci. 16*, 1348–1355.

Freedman, D. J., M. Riesenhuber, T. Poggio, and E. K. Miller (2001). Categorical representation of visual stimuli in the primate prefrontal cortex. *Science 291*, 312–316.

Goudar, V., B. Peysakhovich, D. J. Freedman, E. A. Buffalo, and X.-J. Wang (2023). Schema formation in a neural population subspace underlies learning-to-learn in flexible sensorimotor problem-solving. *Nat. Neurosci. 26*, 879–890.

Lake, B. M., R. Salakhutdinov, and J. B. Tenenbaum (2015). Human-level concept learning through probabilistic program induction. *Science 350*, 1332–1338.

Whittington, J. C., T. H. Muller, S. Mark, G. Chen, C. Barry, N. Burgess, and T. E. Behrens (2020). The Tolman-Eichenbaum machine: Unifying space and relational memory through generalization in the hippocampal formation. *Cell 183*, 1249–1263.

Yang, G. R., M. R. Joglekar, H. F. Song, W. T. Newsome, and X.-J. Wang (2019). Task representations in neural networks trained to perform many cognitive tasks. *Nat. Neurosci. 22*, 297–306.

Intelligence

Hampshire, A., R. R. Highfield, B. L. Parkin, and A. M. Owen (2012). Fractionating human intelligence. *Neuron 76*, 1225–1237.

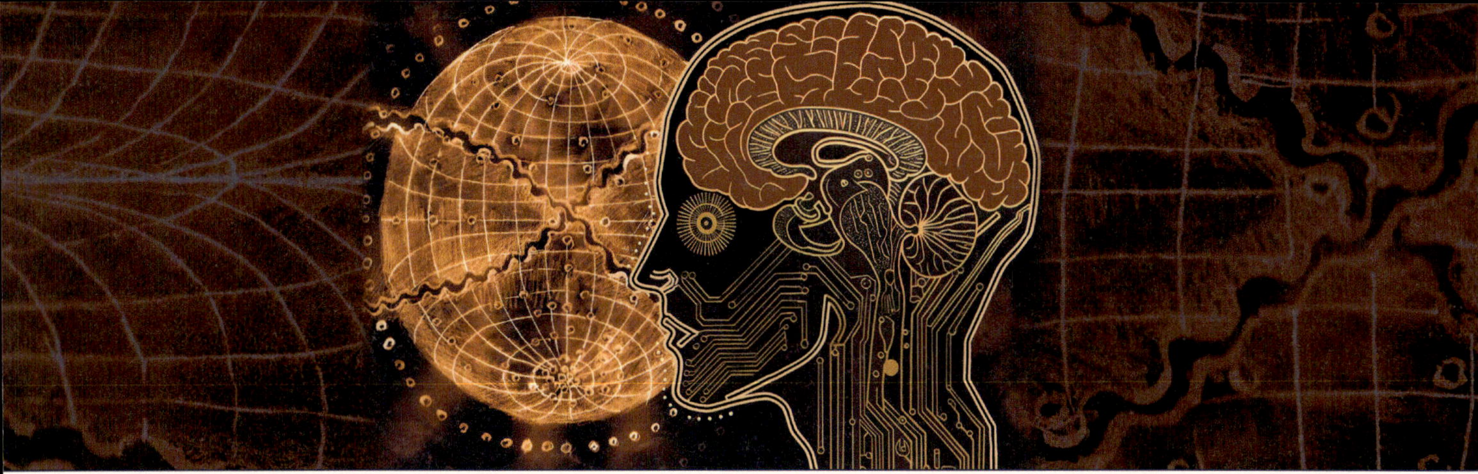

Looking Back and Ahead

12

With your head full of brains and your shoes full of feet, you're too smart to go down any not-so-good street.

Dr. Seuss in *Oh, the Places You'll Go!*

In this book we traveled a wide neuroscientific landscape of cognition. Details matter and, connecting back to the epistemological discussion of Chapter 1, theory is established by solving a range of concrete problems through back-and-forth interactions with experiments. Cognitive processes covered here are limited to the "elemental" variety, in the sense that they are sufficiently simple to be carried out by nonhuman animals in order to dissect the underlying circuit mechanism. We introduced findings, concepts and methods for understanding basic cognitive functions. Taken together, this knowledge forms a basis for the mosaic unity of understanding the brain mechanism of cognition and flexible behavior across levels.

12.1 BUILDING BLOCKS OF BEHAVIOR AND COGNITION

To paraphrase Theodosius Dobzhansky ("Nothing in biology makes sense except in the light of evolution"), nothing in neuroscience makes sense except in the light of behavior. It is thus of cardinal importance for a theorist to pay attention to behavior. In this book, a series of task designs have been introduced progressively, which we recap briefly in **Table 12.1**.

DOI: 10.1201/9781003459361-15

Table 12.1 Behavioral Task List

No.	Task Name	Description	Comment
1	Delayed response task (Section 5.2)	A temporal delay separates a sensory stimulus and a motor response, requiring the former to be held in the brain to guide the latter.	It is commonly used for studying working memory; however, neural signals during the delay period may reflect motor preparation rather than memory of sensory stimulus.
2	Delayed match-to-sample (Section 5.2), ABBA task (Section 6.6)	A sample stimulus (A) and a test stimulus are separated by a time delay, and the subject must respond when it is a match (AA). In the ABBA version of the task, mere repetition (BB) should be ignored.	Since a response cannot be prepared before the test stimulus is shown, delay period activity either stores a memory of the sample or reflects a potential decision conditioned on the upcoming test stimulus.
3	Delayed match-to-rotated sample (Section 5.8)	The decision is whether the test matches the sample rotated by a fixed angle.	This task requires not only the maintenance but also manipulation of information during a delay period.
4	Delayed discrimination (Section 5.2)	Two stimuli shown across a delay are compared, in order to decide whether one is larger or smaller than the other, which involves a different type of computation than match/nonmatch.	Parametric encoding of a graded quantity applies to various types of information such as vibration frequency, reward value.
5	Change detection, delayed estimation (Section 5.1, 5.7)	An array of stimulus items must be remembered across a short time delay.	These tasks probe the capacity of working memory.
6	Detection (Section 6.2)	The subject must judge whether a stimulus is present, which probes the transformation from encoding a physical stimulus to its subjective perception.	Task variants include multi-sensory integration, in which a detection relies on combining inputs from two sensory modalities.
7	Two-alternative forced-choice task (Section 6.2)	Random-dot motion direction discrimination is a well-known example of perceptual decision making.	Task variants include reaction time and fixed stimulation duration versions, and extend to more than two choice alternatives.
8	Two-stage wagering (Section 6.7)	A perception decision is followed by a gamble between an option yielding a large reward only if the first decision is correct versus another that yields a small reward with certainty.	The choice in the second stage probes confidence about the perceptual judgment.
9	Value-based choice (Section 7.2)	A selection is made for one of two or more options that yield different reward types and amounts.	This basic task probes computation of a "common currency" for subjective valuation.
10	Intertemporal choice (Section 7.2)	Two options involve different temporal delays before reward harvest.	A standard way to assess delayed gratification and self-control.
11	Economic choice among bundles (Section 7.4)	A choice must be made between two alternatives, each consisting of two or more good types.	Investigates decision making either using an integrated valuation or separate valuation within each good type.
12	Dynamical foraging task (Section 7.3)	Two reward options are delivered stochastically with different probabilities that may vary over time.	Performance is robustly characterized by the "matching law."
13	Naturalistic foraging (Section 7.2)	A sequential task that involves deciding whether to stay with a choice as return gradually decreases or leaves in search of another option.	Mimics real-life environments studied in behavioral ecology.
14	Two-stage probabilistic decision (Section 7.4)	This task consists of two sequential choices, where each option of the first choice leads to a different gamble in the second stage.	The ultimate reward depends on more than the temporally contingent second choice, probing reinforcement learning beyond action-outcome association.
15	Weather forecasting (Section 7.5)	A number of cues are shown, each associated with the log likelihood odds for predicting which of the two ("rain" or "shine") options will bear out.	Designed to probe probabilistic inference based on the sum of the log likelihood odds.
16	Matching pennies (Section 7.6)	In this two-alternative version of the rock-paper-scissors game, an agent's gain is the other's loss.	The optimal strategy is to play randomly in this zero-sum game.
17	Prisoner's dilemma (Section 7.6)	Two players should cooperate to achieve an optimal outcome.	This task relies on guessing another agent's intention within a social group.

No.	Task Name	Description	Comment
18	Stop-signal task (Section 8.2)	A planned response to a suddenly appearing target must be cancelled upon the presentation of a stop signal.	A simple task for studying executive control of prepotent motor behavior.
19	Pro- versus anti-response (Section 8.2)	The subject must either produce a response towards, or diametrically away from, a stimulus target.	This task engages a combination of inhibitory control (of automated pro-response) and (pro- or anti-) rule-based task switching.
20	Cued rule-dependent decision making (Section 8.5)	Given a stimulus with two (e.g., color and shape) or more features, only one is behaviorally relevant, and the other should be ignored according to a given rule.	A commonly used task for studying selective attention.
21	Uncued rule-dependent sensorimotor mapping (Section 8.6)	A rule currently in play is not signaled but must be inferred from behavioral feedback, exemplified by the Wisconsin Card Sorting Test.	Variants of this task probe internal maintenance of a task rule across trials, behavioral monitoring and rule switching.
22	Momentary happiness test (Section 10.3)	A series of gain and loss continuously update subjective well-being.	Quantifies mood in real time for studies of emotional state.
23	AX continuous performance task (Section 10.4)	While viewing a stream of items such as letters, a response is required to a cue-probe sequence, in which the letter A appears as the cue followed by the letter X as the probe.	This task probes proactive versus reactive control of action.
24	Categorization, delayed match-to-categorization (Section 11.4)	A parametrically varied stimulus feature must be categorized into two or more abstract categories.	Category is a basic form of concepts for generalizable knowledge.
25	Arbitrary sensorimotor mapping (Section 11.5)	Sensory stimuli must be mapped correctly onto corresponding motor responses, and the correspondence must be learned.	This task is used to study learning-to-learn, which denotes acceleration of learning when a number of similar problems are solved sequentially.
26	Raven's matrix (Section 11.6)	A nonverbal puzzle must be solved by filling an empty slot in a matrix of shapes, selecting an item among an array of candidates through reasoning.	Assesses specifically fluid intelligence.

We discussed neural circuit models for the majority, but not all, of the aforementioned tasks. The process of elucidating the underlying neural circuit mechanisms led to the proposal of computational primitives, such as multiple-attractor networks, slow ramping transients, gating by a disinhibitory motif, mixed-selectivity and bifurcation in space. Notable is the fact that these models are biologically realistic, built on components described in Chapters 2–4. Thus, cross-level understanding is achievable for cognitive functions. Thanks to compositionality, these elemental building blocks of cognition can be combined to subserve a variety of new tasks (Yang et al. 2019; Driscoll et al. 2024). There is no obvious reason why this methodological approach cannot be extended to more complex mental abilities such as reasoning, planning and flexible problem solving.

12.2 TAKE-HOME MESSAGES

From the findings covered in this book, what big picture emerges as a whole?

(1) Neural circuits underlying basic cognitive functions cannot be described by purely feedforward models; instead, they are characterized by recurrent network dynamics. Mental processes and behavior in interaction with the world evolve over time that, in the brain, correspond to neural population activity trajectories in the state space, analyzed by novel techniques such as those used for the discovery of

representational manifolds and communication subspaces. The theory of dynamical systems can incorporate neurobiological knowledge on multiple spatial scales and explain emerging brain functions.

(2) A key concept of the theory of nonlinear dynamical systems is bifurcation. This holds the key to explain how a single neural circuit can operate in qualitatively different behavioral regimes, and how two networks with only quantitative differences (e.g., PFC versus V1) can serve radically different brain functions. It also helps solve the riddle of how entirely novel mental abilities can arise from graded and cumulative changes, e.g., through learning, in biological neural circuits.

(3) We discussed at length an "atom" of cognition, namely, working memory. At the core of the multiple attractor model for working memory is recurrent reverberatory dynamics. In contrast to an activity-silent *modus operandi*, neural firing is essential for active internal manipulation of information and executive control.

(4) A "cognitive-type" neural microcircuit is capable of both working memory and decision making, with the prefrontal cortex as a quintessential example. Such a circuit exhibits the duality of, on the one hand, slow transients through reverberation underlying temporal accumulation of information and, on the other hand, attractor dynamics leading to winner-take-all categorical choices.

(5) A microcircuit of the cognitive-type endowed with reinforcement learning can explain value-based decision making in a variety of tasks, such as stochastic foraging, probabilistic inference and interactive games. Therefore, perceptual decisions and economic choices can be explained by a unified model.

(6) Executive control is conceptually understood in terms of the interplay between the parieto-prefrontal circuit and sensory brain regions for selective attention, and top-down signaling into motor systems for volitional control of actions.

(7) At a large and multi-regional scale, the cortex is characterized by directed and weighted inter-areal connections and a quantifiable hierarchy. A new class of cortex models explicitly takes into consideration spatial embedding according to an exponential distance rule.

(8) Area-to-area heterogeneities of neurobiological properties are quantified in the form of low-dimensional macroscopic gradients across the entire cortical mantle. This idea offers an opening for using genomic data in computational models to build a cross-level understanding of the large-scale brain organization.

(9) A hierarchy of temporal integration windows represents a common characteristic of the neocortex. The spatial localization of distinct timescales is realized in spite of dense area-to-area connectivity. This is computationally desirable for a number of functions, such as rapid information processing in early sensory areas and slow temporal integration of information in decision-related areas.

(10) The functional modularity of the neocortex is compatible with its canonical organization. Distinct parts of the brain are capable of diverse functions, not only due to differences in their input and output patterns, but through abrupt transitions mathematically described as bifurcations in space, which lead to the emergence of novel cognitive computational capabilities in distinct subnetworks of the neocortex.

12.3 SHIFTING PERSPECTIVES

We are witnessing drastic changes on multiple fronts, which require collaborations between experimentation and theory on an entirely new level.

First, multiomic data raise new challenges and opportunities for theory building. Connectome, single-cell transcriptome, and massive neurophysiological and imaging measurements demand novel statistical tools for analysis to translate data to discoveries. The deluge of big data provides valuable information for the development of theoretical and biologically based mathematical models of the brain. Computational neuroscientists will benefit from these advances and strive for cross-level understanding from molecules to neural circuits and behavior.

Second, the theory of dynamical systems is coming to the fore. A sea change is underway from time-series analysis of single-cell coding to population analysis in state space (Brette 2019). This is exemplified by recent studies in which vector fields (determining trajectories) and attractors of neural population activity were deduced quantitatively by neurophysiological measurements (Wang et al. 2023; Luo et al. 2023; Genkin et al. 2023). This new perspective has widespread implications for experiments. Consider a brain region where neurons show activity correlated with a categorical decision (Chapter 6). If a manipulation such as optogenetic inactivation of that region yields little effect, it does not mean that the structure is not causally linked to the underlying attractor state. This is because, when an attractor state is strongly stable, it may be hard to perturb; conversely, a system can be more easily perturbed away from a weakly stable attractor. Attractors do not have to be steady states, illustrated by deterministic chaos in a single neuron (Wang 1994) or a recurrent network (van Vreeswijk and Sompolinsky 1996). They can display complicated spatiotemporal patterns (Wang 2008a; Miller 2016; Khona and Fiete 2022), potentially useful for describing neural processes underlying serial organization of behavior.

Third, a current debate is concerned with distributed computation across multiple brain regions versus modular organization. Undoubtedly, the brain is organized in distinct, albeit interactive, modules dedicated to different functions. We are only starting to collect data on neural signals correlated with behavior across multiple brain regions, that are needed to elucidate interpretations and reveal how the extent of activity spread depends on specific tasks or animal species under consideration.

Epistemologically, the dynamical systems approach does not negate the importance of understanding single cell selectivity and circuit biological mechanism; the multi-regional perspective is not incompatible with functional specialization of distinct parts of the brain. Through the interplay between experiments and modeling, ongoing and future work will uncover the general principles of how the large-scale brain underlies cognition.

12.4 LESS CHARTED TERRITORIES

Looking forward, three general open questions stand out:

First, what is the dynamical nature of the restless brain? It is presently unclear what general mathematical framework can account for complex neural activity patterns in the brain. To tackle this challenge, it is important to emphasize that the brain is not an isolated organ and cognition is in fact "embodied" (Shapiro 2019): inputs to the brain include proprioceptive signals due to bodily movements, interoceptive signals from the body, stimulation from sense organs, rewards and punishments, and social communication with others in a society. To survey a vast behavioral repertoire, studies in a natural setting would be desirable in complement to more controlled simple tasks (Krakauer et al. 2017; Datta et al. 2019; Noel et al. 2022). This requires quantification of complex behaviors, which has become feasible with high-resolution video monitoring and machine learning for data analysis (Mathis et al. 2018; Pereira et al. 2019).

Even when the brain is supposedly not directly interacting with the environment, the mind may be engaged in internal mental processes such as mulling over an idea or an emotional feeling. This could be the case in the so-called resting state, when no task performance is required, or during a mnemonic delay period in a working memory-dependent task. Computation through internal mental process would be reflected in complex temporal variations of neural population activity. Study of the neural circuit mechanisms of those internal processes and the associated neural dynamics represents a cutting-edge research area.

Furthermore, experiences and learning lead to plastic changes within the brain. Because those changes occur on multiple timescales, we are never looking at the same brain twice. All of these changes contribute to the observed temporal variations of neural population activities in different parts of the brain. This is not to say that concepts from the theory of dynamical systems such as attractor states and bifurcations are not applicable. Rather, considerations of the action-perception cycle, internal mental processes and plasticity deserve to be taken seriously in our interpretation of observed neural dynamics and theory development.

Second, what constitutes the human cognitive uniqueness? This book has preciously little to say about human language, mathematics, music and visual arts. Winston Churchill (1897), whose leadership in World War II was in no small measure due to his powerful rhetoric, observed when he was 23 years old: "Of all the talents bestowed upon men, none is so precious as the gift of oratory. He who enjoys it wields a power more durable than that of a great king." The writer Tom Wolfe (2016) wrote: "Speech is not one of man's several unique attributes—speech is the attribute of all attributes!"

However, humans display exceptional mental capabilities in multiple dimensions (Laland and Seed 2021). As a matter of fact, many cognitive functions at the core of fluid intelligence engage a cortical circuit that can be dissociated from the brain's language system (Chapter 11). Thanks to the invention of language, humans can communicate with each other far more efficiently than their close primate relatives; it is essential for communication on much larger scales of social groups. Of course, social cognition and communication skills can also be studied without language. Using a battery of tasks performed by 2.5-year-old human children, chimpanzees and orangutans (an ape more distantly related to us than chimpanzee), it has been shown that humans display superior abilities in social learning, communication and theory of mind (Herrmann et al. 2007). An increasing body of evidence shows that language is primarily a tool for communicative function and transmission of cultural knowledge rather than a prerequisite for sophisticated cognition (Fedorenko et al. 2024).

Some opine that "studies of nonhuman animals provide virtually no relevant parallels to human linguistic communication, and none to the underlying biological capacity" (Hauser et al. 2014). However, it is likely that some core computational requirements for language exist in nonhuman animals. There is anatomical evidence for homologs in nonhuman animals to Wernicke's area of speech perception and Broca's area of speech production (Petrides et al. 2005; Roth and Dicke 2005). Do nonhuman animals have the capacity for symbols (Feinman and Lake 2020; Lurz et al. 2022; Sablé-Meyer et al. 2022) or syntactic thinking (Haegeman 2005; Townsend et al. 2018; Dehaene et al. 2022)? A basic property of the language faculty is a generative procedure called recursion to embed phrases within phrases *ad infinitum* (Pinker 1994; Hauser et al. 2002; Penn et al. 2008; Corballis 2014; Roeper and Speas 2014). Recursion is not unique to language; it is essential for generating an infinite number of structures in complex sequences, such as those in mathematical reasoning (Dehaene et al. 2015). Indeed, recursion can be investigated in nonhuman animals performing tasks that require recursive sequences (Gentner et al. 2006; Ferrigno et al. 2020). Cross-species neuroscientific research, combined with computational modeling, promises to further our knowledge about the uniqueness of human intelligence as a result of superior capabilities in symbolic computation, recursion and other cognitive processes (Newell 1990; Laland and Seed 2021; Dehaene et al. 2022).

Third, can brain theory incorporate emotion? Learning and memory are strongly modulated by the emotional tone of our experiences; we remember especially happy or sad experiences. Furthermore, motivation drives actions, and what is optimal decision making depends on emotional context. Without a rigorous science of emotion, it will be impossible to make sense of the ways we behave, the kind of thoughts we harbor or the choices we make (Damasio 1994; Panksepp 2004; Adolphs and Anderson 2018; LeDoux 2020; Mlodinow 2022).

Theory and computational modeling of the emotional brain face the problem that, unlike perception for which inputs can be measured and varied experimentally in psychophysics, feelings are essentially internal body states (e.g., tranquil or agitated, pleasant or unpleasant), which are difficult to quantify and parametrically manipulate. Theorizing is futile without quantification of observed phenomena and testing predictions in controlled conditions. However, manifestations of emotions can be observed, e.g., by studies of facial expressions (Darwin 2015). While facial expressions are highly variable, they can be differentiated in correlation with emotional states (Ekman et al. 1983) and mapped onto the brain during social behavior (Shepherd and Freiwald 2018). Measures using modern video recording and machine learning deployed for the analysis of high-resolution data promise to reveal quantifiable types of facial expressions, furnishing important clues for an empirical data-based theory of emotions.

Presently, studies of emotions are beset by the dichotomy between functional localization of neural representation for an innate emotional state versus emergent subjective experience engaging the large-scale brain (Pessoa 2022). This challenge has been confronted in studies of fear and anxiety: whereas learning and memory of an escape response to a threat depend on the amygdala, the subjective experience of fear involves the interplay between the amygdala and the prefrontal cortex (Sotres-Bayon et al. 2004; Tovote et al. 2015; LeDoux 2020). Future progress along these lines is likely to open the door for a biologically based computational theory for emotion-cognition interplay (affects motivate behavior and influence reason, and executive control mechanisms tame excessive feelings). This enterprise is essential for computational psychiatry, insofar as much suffering of mentally ill individuals is associated with abnormal emotions ranging from debilitating anxiety, unmanageable rage to depression.

Today we are witnessing a watershed moment in neuroscience. Striving for cross-level mechanistic understanding represents a singularly productive epistemology, which hopefully can be achieved with new technologies and big data on multiple fronts, together with theory. Because the brain systems for cognition and executive control are implicated in virtually all major mental disorders, basic research into cognitive neural circuits is crucial for establishing a solid foundation for diagnosis and therapeutic treatments of mental illness in the nascent field of computational psychiatry. Modern artificial intelligence provides the impetus to go beyond the elemental cognition covered in this book. On the other hand, it has been argued that today's AI systems lack the abilities for reasoning, planning and other internally driven mental processes. For this reason, neuroscience of cognition may inspire future AI research. It remains to be seen whether increasingly complex mental processes, including emotions, thoughts and creativity, will yield their secrets through flexible assemblies of neurobiological and computational building blocks. Advances in this endeavor will not only genuinely unify the psychology of the mind and the biology of the brain but also bolster the next-generation of intelligent machines and establish a scientific foundation of psychiatry.

FURTHER READING

Brette, R. (2019). Is coding a relevant metaphor for the brain? *Behav. Brain Sci. 42*, e215.

Corballis, M. C. (2014). *The Recursive Mind: The Origins of Human Language, Thought, and Civilization*. Princeton, NJ: Princeton University Press.

Datta, S. R., D. J. Anderson, K. Branson, P. Perona, and A. Leifer (2019). Computational neuroethology: A call to action. *Neuron 104*, 11–24.

Herrmann, E., J. Call, M. V. Hernàndez-Lloreda, B. Hare, and M. Tomasello (2007). Humans have evolved specialized skills of social cognition: The cultural intelligence hypothesis. *Science 317*, 1360–1366.

Khona, M. and I. R. Fiete (2022). Attractor and integrator networks in the brain. *Nat. Rev. Neurosci. 23*, 744–766.

Panksepp, J. (2004). *Affective Neuroscience: The Foundations of Human and Animal Emotions*. New York: Oxford University Press.

Pinker, S. (1994). *The Language Instinct: How the Mind Creates Language*. New York: HarperPerennial.

Sablé-Meyer, M., K. Ellis, J. Tenenbaum, and S. Dehaene (2022). A language of thought for the mental representation of geometric shapes. *Cogn. Psychol. 139*, 101527.

Wang, S., R. Falcone, B. Richmond, and B. B. Averbeck (2023). Attractor dynamics reflect decision confidence in macaque prefrontal cortex. *Nat. Neurosci. 26*, 1970–1980.

Wang, X.-J. (2008). Attractor network models. In L. Squire, T. Albright, F. Bloom, F. Gage, and N. Spitzer (Eds.), *Encyclopedia of Neuroscience*, pp. 667–679. London: MacMillan Reference Ltd.

REFERENCES

Abbott, L. F. (2006). Where are the switches on this thing? In J. L. van Hemmen and T. J. Sejnowski (Eds.), *23 Problems in Systems Neuroscience*, Chapter 21, pp. 423–431. Oxford: Oxford University Press.

Abbott, L. F. (2008). Theoretical neuroscience rising. *Neuron 60*, 489–495.

Abbott, L. F. and K. I. Blum (1996). Fonctional significance of long-term potentiation for sequence learning and prediction. *Cereb. Cortex 6*, 406–416.

Abbott, L. F., D. D. Bock, E. M. Callaway, W. Denk, C. Dulac, A. L. Fairhall, I. Fiete, K. M. Harris, M. Helmstaedter, V. Jain, N. Kathuri, Y. LeCun, J. W. Lichtman, P. B. Littlewood, L. Luo, J. H. R. Maunsell, C. Reid, B. R. Rosen, G. M. Rubin, T. J. Sejnowski, H. S. Seung, K. Svoboda, D. W. Tank, D. Tsao, and D. van Essen (2020). The mind of a mouse. *Cell 182*, 1372–1376.

Abbott, L. F. and S. B. Nelson (2000). Synaptic plasticity: Taming the beast. *Nat. Neurosci. 3*(Suppl), 1178–1183.

Abbott, L. F. and W. G. Regehr (2004). Synaptic computation. *Nature 431*, 796–803.

Abbott, L. F., J. A. Varela, K. Sen, and S. B. Nelson (1997). Synaptic depression and cortical gain control. *Science 275*, 220–224.

Abeles, M. (1991). *Corticonics*. New York: Cambridge University Press.

Abraham, N., H. Spors, A. Carleton, T. Margrie, T. Kuner, and A. Schaefer (2004). Maintaining accuracy at the expense of speed: Stimulus similarity defines odor discrimination time in mice. *Neuron 44*, 865–876.

Achterberg, J., M. Kadohisa, K. Watanabe, M. Kusunoki, M. J. Buckley, and J. Duncan (2022). A one-shot shift from explore to exploit in monkey prefrontal cortex. *J. Neurosci. 42*, 276–287.

Adam, D. (2013). Mental health: On the spectrum. *Nature 496*, 416–418.

Adelson, E. H. and J. R. Bergen (1985). Spatiotemporal energy models for the perception of motion. *J. Opt. Soc. Am. A 2*, 284–299.

Adelson, E. H. and J. A. Movshon (1982). Phenomenal coherence of moving visual patterns. *Nature 300*, 523–525.

Adesnik, H., W. Bruns, H. Taniguchi, Z. J. Huang, and M. Scanziani (2012). A neural circuit for spatial summation in visual cortex. *Nature 490*, 226–231.

Adolphs, R. and D. J. Anderson (2018). *The Neuroscience of Emotion: A New Synthesis*. Princeton, NJ: Princeton University Press.

Adrian, E. (1950). The electrical activity of the mammalian olfactory bulb. *Electroencephalogr. Clin. Neurophysiol. 2*, 377–388.

Aertsen, A. M. and G. L. Gerstein (1985). Evaluation of neuronal connectivity: Sensitivity of cross-correlation. *Brain Res. 340*, 341–354.

Aggleton, J. P., N. F. Wright, D. L. Rosene, and R. C. Saunders (2015). Complementary patterns of direct amygdala and hippocampal projections to the macaque prefrontal cortex. *Cereb. Cortex 25*, 4351–4373.

Ahmed, B., J. Anderson, R. Douglas, K. Martin, and D. Whitteridge (1998). Estimates of the net excitatory currents evoked by visual stimulation of identified neurons in cat visual cortex. *Cereb. Cortex 8*, 462–476.

Ainslie, G. W. (1974). Impulse control in pigeons. *J. Exp. Anal. Behav. 21*, 485–489.

Ais, J., A. Zylberberg, P. Barttfeld, and M. Sigman (2016). Individual consistency in the accuracy and distribution of confidence judgments. *Cognition 146*, 377–386.

Aksay, E., G. Gamkrelidze, H. S. Seung, R. Baker, and D. W. Tank (2001). In vivo intracellular recording and perturbation of persistent activity in a neural integrator. *Nat. Neurosci. 4*, 184–193.

Albantakis, L. and G. Deco (2009). The encoding of alternatives in multiple-choice decision making. *Proc. Natl. Acad. Sci. U.S.A. 106*, 10308–10313.

Alberini, C. M. (2009). Transcription factors in long-term memory and synaptic plasticity. *Physiol. Rev. 89*, 121–145.

Albert, R. and A. L. Barabási (2002). Statistical mechanics of complex networks. *Rev. Mod. Phys. 74*, 47–97.

Albright, T. D., R. Desimone, and C. G. Gross (1984). Columnar organization of directionally selective cells in visual area mt of the macaque. *J. Neurophysiol. 51*, 16–31.

Albus, J. S. (1971). A theory of cerebellar function. *Math. Biosci. 10*, 25–61.

Alexander, G. E. and M. D. Crutcher (1990). Functional architecture of basal ganglia circuits: Neural substrates of parallel processing. *Trends Neurosci. 13*, 266–271.

Alexander, G. E., M. R. DeLong, and P. L. Strick (1986). Parallel organization of functionally segregated circuits linking basal ganglia and cortex. *Annu. Rev. Neurosci. 9*, 357–381.

Alexander, W. H. and J. W. Brown (2011). Medial prefrontal cortex as an action-outcome predictor. *Nat. Neurosci. 14*, 1338–1344.

Alloway, T. P. and R. G. Alloway (2010). Investigating the predictive roles of working memory and IQ in academic attainment. *J. Exp. Child Psychol. 106*, 20–29.

Allport, A. and G. Wylie (1999). Task-switching: Positive and negative priming of task-set. In G. W. Humphreys, J. Duncan, and A. Treisman (Eds.), *Attention, Space, and Action: Studies in Cognitive Neuroscience*, pp. 273–296. Oxford: Clarendon Press.

Allport, D. A., E. A. Styles, and E. A. Hsieh (1994). Shifting intentional set: Exploring the dynamic control of tasks. In C. Umilta and M. Moscovitch (Eds.), *Attention and Performance* (Vol. XV), pp. 421–452. Cambridge, MA: MIT Press.

Alonso, J., M. Angermeyer, S. Bernert, R. Bruffaerts, T. Brugha, H. Bryson, G. de Girolamo, R. de Graaf, K. Demyttenaere, et al. (2004). Prevalence of mental disorders in europe: Results from the european study of the epidemiology of mental disorders (esemed) project. *Acta Psychiatr. Scand. 109*, 21–27.

Altmann, E. M. (2011). Testing probability matching and episodic retrieval accounts of response repetition effects in task switching. *J. Exp. Psychol. Learn. Mem. Cogn. 37*, 935–951.

Altmann, E. M. and W. D. Gray (2008). An integrated model of cognitive control in task switching. *Psychol. Rev. 115*, 602–639.

Amalric, M., L. Wang, P. Pica, S. Figueira, M. Sigman, and S. Dehaene (2017). The language of geometry: Fast comprehension of geometrical primitives and rules in human adults and preschoolers. *PLoS Comput. Biol. 13*, e1005273.

Amarasingham, A., T. Chen, S. Geman, M. Harrison, and D. Sheinberg (2006). Spike count reliability and the Poisson hypothesis. *J. Neurosci. 26*, 801–809.

Amari, S. (1977). Dynamics of pattern formation in lateral-inhibition type neural fields. *Biol. Cybern. 27*, 77–87.

American Psychiatric Association, D., A. P. Association, et al. (2013). *Diagnostic and Statistical Manual of Mental Disorders: DSM-5*. Washington, DC: American Psychiatric Association.

Amit, D. J. (1989). *Modeling Brain Function*. Cambridge: Cambridge University Press.

Amit, D. J. (1995). The Hebbian paradigm reintegrated: Local reverberations as internal representations. *Behav. Brain Sci. 18*, 617–626.

Amit, D. J. and N. Brunel (1997). Model of global spontaneous activity and local structured activity during delay periods in the cerebral cortex. *Cereb. Cortex 7*, 237–252.

Amit, D. J. and S. Fusi (1994). Dynamic learning in neural networks with material synapses. *Neural Comp. 6*, 957–982.

Amit, D. J., H. Gutfreund, and H. Sompolinsky (1985). Storing infinite numbers of patterns in a spin-glass model of neural networks. *Phys. Rev. Lett. 55*, 1530–1531.

Amunts, K. and T. Lippert (2021). Brain research challenges supercomputing. *Science 374*, 1054–1055.

Amunts, K. and K. Zilles (2015). Architectonic mapping of the human brain beyond Brodmann. *Neuron 88*, 1086–1107.

Andersen, P. and S. A. Andersson (1968). *Physiological Basis of the Alpha Rhythm*. New York: Appleton-Century-Crofts.

Andersen, P., R. Morris, D. Amaral, T. Bliss, and J. O'Keefe (2007). *The Hippocampus Book*. Oxford: Oxford University Press.

Andersen, R. A. (1989). Visual and eye movement functions of the posterior parietal cortex. *Annu. Rev. Neurosci. 12*, 377–403.

Andersen, R. A. and C. A. Buneo (2002). Intentional maps in posterior parietal cortex. *Annu. Rev. Neurosci. 25*, 189–220.

Andersen, R. A., E. J. Hwang, and G. H. Mulliken (2010). Cognitive neural prosthetics. *Annu. Rev. Psychol. 61*, 169–190.

Anderson, J. R. (1983). Retrieval of information from long-term memory. *Science 220*, 25–30.

Anderson, M. C. and C. Green (2001). Suppressing unwanted memories by executive control. *Nature 410*, 366–369.

Anderson, M. C., K. N. Ochsner, B. Kuhl, J. Cooper, E. Robertson, S. W. Gabrieli, G. H. Glover, and J. D. Gabrieli (2004). Neural systems underlying the suppression of unwanted memories. *Science 303*, 232–235.

Anderson, P. W. (1958). Absence of diffusion in certain random lattices. *Phys. Rev. 109*, 1492.

Anderson, P. W. (1963). Plasmons, gauge invariance, and mass. *Phys. Rev. 130*, 439–442.

Anderson, P. W. (1972). More is different. *Science 177*, 393–396.

Anticevic, A., M. W. Cole, G. Repovs, J. D. Murray, M. S. Brumbaugh, A. M. Winkler, A. Savic, J. H. Krystal, G. D. Pearlson, and D. C. Glahn (2014). Characterizing thalamo-cortical disturbances in schizophrenia and bipolar illness. *Cereb. Cortex 24*, 3116–3130.

Anticevic, A., M. Gancsos, J. D. Murray, G. Repovs, N. R. Driesen, D. J. Ennis, M. J. Niciu, P. T. Morgan, T. S. Surti, M. H. Bloch, R. Ramani, M. A. Smith, X.-J. Wang, J. H. Krystal, and P. R. Corlett (2012). NMDA receptor function in large-scale anticorrelated neural systems with implications for cognition and schizophrenia. *Proc. Natl. Acad. Sci. U.S.A. 109*, 16720–16725.

Anticevic, A., G. Repovs, P. R. Corlett, and D. M. Barch (2011). Negative and nonemotional interference with visual working memory in schizophrenia. *Biol. Psychiatry 70*, 1159–1168.

Antoine, M. W., T. Langberg, P. Schnepel, and D. E. Feldman (2019). Increased excitation-inhibition ratio stabilizes synapse and circuit excitability in four autism mouse models. *Neuron 101*, 648–661.

Apšvalka, D., C. S. Ferreira, T. W. Schmitz, J. B. Rowe, and M. C. Anderson (2022). Dynamic targeting enables domain-general inhibitory control over action and thought by the prefrontal cortex. *Nat. Commun. 13*. https://doi.org/10.1038/s41467-021-27926-w.

Aragona, B. J., J. J. Day, M. F. Roitman, N. A. Cleaveland, R. Mark Wightman, and R. M. Carelli (2009). Regional specificity in the real-time development of phasic dopamine transmission patterns during acquisition of a cue–cocaine association in rats. *Eur. J. Neurosci. 30*, 1889–1899.

Ardid, S. and X.-J. Wang (2013). A tweaking principle for executive control: Neuronal circuit mechanism for rule-based task switching and conflict resolution. *J. Neurosci. 33*, 19504–19517.

Ardid, S., X.-J. Wang, and A. Compte (2007). An integrated microcircuit model of attentional processing in the neocortex. *J. Neurosci. 27*, 8486–8495.

Ardid, S., X.-J. Wang, D. Gomez-Cabrero, and A. Compte (2010). Reconciling coherent oscillation with modulation of irregular spiking activity in selective attention: Gamma-range synchronization between sensory and executive cortical areas. *J. Neurosci. 30*, 2856–2870.

Arieli, A., D. Shoham, R. Hildesheim, and A. Grinvald (1995). Coherent spatiotemporal patterns of ongoing activity revealed by real-time optical imaging coupled with single-unit recording in the cat visual cortex. *J. Neurophysiol. 73*, 2072–2093.

Aristieta, A., M. Barresi, S. A. Lindi, G. Barriere, G. Courtand, B. de la Crompe, L. Guilhemsang, S. Gauthier, S. Fioramonti, J. Baufreton, et al. (2021). A disynaptic circuit in the globus pallidus controls locomotion inhibition. *Curr. Biol. 31*, 707–721.

Arnsten, A. F. (1998). Catecholamine modulation of prefrontal cortical cognitive function. *Trends Cogn. Sci. 2*, 436–447.

Arnsten, A. F. (2023). Retrospective: Patricia S. Goldman-Rakic, pioneer in neuroscience and co-founder of the journal, *Cerebral Cortex. Cereb. Cortex.* https://doi.org/10.1093/cercor/bhad159.

Arnsten, A. F., J. Cai, B. Murphy, and P. Goldman-Rakic (1994). Dopamine D1 receptor mechanisms in the cognitive performance of young adult and aged monkeys. *Psychopharmacology 116*, 143–151.

Arnsten, A. F., D. Datta, and M. Wang (2021). The genie in the bottle-magnified calcium signaling in dorsolateral prefrontal cortex. *Mol. Psychiatry 26*, 3684–3700.

Arnsten, A. F., L. E. Jin, N. J. Gamo, B. Ramos, C. D. Paspalas, Y. M. Morozov, A. Kata, N. S. Bamford, M. F. Yeckel, L. K. Kaczmarek, et al. (2019). Role of kcnq potassium channels in stress-induced deficit of working memory. *Neurobiol. Stress 11*, 100187.

Arnsten, A. F., C. D. Paspalas, N. J. Gamo, Y. Yang, and M. Wang (2010). Dynamic network connectivity: A new form of neuroplasticity. *Trends Cogn. Sci. 14*, 365–375.

Arnsten, A. F. and M. Wang (2020). The evolutionary expansion of mGluR3-NAAG-GCPII signaling: Relevance to human intelligence and cognitive disorders. *Am. J. Psychiatry 177*, 1103–1106.

Aron, A. R. (2007). The neural basis of inhibition in cognitive control. *Neuroscientist 13*, 214–228.

Aron, A. R. (2011). From reactive to proactive and selective control: Developing a richer model for stopping inappropriate responses. *Biol. Psychiatry 69*, 55–68.

Aronov, D., R. Nevers, and D. W. Tank (2017). Mapping of a non-spatial dimension by the hippocampal–entorhinal circuit. *Nature 543*, 719–722.

Arrow, K. J. (1959). Rational choice functions and orderings. *Economica 26*, 121–127.

Asaad, W. F., G. Rainer, and E. K. Miller (1998). Neural activity in the primate prefrontal cortex during associative learning. *Neuron 21*, 1399–1407.

Ashby, F. G. and W. T. Maddox (2005). Human category learning. *Annu. Rev. Psychol. 56*, 149–178.

Ashby, W. R. (1956). *An Introduction to Cybernetics*. London: Chapman and Hall.

Ashwood, Z. C., N. A. Roy, I. R. Stone, I. B. Laboratory, A. E. Urai, A. K. Churchland, A. Pouget, and J. W. Pillow (2022). Mice alternate between discrete strategies during perceptual decision-making. *Nat. Neurosci. 25*, 201–212.

Assem, M., M. F. Glasser, D. C. Van Essen, and J. Duncan (2020). A domain-general cognitive core defined in multimodally parcellated human cortex. *Cereb. Cortex 30*, 4361–4380.

Astle, D. E., M. H. Johnson, and D. Akarca (2023). Toward computational neuroconstructivism: A framework for developmental systems neuroscience. *Trends Cogn. Sci. 27*, 726–744.

Atallah, B. V. and M. Scanziani (2009). Instantaneous modulation of gamma oscillation frequency by balancing excitation with inhibition. *Neuron 62*, 566–577.

Atkinson, R. C. and R. M. Shiffrin (1968). Chapter: Human memory: A proposed system and its control processes. In K. W. Spence and J. T. Spence (Eds.), *The Psychology of Learning and Motivation* (Vol. 2), pp. 89–195. New York: Academic Press.

Attwell, D. and S. B. Laughlin (2001). An energy budget for signaling in the grey matter of the brain. *J. Cereb. Blood Flow Metab. 21*, 1133–1145.

Averbeck, B. B. (2022). Pruning recurrent neural networks replicates adolescent changes in working memory and reinforcement learning. *Proc. Natl. Acad. Sci. 119*, e2121331119.

Averbeck, B. B. and M. V. Chafee (2016). Using model systems to understand errant plasticity mechanisms in psychiatric disorders. *Nat. Neurosci. 19*, 1418–1425.

Averbeck, B. B., D. A. Crowe, M. V. Chafee, and A. P. Georgopoulos (2003). Neural activity in prefrontal cortex during copying geometrical shapes. *Exp. Brain Res. 150*, 127–153.

Averbeck, B. B., P. Latham, and A. Pouget (2006). Neural correlations, population coding and computation. *Nat. Rev. Neurosci. 7*, 358–366.

Averbeck, B. B. and M. Seo (2008). The statistical neuroanatomy of frontal networks in the macaque. *PLoS Comput. Biol. 4*, e1000050.

Awh, E. and J. Jonides (2001). Overlapping mechanisms of attention and spatial working memory. *Trends Cogn. Sci. 5*, 119–126.

Axelrod, R. (1984). *The Evoultion of Cooperation* (2nd ed.). New York: Basic Books.

Azevedo, F. A., L. R. Carvalho, L. T. Grinberg, J. M. Farfel, R. E. Ferretti, R. E. Leite, W. J. Filho, R. Lent, and S. Herculano-Houzel (2009). Equal numbers of neuronal and nonneuronal cells make the human brain an isometrically scaled-up primate brain. *J. Comp. Neurol. 513*, 532–541.

Baars, B. J. (1988). *A Cognitive Theory of Consciousness*. Cambridge: Cambridge University Press.

Babadi, B. and H. Sompolinsky (2014). Sparseness and expansion in sensory representations. *Neuron 83*, 1213–1226.

Badcock, J. C., D. R. Badcock, C. Read, and A. Jablensky (2008). Examining encoding imprecision in spatial working memory in schizophrenia. *Schizophr. Res. 100*, 144–152.

Baddeley, A. (2012). Working memory: Theories, models, and controversies. *Annu. Rev. Psychol. 63*, 1–29.

Baddeley, A. and G. J. Hitch (1974). Working memory. In G. A. Bower (Ed.), *The Psychology of Learning and Motivation: Advances in Research and Theory*, pp. 47–89. New York: Academic Press.

Badre, D. (2020). *On Task: How Our Brain Gets Things Done*. Princeton, NJ: Princeton University Press.

Badre, D. (2021). Brain networks for cognitive control: Four unresolved questions. In P. Kalivas and M. P. Paulus (Eds.), *Intrusive Thinking: From Molecules to Free Will*, pp. 203–228. Cambridge, MA: MIT Press.

Baeg, E. H., Y. B. Kim, K. Huh, I. Mook-Jung, H. T. Kim, and M. W. Jung (2003). Dynamics of population code for working memory in the prefrontal cortex. *Neuron 40*, 177–188.

Bahdanau, D., K. Cho, and Y. Bengio (2014). Neural machine translation by jointly learning to align and translate. *arXiv preprint arXiv:1409.0473.*

Bahl, A. and F. Engert (2020). Neural circuits for evidence accumulation and decision making in larval zebrafish. *Nat. Neurosci. 23*, 94–102.

Bair, W., E. Zohary, and W. T. Newsome (2001). Correlated firing in macaque visual area MT: Time scales and relationship to behavior. *J. Neurosci. 21*, 1676–1697.

Bakken, T. E., N. L. Jorstad, Q. Hu, B. B. Lake, W. Tian, B. E. Kalmbach, M. Crow, R. D. Hodge, F. M. Krienen, S. A. Sorensen, et al. (2021). Comparative cellular analysis of motor cortex in human, marmoset and mouse. *Nature 598*, 111–119.

Balan, P. F., J. Oristaglio, D. M. Schneider, and J. Gottlieb (2008). Neuronal correlates of the set-size effect in monkey lateral intraparietal area. *PLoS Biol. 6*, e158.

Balasubramani, P. P., M. C. Pesce, and B. Y. Hayden (2020). Activity in orbitofrontal neuronal ensembles reflects inhibitory control. *Eur. J. Neurosci. 51*, 2033–2051.

Baldassano, C., J. Chen, A. Zadbood, J. W. Pillow, U. Hasson, and K. A. Norman (2017). Discovering event structure in continuous narrative perception and memory. *Neuron 95*, 709–721.

Balleine, B. W. (2019). The meaning of behavior: Discriminating reflex and volition in the brain. *Neuron 104*, 47–62.

Ballesta, S., W. Shi, K. E. Conen, and C. Padoa-Schioppa (2020). Values encoded in orbitofrontal cortex are causally related to economic choices. *Nature 588*, 450–453.

Ballesteros-Yanez, I., R. Benavides-Piccione, J. P. Bourgeois, J. P. Changeux, and J. DeFelipe (2010). Alterations of cortical pyramidal neurons in mice lacking high-affinity nicotinic receptors. *Proc. Natl. Acad. Sci. U.S.A. 107*, 11567–11572.

Balter, M. (2010). Evolution of behavior. Did working memory spark creative culture? *Science 328*, 160–163.

Banca, P., M. D. Vestergaard, V. Rankov, K. Baek, S. Mitchell, T. Lapa, M. Castelo-Branco, and V. Voon (2015). Evidence accumulation in obsessive-compulsive disorder: The role of uncertainty and monetary reward on perceptual decision-making thresholds. *Neuropsychopharmacology 40*, 1192–1202.

Banerjee, A., G. Parente, J. Teutsch, C. Lewis, F. F. Voigt, and F. Helmchen (2020). Value-guided remapping of sensory cortex by lateral orbitofrontal cortex. *Nature 585*, 245–250.

Bang, D. and S. M. Fleming (2018). Distinct encoding of decision confidence in human medial prefrontal cortex. *Proc. Natl. Acad. Sci. U.S.A. 115*, 6082–6087.

Banino, A., C. Barry, B. Uria, C. Blundell, T. Lillicrap, P. Mirowski, A. Pritzel, M. J. Chadwick, T. Degris, J. Modayil, G. Wayne, H. Soyer, F. Viola, B. Zhang, R. Goroshin, N. Rabinowitz, R. Pascanu, C. Beattie, S. Petersen, A. Sadik, S. Gaffney, H. King, K. Kavukcuoglu, D. Hassabis, R. Hadsell, and D. Kumaran (2018). Vector-based navigation using grid-like representations in artificial agents. *Nature 557*, 429–433.

Bao, P., L. She, M. McGill, and D. Y. Tsao (2020). A map of object space in primate inferotemporal cortex. *Nature 583*, 103–108.

Barabási, A.-L. (2016). *Network Science*. New York: Cambridge University Press.

Barack, D. L. and J. W. Krakauer (2021). Two views on the cognitive brain. *Nat. Rev. Neurosci. 22*, 359–371.

Barak, O., D. Sussillo, R. Romo, M. Tsodyks, and L. F. Abbott (2013). From fixed points to chaos: Three models of delayed discrimination. *Prog. Neurobiol. 103*, 214–222.

Barbas, H. (2015). General cortical and special prefrontal connections: Principles from structure to function. *Annu. Rev. Neurosci. 38*, 269–289.

Barbieri, F. and N. Brunel (2008). Can attractor network models account for the statistics of firing during persistent activity in prefrontal cortex? *Front. Neurosci. 2*, 114–122.

Barbosa, J., H. Stein, R. L. Martinez, A. Galan-Gadea, S. Li, J. Dalmau, K. C. Adam, J. Valls-Solé, C. Constantinidis, and A. Compte (2020). Interplay between persistent activity and activity-silent dynamics in the prefrontal cortex underlies serial biases in working memory. *Nat. Neurosci. 23*, 1016–1024.

Barch, D. M. (2005). The cognitive neuroscience of schizophrenia. *Annu. Rev. Clin. Psychol. 1*, 321–353.

Barch, D. M. and A. Ceaser (2012). Cognition in schizophrenia: Core psychological and neural mechanisms. *Trends Cogn. Sci. (Regul. Ed.) 16*, 27–34.

Barch, D. M. and E. Smith (2008). The cognitive neuroscience of working memory: Relevance to CNTRICS and schizophrenia. *Biol. Psychiatry 64*, 11–17.

Bari, A. and T. W. Robbins (2013). Inhibition and impulsivity: Behavioral and neural basis of response control. *Prog. Neurobiol. 108*, 44–79.

Bari, B. A., C. D. Grossman, E. E. Lubin, A. E. Rajagopalan, J. I. Cressy, and J. Y. Cohen (2019). Stable representations of decision variables for flexible behavior. *Neuron 103*, 922–933.

Baria, A. T., B. Maniscalco, and B. J. He (2017). Initial-state-dependent, robust, transient neural dynamics encode conscious visual perception. *PLoS Comput. Biol. 13*, e1005806.

Barkai, N. and S. Leibler (1997). Robustness in simple biochemical networks. *Nature 387*, 913–917.

Barkley, R. A. (1997). Behavioral inhibition, sustained attention, and executive functions: Constructing a unifying theory of ADHD. *Psychol. Bull. 121*, 65–94.

Barlow, H. B. (1961). Possible principles underlying the transformations of sensory messages. In W. A. Rosenblith (Ed.), *Sensory Communication*, pp. 217–234. Cambridge, MA: MIT Press.

Barlow, H. B. (1972). Single units and sensation: A neuron doctrine for perceptual psychology? *Perception 1*, 371–394.

Barlow, H. B. and P. Foldiak (1989). Adaptation and decorrelation in the cortex. In R. Durbin and C. Miall (Eds.), *The Computing Neuron*, pp. 454–472. New York: Addison-Wesley.

Barnett, L. and A. K. Seth (2014). The MVGC multivariate Granger causality toolbox: A new approach to Granger-causal inference. *J. Neurosci. Methods*, 50–68.

Barraclough, D. J., M. L. Conroy, and D. Lee (2004). Prefrontal cortex and decision making in a mixed-strategy game. *Nat. Neurosci. 7*, 404–410.

Barral, J. and A. Reyes (2016). Synaptic scaling rule preserves excitatory-inhibitory balance and salient neuronal network dynamics. *Nat. Neurosci. 19*, 1690–1696.

Barrett, A. B., G. O. Billings, R. G. Morris, and M. C. van Rossum (2009). State based model of long-term potentiation and synaptic tagging and capture. *PLoS Comput. Biol. 5*, e1000259.

Barrett, L. F. (2017). *How Emotions Are Made: The Secret Life of the Brain.* Boston, MA: Houghton Mifflin Harcourt.

Barron, H. C., T. P. Vogels, T. E. Behrens, and M. Ramaswami (2017). Inhibitory engrams in perception and memory. *Proc. Natl. Acad. Sci. 114*, 6666–6674.

Barthélemy, M. (2011). Spatial networks. *Phys. Rep. 499*, 1–101.

Bartlett, F. (1932). *Remembering: A Study in Experimental and Social Psychology* (Reissued in 1995 ed.). Cambridge: Cambridge University Press.

Bartos, M., I. Vida, and P. Jonas (2007). Synaptic mechanisms of synchronized gamma oscillations in inhibitory interneuron networks. *Nat. Rev. Neurosci. 8*, 45–56.

Bassett, D. S., E. Bullmore, B. A. Verchinski, V. S. Mattay, D. R. Weinberger, and A. Meyer-Lindenberg (2008). Hierarchical organization of human cortical networks in health and schizophrenia. *J. Neurosci. 28*, 9239–9248.

Bassett, D. S., A. Meyer-Lindenberg, S. Achard, T. Duke, and E. Bullmore (2006). Adaptive reconfiguration of fractal small-world human brain functional networks. *Proc. Natl. Acad. Sci. U.S.A. 103*, 19518–19523.

Bassett, D. S. and O. Sporns (2017). Network neuroscience. *Nat. Neurosci. 20*, 353–364.

Basso, M. A., M. E. Bickford, and J. Cang (2021). Unraveling circuits of visual perception and cognition through the superior colliculus. *Neuron 109*, 918–937.

Bastos, A. M., R. Loonis, S. Kornblith, M. Lundqvist, and E. K. Miller (2018). Laminar recordings in frontal cortex suggest distinct layers for maintenance and control of working memory. *Proc. Natl. Acad. Sci. 115*, 1117–1122.

Bastos, A. M., W. M. Usrey, R. A. Adams, G. R. Mangun, P. Fries, and K. J. Friston (2012). Canonical microcircuits for predictive coding. *Neuron 76*, 695–711.

Bastos, A. M., J. Vezoli, C. A. Bosman, J. M. Schoffelen, R. Oostenveld, J. R. Dowdall, P. De Weerd, H. Kennedy, and P. Fries (2015). Visual areas exert feedforward and feedback influences through distinct frequency channels. *Neuron 85*, 390–401.

Battaglia, F. P. and C. M. Pennartz (2011). The construction of semantic memory: Grammar-based representations learned from relational episodic information. *Front. Comput. Neurosci. 5*, 36.

Batuev, A., A. Pirogov, and A. Orlov (1979). Unit activity of the prefrontal cortex during delayed alternation performance in monkey. *Acta Physiol. Acad. Sci. Hung. 53*, 345–353.

Bayer, H. and P. Glimcher (2005). Midbrain dopamine neurons encode a quantitative reward prediction error signal. *Neuron 47*, 129–141.

Bayes, T. (1763). An essay towards solving a problem in the doctrine of chances. by the late Rev. Mr. Bayes, FRS communicated by Mr. Price, in a letter to John Canton. *Phil. Trans. Roy. Soc. Lond. 53*, 370–418.

Bays, P. M. (2018). Reassessing the evidence for capacity limits in neural signals related to working memory. *Cereb. Cortex 28*, 1432–1438.

Bays, P. M., R. F. Catalao, and M. Husain (2009). The precision of visual working memory is set by allocation of a shared resource. *J. Vis. 9*, 1–11.

Bays, P. M. and M. Husain (2008). Dynamic shifts of limited working memory resources in human vision. *Science 321*, 851–854.

Bazzari, A. H. and H. R. Parri (2019). Neuromodulators and long-term synaptic plasticity in learning and memory: A steered-glutamatergic perspective. *Brain Sci. 9*, 300.

Beam, E., C. Potts, R. A. Poldrack, and A. Etkin (2021). A data-driven framework for mapping domains of human neurobiology. *Nat. Neurosci.* https://doi.org/10.1038/s41593-021-00948-9.

Bear, M. F. (1996). A synaptic basis for memory storage in the cerebral cortex. *Proc. Natl. Acad. Sci. U.S.A. 93*, 13453–13459.

Bear, M. F., B. W. Connors, and M. A. Paradiso (Eds.) (2015). *Neuroscience: Exploring the Brain* (4th ed.). Alphen aan den Rijn, Netherlands: Wolters Kluwer.

Beauchamp, M. S., K. E. Lee, B. D. Argall, and A. Martin (2004). Integration of auditory and visual information about objects in superior temporal sulcus. *Neuron 41*, 809–823.

Beaulieu-Laroche, L., N. J. Brown, M. Hansen, E. H. Toloza, J. Sharma, Z. M. Williams, M. P. Frosch, G. R. Cosgrove, S. S. Cash, and M. T. Harnett (2021). Allometric rules for mammalian cortical layer 5 neuron biophysics. *Nature*. https://doi.org/10.1038/s41586

Bechara, A., D. Tranel, and H. Damasio (2000). Characterization of the decision-making deficit of patients with ventromedial prefrontal cortex lesions. *Brain 123*, 2189–2202.

Bechtel, W. (2008). *Mental Mechanisms: Philosophical Perspectives on Cognitive Neuroscience*. London: Rutledge.

Bechtel, W. (2019). Resituating cognitive mechanisms within heterarchical networks controlling physiology and behavior. *Theory Psychol. 29*, 620–639.

Bechtel, W. and J. Mundale (1999). Multiple realizability revisited: Linking cognitive and neural states. *Philos. Sci. 66*, 175–207.

Beck, J. M., W. J. Ma, R. Kiani, T. Hanks, A. K. Churchland, J. Roitman, M. N. Shadlen, P. E. Latham, and A. Pouget (2008). Probabilistic population codes for Bayesian decision making. *Neuron 60*, 1142–1152.

Bédard, C. and A. Destexhe (2009). Macroscopic models of local field potentials and the apparent 1/f noise in brain activity. *Biophys. J. 96*, 2589–2603.

Bédard, C., H. Kröger, and A. Destexhe (2006). Does the 1/f frequency scaling of brain signals reflect self-organized critical states? *Phys. Rev. Lett. 97*, 118102.

Behrens, T. E., L. T. Hunt, M. W. Woolrich, and M. F. Rushworth (2008). Associative learning of social value. *Nature 456*, 245–249.

Behrens, T. E., T. H. Muller, J. C. Whittington, S. Mark, A. B. Baram, K. L. Stachenfeld, and Z. Kurth-Nelson (2018). What is a cognitive map? organizing knowledge for flexible behavior. *Neuron 100*, 490–509.

Behrens, T. E., M. Woolrich, M. Walton, and M. Rushworth (2007). Learning the value of information in an uncertain world. *Nat. Neurosci. 10*, 1214–1221.

Bell, C. C., V. Z. Han, Y. Sugawara, and K. Grant (1997). Synaptic plasticity in a cerebellum-like structure depends on temporal order. *Nature 387*, 278–281.

Bell, D. E. (1982). Regret in decision making under uncertainty. *Oper. Res. 30*, 961–981.

Bellman, R. (1957). *Dynamic Programming*. Princeton, NJ: Princeton University Press.

Benna, M. K. and S. Fusi (2016). Computational principles of synaptic memory consolidation. *Nat. Neurosci. 19*, 1697–1706.

Bennett, M. (2023). A brief history of intelligence: evolution, AI, and the five breakthroughs that made our brains. HarperCollins.

Ben-Yishai, R., R. Lev Bar-Or, and H. Sompolinsky (1995). Theory of orientation tuning in visual cortex. *Proc. Natl. Acad. Sci. U.S.A. 92*, 3844–3848.

Berger, B., P. Gaspar, and C. Verney (1991). Dopaminergic innervation of the cerebral cortex: Unexpected differences between rodents and primates. *Trends Neurosci. 14*, 21–27.

Berger, J. (1980). *Statistical Decision Theory and Bayesian Analysis* (2nd ed.). New York: Springer-Verlag.

Berger, M., N. S. Agha, and A. Gail (2020). Wireless recording from unrestrained monkeys reveals motor goal encoding beyond immediate reach in frontoparietal cortex. *Elife 9*, e51322.

Berke, J. D. (2018). What does dopamine mean? *Nat. Neurosci. 21*, 787–793.

Berlemont, K., J.-R. Martin, J. Sackur, and J.-P. Nadal (2020). Nonlinear neural network dynamics accounts for human confidence in a sequence of perceptual decisions. *Sci. Rep. 10*. https://doi.org/10.1038/s41598-020-63582-8.

Bernacchia, A., H. Seo, D. Lee, and X.-J. Wang (2011). A reservoir of time constants for memory traces in cortical neurons. *Nat. Neurosci. 14*, 366–372.

Bernard, C. (1865). *Introduction á l'étude de la médicine expérimentale*. Paris: Baillière.

Bernardi, S., M. K. Benna, M. Rigotti, J. Munuera, S. Fusi, and C. D. Salzman (2020). The geometry of abstraction in the hippocampus and prefrontal cortex. *Cell 183*, 954–967.

Bernhardt, B. C., J. Smallwood, S. Keilholz, and D. S. Margulies (2022). Gradients in brain organization. *NeuroImage 251*, 118987.

Berridge, K. C. (2007). The debate over dopamine's role in reward: The case for incentive salience. *Psychopharmacology (Berl.) 191*, 391–431.

Bertram, R., A. Sherman, and E. F. Stanley (1996). Single-domain/bound calcium hypothesis of transmitter release and facilitation. *J. Neurophysiol. 75*, 1919–1931.

Besson, J. (1999). The neurobiology of pain. *Lancet 353*, 1610–1615.

Beukers, A. O., T. J. Buschman, J. D. Cohen, and K. A. Norman (2021). Is activity silent working memory simply episodic memory? *Trends Cogn. Sci. 25*, 284–293.

Bhalla, U. S. and R. Iyengar (1999). Emergent properties of networks of biological signaling pathways. *Science 283*, 381–387.

Bi, G.-Q. and M.-M. Poo (1998). Synaptic modifications in cultured hippocampal neurons: Dependence on spike timing, synaptic strength, ad postsynaptic cell type. *J. Neurosci. 18*, 10464–10472.

Bi, G.-Q. and M.-M. Poo (2001). Synaptic modification by correlated activity: Hebb's postulate revisited. *Annu. Rev. Neurosci. 24*, 139–166.

Bialek, W. (2012). *Biophysics: Searching for Principles.* Princeton, NJ: Princeton University Press.

Bichot, N. P., A. F. Rossi, and R. Desimone (2005). Parallel and serial neural mechanisms for visual search in macaque area V4. *Science 308*, 529–534.

Bichot, N. P. and J. D. Schall (2002). Priming in macaque frontal cortex during popout visual search: Feature-based facilitation and location-based inhibition of return. *J. Neurosci. 22*, 4675–4685.

Bichot, N. P., R. Xu, A. Ghadooshahy, M. L. Williams, and R. Desimone (2019). The role of prefrontal cortex in the control of feature attention in area V4. *Nat. Commun. 10*, 5727.

Bickel, W. K. and L. A. Marsch (2001). Toward a behavioral economic understanding of drug dependence: Delay discounting processes. *Addiction 96*, 73–86.

Bienenstock, E., L. Cooper, and P. Munro (1982). Theory for the development of neuron selectivity: Orientation specificity and binocular interaction in visual cortex. *J. Neurosci. 2*, 32–48.

Binzegger, T., R. J. Douglas, and K. A. Martin (2004). A quantitative map of the circuit of cat primary visual cortex. *J. Neurosci. 24*, 8441–8453.

Binzegger, T., R. J. Douglas, and K. A. Martin (2009). Topology and dynamics of the canonical circuit of cat V1. *Neural. Netw. 22*, 1071–1078.

Bisley, J. W. and M. E. Goldberg (2003). Neuronal activity in the lateral intraparietal area and spatial attention. *Science 299*, 81–86.

Bittner, K. C., A. D. Milstein, C. Grienberger, S. Romani, and J. C. Magee (2017). Behavioral time scale synaptic plasticity underlies ca1 place fields. *Science 357*, 1033–1036.

Blackman, R. K., D. A. Crowe, A. L. DeNicola, S. Sakellaridi, A. W. MacDonald, and M. V. Chafee (2016). Monkey prefrontal neurons reflect logical operations for cognitive control in a variant of the AX continuous performance task (AX-CPT). *J. Neurosci. 36*, 4067–4079.

Blackman, R. K., A. W. Macdonald, and M. V. Chafee (2013). Effects of ketamine on context-processing performance in monkeys: A new animal model of cognitive deficits in schizophrenia. *Neuropsychopharmacology 38*, 2090–2100.

Blake, R. (1989). A neural theory of binocular rivalry. *Psychol. Rev. 96*, 145–167.

Blakemore, S.-J. and U. Frith (2005). *The Learning Brain: Lessons for Education.* Hoboken, NJ: Blackwell Publishing.

Bland, B. (1986). The physiology and pharmacology of hippocampal formation theta rhythms. *Prog. Neurobiol. 26*, 1–54.

Blazer, D. G., R. C. Kessler, K. A. McGonagle, and M. S. Swartz (1994). The prevalence and distribution of major depression in a national community sample: The national comorbidity survey. *Am. J. Psychiatry 151*, 979–986.

Bliss, D. P., J. J. Sun, and M. D'Esposito (2017). Serial dependence is absent at the time of perception but increases in visual working memory. *Sci. Rep. 7*, 14739.

Bliss, T. V. P. and T. Lomo (1973). Long-lasting potentiation of synaptic transmission in the dentate area of the anaesthetized rabbit following stimulation of the perforant path. *J. Physiol. London 232*, 331–356.

Block, N. (1978). Troubles with functionalism. In C. W. Savage (Ed.), *Perception and Cognition: Issues in the Foundation of Psychology*, pp. 261–325. Minneapolis: University of Minnesota Press.

Block, N. (2005). Two neural correlates of consciousness. *Trends Cogn. Sci. 9*, 46–52.

Block, N. (2019). What is wrong with the no-report paradigm and how to fix it. *Trends Cogn. Sci. 23*, 1003–1013.

Bloom, P. (2002). *How Children Learn the Meanings of Words.* Cambridge, MA: MIT Press.

Boettcher, S. E. P., D. Gresch, A. C. Nobre, and F. van Ede (2021). Output planning at the input stage in visual working memory. *Sci. Adv. 7*, eabe8212.

Boettiger, C. A. and M. D'Esposito (2005). Frontal networks for learning and executing arbitrary stimulus-response associations. *J. Neurosci. 25*, 2723–2732.

Bogacz, R., E. Brown, J. Moehlis, P. Holmes, and J. D. Cohen (2006). The physics of optimal decision making: A formal analysis of models of performance in two-alternative forced-choice tasks. *Psychol. Rev. 113*, 700–765.

Bogacz, R., M. Usher, J. Zhang, and J. McClelland (2007). Extending a biologically inspired model of choice: Multi-alternatives, nonlinearity and value-based multidimensional choice. *Philos. Trans. R. Soc. Lond., B, Biol. Sci. 362*, 1655–1670.

Bogacz, R., E. J. Wagenmakers, B. U. Forstmann, and S. Nieuwenhuis (2010). The neural basis of the speed-accuracy tradeoff. *Trends Neurosci. 33*, 10–16.

Bohland, J. W., C. Wu, H. Barbas, H. Bokil, M. Bota, H. C. Breiter, H. T. Cline, J. C. Doyle, P. J. Freed, R. J. Greenspan, et al. (2009). A proposal for a coordinated effort for the determination of brainwide neuroanatomical connectivity in model organisms at a mesoscopic scale. *PLoS Comput. Biol. 5*, e1000334.

Boldog, E., T. E. Bakken, R. D. Hodge, M. Novotny, B. D. Aevermann, J. Baka, S. Borde, J. L. Close, F. Diez-Fuertes, S. L. Ding, N. Farago, A. K. Kocsis, B. Kovacs, Z. Maltzer, J. M. McCorrison, J. A. Miller, G. Molnar, G. Olah, A. Ozsvar, M. Rozsa, S. I. Shehata, K. A. Smith, S. M. Sunkin, D. N. Tran, P. Venepally, A. Wall, L. G. Puskas, P. Barzo, F. J. Steemers, N. J. Schork, R. H. Scheuermann, R. S. Lasken, E. S. Lein, and G. Tamas (2018). Transcriptomic and morphophysiological evidence for a specialized human cortical GABAergic cell type. *Nat. Neurosci. 21*, 1185–1195.

Bolkan, S. S., J. M. Stujenske, S. Parnaudeau, T. J. Spellman, C. Rauffenbart, A. I. Abbas, A. Z. Harris, J. A. Gordon, and C. Kellendonk (2017). Thalamic projections sustain prefrontal activity during working memory maintenance. *Nat. Neurosci. 20*, 987–996.

Bongioanni, A., D. Folloni, L. Verhagen, J. Sallet, M. Klein-Flugge, and M. Rushworth (2021). Activation and disruption of a neural mechanism for novel choice in monkeys. *Nature.* https://doi.org/10.1038/s41586-020-03115-5.

Bor, D., J. Duncan, R. J. Wiseman, and A. M. Owen (2003). Encoding strategies dissociate prefrontal activity from working memory demand. *Neuron 37*, 361–367.

Börgers, C. and N. Kopell (2003). Synchronization in networks of excitatory and inhibitory neurons with sparse, random connectivity. *Neural Comput. 15*, 509–538.

Born, R. T. and D. C. Bradley (2005). Structure and function of visual area MT. *Annu. Rev. Neurosci. 28*, 157–189.

Borsboom, D. and A. O. Cramer (2013). Network analysis: An integrative approach to the structure of psychopathology. *Annu. Rev. Clin. Psychol. 9*, 91–121.

Borsboom, D., A. O. Cramer, and A. Kalis (2019). Brain disorders? not really: Why network structures block reductionism in psychopathology research. *Behav. Brain Sci. 42*. https://doi.org/10.1017/S0140525X17002266.

Botvinick, M. M., J. D. Cohen, and C. S. Carter (2004). Conflict monitoring and anterior cingulate cortex: An update. *Trends Cogn. Sci. (Regul. Ed.) 8*, 539–546.

Botvinick, M. M., Y. Niv, and A. C. Barto (2009). Hierarchically organized behavior and its neural foundations: A reinforcement learning perspective. *Cognition 113*, 262–280.

Botvinick, M. M., J. X. Wang, W. Dabney, K. J. Miller, and Z. Kurth-Nelson (2020). Deep reinforcement learning and its neuroscientific implications. *Neuron 107*, 603–616.

Bouchacourt, F. and T. J. Buschman (2019). A flexible model of working memory. *Neuron 103*, 147–160.

Bouchacourt, F., S. Palminteri, E. Koechlin, and S. Ostojic (2020). Temporal chunking as a mechanism for unsupervised learning of task-sets. *Elife 9*, e50469.

Boucher, L., T. Palmeri, G. Logan, and J. Schall (2007). Inhibitory control in mind and brain: An interactive race model of countermanding saccades. *Psychol. Rev. 114*, 376–397.

Bowman, C. R., T. Iwashita, and D. Zeithamova (2020). Tracking prototype and exemplar representations in the brain across learning. *Elife 9*, e59360.

Boyden, E. S., F. Zhang, E. Bamberg, G. Nagel, and K. Deisseroth (2005). Millisecond-timescale, genetically targeted optical control of neural activity. *Nat. Neurosci. 8*, 1263–1268.

Bragin, A., G. Jando, Z. Nadasdy, J. Hetke, K. Wise, and G. Buzsáki (1995). Gamma (40–100 Hz) oscillation in the hippocampus of the behaving rat. *J. Neurosci. 15*, 47–60.

Braitenberg, V. and A. Schütz (1991). *Anatomy of the Cortex*. Berlin: Springer-Verlag.

Brandon, N. J., J. K. Millar, C. Korth, H. Sive, K. K. Singh, and A. Sawa (2009). Understanding the role of DISC1 in psychiatric disease and during normal development. *J. Neurosci. 29*, 12768–12775.

Brannon, E. M. and H. S. Terrace (1998). Ordering of the numerosities 1 to 9 by monkeys. *Science 282*, 746–749.

Braun, D. A., C. Mehring, and D. M. Wolpert (2010). Structure learning in action. *Behav. Brain Res. 206*, 157–165.

Braun, U., A. Schaefer, R. F. Betzel, H. Tost, A. Meyer-Lindenberg, and D. S. Bassett (2018). From maps to multi-dimensional network mechanisms of mental disorders. *Neuron 97*, 14–31.

Braver, T. S. (2012). The variable nature of cognitive control: A dual mechanisms framework. *Trends Cogn. Sci. (Regul. Ed.) 16*, 106–113.

Braver, T. S., D. M. Barch, B. A. Keys, C. S. Carter, J. D. Cohen, J. A. Kaye, J. S. Janowsky, S. F. Taylor, J. A. Yesavage, M. S. Mumenthaler, et al. (2001). Context processing in older adults: Evidence for a theory relating cognitive control to neurobiology in healthy aging. *J. Exp. Psychol. Gen. 130*, 746–763.

Brea, J. and W. Gerstner (2016). Does computational neuroscience need new synaptic learning paradigms? *Curr. Opin. Behav. Sci. 11*, 61–66.

Breakspear, M. (2017). Dynamic models of large-scale brain activity. *Nat. Neurosci. 20*, 340–352.

Breese, B. B. (1899). On inhibition. *Psychol. Monogr. Gen. Appl. 3*, 1–65.

Bremer, F. (1958). Cerebral and cerebellar potentials. *Physiol. Rev. 38*, 357–388.

Bressler, S. L. and V. Menon (2010). Large-scale brain networks in cognition: Emerging methods and principles. *Trends Cogn. Sci. 14*, 277–290.

Bressloff, P. C. (2012). Spatiotemporal dynamics of continuum neural fields. *J. Phys. A 45*. https://doi.org/10.1088/1751-8113/45/3/033001.

Brette, R. (2019). Is coding a relevant metaphor for the brain? *Behav. Brain Sci. 42*, e215.

Britten, K. H., W. T. Newsome, M. N. Shadlen, S. Celebrini, and J. A. Movshon (1996). A relationship between behavioral choice and the visual responses of neurons in macaque MT. *Visual Neurosci. 13*, 87–100.

Britten, K. H., M. N. Shadlen, W. T. Newsome, and J. A. Movshon (1992). The analysis of visual motion: A comparison of neuronal and psychophysical performance. *J. Neurosci. 12*, 4745–4765.

Britten, K. H., M. N. Shadlen, W. T. Newsome, and J. A. Movshon (1993). Responses of neurons in macaque MT to stochastic motion signals. *Visual Neurosci. 10*, 1157–1169.

Brodersen, K. H., L. Deserno, F. Schlagenhauf, Z. Lin, W. D. Penny, J. M. Buhmann, and K. E. Stephan (2014). Dissecting psychiatric spectrum disorders by generative embedding. *Neuroimage Clin. 4*, 98–111.

Brodmann, K. (1909). *Vergleichende Lokalisationslehre der Grosshirnrinde in ihren Prinzipien dargestellt auf Grund des Zellenbaues*. Barth.

Brody, C. D. and T. D. Hanks (2016). Neural underpinnings of the evidence accumulator. *Curr. Opin. Neurobiol. 37*, 149–157.

Brody, C. D., A. Hernández, A. Zainos, and R. Romo (2003). Timing and neural encoding of somatosensory parametric working memory in macaque prefrontal cortex. *Cereb. Cortex 13*, 1196–1207.

Brotchie, P. R., R. A. Andersen, L. H. Snyder, and S. J. Goodman (1995). Head position signals used by parietal neurons to encode locations of visual stimuli. *Nature 375*, 232–235.

Brouwer, A. and R. L. Carhart-Harris (2020). Pivotal mental states. *J. Psychopharmacol.*, 0269881120959637.

Brown, J. W., J. R. Reynolds, and T. S. Braver (2007). A computational model of fractionated conflict-control mechanisms in task-switching. *Cogn. Psychol. 55*, 37–85.

Brown, S. D. and A. Heathcote (2008). The simplest complete model of choice response time: Linear ballistic accumulation. *Cogn. Psychol. 57*, 153–178.

Browning, M., C. S. Carter, C. Chatham, H. Den Ouden, C. M. Gillan, J. T. Baker, A. M. Chekroud, R. Cools, P. Dayan, J. Gold, et al. (2020). Realizing the clinical potential of computational psychiatry: Report from the banbury center meeting, february 2019. *Biol. Psychiatry 88*, e5–e10.

Brozoski, T. J., R. M. Brown, H. E. Rosvold, and P. S. Goldman (1979). Cognitive deficit caused by regional depletion of dopamine in prefrontal cortex of rhesus monkey. *Science 205*, 929–932.

Bruce, C. J. and M. E. Goldberg (1985). Primate frontal eye fields. I. Single neurons discharging before saccades. *J. Neurophysiol. 53*, 603–635.

Brunel, N. (2000). Dynamics of sparsely connected networks of excitatory and inhibitory spiking neurons. *J. Comput. Neurosci. 8*, 183–208.

Brunel, N. (2005). Network models of memory. In C. C. Chow, B. Gutkin, D. Hansel, C. Meunier, and J. Dalibard (Eds.), *Methods and Models in Neurophysics*, pp. 407–476. Amsterdam: Elsevier.

Brunel, N. (2016). Is cortical connectivity optimized for storing information? *Nat. Neurosci. 19*, 749–755.

Brunel, N. and V. Hakim (1999). Fast global oscillations in networks of integrate-and-fire neurons with low firing rates. *Neural Comput. 11*, 1621–1671.

Brunel, N. and M. C. van Rossum (2007). Lapicque's 1907 paper: From frogs to integrate-and-fire. *Biol. Cybern. 97*, 337–339.

Brunel, N. and X.-J. Wang (2001). Effects of neuromodulation in a cortical network model of object working memory dominated by recurrent inhibition. *J. Comput. Neurosci. 11*, 63–85.

Brunel, N. and X.-J. Wang (2003). What determines the frequency of fast network oscillations with irregular neural discharges? I. Synaptic dynamics and excitation-inhibition balance. *J. Neurophysiol. 90*, 415–430.

Brunton, B. W., M. M. Botvinick, and C. D. Brody (2013). Rats and humans can optimally accumulate evidence for decision-making. *Science 340*, 95–98.

Buckley, M. J., F. A. Mansouri, H. Hoda, M. Mahboubi, P. G. Browning, S. C. Kwok, A. Phillips, and K. Tanaka (2009). Dissociable components of rule-guided behavior depend on distinct medial and prefrontal regions. *Science 325*, 52–58.

Buckner, R. L. (2012). The serendipitous discovery of the brain's default network. *Neuroimage 2*, 1137–1145.

Buckner, R. L. and D. C. Carroll (2007). Self-projection and the brain. *Trends Cogn. Sci. 11*, 49–57.

Buckner, R. L. and L. M. DiNicola (2019). The brain's default network: Updated anatomy, physiology and evolving insights. *Nat. Rev. Neurosci. 20*, 593–608.

Buffalo, E. A., P. Fries, R. Landman, T. J. Buschman, and R. Desimone (2011). Laminar differences in gamma and alpha coherence in the ventral stream. *Proc. Natl. Acad. Sci. U.S.A. 108*, 11262–11267.

Buhusi, C. V. and W. H. Meck (2005). What makes us tick? functional and neural mechanisms of interval timing. *Nat. Rev. Neurosci. 6*, 755–765.

Bullmore, E. and O. Sporns (2009). Complex brain networks: Graph theoretical analysis of structural and functional systems. *Nat. Rev. Neurosci. 10*, 186–198.

Bundesen, C. (1990). A theory of visual attention. *Psychol. Rev. 97*, 523–547.

Bundesen, C., T. Habekost, and S. Kyllingsbæk (2005). A neural theory of visual attention: Bridging cognition and neurophysiology. *Psychol. Rev. 112*, 291–328.

Bunge, S. A. and J. D. Wallis (Eds.) (2007). *Neuroscience of Rule-Guided Behavior.* New York: Oxford University Press.

Buonomano, D. and R. Laje (2010). Population clocks: Motor timing with neural dynamics. *Trends Cogn. Sci. 14*, 520–527.

Burak, Y. and I. R. Fiete (2009). Accurate path integration in continuous attractor network models of grid cells. *PLoS Comput. Biol. 5*, e1000291.

Burt, J. B., M. Demirta, W. J. Eckner, N. M. Navejar, J. L. Ji, W. J. Martin, A. Bernacchia, A. Anticevic, and J. D. Murray (2018). Hierarchy of transcriptomic specialization across human cortex captured by structural neuroimaging topography. *Nat. Neurosci. 21*, 1251–1259.

Buschman, T. J. and S. Kastner (2015). From behavior to neural dynamics: An integrated theory of attention. *Neuron 88*, 127–144.

Busemeyer, J. and J. Townsend (1993). Decision field theory: A dynamic-cognitive approach to decision making in an uncertain environment. *Psychol. Rev. 100*, 432–459.

Bush, P. C. and T. J. Sejnowski (1993). Reduced compartmental models of neocortical pyramidal cells. *J. Neurosci. Methods 46*, 159–166.

Bush, R. R. and F. Mosteller (1951). A mathematical model for simple learning. *Psychol. Rev. 58*, 313–323.

Buxhoeveden, D. P. and M. F. Casanova (2002). The minicolumn hypothesis in neuroscience. *Brain 125*, 935–951.

Buzsáki, G. (1986). Hippocampal sharp waves: Their origin and significance. *Brain Res. 398*, 242–252.

Buzsaki, G. (1989). Two-stage model of memory trace formation: A role for "noisy" brain states. *Neuroscience 31*, 551–570.

Buzsáki, G. (2002). Theta oscillations in the hippocampus. *Neuron 33*, 325–340.

Buzsáki, G. (2006). *Rhythms of the Brain.* New York: Oxford University Press.

Buzsáki, G. (2019). *The Brain from Inside Out.* New York: Oxford University Press.

Buzsáki, G. (2020). The brain–cognitive behavior problem: A retrospective. *eNeuro 7.* https://doi.org/10.1523/ENEURO.0069-20.2020.

Buzsaki, G., C. A. Anastassiou, and C. Koch (2012). The origin of extracellular fields and currents–EEG, ECoG, LFP and spikes. *Nat. Rev. Neurosci. 13*, 407–420.

Buzsáki, G., D. Buhl, K. Harris, J. Csicsvari, B. Czéh, and A. Morozov (2003). Hippocampal network patterns of activity in the mouse. *Neuroscience 116*, 201–211.

Buzsáki, G., C. Geisler, D. A. Henze, and X.-J. Wang (2004). Interneuron Diversity series: Circuit complexity and axon wiring economy of cortical interneurons. *Trends Neurosci. 27*, 186–193.

Buzsaki, G. and K. Mizuseki (2014). The log-dynamic brain: How skewed distributions affect network operations. *Nat. Rev. Neurosci. 15*, 264–278.

Buzsáki, G., R. Urioste, J. Hetke, and K. Wise (1992). High frequency network oscillation in the hippocampus. *Science 256*, 1025–1027.

Buzsáki, G. and X.-J. Wang (2012). Mechanisms of gamma oscillations. *Annu. Rev. Neurosci. 35*, 203–225.

Bzdok, D. and A. Meyer-Lindenberg (2018). Machine learning for precision psychiatry: Opportunities and challenges. *Biol. Psychiatry Cogn. Neurosci. Neuroimaging 3*, 223–230.

Cadieu, C. F., H. Hong, D. L. Yamins, N. Pinto, D. Ardila, E. A. Solomon, N. J. Majaj, and J. J. DiCarlo (2014). Deep neural networks rival the representation of primate it cortex for core visual object recognition. *PLoS Comput. Biol. 10*, e1003963.

Cai, X. and C. Padoa-Schioppa (2014). Contributions of orbitofrontal and lateral prefrontal cortices to economic choice and the good-to-action transformation. *Neuron 81*, 1140–1151.

Cain, C. K. (2019). Avoidance problems reconsidered. *Curr. Opin. Behav. Sci. 26*, 9–17.

Cajal, S. R. (1911). *Histologie du Système Nerveux de L'Homme et des Vertébrés* (Vol. II). Paris: Maloine, Madrid: Instituto Cajal.

Calhoon, G. G. and K. M. Tye (2015). Resolving the neural circuits of anxiety. *Nat. Neurosci. 18*, 1394–1404.

Callaway, E. M. (2004). Feedforward, feedback and inhibitory connections in primate visual cortex. *Neural Netw. 17*, 625–632.

Calvin, O. L. and A. D. Redish (2021). Global disruption in excitation-inhibition balance can cause localized network dysfunction and schizophrenia-like context-integration deficits. *PLoS Comput. Biol. 17*, e1008985.

Camerer, C., G. Loewenstein, and D. Prelec (2005). Neuroeconomics: How neuroscience can inform economics. *J. Econ. Lit. 43*, 9–64.

Camerer, C. F. (2003). *Behavioral Game Theory: Experiments in Strategic Interaction.* Princeton, NJ: Princeton Unversity Press.

Camille, N., G. Coricelli, J. Sallet, P. Pradat-Diehl, J.-R. Duhamel, and A. Sirigu (2004). The involvement of the orbitofrontal cortex in the experience of regret. *Science 304*, 1167–1170.

Campagnola, L., S. C. Seeman, T. Chartrand, L. Kim, A. Hoggarth, C. Gamlin, S. Ito, J. Trinh, P. Davoudian, C. Radaelli, M. H. Kim, T. Hage, T. Braun, L. Alfiler, J. Andrade, P. Bohn, R. Dalley, A. Henry, S. Kebede, M. Alice, D. Sandman, G. Williams, R. Larsen, C. Teeter, T. L. Daigle, K. Berry, N. Dotson, R. Enstrom, M. Gorham, M. Hupp, S. Dingman Lee, K. Ngo, P. R. Nicovich, L. Potekhina, S. Ransford, A. Gary, J. Goldy, D. McMillen, T. Pham, M. Tieu, L. Siverts, M. Walker, C. Farrell, M. Schroedter, C. Slaughterbeck, C. Cobb, R. Ellenbogen, R. P. Gwinn, C. D. Keene, A. L. Ko, J. G. Ojemann, D. L. Silbergeld, D. Carey, T. Casper, K. Crichton, M. Clark, N. Dee, L. Ellingwood, J. Gloe, M. Kroll, J. Sulc, H. Tung, K. Wadhwani, K. Brouner, T. Egdorf, M. Maxwell, M. McGraw, C. A. Pom, A. Ruiz, J. Bomben, D. Feng, N. Hejazinia, S. Shi,

A. Szafer, W. Wakeman, J. Phillips, A. Bernard, L. Esposito, F. D. D'Orazi, S. Sunkin, K. Smith, B. Tasic, A. Arkhipov, S. Sorensen, E. Lein, C. Koch, G. Murphy, H. Zeng, and T. Jarsky (2022). Local connectivity and synaptic dynamics in mouse and human neocortex. *Science 375*, eabj5861.

Campbell, M. G., S. A. Ocko, C. S. Mallory, I. I. Low, S. Ganguli, and L. M. Giocomo (2018). Principles governing the integration of landmark and self-motion cues in entorhinal cortical codes for navigation. *Nature Neurosci. 21*, 1096–1106.

Camperi, M. and X.-J. Wang (1997). Modeling delay-period activity in the prefrontal cortex during working memory tasks. In J. Bower (Ed.), *Computational Neuroscience: Trends in Research*, pp. 273–279. New York: Plenum Press.

Camperi, M. and X.-J. Wang (1998). A model of visuospatial short-term memory in prefrontal cortex: Recurrent network and cellular bistability. *J. Comput. Neurosci. 5*, 383–405.

Canguilhem, G. (1989). *The Normal and the Pathological.* New York: Zone Books (The original French book was published in 1966).

Cannon, W. B. (1929). Organization for physiological homeostasis. *Physiol. Rev. 9*, 339–431.

Cano-Colino, M., R. Almeida, D. Gomez-Cabrero, F. Artigas, and A. Compte (2014). Serotonin regulates performance nonmonotonically in a spatial working memory network. *Cereb. Cortex 24*, 2449–2463.

Cantlon, J. F., M. L. Platt, and E. M. Brannon (2009). Beyond the number domain. *Trends Cogn. Sci. 13*, 83–91.

Cao, R., A. Pastukhov, S. Aleshin, M. Mattia, and J. Braun (2021). Binocular rivalry reveals an out-of-equilibrium neural dynamics suited for decision-making. *Elife 10*, e61581.

Cao, R., and D. Yamins (2024). Explanatory models in neuroscience, Part 1: Taking mechanistic abstraction seriously. Cognitive Systems Research: 101244.

Caputi, A., A. Rozov, M. Blatow, and H. Monyer (2009). Two calretinin-positive gabaergic cell types in layer 2/3 of the mouse neocortex provide different forms of inhibition. *Cereb. Cortex 19*, 1345–1359.

Carandini, M. (2012). From circuits to behavior: A bridge too far? *Nat. Neurosci. 15*, 507–509.

Carandini, M. and D. J. Heeger (2011). Normalization as a canonical neural computation. *Nat. Rev. Neurosci. 13*, 51–62.

Carandini, M. and D. L. Ringach (1997). Predictions of a recurrent model of orientation selectivity. *Vis. Res. 37*, 3061–3071.

Cardin, J. A. (2018). Inhibitory interneurons regulate temporal precision and correlations in cortical circuits. *Trends Neurosci. 41*, 689–700.

Carew, T. J. and S. H. Magsamen (2010). Neuroscience and education: An ideal partnership for producing evidence-based solutions to guide 21(st) century learning. *Neuron 67*, 685–688.

Carey, S. (2011). *The Origin of Concepts.* New York: Oxford University Press.

Carhart-Harris, R. L., B. Giribaldi, R. Watts, M. Baker-Jones, A. Murphy-Beiner, R. Murphy, J. Martell, A. Blemings, D. Erritzoe, and D. J. Nutt (2021). Trial of psilocybin versus escitalopram for depression. *N. Engl. J. Med. 384*, 1402–1411.

Carhart-Harris, R. L., S. Muthukumaraswamy, L. Roseman, M. Kaelen, W. Droog, K. Murphy, E. Tagliazucchi, E. E. Schenberg, T. Nest, C. Orban, et al. (2016). Neural correlates of the lsd experience revealed by multimodal neuroimaging. *Proc. Natl. Acad. Sci. (USA) 113*, 4853–4858.

Carhart-Harris, R. L. and D. Nutt (2017). Serotonin and brain function: A tale of two receptors. *J. Psychopharmacol. 31*, 1091–1120.

Carland, M. A., D. Thura, and P. Cisek (2019). The urge to decide and act: Implications for brain function and dysfunction. *Neuroscientist 25*, 491–511.

Carlisi, C. O. and O. J. Robinson (2018). The role of prefrontal-subcortical circuitry in negative bias in anxiety: Translational, developmental and treatment perspectives. *Brain Neurosci. Adv. 2*, 1–12.

Caron, S. J., V. Ruta, L. F. Abbott, and R. Axel (2013). Random convergence of olfactory inputs in the Drosophila mushroom body. *Nature 497*, 113–117.

Carpenter, R. H. S. (1981). Oculomotor procrastination. In D. F. Fisher, R. A. Monty, and J. W. Senders (Eds.), *Eye Movements: Cognition and Visual Perception*, pp. 237–246. Hillsdale: Lawrence Erlbaum.

Carr, M. F., S. P. Jadhav, and L. M. Frank (2011). Hippocampal replay in the awake state: A potential substrate for memory consolidation and retrieval. *Nat. Neurosci. 14*, 147–153.

Carroll, J. B. (1993). *Human Cognitive Abilities: A Survey of Factor-Analytic Studies.* New York: Cambridge University Press.

Carter, C. S., D. M. Barch, R. W. Buchanan, E. Bullmore, J. H. Krystal, J. Cohen, M. Geyer, M. Green, K. H. Nuechterlein, T. Robbins, et al. (2008). Identifying cognitive mechanisms targeted for treatment development in schizophrenia: An overview of the first meeting of the cognitive neuroscience treatment research to improve cognition in schizophrenia initiative. *Biol. Psychiatry 64*, 4–10.

Carter, E. and X.-J. Wang (2007). Cannabinoid-mediated disinhibition and working memory: Dynamical interplay of multiple feedback mechanisms in a continuous attractor model of prefrontal cortex. *Cereb. Cortex 17*(Suppl 1), 16–26.

Castellanos, F. X. and R. Tannock (2002). Neuroscience of attention-deficit/hyperactivity disorder: The search for endophenotypes. *Nat. Rev. Neurosci. 3*, 617–628.

Catani, M., R. J. Howard, S. Pajevic, and D. K. Jones (2002). Virtual in vivo interactive dissection of white matter fasciculi in the human brain. *Neuroimage 17*, 77–94.

Cattell, S. J. (1987). *Intelligence: Its Structure, Growth and Action.* Amsterdam: Elsevier.

Cauli, B., E. Audinat, B. Lambolez, M. C. Angulo, N. Ropert, K. Tsuzuki, S. Hestrin, and J. Rossier (1997). Molecular and physiological diversity of cortical nonpyramidal cells. *J. Neurosci. 17*, 3894–3906.

Cavanagh, S. E., N. H. Lam, J. D. Murray, L. T. Hunt, and S. W. Kennerley (2020). A circuit mechanism for decision-making biases and NMDA receptor hypofunction. *eLife 9*, e53664.

Cavanagh, S. E., J. P. Towers, J. D. Wallis, L. T. Hunt, and S. W. Kennerley (2018). Reconciling persistent and dynamic hypotheses of working memory coding in prefrontal cortex. *Nat. Commun. 9*, 3498.

Celada, P., M. Puig, and F. Artigas (2013). Serotonin modulation of cortical neurons and networks. *Front. Integr. Neurosci. 7*, 25.

Cepeda, N. J., H. Pashler, E. Vul, J. T. Wixted, and D. Rohrer (2006). Distributed practice in verbal recall tasks: A review and quantitative synthesis. *Psychol. Bull. 132*, 354–380.

Chafee, M. V. and P. S. Goldman-Rakic (1998). Neuronal activity in macaque prefrontal area 8a and posterior parietal area 7ip related to memory guided saccades. *J. Neurophysiol. 79*, 2919–2940.

Chafee, M. V. and P. S. Goldman-Rakic (2000). Inactivation of parietal and prefrontal cortex reveals interdependence of neural activity during memory-guided saccades. *J. Neurophysiol. 83*, 1550–1566.

Chafee, M. V. and S. R. Heilbronner (2022). Prefrontal cortex. *Curr. Biol. 32*, R346–R351.

Chaisangmongkon, W., S. K. Swaminathan, D. J. Freedman, and X. J. Wang (2017). Computing by robust transience: How the fronto-parietal network performs sequential, category-based decisions. *Neuron 93*, 1504–1517.

Chance, F. S., L. F. Abbott, and A. D. Reyes (2002). Gain modulation from background synaptic input. *Neuron 35*, 773–782.

Chandrasekhar, S. (1943). Stochastic problems in physics and astronomy. *Rev. Mod. Phys. 15*, 1–89.

Chang, L. and D. Y. Tsao (2017). The code for facial identity in the primate brain. *Cell 169*, 1013–1028.

Changeux, J.-P. (1983). *L'homme Neuronal*. Paris, France: Librairie Arthème Fayard.

Changeux, J.-P. and A. Danchin (1976). Selective stabilisation of developing synapses as a mechanism for the specification of neuronal networks. *Nature 264*, 705–712.

Chapman, A. F. and V. S. Störmer (2024). Representational structures as a unifying framework for attention. *Trends Cogn. Sci.* https://doi.org/10.1016/j.tics.2024.01.002.

Chariker, L., R. Shapley, and L. S. Young (2016). Orientation selectivity from very sparse LGN inputs in a comprehensive model of macaque V1 cortex. *J. Neurosci. 36*, 12368–12384.

Charnov, E. L., et al. (1976). Optimal foraging, the marginal value theorem. *Theo. Pop. Biol. 9*, 129–136.

Chau, B. K., N. Kolling, L. T. Hunt, M. E. Walton, and M. F. Rushworth (2014). A neural mechanism underlying failure of optimal choice with multiple alternatives. *Nat. Neurosci. 17*, 463–470.

Chaudhuri, R., A. Bernacchia, and X.-J. Wang (2014). A diversity of localized timescales in network activity. *Elife 3*, e01239.

Chaudhuri, R. and I. Fiete (2016). Computational principles of memory. *Nat. Neurosci. 19*, 394–403.

Chaudhuri, R., B. Gercek, B. Pandey, A. Peyrache, and I. Fiete (2019). The intrinsic attractor manifold and population dynamics of a canonical cognitive circuit across waking and sleep. *Nat. Neurosci. 22*, 1512–1520.

Chaudhuri, R., B. J. He, and X.-J. Wang (2018). Random recurrent networks near criticality capture the broadband power distribution of human ECoG dynamics. *Cereb. Cortex 28*, 3610–3622.

Chaudhuri, R., K. Knoblauch, M. A. Gariel, H. Kennedy, and X.-J. Wang (2015). A large-scale circuit mechanism for hierarchical dynamical processing in the primate cortex. *Neuron 88*, 419–431.

Chekroud, A. M., R. J. Zotti, Z. Shehzad, R. Gueorguieva, M. K. Johnson, M. H. Trivedi, T. D. Cannon, J. H. Krystal, and P. R. Corlett (2016). Cross-trial prediction of treatment outcome in depression: A machine learning approach. *Lancet Psychiatry 3*, 243–250.

Chelazzi, L., E. K. Miller, J. Duncan, and R. Desimone (2001). Responses of neurons in macaque area V4 during memory-guided visual search. *Cereb. Cortex 11*, 761–772.

Chen, A., Y. Sun, Y. Lei, C. Li, S. Liao, J. Meng, . . ., C. Li (2023a). Single-cell spatial transcriptome reveals cell-type organization in the macaque cortex. *Cell 186*, 3726–3743.

Chen, C., T. Takahashi, S. Nakagawa, T. Inoue, and I. Kusumi (2015). Reinforcement learning in depression: A review of computational research. *Neurosci. Biobehav. Rev. 55*, 247–267.

Chen, D. and E. E. Fetz (2005). Characteristic membrane potential trajectories in primate sensorimotor cortex neurons recorded in vivo. *J. Neurophysiol. 94*, 2713–2725.

Chen, G. and P. Gong (2019). Computing by modulating spontaneous cortical activity patterns as a mechanism of active visual processing. *Nat. Commun. 10*, 4915.

Chen, G., P. Greengard, and Z. Yan (2004). Potentiation of NMDA receptor currents by dopamine D1 receptors in prefrontal cortex. *Proc. Natl. Acad. Sci. U.S.A. 101*, 2596–2600.

Chen, S., Y. Liu, Z. Wang, J. Colonell, L. D. Liu, H. Hou, N.-W. Tien, T. Wang, T. Harris, S. Druckmann, et al. (2023b). Brain-wide neural activity underlying memory-guided movement. *BioRxiv*. https://doi.org/10.1101/2023.03.01.530520.

Chen, Z. S., I. R. Galatzer-Levy, B. Bigio, C. Nasca, Y. Zhang, et al. (2022). Modern views of machine learning for precision psychiatry. *Patterns 3*, 100602.

Cheung, C., L. S. Hamiton, K. Johnson, and E. F. Chang (2016). The auditory representation of speech sounds in human motor cortex. *Elife 5*. https://doi.org/10.7554/eLife.12577.

Cheyette, S. J. and S. T. Piantadosi (2020). A unified account of numerosity perception. *Nat. Hum. Behav. 4*, 1265–1272.

Chien, H.-Y. S. and C. J. Honey (2020). Constructing and forgetting temporal context in the human cerebral cortex. *Neuron 106*, 675–686.

Chklovskii, D. B., T. Schikorski, and C. F. Stevens (2002). Wiring optimization in cortical circuits. *Neuron 34*, 341–347.

Cho, K., B. Van Merriënboer, C. Gulcehre, D. Bahdanau, F. Bougares, H. Schwenk, and Y. Bengio (2014). Learning phrase representations using rnn encoder-decoder for statistical machine translation. *arXiv preprint arXiv:1406.1078*.

Chollet, F. (2019) On the measure of intelligence. *arXiv preprint arXiv:1911.01547*

Choudary, P. V., M. Molnar, S. J. Evans, H. Tomita, J. Z. Li, M. P. Vawter, R. M. Myers, W. E. Bunney, H. Akil, S. J. Watson, and E. G. Jones (2005). Altered cortical glutamatergic and GABAergic signal transmission with glial involvement in depression. *Proc. Natl. Acad. Sci. U.S.A. 102*, 15653–15658.

Christophel, T. B., P. C. Klink, B. Spitzer, P. R. Roelfsema, and J. D. Haynes (2017). The distributed nature of working memory. *Trends Cogn. Sci. 21*, 111–124.

Chun, M. M., J. D. Golomb, and N. B. Turk-Browne (2011). A taxonomy of external and internal attention. *Annu. Rev. Psychol. 62*, 73–101.

Chung, J., C. Gulcehre, K. Cho, and Y. Bengio (2014). Empirical evaluation of gated recurrent neural networks on sequence modeling. *arXiv preprint arXiv:1412.3555*.

Churchill, W. S. (1897). *The Scaffolding of Rhetoric*. Unpublished essay.

Churchland, A. K. and L. Abbott (2016). Conceptual and technical advances define a key moment for theoretical neuroscience. *Nat. Neurosci. 19*, 348–349.

Churchland, A. K., R. Kiani, and M. Shadlen (2008). Decision-making with multiple alternatives. *Nat. Neurosci. 11*, 693–702.

Churchland, M. M., J. P. Cunningham, M. T. Kaufman, J. D. Foster, P. Nuyujukian, S. I. Ryu, and K. V. Shenoy (2012). Neural population dynamics during reaching. *Nature 487*, 51.

Churchland, M. M., J. P. Cunningham, M. T. Kaufman, S. I. Ryu, and K. V. Shenoy (2010). Cortical preparatory activity: Representation of movement or first cog in a dynamical machine? *Neuron 68*, 387–400.

Churchland, M. M., B. M. Yu, M. Sahani, and K. V. Shenoy (2007). Techniques for extracting single-trial activity patterns from large-scale neural recordings. *Curr. Opin. Neurobiol. 17*, 609–618.

Churchland, P. (1989). *Neurophilosophy: Toward a Unified Science of the Mind-Brain*. Cambridge, MA: MIT Press.

Churchland, P. (1996). Towards a neurobiology of the mind. In R. Llinás and P. Churchland (Eds.), *The Mind-Brain Continuum*, pp. 281–303. Cambridge, MA: MIT Press.

Churchland, P. and T. Sejnowski (1992). *The Computational Brain*. Cambridge, MA: MIT Press.

Cisek, P. (2019). Resynthesizing behavior through phylogenetic refinement. *Atten. Percept. Psychophys. 81*, 2265–2287.

Cisek, P. and J. F. Kalaska (2005). Neural correlates of reaching decisions in dorsal premotor cortex: Specification of multiple direction choices and final selection of action. *Neuron 45*, 801–814.

Cisek, P. and J. F. Kalaska (2010). Neural mechanisms for interacting with a world full of action choices. *Annu. Rev. Neurosci. 33*, 269–298.

Clark, A. (2013). Whatever next? Predictive brains, situated agents, and the future of cognitive science. *Behav. Brain Sci. 36*(3), 181–204.

Clark, L. A., B. Cuthbert, R. Lewis-Fernandez, W. E. Narrow, and G. M. Reed (2017). Three approaches to understanding and classifying mental disorder: ICD-11, DSM-5, and the National Institute of Mental Health's Research Domain Criteria (RDoC). *Psychol. Sci. Publ. Int. 18*, 72–145.

Clementz, B. A., J. A. Sweeney, J. P. Hamm, E. I. Ivleva, L. E. Ethridge, G. D. Pearlson, M. S. Keshavan, and C. A. Tamminga (2016). Identification of distinct psychosis biotypes using brain-based biomarkers. *Am. J. Psychiatry 173*, 373–384.

Clopath, C. (2012). Synaptic consolidation: An approach to long-term learning. *Cogn. Neurodyn. 6*, 251–257.

Clopath, C., L. Büsing, E. Vasilaki, and W. Gerstner (2010). Connectivity reflects coding: A model of voltage-based stdp with homeostasis. *Nat. Neurosci. 13*, 344.

Cobb, M. (2020). *The Idea of the Brain: The Past and Future of Neuroscience*. New York: Basic Books.

Cochran, W. G. (1937). Problems arising in the analysis of a series of similar experiments. *Suppl. J. R. Stat. Soc. 4*, 102–118.

Cogan, G. B., T. Thesen, C. Carlson, W. Doyle, O. Devinsky, and B. Pesaran (2014). Sensory–motor transformations for speech occur bilaterally. *Nature 507*, 94–98.

Cohen, J. D., D. M. Barch, C. Carter, and D. Servan-Schreiber (1999). Context-processing deficits in schizophrenia: Converging evidence from three theoretically motivated cognitive tasks. *J. Abnorm. Psychol. 108*, 120–133.

Cohen, J. D., K. Dunbar, and J. L. McClelland (1990). On the control of automatic processes: A parallel distributed processing account of the Stroop effect. *Psychol. Rev. 97*, 332–361.

Cohen, J. Y., E. A. Crowder, R. P. Heitz, C. R. Subraveti, K. G. Thompson, G. F. Woodman, and J. D. Schall (2010). Cooperation and competition among frontal eye field neurons during visual target selection. *J. Neurosci. 30*, 3227–3238.

Cohen, J. Y., S. Haesler, L. Vong, B. B. Lowell, and N. Uchida (2012). Neuron-type-specific signals for reward and punishment in the ventral tegmental area. *Nature 482*, 85–88.

Cohen, M. A., K. Ortego, A. Kyroudis, and M. Pitts (2020). Distinguishing the neural correlates of perceptual awareness and postperceptual processing. *J. Neurosci. 40*, 4925–4935.

Cohen, M. R. and A. Kohn (2011). Measuring and interpreting neuronal correlations. *Nat. Neurosci. 14*, 811–819.

Cohen, M. R. and J. H. Maunsell (2009). Attention improves performance primarily by reducing interneuronal correlations. *Nat. Neurosci. 12*, 1594–1600.

Cohen, M. X., N. Axmacher, D. Lenartz, C. E. Elger, V. Sturm, and T. E. Schlaepfer (2009). Nuclei accumbens phase synchrony predicts decision-making reversals following negative feedback. *J. Neurosci. 29*, 7591–7598.

Cohen, S. M., R. W. Tsien, D. C. Goff, and M. M. Halassa (2015). The impact of NMDA receptor hypofunction on GABAergic neurons in the pathophysiology of schizophrenia. *Schizophr. Res. 167*, 98–107.

Coifman, R. R., S. Lafon, A. B. Lee, M. Maggioni, B. Nadler, F. Warner, and S. W. Zucker (2005). Geometric diffusions as a tool for harmonic analysis and structure definition of data: Diffusion maps. *Proc. Natl. Acad. Sci. 102*, 7426–7431.

Cole, M. W., J. A. Etzel, J. M. Zacks, W. Schneider, and T. S. Braver (2011). Rapid transfer of abstract rules to novel contexts in human lateral prefrontal cortex. *Front. Hum. Neurosci. 5*, 142.

Cole, M. W., P. Laurent, and A. Stocco (2013a). Rapid instructed task learning: A new window into the human brain's unique capacity for flexible cognitive control. *Cogn. Affect. Behav. Neurosci. 13*, 1–22.

Cole, M. W., J. R. Reynolds, J. D. Power, G. Repovs, A. Anticevic, and T. S. Braver (2013b). Multi-task connectivity reveals flexible hubs for adaptive task control. *Nat. Neurosci. 16*, 1348–1355.

Coley, A. A., N. Padilla-Coreano, R. Patel, and K. M. Tye (2021). Valence processing in the PFC: Reconciling circuit-level and systems-level views. *Int. Rev. Neurobiol. 158*, 171–212.

Colgin, L. L. (2016). Rhythms of the hippocampal network. *Nat. Rev. Neurosci. 17*, 239–249.

Colgin, L. L., T. Denninger, M. Fyhn, T. Hafting, T. Bonnevie, O. Jensen, M. B. Moser, and E. I. Moser (2009). Frequency of gamma oscillations routes flow of information in the hippocampus. *Nature 462*, 353–357.

Collins, A. G. and J. Cockburn (2020). Beyond dichotomies in reinforcement learning. *Nat. Rev. Neurosci. 21*, 576–586.

Compte, A., N. Brunel, P. S. Goldman-Rakic, and X.-J. Wang (2000). Synaptic mechanisms and network dynamics underlying spatial working memory in a cortical network model. *Cereb. Cortex 10*, 910–923.

Compte, A., C. Constantinidis, J. Tegnér, S. Raghavachari, M. V. Chafee, P. S. Goldman-Rakic, and X.-J. Wang (2003). Temporally irregular mnemonic persistent activity in prefrontal neurons of monkeys during a delayed response task. *J. Neurophysiol. 90*, 3441–3454.

Compte, A. and X.-J. Wang (2006). Tuning curve shift by attention modulation in cortical neurons: A computational study of its mechanisms. *Cereb. Cortex 16*, 761–778.

Condé, F., J. S. Lund, D. M. Jacobowitz, K. G. Baimbridge, and D. A. Lewis (1994). Local circuit neurons immunoreactive for calretinin, calbindin D-28k or parvalbumin in monkey prefrontal cortex: Distribution and morphology. *J. Comp. Neurol. 341*, 95–116.

Connor, C. E., J. L. Gallant, D. C. Preddie, and D. C. Van Essen (1996). Responses in area v4 depend on the spatial relationship between stimulus and attention. *J. Neurophys. 75*, 1306–1308.

Connor, C. E., D. C. Preddie, J. L. Gallant, and D. C. Van Essen (1997). Spatial attention effects in macaque area v4. *J. Neurosci. 17*, 3201–3214.

Constantinescu, A. O., J. X. O'Reilly, and T. E. J. Behrens (2016). Organizing conceptual knowledge in humans with a gridlike code. *Science 352*, 1464–1468.

Constantinidis, C., S. Funahashi, D. Lee, J. D. Murray, X.-L. Qi, M. Wang, and A. F. Arnsten (2018). Persistent spiking activity underlies working memory. *J. Neurosci. 38*, 7020–7028.

Constantinidis, C. and P. S. Goldman-Rakic (2002). Correlated discharges among putative pyramidal neurons and interneurons in the primate prefrontal cortex. *J. Neurophysiol. 88*, 3487–3497.

Constantinidis, C. and T. Klingberg (2016). The neuroscience of working memory capacity and training. *Nat. Rev. Neurosci. 17*, 438–449.

Constantinidis, C. and M. A. Steinmetz (1996). Neuronal activity in posterior parietal area 7a during the delay periods of a spatial memory task. *J. Neurophysiol. 76*, 1352–1355.

Constantinidis, C. and X.-J. Wang (2004). A neural circuit basis for spatial working memory. *Neuroscientist 10*, 553–565.

Constantinople, C. M., A. T. Piet, and C. D. Brody (2019). An analysis of decision under risk in rats. *Curr. Biol. 29*, 2066–2074.

Conway, A. R., M. J. Kane, and R. W. Engle (2003). Working memory capacity and its relation to general intelligence. *Trends Cogn. Sci. (Regul. Ed.) 7*, 547–552.

Cook, E. P. and J. H. R. Maunsell (2002). Dynamics of neuronal responses in macaque MT and VIP during motion detection. *Nat. Neurosci. 5*, 985–994.

Cools, R. and M. D'Esposito (2011). Inverted-u-shaped dopamine actions on human working memory and cognitive control. *Biol. Psychiatry 69*, e113–e125.

Corballis, M. C. (2014). *The Recursive Mind: The Origins of Human Language, Thought, and Civilization*. Prineceton, NJ: Princeton University Pres.

Corbetta, M. and G. Shulman (2002). Control of goal-directed and stimulus-driven attention in the brain. *Nat. Rev. Neurosci. 3*, 201–215.

Corlett, P. R., G. Horga, P. C. Fletcher, B. Alderson-Day, K. Schmack, and A. R. Powers III (2019). Hallucinations and strong priors. *Trends Cogn. Sci. 23*, 114–127.

Corlett, P. R., J. R. Taylor, X. J. Wang, P. C. Fletcher, and J. H. Krystal (2010). Toward a neurobiology of delusions. *Prog. Neurobiol. 92*, 345–369.

Corrado, G., L. Sugrue, H. Seung, and W. Newsome (2005). Linear-Nonlinear-Poisson models of primate choice dynamics. *J. Exp. Anal. Behav. 84*, 581–617.

Cossell, L., M. F. Iacaruso, D. R. Muir, R. Houlton, E. N. Sader, H. Ko, S. B. Hofer, and T. D. Mrsic-Flogel (2015). Functional organization of excitatory synaptic strength in primary visual cortex. *Nature 518*, 399–403.

Costa, R., I. A. Assael, B. Shillingford, N. de Freitas, and T. Vogels (2017). Cortical microcircuits as gated-recurrent neural networks. *Adv. Neural Inf. Process. Syst. 30*, 272–283.

Couey, J. J., A. Witoelar, S.-J. Zhang, K. Zheng, J. Ye, B. Dunn, R. Czajkowski, M.-B. Moser, E. I. Moser, Y. Roudi, et al. (2013). Recurrent inhibitory circuitry as a mechanism for grid formation. *Nat. Neurosci. 16*, 318–326.

Courchesne, E., P. R. Mouton, M. E. Calhoun, K. Semendeferi, C. Ahrens-Barbeau, M. J. Hallet, C. C. Barnes, and K. Pierce (2011). Neuron number and size in prefrontal cortex of children with autism. *JAMA 306*, 2001–2010.

Courtney, S. M., L. Petit, J. M. Maisog, L. G. Ungerleider, and J. V. Haxby (1998). An area specialized for spatial working memory in human frontal cortex. *Science 279*, 1347–1351.

Cowan, N. (2001). The magical number 4 in short-term memory: A reconsideration of mental storage capacity. *Behav. Brain Sci. 24*(1), 87–114.

Cowan, N. (2008). What are the differences between long-term, short-term, and working memory? *Prog. Brain Res. 169*, 323–338.

Cowan, N. (2010). The magical mystery four: How is working memory capacity limited, and why? *Curr. Dir. Psychol. Sci. 19*, 51–57.

Cowan, N. and J. N. Rouder (2009). Comment on "dynamic shifts of limited working memory resources in human vision". *Science 323*, 877–877.

Cox, J. and I. B. Witten (2019). Striatal circuits for reward learning and decision-making. *Nat. Rev. Neurosci. 20*, 482–494.

Coyle, J. T., G. Tsai, and D. Goff (2003). Converging evidence of NMDA receptor hypofunction in the pathophysiology of schizophrenia. *Ann. N. Y. Acad. Sci. 1003*, 318–327.

Craig, A. D. and A. Craig (2009). How do you feel–now? The anterior insula and human awareness. *Nat. Rev. Neurosci. 10*(1).

Crair, M. C., J. C. Horton, A. Antonini, and M. P. Stryker (2001). Emergence of ocular dominance columns in cat visual cortex by 2 weeks of age. *J. Comp. Neurol. 430*, 235–249.

Crandall, S. R., S. J. Cruikshank, and B. W. Connors (2015). A corticothalamic switch: Controlling the thalamus with dynamic synapses. *Neuron 86*, 768–782.

Crapse, T. B., H. Lau, and M. A. Basso (2018). A role for the superior colliculus in decision criteria. *Neuron 97*, 181–194.

Crapse, T. B. and M. A. Sommer (2008). Corollary discharge across the animal kingdom. *Nat. Rev. Neurosci. 9*, 587–600.

Craske, M. G. and M. B. Stein (2016). Anxiety. *Lancet 388*, 3048–3059.

Craver, C. F. (2007). *Explaining the Brain: Mechanisms and the Mosaic Unity of Neuroscience*. Oxford: Clarendon Press.

Craver, C. F. (2015). Levels. In T. Metzinger and J. M. Windt (Eds.), *Open MIND*. Frankfurt am Main: MIND Group. https://doi.org/10.15502/9783958570498.

Crick, F. H. (1984). Neurobiology: Memory and molecular turnover. *Nature 312*, 101–101.

Crick, F. H. (1991). Remembering David Marr. In L. M. Vaina (Ed.), *From the Retina to the Neocortex: Selected Papers of David Marr*, pp. 314–315. Boston, MA: Birkhäuser.

Crick, F. H. and C. Koch (1990). Towards a neurobiological theory of consciousness. *Semin. Neurosci. 2*, 263–275.

Cromer, J. A., M. Machon, and E. K. Miller (2011). Rapid association learning in the primate prefrontal cortex in the absence of behavioral reversals. *J. Cogn. Neurosci. 23*, 1823–1828.

Cromer, J. A., J. E. Roy, and E. K. Miller (2010). Representation of multiple, independent categories in the primate prefrontal cortex. *Neuron 66*, 796–807.

Crook, S., G. Ermentrout, M. Vanier, and J. Bower (1997). The role of axonal delay in the synchronization of networks of coupled cortical oscillators. *J. Comput. Neurosci. 4*, 161–172.

Crowe, D. A., B. B. Averbeck, and M. V. Chafee (2010). Rapid sequences of population activity patterns dynamically encode task critical spatial information in parietal cortex. *J. Neurosci. 30*, 11640–11653.

Cueva, C. J. and X.-X. Wei (2018). Emergence of grid-like representations by training recurrent neural networks to perform spatial localization. *arXiv preprint arXiv:1803.07770.*

Cui, G., S. B. Jun, X. Jin, M. D. Pham, S. S. Vogel, D. M. Lovinger, and R. M. Costa (2013). Concurrent activation of striatal direct and indirect pathways during action initiation. *Nature 494*, 238–242.

Cunningham, J. P. and M. Y. Byron (2014). Dimensionality reduction for large-scale neural recordings. *Nat. Neurosci. 17*, 1500–1509.

Curtis, C. E. and M. D'Esposito (2003). Persistent activity in the prefrontal cortex during working memory. *Trends Cogn. Sci. (Regul. Ed.) 7*, 415–423.

Curtis, C. E. and T. C. Sprague (2021). Persistent activity during working memory from front to back. *Front. Neural Circuits 15*. https://doi.org/10.3389/fncir.2021.696060.

Dalgleish, T. (2004). The emotional brain. *Nat. Rev. Neurosci. 5*, 583–589.

Dalley, J. W., B. J. Everitt, and T. W. Robbins (2011). Impulsivity, compulsivity, and top-down cognitive control. *Neuron 69*, 680–694.

Damasio, A. R. (1994). *Descartes' Error*. New York: G. P. Putnam's Sons.

Damilou, A., S. Apostolakis, E. Thrapsanioti, C. Theleritis, and N. Smyrnis (2016). Shared and distinct oculomotor function deficits in schizophrenia and obsessive compulsive disorder. *Psychophysiology 53*, 796–805.

Dan, Y. and M. M. Poo (2006). Spike timing-dependent plasticity: From synapse to perception. *Physiol. Rev. 86*, 1033–1048.

Dang, W., S. Li, S. Pu, X.-L. Qi, and C. Constantinidis (2022). More prominent non-linear mixed selectivity in the dorsolateral prefrontal than posterior parietal cortex. *Eneuro*. https://doi.org/10.1523/ENEURO.0517-21.2022.

Darwin, C. (2015). *The Expression of the Emotions in Man and Animals*. Chicago: University of Chicago Press. First published in 1872 by London, Engliand: John Murray Publishing House.

Datta, S. R., D. J. Anderson, K. Branson, P. Perona, and A. Leifer (2019). Computational neuroethology: A call to action. *Neuron 104*, 11–24.

Davis, G. W. (2006). Homeostatic control of neural activity: From phenomenology to molecular design. *Annu. Rev. Neurosci. 29*, 307–323.

Davis, Z. W., L. Muller, J. Martinez-Trujillo, T. Sejnowski, and J. H. Reynolds (2020). Spontaneous travelling cortical waves gate perception in behaving primates. *Nature 587*, 432–436.

Daw, N. D. (2016). Reinforcement learning. In M. A. Arbib and J. J. Bonaiuto (Eds.), *From Neuron to Cognition via Computational Neuroscience*, Chapter 10, pp. 299–318. Cambridge, MA: MIT Press.

Daw, N. D., S. J. Gershman, B. Seymour, P. Dayan, and R. J. Dolan (2011). Model-based influences on humans' choices and striatal prediction errors. *Neuron 69*(6), 1204–1215.

Daw, N. D., J. O'Doherty, P. Dayan, B. Seymour, and R. Dolan (2006). Cortical substrates for exploratory decisions in humans. *Nature 441*, 876–879.

Dawkins, R. (1976). Hierarchical organisation: A candidate principle for ethology. In P. P. G. Baterson and R. A. Hinde (Eds.), *Growing Points in Ethology*, pp. 7–54. Cambridge: Cambridge University Press.

Dayan, P. (1993). Improving generalization for temporal difference learning: The successor representation. *Neural Comput. 5*, 613–624.

Dayan, P. and L. F. Abbott (2001). *Theoretical Neuroscience*. Cambridge, MA: MIT Press.

Dayan, P. and H. Nakahara (2018). Models and methods for reinforcement learning. In J. T. Wixted (Ed.), *Stevens' Handbook of Experimental Psychology and Cognitive Neuroscience* (4th ed.). New York: John Wiley & Sons. https://doi.org/10.1002/9781119170174.epcn513.

De, A. and R. Chaudhuri (2023). Common population codes produce extremely nonlinear neural manifolds. *Proc. Natl. Acad. Sci. 120*, e2305853120.

Deco, G. and V. K. Jirsa (2012). Ongoing cortical activity at rest: Criticality, multistability, and ghost attractors. *J. Neurosci. 32*, 3366–3375.

Deco, G., M. Pérez-Sanagustn, V. de Lafuente, and R. Romo (2007). Perceptual detection as a dynamical bistability phenomenon: A neurocomputational correlate of sensation. *Proc. Natl. Acad. Sci. U.S.A. 104*, 20073–20077.

Deco, G., A. Ponce-Alvarez, P. Hagmann, G. L. Romani, D. Mantini, and M. Corbetta (2014). How local excitation-inhibition ratio impacts the whole brain dynamics. *J. Neurosci. 34*, 7886–7898.

Deco, G., E. T. Rolls, L. Albantakis, and R. Romo (2012). Brain mechanisms for perceptual and reward-related decision-making. *Prog. Neurobiol.* https://doi.org/10.1016/j.pneurobio.2012.01.010.

Deco, G., E. T. Rolls, and R. Romo (2009). Stochastic dynamics as a principle of brain function. *Prog. Neurobiol. 88*, 1–16.

Deco, G., G. Tononi, M. Boly, and M. L. Kringelbach (2015). Rethinking segregation and integration: Contributions of whole-brain modelling. *Nat. Rev. Neurosci. 16*, 430–439.

DeFelipe, J. (1997). Types of neurons, synaptic connections and chemical characteristics of cells immunoreactive for calbindin-D28K, parvalbumin and calretinin in the neocortex. *J. Chem. Neuroanat. 14*, 1–19.

DeFelipe, J. (2002). Cortical interneurons: From Cajal to 2001. *Prog. Brain Res. 136*, 215–238.

DeFelipe, J. (2006). Brain plasticity and mental processes: Cajal again. *Nat. Rev. Neurosci. 7*, 811–817.

DeFelipe, J., M. C. Gonzalez-Albo, M. R. Del Rio, and G. N. Elston (1999). Distribution and patterns of connectivity of interneurons containing calbindin, calretinin, and parvalbumin in visual areas of occipital and temporal lobes of the macaque monkeys. *J. Comp. Neurol. 412*, 515–526.

DeFelipe, J., P. L. López-Cruz, R. Benavides-Piccione, C. Bielza, P. Larrañaga, S. Anderson, A. Burkhalter, B. Cauli, A. Fairén, D. Feldmeyer, G. Fishell, D. Fitzpatrick, T. F. Freund, G. González-Burgos, S. Hestrin, S. Hill, P. R. Hof, J. Huang, E. G. Jones, Y. Kawaguchi, Z. Kisvárday, Y. Kubota, D. A. Lewis, O. Marín, H. Markram, C. J. McBain, H. S. Meyer, H. Monyer, S. B. Nelson, K. Rockland, J. Rossier, J. L. Rubenstein, B. Rudy, M. Scanziani, G. M. Shepherd, C. C. Sherwood, J. F. Staiger, G. Tamás, A. Thomson, Y. Wang, R. Yuste, and G. A. Ascoli (2013). New insights into the classification and nomenclature of cortical GABAergic interneurons. *Nat. Rev. Neurosci. 14*, 202–216.

Dehaene, S. (2010). *Reading in the Brain: The New Science of How We Read*. New York: Penguin Books.

Dehaene, S. (2011). *The Number Sense: How the Mind Creates Mathematics*. New York: Oxford University Press.

Dehaene, S. (2014). *Consciousness and the Brain: Deciphering How the Brain Codes Our Thoughts*. New York: Penguin.

Dehaene, S. (2020). *How We Learn: Why Brains Learn Better Than Any Machine . . . for Now*. New York: Viking Press.

Dehaene, S., F. Al Roumi, Y. Lakretz, S. Planton, and M. Sablé-Meyer (2022). Symbols and mental programs: A hypothesis about human singularity. *Trends Cogn. Sci. 26*, 751–766.

Dehaene, S., J. P. Changeux, L. Naccache, J. Sackur, and C. Sergent (2006). Conscious, preconscious, and subliminal processing: A testable taxonomy. *Trends Cogn. Sci. (Regul. Ed.) 10*, 204–211.

Dehaene, S., F. Meyniel, C. Wacongne, L. Wang, and C. Pallier (2015). The neural representation of sequences: From transition probabilities to algebraic patterns and linguistic trees. *Neuron 88*, 1–19.

Dehaene, S., N. Molko, L. Cohen, and A. J. Wilson (2004). Arithmetic and the brain. *Curr. Opin. Neurobiol. 14*, 218–224.

Deisseroth, K. (2021). *Projections: A Story of Human Emotions*. New York: Random House.

de Lafuente, V. and R. Romo (2005). Neuronal correlates of subjective sensory experience. *Nat. Neurosci. 8*, 1698–1703.

Del Cul, A., S. Baillet, and S. Dehaene (2007). Brain dynamics underlying the nonlinear threshold for access to consciousness. *PLoS Biol. 5*, e260.

Dellert, T., M. Müller-Bardorff, I. Schlossmacher, M. Pitts, D. Hofmann, M. Bruchmann, and T. Straube (2021). Dissociating the neural correlates of consciousness and task relevance in face perception using simultaneous EEG-FMRI. *J. Neurosci. 41*, 7864–7875.

DeLong, M. R. (1990). Primate models of movement disorders of basal ganglia origin. *Trends Neurosci. 13*, 281–285.

Demirtas, M., J. B. Burt, M. Helmer, J. L. Ji, B. D. Adkinson, M. F. Glasser, D. C. Van Essen, S. N. Sotiropoulos, A. Anticevic, and J. D. Murray (2019). Hierarchical heterogeneity across human cortex shapes large-scale neural dynamics. *Neuron 101*, 1181–1194.

DeNardo, L. A., D. S. Berns, K. DeLoach, and L. Luo (2015). Connectivity of mouse somatosensory and prefrontal cortex examined with trans-synaptic tracing. *Nat. Neurosci. 18*, 1687–1697.

Deng, J., W. Dong, R. Socher, L.-J. Li, K. Li, and L. Fei-Fei (2009). Imagenet: A large-scale hierarchical image database. In *2009 IEEE Conference on Computer Vision and Pattern Recognition*, pp. 248–255.

Denk, W. and H. Horstmann (2004). Serial block-face scanning electron microscopy to reconstruct three-dimensional tissue nanostructure. *PLoS Biol. 2*, e329.

Dennett, D. C. (1991). *Consciousness Explained*. Boston: Little, Brown and Co.

De Pittà, M. and N. Brunel (2022). Multiple forms of working memory emerge from synapse-astrocyte interactions in a neuron-glia network model. *Proc. Natl. Acad. Sci. U.S.A. 119*, e2207912119.

Derrida, B., E. Gardner, and A. Zippelius (1987). An exactly solvable asymmetric neural network model. *Europhy. Lett. 4*, 167–173.

Descartes, R. (1989). *The Passions of the Soul*. Indiapolis: Hackett Publishing (S. H. Voss, Trans. The original book was published in French in 1649).

Desender, K., A. Boldt, T. Verguts, and T. H. Donner (2019). Confidence predicts speed-accuracy tradeoff for subsequent decisions. *Elife 8*, e43499.

Deserno, L., Q. J. Huys, R. Boehme, R. Buchert, H.-J. Heinze, A. A. Grace, R. J. Dolan, A. Heinz, and F. Schlagenhauf (2015). Ventral striatal dopamine reflects behavioral and neural signatures of model-based control during sequential decision making. *Proc. Natl. Acad. Sci. 112*, 1595–1600.

Deshmukh, S. S. and J. J. Knierim (2011). Representation of non-spatial and spatial information in the lateral entorhinal cortex. *Front. Behav. Neurosci. 5*, https://doi.org/10.3389/fnbeh.2011.00069.

Desimone, R., T. D. Albright, C. G. Gross, and C. Bruce (1984). Stimulus-selective properties of inferior temporal neurons in the macaque. *J. Neurosci. 4*, 2051–2062.

Desimone, R. and J. Duncan (1995). Neural mechanisms of selective visual attention. *Annu. Rev. Neurosci. 18*, 193–222.

D'Esposito, M. (2007). From cognitive to neural models of working memory. *Philos. Trans. R. Soc. Lond., B, Biol. Sci. 362*, 761–772.

D'Esposito, M., J. A. Detre, D. C. Alsop, R. K. Shin, S. Atlas, and M. Grossman (1995). The neural basis of the central executive system of working memory. *Nature 378*, 279–281.

Destexhe, A. (2021). In silico, computer simulations from neurons up to the whole brain. *eNeuro 8*, e0124.

Destexhe, A., D. Contreras, M. Steriade, T. Sejnowski, and J. Huguenard (1996). In vivo, in vitro, and computational analysis of dendritic calcium currents in thalamic reticular neurons. *J. Neurosci. 16*, 169–185.

Destexhe, A., Z. F. Mainen, and T. J. Sejnowski (1998). Kinetic models of synaptic transmission. In C. Koch and I. Segev (Eds.), *Methods in Neuronal Modeling* (2nd ed.)., pp. 1–25. Cambridge, MA: MIT Press.

Destexhe, A., M. Rudolph, and D. Paré (2003). The high-conductance state of neocortical neurons in vivo. *Nat. Rev. Neurosci. 4*, 739–751.

Destexhe, A. and T. J. Sejnowski (2001). *Thalamocortical Assemblies: How Ion Channels, Single Neurons and Large-Scale Networks Organize Sleep Oscillations*. New York: Oxford University Press.

Dezfouli, A., P. Piray, M. M. Keramati, H. Ekhtiari, C. Lucas, and A. Mokri (2009). A neurocomputational model for cocaine addiction. *Neural Comput. 21*, 2869–2893.

Diamond, A. (2013). Executive functions. *Ann. Rev. Psychol. 64*, 135–168.

Diamond, A. and P. S. Goldman-Rakic (1989). Comparison of human infants and rhesus monkeys on Piaget's AB task:

Evidence for dependence on dorsolateral prefrontal cortex. *Exp. Brain Res. 74*, 24–40.

DiCarlo, J. J. and D. D. Cox (2007). Untangling invariant object recognition. *Trends Cogn. Sci. (Regul. Ed.) 11*, 333–341.

Dichter, B. K., J. D. Breshears, M. K. Leonard, and E. F. Chang (2018). The control of vocal pitch in human laryngeal motor cortex. *Cell 174*, 21–31.

Diehl, M. M., J. M. Iravedra-Garcia, J. Morán-Sierra, G. Rojas-Bowe, F. N. Gonzalez-Diaz, V. P. Valentín-Valentín, and G. J. Quirk (2020). Divergent projections of the prelimbic cortex bidirectionally regulate active avoidance. *Elife 9*, e59281.

Diesmann, M., M. Gewaltig, and A. Aertsen (1999). Stable propagation of synchronous spiking in cortical neural networks. *Nature 402*, 529–533.

Ding, L. and J. I. Gold (2013). The basal ganglia's contributions to perceptual decision making. *Neuron 79*, 640–649.

Ding, L. and O. Hikosaka (2006). Comparison of reward modulation in the frontal eye field and caudate of the macaque. *J. Neurosci. 26*, 6695–6703.

Ding, X., S. Froudist-Walsh, J. Jaramillo, J. Jiang, and X.-J. Wang (2024). Cell type-specific connectome predicts distributed working memory activity in the mouse brain. *Elife 13*, e85442.

Ditterich, J. (2006). Evidence for time-variant decision making. *Eur. J. Neurosci. 24*, 3628–3641.

Doebeli, M., Y. Ispolatov, and B. Simon (2017). Towards a mechanistic foundation of evolutionary theory. *Elife 6*, e23804.

Doeller, C. F., C. Barry, and N. Burgess (2010). Evidence for grid cells in a human memory network. *Nature 463*, 657–661.

Doiron, B., A. Litwin-Kumar, R. Rosenbaum, G. K. Ocker, and K. Josić (2016). The mechanics of state-dependent neural correlations. *Nat. Neurosci. 19*, 383–393.

Dolan, R. J. and P. Dayan (2013). Goals and habits in the brain. *Neuron 80*, 312–325.

Doll, B. B., D. A. Simon, and N. D. Daw (2012). The ubiquity of model-based reinforcement learning. *Curr. Opin. Neurobiol. 22*, 1075–1081.

Domenech, P., S. Rheims, and E. Koechlin (2020). Neural mechanisms resolving exploitation-exploration dilemmas in the medial prefrontal cortex. *Science 369*, abb0184.

Donahue, C. H. and D. Lee (2015). Dynamic routing of task-relevant signals for decision making in dorsolateral prefrontal cortex. *Nat. Neurosci. 18*, 295.

Donahue, C. J., S. N. Sotiropoulos, S. Jbabdi, M. Hernandez-Fernandez, T. E. Behrens, T. B. Dyrby, T. Coalson, H. Kennedy, K. Knoblauch, D. C. Van Essen, and M. F. Glasser (2016). Using diffusion tractography to predict cortical connection strength and distance: A quantitative comparison with tracers in the monkey. *J. Neurosci. 36*, 6758–6770.

Donders, F. C. (1969). On the speed of mental processes. *Acta Psychol. 30*, 412–431. Translation of Die Schnelligkeit Psychischer Processe, first published in 1868.

Dong, D., Y. Wang, X. Chang, C. Luo, and D. Yao (2018). Dysfunction of large-scale brain networks in schizophrenia: A meta-analysis of resting-state functional connectivity. *Schizophr. Bull. 44*, 168–181.

Donoghue, J. P. (2008). Bridging the brain to the world: A perspective on neural interface systems. *Neuron 60*, 511–521.

Dorkenwald, S., Matsliah, A., Sterling, A. R., Schlegel, P., Yu, S. C., McKellar, C. E., . . . & Murthy, M. (2024). Neuronal wiring diagram of an adult brain. Nature, *634*, 124–138.

Dorkenwald, S., C. E. McKellar, T. Macrina, N. Kemnitz, K. Lee, R. Lu, J. Wu, S. Popovych, E. Mitchell, B. Nehoran, et al. (2022). Flywire: Online community for whole-brain connectomics. *Nat. Methods 19*, 119–128.

Dorris, M. C. and P. W. Glimcher (2004). Activity in posterior parietal cortex is correlated with the relative subjective desirability of action. *Neuron 44*, 365–378.

Dosher, B. A. and Z. L. Lu (2009). Hebbian reweighting on stable representations in perceptual learning. *Learn Percept. 1*, 37–58.

Dotson, N. M., S. J. Hoffman, B. Goodell, and C. M. Gray (2017). A large-scale semi-chronic microdrive recording system for non-human primates. *Neuron 96*, 769–782.

Dotson, N. M., S. J. Hoffman, B. Goodell, and C. M. Gray (2018). Feature-based visual short-term memory is widely distributed and hierarchically organized. *Neuron 99*, 215–226.

Douglas, R. J., C. Koch, M. Mahowald, K. M. Martin, and H. H. Suarez (1995). Recurrent excitation in neocortical circuits. *Science 269*, 981–985.

Douglas, R. J. and K. A. C. Martin (1991). A functional microcircuit for cat visual cortex. *J. Physiol. (London) 440*, 735–.

Douglas, R. J. and K. A. C. Martin (2004). Neuronal circuits of the neocortex. *Annu. Rev. Neurosci. 27*, 419–451.

Douglas, R. J. and K. A. C. Martin (2012). Behavioral architecture of the cortical sheet. *Curr. Biol. 22*, R1033–1038.

Dowling, J. E. (1987). *The Retina: An Approachable Part of the Brain*. Cambridge, MA: Harvard University Press.

Doya, K. (2000). Complementary roles of basal ganglia and cerebellum in learning and motor control. *Curr. Opin. Neurobiol. 10*, 732–739.

Doya, K., S. Ishii, A. Pouget, and R. P. N. Rao (Eds.) (2011). *Bayesian Brain: Probabilistic Approaches to Neural Coding*. Cambridge, MA: MIT Press.

Drevets, W. C. and M. E. Raichle (1998). Reciprocal suppression of regional cerebral blood flow during emotional versus higher cognitive processes: Implications for interactions between emotion and cognition. *Cogn. Emot. 12*, 353–385.

Driesen, N. R., G. McCarthy, Z. Bhagwagar, M. H. Bloch, V. D. Calhoun, D. C. D'Souza, R. Gueorguieva, G. He, H. C. Leung, R. Ramani, A. Anticevic, R. F. Suckow, P. T. Morgan, and J.

H. Krystal (2013). The impact of NMDA receptor blockade on human working memory-related prefrontal function and connectivity. *Neuropsychopharmacology 38*, 2613–2622.

Driscoll, L. N., Shenoy, K., and Sussillo, D. (2024). Flexible multitask computation in recurrent networks utilizes shared dynamical motifs. *Nat. Neurosci.* 27, 1349–1363..

Druckmann, S. and D. B. Chklovskii (2012). Neuronal circuits underlying persistent representations despite time varying activity. *Curr. Biol. 22*, 2095–2103.

Drugowitsch, J., R. Moreno-Bote, A. K. Churchland, M. N. Shadlen, and A. Pouget (2012). The cost of accumulating evidence in perceptual decision making. *J. Neurosci. 32*, 3612–3628.

Drysdale, A. T., L. Grosenick, J. Downar, K. Dunlop, F. Mansouri, Y. Meng, R. N. Fetcho, B. Zebley, D. J. Oathes, A. Etkin, et al. (2017). Resting-state connectivity biomarkers define neurophysiological subtypes of depression. *Nat. Med. 23*, 28–38.

Duan, C. A., J. C. Erlich, and C. D. Brody (2015). Requirement of prefrontal and midbrain regions for rapid executive control of behavior in the rat. *Neuron 86*, 1491–1503.

Dubreuil, A., A. Valente, M. Beiran, F. Mastrogiuseppe, and S. Ostojic (2022). The role of population structure in computations through neural dynamics. *Nat. Neurosci. 25*, 783–794.

Duncan, J. (2001). An adaptive coding model of neural function in prefrontal cortex. *Nat. Rev. Neurosci. 2*(11), 820–829.

Duncan, J. (2010a). *How Intelligence Happens*. New Haven, CT: Yale University Press.

Duncan, J. (2010b). The multiple-demand (MD) system of the primate brain: Mental programs for intelligent behaviour. *Trends Cogn. Sci. (Regul. Ed.) 14*(4), 172–179.

Duncan, J. (2013). The structure of cognition: Attentional episodes in mind and brain. *Neuron 80*, 35–50.

Duncan, J., P. Burgess, and H. Emslie (1995). Fluid intelligence after frontal lobe lesions. *Neuropsychologia 33*, 261–268.

Duncan, J., D. Chylinski, D. J. Mitchell, and A. Bhandari (2017). Complexity and compositionality in fluid intelligence. *Proc. Natl. Acad. Sci. 114*, 5295–5299.

Duncan, J., R. J. Seitz, J. Kolodny, D. Bor, H. Herzog, A. Ahmed, F. N. Newell, and H. Emslie (2000). A neural basis for general intelligence. *Science 289*, 457–460.

Duncan, K., C. Curtis, and L. Davachi (2009). Distinct memory signatures in the hippocampus: Intentional States distinguish match and mismatch enhancement signals. *J. Neurosci. 29*, 131–139.

Durstewitz, D. and J. K. Seamans (2008). The dual-state theory of prefrontal cortex dopamine function with relevance to catechol-o-methyltransferase genotypes and schizophrenia. *Biol. Psychiatry 64*, 739–749.

Durstewitz, D., J. K. Seamans, and T. J. Sejnowski (2000). Dopamine-mediated stabilization of delay-period activity in a network model of prefrontal cortex. *J. Neurophysiol. 83*, 1733–1750.

Durstewitz, D., N. M. Vittoz, S. B. Floresco, and J. K. Seamans (2010). Abrupt transitions between prefrontal neural ensemble states accompany behavioral transitions during rule learning. *Neuron 66*(3), 438–448.

Dusek, J. A. and H. Eichenbaum (1997). The hippocampus and memory for orderly stimulus relations. *Proc. Natl. Acad. Sci. 94*, 7109–7114.

Ebbinghaus, H. (1885). *Über das Gedächtnis: Untersuchungen zur Experimentellen Psychologie*. Berlin: Duncker & Humblot.

Eccles, J. C., R. Llinas, and K. Sasaki (1966). The mossy fibre-granule cell relay of the cerebellum and its inhibitory control by Golgi cells. *Exp. Brain Res. 1*, 82–101.

Eddy, P. (1982). Probabilistic reasoning in clinical medicine: Problems and opportunities. In P. S. D. Kahneman and A. Tversky (Eds.), *Judgment Under Uncertainty: Heuristics and Biases*, pp. 249–267. New York: Cambridge University Press.

Edelman, G. M. and J. A. Gally (2001). Degeneracy and complexity in biological systems. *Proc. Natl. Acad. Sci. 98*, 13763–13768.

Edin, F., T. Klingberg, P. Johansson, F. McNab, J. Tegnér, and A. Compte (2009). Mechanism for top-down control of working memory capacity. *Proc. Natl. Acad. Sci. U.S.A. 106*, 6802–6807.

Edwards, W. (1965). Optimal strategies for seeking information: Models for statistics, choice reaction times, and human information processing. *J. Math. Psychol. 2*, 312–329.

Egan, M., T. Goldberg, B. Kolachana, J. Callicott, C. Mazzanti, R. Straub, D. Goldman, and D. Weinberger (2001). Effect of COMT Val108/158 Met genotype on frontal lobe function and risk for schizophrenia. *Proc. Natl. Acad. Sci. U.S.A. 98*, 6917–6922.

Egorov, A. V., B. N. Hamam, E. Fransen, M. E. Hasselmo, and A. A. Alonso (2002). Graded persistent activity in entorhinal cortex neurons. *Nature 420*, 173–178.

Ehlers, M. D. (2003). Activity level controls postsynaptic composition and signaling via the ubiquitin-proteasome system. *Nat. Neurosci. 6*, 231–242.

Ehrlich, D. B. and J. D. Murray (2022). Geometry of neural computation unifies working memory and planning. *Proc. Natl. Acad. Sci. 119*, e2115610119.

Eichenbaum, H. (2017). Memory: Organization and control. *Annu. Rev. Psychol. 68*, 19–45.

Eickhoff, S. B., B. T. T. Yeo, and S. Genon (2018). Imaging-based parcellations of the human brain. *Nat. Rev. Neurosci. 19*, 672–686.

Eiselt, A.-K. and A. Nieder (2013). Representation of abstract quantitative rules applied to spatial and numerical magnitudes in primate prefrontal cortex. *J. Neurosci. 33*, 7526–7534.

Eisenberg, I. W., P. G. Bissett, A. Zeynep Enkavi, J. Li, D. P. MacKinnon, L. A. Marsch, and R. A. Poldrack (2019). Uncovering the structure of self-regulation through data-driven ontology discovery. *Nat. Commun. 10*, 2319.

Ekman, P., R. W. Levenson, and W. V. Friesen (1983). Autonomic nervous system activity distinguishes among emotions. *Science 221*, 1208–1210.

Eliades, S. J. and X. Wang (2008). Neural substrates of vocalization feedback monitoring in primate auditory cortex. *Nature 453*, 1102–1106.

Eliasmith, C. (2013). *How to Build a Brain: A Neural Architecture for Biological Cognition*. Oxford: Oxford University Press.

Eliasmith, C., T. C. Stewart, X. Choo, T. Bekolay, T. DeWolf, Y. Tang, and D. Rasmussen (2012). A large-scale model of the functioning brain. *Science 338*, 1202–1205.

Elsayed, G. F., A. H. Lara, M. T. Kaufman, M. M. Churchland, and J. P. Cunningham (2016). Reorganization between preparatory and movement population responses in motor cortex. *Nat. Commun. 7*, 13239.

Elston, G. H. (2000). Pyramidal cells of the frontal lobe: All the more spinous to think with. *J. Neurosci. 20*, RC95.

Elston, G. H. (2007). Specialization of the neocortical pyramidal cell during primate evolution. In J. H. Kaass and T. M. Preuss (Eds.), *Evolution of the Nervous Systems: A Comprehensive Reference, Vol. 4 Primates*, pp. 191–242. Amsterdam: Elsevier.

Elston, G. N., R. Benavides-Piccione, and J. DeFelipe (2001). The pyramidal cell in cognition: A comparative study in human and monkey. *J. Neurosci. 21*, RC163–RC163.

Emmons, E. B., B. J. De Corte, Y. Kim, K. L. Parker, M. S. Matell, and N. S. Narayanan (2017). Rodent medial frontal control of temporal processing in the dorsomedial striatum. *J. Neurosci. 37*, 8718–8733.

Enel, P., J. D. Wallis, and E. L. Rich (2020). Stable and dynamic representations of value in the prefrontal cortex. *eLife 9*, e54313.

Engel, T. A., W. Chaisangmongkon, D. J. Freedman, and X. J. Wang (2015). Choice-correlated activity fluctuations underlie learning of neuronal category representation. *Nat. Commun. 6*, 6454.

Engel, T. A. and X.-J. Wang (2011). Same or different? A neural circuit mechanism of similarity-based pattern match decision making. *J. Neurosci. 31*, 6982–6996.

Engelhard, B., J. Finkelstein, J. Cox, W. Fleming, H. J. Jang, S. Ornelas, S. A. Koay, S. Y. Thiberge, N. D. Daw, D. W. Tank, et al. (2019). Specialized coding of sensory, motor and cognitive variables in vta dopamine neurons. *Nature 570*, 509–513.

Ercsey-Ravasz, M., N. T. Markov, C. Lamy, D. C. Van Essen, K. Knoblauch, Z. Toroczkai, and H. Kennedy (2013). A predictive network model of cerebral cortical connectivity based on a distance rule. *Neuron 80*, 184–197.

Erdos, P. and A. Renyi (1960). On the evolution of random graphs. *Publ. Math. Inst. Hungari. Acad. Sci. 5*, 17–61.

Erlich, J. C., B. W. Brunton, C. A. Duan, T. D. Hanks, and C. D. Brody (2015). Distinct effects of prefrontal and parietal cortex inactivations on an accumulation of evidence task in the rat. *Elife 4*, e05457.

Ermentrout, B. and D. H. Terman (2010). *Mathematical Foundations of Neuroscience*. New York: Springer.

Ermentrout, G. B. (1998). Neural networks as spatio-temporal pattern-forming systems. *Rep. Prog. Phys. 61*, 353–430.

Ermentrout, G. B. and D. Kleinfeld (2001). Traveling electrical waves in cortex: Insights from phase dynamics and speculation on a computational role. *Neuron 29*, 33–44.

Ernst, M. and M. Banks (2002). Humans integrate visual and haptic information in a statistically optimal fashion. *Nature 415*, 429–433.

Etter, W. L. (1974). Benjamin Franklin and prudential algebra. *Decis. Sci. 5*, 145–147.

Evans, D. A., A. V. Stempel, R. Vale, S. Ruehle, Y. Lefler, and T. Branco (2018). A synaptic threshold mechanism for computing escape decisions. *Nature 558*, 590–594.

Evenden, J. L. (1999). Varieties of impulsivity. *Psychopharmacology 146*, 348–361.

Everitt, B. J. and T. W. Robbins (2005). Neural systems of reinforcement for drug addiction: From actions to habits to compulsion. *Nat. Neurosci. 8*, 1481–1489.

Everitt, B. J. and T. W. Robbins (2016). Drug addiction: Updating actions to habits to compulsions ten years on. *Annu. Rev. Psychol. 67*, 23–50.

Everling, S. and J. F. DeSouza (2005). Rule-dependent activity for prosaccades and antisaccades in the primate prefrontal cortex. *J. Cogn. Neurosci. 17*, 1483–1496.

Everling, S., M. C. Dorris, R. M. Klein, and D. P. Munoz (1999). Role of primate superior colliculus in preparation and execution of anti-saccades and pro-saccades. *J. Neurosci. 19*, 2740–2754.

Everling, S. and B. Fischer (1998). The antisaccade: A review of basic research and clinical studies. *Neuropsychologia 36*, 885–899.

Ewert, J. (1997). Neural correlates of key stimulus and releasing mechanism: A case study and two concepts. *Trends Neurosci. 20*, 332–339.

Fairhall, A. (2021). In silico: Where next? *eNeuro 8*, e0131.

Farashahi, S., C. H. Donahue, B. Y. Hayden, D. Lee, and A. Soltani (2019). Flexible combination of reward information across primates. *Nat. Hum. Behav. 3*, 1215–1224.

Farashahi, S., C. H. Donahue, P. Khorsand, H. Seo, D. Lee, and A. Soltani (2017). Metaplasticity as a neural substrate for adaptive learning and choice under uncertainty. *Neuron 94*, 401–414.

Fechner, G. T. (1860). *Element der Psychophysik*. Leipzig: Breitkopf and Härterl.

Fedorenko, E., M. K. Behr, and N. Kanwisher (2011). Functional specificity for high-level linguistic processing in the human brain. *Proc. Natl. Acad. Sci. 108*, 16428–16433.

Fedorenko, E., J. Duncan, and N. Kanwisher (2012). Language-selective and domain-general regions lie side by side within broca's area. *Curr. Biol. 22*, 2059–2062.

Fedorenko, E., Piantadosi, S. T., and Gibson, E. A. (2024). Language is primarily a tool for communication rather than thought. *Nature 630*, 575–586.

Feinman, R. and B. M. Lake (2020). Learning task-general representations with generative neuro-symbolic modeling. *arXiv preprint arXiv:2006.14448*.

Feldmeyer, D., V. Egger, J. Lubke, and B. Sakmann (1999). Reliable synaptic connections between pairs of excitatory layer 4 neurones within a single 'barrel' of developing rat somatosensory cort e. *J. Physiol. 521*, 169–190.

Feldmeyer, D., G. Qi, V. Emmenegger, and J. F. Staiger (2018). Inhibitory interneurons and their circuit motifs in the many layers of the barrel cortex. *Neuroscience 368*, 132–151.

Felleman, D. J. and D. C. Van Essen (1991). Distributed hierarchical processing in the primate cerebral cortex. *Cereb. Cortex 1*, 1–47.

Fellows, L. K. (2006). Deciding how to decide: Ventromedial frontal lobe damage affects information acquisition in multi-attribute decision making. *Brain 129*, 944–952.

Fernandez-Ruiz, A., A. Oliva, E. Fermino de Oliveira, F. Rocha-Almeida, D. Tingley, and G. Buzsaki (2019). Long-duration hippocampal sharp wave ripples improve memory. *Science 364*, 1082–1086.

Ferrigno, S., S. J. Cheyette, S. T. Piantadosi, and J. F. Cantlon (2020). Recursive sequence generation in monkeys, children, us adults, and native amazonians. *Sci. Adv. 6*, eaaz1002.

Ferster, D. and K. D. Miller (2000). Neural mechanisms of orientation selectivity in the visual cortex. *Annu. Rev. Neurosci. 23*, 441–471.

Fiebig, F., P. Herman, and A. Lansner (2020). An indexing theory for working memory based on fast hebbian plasticity. *eNeuro 7*. https://doi.org/10.1523/ENEURO.0374-19.2020.

Fiete, I. R., Y. Burak, and T. Brookings (2008). What grid cells convey about rat location. *J. Neurosci. 28*, 6858–6871.

Finkelstein, A., L. Fontolan, M. N. Economo, N. Li, S. Romani, and K. Svoboda (2021). Attractor dynamics gate cortical information flow during decision-making. *Nat. Neurosci.* https://doi.org/10.1038/s41593-021-00840-6.

Finn, C., P. Abbeel, and S. Levine (2017). Model-agnostic meta-learning for fast adaptation of deep networks. In *International Conference on Machine Learning*, pp. 1126–1135.

Finn, E. S., L. Huber, D. C. Jangraw, P. J. Molfese, and P. A. Bandettini (2019). Layer-dependent activity in human prefrontal cortex during working memory. *Nat. Neurosci. 22*, 1687–1695.

Fiorillo, C., P. Tobler, and W. Schultz (2003). Discrete coding of reward probability and uncertainty by dopamine neurons. *Science 299*, 1898–1902.

Fisahn, A., F. G. Pike, E. H. Buhl, and O. Paulsen (1998). Cholinergic induction of network oscillations at 40hz in the hippocampus *in vitro*. *Nature 394*, 186–189.

Fiser, A., D. Mahringer, H. K. Oyibo, A. V. Petersen, M. Leinweber, and G. B. Keller (2016). Experience-dependent spatial expectations in mouse visual cortex. *Nat. Neurosci. 19*, 1658–1664.

Fisher, S. A., T. M. Fischer, and T. J. Carew (1997). Multiple overlapping processes underlying short-term synaptic enhancement. *Trends Neurosci. 20*, 170–177.

Flavell, J. H. (1979). Metacognition and cognitive monitoring: A new area of cognitive-developmental inquiry. *Am. Psychol. 34*, 906–911.

Fleming, S. M., J. Huijgen, and R. J. Dolan (2012). Prefrontal contributions to metacognition in perceptual decision making. *J. Neurosci. 32*, 6117–6125.

Fleming, S. M., J. Ryu, J. G. Golfinos, and K. E. Blackmon (2014). Domain-specific impairment in metacognitive accuracy following anterior prefrontal lesions. *Brain 137*, 2811–2822.

Fodor, J. A. (1975). *The Language of Thought* (Vol. 5). Cambridge, MA: Harvard University Press.

Fodor, J. A. (1983). *The Modularity of Mind: An Essay on Faculty Psychology*. Cambridge, MA: MIT Press.

Fornito, A., A. Arnatkevičiūtė, and B. D. Fulcher (2019). Bridging the gap between connectome and transcriptome. *Trends Cogn. Sci. 23*, 34–50.

Fornito, A., A. Zalesky, and M. Breakspear (2015). The connectomics of brain disorders. *Nat. Rev. Neurosci. 16*, 159–172.

Fornito, A., A. Zalesky, and E. Bullmore (2016). *Fundamentals of Brain Network Analysis*. London: Academic Press.

Forstmann, B. U., G. Dutilh, S. Brown, J. Neumann, D. Y. von Cramon, K. R. Ridderinkhof, and E. J. Wagenmakers (2008). Striatum and pre-SMA facilitate decision-making under time pressure. *Proc. Natl. Acad. Sci. U.S.A. 105*, 17538–17542.

Forstmann, B. U., R. Ratcliff, and E. J. Wagenmakers (2016). Sequential sampling models in cognitive neuroscience: Advantages, applications, and extensions. *Annu. Rev. Psychol. 67*, 641–666.

Foss-Feig, J. H., B. D. Adkinson, J. L. Ji, G. Yang, V. H. Srihari, J. C. McPartland, J. H. Krystal, J. D. Murray, and A. Anticevic (2017). Searching for cross-diagnostic convergence: neural mechanisms governing excitation and inhibition balance in schizophrenia and autism spectrum disorders. *Biol. Psychiatry 81*, 848–861.

Foster, D. J. (2017). Replay comes of age. *Annu. Rev. Neurosci. 40*, 581–602.

Foster, D. J. and M. A. Wilson (2006). Reverse replay of behavioural sequences in hippocampal place cells during the awake state. *Nature 440*, 680–683.

Foster, N. N., J. Barry, L. Korobkova, L. Garcia, L. Gao, M. Becerra, Y. Sherafat, B. Peng, X. Li, J.-H. Choi, et al. (2021). The mouse cortico–basal ganglia–thalamic network. *Nature 598*, 188–194.

Fourcaud-Trocme, N., D. Hansel, C. van Vreeswijk, and N. Brunel (2003). How spike generation mechanisms determine the neuronal response to fluctuating inputs. *J. Neurosci. 23*, 11628–11640.

Fox, M. D., A. Z. Snyder, J. L. Vincent, M. Corbetta, D. C. Van Essen, and M. E. Raichle (2005). The human brain is intrinsically organized into dynamic, anticorrelated functional networks. *Proc. Natl. Acad. Sci. U.S.A. 102*, 9673–9678.

Frank, M. C. (2023). Baby steps in evaluating the capacities of large language models. *Nat. Rev. Psychol.* https://doi.org/10.1038/s44159-023-00211-x.

Frank, M. J. (2006). Hold your horses: A dynamic computational role for the subthalamic nucleus in decision making. *Neural Netw. 19*, 1120–1136.

Frank, M. J., B. Loughry, and R. C. O'Reilly (2001). Interactions between frontal cortex and basal ganglia in working memory: A computational model. *Cogn. Affect. Behav. Neurosci. 1*, 137–160.

Frank, M. J., J. Samanta, A. A. Moustafa, and S. J. Sherman (2007). Hold your horses: Impulsivity, deep brain stimulation, and medication in parkinsonism. *Science 318*, 1309–1312.

Frank, M. J., L. C. Seeberger, and R. C. O'Reilly (2004). By carrot or by stick: Cognitive reinforcement learning in parkinsonism. *Science 306*, 1940–1943.

Frank, R. H. (2003). *Microenomics and Behavior* (5th ed.). New York: McGraw-Hill.

Frankland, P. W. and B. Bontempi (2005). The organization of recent and remote memories. *Nat. Rev. Neurosci. 6*, 119–130.

Frankland, S. M. and J. D. Greene (2020). Concepts and compositionality: In search of the brain's language of thought. *Annu. Rev. Psychol. 71*, 273–303.

Fransén, E., B. Tahvildari, A. V. Egorov, M. E. Hasselmo, and A. A. Alonso (2006). Mechanism of graded persistent cellular activity of entorhinal cortex layer v neurons. *Neuron 49*, 735–746.

Frässle, S., A. F. Marquand, L. Schmaal, R. Dinga, D. J. Veltman, N. J. Van der Wee, M.-J. van Tol, D. Schöbi, B. W. Penninx, and K. E. Stephan (2020). Predicting individual clinical trajectories of depression with generative embedding. *NeuroImage Clin. 26*, 102213.

Frederick, S., G. Loewenstein, and T. O'donoghue (2002). Time discounting and time preference: A critical review. *J. Econ. Lit. 40*, 351–401.

Freedman, D. J. and J. A. Assad (2006). Experience-dependent representation of visual categories in parietal cortex. *Nature 443*, 85–88.

Freedman, D. J. and J. A. Assad (2016). Neuronal mechanisms of visual categorization: An abstract view on decision making. *Annu. Rev. Neurosci. 39*, 129–147.

Freedman, D. J. and E. K. Miller (2008). Neural mechanisms of visual categorization: Insights from neurophysiology. *Neurosci. Biobehav. Rev. 32*, 311–329.

Freedman, D. J., M. Riesenhuber, T. Poggio, and E. K. Miller (2001). Categorical representation of visual stimuli in the primate prefrontal cortex. *Science 291*, 312–316.

Freedman, R., D. A. Lewis, R. Michels, D. S. Pine, S. K. Schultz, C. A. Tamminga, G. O. Gabbard, S. S.-F. Gau, D. C. Javitt, M. A. Oquendo, et al. (2013). The initial field trials of dsm-5: New blooms and old thorns. *Am. J. Psychiatry* 170, 1–5.

Freeman, W. J. (1975). *Mass Action in the Nervous System*. New York: Academic Press.

Freeman, W. J. and J. Barrie (2000). Analysis of spatial patterns of phase in neocortical gamma EEGs in rabbit. *J. Neurophysiol. 84*, 1266–1278.

Frégnac, Y. (2017). Big data and the industrialization of neuroscience: A safe roadmap for understanding the brain? *Science*, 470–477.

Frégnac, Y. (2021). How blue is the sky. *eNeuro 8*, e0130.

Frémaux, N., H. Sprekeler, and W. Gerstner (2010). Functional requirements for reward-modulated spike-timing-dependent plasticity. *J. Neurosci. 30*, 13326–13337.

French, R. M. (1999). Catastrophic forgetting in connectionist networks. *Trends Cogn. Sci. 3*, 128–135.

Freund, T. F. and G. Buzsáki (1996). Interneurons of the hippocampus. *Hippocampus 6*, 347–470.

Freund, T. F., I. Katona, and D. Piomelli (2003). Role of endogenous cannabinoids in synaptic signaling. *Physiol. Rev. 83*, 1017–1066.

Freund, T. F., J. F. Powell, and A. D. Smith (1984). Tyrosine hydroxylase-immunoreactive boutons in synaptic contact with identified striatonigral neurons, with particular reference to dendritic spines. *Neuroscience 13*, 1189–1215.

Frey, U. and R. G. Morris (1997). Synaptic tagging and long-term potentiation. *Nature 385*, 533–536.

Fries, P. (2005). A mechanism for cognitive dynamics: Neuronal communication through neuronal coherence. *Trends Cogn. Sci. 9*, 474–480.

Fries, P., J. Reynolds, A. Rorie, and R. Desimone (2001). Modulation of oscillatory neuronal synchronization by selective visual attention. *Science 291*, 1560–1563.

Fries, P., R. Scheeringa, and R. Oostenveld (2008). Finding gamma. *Neuron 58*, 303–305.

Fries, P., T. Womelsdorf, R. Oostenveld, and R. Desimone (2008). The effects of visual stimulation and selective visual attention on rhythmic neuronal synchronization in macaque area V4. *J. Neurosci. 28*, 4823–4835.

Friston, K. J., L. Harrison, and W. Penny (2003). Dynamic causal modelling. *Neuroimage 19*, 1273–1302.

Friston, K. J., A. D. Redish, and J. A. Gordon (2017). Computational nosology and pecision psychiatry. *Comput. Psychiatr. 1*, 2–23.

Frith, C. (2005). The neural basis of hallucinations and delusions. *C. R. Biol. 328*, 169–175.

Froemke, R. C. (2015). Plasticity of cortical excitatory-inhibitory balance. *Annu. Rev. Neurosci. 38*, 195–219.

Froudist-Walsh, S., D. P. Bliss, X. Ding, L. Rapan, M. Niu, K. Knoblauch, K. Zilles, H. Kennedy, N. Palomero-Gallagher, and X.-J. Wang (2021). A dopamine gradient controls access to distributed working memory in the large-scale monkey cortex. *Neuron 109*, 3500–3520.

Froudist-Walsh, S., T. Xu, M. Niu, L. Rapan, L. Zhao, D. S. Margulies, K. Zilles, X. J. Wang, and N. Palomero-Gallagher (2023). Gradients of neurotransmitter receptor expression in the macaque cortex. *Nat. Neurosci. 26*, 1281–1294.

Fu, Y., J. M. Tucciarone, J. S. Espinosa, N. Sheng, D. P. Darcy, R. A. Nicoll, Z. J. Huang, and M. P. Stryker (2014). A cortical circuit for gain control by behavioral state. *Cell 156*, 1139–1152.

Fu, Z., A. Sajad, S. P. Errington, J. D. Schall, and U. Rutishauser (2023). Neurophysiological mechanisms of error monitoring in human and non-human primates. *Nat. Rev. Neurosci. 24*, 153–172.

Fuhs, M. C. and D. S. Touretzky (2006). A spin glass model of path integration in rat medial entorhinal cortex. *J. Neurosci. 26*, 4266–4276.

Fukushima, K. (1988). Neocognitron: A hierarchical neural network capable of visual pattern recognition. *Neural Netw. 1*, 119–130.

Fukushima, K. and S. Miyake (1982). Neocognitron: A new algorithm for pattern recognition tolerant of deformations and shifts in position. *Pattern Recognit. 15*, 455–469.

Fulcher, B. D., J. D. Murray, V. Zerbi, and X. J. Wang (2019). Multimodal gradients across mouse cortex. *Proc. Natl. Acad. Sci. U.S.A. 116*, 4689–4695.

Funahashi, S. (2013). Thalamic mediodorsal nucleus and its participation in spatial working memory processes: Comparison with the prefrontal cortex. *Front. Syst. Neurosci. 7*, 36.

Funahashi, S., C. J. Bruce, and P. S. Goldman-Rakic (1989). Mnemonic coding of visual space in the monkey's dorsolateral prefrontal cortex. *J. Neurophysiol. 61*, 331–349.

Funahashi, S., M. V. Chafee, and P. S. Goldman-Rakic (1993). Prefrontal neuronal activity in rhesus monkeys performing a delayed anti-saccade task. *Nature 365*, 753–756.

Funahashi, S., K. Takeda, and Y. Watanabe (2004). Neural mechanisms of spatial working memory: Contributions of the dorsolateral prefrontal cortex and the thalamic mediodorsal nucleus. *Cogn. Affect. Behav. Neurosci. 4*, 409–420.

Furman, M. and X.-J. Wang (2008). Similarity effect and optimal control of multiple-choice decision making. *Neuron 60*, 1153–1168.

Fusi, S. (2002). Hebbian spike-driven synaptic plasticity for learning patterns of mean firing rates. *Biol. Cybern. 87*, 459–470.

Fusi, S., W. F. Asaad, E. K. Miller, and X.-J. Wang (2007). A neural circuit model of flexible sensorimotor mapping: Learning and forgetting on multiple timescales. *Neuron 54*, 319–333.

Fusi, S., P. J. Drew, and L. F. Abbott (2005). Cascade models of synaptically stored memories. *Neuron 45*, 599–611.

Fusi, S., E. K. Miller, and M. Rigotti (2016). Why neurons mix: High dimensionality for higher cognition. *Curr. Opin. Neurobiol. 37*, 66–74.

Fusi, S. and X.-J. Wang (2016). Short-term, long-term, and working memory. In M. A. Arbib and J. J. Bonaiuto (Eds.), *From Neuron to Cognition via Computational Neuroscience*, Chapter 11, pp. 319–344. Cambridge, MA: MIT Press.

Fuster, J. M. (1985). The prefrontal cortex, mediator of cross-temporal contingencies. *Hum. Neurobiol. 4*, 169–179.

Fuster, J. M. (1989). *The Prefrontal Cortex* (2nd ed.). New York: Academic Press.

Fuster, J. M. (2013). *The Neuroscience of Freedom and Creativity: Our Predictive Brain*. Cambridge: Cambridge University Press.

Fuster, J. M. and G. E. Alexander (1971). Neuron activity related to short-term memory. *Science 173*, 652–654.

Fuster, J. M. and G. E. Alexander (1973). Firing changes in cells of the nucleus medialis dorsalis associated with delayed response behavior. *Brain Res. 61*, 79–91.

Fuster, J. M. and J. P. Jervey (1981). Inferotemporal neurons distinguish and retain behaviourally relevant features of visual stimuli. *Science 212*, 952–955.

Gabbiani, F. and H. G. Krapp (2006). Spike-frequency adaptation and intrinsic properties of an identified, looming-sensitive neuron. *J. Neurophysiol. 96*, 2951–2962.

Gabbott, P. L. A. and S. J. Bacon (1996). Local circuit neurons in the medial prefrontal cortex (areas 24a,b,c, 25 and 32) in the monkey. *J. Comp. Neurol. 364*, 567–636.

Galakhova, A., S. Hunt, R. Wilbers, D. Heyer, C. de Kock, H. Mansvelder, and N. Goriounova (2022). Evolution of cortical neurons supporting human cognition. *Trends Cogn. Sci. 26*, 909–922.

Gallego, J. A., M. G. Perich, L. E. Miller, and S. A. Solla (2017). Neural manifolds for the control of movement. *Neuron 94*, 978–984.

Gallero-Salas, Y., S. Han, Y. Sych, F. F. Voigt, B. Laurenczy, A. Gilad, and F. Helmchen (2021). Sensory and behavioral components of neocortical signal flow in discrimination tasks with short-term memory. *Neuron 109*, 135–148.

Gallistel, C. R. (1990). *The Organization of Learning (Learning, Development, and Conceptual Change)*. Cambridge, MA: MIT Press.

Gallistel, C. R. and I. Gelman (2000). Non-verbal numerical cognition: From reals to integers. *Trends Cogn. Sci. 4*, 59–65.

Gămănuţ, R., H. Kennedy, Z. Toroczkai, M. Ercsey-Ravasz, D. C. Van Essen, K. Knoblauch, and A. Burkhalter (2018). The mouse cortical connectome, characterized by an ultra-dense cortical graph, maintains specificity by distinct connectivity profiles. *Neuron 97*, 698–715.

Gandal, M. J., J. R. Haney, N. N. Parikshak, V. Leppa, G. Ramaswami, C. Hartl, A. J. Schork, V. Appadurai, A. Buil, T. M. Werge, C. Liu, K. P. White, S. Horvath, D. H. Geschwind, N. Sestan, F. Vaccarino, M. Gerstein, S. Weissman, S. Pochareddy, M. State, J. Knowles, P. Farnham, S. Akbarian, D. Pinto, H. Van Baekl, S. Dracheva, A. Jaffe, T. Hyde, P. Zandi, G. Crawford, P. Sullivan, W. K. Thompson, P. B. Mortensen, E. Agerbo, M. G. Pedersen, C. B. Pedersen, O. Mors, A. D. Børglum, M. Nordentoft, D. M. Hougaard, J. Bybjerg-Grauholm, M. Bækvad-Hansen, A. R. Martin, A. Dumont, C. Stevens, C. Churchhouse, D. P. Howrigan, D. S. Palmer, E. B. Robinson, K. F. Satterstrom, F. Cerrato, H. Huang, J. Goldstein, J. Moran, J. Martin, J. Maller, K. Chambert, P. Turley, R. Walters, R. Belliveau, S. Ripke, T. Poterba, M. J. Daly, B. Neale, M. Fromer, P. Roussos, J. S. Johnson, H. R. Shah, M. C. Mahajan, E. Schadt, V. Haroutunian, D. M. Ruderfer, J. D. Buxbaum, S. K. Sieberts, K. Dang, B. Logsdon, L. M. Mangravite, M. Peters, R. E. Gur, C. G. Hahn, B. Devlin, L. L. Klei, D. Lewis, B. Lipska, K. Hirai, H. Toyoshiba, E. Domenici, and P. Sklar (2018). Shared molecular neuropathology across major psychiatric disorders parallels polygenic overlap. *Science 359*, 693–697.

Ganguli, S., J. Bisley, J. Roitman, M. Shadlen, M. Goldberg, and K. Miller (2008). One-dimensional dynamics of attention and decision making in LIP. *Neuron 58*, 15–25.

Ganguli, S. and H. Sompolinsky (2012). Compressed sensing, sparsity, and dimensionality in neuronal information processing and data analysis. *Annu. Rev. Neurosci. 35*, 485–508.

Gao, P. and S. Ganguli (2015). On simplicity and complexity in the brave new world of large-scale neuroscience. *Curr. Opin. Neurobiol. 32*, 148–155.

Gao, R., R. L. van den Brink, T. Pfeffer, and B. Voytek (2020). Neuronal timescales are functionally dynamic and shaped by cortical microarchitecture. *Elife 9*, e61277.

Gao, W.-J., Y. Wang, and P. S. Goldman-Rakic (2003). Dopamine modulation of perisomatic and peridendritic inhibition in prefrontal cortex. *J. Neurosci. 23*, 1622–1630.

Gao, Z., H. Wang, C. Lu, T. Lu, S. Froudist-Walsh, M. Chen, X.-J. Wang, J. Hu, and W. Sun (2021). The neural basis of delayed gratification. *Sci. Adv. 7*, eabg6611.

Garcia Del Molino, L. C., G. R. Yang, J. F. Mejias, and X. J. Wang (2017). Paradoxical response reversal of top-down modulation in cortical circuits with three interneuron types. *Elife 6*, e29742.

Gardiner, G. W. (Ed.) (1985). *Handbook of Stochastic Methods* (2nd ed.). New York: Springer.

Gardner, H. E. (2011). *Frames of Mind: The Theory of Multiple Intelligences*. New York: Basic Books.

Garrido, M. I., J. M. Kilner, S. J. Kiebel, and K. J. Friston (2007). Evoked brain responses are generated by feedback loops. *Proc. Natl. Acad. Sci. U.S.A. 104*, 20961–20966.

Garrido, M. I., J. M. Kilner, K. E. Stephan, and K. J. Friston (2009). The mismatch negativity: A review of underlying mechanisms. *Clin. Neurophysiol. 120*, 453–463.

Garvert, M. M., R. J. Dolan, and T. E. Behrens (2017). A map of abstract relational knowledge in the human hippocampal–entorhinal cortex. *Elife 6*, e17086.

Gazzaley, A. and A. C. Nobre (2012). Top-down modulation: Bridging selective attention and working memory. *Trends Cogn. Sci. 16*, 129–135.

Gazzaniga, M. S., R. B. Ivry, and G. R. Mangun (2013). *Cognitive Neuroscience: The Biology of the Mind* (4th ed.). New York: W. W. Norton & Company.

Gehring, W. J., B. Goss, M. G. Coles, D. E. Meyer, and E. Donchin (1993). A neural system for error detection and compensation. *Psychol. Sci. 4*, 385–390.

Geisler, C., N. Brunel, and X.-J. Wang (2005). Contributions of intrinsic membrane dynamics to fast network oscillations with irregular neuronal discharges. *J. Neurophysiol. 94*, 4344–4361.

Genkin, M., K. V. Shenoy, C. Chandrasekaran, and T. A. Engel (2023). The dynamics and geometry of choice in premotor cortex. *BioRxiv*. https://doi.org/10.1101/2023.07.22.550183.

Gentner, T. Q., K. M. Fenn, D. Margoliash, and H. C. Nusbaum (2006). Recursive syntactic pattern learning by songbirds. *Nature 440*, 1204–1207.

Georgopoulos, A. P., J. Kalaska, and R. Caminiti (1982). On the relations between the direction of two-dimensional arm movements and cell discharge in primate motor cortex. *J. Neurosci. 2*, 1527–1537.

Georgopoulos, A. P., A. B. Schwartz, and R. E. Kettner (1986). Neuronal population coding of movement direction. *Science 233*, 1416–1419.

Gers, F. A., J. Schmidhuber, and F. Cummins (2000). Learning to forget: Continual prediction with LSTM. *Neural Comput. 12*(10), 2451–2471.

Gerstein, G. L. and D. H. Perkel (1972). Mutual temporal relationships among neuronal spike trains. Statistical techniques for display and analysis. *Biophys. J. 12*, 453–473.

Gerstner, W., R. Kempter, J. van Hemmen, and H. Wagner (1996). A neuronal learning rule for sub-millisecond temporal coding. *Nature 383*, 76–81.

Gerstner, W., W. M. Kistler, R. Naud, and L. Paninski (2014). *Neuronal Dynamics: From Single Neurons to Networks and Models of Cognition*. Cambridge: Cambridge University Press.

Getting, P. A. (1989). Emerging principles governing the operation of neural networks. *Annu. Rev. Neurosci. 12*, 185–204.

Gibbon, J. (1977). Scalar expectancy theory and Weber's law in animal timing. *Psychol. Rev. 84*, 279–325.

Gibbon, J., C. Malapani, C. L. Dale, and C. R. Gallistel (1997). Toward a neurobiology of temporal cognition: Advances and challenges. *Curr. Opinion. Neurobiol. 7*, 170–184.

Gibson, J. J. (1979). *The Ecological Approach to Visual Perception*. Boston, MA: Houghton Mifflin.

Gibson, J. J., M. Beierlein, and B. Connors (1999). Two networks of electrically coupled inhibitory neurons in neocortex. *Nature 402*, 75–79.

Gidon, A., T. A. Zolnik, P. Fidzinski, F. Bolduan, A. Papoutsi, P. Poirazi, M. Holtkamp, I. Vida, and M. E. Larkum (2020). Dendritic action potentials and computation in human layer 2/3 cortical neurons. *Science 367*, 83–87.

Gigante, G., M. Mattia, J. Braun, and P. Del Giudice (2009). Bistable perception modeled as competing stochastic integrations at two levels. *PLoS Comput. Biol. 5*, e1000430.

Gigerenzer, G. and W. Gaissmaier (2011). Heuristic decision making. *Annu. Rev. Psychol. 62*, 451–482.

Gigerenzer, G. and U. Hoffrage (1995). How to improve bayesian reasoning without instruction: Frequency formats. *Psychol. Rev. 102*, 684–704.

Gigerenzer, G., U. Hoffrage, and H. Kleinbolting (1991). Probabilistic mental models: A Brunswikian theory of confidence. *Psychol. Rev. 98*, 506–528.

Gigerenzer, G., P. M. Todd, and A. R. Group (2000). *Simple Heuristics That Make Us Smart*. Oxford: Oxford University Press.

Gilbert, C. D. (1983). Microcircuitry of the visual cortex. *Annu. Rev. Neurosci. 6*, 217–247.

Gilbert, C. D. and W. Li (2013). Top-down influences on visual processing. *Nat. Rev. Neurosci. 14*, 350–363.

Gilboa, A. and H. Marlatte (2017). Neurobiology of schemas and schema-mediated memory. *Trends Cogn. Sci. 21*, 618–631.

Gil-da Costa, R., G. R. Stoner, R. Fung, and T. D. Albright (2013). Nonhuman primate model of schizophrenia using a noninvasive EEG method. *Proc. Natl. Acad. Sci. U.S.A. 110*, 15425–15430.

Gillan, C. M., M. Kosinski, R. Whelan, E. A. Phelps, and N. D. Daw (2016). Characterizing a psychiatric symptom dimension related to deficits in goal-directed control. *Elife 5*, e11305.

Gillan, C. M., S. Morein-Zamir, G. P. Urcelay, A. Sule, V. Voon, A. M. Apergis-Schoute, N. A. Fineberg, B. J. Sahakian, and T. W. Robbins (2014). Enhanced avoidance habits in obsessive-compulsive disorder. *Biol. Psychiatry 75*, 631–638.

Gilman, J. P., M. Medalla, and J. I. Luebke (2017). Area-specific features of pyramidal neurons-a comparative study in mouse and rhesus monkey. *Cereb. Cortex 27*, 2078–2094.

Girshick, A. R., M. S. Landy, and E. P. Simoncelli (2011). Cardinal rules: Visual orientation perception reflects knowledge of environmental statistics. *Nat. Neurosci. 14*, 926–932.

Glantz, L. A. and D. A. Lewis (2000). Decreased dendritic spine density on prefrontal cortical pyramidal neurons in schizophrenia. *Arch. Gen. Psychiatry 57*, 65–73.

Glasser, M. F., T. S. Coalson, E. C. Robinson, C. D. Hacker, J. Harwell, E. Yacoub, K. Ugurbil, J. Andersson, C. F. Beckmann, M. Jenkinson, S. M. Smith, and D. C. Van Essen (2016). A multi-modal parcellation of human cerebral cortex. *Nature 536*, 171–178.

Glasser, M. F. and D. C. Van Essen (2011). Mapping human cortical areas in vivo based on myelin content as revealed by T1- and T2-weighted MRI. *J. Neurosci. 31*, 11597–11616.

Gleeson, P., S. Crook, D. Turner, K. Mantel, M. Raunak, T. Willke, and J. D. Cohen (2023). Integrating model development across computational neuroscience, cognitive science, and machine learning. *Neuron 111*, 1526–1530.

Glimcher, P. W. (2003). *Decisions, Uncertainty, and the Brain: The Science of Neuroeconomics*. Cambridge, MA: MIT Press.

Glimcher, P. W. (2005). Indeterminacy in brain and behavior. *Annu. Rev. Psychol. 56*, 25–56.

Glimcher, P. W. (2011). Understanding dopamine and reinforcement learning: The dopamine reward prediction error hypothesis. *Proc. Natl. Acad. Sci. U.S.A. 108*(Suppl 3), 15647–15654.

Glimcher, P. W. (2022). Efficiently irrational: Deciphering the riddle of human choice. *Trends Cogn. Sci. 26*, 669–687.

Gluck, M. A. and G. H. Bower (1988). From conditioning to category learning: An adaptive network model. *J. Exp. Psychol. Gen. 117*, 227–247.

Gluck, M. A., D. Shohamy, and C. Myers (2002). How do people solve the "weather prediction" task?: Individual variability in strategies for probabilistic category learning. *Learn. Mem. 9*, 408–418.

Gluth, S., N. Kern, M. Kortmann, and C. L. Vitali (2020). Value-based attention but not divisive normalization influences decisions with multiple alternatives. *Nat. Hum. Behav. 4*, 634–645.

Gnadt, J. W. and R. A. Andersen (1988). Memory related motor planning activity in posterior parietal cortex of macaque. *Exp. Brain Res. 70*, 216–220.

Goetz, L., A. Roth, and M. Häusser (2021). Active dendrites enable strong but sparse inputs to determine orientation selectivity. *Proc. Natl. Acad. Sci. 118*. https://doi.org/10.1073/pnas.2017339118.

Gold, J. I. and M. N. Shadlen (2001). Neural computations that underlie decisions about sensory stimuli. *Trends Cogn. Sci. 5*, 10–16.

Gold, J. I. and M. N. Shadlen (2002). Banburismus and the brain: Decoding the relationship between sensory stimuli, decisions, and reward. *Neuron 36*, 299–308.

Gold, J. I. and M. N. Shadlen (2007). The neural basis of decision making. *Annu. Rev. Neurosci. 30*, 535–574.

Goldman, M. S. (2009). Memory without feedback in a neural network. *Neuron 61*, 621–634.

Goldman, M. S., A. Compte, and X.-J. Wang (2009). Neural integrators: Recurrent mechanisms and models. In L. Squire, T. Albright, F. Bloom, F. Gage, and N. Spitzer (Eds.), *Encyclopedia of Neuroscience*, pp. 165–178. London: MacMillan Reference Ltd.

Goldman, M. S., J. H. Levine, G. Major, D. W. Tank, and H. S. Seung (2003). Robust persistent neural activity in a model integrator with multiple hysteretic dendrites per neuron. *Cereb. Cortex 13*, 1185–1195.

Goldman-Rakic, P. S. (1987a). Circuitry of primate prefrontal cortex and regulation of behavior by representational memory. In F. Plum and V. Mountcastle (Eds.), *Handbook of Physiology—The Nervous System V*, Chapter 9, pp. 373–417. Bethesda, MD: American Physiological Society.

Goldman-Rakic, P. S. (1987b). Development of cortical circuitry and cognitive function. *Child Dev. 58*, 601–622.

Goldman-Rakic, P. S. (1992). Working memory and the mind. *Sci. Am. 267*, 110–117.

Goldman-Rakic, P. S. (1994). Working memory dysfunction in schizophrenia. *J. Neuropsych. Clin. Neurosci. 6*, 348–357.

Goldman-Rakic, P. S. (1995). Cellular basis of working memory. *Neuron 14*, 477–485.

Goldstein, R. Z. and N. D. Volkow (2011). Dysfunction of the prefrontal cortex in addiction: Neuroimaging findings and clinical implications. *Nat. Rev. Neurosci. 12*, 652–669.

Goldstone, R. L. and A. T. Hendrickson (2010). Categorical perception. *Wiley Interdiscip. Rev. Cogn. Sci. 1*, 69–78.

Goleman, D. (1996). *Emotional Intelligence: Why It Can Matter More Than IQ*. London: Bloomsbury Publishing.

Golomb, D. and D. Hansel (2000). The number of synaptic inputs and the synchrony of large sparse neuronal networks. *Neural Comput. 12*, 1095–1139.

Gomez-Marin, A. and A. A. Ghazanfar (2019). The life of behavior. *Neuron 104*, 25–36.

Gonchar, Y. and A. Burkhalter (1999). Connectivity of GABAergic calretinin-immunoreactive neurons in rat primary visual cortex. *Cereb. Cortex 9*, 683–696.

González-Burgos, G., G. Barrionuevo, and D. A. Lewis (2000). Horizontal synaptic connections in monkey prefrontal cortex: An in vitro electrophysiological study. *Cereb. Cortex 10*, 82–92.

Gonzalez-Burgos, G. and D. Lewis (2008). GABA neurons and the mechanisms of network oscillations: Implications for understanding cortical dysfunction in schizophrenia. *Schizophr. Bull. 34*, 944–961.

Gonzelez-Burgos, G., T. Miyamae, Y. Krimer, Y. Gulchina, D. E. Pafundo, O. Krimer, H. Bazmi, D. Arion, J. F. Enwright, K. N. Fish, and D. A. Lewis (2019). Distinct properties of layer 3 pyramidal neurons from prefrontal and parietal areas of the monkey neocortex. *J. Neurosci. 39*, 7277–7290.

Goodfellow, I., Y. Bengio, and A. Courville (2016). *Deep Learning*. Cambridge, MA: MIT Press.

Goodnow, J. J. and R. A. Levine (1974). 'The grammar of action': Sequence and syntax in children's copying. *Cogn. Psych. 4*, 82–98.

Goriounova, N. A., D. B. Heyer, R. Wilbers, M. B. Verhoog, M. Giugliano, C. Verbist, J. Obermayer, A. Kerkhofs, H. Smeding, M. Verberne, et al. (2018). Large and fast human pyramidal neurons associate with intelligence. *Elife 7*, e41714.

Goris, R. L., J. A. Movshon, and E. P. Simoncelli (2014). Partitioning neuronal variability. *Nat. Neurosci. 17*, 858–865.

Gottesman, I. I. and T. D. Gould (2003). The endophenotype concept in psychiatry: Etymology and strategic intentions. *Am. J. Psychiatry 160*, 636–645.

Gottfredson, L. S. (1997). Mainstream science on intelligence: An editorial with 52 signatories, history, and bibliography. *Intelligence* 24, 13–23.

Goudar, V. and D. V. Buonomano (2018). Encoding sensory and motor patterns as time-invariant trajectories in recurrent neural networks. *eLife 7*, e31134.

Goudar, V., B. Peysakhovich, D. J. Freedman, E. A. Buffalo, and X.-J. Wang (2023). Schema formation in a neural population subspace underlies learning-to-learn in flexible sensorimotor problem-solving. *Nat. Neurosci. 26*, 879–890.

Goulas, A., K. Zilles, and C. C. Hilgetag (2018). Cortical gradients and laminar projections in mammals. *Trends Neurosci. 41*, 775–788.

Gould, S. J. and N. Eldredge (1993). Punctuated equilibrium comes of age. *Nature 366*, 223–227.

Gouwens, N. W., S. A. Sorensen, F. Baftizadeh, A. Budzillo, B. R. Lee, T. Jarsky, L. Alfiler, K. Baker, E. Barkan, K. Berry, et al. (2020). Integrated morphoelectric and transcriptomic classification of cortical gabaergic cells. *Cell 183*, 935–953.

Grace, A. A. (1991). Phasic versus tonic dopamine release and the modulation of dopamine system responsivity: A hypothesis for the etiology of schizophrenia. *Neuroscience 41*, 1–24.

Grange, J. A. and G. Houghton (2014). *Task Switching and Cognitive Control*. New York: Oxford University Press.

Granger, C. W. J. (1969). Investigating causal relations by econometric models and cross-spectral methods. *Econometrics 37*, 424–438.

Gratton, G., M. Coles, E. Sirevaag, C. Eriksen, and E. Donchin (1988). Pre- and poststimulus activation of response channels: A psychophysiological analysis. *J. Exp. Psychol. Hum. Percept. Perform. 14*, 331–344.

Graupner, M. and N. Brunel (2012). Calcium-based plasticity model explains sensitivity of synaptic changes to spike pattern, rate, and dendritic location. *Proc. Natl. Acad. Sci. U.S.A. 109*, 3991–3996.

Gray, C. M. (1994). Synchronous oscillations in neuronal systems: Mechanisms and functions. *J. Comput. Neurosci. 1*, 11–38.

Gray, C. M. and W. Singer (1989). Stimulus-specific neuronal oscillations in orientation columns of cat visual cortex. *Proc. Natl. Acad. Sci. U.S.A. 86*, 1698–1702.

Gray, J. R., C. F. Chabris, and T. S. Braver (2003). Neural mechanisms of general fluid intelligence. *Nat. Neurosci. 6*, 316–322.

Graybiel, A. M. (2008). Habits, rituals, and the evaluative brain. *Annu. Rev. Neurosci. 31*, 359–387.

Graybiel, A. M. and S. L. Rauch (2000). Toward a neurobiology of obsessive-compulsive disorder. *Neuron 28*, 343–347.

Green, D. M. and J. A. Swets (1966). *Signal Detection Theory and Psychophysics*. New York: John Wiley and Sons.

Gregoriou, G. G., S. J. Gotts, H. Zhou, and R. Desimone (2009). High-frequency, long-range coupling between prefrontal and visual cortex during attention. *Science 324*, 1207–1210.

Greicius, M. D., K. Supekar, V. Menon, and R. F. Dougherty (2009). Resting-state functional connectivity reflects structural connectivity in the default mode network. *Cereb. Cortex 19*, 72–78.

Grienberger, C. and A. Konnerth (2012). Imaging calcium in neurons. *Neuron 73*, 862–885.

Grienberger, C. and J. C. Magee (2022). Entorhinal cortex directs learning-related changes in ca1 representations. *Nature 611*, 554–562.

Grillner, S. and B. Robertson (2016). The basal ganglia over 500 million years. *Curr. Biol. 26*, R1088–R1100.

Grillon, C., O. J. Robinson, K. O'Connell, A. Davis, G. Alvarez, D. S. Pine, and M. Ernst (2017). Clinical anxiety promotes excessive response inhibition. *Psychol. Med. 47*, 484–494.

Grimaldi, P., H. Lau, and M. A. Basso (2015). There are things that we know that we know, and there are things that we do not know we do not know: Confidence in decision-making. *Neurosci. Biobehav. Rev. 55*, 88–97.

Guckenheimer, J. and P. Holmes (1983). *Nonlinear Oscillations, Dynamical Systems, and Bifurcations of Vector Fields*. New York: Springer Verlag.

Guitart-Masip, M., Q. J. Huys, L. Fuentemilla, P. Dayan, E. Duzel, and R. J. Dolan (2012). Go and no-go learning in reward and punishment: Interactions between affect and effect. *Neuroimage 62*, 154–166.

Gulyás, A. I., N. Hájos, and T. Freund (1996). Interneurons containing calretinin are specialized to control other interneurons in the rat hippocampus. *J. Neurosci. 16*, 3397–3411.

Guo, Z. V., H. K. Inagaki, K. Daie, S. Druckmann, C. R. Gerfen, and K. Svoboda (2017). Maintenance of persistent activity in a frontal thalamocortical loop. *Nature 545*, 181–186.

Guo, Z. V., N. Li, D. Huber, E. Ophir, D. Gutnisky, J. T. Ting, G. Feng, and K. Svoboda (2014). Flow of cortical activity underlying a tactile decision in mice. *Neuron 81*, 179–194.

Guru, A., C. Seo, R. J. Post, D. S. Kullakanda, J. A. Schaffer, and M. R. Warden (2020). Ramping activity in midbrain dopamine neurons signifies the use of a cognitive map. *bioRxiv*. https://doi.org/10.1101/2020.05.21.108886.

Gutenkunst, R. N., J. J. Waterfall, F. P. Casey, K. S. Brown, C. R. Myers, and J. P. Sethna (2007). Universally sloppy parameter sensitivities in systems biology models. *PLoS Comput. Biol. 3*, 1871–1878.

Haber, S. N., J. L. Fudge, and N. R. McFarland (2000). Striatonigrostriatal pathways in primates form an ascending spiral from the shell to the dorsolateral striatum. *J. Neurosci. 20*, 2369–2382.

Haegeman, L. (2005). *Thinking Syntactically: A Guide to Argumentation and Analysis*. Hoboken, NJ: Wiley-Blackwell.

Hafting, T., M. Fyhn, S. Molden, M. B. Moser, and E. I. Moser (2005). Microstructure of a spatial map in the entorhinal cortex. *Nature 436*, 801–806.

Hagan, M. A. and B. Pesaran (2022). Modulation of inhibitory communication coordinates looking and reaching. *Nature 604*, 708–713.

Hagmann, P., L. Cammoun, X. Gigandet, R. Meuli, C. Honey, V. Wedeen, and O. Sporns (2008). Mapping the structural core of human cerebral cortex. *PLoS Biol. 6*, e159.

Haider, B., M. R. Krause, A. Duque, Y. Yu, J. Touryan, J. A. Mazer, and D. A. McCormick (2010). Synaptic and network mechanisms of sparse and reliable visual cortical activity during nonclassical receptive field stimulation. *Neuron 65*, 107–121.

Haier, R. J. (2017). *The Neuroscience of Intelligence*. Cambridge: Cambridge University Press.

Haj-Dahmane, S. and R. Andrade (1998). Ionic mechanism of the slow afterdepolarization induced by muscarinic receptor activation in rat prefrontal cortex. *J. Neurophysiol. 80*, 1197–1210.

Hall, W. C. and A. Moschovakis (Eds.) (2003). *The Superior Colliculus: New Approaches for Studying Sensorimotor Integration*. New York: CRC Press.

Hallett, P. E. (1978). Primary and secondary saccades to goals defined by instructions. *Vision Res. 18*, 1279–1296.

Hall-McMaster, S., P. S. Muhle-Karbe, N. E. Myers, and M. G. Stokes (2019). Reward boosts neural coding of task rules to optimize cognitive flexibility. *J. Neurosci. 39*, 8549–8561.

Hampshire, A., R. R. Highfield, B. L. Parkin, and A. M. Owen (2012). Fractionating human intelligence. *Neuron 76*, 1225–1237.

Hanes, D. P., W. F. Patterson, and J. D. Schall (1998). Role of frontal eye fields in countermanding saccades: Visual, movement, and fixation activity. *J. Neurophysiol. 79*, 817–834.

Hanes, D. P. and J. D. Schall (1996). Neural control of voluntary movement initiation. *Science 274*, 427–430.

Hangya, B., J. L. Sanders, and A. Kepecs (2016). A mathematical framework for statistical decision confidence. *Neural Comput.*, 1–18.

Hanks, T. D., J. Ditterich, and M. N. Shadlen (2006). Microstimulation of macaque area LIP affects decision-making in a motion discrimination task. *Nat. Neurosci. 9*, 682–689.

Hanks, T. D., R. Kiani, R. Kira, and M. N. Shadlen (2007). Evidence for the incorporation of psychological urgency in decision making. *Soc. Neurosci. Abstr. 38*, 507.4.

Hanks, T. D., R. Kiani, and M. N. Shadlen (2014). A neural mechanism of speed-accuracy tradeoff in macaque area LIP. *Elife 3*, e02260.

Hanks, T. D., C. D. Kopec, B. W. Brunton, C. A. Duan, J. C. Erlich, and C. D. Brody (2015). Distinct relationships of parietal and prefrontal cortices to evidence accumulation. *Nature 520*, 220–223.

Hansel, D. and G. Mato (2003). Asynchronous states and the emergence of synchrony in large networks of interacting excitatory and inhibitory neurons. *Neural Comput. 15*, 1–56.

Hansel, D. and G. Mato (2013). Short-term plasticity explains irregular persistent activity in working memory tasks. *J. Neurosci. 33*, 133–149.

Hansel, D., G. Mato, and C. Meunier (1995). Synchrony in excitatory neural networks. *Neural Comput. 7*, 307–337.

Hansel, D. and H. Sompolinsky (1998). Modeling feature selectivity in local cortical circuits. In C. Koch and I. Segev (Eds.), *Methods in Neuronal Modeling* (2nd ed.), pp. 499–567. Cambridge, MA: MIT Press.

Harlow, H. F. (1949). The formation of learning sets. *Psychol. Rev. 56*, 51–65.

Harlow, J. M. (1868). Recovery from the passage of an iron bar through the head. *Publ. Mass. Med. Soc. 2*, 327–346.

Harnard, S. (1987). *Categorical Perception: The Groundwork of Cognition*. New York: Cambridge University Press.

Haroush, K. and Z. M. Williams (2015). Neuronal prediction of opponent's behavior during cooperative social interchange in primates. *Cell 160*, 1233–1245.

Harris, J. A., S. Mihalas, K. E. Hirokawa, J. D. Whitesell, H. Choi, A. Bernard, P. Bohn, S. Caldejon, L. Casal, A. Cho, et al. (2019). Hierarchical organization of cortical and thalamic connectivity. *Nature 575*, 195–202.

Harris, K. D. and T. D. Mrsic-Flogel (2013). Cortical connectivity and sensory coding. *Nature 503*, 51–58.

Harris, K. D. and G. M. Shepherd (2015). The neocortical circuit: Themes and variations. *Nat. Neurosci. 18*, 170–181.

Harrison, S. A. and F. Tong (2009). Decoding reveals the contents of visual working memory in early visual areas. *Nature 458*, 632–635.

Harvey, C. D., P. Coen, and D. W. Tank (2012). Choice-specific sequences in parietal cortex during a virtual-navigation decision task. *Nature 484*, 62–68.

Hassabis, D., D. Kumaran, C. Summerfield, and M. Botvinick (2017). Neuroscience-inspired artificial intelligence. *Neuron 95*, 245–258.

Hasselmo, M. E. and C. E. Stern (2018). A network model of behavioural performance in a rule learning task. *Philos. Trans. R. Soc. B: Biol. Sci. 373*, 20170275.

Hasson, U., J. Chen, and C. J. Honey (2015). Hierarchical process memory: Memory as an integral component of information processing. *Trends Cogn. Sci. (Regul. Ed.) 19*, 304–313.

Hasson, U., E. Yang, I. Vallines, D. J. Heeger, and N. Rubin (2008). A hierarchy of temporal receptive windows in human cortex. *J. Neurosci. 28*, 2539–2550.

Hattori, R., B. Danskin, Z. Babic, N. Mlynaryk, and T. Komiyama (2019). Area-specificity and plasticity of history-dependent value coding during learning. *Cell 177*, 1858–1872.

Hauser, M. D., N. Chomsky, and W. T. Fitch (2002). The faculty of language: What is it, who has it, and how did it evolve? *Science 298*, 1569–1579.

Hauser, M. D., C. Yang, R. C. Berwick, I. Tattersall, M. J. Ryan, J. Watumull, N. Chomsky, and R. C. Lewontin (2014). The mystery of language evolution. *Front. Psychol. 5*, 401.

Hauser, T. U., G.-J. Will, M. Dubois, and R. J. Dolan (2019). Annual research review: Developmental computational psychiatry. *J. Child Psychol. Psychiatry 60*, 412–426.

Hawkins, G. E., B. U. Forstmann, E. J. Wagenmakers, R. Ratcliff, and S. D. Brown (2015). Revisiting the evidence for collapsing boundaries and urgency signals in perceptual decision-making. *J. Neurosci. 35*, 2476–2484.

Hawrylycz, M., J. A. Miller, V. Menon, D. Feng, T. Dolbeare, A. L. Guillozet-Bongaarts, A. G. Jegga, B. J. Aronow, C.-K. Lee, A. Bernard, M. F. Glasser, D. L. Dierker, J. Menche, A. Szafer, F. Collman, P. Grange, K. A. Berman, S. Mihalas, Z. Yao, L. Stewart, A.-L. Barabási, J. Schulkin, J. Phillips, L. Ng, C. Dang, D. R. Haynor, A. Jones, D. C. Van Essen, C. Koch, and E. Lein (2015). Canonical genetic signatures of the adult human brain. *Nat. Neurosci. 18*(12), 1832–1844.

Haykin, S. S. (2002). *Adaptive Filter Theory*. Upper Saddle River, NJ: Prentice Hall.

He, B. J. (2014). Scale-free brain activity: Past, present, and future. *Trends Cogn. Sci. (Regul. Ed.) 18*, 480–487.

He, B. J., J. M. Zempel, A. Z. Snyder, and M. E. Raichle (2010). The temporal structures and functional significance of scale-free brain activity. *Neuron 66*, 353–369.

Hebb, D. O. (1949). *Organization of Behavior*. New York: Wiley.

Heeger, D. J. (1992). Normalization of cell responses in cat striate cortex. *Vis. Neurosci. 9*, 181–197.

Heeger, D. J., E. P. Simoncelli, and J. A. Movshon (1996). Computational models of cortical visual processing. *Proc. Natl. Acad. Sci. U.S.A. 93*, 623–627.

Heekeren, H. R., S. Marrett, P. A. Bandettini, and L. G. Ungerleider (2004). A general mechanism for perceptual decision-making in the human brain. *Nature 431*, 859–862.

Heekeren, H. R., S. Marrett, and L. Ungerleider (2008). The neural systems that mediate human perceptual decision making. *Nat. Rev. Neurosci. 9*, 467–479.

Heilbron, M. and M. Chait (2018). Great expectations: Is there evidence for predictive coding in auditory cortex? *Neuroscience 389*, 54–73.

Hein, G. and R. T. Knight (2008). Superior temporal sulcus-it's my area: Or is it? *J. Cogn. Neurosci. 20*, 2125–2136.

Heinbockel, T., D. H. Brager, C. G. Reich, J. Zhao, S. Muralidharan, B. E. Alger, and J. P. Kao (2005). Endocannabinoid signaling dynamics probed with optical tools. *J. Neurosci. 25*, 9449–9459.

Heinz, A. (2017). *A New Understanding of Mental Disorders: Computational Models for Dimensional Psychiatry*. Cambridge, MA: MIT Press.

Heinzle, J., K. Hepp, and K. Martin (2007). A microcircuit model of the frontal eye fields. *J. Neurosci. 27*, 9341–9353.

Heisenberg, M. (2009). Is free will an illusion? *Nature 459*, 164–165.

Heitz, R. P., J. Y. Cohen, G. F. Woodman, and J. D. Schall (2010). Neural correlates of correct and errant attentional selection revealed through N2pc and frontal eye field activity. *J. Neurophysiol. 104*, 2433–2441.

Heitz, R. P. and J. D. Schall (2012). Neural mechanisms of speed-accuracy tradeoff. *Neuron 76*, 616–628.

Helfrich, R. F., I. C. Fiebelkorn, S. M. Szczepanski, J. J. Lin, J. Parvizi, R. T. Knight, and S. Kastner (2018). Neural Mechanisms of Sustained Attention Are Rhythmic. *Neuron 99*, 854–865.

Helmchen, F. and W. Denk (2005). Deep tissue two-photon microscopy. *Nat. Methods 2*, 932–940.

Helmchen, F., K. Imoto, and B. Sakmann (1996). Ca2+ buffering and action potential-evoked Ca2+ signaling in dendrites of pyramidal neurons. *Biophys. J. 70*, 1069–1081.

Helmstaedter, M., K. L. Briggman, and W. Denk (2008). 3D structural imaging of the brain with photons and electrons. *Curr. Opin. Neurobiol. 18*, 633–641.

Helmstaedter, M., K. L. Briggman, S. C. Turaga, V. Jain, H. S. Seung, and W. Denk (2013). Connectomic reconstruction of the inner plexiform layer in the mouse retina. *Nature 500*, 168–174.

Hempel, C. M., K. H. Hartman, X.-J. Wang, G. Turrigiano, and S. B. Nelson (2000). Multiple forms of short-term plasticity at excitatory synapses in rat medial prefrontal cortex. *J. Neurophysiol. 83*, 3031–3041.

Henderson, M. M., R. L. Rademaker, and J. T. Serences (2022). Flexible utilization of spatial- and motor-based codes for the storage of visuo-spatial information. *Elife 11*, e75688.

Hennequin, G., E. J. Agnes, and T. P. Vogels (2017). Inhibitory plasticity: Balance, control, and codependence. *Annu. Rev. Neurosci. 40*, 557–579.

Hennequin, G., T. P. Vogels, and W. Gerstner (2014). Optimal control of transient dynamics in balanced networks supports generation of complex movements. *Neuron 82*, 1394–1406.

Henry, G. H., B. Dreher, and P. Bishop (1974). Orientation specificity of cells in cat striate cortex. *J. Neurophysiol. 37*, 1394–1409.

Hensch, T. and M. Fagiolini (2004). *Excitatory-Inhibitory Balance: Synapses, Circuits, Systems*. New York: Springer.

Hernández, A., A. Zainos, and R. Romo (2002). Temporal evolution of a decision making process in medial premotor cortex. *Neuron 33*, 950–972.

Herrmann, E., J. Call, M. V. Hernández-Lloreda, B. Hare, and M. Tomasello (2007). Humans have evolved specialized skills of social cognition: The cultural intelligence hypothesis. *Science 317*, 1360–1366.

Herrnstein, R. J. (1961). Relative and absolute strength of response as a function of frequency of reinforcement. *J. Exp. Anal. Behav. 4*, 267–272.

Herrnstein, R. J. (1990). Rational choice theory: Necessary but not sufficient. *Am. Psychol. 45*, 356.

Herrnstein, R. J. and D. Prelec (1991). Melioration: A theory of distributed choice. *J. Econ. Perspect. 5*, 137–156.

Hersche, M., M. Zeqiri, L. Benini, A. Sebastian, and A. Rahimi (2023). A neuro-vector-symbolic architecture for solving Raven's progressive matrices. *Nat. Mach. Intell. 5*, 363–375.

Hertäg, L. and H. Sprekeler (2019). Amplifying the redistribution of somato-dendritic inhibition by the interplay of three interneuron types. *PLoS Comput. Biol. 15*, e1006999.

Herwig, A., M. Beisert, and W. X. Schneider (2010). On the spatial interaction of visual working memory and attention: Evidence for a global effect from memory-guided saccades. *J. Vis. 10*, 8.

Herz, J. A., A. S. Krogh, and R. G. Palmer (1991). *Introduction to the Theory of Neural Computation*. Boston: Addison-Wesley.

Herzfeld, D. J., Y. Kojima, R. Soetedjo, and R. Shadmehr (2018). Encoding of error and learning to correct that error by the purkinje cells of the cerebellum. *Nat. Neurosci. 21*, 736–743.

Hestrin, S., D. J. Perkel, P. Sah, T. Manabe, P. Renner, and R. A. Nicoll (1990a). Physiological properties of excitatory synaptic transmission in the central nervous system. *Cold Spring Harbor. Symp. Quant. Biol. 55*, 87–93.

Hestrin, S., P. Sah, and R. Nicoll (1990b). Mechanisms generating the time course of dual component excitatory synaptic currents recorded in hippocampal slices. *Neuron 5*, 247–253.

Hick, W. E. (1952). On the rate of gain of information. *Quart. J. Exp. Psychol. 4*, 11–26.

Hickok, G. and D. Poeppel (2007). The cortical organization of speech processing. *Nat. Rev. Neurosci. 8*, 393–402.

Higgins, I., L. Matthey, A. Pal, C. Burgess, X. Glorot, M. Botvinick, S. Mohamed, and A. Lerchner (2017). beta-VAE: Learning basic visual concepts with a constrained variational framework. In *International Conference on Learning Representations* (poster).

Hikosaka, O. and M. Isoda (2010). Switching from automatic to controlled behavior: Cortico-basal ganglia mechanisms. *Trends Cogn. Sci. 14*, 154–161.

Hikosaka, O., M. Sakamoto, and S. Usui (1989). Functional properties of monkey caudate neurons. iii. activities related to expectation of target and reward. *J. Neurophysiol. 61*, 814–832.

Hikosaka, O., Y. Takikawa, and R. Kawagoe (2000). Role of the basal ganglia in the control of purposive saccadic eye movements. *Physiol. Rev. 80*, 953–978.

Hikosaka, O. and R. H. Wurtz (1983). Visual and oculomotor functions of monkey substantia nigra pars reticulata. III. memory-contingent visual and saccade responses. *J. Neurophysiol. 49*, 1268–1284.

Hildebrand, D. G. C., M. Cicconet, R. M. Torres, W. Choi, T. M. Quan, J. Moon, A. W. Wetzel, A. S. Champion, B. J. Graham, O. Randlett, et al. (2017). Whole-brain serial-section electron microscopy in larval zebrafish. *Nature 545*, 345–349.

Hille, B. (2001). *Ion Channels of Excitable Membranes* (3rd ed.). Sunderland, MA: Sinauer Associates.

Hinde, R. A. (1966). *Animal Behavior: A Synthesis of Ethology and Comparative Psychology*. New York: McGraw-Hill.

Hinton, S. C. and W. H. Meck (2004). Frontal–striatal circuitry activated by human peak-interval timing in the supra-seconds range. *Cogn. Brain Res. 21*, 171–182.

Hirokawa, J., A. Vaughan, P. Masset, T. Ott, and A. Kepecs (2019). Frontal cortex neuron types categorically encode single decision variables. *Nature 576*, 446–451.

Hiser, J. and M. Koenigs (2018). The multifaceted role of the ventromedial prefrontal cortex in emotion, decision making, social cognition, and psychopathology. *Biol. Psychiatry 83*, 638–647.

Hitchcock, P. F., E. I. Fried, and M. J. Frank (2022). Computational psychiatry needs time and context. *Annu. Rev. Psychol. 73*, 243–270.

Hochreiter, S. and J. Schmidhuber (1997). Long short-term memory. *Neural Comput. 9*, 1735–1780.

Hochreiter, S., A. S. Younger, and P. R. Conwell (2001). Learning to learn using gradient descent. In G. Dorffner, H. Bischof, and K. Hornik (Eds.), *Artificial Neural Networks—ICANN 2001*, pp. 87–94. Berlin, Heidelberg: Springer.

Hodge, R. D., T. E. Bakken, J. A. Miller, K. A. Smith, E. R. Barkan, L. T. Graybuck, J. L. Close, B. Long, N. Johansen, O. Penn, Z. Yao, J. Eggermont, T. Hollt, B. P. Levi, S. I. Shehata, B. Aevermann, A. Beller, D. Bertagnolli, K. Brouner, T. Casper, C. Cobbs, R. Dalley, N. Dee, S. L. Ding, R. G. Ellenbogen, O. Fong, E. Garren, J. Goldy, R. P. Gwinn, D. Hirschstein, C. D. Keene, M. Keshk, A. L. Ko, K. Lathia, A. Mahfouz, Z. Maltzer, M. McGraw, T. N. Nguyen, J. Nyhus, J. G. Ojemann, A. Oldre, S. Parry, S. Reynolds, C. Rimorin, N. V. Shapovalova, S. Somasundaram, A. Szafer, E. R. Thomsen, M. Tieu, G. Quon, R. H. Scheuermann, R. Yuste, S. M. Sunkin, B. Lelieveldt, D. Feng, L. Ng, A. Bernard, M. Hawrylycz, J. W. Phillips, B. Tasic, H. Zeng, A. R. Jones, C. Koch, and E. S. Lein (2019). Conserved cell types with divergent features in human versus mouse cortex. *Nature 573*, 61–68.

Hodgkin, A. L. (1948). The local electric changes associated with repetitive action in a non-medullated axon. *J. Physiol. (Lond.) 107*, 165–181.

Hodgkin, A. L. and A. F. Huxley (1952). A quantitative description of membrane current and its application to conductance and excitation in nerve. *J. Physiol. 117*, 500–544.

Hofer, S. B., H. Ko, B. Pichler, J. Vogelstein, H. Ros, H. Zeng, E. Lein, N. A. Lesica, and T. D. Mrsic-Flogel (2011). Differential connectivity and response dynamics of excitatory and inhibitory neurons in visual cortex. *Nat. Neurosci. 14*, 1045–1052.

Hoftman, G. D., S. J. Dienel, H. H. Bazmi, Y. Zhang, K. Chen, and D. A. Lewis (2018). Altered gradients of glutamate and Gamma-aminobutyric acid transcripts in the cortical visuospatial working memory network in schizophrenia. *Biol. Psychiatry*, 670–679.

Hol, K. and S. Treue (2001). Different populations of neurons contribute to the detection and discrimination of visual motion. *Vis. Res. 41*, 685–689.

Holland, J. H., K. J. Holyoak, R. E. Nisbett, and P. R. Thagard (1986). *Induction: Processes of Inference, Learning, and Discovery*. Cambridge, MA: MIT Press.

Holmgren, C., T. Harkany, B. Svennenfors, and Y. Zilberter (2003). Pyramidal cell communication within local networks in layer 2/3 of rat neocortex. *J. Physiol. (Lond.) 551*, 139–153.

Holt, G. R. and C. Koch (1997). Shunting inhibition does not have a divisive effect on firing rates. *Neural Comput. 9*, 1001–1013.

Holtmaat, A. and P. Caroni (2016). Functional and structural underpinnings of neuronal assembly formation in learning. *Nat. Neurosci. 19*, 1553–1562.

Homayoun, H. and B. Moghaddam (2007). NMDA receptor hypofunction produces opposite effects on prefrontal cortex interneurons and pyramidal neurons. *J. Neurosci. 27*, 11496–11500.

Honey, C. J., R. Kötter, M. Breakspear, and O. Sporns (2007). Network structure of cerebral cortex shapes functional connectivity on multiple time scales. *Proc. Natl. Acad. Sci. 104*, 10240–10245.

Honey, C. J., T. Thesen, T. H. Donner, L. J. Silbert, C. E. Carlson, O. Devinsky, W. K. Doyle, N. Rubin, D. J. Heeger, and U. Hasson (2012). Slow cortical dynamics and the accumulation of information over long timescales. *Neuron 76*, 423–434.

Hong, H., D. L. Yamins, N. J. Majaj, and J. J. DiCarlo (2016). Explicit information for category-orthogonal object properties increases along the ventral stream. *Nat. Neurosci. 19*, 613.

Hong, S.-J., R. V. De Wael, R. A. Bethlehem, S. Lariviere, C. Paquola, S. L. Valk, M. P. Milham, A. Di Martino, D. S. Margulies, J. Smallwood, et al. (2019). Atypical functional connectome hierarchy in autism. *Nat. Commun. 10*, 1–13.

Hong, T. and W. R. Stauffer (2023). Computational complexity drives sustained deliberation. *Nat. Neurosci.* https://doi.org/10.1038/s41593-023-01307-6.

Honig, A., M. A. Romme, B. J. Ensink, S. D. Escher, M. H. Pennings, and M. Devries (1998). Auditory hallucinations: A comparison between patients and nonpatients. *J. Nerv. Ment. Dis. 186*, 646–651.

Honig, W. K. and R. K. Thompson (1982). Retrospective and prospective processing in animal working memory. In G. H. Bower (Ed.), *Psychology of Learning and Motivation*, pp. 239–283. New York: Academic.

Hopfield, J. J. (1982). Neural networks and physical systems with emergent collective computational abilities. *Proc. Natl. Acad. Sci. U.S.A. 79*, 2554–2558.

Hopfield, J. J. (1984). Neurons with graded response have collective computational properties like those of two-state neurons. *Proc. Natl. Acad. Sci. U.S.A. 81*, 3088–3092.

Horga, G. and A. Abi-Dargham (2019). An integrative framework for perceptual disturbances in psychosis. *Nat. Rev. Neurosci. 20*, 763–778.

Horvát, S., R. Gămănuţ, M. Ercsey-Ravasz, L. Magrou, B. Gămănuţ, D. C. Van Essen, A. Burkhalter, K. Knoblauch, Z. Toroczkai, and H. Kennedy (2016). Spatial embedding and wiring cost constrain the functional layout of the cortical network of rodents and primates. *PLoS Biol. 14*, e1002512.

Horwitz, G. D., A. P. Batista, and W. T. Newsome (2004). Representation of an abstract perceptual decision in macaque superior colliculus. *J. Neurophysiol. 91*, 2281–2296.

Horwitz, G. D. and W. T. Newsome (2001). Target selection for saccadic eye movements: Prelude activity in the superior colliculus during a direction-discrimination task. *J. Neurophysiol. 86*, 2543–2558.

Howard, D. M., M. J. Adams, T.-K. Clarke, J. D. Hafferty, J. Gibson, M. Shirali, J. R. Coleman, S. P. Hagenaars, J. Ward, E. M. Wigmore, et al. (2019). Genome-wide meta-analysis of depression identifies 102 independent variants and highlights the importance of the prefrontal brain regions. *Nat. Neurosci. 22*, 343–352.

Howe, M. W., P. L. Tierney, S. G. Sandberg, P. E. Phillips, and A. M. Graybiel (2013). Prolonged dopamine signalling in striatum signals proximity and value of distant rewards. *Nature 500*, 575–579.

Howie, D. (2002). *Interpreting Probability: Controversies and Developments in the Early Twentieth Century*. Cambridge: Cambridge University Press.

Hu, J. M., M. Z. Qian, H. Tanigawa, X. M. Song, and A. W. Roe (2020). Focal electrical stimulation of cortical functional networks. *Cereb. Cortex 30*, 5532–5543.

Huang, C., D. A. Ruff, R. Pyle, R. Rosenbaum, M. R. Cohen, and B. Doiron (2019). Circuit models of low-dimensional shared variability in cortical networks. *Neuron 101*, 337–348.

Huang, X., W. C. Troy, Q. Yang, H. Ma, C. R. Laing, S. J. Schiff, and J. Y. Wu (2004). Spiral waves in disinhibited mammalian neocortex. *J. Neurosci. 24*, 9897–9902.

Hubel, D. H. and T. N. Wiesel (1962). Receptive fields, binocular interaction and functional architecture in the cat's visual cortex. *J. Physiol. (Lond.) 160*, 106–154.

Hubel, D. H. and T. N. Wiesel (1968). Receptive fields and functional architecture of monkey striate cortex. *J. Physiol. (Lond.) 195*, 215–243.

Hubel, D. H. and T. N. Wiesel (1977). Ferrier lecture-functional architecture of macaque monkey visual cortex. *Philos. Trans. R. Soc. B: Biol. Sci. 198*, 1–59.

Huh, D. and T. J. Sejnowski (2017). Gradient descent for spiking neural networks. *arXiv preprint arXiv:1706.04698*.

Huk, A. C. and M. N. Shadlen (2005). Neural activity in macaque parietal cortex reflects temporal integration of visual motion signals during perceptual decision making. *J. Neurosci. 25*, 10420–10436.

Hunt, L. T., N. Daw, P. Kaanders, M. MacIver, U. Mugan, E. Procyk, A. Redish, E. Russo, J. Scholl, K. Stachenfeld, et al. (2021). Formalizing planning and information search in naturalistic decision-making. *Nat. Neurosci. 24*, 1051–1064.

Hunt, L. T. and B. Y. Hayden (2017). A distributed, hierarchical and recurrent framework for reward-based choice. *Nat. Rev. Neurosci. 18*, 172–182.

Hunt, L. T., N. Kolling, A. Soltani, M. W. Woolrich, M. F. Rushworth, and T. E. Behrens (2012). Mechanisms underlying cortical activity during value-guided choice. *Nat. Neurosci. 15*, 470–476.

Hunt, L. T., W. N. Malalasekera, A. O. de Berker, B. Miranda, S. F. Farmer, T. E. Behrens, and S. W. Kennerley (2018). Triple dissociation of attention and decision computations across prefrontal cortex. *Nat. Neurosci. 21*, 1471–1481.

Huntenburg, J. M., P. L. Bazin, A. Goulas, C. L. Tardif, A. Villringer, and D. S. Margulies (2017). A systematic relationship between functional connectivity and intracortical myelin in the human cerebral cortex. *Cereb. Cortex 27*, 981–997.

Hunter, W. S. (1913). The delayed reactions in animals and children. *Behav. Monogr. 2*, 1–86.

Husain, M. and J. P. Roiser (2018). Neuroscience of apathy and anhedonia: A transdiagnostic approach. *Nat. Rev. Neurosci. 19*, 470–484.

Hussar, C. R. and T. Pasternak (2013). Common rules guide comparisons of speed and direction of motion in the dorsolateral prefrontal cortex. *J. Neurosci. 33*, 972–986.

Hutt, S. J. and C. Hutt (1970). *Direct Observation and Measurement of Behavior*. Springfield, IL: CC Thomas Publisher.

Huttenlocher, P. R. and A. S. Dabholkar (1997). Regional differences in synaptogenesis in human cerebral cortex. *J. Comp. Neurol. 387*, 167–178.

Huxter, J., N. Burgess, and J. O'Keefe (2003). Independent rate and temporal coding in hippocampal pyramidal cells. *Nature 425*, 828–832.

Huys, Q. J., M. Browning, M. P. Paulus, and M. J. Frank (2021). Advances in the computational understanding of mental illness. *Neuropsychopharmacology 46*, 3–19.

Huys, Q. J., N. D. Daw, and P. Dayan (2015). Depression: A decision-theoretic analysis. *Annu. Rev. Neurosci. 38*, 1–23.

Huys, Q. J., T. V. Maia, and M. J. Frank (2016). Computational psychiatry as a bridge from neuroscience to clinical applications. *Nat. Neurosci. 19*, 404–413.

Huys, Q. J., D. A. Pizzagalli, R. Bogdan, and P. Dayan (2013). Mapping anhedonia onto reinforcement learning: A behavioural meta-analysis. *Biol. Mood. Anxiety Disord. 3*, https://doi.org/10.1186/2045-5380-3-12.

Iigaya, K., Y. Ahmadian, L. P. Sugrue, G. S. Corrado, Y. Loewenstein, W. T. Newsome, and S. Fusi (2019). Deviation from the matching law reflects an optimal strategy involving learning over multiple timescales. *Nat. Commun. 10*. https://doi.org/10.1038/s41467-019-09388-3.

Inagaki, H. K., L. Fontolan, S. Romani, and K. Svoboda (2019). Discrete attractor dynamics underlies persistent activity in the frontal cortex. *Nature 566*, 212–217.

Inglebert, Y., J. Aljadeff, N. Brunel, and D. Debanne (2020). Synaptic plasticity rules with physiological calcium levels. *Proc. Natl. Acad. Sci. 117*, 33639–33648.

Insel, T. R. (2010). Rethinking schizophrenia. *Nature 468*, 187–193.

Insel, T. R. (2014). The NIMH Research Domain Criteria (RDoC) Project: Precision medicine for psychiatry. *Am. J. Psychiatry 171*, 395–397.

Insel, T. R., B. Cuthbert, M. Garvey, R. Heinssen, D. S. Pine, K. Quinn, C. Sanislow, and P. Wang (2010). Research domain criteria (RDoC): Toward a new classification framework for research on mental disorders. *Am. J. Psychiatry 167*, 748–751.

International Brain Lab, B. Benson, J. Benson, D. Birman, N. Bonacchi, M. Carandini, J. A. Catarino, G. A. Chapuis, A. K. Churchland, Y. Dan, et al. (2023). A brain-wide map of neural activity during complex behaviour. *bioRxiv*. https://doi.org/10.1101/2023.07.04.547681.

Ipata, A. E., A. L. Gee, M. E. Goldberg, and J. W. Bisley (2006). Activity in the lateral intraparietal area predicts the goal and latency of saccades in a free-viewing visual search task. *J. Neurosci. 26*, 3656–3661.

Isoda, M. and O. Hikosaka (2007). Switching from automatic to controlled action by monkey medial frontal cortex. *Nat. Neurosci. 10*, 240–248.

Issa, J. B., G. Tocker, M. E. Hasselmo, J. G. Heys, and D. A. Dombeck (2020). Navigating through time: A spatial navigation perspective on how the brain may encode time. *Ann. Rev. Neurosci. 43*. https://doi.org/10.1146/annurev-neuro-101419-011117.

Ito, M. (2000). Mechanisms of motor learning in the cerebellum. *Brain Res. 886*, 237–245.

Itskov, V., D. Hansel, and M. Tsodyks (2011). Short-term facilitation may stabilize parametric working memory trace. *Front. Comput. Neurosci. 5*, 40.

Itti, L. and C. Koch (2001). Computational modelling of visual attention. *Nat. Rev. Neurosci. 2*, 194–203.

Iturria-Medina, Y., E. J. Canales-Rodrguez, L. Melie-Garca, P. A. Valdés-Hernández, E. Martnez-Montes, Y. Alemán-Gómez, and J. Sánchez-Bornot (2007). Characterizing brain anatomical connections using diffusion weighted mri and graph theory. *Neuroimage 36*, 645–660.

Ivanova, A. A., S. Srikant, Y. Sueoka, H. H. Kean, R. Dhamala, U.-M. O'reilly, M. U. Bers, and E. Fedorenko (2020). Comprehension of computer code relies primarily on domain-general executive brain regions. *eLife 9*, e58906.

Izhikevich, E. M. (2007). Solving the distal reward problem through linkage of STDP and dopamine signaling. *Cereb. Cortex 17*, 2443–2452.

Izhikevich, E. M. and G. M. Edelman (2008). Large-scale model of mammalian thalamocortical systems. *Proc. Natl. Acad. Sci. U.S.A. 105*, 3593–3598.

Jacobs, B. L. and E. C. Azmitia (1992). Structure and function of the brain serotonin system. *Physiol. Rev. 72*, 165–229.

Jacobsen, C. F. (1936). Studies of cerebral function in primates: I. The functions of the frontal association areas in monkeys. *Comp. Psychol. Monogr. 13*, 1–68.

Jaeger, H. and H. Haas (2004). Harnessing nonlinearity: Predicting chaotic systems and saving energy in wireless communication. *Science 304*, 78–80.

Jaeger, H., M. Lukosevicius, D. Popovici, and U. Siewert (2007). Optimization and applications of echo state networks with leaky-integrator neurons. *Neural Netw. 20*, 335–352.

Jahr, C. E. and C. F. Stevens (1990). Voltage dependence of NMDA-activated macroscopic conductances predicted by single-channel kinetics. *J. Neurosci. 10*, 3178–3182.

James, W. (1890). *The Principles of Psychology*. New York: Henry Holt & Company.

Janacsek, K., T. M. Evans, M. Kiss, L. Shah, H. Blumenfeld, and M. T. Ullman (2022). Subcortical cognition: The fruit below the rind. *Annu. Rev. Neurosci. 45*. https://doi.org/10.1146/annurev-neuro-110920-013544.

Jaramillo, J., J. F. Mejias, and X. J. Wang (2019). Engagement of pulvino-cortical feedforward and feedback pathways in cognitive computations. *Neuron 101*, 321–336.

Javitt, D. C., M. Steinschneider, C. E. Schroeder, and J. C. Arezzo (1996). Role of cortical N-methyl-D-aspartate receptors in auditory sensory memory and mismatch negativity generation: Implications for schizophrenia. *Proc. Natl. Acad. Sci. U.S.A. 93*, 11962–11967.

Javitt, D. C. and R. A. Sweet (2015). Auditory dysfunction in schizophrenia: Integrating clinical and basic features. *Nat. Rev. Neurosci. 16*, 535–550.

Jazayeri, M. and J. A. Movshon (2006). Optimal representation of sensory information by neural populations. *Nat. Neurosci. 9*, 690–696.

Jazayeri, M. and S. Ostojic (2021). Interpreting neural computations by examining intrinsic and embedding dimensionality of neural activity. *Curr. Opin. Neurobiol. 70*, 113–120.

Jensen, K. T., G. Hennequin, and M. G. Mattar (2023). A recurrent network model of planning explains hippocampal replay and human behavior. *bioRxiv*. https://doi.org/10.1101/2023.01.16.523429.

Jensen, O. and L. Colgin (2007). Cross-frequency coupling between neuronal oscillations. *Trends Cogn. Sci. 11*, 267–269.

Ji, J. L., M. Spronk, K. Kulkarni, G. Repovš, A. Anticevic, and M. W. Cole (2019). Mapping the human brain's cortical-subcortical functional network organization. *Neuroimage 185*, 35–57.

Jiang, X., E. Bradley, R. A. Rini, T. Zeffiro, J. VanMeter, and M. Riesenhuber (2007). Categorization training results in shape- and category-selective human neural plasticity. *Neuron 53*, 891–903.

Jiang, X., S. Shen, C. R. Cadwell, P. Berens, F. Sinz, A. S. Ecker, S. Patel, and A. S. Tolias (2015). Principles of connectivity among morphologically defined cell types in adult neocortex. *Science 350*, aac9462.

Jin, D. Z., N. Fujii, and A. M. Graybiel (2009). Neural representation of time in cortico-basal ganglia circuits. *Proc. Natl. Acad. Sci. U.S.A. 106*(45), 19156–19161.

Jin, L., M. Wang, V. Galvin, T. Lightbourne, P. Conn, A. Arnsten, and C. Paspalas (2018). mglur2 vs. mglur3 in primate prefrontal cortex: Postsynaptic mglur3 strengthen cognitive networks. *Cereb. Cortex 28*, 974–987.

Jirsa, V. (2021). Say what was not said. *eNeuro 8*, e0128.

Jocham, G., L. T. Hunt, J. Near, and T. E. Behrens (2012). A mechanism for value-guided choice based on the excitation-inhibition balance in prefrontal cortex. *Nat. Neurosci. 15*, 960.

Joelving, F., A. Compte, and C. Constantinidis (2007). Temporal properties of posterior parietal neuron discharges during working memory and passive viewing. *J. Neurophysiol. 97*, 2254–2266.

Joglekar, M. R., J. F. Mejias, G. R. Yang, and X.-J. Wang (2018). Inter-areal balanced amplification enhances signal propagation in a large-scale circuit model of the primate cortex. *Neuron 98*, 222 234.

Johansen-Berg, H. and M. F. Rushworth (2009). Using diffusion imaging to study human connectional anatomy. *Annu. Rev. Neurosci. 32*, 75–94.

John, Y. J., B. Zikopoulos, D. Bullock, and H. Barbas (2018). Visual attention deficits in schizophrenia can arise from inhibitory dysfunction in thalamus or cortex. *Comput. Psychiatr. 2*, 223–257.

Johnson, M. K., R. P. McMahon, B. M. Robinson, A. N. Harvey, B. Hahn, C. J. Leonard, S. J. Luck, and J. M. Gold (2013). The relationship between working memory capacity and broad measures of cognitive ability in healthy adults and people with schizophrenia. *Neuropsychology 27*, 220–229.

Johnston, K., H. M. Levin, M. J. Koval, and S. Everling (2007). Top-down control-signal dynamics in anterior cingulate and prefrontal cortex neurons following task switching. *Neuron 53*, 453–462.

Jones, M. and B. C. Love (2011). Bayesian Fundamentalism or Enlightenment? On the explanatory status and theoretical contributions of Bayesian models of cognition. *Behav. Brain Sci. 34*, 169–188.

Jones, S. R., D. L. Pritchett, M. A. Sikora, S. M. Stufflebeam, M. Hamalainen, and C. I. Moore (2009). Quantitative analysis and biophysically realistic neural modeling of the MEG mu rhythm: Rhythmogenesis and modulation of sensory-evoked responses. *J. Neurophysiol. 102*, 3554–3572.

Jonides, J., R. L. Lewis, D. E. Nee, C. A. Lustig, M. G. Berman, and K. S. Moore (2008). The mind and brain of short-term memory. *Annu. Rev. Psychol. 59*, 193–224.

Josselyn, S. A., S. Kohler, and P. W. Frankland (2015). Finding the engram. *Nat. Rev. Neurosci. 16*, 521–534.

Jouravlev, O., D. Zheng, Z. Balewski, A. L. A. Pongos, Z. Levan, S. Goldin-Meadow, and E. Fedorenko (2019). Speech-accompanying gestures are not processed by the language-processing mechanisms. *Neuropsychologia 132*, 107132.

Jumper, J., R. Evans, A. Pritzel, T. Green, M. Figurnov, O. Ronneberger, K. Tunyasuvunakool, R. Bates, A. Žídek, A. Potapenko, et al. (2021). Highly accurate protein structure prediction with alphafold. *Nature 596*, 583–589.

Jun, J. J., N. A. Steinmetz, J. H. Siegle, D. J. Denman, M. Bauza, B. Barbarits, A. K. Lee, C. A. Anastassiou, A. Andrei, C. Aydin, M. Barbic, T. J. Blanche, V. Bonin, J. Couto, B. Dutta, S. L. Gratiy, D. A. Gutnisky, M. Hausser, B. Karsh, P. Ledochowitsch, C. M. Lopez, C. Mitelut, S. Musa, M. Okun, M. Pachitariu, J. Putzeys, P. D. Rich, C. Rossant, W. L. Sun, K. Svoboda, M. Carandini, K. D. Harris, C. Koch, J. O'Keefe, and T. D. Harris (2017). Fully integrated silicon probes for high-density recording of neural activity. *Nature 551*, 232–236.

Kaanders, P., H. Nili, J. X. O'Reilly, and L. Hunt (2021). Medial frontal cortex activity predicts information sampling in economic choice. *J. Neurosci. 41*, 8403–8413.

Kacelnik, A. (1984). Central place foraging in starlings (sturnus vulgaris). i. patch residence time. *J. Anim. Ecol. 53*, 283–299.

Kahneman, D. (2000). Experienced utility and objective happiness: a moment-based approach. In D. Kahneman and A. Tversky (Eds.), *Choices, Values and Frames*, pp. 673–692. New York: Cambridge University Press.

Kahneman, D. (2011). *Thinking, Fast and Slow*. New York: Farrar, Straus and Giroux.

Kahneman, D., P. Slovic, and A. Tversky (1982). *Judgment under Uncertainty: Heuristics and Biases*. New York: Cambridge University Press.

Kahneman, D. and A. Tversky (1979). Prospect theory: An analysis of decision under risk. *Econometrica 47*, 263–292.

Kalman, R. E. and R. S. Bucy (1961). New results in linear filtering and prediction theory. *J. Basic Eng. Series B 83*, 95–108.

Kamin, L. J. (1967). Predictability, surprise, attention, and conditioning. In B. A. Campbell and R. M. Church (Eds.), *Punishment and Aversive Behavior*, pp. 279–296. New York: Appelton-Century-Crofts.

Kaminski, J. and U. Rutishauser (2019). Between persistently active and activity-silent frameworks: Novel vistas on the cellular basis of working memory. *Ann. N. Y. Acad. Sci. 1459*. https://doi.org/10.1111/nyas.14213.

Kaminski, J., S. Sullivan, J. M. Chung, I. B. Ross, A. N. Mamelak, and U. Rutishauser (2017). Persistently active neurons in human medial frontal and medial temporal lobe support working memory. *Nat. Neurosci. 20*, 590–601.

Kampen, N. G. V. (2007). *Stochastic Processes in Physics and Chemistry* (3rd ed.). Amsterdam: North Holland.

Kandel, E. R. (1979). Psychotherapy and the single synapse. *N. Engl. J. Med. 301*, 1028–1037.

Kandel, E. R. (2001). The molecular biology of memory storage: A dialogue between genes and synapses. *Science 294*, 1030–1038.

Kandel, E. R. (2005). *Psychiatry, Psychoanalysis, and the New Biology of Mind*. Washington, DC: American Psychiatry Publishing, Inc.

Kang, B. and S. Druckmann (2020). Object recognition under occlusion revisited: Elucidating algorithmic advantages of recurrent computation. *bioRxiv*. https://doi.org/10.1101/2020.12.16.422991.

Kant, I. (1900). *Critique of Pure Reason*. New York: P. F. Collier & Son (English translation by J. M. D. Meiklejohn, First published in German in 1781).

Kanwisher, N. (2010). Functional specificity in the human brain: A window into the functional architecture of the mind. *Proc. Natl. Acad. Sci. 107*, 11163–11170.

Kanwisher, N., J. McDermott, and M. M. Chun (1997). The fusiform face area: A module in human extrastriate cortex specialized for face perception. *J. Neurosci. 17*, 4302–4311.

Kaplan, D. M. (2011). Explanation and description in computational neuroscience. *Synthese 183*, 339–373.

Kaplanis, C., C. Clopath, and M. Shanahan (2020). Continual reinforcement learning with multi-timescale replay. *arXiv preprint arXiv:2004.07530.*

Kaplanis, C., M. Shanahan, and C. Clopath (2018). Continual reinforcement learning with complex synapses. *arXiv preprint arXiv:1802.07239.*

Kaplanis, C., M. Shanahan, and C. Clopath (2019). Policy consolidation for continual reinforcement learning. *arXiv preprint arXiv:1902.00255.*

Kapoor, V., A. Dwarakanath, S. Safavi, J. Werner, M. Besserve, T. I. Panagiotaropoulos, and N. K. Logothetis (2022). Decoding internally generated transitions of conscious contents in the prefrontal cortex without subjective reports. *Nat. Commun. 13*, 1535.

Kar, K. and J. J. DiCarlo (2021). Fast recurrent processing via ventrolateral prefrontal cortex is needed by the primate ventral stream for robust core visual object recognition. *Neuron 109*, 164–176.

Kaschube, M., M. Schnabel, S. Löwel, D. M. Coppola, L. E. White, and F. Wolf (2010). Universality in the evolution of orientation columns in the visual cortex. *Science 330*, 1113–1116.

Kass, R. E., S.-I. Amari, K. Arai, E. N. Brown, C. O. Diekman, M. Diesmann, B. Doiron, U. T. Eden, A. L. Fairhall, G. M. Fiddyment, et al. (2018). Computational neuroscience: Mathematical and statistical perspectives. *Ann. Rev. Stat. Appl. 5*, 183–214.

Kastellakis, G., D. J. Cai, S. C. Mednick, A. J. Silva, and P. Poirazi (2015). Synaptic clustering within dendrites: An emerging theory of memory formation. *Prog. Neurobiol. 126*, 19–35.

Kastner, S., M. A. Pinsk, P. De Weerd, R. Desimone, and L. G. Ungerleider (1999). Increased activity in human visual cortex during directed attention in the absence of visual stimulation. *Neuron 22*, 751–761.

Kastner, S. and L. G. Ungerleider (2000). Mechanisms of visual attention in the human cortex. *Annu. Rev. Neurosci. 23*, 315–341.

Katona, I., B. Sperlagh, A. Sk, A. Käfalvi, E. S. Vizi, K. Mackie, and T. F. Freund (1999). Presynaptically located cb1 cannabinoid receptors regulate gaba release from axon terminals of specific hippocampal interneurons. *J. Neurosci. 19*, 4544–4558.

Katsuki, F. and C. Constantinidis (2012). Unique and shared roles of the posterior parietal and dorsolateral prefrontal cortex in cognitive functions. *Front. Integr. Neurosci. 6*, 17.

Katz, B. (1966). *Nerve, Muscle, and Synapse*. New York: McGraw-Hill.

Katz, B. and R. Miledi (1968). The role of calcium in neuromuscular facilitation. *J. Physiol. 195*, 481–492.

Katzner, S., I. Nauhaus, A. Benucci, V. Bonin, D. L. Ringach, and M. Carandini (2009). Local origin of field potentials in visual cortex. *Neuron 61*, 35–41.

Kauer, J. A. and R. C. Malenka (2007). Synaptic plasticity and addiction. *Nat. Rev. Neurosci. 8*, 844–858.

Kaufman, M. T., M. M. Churchland, S. I. Ryu, and K. V. Shenoy (2014). Cortical activity in the null space: Permitting preparation without movement. *Nat. Neurosci. 17*, 440–448.

Kaufman, M. T., J. S. Seely, D. Sussillo, S. I. Ryu, K. V. Shenoy, and M. M. Churchland (2016). The largest response component in the motor cortex reflects movement timing but not movement type. *eNeuro 3*. https://doi.org/10.1523/ENEURO.0085-16.2016.

Kawaguchi, Y. (1997). Selective cholinergic modulation of cortical GABAergic cell subtypes. *J. Neurophysiol. 78*, 1743–1747.

Kawaguchi, Y. and Y. Kubota (1993). Correlation of physiological subgroupings of nonpyramidal cells with parvalbumin- and calbindinD28k-immunoreactive neurons in layer V of rat frontal cortex. *J. Neurophysiol. 70*, 387–396.

Kawaguchi, Y. and Y. Kubota (1997). GABAergic cell subtypes and their synaptic connections in rat frontal cortex. *Cereb. Cortex 7*, 476–486.

Kawai, T., H. Yamada, N. Sato, M. Takada, and M. Matsumoto (2019). Preferential representation of past outcome information and future choice behavior by putative inhibitory interneurons rather than putative pyramidal neurons in the primate dorsal anterior cingulate cortex. *Cereb. Cortex 29*, 2339–2352.

Kay, S. R., A. Fiszbein, and L. A. Opler (1987). The positive and negative syndrome scale (panss) for schizophrenia. *Schizophr. Bull. 13*, 261–276.

Keemink, S. W. and C. K. Machens (2019). Decoding and encoding (de)mixed population responses. *Curr. Opin. Neurobiol. 58*, 112–121.

Keller, G. B. and T. D. Mrsic-Flogel (2018). Predictive processing: A canonical cortical computation. *Neuron 100*, 424–435.

Kelly, S. P. and R. G. O'Connell (2013). Internal and external influences on the rate of sensory evidence accumulation in the human brain. *J. Neurosci. 33*, 19434–19441.

Kelty-Stephen, D. G., P. E. Cisek, B. De Bari, J. Dixon, L. H. Favela, F. Hasselman, F. Keijzer, V. Raja, J. B. Wagman, B. J. Thomas, et al. (2022). In search for an alternative to the computer metaphor of the mind and brain. *arXiv preprint arXiv:2206.04603.*

Kemp, C. and J. B. Tenenbaum (2008). The discovery of structural form. *Proc. Natl. Acad. Sci. U.S.A. 105*, 10687–10692.

Kempter, R., W. Gerstner, and J. van Hemmen (1999). Hebbian learning and spiking neurons. *Phys. Rev. E 59*, 4498–4515.

Kennedy, A., G. Wayne, P. Kaifosh, K. Alviña, L. Abbott, and N. B. Sawtell (2014). A temporal basis for predicting the sensory consequences of motor commands in an electric fish. *Nat. Neurosci. 17*, 416–422.

Kennerley, S. W., A. F. Dahmubed, A. H. Lara, and J. D. Wallis (2009). Neurons in the frontal lobe encode the value of multiple decision variables. *J. Cogn. Neurosci. 21*, 1162–1178.

Kepecs, A. and G. Fishell (2014). Interneuron cell types are fit to function. *Nature 505*, 318–326.

Kepecs, A., N. Uchida, H. Zariwala, and Z. F. Mainen (2008). Neural correlations, computation and behavioral impact of decision confidence. *Nature 455*, 227–231.

Kepecs, A., X.-J. Wang, and J. Lisman (2002). Bursting neurons signal input slope. *J. Neurosci. 22*, 9053–9062.

Kepple, D. R., R. Engelken, and K. Rajan (2022). Curriculum learning as a tool to uncover learning principles in the brain. In *International Conference on Learning Representations* (poster).

Keramati, M., A. Durand, P. Girardeau, B. Gutkin, and S. H. Ahmed (2017). Cocaine addiction as a homeostatic reinforcement learning disorder. *Psychol. Rev. 124*, 130–153.

Kessler, R. C., P. Berglund, O. Demler, R. Jin, D. Koretz, K. R. Merikangas, A. J. Rush, E. E. Walters, and P. S. Wang (2003). The epidemiology of major depressive disorder: Results from the national comorbidity survey replication (ncs-r). *JAMA 289*, 3095–3105.

Kessler, R. C., S. Heeringa, M. D. Lakoma, M. Petukhova, A. E. Rupp, M. Schoenbaum, P. S. Wang, and A. M. Zaslavsky (2008). Individual and societal effects of mental disorders on earnings in the united states: Results from the national comorbidity survey replication. *A. J. Psychiatry 165*, 703–711.

Khaligh-Razavi, S.-M. and N. Kriegeskorte (2014). Deep supervised, but not unsupervised, models may explain IT cortical representation. *PLoS Comput. Biol. 10*, e1003915.

Khlestova, E., J. W. Johnson, J. H. Krystal, and J. Lisman (2016). The role of GluN2C-containing NMDA receptors in Ketamine's psychotogenic action and in schizophrenia models. *J. Neurosci. 36*, 11151–11157.

Khona, M. and I. R. Fiete (2022). Attractor and integrator networks in the brain. *Nat. Rev. Neurosci. 23*, 744–766.

Kiani, R., A. K. Churchland, and M. N. Shadlen (2013). Integration of direction cues is invariant to the temporal gap between them. *J. Neurosci. 33*, 16483–16489.

Kiani, R., L. Corthell, and M. N. Shadlen (2014a). Choice certainty is informed by both evidence and decision time. *Neuron 84*, 1329–1342.

Kiani, R., C. J. Cueva, J. B. Reppas, and W. T. Newsome (2014b). Dynamics of neural population responses in prefrontal cortex indicate changes of mind on single trials. *Curr. Biol. 24*, 1542–1547.

Kiani, R., T. Hanks, and M. Shadlen (2008). Bounded integration in parietal cortex underlies decisions even when viewing duration is dictated by the environment. *J. Neurosci. 28*, 3017–3029.

Kiani, R. and M. N. Shadlen (2009). Representation of confidence associated with a decision by neurons in the parietal cortex. *Science 324*, 759–764.

Kiebel, S. J., J. Daunizeau, and K. J. Friston (2008). A hierarchy of time-scales and the brain. *PLoS Comput. Biol. 4*, e1000209.

Killeen, P. R. and J. G. Fetterman (1988). A behavioral theory of timing. *Psychol. Rev. 95*, 274–295.

Kilpatrick, Z. P. (2018). Synaptic mechanisms of interference in working memory. *Sci. Rep. 8*, 7879.

Kilpatrick, Z. P. and P. C. Bressloff (2009). Spatially structured oscillations in a two-dimensional excitatory neuronal network with synaptic depression. *J. Comput. Neurosci. 28*, 193–209.

Kim, B. and M. A. Basso (2008). Saccade target selection in the superior colliculus: A signal detection theory approach. *J. Neurosci. 28*, 2991–3007.

Kim, E. and M. Sheng (2004). PDZ domain proteins of synapses. *Nat. Rev. Neurosci. 5*, 771–781.

Kim, H. R., A. N. Malik, J. G. Mikhael, P. Bech, I. Tsutsui-Kimura, F. Sun, Y. Zhang, Y. Li, M. Watabe-Uchida, S. J. Gershman, et al. (2020). A unified framework for dopamine signals across timescales. *Cell 183*, 1600–1616.

Kim, J.-N. and M. N. Shadlen (1999). Neural correlates of a decision in the dorsolateral prefrontal cortex of the macaque. *Nat. Neurosci. 2*, 176–183.

Kim, S. S., A. M. Hermundstad, S. Romani, L. Abbott, and V. Jayaraman (2019). Generation of stable heading representations in diverse visual scenes. *Nature 576*, 126–131.

Kim, Y., G. R. Yang, K. Pradhan, K. U. Venkataraju, M. Bota, L. C. Garcia Del Molino, G. Fitzgerald, K. Ram, M. He, J. M. Levine, P. Mitra, Z. J. Huang, X.-J. Wang, and P. Osten (2017). Brain-wide maps reveal stereotyped cell-type-based cortical architecture and subcortical sexual dimorphism. *Cell 171*, 456–469.

Kimberg, D. Y. and M. J. Farah (1993). A unified account of cognitive impairments following frontal lobe damage: The role of working memory in complex, organized behavior. *J. Exp. Psychol. Gen. 122*, 411.

King, J. R. and S. Dehaene (2014). Characterizing the dynamics of mental representations: The temporal generalization method. *Trends Cogn. Sci. (Regul. Ed.) 18*, 203–210.

Kira, S., T. Yang, and M. N. Shadlen (2015). A neural implementation of Wald's sequential probability ratio test. *Neuron 85*, 861–873.

Kirchner, W. K. (1958). Age differences in short-term retention of rapidly changing information. *J. Exp. Psychol. 55*, 352.

Kirkpatrick, J., R. Pascanu, N. Rabinowitz, J. Veness, G. Desjardins, A. A. Rusu, K. Milan, J. Quan, T. Ramalho, A. Grabska-Barwinska, D. Hassabis, C. Clopath, D. Kumaran, and R. Hadsell (2017). Overcoming catastrophic forgetting in neural networks. *Proc. Natl. Acad. Sci. U.S.A. 114*, 3521–3526.

Kisvarday, Z. F., A. S. Ferecska, K. Kovaics, P. Buzas, J. M. L. Budd, and U. T. Eysel (2003). One axon-multiple functions: Specificity of lateral inhibitory connections by large basket cells. *J. Neurocytol. 31*, 255–264.

Klatzmann, U., S. Froudist-Walsh, D. P. Bliss, P. Theodoni, J. Mejas, M. Niu, L. Rapan, N. Palomero-Gallagher, C. Sergent, S. Dehaene, and X.-J. Wang (2024). A dynamic bifurcation mechanism explains cortex-wide neural correlates of conscious access. *Cell Reports*, in press. *bioRxiv*. https://doi.org/10.1101/2022.02.20.481230.

Klein, S. B. (2013). The complex act of projecting oneself into the future. *Wiley Interdiscip. Rev. Cogn. Sci. 4*, 63–79.

Klein-Flügge, M. C., A. Bongioanni, and M. F. Rushworth (2022). Medial and orbital frontal cortex in decision-making and flexible behavior. *Neuron*, 2743–2770.

Kleinman, M., C. Chandrasekaran, and J. Kao (2021). A mechanistic multi-area recurrent network model of decision-making. *Adv. Neural Inf. Process. Syst. 34*, 23152–23165.

Klingberg, T. (2009). *The Overflowing Brain: Information Overload and the Limits of Working Memory*. New York: Oxford University Press.

Klukas, M., Lewis, M., and Fiete, I. (2020). Efficient and flexible representation of higher-dimensional cognitive variables with grid cells. *PLoS Comput. Biol. 16*, e1007796.

Knill, D. C. and A. Pouget (2004). The Bayesian brain: The role of uncertainty in neural coding and computation. *Trends Neurosci. 27*, 712–719.

Knill, D. C. and W. Richards (Eds.) (1996). *Perception as Bayesian Inference*. New York: Cambridge University Press.

Knudsen, E. B. and J. D. Wallis (2020). Closed-loop theta stimulation in the orbitofrontal cortex prevents reward-based learning. *Neuron 106*, 537–547.

Knudsen, E. I. (1994). Supervised learning in the brain. *J. Neurosci. 14*, 3985–3997.

Knutson, B., J. Taylor, M. Kaufman, R. Peterson, and G. Glover (2005). Distributed neural representation of expected value. *J. Neurosci. 25*, 4806–4812.

Kobak, D., W. Brendel, C. Constantinidis, C. E. Feierstein, A. Kepecs, Z. F. Mainen, X. L. Qi, R. Romo, N. Uchida, and C. K. Machens (2016). Demixed principal component analysis of neural population data. *Elife 5*. https://doi.org/10.7554/eLife.10989.

Kobak, D., J. L. Pardo-Vazquez, M. Valente, C. K. Machens, and A. Renart (2019). State-dependent geometry of population activity in rat auditory cortex. *Elife 8*.

Koch, C. (1999). *Biophysics of Computation: Information Processing in Single Neurons*. New York: Oxford University Press.

Koch, C. (2004). *The Quest for Consciousness: A Neurobiological Approach*. Englewood, CO: Roberts and Company Publishers.

Kohn, A., A. I. Jasper, J. D. Semedo, E. Gokcen, C. K. Machens, and M. Y. Byron (2020). Principles of corticocortical communication: Proposed schemes and design considerations. *Trends in Neurosciences 43*, 725–737.

Koldewyn, K., D. Whitney, and S. M. Rivera (2010). The psychophysics of visual motion and global form processing in autism. *Brain 133*, 599–610.

Kolling, N., T. E. Behrens, R. B. Mars, and M. F. Rushworth (2012). Neural mechanisms of foraging. *Science 336*, 95–98.

Komura, Y., A. Nikkuni, N. Hirashima, T. Uetake, and A. Miyamoto (2013). Responses of pulvinar neurons reflect a subject's confidence in visual categorization. *Nat. Neurosci. 16*, 749–755.

Komura, Y., R. Tamura, T. Uwano, H. Nishijo, K. Kaga, and T. Ono (2001). Retrospective and prospective coding for predicted reward in the sensory thalamus. *Nature 412*, 546–549.

Kondo, H., K. Tanaka, T. Hashikawa, and E. G. Jones (1999). Neurochemical gradients along monkey sensory cortical pathways: Calbindin-immunoreactive pyramidal neurons in layers II and III. *Euro. J. Neurosci. 11*, 4197–4203.

Kong, X., R. Kong, C. Orban, P. Wang, S. Zhang, K. Anderson, A. Holmes, J. D. Murray, G. Deco, M. van den Heuvel, et al. (2021). Sensory-motor cortices shape functional connectivity dynamics in the human brain. *Nat. Commun. 12*. https://doi.org/10.1038/s41467-021-26704-y.

Konidaris, G. (2019). On the necessity of abstraction. *Curr. Opin. Behav. Sci. 29*, 1–7.

Koob, G. F. and M. Le Moal (1997). Drug abuse: Hedonic homeostatic dysregulation. *Science 278*, 52–58.

Kooijmans, R. N., W. Sierhuis, M. W. Self, and P. R. Roelfsema (2020). A quantitative comparison of inhibitory interneuron size and distribution between mouse and macaque v1, using calcium-binding proteins. *Cereb. Cortex Commun. 1*, tgaa068.

Kopec, C. D., J. C. Erlich, B. W. Brunton, K. Deisseroth, and C. D. Brody (2015). Cortical and subcortical contributions to short-term memory for orienting movements. *Neuron 88*, 367–377.

Kornell, N., L. K. Son, and H. S. Terrace (2007). Transfer of metacognitive skills and hint seeking in monkeys. *Psychol. Sci. 18*, 64–71.

Korzeniowska, M., A. Theodorsen, M. Rypdal, and O. Garcia (2023). Apparent universality of 1/*f* spectra as an artifact of finite-size effects. *arXiv preprint arXiv:2304.08371*.

Kotermanski, S. E. and J. W. Johnson (2009). Mg2+ imparts NMDA receptor subtype selectivity to the Alzheimer's drug memantine. *J. Neurosci. 29*, 2774–2779.

Koulakov, A. A., S. Raghavachari, A. Kepecs, and J. E. Lisman (2002). Model for a robust neural integrator. *Nat. Neurosci. 5*, 775–782.

Koyluoglu, O. O., Y. Pertzov, S. Manohar, M. Husain, and I. R. Fiete (2017). Fundamental bound on the persistence and capacity of short-term memory stored as graded persistent activity. *Elife 6*.

Kraepelin, E. (1904). *Lectures on Clinical Psychiatry*. London: Baillière, Tindall and Cox.

Krajbich, I., C. Armel, and A. Rangel (2010). Visual fixations and the computation and comparison of value in simple choice. *Nat. Neurosci. 13*, 1292–1298.

Krajbich, I. and A. Rangel (2011). Multialternative drift-diffusion model predicts the relationship between visual fixations and choice in value-based decisions. *Proc. Natl. Acad. Sci. U.S.A. 108*, 13852–13857.

Krakauer, J. W., A. A. Ghazanfar, A. Gomez-Marin, M. A. MacIver, and D. Poeppel (2017). Neuroscience needs behavior: Correcting a reductionist bias. *Neuron 93*, 480–490.

Kravitz, A. V., B. S. Freeze, P. R. Parker, K. Kay, M. T. Thwin, K. Deisseroth, and A. C. Kreitzer (2010). Regulation of parkinsonian motor behaviours by optogenetic control of basal ganglia circuitry. *Nature 466*, 622–626.

Kreiman, G. and T. Serre (2020). Beyond the feedforward sweep: Feedback computations in the visual cortex. *Ann. N. Y. Acad. Sci. 1464*, 222–241.

Kriegeskorte, N. (2015). Deep neural networks: A new framework for modeling biological vision and brain information processing. *Ann. Rev. Vis. Sci. 1*, 417–446.

Kriegeskorte, N., M. Mur, D. A. Ruff, R. Kiani, J. Bodurka, H. Esteky, K. Tanaka, and P. A. Bandettini (2008). Matching categorical object representations in inferior temporal cortex of man and monkey. *Neuron 60*, 1126–1141.

Krienen, F. M., M. Goldman, Q. Zhang, R. C. Del Rosario, M. Florio, R. Machold, A. Saunders, K. Levandowski, H. Zaniewski, B. Schuman, et al. (2020). Innovations present in the primate interneuron repertoire. *Nature 586*, 262–269.

Kriener, B., R. Chaudhuri, and I. R. Fiete (2020). Robust parallel decision-making in neural circuits with nonlinear inhibition. *Proc. Natl. Acad. Sci. 117*, 25505–25516.

Krimer, L. S. and P. S. Goldman-Rakic (2001). Prefrontal microcircuits: Membrane properties and excitatory input of local, medium, and wide arbor interneurons. *J. Neurosci. 21*, 3788–3796.

Krishnamurthy, K., T. Can, and D. J. Schwab (2022). Theory of gating in recurrent neural networks. *Phy. Rev. X 12*, 011011.

Kristan, W. B., and Katz, P. (2006). Form and function in systems neuroscience. *Curr. Biol.* 16, R828–R831.

Kritzer, M. F. and P. S. Goldman-Rakic (1995). Intrinsic circuit organization of the major layers and sublayers of the dorsolateral prefrontal cortex in the rhesus monkey. *J. Comp. Neurol. 359*, 131–143.

Krizhevsky, A., I. Sutskever, and G. E. Hinton (2012). Imagenet classification with deep convolutional neural networks. *Adv. Neural Inf. Process. Syst.* 1097–1105.

Krueger, R. F. (1999). The structure of common mental disorders. *Arch. Gen. Psychiatry 56*, 921–926.

Krumin, M., J. J. Lee, K. D. Harris, and M. Carandini (2018). Decision and navigation in mouse parietal cortex. *eLife 7*, e42583.

Krystal, J. H., A. Anticevic, G. J. Yang, G. Dragoi, N. R. Driesen, X. J. Wang, and J. D. Murray (2017). Impaired tuning of neural ensembles and the pathophysiology of Schizophrenia: A translational and computational neuroscience perspective. *Biol. Psychiatry 81*, 874–885.

Krystal, J. H., D. C. D'Souza, D. Mathalon, E. Perry, A. Belger, and R. Hoffman (2003). NMDA receptor antagonist effects, cortical glutamatergic function, and schizophrenia: Toward a paradigm shift in medication development. *Psychopharmacology 169*, 215–233.

Krystal, J. H., L. P. Karper, J. P. Seibyl, G. K. Freeman, R. Delaney, J. D. Bremner, G. R. Heninger, M. B. Bowers, and D. S. Charney (1994). Subanesthetic effects of the noncompetitive NMDA antagonist, ketamine, in humans.

psychotomimetic, perceptual, cognitive, and neuroendocrine responses. *Arch. Gen. Psychiatry 51*, 199–214.

Krystal, J. H. and M. W. State (2014). Psychiatric disorders: Diagnosis to therapy. *Cell 157*, 201–214.

Kubanek, J. (2017). Optimal decision making and matching are tied through diminishing returns. *Proc. Natl. Acad. Sci. 114*, 8499–8504.

Kubota, K. and H. Niki (1971). Prefrontal cortical unit activity and delayed alternation performance in monkeys. *J. Neurophysiol. 34*, 337–347.

Kulkarni, A., J. Ranft, and V. Hakim (2020). Synchronization, stochasticity, and phase waves in neuronal networks with spatially-structured connectivity. *Front. Comput. Neurosci. 14*, 569644.

Kumar, A., S. Rotter, and A. Aertsen (2010). Spiking activity propagation in neuronal networks: Reconciling different perspectives on neural coding. *Nat. Rev. Neurosci. 11*, 615–627.

Kumaran, D., J. J. Summerfield, D. Hassabis, and E. A. Maguire (2009). Tracking the emergence of conceptual knowledge during human decision making. *Neuron 63*, 889–901.

Kummerfeld, E., S. Ma, R. K. Blackman, A. L. DeNicola, A. D. Redish, S. Vinogradov, D. A. Crowe, and M. V. Chafee (2020). Cognitive control errors in nonhuman primates resembling those in schizophrenia reflect opposing effects of nmda receptor blockade on causal interactions between cells and circuits in prefrontal and parietal cortices. *Biol. Psychiatry Cogn. Neurosci. Neuroimaging 5*, 705–714.

Kurth-Nelson, Z., T. Behrens, G. Wayne, K. Miller, L. Luettgau, R. Dolan, Y. Liu, and P. Schwartenbeck (2023). Replay and compositional computation. *Neuron 111*, 454–469.

Kuśmierz, Ł., T. Isomura, and T. Toyoizumi (2017). Learning with three factors: Modulating hebbian plasticity with errors. *Curr. Opin. Neurobiol. 46*, 170–177.

Kutas, M., G. McCarthy, and E. Donchin (1977). Augmenting mental chronometry: The P300 as a measure of stimulus evaluation time. *Science 197*, 792–795.

Kuwabara, M., N. Kang, T. E. Holy, and C. Padoa-Schioppa (2020). Neural mechanisms of economic choices in mice. *Elife 9*, e49669.

Kyzar, E. J. and G. H. Denfield (2023). Taking subjectivity seriously: Towards a unification of phenomenology, psychiatry, and neuroscience. *Mol. Psychiatry 28*, 10–16.

Laibson, D. (1997). Golden eggs and hyperbolic discounting. *Q. J. Econ. 112*, 443–478.

Laje, R. and D. V. Buonomano (2013). Robust timing and motor patterns by taming chaos in recurrent neural networks. *Nat. Neurosci. 16*, 925–933.

Lak, A., M. Okun, M. M. Moss, H. Gurnani, K. Farrell, M. J. Wells, C. B. Reddy, A. Kepecs, K. D. Harris, and M. Carandini (2020). Dopaminergic and prefrontal basis of learning from sensory confidence and reward value. *Neuron 105*, 700–711.

Lake, B. M., R. Salakhutdinov, and J. B. Tenenbaum (2015). Human-level concept learning through probabilistic program induction. *Science 350*, 1332–1338.

Lake, B. M., T. D. Ullman, J. B. Tenenbaum, and S. J. Gershman (2017). Building machines that learn and think like people. *Behav. Brain Sci. 40*, 1–72.

Lakshminarayanan, V. R., M. K. Chen, and L. R. Santos (2011). The evolution of decision-making under risk: Framing effects in monkey risk preferences. *J. Exp. Soc. Psychol. 47*, 689–693.

Laland, K. and A. Seed (2021). Understanding human cognitive uniqueness. *Annu. Rev. Psychol. 72*, 689–716.

Lalazar, H., L. F. Abbott, and E. Vaadia (2016). Tuning curves for arm posture control in motor cortex are consistent with random connectivity. *PLoS Comput. Biol. 12*, e1004910.

Lam, N. H., T. Borduqui, J. Hallak, A. Roque, A. Anticevic, J. H. Krystal, X. J. Wang, and J. D. Murray (2022). Effects of altered excitation-inhibition balance on decision making in a cortical circuit model. *J. Neurosci. 42*, 1035–1053.

Lamb, T. D. (2013). Evolution of phototransduction, vertebrate photoreceptors and retina. *Prog. Retin. Eye Res. 36*, 52–119.

Laming, D. (1968). *Information Theory of Choice Reaction Times*. New York: Academic Press.

Landauer, T. K. (1969). Reinforcement as consolidation. *Psychol. Rev. 76*, 82–96.

Landauer, T. K. (1986). How much do people remember? Some estimates of the quantity of learned information in long-term memory. *Cogn. Sci. 10*, 477–493.

Lapish, C. C., D. Durstewitz, L. J. Chandler, and J. K. Seamans (2008). Successful choice behavior is associated with distinct and coherent network states in anterior cingulate cortex. *Proc. Natl. Acad. Sci. U.S.A. 105*, 11963–11968.

Lappin, J. S. and C. W. Eriksen (1966). Use of a delayed signal to stop a visual reaction-time response. *J. Exp. Psychol. 72*, 805–811.

Larkum, M. E. (2013). A cellular mechanism for cortical associations: An organizing principle for the cerebral cortex. *Trends Neurosci. 36*, 141–151.

Larkum, M. E., T. Nevian, M. Sandler, A. Polsky, and J. Schiller (2009). Synaptic integration in tuft dendrites of layer 5 pyramidal neurons: A new unifying principle. *Science 325*, 756–760.

Larkum, M. E., J. J. Zhu, and B. Sakmann (1999). A new cellular mechanism for coupling inputs arriving at different cortical layers. *Nature 398*, 338–341.

Larson, R. and D. C. Falvo (2009). *Elementary Linear Algebra*. Boston: Cengage Learning.

Laruelle, M. and A. Abi-Dargham (1999). Dopamine as the wind of the psychotic fire: New evidence from brain imaging studies. *J. Psychopharmacol. 13*, 358–371.

Lashley, K. S. (1951). The problem of serial order in behavior. In L. A. Jeffress (Ed.), *Cerebral Mechanisms in Behavior: The Hixon Symposium*, pp. 112–146. New York: Wiley.

Latimer, K. W., J. L. Yates, M. L. Meister, A. C. Huk, and J. W. Pillow (2015). Single-trial spike trains in parietal cortex reveal discrete steps during decision-making. *Science 349*, 184–187.

Lau, B. and P. W. Glimcher (2005). Dynamic response-by-response models of matching behavior in rhesus monkeys. *J. Exp. Anal. Behav. 84*, 555–579.

Lau, B. and P. W. Glimcher (2007). Action and outcome encoding in the primate caudate nucleus. *J. Neurosci. 27*, 14502–14514.

Laubach, M., L. M. Amarante, K. Swanson, and S. R. White (2018). What, if anything, is rodent prefrontal cortex? *Eneuro*. https://doi.org/10.1523/ENEURO.0315-18.2018.

Lavi, A., M. Sehgal, A. F. de Sousa, D. Ter-Mkrtchyan, F. Sisan, A. Luchetti, A. Okabe, C. Bear, and A. J. Silva (2023). Local memory allocation recruits memory ensembles across brain regions. *Neuron 111*, 470–480.

Law, C. and J. Gold (2008). Neural correlates of perceptual learning in a sensory-motor, but not a sensory, cortical area. *Nat. Neurosci. 11*, 505–513.

Leavitt, M. L., D. Mendoza-Halliday, and J. C. Martinez-Trujillo (2017). Sustained activity encoding working memories: Not fully distributed. *Trends Neurosci. 40*, 328–346.

LeCun, Y. (1988). A theoretical framework for back-propagation. In D. Touretzky, G. Hinton, and T. Sejnowski (Eds.), *Proceedings of the 1988 Connectionist Models Summer School*, pp. 21–28. Burlington, MA: Morgan Kaufmann.

LeCun, Y. and Y. Bengio (1995). Convolutional networks for images, speech, and time series. In M. A. Arbib (Ed.), *The Handbook of Brain Theory and Neural Networks*, pp. 255–258. Cambridge, MA: MIT Press.

LeCun, Y., Y. Bengio, and G. Hinton (2015). Deep learning. *Nature 521*, 436–444.

LeCun, Y., L. Bottou, Y. Bengio, and P. Haffner (1998). Gradient-based learning applied to document recognition. *Proc. IEEE 86*, 2278–2324.

LeDoux, J. E. (2000). Emotion circuits in the brain. *Annu. Rev. Neurosci. 23*, 155–184.

LeDoux, J. E. (2015). *Anxious: Using the Brain to Understand and Treat Fear and Anxiety*. New York: Penguin Books.

LeDoux, J. E. (2020). *The Deep History of Ourselves: The Four-Billion-Year Story of How We Got Conscious Brains*. New York: Penguin.

LeDoux, J. E., J. Moscarello, R. Sears, and V. Campese (2017). The birth, death and resurrection of avoidance: A reconceptualization of a troubled paradigm. *Mol. Psychiatry 22*, 24–36.

Lee, A. K. and M. A. Wilson (2002). Memory of sequential experience in the hippocampus during slow wave sleep. *Neuron 36*, 1183–1194.

Lee, C. S., M. Aly, and C. Baldassano (2021). Anticipation of temporally structured events in the brain. *eLife 10*, e64972.

Lee, D. (2008). Game theory and neural basis of social decision making. *Nat. Neurosci. 11*, 404–409.

Lee, D. (2013). Decision making: From neuroscience to psychiatry. *Neuron 78*, 233–248.

Lee, D. (2020). *Birth of Intelligence: From RNA to Artificial Intelligence*. Oxford: Oxford University Press.

Lee, D. and H. Seo (2016). Neural basis of strategic decision making. *Trends Neurosci. 39*, 40–48.

Lee, J. H., C. Koch, and S. Mihalas (2017). A computational analysis of the function of three inhibitory cell types in contextual visual processing. *Front. Comput. Neurosci. 11*, 28.

Lee, J. H. and S. Park (2005). Working memory impairments in schizophrenia: A meta-analysis. *J. Abnorm. Psychol. 114*, 599.

Lee, S., I. Kruglikov, Z. J. Huang, G. Fishell, and B. Rudy (2013). A disinhibitory circuit mediates motor integration in the somatosensory cortex. *Nat. Neurosci. 16*, 1662–1670.

Lefort, S., C. Tomm, J.-C. F. Sarria, and C. C. Petersen (2009). The excitatory neuronal network of the c2 barrel column in mouse primary somatosensory cortex. *Neuron 61*, 301–316.

Legenstein, R., S. M. Chase, A. B. Schwartz, and W. Maass (2010). A reward-modulated hebbian learning rule can explain experimentally observed network reorganization in a brain control task. *J. Neurosci. 30*, 8400–8410.

Legg, S., M. Hutter, et al. (2007). A collection of definitions of intelligence. *Front. Artif. Intell. Appl. 157*, 17.

Lein, E. S., M. J. Hawrylycz, N. Ao, M. Ayres, A. Bensinger, A. Bernard, A. F. Boe, M. S. Boguski, K. S. Brockway, E. J. Byrnes, L. Chen, L. Chen, T. M. Chen, M. C. Chin, J. Chong, B. E. Crook, A. Czaplinska, C. N. Dang, S. Datta, N. R. Dee, A. L. Desaki, T. Desta, E. Diep, T. A. Dolbeare, M. J. Donelan, H. W. Dong, J. G. Dougherty, B. J. Duncan, A. J. Ebbert, G. Eichele, L. K. Estin, C. Faber, B. A. Facer, R. Fields, S. R. Fischer, T. P. Fliss, C. Frensley, S. N. Gates, K. J. Glattfelder, K. R. Halverson, M. R. Hart, J. G. Hohmann, M. P. Howell, D. P. Jeung, R. A. Johnson, P. T. Karr, R. Kawal, J. M. Kidney, R. H. Knapik, C. L. Kuan, J. H. Lake, A. R. Laramee, K. D. Larsen, C. Lau, T. A. Lemon, A. J. Liang, Y. Liu, L. T. Luong, J. Michaels, J. J. Morgan, R. J. Morgan, M. T. Mortrud, N. F. Mosqueda, L. L. Ng, R. Ng, G. J. Orta, C. C. Overly, T. H. Pak, S. E. Parry, S. D. Pathak, O. C. Pearson, R. B. Puchalski, Z. L. Riley, H. R. Rockett, S. A. Rowland, J. J. Royall, M. J. Ruiz, N. R. Sarno, K. Schaffnit, N. V. Shapovalova, T. Sivisay, C. R. Slaughterbeck, S. C. Smith, K. A. Smith, B. I. Smith, A. J. Sodt, N. N. Stewart, K. R. Stumpf, S. M. Sunkin, M. Sutram, A. Tam, C. D. Teemer, C. Thaller, C. L. Thompson, L. R. Varnam, A. Visel, R. M. Whitlock, P. E. Wohnoutka, C. K. Wolkey, V. Y. Wong, M. Wood, M. B. Yaylaoglu, R. C. Young, B. L. Youngstrom, X. F. Yuan, B. Zhang, T. A. Zwingman, and A. R. Jones (2007). Genome-wide atlas of gene expression in the adult mouse brain. *Nature 445*, 168–176.

Le Merre, P., S. Ährlund-Richter, and M. Carlén (2021). The mouse prefrontal cortex: Unity in diversity. *Neuron*. https://doi.org/10.1016/j.neuron.2021.03.035.

Lempert, K. M. and E. A. Phelps (2014). Neuroeconomics of emotion and decision making. In P. Glimcher and E. Fehr (Eds.), *Neuroeconomics* (2nd ed.), pp. 219–236. Cambridge, MA: MIT Press.

Lempert, K. M., J. E. Steinglass, A. Pinto, J. W. Kable, and H. B. Simpson (2019). Can delay discounting deliver on the promise of rdoc? *Psychol. Med. 49*, 190–199.

Lemus, L., A. Hernández, R. Luna, A. Zainos, V. Nácher, and R. Romo (2007). Neural correlates of a postponed decision report. *Proc. Natl. Acad. Sci. U.S.A. 104*, 17174–17179.

Leopold, D. A. and N. K. Logothetis (1996). Activity changes in early visual cortex reflect monkeys' percepts during binocular rivalry. *Nature 379*, 549–553.

Leopold, D. A., Y. Murayama, and N. Logothetis (2003). Very slow activity fluctuations in monkey visual cortex: Implications for functional brain imaging. *Cereb. Cortex 13*, 422–433.

Lerner, Y., C. J. Honey, L. J. Silbert, and U. Hasson (2011). Topographic mapping of a hierarchy of temporal receptive windows using a narrated story. *J. Neurosci. 31*, 2906–2915.

Letzkus, J. J., S. B. Wolff, E. M. Meyer, P. Tovote, J. Courtin, C. Herry, and A. Lüthi (2011). A disinhibitory microcircuit for associative fear learning in the auditory cortex. *Nature 480*, 331–335.

Levelt, W. J. (1965). *On Binocular Rivalry*. Ph.D. thesis. Van Gorcum Assen.

Levitin, D. J. (2006). *This Is Your Brain on Music: The Science of a Human Obsession*. New York: Penguin.

Levitt, B., D. A. Lewis, T. Yoshioka, and J. Lund (1993). Topography of pyramidal neuron intrinsic connections in macaque monkey prefrontal cortex (areas 9 and 46). *J. Comp. Neurol. 338*, 360–376.

Levy, R. B. (2024). The prefrontal cortex: From monkey to man. *Brain 147*, 794–815.

Levy, R. B. and A. D. Reyes (2012). Spatial profile of excitatory and inhibitory synaptic connectivity in mouse primary auditory cortex. *J. Neurosci. 32*, 5609–5619.

Lewis, D., T. Hashimoto, and D. Volk (2005). Cortical inhibitory neurons and schizophrenia. *Nat. Rev. Neurosci. 6*, 312–324.

Li, B., B. N. Routh, D. Johnston, E. Seidemann, and N. J. Priebe (2020). Voltage-gated intrinsic conductances shape the input-output relationship of cortical neurons in behaving primate V1. *Neuron 107*, 185–196.

Li, C.-S. R. (2020). Response inhibition. In A. W. Toga (Ed.), *Brain Mapping: An Encyclopaedic Reference*, Chapter 248. New York: Academic Press.

Li, D., C. Constantinidis, and J. D. Murray (2021a). Trial-to-trial variability of spiking delay activity in prefrontal cortex constrains burst-coding models of working memory. *J. Neurosci. 41*, 8928–8945.

Li, H. H., T. C. Sprague, A. H. Yoo, W. J. Ma, and C. E. Curtis (2021b). Joint representation of working memory and uncertainty in human cortex. *Neuron 109*, 3699–3712.

Li, N., T. W. Chen, Z. V. Guo, C. R. Gerfen, and K. Svoboda (2015). A motor cortex circuit for motor planning and movement. *Nature 519*, 51–56.

Li, N., K. Daie, K. Svoboda, and S. Druckmann (2016). Robust neuronal dynamics in premotor cortex during motor planning. *Nature 532*, 459–464.

Li, Q. and H. Sompolinsky (2020). Statistical mechanics of deep linear neural networks: The back-propagating renormalization group. *arXiv e-prints*, arXiv:2012.04030.

Li, S. and X.-J. Wang (2022). Hierarchical timescales in the neocortex: Mathematical mechanism and biological insights. *Pro. Natl. Acad. Sci. U.S.A. 119*, e2110274119.

Li, Z., F. Zhou, F. Chen, and H. Li (2017). Meta-sgd: Learning to learn quickly for few shot learning. *arXiv preprint arXiv:1707.09835*.

Lichtman, J. W., J. Livet, and J. R. Sanes (2008). A technicolour approach to the connectome. *Nat. Rev. Neurosci. 9*, 417–422.

Lidow, M. S., G. V. Williams, and P. S. Goldman-Rakic (1998). The cerebral cortex: A case for a common site of action of antipsychotics. *Trends Pharmacol. Sci. 19*, 136–140.

Lieder, F. and T. L. Griffiths (2020). Resource-rational analysis: Understanding human cognition as the optimal use of limited computational resources. *Behav. Brain Sci. 43*, https://doi.org/10.1017/S0140525X1900061X.

Lim, S. and M. S. Goldman (2013). Balanced cortical microcircuitry for maintaining information in working memory. *Nat. Neurosci. 16*, 1306–1314.

Lim, S. and M. S. Goldman (2014). Balanced cortical microcircuitry for spatial working memory based on corrective feedback control. *J. Neurosci. 34*, 6790–6806.

Lin, I.-C., M. Okun, M. Carandini, and K. D. Harris (2020). Equations governing dynamics of excitation and inhibition in the mouse corticothalamic network. *bioRxiv*. https://doi.org/10.1101/2020.06.03.132688.

Link, S. W. (1992). *The Wave Theory of Difference and Similarity*. Mahwah, NJ: Lawrence Erlbaum Associates.

Lisman, J. E. (1985). A mechanism for memory storage insensitive to molecular turnover: A bistable autophosphorylating kinase. *Proc. Natl. Acad. Sci. U.S.A. 82*, 3055–3057.

Lisman, J. E. (1997). Bursts as a unit of neural information: Making unreliable synapses reliable. *Trends Neurosci. 20*, 38–43.

Lisman, J. E., J. Coyle, R. Green, D. Javitt, F. Benes, S. Heckers, and A. Grace (2008). Circuit-based framework for understanding neurotransmitter and risk gene interactions in schizophrenia. *Trends Neurosci. 31*, 234–242.

Lisman, J. E. and M. A. Idiart (1995). Storage of 7 ± 2 short-term memories in oscillatory subcycles. *Science 267*, 1512–1515.

Lisman, J. E., R. Yasuda, and S. Raghavachari (2012). Mechanisms of camkii action in long-term potentiation. *Nat. Rev. Neurosci. 13*, 169–182.

Lisman, J. E. and A. M. Zhabotinsky (2001). A model of synaptic memory: A CaMKII/PP1 switch that potentiates transmission by organizing an AMPA receptor anchoring assembly. *Neuron 31*, 191–201.

Litvak, V., H. Sompolinsky, I. Segev, and M. Abeles (2003). On the transmission of rate code in long feedforward networks with excitatory-inhibitory balance. *J. Neurosci. 23*, 3006–3015.

Litwin-Kumar, A. and B. Doiron (2014). Formation and maintenance of neuronal assemblies through synaptic plasticity. *Nat. Commun. 5*, 1–12.

Litwin-Kumar, A., K. D. Harris, R. Axel, H. Sompolinsky, and L. F. Abbott (2017). Optimal degrees of synaptic connectivity. *Neuron 93*, 1153–1164.

Litwin-Kumar, A., R. Rosenbaum, and B. Doiron (2016). Inhibitory stabilization and visual coding in cortical circuits with multiple interneuron subtypes. *J. Neurophysiol. 115*, 1399–1409.

Liu, D., X. Gu, J. Zhu, X. Zhang, Z. Han, W. Yan, Q. Cheng, J. Hao, H. Fan, R. Hou, Z. Chen, Y. Chen, and C. T. Li (2014). Medial prefrontal activity during delay period contributes to learning of a working memory task. *Science 346*, 458–463.

Liu, F. and X.-J. Wang (2008). A common cortical circuit mechanism for perceptual categorical discrimination and veridical judgment. *PLoS Comput. Biol. 4*, e1000253.

Liu, Y. H., R. J. Dolan, Z. Kurth-Nelson, and T. E. J. Behrens (2019). Human replay spontaneously reorganizes experience. *Cell 178*, 640–652.

Liu, Y. H. and X.-J. Wang (2001). Spike-frequency adaptation of a generalized leaky integrate-and-fire model neuron. *J. Comput. Neurosci. 10*, 25–45.

Liu, Y. and X.-J. Wang (2024). Flexible gating between subspaces: A neural network model of internally guided task switching. *Nat. Commun. 15*, 6497.

Liu, Y. H., J. Zhu, C. Constantinidis, and X. Zhou (2021). Emergence of prefrontal neuron maturation properties by training recurrent neural networks in cognitive tasks. *Iscience 24*, 103178.

Liu, Z., J. Golowasch, E. Marder, and L. Abbott (1998). A model neuron with activity-dependent conductances regulated by multiple calcium sensors. *J. Neurosci. 18*, 2309–2320.

Liu, Z., R. Lin, and M. Luo (2020). Reward contributions to serotonergic functions. *Annu. Rev. Neurosci. 43*, 141–162.

Llinás, R. R. (1982). Calcium in synaptic transmission. *Sci. Am. 247*, 56–65.

Llinás, R. R. (1988). The intrinsic electrophysiological properties of mammalian neurons: Insights into central nervous system function. *Science 242*, 1654–1664.

Llinás, R. R. (2002). *I of the Vortex: From Neurons to Self.* Cambridge, MA: MIT Press.

Lo, C. C., L. Boucher, M. Pare, J. D. Schall, and X.-J. Wang (2009). Proactive inhibitory control and attractor dynamics in countermanding action: A spiking neural circuit model. *J. Neurosci. 29*, 9059–9071.

Lo, C. C., C. T. Wang, and X.-J. Wang (2015). Speed-accuracy tradeoff by a control signal with balanced excitation and inhibition. *J. Neurophysiol. 114*, 650–661.

Lo, C. C. and X.-J. Wang (2006). Cortico-basal ganglia circuit mechanism for a decision threshold in reaction time tasks. *Nat. Neurosci. 9*, 956–963.

Lo, C. C. and X.-J. Wang (2016). Conflict resolution as near-threshold decision-making: A spiking neural circuit model with two-stage competition for antisaccadic task. *PLoS Comput. Biol. 12*, e1005081.

Locke, J. (1706). *An Essay Concerning Humane Understanding: In Four Books.* London: Awnsham and John Churchill.

Lockwood, P. L., M. A. Apps, and S. W. Chang (2020). Is there a 'social' brain? Implementations and algorithms. *Trends Cogn. Sci. 24*, 802–813.

Loewenstein, Y. and H. Seung (2006). Operant matching is a generic outcome of synaptic plasticity based on the covariance between reward and neural activity. *Proc. Natl. Acad. Sci. U.S.A. 103*, 15224–15229.

Loewenstein, Y. and H. Sompolinsky (2003). Temporal integration by calcium dynamics in a model neuron. *Nat. Neurosci. 6*, 961–967.

Logan, G. D. (1988). Toward an instance theory of automatization. *Psychol. Rev. 95*, 492–527.

Logan, G. D. and W. B. Cowan (1984). On the ability to inhibit thought and action: A theory of an act of control. *Psych. Rev. 91*, 295–327.

Logothetis, N. and B. Wandell (2004). Interpreting the BOLD signal. *Annu. Rev. Physiol. 66*, 735–769.

London, M., A. Roth, L. Beeren, M. Haüsser, and P. E. Latham (2010). Sensitivity to perturbations in vivo implies high noise and suggests rate coding in cortex. *Nature 466*, 123–127.

Lord, C., S. Risi, L. Lambrecht, E. H. Cook, B. L. Leventhal, P. C. DiLavore, A. Pickles, and M. Rutter (2000). The autism diagnostic observation schedule—generic: A standard measure of social and communication deficits associated with the spectrum of autism. *J. Autism Dev. Disord. 30*, 205–223.

Lorente de Nó, R. (1933). Vestibulo-ocular reflex arc. *Arch. Neurol. Psych. 30*, 245–291.

Lorenz, C. (1961). *King Solomon's Ring.* London: Methuen.

Lorenzon, N. M. and R. C. Foehring (1992). Relationship between repetitive firing and afterhyperpolarizations in human neocortical neurons. *J. Neurophysiol. 67*, 350–363.

Louie, K. and P. W. Glimcher (2010). Separating value from choice: Delay discounting activity in the lateral intraparietal area. *J. Neurosci. 30*, 5498–5507.

Louie, K., L. E. Grattan, and P. W. Glimcher (2011). Reward value-based gain control: Divisive normalization in parietal cortex. *J. Neurosci. 31*, 10627–10639.

Louie, K., T. LoFaro, R. Webb, and P. W. Glimcher (2014). Dynamic divisive normalization predicts time-varying value coding in decision-related circuits. *J. Neurosci. 34*, 16046–16057.

Louie, K. and M. A. Wilson (2001). Temporally structured replay of awake hippocampal ensemble activity during rapid eye movement sleep. *Neuron 29*, 145–156.

Low, R. J., S. Lewallen, D. Aronov, R. Nevers, and D. W. Tank (2018). Probing variability in a cognitive map using manifold inference from neural dynamics. *bioRxiv*, 418939.

Lu, J., J. Tucciarone, N. Padilla-Coreano, M. He, J. A. Gordon, and Z. J. Huang (2017). Selective inhibitory control of pyramidal neuron ensembles and cortical subnetworks by chandelier cells. *Nat. Neurosci. 20*, 1377–1383.

Lubenov, E. V. and A. G. Siapas (2009). Hippocampal theta oscillations are travelling waves. *Nature 459*, 534–539.

Luce, R. D. (1986). *Response Time: Their Role in Inferring Elementary Mental Organization*. New York: Oxford University Press.

Luce, R. D. (2005). *Individual Choice Behavior: A Theoretical Analysis*. New York: Dover Publications.

Luck, S. J. and E. K. Vogel (1997). The capacity of visual working memory for features and conjunctions. *Nature 390*, 279–281.

Luck, S. J. and E. K. Vogel (2013). Visual working memory capacity: From psychophysics and neurobiology to individual differences. *Trends Cogn. Sci. (Regul. Ed.) 17*, 391–400.

Luhmann, C. C., M. M. Chun, D.-J. Yi, D. Lee, and X.-J. Wang (2008). Neural dissociation of delay and uncertainty in intertemporal choice. *J. Neurosci. 28*, 14459–14466.

Lukosevicius, M. and H. Jaeger (2009). Reservoir computing approaches to recurrent neural network training. *Comput. Sci. Rev. 3*, 127–149.

Lundqvist, M., P. Herman, and E. K. Miller (2018). Working memory: Delay activity, yes! Persistent activity? Maybe not. *J. Neurosci. 38*, 7013–7019.

Lundqvist, M., J. Rose, P. Herman, S. L. Brincat, T. J. Buschman, and E. K. Miller (2016). Gamma and beta bursts underlie working memory. *Neuron 90*, 152–164.

Luo, L., E. Callaway, and K. Svoboda (2008). Genetic dissection of neural circuits. *Neuron 57*, 634–660.

Luo, T. Z., T. D. Kim, D. Gupta, A. G. Bondy, C. D. Kopec, V. A. Elliot, B. DePasquale, and C. D. Brody (2023). Transitions in dynamical regime and neural mode underlie perceptual decision-making. *bioRxiv*. https://doi.org/10.1101/2023.10.15.562427.

Luria, A. R. (1980). *Higher Cortical Functions in Man* (2nd ed.). New York: Plenum Publishing (The original book in Russian was first published in 1962).

Lurz, R., C. Krachun, W. D. Hopkins, and J. Taglialatela (2022). Evidence of psychological essentialism in a symbol-trained bonobo (pan paniscus). *Int. J. Primatol. 43*, 235–252.

Lyu, C., Abbott, L. F., and Maimon, G. (2022). Building an allocentric travelling direction signal via vector computation. *Nature 601*, 92–97.

Ma, W. J. (2010). Signal detection theory, uncertainty, and poisson-like population codes. *Vis. Res. 50*, 2308–2319.

Ma, W. J. (2019). Bayesian decision models: A primer. *Neuron 104*, 164–175.

Ma, W. J., J. Beck, P. Latham, and A. Pouget (2006). Bayesian inference with probabilistic population codes. *Nat. Neurosci. 9*, 1432–1438.

Ma, W. J., M. Husain, and P. M. Bays (2014). Changing concepts of working memory. *Nat. Neurosci. 17*, 347–356.

Ma, Y., H. Hu, and A. Agmon (2012). Short-term plasticity of unitary inhibitory-to-inhibitory synapses depends on the presynaptic interneuron subtype. *J. Neurosci. 32*, 983–988.

Maass, W., T. Natschläger, and H. Markram (2002). Real-time computing without stable states: A new framework for neural computation based on perturbations. *Neural Comput. 14*, 2531–2560.

MacDonald, A. W., C. S. Carter, J. G. Kerns, S. Ursu, D. M. Barch, A. J. Holmes, V. A. Stenger, and J. D. Cohen (2005). Specificity of prefrontal dysfunction and context processing deficits to schizophrenia in never-medicated patients with first-episode psychosis. *Am. J. Psychiatry 162*, 475–484.

Machens, C. K., R. Romo, and C. D. Brody (2005). Flexible control of mutual inhibition: A neural model of two-interval discrimination. *Science 18*, 1121–1124.

Machens, C. K., R. Romo, and C. D. Brody (2010). Functional, but not anatomical, separation of 'what' and 'when' in prefrontal cortex. *J. Neurosci. 30*, 350–360.

Machizawa, M. G. and J. Driver (2011). Principal component analysis of behavioural individual differences suggests that particular aspects of visual working memory may relate to specific aspects of attention. *Neuropsychologia 49*, 1518–1526.

Machold, R., S. Dellal, M. Valero, H. Zurita, I. Kruglikov, J. H. Meng, . . . Rudy, B. (2023). Id2 GABAergic interneurons comprise a neglected fourth major group of cortical inhibitory cells. *Elife 12*, e85893.

Mackintosh, N. J. (1974). *The Psychology of Animal Learning*. London: Acdemic Press.

Maclean, E. L. (2016). Unraveling the evolution of uniquely human cognition. *Proc. Natl. Acad. Sci. 113*, 6348–6354.

MacLeod, C. M. (1991). Half a century of research on the Stroop effect: An integrative review. *Psychol. Bull. 109*, 163–203.

Macy, M. W. and A. Flache (2002). Learning dynamics in social dilemmas. *Proc. Natl. Acad. Sci. 99*, 7229–7236.

Magee, J. C. and C. Grienberger (2020). Synaptic plasticity forms and functions. *Annu. Rev. Neurosci. 43*, 96–117.

Maia, T. V. (2010). Two-factor theory, the actor-critic model, and conditioned avoidance. *Learn. Behav. 38*, 50–67.

Maia, T. V. and M. Cano-Colino (2015). The role of serotonin in orbitofrontal function and obsessive-compulsive disorder. *Clin. Psychol. Sci. 3*, 460–482.

Mainen, Z. F., R. Malinow, and K. Svoboda (1999). Synaptic calcium transients in single spines indicate that nmda receptors are not saturated. *Nature 399*, 151–155.

Major, G. and D. Tank (2004). Persistent neural activity: Prevalence and mechanisms. *Curr. Opin. Neurobiol. 14*, 675–684.

Malenka, R. C. and M. F. Bear (2004). LTP and LTD: An embarrassment of riches. *Neuron 44*, 5–21.

Malenka, R. C. and R. A. Nicoll (1999). Long-term potentiation–a decade of progress? *Science 285*, 1870–1874.

Malerba, P. and M. Bazhenov (2019). Circuit mechanisms of hippocampal reactivation during sleep. *Neurobiol. Learn. Mem. 160*, 98–107.

Mallet, N., R. Schmidt, D. Leventhal, F. Chen, N. Amer, T. Boraud, and J. D. Berke (2016). Arkypallidal cells send a stop signal to striatum. *Neuron 89*, 308–316.

Mamassian, P. (2016). Visual confidence. *Annu. Rev. Vis. Sci. 2*, 459–481.

Manning, J. R., J. Jacobs, I. Fried, and M. J. Kahana (2009). Broadband shifts in local field potential power spectra are correlated with single-neuron spiking in humans. *J. Neurosci. 29*, 13613–13620.

Manns, J. R. and H. Eichenbaum (2006). Evolution of declarative memory. *Hippocampus 16*, 795–808.

Mansouri, F. A., M. J. Buckley, and K. Tanaka (2007). Mnemonic function of the dorsolateral prefrontal cortex in conflict-induced behavioral adjustment. *Science 318*, 987–990.

Mansouri, F. A., D. J. Freedman, and M. J. Buckley (2020). Emergence of abstract rules in the primate brain. *Nat. Rev. Neurosci. 21*, 595–610.

Mansouri, F. A., K. Matsumoto, and K. Tanaka (2006). Prefrontal cell activities related to monkeys' success and failure in adapting to rule changes in a Wisconsin Card Sorting Test analog. *J. Neurosci. 26*, 2745–2756.

Mansouri, F. A., K. Tanaka, and M. J. Buckley (2009). Conflict-induced behavioural adjustment: A clue to the executive functions of the prefrontal cortex. *Nat. Rev. Neurosci. 10*, 141–152.

Mante, V., D. Sussillo, K. V. Shenoy, and W. T. Newsome (2013). Context-dependent computation by recurrent dynamics in prefrontal cortex. *Nature 503*, 78–84.

Mao, J., C. Gan, P. Kohli, J. B. Tenenbaum, and J. Wu (2019). The neuro-symbolic concept learner: Interpreting scenes, words, and sentences from natural supervision. *arXiv preprint arXiv:1904.12584*.

Marcus, G. F. (2003). *The Algebraic Mind: Integrating Connectionism and Cognitive Science*. Cambridge, MA: MIT Press.

Marder, E., L. Abbott, G. Turrigiano, Z. Liu, and J. Golowash (1996). Memory from the dynamics of intrinsic membrane currents. *Proc. Natl. Acad. Sci. U.S.A. 93*, 13482–13486.

Marder, E. and R. L. Calabrese (1996). Principles of rhythmic motor pattern generation. *Physiol. Rev. 76*, 687–717.

Marder, E. and J.-M. Goaillard (2006). Variability, compensation and homeostasis in neuron and network function. *Nat. Rev. Neurosci. 7*, 563–574.

Marder, E. and A. A. Prinz (2002). Modeling stability in neuron and network function: The role of activity in homeostasis. *Bioessays 24*, 1145–1154.

Marder, E. and M. C. Rue (2021). From the neuroscience of individual variability to climate change. *J. Neurosci. 41*, 10213–10221.

Marek, S., B. Tervo-Clemmens, F. J. Calabro, D. F. Montez, B. P. Kay, A. S. Hatoum, M. R. Donohue, W. Foran, R. L. Miller, T. J. Hendrickson, et al. (2022). Reproducible brain-wide association studies require thousands of individuals. *Nature 603*, 654–660.

Maren, S. (2001). Neurobiology of Pavlovian fear conditioning. *Annu. Rev. Neurosci. 24*, 897–931.

Margulies, D. S., S. S. Ghosh, A. Goulas, M. Falkiewicz, J. M. Huntenburg, G. Langs, G. Bezgin, S. B. Eickhoff, F. X. Castellanos, M. Petrides, E. Jefferies, and J. Smallwood (2016). Situating the default-mode network along a principal gradient of macroscale cortical organization. *Proc. Natl. Acad. Sci. U.S.A. 113*, 12574–12579.

Markov, N. T., M. M. Ercsey-Ravasz, D. C. Van Essen, K. Knoblauch, Z. Toroczkai, and H. Kennedy (2013). Cortical high-density counterstream architectures. *Science 342*, 1238406.

Markov, N. T., M. M. Ercsey-Ravasz, A. R. Ribeiro Gomes, C. Lamy, L. Magrou, J. Vezoli, P. Misery, A. Falchier, R. Quilodran, M. A. Gariel, J. Sallet, R. Gamanut, C. Huissoud, S. Clavagnier, P. Giroud, D. Sappey-Marinier, P. Barone, C. Dehay, Z. Toroczkai, K. Knoblauch, D. C. Van Essen, and H. Kennedy (2014a). A weighted and directed interareal connectivity matrix for macaque cerebral cortex. *Cereb. Cortex 24*, 17–36.

Markov, N. T., J. Vezoli, P. Chameau, A. Falchier, R. Quilodran, C. Huissoud, C. Lamy, P. Misery, P. Giroud, S. Ullman, P. Barone, C. Dehay, K. Knoblauch, and H. Kennedy (2014b). Anatomy of hierarchy: Feedforward and feedback pathways in macaque visual cortex. *J. Comp. Neurol. 522*, 225–259.

Markowitz, J. E., W. F. Gillis, C. C. Beron, S. Q. Neufeld, K. Robertson, N. D. Bhagat, R. E. Peterson, E. Peterson, M. Hyun, S. W. Linderman, B. L. Sabatini, and S. R. Datta (2018). The striatum organizes 3D behavior via moment-to-moment action selection. *Cell 174*, 44–58.

Markram, H., J. Lubke, M. Frotscher, A. Roth, and B. Sakmann (1997). Physiology and anatomy of synaptic connections between thick tufted pyramidal neurones in the developing rat neocortex. *J. Physiol. (London) 500*, 409–440.

Markram, H., M. Toledo-Rodriguez, Y. Wang, A. Gupta, G. Silberberg, and C. Wu (2004). Interneurons of the neocortical inhibitory system. *Nat. Rev. Neurosci. 5*, 793–807.

Markram, H. and M. Tsodyks (1996). Redistribution of synaptic efficacy between neocortical pyramidal neurons. *Nature 382*, 807–810.

Markram, H., Y. Wang, and M. Tsodyks (1998). Differential signaling via the same axon of neocortical pyramidal neurons. *Proc. Natl. Acad. Sci. U.S.A. 95*, 5323–5328.

Marom, S. (2010). Neural timescales or lack thereof. *Prog. Neurobiol. 90*, 16–28.

Marr, D. (1969). A theory of cerebellar cortex. *J. Physiol. (Lond.) 202*, 437–470.

Marr, D. (1970). A theory for cerebral neocortex. *Proc. Roy. Soc. Lond. B. 176*, 161–234.

Marr, D. (1971). Simple memory: A theory for archicortex. *Philos. Trans. R. Soc. London B 262*, 23–81.

Marr, D. (1982). *Vision*. New York: W.H. Freeman & Co.

Marr, D. and T. Poggio (1977). From understanding computation to understanding neural circuitry. In E. Poppel, R. Held, and J. E. Dowling (Eds.), *Neurosciences Research Program Bulletin* (Vol. 15), pp. 470–488. Boca Raton: CRC Press.

Martens, J. and I. Sutskever (2011). Learning recurrent neural networks with hessian-free optimization. In *Proceedings of the 28th International Conference on Machine Learning (ICML-11)*, pp. 1033–1040.

Martinez-Trujillo, J. and S. Treue (2004). Feature-based attention increases the selectivity of population responses in primate visual cortex. *Curr. Biol. 14*, 744–751.

Mashour, G. A., P. Roelfsema, J. P. Changeux, and S. Dehaene (2020). Conscious processing and the global neuronal workspace hypothesis. *Neuron 105*, 776–798.

Masse, N. Y., M. C. Rosen, and D. J. Freedman (2020). Reevaluating the role of persistent neural activity in short-term memory. *Trends Cogn. Sci. 24*, 242–258.

Masse, N. Y., G. R. Yang, H. F. Song, X.-J. Wang, and D. J. Freedman (2019). Circuit mechanisms for the maintenance and manipulation of information in working memory. *Nat. Neurosci. 22*, 1159–1167.

Massimini, M., R. Huber, F. Ferrarelli, S. Hill, and G. Tononi (2004). The sleep slow oscillation as a traveling wave. *J. Neurosci. 24*, 6862–6870.

Mastrogiuseppe, F. and S. Ostojic (2018). Linking connectivity, dynamics, and computations in low-rank recurrent neural networks. *Neuron 99*, 609–623.

Mathis, A., P. Mamidanna, K. M. Cury, T. Abe, V. N. Murthy, M. W. Mathis, and M. Bethge (2018). Deeplabcut: Markerless pose estimation of user-defined body parts with deep learning. *Nat. Neurosci. 21*, 1281.

Mathy, F. and J. Feldman (2012). What's magic about magic numbers? Chunking and data compression in short-term memory. *Cognition 122*, 346–362.

Matsuda, Y., A. Marzo, and S. Otani (2006). The presence of background dopamine signal converts long-term synaptic depression to potentiation in rat prefrontal cortex. *J. Neurosci. 26*, 4803–4810.

Mattar, M. G. and M. Lengyel (2022). Planning in the brain. *Neuron 110*, 914–934.

Matveev, V. and X.-J. Wang (2000). Differential short-term synaptic plasticity and transmission of complex spike trains: To depress or to facilitate? *Cereb. Cortex 10*, 1143–1153.

Maunsell, J. H. R. (2015). Neuronal mechanisms of visual attention. *Annu. Rev. Vis. Sci. 1*, 373–391.

Maunsell, J. H. R. and S. Treue (2006). Feature-based attention in visual cortex. *Trends Neurosci. 29*, 317–322.

Mayberg, H. S., A. M. Lozano, V. Voon, H. E. McNeely, D. Seminowicz, C. Hamani, J. M. Schwalb, and S. H. Kennedy (2005). Deep brain stimulation for treatment-resistant depression. *Neuron 45*, 651–660.

Mayer, J. D., D. R. Caruso, and P. Salovey (1999). Emotional intelligence meets traditional standards for an intelligence. *Intelligence 27*, 267–298.

Mayer, J. S. and S. Park (2012). Working memory encoding and false memory in schizophrenia and bipolar disorder in a spatial delayed response task. *J. Abnorm. Psychol. 121*, 784–794.

Mazor, O. and G. Laurent (2005). Transient dynamics versus fixed points in odor representations by locust antennal lobe projection neurons. *Neuron 48*, 661–673.

Mazur, J. E. (1981). Optimization theory fails to predict performance of pigeons in a two-response situation. *Science 214*, 823–825.

Mazur, J. E. (1987). An adjusting procedure for studying delayed reinforcement. In M. L. Commons, J. E. Mazur, J. A. Nevin, and H. Rachlin (Eds.), *Quantitative Analysis of Behavior: The Effects of Delay and Intervening Events on Reinforcement*, pp. 55–73. Hillsdale, NJ: Lawrence Erlbaum Associates, Publishers.

Mazurek, M. E., J. D. Roitman, J. Ditterich, and M. N. Shadlen (2003). A role for neural integrators in perceptual decision making. *Cereb. Cortex. 13*, 1257–1269.

Mazzoni, A., N. Brunel, S. Cavallari, N. K. Logothetis, and S. Panzeri (2011). Cortical dynamics during naturalistic sensory stimulations: Experiments and models. *J. Physiol. Paris 105*, 2–15.

Mazzoni, A., K. Whittingstall, N. Brunel, N. K. Logothetis, and S. Panzeri (2009). Understanding the relationships between spike rate and delta/gamma frequency bands of LFPs and EEGs using a local cortical network model. *Neuroimage 52*, 956–972.

Mazzoni, A., K. Whittingstall, N. Brunel, N. K. Logothetis, and S. Panzeri (2010). Understanding the relationships between spike rate and delta/gamma frequency bands of lfps and eegs using a local cortical network model. *Neuroimage 52*, 956–972.

McAdams, C. J. and J. H. R. Maunsell (1999). Effects of attention on orientation-tuning functions of single neurons in macaque cortical area V4. *J. Neurosci. 19*, 431–441.

McCarthy, G. and E. Donchin (1981). A metric for thought: A comparison of P300 latency and reaction time. *Science 211*, 77–80.

McClelland, J. L., B. L. McNaughton, and R. C. O'Reilly (1995). Why there are complementary learning systems in the hippocampus and neocortex: Insights from the successes and failures of connectionist models of learning and memory. *Psychol. Rev. 102*, 419–457.

McCloskey, M. and N. J. Cohen (1989). Catastrophic interference in connectionist networks: The sequential learning problem. *Psychology of Learning and Motivation 24*, 109–165.

McClure, S. M., D. I. Laibson, G. Loewenstein, and J. D. Cohen (2004). Separate neural systems value immediate and delayed monetary rewards. *Science 306*, 503–507.

McCormick, D. A., B. Connors, J. Lighthall, and D. Prince (1985). Comparative electrophysiology of pyramidal and sparsely spiny stellate neurons in the neocortex. *J. Neurophysiol. 54*, 782–806.

McCormick, D. A. and J. R. Huguenard (1992). A model of the electrophysiological properties of thalamocortical relay neurons. *J. Neurophysiol. 68*, 1384–1400.

McCulloch, W. S. and W. Pitts (1943). A logical calculus of the ideas immanent in nervous activity. *Bull. Math. Biophys. 5*, 115–133.

McEwen, B. S. (1998). Stress, adaptation, and disease: Allostasis and allostatic load. *Ann. N. Y. Acad. Sci. 840*, 33–44.

McEwen, B. S. and J. C. Wingfield (2003). The concept of allostasis in biology and biomedicine. *Horm. Behav. 43*, 2–15.

McGonigle, B. O. and M. Chalmers (1977). Are monkeys logical? *Nature 267*, 694–696.

McMillen, T. and P. Holmes (2006). The dynamics of choice among multiple alternatives. *J. Math. Psych. 50*, 30–57.

McNaughton, B. L., F. P. Battaglia, O. Jensen, E. I. Moser, and M. B. Moser (2006). Path integration and the neural basis of the 'cognitive map'. *Nat. Rev. Neurosci. 7*, 663–678.

McPeek, R. and E. Keller (2002). Saccade target selection in the superior colliculus during a visual search task. *J. Neurophysiol. 88*, 2019–2034.

Meck, W. H. (2006). Neuroanatomical localization of an internal clock: A functional link between mesolimbic, nigrostriatal, and mesocortical dopaminergic systems. *Brain Res. 1109*, 93–107.

Mehta, M., M. Quirk, and M. Wilson (2000). Experience-dependent asymmetric shape of hippocampal receptive fields. *Neuron 25*, 707–715.

Meiran, N. (2000). Modeling cognitive control in task-switching. *Psychol. Res. 63*, 234–249.

Mejias, J. F., J. D. Murray, H. Kennedy, and X. J. Wang (2016). Feedforward and feedback frequency-dependent interactions in a large-scale laminar network of the primate cortex. *Sci. Adv. 2*, e1601335.

Mejias, J. F. and X.-J. Wang (2022). Mechanisms of distributed working memory in a large-scale model of the macaque neocortex. *eLife 11*, e72136.

Mello, G. B., S. Soares, and J. J. Paton (2015). A scalable population code for time in the striatum. *Curr. Biol. 25*, 1113–1122.

Mendona, P. R., M. Vargas-Caballero, F. Erdélyi, S. G., O. Paulsen, and H. P. Robinson (2016). Stochastic and deterministic dynamics of intrinsically irregular firing in cortical inhibitory interneurons. *Elife 5*, e16475.

Mendoza-Halliday, D., S. Torres, and J. C. Martinez-Trujillo (2014). Sharp emergence of feature-selective sustained activity along the dorsal visual pathway. *Nat. Neurosci. 17*, 1255–1262.

Mendoza-Halliday, D., H. Xu, F. A. Azevedo, and R. Desimone (2024). Dissociable neuronal substrates of visual feature attention and working memory. *Neuron 112*, 850–863.

Meng, J. H., B. Schuman, B. Rudy, and X.-J. Wang (2023). Mechanisms of dominant electrophysiological features of four subtypes of layer 1 interneurons. *J. Neurosci. 43*, 3202–3218.

Merchant, H., D. L. Harrington, and W. H. Meck (2013). Neural basis of the perception and estimation of time. *Ann. Rev. Neurosci. 36*, 313–336.

Meskenaite, V. (1997). Calretinin-immunoreactive local circuit neurons in the area 17 of the cynomolgus monkey, Macaca fascicularis. *J. Comp. Neurol. 379*, 113–132.

Mesulam, M.-M. (2000). *Principles of Behavioral and Cognitive Neurology* (2nd ed.). New York: Oxford University Press.

Meyer, D. E., A. M. Osman, D. E. Irwin, and S. Yantis (1988). Modern mental chronometry. *Biol. Psychol. 26*, 3–67.

Meyer-Lindenberg, A. and D. R. Weinberger (2006). Intermediate phenotypes and genetic mechanisms of psychiatric disorders. *Nat. Rev. Neurosci. 7*, 818–827.

Meyers, E. M. (2018). Dynamic population coding and its relationship to working memory. *J. Neurophysiol. 120*, 2260–2268.

Meyers, E. M., D. J. Freedman, G. Kreiman, E. K. Miller, and T. Poggio (2008). Dynamic population coding of category information in inferior temporal and prefrontal cortex. *J. Neurophysiol. 100*, 1407–1419.

Meyers, E. M., X. L. Qi, and C. Constantinidis (2012). Incorporation of new information into prefrontal cortical activity after learning working memory tasks. *Proc. Natl. Acad. Sci. U.S.A. 109*, 4651–4656.

Meyniel, F., M. Sigman, and Z. F. Mainen (2015). Confidence as bayesian probability: From neural origins to behavior. *Neuron 88*, 78–92.

Mi, Y., M. Katkov, and M. Tsodyks (2017). Synaptic correlates of working memory capacity. *Neuron 93*, 323–330.

Miall, C. (1989). The storage of time intervals using oscillating neurons. *Neural Comput. 1*, 359–371.

Michalareas, G., J. Vezoli, S. van Pelt, J. M. Schoffelen, H. Kennedy, and P. Fries (2016). Alpha-beta and gamma rhythms subserve feedback and feedforward influences among human visual cortical areas. *Neuron 89*, 384–397.

MICrONS Consortium, J. A. Bae, M. Baptiste, C. A. Bishop, A. L. Bodor, D. Brittain, J. Buchanan, D. J. Bumbarger, M. A. Castro, B. Celii, et al. (2021). Functional connectomics spanning multiple areas of mouse visual cortex. *BioRxiv*. https://doi.org/10.1101/2021.07.28.454025.

Middlebrooks, P. G. and J. D. Schall (2014). Response inhibition during perceptual decision making in humans and macaques. *Atten. Percept. Psychophys 76*, 353–366.

Middlebrooks, P. G. and M. A. Sommer (2012). Neuronal correlates of metacognition in primate frontal cortex. *Neuron 75*, 517–530.

Miller, E. K. and J. D. Cohen (2001). An integrative theory of prefrontal cortex function. *Annu. Rev. Neurosci. 24*, 167–202.

Miller, E. K., C. A. Erickson, and R. Desimone (1996). Neural mechanisms of visual working memory in prefrontal cortex of the macaque. *J. Neurosci. 16*, 5154–5167.

Miller, E. K., L. Li, and R. Desimone (1993). Activity of neurons in anterior inferior temporal cortex during a short-term memory task. *J. Neurosci. 13*, 1460–1478.

Miller, E. K., A. Nieder, D. J. Freedman, and J. D. Wallis (2003a). Neural correlates of categories and concepts. *Curr. Opin. Neurobiol. 13*, 198–203.

Miller, G. A. (1956). The magical number seven, plus or minus two: Some limits on our capacity for processing information. *Psychol. Rev. 63*, 81.

Miller, J. A., A. Tambini, A. Kiyonaga, and M. D'Esposito (2022). Long-term learning transforms prefrontal cortex representations during working memory. *Neuron 110*, 3805–3819.

Miller, K. D. (1990). Correlation-based models of neural development. In D. E. R. M. A. Gluck (Ed.), *Neuroscience and Connectionist Theory*, pp. 267–363. Hillsdale, NJ: Lawrence Erlbaum Associates.

Miller, K. D., J. B. Keller, and M. P. Stryker (1989). Ocular dominance column development: Analysis and simulation. *Science 245*, 605–615.

Miller, K. J., L. B. Sorensen, J. G. Ojemann, and M. den Nijs (2009). Power-law scaling in the brain surface electric potential. *PLoS Comput. Biol. 5*, e1000609.

Miller, P. (2016). Itinerancy between attractor states in neural systems. *Curr. Opin. Neurobiol. 40*, 14–22.

Miller, P., C. D. Brody, R. Romo, and X.-J. Wang (2003b). A recurrent network model of somatosensory parametric working memory in the prefrontal cortex. *Cereb. Cortex 13*, 1208–1218.

Miller, P. and D. B. Katz (2010). Stochastic transitions between neural states in taste processing and decision-making. *J. Neurosci. 30*, 2559–2570.

Miller, P. and X.-J. Wang (2006). Inhibitory control by an integral feedback signal in prefrontal cortex: A model of discrimination between sequential stimuli. *Proc. Natl. Acad. Sci. U.S.A. 103*, 201–206.

Miller, P., A. M. Zhabotinsky, J. E. Lisman, and X.-J. Wang (2005). The stability of a stochastic CaMKII switch: Dependence on the number of enzyme molecules and protein turnover. *PLoS Biol. 3*, e107.

Milne, E., A. Scope, O. Pascalis, D. Buckley, and S. Makeig (2009). Independent component analysis reveals atypical electroencephalographic activity during visual perception in individuals with autism. *Biol. Psychiatry 65*, 22–30.

Milne, E., J. Swettenham, P. Hansen, R. Campbell, H. Jeffries, and K. Plaisted (2002, Feb). High motion coherence thresholds in children with autism. *J. Child Psychol. Psychiatry 43*, 255–263.

Milner, B. (1963). Effects of different brain lesions on card sorting. *Arch. Neurol. 9*, 90–100.

Milner, G. A., E. Galanter, and K. H. Pribram (1960). *Plans and the Structure of Behavior*. New York: Adams Bannister Cox.

Milstein, A. D., Y. Li, K. C. Bittner, C. Grienberger, I. Soltesz, J. C. Magee, and S. Romani (2021). Bidirectional synaptic plasticity rapidly modifies hippocampal representations. *Elife 10*, e73046.

Milton, J. G., P. H. Chu, and J. D. Cowan (1993). Spiral waves in integrate-and-fire neural network. In C. L. G. S. J. Hanson and J. D. Cowan (Eds.), *Advances in Neural Information Processing Systems*, pp. 1001–1007. San Mateo: Morgan Kaufmann.

Min, B., D. P. Bliss, A. Sarma, D. J. Freedman, and X.-J. Wang (2020). A neural circuit mechanism of categorical perception: Top-down signaling in the primate cortex. *bioRxiv*. https://doi.org/10.1101/2020.06.15.151506.

Mindell, D. A. and D. A. Mindell (2002). *Between Human and Machine: Feedback, Control, and Computing before Cybernetics*. Baltimore, MA: Johns Hopkins University Press.

Minsky, M. (1988). *Society of Mind*. New York: Simon and Schuster.

Minsky, M. and S. Papert (1969). *Perceptrons*. Cambridge, MA: MIT Press.

Minxha, J., R. Adolphs, S. Fusi, A. N. Mamelak, and U. Rutishauser (2020). Flexible recruitment of memory-based choice representations by the human medial frontal cortex. *Science 368*(6498).

Mirabella, G., G. Bertini, I. Samengo, B. Kilavik, D. Frilli, C. Della Libera, and L. Chelazzi (2007). Neurons in area V4 of the macaque translate attended visual features into behaviorally relevant categories. *Neuron 54*, 303–318.

Mischel, W., Y. Shoda, and M. I. Rodriguez (1989). Delay of gratification in children. *Science 244*, 933–938.

Mitchell, J., K. Sundberg, and J. Reynolds (2007). Differential attention-dependent response modulation across cell classes in macaque visual area V4. *Neuron 55*, 131–141.

Mitchell, M. (2023). How do we know how smart AI systems are? *Science 381*, adj5957.

Miyashita, Y. and H. S. Chang (1988). Neuronal correlate of pictorial short-term memory in the primate temporal cortex. *Nature 331*, 68–70.

Mkrtchian, A., J. Aylward, P. Dayan, J. P. Roiser, and O. J. Robinson (2017). Modeling avoidance in mood and anxiety disorders using reinforcement learning. *Biol. Psychiatry 82*, 532–539.

Mlodinow, L. (2022). *Emotional: How Feelings Shape Our Thinking*. New York: Pantheon.

Mnih, V., A. P. Badia, M. Mirza, A. Graves, T. Lillicrap, T. Harley, D. Silver, and K. Kavukcuoglu (2016). Asynchronous methods for deep reinforcement learning. *PMLR*. 1928–1937.

Mnih, V., K. Kavukcuoglu, D. Silver, A. A. Rusu, J. Veness, M. G. Bellemare, A. Graves, M. Riedmiller, A. K. Fidjeland, G. Ostrovski, et al. (2015). Human-level control through deep reinforcement learning. *Nature 518*, 529–533.

Mobbs, D. and J. J. Kim (2015). Neuroethological studies of fear, anxiety, and risky decision-making in rodents and humans. *Curr. Opin. Behav. Sci. 5*, 8–15.

Mogenson, G. J., D. L. Jones, and C. Y. Yim (1980). From motivation to action: Functional interface between the limbic system and the motor system. *Prog. Neurobiol. 14*, 69–97.

Momennejad, I. (2020). Learning structures: Predictive representations, replay, and generalization. *Curr. Opin. Behav. Sci. 32*, 155–166.

Mongillo, G., O. Barak, and M. Tsodyks (2008). Synaptic theory of working memory. *Science 319*, 1543–1546.

Montague, P. R., P. Dayan, and T. Sejnowski (1996). A framework for mesencephalic dopamine systems based on predictive Hebbian learning. *J. Neurosci. 16*, 1936–1947.

Montague, P. R., R. J. Dolan, K. J. Friston, and P. Dayan (2012). Computational psychiatry. *Trends Cogn. Sci. 16*, 72–80.

Moore, T. (2006). The neurobiology of visual attention: Finding sources. *Curr. Opin. Neurobiol. 16*, 159–165.

Moore, T. and K. M. Armstrong (2003). Selective gating of visual signals by microstimulation of frontal cortex. *Nature 421*, 370–373.

Moore, T. and M. Zirnsak (2017). Neural mechanisms of selective visual attention. *Annu. Rev. Psychol. 68*, 47–72.

Moran, J. and R. Desimone (1985). Selective attention gates visual processing in the extrastriate cortex. *Science 229*, 782–784.

Moravec, H. (1988). *Mind Children: The Future of Robot and Human Intelligence*. Cambridge, MA: Harvard University Press.

Moreno-Bote, R. (2010). Decision confidence and uncertainty in diffusion models with partially correlated neuronal integrators. *Neural Comput. 22*, 1786–1811.

Moreno-Bote, R., J. Rinzel, and N. Rubin (2007). Noise-induced alternations in an attractor network model of perceptual bistability. *J. Neurophysiol. 98*, 1125–1139.

Mori, S., B. J. Crain, V. P. Chacko, and P. C. Van Zijl (1999). Three-dimensional tracking of axonal projections in the brain by magnetic resonance imaging. *Ann. Neurol. 45*, 265–269.

Morris, R. G. (2006). Elements of a neurobiological theory of hippocampal function: The role of synaptic plasticity, synaptic tagging and schemas. *Eur. J. Neurosci. 23*, 2829–2846.

Morton, N. W., M. L. Schlichting, and A. R. Preston (2020). Representations of common event structure in medial temporal lobe and frontoparietal cortex support efficient inference. *Proc. Natl. Acad. Sci. U.S.A. 117*, 29338–29345.

Moschovakis, A. K., C. A. Scudder, and S. M. Highstein (1996). The microscopic anatomy and physiology of the mammalian saccadic system. *Prog. Neurobiol. 50*, 133–254.

Moser, E. I., Y. Roudi, M. P. Witter, C. Kentros, T. Bonhoeffer, and M. B. Moser (2014). Grid cells and cortical representation. *Nat. Rev. Neurosci. 15*, 466–481.

Mountcastle, V. B. (1997). The columnar organization of the neocortex. *Brain 120*, 701–722.

Mountcastle, V. B., M. Steinmetz, and R. Romo (1990). Frequency discrimination in the sense of flutter: Psychophysical measurements correlated with postcentral events in behaving monkeys. *J. Neurosci. 10*, 3032–3044.

Mountcastle, V. B., W. Talbot, H. Sakata, and J. Hyvärinen (1969). Cortical neuronal mechanisms in flutter-vibration studied in unasthetized monkeys. neuronal periodicity and frequency discrimination. *J. Neurophysiol. 32*, 452–.

Moutoussis, M., N. Shahar, T. U. Hauser, and R. J. Dolan (2018). Computation in psychotherapy, or how computational psychiatry can aid learning-based psychological therapies. *Comput. Psychiatr. 2*, 50–73.

Movshon, J. A., T. D. Albright, G. R. Stoner, N. J. Majaj, and M. A. Smith (2003). Cortical responses to visual motion in alert and anesthetized monkeys. *Nat. Neurosci. 6*, 3.

Mueller, A., R. M. Krock, S. Shepard, and T. Moore (2020). Dopamine receptor expression among local and visual cortex-projecting frontal eye field neurons. *Cereb. Cortex 30*, 148–164.

Mullally, S. L. and E. A. Maguire (2014). Memory, imagination, and predicting the future: A common brain mechanism? *Neuroscientist 20*, 220–234.

Muller, L., F. Chavane, J. Reynolds, and T. J. Sejnowski (2018). Cortical travelling waves: Mechanisms and computational principles. *Nat. Rev. Neurosci. 19*, 255–268.

Muller, L., G. Piantoni, D. Koller, S. S. Cash, E. Halgren, and T. J. Sejnowski (2016). Rotating waves during human sleep spindles organize global patterns of activity that repeat precisely through the night. *Elife 5*. https://doi.org/10.7554/eLife.09868.

Muller, S. Z., A. N. Zadina, L. Abbott, and N. B. Sawtell (2019). Continual learning in a multi-layer network of an electric fish. *Cell 179*, 1382–1392.

Muly, E. C., K. Szigeti, and P. S. Goldman-Rakic (1998). D1 receptor in interneurons of macaque prefrontal cortex: Distribution and subcellular localization. *J. Neurosci. 18*, 10553–10565.

Mumford, D. (1992). On the computational architecture of the neocortex. II. The role of cortico-cortical loops. *Biol. Cybern 66*, 241–251.

Munn, R. G. and L. M. Giocomo (2020). Multiple head direction signals within entorhinal cortex: Origin and function. *Curr. Opin. Neurobiol. 64*, 32–40.

Munoz, D. P. and S. Everling (2004). Look away: The anti-saccade task and the voluntary control of eye movement. *Nat. Rev. Neurosci. 5*, 218–228.

Munoz, D. P. and J. H. Fecteau (2002). Vying for dominance: Dynamic interactions control visual fixation and saccadic initiation in the superior colliculus. *Prog. Brain Res. 140*, 3–19.

Munoz, D. P. and R. H. Wurtz (1993). Fixation cells in monkey superior colliculus. I. Characteristics of cell discharge. *J. Neurophysiol. 70*, 559–575.

Munoz, D. P. and R. H. Wurtz (1995). Saccade-related activity in monkey superior colliculus. 1. characteristics of burst and buildup cells. *J. Neurophysiol. 73*, 2313–2333.

Munoz, W., R. Tremblay, D. Levenstein, and B. Rudy (2017). Layer-specific modulation of neocortical dendritic inhibition during active wakefulness. *Science 355*, 954–959.

Murphy, B. K. and K. D. Miller (2003). Multiplicative gain changes are induced by excitation or inhibition alone. *J. Neurosci. 23*, 10040–10051.

Murphy, B. K. and K. D. Miller (2009). Balanced amplification: A new mechanism of selective amplification of neural activity patterns. *Neuron 61*, 635–648.

Murphy, G. L. (2004). *The Big Book of Concepts*. Cambridge, MA: MIT Pres.

Murray, E. A., S. P. Wise, and K. S. Graham (2017). *The Evolution of Memory Systems: Ancestors, Anatomy, ana Adaptations*. Oxford: Oxford University Press.

Murray, J. D. (1993). *Mathematical Biology*. Berlin: Springer-Verlag.

Murray, J. D., A. Anticevic, M. Gancsos, M. Ichinose, P. R. Corlett, J. H. Krystal, and X.-J. Wang (2014a). Linking microcircuit dysfunction to cognitive impairment: Effects of disinhibition associated with schizophrenia in a cortical working memory model. *Cereb. Cortex 24*, 859–872.

Murray, J. D., A. Bernacchia, D. J. Freedman, R. Romo, J. D. Wallis, X. Cai, C. Padoa-Schioppa, T. Pasternak, H. Seo, D. Lee, and X. J. Wang (2014b). A hierarchy of intrinsic timescales across primate cortex. *Nat. Neurosci. 17*, 1661–1663.

Murray, J. D., A. Bernacchia, N. A. Roy, C. Constantinidis, R. Romo, and X. J. Wang (2017a). Stable population coding for working memory coexists with heterogeneous neural dynamics in prefrontal cortex. *Proc. Natl. Acad. Sci. U.S.A. 114*, 394–399.

Murray, J. D., J. Jaramillo, and X. J. Wang (2017b). Working memory and decision-making in a frontoparietal circuit model. *J. Neurosci. 37*, 12167–12186.

Murray, J. D. and X.-J. Wang (2017). Cortical circuit models in psychiatry: Linking disrupted excitation-inhibition balance to cognitive deficits associated with schizophrenia. In A. Anticevic and J. Murray (Eds.), *Computational Psychiatry: Mathematical Modeling of Mental Illness*. Cambridge, MA: Academic Press, pp. 3–25.

Musall, S., M. T. Kaufman, A. L. Juavinett, S. Gluf, and A. K. Churchland (2019). Single-trial neural dynamics are dominated by richly varied movements. *Nat. Neurosci. 22*, 1677–1686.

Mushiake, H., N. Saito, K. Sakamoto, Y. Itoyama, and J. Tanji (2006). Activity in the lateral prefrontal cortex reflects multiple steps of future events in action plans. *Neuron 50*, 631–641.

Myers, N. E., M. G. Stokes, and A. C. Nobre (2017). Prioritizing information during working memory: Beyond sustained internal attention. *Trends Cogn. Sci. (Regul. Ed.) 21*, 449–461.

Nadasdy, Z., H. Hirase, A. Czurko, J. Csicsvari, and G. Buzsaki (1999). Replay and time compression of recurring spike sequences in the hippocampus. *J. Neurosci. 19*, 9497–9507.

Najafi, F. and A. K. Churchland (2018). Perceptual decision-making: A field in the midst of a transformation. *Neuron 100*, 453–462.

Najafi, F., G. F. Elsayed, R. Cao, E. Pnevmatikakis, P. E. Latham, J. P. Cunningham, and A. K. Churchland (2020). Excitatory and inhibitory subnetworks are equally selective during decision-making and emerge simultaneously during learning. *Neuron 105*, 165–179.

Nambu, A., H. Tokuno, I. Hamada, H. Kita, M. Imanishi, T. Akazawa, Y. Ikeuchi, and N. Hasegawa (2000). Excitatory cortical inputs to pallidal neurons via the subthalamic nucleus in the monkey. *J. Neurophysiol. 84*, 289–300.

Nambu, Y. (2009). Nobel lecture: Spontaneous symmetry breaking in particle physics: A case of cross fertilization. *Rev. Mod. Phys. 81*, 1015–1018.

Nash, J. F., et al. (1950). Equilibrium points in n-person games. *Proc. Natl. Acad. Sci. 36*, 48–49.

Navajas, J., C. Hindocha, H. Foda, M. Keramati, P. E. Latham, and B. Bahrami (2017). The idiosyncratic nature of confidence. *Nat. Hum. Behav. 1*, 810–818.

Neisser, U., G. Boodoo, T. J. Bouchard Jr, A. W. Boykin, N. Brody, S. J. Ceci, D. F. Halpern, J. C. Loehlin, R. Perloff, R. J. Sternberg, et al. (1996). Intelligence: Knowns and unknowns. *Am. Psychol. 51*, 77–101.

Nelson, S. B. (2002). Cortical microcircuits: Diverse or canonical? *Neuron 36*, 19–27.

Nestler, E. J. and W. A. Carlezon (2006). The mesolimbic dopamine reward circuit in depression. *Biol. Psychiatry 59*, 1151–1159.

Neubert, F.-X., R. B. Mars, J. Sallet, and M. F. Rushworth (2015). Connectivity reveals relationship of brain areas for reward-guided learning and decision making in human and monkey frontal cortex. *Proc. Natl. Acad. Sci. 112*, E2695–E2704.

Neupane, S., Fiete, I., & Jazayeri. M. (2004) Mental navigation in the primate entorhinal cortex. Nature *630*, 704–711.

Nevian, T. and B. Sakmann (2006). Spine ca2+ signaling in spike-timing-dependent plasticity. *J. Neurosci. 26*, 11001–11013.

Newell, A. (1990). *Unified Theories of Cognition*. Cambridge, MA: Harvard University Press.

Newman, M. (2010). *Networks: An Introduction*. New York: Oxford University Press.

Newsome, W. T., K. H. Britten, and J. A. Movshon (1989). Neuronal correlates of a perceptual decision. *Nature 341*, 52–54.

Newsome, W. T. and E. B. Paré (1988). A selective impairment of motion perception following lesions of the middle temporal visual area (MT). *J. Neurosci. 8*, 2201–2211.

Ni, A. M., B. S. Bowes, D. A. Ruff, and M. R. Cohen (2022). Methylphenidate as a causal test of translational and basic neural coding hypotheses. *Proc. Natl. Acad. Sci. 119*, e2120529119.

Ni, A. M., D. A. Ruff, J. J. Alberts, J. Symmonds, and M. R. Cohen (2018). Learning and attention reveal a general relationship between population activity and behavior. *Science 359*, 463–465.

Nichols, J. G., A. R. Martin, and B. G. Wallace (1992). *From Neuron to Brain* (3rd ed.). Sunderland, MA: Sinauer Associates.

Nichols, M. J. and W. T. Newsome (2002). Middle temporal visual area microstimulation influences veridical judgments of motion direction. *J. Neurosci. 22*, 9530–9540.

Nicholson, D. A., R. Trana, Y. Katz, W. L. Kath, N. Spruston, and Y. Geinisman (2006). Distance-dependent differences in synapse number and AMPA receptor expression in hippocampal CA1 pyramidal neurons. *Neuron 50*, 431–442.

Nicolis, G. and I. Prigogine (1977). *Self-Organization in Nonequilibrium Systems: From Dissipative Structure to Order Through Fluctuations*. New York: John Wiley Sons.

Nicoll, R. A. (2017). A brief history of long-term potentiation. *Neuron 93*, 281–290.

Nieder, A. (2019). *A Brain for Numbers: The Biology of the Number Instinct*. Cambridge, MA: MIT Press.

Nieder, A., D. J. Freedman, and E. K. Miller (2002). Representation of the quantity of visual items in the primate prefrontal cortex. *Science 297*, 1708–1711.

Nieder, A. and E. K. Miller (2003). Coding of cognitive magnitude: Compressed scaling of numerical information in the primate prefrontal cortex. *Neuron 37*, 149–157.

Nienborg, H. and B. G. Cumming (2009). Decision-related activity in sensory neurons reflects more than a neuron's causal effect. *Nature 459*, 89–92.

Niv, Y. (2021). The primacy of behavioral research for understanding the brain. *Behav. Neurosci. 135*, 601–609.

Niv, Y., N. D. Daw, D. Joel, and P. Dayan (2007). Tonic dopamine: Opportunity costs and the control of response vigor. *Psychopharmacology (Berl.) 191*, 507–520.

Niwa, M. and J. Ditterich (2008). Perceptual decisions between multiple directions of visual motion. *J. Neurosci. 28*, 4435–4445.

Nobre, A. C. and S. Kastner (Eds.) (2014). *The Oxford Handbook of Attention*. Oxford: Oxford University Press.

Nobre, A. C. and M. G. Stokes (2019). Premembering experience: A hierarchy of time-scales for proactive attention. *Neuron 104*, 132–146.

Noel, J.-P., E. Balzani, E. Avila, K. J. Lakshminarasimhan, S. Bruni, P. Alefantis, C. Savin, and D. E. Angelaki (2022). Coding of latent variables in sensory, parietal, and frontal cortices during closed-loop virtual navigation. *eLife 11*, e80280.

Noori, H. R., J. Schöttler, M. Ercsey-Ravasz, A. Cosa-Linan, M. Varga, Z. Toroczkai, and R. Spanagel (2017). A multiscale cerebral neurochemical connectome of the rat brain. *PLoS Biol. 15*, e2002612.

Norman, D. A. and T. Shallice (1986). Attention to action: willed and automatic control of behavior. In S. R. S. D. Davidson (Ed.), *Consciousness and Self-Regulation*. Berlin: Plenum Press, pp. 1–18.

Nosofsky, R. M. (1986). Attention, similarity, and the identification-categorization relationship. *J. Exp. Psychol. Gen. 115*, 39–61.

Nozari, E., M. A. Bertolero, J. Stiso, L. Caciagli, E. J. Cornblath, X. He, A. S. Mahadevan, G. J. Pappas, and D. S. Bassett (2020). Is the brain macroscopically linear? A system identification of resting state dynamics. *arXiv preprint arXiv:2012.12351*.

Oberauer, K. (2009). Design for a working memory. *Psychol. Learn. Motiv. 51*, 45–100.

O'Connell, R. G., P. M. Dockree, and S. P. Kelly (2012). A supramodal accumulation-to-bound signal that determines perceptual decisions in humans. *Nat. Neurosci. 15*, 1729–1735.

O'Connell, R. G., M. N. Shadlen, K. Wong-Lin, and S. P. Kelly (2018). Bridging neural and computational viewpoints on perceptual decision-making. *Trends Neurosci. 41*, 838–852.

O'Connor, D. H., G. M. Wittenberg, and S. S.-H. Wang (2005). Graded bidirectional synaptic plasticity is composed of switch-like unitary events. *Proc. Natl. Acad. Sci. U.S.A. 102*, 9679–9684.

O'doherty, J. P. (2004). Reward representations and reward-related learning in the human brain: Insights from neuroimaging. *Curr. Opin. Neurobiol. 14*(6), 769–776.

Ogawa, S., T.-M. Lee, A. R. Kay, and D. W. Tank (1990). Brain magnetic resonance imaging with contrast dependent on blood oxygenation. *Proc. Natl. Acad. Sci. 87*, 9868–9872.

Ogawa, T. and H. Komatsu (2010). Differential temporal storage capacity in the baseline activity of neurons in macaque frontal eye field and area V4. *J. Neurophysiol. 103*, 2433–2445.

Oh, S. W., J. A. Harris, L. Ng, B. Winslow, N. Cain, S. Mihalas, Q. Wang, C. Lau, L. Kuan, A. M. Henry, M. T. Mortrud, B. Ouellette, T. N. Nguyen, S. A. Sorensen, C. R. Slaughterbeck, W. Wakeman, Y. Li, D. Feng, A. Ho, E. Nicholas, K. E. Hirokawa, P. Bohn, K. M. Joines, H. Peng, M. J. Hawrylycz, J. W. Phillips, J. G. Hohmann, P. Wohnoutka, C. R. Gerfen, C. Koch, A. Bernard, C. Dang, A. R. Jones, and H. Zeng (2014). A mesoscale connectome of the mouse brain. *Nature 508*, 207–214.

Ohki, K., S. Chung, P. Kara, M. Hübener, T. Bonhoeffer, and R. C. Reid (2006). Highly ordered arrangement of single neurons in orientation pinwheels. *Nature 442*, 925–928.

Ohzawa, I., G. Sclar, and R. D. Freeman (1985). Contrast gain control in the cat's visual system. *J. Neurophysiol. 54*, 651–667.

Okabe, A., B. Boots, K. Sugihara, and S. N. Chiu (2000). *Spatial Tessellations: Concepts and Applications of Voronoi Diagrams* (2nd ed.). Chichester: John Wiley & Sons.

Okamoto, H., Y. Isomura, M. Takada, and T. Fukai (2007). Temporal integration by stochastic recurrent network dynamics with bimodal neurons. *J. Neurophysiol. 97*, 3859–3867.

Okazawa, G., C. E. Hatch, A. Mancoo, C. K. Machens, and R. Kiani (2021). Representational geometry of perceptual decisions in the monkey parietal cortex. *Cell 184*, 3748–3761.

O'Keefe, J. and J. Dostrovsky (1971). The hippocampus as a spatial map. Preliminary evidence from unit activity in the freely moving rat. *Exp. Brain Res. 34*, 171–175.

O'Keefe, J. and L. Nadel (1978). *The Hippocampus as a Cognitive Map*. Oxford: Clarendon Press.

O'Keefe, J. and M. L. Recce (1993). Phase relationship between hippocampal place units and the EEG theta rhythm. *Hippocampus 3*, 317–330.

Olsen, S. R., V. Bhandawat, and R. I. Wilson (2010). Divisive normalization in olfactory population codes. *Neuron 66*, 287–299.

Olshausen, B. A., C. H. Anderson, and D. C. Van Essen (1993). A neurobiological model of visual attention and invariant pattern recognition based on dynamic routing of information. *J. Neurosci. 13*, 4700–4719.

Olshausen, B. A. and D. J. Field (1996). Emergence of simple-cell receptive field properties by learning a sparse code for natural images. *Nature 381*, 607–609.

O'Reilly, R. C., R. Bhattacharyya, M. D. Howard, and N. Ketz (2014). Complementary learning systems. *Cogn. Sci. 38*, 1229–1248.

O'Reilly, R. C., J. L. Russin, M. Zolfaghar, and J. Rohrlich (2021). Deep predictive learning in neocortex and pulvinar. *J. Cogn. Neurosci. 33*, 1158–1196.

Ostojic, S. (2014). Two types of asynchronous activity in networks of excitatory and inhibitory spiking neurons. *Nat. Neurosci. 17*, 594–600.

Otor, Y., S. Achvat, N. Cermak, H. Benisty, M. Abboud, O. Barak, Y. Schiller, A. Poleg-Polsky, and J. Schiller (2022). Dynamic compartmental computations in tuft dendrites of layer 5 neurons during motor behavior. *Science 376*, 267–275.

Ottenheimer, D. J., B. A. Bari, E. Sutlief, K. M. Fraser, T. H. Kim, J. M. Richard, J. Y. Cohen, and P. H. Janak (2020). A quantitative reward prediction error signal in the ventral pallidum. *Nat. Neurosci. 23*. https://doi.org/10.1038/s41593-020-0688-5.

Owen, A. M., K. M. McMillan, A. R. Laird, and E. Bullmore (2005). N-back working memory paradigm: A meta-analysis of normative functional neuroimaging studies. *Hum. Brain Mapp. 25*, 46–59.

Ozeki, H., I. M. Finn, E. S. Schaffer, K. D. Miller, and D. Ferster (2009). Inhibitory stabilization of the cortical network underlies visual surround suppression. *Neuron 62*, 578–592.

Padoa-Schioppa, C. (2011). Neurobiology of economic choice: A good-based model. *Annu. Rev. Neurosci. 34*, 333–359.

Padoa-Schioppa, C. (2013). Neuronal origins of choice variability in economic decisions. *Neuron 80*, 1322–1336.

Padoa-Schioppa, C. and J. Assad (2006). Neurons in the orbitofrontal cortex encode economic value. *Nature 441*, 223–226.

Padoa-Schioppa, C. and K. E. Conen (2017). Orbitofrontal cortex: A neural circuit for economic decisions. *Neuron 96*, 736–754.

Pagan, M., V. D. Tang, M. C. Aoi, J. W. Pillow, V. Mante, D. Sussillo, and C. D. Brody (2022). A new theoretical framework jointly explains behavioral and neural variability across subjects performing flexible decision-making. *bioRxiv*. https://doi.org/10.1101/2022.11.28.518207.

Palmer, J., A. C. Huk, and M. N. Shadlen (2005). The effect of stimulus strength on the speed and accuracy of a perceptual decision. *J. Vis. 5*, 376–404.

Palmigiano, A., F. Fumarola, D. P. Mossing, N. Kraynyukova, H. Adesnik, and K. D. Miller (2020). Structure and variability of optogenetic responses identify the operating regime of cortex. *bioRxiv*. https://doi.org/10.1101/2020.11.11.378729.

Palmigiano, A., T. Geisel, F. Wolf, and D. Battaglia (2017). Flexible information routing by transient synchrony. *Nat. Neurosci. 20*, 1014–1022.

Pandya, N. D., B. Seltzer, and H. Barbas (1988). Input-output organization of the primate cerebral cortex. In H. D. Steklis and J. Erwin (Eds.), *Comparative Primate Biology* (Vol. 4), pp. 39–80. New York: Alan R. Liss.

Pang, J. C., K. M. Aquino, M. Oldehinkel, P. A. Robinson, B. D. Fulcher, M. Breakspear, and A. Fornito (2023). Geometric constraints on human brain function. *Nature*. https://doi.org/10.1038/s41586-023-06098-1.

Panichello, M. F. and T. J. Buschman (2021). Shared mechanisms underlie the control of working memory and attention. *Nature*. https://doi.org/10.1038/s41586-021-03390-w.

Panichello, M. F., Jonikaitis, D., Oh, Y. J., Zhu, S., Trepka, E. B., & Moore, T. (2024). Intermittent rate coding and cue-specific ensembles support working memory. Nature, https://doi.org/10.1038/s41586-024-08139-9

Panksepp, J. (2004). *Affective Neuroscience: The Foundations of Human and Animal Emotions*. New York: Oxford University Press.

Panlilio, L. V., E. B. Thorndike, and C. W. Schindler (2007). Blocking of conditioning to a cocaine-paired stimulus: Testing the hypothesis that cocaine perpetually produces a signal of larger-than-expected reward. *Pharmacol. Biochem. Behav. 86*, 774–777.

Papadimitriou, C., A. Ferdoash, and L. H. Snyder (2015). Ghosts in the machine: Memory interference from the previous trial. *J. Neurophysiol. 113*, 567–577.

Papoutsi, A., K. Sidiropoulou, V. Cutsuridis, and P. Poirazi (2013). Induction and modulation of persistent activity in a layer V PFC microcircuit model. *Front. Neural Circuits 7*. https://doi.org/10.3389/fncir.2013.00161.

Papyan, V., X. Han, and D. L. Donoho (2020). Prevalence of neural collapse during the terminal phase of deep learning training. *Proc. Natl. Acad. Sci. 117*, 24652–24663.

Pardo-Vazquez, J. L., J. R. Castiñeiras-de Saa, M. Valente, I. Damião, T. Costa, M. I. Vicente, A. G. Mendon ca, Z. F. Mainen, and A. Renart (2019). The mechanistic foundation of Weber's law. *Nat. Neurosci. 22*, 1493–1502.

Paré, M. and D. Hanes (2003). Controlled movement processing: Superior colliculus activity associated with countermanded saccades. *J. Neurosci. 23*, 6480–6489.

Parisi, G. I. (1979). Infinite number of order parameters for spin-glasses. *Phys. Rev. Lett. 43*, 1754–1756.

Parisi, G. I., R. Kemker, J. L. Part, C. Kanan, and S. Wermter (2019). Continual lifelong learning with neural networks: A review. *Neural Netw. 113*, 54–71.

Park, S. and P. S. Holzman (1992). Schizophrenics show spatial working memory deficits. *Arch. Gen. Psychiatry 49*, 975–982.

Parker, A. J. and K. Krug (2003). Neuronal mechanisms for the perception of ambiguous stimuli. *Curr. Opin. Neurobiol. 13*, 433–439.

Parker, A. J. and W. T. Newsome (1998). Sense and the single neuron: Probing the physiology of perception. *Ann. Rev. Neurosci. 21*, 227–277.

Parker, N. F., A. Baidya, J. Cox, L. M. Haetzel, A. Zhukovskaya, M. Murugan, . . . I. B. Witten (2022). Choice-selective sequences dominate in cortical relative to thalamic inputs to NAc to support reinforcement learning. *Cell Rep. 39*, 110756.

Parr, T. (2019). *The Computational Neurology of Active Vision*. Ph.D. thesis, UCL (University College London).

Parr, T., J. Limanowski, V. Rawji, and K. Friston (2021). The computational neurology of movement under active inference. *Brain 144*, 1799–1818.

Parras, G. G., L. Casado-Román, E. Schröger, and M. S. Malmierca (2021). The posterior auditory field is the chief generator of prediction error signals in the auditory cortex. *NeuroImage 242*, 118446.

Parthasarathy, A., R. Herikstad, J. H. Bong, F. S. Medina, C. Libedinsky, and S.-C. Yen (2017). Mixed selectivity morphs population codes in prefrontal cortex. *Nat. Neurosci. 20*, 1770–1779.

Pascanu, R., T. Mikolov, and Y. Bengio (2013). On the difficulty of training recurrent neural networks. In *International Conference on Machine Learning*, pp. 1310–1318.

Pashler, H. (1988). Familiarity and visual change detection. *Percept. Psychophys. 44*, 369–378.

Passingham, R. E. (1995). *The Frontal Lobes and Voluntary Action*. New York: Oxford University Press.

Passingham, R. E. (2008). *What Is Special About the Human Brain?* Oxford: Oxford University Press.

Passingham, R. E. (2021). *Understanding the Prefrontal Cortex: Selective Advantage, Connectivity, and Neural Operations*. Oxford: Oxford University Press.

Passingham, R. E. and S. P. Wise (2012). *The Neurobiology of the Prefrontal Cortex: Anatomy, Evolution, and the Origin of Insight*. Oxford: Oxford University Press.

Pasternak, T. and M. Greenlee (2005). Working memory in primate sensory systems. *Nat. Rev. Neurosci. 6*, 97–107.

Pastor-Bernier, A., C. R. Plott, and W. Schultz (2017). Monkeys choose as if maximizing utility compatible with basic principles of revealed preference theory. *Proc. Natl. Acad. Sci. 114*, E1766–E1775.

Pastor-Bernier, A., A. Stasiak, and W. Schultz (2019). Orbitofrontal signals for two-component choice options comply with indifference curves of revealed preference theory. *Nat. Commun. 10*, 1–19.

Pasupathy, A. and E. K. Miller (2005). Different time courses of learning-related activity in the prefrontal cortex and striatum. *Nature 433*, 873–876.

Paszke, A., S. Gross, F. Massa, A. Lerer, J. Bradbury, G. Chanan, T. Killeen, Z. Lin, N. Gimelshein, L. Antiga, et al. (2019). Pytorch: An imperative style, high-performance deep learning library. *Adv. Neural Inf. Process. Syst. 32*, 8026–8037.

Paton, J. J. and D. V. Buonomano (2018). The neural basis of timing: Distributed mechanisms for diverse functions. *Neuron 98*, 687–705.

Patros, C. H., R. M. Alderson, L. J. Kasper, S. J. Tarle, S. E. Lea, and K. L. Hudec (2016). Choice-impulsivity in children and adolescents with attention-deficit/hyperactivity disorder (adhd): A meta-analytic review. *Clin. Psychol. Rev. 43*, 162–174.

Patterson, K., P. J. Nestor, and T. T. Rogers (2007). Where do you know what you know? The representation of semantic knowledge in the human brain. *Nat. Rev. Neurosci. 8*, 976–987.

Paul, A., M. Crow, R. Raudales, M. He, J. Gillis, and Z. J. Huang (2017). Transcriptional architecture of synaptic communication delineates GABAergic neuron identity. *Cell 171*, 522–539.

Paulk, A. C., Y. Kfir, A. R. Khanna, M. L. Mustroph, E. M. Trautmann, D. J. Soper, S. D. Stavisky, M. Welkenhuysen, B. Dutta, K. V. Shenoy, et al. (2022). Large-scale neural recordings with single neuron resolution using neuropixels probes in human cortex. *Nat. Neurosci. 25*, 52–263.

Paulus, M. P., Q. J. Huys, and T. V. Maia (2016). A roadmap for the development of applied computational psychiatry. *Biol. Psychiatry Cogn. Neurosci. Neuroimaging 1*, 386–392.

Pavlov, I. P. (2003). *Conditioned Reflex*. New York: Dover (The English translation from Russian was first published in 1927).

Pawlak, V. and J. Kerr (2008). Dopamine receptor activation is required for corticostriatal spike-timing-dependent plasticity. *J. Neurosci. 28*, 2435–2446.

Pawlak, V., J. R. Wickens, A. Kirkwood, and J. N. Kerr (2010). Timing is not everything: Neuromodulation opens the STDP gate. *Front. Synaptic Neurosci. 2*, 146.

Pearl, J. (2009). *Causality: Models, Reasoning, and Inference* (2nd ed.). New York: Cambridge University Press.

Pearson, J. M., K. K. Watson, and M. L. Platt (2014). Decision making: The neuroethological turn. *Neuron 82*, 950–965.

Peixoto, D., J. R. Verhein, R. Kiani, J. C. Kao, P. Nuyujukian, C. Chandrasekaran, J. Brown, S. Fong, S. I. Ryu, K. V. Shenoy, et al. (2021). Decoding and perturbing decision states in real time. *Nature 591*, 604–609.

Penn, D. C., K. J. Holyoak, and D. J. Povinelli (2008). Darwin's mistake: Explaining the discontinuity between human and nonhuman minds. *Behav. Brain Sci. 31*, 109–130.

Penzes, P., M. E. Cahill, K. A. Jones, J.-E. VanLeeuwen, and K. M. Woolfrey (2011). Dendritic spine pathology in neuropsychiatric disorders. *Nat. Neurosci. 14*, 285–293.

Pereira, J. and X. J. Wang (2015). A tradeoff between accuracy and flexibility in a working memory circuit endowed with slow feedback mechanisms. *Cereb. Cortex 25*, 3586–3601.

Pereira, T. D., D. E. Aldarondo, L. Willmore, M. Kislin, S. S.-H. Wang, M. Murthy, and J. W. Shaevitz (2019). Fast animal pose estimation using deep neural networks. *Nat. Methods 16*, 117–125.

Pereira-Obilinovic, U., H. Hou, K. Svoboda, and X.-J. Wang (2024). Brain mechanism of foraging: Reward-dependent synaptic plasticity versus neural integration of values. *Proc. Natl. Acad. Sci. U.S.A. 121*, e2318521121.

Pérez-Otaño, I., R. S. Larsen, and J. F. Wesseling (2016). Emerging roles of GluN3-containing NMDA receptors in the CNS. *Nat. Rev. Neurosci. 17*, 623–635.

Perich, M. G., C. Arlt, S. Soares, M. E. Young, C. P. Mosher, J. Minxha, E. Carter, U. Rutishauser, P. H. Rudebeck, C. D. Harvey, et al. (2021). Inferring brain-wide interactions using data-constrained recurrent neural network models. *bioRxiv*. https://doi.org/10.1101/2020.12.18.423348.

Perkel, D. H. and T. H. Bullock (1968). Neural coding. *Neurosci. Res. Program Bull. 6*, 219–349.

Perkel, D. H. and B. Mulloney (1974). Motor pattern production in reciprocally inhibitory neurons exhibiting postinhibitory rebound. *Science 185*, 181–183.

Pesaran, B., J. S. Pezaris, M. Sahani, P. P. Mitra, and R. A. Andersen (2002). Temporal structure in neuronal activity during working memory in marcaque pariental cortex. *Nat. Neurosci. 5*, 805–811.

Pessoa, L. (2022). *The Entangled Brain: How Perception, Cognition, and Emotion Are Woven Together*. Cambridge, MA: MIT Press.

Pessoa, L., E. Gutierrez, P. A. Bandettini, and L. G. Ungerleider (2002). Neural correlates of visual working memory: Fmri amplitude predicts task performance. *Neuron 35*, 975–987.

Pessoa, L., L. Medina, and E. Desfilis (2022). Refocusing neuroscience: Moving away from mental categories and towards complex behaviours. *Philos. Trans. R. Soc. Lond. B Biol. Sci. 377*, 20200534.

Petanjek, Z., M. Judaš, G. Šimic, M. R. Rašin, H. B. Uylings, P. Rakic, and I. Kostovic (2011). Extraordinary neoteny of synaptic spines in the human prefrontal cortex. *Proc. Natl. Acad. Sci. 108*, 13281–13286.

Petersen, C. C., R. C. Malenka, R. A. Nicoll, and J. J. Hopfield (1998). All-or-none potentiation at CA3-CA1 synapses. *Proc. Natl. Acad. Sci. U.S.A. 95*, 4732–4737.

Petersen, S. E. and M. I. Posner (2012). The attention system of the human brain: 20 years after. *Annu. Rev. Neurosci. 35*, 73–89.

Petreanu, L., T. Mao, S. M. Sternson, and K. Svoboda (2009). The subcellular organization of neocortical excitatory connections. *Nature 457*, 1142–1145.

Petrides, M. E., G. Cadoret, and S. Mackey (2005). Orofacial somatomotor responses in the macaque monkey homologue of Broca's area. *Nature 435*, 1235–1238.

Petrides, M. E. and D. N. Pandya (2002). Association pathways of the prefrontal cortex and functional observations. In D. T. Stuss and R. T. Knight (Eds.), *Principles of Frontal Lobe Function*, pp. 31–50. New York: Oxford University Press.

Pettine, W. W., K. Louie, J. D. Murray, and X.-J. Wang (2021). Excitatory-inhibitory tone shapes decision strategies in a hierarchical neural network model of multi-attribute choice. *PLoS Comput. Biol. 17*, e1008791.

Peyrache, A., M. M. Lacroix, P. C. Petersen, and G. Buzsáki (2015). Internally organized mechanisms of the head direction sense. *Nat. Neurosci. 18*, 569–575.

Pfeiffer, M., B. Nessler, R. J. Douglas, and W. Maass (2010). Reward-modulated Hebbian learning of decision making. *Neural Comput. 22*, 1399–1444.

Pfister, J.-P. and W. Gerstner (2006). Triplets of spikes in a model of spike timing-dependent plasticity. *J. Neurosci. 26*, 9673–9682.

Pi, H. J., B. Hangya, D. Kvitsiani, J. I. Sanders, Z. J. Huang, and A. Kepecs (2013). Cortical interneurons that specialize in disinhibitory control. *Nature 503*, 521–524.

Piaget, J. (1923). *Le Langage et la Pensée chez L'Enfant.* Neuchâtel, Switzerland: Delachaux et Niestlé.

Pickering, A. (2010). *The Cybernetic Brain.* Chicago: University of Chicago Press.

Pickover, C. (1998). *Strange Brains and Genius: The Secret Lives of Eccentric Scientists and Madmen.* New York: HarpherCollins Publishers Inc.

Piet, A. T., J. C. Erlich, C. D. Kopec, and C. D. Brody (2017). Rat prefrontal cortex inactivations during decision making are explained by bistable attractor dynamics. *Neural Comput. 29*, 2861–2886.

Pinker, S. (1994). *The Language Instinct: How the Mind Creates Language.* New York: HarperPerennial.

Pinsky, P. and J. Rinzel (1994). Intrinsic and network rhythmogenesis in a reduced Traub model for CA3 neurons. *J. Comput. Neurosci. 1*, 39–60.

Pinto, L., K. Rajan, B. DePasquale, S. Y. Thiberge, D. W. Tank, and C. D. Brody (2019). Task-dependent changes in the large-scale dynamics and necessity of cortical regions. *Neuron 104*, 810–824.

Pizzagalli, D. A., A. L. Jahn, and J. P. O'Shea (2005). Toward an objective characterization of an anhedonic phenotype: A signal-detection approach. *Biol. Psychiatry 57*, 319–327.

Platt, J. R. (1964). Strong Inference: Certain systematic methods of scientific thinking may produce much more rapid progress than others. *Science 146*, 347–353.

Platt, M. L. and P. W. Glimcher (1999). Neural correlates of decision variables in parietal cortex. *Nature 400*, 233–238.

Pliszka, S. R., M. Liotti, and M. G. Woldorff (2000). Inhibitory control in children with attention-deficit/hyperactivity disorder: Event-related potentials identify the processing component and timing of an impaired right-frontal response-inhibition mechanism. *Biol. Psychiatry 48*, 238–246.

Ploner, C. J., A. Tschirch, F. Ostendorf, S. Dick, B. M. Gaymard, S. Rivaud-Pechoux, F. Sporkert, F. Pragst, and A. M. Stadelmann (2002). Oculomotor effects of delta-9-tetrahydrocannabinol in humans: Implications for the functional neuroanatomy of the brain cannabinoid system. *Cereb. Cortex 12*, 1016–1023.

Poeppel, D. (2003). The analysis of speech in different temporal integration windows: Cerebral lateralization as 'asymmetric sampling in time'. *Speech Commun. 41*, 245–255.

Poeppel, D. and F. Adolfi (2020). Against the epistemological primacy of the hardware: The brain from inside out, turned upside down. *eNeuro 7.* https://doi.org/10.1523/ENEURO.0215-20.2020.

Poeppel, D., K. Emmorey, G. Hickok, and L. Pylkkänen (2012). Towards a new neurobiology of language. *J. Neurosci. 32*, 14125–14131.

Poggio, T. (2012). The levels of understanding framework, revised. *Perception 41*, 1017–1023.

Poirazi, P., T. Brannon, and B. W. Mel (2003). Pyramidal neuron as two-layer neural network. *Neuron 37*, 989–999.

Poirazi, P. and A. Papoutsi (2020). Illuminating dendritic function with computational models. *Nat. Rev. Neurosci. 21*, 303–321.

Polsky, A., B. W. Mel, and J. Schiller (2004). Computational subunits in thin dendrites of pyramidal cells. *Nat. Neurosci. 7*, 621–627.

Posner, M. I. (1978). *Chronometric Exploration of Mind.* Hillsdale, NJ: Lawrence Erlbaum Associates.

Posner, M. I. and S. W. Keele (1968). On the genesis of abstract ideas. *J. Exp. Psychol. 77*, 353–363.

Posner, M. I. and M. E. Raichle (1994). *Images of Mind.* New York: W. H. Freeman.

Pouget, A., P. Dayan, and R. Zemel (2003). Inference and computation with population codes. *Annu. Rev. Neurosci. 26*, 381–410.

Pouget, A., J. Drugowitsch, and A. Kepecs (2016). Confidence and certainty: Distinct probabilistic quantities for different goals. *Nat. Neurosci. 19*, 366–374.

Pouget, A. and T. J. Sejnowski (1997a). A new view of hemineglect based on the response properties of parietal neurones. *Phil. Trans. R. Soc. London. Series B 352*(1360), 1449–1459.

Pouget, A. and T. J. Sejnowski (1997b). Spatial transformations in the parietal cortex using basis functions. *J. Cogn. Neursci. 9*(2), 222–237.

Poundstone, W. (1992). *Prisoner's Dilemma.* New York: Random House.

Povysheva, N. V., G. Gonzalez-Burgos, A. V. Zaitsev, S. Kroner, G. Barrionuevo, D. A. Lewis, and L. S. Krimer (2006). Properties of excitatory synaptic responses in fast-spiking interneurons and pyramidal cells from monkey and rat prefrontal cortex. *Cereb. Cortex 16*, 541–552.

Powers, A. R., C. Mathys, and P. R. Corlett (2017). Pavlovian conditioning-induced hallucinations result from overweighting of perceptual priors. *Science 357*, 596–600.

Prechtl, J., L. Cohen, B. Pesaran, P. Mitra, and D. Kleinfeld (1997). Visual stimuli induce waves of electrical activity in turtle cortex. *Proc. Natl. Acad. Sci. U.S.A. 94*, 7621–7626.

Preuss, T. M. (1995). Do rats have prefrontal cortex? the rose-woolsey-akert program reconsidered. *J. Cogn. Neurosci. 7*, 1–24.

Preuss, T. M. and S. P. Wise (2022). Evolution of prefrontal cortex. *Neuropsychopharmacology 47*, 3–19.

Price, D. J., H. Kennedy, C. Dehay, L. Zhou, M. Mercier, Y. Jossin, A. M. Goffinet, F. Tissir, D. Blakey, and Z. Molnár (2006). The development of cortical connections. *Eur. J. Neurosci. 23*, 910–920.

Prinz, A. A., D. Bucher, and E. Marder (2004). Similar network activity from disparate circuit parameters. *Nat. Neurosci. 7*, 1345–1352.

Prinz, W. (1997). Perception and action planning. *Eur. J. Cogn. Psychol. 9*, 129–154.

Pritchett, B. L., C. Hoeflin, K. Koldewyn, E. Dechter, and E. Fedorenko (2018). High-level language processing regions are not engaged in action observation or imitation. *J. Neurophysiol. 120*, 2555–2570.

Procyk, E., Y. L. Tanaka, and J. P. Joseph (2000). Anterior cingulate activity during routine and non-routine sequential behaviors in macaques. *Nat. Neurosci. 3*(5), 502–508.

Proske, U. and S. C. Gandevia (2012). The proprioceptive senses: Their roles in signaling body shape, body position and movement, and muscle force. *Physiol. Rev. 92*, 1651–1697.

Proust, M. (1913). *À la Recherche du Temps Perdu*. Paris, France: Grasset and Gallimard.

Prut, Y. and E. E. Fetz (1999). Primate spinal interneurons show pre-movement instructed delay activity. *Nature 401*, 590–594.

Pulcu, E. and M. Browning (2017). Affective bias as a rational response to the statistics of rewards and punishments. *Elife 6*, e27879.

Purushothaman, G. and D. C. Bradley (2005). Neural population code for fine perceptual decisions in area MT. *Nat. Neurosci. 8*, 99–106.

Putnam, H. (1960). Minds and machines. In S. Hook (Ed.), *Dimensions of Mind*, pp. 138–164. New York: New York University Press.

Quinlan, E. M., D. H. Olstein, and M. F. Bear (1999). Bidirectional, experience-dependent regulation of N-methyl-D-aspartate receptor subunit composition in the rat visual cortex during postnatal development. *Proc. Natl. Acad. Sci. U.S.A. 96*, 12876–12880.

Quiroga, R. Q., L. Reddy, G. Kreiman, C. Koch, and I. Fried (2005). Invariant visual representation by single neurons in the human brain. *Nature 435*, 1102–1107.

Rabinovich, M., R. Huerta, and G. Laurent (2008). Transient dynamics for neural processing. *Science 321*, 48–50.

Rachlin, H. and D. I. Laibson (1997). *Richard J. Herrnstein, the Matching Law: Papers in Psychology and Economics*. Cambridge, MA: Harvard University Press.

Rachlin, H., A. Raineri, and D. Cross (1991). Subjective probability and delay. *J. Exp. Anal. Behav. 55*, 233–244.

Radford, A., K. Narasimhan, T. Salimans, I. Sutskever, et al. (2018). *Improving Language Understanding by Generative Pre-Training*. OpenAI.

Rahnev, D. and R. N. Denison (2018). Suboptimality in perceptual decision making. *Behav. Brain Sci.*, 1–17. Discussion 17–66.

Raichle, M. E. (2009). A brief history of human brain mapping. *Trends Neurosci. 32*, 118–126.

Raichle, M. E. (2015). The brain's default mode network. *Annu. Rev. Neurosci. 38*, 433–447.

Raichle, M. E., A. M. MacLeod, A. Z. Snyder, W. J. Powers, D. A. Gusnard, and G. L. Shulman (2001). A default mode of brain function. *Proc. Natl. Acad. Sci. U.S.A. 98*, 676–682.

Rainer, G., W. F. Assad, and E. K. Miller (1998). Memory fields of neurons in the primate prefrontal cortex. *Proc. Natl. Acad. Sci. U.S.A. 95*, 15008–15013.

Rainer, G., S. C. Rao, and E. K. Miller (1999). Prospective coding for objects in primate prefrontal cortex. *J. Neurosci. 19*, 5493–5505.

Rajan, K. and L. F. Abbott (2006). Eigenvalue spectra of random matrices for neural networks. *Phys. Rev. Lett. 97*, 188104.

Rajan, K., C. D. Harvey, and D. W. Tank (2016). Recurrent network models of sequence generation and memory. *Neuron 90*, 128–142.

Rakic, P. (2009). Evolution of the neocortex: A perspective from developmental biology. *Nat. Rev. Neurosci. 10*, 724–735.

Rall, W. (1964). Theoretical significance of dendritic trees for neuronal input-output relations. In R. F. Reiss (Ed.), *Neural Theory and Modeling*, pp. 73–97. Stanford: Stanford University Press.

Rall, W. and G. Shepherd (1968). Theoretical reconstruction of field potentials and dendrodendritic synaptic interactions in olfactory bulb. *J. Neurophysiol. 31*, 884–915.

Ramcharan, E. J., J. W. Gnadt, and S. M. Sherman (2005). Higher-order thalamic relays burst more than first-order relays. *Proc. Natl. Acad. Sci. U.S.A. 102*, 12236–12241.

Ramirez-Mahaluf, J. P., A. Roxin, H. S. Mayberg, and A. Compte (2017). A computational model of major depression: The role of glutamate dysfunction on cingulo-frontal network dynamics. *Cereb. Cortex 27*, 660–679.

Rangel, A., C. Camerer, and P. R. Montague (2008). A framework for studying the neurobiology of value-based decision making. *Nat. Rev. Neurosci. 9*, 545–556.

Rao, R. P. and D. H. Ballard (1999). Predictive coding in the visual cortex: A functional interpretation of some extra-classical receptive-field effects. *Nat. Neurosci. 2*, 79–87.

Rapoport, A. and D. V. Budescu (1992). Generation of random series in two-person strictly competitive games. *J. Exp. Psychol. Gen. 121*, 352–363.

Rapoport, A. and A. Chammah (1970). *Prisoner's Dilemma* (2nd ed.). Ann Arbor, MI: The University of Michigan Press.

Rascovsky, K., M. E. Growdon, I. R. Pardo, S. Grossman, and B. L. Miller (2009). 'The quicksand of forgetfulness': Semantic dementia in one hundred years of solitude. *Brain 132*, 2609–2616.

Ratcliff, R. (1978). A theory of memory retrieval. *Psychol. Rev. 85*, 59–108.

Ratcliff, R. and G. McKoon (2008). The diffusion decision model: Theory and data for two-choice decision tasks. *Neural Comput. 20*, 873–922.

Ratcliff, R. and J. F. Rouder (1998). Modeling response times for two-choice decisions. *Psychol. Sci. 9*, 347–356.

Ratcliff, R., P. L. Smith, S. D. Brown, and G. McKoon (2016). Diffusion decision model: Current issues and history. *Trends Cogn. Sci. (Regul. Ed.) 20*, 260–281.

Raut, R. V., A. Z. Snyder, and M. E. Raichle (2020). Hierarchical dynamics as a macroscopic organizing principle of the human brain. *Proc. Natl. Acad. Sci. 117*, 20890–20897.

Raven, J. C., J. H. Court, and J. Raven (1988). *Manual for Raven's Progressive Matrices and Vocabulary Scales*. London: H. K. Lewis & Co.

Rawley, J. B. and C. Constantinidis (2010). Effects of task and coordinate frame of attention in area 7a of the primate posterior parietal cortex. *J. Vis. 10*, 1–16.

Ray, S. and J. H. Maunsell (2015). Do gamma oscillations play a role in cerebral cortex? *Trends Cogn. Sci. (Regul. Ed.) 19*, 78–85.

Raymond, J. L. and J. F. Medina (2018). Computational principles of supervised learning in the cerebellum. *Annu. Rev. Neurosci. 41*, 233–253.

Real, L. (1991). Animal choice behavior and the evolution of cognitive architecture. *Science 253*, 980–986.

Reddi, B. and R. H. Carpenter (2000). The influence of urgency on decision time. *Nat. Neurosci. 3*, 827–831.

Redish, A. D. (2004). Addiction as a computational process gone awry. *Science 306*, 1944–1947.

Redish, A. D. (2013). *The Mind Within the Brain: How We Make Decisions and How Those Decisions Go Wrong*. New York: Oxford University Press.

Redish, A. D. and J. A. Gordon (Eds.) (2016). *Computational Psychiatry: New Perspectives on Mental Illness*. Cambridge, MA: MIT Press.

Redish, A. D., S. Jensen, and A. Johnson (2008). Addiction as vulnerabilities in the decision process. *Behav. Brain Sci. 31*, 461–487.

Reid, R. C. and J. M. Alonso (1995). Specificity of monosynaptic connections from thalamus to visual cortex. *Nature 378*, 281–284.

Remington, E. D., D. Narain, E. A. Hosseini, and M. Jazayeri (2018). Flexible sensorimotor computations through rapid reconfiguration of cortical dynamics. *Neuron 98*, 1005–1019.

Renart, A., N. Brunel, and X.-J. Wang (2003). Mean-field theory of recurrent cortical networks: Working memory circuits with irregularly spiking neurons. In J. Feng (Ed.), *Computational Neuroscience: A Comprehensive Approach*, pp. 432–490. Boca Raton: CRC Press.

Renart, A., J. de la Rocha, P. Bartho, L. Hollender, N. Parga, A. Reyes, and K. D. Harris (2010). The asynchronous state in cortical circuits. *Science 327*, 587–590.

Renart, A., R. Moreno-Bote, X.-J. Wang, and N. Parga (2007). Mean-driven and fluctuation-driven persistent activity in recurrent networks. *Neural Comput. 19*, 1–46.

Renart, A., P. Song, and X.-J. Wang (2003). Robust spatial working memory through homeostatic synaptic scaling in heterogeneous cortical networks. *Neuron 38*, 473–485.

Rescorla, R. and A. Wagner (1972). A theory of pavlovian conditioning: variations in the effectiveness of reinforcement and nonreinforcement. In A. H. Black and W. F. Prokasy (Eds.), *Classical Conditioning: Vol. 2. Current Research and Theory*, Chapter 9, pp. 64–99. New York: Appleton-Century-Crofts.

Resulaj, A., R. Kiani, D. M. Wolpert, and M. N. Shadlen (2009). Changes of mind in decision-making. *Nature 461*, 263–266.

Reynolds, J. H. and L. Chelazzi (2004). Attentional modulation of visual processing. *Annu. Rev. Neurosci. 27*, 611–647.

Reynolds, J. H., L. Chelazzi, and R. Desimone (1999). Competitive mechanisms subserve attention in macaque areas V2 and V4. *J. Neurosci. 19*, 1736–1753.

Reynolds, J. H. and D. J. Heeger (2009). The normalization model of attention. *Neuron 61*, 168–185.

Reynolds, J. N. J., B. I. Hyland, and J. R. Wickens (2001). A cellular mechanism of reward-related learning. *Nature 413*, 67–70.

Ribary, U., A. Ioannides, K. Singh, R. Hasson, J. Bolton, F. Lado, A. Mogilner, and R. Llinás (1991). Magnetic field tomography of coherent thalamocortical 40-Hz oscillations in humans. *Proc. Natl. Acad. Sci. U.S.A. 88*, 11037–11041.

Ribot, T.-A. (1881). *Les Maladies de la Mémoire*. Paris: Germer Baillere.

Ricciardi, L. M. (1977). *Diffusion Processes and Related Topics on Biology*. Berlin: Springer-Verlag.

Rich, E. L. and J. D. Wallis (2016). Decoding subjective decisions from orbitofrontal cortex. *Nat. Neurosci. 19*, 973–980.

Richards, J. B., S. H. Mitchell, H. De Wit, and L. S. Seiden (1997). Determination of discount functions in rats with an adjusting-amount procedure. *J. Exp. Anal. Behav. 67*, 353–366.

Richardson, M. J. E., N. Brunel, and V. Hakim (2003). From subthreshold to firing-rate resonance. *J. Neurophysiol. 89*, 2538–2554.

Ridderinkhof, K. R., M. Ullsperger, E. A. Crone, and S. Nieuwenhuis (2004). The role of the medial frontal cortex in cognitive control. *Science 306*, 443–447.

Rieke, F., D. Warland, R. V. Steveninck, and W. Bialek (1999). *Spikes: Exploring the Neural Code*. Cambridge, MA: MIT Press.

Riesenhuber, M. and T. Poggio (1999). Hierarchical models of object recognition in cortex. *Nat. Neurosci. 2*, 1019–1025.

Rigotti, M., O. Barak, M. R. Warden, X. J. Wang, N. D. Daw, E. K. Miller, and S. Fusi (2013). The importance of mixed selectivity in complex cognitive tasks. *Nature 497*, 585–590.

Rigotti, M., D. B. Rubin, X.-J. Wang, and S. Fusi (2010). Internal representation of task rules by recurrent dynamics: The importance of the diversity of neural responses. *Front. Comput. Neurosci. 4*, 24.

Rinzel, J. (1987). A formal classification of bursting mechanisms in excitable systems. In A. M. Gleason (Ed.), *Proceedings of the International Congress of Mathematicians*, pp. 1578–1594. Providence, RI: American Mathematical Society.

Rinzel, J. and G. B. Ermentrout (1989). Analysis of neural excitability and oscillations. In C. Koch and I. Segev (Eds.), *Methods in Neuronal Modeling: From Synapses to Networks*, pp. 135–169. Cambridge, MA: MIT Press.

Ripke, S., B. M. Neale, A. Corvin, J. T. Walters, K.-H. Farh, P. A. Holmans, P. Lee, B. Bulik-Sullivan, D. A. Collier, H. Huang, et al. (2014). Biological insights from 108 schizophrenia-associated genetic loci. *Nature 511*, 421–427.

Riveland, R. and A. Pouget (2022). A neural model of task compositionality with natural language instructions. *bioRxiv*. https://doi.org/10.1101/2022.02.22.481293.

Robbins, T. W., C. M. Gillan, D. G. Smith, S. de Wit, and K. D. Ersche (2012). Neurocognitive endophenotypes of impulsivity and compulsivity: Towards dimensional psychiatry. *Trends Cogn. Sci. 16*, 81–91.

Roberts, A. C. (2020). Prefrontal regulation of threat-elicited behaviors: A pathway to translation. *Annu. Rev. Psychol. 71*, 357–387.

Roberts, J. A., T. W. Boonstra, and M. Breakspear (2015). The heavy tail of the human brain. *Curr. Opin. Neurobiol. 31*, 164–172.

Roberts, P. D. and C. C. Bell (2000). Computational consequences of temporally asymmetric learning rules: Ii. sensory image cancellation. *J. Comput. Neurosci. 9*, 67–83.

Roelfsema, P. R. and A. Holtmaat (2018). Control of synaptic plasticity in deep cortical networks. *Nat. Rev. Neurosci. 19*, 166–180.

Roelfsema, P. R. and A. van Ooyen (2005). Attention-gated reinforcement learning of internal representations for classification. *Neural Comput. 17*, 2176–2214.

Roeper, T. and M. Speas (Eds.) (2014). *Recursion: Complexitiy in Cognition*. Amherst, MA: Springer.

Roesch, M. R. and C. R. Olson (2005). Neuronal activity in primate orbitofrontal cortex reflects the value of time. *J. Neurophysiol. 94*, 2457–2471.

Rogers, R. D. and S. Monsell (1995). Costs of a predictible switch between simple cognitive tasks. *J. Exp. Psych. Gen. 124*, 207–231.

Rogers, T. T. and J. L. McClelland (2006). *Semantic Cognition: A Parallel Distributed Processing Approach*. Camrbidge, MA: MIT Press.

Rogers, T. T. and J. L. McClelland (2014). Parallel distributed processing at 25: Further explorations in the microstructure of cognition. *Cogn. Sci. 38*, 1024–1077.

Roiser, J. (2012). What has neuroscience ever done for us? *Psychologist 28*, 284–287.

Roitblat, H. L. (1980). Codes and coding processes in pigeon short-term memory. *Anim. Learn. Behav. 8*, 341–251.

Roitman, J. D. and M. N. Shadlen (2002). Response of neurons in the lateral intraparietal area during a combined visual discrimination reaction time task. *J. Neurosci. 22*, 9475–9489.

Rolls, E., M. Loh, G. Deco, and G. Winterer (2008). Computational models of schizophrenia and dopamine modulation in the prefrontal cortex. *Nat. Rev. Neurosci. 9*, 696–709.

Romo, R., C. D. Brody, A. Hernández, and L. Lemus (1999). Neuronal correlates of parametric working memory in the prefrontal cortex. *Nature 399*, 470–474.

Romo, R., A. Hernandez, and A. Zainos (2004). Neuronal correlates of a perceptual decision in ventral premotor cortex. *Neuron 41*, 165–173.

Romo, R., A. Hernández, A. Zainos, L. Lemus, and C. Brody (2002). Neuronal correlates of decision-making in secondary somatosensory cortex. *Nat. Neurosci. 5*, 1217–1225.

Romo, R. and R. Rossi-Pool (2020). Turning touch into perception. *Neuron 105*, 16–33.

Romo, R. and E. Salinas (2001). Touch and go: Decision-making mechanisms in somatosensation. *Ann. Rev. Neurosci. 24*, 107–137.

Rosch, E. (1978). Principles of categorization. In E. Rosch and B. B. Lloyd (Eds.), *Cognition and Categorization*, pp. 28–49. Hillsdale, NJ: Lawrence Erlbaum Associates.

Rosch, E. (1999). Principles of categorization. In E. Margolis and S. Laurence (Eds.), *Concepts: Core Readings*, pp. 189–206. Cambridge, MA: MIT Press.

Rosch, E. and C. B. Mervis (1975). Family resemblances: Studies in the internal structure of categories. *Cogn. Psychol. 7*, 573–605.

Rose, N. S., J. J. LaRocque, A. C. Riggall, O. Gosseries, M. J. Starrett, E. E. Meyering, and B. R. Postle (2016). Reactivation of latent working memories with transcranial magnetic stimulation. *Science 354*, 1136–1139.

Rosenbaum, R., M. A. Smith, A. Kohn, J. E. Rubin, and B. Doiron (2017). The spatial structure of correlated neuronal variability. *Nat. Neurosci. 20*, 107–114.

Rosenberg, M. D., S. A. Martinez, K. M. Rapuano, M. I. Conley, A. O. Cohen, M. D. Cornejo, D. J. Hagler, W. J. Meredith, K. M. Anderson, T. D. Wager, et al. (2020). Behavioral and neural signatures of working memory in childhood. *J. Neurosci. 40*, 5090–5104.

Rosenberg, M. D., T. Zhang, P. Perona, and M. Meister (2021). Mice in a labyrinth show rapid learning, sudden insight, and efficient exploration. *Elife 10*, e66175.

Rosenblatt, F. (1962). *Principles of Neurodynamics*. New York: Spartan.

Ross, J. M. and J. P. Hamm (2020). Cortical microcircuit mechanisms of mismatch negativity and its uuderlying subcomponents. *Front. Neural Circuits 14*, 13.

Rossi, L. F., K. D. Harris, and M. Carandini (2020). Spatial connectivity matches direction selectivity in visual cortex. *Nature 588*, 648–652.

Rotaru, D. C., H. Yoshino, D. A. Lewis, G. B. Ermentrout, and G. Gonzalez-Burgos (2011). Glutamate receptor subtypes mediating synaptic activation of prefrontal cortex neurons: Relevance for schizophrenia. *J. Neurosci. 31*, 142–156.

Roth, G. and U. Dicke (2005). Evolution of the brain and intelligence. *Trends Cogn. Sci. 9*, 250–257.

Roxin, A., N. Brunel, D. Hansel, G. Mongillo, and C. van Vreeswijk (2011). On the distribution of firing rates in networks of cortical neurons. *J. Neurosci. 31*, 16217–16226.

Roxin, A. and A. Ledberg (2008). Neurobiological models of two-choice decision making can be reduced to a one-dimensional nonlinear diffusion equation. *PLoS Comput. Biol. 4*, e1000046.

Roy, D. S., Y.-G. Park, M. E. Kim, Y. Zhang, S. K. Ogawa, N. DiNapoli, X. Gu, J. H. Cho, H. Choi, L. Kamentsky, et al. (2022). Brain-wide mapping reveals that engrams for a single memory are distributed across multiple brain regions. *Nat. Commun. 13*, 1799.

Rubenstein, J. L. and M. M. Merzenich (2003). Model of autism: Increased ratio of excitation/inhibition in key neural systems. *Genes Brain Behav. 2*, 255–267.

Rubia, K., T. Russell, S. Overmeyer, M. J. Brammer, E. T. Bullmore, T. Sharma, A. Simmons, S. C. Williams, V. Giampietro, C. M. Andrew, et al. (2001). Mapping motor inhibition: Conjunctive brain activations across different versions of go/no-go and stop tasks. *Neuroimage 13*, 250–261.

Rubinov, M. and E. Bullmore (2013). Fledgling pathoconnectomics of psychiatric disorders. *Trends Cogn. Sci. (Regul. Ed.) 17*, 641–647.

Rudebeck, P. H., M. J. Buckley, M. E. Walton, and M. F. Rushworth (2006). A role for the macaque anterior cingulate gyrus in social valuation. *Science 313*, 1310–1312.

Rudebeck, P. H. and E. A. Murray (2014). The orbitofrontal oracle: Cortical mechanisms for the prediction and evaluation of specific behavioral outcomes. *Neuron 84*, 1143–1156.

Rudy, B. (1988). Diversity and ubiquity of k channels. *Neuroscience 25*, 729–749.

Rumelhart, D. E. (1980). Schemata: The building blocks of cognition. In R. J. Spiro et al. (Ed.), *Theoretical Issues in Reading Comprehension*, pp. 33–58. Hillsdale, NJ: Lawrence Erlbaum.

Rumelhart, D. E., G. E. Hinton, and R. J. Williams (1986a). Learning representations by back-propagating errors. *Nature 323*, 533–536.

Rumelhart, D. E., J. L. McClelland, and the PDP Group (Eds.) (1986b). *Parallel Distributed Processing: Explorations in the Microstructure of Cognition*. Cambridge, MA: MIT Press.

Runyan, C. A., E. Piasini, S. Panzeri, and C. D. Harvey (2017). Distinct timescales of population coding across cortex. *Nature 548*, 92–96.

Rushworth, M. F. and T. Behrens (2008). Choice, uncertainty and value in prefrontal and cingulate cortex. *Nat. Neurosci. 11*, 389–397.

Rushworth, M. F., N. Kolling, J. Sallet, and R. B. Mars (2012). Valuation and decision-making in frontal cortex: One or many serial or parallel systems? *Curr. Opin. Neurobiol. 22*, 946–955.

Russek, E. M., I. Momennejad, M. M. Botvinick, S. J. Gershman, and N. D. Daw (2017). Predictive representations can link model-based reinforcement learning to model-free mechanisms. *PLoS Comput. Biol. 13*, e1005768.

Rust, N. C., V. Mante, E. P. Simoncelli, and J. A. Movshon (2006). How MT cells analyze the motion of visual patterns. *Nat. Neurosci. 9*, 1421–1431.

Rustichini, A. and C. Padoa-Schioppa (2015). A neuro-computational model of economic decisions. *J. Neurophysiol. 114*, 1382 1398.

Rutledge, R. B., M. Moutoussis, P. Smittenaar, P. Zeidman, T. Taylor, L. Hrynkiewicz, J. Lam, N. Skandali, J. Z. Siegel, O. T. Ousdal, G. Prabhu, P. Dayan, P. Fonagy, and R. J. Dolan (2017). Association of neural and emotional impacts of reward prediction errors with major depression. *JAMA Psychiat. 74*, 790–797.

Rutledge, R. B., N. Skandali, P. Dayan, and R. J. Dolan (2014). A computational and neural model of momentary subjective well-being. *Proc. Natl. Acad. Sci. U.S.A. 111*, 12252–12257.

Saalmann, Y. B., M. A. Pinsk, L. Wang, X. Li, and S. Kastner (2012). The pulvinar regulates information transmission between cortical areas based on attention demands. *Science 337*, 753–756.

Sabatini, B. L., T. G. Oertner, and K. Svoboda (2002). The life cycle of Ca(2+) ions in dendritic spines. *Neuron 33*, 439–452.

Sablé-Meyer, M., K. Ellis, J. Tenenbaum, and S. Dehaene (2022). A language of thought for the mental representation of geometric shapes. *Cogn. Psychol. 139*, 101527.

Sacks, O. (2010). *Musicophilia: Tales of Music and the Brain.* New York, NY: Knopf Doubleday Publishing.

Sacktor, T. C. (2011). How does PKMζ maintain long-term memory? *Nat. Rev. Neurosci. 12*, 9–15.

Sadacca, B. F., J. L. Jones, and G. Schoenbaum (2016). Midbrain dopamine neurons compute inferred and cached value prediction errors in a common framework. *Elife 5*, e13665.

Sah, P. (1996). Ca(2+)-activated K+ currents in neurones: Types, physiological roles and modulation. *Trends Neurosci. 19*, 150–154.

Sajad, A., D. C. Godlove, and J. D. Schall (2019). Cortical microcircuitry of performance monitoring. *Nat. Neurosci. 22*, 265–274.

Sakai, K. (2008). Task set and prefrontal cortex. *Annu. Rev. Neurosci. 31*, 219–245.

Salin, P. A. and D. A. Prince (1996). Spontaneous GABAA receptor mediated inhibitory currents in adult rat somatosensory cortex. *J. Neurophysiol. 75*, 1573–1588.

Salinas, E. (2004). Context-dependent selection of visuomotor maps. *BMC Neurosci 5*, 47.

Salinas, E. and L. F. Abbott (1995). Transfer of coded information from sensory to motor networks. *J. Neurosci. 15*, 6461–6474.

Salinas, E. and L. F. Abbott (1996). A model of multiplicative neural responses in parietal cortex. *Proc. Natl. Acad. Sci. U.S.A. 93*, 11956–11961.

Salinas, E., A. Hernandez, A. Zainos, and R. Romo (2000). Periodicity and firing rate as candidate neural codes for the frequency of vibrotactile stimuli. *J. Neurosci. 20*, 5503–5515.

Salinas, E., S. Shankar, M. G. Costello, D. Zhu, and T. R. Stanford (2010). Waiting is the hardest part: Comparison of two computational strategies for performing a compelled-response task. *Front. Comput. Neurosci. 4*, 153.

Salzman, C. D., K. H. Britten, and W. T. Newsome (1990). Cortical microstimulation influences perceptual judgments of motion direction. *Nature 346*, 174–177.

Samejima, K., Y. Ueda, K. Doya, and M. Kimura (2005). Representation of action-specific reward values in the striatum. *Science 310*, 1337–1340.

Samuelson, P. A. (1937). A note on measurement of utility. *Rev. Econ. Stud. 4*, 155–161.

Sanborn, A. N. and N. Chater (2016). Bayesian brains without probabilities. *Trends Cogn. Sci. 20*, 883–893.

Sanchez-Vives, M. V. and D. A. McCormick (2000). Cellular and network mechanisms of rhythmic recurrent activity in neocortex. *Nat. Neurosci. 3*, 1027–1034.

Sanchez-Vives, M. V., L. G. Nowak, and D. A. McCormick (2000). Cellular mechanisms of long-lasting adaptation in visual cortical neurons in vitro. *J. Neurosci. 20*, 4286–4299.

Sanders, H., M. Berends, G. Major, M. S. Goldman, and J. E. Lisman (2013). Nmda and gabab (kir) conductances: The 'perfect couple' for bistability. *J. Neurosci. 33*, 424–429.

Sanides, F. (1972). Representation in the cerebral cortex and its area lamination patterns. In G. H. Bourne (Ed.), *The Structure and Function of the Nervous System*, Chapter 7, pp. 329–453. New York: Academic Press.

Sapolsky, R. M. (2017). *Behave: The Biology of Humans at Our Best and Worst.* London: Penguin Press.

Sarma, A., N. Y. Masse, X.-J. Wang, and D. J. Freedman (2016). Task-specific versus generalized mnemonic representations in parietal and prefrontal cortices. *Nat. Neurosci. 19*, 143–149.

Sasaki, K. and H. Gemba (1986). Electrical activity in the prefrontal cortex specific to no-go reaction of conditioned hand movement with colour discrimination in the monkey. *Exp. Brain Res. 64*, 603–606.

Sato, T. R., A. Murthy, K. G. Thompson, and J. D. Schall (2001). Search efficiency but not response interference affects visual selection in frontal eye field. *Neuron 30*, 583–591.

Sato, T. R. and J. D. Schall (2003). Effects of stimulus-response compatibility on neural selection in frontal eye field. *Neuron 38*, 637–648.

Satterthwaite, T. D., D. H. Wolf, G. Erus, K. Ruparel, M. A. Elliott, E. D. Gennatas, R. Hopson, C. Jackson, K. Prabhakaran, W. B. Bilker, M. E. Calkins, J. Loughead, A. Smith, D. R. Roalf, H. Hakonarson, R. Verma, C. Davatzikos, R. C. Gur, and R. E. Gur (2013). Functional maturation of the executive system during adolescence. *J. Neurosci. 33*, 16249–16261.

Sawaguchi, T. and P. S. Goldman-Rakic (1991). D1 dopamine receptors in prefrontal cortex: Involvement in working memory. *Science 251*, 947–950.

Saxe, A., S. Nelli, and C. Summerfield (2021). If deep learning is the answer, what is the question? *Nat. Rev. Neurosci. 22*, 55–67.

Saxena, S., A. L. Brody, J. M. Schwartz, and L. R. Baxter (1998). Neuroimaging and frontal-subcortical circuitry in obsessive-compulsive disorder. *Br. J. Psychiatry Suppl. 35*, 26–37.

Schaffer, E. S., N. Mishra, M. R. Whiteway, W. Li, M. B. Vancura, J. Freedman . . . R. Axel. (2023). The spatial and temporal structure of neural activity across the fly brain. *Nat. Commun. 14*, 5572.

Schall, J. D. (2001). Neural basis of deciding, choosing and acting. *Nat. Neurosci. 2*, 33–42.

Schall, J. D. (2004). On building a bridge between brain and behavior. *Annu. Rev. Psychol. 55*, 23–50.

Schall, J. D. (2019). Accumulators, neurons, and response time. *Trends Neurosci. 42*, 848–860.

Schall, J. D. and D. P. Hanes (1993). Neural basis of saccade target selection in frontal eye field during visual search. *Nature 366*, 467–469.

Schall, J. D., T. J. Palmeri, and G. D. Logan (2017). Models of inhibitory control. *Philos. Trans. R. Soc. B: Biol. Sci. 372*, 20160193.

Schapiro, A. C., N. B. Turk-Browne, K. A. Norman, and M. M. Botvinick (2016). Statistical learning of temporal community structure in the hippocampus. *Hippocampus 26*, 3–8.

Schecklmann, M., A.-C. Ehlis, M. M. Plichta, T. Dresler, M. Heine, A. Boreatti-Hümmer, M. Romanos, C. Jacob, P. Pauli, and A. J. Fallgatter (2013). Working memory and response inhibition as one integral phenotype of adult ADHD? A behavioral and imaging correlational investigation. *J. Atten. Disord. 17*, 470–482.

Scheffer, L. K., C. S. Xu, M. Januszewski, Z. Lu, S.-y. Takemura, K. J. Hayworth, G. Huang, K. Shinomiya, J. Maitlin-Shepard, S. Berg, et al. (2020). A connectome and analysis of the adult Drosophila central brain. *BioRxiv*. https://doi.org/10.1101/2020.04.07.030213.

Scheffer, M. (2009). *Critical Transitions in Nature and Society*. Princeton, NJ: Princeton University Press.

Scherberger, H. and R. Andersen (2007). Target selection signals for arm reaching in the posterior parietal cortex. *J. Neurosci. 27*, 2001–2012.

Scherzer, C. R., G. B. Landwehrmeyer, J. A. Kerner, T. J. Counihan, C. M. Kosinski, D. G. Standaert, L. P. Daggett, G. Veliçelebi, J. B. Penney, and A. B. Young (1998). Expression of N-Methyl-D-Aspartate receptor subunit mRNAs in the human brain: Hippocampus and cortex. *J. Comp. Neurol. 390*, 75–90.

Schiller, J., G. Major, H. J. Koester, and Y. Schiller (2000). NMDA spikes in basal dendrites of cortical pyramidal neurons. *Nature 404*, 285–289.

Schlag-Rey, M., N. Amador, H. Sanchez, and J. Schlag (1997). Antisaccade performance predicted by neuronal activity in the supplementary eye field. *Nature 390*, 398–401.

Schlichting, M. L., J. A. Mumford, and A. R. Preston (2015). Learning-related representational changes reveal dissociable integration and separation signatures in the hippocampus and prefrontal cortex. *Nat. Commun. 6*, 1–10.

Schmack, K., M. Bosc, T. Ott, J. Sturgill, and A. Kepecs (2021). Striatal dopamine mediates hallucination-like perception in mice. *Science 372*. https://doi.org/10.1126/science.abf4740.

Schmidt, E. R., H. T. Zhao, J. M. Park, M. Dipoppa, M. M. Monsalve-Mercado, J. B. Dahan, C. C. Rodgers, A. Lejeune, E. Hillman, K. D. Miller, et al. (2021). A human-specific modifier of cortical connectivity and circuit function. *Nature 599*, 640–644.

Schmidt, R., D. K. Leventhal, N. Mallet, F. Chen, and J. D. Berke (2013). Canceling actions involves a race between basal ganglia pathways. *Nat. Neurosci. 16*, 1118–1124.

Schmitt, L. I., R. D. Wimmer, M. Nakajima, M. Happ, S. Mofakham, and M. M. Halassa (2017). Thalamic amplification of cortical connectivity sustains attentional control. *Nature 545*, 219–223.

Schmolesky, M. T., Y. Wang, D. P. Hanes, K. G. Thompson, S. Leutgeb, J. D. Schall, and A. G. Leventhal (1998). Signal timing across the macaque visual system. *J. Neurophysiol. 79*, 3272–3278.

Schneider, M., A. C. Broggini, B. Dann, A. Tzanou, C. Uran, S. Sheshadri, H. Scherberger, and M. Vinck (2021). A mechanism for inter-areal coherence through communication based on connectivity and oscillatory power. *Neuron 109*, 4050–4067.

Schoenbaum, G., A. Chiba, and M. Gallagher (1998). Orbitofrontal cortex and basolateral amygdala encode expected outcomes during learning. *Nat. Neurosci. 1*, 155–159.

Schoenbaum, G., M. R. Roesch, T. A. Stalnaker, and Y. K. Takahashi (2009). A new perspective on the role of the orbitofrontal cortex in adaptive behaviour. *Nat. Rev. Neurosci. 10*, 885–892.

Scholtens, L. H., R. Schmidt, M. A. de Reus, and M. P. van den Heuvel (2014). Linking macroscale graph analytical organization to microscale neuroarchitectonics in the macaque connectome. *J. Neurosci. 34*, 12192–12205.

Schork, A. J., H. Won, V. Appadurai, R. Nudel, M. Gandal, O. Delaneau, M. R. Christiansen, D. M. Hougaard, M. Bækved-Hansen, J. Bybjerg-Grauholm, et al. (2019). A genome-wide association study of shared risk across psychiatric disorders implicates gene regulation during fetal neurodevelopment. *Nat. Neurosci. 22*, 353–361.

Schrimpf, M., J. Kubilius, H. Hong, N. J. Majaj, R. Rajalingham, E. B. Issa, K. Kar, P. Bashivan, J. Prescott-Roy, K. Schmidt, et al. (2018). Brain-score: Which artificial neural network for object recognition is most brain-like? *BioRxiv*, 407007.

Schrimpf, M., J. Kubilius, M. J. Lee, N. A. Ratan Murty, R. Ajemian, and J. J. DiCarlo (2020). Integrative benchmarking to advance neurally mechanistic models of human intelligence. *Neuron 108*, 413–423.

Schuck, N. W., M. B. Cai, R. C. Wilson, and Y. Niv (2016). Human orbitofrontal cortex represents a cognitive map of state space. *Neuron 91*, 1402–1412.

Schuck, N. W. and Y. Niv (2019). Sequential replay of nonspatial task states in the human hippocampus. *Science 364*, eaaw5181.

Schulman, J., F. Wolski, P. Dhariwal, A. Radford, and O. Klimov (2017). Proximal policy optimization algorithms. *arXiv preprint arXiv:1707.06347*.

Schultz, W. (1998). Predictive reward signal of dopamine neurons. *J. Neurophysiol. 80*, 1–27.

Schultz, W., P. Dayan, and P. R. Montague (1997). A neural substrate of prediction and reward. *Science 275*, 1593–1599.

Schulz, D. J., J.-M. Goaillard, and E. E. Marder (2006). Variable channel expression in identified single and electrically coupled neurons in different animals. *Nat. Neurosci. 9*, 356–362.

Schulz, D. J., J.-M. Goaillard, and E. E. Marder (2007). Quantitative expression profiling of identified neurons reveals cell-specific constraints on highly variable levels of gene expression. *Proc. Natl. Acad. Sci. 104*, 13187–13191.

Schwartz, O., J. W. Pillow, N. C. Rust, and E. P. Simoncelli (2006). Spike-triggered neural characterization. *J. Vis. 6*, 484–507.

Schwartz-Ziv, R. and N. Tishby (2017). Opening the black box of deep neural networks via information. *arXiv preprint arXiv:1703.00810.*

Schweighofer, N., M. Bertin, K. Shishida, Y. Okamoto, S. C. Tanaka, S. Yamawaki, and K. Doya (2008). Low-serotonin levels increase delayed reward discounting in humans. *J. Neurosci. 28*, 4528–4532.

Schweighofer, N. and K. Doya (2003). Meta-learning in reinforcement learning. *Neural Netw. 16*, 5–9.

Sclar, G. and R. D. Freeman (1982). Orientation selectivity in the cat's striate cortex is invariant with stimulus contrast. *Exp. Brain Res. 46*, 457–461.

Scoville, W. B. and B. Milner (1957). Loss of recent memory after bilateral hippocampal lesions. *J. Neurol. Neurosurg. Psychiatr. 20*, 11–21.

Seamans, J. K., D. Durstewitz, B. R. Christie, C. F. Stevens, and T. J. Sejnowski (2000). Dopamine D1/D5 receptor modulation of excitatory synaptic inputs to layer V prefrontal cortex neurons. *Proc. Natl. Acad. Sci. U.S.A. 98*, 301–306.

Seamans, J. K. and C. Yang (2004). The principal features and mechanisms of dopamine modulation in the prefrontal cortex. *Prog. Neurobiol. 74*, 1–58.

Seeholzer, A., M. Deger, and W. Gerstner (2019). Stability of working memory in continuous attractor networks under the control of short-term plasticity. *PLoS Comput. Biol. 15*, e1006928.

Seger, C. A. and E. K. Miller (2010). Category learning in the brain. *Annu. Rev. Neurosci. 33*, 203–219.

Segev, I., J. Rinzel, and G. M. Shepherd (1995). *The Theoretical Foundation of Dendritic Function: Selected Papers of Wilfrid Ralll.* Cambridge, MA: MIT Press.

Segrè, E. (2007). *From X-rays to Quarks: Modern Physicists and Their Discoveries.* New York: Dover Publications.

Sejnowski, T. J. (1977). Storing covariance with nonlinearly interacting neurons. *J. Math. Biol. 4*, 303.

Sejnowski, T. J., C. Koch, and P. S. Churchland (1988). Computational neuroscience. *Science 241*, 1299–1306.

Selemon, L. D. (2013). A role for synaptic plasticity in the adolescent development of executive function. *Transl. Psychiatry 3*, e238–e238.

Semedo, J. D., A. I. Jasper, A. Zandvakili, A. Aschner, C. K. Machens, A. Kohn, and M. Y. Byron (2022). Feedforward and feedback interactions between visual cortical areas use different population activity patterns. *Nat. Commun.*, 1099.

Semedo, J. D., A. Zandvakili, C. K. Machens, M. Y. Byron, and A. Kohn (2019). Cortical areas interact through a communication subspace. *Neuron 102*, 249–259.

Seo, H. and D. Lee (2008). Cortical mechanisms for reinforcement learning in competitive games. *Philos. Trans. R. Soc. B: Biol. Sci. 363*, 3845–3857.

Seo, H., T. J. Vickery, and D. Lee (2011). Game theory in neuroscience. *Cogn. Critique 4*, 87–120.

Seol, G. H., J. Ziburkus, S. Huang, L. Song, I. T. Kim, K. Takamiya, R. L. Huganir, H. K. Lee, and A. Kirkwood (2007). Neuromodulators control the polarity of spike-timing-dependent synaptic plasticity. *Neuron 55*, 919–929.

Serafin, M., S. Williams, A. Khateb, P. Fort, and M. Muhlethaler (1996). Rhythmic firing of medial septum non-cholinergic neurons. *Neuroscience 75*, 671–675.

Sereno, A. and S. Amador (2006). Attention and memory-related responses of neurons in the lateral intraparietal area during spatial and shape-delayed match-to-sample tasks. *J. Neurophysiol. 95*, 1078–1098.

Sergent, C., M. Corazzol, G. Labouret, F. Stockart, M. Wexler, J.-R. King, F. Meyniel, and D. Pressnitzer (2021). Bifurcation in brain dynamics reveals a signature of conscious processing independent of report. *Nat. Commun. 12.* https://doi.org/10.1038/s41467-021-21393-z.

Seriès, P. (Ed.) (2020). *Computational Psychiatry: A Primer.* Cambridge, MA: MIT Press.

Serre, T., A. Oliva, and T. Poggio (2007a). A feedforward architecture accounts for rapid categorization. *Proc. Natl. Acad. Sci. 104*, 6424–6429.

Serre, T., L. Wolf, S. Bileschi, M. Riesenhuber, and T. Poggio (2007b). Robust object recognition with cortex-like mechanisms. *IEEE Trans. Pattern Anal. Mach. Intell. 29*, 411–426.

Seung, H. S. (1996). How the brain keeps the eyes still. *Proc. Natl. Acad. Sci. U.S.A. 93*, 13339–13344.

Seung, H. S. (2012). *Connectome: How the Brain's Wiring Makes Who We Are.* Boston, MA: Houghton Mifflin Harcourt.

Seung, H. S., D. D. Lee, B. Y. Reis, and D. W. Tank (2000). Stability of the memory of eye position in a recurrent network of conductance-based model neurons. *Neuron 26*, 259–271.

Seung, H. S. and H. Sompolinsky (1993). Simple models for reading neuronal population codes. *Proc. Natl. Acad. Sci. 90*, 10749–10753.

Shadlen, M. N., K. H. Britten, W. T. Newsome, and J. A. Movshon (1996). A computational analysis of the relationship between neuronal and behavioral responses to visual motion. *J. Neurosci. 16*, 1486–1510.

Shadlen, M. N., T. D. Hanks, A. K. Churchland, R. Kiani, and T. Yang (2007). The speed and accuracy of a simple perceptual decision: A mathematical primer. In K. Doya, A. Pouget, and R. P. N. Rao (Eds.), *Bayesian Brain: Probabilistic Approaches to Neural Coding*, pp. 209–237. Cambridge, MA: MIT Press.

Shadlen, M. N. and E. R. Kandel (2021). Decision-making and consciousness. In E. R. Kandel, J. D. Koester, S. H. Mack, and S. A. Siegelbaum (Eds.), *Principles of Neural Science* (6th ed.), pp. 1392–1416. New York: McGraw-hill.

Shadlen, M. N. and R. Kiani (2013). Decision making as a window on cognition. *Neuron 80*, 791–806.

Shadlen, M. N. and W. T. Newsome (1994). Noise, neural codes and cortical organization. *Curr. Opin. Neurobiol. 4*, 569–579.

Shadlen, M. N. and W. T. Newsome (1996). Motion perception: Seeing and deciding. *Proc. Natl. Acad. Sci. U.S.A. 93*, 628–633.

Shadlen, M. N. and W. T. Newsome (1998). The variable discharge of cortical neurons: Implications for connectivity, computation, and information coding. *J. Neurosci. 18*, 3870–3896.

Shadlen, M. N. and W. T. Newsome (2001). Neural basis of a perceptual decision in the parietal cortex (area LIP) of the rhesus monkey. *J. Neurophysiol. 86*, 1916–1936.

Shadlen, M. N. and D. Shohamy (2016). Decision making and sequential sampling from memory. *Neuron 90*, 927–939.

Shadmehr, R. and S. Mussa-Ivaldi (2012). *Biological Learning and Control: How the Brain Builds Representations, Predicts Events, and Makes Decisions*. Cambridge, MA: MIT Press.

Shafiei, G., R. D. Markello, R. V. De Wael, B. C. Bernhardt, B. D. Fulcher, and B. Misic (2020). Topographic gradients of intrinsic dynamics across neocortex. *Elife 9*, e62116.

Shallice, T. (1988). *From Neuropsychology to Mental Structure*. Cambridge: Cambridge University Press.

Shapiro, L. (2019). *Embodied Cognition* (2nd ed.). Milton Park, UK: Routledge.

Shatz, C. J. (1990). Impulse activity and the patterning of connections during cns development. *Neuron 5*, 745–756.

Shen, W., M. Flajolet, P. Greengard, and D. Surmeier (2008). Dichotomous dopaminergic control of striatal synaptic plasticity. *Science 321*, 848–851.

Shenhav, A., M. M. Botvinick, and J. D. Cohen (2013). The expected value of control: An integrative theory of anterior cingulate cortex function. *Neuron 79*, 217–240.

Shenoy, K. V., M. T. Kaufman, M. Sahani, and M. M. Churchland (2011). A dynamical systems view of motor preparation: Implications for neural prosthetic system design. *Prog. Brain Res. 192*, 33–58.

Shenoy, K. V., M. Sahani, and M. M. Churchland (2013). Cortical control of arm movements: A dynamical systems perspective. *Annu. Rev. Neurosci. 36*, 337–359.

Shepard, R. N. (1980). Multidimensional scaling, tree-fitting, and clustering. *Science 210*, 390–398.

Shepherd, G. M. (Ed.) (1998). *The Synaptic Organization of the Brain* (4th ed.). New York: Oxford University Press.

Shepherd, G. M. and S. Grillner (Eds.) (2018). *Handbook od Brain Microcircuits* (2nd ed.). New York: Oxford University Press.

Shepherd, S. V. and W. A. Freiwald (2018). Functional networks for social communication in the macaque monkey. *Neuron 99*, 413–420.

Sherman, M. A., S. Lee, R. Law, S. Haegens, C. A. Thorn, M. S. Hamalainen, C. I. Moore, and S. R. Jones (2016). Neural mechanisms of transient neocortical beta rhythms: Converging evidence from humans, computational modeling, monkeys, and mice. *Proc. Natl. Acad. Sci. U.S.A. 113*, E4885–4894.

Sherrington, C. (1906). *The integrative Action of the Nervous System*. New Haven, CT: Yale University Press.

Shi, L., T. L. Griffiths, N. H. Feldman, and A. N. Sanborn (2010). Exemplar models as a mechanism for performing Bayesian inference. *Psychon. Bull. Rev. 17*, 443–464.

Shields, W., J. Smith, K. Guttmannova, and D. Washburn (2005). Confidence judgments by humans and rhesus monkeys. *J. Gen. Psychol. 132*, 165–186.

Shin, E. J., Y. Jang, S. Kim, H. Kim, X. Cai, H. Lee, J. H. Sul, S. H. Lee, Y. Chung, D. Lee, and M. W. Jung (2021). Robust and distributed neural representation of action values. *Elife 10*, e53045.

Shouval, H. Z., M. F. Bear, and L. N. Cooper (2002). A unified model of nmda receptor-dependent bidirectional synaptic plasticity. *Proc. Natl. Acad. Sci. 99*, 10831–10836.

Siegel, M., T. J. Buschman, and E. K. Miller (2015). Cortical information flow during flexible sensorimotor decisions. *Science 348*, 1352–1355.

Siegle, J. H., X. Jia, S. Durand, S. Gale, C. Bennett, N. Graddis, G. Heller, T. K. Ramirez, H. Choi, J. A. Luviano, et al. (2021). Survey of spiking in the mouse visual system reveals functional hierarchy. *Nature 592*, 86–92.

Sigala, N., M. Kusunoki, I. Nimmo-Smith, D. Gaffan, and J. Duncan (2008). Hierarchical coding for sequential task events in the monkey prefrontal cortex. *Proc. Natl. Acad. Sci. U.S.A. 105*, 11969–11974.

Sigman, M. and S. Dehaene (2005). Parsing a cognitive task: A characterization of the mind's bottleneck. *PLoS Biol. 3*, e37.

Sik, A., M. Penttonen, A. Ylinen, and G. Buzsáki (1995). Hippocampal CA1 interneurons: An in vivo intracellular labeling study. *J. Neurosci. 15*, 6651–6665.

Silberberg, G. and H. Markram (2007). Disynaptic inhibition between neocortical pyramidal cells mediated by Martinotti cells. *Neuron 53*, 735–746.

Silva, L., Y. Amitai, and B. Connors (1991). Intrinsic oscillations of neocortex generated by layer 5 pyramidal neurons. *Science 251*, 432–435.

Silver, D., J. Schrittwieser, K. Simonyan, I. Antonoglou, A. Huang, A. Guez, T. Hubert, L. Baker, M. Lai, A. Bolton, et al. (2017). Mastering the game of go without human knowledge. *Nature 550*, 354–359.

Simen, P., F. Balci, L. de Souza, J. D. Cohen, and P. Holmes (2011). A model of interval timing by neural integration. *J. Neurosci. 31*, 9238–9253.

Simon, H. A. (1997). *Models of Bounded Rationality: Empirically Grounded Economic Reason*. Cambridge, MA: MIT Press.

Simons, D. J. and C. F. Chabris (1999). Gorillas in our midst: Sustained inattentional blindness for dynamic events. *Perception 28*, 1059–1074.

Sjöström, P. J., G. G. Turrigiano, and S. B. Nelson (2001). Rate, timing, and cooperativity jointly determine cortical synaptic plasticity. *Neuron 32*, 1149–1164.

Skaggs, W. E., J. J. Knierim, H. S. Kudrimoti, and B. L. McNaughton (1995). A model of the neural basis of the rat's sense of direction. In D. S. T. G. Tesauro and T. K. Leen (Eds.), *Advances in Neural Information Processing Systems 7*. Cambridge, MA: MIT Press.

Skaggs, W. E., B. McNaughton, M. Wilson, and C. Barnes (1996). Theta phase precession in hippocampal neuronal populations and the compression of temporal sequences. *Hippocampus 6*, 149–172.

Smaers, J. B., A. Gómez-Robles, A. N. Parks, and C. C. Sherwood (2017). Exceptional evolutionary expansion of prefrontal cortex in great apes and humans. *Curr. Biol. 27*, 714–720.

Smith, G. D., C. L. Cox, S. M. Sherman, and J. Rinzel (2000). Fourier analysis of sinusoidally driven thalamocortical relay neurons and a minimal integrate-and-fire-or-burst model. *J. Neurophysiol. 83*, 588–610.

Smith, J. D., J. J. Couchman, and M. J. Beran (2014). Animal metacognition: A tale of two comparative psychologies. *J. Comp. Psychol. 128*, 115–131.

Smith, P. L. (2016). Diffusion theory of decision making in continuous report. *Psychol. Rev. 123*, 425–451.

Smith, S. M., P. T. Fox, K. L. Miller, D. C. Glahn, P. M. Fox, C. E. Mackay, N. Filippini, K. E. Watkins, R. Toro, A. R. Laird, et al. (2009). Correspondence of the brain's functional architecture during activation and rest. *Proc. Natl. Acad. Sci. 106*, 13040–13045.

Smolen, P., D. A. Baxter, and J. H. Byrne (2006). A model of the roles of essential kinases in the induction and expression of late long-term potentiation. *Biophys. J. 90*, 2760–2775.

Smolen, P., D. A. Baxter, and J. H. Byrne (2019). How can memories last for days, years, or a lifetime? Proposed mechanisms for maintaining synaptic potentiation and memory. *Learn. Mem. 26*, 133–150.

Smolen, P., Y. Zhang, and J. H. Byrne (2016). The right time to learn: Mechanisms and optimization of spaced learning. *Nat. Rev. Neurosci. 17*, 77–88.

Smolensky, P. (1990). Tensor product variable binding and the representation of symbolic structures in connectionist systems. *Artif. Intell. 46*, 159–216.

Smucny, J., S. J. Dienel, D. A. Lewis, and C. S. Carter (2022). Mechanisms underlying dorsolateral prefrontal cortex contributions to cognitive dysfunction in schizophrenia. *Neuropsychopharmacology 47*, 292–308.

Snyder, L. H., A. P. Batista, and R. A. Andersen (1997). Coding of intention in the posterior parietal cortex. *Nature 386*, 167–170.

Sofroniew, N. J., D. Flickinger, J. King, and K. Svoboda (2016). A large field of view two-photon mesoscope with subcellular resolution for in vivo imaging. *eLife 5*, e14472.

Softky, W. R. (1994). Sub-millisecond coincidence detection in active dentritic trees. *Neuroscience 58*, 13–41.

Softky, W. R. and C. Koch (1993). The highly irregular firing of cortical cells is inconsistent with temporal integration of random EPSPs. *J. Neurosci. 13*, 334–350.

Sohn, H., N. Meirhaeghe, R. Rajalingham, and M. Jazayeri (2020). A network perspective on sensorimotor learning. *Trends Neurosci.* https://doi.org/10.1016/j.tins.2020.11.007.

Solso, R. L. (1994). *Cognition and the Visual Arts*. Cambridge, MA: MIT Press.

Soltani, A., B. De Martino, and C. Camerer (2012). A range-normalization model of context-dependent choice: A new model and evidence. *PLoS Comput. Biol. 8*, e1002607.

Soltani, A., P. Khorsand, C. Guo, S. Farashahi, and J. Liu (2016). Neural substrates of cognitive biases during probabilistic inference. *Nat. Commun. 7*, 1–14.

Soltani, A., D. Lee, and X.-J. Wang (2006). Neural mechanism for stochastic behaviour during a competitive game. *Neural Netw. 19*, 1075–1090.

Soltani, A. and X.-J. Wang (2006). A biophysically based neural model of matching law behavior: Melioration by stochastic synapses. *J. Neurosci. 26*, 3731–3744.

Soltani, A. and X.-J. Wang (2010). Synaptic computation underlying probabilistic inference. *Nat. Neurosci. 13*, 112–119.

Soltesz, I. and M. Deschênes (1993, Jul). Low- and high-frequency membrane potential oscillations during theta activity in CA1 and CA3 pyramidal neurons of the rat hippocampus under ketamine-xylazine anesthesia. *J. Neurophysiol. 70*, 97–116.

Somers, D. C., S. B. Nelson, and M. Sur (1995). An emergent model of orientation selectivity in cat visual cortical simple cells. *J. Neurosci. 15*, 5448–5465.

Sommer, M. A. and R. H. Wurtz (2002). A pathway in primate brain for internal monitoring of movements. *Science 296*, 1480–1482.

Sommer, M. A. and R. H. Wurtz (2008). Brain circuits for the internal monitoring of movements. *Annu. Rev. Neurosci. 31*, 317–338.

Sommer, T. (2017). The emergence of knowledge and how it supports the memory for novel related information. *Cereb. Cortex 27*, 1906–1921.

Somogyi, P. (1977). A specific 'axo-axonal'interneuron in the visual cortex of the rat. *Brain Res. 136*, 345–350.

Somogyi, P., G. Tamás, R. Lujan, and E. H. Buhl (1998). Salient features of synaptic organization in the cerebral cortex. *Brain Res. Rev. 26*, 113–135.

Sompolinsky, H., A. Crisanti, and H. J. Sommers (1988). Chaos in random neural networks. *Phys. Rev. Lett. 61*, 259–262.

Sompolinsky, H. and R. Shapley (1997). New perspectives on the mechanisms for orientation selectivity. *Curr. Opin. Neurobiol. 7*, 514–522.

Song, H. F., H. Kennedy, and X.-J. Wang (2014). Spatial embedding of similarity structure in the cerebral cortex. *Proc. Natl. Acad. Sci. U.S.A. 111*, 16580–16585.

Song, H. F. and X.-J. Wang (2014). Simple, distance-dependent formulation of the Watts-Strogatz model for directed and undirected small-world networks. *Phys. Rev. E Stat. Nonlin. Soft. Matter Phys. 90*, 062801.

Song, H. F., G. R. Yang, and X.-J. Wang (2016). Training excitatory-inhibitory recurrent neural networks for cognitive tasks: A simple and flexible framework. *PLoS Comput. Biol. 12*, e1004792.

Song, H. F., G. R. Yang, and X.-J. Wang (2017). Reward-based training of recurrent neural networks for cognitive and value-based tasks. *Elife 6*. https://doi.org/10.7554/eLife.21492.

Song, M., E. J. Shin, H. Seo, A. Soltani, N. A. Steinmetz, D. Lee, M. W. Jung, and S.-B. Paik (2023). Hierarchical gradient of timescales in the mammalian forebrain. *bioRxiv*. https://doi.org/10.1101/2023.05.12.540610.

Song, P. and X.-J. Wang (2005). Angular path integration by moving 'hill of activity': A spiking neuron model without recurrent excitation of the head-direction system. *J. Neurosci. 25*, 1002–1014.

Song, S., K. D. Miller, and L. F. Abbott (2000). Competitive hebbian learning through spike-time-dependent synaptic plasticity. *Nat. Neurosci. 3*, 919–926.

Song, S., P. Sjöström, M. Reigl, S. Nelson, and D. Chklovskii (2005). Highly nonrandom features of synaptic connectivity in local cortical circuits. *PLoS Biol. 3*, e68.

Sorscher, B., G. C. Mel, S. A. Ocko, L. M. Giocomo, and S. Ganguli (2023). A unified theory for the computational and mechanistic origins of grid cells. *Neuron 111*, 121–137.

Soto, D., J. Hodsoll, P. Rotshtein, and G. W. Humphreys (2008). Automatic guidance of attention from working memory. *Trends Cogn. Sci. 12*, 342–348.

Sotres-Bayon, F., D. E. Bush, and J. E. LeDoux (2004). Emotional perseveration: An update on prefrontal-amygdala interactions in fear extinction. *Learn. Mem. 11*, 525–535.

Spaak, E., M. Bonnefond, A. Maier, D. A. Leopold, and O. Jensen (2012). Layer-specific entrainment of gamma-band neural activity by the alpha rhythm in monkey visual cortex. *Curr. Biol. 22*, 2313–2318.

Sparks, D. L. (2002). The brainstem control of saccadic eye movements. *Nat. Rev. Neurosci. 3*, 952–964.

Spearman, C. (1927). *The Abilities of Man*. New York: McMillan.

Spencer, K., P. Nestor, M. Niznikiewicz, D. Salisbury, M. Shenton, and R. McCarley (2003). Abnormal neural synchrony in schizophrenia. *J. Neurosci. 23*, 7407–7411.

Spitmaan, M., H. Seo, D. Lee, and A. Soltani (2020). Multiple timescales of neural dynamics and integration of task-relevant signals across cortex. *Proc. Natl. Acad. Sci. 117*, 22522–22531.

Sporns, O. (2009). *Networks of the Brain*. Cambridge, MA: MIT Press.

Sporns, O. (2014). Contributions and challenges for network models in cognitive neuroscience. *Nat. Neurosci. 17*, 652–660.

Sporns, O., G. Tononi, and R. Kötter (2005). The human connectome: A structural description of the human brain. *PLoS Comput. Biol. 1*, e42.

Sporns, O. and J. D. Zwi (2004). The small world of the cerebral cortex. *Neuroinformatics 2*, 145–162.

Spreng, R. N., W. D. Stevens, J. P. Chamberlain, A. W. Gilmore, and D. L. Schacter (2010). Default network activity, coupled with the frontoparietal control network, supports goal-directed cognition. *Neuroimage 53*, 303–317.

Spruston, N. (2008). Pyramidal neurons: Dendritic structure and synaptic integration. *Nat. Rev. Neurosci. 9*, 206–221.

Spruston, N., P. Jonas, and B. Sakmann (1995). Dendritic glutamate receptor channel in rat hippocampal CA3 and CA1 pyramidal neurons. *J. Physiol. 482*, 325–352.

Squire, L. R. (1987). *Memory and Brain*. New York: Oxford University Press.

Squire, L. R. (2009). The legacy of patient H.M. for neuroscience. *Neuron 61*, 6–9.

Squire, L. R. and E. Kandel (1999). *Memory: From Mind to Molecules*. Scientific American Library.

Sreenivasan, K. K. and M. D'Esposito (2019). The what, where and how of delay activity. *Nat. Rev. Neurosci. 20*, 466–481.

Starc, M., J. D. Murray, N. Santamauro, A. Savic, C. Diehl, Y. T. Cho, V. Srihari, P. T. Morgan, J. H. Krystal, X.-J. Wang, G. Repovs, and A. Anticevic (2017). Schizophrenia is associated with a pattern of spatial working memory deficits consistent with cortical disinhibition. *Schizophr. Res. 181*, 107–116.

Stein, H., J. Barbosa, and A. Compte (2021). Towards biologically constrained attractor models of schizophrenia. *Curr. Opin. Neurobiol. 70*, 171–181.

Steinemann, N. A., G. M. Stine, E. M. Trautmann, A. Zylberberg, D. M. Wolpert, and M. N. Shadlen (2022). Direct observation of the neural computations underlying a single decision. *bioRxiv*. https://doi.org/10.1101/2022.05.02.490321.

Steiner, A. P. and A. D. Redish (2014). Behavioral and neurophysiological correlates of regret in rat decision-making on a neuroeconomic task. *Nat. Neurosci. 17*, 995.

Steinmetz, N. A., C. Aydin, A. Lebedeva, M. Okun, M. Pachitariu, M. Bauza, M. Beau, J. Bhagat, C. Böhm, M. Broux, et al. (2021). Neuropixels 2.0: A miniaturized high-density probe for stable, long-term brain recordings. *Science 372*, eabf4588.

Steinmetz, N. A., P. Zatka-Haas, M. Carandini, and K. D. Harris (2019). Distributed coding of choice, action and engagement across the mouse brain. *Nature 576*, 266–273.

Stent, G. S. (1973). A physiological mechanism for Hebb's postulate of learning. *Proc. Natl. Acad. Sci. U.S.A. 70*, 997–1001.

Stephan, K. E., D. R. Bach, P. C. Fletcher, J. Flint, M. J. Frank, K. J. Friston, A. Heinz, Q. J. Huys, M. J. Owen, E. B. Binder, P. Dayan, E. C. Johnstone, A. Meyer-Lindenberg, P. R. Montague, U. Schnyder, X. J. Wang, and M. Breakspear (2016). Charting the landscape of priority problems in psychiatry, part 1: Classification and diagnosis. *Lancet Psychiatry 3*, 77–83.

Stephan, K. E., K. J. Friston, and C. D. Frith (2009). Dysconnection in schizophrenia: From abnormal synaptic plasticity to failures of self-monitoring. *Schizophr. Bull. 35*, 509–527.

Stephan, K. E., L. M. Harrison, S. J. Kiebel, O. David, W. D. Penny, and K. J. Friston (2007). Dynamic causal models of neural system dynamics: current state and future extensions. *J. Biosci. 32*, 129–144.

Stephan, K. E., L. Kamper, A. Bozkurt, G. A. Burns, M. P. Young, and R. Kötter (2001). Advanced database methodology for the collation of connectivity data on the macaque brain (CoCoMac). *Philos. Trans. R. Soc. Lond. Series B Biol. Sci. 356*, 1159–1186.

Stephan, K. E., F. Schlagenhauf, Q. J. Huys, S. Raman, E. A. Aponte, K. H. Brodersen, L. Rigoux, R. J. Moran, J. Daunizeau, R. J. Dolan, et al. (2017). Computational neuroimaging strategies for single patient predictions. *Neuroimage 145*, 180–199.

Stephens, D. W. and J. R. Krebs (1986). *Foraging Theory*. Prineceton, NJ: Prineceton University Press.

Stephens, G. J., C. J. Honey, and U. Hasson (2013). A place for time: The spatiotemporal structure of neural dynamics during natural audition. *J. Neurophysiol. 110*, 2019–2026.

Steriade, M. and R. Llinas (1988). The functional states of the thalamus and the associated neuronal interplay. *Physiol. Rev. 68*, 649–742.

Steriade, M., D. McCormick, and T. Sejnowski (1993). Thalamocortical oscillations in the sleeping and aroused brain. *Science 262*, 679–685.

Sternberg, R. J. and D. K. Detterman (Eds.) (1986). *What Is Intelligence? Comtemporary Views on Its Nature and Definition*. Norwood, NJ: Ablex.

Stewart, M. and S. Fox (1990). Do septal neurons pace the hippocampal theta rhythm? *Trends Neurosci. 13*, 163–168.

Stiefel, K. M., B. Englitz, and T. J. Sejnowski (2013). Origin of intrinsic irregular firing in cortical interneurons. *Proc. Natl. Acad. Sci. 110*, 7886–7891.

Stiles, J. and T. L. Jernigan (2010). The basics of brain development. *Neuropsychol. Rev. 20*, 327–348.

Stine, G. M., E. M. Trautmann, D. Jeurissen, and M. N. Shadlen (2023). A neural mechanism for terminating decisions. *Neuron 111*, 2601–2613.

Stoet, G. and L. H. Snyder (2007). Extensive practice does not eliminate human switch costs. *Cogn. Affect Behav. Neurosci. 7*, 192–197.

Stoet, G. and L. H. Snyder (2009). Neural correlates of executive control functions in the monkey. *Trends Cogn. Sci. (Regul. Ed.) 13*, 228–234.

Stokes, M. G. (2015). 'Activity-silent' working memory in prefrontal cortex: A dynamic coding framework. *Trends Cogn. Sci. (Regul. Ed.) 19*, 394–405.

Stokes, M. G., M. Kusunoki, N. Sigala, H. Nili, D. Gaffan, and J. Duncan (2013). Dynamic coding for cognitive control in prefrontal cortex. *Neuron 78*, 364–375.

Stringer, C., M. Pachitariu, N. Steinmetz, C. B. Reddy, M. Carandini, and K. D. Harris (2019). Spontaneous behaviors drive multidimensional, brainwide activity. *Science 364*, 255.

Strogatz, S. H. (2016). *Nonlinear Dynamics and Chaos: With Applications to Physics, Biology, Chemistry and Engineering* (2nd ed.). Oxford: Taylor & Francis Group.

Stroop, J. R. (1935). Studies of interference in serial verbal reactions. *J. Exp. Psychol. 18*, 643–662.

Stuart, G., N. Spruston, and M. Hausser (2016). *Dendrite* (3rd ed.). Cambridge, MA: MIT Press.

Stuart, G., N. Spruston, B. Sakmann, and M. Hausser (1997). Action potential initiation and backpropagation in neurons of the mammalian CNS. *Trends Neurosci. 20*, 125–131.

Stumpf, C. (1965, Apr). The fast component in the electrical activity of rabbit's hippocampus. *Electroencephalogr. Clin. Neurophysiol. 18*, 477–486.

Stuss, D. T. and R. T. Knight (2002). *Principles of Frontal Lobe Function*. New York: Oxford University Press.

Suddendorf, T. and J. Busby (2003). Mental time travel in animals? *Trends Cogn. Sci. 7*, 391–396.

Sugrue, L. P., G. C. Corrado, and W. T. Newsome (2004). Matching behavior and representation of value in parietal cortex. *Science 304*, 1782–1787.

Sugrue, L. P., G. S. Corrado, and W. T. Newsome (2005). Choosing the greater of two goods: Neural currencies for valuation and decision making. *Nat. Rev. Neurosci. 6*, 363–375.

Summerfield, C., T. Egner, M. Greene, E. Koechlin, J. Mangels, and J. Hirsch (2006). Predictive codes for forthcoming perception in the frontal cortex. *Science 314*, 1311–1314.

Summerfield, C. and P. Parpart (2021). Normative principles for decision-making in natural environments. *Ann. Rev. Psychhol. 73*, https://doi.org/10.1146/annurev-psych-020821-104057.

Surmeier, D. J., W. Shen, M. Day, T. Gertler, S. Chan, X. Tian, and J. L. Plotkin (2010). The role of dopamine in modulating the structure and function of striatal circuits. *Prog. Brain Res. 183*, 149–167.

Sussillo, D. and L. F. Abbott (2009). Generating coherent patterns of activity from chaotic neural networks. *Neuron 63*, 544–557.

Sussillo, D. and O. Barak (2013). Opening the black box: Low-dimensional dynamics in high-dimensional recurrent neural networks. *Neural Comput. 25*, 626–649.

Sussillo, D., M. M. Churchland, M. T. Kaufman, and K. V. Shenoy (2015). A neural network that finds a naturalistic solution for the production of muscle activity. *Nat. Neurosci. 18*, 1025–1033.

Sutskever, I., O. Vinyals, and Q. V. Le (2014). Sequence to sequence learning with neural networks. *Adv. Neural Inf. Process. Syst. 27*, 3104–3112.

Sutton, R. S. and A. G. Barto (2018). *Reinforcement Learning: An Introduction* (2nd ed.). Cambridge, MA: MIT Press.

Suzuki, M. and J. Gottlieb (2013). Distinct neural mechanisms of distractor suppression in the frontal and parietal lobe. *Nat. Neurosci. 16*, 98–104.

Suzuki, M., C. M. Pennartz, and J. Aru (2023). How deep is the brain? The shallow brain hypothesis. *Nat. Rev. Neurosci. 24*, 778–791.

Svoboda, K. and N. Li (2018). Neural mechanisms of movement planning: Motor cortex and beyond. *Curr. Opin. Neurobiol. 49*, 33–41.

Swadlow, H. A. (1990). Efferent neurons and suspected interneurons in s-1 forelimb representation of the awake rabbit: Receptive fields and axonal properties. *J. Neurophysiol. 63*, 1477–1498.

Swanson, L. W. and J. W. Lichtman (2016). From cajal to connectome and beyond. *Annu. Rev. Neurosci. 39*, 197–216.

Swindale, N. V. (1980). A model for the formation of ocular dominance stripes. *Proc. R. Soc. London. Series B. Biol. Sci. 208*, 243–264.

Swindale, N. V. (1996). The development of topography in the visual cortex: A review of models. *Netw. Comput. Neural Syst. 7*, 161–247.

Szczepanski, S. M. and R. T. Knight (2014). Insights into human behavior from lesions to the prefrontal cortex. *Neuron 83*, 1002–1018.

Szpunar, K. K., R. N. Spreng, and D. L. Schacter (2014). A taxonomy of prospection: Introducing an organizational framework for future-oriented cognition. *Proc. Natl. Acad. Sci. 111*, 18414–18421.

Taine, H. (1875). *On Intelligence*. New York: Henry Holt and Company (Translation of French by T. D. Haye, from D'Intelligence, Libraire Hachette et Cie).

Takahashi, Y. K., M. R. Roesch, R. C. Wilson, K. Toreson, P. O'Donnell, Y. Niv, and G. Schoenbaum (2011). Expectancy-related changes in firing of dopamine neurons depend on orbitofrontal cortex. *Nat. Neurosci. 14*, 1590–1597.

Takeda, K. and S. Funahashi (2002). Prefrontal task-related activity representing visual cue location or saccade direction in spatial working memory tasks. *J. Neurophysiol. 87*, 567–588.

Takeuchi, T., M. Tamura, D. Tse, Y. Kajii, G. Fernández, and R. G. Morris (2022). Brain region networks for the assimilation of new associative memory into a schema. *Mol. Brain 15*. https://doi.org/10.1186/s13041-022-00908-9.

Tamás, G., A. Lorincz, A. Simon, and J. Szabadics (2003). Identified sources and targets of slow inhibition in the neocortex. *Science 299*, 1902–1905.

Tanaka, S. (2006). Dopaminergic control of working memory and its relevance to schizophrenia: A circuit dynamics perspective. *Neuroscience 139*, 153–171.

Tang, H., X.-L. Qi, M. R. Riley, and C. Constantinidis (2019). Working memory capacity is enhanced by distributed prefrontal activation and invariant temporal dynamics. *Proc. Natl. Acad. Sci. 116*, 7095–7100.

Tanigawa, H., H. D. Lu, and A. W. Roe (2010). Functional organization for color and orientation in macaque V4. *Nat. Neurosci. 13*, 1542–1548.

Tanji, J. and K. Shima (1994). Role for supplementary motor area cells in planning several movements ahead. *Nature 371*, 413–416.

Tank, D. W. and J. Hopfield (1987). Collective computation in neuronlike circuits. *Sci. Am. 257*, 104–114.

Tao, T. (2011). Outliers in the spectrum of iid matrices with bounded rank perturbations. *Probab. Theory Rel.*, 1–33.

Tasic, B., Z. Yao, L. T. Graybuck, K. A. Smith, T. N. Nguyen, D. Bertagnolli, J. Goldy, E. Garren, M. N. Economo, S. Viswanathan, et al. (2018). Shared and distinct transcriptomic cell types across neocortical areas. *Nature 563*, 72–78.

Taube, J. S. and J. P. Bassett (2003). Persistent neural activity in head direction cells. *Cereb. Cortex 13*, 1162–1172.

Taube, J. S., R. U. Muller, and J. B. Ranck (1990). Head-direction cells recorded from the postsubiculum in freely moving rats. I. description and quantitative analysis. *J. Neurosci. 10*, 420–435.

Tegnér, J., A. Compte, and X.-J. Wang (2002). The dynamical stability of reverberatory neural circuits. *Biol. Cybern. 87*, 471–481.

Terrace, H. S. and L. K. Son (2009). Comparative metacognition. *Curr. Opin. Neurobiol. 19*, 67–74.

Testard, C., S. Tremblay, F. Parodi, R. W. DiTullio, A. Acevedo-Ithier, K. L. Gardiner, K. Kording, and M. L. Platt (2024). Neural signatures of natural behaviour in socializing macaques. *Nature*. https://doi.org/10.1038/s41586-024-07178-6.

Theodoni, P., P. Majka, D. H. Reser, D. K. Wójcik, M. G. Rosa, and X.-J. Wang (2022). Structural attributes and principles of the neocortical connectome in the marmoset monkey. *Cereb. Cortex 32*, 15–28.

Theodoni, P., T. I. Panagiotaropoulos, V. Kapoor, N. K. Logothetis, and G. Deco (2011). Cortical microcircuit dynamics mediating binocular rivalry: The role of adaptation in inhibition. *Front. Hum. Neurosci. 5*. https://doi.org/10.3389/fnhum.2011.00145.

Thomas, N. W. and M. Paré (2007). Temporal processing of saccade targets in parietal cortex area LIP during visual search. *J. Neurophysiol. 97*, 942–947.

Thompson, K. G. and J. D. Schall (1999). The detection of visual signals by macaque frontal eye field during masking. *Nat. Neurosci. 2*, 283–288.

Thomson, A. M., A. P. Bannister, A. Mercer, and O. T. Morris (2002). Target and temporal pattern selection at neocortical synapses. *Philosophical Trans. R. Soc. London. Series B: Biol. Sci. 357*, 1781–1791.

Thorpe, S., D. Fize, and C. Marlot (1996). Speed of processing in the human visual system. *Nature 381*, 520–522.

Thorpe, S., E. Rolls, and S. Maddison (1983). The orbitofrontal cortex: Neuronal activity in the behaving monkey. *Exp. Brain Res. 49*, 93–115.

Thrun, S. and L. Pratt (Eds.) (1998). *Learning to Learn*. New York: Springer Science & Business Media.

Thura, D. and P. Cisek (2016). Modulation of premotor and primary motor cortical activity during volitional adjustments of speed-accuracy trade-offs. *J. Neurosci. 36*, 938–956.

Tinbergen, N. (1951). *The Study of Instinct*. Oxford: Clarendon Press.

Titley, H. K., N. Brunel, and C. Hansel (2017). Toward a neurocentric view of learning. *Neuron 95*, 19–32.

Todd, J. J. and R. Marois (2004). Capacity limit of visual short-term memory in human posterior parietal cortex. *Nature 428*, 751–754.

Todd, P. M. and G. E. Gigerenzer (2012). *Ecological Rationality: Intelligence in the World*. Oxford: Oxford University Press.

Todorov, E. (2004). Optimality principles in sensorimotor control. *Nat. Neurosci. 7*, 907–915.

Tolhurst, D., J. Movshon, and A. Dean (1983). The statistical reliability of signals in single neurons in cat and monkey visual cortex. *Vis. Res. 23*, 775–785.

Tolman, E. C. (1948). Cognitive maps in rats and men. *Psychol. Rev. 55*, 189–208.

Tomé, D. F., Y. Zhang, T. Aida, S. Sadeh, D. S. Roy, and C. Clopath (2024). Dynamic and selective engrams emerge with memory consolidation. *Nat. Neurosci.* https://doi.org/10.1038/s41593-023-01551-w.

Tonegawa, S., M. Pignatelli, D. S. Roy, and T. J. Ryan (2015). Memory engram storage and retrieval. *Curr. Opin. Neurobiol. 35*, 101–109.

Tononi, G., M. Boly, M. Massimini, and C. Koch (2016). Integrated information theory: From consciousness to its physical substrate. *Nat. Rev. Neurosci. 17*, 450–461.

Torres-Gomez, S., J. D. Blonde, D. Mendoza-Halliday, E. Kuebler, M. Everest, X. J. Wang, W. Inoue, M. O. Poulter, and J. Martinez-Trujillo (2020). Changes in the proportion of inhibitory interneuron types from sensory to executive areas of the primate neocortex: Implications for the origins of working memory representations. *Cereb. Cortex 30*, 4544–4562.

Tovote, P., J. P. Fadok, and A. Luthi (2015). Neuronal circuits for fear and anxiety. *Nat. Rev. Neurosci. 16*, 317–331.

Townsend, S. W., S. Engesser, S. Stoll, K. Zuberbühler, and B. Bickel (2018). Compositionality in animals and humans. *PLoS Biol. 16*, e2006425.

Toyoizumi, T., M. Kaneko, M. P. Stryker, and K. D. Miller (2014). Modeling the dynamic interaction of hebbian and homeostatic plasticity. *Neuron 84*, 497–510.

Tozzi, L., Zhang, X., Pines, A., Olmsted, A. M., Zhai, E. S., Anene, E. T., ... and Williams, L. M. (2024). Personalized brain circuit scores identify clinically distinct biotypes in depression and anxiety. *Nature Medicine*, https://doi.org/10.1038/s41591-024-03057-9

Traub, R. D., R. K. Wong, R. Miles, and H. Michelson (1991). A model of a ca3 hippocampal pyramidal neuron incorporating voltage-clamp data on intrinsic conductances. *J. Neurophysiol. 66*, 635–650.

Trautmann, E. M., J. K. Hesse, G. M. Stine, R. Xia, S. Zhu, D. J. O'Shea, B. Karsh, J. Colonell, F. F. Lanfranchi, S. Vyas, et al. (2023). Large-scale high-density brain-wide neural recording in nonhuman primates. *bioRxiv*. https://doi.org/10.1101/2023.02.01.526664.

Treisman, M. (1963). Temporal discrimination and the indifference interval: Implications for a model of the 'internal clock'. *Psychol. Monogr. 77*(13), 1–31.

Tremblay, R., S. Lee, and B. Rudy (2016). GABAergic interneurons in the neocortex: From cellular properties to circuits. *Neuron 91*, 260–292.

Treue, S., K. Hol, and H. Rauber (2000). Seeing multiple directions of motion-physiology and psychophysics. *Nat. Neurosci. 3*, 270–276.

Treue, S. and J. C. Mart nez Trujillo (1999). Feature-based attention influences motion processing gain in macaque visual cortex. *Nature 399*, 575–579.

Treue, S. and J. H. R. Maunsell (1999). Effects of attention on the processing of motion in macaque middle temporal and medial superior temporal visual cortical areas. *J. Neurosci. 19*, 7591–7602.

Treves, A. (1993). Mean-field analysis of neuronal spike dynamics. *Network 4*, 259–284.

Tritsch, N. X. and B. L. Sabatini (2012). Dopaminergic modulation of synaptic transmission in cortex and striatum. *Neuron 76*, 33–50.

Trübutschek, D., S. Marti, H. Ueberschär, and S. Dehaene (2019). Probing the limits of activity-silent non-conscious working memory. *Proc. Natl. Acad. Sci. 116*, 14358–14367.

Tsao, D. Y., W. A. Freiwald, R. B. Tootell, and M. S. Livingstone (2006). A cortical region consisting entirely of face-selective cells. *Science 311*, 670–674.

Tsao, D. Y. and M. S. Livingstone (2008). Mechanisms of face perception. *Annu. Rev. Neurosci. 31*, 411–437.

Tse, D., R. F. Langston, M. Kakeyama, I. Bethus, P. A. Spooner, E. R. Wood, M. P. Witter, and R. G. Morris (2007). Schemas and memory consolidation. *Science 316*, 76–82.

Tsien, R., A. Fox, P. Hess, E. McCleskey, B. Nilius, M. Nowycky, and R. Rosenberg (1987). Multiple types of calcium channel in excitable cells. *Soc. Gen. Physiol. Ser. 41*, 167–187.

Tsodyks, M. V. (1990). Associative memory in neural networks with binary synapses. *Mod. Phys. Lett. B4*, 713–716.

Tsodyks, M. V., T. Kenet, A. Grinvald, and A. Arieli (1999, Dec). Linking spontaneous activity of single cortical neurons and the underlying functional architecture. *Science 286*(5446), 1943–1946.

Tsodyks, M. V. and H. Markram (1997). The neural code between neocortical pyramidal neurons depends on neurotransmitter release probability. *Proc. Natl. Acad. Sci. U.S.A. 94*, 719–723.

Tsodyks, M. V., K. Pawelzik, and H. Markram (1998). Neural networks with dynamic synapses. *Neural Comput. 10*, 821–835.

Tsodyks, M. V., W. Skaggs, T. Sejnowski, and B. McNaughton (1997). Paradoxical effects of external modulation of inhibitory interneurons. *J. Neurosci. 17*, 4382–4388.

Tuckwell, H. C. (1988). *Introduction to Theoretical Neurobiology*. Cambridge: Cambridge University Press.

Tulving, E. (1995). *Organization of Memory: Quo vadis?* Cambridge, MA: MIT Press.

Turing, A. M. (1950). Computing machinery and intelligence. *Mind LIX*, 433–460.

Turing, A. M. (1952). The chemical basis of morphogenesis. *Phil. Trans. R. Soc. Lond. B 237*, 37–72.

Turk-Browne, N. B., B. J. Scholl, M. K. Johnson, and M. M. Chun (2010). Implicit perceptual anticipation triggered by statistical learning. *J. Neurosci. 30*, 11177–11187.

Turner-Evans, D. B., K. T. Jensen, S. Ali, T. Paterson, A. Sheridan, R. P. Ray, T. Wolff, J. S. Lauritzen, G. M. Rubin, D. D. Bock, et al. (2020). The neuroanatomical ultrastructure and function of a biological ring attractor. *Neuron 108*, 145–163.

Turrigiano, G. G., K. R. Leslie, N. S. Desai, L. C. Rutherford, and S. B. Nelson (1998). Activity-dependent scaling of quantal amplitude in neocortical neurons. *Nature 391*, 892–896.

Tversky, A. and D. Kahneman (1981). The framing of decisions and the psychology of choice. *Science 211*, 453–458.

Uchida, A., A. Kepecs, and Z. F. Mainen (2006). Seeing at a glance, smelling in a whiff: Rapid forms of perceptual decision making. *Nat. Rev. Neurosci. 7*, 485–491.

Uchida, N. and Z. Mainen (2003). Speed and accuracy of olfactory discrimination in the rat. *Nat. Neurosci. 6*, 1224–1229.

Uhlhaas, P. J. and W. Singer (2010). Abnormal neural oscillations and synchrony in schizophrenia. *Nat. Rev. Neurosci. 11*, 100–113.

Ullman, S. (1995). Sequence seeking and counter streams: A computational model for bidirectional information flow in the visual cortex. *Cereb. Cortex 5*, 1–11.

Ullsperger, M., C. Danielmeier, and G. Jocham (2014). Neurophysiology of performance monitoring and adaptive behavior. *Physiol. Rev. 94*, 35–79.

Urai, A. E., B. Doiron, A. M. Leifer, and A. K. Churchland (2021). Large-scale neural recordings call for new insights to link brain and behavior. *arXiv preprint arXiv:2103.14662*.

Urban-Ciecko, J. and A. L. Barth (2016). Somatostatin-expressing neurons in cortical networks. *Nat. Rev. Neurosci. 17*, 401–409.

Usher, M. and J. McClelland (2001). On the time course of perceptual choice: The leaky competing accumulator model. *Psychol. Rev. 108*, 550–592.

van Bergen, R. S. and N. Kriegeskorte (2020). Going in circles is the way forward: The role of recurrence in visual inference. *Curr. Opin. Neurobiol. 65*, 176–193.

Van den Berg, R., H. Shin, W.-C. Chou, R. George, and W. J. Ma (2012). Variability in encoding precision accounts for visual short-term memory limitations. *Proc. Natl. Acad. Sci. 109*, 8780–8785.

van den Berg, R., A. Zylberberg, R. Kiani, M. N. Shadlen, and D. M. Wolpert (2016). Confidence is the bridge between multi-stage decisions. *Curr. Biol. 26*, 3157–3168.

van der Stelt, O. and A. Belger (2007). Application of electroencephalography to the study of cognitive and brain functions in schizophrenia. *Schizophr. Bull. 33*, 955–970.

Vanderwolf, C. (1969). Hippocampal electrical activity and voluntary movement in the rat. *Electroencephalogr. Clin. Neurophysiol. 26*, 407–418.

Van Essen, D. C. (2013). Cartography and connectomes. *Neuron 80*, 775–790.

Van Gelder, T. (1995). What might cognition be, if not computation? *J. Philos. 92*, 345–381.

van Hatern, J. H. and A. van der Schaaf (1996). Temporal properties of natural scenes. In *Human Vision and Electronic Imaging*, vol. 2657, pp. 139–143.

van Kerkoerle, T., M. W. Self, B. Dagnino, M. A. Gariel-Mathis, J. Poort, C. van der Togt, and P. R. Roelfsema (2014). Alpha and gamma oscillations characterize feedback and feedforward processing in monkey visual cortex. *Proc. Natl. Acad. Sci. U.S.A. 111*, 14332–14341.

van Opheusden, B., I. Kuperwajs, G. Galbiati, Z. Bnaya, Y. Li, and W. J. Ma (2023). Expertise increases planning depth in human gameplay. *Nature*. https://doi.org/10.1038/s41586-023-06124-2.

Van Os, J., Linscott, R. J., Myin-Germeys, I., Delespaul, P., and Krabbendam, L. J. P. M. (2009). A systematic review and meta-analysis of the psychosis continuum: evidence for a psychosis proneness–persistence–impairment model of psychotic disorder. *Psychol. Med. 39*, 179–195.

van Rossum, M. C. W., G. Q. Bi, and G. G. Turrigiano (2000). Stable hebbian learning from spike timing-dependent plasticity. *J. Neurosci. 20*, 8812–8821.

van Vreeswijk, C., L. Abbott, and G. B. Ermentrout (1994). When inhibition not excitation synchronizes neural firing. *J. Comput. Neurosci. 1*, 313–321.

van Vreeswijk, C. and H. Sompolinsky (1996). Chaos in neuronal networks with balanced excitatory and inhibitory activity. *Science 274*, 1724–1726.

van Vreeswijk, C. and H. Sompolinsky (1998). Chaotic balanced state in a model of cortical circuits. *Neural Comp. 10*, 1321–1371.

van Vugt, B., B. Dagnino, D. Vartak, H. Safaai, S. Panzeri, S. Dehaene, and P. R. Roelfsema (2018). The threshold for conscious report: Signal loss and response bias in visual and frontal cortex. *Science 360*, 537–542.

van Vugt, B., T. van Kerkoerle, D. Vartak, and P. R. Roelfsema (2020). The contribution of ampa and nmda receptors to persistent firing in the dorsolateral prefrontal cortex in working memory. *J. Neurosci. 40*, 2458–2470.

van Vugt, M. K., M. A. Beulen, and N. A. Taatgen (2019). Relation between centro-parietal positivity and diffusion model parameters in both perceptual and memory-based decision making. *Brain Res. 1715*, 1–12.

Vapnik, V. (2013). *The Nature of Statistical Learning Theory.* Berlin: Springer Science & Business Media.

Vaswani, A., N. Shazeer, N. Parmar, J. Uszkoreit, L. Jones, A. N. Gomez, Ł. Kaiser, and I. Polosukhin (2017). Attention is all you need. *Adv. Neural Inf. Process. Syst. 30.*

Vaz, A. P., J. H. Wittig Jr, S. K. Inati, and K. A. Zaghloul (2020). Replay of cortical spiking sequences during human memory retrieval. *Science 367*, 1131–1134.

Vázquez-Rodriguez, B., L. E. Suárez, R. D. Markello, G. Shafiei, C. Paquola, P. Hagmann, M. P. Van Den Heuvel, B. C. Bernhardt, R. N. Spreng, and B. Misic (2019). Gradients of structure–function tethering across neocortex. *Proc. Natl. Acad. Sci. 116*, 21219–21227.

Verbruggen, F., A. R. Aron, G. P. Band, C. Beste, P. G. Bissett, A. T. Brockett, J. W. Brown, S. R. Chamberlain, C. D. Chambers, H. Colonius, et al. (2019). A consensus guide to capturing the ability to inhibit actions and impulsive behaviors in the stop-signal task. *Elife 8*, e46323.

Verhoef, B. E. and J. H. R. Maunsell (2017). Attention-related changes in correlated neuronal activity arise from normalization mechanisms. *Nat. Neurosci. 20*, 969–977.

Vezoli, J., L. Magrou, R. Goebel, X.-J. Wang, K. Knoblauch, M. Vinck, and H. Kennedy (2021a). Cortical hierarchy, dual counterstream architecture and the importance of top-down generative networks. *NeuroImage 225*, 117479.

Vezoli, J., M. Vinck, C. A. Bosman, A. M. Bastos, C. M. Lewis, H. Kennedy, and P. Fries (2021b). Brain rhythms define distinct interaction networks with differential dependence on anatomy. *Neuron 109*, 3862–3878.

Vickers, D. (1979). *Decision Processes in Visual Perception.* London: Academic Press.

Vickery, T. J., M. M. Chun, and D. Lee (2011). Ubiquity and specificity of reinforcement signals throughout the human brain. *Neuron 72*, 166–177.

Vierling-Claassen, D., P. Siekmeier, S. Stufflebeam, and N. Kopell (2008). Modeling GABA alterations in schizophrenia: A link between impaired inhibition and altered gamma and beta range auditory entrainment. *J. Neurophysiol. 99*, 2656–2671.

Vigo, D., G. Thornicroft, and R. Atun (2016). Estimating the true global burden of mental illness. *Lancet Psychiatry 3*, 171–178.

Vijayraghavan, S., M. Wang, S. G. Birnbaum, G. V. Williams, and A. F. Arnsten (2007). Inverted-U dopamine D1 receptor actions on prefrontal neurons engaged in working memory. *Nat. Neurosci. 10*(3), 376–384.

Vinje, W. E. and J. L. Gallant (2000). Sparse coding and decorrelation in primary visual cortex during natural vision. *Science 287*, 1273–1276.

Vinograd, A., Nair, A., Kim, J. H., Linderman, S. W., & Anderson, D. J. (2024). Causal evidence of a line attractor encoding an affective state. Nature, *634*, 910–918.

Vogel, E. K. and M. G. Machizawa (2004). Neural activity predicts individual differences in visual working memory capacity. *Nature 428*, 748–751.

Vogels, T. P. and L. F. Abbott (2009). Gating multiple signals through detailed balance of excitation and inhibition in spiking networks. *Nat. Neurosci. 12*, 483–491.

Vogels, T. P., H. Sprekeler, F. Zenke, C. Clopath, and W. Gerstner (2011). Inhibitory plasticity balances excitation and inhibition in sensory pathways and memory networks. *Science 334*, 1569–1573.

Volkow, N. D., J. S. Fowler, G.-J. Wang, R. Hitzemann, J. Logan, D. J. Schlyer, S. L. Dewey, and A. P. Wolf (1993). Decreased dopamine D2 receptor availability is associated with reduced frontal metabolism in cocaine abusers. *Synapse 14*, 169–177.

Volkow, N. D. and T. K. Li (2004). Drug addiction: The neurobiology of behaviour gone awry. *Nat. Rev. Neurosci. 5*, 963–970.

Volkow, N. D. and M. Morales (2015). The brain on drugs: From reward to addiction. *Cell 162*, 712–725.

Volkow, N. D., A. P. Wolf, J. D. Brodie, R. Cancro, J. E. Overall, H. Rhoades, and P. Van Gelder (1988). Brain interactions in chronic schizophrenics under resting and activation conditions. *Schizophr. Res. 1*, 47–53.

Volman, V., M. M. Behrens, and T. J. Sejnowski (2011). Downregulation of parvalbumin at cortical GABA synapses reduces network gamma oscillatory activity. *J. Neurosci. 31*, 18137–18148.

von Economo, C. (1929). *The Cytoarchitectonics of the Human Cerebral Cortex.* London: Oxford University Press.

von Frisch, K. (1974). *Animal Architecture.* New York: Helen and Kurt Wolff.

von Neumann, J. (1958). *The Computer and the Brain.* New Haven, CT: Yale University Press.

von Neumann, J. and O. Morgenstern (1944). *Theory of Games and Economic Behavior*. Prineceton, NJ: Prineceton University Press.

Von Uexküll, J. (2010). *A Foray into the Worlds of Animals and Humans: With a Theory of Meaning*. Minneapolis: University of Minnesota Press (First edition published in 1934).

Voon, V., K. Derbyshire, C. Ruck, M. A. Irvine, Y. Worbe, J. Enander, L. R. Schreiber, C. Gillan, N. A. Fineberg, B. J. Sahakian, T. W. Robbins, N. A. Harrison, J. Wood, N. D. Daw, P. Dayan, J. E. Grant, and E. T. Bullmore (2015). Disorders of compulsivity: A common bias towards learning habits. *Mol. Psychiatry 20*, 345–352.

Vyas, S., M. D. Golub, D. Sussillo, and K. V. Shenoy (2020). Computation through neural population dynamics. *Annu. Rev. Neurosci. 43*, 249–275.

Wald, A. (1948). On cumulative sums of random variables. *Ann. Math. Stat. 15*, 283–296.

Wall, N. R., M. De La Parra, J. M. Sorokin, H. Taniguchi, Z. J. Huang, and E. M. Callaway (2016). Brain-wide maps of synaptic input to cortical interneurons. *J. Neurosci. 36*, 4000–4009.

Wallis, J. D. (2007). Orbitofrontal cortex and its contribution to decision-making. *Annu. Rev. Neurosci. 30*, 31–56.

Wallis, J. D., K. Anderson, and E. Miller (2001). Single neurons in prefrontal cortex encode abstract rules. *Nature 411*, 953–956.

Wang, B., L. Yin, X. Zou, M. Ye, Y. Liu, T. He, S. Deng, Y. Jiang, R. Zheng, Y. Wang, M. Yang, H. Lu, S. Wu, and Y. Shu (2015). A subtype of inhibitory interneuron with intrinsic persistent activity in human and monkey neocortex. *Cell Rep. 10*. https://doi.org/10.1016/j.celrep.2015.02.018.

Wang, H., G. G. Stradtman, X.-J. Wang, and W. J. Gao (2008). A specialized NMDA receptor function in layer 5 recurrent microcircuitry of the adult rat prefrontal cortex. *Proc. Natl. Acad. Sci. U.S.A. 105*, 16791–16796.

Wang, J., D. Narain, E. A. Hosseini, and M. Jazayeri (2018a). Flexible timing by temporal scaling of cortical responses. *Nat. Neurosci. 21*, 102–110.

Wang, J. and R. S. Zucker (2001). Photolysis-induced suppression of inhibition in rat hippocampal CA1 pyramidal neurons. *J. Physiol. 533*, 757–763.

Wang, J. X., Z. Kurth-Nelson, D. Kumaran, D. Tirumala, H. Soyer, J. Z. Leibo, D. Hassabis, and M. Botvinick (2018b). Prefrontal cortex as a meta-reinforcement learning system. *Nat. Neurosci. 21*, 860–868.

Wang, M., S. Vijayraghavan, and P. S. Goldman-Rakic (2004). Selective D2 receptor actions on the functional circuitry of working memory. *Science 303*, 853–856.

Wang, M., Y. Yang, C. J. Wang, N. J. Gamo, L. E. Jin, J. A. Mazer, J. H. Morrison, X. J. Wang, and A. F. Arnsten (2013). NMDA receptors subserve persistent neuronal firing during working memory in dorsolateral prefrontal cortex. *Neuron 77*, 736–749.

Wang, P., R. Kong, X. Kong, R. Liegeois, C. Orban, G. Deco, M. P. van den Heuvel, and B. T. Thomas Yeo (2019). Inversion of a large-scale circuit model reveals a cortical hierarchy in the dynamic resting human brain. *Sci. Adv. 5*, eaat7854.

Wang, S., R. Falcone, B. Richmond, and B. B. Averbeck (2023). Attractor dynamics reflect decision confidence in macaque prefrontal cortex. *Nat. Neurosci. 26*, 1970–1980.

Wang, X., Y. Chen, and W. Zhu (2021). A survey on curriculum learning. *IEEE Trans. Pattern Anal. Mach. Intell.* https://doi.org/10.1109/TPAMI.2021.3069908.

Wang, X.-J. (1993). Ionic basis for intrinsic 40 Hz neuronal oscillations. *Neuroreport 5*, 221–224.

Wang, X.-J. (1994). Multiple dynamical modes of thalamic relay neurons: Rhythmic bursting and intermittent phase-locking. *Neuroscience 59*, 21–31.

Wang, X.-J. (1998). Calcium coding and adaptive temporal computation in cortical pyramidal neurons. *J. Neurophysiol. 79*, 1549–1566.

Wang, X.-J. (1999a). Fast burst firing and short-term synaptic plasticity: A model of neocortical chattering neurons. *Neuroscience 89*, 347–362.

Wang, X.-J. (1999b). Synaptic basis of cortical persistent activity: The importance of NMDA receptors to working memory. *J. Neurosci. 19*, 9587–9603.

Wang, X.-J. (2001). Synaptic reverberation underlying mnemonic persistent activity. *Trends Neurosci. 24*, 455–463.

Wang, X.-J. (2002a). Pacemaker neurons for the theta rhythm and their synchronization in the septohippocampal reciprocal loop. *J. Neurophysiol. 87*, 889–900.

Wang, X.-J. (2002b). Probabilistic decision making by slow reverberation in cortical circuits. *Neuron 36*, 955–968.

Wang, X.-J. (2005). Discovering spatial working memory fields in prefrontal cortex. *J. Neurophysiol. 93*, 3027–3028.

Wang, X.-J. (2006a). A microcircuit model of prefrontal functions: Yin and yang of reverberatory neurodynamics in cognition. In J. Risberg, J. Grafman, and F. Boller (Eds.), *The Prefrontal Lobes: Development, Function and Pathology*, pp. 92–127. Cambridge: Cambridge University Press.

Wang, X.-J. (2006b). Scale invariance of reaction time statistics and its neuronal mechanism in a cortical network model of simple decisions. *Soc. Neurosci. Abstr. 37*, 605.10.

Wang, X.-J. (2006c). Toward a prefrontal microcircuit model for cognitive deficits in schizophrenia. *Pharmacopsychiatry 39*(Suppl 1), 80–87.

Wang, X.-J. (2008a). Attractor network models. In L. Squire, T. Albright, F. Bloom, F. Gage, and N. Spitzer (Eds.), *Encyclopedia of Neuroscience*, pp. 667–679. London: MacMillan Reference Ltd.

Wang, X.-J. (2008b). Decision making in recurrent neuronal circuits. *Neuron 60*, 215–234.

Wang, X.-J. (2010). Neurophysiological and computational principles of cortical rhythms in cognition. *Physiol. Rev. 90*, 1195–1268.

Wang, X.-J. (2012). Neural dynamics and circuit mechanisms of decision-making. *Curr. Opin. Neurobiol. 22*, 1039–1046.

Wang, X.-J. (2013). The prefrontal cortex as a quintessential 'cognitive-type' neural circuit: Working memory and decision making. In D. T. Stuss and R. T. Knight (Eds.), *Principles of Frontal Lobe Function* (2nd ed.), pp. 226–248. New York: Cambridge University Press.

Wang, X.-J. (2014). Neuronal circuit computation of choice. In P. Glimcher and E. Fehr (Eds.), *Neuroeconomics* (2nd ed.), pp. 435–453. Cambridge, MA: MIT Press.

Wang, X.-J. (2020). Macroscopic gradients of synaptic excitation and inhibition in the neocortex. *Nat. Rev. Neurosci. 21*, 169–178.

Wang, X.-J. (2021). 50 years of mnemonic persistent activity: Quo vadis? *Trends Neurosci. 44*, 888–902.

Wang, X.-J. (2022). Theory of the multiregional neocortex: Large-scale neural dynamics and distributed cognition. *Ann. Rev. Neurosci. 45*, 533–560.

Wang, X.-J. and G. Buzsáki (1996). Gamma oscillation by synaptic inhibition in a hippocampal interneuronal network model. *J. Neurosci. 16*, 6402–6413.

Wang, X.-J., D. Golomb, and J. Rinzel (1995). Emergent spindle oscillations and intermittent burst firing in a thalamic model: Specific neuronal mechanisms. *Proc. Natl. Acad. Sci. U.S.A. 92*, 5577–5581.

Wang, X.-J., J. Jiang, and U. Pereira-Obilinovic (2023a). Bifurcation in space: Emergence of functional modularity in the neocortex. *bioRxiv*. https://doi.org/10.1101/2023.06.04.543639.

Wang, X.-J. and H. Kennedy (2016). Brain structure and dynamics across scales: In search of rules. *Curr. Opin. Neurobiol. 37*, 92–98.

Wang, X.-J. and J. H. Krystal (2014). Computational psychiatry. *Neuron 84*, 638–654.

Wang, X.-J., Y. Liu, M. V. Sanchez-Vives, and D. A. McCormick (2003). Adaptation and temporal decorrelation by single neurons in the primary visual cortex. *J. Neurophysiol. 89*, 3279–3293.

Wang, X.-J. and J. Murray (2020). Biophysically-based neural circuit modeling of working memory, decision making and related psychiatric deficits. In P. Seriès (Ed.), *Computational Psychiatry: A Primer*, pp. 66–84. Cambridge, MA: MIT Press.

Wang, X.-J., U. Pereira, M. G. Rosa, and H. Kennedy (2020). Brain connectomes come of age. *Curr. Opin. Neurobiol. 65*, 152–161.

Wang, X.-J. and J. Rinzel (1992). Alternating and synchronous rhythms in reciprocally inhibitory model neurons. *Neural Comput. 4*, 84–97.

Wang, X.-J. and J. Rinzel (1993). Spindle rhythmicity in the reticularis thalami nucleus: Synchronization among mutually inhibitory neurons. *Neuroscience 53*, 899–904.

Wang, X.-J. and J. Rinzel (1995). Oscillatory and bursting properties of neurons. In M. Arbib (Ed.), *Handbook of Brain Theory and Neural Networks*, pp. 686–691. Cambridge, MA: MIT Press.

Wang, X.-J., J. Rinzel, and M. A. Rogawski (1991). A model of the T-type calcium current and the low-threshold spike in thalamic neurons. *J. Neurophysiol. 66*, 839–850.

Wang, X.-J., J. Tegnér, C. Constantinidis, and P. S. Goldman-Rakic (2004). Division of labor among distinct subtypes of inhibitory neurons in a cortical microcircuit of working memory. *Proc. Natl. Acad. Sci. U.S.A. 101*, 1368–1373.

Wang, X.-J. and G. R. Yang (2018). A disinhibitory circuit motif and flexible information routing in the brain. *Curr. Opin. Neurobiol. 49*, 75–83.

Wang, Y., H. Markram, P. H. Goodman, T. K. Berger, J. Ma, and P. S. Goldman-Rakic (2006). Heterogeneity in the pyramidal network of the medial prefrontal cortex. *Nat. Neurosci. 9*, 534–542.

Wang, Z. A., S. Chen, Y. Liu, L. D. Liu, K. Svoboda, N. Li, and S. Druckmann (2023b). Not everything, not everywhere, not all at once: A study of brain-wide encoding of movement. *bioRxiv*. https://doi.org/10.1101/2023.06.08.544257.

Warden, M. R. and E. K. Miller (2010). Task-dependent changes in short-term memory in the prefrontal cortex. *J. Neurosci. 30*, 15801–15810.

Warren, J. M. (1974). Possibly unique characteristics of learning by primates. *J. Hum. Evol. 3*, 445–454.

Warren, W. H. (2006). The dynamics of perception and action. *Psychol. Rev. 113*, 358–389.

Wasmuht, D. F., E. Spaak, T. J. Buschman, E. K. Miller, and M. G. Stokes (2018). Intrinsic neuronal dynamics predict distinct functional roles during working memory. *Nat. Commun. 9*, 3499.

Watanabe, M. and D. P. Munoz (2009). Neural correlates of conflict resolution between automatic and volitional actions by basal ganglia. *Eur. J. Neurosci. 30*, 2165–2176.

Watkins, C. J. C. H. and P. Dayan (2012). Q-learning. *Mach. Learn. 8*, 279–292.

Watts, D. J. (2004). *Six Degrees: The Science of a Connected Age*. New York: W. W. Norton & Company.

Watts, D. J. and S. Strogatz (1998). Collective dynamics of 'small-world' networks. *Nature 393*, 440–442.

Wax, N. (1954). *Selected Papers on Noise and Stochastic Processes*. New York: Dover.

Webb, C. A., M. H. Trivedi, Z. D. Cohen, D. G. Dillon, J. C. Fournier, F. Goer, M. Fava, P. J. McGrath, M. Weissman, R. Parsey, P. Adams, J. M. Trombello, C. Cooper, P. Deldin, M. A. Oquendo, M. G. McInnis, Q. Huys, G. Bruder, B. T. Kurian, M. Jha, R. J. DeRubeis, and D. A. Pizzagalli (2018). Personalized prediction of antidepressant v. placebo response: Evidence from the EMBARC study. *Psychol. Med.*, 1–10.

Webb, T., S. S. Mondal, C. Wang, B. Krabach, and I. Momennejad (2023a). A prefrontal cortex-inspired architecture for planning in large language models. *arXiv preprint arXiv:2310.00194.*

Webb, T., Holyoak, K. J., and Lu, H. (2023b). Emergent analogical reasoning in large language models. *Nat. Hum. Behav.* 7, 1526–1541.

Weber, A. I., K. Krishnamurthy, and A. L. Fairhall (2019). Coding principles in adaptation. *Annu. Rev. Vis. Sci. 5*, 427–449.

Weber, E. H. (1834). *De Pulsu, Resorptione, Auditu et Tactu: Annotationes Anatomicae et Physiologicae, Auctore.* Prostat Apud CF Koehler.

Wei, W., J. E. Rubin, and X.-J. Wang (2015). Role of the indirect pathway of the basal ganglia in perceptual decision making. *J. Neurosci. 35*, 4052–4064.

Wei, W. and X. J. Wang (2016). Inhibitory control in the cortico-basal ganglia-thalamocortical loop: Complex regulation and interplay with memory and decision processes. *Neuron 92*, 1093–1105.

Wei, Z. and X.-J. Wang (2015). Confidence estimation as a stochastic process in a neurodynamical system of decision making. *J. Neurophysiol. 114*, 99–113.

Wei, Z., X. J. Wang, and D. H. Wang (2012). From distributed resources to limited slots in multiple-item working memory: A spiking network model with normalization. *J. Neurosci. 32*, 11228–11240.

Werbos, P. J. (1990). Backpropagation through time: What it does and how to do it. *Proc. IEEE 78*, 1550–1560.

Whitacre, J. and A. Bender (2010). Degeneracy: A design principle for achieving robustness and evolvability. *J. Theor. Biol. 263*, 143–153.

White, J., J. Rubinstein, and A. Kay (2000). Channel noise in neurons. *Trends Neurosci. 23*, 131–137.

White, J. G., E. Southgate, J. N. Thomson, and S. Brenner (1986). The structure of the nervous system of the nematode caenorhabditis elegans. *Philos. Trans. R. Soc. Lond. B Biol. Sci. 314*, 1–340.

White, J. M., D. L. Sparks, and T. R. Stanford (1994). Saccades to remembered target locations: And analysis of systematic and variable errors. *Vis. Res. 34*, 79–92.

Whiteford, H. A., A. J. Ferrari, L. Degenhardt, V. Feigin, and T. Vos (2015). The global burden of mental, neurological and substance use disorders: An analysis from the global burden of disease study 2010. *PLoS ONE 10*, e0116820.

Whittington, J. C., D. McCaffary, J. J. Bakermans, and T. E. Behrens (2022). How to build a cognitive map: Insights from models of the hippocampal formation. *arXiv preprint arXiv:2202.01682.*

Whittington, J. C., T. H. Muller, S. Mark, G. Chen, C. Barry, N. Burgess, and T. E. Behrens (2020). The tolman-eichenbaum machine: Unifying space and relational memory through generalization in the hippocampal formation. *Cell 183*, 1249–1263.

Whittington, M. A., R. D. Traub, and J. G. R. Jefferys (1995). Synchronized oscillations in interneuron networks driven by metabotropic glutamate receptor activation. *Nature 373*, 612–615.

Wickelgren, W. A. (1977). Speed-accuracy tradeoff and information processing dynamics. *Acta Psychol. 41*, 67–85.

Wickens, J. R., J. N. J. Reynolds, and B. I. Hyland (2003). Neural mechanisms of reward-related motor learning. *Curr. Opin. Neurobiol. 13*, 685–690.

Widrow, B. and M. E. J. Hoff (1960). Adaptive switching circuits. In *1960 IRE WSCON Covention Record, Part 4*, pp. 96–104. New York: IRE.

Wiecki, T. V. and M. J. Frank (2013). A computational model of inhibitory control in frontal cortex and basal ganglia. *Psychol. Rev. 120*, 329–355.

Wiecki, T. V., J. S. Poland, and M. J. Frank (2015). Model-based cognitive neuroscience approaches to computational psychiatry: Clustering and classification. *Clin. Psychol. Sci. 120*, 329–355.

Wiener, N. (1948). *Cybernetics or Control and Communication in the Animal and the Machine.* Cambridge, MA: MIT Press.

Wiesel, T. N. and D. H. Hubel (1965). Comparison of the effects of unilateral and bilateral eye closure on cortical unit responses in kittens. *J. Neurophysiol. 28*, 1029–1040.

Wildenberg, G. A., M. R. Rosen, J. Lundell, D. Paukner, D. J. Freedman, and N. Kasthuri (2021). Primate neuronal connections are sparse in cortex as compared to mouse. *Cell Rep. 36*, 109709.

Williams, A. H., T. H. Kim, F. Wang, S. Vyas, S. I. Ryu, K. V. Shenoy, M. Schnitzer, T. G. Kolda, and S. Ganguli (2018). Unsupervised discovery of demixed, low-dimensional neural dynamics across multiple timescales through tensor component analysis. *Neuron 98*, 1099–1115.

Williams, G. V. and P. S. Goldman-Rakic (1995). Modulation of memory fields by dopamine D1 receptors in prefrontal cortex. *Nature 376*, 572–575.

Williams, R. J. (1992). Simple statistical gradient-following algorithms for connectionist reinforcement learning. *Mach. Learn. 8*, 229–256.

Williford, T. and J. H. Maunsell (2006). Effects of spatial attention on contrast response functions in macaque area v4. *J. Neurophysiol. 96*, 40–54.

Wilming, N., P. R. Murphy, F. Meyniel, and T. H. Donner (2020). Large-scale dynamics of perceptual decision information across human cortex. *Nat. Commun. 11*, 5109.

Wilson, H. R. (2003). Computational evidence for a rivalry hierarchy in vision. *Proc. Natl. Acad. Sci. 100*, 14499–14503.

Wilson, H. R. and J. D. Cowan (1972). Excitatory and inhibitory interactions in localized populations of model neurons. *Biophys. J. 12*, 1–24.

Wilson, H. R. and J. D. Cowan (1973). A mathematical theory of the functional dynamics of cortical and thalamic nervous tissue. *Kybernetik 13*, 55–80.

Wilson, M. A. and B. L. McNaughton (1994). Reactivation of hippocampal ensemble memories during sleep. *Science 265*, 676–679.

Wilson, R. I. and R. A. Nicoll (2001). Endogenous cannabinoids mediate retrograde signalling at hippocampal synapses. *Nature 410*, 588–592.

Wilson, R. I. and R. A. Nicoll (2002). Endocannabinoid signaling in the brain. *Science 296*, 678–682.

Wiltschko, A. B., T. Tsukahara, A. Zeine, R. Anyoha, W. F. Gillis, J. E. Markowitz, R. E. Peterson, J. Katon, M. J. Johnson, and S. R. Datta (2020). Revealing the structure of pharmacobehavioral space through motion sequencing. *Nat. Neurosci. 23*, 1433–1443.

Wimmer, K., A. Compte, A. Roxin, D. Peixoto, A. Renart, and J. de la Rocha (2015). Sensory integration dynamics in a hierarchical network explains choice probabilities in cortical area MT. *Nat. Commun. 6*, 6177.

Wimmer, K., D. Q. Nykamp, C. Constantinidis, and A. Compte (2014). Bump attractor dynamics in prefrontal cortex explains behavioral precision in spatial working memory. *Nat. Neurosci. 17*, 431–439.

Wimsatt, W. C. (1976). Reductionism, levels of organization, and the mind-body problem. In G. G. Globus, G. Maxwell, and I. Savodnik (Eds.), *Consciousness and the Brain*, pp. 205–267. Springer.

Wise, R. A. and P. P. Rompre (1989). Brain dopamine and reward. *Annu. Rev. Psychol. 40*, 191–225.

Wise, S. P. (2008). Forward frontal fields: Phylogeny and fundamental function. *Trends Neurosci. 31*, 599–608.

Wise, S. P. and E. A. Murray (2000). Arbitrary associations between antecedents and actions. *Trends Neurosci. 23*(6), 271–276.

Wittmann, M. K., P. L. Lockwood, and M. F. Rushworth (2018). Neural mechanisms of social cognition in primates. *Annu. Rev. Neurosci. 41*, 99–118.

Wixted, J. T. (2004). The psychology and neuroscience of forgetting. *Annu. Rev. Psychol. 55*, 235–269.

Wixted, J. T. and E. B. Ebbesen (1997). Genuine power curves in forgetting: A quantitative analysis of individual subject forgetting functions. *Mem. Cognit. 25*, 731–739.

Wolfe, J. M. (1994). Guided search 2.0: A revised model of visual search. *Psychon. Bull. Rev. 1*, 202–238.

Wolfe, T. (2016). *The Kingdom of Speech*. New York: Random House.

Wolff, M. J., J. Jochim, E. G. Akyürek, and M. G. Stokes (2017). Dynamic hidden states underlying working-memory-guided behavior. *Nat. Neurosci. 20*, 864–871.

Wolpert, D. M. and Z. Ghahramani (2000). Computational principles of movement neuroscience. *Nat. Neurosci. 3*, 1212–1217.

Wong, K. F., A. C. Huk, M. N. Shadlen, and X.-J. Wang (2007). Neural circuit dynamics underlying accumulation of time-varying evidence during perceptual decision-making. *Front. Comput. Neurosci. 1*, 6. https://doi.org/10.3389/neuro.10/006.2007.

Wong, K. F. and X.-J. Wang (2006). A recurrent network mechanism of time integration in perceptual decisions. *J. Neurosci. 26*, 1314–1328.

Wong, R. K., D. A. Prince, and A. I. Basbaum (1979). Intradendritic recordings from hippocampal neurons. *Proc. Natl. Acad. Sci. U.S.A. 76*, 986–990.

Wong, R. K., J. Selvanayagam, K. Johnston, and S. Everling (2022). Delay-related activity in marmoset prefrontal cortex. *bioRxiv*. https://doi.org/10.1101/2022.04.04.487026.

Wong-Lin, K., J. M. Sanchez-Bornot, N. McCombe, D. Kaur, P. L. McClean, X. Zou, V. Youssofzadeh, X. Ding, M. Bucholc, S. Yang, et al. (2020). Computational neurology: Computational modeling approaches in dementia. *arXiv preprint arXiv:2005.02214*.

Wong-Lin, K., D.-H. Wang, A. A. Moustafa, J. Y. Cohen, and K. Nakamura (2017). Toward a multiscale modeling framework for understanding serotonergic function. *J. Psychopharmacol. 31*, 1121–1136.

Woodin, M. A., K. Ganguly, and M. M. Poo (2003). Coincident pre- and postsynaptic activity modifies GABAergic synapses by postsynaptic changes in Cl- transporter activity. *Neuron 39*, 807–820.

Woodman, G. F., M. S. Kang, K. Thompson, and J. D. Schall (2008). The effect of visual search efficiency on response preparation: Neurophysiological evidence for discrete flow. *Psychol. Sci. 19*, 128–136.

Woolgar, A., J. Duncan, F. Manes, and E. Fedorenko (2018). The multiple-demand system but not the language system supports fluid intelligence. *Nat. Hum. Behav. 2*, 200–204.

Wu, S., K. Hamaguchi, and S. Amari (2008). Dynamics and computation of continuous attractors. *Neural Comput. 20*, 994–1025.

Wu, Y. K. and F. Zenke (2021). Nonlinear transient amplification in recurrent neural networks with short-term plasticity. *Elife 10*, e71263.

Wu, Z., A. Litwin-Kumar, P. Shamash, A. Taylor, R. Axel, and M. N. Shadlen (2020). Context-dependent decision making in a premotor circuit. *Neuron 106*, 316–328.

Wurtz, R. H. (1969). Visual receptive fields of striate cortex neurons in awake monkeys. *J. Neurophysiol. 32*, 727–742.

Wurtz, R. H. and M. E. Goldberg (1971). Superior colliculus cell responses related to eye movements in awake monkeys. *Science 171*, 82–84.

Wynn, R. M. and F. L. Coolidge (2008). A stone-age meeting of minds. *Amer. Scientist. 96*, 44–51.

Wynn, T. and F. L. Coolidge (2010). Beyond symbolism and language: An introduction to supplement 1, working memory. *Curr. Anthropol. 51*, S5–S16.

Wynn, T. and F. L. Coolidge (2011). The implications of the working memory model for the evolution of modern cognition. *Int. J. Evol. Biol. 2011*, 741357.

Xiang, Z., J. R. Huguenard, and D. A. Prince (1998). GABAA receptor mediated currents in interneurons and pyramidal cells of rat visual cortex. *J. Physiol. 506*, 715–730.

Xie, X., R. H. R. Hahnloser, and H. S. Seung (2002). Double-ring network model of the head-direction system. *Phy. Rev. E 66*, 041902.

Xing, D., C.-I. Yeh, and R. M. Shapley (2009). Spatial spread of the local field potential and its laminar variation in visual cortex. *J. Neurosci. 29*, 11540–11549.

Xu, C. S., M. Januszewski, Z. Lu, S.-y. Takemura, K. Hayworth, G. Huang, K. Shinomiya, J. Maitin-Shepard, D. Ackerman, S. Berg, et al. (2020). A connectome of the adult Drosophila central brain. *BioRxiv*. https://doi.org/10.1101/2020.01.21.911859.

Xu, F., Y. Shen, L. Ding, C.-Y. Yang, H. Tan, H. Wang, Q. Zhu, R. Xu, F. Wu, Y. Xiao, et al. (2021). High-throughput mapping of a whole rhesus monkey brain at micrometer resolution. *Nat. Biotechnol. 39*, 1521–1528.

Xu, K., J. Ba, R. Kiros, K. Cho, A. Courville, R. Salakhudinov, R. Zemel, and Y. Bengio (2015). Show, attend and tell: Neural image caption generation with visual attention. In *International Conference on Machine Learning*, pp. 2048–2057.

Xu, R., N. P. Bichot, A. Takahashi, and R. Desimone (2022). The cortical connectome of primate lateral prefrontal cortex. *Neuron 110*, 312–327.

Xu, T. X. and W. D. Yao (2010). D1 and D2 dopamine receptors in separate circuits cooperate to drive associative long-term potentiation in the prefrontal cortex. *Proc. Natl. Acad. Sci. U.S.A. 107*, 16366–16371.

Xu, X., N. D. Olivas, T. Ikrar, T. Peng, T. C. Holmes, Q. Nie, and Y. Shi (2016). Primary visual cortex shows laminar-specific and balanced circuit organization of excitatory and inhibitory synaptic connectivity. *J. Physiol. 594*, 1891–1910.

Xu, Y. and M. M. Chun (2006). Dissociable neural mechanisms supporting visual short-term memory for objects. *Nature 440*, 91–95.

Yadav, N., C. Noble, J. E. Niemeyer, A. Terceros, J. Victor, C. Liston, and P. Rajasethupathy (2022). Prefrontal feature representations drive memory recall. *Nature 608*, 153–160.

Yamada, H., A. Tymula, K. Louie, and P. W. Glimcher (2013). Thirst-dependent risk preferences in monkeys identify a primitive form of wealth. *Proc. Natl. Acad. Sci. 110*, 15788–15793.

Yamada, M., M. C. Pita, T. Iijima, and K. Tsutsui (2010). Rule-dependent anticipatory activity in prefrontal neurons. *Neurosci. Res. 67*, 162–171.

Yamins, D. L. and J. J. DiCarlo (2016). Using goal-driven deep learning models to understand sensory cortex. *Nat. Neurosci. 19*, 356–365.

Yamins, D. L., H. Hong, C. F. Cadieu, E. A. Solomon, D. Seibert, and J. J. DiCarlo (2014). Performance-optimized hierarchical models predict neural responses in higher visual cortex. *Proc. Natl. Acad. Sci. U.S.A. 111*, 8619–8624.

Yang, G. J., C. S. Lai, J. Cichon, L. Ma, W. Li, and W. B. Gan (2014). Sleep promotes branch-specific formation of dendritic spines after learning. *Science 344*, 1173–1178.

Yang, G. J., J. D. Murray, X. J. Wang, D. C. Glahn, G. D. Pearlson, G. Repovs, J. H. Krystal, and A. Anticevic (2016a). Functional hierarchy underlies preferential connectivity disturbances in schizophrenia. *Proc. Natl. Acad. Sci. U.S.A. 113*, E219–228.

Yang, G. R., M. R. Joglekar, H. F. Song, W. T. Newsome, and X.-J. Wang (2019). Task representations in neural networks trained to perform many cognitive tasks. *Nat. Neurosci. 22*, 297–306.

Yang, G. R. and M. Molano-Mazón (2021). Towards the next generation of recurrent network models for cognitive neuroscience. *Curr. Opin. Neurobiol. 70*, 182–192.

Yang, G. R., J. D. Murray, and X.-J. Wang (2016b). A dendritic disinhibitory circuit mechanism for pathway-specific gating. *Nat. Commun. 7*, 12815.

Yang, G. R. and X.-J. Wang (2020). Artificial neural networks for neuroscientists: A primer. *Neuron 107*, 1048–1070.

Yang, S., H. Seo, M. Wang, and A. F. Arnsten (2021). NMDAR neurotransmission needed for persistent neuronal firing: Potential roles in mental disorders. *Front. Psychiatry 12*, 337.

Yang, T. and M. Shadlen (2007). Probabilistic reasoning by neurons. *Nature 447*, 1075–1080.

Yang, Y.-Y. and S.-W. Wu (2020). Base rate neglect and neural computations for subjective weight in decision under uncertainty. *Proc. Natl. Acad. Sci. 117*, 16908–16919.

Yao, Z., C. T. van Velthoven, M. Kunst, M. Zhang, D. McMillen, C. Lee, W. Jung, J. Goldy, A. Abdelhak, P. Baker, et al. (2023). A high-resolution transcriptomic and spatial atlas of cell types in the whole mouse brain. *bioRxiv*. https://doi.org/10.1101/2023.03.06.531121.

Yarbus, A. L. (1967). *Eye Movements and Vision*. New York: Plenum Press (Translated by B. Haigh. The original book was published in Russian in 1965).

Yartsev, M. M., T. D. Hanks, A. M. Yoon, and C. D. Brody (2018). Causal contribution and dynamical encoding in the striatum during evidence accumulation. *Elife 7*, e34929.

Yeo, B. T., F. M. Krienen, J. Sepulcre, M. R. Sabuncu, D. Lashkari, M. Hollinshead, J. L. Roffman, J. W. Smoller, L. Zollei, J. R. Polimeni, B. Fischl, H. Liu, and R. L. Buckner (2011). The organization of the human cerebral cortex estimated by intrinsic functional connectivity. *J. Neurophysiol. 106*, 1125–1165.

Yi, H. G., M. K. Leonard, and E. F. Chang (2019). The encoding of speech sounds in the superior temporal gyrus. *Neuron 102*, 1096–1110.

Yi, T. M., Y. Huang, M. I. Simon, and J. Doyle (2000). Robust perfect adaptation in bacterial chemotaxis through integral feedback control. *Proc. Natl. Acad. Sci. U.S.A. 97*, 4649–4653.

Yim, M. Y., X. Cai, and X.-J. Wang (2019). Transforming the choice outcome to an action plan in monkey lateral prefrontal cortex: A neural circuit model. *Neuron 103*, 520–532.

Yizhar, O., L. E. Fenno, M. Prigge, F. Schneider, T. J. Davidson, D. J. O'shea, V. S. Sohal, I. Goshen, J. Finkelstein, J. T. Paz, et al. (2011). Neocortical excitation/inhibition balance in information processing and social dysfunction. *Nature 477*, 171–178.

Yoo, S.-A., J. C. Martinez-Trujillo, S. Treue, J. K. Tsotsos, and M. Fallah (2022). Attention to visual motion suppresses neuronal and behavioral sensitivity in nearby feature space. *BMC Biol. 20*. https://doi.org/10.1186/s12915-022-01428-7.

Yoo, S. B. M. and B. Y. Hayden (2018). Economic choice as an untangling of options into actions. *Neuron 99*, 434–447.

Zador, A., S. Escola, B. Richards, B. Ölveczky, Y. Bengio, K. Boahen, M. Botvinick, D. Chklovskii, A. Churchland, C. Clopath, et al. (2023). Catalyzing next-generation artificial intelligence through neuroai. *Nat. Commun. 14*, 1597.

Zagha, E., J. C. Erlich, S. Lee, G. Lur, D. H. O'Connor, N. A. Steinmetz, C. Stringer, and H. Yang (2022). The importance of accounting for movement when relating neuronal activity to sensory and cognitive processes. *J. Neurosci. 42*, 1375–1382.

Zagha, E., X. Ge, and D. A. McCormick (2015). Competing neural ensembles in motor cortex gate goal-directed motor output. *Neuron 88*, 565–577.

Zahrt, J., J. R. Taylor, R. G. Mathew, and A. F. Arnsten (1997). Supranormal stimulation of D1 dopamine receptors in the rodent prefrontal cortex impairs spatial working memory performance. *J. Neurosci. 17*, 8528–8535.

Zaitsev, A. V., G. Gonzalez-Burgos, N. V. Povysheva, S. Kroner, D. A. Lewis, and L. S. Krimer (2005). Localization of calcium-binding proteins in physiologically and morphologically characterized interneurons of monkey dorsolateral prefrontal cortex. *Cereb. Cortex 15*, 1178–1186.

Zaksas, D. and T. Pasternak (2006). Directional signals in the prefrontal cortex and in area MT during a working memory for visual motion task. *J. Neurosci. 26*, 11726–11742.

Zatorre, R. J., P. Belin, and V. B. Penhune (2002). Structure and function of auditory cortex: Music and speech. *Trends Cogn. Sci. 6*, 37–46.

Zeisel, A., A. B. Muñoz-Manchado, S. Codeluppi, P. Lönnerberg, G. La Manno, A. Juréus, S. Marques, H. Munguba, L. He, C. Betsholtz, C. Rolny, G. Castelo-Branco, J. Hjerling-Leffler, and S. Linnarsson (2015). Cell types in the mouse cortex and hippocampus revealed by single-cell RNA-seq. *Science 347*, 1138–1142.

Zeithamova, D., M. L. Schlichting, and A. R. Preston (2012). The hippocampus and inferential reasoning: Building memories to navigate future decisions. *Front. Hum. Neurosci. 6*, 70.

Zeki, S. (2001). Artistic creativity and the brain. *Science 293*, 51–52.

Zemelman, B. V., G. A. Lee, M. Ng, and G. Miesenböck (2002). Selective photostimulation of genetically chARGed neurons. *Neuron 33*, 15–22.

Zeng, G., Y. Chen, B. Cui, and S. Yu (2019). Continual learning of context-dependent processing in neural networks. *Nat. Mach. Intell. 1*, 364–372.

Zenke, F., B. Poole, and S. Ganguli (2017). Continual learning through synaptic intelligence. *Proc. Mach. Learn. Res. 70*, 3987–3995.

Zhang, J. C., K. C. Berridge, A. J. Tindell, K. S. Smith, and J. W. Aldridge (2009a). A neural computational model of incentive salience. *PLoS Comput. Biol. 5*, e1000437.

Zhang, J. C., P. M. Lau, and G. Q. Bi (2009b). Gain in sensitivity and loss in temporal contrast of STDP by dopaminergic modulation at hippocampal synapses. *Proc. Natl. Acad. Sci. U.S.A. 106*, 13028–13033.

Zhang, K. (1996). Representation of spatial orientation by the intrinsic dynamics of the head-direction cell ensembles: A theory. *J. Neurosci. 16*, 2112–2126.

Zhang, K. and T. J. Sejnowski (2000). A universal scaling law between gray matter and white matter of cerebral cortex. *Proc. Natl. Acad. Sci. 97*, 5621–5626.

Zhang, M. and S. Barash (2000). Neuronal switching of sensorimotor transformations for antisaccades. *Nature 408*, 971–975.

Zhang, S., M. Xu, T. Kamigaki, J. P. Hoang Do, W. C. Chang, S. Jenvay, K. Miyamichi, L. Luo, and Y. Dan (2014). Selective attention. Long-range and local circuits for top-down modulation of visual cortex processing. *Science 345*, 660–665.

Zhang, W. and S. J. Luck (2008). Discrete fixed-resolution representations in visual working memory. *Nature 453*, 233–235.

Zhang, W., S. Wu, K. Josic, and B. Doiron (2022). Sampling-based bayesian inference in recurrent circuits of stochastic spiking neurons. *bioRxiv*. https://doi.org/10.1101/2022.01.26.477877.

Zhang, Y., R. Y. Liu, G. A. Heberton, P. Smolen, D. A. Baxter, L. J. Cleary, and J. H. Byrne (2011). Computational design of enhanced learning protocols. *Nat. Neurosci. 15*, 294–297.

Zhang, Z. and M. Sabuncu (2018). Generalized cross entropy loss for training deep neural networks with noisy labels. *Adv. Neural Inf. Process. Syst. 31*.

Zhao, Y.-J., T. Ma, L. Zhang, X. Ran, R.-Y. Zhang, and Y. Ku (2021). Atypically larger variability of resource allocation accounts for visual working memory deficits in schizophrenia. *PLoS Comput. Biol. 17*, e1009544.

Zheng, Z., J. S. Lauritzen, E. Perlman, C. G. Robinson, M. Nichols, D. Milkie, O. Torrens, J. Price, C. B. Fisher, N. Sharifi, et al. (2018). A complete electron microscopy volume of the brain of adult Drosophila melanogaster. *Cell 174*, 730–743.

Zhou, J., M. Jia, C.and Montesinos-Cartagena, M. P. H. Gardner, W. Zong, and G. Schoenbaum (2020). Evolving schema representations in orbitofrontal ensembles during learning. *Nature 589*. https://doi.org/10.1038/s41586-020-03061-2.

Zhou, X., F. Katsuki, X. L. Qi, and C. Constantinidis (2012). Neurons with inverted tuning during the delay periods of working memory tasks in the dorsal prefrontal and posterior parietal cortex. *J. Neurophysiol. 108*, 31–38.

Zhou, Y., J. Sharma, Q. Ke, R. Landman, J. Yuan, H. Chen, D. S. Hayden, J. W. Fisher, M. Jiang, W. Menegas, et al. (2019). Atypical behaviour and connectivity in shank3-mutant macaques. *Nature 570*, 326–331.

Zhu, F., M. Cizeron, Z. Qiu, R. Benavides-Piccione, M. V. Kopanitsa, N. G. Skene, B. Koniaris, J. DeFelipe, E. Fransén, N. H. Komiyama, et al. (2018). Architecture of the mouse brain synaptome. *Neuron 99*, 781–799.

Zhu, J., Q. Cheng, Y. Chen, H. Fan, Z. Han, R. Hou, Z. Chen, and C. T. Li (2020). Transient delay-period activity of agranular insular cortex controls working memory maintenance in learning novel tasks. *Neuron 105*, 934–946.

Ziegler, G., T. U. Hauser, M. Moutoussis, E. T. Bullmore, I. M. Goodyer, P. Fonagy, P. B. Jones, U. Lindenberger, and R. J. Dolan (2019). Compulsivity and impulsivity traits linked to attenuated developmental frontostriatal myelination trajectories. *Nat. Neurosci. 22*, 992–999.

Zilles, K. and N. Palomero-Gallagher (2017). Multiple transmitter receptors in regions and layers of the human cerebral cortex. *Front. Neuroanat. 11*, 78.

Zingg, B., H. Hintiryan, L. Gou, M. Y. Song, M. Bay, M. S. Bienkowski, N. N. Foster, S. Yamashita, I. Bowman, A. W. Toga, and H. W. Dong (2014). Neural networks of the mouse neocortex. *Cell 156*, 1096–1111.

Zink, C. F., P. B. Barker, A. Sawa, D. R. Weinberger, M. Wang, H. Quillian, W. S. Ulrich, Q. Chen, A. E. Jaffe, J. E. Kleinman, et al. (2020). Association of missense mutation in folh1 with decreased naag levels and impaired working memory circuitry and cognition. *Am. J. Psychiatry 177*, 1129–1139.

Zipser, D. and R. A. Andersen (1988). A back-propagation programmed network that simulates response properties of a subset of posterior parietal neurons. *Nature 331*, 679.

Zohary, E., M. N. Shadlen, and W. T. Newsome (1994). Correlated neuronal discharge rate and its implications for psychophysical performance. *Nature 370*, 140–143.

Zorumski, C. F., Y. Izumi, and S. Mennerick (2016). Ketamine: NMDA receptors and beyond. *J. Neurosci. 36*, 11158–11164.

Zucker, R. S. (1989). Short-term synaptic plasticity. *Ann. Rev. Neurosci. 12*, 13–21.

Zuo, X.-N., Y. He, R. F. Betzel, S. Colcombe, O. Sporns, and M. P. Milham (2017). Human connectomics across the life span. *Trends Cogn. Sci. 21*, 32–45.

Zylberberg, A., P. Barttfeld, and M. Sigman (2012). The construction of confidence in a perceptual decision. *Front. Integr. Neurosci. 6*, 79.

Zylberberg, J. and B. W. Strowbridge (2017). Mechanisms of persistent activity in cortical circuits: Possible neural substrates for working memory. *Annu. Rev. Neurosci. 40*, 603–627.

INDEX

Note: Page numbers in *italics* indicate a figure and page numbers in **bold** indicate a table on the corresponding page.

I

K

L

M

R

S

T